A BRIEF TABLE OF INTEGRALS* (continued)

$$\int u \sin u \, du = \sin u - u \cos u.$$

$$\int u^n \sin u \, du = -u^n \cos u + n \int u^{n-1} \cos u \, du.$$

$$\int u \cos u \, du = \cos u + u \sin u.$$

$$\int u^n \cos u \, du = u^n \sin u - n \int u^{n-1} \sin u \, du.$$

$$\int e^{au} \sin nu \, du = \frac{e^{au}(a \sin nu - n \cos nu)}{a^2 + n^2}.$$

$$\int e^{au} \cos nu \, du = \frac{e^{au}(a \cos nu + n \sin nu)}{a^2 + n^2}.$$

$$\int \sin au \sin bu \, du = -\frac{\sin(a + b)u}{2(a + b)} + \frac{\sin(a - b)u}{2(a - b)}, \qquad a^2 \neq b^2.$$

$$\int \cos au \cos bu \, du = \frac{\sin(a + b)u}{2(a + b)} + \frac{\sin(a - b)u}{2(a - b)}, \qquad a^2 \neq b^2.$$

$$\int \sin au \cos bu \, du = -\frac{\cos(a + b)u}{2(a + b)} - \frac{\cos(a - b)u}{2(a - b)}, \qquad a^2 \neq b^2.$$

$$\int \sinh u \, du = \cosh u. \qquad \int \cosh u \, du = \sinh u.$$

$$\Gamma(t) = \int_0^\infty e^{-u} u^{t-1} \, du, \quad t > 0; \qquad \Gamma(\tfrac{1}{2}) = \sqrt{\pi}; \quad \text{and} \quad \Gamma(n + 1) = n!, \text{ if } n \text{ is a positive integer.}$$

SOME POWER SERIES EXPANSIONS

$$f(x) = f(a) + (x - a)f'(a) + \frac{(x - a)^2}{2!} f''(a) + \cdots + \frac{(x - a)^n}{n!} f^{(n)}(a) + \cdots \qquad \text{(Taylor series)}$$

$$e^x = \sum_{n=0}^\infty \frac{x^n}{n!} \qquad \sin x = \sum_{n=0}^\infty \frac{(-1)^n x^{2n+1}}{(2n + 1)!} \qquad \cos x = \sum_{n=0}^\infty \frac{(-1)^n x^{2n}}{(2n)!}$$

$$(1 - x)^{-1} = \sum_{n=0}^\infty x^n \qquad (1 - x)^{-2} = \sum_{n=0}^\infty (n + 1)x^n \qquad \ln(1 - x) = -\sum_{n=1}^\infty \frac{x^n}{n}$$

$$\tan x = x + \tfrac{1}{3}x^3 + \tfrac{2}{15}x^5 + \tfrac{17}{315}x^7 + \tfrac{62}{2835}x^9 + \cdots$$

$$\arcsin x = x + \frac{x^3}{2 \cdot 3} + \frac{1 \cdot 3}{2 \cdot 4 \cdot 5} x^5 + \frac{1 \cdot 3 \cdot 5}{2 \cdot 4 \cdot 6 \cdot 7} x^7 + \cdots$$

$$\arctan x = \sum_{n=0}^\infty (-1)^n \frac{x^{2n+1}}{2n + 1}$$

$$J_0(x) = \sum_{k=0}^\infty \frac{(-1)^k x^{2k}}{(k!)^2 2^{2k}} \qquad J_1(x) = \sum_{k=0}^\infty \frac{(-1)^k x^{2k+1}}{k!(k + 1)! 2^{2k+1}} \qquad J_n(x) = \sum_{k=0}^\infty \frac{(-1)^k x^{2k+n}}{k! \Gamma(n + k + 1) 2^{2k+n}}$$

* *Note:* An arbitrary constant is to be added to each formula.

Fundamentals of Differential Equations and Boundary Value Problems

Second Edition

R. Kent Nagle & Edward B. Saff

UNIVERSITY OF SOUTH FLORIDA

with contributions by
A. D. Snider

UNIVERSITY OF SOUTH FLORIDA

Addison-Wesley Publishing Company

READING, MASSACHUSETTS MENLO PARK, CALIFORNIA NEW YORK
DON MILLS, ONTARIO WOKINGHAM, ENGLAND AMSTERDAM BONN SYDNEY
SINGAPORE TOKYO MADRID SAN JUAN MILAN PARIS

Acquisitions Editor Laurie Rosatone
Senior Production Supervisor Karen Wernholm
Senior Marketing Manager Andrew Fisher
Prepress Buyer Caroline Fell
Art Buyer Joseph Vetere
Senior Manufacturing Manager Roy Logan
Editorial Production Services Barbara Pendergast
Art Editor Alena Konecny
Text Design Ron Kosciak, Dragonfly Design
Cover Design Diana C. Coe
Composition EPS Group, Beacon Graphics
Technical Illustration Academy Artworks

Library of Congress Cataloging-in-Publication Data

Nagle, R. Kent.
 Fundamentals of differential equations and boundary value problems
 / R. Kent Nagle, Edward B. Saff ; with contributions by Arthur David
Snider, —2nd ed.
 p. cm.
 Includes bibliographical references and index.
 ISBN 0-201-80879-X
 1. Differential equations. 2. Boundary value problems. I. Saff,
E.B. II. Snider, Arthur David. III. Title.
QA371.N243 1996 95-10162
515'.35—dc20 CIP

Reprinted with corrections, February 1999

7 8 9 10-DOC-01 00 99

In Memoriam: R. Kent Nagle

He has left his imprint not only on these pages but upon all who knew him. He was that rare mathematician who could effectively communicate at all levels, imparting his love for the subject with the same ease to undergraduates, graduates, precollege students, public school teachers, and his colleagues at the University of South Florida.

Kent was at peace in life—a peace that emanated from the depth of his understanding of the human condition and the strength of his beliefs in the institutions of family, religion, and education. He was a research mathematician, an accomplished author, a Sunday school teacher, and a devoted husband and father.

Kent was also my dear friend and my jogging partner who has left me behind still struggling to keep pace with his high ideals.

E. B. Saff

Preface

OUR GOAL

Fundamentals of Differential Equations and Boundary Value Problems is designed to serve the needs of a one- or two-semester course in basic theory as well as applications of differential equations. For this purpose, we have augmented our original text by including chapters on Eigenvalue Problems and Sturm-Liouville Equations, Stability of Autonomous Systems, and Existence and Uniqueness Theory. We have striven to create a flexible text that affords the instructor substantial latitude in designing a syllabus (we provide some sample syllabi later in this preface), in course emphasis (theory, applications and techniques, numerical methods), and in using commercially available computer software and/or the software utility programs specifically designed for this text.

IMPROVEMENTS IN THIS EDITION

In response to requests of users and reviewers, and recognizing recent developments in teaching and learning, we offer the following:

Focus on Analytical, Graphical, and Numerical Methods
Certainly these three aspects strengthen conceptual learning. Moreover, current trends in education increasingly reflect the ideology that all three are essential pedagogical ingredients that are best utilized in combination with one another. With this in mind, we have substantially augmented our discussions of theory and methods with many new graphical illustrations and have enhanced both graphical and numerical techniques by providing the reader with a software utility package that is keyed to the text. With these new features, students are free of the burden of computations and graphing and can focus more immediately on interpretative aspects. Also, more Group Projects have been included (see discussion below) since they provide the student with a vehicle for blending analytical, graphical, and numerical methods in the framework of a challenging joint problem-solving activity. At the same time, we have retained a substantial number of practice exercises that are essential for building basic skills.

New Group Projects
Twenty percent of the group projects are new. As a result of a national search sponsored by Addison-Wesley, instructors contributed their own most successful classroom-tested group projects (see acknowledgments below). Group projects appear at the end of each chapter and

relate to the material covered in the chapter. They often involve a more challenging application, delve deeper into theory, or introduce more advanced topics. New projects include:

Chapter 2	*Torricelli's Law of Fluid Flow*
	Equilibrium Points of Autonomous First Order Equations
Chapter 3	*Aircraft Guidance in a Crosswind*
	Bang-Bang Controls
Chapter 5	*Designing a Landing System for Interplanetary Travel*
	Things That Bob
	A Growth Model for Phytoplankton—Part I
Chapter 9	*Undamped Second Order Systems*
Chapter 12	*Ecosystem on Planet Glia-2*
	A Growth Model for Phytoplankton—Part II

Revised Exercises Twenty percent of the exercises are new, including computer implementation problems as well as exercises that require students to interpret solutions, building conceptual knowledge.

New Figures To help students visualize concepts, dozens of new illustrations have been added, including direction field plots (Section 1.3), graphs of power series solutions (Chapter 8), and three-dimensional plots of solutions to partial differential equations (Chapter 10).

Software New exploratory software that is pedagogically tied to the text is available. It is student friendly, functional, and easy to use. A complete description appears under "Supplements" later in this preface. Also available is a MAPLE based instructor's guide that includes sample worksheets and additional group projects.

World Wide Web The new exploratory software is also available for downloading from the World Wide Web. Point your web browser to http://hepg.awl.com, keyword: Nagle-Saff.

CONTENT CHANGES

- **Direction Fields** Now a separate section, 1.3, that includes analytical examples and exercises dealing with population growth and motion of particles. Graphs illustrate the basic existence-uniqueness theorem.

- **Introduction to Systems of First Order Equations** The concept is introduced much earlier, in Section 4.10, following the basic theory and techniques for second order linear equations. The material is presented in a manner that does not require background in linear algebra.

- **Introduction to Phase Plane Analysis** This section, 5.7, is presented earlier. It introduces the study of trajectories of autonomous systems, critical points, and stability. Chapter 12 contains an in-depth treatment of these topics.

- **Introduction to Laplace Transforms: A Mixing Problem** This new section, 7.1, emphasizes advantages of the Laplace method in solving linear equations when the forcing function is discontinuous. It includes a schematic comparing the Laplace method with the standard method.

- **Chapter 2** The section on linear equations is now presented before exact equations.

PREREQUISITES

While some universities make linear algebra a prerequisite for differential equations, many schools (especially engineering) only require calculus. With this in mind, we have designed the text so that only Chapter 6 (Theory of Higher-Order Linear Differential Equations) and Chapter 9 (Matrix Methods for Linear Systems) require more than high school linear algebra. Moreover, Chapter 9 contains a review section on matrices and vectors as well as specific references for the deeper results used from the theory of linear algebra. We have also written Section 4.10 and Chapter 5 so as to give an introduction to systems of differential equations—including methods of solving, applications, numerical procedures, phase plane analysis, and Poincaré maps—that does not require a background in linear algebra.

SAMPLE SYLLABI

As a rough guide in designing a two-semester syllabus related to this text, we provide two samples that can be used for a sequence of two 15-week courses that meet three hours per week: the first emphasizes applications and computations including phase plane analysis; the second is designed for courses that place more emphasis on theory. As in the original text, Chapters 1, 2, and 4 provide the core for the first-semester course. The rest of the chapters are, for the most part, independent of each other.

	Methods, Computations, and Applications	Theory and Methods		
	SEMESTER 1			**SEMESTER 2**
Week	Sections	Sections	Week	Sections
1	1.1, 1.2, 1.3	1.1, 1.2, 1.3	1	9.2, 9.3
2	1.4, 2.2	2.2, 2.3	2	9.4, 9.5
3	2.3, 2.4	2.4, 3.2, 3.4	3	9.6, 9.7
4	3.2, 3.4, 3.5	4.2, 4.3	4	12.1, 12.2
5	3.6, 4.2	4.4, 4.5, 4.6	5	12.3, 12.4
6	4.3, 4.5	4.7, 4.8	6	12.5, 12.6

(continued)

| | SEMESTER 1 | | | | SEMESTER 2 | |
|---|---|---|---|---|---|
| Week | Sections | Sections | | Week | Sections |
| 7 | 4.6, 4.7, 4.8 | 4.9, 4.10 | | 7 | 12.7, 10.2 |
| 8 | 4.9, 4.10 | 5.7, 6.1 | | 8 | 10.3, 10.4 |
| 9 | 5.1, 5.2, 5.3 | 6.2, 6.3 | | 9 | 10.5, 10.6 |
| 10 | 5.4, 5.6, 5.7 | 7.2, 7.3 | | 10 | 10.7, 11.2 |
| 11 | 7.2, 7.3 | 7.4, 7.5 | | 11 | 11.3, 11.4 |
| 12 | 7.4, 7.5 | 7.6, 7.7 | | 12 | 11.5, 11.6 |
| 13 | 7.6, 7.7 | 8.2, 8.3 | | 13 | 11.7, 11.8 |
| 14 | 8.2, 8.3 | 8.5, 8.6, 8.7 | | 14 | 13.1, 13.2 |
| 15 | 8.5, 8.6 | 8.8, 8.9 | | 15 | 13.3, 13.4 |

RETAINED FEATURES

Flexible Organization

Most of the material is modular in nature to allow for various course configurations and emphasis (theory, applications and techniques, and concepts).

Optional Use of Computer Software

The availability of computer packages such as MATHEMATICA®, DERIVE®, and MAPLE® provides an opportunity for the student to conduct numerical experiments and tackle realistic applications that give additional insights into the subject. Consequently, we have added several exercises and projects throughout the text that are designed for the student to employ available software in phase plane analysis, eigenvalue computations, and the numerical solutions of various equations.

Choice of Applications

Because of syllabus constraints, some courses will have little or no time for sections (such as those in Chapters 3 and 5) that exclusively deal with applications. Therefore, we have made the sections in these chapters almost completely independent of each other. To afford the instructor even greater flexibility, we have built in a variety of applications in the exercises for the theoretical sections. In addition, we have included many projects that deal with such applications.

Technical Writing Exercises

Communication skills are, of course, an essential aspect of professional activities. Yet few texts provide opportunities for the reader to develop these skills. Thus we have added at the end of most chapters a set of clearly marked technical writing exercises which invite students to make documented responses to questions dealing with the concepts in the chapter. In so doing, students are encouraged to make comparisons between various methods and to present examples that support their analysis.

Historical Footnotes

Throughout the text historical footnotes are set off by colored daggers ($\dagger$). These footnotes typically provide the name of the person who developed the technique, the date, and the context of the original research.

Motivating Problem Most chapters begin with a discussion of a problem from physics or engineering that motivates the topic presented and illustrates the methodology.

Chapter Summary and Review Problems All of the main chapters contain a set of review problems along with a synopsis of the major concepts presented.

Computer Graphics Most of the figures in the text were generated on a microcomputer. Computer graphics not only ensure greater accuracy in the illustrations, they demonstrate the use of numerical experimentation in studying the behavior of solutions.

Proofs While more pragmatic students may balk at proofs, most instructors regard these justifications as an essential ingredient in a textbook on differential equations. As with any text at this level, certain details in the proofs must be omitted. When this occurs, we flag the instance and refer readers either to a problem in the exercises or to another text. For convenience, the end of a proof is marked by the symbol ■ End.

Linear Theory We have developed the theory of linear differential equations in a gradual manner. In Chapter 4 (Linear Second Order Equations) we present the basic theory for linear second order equations and discuss various techniques for solving these equations. Higher-order equations are briefly mentioned in this chapter. A more detailed discussion of linear higher-order differential equations is given in Chapter 6 (Theory of Higher-Order Linear Differential Equations). For a beginning course emphasizing methods of solution, the presentation in Chapter 4 is sufficient and Chapter 6 can be skipped.

Numerical Algorithms Several numerical methods for approximating solutions to differential equations are presented along with program outlines that are easily implemented on a microcomputer. These methods are introduced early in the text so that teachers and/or students can use them for numerical experimentation and for tackling complicated applications. All algorithms discussed in the text are implemented on the software package available with this text.

Exercises An abundance of exercises is graduated in difficulty from straightforward, routine problems to more challenging ones. Deeper theoretical questions, along with applications, usually occur toward the end of the exercise sets. Throughout the text we have included problems and projects that require the use of a microcomputer. These exercises are denoted by the symbol ▄. The software specifically designed for use with this text greatly facilitates the solutions to these numerical problems.

Optional Sections These sections can be omitted without affecting the logical development of the material. They are marked with an asterisk in the table of contents. As mentioned earlier, the sections in Chapters 3 and 5 are almost completely independent of each other.

Laplace Transforms We provide a detailed chapter on Laplace transforms since this is a recurring topic for engineers. Our treatment emphasizes discontinuous forcing terms and includes a section on the Dirac delta function.

Power Series Power series solutions is a topic that occasionally causes student anxiety. Possibly, this is due to inadequate preparation in calculus where the more subtle subject of convergent series is (not infrequently) covered at a rapid pace. Our solution has been to provide a thorough treatment of power series solutions that also includes a review of their properties as well as a discussion of real analytic functions. Unlike many texts, this one provides an extensive section on the *method of Frobenius* (Section 8.6) and two sections on the various methods for finding a second linearly independent solution.

While we have given considerable space to power series solutions, we have also taken great care to accommodate the instructor who only wishes to give a basic introduction to the topic. *An introduction to solving differential equations using power series and the method of Frobenius can be accomplished by covering the materials in Section 8.2, Section 8.3, and Section 8.6.*

Partial Differential Equations An introduction to this subject is provided in Chapter 10, which covers the method of separation of variables, Fourier series, the heat equation, the wave equation, and Laplace's equation. Examples in two and three dimensions are included.

SUPPLEMENTS

Student Solutions Manual By Anne Kusmierczyk and E. B. Saff. Containing complete, worked-out solutions to selected exercises, this provides students with an excellent study tool. ISBN 0-201-80877-3

Instructor's Resource Guide Contains short answers to all exercises and additional group projects. ISBN 0-201-80876-5

MAPLE Based Instructor's Guide By Douglas B. Meade. Contains MAPLE demonstrations for ODEs, sample MAPLE worksheets for students, and additional group projects designed for using MAPLE. ISBN 0-201-96429-5

Software Available free to adopters, this DOS-based software is student-friendly, requiring little learning time. It is designed not only to aid the student in solving numerical problems in the text, but also as a convenient utility to encourage students to experiment. The software includes direction field plots, improved Euler's method, Runge-Kutta methods (also for systems), phase plane diagrams, eigenvalues and eigenvectors, computational methods such as Newton's method for solving equations, Simpson's rule, plots and tabulation of one or several functions, and many other useful algorithms. ISBN 0-201-82688-7

World Wide Web The new exploratory software is also available for downloading from the World Wide Web. Point your web browser to http://hepg.awl.com, keyword: Nagle-Saff.

ACKNOWLEDGMENTS

The staging of the original text involved considerable behind-the-scenes activity. We want to thank Frank Glaser (California State Polytechnic University, Pomona) for many of the historical footnotes. We are indebted to Herbert E. Rauch (Lockheed Research Laboratory) for help with Section 3.3 on heating and cooling of buildings, Project B in Chapter 3 on aquaculture, and other application problems. Our appreciation goes to George Fix and R. Kannan (University of Texas, Arlington) for their useful suggestions concerning Section 3.7. We give special thanks to Richard H. Elderkin (Pomona College), Jerrold Marsden (University of California, Berkeley), T. G. Proctor (Clemson University), and Philip W. Schaefer (University of Tennessee), who read and reread the manuscript for the original text, making numerous suggestions that greatly improved the book. We are also indebted to the many people who reviewed the manuscript for this new edition:

Dean Brown, YOUNGSTOWN STATE UNIVERSITY
Alfred Clark, Jr., UNIVERSITY OF ROCHESTER
Don Hartig, CALIFORNIA POLYTECHNIC STATE UNIVERSITY
D. J. Kaup, CLARKSON UNIVERSITY
Paul Kumpel, STATE UNIVERSITY OF NEW YORK AT STONY BROOK
Glenn Ledder, UNIVERSITY OF NEBRASKA—LINCOLN
Benedict Pollina, UNIVERSITY OF HARTFORD
Mark W. Winstead, UNIVERSITY OF CALIFORNIA—SAN DIEGO

In addition, we would like to offer our sincere thanks to all of the instructors who shared their own group projects with us for inclusion in this edition:

Oliver Bernard, STATION ZOOLOGIQUE DE VILLEFRANCHE-SUR-MER
Richard Bernatz, LUTHER COLLEGE
Randall K. Campbell-Wright, UNIVERSITY OF TAMPA
Alfred Clark, Jr., UNIVERSITY OF ROCHESTER
Leopoldo P. Franca, UNIVERSITY OF COLORADO—DENVER
Jean-Luc Gouzé, INRIA, SOPHIA-ANTIPOLIS
T. L. Pearson, ACADIA UNIVERSITY
David Stapleton, UNIVERSITY OF CENTRAL OKLAHOMA

The timely completion of this edition would not have been possible without the contributions and dedication of my colleague and collaborator, Dr. Dave Snider. He not only lent his mathematical expertise, his engineering perspective, and his lively writing style, but also provided the entire project with a reassuring calm during the difficult period surrounding the passing of Kent Nagle.

The newly developed software that accompanies this text was expertly developed by Emil Moskona and Spas Tashev of the Bulgarian Academy of Sciences. Their efforts not only enhance the usefulness of this text, but also provide a convenient utility for calculus/differential equations problem solving.

We are grateful for the encouragement we received from our students and staff here at the University of South Florida. Maria Carvalho, through her untiring efforts and devotion

to duty, somehow lent order to an otherwise chaotic existence during the creation of this edition.

Finally, we want to thank the staff at Addison-Wesley for their dedicated assistance in bringing this edition to fruition. Special kudos go to mathematics editor Laurie Rosatone for her help, encouragement, and sensitivity. The production phase was carefully guided by Karen Wernholm and Barbara Pendergast, who admirably put up with my frequent outbursts of anxiety.

E. B. SAFF
TAMPA, FLORIDA

Contents

*Denotes optional sections that can be omitted without compromising the logical flow.

Chapter 3 MATHEMATICAL MODELS AND NUMERICAL METHODS INVOLVING FIRST ORDER EQUATIONS

Chapter 4 LINEAR SECOND ORDER EQUATIONS

Chapter 5 APPLICATIONS AND NUMERICAL METHODS FOR SECOND ORDER EQUATIONS AND SYSTEMS 240

Chapter 6 THEORY OF HIGHER-ORDER LINEAR DIFFERENTIAL EQUATIONS

Chapter 7 LAPLACE TRANSFORMS

Chapter 10 PARTIAL DIFFERENTIAL EQUATIONS 573

Chapter 11 EIGENVALUE PROBLEMS AND STURM-LIOUVILLE EQUATIONS 658

Chapter 12 STABILITY OF AUTONOMOUS SYSTEMS 734

Chapter 13 EXISTENCE AND UNIQUENESS THEORY 805

Chapter 1

Introduction

1.1 BACKGROUND

In the sciences and engineering, mathematical models are developed to aid in the understanding of physical phenomena. These models often yield an equation that contains some derivatives of an unknown function. Such an equation is called a **differential equation**. Two examples of models developed in calculus are the free fall of a body and the decay of a radioactive substance.

In the case of free fall, an object is dropped from a certain height above the ground and falls under the force of gravity.[†] Newton's second law, which states that an object's mass times its acceleration equals the total force acting on it, can be applied to the falling object. This leads to the equation (see Figure 1.1)

$$m \frac{d^2 h}{dt^2} = -mg,$$

where m is the mass of the object, h is its height above the ground, d^2h/dt^2 is its acceleration, g is the (constant) gravitational acceleration, and $-mg$ is the force due to gravity. This is a differential equation containing the second derivative of the unknown height h as a function of time.

Figure 1.1 Apple in free fall

[†] We are assuming here that gravity is the *only* force acting on the object and that this force is constant. More general models would take into account other forces, such as air resistance.

Fortunately, the above equation is easy to solve for h. All we have to do is divide by m and integrate twice with respect to t. That is,

$$\frac{d^2h}{dt^2} = -g,$$

so

$$\frac{dh}{dt} = -gt + c_1$$

and

$$h = h(t) = \frac{-gt^2}{2} + c_1 t + c_2.$$

We shall see that the constants of integration, c_1 and c_2, are determined if we know the *initial* height and the *initial* velocity of the object. We then have a formula for the height of the object at time t.

In the case of radioactive decay (Figure 1.2), we begin from the premise that the rate of decay is proportional to the amount of radioactive substance present. This leads to the equation

$$\frac{dA}{dt} = -kA, \qquad k > 0,$$

where $A(>0)$ is the unknown amount of radioactive substance present at time t and k is the proportionality constant. To solve this differential equation, we rewrite it in the form

$$\frac{1}{A} dA = -k \, dt$$

and integrate to obtain

$$\int \frac{1}{A} dA = \int -k \, dt$$

$$\ln A + C_1 = -kt + C_2.$$

Figure 1.2 Radioactive decay

Solving for A yields

$$A = A(t) = e^{\ln A} = e^{-kt}e^{C_2 - C_1} = Ce^{-kt},$$

where C is the combination of integration constants $e^{C_2 - C_1}$. The value of C, as we shall see later, is determined if the *initial* amount of radioactive substance is given. We then have a formula for the amount of radioactive substance at any future time t.

Even though the above examples were easily solved by methods learned in calculus, they do give us some insight into the study of differential equations in general. First, notice that the solution of a differential equation is a *function*, like $h(t)$ or $A(t)$, not merely a number. Second, integration is an important tool in solving differential equations (not surprisingly!). Third, we cannot expect to get a unique solution to a differential equation, since there will be arbitrary "constants of integration." The second derivative d^2h/dt^2 in the free fall equation gave rise to two constants, c_1 and c_2, and the first derivative in the decay equation gave rise, ultimately, to one constant, C.

Whenever a mathematical model involves the **rate of change** of one variable with respect to another, a differential equation is apt to appear. Unfortunately, in contrast to the examples for free fall and radioactive decay, the differential equation may be very complicated and difficult to analyze.

Differential equations arise in a variety of subject areas, including not only the physical sciences but also such diverse fields as economics, medicine, psychology, and operations research. We now list a few specific examples.

1. A classic application of differential equations is found in the study of an electric circuit consisting of a resistor, an inductor, and a capacitor driven by an electromotive force (see Figure 1.3). Here an application of Kirchhoff's laws[†] leads to the equation

 (1) $$L\frac{d^2q}{dt^2} + R\frac{dq}{dt} + \frac{1}{C}q = E(t),$$

 where L is the inductance, R is the resistance, C is the capacitance, $E(t)$ is the electromotive force, $q(t)$ is the charge, and t is the time.

Figure 1.3 Schematic for a series RLC circuit

[†] We will discuss the applications of Kirchhoff's laws in Section 5.5.

2. In the study of the gravitational equilibrium of a star, an application of Newton's law of gravity and of the Stefan-Boltzmann law for gases leads to the equilibrium equation

(2) $$\frac{1}{r^2}\frac{d}{dr}\left(\frac{r^2}{\rho}\frac{dP}{dr}\right) = -4\pi\rho G,$$

where P is the sum of the gas kinetic pressure and the radiation pressure, r is the distance from the center of the star, ρ is the density of matter, and G is the gravitational constant.

3. In psychology, one model of the learning of a task involves the equation

(3) $$\frac{dy/dt}{y^{3/2}(1-y)^{3/2}} = \frac{2p}{\sqrt{n}}.$$

Here the variable y represents the learner's skill level as a function of time t. The constants p and n depend on the individual learner and the nature of the task.

4. In the study of vibrating strings and the propagation of waves, we find the *partial* differential equation

(4) $$\frac{\partial^2 u}{\partial t^2} - c^2\frac{\partial^2 u}{\partial x^2} = 0,^{\dagger}$$

where t represents time, x the location along the string, c the wave speed, and u the displacement of the string, *which is a function of time and location.*

To begin our study of differential equations we need some common terminology. If an equation involves the derivative of one variable with respect to another, then the former is called a **dependent variable** and the latter an **independent variable**. Thus in the equation

(5) $$\frac{d^2x}{dt^2} + a\frac{dx}{dt} + kx = 0,$$

t is the independent variable and x is the dependent variable. We refer to a and k as **coefficients** in equation (5). In the equation

(6) $$\frac{\partial u}{\partial x} - \frac{\partial u}{\partial y} = x - 2y,$$

x and y are independent variables and u is a dependent variable.

A differential equation involving ordinary derivatives with respect to a single independent variable is called an **ordinary differential equation**. A differential equation involving

†*Historical Footnote:* This partial differential equation was first discovered by Jean le Rond d'Alembert (1717–1783) in 1747.

partial derivatives with respect to more than one independent variable is a **partial differential equation**. Equation (5) is an ordinary differential equation, and equation (6) is a partial differential equation.

The **order** of a differential equation is the order of the highest-order derivatives present in the equation. Equation (5) is a second order equation because d^2x/dt^2 is the highest-order derivative present. Equation (6) is a first order equation because only first order partial derivatives occur.

It will be useful to classify ordinary differential equations as being either linear or nonlinear. A **linear** differential equation is one in which the occurrence of the dependent variable y has the format

$$(7) \qquad a_n(x)\frac{d^ny}{dx^n} + a_{n-1}(x)\frac{d^{n-1}y}{dx^{n-1}} + \cdots + a_1(x)\frac{dy}{dx} + a_0(x)y = F(x),$$

where $a_n(x)$, $a_{n-1}(x)$, $\ldots$, $a_0(x)$, and $F(x)$ depend only on the independent variable x. No restriction is imposed on the nature of the x-dependence. If an ordinary differential equation is not linear, then we call it **nonlinear**. For example,

$$\frac{d^2y}{dx^2} + y^3 = 0$$

is a nonlinear second order ordinary differential equation because of the y^3 term, whereas

$$\frac{d^2y}{dx^2} + y = x^3$$

is linear (despite the x^3 term). The equation

$$\frac{d^2y}{dx^2} - y\frac{dy}{dx} = \cos x$$

is nonlinear because of the $y\ dy/dx$ term. As we shall see in Chapter 4, linear equations are more amenable to analytic attack than nonlinear ones are.

EXERCISES 1.1

In Problems 1 through 14, a differential equation is given along with the field or problem area in which it arises. Classify each as an ordinary differential equation (ODE) or a partial differential equation (PDE), give the order, and indicate the independent and dependent variables. If the equation is an ordinary differential equation, indicate whether the equation is linear or nonlinear.

1. $3\dfrac{d^2x}{dt^2} + 4\dfrac{dx}{dt} + 9x = 2\cos 3t$

(mechanical vibrations, electrical circuits, seismology).

2. $\dfrac{d^2y}{dx^2} - 2x\dfrac{dy}{dx} + 2y = 0$

(Hermite's equation, quantum mechanical harmonic oscillator).

3. $\dfrac{dy}{dx} = \dfrac{y(2 - 3x)}{x(1 + 3y)}$

(competition between two species, ecology).

4. $\dfrac{\partial^2 u}{\partial x^2} + \dfrac{\partial^2 u}{\partial y^2} = 0$

(Laplace's equation, potential theory, electricity, heat, aerodynamics).

5. $\dfrac{dp}{dt} = kp(P - p)$, where k and P are constants

(logistic curve, epidemiology, economics).

6. $\dfrac{dx}{dt} = (4 - x)(1 - x)$

(chemical reaction rates).

7. $y\left[1 + \left(\dfrac{dy}{dx}\right)^2\right] = C$, where C is a constant

(brachistochrone problem,[†] calculus of variations).

8. $\sqrt{1 - y}\,\dfrac{d^2y}{dx^2} + 2x\dfrac{dy}{dx} = 0$

(Kidder's equation, flow of gases through a porous medium).

9. $x\dfrac{d^2y}{dx^2} + \dfrac{dy}{dx} + xy = 0$

(aerodynamics, stress analysis).

10. $8\dfrac{d^4y}{dx^4} = x(1 - x)$

(deflection of beams).

11. $\dfrac{\partial N}{\partial t} = \dfrac{\partial^2 N}{\partial r^2} + \dfrac{1}{r}\dfrac{\partial N}{\partial r} + kN$, where k is a constant

(nuclear fission).

12. $\dfrac{d^2y}{dx^2} - 0.1(1 - y^2)\dfrac{dy}{dx} + 9y = 0$

(Van der Pol's equation, triode vacuum tube).

13. $x\dfrac{d^2y}{dx^2} + 2\dfrac{dy}{dx} + x(y^2 - 1)^{3/2} = 0$

(white dwarf equation, gravitational potential in degenerate stars).

14. $(y \tan \alpha)^2\dfrac{dy}{dt} = (gy)^{1/2}$, where α and g are constants

(liquid flow from a container).

In Problems 15 and 16, write a differential equation that fits the given description.

15. The rate of change in the temperature T of coffee at time t is proportional to the difference between the temperature M of the air at time t and the temperature of the coffee at time t.

16. The rate of change of the mass A of salt at time t is proportional to the square of the mass of salt present at time t.

17. Drag Race. Two drivers, Alison and Kevin, are participating in a drag race. Beginning from a standing start, they each proceed with a constant acceleration. Kevin covers the last 1/4 of the distance in 3 seconds, whereas Alison covers the last 1/3 of the distance in 4 seconds. Who wins and by how much time?

1.2 SOLUTIONS AND INITIAL VALUE PROBLEMS

An nth order ordinary differential equation is an equality involving the independent variable x, the dependent variable y, and the first n derivatives of y. Examples are

$$x^2\dfrac{d^2y}{dx^2} + x\dfrac{dy}{dx} + y = x^3 \quad \text{(second order)}$$

$$\sqrt{1 - \left(\dfrac{d^2y}{dx^2}\right)^2} - y = 0 \quad \text{(second order)}$$

$$\dfrac{d^4y}{dx^4} = xy \quad \text{(fourth order)}.$$

[†] *Historical Footnote:* In 1630 Galileo formulated the brachistochrone problem ($\beta\rho\acute{\alpha}\chi\acute{\iota}\sigma\tau o\varsigma$ = shortest, $\chi\rho\acute{o}\nu o\varsigma$ = time), that is, to determine a path down which a particle will fall from one given point to another in the shortest time. It was reproposed by John Bernoulli in 1696 and solved by him the following year.

Thus a general form for an *n*th order equation would be

(1) $$F\left(x, y, \frac{dy}{dx}, \ldots, \frac{d^n y}{dx^n}\right) = 0,$$

where *F* is a function of the independent variable *x*, the dependent variable *y*, and the derivatives of *y* up to order *n*; that is, $x, y, \ldots, d^n y/dx^n$. We assume that the equation holds for all *x* in an interval *I* (which may or may not include its endpoints: $a \le x \le b$, $a < x \le b$, etc.). In many cases, we can isolate the highest-order term $d^n y/dx^n$, and write (1) as

(2) $$\frac{d^n y}{dx^n} = f\left(x, y, \frac{dy}{dx}, \ldots, \frac{d^{n-1} y}{dx^{n-1}}\right),$$

which is often preferable to (1) for theoretical and computational purposes.

EXPLICIT SOLUTION

Definition 1. A function $\phi(x)$ that when substituted for *y* in equation (1) [or (2)] satisfies the equation for all *x* in the interval *I* is called an **explicit solution** to the equation on *I*.

EXAMPLE 1 Show that $\phi(x) = x^2 - x^{-1}$ is an explicit solution to

(3) $$\frac{d^2 y}{dx^2} - \frac{2}{x^2} y = 0.$$

Solution The functions $\phi(x) = x^2 - x^{-1}$, $\phi'(x) = 2x + x^{-2}$, and $\phi''(x) = 2 - 2x^{-3}$ are defined for all $x \neq 0$. Substitution of $\phi(x)$ for *y* in equation (3) gives

$$(2 - 2x^{-3}) - \frac{2}{x^2}(x^2 - x^{-1}) = (2 - 2x^{-3}) - (2 - 2x^{-3}) = 0.$$

Since this is valid for any $x \neq 0$, the function $\phi(x) = x^2 - x^{-1}$ is an explicit solution to (3) on $(-\infty, 0)$ and also on $(0, \infty)$. ■ End

EXAMPLE 2 Show that for *any* choice of the constants c_1 and c_2, the function

$$\phi(x) = c_1 e^{-x} + c_2 e^{2x}$$

is an explicit solution to

(4) $$y'' - y' - 2y = 0.$$

Solution We compute $\phi'(x) = -c_1 e^{-x} + 2c_2 e^{2x}$ and $\phi''(x) = c_1 e^{-x} + 4c_2 e^{2x}$. Substitution of ϕ, ϕ', and ϕ'' for *y*, y', and y'' in equation (4) yields

$$(c_1 e^{-x} + 4c_2 e^{2x}) - (-c_1 e^{-x} + 2c_2 e^{2x}) - 2(c_1 e^{-x} + c_2 e^{2x})$$

$$= (c_1 + c_1 - 2c_1)e^{-x} + (4c_2 - 2c_2 - 2c_2)e^{2x} = 0.$$

Since equality holds for all x in $(-\infty, \infty)$, then $\phi(x) = c_1 e^{-x} + c_2 e^{2x}$ is an explicit solution to (4) on the interval $(-\infty, \infty)$ for any choice of the constants c_1 and c_2. ∎ End

As we shall see in Chapter 2, the methods for solving differential equations do not always yield an explicit solution for the equation. We may have to settle for a solution that is defined implicitly. Consider the following example.

EXAMPLE 3 Show that the relation

(5) $y^2 - x^3 + 8 = 0$

implicitly defines a solution to

(6) $\dfrac{dy}{dx} = \dfrac{3x^2}{2y}$

on the interval $(2, \infty)$.

Solution When we solve (5) for y, we obtain $y = \pm\sqrt{x^3 - 8}$. Let's try $\phi(x) = \sqrt{x^3 - 8}$ to see if it is an explicit solution. Since $d\phi/dx = 3x^2/2\sqrt{x^3 - 8}$, both ϕ and $d\phi/dx$ are defined on $(2, \infty)$. Substituting them into (6) yields

$$\frac{3x^2}{2\sqrt{x^3 - 8}} = \frac{3x^2}{2(\sqrt{x^3 - 8})},$$

which is valid for all x in $(2, \infty)$. Hence relation (5) is an implicit solution to (6) on $(2, \infty)$. (The reader can check that $\psi(x) = -\sqrt{x^3 - 8}$ is also an explicit solution to (6).) ∎ End

IMPLICIT SOLUTION

Definition 2. A relation $G(x, y) = 0$ is said to be an **implicit solution** to equation (1) on the interval I if it defines one or more explicit solutions on I.

EXAMPLE 4 Show that

(7) $x + y + e^{xy} = 0$

is an implicit solution to

(8) $(1 + xe^{xy})\dfrac{dy}{dx} + 1 + ye^{xy} = 0.$

Solution First, we observe that we are unable to solve (7) directly for y in terms of x alone. However, for (7) to hold, we realize that any change in x requires a change in y, so we expect the relation (7) to define implicitly at least one function $y(x)$. This is difficult to show directly but can be rigorously verified using the **implicit function theorem**[†] of advanced calculus, which guarantees that such a function $y(x)$ exists that is also differentiable (see Problem 30).

[†] See *Vector Calculus*, Third Edition, by J. E. Marsden and A. J. Tromba (W. H. Freeman and Company, San Francisco, 1988), p. 281.

Once we know that y is a differentiable function of x, we can use the technique of implicit differentiation. Indeed, from (7) we obtain on differentiating with respect to x,

$$1 + \frac{dy}{dx} + e^{xy}\left(y + x\frac{dy}{dx}\right) = 0$$

or

$$(1 + xe^{xy})\frac{dy}{dx} + 1 + ye^{xy} = 0,$$

which is identical to the differential equation (8). Thus relation (7) is an implicit solution on some interval guaranteed by the implicit function theorem. ▪ End

EXAMPLE 5 Verify that $4x^2 - y^2 = C$, where C is an arbitrary constant, gives a one-parameter family of implicit solutions to

(9) $$y\frac{dy}{dx} - 4x = 0$$

and graph several of these solution curves.

Solution When we implicitly differentiate the equation $4x^2 - y^2 = C$ with respect to x, we find

$$8x - 2y\frac{dy}{dx} = 0,$$

which is equivalent to (9). In Figure 1.4 we have sketched the implicit solutions for $C = 0$, ±1, ±4. The curves are hyperbolas with common asymptotes $y = \pm2x$. Notice that the implicit solution curves (with C arbitrary) fill the entire plane and are nonintersecting for $C \neq 0$. For $C = 0$, the implicit solution gives rise to the two explicit solutions $y = 2x$ and $y = -2x$, both of which pass through the origin. ▪ End

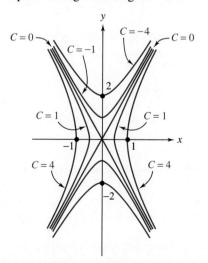

Figure 1.4 Implicit solutions $4x^2 - y^2 = C$

For brevity, we hereafter use the term *solution* to mean either an explicit or an implicit solution.

In the beginning of Section 1.1, we saw that the solution of the *second* order free fall equation invoked two arbitrary constants of integration c_1, c_2:

$$h(t) = \frac{-gt^2}{2} + c_1 t + c_2,$$

while the solution of the *first* order radioactive decay equation contained a single constant C:

$$A(t) = Ce^{-kt}.$$

It is clear that integration of the simple *fourth* order equation

$$\frac{d^4 y}{dx^4} = 0$$

brings in four undetermined constants:

$$y(x) = c_1 x^3 + c_2 x^2 + c_3 x + c_4.$$

It will be shown later in the text that, in general, the methods for solving nth order differential equations evoke n arbitrary constants. In most cases, we will be able to evaluate these constants if we know n initial values $y(x_0)$, $y'(x_0)$, ..., $y^{(n-1)}(x_0)$.

INITIAL VALUE PROBLEM

Definition 3. By an **initial value problem** for an nth order differential equation

$$F\left(x, y, \frac{dy}{dx}, \ldots, \frac{d^n y}{dx^n}\right) = 0,$$

we mean: Find a solution to the differential equation on an interval I that satisfies at x_0 the n initial conditions

$$y(x_0) = y_0,$$

$$\frac{dy}{dx}(x_0) = y_1,$$

$$\vdots$$

$$\frac{d^{n-1} y}{dx^{n-1}}(x_0) = y_{n-1},$$

where $x_0 \in I$ and $y_0, y_1, \ldots, y_{n-1}$ are given constants.

In the case of a first order equation, the initial conditions reduce to the single requirement

$$y(x_0) = y_0,$$

and in the case of a second order equation, the initial conditions have the form

$$y(x_0) = y_0, \qquad \frac{dy}{dx}(x_0) = y_1.$$

The terminology *initial conditions* comes from mechanics, where the independent variable x represents *time* and is customarily symbolized as t. Then if t_0 is the starting time, $y(t_0) = y_0$ represents the initial location of an object and $dy(t_0)/dt$ gives its initial velocity.

EXAMPLE 6 Show that $\phi(x) = \sin x - \cos x$ is a solution to the initial value problem

(10) $$\frac{d^2y}{dx^2} + y = 0; \qquad y(0) = -1, \qquad \frac{dy}{dx}(0) = 1.$$

Solution Observe that $\phi(x) = \sin x - \cos x$, $d\phi/dx = \cos x + \sin x$, and $d^2\phi/dx^2 = -\sin x + \cos x$ are all defined on $(-\infty, \infty)$. Substituting into the differential equation gives

$$(-\sin x + \cos x) + (\sin x - \cos x) = 0,$$

which holds for all $x \in (-\infty, \infty)$. Hence $\phi(x)$ is a solution to the differential equation in (10) on $(-\infty, \infty)$. When we check the initial conditions, we find

$$\phi(0) = \sin 0 - \cos 0 = -1,$$

$$\frac{d\phi}{dx}(0) = \cos 0 + \sin 0 = 1,$$

which meets the requirements of (10). Therefore $\phi(x)$ is a solution to the given initial value problem. ■ End

EXAMPLE 7 As shown in Example 2, the function $\phi(x) = c_1 e^{-x} + c_2 e^{2x}$ is a solution to

$$\frac{d^2y}{dx^2} - \frac{dy}{dx} - 2y = 0$$

for any choice of the constants c_1 and c_2. Determine c_1 and c_2 so that the initial conditions

$$y(0) = 2 \quad \text{and} \quad \frac{dy}{dx}(0) = -3$$

are satisfied.

Solution To determine the constants c_1 and c_2, we first compute $d\phi/dx$ to get $d\phi/dx = -c_1 e^{-x} + 2c_2 e^{2x}$. Substituting in our initial conditions gives the following system of equations:

$$\begin{cases} \phi(0) = c_1 e^0 + c_2 e^0 = 2, \\ \dfrac{d\phi}{dx}(0) = -c_1 e^0 + 2c_2 e^0 = -3, \end{cases} \quad \text{or} \quad \begin{cases} c_1 + c_2 = 2, \\ -c_1 + 2c_2 = -3. \end{cases}$$

Adding the last two equations yields $3c_2 = -1$, so $c_2 = -1/3$. Since $c_1 + c_2 = 2$, we find $c_1 = 7/3$. Hence the solution to the initial value problem is $\phi(x) = (7/3)\,e^{-x} - (1/3)\,e^{2x}$. ■ End

We now state an existence and uniqueness theorem for first order initial value problems. We presume the differential equation has been cast into the format

$$\frac{dy}{dx} = f(x, y).$$

Of course the right-hand side, $f(x, y)$, must be well defined at the starting value x_0 for x and at the stipulated initial value $y_0 = y(x_0)$ for y. The theorem, in fact, requires some continuity hypotheses in some *interval $a < x < b$* containing x_0, and some interval $c < y < d$ containing y_0. Notice that the set of points in the (x, y) plane that satisfy $a < x < b$ and $c < y < d$ constitutes a *rectangle*. Figure 1.5 depicts this ''rectangle of continuity'' with the initial point (x_0, y_0) in its interior and a sketch of a portion of the solution curve contained therein.

EXISTENCE AND UNIQUENESS OF SOLUTION

Theorem 1. Given the initial value problem

$$\frac{dy}{dx} = f(x, y), \qquad y(x_0) = y_0,$$

assume that f and $\partial f / \partial y$ are continuous functions in a rectangle

$$R = \{(x, y): a < x < b, \; c < y < d\}$$

that contains the point (x_0, y_0). Then the initial value problem has a unique solution $\phi(x)$ in some interval $x_0 - h < x < x_0 + h$, where h is a positive number.

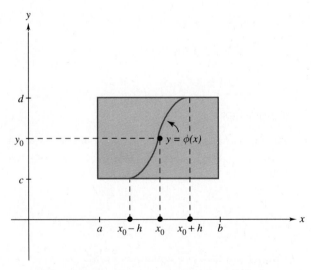

Figure 1.5 Layout for the existence-uniqueness theorem

The preceding theorem tells us two things. First, when an equation satisfies the hypotheses of Theorem 1, we are assured that a solution to the initial value problem exists. Naturally it is desirable to know whether the equation we are trying to solve actually has a solution before we spend too much time trying to solve it. Second, when the hypotheses are satisfied, there is a **unique** solution to the initial value problem. This uniqueness tells us that if we can find a solution, then it is the *only* solution for the initial value problem. Graphically, the theorem says that there is only one solution curve that passes through the point (x_0, y_0). In fact, for this first order equation, two solutions can't cross anywhere in the rectangle. Notice that the existence and uniqueness of the solution holds only in *some* neighborhood $(x_0 - h, x_0 + h)$. Unfortunately, the theorem does not tell us the span $(2h)$ of this neighborhood (merely that it is not zero). Problem 18 elaborates on this feature.

Problem 19 gives an equation with no solution. Problem 29 displays an initial value problem for which the solution is not unique. Of course, the hypotheses of Theorem 1 are not met for these cases.

When initial value problems are used to model physical phenomena, many practitioners tacitly presume the conclusions of Theorem 1 to be valid. Indeed, for the initial value problem to be a reasonable model, we certainly expect it to have a solution, since physically "something does happen." Moreover, the solution should be unique in those cases when repetition of the experiment under identical conditions yields the same results.[†]

The proof of Theorem 1 involves converting the initial value problem into an integral equation and then using Picard's method to generate a sequence of successive approximations that converge to the solution. The conversion to an integral equation and Picard's method are discussed in Project B at the end of this chapter. A detailed discussion and proof of the theorem are given in Chapter 13.[††]

EXAMPLE 8 For the initial value problem

(11) $$\dfrac{dy}{dx} = x^2 - xy^3, \qquad y(1) = 6,$$

does Theorem 1 imply the existence of a unique solution?

Solution Here $f(x, y) = x^2 - xy^3$ and $\partial f/\partial y = -3xy^2$. Both of these functions are continuous in any rectangle containing the point $(1, 6)$, so the hypotheses of Theorem 1 are satisfied. It then follows from Theorem 1 that the initial value problem (11) has a unique solution in an interval about $x = 1$ of the form $(1 - h, 1 + h)$, where h is some positive number. ∎ End

EXAMPLE 9 For the initial value problem

(12) $$\dfrac{dy}{dx} = 3y^{2/3}, \qquad y(2) = 0,$$

does Theorem 1 imply the existence of a unique solution?

[†] At least this is the case when we are considering a deterministic model, as opposed to a probabilistic model.
[††] All references to Chapters 11–13 refer to the expanded text *Fundamentals of Differential Equations and Boundary Value Problems*, 2nd edition.

Solution Here $f(x, y) = 3y^{2/3}$ and $\partial f/\partial y = 2y^{-1/3}$. Unfortunately $\partial f/\partial y$ is not continuous or even defined when $y = 0$. Consequently there is no rectangle containing $(2, 0)$ in which both f and $\partial f/\partial y$ are continuous. Because the hypotheses of Theorem 1 do not hold, we cannot use Theorem 1 to determine whether the initial value problem does or does not have a unique solution. It turns out that this initial value problem does *not* have a unique solution. We refer the reader to Problem 29 for the details. ∎ End

 In Example 9, suppose the initial condition is changed to $y(2) = 1$. Then, since f and $\partial f/\partial y$ are continuous in any rectangle that contains the point $(2, 1)$ but does not intersect the x-axis, say $R = \{(x, y): 0 < x < 10, 0 < y < 5\}$, it follows from Theorem 1 that this *new* initial value problem has a unique solution in some interval about $x = 2$.

EXERCISES 1.2

1. (a) Show that $\phi(x) = 2x^3$ is an explicit solution to

$$x\frac{dy}{dx} = 3y$$

on the interval $(-\infty, \infty)$.
(b) Show that $\phi(x) = e^x - x$ is an explicit solution to

$$\frac{dy}{dx} + y^2 = e^{2x} + (1 - 2x)e^x + x^2 - 1$$

on the interval $(-\infty, \infty)$.
(c) Show that $\phi(x) = x^2 - x^{-1}$ is an explicit solution to $x^2 d^2y/dx^2 = 2y$ on the interval $(0, \infty)$.

2. (a) Show that $y^2 + x - 3 = 0$ is an implicit solution to $dy/dx = -1/(2y)$ on the interval $(-\infty, 3)$.
(b) Show that $xy^3 - xy^3\sin x = 1$ is an implicit solution to

$$\frac{dy}{dx} = \frac{(x\cos x + \sin x - 1)y}{3(x - x\sin x)}$$

on the interval $(0, \pi/2)$.

In Problems 3 through 8, determine whether the given function is a solution to the given differential equation.

3. $y = \sin x + x^2$, $\dfrac{d^2y}{dx^2} + y = x^2 + 2$.

4. $x = 3\cos t - 5\sin t$, $x'' + x = 0$.

5. $x = \cos 2t$, $\dfrac{dx}{dt} + tx = \sin 2t$.

6. $\theta = 2e^{3t} - e^{2t}$, $\dfrac{d^2\theta}{dt^2} - \theta\dfrac{d\theta}{dt} + 3\theta = -2e^{2t}$.

7. $y = e^{2x} - 3e^{-x}$, $\dfrac{d^2y}{dx^2} - \dfrac{dy}{dx} - 2y = 0$.

8. $y = 3\sin 2x + e^{-x}$, $y'' + 4y = 5e^{-x}$.

In Problems 9 through 13, determine whether the given relation is an implicit solution to the differential equation. Assume that the relationship does define y implicitly as a function of x and use implicit differentiation.

9. $x^2 + y^2 = 6$, $\dfrac{dy}{dx} = \dfrac{x}{y}$.

10. $y - \ln y = x^2 + 1$, $\dfrac{dy}{dx} = \dfrac{2xy}{y - 1}$.

11. $e^{xy} + y = x - 1$, $\dfrac{dy}{dx} = \dfrac{e^{-xy} - y}{e^{-xy} + x}$.

12. $x^2 - \sin(x + y) = 1$, $\dfrac{dy}{dx} = 2x\sec(x + y) - 1$.

13. $\sin y + xy - x^3 = 2$, $y'' = \dfrac{6xy' + (y')^3 \sin y - 2(y')^2}{3x^2 - y}$.

14. Show that $\phi(x) = c_1 \sin x + c_2 \cos x$ is a solution to $d^2y/dx^2 + y = 0$ for any choice of the constants c_1 and c_2. Thus $c_1 \sin x + c_2 \cos x$ is a two-parameter family of solutions to the differential equation.

15. Show that $\phi(x) = Ce^{3x} + 1$ is a solution to $dy/dx - 3y = -3$ for any choice of the constant C. Thus $Ce^{3x} + 1$ is a one-parameter family of solutions to the differential equation. Graph several of the solution curves using the same coordinate axes.

16. Verify that $x^2 + cy^2 = 1$, where c is an arbitrary non-zero constant, is a one-parameter family of implicit solutions to

$$\frac{dy}{dx} = \frac{xy}{x^2 - 1}$$

and graph several of the solution curves using the same coordinate axes.

17. Verify that $\phi(x) = 2/(1 - ce^x)$, where c is an arbitrary constant, is a one-parameter family of solutions to

$$\frac{dy}{dx} = \frac{y(y - 2)}{2}.$$

Graph the solution curves corresponding to $c = 0, \pm 1, \pm 2$ using the same coordinate axes.

18. Show that for any choice of the constant C, the function $\phi(x) = (C - x)^{-1}$ is a solution to $dy/dx = y^2$ on the interval (C, ∞) and on $(-\infty, C)$. Moreover, this solution becomes unbounded as x approaches C. (Observe that there is no clue from the differential equation $dy/dx = y^2$ itself, nor from the initial value $y(0) = C^{-1}$, that the solution $\phi(x)$ "blows up" at $x = C$.)

19. Show that the equation $(dy/dx)^2 + y^2 + 3 = 0$ has no (real-valued) solution.

20. Determine for which values of m the function $\phi(x) = e^{mx}$ is a solution to the given equation.

(a) $\dfrac{d^2y}{dx^2} + 6\dfrac{dy}{dx} + 5y = 0.$

(b) $\dfrac{d^3y}{dx^3} + 3\dfrac{d^2y}{dx^2} + 2\dfrac{dy}{dx} = 0.$

21. Determine for which values of m the function $\phi(x) = x^m$ is a solution to the given equation.

(a) $3x^2\dfrac{d^2y}{dx^2} + 11x\dfrac{dy}{dx} - 3y = 0.$

(b) $x^2\dfrac{d^2y}{dx^2} - x\dfrac{dy}{dx} - 5y = 0.$

22. Verify that the function $\phi(x) = c_1e^{-x} + c_2e^{2x}$ is a solution to

$$\frac{d^2y}{dx^2} - \frac{dy}{dx} - 2y = 0$$

for any choice of the constants c_1 and c_2. Determine c_1 and c_2 so that each of the following initial conditions is satisfied.

(a) $y(0) = 2, \quad y'(0) = 1.$

(b) $y(1) = 1, \quad y'(1) = 0.$

In Problems 23 through 28, determine whether Theorem 1 implies that the given initial value problem has a unique solution.

23. $\dfrac{dy}{dx} = x^3 - y^3, \qquad\qquad y(0) = 6.$

24. $\dfrac{dy}{d\theta} - \theta y = \sin^2\theta, \qquad\quad y(\pi) = 5.$

25. $x\dfrac{dx}{dt} + 4t = 0, \qquad\qquad x(2) = -\pi.$

26. $\dfrac{dy}{dx} = 3x - \sqrt[3]{y - 1}, \qquad y(2) = 1.$

27. $y\dfrac{dy}{dx} = x, \qquad\qquad\qquad y(1) = 0.$

28. $\dfrac{dx}{dt} + \cos x = \sin t, \qquad x(\pi) = 0.$

29. For the initial value problem (12) of Example 9 show that $\phi_1(x) \equiv 0$ and $\phi_2(x) = (x - 2)^3$ are solutions. Hence the initial value problem does not have a *unique* solution.

30. Implicit Function Theorem. *Let $G(x, y)$ have continuous first partial derivatives in the rectangle $R = \{(x, y): a < x < b, c < y < d\}$ containing the point (x_0, y_0). If $G(x_0, y_0) = 0$ and $G_y(x_0, y_0) \neq 0$, then there exists a differentiable function $y = \phi(x)$, defined in some interval $I = (x_0 - h, x_0 + h)$, that satisfies $G(x, \phi(x)) = 0$ for all $x \in I$.*

The implicit function theorem gives conditions under which the relationship $G(x, y) = 0$ defines y implicitly as a function of x. Use the implicit function theorem to show that the relationship $x + y + e^{xy} = 0$, given in Example 4, defines y implicitly as a function of x near the point $(0, -1)$.

31. Consider the equation of Example 5,

$$(13) \quad y\frac{dy}{dx} - 4x = 0.$$

(a) Does Theorem 1 imply the existence of a unique solution to (13) that satisfies $y(x_0) = 0$?

(b) Show that when $x_0 \neq 0$, equation (13) can't possibly have a solution in a neighborhood of $x = x_0$ that satisfies $y(x_0) = 0$.

(c) Show that there are two distinct solutions to (13) satisfying $y(0) = 0$ (see Figure 1.4 on page 9).

1.3 DIRECTION FIELDS

The existence and uniqueness theorem discussed in the preceding section certainly has great value, but it stops short of telling us anything about the *nature* of the solution to a differential equation. For practical reasons, we may need to know the value of the solution at a certain point or the intervals where the solution is increasing or the points where the solution attains a maximum value. Certainly knowing an explicit representation (a formula) for the solution would be a considerable help in answering these questions. However, for many of the differential equations that the reader is likely to encounter in "real-world" applications, it will be impossible to find such a formula. Moreover, even if we are lucky enough to obtain an implicit solution, it may be difficult to use this relationship to determine an explicit form. Thus we must rely on other methods to analyze and approximate the solution.

One technique that is useful in visualizing (graphing) the solutions to a first order differential equation is to sketch the direction field for the equation. To describe this method, we need to make a general observation. Namely, a first order equation

$$\frac{dy}{dx} = f(x, y)$$

specifies a slope at each point in the *xy*-plane where *f* is defined. In other words, it gives the direction that a solution to the equation must have at each point. Consider, for example, the equation

(1) $\frac{dy}{dx} = x^2 - y.$

The graph of a solution to (1) that passes through the point $(-2, 1)$ must have slope $(-2)^2 - 1 = 3$ at that point, and a solution through $(-1, 1)$ has zero slope at that point.

A plot of short line segments drawn at various points in the *xy*-plane showing the slope of the solution curve there is called a **direction field** for the differential equation. Because the direction field gives the "flow of solutions," it facilitates the drawing of any particular solution (such as the solution to an initial value problem). In Figure 1.6(a) we have sketched the direction field for the equation (1) and in Figure 1.6(b) drawn several solution curves in color.

Some other interesting direction field patterns are displayed in Figure 1.7. Figure 1.7(a) depicts the pattern for the radioactive decay equation $dy/dx = -2y$ (recall that in Section 1.1, we analyzed this equation in the form $dA/dt = -kA$). From the flow patterns, we can see that all solutions tend asymptotically to the positive *x*-axis as *x* gets larger. In other words, any material decaying according to this law eventually dwindles to practically nothing. This is consistent with the solution formula we derived earlier,

$$A = Ce^{-kt}, \text{ or } y = Ce^{-2x}.$$

From the direction field in Figure 1.7(b), we can anticipate that all solutions to $dy/dx = -y/x$ also approach the *x*-axis as *x* approaches infinity (plus or minus infinity, in fact). But more interesting is the observation that no solution can make it across the *y*-axis;

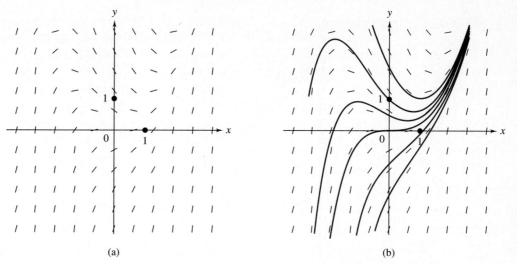

(a)

(b)

Figure 1.6 (a) Direction field for $dy/dx = x^2 - y$ (b) Solutions to $dy/dx = x^2 - y$

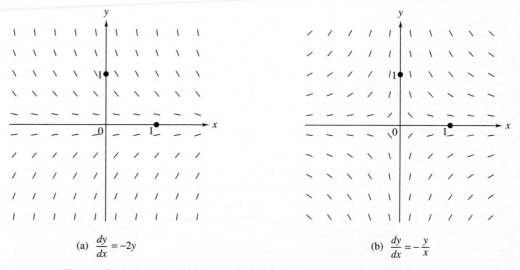

(a) $\dfrac{dy}{dx} = -2y$

(b) $\dfrac{dy}{dx} = -\dfrac{y}{x}$

Figure 1.7 (a) Direction field for $dy/dx = -2y$ (b) Direction field for $dy/dx = -y/x$

$|y(x)|$ ''blows up'' as x goes to zero from either direction. Exception: On close examination, it appears the function $y(x) \equiv 0$ might just make it through this barrier. As a matter of fact, in Problem 19 the reader is invited to show that the solutions to this differential equation are given by $y = C/x$, with C an arbitrary constant. So they do diverge at $x = 0$, unless $C = 0$.

Let's interpret the existence-uniqueness theorem of Section 1.2 for these direction fields. For Figure 1.7(a), where $dy/dx = f(x, y) = -2y$, we select a starting point x_0 and an

initial value $y(x_0) = y_0$, as in Figure 1.8(a). Because the right-hand side $f(x, y) = -2y$ is continuously differentiable for all x and y, we can enclose any initial point (x_0, y_0) in a "rectangle of continuity." We conclude that the equation has one and only one solution curve passing through (x_0, y_0), as depicted in the figure.

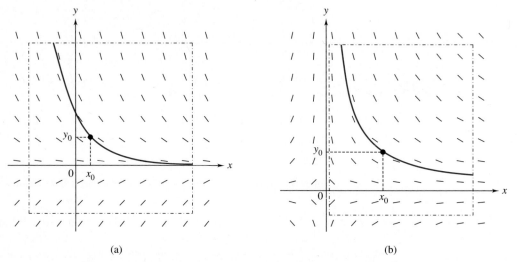

(a)　　　　　　　　　　　　　　　　　(b)

Figure 1.8 (a) A solution for $dy/dx = -2y$　(b) A solution for $dy/dx = -y/x$

For the equation

$$\frac{dy}{dx} = f(x, y) = -\frac{y}{x},$$

the right-hand side $f(x, y) = -y/x$ does not meet the continuity conditions when $x = 0$ (i.e., for points on the y-axis). However, for any *nonzero* starting value x_0 and *any* initial value $y(x_0) = y_0$, we *can* enclose (x_0, y_0) in a rectangle of continuity that excludes the y-axis, as in Figure 1.8(b). Thus we can be assured of one and only one solution curve passing through such a point.

The direction field for the equation

$$\frac{dy}{dx} = 3y^{2/3}$$

is intriguing because Example 9 of the preceding section showed that the hypotheses of Theorem 1 do *not* hold in any rectangle enclosing the initial point $x_0 = 2$, $y_0 = 0$. Indeed Problem 29 of that section demonstrated the violation of uniqueness by exhibiting *two* solutions, $y(x) \equiv 0$ and $y(x) = (x - 2)^3$, passing through $(2, 0)$. Figure 1.9(a) displays this direction field, and Figure 1.9(b) demonstrates how both solution curves can successfully "negotiate" this flow pattern.

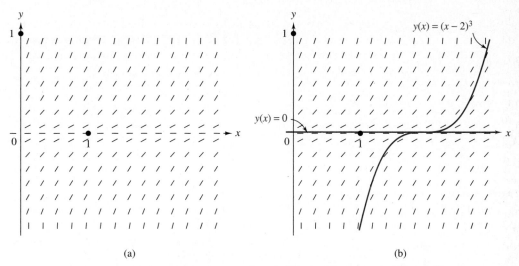

Figure 1.9 (a) Direction field for $dy/dx = 3y^{2/3}$ (b) Solutions for $dy/dx = 3y^{2/3}$

Clearly a sketch of the direction field of a first order differential equation can be helpful in visualizing the solutions. However, such a sketch is not sufficient to enable us to trace, unambiguously, the solution curve passing through a given initial point (x_0, y_0). If we tried to trace one of the solution curves in Figure 1.6(b) on page 17, for example, we could easily "slip" over to an adjacent curve. For nonunique situations like that in Figure 1.9(b), as one negotiates the flow along the curve $y = (x - 2)^3$, and reaches the inflection point, one cannot decide whether to turn or to – literally – go off on the tangent ($y = 0$).

EXAMPLE 1 The logistic equation for the population p (in thousands) at time t of a certain species is given by

(2) $$\frac{dp}{dt} = p(2 - p).$$

(Of course p is nonnegative. The interpretation of the terms in the logistic equation are discussed in Section 3.2.) From the direction field sketched in Figure 1.10 on page 20, answer the following.

(a) If the initial population is 3000 (that is, $p(0) = 3$), what can you say about the limiting population $\lim_{t\to\infty} p(t)$?

(b) Can a population of 1000 ever decline to 500?

(c) Can a population of 1000 ever increase to 3000?

Solution

(a) The direction field indicates that all solution curves (other than $p(t) \equiv 0$) will approach the horizontal line $p = 2$ as $t \to \infty$, that is, this line is an asymptote for all positive solutions. Thus $\lim_{t\to\infty} p(t) = 2$.

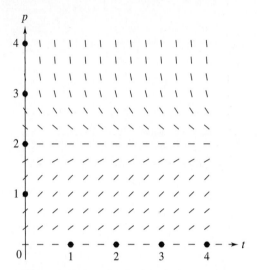

Figure 1.10 Direction field for logistic equation

(b) The direction field further shows that populations greater than 2000 will steadily decrease, while those less than 2000 will increase. In particular, a population of 1000 can never decline to 500.

(c) As mentioned in part (b), a population of 1000 will increase with time. But the direction field indicates it can never reach 2000 or any larger value, i.e., the solution curve cannot cross the line $p = 2$. Indeed the constant function $p(t) \equiv 2$ is a solution to (2), and the uniqueness part of Theorem 1, page 12, precludes intersections of solution curves. ■ End

Notice that the direction field in Figure 1.10 has the nice feature that the slopes do not depend on t, that is, the pattern of slopes is the same along each vertical line. The same is true for Figures 1.8(a) and 1.9. This is the key property of so-called *autonomous equations* $y' = f(y)$, where the right-hand side is a function of the dependent variable only. (See also Project E, Chapter 2.)

Hand sketching the direction field for a differential equation is often tedious. Fortunately several software programs have been developed to obviate this task.[†] When hand sketching is necessary, however, the **method of isoclines** can be helpful in reducing the work.

The Method of Isoclines

An isocline for the differential equation

$$y' = f(x, y)$$

[†] There are several microcomputer programs available for drawing direction fields, such as MACMATH by John H. Hubbard and Beverly H. West, available from Springer-Verlag, New York, and PHASEPLANE by Brad Ermentrout, available from Prindle, Weber & Schmidt, Boston.

is a set of points in the xy-plane where all the solutions have the same slope dy/dx; thus it is a level curve for the function $f(x, y)$. For example, if

(3) $y' = f(x, y) = x + y,$

the isoclines are simply the curves (straight lines) $x + y = c$ or $y = -x + c$. Here c is an arbitrary constant. But c can be interpreted as the numerical value of the slope dy/dx of every solution curve as it crosses the isocline. (Note that c is not the slope of the isocline itself; the latter is, obviously, -1.) Figure 1.11(a) depicts the isoclines for equation (3).

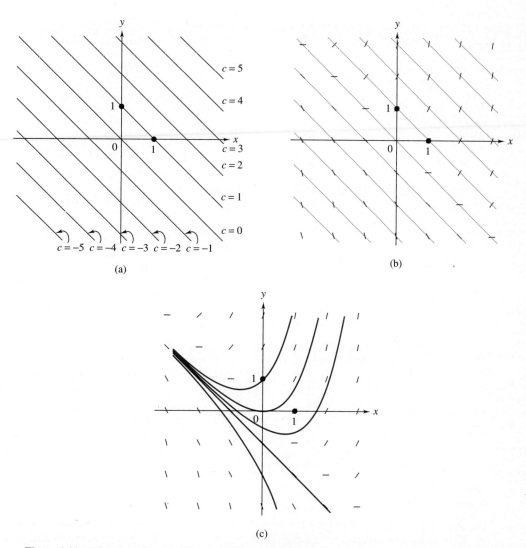

(a)

(b)

(c)

Figure 1.11 (a) Isoclines for $y' = x + y$ (b) Direction field for $y' = x + y$ (c) Solutions to $y' = x + y$

To implement the method of isoclines for sketching direction fields, we draw hash marks with slope c along the isocline $f(x, y) = c$ for a few selected values of c. If we then erase the underlying isocline curves, the hash marks constitute a part of the direction field for the differential equation. Figure 1.11(b) depicts this process for the isoclines shown in Figure 1.11(a), and Figure 1.11(c) displays some solution curves.

Remark. The isoclines themselves are not always straight lines. For equation (1) at the beginning of this section (page 16), they are parabolas $x^2 - y = c$. When the isocline curves are complicated, this method is not recommended.

EXERCISES 1.3

1. The direction field for $dy/dx = 4x/y$ is shown in Figure 1.12.

 (a) Verify that the straight lines $y = \pm 2x$ are solution curves, provided $x \neq 0$.
 (b) Sketch the solution curve with initial condition $y(0) = 2$.
 (c) Sketch the solution curve with initial condition $y(2) = 1$.
 (d) What can you say about the behavior of the above solutions as $x \to \infty$? How about $x \to -\infty$?

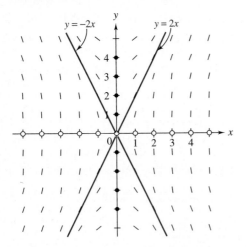

Figure 1.12 Direction field for $dy/dx = 4x/y$

2. The direction field for $dy/dx = 2x + y$ is shown in Figure 1.13.

 (a) Sketch the solution curve that passes through $(0, -2)$. From this sketch write the equation for the solution.

 (b) Sketch the solution curve that passes through $(-1, 3)$.
 (c) What can you say about the solution in part (b) as $x \to \infty$? How about $x \to -\infty$?

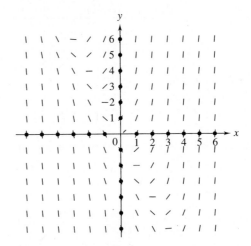

Figure 1.13 Direction field for $dy/dx = 2x + y$

3. A model for the velocity v at time t of a certain object falling under the influence of gravity in a viscous medium is given by the equation

$$\frac{dv}{dt} = 1 - \frac{v}{8}.$$

From the direction field shown in Figure 1.14, sketch the solutions with the initial conditions $v(0) = 5, 8$, and 15. Why is the value $v = 8$ called the "terminal velocity"?

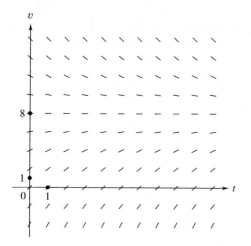

Figure 1.14 Direction field for $\dfrac{dv}{dt} = 1 - \dfrac{v}{8}$.

4. If the viscous force in Problem 3 is nonlinear, a possible model would be provided by the differential equation

$$\frac{dv}{dt} = 1 - \frac{v^3}{8}.$$

Redraw the direction field in Figure 1.14 to incorporate this v^3 dependence. Sketch the solutions with initial conditions $v(0) = 0, 1, 2, 3$. What is the terminal velocity in this case?

5. The logistic equation for the population (in thousands) of a certain species is given by

$$\frac{dp}{dt} = 3p - 2p^2.$$

(a) Sketch the direction field by using either a computer software package or the method of isoclines.
(b) If the initial population is 2000 (that is, $p(0) = 2$), what can you say about the limiting population $\lim_{t \to \infty} p(t)$?
(c) If $p(0) = 0.5$, what is $\lim_{t \to \infty} p(t)$?
(d) Can a population of 3000 ever decline to 500?

6. Consider the differential equation

$$\frac{dy}{dx} = x + \sin y.$$

(a) A solution curve passes through the point $(1, \pi/2)$. What is its slope at this point?
(b) Argue that every solution curve is increasing for $x > 1$.

(c) Show that the second derivative of every solution satisfies

$$\frac{d^2y}{dx^2} = 1 + x \cos y + \frac{1}{2} \sin 2y.$$

(d) A solution curve passes through $(0, 0)$. Prove that this curve has a relative minimum at $(0, 0)$.

7. Consider the differential equation

$$\frac{dp}{dt} = p(p - 1)(2 - p)$$

for the population p (in thousands) of a certain species at time t.

(a) Sketch the direction field by using either a computer software package or the method of isoclines.
(b) If the initial population is 3000 (that is, $p(0) = 3$), what can you say about the limiting population $\lim_{t \to \infty} p(t)$?
(c) If $p(0) = 1.5$, what is $\lim_{t \to \infty} p(t)$?
(d) If $p(0) = 0.5$, what is $\lim_{t \to \infty} p(t)$?
(e) Can a population of 900 ever increase to 1100?

8. The motion of a set of particles moving along the x-axis is governed by the differential equation

$$\frac{dx}{dt} = t^3 - x^3,$$

where $x(t)$ denotes the position at time t of the particle.

(a) If a particle resides at $x = 1$ when $t = 2$, what is its velocity at this time?
(b) Show that the acceleration of a particle is given by

$$\frac{d^2x}{dt^2} = 3t^2 - 3t^3x^2 + 3x^5.$$

(c) If a particle is located at $x = 2$ when $t = 2.5$, can it reach the location $x = 1$ at any later time? [*Hint:* $t^3 - x^3 = (t - x)(t^2 + xt + x^2)$.]

9. Let $\phi(x)$ denote the solution to the initial value problem

$$\frac{dy}{dx} = x - y, \qquad y(0) = 1.$$

(a) Show that $\phi''(x) = 1 - \phi'(x) = 1 - x + \phi(x)$.
(b) Argue that the graph of ϕ is decreasing for x near zero and that as x increases from zero, $\phi(x)$ decreases until it crosses the line $y = x$, where its derivative is zero.
(c) Let x^* be the abscissa of the point where the solution curve $y = \phi(x)$ crosses the line $y = x$. Consider the sign of $\phi''(x^*)$ and argue that ϕ has a relative minimum at x^*.

(d) What can you say about the graph of $y = \phi(x)$ for $x > x^*$?

(e) Verify that $y = x - 1$ is a solution to $dy/dx = x - y$ and explain why the graph of $\phi(x)$ always stays above the line $y = x - 1$.

(f) Sketch the direction field for $dy/dx = x - y$ by using the method of isoclines or a computer software package.

(g) Sketch the solution $y = \phi(x)$ using the direction field in part (f).

10. Use a computer software package to sketch the direction field for the following differential equations. Sketch some of the solution curves.

 (a) $dy/dx = \sin x$.
 (b) $dy/dx = \sin y$.
 (c) $dy/dx = \sin x \sin y$.
 (d) $dy/dx = x^2 + 2y^2$.
 (e) $dy/dx = x^2 - 2y^2$.

In Problems 11 through 16, draw the isoclines with their direction markers and sketch several solution curves, including the curve satisfying the given initial conditions.

11. $dy/dx = 2x$, $y(0) = -1$.

12. $dy/dx = y$, $y(0) = 1$.

13. $dy/dx = -x/y$, $y(0) = 4$.

14. $dy/dx = x/y$, $y(0) = -1$.

15. $dy/dx = 2x^2 - y$, $y(0) = 0$.

16. $dy/dx = x + 2y$, $y(0) = 1$.

17. From a sketch of the direction field, what can one say about the behavior as x approaches ∞ of a solution to

$$\frac{dy}{dx} = 3 - y + \frac{1}{x}?$$

18. From a sketch of the direction field, what can one say about the behavior as x approaches ∞ of a solution to

$$\frac{dy}{dx} = -y?$$

19. By rewriting the differential equation $dy/dx = -y/x$ in the form

$$\frac{1}{y}\,dy = \frac{-1}{x}\,dx,$$

integrate both sides to obtain the solution $y = C/x$ for an arbitrary constant C.

1.4 THE APPROXIMATION METHOD OF EULER

Euler's method (or the tangent line method) is a procedure for constructing approximate solutions to an initial value problem for a first order differential equation

(1) $y' = f(x, y)$, $y(x_0) = y_0$.

It could be described as a "mechanical" or "computerized" implementation of the informal procedure for hand sketching the solution curve from a picture of the direction field. As such, it remains subject to the failing that it may skip across solution curves. However, under fairly general conditions, iterations of the procedure do converge to true solutions.

The method is illustrated in Figure 1.15. Starting at the initial point (x_0, y_0), we follow the straight line with slope $f(x_0, y_0)$, the tangent line, for some distance to the point (x_1, y_1). Then we reset the slope to the value $f(x_1, y_1)$ and follow this line to (x_2, y_2). In this way, we construct polygonal (broken line) approximations to the solution. As we take smaller spacings between points (and thus employ more points), we may expect to converge to the true solution.

To be more precise, assume the initial value problem (1) has a unique solution $\phi(x)$ in some interval centered at x_0. Let h be a fixed positive number (called the *step size*) and consider the equally spaced points[†]

$$x_n := x_0 + nh, \qquad n = 0, 1, 2, \ldots.$$

[†] The symbol $:=$ means "is defined to be."

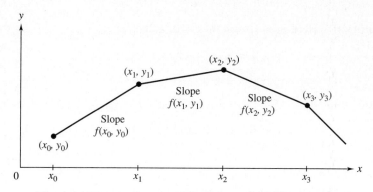

Figure 1.15 Polygonal line approximation given by Euler's method

The construction of values y_n that approximate the solution values $\phi(x_n)$ proceeds as follows. At the point (x_0, y_0), the slope of the solution to (3) is given by $dy/dx = f(x_0, y_0)$. Hence the tangent line to the solution curve at the initial point (x_0, y_0) is

$$y = y_0 + (x - x_0)f(x_0, y_0).$$

Using this tangent line to approximate $\phi(x)$, we find that for the point $x_1 = x_0 + h$

$$\phi(x_1) \approx y_1 := y_0 + hf(x_0, y_0).$$

Next, starting at the point (x_1, y_1), we construct the line with slope given by the direction field at the point (x_1, y_1); that is, with slope equal to $f(x_1, y_1)$. If we follow this line[†] (namely, $y = y_1 + (x - x_1)f(x_1, y_1)$) in stepping from x_1 to $x_2 = x_1 + h$, we arrive at the approximation

$$\phi(x_2) \approx y_2 := y_1 + hf(x_1, y_1).$$

Repeating the process (as illustrated in Figure 1.15), we get

$$\phi(x_3) \approx y_3 := y_2 + hf(x_2, y_2),$$
$$\phi(x_4) \approx y_4 := y_3 + hf(x_3, y_3), \qquad \text{etc.}$$

This simple procedure is **Euler's method** and can be summarized by the recursive formulas

(2) $x_{n+1} = x_n + h,$

(3) $y_{n+1} = y_n + hf(x_n, y_n), \qquad n = 0, 1, 2, \ldots.$

EXAMPLE 1 Use Euler's method with step size $h = 0.1$ to approximate the solution to the initial value problem

(4) $y' = x\sqrt{y}, \qquad y(1) = 4$

at the points $x = 1.1, 1.2, 1.3, 1.4,$ and $1.5.$

[†] Because y_1 is an approximation to $\phi(x_1)$, we cannot assert that this line is tangent to the solution curve $y = \phi(x)$.

Solution Here $x_0 = 1$, $y_0 = 4$, $h = 0.1$, and $f(x, y) = x\sqrt{y}$. Thus the recursive formula (3) for y_n is

$$y_{n+1} = y_n + hf(x_n, y_n) = y_n + (0.1)x_n\sqrt{y_n}.$$

Substituting $n = 0$, we get

$$x_1 = x_0 + 0.1 = 1 + 0.1 = 1.1,$$
$$y_1 = y_0 + (0.1)x_0\sqrt{y_0} = 4 + (0.1)(1)\sqrt{4} = 4.2.$$

Putting $n = 1$ yields

$$x_2 = x_1 + 0.1 = 1.1 + 0.1 = 1.2,$$
$$y_2 = y_1 + (0.1)x_1\sqrt{y_1} = 4.2 + (0.1)(1.1)\sqrt{4.2} = 4.42543.$$

Continuing in this manner, we obtain the results listed in Table 1.1. For comparison, we have included the exact value (to five decimal places) of the solution $\phi(x) = (x^2 + 7)^2/16$ to (4), which can be obtained using separation of variables (see Section 2.2). As one might expect, the approximation deteriorates as x moves further away from 1. ■ End

TABLE 1.1	COMPUTATIONS FOR		
$y' = x\sqrt{y}$, $y(1) = 4$			
n	x_n	Euler's Method	Exact Value
0	1	4	4
1	1.1	4.2	4.21276
2	1.2	4.42543	4.45210
3	1.3	4.67787	4.71976
4	1.4	4.95904	5.01760
5	1.5	5.27081	5.34766

Given the initial value problem (1) and a specific point x, how can Euler's method be used to approximate $\phi(x)$? Starting at x_0, we can take one giant step that lands on x, or we can take several smaller steps to arrive at x. If we wish to take N steps, then we set $h = (x - x_0)/N$ so that the step size h and the number of steps N are related in a specific way. For example, if $x_0 = 1.5$ and we wish to approximate $\phi(2)$ using 10 steps, then we would take $h = (2 - 1.5)/10 = 0.05$. It is expected that the more steps we take, the better will be the approximation. (But keep in mind that more steps mean more computations and hence greater accumulated roundoff error.)

EXAMPLE 2 Use Euler's method to find approximates to the solution of the initial value problem

(5) $y' = y$, $y(0) = 1$

at $x = 1$, taking 1, 2, 4, 8, and 16 steps.

Remark. Observe that the solution to (5) is just $\phi(x) = e^x$, so Euler's method will generate algebraic approximations to the transcendental number $e = 2.71828 \ldots .$

Solution Here $f(x, y) = y$, $x_0 = 0$, and $y_0 = 1$. The recursive formula for Euler's method is

$$y_{n+1} = y_n + hy_n = (1 + h)y_n.$$

To obtain approximations at $x = 1$ with N steps, we take the step size $h = 1/N$. For $N = 1$, we have

$$\phi(1) \approx y_1 = (1 + 1)(1) = 2.$$

For $N = 2$, $\phi(x_2 = 1) \approx y_2$. In this case, we get

$$y_1 = (1 + 0.5)(1) = 1.5,$$
$$\phi(1) \approx y_2 = (1 + 0.5)(1.5) = 2.25.$$

For $N = 4$, $\phi(x_4 = 1) \approx y_4$, where

$$y_1 = (1 + 0.25)(1) = 1.25,$$
$$y_2 = (1 + 0.25)(1.25) = 1.5625,$$
$$y_3 = (1 + 0.25)(1.5625) = 1.95313,$$
$$\phi(1) \approx y_4 = (1 + 0.25)(1.95313) = 2.44141.$$

(In the above computations, we have rounded to five decimal places.) Similarly, taking $N = 8$ and 16, we obtain even better estimates for $\phi(1)$. These approximations are shown in Table 1.2. For comparison, Figure 1.16 on page 28 displays the polygonal line approximations to e^x using Euler's method with $h = 1/4$ ($N = 4$) and $h = 1/8$ ($N = 8$). Notice that the smaller step size yields the better approximation. ■ End

TABLE 1.2 EULER'S METHOD FOR $y' = y$, $y(0) = 1$		
N	h	Approximation for $\phi(1) = e$
1	1.0	2.0
2	0.5	2.25
4	0.25	2.44141
8	0.125	2.56578
16	0.0625	2.63793

How good (or bad) is Euler's method? In judging a numerical scheme, we must begin with two fundamental questions. Does the method converge? And, if so, what is the rate of convergence? These important issues are discussed in Section 3.5, where improvements in Euler's method are introduced (see also Problems 12 and 13).

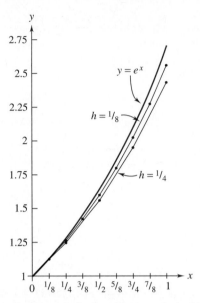

Figure 1.16 Approximations of e^x using Euler's method with $h = 1/4$ and $1/8$

EXERCISES 1.4

In many of the problems below, it will be helpful to have a calculator or microcomputer available. The reader may also find it convenient to write a program for solving initial value problems using Euler's method. (Remember, all trigonometric calculations are done in radians.)

In Problems 1 through 4, use Euler's method to approximate the solution to the given initial value problem at the points $x = 0.1, 0.2, 0.3, 0.4,$ and 0.5 using steps of size 0.1 $(h = 0.1)$.

1. $dy/dx = x/y$, $y(0) = -1$.

2. $dy/dx = -x/y$, $y(0) = 4$.

3. $dy/dx = y(2 - y)$, $y(0) = 3$.

4. $dy/dx = x + y$, $y(0) = 1$.

5. Use Euler's method with step size $h = 0.2$ to approximate the solution to the initial value problem

$$y' = \frac{1}{x}(y^2 + y), \qquad y(1) = 1$$

at the points $x = 1.2, 1.4, 1.6,$ and 1.8.

6. Use Euler's method with step size $h = 0.1$ to approximate the solution to the initial value problem

$$y' = x - y^2, \qquad y(1) = 0$$

at the points $x = 1.1, 1.2, 1.3, 1.4,$ and 1.5.

7. Use Euler's method to find approximates to the solution of the initial value problem

$$\frac{dx}{dt} = 1 + t \sin(tx), \qquad x(0) = 0$$

at $t = 1$, taking 1, 2, 4, and 8 steps.

8. Use Euler's method to find approximates to the solution of the initial value problem

$$y' = 1 - \sin y, \qquad y(0) = 0$$

at $x = \pi$, taking 1, 2, 4, and 8 steps.

9. Use Euler's method with $h = 0.1$ to approximate the solution to the initial value problem

$$y' = \frac{1}{x^2} - \frac{y}{x} - y^2, \qquad y(1) = -1$$

on the interval $1 \le x \le 2$. Compare these approximations with the actual solution $y = -1/x$ (verify!) by

graphing the polygonal line approximation and the actual solution on the same coordinate system.

10. Use Euler's method with $h = 0.1$ to approximate the solution to the initial value problem

$$y' = x - y, \qquad y(0) = 0$$

on the interval $0 \leq x \leq 1$. Compare these approximations with the actual solution $y = e^{-x} + x - 1$ (verify!) by graphing the polygonal line approximation and the actual solution on the same coordinate system.

11. Use Euler's method with 20 steps to approximate the solution to the initial value problem

$$\frac{dx}{dt} = 1 + x^2, \qquad x(0) = 0$$

at $t = 1$. Compare the approximation with the actual solution $x = \tan t$ (verify!) evaluated at $t = 1$.

12. In Example 2, we approximated the transcendental number e by using Euler's method to solve the initial value problem

$$y' = y, \qquad y(0) = 1.$$

Show that the Euler approximation y_n obtained by using the step size $1/n$ is given by the formula

$$y_n = \left(1 + \frac{1}{n}\right)^n, \qquad n = 1, 2, \ldots.$$

Recall from calculus that

$$\lim_{n \to \infty} \left(1 + \frac{1}{n}\right)^n = e,$$

and hence Euler's method converges (theoretically) to the correct value.

13. Prove that the "rate of convergence" for Euler's method in Problem 12 is comparable to $1/n$ by showing that

$$\lim_{n \to \infty} \frac{e - y_n}{1/n} = \frac{e}{2}.$$

[*Hint*: Use L'Hôpital's rule and the Maclaurin expansion for $\ln(1 + t)$.]

14. Use Euler's method with the spacings $h = 0.5$, 0.1, 0.05, 0.01 to approximate the solution to the initial value problem

$$y' = y^2, \qquad y(0) = 1$$

on the interval $0 \leq x \leq 2$. (The explanation for the erratic results lies in Problem 18 of Exercises 1.2.)

Heat Exchange. *There are basically two mechanisms by which a physical body exchanges heat with its environment. The* **contact** *heat transfer across the body's surface is driven by the difference in the body's temperature and that of the environment; this is known as Newton's law of cooling. However, heat transfer also occurs due to* **thermal radiation**, *which according to Stefan's law of radiation is governed by the difference of the fourth powers of these temperatures. In most cases, one of these modes dominates the other. Problems 15 and 16 invite the reader to simulate each mode numerically for a given set of initial conditions.*

15. Newton's Law of Cooling. Newton's law of cooling states that the rate of change in the temperature $T(t)$ of a body is proportional to the difference between the temperature of the medium $M(t)$ and the temperature of the body. That is,

$$\frac{dT}{dt} = K[M(t) - T(t)],$$

where K is a constant. Let $K = 1$ $(\text{min})^{-1}$ and the temperature of the medium be constant, $M(t) \equiv 70°$. If the body is initially at $100°$, use Euler's method with $h = 0.1$ to approximate the temperature of the body after

(a) 1 minute.

(b) 2 minutes.

16. Stefan's Law of Radiation. Stefan's law of radiation states that the rate of change in temperature of a body at $T(t)$ degrees in a medium at $M(t)$ degrees is proportional to $M^4 - T^4$. That is,

$$\frac{dT}{dt} = K(M^4 - T^4),$$

where K is a constant. Let $K = (40)^{-4}$ and assume the medium temperature is constant, $M(t) \equiv 70°$. If $T(0) = 100°$, use Euler's method with $h = 0.1$ to approximate $T(1)$ and $T(2)$.

CHAPTER SUMMARY

In this chapter, we introduced some basic terminology for differential equations. The **order** of a differential equation is the order of the highest derivative present. The subject of this text is **ordinary** differential equations, which involve derivatives with respect to a single independent variable. Such equations are classified as **linear** or **nonlinear**.

An **explicit solution** of a differential equation is a function that satisfies the equation on some interval. An **implicit solution** is a relation that implicitly defines a function that is an explicit solution. A differential equation typically has infinitely many solutions. In contrast, there are theorems that ensure a unique solution exists for certain **initial value problems** in which one must find a solution to the differential equation that also satisfies given initial conditions. For an nth order equation, these conditions refer to the values of the solution and its first $n - 1$ derivatives at some point.

Even if one is not successful in finding explicit solutions to a differential equation, several techniques can be used to help analyze the solutions. One such method for first order equations views the differential equation $dy/dx = f(x, y)$ as specifying directions (slopes) at points on the plane. The conglomerate of such slopes is the **direction field** for the equation. Knowing the "flow of solutions" is helpful in sketching the solution to an initial value problem. Furthermore, carrying out this method algebraically leads to numerical approximations to the desired solution. This numerical process is called **Euler's method**.

TECHNICAL WRITING EXERCISES

1. Select four fields (for example, astronomy, geology, biology, and economics) and for each field discuss a situation in which differential equations are used to solve a problem. Select examples that are not covered in Section 1.1.

2. Compare the different types of solutions discussed in this chapter—explicit, implicit, graphical, and numerical. What are advantages and disadvantages of each?

GROUP PROJECTS FOR CHAPTER 1

A. Taylor Series Method

Euler's method is based on the fact that the tangent line gives a good local approximation for the function. But why restrict ourselves to linear approximants when higher-degree polynomial approximants are available? For example, we can use the **Taylor polynomial** of degree n about $x = x_0$, which is defined by

$$P_n(x) := y(x_0) + y'(x_0)(x - x_0) + \frac{y''(x_0)}{2!}(x - x_0)^2 + \cdots + \frac{y^{(n)}(x_0)}{n!}(x - x_0)^n.$$

This polynomial is the nth partial sum of the Taylor series representation

$$\sum_{k=0}^{\infty} \frac{y^{(k)}(x_0)}{k!}(x - x_0)^k.$$

To determine the Taylor series for the solution $\phi(x)$ to the initial value problem

$$dy/dx = f(x, y), \qquad y(x_0) = y_0,$$

we need only determine the values of the derivatives of ϕ (assuming they exist) at x_0; that is, $\phi(x_0)$, $\phi'(x_0), \ldots$. The initial condition gives the first value $\phi(x_0) = y_0$. Using the equation $y' = f(x, y)$, we find $\phi'(x_0) = f(x_0, y_0)$. To determine $\phi''(x_0)$, we differentiate the equation $y' = f(x, y)$ implicitly with respect to x to obtain

$$y'' = \frac{\partial f}{\partial x} + \frac{\partial f}{\partial y}\frac{dy}{dx} = \frac{\partial f}{\partial x} + \frac{\partial f}{\partial y}f$$

and thereby we can compute $\phi''(x_0)$.

(a) Compute the Taylor polynomials of degree 4 for the solutions to the given initial value problems. Use these Taylor polynomials to approximate the solution at $x = 1$.

(i) $\dfrac{dy}{dx} = x - 2y; \quad y(0) = 1.$ (ii) $\dfrac{dy}{dx} = y(2 - y); \quad y(0) = 4.$

(b) Compare the use of Euler's method with that of Taylor series to approximate the solution $\phi(x)$ to the initial value problem

$$\frac{dy}{dx} + y = \cos x - \sin x, \qquad y(0) = 0.$$

Do this by completing Table 1.3 on page 32. Give the approximations for $\phi(1)$ and $\phi(3)$ to the nearest thousandth. Verify that $\phi(x) = \cos x - e^{-x}$ and use this formula together with a calculator or tables to find the exact values of $\phi(x)$ to the nearest thousandth. Finally, decide which of the first four methods in Table 1.3 will yield the closest approximation to $\phi(10)$ and give the reasons for your choice. (Remember that the computation of trigonometric functions must be done in the radian mode.)

(c) Compute the Taylor polynomial of degree 6 for the solution to the Airy equation

$$\frac{d^2y}{dx^2} = xy$$

with the initial conditions $y(0) = 1$, $y'(0) = 0$. Do you see how, in general, the Taylor series method for an nth order differential equation will employ each of the n initial conditions mentioned in Definition 3, Section 1.2?

TABLE 1.3

Method	Approximation of $\phi(1)$	Approximation of $\phi(3)$
Euler's method using steps of size 0.1		
Euler's method using steps of size 0.01		
Taylor polynomial of degree 2		
Taylor polynomial of degree 5		
Exact value of $\phi(x)$ to nearest thousandth		

B. Picard's Method

The initial value problem

(1) $y'(x) = f(x, y), \qquad y(x_0) = y_0$

can be rewritten as an **integral** equation. This is obtained by integrating both sides of (1) with respect to x from $x = x_0$ to $x = x_1$:

(2) $\displaystyle\int_{x_0}^{x_1} y'(x)\,dx = y(x_1) - y(x_0) = \int_{x_0}^{x_1} f(x, y(x))\,dx.$

Substituting $y(x_0) = y_0$ and solving for $y(x_1)$ gives

(3) $\displaystyle y(x_1) = y_0 + \int_{x_0}^{x_1} f(x, y(x))\,dx.$

If we use t instead of x as the variable of integration, we can let $x = x_1$ be the upper limit of integration. Equation (3) then becomes

(4) $\displaystyle y(x) = y_0 + \int_{x_0}^{x} f(t, y(t))\,dt.$

Equation (4) can be used to generate successive approximations of a solution to (1). Let the function $\phi_0(x)$ be an initial guess or approximation of a solution to (1). Then a new approximation

function is given by

$$\phi_1(x) := y_0 + \int_{x_0}^x f(t, \phi_0(t))\,dt,$$

where we have replaced $y(t)$ by the approximation $\phi_0(t)$ in the argument of f. In a similar fashion, we can use $\phi_1(x)$ to generate a new approximation $\phi_2(x)$, and so on. In general, we obtain the $(n + 1)$st approximation from the relation

(5) $$\phi_{n+1}(x) := y_0 + \int_{x_0}^x f(t, \phi_n(t))\,dt.$$

This procedure is called **Picard's method**.[†] Under certain assumptions on f and $\phi_0(x)$, the sequence $\{\phi_n(x)\}$ is known to converge to a solution to (1). These assumptions and the proof of convergence are given in Chapter 13.[††]

Without further information about the solution to (1), it is common practice to take $\phi_0(x) \equiv y_0$.

(a) Use Picard's method with $\phi_0(x) \equiv 1$ to obtain the next four successive approximations of the solution to

(6) $y'(x) = y(x), \qquad y(0) = 1.$

Show that these approximations are just the partial sums of the Maclaurin series for the actual solution e^x.

(b) Use Picard's method with $\phi_0(x) \equiv 0$ to obtain the next three successive approximations of the solution to the nonlinear problem

(7) $y'(x) = 2x - [y(x)]^2, \qquad y(0) = 0.$

Graph these approximations for $0 \le x \le 1$.

(c) In Problem 29 in Exercises 1.2, we showed that the initial value problem

(8) $y'(x) = 3[y(x)]^{2/3}, \qquad y(2) = 0$

does not have a unique solution. Show that Picard's method beginning with $\phi_0(x) \equiv 0$ converges to the solution $y(x) \equiv 0$, whereas Picard's method beginning with $\phi_0(x) = x - 2$ converges to the second solution $y(x) = (x - 2)^3$. [*Hint:* For the guess $\phi_0(x) = x - 2$, show that $\phi_n(x)$ has the form $c_n(x - 2)^{r_n}$, where $c_n \to 1$ and $r_n \to 3$ as $n \to \infty$.]

C. *Magnetic "Dipole"*

A bar magnet is often modeled as a magnetic dipole with one end labeled the north pole N and the opposite end labeled the south pole S. The magnetic field for the magnetic dipole is symmetric with respect to rotation about the axis passing lengthwise through the center of the bar. Hence we can study the magnetic field by restricting ourselves to a plane containing this axis of symmetry with the origin at the center of the bar (see Figure 1.17 on page 34).

[†] *Historical Footnote*: This approximation method is a by-product of the famous Picard-Lindelöf existence theorem formulated at the end of the nineteenth century.
[††] All references to Chapters 11–13 refer to the expanded text *Fundamentals of Differential Equations and Boundary Value Problems*, 2nd edition.

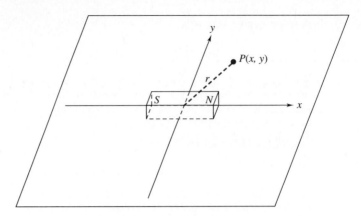

Figure 1.17 Magnetic "dipole"

For a point P that is located a distance r from the origin, where r is much greater than the length of the magnet, the **magnetic field lines** satisfy the differential equation

(9) $$\frac{dy}{dx} = \frac{3xy}{2x^2 - y^2}$$

and the **equipotential lines** satisfy the equation

(10) $$\frac{dy}{dx} = \frac{y^2 - 2x^2}{3xy}.$$

(a) Show that the two families of curves are perpendicular where they intersect. [*Hint*: Consider the slopes of the tangent lines of the two curves at a point of intersection.]

(b) Sketch the direction field for equation (9) for $-5 \leq x, y \leq 5$. You can use a microcomputer to generate the direction field or use the method of isoclines discussed in Section 1.3. The direction field should remind you of the experiment where iron filings are sprinkled on a sheet of paper that is held above a bar magnet. The iron filings correspond to the hash marks.

(c) Use the direction field found in part (b) to help sketch the magnetic field lines that are solutions to (9).

(d) Apply the statement of part (a) to the curves in part (c) to sketch the equipotential lines that are solutions to (10). The magnetic field lines and the equipotential lines are examples of *orthogonal trajectories*. (See Problem 32 in Exercises 2.4, page 64.)

Chapter 2

First Order Differential Equations

2.1 INTRODUCTION: MOTION OF A FALLING BODY

> **An object falls through the air toward Earth. Assuming that the only forces acting on the object are gravity and air resistance, determine the velocity of the object as a function of time.**

Newton's second law states that force is equal to mass times acceleration. We can express this by the equation

$$m\frac{dv}{dt} = F,$$

where F represents the total force on the object, m is the mass of the object, and dv/dt is the acceleration, expressed as the derivative of velocity with respect to time.

Near Earth's surface, the force due to gravity is just the weight of the object. This force can be expressed by mg, where g is the acceleration due to gravity. There is no general law that precisely models the air resistance acting on the object, since this force seems to depend on the velocity of the object, the density of the air, and the shape of the object, among other things. However, in some instances air resistance can be reasonably represented by $-kv$, where k is a positive constant depending on the density of the air and the shape of the object. We use the negative sign because air resistance is a force that opposes the motion. The forces acting on the object are depicted in Figure 2.1 on page 36. (Note that we have generalized the free fall model in Section 1.1 by including air resistance.)

Applying Newton's law, we obtain the first order differential equation

(1) $$m\frac{dv}{dt} = mg - kv.$$

Figure 2.1 Forces on falling object

To solve this equation, we exploit a technique called **separation of variables**, which was used to analyze the radioactive decay model in Section 1.1 and will be developed in full detail in Section 2.2. Treating dv and dt as differentials, we rewrite equation (1) so as to isolate the variables v and t on opposite sides of the equation:

$$\frac{dv}{mg - kv} = \frac{dt}{m}.$$

(Hence the nomenclature "separation of variables.")

Next we integrate the separated equation

$$\int \frac{dv}{mg - kv} = \int \frac{dt}{m}$$

and derive

$$-\frac{1}{k} \ln |mg - kv| = \frac{t}{m} + c.$$

Therefore

$$|mg - kv| = e^{-kc} e^{-kt/m}$$

or

$$mg - kv = Ae^{-kt/m},$$

where the new constant A has magnitude e^{-kc} and the same sign ($\pm$) as $(mg - kv)$. Solving for v, we obtain

(2) $$v = \frac{mg}{k} - \frac{A}{k} e^{-kt/m},$$

which is called a **general solution** to the differential equation because, as we shall see in Section 2.3, every solution to (1) can be expressed in the form given in (2).

In a specific case, we would be given the values of m, g, and k. To determine the constant A in the general solution, we can use the initial velocity of the object v_0. That is,

we solve the **initial value problem**

$$m \frac{dv}{dt} = mg - kv, \qquad v(0) = v_0.$$

Substituting $v = v_0$ and $t = 0$ into the general solution to the differential equation, we can solve for A. With this value for A, the solution to the initial value problem is

$$(3) \qquad v = \frac{mg}{k} + \left(v_0 - \frac{mg}{k} \right) e^{-kt/m}.$$

The preceding formula gives the velocity of the object falling through the air as a function of time if the initial velocity of the object is v_0. In Figure 2.2, we have sketched the graph of $v(t)$ for various values of v_0. It appears from Figure 2.2 that the velocity $v(t)$ approaches *mg/k regardless of the initial velocity* v_0. (This is easy to see from formula (3) by letting $t \to \infty$.) The constant mg/k is referred to as the **limiting** or **terminal velocity** of the object.

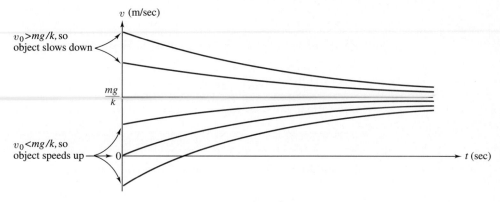

Figure 2.2 Graph of $v(t)$ for six different initial velocities v_0. ($g = 9.8$ m/sec^2, $m/k = 5$ sec)

From this model for a falling body, we can make certain observations. Because $e^{-kt/m}$ rapidly tends to zero, the velocity is approximately the weight, mg, divided by the coefficient of air resistance, k. Thus in the presence of air resistance, the heavier the object, the faster it will fall, assuming shapes and sizes are the same. Also, when air resistance is lessened (k is made smaller), the object will fall faster. These observations certainly agree with our experience.

Many other physical problems,[†] when formulated mathematically, lead to first order differential equations or initial value problems. Several of these are discussed in Chapter 3. In this chapter, we shall learn how to recognize and obtain solutions for some special types of first order equations. We begin by studying separable equations, then linear equa-

[†] The physical problem just discussed has other mathematical models. For example, one could take into account the variations in the gravitational field of Earth and the more general equations for air resistance.

tions, and then exact equations. The methods for solving these are the most basic. In the last two sections, we illustrate how devices such as integrating factors, substitutions, and transformations can be used to transform certain equations into either separable, exact, or linear equations that we can solve. Through our discussion of these special types of equations, the reader will gain insight into the behavior of solutions to more general equations and the possible difficulties in finding these solutions.

2.2 SEPARABLE EQUATIONS

A simple class of first order differential equations that can be solved using integration is the class of **separable equations**. These are equations

(1) $\dfrac{dy}{dx} = f(x, y),$

which can be rewritten so as to isolate the variables x and y (together with their differentials dx and dy) on opposite sides of the equation, as in

$$h(y)dy = g(x)dx.$$

So the original right-hand side $f(x, y)$ must have the factored form

$$f(x, y) = g(x) \cdot \dfrac{1}{h(y)}.$$

More formally, we write $p(y) = h(y)^{-1}$ and present the following definition.

SEPARABLE EQUATION

Definition 1. If the right-hand side of the equation

$$\dfrac{dy}{dx} = f(x, y)$$

can be expressed as a function $g(x)$ that depends only on x times a function $p(y)$ that depends only on y, then the differential equation is called **separable.**[†]

In other words, a first order equation is separable if it can be written in the form

(2) $\dfrac{dy}{dx} = g(x)p(y).$

For example, the equation

$$\dfrac{dy}{dx} = \dfrac{2x + xy}{y^2 + 1}$$

[†] ***Historical Footnote:*** A procedure for solving separable equations was discovered implicitly by Gottfried Leibniz in 1691. The explicit technique called separation of variables was formalized by John Bernoulli in 1694.

is separable, since (if one is sufficiently alert to detect the factorization)

$$\frac{2x + xy}{y^2 + 1} = x\frac{2 + y}{y^2 + 1} = g(x)p(y).$$

However, the equation

$$\frac{dy}{dx} = 1 + xy$$

admits no such factorization of the right-hand side and so is not separable.

Informally speaking, one solves separable equations by performing the separation and then integrating each side.

METHOD FOR SOLVING SEPARABLE EQUATIONS

To solve the equation

$$\frac{dy}{dx} = g(x)p(y)$$

multiply by dx and by $h(y) := p(y)^{-1}$ to obtain

$$h(y)dy = g(x)dx.$$

Then integrate both sides:

$$\int h(y)dy = \int g(x)dx,$$

(3) $H(y) = G(x) + C.$

The last equation gives an implicit solution to the differential equation.

We will look at the mathematical justification of this "streamlined" procedure shortly, but first we study some examples.

EXAMPLE 1 Solve

$$\frac{dy}{dx} = \frac{x - 5}{y^2}.$$

Solution Following the streamlined approach, we separate the variables and rewrite the equation in the form

$$y^2\,dy = (x - 5)dx.$$

Integrating, we have

$$\int y^2 \, dy = \int (x - 5) \, dx$$

$$\frac{y^3}{3} = \frac{x^2}{2} - 5x + C,$$

and solving this last equation for y gives

$$y = \left(\frac{3x^2}{2} - 15x + 3C \right)^{1/3}.$$

Since C is a constant of integration that can be any real number, $3C$ can also be any real number. Replacing $3C$ by the single symbol K, we then have

$$y = \left(\frac{3x^2}{2} - 15x + K \right)^{1/3}.$$

If we wish to abide by the custom of letting C represent an arbitrary constant, we can go one step further and use C instead of K in the final answer. This solution family is graphed in Figure 2.3. ■ End

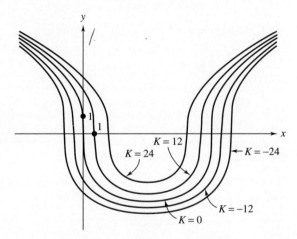

Figure 2.3 Family of solutions for Example 1

As Example 1 attests, separable equations are among the easiest to solve. However, the procedure does require a facility for computing integrals. Many of the procedures to be discussed later in the text also require a familiarity with the techniques of integration. For the reader's convenience, a brief table of integrals can be found on the inside front cover.

EXAMPLE 2 Solve the initial value problem

(4) $\dfrac{dy}{dx} = \dfrac{y-1}{x+3}$, $y(-1) = 0$.

Solution Separating the variables and integrating gives

$$\frac{dy}{y-1} = \frac{dx}{x+3},$$

$$\int \frac{dy}{y-1} = \int \frac{dx}{x+3},$$

(5) $\ln|y-1| = \ln|x+3| + C$.

At this point, we can either solve for y explicitly (retaining the constant C) or use the initial condition to determine C and then solve explicitly for y. Let's try the first approach.
 Exponentiating equation (5), we have

$$e^{\ln|y-1|} = e^{\ln|x+3|+C} = e^{C}e^{\ln|x+3|},$$

(6) $|y-1| = e^{C}|x+3| = C_1|x+3|$,

where $C_1 := e^{C}$.[†] Now, depending on the values of y, we have $|y-1| = \pm(y-1)$; and similarly, $|x+3| = \pm(x+3)$. Thus (6) can be written as

$$y - 1 = \pm C_1(x+3) \quad \text{or} \quad y = 1 \pm C_1(x+3),$$

where the choice of sign depends (as we said), on the values of x and y. Because C_1 is a *positive* constant (recall that $C_1 = e^{C} > 0$), we can replace $\pm C_1$ by K, where K now represents an *arbitrary* nonzero constant. We then obtain

(7) $y = 1 + K(x+3)$.

Finally, we determine K such that the initial condition $y(-1) = 0$ is satisfied. Putting $x = -1$ and $y = 0$ in equation (7) gives

$$0 = 1 + K(-1+3) = 1 + 2K,$$

and so $K = -1/2$. Thus the solution to the initial value problem is

(8) $y = 1 - \dfrac{1}{2}(x+3) = -\dfrac{1}{2}(x+1)$.

Alternative Approach. The second approach is to first set $x = -1$ and $y = 0$ in equation (5) and solve for C. In this case, we obtain

$$\ln|0-1| = \ln|-1+3| + C,$$
$$0 = \ln 1 = \ln 2 + C,$$

[†] The symbol := means ''is defined to be.''

and so $C = -\ln 2$. Thus, from (5), the solution y is given implicitly by

$$\ln(1 - y) = \ln(x + 3) - \ln 2.$$

Here we have replaced $|y - 1|$ by $1 - y$ and $|x + 3|$ by $x + 3$, since we are interested in x and y near the initial values $x = -1$, $y = 0$ (for such values, $y - 1 < 0$ and $x + 3 > 0$). Solving for y, we find

$$\ln(1 - y) = \ln(x + 3) - \ln 2 = \ln\left(\frac{x + 3}{2}\right),$$

$$1 - y = \frac{x + 3}{2},$$

$$y = 1 - \frac{1}{2}(x + 3) = -\frac{1}{2}(x + 1),$$

which agrees with the solution (8) found by the first method. ■ End

EXAMPLE 3 Solve

(9) $$\frac{dy}{dx} = \frac{6x^5 - 2x + 1}{\cos y + e^y}.$$

Solution Separating variables and integrating, we find

$$(\cos y + e^y)dy = (6x^5 - 2x + 1)dx,$$

$$\int (\cos y + e^y)dy = \int (6x^5 - 2x + 1)dx,$$

$$\sin y + e^y = x^6 - x^2 + x + C.$$

At this point, we reach an impasse. We would like to solve for y explicitly, but we cannot. This is often the case in solving nonlinear first order equations. Consequently, when we say "solve the equation," we must on occasion be content if only an implicit form of the solution has been found. ■ End

The separation-of-variables technique, as well as several other techniques discussed in this book, entails rewriting a differential equation by performing certain algebraic operations on it. "Rewriting $dy/dx = g(x)p(y)$ as $h(y)dy = g(x)dx$" amounts to dividing both sides by $p(y)$. The reader may recall from his or her algebra days that doing this can be treacherous. For example, the equation $x(x - 2) = 4(x - 2)$ has two solutions: $x = 2$ and $x = 4$. But if we "rewrite" the equation as $x = 4$ by dividing both sides by $(x - 2)$, we lose track of the root $x = 2$. Thus we should record the roots of $(x - 2)$ itself before dividing by this factor.

By the same token we must take note of the zeros of $p(y)$ in the separable equation $dy/dx = g(x)p(y)$ prior to dividing. After all, if (say) $g(x)p(y) = (x - 2)^2(y - 13)$, then

observe that the constant function $y(x) \equiv 13$ solves the differential equation $dy/dx = g(x)p(y)$:

$$\frac{dy}{dx} = \frac{d(13)}{dx} = 0,$$

$$g(x)p(y) = (x - 2)^2(13 - 13) = 0.$$

Indeed, in solving the equation of Example 2,

$$\frac{dy}{dx} = \frac{y - 1}{x + 3},$$

we obtained $y = 1 + K(x + 3)$ as the set of solutions, where K was a *nonzero* constant (since K replaced $\pm e^C$). But notice that the constant function $y \equiv 1$ (which in this case corresponds to $K = 0$) is also a solution to the differential equation. The reason we lost this solution can be traced back to a division by $y - 1$ in the separation process. (See Problem 31 for an example of where a solution is lost and cannot be retrieved by setting the constant $K = 0$.)

Formal Justification of Method

We close this section by reviewing the separation-of-variables procedure in a more rigorous framework. The original differential equation (2) is rewritten in the form

(10) $$h(y)\frac{dy}{dx} = g(x),$$

where $h(y) := 1/p(y)$. Letting $H(y)$ and $G(x)$ denote antiderivatives (indefinite integrals) of $h(y)$ and $g(x)$, respectively, that is

$$H'(y) = h(y), \qquad G'(x) = g(x),$$

we recast equation (10) as

$$H'(y)\frac{dy}{dx} = G'(x).$$

By the chain rule for differentiation, the left-hand side is the derivative of the composite function $H(y(x))$:

$$\frac{d}{dx}H(y(x)) = H'(y(x))\frac{dy}{dx}.$$

Thus if $y(x)$ is a solution to equation (2), then $H(y(x))$ and $G(x)$ are two functions of x that have the same derivative. Therefore they differ by a constant:

(11) $$H(y(x)) = G(x) + C.$$

Equation (11) agrees with equation (3), which was derived informally, and thus we have verified that the latter can be used to construct implicit solutions.

EXERCISES 2.2

In Problems 1 through 6, determine whether the given differential equation is separable.

1. $\dfrac{dy}{dx} = 2y^3 + y + 4.$ **2.** $\dfrac{dy}{dx} = \sin(x + y).$

3. $\dfrac{dy}{dx} = \dfrac{ye^{x+y}}{x^2 + 2}.$ **4.** $\dfrac{ds}{dt} = t \ln(s^{2t}) + 8t^2.$

5. $s^2 + \dfrac{ds}{dt} = \dfrac{s + 1}{st}.$

6. $(xy^2 + 3y^2)\,dy - 2x\,dx = 0.$

In Problems 7 through 16, solve the equation.

7. $\dfrac{dy}{dx} = \dfrac{1 - x^2}{y^2}.$ **8.** $\dfrac{dy}{dx} = \dfrac{1}{xy^3}.$

9. $\dfrac{dx}{dt} = 3xt^2.$ **10.** $\dfrac{dy}{dx} = y(2 + \sin x).$

11. $\dfrac{dy}{dx} = \dfrac{\sec^2 y}{1 + x^2}.$ **12.** $x\dfrac{dv}{dx} = \dfrac{1 - 4v^2}{3v}.$

13. $\dfrac{dx}{dt} + x^2 = x.$ **14.** $\dfrac{dy}{dx} = 3x^2(1 + y^2).$

15. $y^{-1}\,dy + ye^{\cos x}\sin x\,dx = 0.$

16. $(x + xy^2)\,dx + e^{x^2}y\,dy = 0.$

In Problems 17 through 26, solve the initial value problem.

17. $y' = x^3(1 - y), \qquad y(0) = 3.$

18. $\dfrac{dy}{dx} = (1 + y^2)\tan x, \qquad y(0) = \sqrt{3}.$

19. $\dfrac{dy}{d\theta} = y \sin \theta, \qquad y(\pi) = -3.$

20. $\dfrac{dy}{dx} = \dfrac{3x^2 + 4x + 2}{2y + 1}, \qquad y(0) = -1.$

21. $\dfrac{dy}{dx} = 2\sqrt{y + 1}\cos x, \qquad y(\pi) = 0.$

22. $x^2\,dx + 2y\,dy = 0, \qquad y(0) = 2.$

23. $\dfrac{dy}{dx} = 2x\cos^2 y, \qquad y(0) = \pi/4.$

24. $\dfrac{dy}{dx} = 8x^3 e^{-2y}, \qquad y(1) = 0.$

25. $\dfrac{dy}{dx} = x^2(1 + y), \qquad y(0) = 3.$

26. $\sqrt{y}\,dx + (1 + x)\,dy = 0, \qquad y(0) = 1.$

27. Solutions Not Expressible in Terms of Elementary Functions. As discussed in calculus, certain indefinite integrals (antiderivatives) such as $\int e^{x^2}\,dx$ cannot be expressed in finite terms using elementary functions. When such an integral is encountered while solving a differential equation, it is often helpful to use definite integration (integrals with variable upper limit). For example, consider the initial value problem

$$\dfrac{dy}{dx} = e^{x^2}y^2, \qquad y(2) = 1.$$

The differential equation separates if we divide by y^2 and multiply by dx. We integrate the separated equation from $x = 2$ to $x = x_1$ and find

$$\int_{x=2}^{x=x_1} e^{x^2}\,dx = \int_{x=2}^{x=x_1} \dfrac{dy}{y^2}$$

$$= -\dfrac{1}{y}\bigg|_{x=2}^{x=x_1}$$

$$= -\dfrac{1}{y(x_1)} + \dfrac{1}{y(2)}.$$

If we let t be the variable of integration and replace x_1 by x and $y(2)$ by 1, then we can express the solution to the initial value problem by

$$y(x) = \left\{ 1 - \int_2^x e^{t^2}\,dt \right\}^{-1}.$$

Use definite integration to find an explicit solution to the initial value problems in (a)–(c).

(a) $dy/dx = e^{x^2}, \qquad y(0) = 0.$

(b) $dy/dx = e^{x^2}y^{-2}, \qquad y(0) = 1.$

(c) $dy/dx = \sqrt{1 + \sin x}\,(1 + y^2), \qquad y(0) = 1.$

(d) Use a numerical integration algorithm (such as Simpson's rule, described in Appendix B) to approximate the solution to part (b) at $x = 0.5$ to three decimal places.

28. Sketch the solution to the initial value problem

$$\dfrac{dy}{dt} = 2y - 2yt, \qquad y(0) = 3$$

and determine its maximum value.

29. Uniqueness Questions. In Chapter 1, we indicated that in applications most *initial value problems* will have a unique solution. In fact, the existence of unique solutions was so important that we stated an existence

and uniqueness theorem, Theorem 1, page 12. The method for separable equations will give us a solution, but it may not give us all the solutions (also see Problem 31). To illustrate this, consider the equation $dy/dx = y^{1/3}$.

(a) Use the method of separation of variables to show that

$$y = \left(\frac{2x}{3} + C\right)^{3/2}$$

is a solution.

(b) Show that the initial value problem $dy/dx = y^{1/3}$ with $y(0) = 0$ is satisfied for $C = 0$ by $y = (2x/3)^{3/2}$.

(c) Now show that the constant function $y \equiv 0$ also satisfies the initial value problem given in part (b). Hence this initial value problem does not have a unique solution.

(d) Finally, show that the conditions of Theorem 1 on page 12 are not satisfied.

(The solution $y \equiv 0$ was lost because of the division by zero in the separation process.)

30. Formal Solutions. The method of separation of variables usually produces an implicit solution to the equation, which in certain cases can be solved to obtain an explicit solution. This problem illustrates that the method actually gives expressions that "formally" satisfy the equation, but some care is required when choosing the arbitrary constant if one is, in fact, going to obtain an implicit solution.

(a) Solve the equation $x + y\, dy/dx = 0$ using separation of variables to get an expression of the form $x^2 + y^2 = C$.

(b) Show that when $C = -1$, there are no real values for x and y that satisfy $x^2 + y^2 = C$.

(c) While C was an arbitrary constant of integration, show that it is only for $C > 0$ that we obtain an implicit relationship between x and y.

31. As stated in this section, the separation of equation (2) on page 38 requires division by $p(y)$, and this may disguise the fact that the roots of the equation $p(y) = 0$ are actually constant solutions to the differential equation.

(a) To explore this further, separate the equation

$$\frac{dy}{dx} = (x - 3)(y + 1)^{2/3}$$

to derive the solution.

$$y = -1 + (x^2/6 - x + C)^3.$$

(b) Show that $y \equiv -1$ satisfies the original equation $dy/dx = (x - 3)(y + 1)^{2/3}$.

(c) Show that there is no choice of the constant C that will make the solution in part (a) yield the solution $y \equiv -1$. Thus we lost the solution $y \equiv -1$ when we divided by $(y + 1)^{2/3}$.

32. Interval of Definition. By looking at an initial value problem $dy/dx = f(x, y)$ with $y(x_0) = y_0$, it is not always possible to determine the domain of the solution $y(x)$ or the interval over which the function $y(x)$ satisfies the differential equation.

(a) Solve the equation $dy/dx = xy^3$.

(b) Give explicitly the solutions to the initial value problem with $y(0) = 1$; $y(0) = 1/2$; $y(0) = 2$.

(c) Determine the domains of the solutions in part (b).

(d) As found in part (c), the domains of the solutions depend on the initial conditions. For the initial value problem $dy/dx = xy^3$ with $y(0) = a$, $a > 0$, show that *as* $a \to 0^+$, the domain approaches the whole real line $(-\infty, \infty)$, and as $a \to +\infty$, the domain shrinks to a single point.

(e) Sketch the solutions to the initial value problem $dy/dx = xy^3$ with $y(0) = a$ for $a = \pm 1/2$, ± 1, and ± 2.

33. Mixing. Suppose a brine containing 3 kg of salt per liter (L) runs into a tank initially filled with 400 L of water containing 20 kg of salt. If the brine enters at 10 L/min, the mixture is kept uniform by stirring, and the mixture flows out at the same rate, find the mass of salt in the tank after 10 min (see Figure 2.4). [*Hint*: Let A denote the number of kilograms of salt in the tank at t minutes after the process begins, and use the fact that

rate of increase in A = **rate of input** $-$ **rate of exit**.

A further discussion of mixing problems is given in Section 3.2.]

Figure 2.4 Schematic representation of a mixing problem

34. Newton's Law of Cooling. According to Newton's law of cooling, if an object at temperature T is

immersed in a medium having the constant temperature M, then the rate of change of T is proportional to the difference of temperature $M - T$. This gives the differential equation

$$dT/dt = k(M - T).$$

(a) Solve the differential equation for T.

(b) A thermometer reading $100°$ is placed in a medium having a constant temperature of $70°$. After 6 min, the thermometer reads $80°$. What is the reading after 20 min?

(Further applications of Newton's law of cooling appear in Section 3.3.)

35. Blood plasma is stored at $40°$. Before the plasma can be used, it must be at $90°$. When the plasma is placed in an oven at $120°$, it takes 45 min for the plasma to warm to $90°$. Assume Newton's law of cooling (Problem 34) applies. How long will it take the plasma to warm to $90°$ if the oven temperature is set at **(a)** $100°$, **(b)** $140°$, and **(c)** $80°$?

36. Free Fall. In Section 2.1, we discussed a model for an object falling toward Earth. Assuming only air resis-

tance and gravity are acting on the object, we found that the velocity v must satisfy the equation

$$m\frac{dv}{dt} = mg - kv,$$

where m is the mass, g is the acceleration due to gravity, and $k > 0$ is a constant (see Figure 2.1). If $m = 100$ kg, $g = 9.8$ m/sec^2, $k = 5$ kg/sec, and $v(0) = 10$ m/sec, solve for $v(t)$. What is the limiting (or terminal) velocity of the object?

37. Compound Interest. If $P(t)$ is the amount of dollars in a savings bank account that pays a yearly interest rate of $r\%$ *compounded continuously*, then

$$\frac{dP}{dt} = \frac{r}{100}P, \qquad t \text{ in years.}$$

Assume the interest is 5% annually, $P(0) = \$1000$, and no monies are withdrawn.

(a) How much will be in the account after 2 yrs?

(b) When will the account reach \$4000?

(c) If \$1000 is added to the account every 12 months, how much will be in the account after $3\frac{1}{2}$ yrs?

2.3 LINEAR EQUATIONS

A type of first order differential equation that occurs frequently in applications is the linear equation. Recall from Section 1.1 that a **linear first order equation** is an equation that can be expressed in the form

$$(1) \qquad a_1(x)\frac{dy}{dx} + a_0(x)y = b(x),$$

where $a_1(x)$, $a_0(x)$, and $b(x)$ depend only on the independent variable x, not on y.

For example, the equation

$$x^2 \sin x - (\cos x)y = (\sin x)\frac{dy}{dx}$$

is linear, since it can be rewritten in the form

$$(\sin x)\frac{dy}{dx} + (\cos x)y = x^2 \sin x.$$

However, the equation

$$y\frac{dy}{dx} + (\sin x)y^3 = e^x + 1$$

is not linear; it cannot be put in the form of equation (1) due to the presence of the y^3 and $y\,dy/dx$ terms.

There are two situations for which the solution of a linear differential equation is quite immediate. The first arises if the coefficient $a_0(x)$ is zero, for then equation (1) reduces to

$$(2) \qquad a_1(x)\frac{dy}{dx} = b(x),$$

which is equivalent to

$$y(x) = \int \frac{b(x)}{a_1(x)}\,dx + C$$

(as long as $a_1(x)$ is not zero).

The second is less trivial. Note that if $a_0(x)$ happens to equal the derivative of $a_1(x)$, that is, $a_0(x) = a_1'(x)$, then the two terms on the left-hand side of equation (1) simply comprise the derivative of the product $a_1(x)y$:

$$a_1(x)y' + a_0(x)y = a_1(x)y' + a_1'(x)y = \frac{d}{dx}[a_1(x)y].$$

Therefore equation (1) becomes

$$(3) \qquad \frac{d}{dx}[a_1(x)y] = b(x)$$

and the solution is again elementary:

$$a_1(x)y = \int b(x)\,dx + C,$$

$$y(x) = \frac{1}{a_1(x)}\left\{\int b(x)\,dx + C\right\}.$$

One can seldom rewrite a linear differential equation so that it reduces to a form as simple as (2). However, the form (3) can be achieved through multiplication of the original equation (1) by a well-chosen function $\mu(x)$. The function $\mu(x)$ is then called an ''integrating factor'' for equation (1). The easiest way to see this is first to divide the original equation (1) by $a_1(x)$ and put it into **standard form**

$$(4) \qquad \frac{dy}{dx} + P(x)y = Q(x).$$

Next we wish to determine $\mu(x)$ so that the left-hand side of the multiplied equation

$$(5) \qquad \mu(x)\frac{dy}{dx} + \mu(x)P(x)y = \mu(x)Q(x)$$

is just the derivative of the product $\mu(x)y$:

$$\mu(x)\frac{dy}{dx} + \mu(x)P(x)y = \frac{d}{dx}[\mu(x)y] = \mu(x)\frac{dy}{dx} + \mu'(x)y.$$

Clearly this requires that μ satisfy

(6) $\mu' = \mu P.$

To find such a function, we recognize that equation (6) is a separable differential equation for which one solution is given by $\int d\mu/\mu = \int P(x)\,dx$, or

(7) $\mu(x) = e^{\int P(x)\,dx}.$

With this choice for $\mu(x)$, equation (5) becomes

$$\frac{d}{dx}[\mu(x)y] = \mu(x)Q(x),$$

which has the solution

(8) $y(x) = \dfrac{1}{\mu(x)}\left\{ \displaystyle\int \mu(x)Q(x)\,dx + C \right\}.$

Here C is an arbitrary constant, so (8) gives a one-parameter family of solutions to (4). This form is known as the **general solution** to (4).

We can summarize the method for solving linear equations as follows.

METHOD FOR SOLVING LINEAR EQUATIONS

(a) Write the equation in the standard form

$$\frac{dy}{dx} + P(x)y = Q(x).$$

(b) Calculate the integrating factor $\mu(x)$ by the formula

$$\mu(x) = \exp\left(\int P(x)\,dx \right).$$

(c) Multiply the equation in standard form by $\mu(x)$ and, recalling that the left-hand side is just $\dfrac{d}{dx}(\mu(x)y)$, obtain

$$\underbrace{\mu(x)\frac{dy}{dx} + P(x)\mu(x)y}_{\dfrac{d}{dx}(\mu(x)y)} = \mu(x)Q(x),$$

$$\frac{d}{dx}(\mu(x)y) = \mu(x)Q(x).$$

(d) Integrate the last equation and solve for y by dividing by $\mu(x)$ to obtain (8).

EXAMPLE 1 Find the general solution to

(9) $\dfrac{1}{x}\dfrac{dy}{dx} - \dfrac{2y}{x^2} = x\cos x,\qquad x > 0.$

Solution To put this linear equation in standard form, we multiply by x to obtain

(10) $$\frac{dy}{dx} - \frac{2}{x}y = x^2 \cos x.$$

Here $P(x) = -2/x$, so

$$\int P(x)\,dx = \int \frac{-2}{x}\,dx = -2 \ln |x|.$$

Thus an integrating factor is

$$\mu(x) = e^{-2 \ln |x|} = e^{\ln(x^{-2})} = x^{-2}.$$

Multiplying equation (10) by $\mu(x)$ yields

$$\underbrace{x^{-2}\frac{dy}{dx} - 2x^{-3}y}_{\frac{d}{dx}(x^{-2}y)} = \cos x,$$

$$\frac{d}{dx}(x^{-2}y) = \cos x.$$

We now integrate both sides and solve for y to find

$$x^{-2}y = \int \cos x\,dx = \sin x + C.$$

(11) $$y = x^2 \sin x + Cx^2.$$

It is easily checked that this solution is valid for all $x > 0$. In Figure 2.5 we have sketched solutions for various values of the constant C in (11). ∎ End

Figure 2.5 Graph of $y = x^2 \sin x + Cx^2$ for five values of the constant C

In the next example, we encounter a linear equation that arises in the study of the radioactive decay of an isotope.

EXAMPLE 2 A rock contains two radioactive isotopes, RA_1 and RA_2, that belong to the same radioactive series; that is, RA_1 decays into RA_2, which then decays into stable atoms. Assume that the

rate that RA_1 decays into RA_2 is $50e^{-10t}$ kg/sec. Because the rate of decay of RA_2 is proportional to the mass $y(t)$ of RA_2 present, the rate of change in RA_2 is

$$\frac{dy}{dt} = \text{rate of creation} - \text{rate of decay,}$$

(12) $$\frac{dy}{dt} = 50e^{-10t} - ky,$$

where $k > 0$ is the decay constant. If $k = 2$/sec and initially $y(0) = 40$ kg, find the mass $y(t)$ of RA_2 for $t \geq 0$.

Solution Equation (12) is linear, so we begin by writing it in standard form

(13) $$\frac{dy}{dt} + 2y = 50e^{-10t}, \qquad y(0) = 40,$$

where we have substituted $k = 2$ and displayed the initial condition. We now see that $P(t) = 2$, so $\int P(t)\,dt = \int 2\ dt = 2t$. Thus an integrating factor is $\mu(t) = e^{2t}$. Multiplying equation (13) by $\mu(t)$ yields

$$e^{2t}\frac{dy}{dt} + 2e^{2t}y = 50e^{-10t+2t} = 50e^{-8t},$$

$$\frac{d}{dt}(e^{2t}y) = 50e^{-8t}.$$

Integrating both sides and solving for y, we find

$$e^{2t}y = -\frac{25}{4}e^{-8t} + C,$$

$$y = -\frac{25}{4}e^{-10t} + Ce^{-2t}.$$

Substituting $t = 0$ and $y(0) = 40$ gives

$$40 = -\frac{25}{4}e^0 + Ce^0 = -\frac{25}{4} + C,$$

so $C = 40 + 25/4 = 185/4$. Thus the mass $y(t)$ of RA_2 at time t is given by

(14) $$y(t) = \left(\frac{185}{4}\right)e^{-2t} - \left(\frac{25}{4}\right)e^{-10t}, \qquad t \geq 0. \quad \blacksquare \text{ End}$$

EXAMPLE 3 For the initial value problem

$$y' + y = \sqrt{1 + \cos^2 x}, \qquad y(1) = 4,$$

find the value of $y(2)$.

Solution The integrating factor for the differential equation is, from equation (7),

$$\mu(x) = e^{\int 1\,dx} = e^x.$$

The general solution form (8) thus reads

$$y(x) = e^{-x} \left[\int e^x \sqrt{1 + \cos^2 x} \, dx + C \right].$$

However, this indefinite integral cannot be expressed in finite terms with elementary functions (recall a similar situation in Problem 27 of Exercises 2.2). Because we *can* use numerical algorithms such as Simpson's rule (Appendix B) to perform *definite* integration, we revert to the form (5), which in this case reads

$$\frac{d}{dx} [e^x y] = e^x \sqrt{1 + \cos^2 x},$$

and take the definite integral from the initial value $x = 1$ to the desired value $x = 2$:

$$e^x y \Big|_{x=1}^{x=2} = e^2 y(2) - e^1 y(1) = \int_{x=1}^{x=2} e^x \sqrt{1 + \cos^2 x} \, dx.$$

Inserting the given value of $y(1)$ and solving, we express

$$y(2) = e^{-2+1}(4) + e^{-2} \int_1^2 e^x \sqrt{1 + \cos^2 x} \, dx.$$

Using Simpson's rule, we find that the definite integral is approximately 4.841, so

$$y(2) \approx 4e^{-1} + 4.841e^{-2} \approx 2.127. \quad \blacksquare \text{ End}$$

In Example 3, we had no difficulty expressing the integral for the integrating factor $\mu(x) = e^{\int 1 \, dx} = e^x$. Clearly, situations will arise where this integral, too, cannot be expressed with elementary functions. In such cases, we must again resort to a numerical procedure such as Euler's method (Section 1.4) or to a ''nested loop'' implementation of Simpson's rule. The reader is invited to explore such a possibility in Problem 27.

Because we have established explicit formulas for the solutions to *linear* first order differential equations, we get as a dividend a direct proof of the following theorem.

EXISTENCE AND UNIQUENESS OF SOLUTION

Theorem 1. Suppose $P(x)$ and $Q(x)$ are continuous on an interval (a, b) that contains the point x_0. Then for any choice of initial value y_0, there exists a unique solution $y(x)$ on (a, b) to the initial value problem

$$(15) \qquad \frac{dy}{dx} + P(x)y = Q(x), \qquad y(x_0) = y_0.$$

In fact, the solution is given by (8) for a suitable value of C.

The essentials of the proof of Theorem 1 are contained in the deliberations leading to equation (8); Problem 34 provides the details. This theorem differs from Theorem 1 on

page 12 in that for the *linear* initial value problem (15), we have the existence and uniqueness of the solution on the *whole* interval (a, b), rather than on some smaller unspecified interval about x_0.

The theory of linear differential equations is an important branch of mathematics not only because these equations occur in applications, but also because of the elegant structure associated with them. For example, first order linear equations always have a general solution given by equation (8). Some further properties of first order linear equations are described in Problems 28 and 36. Higher-order linear equations are treated in Chapters 4, 6, and 8.

EXERCISES 2.3

In Problems 1 through 6, determine whether the given equation is separable, linear, neither, or both.

1. $x^2 \dfrac{dy}{dx} + \cos x = y.$

2. $\dfrac{dx}{dt} + xt = e^x.$

3. $x\dfrac{dx}{dt} + t^2 x = \sin t.$

4. $3t = e^t \dfrac{dy}{dt} + y \ln t.$

5. $(t^2 + 1)\dfrac{dy}{dt} = yt - y.$

6. $3r = \dfrac{dr}{d\theta} - \theta^3.$

In Problems 7 through 16, obtain the general solution to the equation.

7. $\dfrac{dy}{dx} - y = e^{3x}.$

8. $\dfrac{dy}{dx} = \dfrac{y}{x} + 2x + 1.$

9. $\dfrac{dr}{d\theta} + r \tan \theta = \sec \theta.$

10. $x\dfrac{dy}{dx} + 2y = x^{-3}.$

11. $(t + y + 1)dt - dy = 0.$

12. $\dfrac{dy}{dx} = x^2 e^{-4x} - 4y.$

13. $y\dfrac{dx}{dy} + 2x = 5y^3.$

14. $x\dfrac{dy}{dx} + 3y + 2x^2 = x^3 + 4x.$

15. $(x^2 + 1)\dfrac{dy}{dx} + xy = x.$

16. $(x^2 + 1)\dfrac{dy}{dx} = x^2 + 2x - 1 - 4xy.$

In Problems 17 through 22, solve the initial value problem.

17. $\dfrac{dy}{dx} - \dfrac{y}{x} = xe^x,$ $y(1) = e - 1.$

18. $\dfrac{dy}{dx} + 4y - e^{-x} = 0,$ $y(0) = \dfrac{4}{3}.$

19. $t^3 \dfrac{dx}{dt} + 3t^2 x = t,$ $x(2) = 0.$

20. $\dfrac{dy}{dx} + \dfrac{3y}{x} + 2 = 3x,$ $y(1) = 1.$

21. $\cos x \dfrac{dy}{dx} + y \sin x = 2x \cos^2 x,$ $y\left(\dfrac{\pi}{4}\right) = \dfrac{-15\sqrt{2}\pi^2}{32}.$

22. $\sin x \dfrac{dy}{dx} + y \cos x = x \sin x,$ $y\left(\dfrac{\pi}{2}\right) = 2.$

23. Radioactive Decay. In Example 2, assume the rate at which RA_1 decays into RA_2 is $40e^{-20t}$ kg/sec and the decay constant for RA_2 is $k = 5$/sec. Find the mass $y(t)$ of RA_2 for $t \geq 0$ if initially $y(0) = 10$ kg.

24. In Example 2, the decay constant for isotope RA_1 was 10/sec, which expresses itself in the exponent of the rate term $50e^{-10t}$ kg/sec. When the decay constant for RA_2 is $k = 2$/sec, we see that in formula (14) for y the term $(185/4)e^{-2t}$ eventually dominates (has greater magnitude for t large).

(a) Redo Example 2 taking $k = 20$/sec. Now which term in the solution eventually dominates?

(b) Redo Example 2 taking $k = 10$/sec.

25. (a) Using definite integration, show that the solution to the initial value problem

$$\frac{dy}{dx} + 2xy = 1, \qquad y(2) = 1,$$

can be expressed as

$$y(x) = e^{-x^2}\left\{ e^4 + \int_2^x e^{t^2}\, dt \right\}.$$

(b) Use numerical integration (such as Simpson's rule, Appendix B) to approximate the solution at $x = 3$.

26. Use numerical integration (such as Simpson's rule, Appendix B) to approximate the solution, at $x = 1$, to the initial value problem

$$\frac{dy}{dx} + \frac{\sin 2x}{2(1 + \sin^2 x)} y = 1, \qquad y(0) = 0.$$

Ensure your approximation is accurate to three decimal places.

27. Consider the initial value problem

$$\frac{dy}{dx} + \sqrt{1 + \sin^2 x}\; y = x, \qquad y(0) = 2.$$

(a) Using definite integration, show that the integrating factor for the differential equation can be written

$$\mu(x) = \exp\left(\int_0^x \sqrt{1 + \sin^2 t}\; dt \right)$$

and that the solution to the initial value problem is

$$y(x) = \frac{1}{\mu(x)} \int_0^x \mu(s)s\; ds + \frac{2}{\mu(x)}.$$

(b) Obtain an approximation to the solution at $x = 1$ by using numerical integration (such as Simpson's rule, Appendix B) in a nested loop to estimate values of $\mu(x)$ and, thereby, the value of

$$\int_0^1 \mu(s)\, s\; ds.$$

[*Hint:* First, use Simpson's rule to approximate $\mu(x)$ at $x = 0.1, 0.2, \ldots, 1$. Then use these values and apply Simpson's rule again to approximate $\int_0^1 \mu(s)s\; ds$.]

(c) Use Euler's method (Section 1.4) to approximate the solution at $x = 1$, with step sizes $h = 0.1$ and 0.05.

(A direct comparison of the merits of the two numerical schemes in (b) and (c) is very complicated, since it should take into account the number of functional evaluations in each algorithm as well as the inherent accuracies.)

28. Constant Multiples of Solutions

(a) Show that $y = e^{-x}$ is a solution of the linear equation

(16) $$\frac{dy}{dx} + y = 0,$$

and $y = x^{-1}$ is a solution of the nonlinear equation

(17) $$\frac{dy}{dx} + y^2 = 0.$$

(b) Show that for any constant C, Ce^{-x} is a solution of equation (16), while Cx^{-1} is a solution of equation (17) only when $C = 0$ or 1.

(c) Show that for any linear equation of the form

$$\frac{dy}{dx} + P(x)y = 0,$$

if $\hat{y}(x)$ is a solution, then for any constant C, the function $C\hat{y}(x)$ is also a solution.

29. Use your ingenuity to solve the equation

$$\frac{dy}{dx} = \frac{1}{e^{4y} + 2x}.$$

30. Bernoulli Equations. The equation

(18) $$\frac{dy}{dx} + 2y = xy^{-2}$$

is an example of a Bernoulli equation. (Further discussion of Bernoulli equations is in Section 2.6.)

(a) Show that the substitution $v = y^3$ reduces equation (18) to the equation

(19) $$\frac{dv}{dx} + 6v = 3x.$$

(b) Solve equation (19) for v. Then make the substitution $v = y^3$ to obtain the solution to equation (18).

31. Discontinuous Coefficients. As we will see in Chapter 3, there are times when the coefficient $P(x)$ in a linear equation may not be continuous but may have a jump discontinuity. Fortunately we may still obtain a "reasonable" solution. For example, consider the initial value problem

$$\frac{dy}{dx} + P(x)y = x, \qquad y(0) = 1,$$

where

$$P(x) := \begin{cases} 1, & 0 \le x \le 2, \\ 3, & x > 2. \end{cases}$$

(a) Find the general solution for $0 \le x \le 2$.

(b) Choose the constant in the solution of part (a) so that the initial condition is satisfied.

(c) Find the general solution for $x > 2$.

(d) Now choose the constant in the general solution from part (c) so that the solution from part (b) and the solution from part (c) agree at $x = 2$. By patching the two solutions together, we are able to obtain a continuous function that satisfies the differential equation except at $x = 2$, where its derivative is undefined.

(e) Sketch the graph of the solution from $x = 0$ to $x = 5$.

32. Discontinuous Forcing Terms. There are times when the forcing term $Q(x)$ in a linear equation may not be continuous, but may have a jump discontinuity. Fortunately, we may still obtain a reasonable solution imitating the procedure discussed in Problem 31. Use this procedure to find the continuous solution to the initial value problem

$$\frac{dy}{dx} + 2y = Q(x), \qquad y(0) = 0,$$

where

$$Q(x) := \begin{cases} 2, & 0 \le x \le 3, \\ -2, & x > 3. \end{cases}$$

Sketch the graph of the solution from $x = 0$ to $x = 7$.

33. Singular Points. Those values of x for which $P(x)$ in equation (4) is not defined are called **singular points** of the equation. For example, $x = 0$ is a singular point of the equation $xy' + 2y = 3x$, since when the equation is written in the standard form $y' + (2/x)y = 3$, we see that $P(x) = 2/x$ is not defined at $x = 0$. On an interval containing a singular point, the questions of the existence and uniqueness of a solution are left unanswered, since Theorem 1 does not apply. To show the possible behavior of solutions near a singular point, consider the following equations.

(a) Show that $xy' + 2y = 3x$ has only one solution defined at $x = 0$. Then show that the initial value problem for this equation with initial condition $y(0) = y_0$ has a unique solution when $y_0 = 0$ and no solution when $y_0 \ne 0$.

(b) Show that $xy' - 2y = 3x$ has an infinite number of solutions defined at $x = 0$. Then show that the initial value problem for this equation with initial condition $y(0) = 0$ has an infinite number of solutions.

34. Existence and Uniqueness. Under the assumptions of Theorem 1, we will prove that equation (8) gives a solution to equation (4) on (a, b). We can then choose the constant C in equation (8) so that the initial value problem (15) is solved.

(a) Show that since $P(x)$ is continuous on (a, b), then $\mu(x)$ defined in (7) is a positive, continuous function satisfying $d\mu/dx = P(x)\mu(x)$ on (a, b).

(b) Since

$$\frac{d}{dx} \int \mu(x)Q(x)\,dx = \mu(x)Q(x),$$

verify that y given in equation (8) satisfies equation (4) by differentiating both sides of equation (8).

(c) Show that when we let $\int \mu(x)Q(x)\,dx$ be the antiderivative whose value at x_0 is 0 (i.e., $\int_{x_0}^{x} \mu(t)Q(t)\,dt$) and choose C to be $y_0\mu(x_0)$, the initial condition $y(x_0) = y_0$ is satisfied.

(d) Start with the assumption that $y(x)$ is a solution to the initial value problem (15) and argue that the discussion leading to equation (8) implies that $y(x)$ must obey equation (8). Then argue that the initial condition in (15) determines the constant C uniquely.

35. Mixing. Suppose a brine containing 2 kg of salt per liter runs into a tank initially filled with 500 L of water containing 50 kg of salt. The brine enters the tank at a rate of 5 L/min. The mixture, kept uniform by stirring, is flowing out at the rate of 5 L/min (see Figure 2.6).

(a) Find the concentration, in kilograms per liter, of salt in the tank after 10 min. [*Hint:* Let A denote the number of kilograms of salt in the tank at t minutes after the process begins and use the fact that

rate of increase in A = rate of input − rate of exit.

A further discussion of mixing problems is given in Section 3.2.]

5 L/min
2 kg/L

$A(t)$

500 L

$A(0) = 50$ kg

5 L/min

Figure 2.6 Mixing problem with equal flow rates

(b) After 10 min, a leak develops in the tank and an additional liter per minute of mixture flows out of the tank (see Figure 2.7). What will be the concentration, in kilograms per liter, of salt in the tank 20 min after the leak develops? [*Hint:* Use the method discussed in Problems 31 and 32.]

Figure 2.7 Mixing problem with unequal flow rates

36. Variation of Parameters. Here is another procedure for solving linear equations that is particularly useful for higher-order linear equations. This method is called **variation of parameters**. It is based on the idea that just by knowing the *form* of the solution, we can substitute into the given equation and solve for any unknowns. Here we illustrate the method for first order equations (see Sections 4.9 and 6.4 for the generalization to higher-order equations).

(a) Show that the general solution to

(20) $\dfrac{dy}{dx} + P(x)y = Q(x)$

has the form

$y(x) = Cy_h(x) + y_p(x),$

where y_h is a solution to equation (20) when $Q(x) \equiv 0$, C is a constant, and $y_p(x) = v(x)y_h(x)$ for a suitable function $v(x)$. [*Hint:* Show that we can take $y_h = \mu^{-1}(x)$ and then use equation (8).]

Knowing this form, we can determine the unknown function y_h by solving a separable equation. Then direct substitution into the original equation will give a simple equation that can be solved for v.

Use this procedure to find the general solution to

(21) $\dfrac{dy}{dx} + \dfrac{3}{x}y = x^2,$ $x > 0,$

by completing the following steps:

(b) Find a nontrivial solution y_h to the separable equation

(22) $\dfrac{dy}{dx} + \dfrac{3}{x}y = 0,$ $x > 0.$

(c) Assuming (21) has a solution of the form $y_p(x) = v(x)y_h(x)$, substitute this into equation (21), and simplify to obtain $v'(x) = x^2/y_h(x)$.

(d) Now integrate to get $v(x)$.

(e) Verify that $y(x) = Cy_h(x) + v(x)y_p(x)$ is a general solution to (21).

37. Secretion of Hormones. The secretion of hormones into the blood is often a periodic activity. If a hormone is secreted on a 24-hr cycle, then the rate of change of the level of the hormone in the blood may be represented by the initial value problem

$$\frac{dx}{dt} = \alpha - \beta \cos\frac{\pi t}{12} - kx, x(0) = x_0,$$

where $x(t)$ is the amount of the hormone in the blood at time t, α is the average secretion rate, β is the amount of daily variation in the secretion, and k is a positive constant reflecting the rate at which the body removes the hormone from the blood. If $\alpha = \beta = 1$, $k = 2$, and $x_0 = 10$, solve for $x(t)$.

38. Use the separation-of-variables technique to derive the solution (7) to the differential equation (6).

2.4 EXACT EQUATIONS

Suppose the mathematical function $F(x, y)$ represents some physical quantity, such as temperature, in a region of the xy-plane. Then the level curves of F, where $F(x, y) = constant$, could be interpreted as isotherms on a weather map, as depicted in Figure 2.8 on page 56.

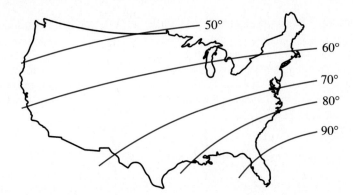

Figure 2.8 Level curves of $F(x, y)$

How does one calculate the slope of the tangent to a level curve? It's easy. Because $F(x, y) = constant$ on the level curve, the total differential of F along the curve is zero:

(1) $$dF(x, y) := \frac{\partial F}{\partial x} dx + \frac{\partial F}{\partial y} dy = 0.$$

Thus the equation for its slope $f(x, y)$ at the point (x, y) is given by

(2) $$\frac{dy}{dx} = f(x, y) = - \frac{\partial F/\partial x}{\partial F/\partial y}.$$

As equation (2) has the form of a differential equation, we should be able to reverse this logic and come up with a very easy technique for solving some differential equations. After all, any first order differential equation $dy/dx = f(x, y)$ can be rewritten in the (differential) form

(3) $$M(x, y) dx + N(x, y) dy = 0$$

(in a variety of ways). Now, if the left-hand side of equation (3) can be identified as a total differential:

$$M(x, y) dx + N(x, y) dy = \frac{\partial F}{\partial x} dx + \frac{\partial F}{\partial y} dy = dF(x, y),$$

then its solutions are given (implicitly) by the level curves

$$F(x, y) = C$$

for an arbitrary constant C.

EXAMPLE 1 Solve the differential equation

$$\frac{dy}{dx} = - \frac{2xy^2 + 1}{2x^2 y}.$$

Solution Some of the choices of differential forms corresponding to this equation are

$$(2xy^2 + 1)\,dx + 2x^2y\,dy = 0,$$

$$\frac{2xy^2 + 1}{2x^2y}\,dx + dy = 0,$$

$$dx + \frac{2x^2y}{2xy^2 + 1}\,dy = 0, \quad \text{etc.}$$

However, the first form is best for our purposes because it is a total differential of the function $F(x, y) = x^2y^2 + x$:

$$(2xy^2 + 1)\,dx + 2x^2y\,dy = d[x^2y^2 + x]$$

$$= \frac{\partial}{\partial x}(x^2y^2 + x)\,dx + \frac{\partial}{\partial y}(x^2y^2 + x)\,dy.$$

Thus the solutions are given implicitly by the formula

$$x^2y^2 + x = C.$$

See Figure 2.9. ■ End

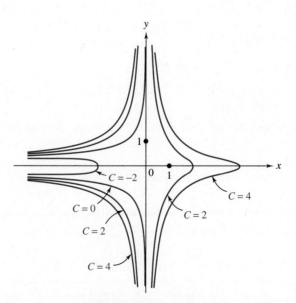

Figure 2.9 Solutions of Example 1

Next we introduce some terminology.

EXACT DIFFERENTIAL FORM

Definition 2. The differential form

$$M(x, y)\,dx + N(x, y)\,dy$$

is said to be **exact** in a rectangle R if there is a function $F(x, y)$ such that

(4) $\qquad \dfrac{\partial F}{\partial x}(x, y) = M(x, y) \quad$ **and** $\quad \dfrac{\partial F}{\partial y}(x, y) = N(x, y)$

for all (x, y) in R. That is, the total differential of $F(x, y)$ satisfies

$$dF(x, y) = M(x, y)\,dx + N(x, y)\,dy.$$

If $M(x, y)\,dx + N(x, y)\,dy$ is an exact differential form, then the equation

$$M(x, y)\,dx + N(x, y)\,dy = 0$$

is called an **exact equation**.

As you might suspect, in applications a differential equation is rarely given to us in exact differential form. However, the solution procedure is so quick and simple for such equations that we devote this section to it. From Example 1, we see that what is needed is (i) a test to determine if a differential form $M(x, y)\,dx + N(x, y)\,dy$ *is* exact and, if so, (ii) a procedure for finding the function $F(x, y)$ itself.

The test for exactness arises from the following observation. If

$$M(x, y)\,dx + N(x, y)\,dy = \frac{\partial F}{\partial x}\,dx + \frac{\partial F}{\partial y}\,dy,$$

then the calculus theorem concerning the equality of continuous mixed partial derivatives

$$\frac{\partial}{\partial y}\frac{\partial F}{\partial x} = \frac{\partial}{\partial x}\frac{\partial F}{\partial y}$$

would dictate a "compatibility condition" on the functions M and N:

$$\frac{\partial}{\partial y}M(x, y) = \frac{\partial}{\partial x}N(x, y).$$

In fact, the following Theorem 2 states that the compatibility condition is also *sufficient* for the differential form to be exact.

TEST FOR EXACTNESS

Theorem 2. Suppose the first partial derivatives of $M(x, y)$ and $N(x, y)$ are continuous in a rectangle R. Then

$$M(x, y)\,dx + N(x, y)\,dy = 0$$

is an exact equation in R if and only if the compatibility condition

$$(5) \qquad \frac{\partial M}{\partial y}(x, y) = \frac{\partial N}{\partial x}(x, y)$$

holds for all (x, y) in R.[†]

Before we address the proof of Theorem 2, note that in Example 1 the differential form that led to the total differential was

$$(2xy^2 + 1)\,dx + (2x^2 y)\,dy = 0.$$

The compatibility conditions are easily confirmed:

$$\frac{\partial M}{\partial y} = \frac{\partial}{\partial y}(2xy^2 + 1) = 4xy,$$

$$\frac{\partial N}{\partial x} = \frac{\partial}{\partial x}(2x^2 y) = 4xy.$$

Also clear is the fact that the other differential forms considered,

$$\frac{2xy^2 + 1}{2x^2 y}\,dx + dy = 0, \qquad dx + \frac{2x^2 y}{2xy^2 + 1}\,dy = 0,$$

do *not* meet the compatibility conditions.

Proof of Theorem 2. There are two parts to the theorem: Exactness implies compatibility, and compatibility implies exactness. First, we have seen that if the differential equation is exact, then the two members of equation (5) are simply the mixed second partials of a function $F(x, y)$. As such, their equality is assured by the theorem of calculus that states that mixed second partials are equal if they are continuous. Because the hypothesis of Theorem 2 guarantees the latter condition, equation (5) is validated.

Rather than proceed directly with the proof of the second part of the theorem, let's derive a formula for a function $F(x, y)$ that satisfies $\partial F/\partial x = M$ and $\partial F/\partial y = N$. Integrating the first equation with respect to x yields

$$(6) \qquad F(x, y) = \int M(x, y)\,dx + g(y).$$

Notice that instead of using C to represent the constant of integration, we have written $g(y)$. This is because y is held fixed while integrating with respect to x, and so our

[†] ***Historical Footnote:*** This theorem was proven by Leonhard Euler in 1734.

"constant" may well depend on y. To determine $g(y)$, we differentiate both sides of (6) with respect to y to obtain

(7) $\qquad \dfrac{\partial F}{\partial y}(x, y) = \dfrac{\partial}{\partial y} \displaystyle\int M(x, y)\,dx + \dfrac{\partial}{\partial y} g(y).$

As g is a function of y alone, we can write $\partial g/\partial y = g'(y)$, and solving (7) for $g'(y)$ gives

$$g'(y) = \dfrac{\partial F}{\partial y}(x, y) - \dfrac{\partial}{\partial y} \int M(x, y)\,dx.$$

Since $\partial F/\partial y = N$, this last equation becomes

(8) $\qquad g'(y) = N(x, y) - \dfrac{\partial}{\partial y} \displaystyle\int M(x, y)\,dx.$

Notice that although the right-hand side of (8) indicates a possible dependence on x, the appearances of this variable must cancel because the left-hand side, $g'(y)$, depends only on y. By integrating (8), we can determine $g(y)$ up to a numerical constant, and therefore we can determine the function $F(x, y)$ up to a numerical constant from the functions $M(x, y)$ and $N(x, y)$.

To finish the proof of Theorem 2, we need to show that the condition (5) implies that $M\,dx + N\,dy = 0$ is an exact equation. This we do by actually exhibiting a function $F(x, y)$ that satisfies $\partial F/\partial x = M$ and $\partial F/\partial y = N$. Fortunately we needn't look too far for such a function. The discussion in the first part of the proof suggests (6) as a candidate, where $g'(y)$ is given by (8). Namely, we *define* $F(x, y)$ by

(9) $\qquad F(x, y) := \displaystyle\int_{x_0}^{x} M(t, y)\,dt + g(y),$

where (x_0, y_0) is a fixed point in the rectangle R and $g(y)$ is determined, up to a numerical constant, by the equation

(10) $\qquad g'(y) := N(x, y) - \dfrac{\partial}{\partial y} \displaystyle\int_{x_0}^{x} M(t, y)\,dt.$

Before proceeding, we must address an extremely important question concerning the definition of $F(x, y)$. That is, how can we be sure (in this portion of the proof) that $g'(y)$, as given in equation (10), is really a function of just y alone? To show that the right-hand side of (10) is independent of x—that is, that the appearances of the variable x cancel— all we need to do is show that its partial derivative with respect to x is zero. This is where condition (5) is utilized. We leave to the reader this computation and the verification that $F(x, y)$ satisfies conditions (4) (see Problems 35 and 36). ■ End

The construction in the proof of Theorem 2 actually provides an explicit procedure for solving exact equations. Let's recap and look at some examples.

METHOD FOR SOLVING EXACT EQUATIONS

(a) If $M\,dx + N\,dy = 0$ is exact, then $\partial F/\partial x = M$. Integrate this last equation with respect to x to get

(11) $$F(x, y) = \int M(x, y)\,dx + g(y).$$

(b) To determine $g(y)$, take the partial derivative with respect to y of both sides of equation (11) and substitute N for $\partial F/\partial y$. We can now solve for $g'(y)$.

(c) Integrate $g'(y)$ to obtain $g(y)$ up to a numerical constant. Substituting $g(y)$ into equation (11) gives $F(x, y)$.

(d) The solution to $M\,dx + N\,dy = 0$ is given implicitly by

$$F(x, y) = C.$$

(Alternatively, starting with $\partial F/\partial y = N$, the implicit solution can be found by first integrating with respect to y; see Example 3.)

EXAMPLE 2 Solve

(12) $$(2xy - \sec^2 x)\,dx + (x^2 + 2y)\,dy = 0.$$

Solution Here $M(x, y) = 2xy - \sec^2 x$ and $N(x, y) = x^2 + 2y$. Because

$$\frac{\partial M}{\partial y} = 2x = \frac{\partial N}{\partial x},$$

equation (12) is exact. To find $F(x, y)$, we begin by integrating M with respect to x:

(13) $$\begin{aligned} F(x, y) &= \int (2xy - \sec^2 x)\,dx + g(y) \\ &= x^2 y - \tan x + g(y). \end{aligned}$$

Next we take the partial derivative of (13) with respect to y and substitute $x^2 + 2y$ for N:

$$\frac{\partial F}{\partial y}(x, y) = N(x, y),$$

$$x^2 + g'(y) = x^2 + 2y.$$

Thus $g'(y) = 2y$, and since the choice of the constant of integration is not important, we can take $g(y) = y^2$. Hence, from (13), we have $F(x, y) = x^2 y - \tan x + y^2$, and the solution to equation (12) is given implicitly by $x^2 y - \tan x + y^2 = C$. ∎ End

Remark. The procedure for solving exact equations requires several steps. As a check on our work, we observe that when we solve for $g'(y)$, we must obtain a function that is independent of x. If this is not the case, then we have erred either in our computation of $F(x, y)$ or in computing $\partial M/\partial y$ or $\partial N/\partial x$.

In the construction of $F(x, y)$, we can first integrate $N(x, y)$ with respect to y to get

(14) $$F(x, y) = \int N(x, y)\,dy + h(x)$$

and then proceed to find $h(x)$. We illustrate this alternative method in the next example.

EXAMPLE 3 Solve

(15) $(1 + e^x y + xe^x y)\,dx + (xe^x + 2)\,dy = 0.$

Solution Here $M = 1 + e^x y + xe^x y$ and $N = xe^x + 2$. Because

$$\frac{\partial M}{\partial y} = e^x + xe^x = \frac{\partial N}{\partial x},$$

equation (15) is exact. If we now integrate $N(x, y)$ with respect to y, we obtain

$$F(x, y) = \int (xe^x + 2)\,dy + h(x) = xe^x y + 2y + h(x).$$

When we take the partial derivative with respect to x and substitute for M, we get

$$\frac{\partial F}{\partial x}(x, y) = M(x, y)$$

$$e^x y + xe^x y + h'(x) = 1 + e^x y + xe^x y.$$

Thus $h'(x) = 1$, so we take $h(x) = x$. Hence $F(x, y) = xe^x y + 2y + x$, and the solution to equation (15) is given implicitly by $xe^x y + 2y + x = C$. In this case, we can solve explicitly for y to obtain $y = (C - x)/(2 + xe^x)$. ∎ End

Remark. Since we can use either procedure for finding $F(x, y)$, it may be worthwhile to consider each of the integrals $\int M(x, y)\,dx$ and $\int N(x, y)\,dy$. If one is easier to evaluate than the other, this would be sufficient reason for us to use one method over the other. (The skeptical reader should try solving equation (15) by first integrating $M(x, y)$.)

EXAMPLE 4 Show that

(16) $(x + 3x^3 \sin y)\,dx + (x^4 \cos y)\,dy = 0$

is *not* exact but that multiplying this equation by the factor x^{-1} yields an exact equation. Use this fact to solve (16).

Solution In equation (16), $M = x + 3x^3 \sin y$ and $N = x^4 \cos y$. Because

$$\frac{\partial M}{\partial y} = 3x^3 \cos y \neq 4x^3 \cos y = \frac{\partial N}{\partial x},$$

equation (16) is not exact. When we multiply (16) by the factor x^{-1}, we obtain

(17) $(1 + 3x^2 \sin y)\,dx + (x^3 \cos y)\,dy = 0.$

For this new equation, $M = 1 + 3x^2 \sin y$ and $N = x^3 \cos y$. If we test for exactness, we now find that

$$\frac{\partial M}{\partial y} = 3x^2 \cos y = \frac{\partial N}{\partial x},$$

and hence (17) is exact. Upon solving (17), we find that the solution is given implicitly by $x + x^3 \sin y = C$. Since equations (16) and (17) differ only by a factor of x, then any solution to one will be a solution for the other whenever $x \neq 0$. Hence the solution to equation (16) is given implicitly by $x + x^3 \sin y = C$. ■ End

In Section 2.5, we discuss methods for finding factors that, like x^{-1} in Example 4, change inexact equations into exact equations.

EXERCISES 2.4

In Problems 1 through 8, classify the equation as separable, linear, exact, or none of these. Notice that some equations may have more than one classification.

1. $(x^2 y + x^4 \cos x) dx - x^3 \, dy = 0$.

2. $(x^{10/3} - 2y) dx + x \, dy = 0$.

3. $(ye^{xy} + 2x) dx + (xe^{xy} - 2y) dy = 0$.

4. $\sqrt{-2y - y^2} \, dx + (3 + 2x - x^2) dy = 0$.

5. $y^2 \, dx + (2xy + \cos y) dy = 0$.

6. $xy \, dx + dy = 0$.

7. $\theta \, dr + (3r - \theta - 1) d\theta = 0$.

8. $[2x + y \cos(xy)] dx + [x \cos(xy) - 2y] dy = 0$.

In Problems 9 through 20, determine whether the equation is exact. If it is, then solve it.

9. $(2xy + 3) dx + (x^2 - 1) dy = 0$.

10. $(2x + y) dx + (x - 2y) dy = 0$.

11. $(\cos x \cos y + 2x) dx - (\sin x \sin y + 2y) \, dy = 0$.

12. $(e^x \sin y - 3x^2) dx + (e^x \cos y + y^{-2/3}/3) dy = 0$.

13. $(t/y) dy + (1 + \ln y) dt = 0$.

14. $e^t(y - t) dt + (1 + e^t) dy = 0$.

15. $\cos \theta \, dr - (r \sin \theta - e^\theta) d\theta = 0$.

16. $(ye^{xy} - 1/y) dx + (xe^{xy} + x/y^2) dy = 0$.

17. $(1/y) dx - (3y - x/y^2) dy = 0$.

18. $[2x + y^2 - \cos(x + y)] dx$
$\qquad + [2xy - \cos(x + y) - e^y] dy = 0$.

19. $\left(2x + \dfrac{y}{1 + x^2 y^2}\right) dx + \left(\dfrac{x}{1 + x^2 y^2} - 2y\right) dy = 0$.

20. $\left[\dfrac{2}{\sqrt{1 - x^2}} + y \cos(xy)\right] dx + [x \cos(xy) - y^{-1/3}] dy = 0$.

In Problems 21 through 26, solve the initial value problem.

21. $(1/x + 2y^2 x) dx + (2yx^2 - \cos y) dy = 0,$ $y(1) = \pi$.

22. $(ye^{xy} - 1/y) dx + (xe^{xy} + x/y^2) dy = 0,$ $y(1) = 1$.

23. $(e^t y + te^t y) dt + (te^t + 2) dy = 0,$ $y(0) = -1$.

24. $(e^t x + 1) dt + (e^t - 1) dx = 0,$ $x(1) = 1$.

25. $(y^2 \sin x) dx + (1/x - y/x) dy = 0,$ $y(\pi) = 1$.

26. $(\tan y - 2) dx + (x \sec^2 y + 1/y) dy = 0,$ $y(0) = 1$.

27. For each of the following equations, find the most general function $M(x, y)$ so that the equation is exact:

 (a) $M(x, y) dx + (\sec^2 y - x/y) dy = 0$.

 (b) $M(x, y) dx + (\sin x \cos y - xy - e^{-y}) dy = 0$.

28. For each of the following equations, find the most general function $N(x, y)$ so that the equation is exact:

 (a) $[y \cos(xy) + e^x] dx + N(x, y) dy = 0$.

 (b) $[ye^{xy} - 4x^3 y + 2] dx + N(x, y) dy = 0$.

29. Consider the equation

 $$(y^2 + 2xy) dx - x^2 \, dy = 0.$$

 (a) Show that this equation is not exact.

 (b) Show that multiplying both sides of the equation by y^{-2} yields a new equation that is exact.

(c) Use the solution of the resulting exact equation to solve the original equation.

(d) Were any solutions ''lost'' in the process?

30. Consider the equation

$$(5x^2y + 6x^3y^2 + 4xy^2)\,dx$$
$$+ (2x^3 + 3x^4y + 3x^2y)\,dy = 0.$$

(a) Show that the equation is not exact.

(b) Multiply the equation by $x^n y^m$ and determine values for n and m that make the resulting equation exact.

(c) Use the solution of the resulting exact equation to solve the original equation.

31. Argue that in the proof of Theorem 2, the function g can be taken as

$$g(y) = \int_{y_0}^{y} N(x, t)\,dt - \int_{y_0}^{y} \left[\frac{\partial}{\partial t} \int_{x_0}^{x} M(s, t)\,ds \right] dt,$$

which can be expressed as

$$g(y) = \int_{y_0}^{y} N(x, t)\,dt - \int_{x_0}^{x} M(s, y)\,ds + \int_{x_0}^{x} M(s, y_0)\,ds.$$

This leads ultimately to the representation

(18) $F(x, y) = \displaystyle\int_{y_0}^{y} N(x, t)\,dt + \int_{x_0}^{x} M(s, y_0)\,ds.$

Evaluate this formula directly with $x_0 = 0$, $y_0 = 0$ to rework

(a) Example 1.

(b) Example 2.

(c) Example 3.

32. Orthogonal Trajectories. A geometric problem occurring often in engineering is that of finding a family of curves (orthogonal trajectories) that intersects a given family of curves orthogonally at each point. For example, we may be given the lines of force of an electric field and want to find the equation for the equipotential curves. Consider the family of curves described by $F(x, y) = k$, where k is a parameter. Recall from the discussion of equation (2) that for each curve in the family, the slope is given by

$$\frac{dy}{dx} = - \frac{\partial F}{\partial x} \bigg/ \frac{\partial F}{\partial y}.$$

(a) Recall that the slope of a curve that is orthogonal (perpendicular) to a given curve is just the negative reciprocal of the slope of the given curve. Using

this fact, show that the curves orthogonal to the family $F(x, y) = k$ satisfy the differential equation

$$\frac{\partial F}{\partial y}(x, y)\,dx - \frac{\partial F}{\partial x}(x, y)\,dy = 0.$$

(b) Using the preceding differential equation, show that the orthogonal trajectories to the family of circles $x^2 + y^2 = k$ are just straight lines through the origin (see Figure 2.10).

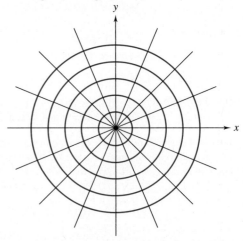

Figure 2.10 Orthogonal trajectories for concentric circles are lines through the center

(c) Show that the orthogonal trajectories to the family of hyperbolas $xy = k$ are the hyperbolas $x^2 - y^2 = k$ (see Figure 2.11).

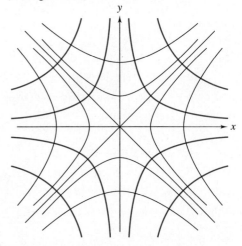

Figure 2.11 Families of orthogonal hyperbolas

33. Use the method in Problem 32 to find the orthogonal trajectories for each of the given families of curves.

(a) $2x^2 + y^2 = k.$ (b) $y = kx^4.$

(c) $y = e^{kx}.$ (d) $y^2 = kx.$

[*Hint:* First express the family in the form $F(x, y) = k.$]

34. Use the method described in Problem 32 to show that the orthogonal trajectories to the family of curves $x^2 + y^2 = kx$, k a parameter, satisfy

$$(2yx^{-1})dx + (y^2x^{-2} - 1)dy = 0.$$

Find the orthogonal trajectories by solving the above equation. Sketch the family of curves, along with their orthogonal trajectories. [*Hint:* Try multiplying the equation by $x^m y^n$ as in Problem 30.]

35. Using condition (5), show that the right-hand side of (10) is independent of x by showing that its partial derivative with respect to x is zero. [*Hint:* Since the partial derivatives of M are continuous, Leibniz's theorem allows you to interchange the operations of integration and differentiation.]

36. Verify that $F(x, y)$ as defined by (9) and (10) satisfies conditions (4).

2.5 SPECIAL INTEGRATING FACTORS

If we take the standard form for the linear differential equation of Section 2.3,

$$\frac{dy}{dx} + P(x)y = Q(x),$$

and rewrite it in differential form by multiplying through by dx, we obtain

$$[P(x)y - Q(x)]dx + dy = 0.$$

This form is certainly not exact, but it becomes exact upon multiplication by the integrating factor $\mu(x) = e^{\int P(x)dx}$. We have

$$[\mu(x)P(x)y - \mu(x)Q(x)]dx + \mu(x)dy$$

as the form, and the compatibility condition is precisely the identity $\mu(x)P(x) = \mu'(x)$ (see Problem 18).

This leads us to generalize the notion of an integrating factor.

INTEGRATING FACTOR

Definition 3. If the equation

(1) $M(x, y)dx + N(x, y)dy = 0$

is not exact, but the equation

(2) $\mu(x, y)M(x, y)dx + \mu(x, y)N(x, y)dy = 0,$

which results from multiplying equation (1) by the function $\mu(x, y)$, is exact, then $\mu(x, y)$ is called an **integrating factor**[†] of the equation (1).

[†] *Historical Footnote:* A general theory of integrating factors was developed by Alexis Clairaut in 1739. Leonhard Euler also studied classes of equations that could be solved using a specific integrating factor.

EXAMPLE 1 Show that $\mu(x, y) = xy^2$ is an integrating factor for

(3) $(2y - 6x)dx + (3x - 4x^2y^{-1})dy = 0.$

Use this integrating factor to solve the equation.

Solution We leave it for the reader to show that (3) is not exact. Multiplying (3) by $\mu(x, y) = xy^2$, we obtain

(4) $(2xy^3 - 6x^2y^2)dx + (3x^2y^2 - 4x^3y)dy = 0.$

For this equation we have $M = 2xy^3 - 6x^2y^2$ and $N = 3x^2y^2 - 4x^3y$. Because

$$\frac{\partial M}{\partial y}(x, y) = 6xy^2 - 12x^2y = \frac{\partial N}{\partial x}(x, y),$$

equation (4) is exact. Hence $\mu(x, y) = xy^2$ is indeed an integrating factor of equation (3).

Let's now solve equation (4) using the procedure of Section 2.4. To find $F(x, y)$, we begin by integrating M with respect to x:

$$F(x, y) = \int (2xy^3 - 6x^2y^2)dx + g(y) = x^2y^3 - 2x^3y^2 + g(y).$$

When we take the partial derivative with respect to y and substitute for N, we find

$$\frac{\partial F}{\partial y}(x, y) = N(x, y)$$

$$3x^2y^2 - 4x^3y + g'(y) = 3x^2y^2 - 4x^3y.$$

Thus $g'(y) = 0$, so we can take $g(y) \equiv 0$. Hence $F(x, y) = x^2y^3 - 2x^3y^2$, and the solution to equation (4) is given implicitly by

$$x^2y^3 - 2x^3y^2 = C.$$

While equations (3) and (4) have essentially the same solutions, *it is possible to lose or gain solutions when multiplying by* $\mu(x, y)$. In this case, $y \equiv 0$ is a solution of equation (4) but not of equation (3). The extraneous solution arises because when we multiply (3) by $\mu = xy^2$ to obtain (4), we are actually multiplying both sides of (3) by zero if $y \equiv 0$. This gives us $y \equiv 0$ as a solution to (4), while it is not a solution to (3). ■ End

How do we find an integrating factor? If $\mu(x, y)$ is an integrating factor of (1) with continuous first partial derivatives, then testing (2) for exactness, we must have

$$\frac{\partial}{\partial y}[\mu(x, y)M(x, y)] = \frac{\partial}{\partial x}[\mu(x, y)N(x, y)].$$

By use of the product rule, this reduces to the equation

$$(5) \qquad M\frac{\partial \mu}{\partial y} - N\frac{\partial \mu}{\partial x} = \left(\frac{\partial N}{\partial x} - \frac{\partial M}{\partial y}\right)\mu.$$

But solving the partial differential equation (5) for μ is usually more difficult than solving the original equation (1). There are, however, two important exceptions.

Let's assume equation (1) has an integrating factor that depends only on x; that is, $\mu = \mu(x)$. In this case, equation (5) reduces to the separable equation

$$(6) \qquad \frac{d\mu}{dx} = \left\{\frac{\partial M/\partial y - \partial N/\partial x}{N}\right\}\mu,$$

where $(\partial M/\partial y - \partial N/\partial x)/N$ is (presumably) just a function of x. In a similar fashion, if equation (1) has an integrating factor that depends only on y, then equation (5) reduces to the separable equation

$$(7) \qquad \frac{d\mu}{dy} = \left\{\frac{\partial N/\partial x - \partial M/\partial y}{M}\right\}\mu,$$

where $(\partial N/\partial x - \partial M/\partial y)/M$ is just a function of y.

We can reverse the above argument. In particular, if $(\partial M/\partial y - \partial N/\partial x)/N$ is a function that depends only on x, then we can solve the separable equation (6) to obtain the integrating factor

$$\mu(x) = \exp\left(\int \left\{\frac{\partial M/\partial y - \partial N/\partial x}{N}\right\}dx\right)$$

for equation (1). We summarize these observations in the following theorem.

SPECIAL INTEGRATING FACTORS

Theorem 3. If $(\partial M/\partial y - \partial N/\partial x)/N$ is continuous and depends only on x, then

$$(8) \qquad \mu(x) = \exp\left(\int \left\{\frac{\partial M/\partial y - \partial N/\partial x}{N}\right\}dx\right)$$

is an integrating factor for equation (1).

If $(\partial N/\partial x - \partial M/\partial y)/M$ is continuous and depends only on y, then

$$(9) \qquad \mu(y) = \exp\left(\int \left\{\frac{\partial N/\partial x - \partial M/\partial y}{M}\right\}dy\right)$$

is an integrating factor for equation (1).

Theorem 3 suggests the following procedure.

METHOD FOR FINDING SPECIAL INTEGRATING FACTORS

If $M\,dx + N\,dy = 0$ is neither separable nor linear, compute $\partial M/\partial y$ and $\partial N/\partial x$. If $\partial M/\partial y = \partial N/\partial x$, then the equation is exact. If it is not exact, consider

(10) $\qquad \dfrac{\partial M/\partial y - \partial N/\partial x}{N}.$

If (10) is a function of just x, then an integrating factor is given by formula (8). If not, consider

(11) $\qquad \dfrac{\partial N/\partial x - \partial M/\partial y}{M}.$

If (11) is a function of just y, then an integrating factor is given by formula (9).

EXAMPLE 2 Solve

(12) $\qquad (2x^2 + y)\,dx + (x^2 y - x)\,dy = 0.$

Solution A quick inspection shows that equation (12) is neither separable nor linear. We also note that

$$\frac{\partial M}{\partial y} = 1 \neq (2xy - 1) = \frac{\partial N}{\partial x}.$$

Because (12) is not exact, we compute

$$\frac{\partial M/\partial y - \partial N/\partial x}{N} = \frac{1 - (2xy - 1)}{x^2 y - x} = \frac{2(1 - xy)}{-x(1 - xy)} = \frac{-2}{x}.$$

We obtain a function of only x, so an integrating factor for (12) is given by formula (8). That is,

$$\mu(x) = \exp\!\left(\int \frac{-2}{x}\,dx \right) = x^{-2}.$$

When we multiply (12) by $\mu = x^{-2}$, we get the exact equation

$$(2 + yx^{-2})\,dx + (y - x^{-1})\,dy = 0.$$

Solving this equation, we ultimately derive the implicit solution

(13) $\qquad 2x - yx^{-1} + \dfrac{y^2}{2} = C.$

Notice that the solution $x \equiv 0$ was lost in multiplying by $\mu = x^{-2}$. Hence (13) and $x \equiv 0$ are solutions to equation (12). ■ End

There are many differential equations that are not covered by Theorem 3 but for which an integrating factor nevertheless exists. The major difficulty, however, is in finding an explicit formula for these integrating factors, which in general will depend on both x and y.

EXERCISES 2.5

In Problems 1 through 6, identify the equation as separable, linear, exact, or having an integrating factor that is a function of either x alone or y alone.

1. $(2y^3 + 2y^2)\,dx + (3y^2x + 2xy)\,dy = 0$.

2. $(2x + yx^{-1})\,dx + (xy - 1)\,dy = 0$.

3. $(y^2 + 2xy)\,dx - x^2\,dy = 0$.

4. $(2x + y)\,dx + (x - 2y)\,dy = 0$.

5. $(2y^2x - y)\,dx + x\,dy = 0$.

6. $(x^2 \sin x + 4y)\,dx + x\,dy = 0$.

In Problems 7 through 12, solve the equation.

7. $(3x^2 + y)\,dx + (x^2y - x)\,dy = 0$.

8. $(2xy)\,dx + (y^2 - 3x^2)\,dy = 0$.

9. $(2y^2 + 2y + 4x^2)\,dx + (2xy + x)\,dy = 0$.

10. $(x^4 - x + y)\,dx - x\,dy = 0$.

11. $(y^2 + 2xy)\,dx - x^2\,dy = 0$.

12. $(2xy^3 + 1)\,dx + (3x^2y^2 - y^{-1})\,dy = 0$.

In Problems 13 and 14, find an integrating factor of the form $x^n y^m$ and solve the equation.

13. $(2y^2 - 6xy)\,dx + (3xy - 4x^2)\,dy = 0$.

14. $(12 + 5xy)\,dx + (6xy^{-1} + 3x^2)\,dy = 0$.

15. Show that if $(\partial N/\partial x - \partial M/\partial y)/(xM - yN)$ depends only on the product xy, that is

$$\frac{\partial N/\partial x - \partial M/\partial y}{xM - yN} = H(xy),$$

then the equation $M(x, y)\,dx + N(x, y)\,dy = 0$ has an integrating factor of the form $\mu(xy)$. Give the general formula for $\mu(xy)$.

16. If $xM(x, y) + yN(x, y) \equiv 0$, find the solution to the equation

$$M(x, y)\,dx + N(x, y)\,dy = 0.$$

17. Fluid Flow. The streamlines associated with a certain fluid flow are represented by the family of curves $y = x - 1 + ke^{-x}$. The velocity potentials of the flow are just the orthogonal trajectories of this family.

(a) Use the method described in Problem 32 of Exercises 2.4 to show that the velocity potentials satisfy

$$dx + (x - y)\,dy = 0.$$

[*Hint:* First express the family $y = x - 1 + ke^{-x}$ in the form $F(x, y) = k$.]

(b) Find the velocity potentials by solving the equation obtained in part (a).

18. Verify that when the linear differential equation $[P(x)y - Q(x)]\,dx + dy = 0$ is multiplied by $\mu(x) = e^{\int P(x)\,dx}$, the result is exact.

2.6 SUBSTITUTIONS AND TRANSFORMATIONS

When the equation

$$M(x, y)\,dx + N(x, y)\,dy = 0$$

is not a separable, exact, or linear equation, it may still be possible to transform it into one that we know how to solve. This, in fact, was our approach in Section 2.5, where we used an integrating factor to transform our original equation into an exact equation.

In this section, we study four types of equations that can be transformed into either a separable or linear equation by means of a suitable substitution or transformation.

SUBSTITUTION PROCEDURE

(a) Identify the type of equation and determine the appropriate substitution or transformation.
(b) Rewrite the original equation in terms of new variables.
(c) Solve the transformed equation.
(d) Express the solution in terms of the original variables.

Homogeneous Equations

HOMOGENEOUS EQUATION

Definition 4. If the right-hand side of the equation

(1) $\dfrac{dy}{dx} = f(x, y)$

can be expressed as a function of the ratio y/x alone, then we say the equation is **homogeneous.**

For example, the equation

(2) $(x - y)dx + x\,dy = 0$

can be written in the form

$$\frac{dy}{dx} = \frac{y - x}{x} = \frac{y}{x} - 1.$$

Since we have expressed $(y - x)/x$ as a function of the ratio y/x; that is, $(y - x)/x = G(y/x)$, where $G(v) := v - 1$, then equation (2) is homogeneous.

The equation

(3) $(x - 2y + 1)dx + (x - y)dy = 0$

can be written in the form

$$\frac{dy}{dx} = \frac{x - 2y + 1}{y - x} = \frac{1 - 2(y/x) + (1/x)}{(y/x) - 1}.$$

Here the right-hand side cannot be expressed as a function of y/x alone because of the term $1/x$ in the numerator. Hence equation (3) is not homogeneous.

One test for the homogeneity of equation (1) is to replace x by tx and y by ty. Then (1) is homogeneous if and only if

$$f(tx, ty) = f(x, y)$$

for all $t \neq 0$ (see Problem 43(a)).

To solve a homogeneous equation, we make a rather obvious substitution. Let

$$v = \frac{y}{x}.$$

Our homogeneous equation now has the form

(4) $\dfrac{dy}{dx} = G(v),$

and all we need is to express dy/dx in terms of x and v. Since $v = y/x$, then $y = vx$. Keeping in mind that both v and y are functions of x, we use the product rule for differentiation to deduce from $y = vx$ that

$$\frac{dy}{dx} = v + x\frac{dv}{dx}.$$

We then substitute the above expression for dy/dx into equation (4) to obtain

(5) $v + x\dfrac{dv}{dx} = G(v).$

The new equation (5) is separable, and we can obtain its implicit solution from

$$\int \frac{1}{G(v) - v}\,dv = \int \frac{1}{x}\,dx.$$

All that remains to do is to express the solution in terms of the original variables x and y.

EXAMPLE 1 Solve

(6) $(xy + y^2 + x^2)dx - x^2\,dy = 0.$

Solution A check will show that equation (6) is not separable, exact, or linear. If we express (6) in the derivative form

(7) $\dfrac{dy}{dx} = \dfrac{xy + y^2 + x^2}{x^2} = \dfrac{y}{x} + \left(\dfrac{y}{x}\right)^2 + 1,$

then we see that the right-hand side of (7) is a function of just y/x. Thus equation (6) is homogeneous.

Now let $v = y/x$ and recall that $dy/dx = v + x(dv/dx)$. With these substitutions, equation (7) becomes

$$v + x\frac{dv}{dx} = v + v^2 + 1.$$

The above equation is separable, and, on separating the variables and integrating, we obtain

$$\int \frac{1}{v^2 + 1}\,dv = \int \frac{1}{x}\,dx,$$

$$\arctan v = \ln|x| + C.$$

Hence

$$v = \tan(\ln|x| + C).$$

Finally, we substitute y/x for v and solve for y to get

$$y = x \tan(\ln|x| + C)$$

as an explicit solution to equation (6). ∎ End

Equations of the Form $dy/dx = G(ax + by)$

When the right-hand side of the equation $dy/dx = f(x, y)$ can be expressed as a function of the combination $ax + by$, where a and b are constants, that is,

$$\frac{dy}{dx} = G(ax + by),$$

then the substitution

$$z = ax + by$$

transforms the equation into a separable one. The method is illustrated in the next example.

EXAMPLE 2 Solve

(8) $$\frac{dy}{dx} = y - x - 1 + (x - y + 2)^{-1}.$$

Solution The right-hand side can be expressed as a function of $x - y$, that is,

$$y - x - 1 + (x - y + 2)^{-1} = -(x - y) - 1 + [(x - y) + 2]^{-1},$$

so let $z = x - y$. To solve for dy/dx, we differentiate $z = x - y$ with respect to x to obtain $dz/dx = 1 - dy/dx$, and so $dy/dx = 1 - dz/dx$. Substituting into (8) yields

$$1 - \frac{dz}{dx} = -z - 1 + (z + 2)^{-1},$$

or

$$\frac{dz}{dx} = (z + 2) - (z + 2)^{-1}.$$

Solving this separable equation, we obtain

$$\int \frac{z + 2}{(z + 2)^2 - 1} \, dz = \int dx,$$

$$\frac{1}{2} \ln|(z + 2)^2 - 1| = x + C_1,$$

from which it follows that

$$(z + 2)^2 = Ce^{2x} + 1.$$

Finally replacing z by $x - y$ yields

$$(x - y + 2)^2 = Ce^{2x} + 1$$

as an implicit solution to equation (8). ∎ End

Bernoulli Equations

BERNOULLI EQUATION

Definition 5. A first order equation that can be written in the form

(9) $$\frac{dy}{dx} + P(x)y = Q(x)y^n,$$

where $P(x)$ and $Q(x)$ are continuous on an interval (a, b) and n is a real number, is called a **Bernoulli equation.**[†]

Notice that when $n = 0$ or 1, equation (9) is also a linear equation and can be solved by the method discussed in Section 2.3. For other values of n, the substitution

$$v = y^{1-n}$$

transforms the Bernoulli equation into a linear equation, as we now show.

Dividing equation (9) by y^n yields

(10) $$y^{-n}\frac{dy}{dx} + P(x)y^{1-n} = Q(x).$$

Taking $v = y^{1-n}$, we find via the chain rule that

$$\frac{dv}{dx} = (1 - n)y^{-n}\frac{dy}{dx},$$

and so equation (10) becomes

$$\frac{1}{1 - n}\frac{dv}{dx} + P(x)v = Q(x).$$

Because $1/(1 - n)$ is just a constant, the last equation is indeed linear.

EXAMPLE 3 Solve

(11) $$\frac{dy}{dx} - 5y = -\frac{5}{2}xy^3.$$

[†] *Historical Footnote:* This equation was proposed for solution by James Bernoulli in 1695. It was solved by his brother John Bernoulli. (James and John were two of eight mathematicians in the Bernoulli family.) In 1696 Gottfried Leibniz showed that the Bernoulli equation can be reduced to a linear equation by making the substitution $v = y^{1-n}$.

Solution This is a Bernoulli equation with $n = 3$, $P(x) = -5$, and $Q(x) = -5x/2$. To transform (11) into a linear equation, we first divide by y^3 to obtain

$$y^{-3}\frac{dy}{dx} - 5y^{-2} = -\frac{5}{2}x.$$

Next we make the substitution $v = y^{-2}$. Since $dv/dx = -2y^{-3}\,dy/dx$, the transformed equation is

$$-\frac{1}{2}\frac{dv}{dx} - 5v = -\frac{5}{2}x,$$

(12) $$\frac{dv}{dx} + 10v = 5x.$$

Equation (12) is linear, so we can solve it for v using the method discussed in Section 2.3. When we do this, it turns out that

$$v = \frac{x}{2} - \frac{1}{20} + Ce^{-10x}.$$

Substituting $v = y^{-2}$ gives the solution

$$y^{-2} = \frac{x}{2} - \frac{1}{20} + Ce^{-10x}.$$

Not included in the last equation is the solution $y \equiv 0$ that was lost in the process of dividing (11) by y^3. ■ End

Equations with Linear Coefficients

We have used various substitutions for y to transform the original equation into a new equation that we could solve. In some cases, we must transform *both* x and y into new variables, say u and v. This is the situation for **equations with linear coefficients**, that is, equations of the form

(13) $(a_1x + b_1y + c_1)dx + (a_2x + b_2y + c_2)dy = 0,$

where the a_i's, b_i's and c_i's are constants. We leave it as an exercise to show that when $a_1b_2 = a_2b_1$, equation (13) can be put in the form $dy/dx = G(ax + by)$, which we solved via the substitution $z = ax + by$.

Before considering the general case when $a_1b_2 \neq a_2b_1$, let's first look at the special situation when $c_1 = c_2 = 0$. Equation (13) then becomes

$$(a_1x + b_1y)dx + (a_2x + b_2y)dy = 0,$$

which can be rewritten in the form

$$\frac{dy}{dx} = -\frac{a_1x + b_1y}{a_2x + b_2y} = -\frac{a_1 + b_1(y/x)}{a_2 + b_2(y/x)}.$$

This equation is homogeneous, so we can solve it using the method discussed earlier in this section.

The above discussion suggests the following procedure for solving (13). If $a_1b_2 \neq a_2b_1$, then we seek a translation of axes of the form

$$x = u + h \quad \text{and} \quad y = v + k,$$

where h, k are constants, that will change $a_1x + b_1y + c_1$ into $a_1u + b_1v$ and change $a_2x + b_2y + c_2$ into $a_2u + b_2v$. Some elementary algebra shows that such a transformation exists if the system of equations

(14)
$$a_1h + b_1k + c_1 = 0,$$
$$a_2h + b_2k + c_2 = 0$$

has a solution. This is assured by the assumption $a_1b_2 \neq a_2b_1$, which is geometrically equivalent to assuming that the two lines described by the system (14) intersect. Now if (h, k) satisfies (14), then the substitutions $x = u + h$ and $y = v + k$ transform equation (13) into the homogeneous equation

(15)
$$\frac{dv}{du} = -\frac{a_1u + b_1v}{a_2u + b_2v} = -\frac{a_1 + b_1(v/u)}{a_2 + b_2(v/u)},$$

which we know how to solve.

EXAMPLE 4 Solve

(16) $(-3x + y + 6)dx + (x + y + 2)dy = 0.$

Solution Since $a_1b_2 = (-3)(1) \neq (1)(1) = a_2b_1$, we will use the translation of axes $x = u + h$, $y = v + k$, where h and k satisfy the system

$$-3h + k + 6 = 0,$$
$$h + k + 2 = 0.$$

Solving the above system for h and k gives $h = 1$, $k = -3$. Hence we let $x = u + 1$ and $y = v - 3$. As $dy = dv$ and $dx = du$, substituting in equation (16) for x and y yields

$$(-3u + v)du + (u + v)dv = 0$$

or

$$\frac{dv}{du} = \frac{3 - (v/u)}{1 + (v/u)}.$$

The above equation is homogeneous, so we let $z = v/u$. Then $dv/du = z + u(dz/du)$, and, substituting for v/u, we obtain

$$z + u\frac{dz}{du} = \frac{3 - z}{1 + z}.$$

Separating variables gives

$$\int \frac{z + 1}{z^2 + 2z - 3} dz = -\int \frac{1}{u} du,$$

$$\frac{1}{2}\ln|z^2 + 2z - 3| = -\ln|u| + C_1,$$

from which it follows that

$$z^2 + 2z - 3 = Cu^{-2}.$$

When we substitute back in for z, u, and v, we find

$$(v/u)^2 + 2(v/u) - 3 = Cu^{-2},$$
$$v^2 + 2uv - 3u^2 = C,$$
$$(y + 3)^2 + 2(x - 1)(y + 3) - 3(x - 1)^2 = C.$$

This last equation gives an implicit solution to (16). ■ End

EXERCISES 2.6

In Problems 1 through 8, identify the equation as homogeneous, Bernoulli, linear coefficients, or of the form $y' = G(ax + by)$.

1. $(y - 4x - 1)^2 \, dx - dy = 0.$

2. $2tx \, dx + (t^2 - x^2) \, dt = 0.$

3. $(t + x + 2) \, dx + (3t - x - 6) \, dt = 0.$

4. $dy/dx + y/x = x^3 y^2.$

5. $(ye^{-2x} + y^3) \, dx - e^{-2x} \, dy = 0.$

6. $\theta \, dy - y \, d\theta = \sqrt{\theta y} \, d\theta.$

7. $(y^3 - \theta y^2) \, d\theta + 2\theta^2 y \, dy = 0.$

8. $\cos(x + y) \, dy = \sin(x + y) \, dx.$

Use the method discussed under "Homogeneous Equations" to solve Problems 9 through 16.

9. $(3x^2 - y^2) \, dx + (xy - x^3 y^{-1}) \, dy = 0.$

10. $(xy + y^2) \, dx - x^2 \, dy = 0.$

11. $(y^2 - xy) \, dx + x^2 \, dy = 0.$

12. $(x^2 + y^2) \, dx + 2xy \, dy = 0.$

13. $\dfrac{dx}{dt} = \dfrac{x^2 + t\sqrt{t^2 + x^2}}{tx}.$ **14.** $\dfrac{dy}{d\theta} = \dfrac{\theta \sec(y/\theta) + y}{\theta}.$

15. $\dfrac{dy}{dx} = \dfrac{x^2 - y^2}{3xy}.$ **16.** $\dfrac{dy}{dx} = \dfrac{y(\ln y - \ln x + 1)}{x}.$

Use the method discussed under "Equations of the Form $dy/dx = G(ax + by)$" to solve Problems 17 through 20.

17. $dy/dx = \sqrt{x + y} - 1.$ **18.** $dy/dx = (x + y + 2)^2.$

19. $dy/dx = (x - y + 5)^2.$ **20.** $dy/dx = \sin(x - y).$

Use the method discussed under "Bernoulli Equations" to solve Problems 21 through 28.

21. $\dfrac{dy}{dx} + \dfrac{y}{x} = x^2 y^2.$

22. $\dfrac{dy}{dx} - y = e^{2x} y^3.$

23. $\dfrac{dy}{dx} = \dfrac{2y}{x} - x^2 y^2.$

24. $\dfrac{dy}{dx} + \dfrac{y}{(x - 2)} = 5(x - 2)y^{1/2}.$

25. $\dfrac{dx}{dt} + tx^3 + \dfrac{x}{t} = 0.$ **26.** $\dfrac{dy}{dx} + y = e^x y^{-2}.$

27. $\dfrac{dr}{d\theta} = \dfrac{r^2 + 2r\theta}{\theta^2}.$ **28.** $\dfrac{dy}{dx} + y^3 x + y = 0.$

Use the method discussed under "Equations with Linear Coefficients" to solve Problems 29 through 32.

29. $(-3x + y - 1) \, dx + (x + y + 3) \, dy = 0.$

30. $(x + y - 1) \, dx + (y - x - 5) \, dy = 0.$

31. $(2x - y) \, dy + (4x + y - 3) \, dy = 0.$

32. $(2x + y + 4) \, dx + (x - 2y - 2) \, dy = 0.$

In Problems 33 through 40, solve the equation given in:

33. Problem 1. **34.** Problem 2.

35. Problem 3. **36.** Problem 4.

37. Problem 5. **38.** Problem 6.

39. Problem 7. **40.** Problem 8.

41. Use the substitution $v = x - y + 2$ to solve equation (8).

42. Use the substitution $y = vx^2$ to solve

$$\frac{dy}{dx} = \frac{2y}{x} + x \cos(y/x^2).$$

43. (a) Show that the equation $dy/dx = f(x, y)$ is homogeneous if and only if $f(tx, ty) = f(x, y)$. [*Hint:* Let $t = 1/x$.]

(b) A function $H(x, y)$ is called **homogeneous of order** n if $H(tx, ty) = t^n H(x, y)$. Show that the equation

$$M(x, y)\,dx + N(x, y)\,dy = 0$$

is homogeneous if $M(x, y)$ and $N(x, y)$ are both homogeneous of the same order.

44. Show that equation (13) reduces to an equation of the form

$$\frac{dy}{dx} = G(ax + by),$$

when $a_1 b_2 = a_2 b_1$. [*Hint:* If $a_1 b_2 = a_2 b_1$, then $a_2/a_1 = b_2/b_1 = k$, so that $a_2 = ka_1$ and $b_2 = kb_1$.]

45. Ecological Systems. In modeling ecological systems, there often arise coupled equations of the form

$$\frac{dy}{dt} = ax + by,$$

$$\frac{dx}{dt} = \alpha x + \beta y,$$

where a, b, α, and β are constants. The quantities x and y represent the populations at time t of two competing species. In this problem, we wish to determine the relationship between x and y rather than the individual populations as functions of t. For this purpose, divide the first equation by the second to obtain

$$(17)\qquad \frac{dy}{dx} = \frac{ax + by}{\alpha x + \beta y}.$$

This new equation is homogeneous, so we can solve it via the substitution $v = y/x$. We refer to the solutions of (17) as **integral curves**. Determine the integral curves for the system

$$\frac{dy}{dt} = -4x - y,$$

$$\frac{dx}{dt} = 2x - y.$$

46. Riccati Equation. An equation of the form

$$(18)\qquad \frac{dy}{dx} = P(x)y^2 + Q(x)y + R(x)$$

is called a generalized Riccati equation.[†]

(a) If one solution, say $u(x)$, of (18) is known, show that the substitution $y = u + 1/v$ reduces (18) to a linear equation in v.

(b) Given that $u(x) = x$ is a solution to

$$\frac{dy}{dx} = x^3(y - x)^2 + \frac{y}{x},$$

use the result of part (a) to find all the other solutions to this equation. (The particular solution $u(x) = x$ can be found by inspection or by using a Taylor series method; see Section 8.2.)

47. Guerrilla Combat Model.[††] In modeling a pair of guerrilla forces in combat, the following system arises

$$\frac{dy}{dt} = -axy,$$

$$\frac{dx}{dt} = -bxy,$$

where $x(t)$ and $y(t)$ are the strengths of opposing forces at time t and a and b are positive constants. The terms $-axy$ and $-bxy$ represent the *combat loss rate* for the troops y and x, respectively. This model assumes no reinforcements.

(a) Show that x and y satisfy the equation

$$\frac{dy}{dx} = \frac{a}{b}.$$

(b) Let $y(0) = y_0$ and $x(0) = x_0$. Solving the equation in part (a), derive the **linear combat law**

$$by - ax = c,$$

where $c = by_0 - ax_0$.

(c) Use the linear combat law to show that the y troops win if $c > 0$ and lose if $c < 0$.

48. Mixed Combat Model. In modeling a conflict between a guerrilla force and a conventional force, the

[†] *Historical Footnote:* Count Jacopo Riccati studied a particular case of this equation in 1724 during his investigation of curves whose radii of curvature depend only on the variable y and not the variable x.

[††] For a discussion of combat models, see *Differential Equation Models*, M. Braun, C. S. Coleman, and D. A. Drew, eds. (Springer-Verlag, New York, 1983), Chapter 8.

following system arises

$$\frac{dy}{dt} = -ax,$$

$$\frac{dx}{dt} = -bxy,$$

where $x(t)$ is the strength of the guerrilla force at time t, $y(t)$ is the strength of the conventional force, and a and b are positive constants. The terms $-ax$ and $-bxy$ represent the *combat loss rate* for troops y and x, respectively. This model assumes no reinforcements.

(a) Show that x and y satisfy the equation

$$\frac{dy}{dx} = \frac{a}{by}.$$

(b) Let $y(0) = y_0$ and $x(0) = x_0$. Solving the equation in part (a), derive the **parabolic combat law**

$$by^2 - 2ax = c,$$

where $c = by_0^2 - 2ax_0$.

(c) Use the parabolic combat law to show that the y troops win if $c > 0$ and lose if $c < 0$.

CHAPTER SUMMARY

In this chapter, we have discussed various types of first order differential equations. The most important were the separable, linear, and exact equations. Their principal features and method of solution are outlined below.

Separable Equations: $dy/dx = g(x)p(y)$. Separate the variables and integrate.

Linear Equations: $dy/dx + P(x)y = Q(x)$. The integrating factor $\mu = \exp(\int P(x)\,dx)$ reduces the equation to $d(\mu y)/dx = \mu Q$, so that $\mu y = \int \mu Q\, dx + C$.

Exact Equations: $dF(x, y) = 0$. Solutions are given implicitly by $F(x, y) = C$. If $\partial M/\partial y = \partial N/\partial x$, then $M\, dx + N\, dy = 0$ is exact and F is given by

$$F = \int M\, dx + g(y), \quad \text{where} \quad g'(y) = N - \frac{\partial}{\partial y}\int M\, dx$$

or

$$F = \int N\, dy + h(x), \quad \text{where} \quad h'(x) = M - \frac{\partial}{\partial x}\int N\, dy.$$

When an equation is not separable, linear, or exact, it may be possible to find an integrating factor or perform a substitution that will enable us to solve the equation.

Special Integrating Factors: $\mu M dx + \mu N dy = 0$ is exact. If $(\partial M/\partial y - \partial N/\partial x)/N$ depends only on x, then

$$\mu(x) = \exp\left(\int \left\{\frac{\partial M/\partial y - \partial N/\partial x}{N}\right\} dx\right)$$

is an integrating factor. If $(\partial N/\partial x - \partial M/\partial y)/M$ depends only on y, then

$$\mu(y) = \exp\left(\int \left\{\frac{\partial N/\partial x - \partial M/\partial y}{M}\right\} dy\right)$$

is an integrating factor.

Homogeneous Equations: $dy/dx = G(y/x)$. Let $v = y/x$. Then $dy/dx = v + x(dv/dx)$, and the transformed equation in the variables v and x is separable.

Equations of the Form: $dy/dx = G(ax + by)$. Let $z = ax + by$. Then $dz/dx = a + b(dy/dx)$, and the transformed equation in the variables z and x is separable.

Bernoulli Equations: $dy/dx + P(x)y = Q(x)y^n$. For $n \neq 0$ or 1, let $v = y^{1-n}$. Then $dv/dx = (1 - n)y^{-n}(dy/dx)$, and the transformed equation in the variables v and x is linear.

Linear Coefficients: $(a_1 x + b_1 y + c_1)dx + (a_2 x + b_2 y + c_2)dy = 0$. For $a_1 b_2 \neq a_2 b_1$, let $x = u + h$ and $y = v + k$, where h and k satisfy

$$a_1 h + b_1 k + c_1 = 0,$$
$$a_2 h + b_2 k + c_2 = 0.$$

Then the transformed equation in the variables u and v is homogeneous.

REVIEW PROBLEMS

In Problems 1 through 30, solve the equation.

1. $\dfrac{dy}{dx} = \dfrac{e^{x+y}}{y - 1}.$

2. $\dfrac{dy}{dx} - 4y = 32x^2.$

3. $(x^2 - 2y^{-3})dy + (2xy - 3x^2)dx = 0.$

4. $\dfrac{dy}{dx} + \dfrac{3y}{x} = x^2 - 4x + 3.$

5. $[\sin(xy) + xy\cos(xy)]dx + [1 + x^2\cos(xy)]dy = 0.$

6. $2xy^3\,dx - (1 - x^2)dy = 0.$

7. $t^3y^2\,dt + t^4y^{-6}\,dy = 0.$

8. $\dfrac{dy}{dx} + \dfrac{2y}{x} = 2x^2 y^2.$

9. $(x^2 + y^2)dx + 3xy\,dy = 0.$

10. $[1 + (1 + x^2 + 2xy + y^2)^{-1}]dx$
$\qquad + [y^{-1/2} + (1 + x^2 + 2xy + y^2)^{-1}]dy = 0.$

11. $\dfrac{dx}{dt} = 1 + \cos^2(t - x).$

12. $(y^3 + 4e^x y)dx + (2e^x + 3y^2)dy = 0.$

13. $\dfrac{dy}{dx} - \dfrac{y}{x} = x^2 \sin 2x.$

14. $\dfrac{dx}{dt} - \dfrac{x}{t - 1} = t^2 + 2.$

15. $\dfrac{dy}{dx} = 2 - \sqrt{2x - y + 3}.$

16. $\dfrac{dy}{dx} + y\tan x + \sin x = 0.$

17. $\dfrac{dy}{d\theta} + 2y = y^2.$

18. $\dfrac{dy}{dx} = (2x + y - 1)^2.$

19. $(x^2 - 3y^2)dx + 2xy\,dy = 0.$

20. $\dfrac{dy}{d\theta} + \dfrac{y}{\theta} = -4\theta y^{-2}.$

21. $(y - 2x - 1)dx + (x + y - 4)dy = 0.$

22. $(2x - 2y - 8)dx + (x - 3y - 6)dy = 0.$

23. $(y - x)dx + (x + y)dy = 0.$

24. $(\sqrt{y/x} + \cos x)dx + (\sqrt{x/y} + \sin y)dy = 0.$

25. $y(x - y - 2)dx + x(y - x + 4)dy = 0.$

26. $\dfrac{dy}{dx} + xy = 0.$

27. $(3x - y - 5)dx + (x - y + 1)dy = 0.$

28. $\dfrac{dy}{dx} = \dfrac{x - y - 1}{x + y + 5}.$

29. $(4xy^3 - 9y^2 + 4xy^2)dx + (3x^2 y^2 - 6xy + 2x^2 y)dy = 0.$

30. $\dfrac{dy}{dx} = (x + y + 1)^2 - (x + y - 1)^2.$

In Problems 31 through 40, solve the initial value problem.

31. $(x^3 - y)dx + x\,dy = 0,$ $y(1) = 3.$

32. $\dfrac{dy}{dx} = \left(\dfrac{x}{y} + \dfrac{y}{x}\right),$ $y(1) = -4.$

33. $(t + x + 3)dt + dx = 0,$ $x(0) = 1.$

34. $\dfrac{dy}{dx} - \dfrac{2y}{x} = x^2 \cos x,$ $y(\pi) = 2.$

35. $(2y^2 + 4x^2)dx - xy\,dy = 0,$ $y(1) = -2.$

36. $[2\cos(2x + y) - x^2]dx + [\cos(2x + y) + e^y]dy = 0,$ $y(1) = 0.$

37. $(2x - y)dx + (x + y - 3)dy = 0,$ $y(0) = 2.$

38. $\sqrt{y}\,dx + (x^2 + 4)dy = 0,$ $y(0) = 4.$

39. $\dfrac{dy}{dx} - \dfrac{2y}{x} = x^{-1}y^{-1},$ $y(1) = 3.$

40. $\dfrac{dy}{dx} - 4y = 2xy^2,$ $y(0) = -4.$

TECHNICAL WRITING EXERCISES

1. An instructor at Ivey U. asserted: "All you need to know about first order differential equations is how to solve those that are exact." Give arguments that support and arguments that refute the instructor's claim.

2. What properties do solutions to linear equations have that are not shared by solutions to either separable or exact equations? Give some specific examples to support your conclusions.

3. Consider the differential equation

$$\frac{dy}{dx} = ay + be^{-x}, \qquad y(0) = c,$$

where a, b, and c are constants. Describe what happens to the asymptotic behavior as $x \to +\infty$ of the solution as the constants a, b, and c are varied. Illustrate with figures and/or graphs.

GROUP PROJECTS FOR CHAPTER 2

A. The Snowplow Problem

To apply the techniques discussed in this chapter to real-world problems, it is necessary to "translate" these problems into questions that can be answered mathematically. The process of reformulating a real-world problem as a mathematical one often requires making certain simplifying assumptions. To illustrate this, consider the following snowplow problem:

> One morning it began to snow very hard and continued snowing steadily throughout the day. A snowplow set out at 8:00 A.M. to clear a road, clearing 3 mi by 11:00 A.M. and an additional mile by 1:00 P.M. *At what time did it start snowing?*

To solve this problem, you can make two physical assumptions concerning the rate at which it is snowing and the rate at which the snowplow can clear the road. Because it is snowing steadily, it is reasonable to assume it is snowing at a constant rate. From the data given (and from our experience), the deeper the snow, the slower the snowplow moves. With this in mind, assume the rate (in mi/hr) at which a snowplow can clear a road is inversely proportional to the depth of the snow.

B. Asymptotic Behavior of Solutions to Linear Equations

To illustrate how the asymptotic behavior of the forcing term $Q(x)$ affects the solution to a linear equation, consider the equation

(1) $\quad \dfrac{dy}{dx} + ay = Q(x),$

where the constant a is positive and $Q(x)$ is continuous on $(0, \infty)$.

 (a) Show that the general solution to equation (1) can be written in the form

(2) $\quad y(x) = y(x_0)e^{-a(x-x_0)} + e^{-ax}\displaystyle\int_{x_0}^{x} e^{at}Q(t)\,dt,$

 where x_0 is a nonnegative constant.

 (b) If $|Q(x)| \le k$ for $x \ge x_0$, where k and x_0 are nonnegative constants, show that

$$|y(x)| \le |y(x_0)|e^{-a(x-x_0)} + \frac{k}{a}[1 - e^{-a(x-x_0)}] \qquad \text{for } x \ge x_0.$$

 (c) Let $z(x)$ satisfy the same equation as (1) but with forcing function $\tilde{Q}(x)$. That is,

$$\frac{dz}{dx} + az = \tilde{Q}(x),$$

 where $\tilde{Q}(x)$ is continuous on $(0, \infty)$. Show that if

$$|\tilde{Q}(x) - Q(x)| \le K \qquad \text{for} \qquad x \ge x_0,$$

then

$$|z(x) - y(x)| \leq |z(x_0) - y(x_0)|e^{-a(x-x_0)} + \frac{K}{a}[1 - e^{-a(x-x_0)}] \qquad \text{for } x \geq x_0.$$

(d) Now show that if $Q(x) \to \beta$ as $x \to \infty$, then any solution $y(x)$ of equation (1) satisfies $y(x) \to \beta/a$ as $x \to \infty$. [*Hint:* Take $\tilde{Q}(x) \equiv \beta$ and $z(x) \equiv \beta/a$ in part (c).]

(e) As an application of part (d), suppose that a brine solution containing $q(t)$ kilograms of salt per liter at time t runs into a tank of water at a fixed rate and that the mixture, kept uniform by stirring, flows out at the same rate. Given that $q(t) \to \beta$ as $t \to \infty$, use the result of part (d) to determine the limiting concentration of the salt in the tank as $t \to \infty$ (see Exercises 2.3, Problem 35).

C. Torricelli's Law of Fluid Flow

Randall K. Campbell-Wright, University of Tampa

How long does it take for water to drain through a hole in the bottom of a tank? Consider the tank pictured in Figure 2.12, which drains through a small, round hole. **Torricelli's law**[†] states that when the surface of the water is at a height h, the water drains with the velocity it would have if it fell freely from a height h (ignoring various forms of friction).

(a) Show that the standard gravity differential equation

$$\frac{d^2h}{dt^2} = -g$$

leads to the conclusion that an object that falls from a height h will land with a velocity of $-\sqrt{2gh}$.

(b) Let $A(h)$ be the cross-sectional area of the water in the tank at height h and a the area of the drain hole. The rate at which water is flowing out of the tank at time t can be expressed as the cross-sectional area at height h times the rate at which the height of the

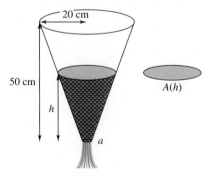

Figure 2.12 Conical tank

[†] **Historical Footnote:** Evangelista Torricelli (1608–1647) invented the barometer and worked on computing the value of the acceleration of gravity as well as observing this principle of fluid flow.

water is changing. Alternatively, the rate at which water flows out of the hole can be expressed as the area of the hole times the velocity of the draining water. Set these two equal to each other to derive the differential equation

(3) $$A(h)\frac{dh}{dt} = -a\sqrt{2gh}.$$

(c) The conical tank of Figure 2.12 has a radius of 20 cm when it is filled to an initial depth of 50 cm. A small round hole at the bottom has a diameter of one cm. Determine $A(h)$ and a and then solve the differential equation in (3), thus explicitly describing the height of the water in this tank as a function of time.

(d) Use your solution to (c) to predict how long it will take for the tank to drain entirely.

(e) Which would drain faster, the tank pictured or an upside down conical tank of the same dimensions draining through a hole of the same size? How long would it take to drain the upside down tank?

(f) Find a water tank and time how long it takes to drain. (You may be able to borrow a ''separatory funnel'' from your chemistry department or use a large water cooler.) The tank should be large enough to take several minutes to drain, and the drain hole should be large enough to allow water to flow freely. The top of the tank should be open (so that the water will not ''glug''). Repeat steps (c) and (d) for your tank and compare the prediction of Torricelli's law to your experimental results.

D. Clairaut Equations and Singular Solutions

An equation of the form

(4) $$y = x\frac{dy}{dx} + f(dy/dx),$$

where the continuously differentiable function $f(t)$ is evaluated at $t = dy/dx$, is called a **Clairaut equation.**[†] Interest in these equations is due to the fact that (4) has a one-parameter family of solutions that consist of *straight lines*. Further, the **envelope** of this family—that is, the curve whose tangent lines are given by the family—is also a solution to (4) and is called the **singular solution**.
 To solve a Clairaut equation:

(a) Differentiate equation (4) with respect to x and simplify to show that

(5) $$[x + f'(dy/dx)]\frac{d^2y}{dx^2} = 0, \qquad \text{where} \qquad f'(t) = \frac{d}{dt}f(t).$$

(b) From (5), conclude that $dy/dx = c$ or $f'(dy/dx) = -x$. Assume $dy/dx = c$ and substitute back into (4) to obtain the family of *straight-line solutions*

$$y = cx + f(c).$$

(c) Show that another solution to (4) is given parametrically by

$$x = -f'(p),$$
$$y = f(p) - pf'(p),$$

where the parameter $p = dy/dx$. This solution is the *singular solution*.

[†] ***Historical Footnote:*** These equations were studied by Alexis Clairaut in 1734.

(d) Use the above method to find the family of straight-line solutions and the singular solution to the equation

$$y = x\left(\frac{dy}{dx}\right) + 3\left(\frac{dy}{dx}\right)^2.$$

Here $f(t) = 3t^2$. Sketch several of the straight-line solutions along with the singular solution on the same coordinate system. Observe that the straight-line solutions are all tangent to the singular solution.

(e) Repeat part (d) for the equation

$$2x\left(\frac{dy}{dx}\right)^3 - y\left(\frac{dy}{dx}\right)^2 + 1 = 0.$$

E. Equilibrium Points of Autonomous First Order Equations

A first order equation of the form

$$y' = f(y),$$

where $f(y)$ is independent of x, is called an **autonomous equation**. Autonomous equations are separable, and in fact, the rearrangement

$$dy/f(y) = dx,$$

when integrated, renders the solution implicitly. However, there are certain qualitative features of autonomous equations that can be established without explicit integration of $1/f(y)$. In particular, we have already noted that if y_0 is a zero of $f(y)$,

$$f(y_0) = 0,$$

then the constant function $y(x) \equiv y_0$ is a solution to the autonomous equation:

$$y_0' = 0 = f(y_0).$$

Such (constant) solutions are called **equilibrium points** for the equation (by analogy with physics, where a system that does not change in time is said to be in equilibrium).

(a) Discuss how the direction field depicted in Figure 2.13 exhibits the key feature of autonomous equations, namely, the nondependence of the slopes $f(y)$ on x.

(b) Identify the equilibrium solutions from the direction field in Figure 2.13. It may be helpful to sketch a few of the solutions.

(c) Sketch some of the solutions that start near the equilibrium solution $y \equiv y_1$ in Figure 2.13. The latter is known as a **stable equilibrium**. Explain this nomenclature. In contrast, the equilibrium solution $y \equiv y_2$ is an **unstable equilibrium**. Explain.

(d) In Figure 2.14, $f(y)$ is graphed as a function of y. Explain how the nature of the ''zero crossings'' at y_1 and y_2 in Figure 2.14 are related to the stability of the corresponding equilibrium solutions.

(e) Suppose that for some function $f(y)$, we have $f(y^*) = 0$ and that f has a relative minimum point at y^*. Sketch the direction field. How would you describe the state of stabil-

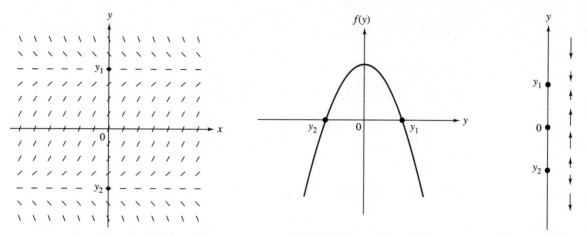

Figure 2.13 Direction field for autonomous equation **Figure 2.14** Graph of $f(y)$ versus y **Figure 2.15** Phase line plot

ity of the equilibrium solution $y \equiv y^*$? Answer the same question for a relative maximum.

(f) Because $f(y)$ for an autonomous equation is independent of x, much of the information in its direction field graph can be displayed in the condensed "phase line" plot. Figure 2.15 depicts the phase line plot for the direction field of Figure 2.13. Explain how the direction of the arrows in the phase line plot corresponds to the sign of $f(y)$. How does one identify a stable equilibrium from a phase line plot?

(g) Sketch the phase line plot for the equation

$$y' = (y - 1)(y - 2)(y - 3).$$

What are the equilibrium solutions? Which are stable?

Chapter 3

Mathematical Models and Numerical Methods Involving First Order Equations

3.1 MATHEMATICAL MODELING

Adopting the Babylonian practices of careful measurement and detailed observations, the ancient Greeks sought to comprehend nature by logical analysis. Aristotle's convincing arguments that the world was not flat, but spherical, led the intellectuals of that day to ponder the question: What is the circumference of Earth? And it was astonishing that Eratosthenes managed to obtain a fairly accurate answer to this problem without having to set foot beyond the ancient city of Alexandria. His method involved certain assumptions and simplifications: Earth is a perfect sphere, the sun's rays travel parallel paths, the city of Syene was 5000 stadia due south of Alexandria, and so on. With these idealizations, Eratosthenes created a mathematical context in which the principles of geometry could be applied.[†]

Today, as scientists seek to further our understanding of nature and as engineers seek, on a more pragmatic level, to find answers to technical problems, the technique of representing our "real world" in mathematical terms has become an invaluable tool. This process of mimicking reality by using the language of mathematics is known as **mathematical modeling**.

Formulating problems in mathematical terms has several benefits. First, it requires that we clearly state our premises. Real-world problems are often complex, involving several different and possibly interrelated processes. Before mathematical treatment can proceed, one must determine which variables are significant and which can be ignored. Often, for the relevant variables, relationships are postulated in the form of laws, formulas, theories, etc. These assumptions constitute the **idealizations** of the model.

Mathematics contains a wealth of theorems and techniques for making logical deductions and manipulating equations. Hence it provides a context in which analysis can take place free of any preconceived notions of the outcome. It is also of great practical importance that mathematics provides a format for obtaining numerical answers via a computer.

[†]For further reading, see, for example, *The Mapmakers*, by John Noble Wilford (Vintage Books, New York, 1982), Chapter 2.

The process of building an effective mathematical model takes skill, imagination, and objective evaluation. Certainly an exposure to several existing models that illustrate various aspects of modeling can lead to a better feel for the process. Several excellent books and articles are devoted exclusively to the subject.[†] In this chapter, we concentrate on examples of models that involve first order differential equations. In studying these and in building your own models, the following broad outline of the process may be helpful.

Formulate the Problem

Here you must pose the problem in such a way that it can be ''answered'' mathematically. This requires an understanding of the problem area as well as the mathematics. At this stage, you may need to spend time talking with nonmathematicians and reading the relevant literature.

Develop the Model

There are two things to be done here. First you must decide which variables are important and which are not. The former are then classified as independent variables or dependent variables. The unimportant variables are those that have very little or no effect on the process. (For example, in studying the motion of a falling body, its color is usually of little interest.) The independent variables are those whose effect is significant and that will serve as input for the model.[††] For the falling body, its shape, mass, initial position, initial velocity, and time from release are possible independent variables. The dependent variables are those that are affected by the independent variables and that are important to solving the problem. Again, for a falling body, its velocity, location, and time of impact are all possible dependent variables.

Second, you must determine or specify the relationships (for example, a differential equation) that exist among the relevant variables. This requires a good background in the area and insight into the problem. You may begin with a crude model and then, based upon testing, refine the model as needed. For example, you might begin by ignoring any friction acting on the falling body. Then, if necessary to obtain a more acceptable answer, try to take into account any frictional forces that may affect the motion.

Test the Model

Before attempting to ''verify'' a model by comparing its output with experimental data, the following questions should be considered:

Are the assumptions reasonable?

Are the equations dimensionally consistent? (For example, we don't want to add units of force to units of velocity.)

[†]See, for example, *A First Course in Mathematical Modeling*, by F. R. Giordano and M. D. Weir (Brooks/Cole Publishing Company, Monterey, California, 1985) or *Concepts of Mathematical Modeling*, by W. J. Meyer (McGraw-Hill Book Company, New York, 1984).

[††]In the mathematical formulation of the model, certain of the independent variables may be called **parameters**.

Is the model internally consistent in the sense that equations do not contradict one another?

Do the relevant equations have solutions?

Are the solutions unique?

How difficult is it to obtain the solutions?

Do the solutions provide an answer for the problem being studied?

When possible, try to validate the model by comparing its predictions with any experimental data. Begin with rather simple predictions that involve little computation or analysis. Then, as the model is refined, check to see that the accuracy of the model's predictions is acceptable to you. In some cases, validation is impossible or socially, politically, economically, or morally unreasonable. For example, how does one validate a model that predicts when our sun will die out?

Each time the model is used to predict the outcome of a process and hence solve a problem, it provides a test of the model that may lead to further refinements or simplifications. In many cases, a model is simplified to give a quicker or less expensive answer—provided, of course, that sufficient accuracy is maintained.

One should always keep in mind that a model is *not* reality but only a representation of reality. The more refined models *may* provide an understanding of the underlying processes of nature. For this reason, applied mathematicians strive for better, more refined models. Still, the real test of a model is its ability to find an acceptable answer for the posed problem.

In this chapter, we discuss various models that involve differential equations. In Section 3.2, Compartmental Analysis, and Section 3.4, Newtonian Mechanics, we consider two general models that result from interpreting the derivative as a rate of change of one variable with respect to another. Section 3.3 describes a model for the heating and cooling of a building. Sections 3.5 and 3.6 contain a brief introduction to some numerical techniques for solving first order initial value problems. This will enable us to study more realistic models that cannot be solved using the techniques of Chapter 2.

3.2 COMPARTMENTAL ANALYSIS

Many complicated processes can be broken down into distinct stages and the entire system modeled by describing the interactions between the various stages. Such systems are called **compartmental** and are graphically depicted by **block diagrams**. In this section, we study the basic unit of these systems, a single compartment, and analyze some simple processes that can be handled by such a model.

The basic one-compartment system consists of a function $x(t)$ that represents the amount of a substance in the compartment at time t, an input rate at which the substance enters the compartment, and an output rate at which the substance leaves the compartment (see Figure 3.1).

Because the derivative of x with respect to t can be interpreted as the rate of change in the amount of the substance in the compartment with respect to time, the one-compart-

Figure 3.1 Schematic representation of a one-compartment system

ment system suggests

(1) $\dfrac{dx}{dt}$ **= input rate − output rate**

as a mathematical model for the process.

Mixing Problems

A problem for which the one-compartment system provides a useful representation is the mixing of fluids in a tank. Let $x(t)$ represent the amount of a substance in a tank (compartment) at time t. To use the compartmental analysis model, we must be able to determine the rates at which this substance enters and leaves the tank. In mixing problems, one is often given the rate at which a fluid containing the substance flows into the tank, along with the concentration of the substance in that fluid. Hence multiplying the flow rate (volume/time) by the concentration (amount/volume) yields the input rate (amount/time).

The output rate of the substance is usually more difficult to determine. If we are given the exit rate of the mixture of fluids in the tank, then how do we determine the concentration of the substance in this mixture? One simplifying assumption that we might make is that the concentration is kept uniform in the mixture. Then we can compute the concentration of the substance in the mixture by dividing the amount $x(t)$ by the volume of the mixture in the tank at time t. Multiplying this concentration by the exit rate of the mixture then gives the desired output rate of the substance. This model is used in the next two examples.

EXAMPLE 1 Consider a large tank holding 1000 L of water into which a brine solution of salt begins to flow at a constant rate of 6 L/min. The solution inside the tank is kept well stirred and is flowing out of the tank at a rate of 6 L/min. If the concentration of salt in the brine entering the tank is 1 kg/L, determine when the concentration of salt in the tank will reach 1/2 kg/L (see Figure 3.2).

Figure 3.2 Mixing problem with equal flow rates

Solution We can view the tank as a compartment containing salt. If we let $x(t)$ denote the mass of salt in the tank at time t, we can determine the concentration of salt in the tank by dividing $x(t)$ by the volume of fluid in the tank at time t. We use the mathematical model described by equation (1) to solve for $x(t)$.

First we must determine the rate at which salt enters the tank. We are given that brine flows into the tank at a rate of 6 L/min. Since the concentration is 1 kg/L, we conclude that the input rate of salt into the tank is

(2) (6 L/min)(1 kg/L) = 6 kg/min.

We must now determine the output rate of salt from the tank. The brine solution in the tank is kept well stirred, so let's assume the concentration of salt in the tank is uniform. That is, the concentration of salt in any part of the tank at time t is just $x(t)$ divided by the volume of fluid in the tank. Because the tank initially contains 1000 L and the rate of flow into the tank is the same as the rate of flow out, the volume is a constant 1000 L. Hence the output rate of salt is

(3) $(6 \text{ L/min}) \left(\dfrac{x(t)}{1000} \text{ kg/L} \right) = \dfrac{3x(t)}{500} \text{ kg/min.}$

The tank initially contained just water, so we set $x(0) = 0$. Substituting the rates in (2) and (3) into equation (1) then gives the initial value problem

(4) $\dfrac{dx}{dt} = 6 - \dfrac{3x}{500}, \qquad x(0) = 0,$

as a mathematical model for the mixing problem.

The equation in (4) is separable (and linear) and easy to solve. Using the initial condition $x(0) = 0$ to evaluate the arbitrary constant, we obtain

(5) $x(t) = 1000(1 - e^{-3t/500}).$

Thus the concentration of salt in the tank at time t is

$$\frac{x(t)}{1000} = 1 - e^{-3t/500} \text{ kg/L.}$$

To determine when the concentration of salt is 1/2 kg/L, we set the right-hand side equal to 1/2 and solve for t. This gives

$$1 - e^{-3t/500} = \frac{1}{2} \quad \text{or} \quad e^{-3t/500} = \frac{1}{2},$$

and hence

$$t = \frac{500 \ln 2}{3} \approx 115.52 \text{ min.}$$

Consequently the concentration of salt in the tank will be 1/2 kg/L after 115.52 min.

◼ End

From equation (5), we observe that the mass of salt in the tank steadily increases and has the limiting value

$$\lim_{t \to \infty} x(t) = \lim_{t \to \infty} 1000(1 - e^{-3t/500}) = 1000 \text{ kg}.$$

Thus the limiting concentration of salt in the tank is 1 kg/L, which is the same as the concentration of salt in the brine flowing into the tank. This certainly agrees with our expectations!

It might be interesting to see what would happen to the concentration if the flow rate into the tank is greater than the flow rate out.

EXAMPLE 2 For the mixing problem described in Example 1, assume now that the brine leaves the tank at a rate of 5 L/min instead of 6 L/min, with all else being the same (see Figure 3.3). Determine the concentration of salt in the tank as a function of time.

6 L/min
1 kg/L

$x(t)$

? L

$x(0) = 0$ kg

5 L/min

Figure 3.3 Mixing problem with unequal flow rates

Solution The difference between the rate of flow into the tank and the rate of flow out is $6 - 5 = 1$ L/min, so the volume of fluid in the tank after t minutes is $(1000 + t)$ L. Hence the rate at which salt leaves the tank is

$$(5 \text{ L/min}) \left[\frac{x(t)}{1000 + t} \text{ kg/L} \right] = \frac{5x(t)}{1000 + t} \text{ kg/min}.$$

Using this in place of (3) for the output rate gives the initial value problem

(6) $$\frac{dx}{dt} = 6 - \frac{5x}{1000 + t}, \qquad x(0) = 0,$$

as a mathematical model for the mixing problem.

The differential equation in (6) is linear, so we can use the procedure outlined on page 48 to solve for $x(t)$. The integrating factor is $\mu(t) = (1000 + t)^5$. Thus

$$\frac{d}{dt}[(1000 + t)^5 x] = 6(1000 + t)^5,$$

$$(1000 + t)^5 x = (1000 + t)^6 + c,$$

$$x(t) = (1000 + t) + c(1000 + t)^{-5}.$$

Using the initial condition $x(0) = 0$, we find $c = -(1000)^6$, and thus the solution to (6) is

$$x(t) = (1000 + t) - (1000)^6(1000 + t)^{-5}.$$

Hence the concentration of salt in the tank at time t is

(7) $$\frac{x(t)}{1000 + t} = 1 - (1000)^6(1000 + t)^{-6} \text{ kg/L.}$$ ■ End

As in Example 1, the concentration given by (7) approaches 1 kg/L as $t \to \infty$. However, in Example 2 the volume of fluid in the tank becomes unbounded, and when the tank begins to overflow, the model in (6) is no longer appropriate.

For mixing problems, the one-compartment model is extremely useful. The only assumption made that is the least bit questionable is that the well-stirred mixture in the tank has a uniform concentration. Even with stirring, the salt entering the tank at the top has a delayed effect on the concentration of the brine leaving at the bottom. It is more realistic to assume that the concentration of brine leaving the tank at time t is equal to the average concentration at an earlier instant, say at time $t - t_0$, where t_0 is a positive constant. The original problem of Example 1 would now be described by

$$x'(t) = 6 - \frac{3}{500}x(t - t_0), \qquad x(t) = 0 \quad \text{for} \quad t \in [-t_0, 0].$$

Such equations are called **delay differential equations** due to the "delay" appearing in the argument of $x(t - t_0)$. These equations are discussed further in the projects at the end of this chapter (see Project A).

Population Models

How does one predict the growth of a population? If we are interested in a single population, we can think of the species as being contained in a compartment (a petri dish, an island, a country, etc.) and study the growth process as a one-compartment system.

Let $p(t)$ be the population at time t. While the population is always an integer, it is usually large enough so that very little error is introduced in assuming that $p(t)$ is a continuous function. We now need to determine the growth (input) rate and the death (output) rate for the population.

Let's consider a population of bacteria that reproduce by simple cell division. In our model, we assume that the growth rate is proportional to the population present. This assumption is consistent with observations of bacterial growth. As long as there are sufficient space and ample food supply for the bacteria, we can also assume that the death rate is zero. (Remember that in cell division, the parent cell does not die, but becomes two new cells.) Hence a mathematical model for a population of bacteria is

(8) $$\frac{dp}{dt} = k_1p, \qquad p(0) = p_0,$$

where $k_1 > 0$ is the proportionality constant for the growth rate and p_0 is the population at time $t = 0$. For human populations, the assumption that the death rate is zero is certainly

wrong! However, if we assume that people die only of natural causes, we might expect the death rate also to be proportional to the size of the population. That is, we revise (8) to read

$$(9) \qquad \frac{dp}{dt} = k_1 p - k_2 p = (k_1 - k_2)p = kp,$$

where $k := k_1 - k_2$ and k_2 is the proportionality constant for the death rate. Let's assume $k_1 > k_2$ so that $k > 0$. This gives the mathematical model

$$(10) \qquad \frac{dp}{dt} = kp, \qquad p(0) = p_0,$$

which is called the **Malthusian,** or **exponential, law** of population growth. This equation is separable, and solving the initial value problem for $p(t)$ gives

$$(11) \qquad p(t) = p_0 e^{kt}.$$

To test the Malthusian model, let's apply it to the demographic history of the United States.

EXAMPLE 3 In 1790 the population of the United States was 3.93 million, and in 1890 it was 62.95 million. Using the Malthusian model, estimate the U.S. population as a function of time.

Solution If we set $t = 0$ to be the year 1790, then by formula (11) we have

$$(12) \qquad p(t) = (3.93)e^{kt},$$

where $p(t)$ is the population in millions. To determine k, we observe that in 1890, when $t = 100$ years, we have

$$p(100) = 62.95 = (3.93)e^{100k}.$$

Solving for k yields

$$k = \frac{\ln(62.95) - \ln(3.93)}{100} \approx 0.027737.$$

Substituting this value in equation (12), we find

$$(13) \qquad p(t) = (3.93)e^{(0.027737)t}. \qquad \blacksquare \text{ End}$$

In Table 3.1 on page 94, we list the United States population as given by the U.S. Bureau of the Census, and the population predicted by the Malthusian model using equation (13). From Table 3.1, we see that the predictions based on the Malthusian model are in reasonable agreement with the census data until about 1900. After 1900, the predicted population is too large, and so the Malthusian model is unacceptable.

We remark that a Malthusian model can be generated using the census data for any two different years. We selected 1790 and 1890 for purposes of comparison with the logistic model that we now describe.

		Malthusian	Logistic	$\dfrac{1}{p}\dfrac{dp}{dt}$	Logistic
Year	U.S. Census	(Example 3)	(Example 4)		(Least Squares)
1790	3.93	3.93	3.93		4.01
1800	5.31	5.19	5.30	0.0312	5.34
1810	7.24	6.84	7.13	0.0299	7.10
1820	9.64	9.03	9.58	0.0292	9.42
1830	12.87	11.92	12.82	0.0289	12.47
1840	17.07	15.73	17.07	0.0302	16.44
1850	23.19	20.76	22.60	0.0310	21.58
1860	31.44	27.39	29.70	0.0264	28.16
1870	39.82	36.15	38.65	0.0235	36.47
1880	50.16	47.70	49.69	0.0231	46.80
1890	62.95	62.95	62.95	0.0205	59.37
1900	75.99	83.07	78.37	0.0191	74.28
1910	91.97	109.63	95.64	0.0162	91.43
1920	105.71	144.67	114.21	0.0146	110.52
1930	122.78	190.91	133.28	0.0106	130.94
1940	131.67	251.94	152.00	0.0108	151.93
1950	151.33	332.47	169.56	0.0157	172.61
1960	179.32	438.75	185.35	0.0145	192.16
1970	203.21	579.00	199.01	0.0116	209.93
1980	226.50	764.08	210.46	0.0102	225.52
1990	249.63	1008.32	219.77		238.77
2000	?	1330.63	227.19		249.74

TABLE 3.1 A COMPARISON OF THE MALTHUSIAN AND LOGISTIC MODELS WITH U.S. CENSUS DATA (POPULATION IS GIVEN IN MILLIONS)

The Malthusian model considered only death by natural causes. What about premature deaths due to malnutrition, inadequate medical supplies, communicable diseases, violent crimes, etc.? These factors involve a competition within the population, so we might assume there is another component of the death rate that is proportional to the number of two-party interactions. For a population of size p, there are $p(p-1)/2$ such interactions. Thus we will assume the death rate due to intraspecies competition is

$$k_3 \frac{p(p-1)}{2},$$

where k_3 is the proportionality constant. Together with equation (9), this leads to the **logistic model**

$$(14) \qquad \frac{dp}{dt} = ap - bp^2, \qquad p(0) = p_0,$$

where $a = k_1 - k_2 + k_3/2$ and $b = k_3/2$.

The equation (14) is separable and can be solved using a partial fractions expansion:

$$\int \frac{1}{p(a - bp)}\,dp = \int dt,$$

$$\frac{1}{a}\int \frac{1}{p}\,dp - \frac{1}{a}\int \frac{-b}{a - bp}\,dp = \int dt,$$

$$\frac{1}{a}\ln|p| - \frac{1}{a}\ln|a - bp| = t + c_1,$$

$$\ln\left|\frac{p}{a - bp}\right| = at + c_2,$$

$$\frac{p}{a - bp} = c_3 e^{at}.$$

Solving for p, we have

$$p(t) = \frac{ac_3 e^{at}}{bc_3 e^{at} + 1} = \frac{ac_3}{bc_3 + e^{-at}}.$$

Using the initial condition in (14), we find

$$c_3 = \frac{p_0}{a - bp_0}, \quad \text{if} \quad p_0 \neq a/b \quad \text{(or 0)}.$$

Thus, on substituting for c_3 and simplifying, we get

(15) $$p(t) = \frac{ap_0}{bp_0 + (a - bp_0)e^{-at}}.$$

The function $p(t)$ given in (15) is called the **logistic function**. In Figure 3.4 we have sketched the graph of this function, which is called the **logistic curve**. The logistic model for population growth was first developed by P. F. Verhulst around 1840.

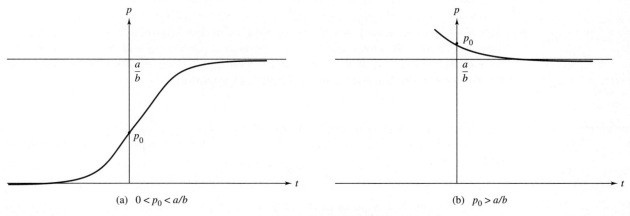

(a) $0 < p_0 < a/b$ (b) $p_0 > a/b$

Figure 3.4 The logistic curves

An important property of the logistic function is that

$$\lim_{t\to\infty} p(t) = \frac{a}{b} \quad \text{if} \quad p_0 > 0.$$

Consequently we see that any population following the logistic model will remain **bounded**. Moreover, when the initial population p_0 is greater than the ratio a/b, the population will decrease toward a/b and when the initial population is less than a/b, it will increase toward a/b as illustrated in Figure 3.4. This behavior is also evident from a sketch of the direction field (see Fig. 1.10, page 20).

In deriving the solution given in (15), we excluded the cases when $p_0 = a/b$ or 0. These two populations are referred to as **equilibrium populations**. This terminology follows from the observation that when $p(t) \equiv a/b$ (or 0), then

$$0 \equiv \frac{dp}{dt} \quad \text{and} \quad ap - bp^2 = p(a - bp) \equiv 0.$$

Hence $p(t) \equiv a/b$ (or 0) are constant solutions to (14). Notice that the graphs of these equilibrium populations are the horizontal asymptotes of the logistic curves in Figure 3.4.

Let's test the logistic model on the population growth of the United States.

EXAMPLE 4 Taking the 1790 population of 3.93 million as the initial population and given the 1840 and 1890 populations of 17.07 and 62.95 million, respectively, use the logistic model to estimate the population at time t.

Solution With $t = 0$ corresponding to the year 1790, we know that $p_0 = 3.93$. We must now determine the parameters a, b in equation (15). For this purpose, we use the given facts that $p(50) = 17.07$ and $p(100) = 62.95$; that is,

(16) $17.07 = \dfrac{3.93a}{3.93b + (a - 3.93b)e^{-50a}}$,

(17) $62.95 = \dfrac{3.93a}{3.93b + (a - 3.93b)e^{-100a}}$.

Equations (16) and (17) are two nonlinear equations in the two unknowns a, b. To solve such a system, we would usually resort to a numerical approximation scheme such as Newton's method. However, for the case at hand, it is possible to find the solutions directly because the data are given at times t_1 and t_2 with $t_2 = 2t_1$ (see Problem 12). Carrying out the algebra described in Problem 12, we ultimately find that

(18) $a = 0.0304667 \quad \text{and} \quad b = 0.0001214.$

Thus the logistic model for the given data is

(19) $p(t) = \dfrac{0.1197340}{0.0004771 + (0.0299896)e^{-(0.0304667)t}}.$ ■ End

The population data predicted by (19) are displayed in column 4 of Table 3.1. As you can see, these predictions are in better agreement with the census data than the Malthusian

model is. And, of course, the agreement is perfect in the years 1790, 1840, and 1890. However, the choice of these particular years for estimating the parameters p_0, a, and b is quite arbitrary, and we would expect that a more robust model would use *all* of the data, in some way, for the estimation. One way to implement this idea is the following.

Observe from equation (14) that the logistic model predicts a linear relationship between $(dp/dt)/p$ and p:

$$\frac{1}{p}\frac{dp}{dt} = a - bp,$$

with a as the intercept and $-b$ as the slope. In column five of Table 3.1, we list values of $(dp/dt)/p$, which are estimated from centered differences according to

(20) $$\frac{1}{p(t)}\frac{dp}{dt}(t) \approx \frac{1}{p(t)}\frac{p(t+10) - p(t-10)}{20}$$

(see Problem 16). In Figure 3.5, these estimated values of $(dp/dt)/p$ are plotted against p in what is called a *scatter diagram*. The linear relationship predicted by the logistic model suggests that we approximate the plot by a straight line. A standard technique for doing this is the so-called *least-squares linear fit* which is discussed in Appendix D. This yields the straight line

$$\frac{1}{p}\frac{dp}{dt} = 0.0291108 - 0.0001007p,$$

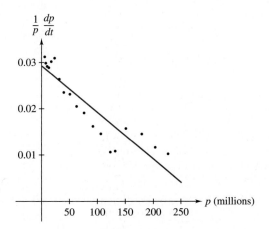

Figure 3.5 Scatter data and straight line fit

which is also depicted in Figure 3.5. Now with $a = 0.0291108$ and $b = 0.0001007$, we can solve equation (15) for p_0:

(21) $$p_0 = \frac{ap(t)e^{-at}}{a - bp(t)(1 - e^{-at})}.$$

By averaging the right-hand side of (21) over all the data, we obtain the estimate $p_0 \approx 4.009$. Finally the insertion of these estimates for the parameters in equation (15) leads to the predictions listed in column six of Table 3.1.

Note that this model yields the following limit on the future population of the United States:

$$\lim_{t \to \infty} p(t) = \frac{a}{b} = \frac{0.0291108}{0.0001007} \approx 289 \text{ million.}$$

EXERCISES 3.2

1. A brine solution of salt flows at a constant rate of 8 L/min into a large tank that initially held 100 L of brine solution in which was dissolved 5 kg of salt. The solution inside the tank is kept well stirred and flows out of the tank at the same rate. If the concentration of salt in the brine entering the tank is 0.5 kg/L, determine the mass of salt in the tank after t minutes. When will the concentration of salt in the tank reach 0.2 kg/L?

2. A brine solution of salt flows at a constant rate of 6 L/min into a large tank that initially held 50 L of brine solution in which was dissolved 5 kg of salt. The solution inside the tank is kept well stirred and flows out of the tank at the same rate. If the concentration of salt in the brine entering the tank is 0.5 kg/L, determine the mass of salt in the tank after t minutes. When will the concentration of salt in the tank reach 0.3 kg/L?

3. A nitric acid solution flows at a constant rate of 6 L/min into a large tank that initially held 200 L of a 0.5% nitric acid solution. The solution inside the tank is kept well stirred and flows out of the tank at a rate of 8 L/min. If the solution entering the tank is 20% nitric acid, determine the volume of nitric acid in the tank after t minutes. When will the percentage of nitric acid in the tank reach 10%?

4. A brine solution of salt flows at a constant rate of 4 L/min into a large tank that initially held 100 L of water. The solution inside the tank is kept well stirred and flows out of the tank at a rate of 3 L/min. If the concentration of salt in the brine entering the tank is 0.2 kg/L, determine the mass of salt in the tank after t minutes. When will the concentration of salt in the tank reach 0.1 kg/L?

5. A swimming pool whose volume is 10,000 gal contains water that is 0.01% chlorine. Starting at $t = 0$, city water containing 0.001% chlorine is pumped into the pool at a rate of 5 gal/min. The pool water flows out at the same rate. What is the percentage of chlorine in the pool after 1 hr? When will the pool water be 0.002% chlorine?

6. The air in a small room 12 ft by 8 ft by 8 ft is 3% carbon monoxide. Starting at $t = 0$, fresh air containing no carbon monoxide is blown into the room at a rate of 100 ft^3/min. If air in the room flows out through a vent at the same rate, when will the air in the room be 0.01% carbon monoxide?

7. Blood carries a drug into an organ at a rate of 3 cm^3/sec and leaves at the same rate. The organ has a liquid volume of 125 cm^3. If the concentration of the drug in the blood entering the organ is 0.2 g/cm^3, what is the concentration of the drug in the organ at time t if there was no trace of the drug initially? When will the concentration of the drug in the organ reach 0.1 g/cm^3?

8. Water flows into Lake Magdalene from Sweetwater Creek at a rate of 300 gal/min. Lake Magdalene contains about 100 million gal of water. The spraying of nearby orange groves has caused the concentration of pesticides in the lake to reach 0.000035, or 35 parts per million. If pesticides are banned, how long will it be before the concentration of pesticides in Lake Magdalene is below 10 parts per million? (Assume Sweetwater Creek contains no pesticides and the volume of Lake Magdalene stays constant.)

9. In 1980 the Department of Natural Resources released 1000 splake (a crossbreed of fish) into a lake. In 1987 the population of splake in the lake was estimated to be 3000. Using the Malthusian law for population growth, estimate the population of splake in the lake

in 1990. What would the estimate be for the year 2001 using the Malthusian law?

10. Use a sketch of the direction field (see Section 1.3) to argue that any solution to the mixing problem model

$$\frac{dx}{dt} = a - bx; \qquad a, b > 0,$$

approaches the equilibrium solution $x(t) \equiv a/b$ as t approaches ∞.

11. Use a sketch of the direction field (see Section 1.3) to argue that any solution to the logistic model

$$\frac{dp}{dt} = (a - bp)p; \quad p(t_0) = p_0,$$

where a, b, and p_0 are positive constants, approaches the equilibrium solution $p(t) \equiv a/b$ as t approaches ∞.

12. For the logistic curve (15), assume $p_1 := p(t_1)$ and $p_2 := p(t_2)$ are given with $t_2 = 2t_1 (t_1 > 0)$. Show that

$$a = \frac{1}{t_1} \ln \left[\frac{p_2(p_1 - p_0)}{p_0(p_2 - p_1)} \right],$$

$$b = \frac{a}{p_1} \left[\frac{p_1^2 - p_0 p_2}{p_1 p_2 - 2p_0 p_2 + p_0 p_1} \right].$$

[*Hint:* Solve separately the equations $p_1 = p(t_1)$ and $p_2 = p(t_2)$ for bp_0 and set the answers equal.] Use these formulas to verify the numerical values given in (18).

13. In Problem 9, suppose we have the additional information that the population of splake in 1994 was estimated to be 5000. Use a logistic model to estimate the population of splake in the year 2001. What is the predicted limiting population? [*Hint:* Use the formulas in Problem 12.]

14. In 1970 the population of alligators on the Kennedy Space Center grounds was estimated to be 300. In 1980 the population had grown to an estimated 1500. Using the Malthusian law for population growth, estimate the alligator population on the Kennedy Space Center grounds in the year 2000.

15. In Problem 14, suppose we have the additional information that the population of alligators on the grounds of the Kennedy Space Center in 1975 was estimated to be 1200. Use a logistic model to estimate the population of alligators in the year 2000. What is the predicted limiting population? [*Hint:* Use the formulas in Problem 12.]

16. Show that for a differentiable function $p(t)$, we have

$$\lim_{h \to 0} \frac{p(t + h) - p(t - h)}{2h} = p'(t),$$

which is the basis of the centered difference approximation used in (20).

17. (a) For the U.S. census data, use the forward difference approximation to the derivative, that is,

$$\frac{1}{p(t)} \frac{dp}{dt}(t) \approx \frac{1}{p(t)} \frac{p(t + 10) - p(t)}{10},$$

to recompute column 5 of Table 3.1.

 (b) Using the data from part (a), determine the constants a, b in the least-squares fit

$$\frac{1}{p} \frac{dp}{dt} \approx a - bp$$

(see Appendix D).

 (c) With the values for a and b found in part (b), determine p_0 by averaging formula (21) over the data.

 (d) Substitute a, b, and p_0 as determined above into the logistic formula (15) and calculate the populations predicted for each of the years listed in Table 3.1.

 (e) Compare this model with that of the centered difference-based model in column 6 of Table 3.1.

18. A population model used in actuarial predictions is based on the **Gompertz equation**

$$\frac{dP}{dt} = P(a - b \ln P),$$

where a and b are constants.

 (a) Solve the Gompertz equation for $P(t)$.

 (b) If $P(0) = P_0 > 0$, give a formula for $P(t)$ in terms of a, b, P_0, and t.

 (c) Describe the behavior of $P(t)$ as $t \to \infty$. [*Hint:* Consider cases for $b > 0$ and $b < 0$.]

19. The initial mass of a certain species of fish is 7 million tons. The mass of fish, if left alone, would increase at a rate proportional to the mass, with a proportionality constant of 2/yr. However, commercial fishing removes fish mass at a constant rate of 15 million tons per year. When will all the fish be gone? If the fishing rate is changed so that the mass of fish remains constant, what should that rate be?

20. From theoretical considerations, it is known that light from a certain star should reach Earth with intensity I_0. However, the path taken by the light from the star to Earth passes through a dust cloud, with absorption

coefficient 0.1/light year. The light reaching Earth has intensity $1/2\,I_0$. How thick is the dust cloud? (The rate of change of light intensity with respect to thickness is proportional to the intensity. One light year is the distance traveled by light during 1 yr.)

21. A snowball melts in such a way that the rate of change in its volume is proportional to its surface area. If the snowball was initially 4 in. in diameter and after 30 min its diameter is 3 in., when will its diameter be 2 in.? Mathematically speaking, when will the snowball disappear?

22. Suppose the snowball in Problem 21 melts so that the rate of change in its *diameter* is proportional to its surface area. Using the same given data, determine when its diameter will be 2 in. Mathematically speaking, when will the snowball disappear?

In Problems 23 through 27, assume the rate of decay of a radioactive substance is proportional to the amount of the substance present. The half-life of a radioactive substance is the time it takes for one-half of the substance to disintegrate.

23. If initially there are 50 g of a radioactive substance and after 3 days there are only 10 g remaining, what percentage of the original amount remains after 4 days?

24. If initially there are 300 g of a radioactive substance and after 5 years there are 200 g remaining, how much time must elapse before only 10 g remain?

25. Carbon dating is often used to determine the age of a fossil. For example, a humanoid skull was found in a cave in South Africa along with the remains of a campfire. Archaeologists believe the age of the skull to be the same age as the campfire. It is determined that only 2% of the original amount of carbon-14 remains in the burnt wood of the campfire. Estimate the age of the skull if the half-life of carbon-14 is about 5600 years.

26. To see how sensitive the technique of carbon dating of Problem 25 is,

(a) Redo Problem 25 assuming the half-life of carbon-14 is 5550 years.

(b) Redo Problem 25 assuming 3% of the original mass remains.

(c) If each of the figures in parts (a) and (b) represents a 1% error in measuring the two parameters of half-life and percent of mass remaining, to which parameter is the model more sensitive?

27. The only undiscovered isotopes of the two unknown elements hohum and inertium (symbols Hh and It) are radioactive. Hohum decays into inertium with a decay constant of 2/yr, and inertium decays into the nonradioactive isotope of bunkum (symbol Bu) with a decay constant of 1/yr. An initial mass of 1 kg of hohum is put into a nonradioactive container, with no other source of hohum, inertium, or bunkum. How much of each of the three elements is in the container after t years? (The **decay constant** is the constant of proportionality in the statement that the rate of loss of mass of the element at any time is proportional to the mass of the element at that time.)

3.3 HEATING AND COOLING OF BUILDINGS

Our goal is to formulate a mathematical model that describes the 24-hr temperature profile inside a building as a function of the outside temperature, the heat generated inside the building, and the furnace heating or air conditioning cooling. From this model, we would like to answer the following three questions:

(a) How long does it take to change the building temperature substantially?

(b) How does the building temperature vary during spring and fall when there is no furnace heating or air conditioning?

(c) How does the building temperature vary in summer when there is air conditioning or in the winter when there is furnace heating?

A natural approach to modeling the temperature inside a building is to use compartmental analysis. Let $T(t)$ represent the temperature inside the building at time t and view

the building as a single compartment. Then the rate of change in the temperature is determined by all the factors that generate or dissipate heat.

We will consider three main factors that affect the temperature inside the building. First is the heat produced by people, lights, and machines inside the building. This causes a rate of increase in temperature that we will denote by $H(t)$. Second is the heating (or cooling) supplied by the furnace (or air conditioning). This rate of increase (or decrease) in temperature will be represented by $U(t)$. In general, the additional heating rate $H(t)$ and the furnace (or air conditioning) rate $U(t)$ are described in terms of energy per unit time (such as British thermal units per hour). However, by multiplying by the heat capacity of the building (in units of degrees temperature change per unit heat energy), we can express the two quantities $H(t)$ and $U(t)$ in terms of temperature per unit time.

The third factor is the effect of the outside temperature $M(t)$ on the temperature inside the building. Experimental evidence has shown that this factor can be modeled using **Newton's law of cooling**. This law states that there is a rate of change in the temperature $T(t)$ that is proportional to the difference between the outside temperature $M(t)$ and the inside temperature $T(t)$. That is, the rate of change in the building temperature due to $M(t)$ is

$$K[M(t) - T(t)].$$

The positive constant K depends on the physical properties of the building, such as the number of doors and windows and the type of insulation, but K does not depend on M, T, or t. Hence, when the outside temperature is greater than the inside temperature, $M(t) - T(t) > 0$ and there is an increase in the rate of change of the building temperature due to $M(t)$. On the other hand, when the outside temperature is less than the inside temperature, then $M(t) - T(t) < 0$ and there is a decrease in this rate of change.

Summarizing we find

(1) $$\frac{dT}{dt} = K[M(t) - T(t)] + H(t) + U(t),$$

where the additional heating rate $H(t)$ is always nonnegative and $U(t)$ is positive for furnace heating and negative for air conditioner cooling. A more detailed model of the temperature dynamics of the building could involve more variables to represent different temperatures in different rooms or zones. Such an approach would use compartmental analysis, with the rooms as different compartments (see Problems 33–36, Exercises 4.10).

Because equation (1) is linear, it can be solved using the method discussed in Section 2.3. Rewriting (1) in the standard form

(2) $$\frac{dT}{dt}(t) + P(t)T(t) = Q(t),$$

where

$$P(t) := K,$$

(3) $$Q(t) := KM(t) + H(t) + U(t),$$

we find that the integrating factor is

$$\mu(t) = \exp\left(\int K\,dt\right) = e^{Kt}.$$

To solve (2), multiply each side by e^{Kt} and integrate:

$$e^{Kt}\frac{dT}{dt}(t) + Ke^{Kt}T(t) = e^{Kt}Q(t),$$

$$e^{Kt}T(t) = \int e^{Kt}Q(t)\,dt + C.$$

Solving for $T(t)$ gives

(4) $$T(t) = e^{-Kt}\int e^{Kt}Q(t)\,dt + Ce^{-Kt}$$

$$= e^{-Kt}\left\{\int e^{Kt}[KM(t) + H(t) + U(t)]\,dt + C\right\}.$$

EXAMPLE 1 Suppose that at the end of the day (at time t_0) when people leave the building, the outside temperature stays constant at M_0, the additional heating rate H inside the building is zero, and the furnace/air conditioner rate U is zero. Determine $T(t)$, given the initial condition $T(t_0) = T_0$.

Solution With $M = M_0$, $H = 0$, and $U = 0$, equation (4) becomes

$$T(t) = e^{-Kt}\left\{\int e^{Kt}KM_0\,dt + C\right\} = e^{-Kt}[M_0e^{Kt} + C]$$

$$= M_0 + Ce^{-Kt}.$$

Setting $t = t_0$ and using the initial value T_0 of the temperature, we find that the constant C is $(T_0 - M_0)\exp(KT_0)$. Hence,

(5) $$T(t) = M_0 + (T_0 - M_0)e^{-K(t-t_0)}. \quad \blacksquare \text{ End}$$

When $M_0 < T_0$, the solution in (5) decreases exponentially from the initial temperature T_0 to the final temperature M_0. To determine a measure of the time it takes for the temperature to change "substantially," consider the simple linear equation $dA/dt = -\alpha A$, whose solutions have the form $A(t) = A(0)e^{-\alpha t}$. Now as $t \to \infty$, the function $A(t)$ either decays exponentially ($\alpha > 0$) or grows exponentially ($\alpha < 0$). In either case, the time it takes for $A(t)$ to change from $A(0)$ to $A(0)/e$ ($\approx 0.368\,A(0)$) is just $1/\alpha$ because

$$A\left(\frac{1}{\alpha}\right) = A(0)e^{-\alpha(1/\alpha)} = \frac{A(0)}{e}.$$

The quantity $1/|\alpha|$, which is independent of $A(0)$, is called the **time constant** for the equation. For linear equations of the more general form $dA/dt = -\alpha A + g(t)$, we again refer to $1/|\alpha|$ as the time constant.

Returning to Example 1, we see that the temperature $T(t)$ satisfies the equations

$$\frac{dT}{dt}(t) = -KT(t) + KM_0, \qquad \frac{d(T - M_0)}{dt}(t) = -K[T(t) - M_0],$$

for M_0 a constant. In either case, the time constant is just $1/K$, which represents the time it takes for the temperature difference $T - M_0$ to change from $T_0 - M_0$ to $(T_0 - M_0)/e$. We also call $1/K$ the **time constant for the building** (without heating or air conditioning). A typical value for the time constant of a building is 2 to 4 hr, but the time constant can be much shorter if windows are open or if there is a fan circulating air. Or it can be much longer if the building is well insulated.

In the context of Example 1, we can use the notion of time constant to answer our initial question (a): The building temperature changes exponentially with a time constant of $1/K$. An answer to question (b) about the temperature inside the building during spring and fall is given in the next example.

EXAMPLE 2 Find the building temperature $T(t)$ if the additional heating rate $H(t)$ is equal to the constant H_0, there is no heating or cooling ($U(t) \equiv 0$), and the outside temperature M varies as a sine wave over a 24-hr period, with its minimum at $t = 0$ (midnight) and its maximum at $t = 12$ (noon). That is,

$$M(t) = M_0 - B \cos \omega t,$$

where B is a positive constant, M_0 is the average outside temperature, and $\omega = 2\pi/24 = \pi/12$ radians/hr. (This could be the situation during the spring or fall when there is neither furnace heating nor air conditioning.)

Solution The function $Q(t)$ in (3) is now

$$Q(t) = K(M_0 - B \cos \omega t) + H_0.$$

Setting $B_0 := M_0 + H_0/K$, we can rewrite Q as

(6) $$Q(t) = K(B_0 - B \cos \omega t),$$

where KB_0 represents the daily average value of $Q(t)$; that is,

$$KB_0 = \frac{1}{24} \int_0^{24} Q(t)\,dt.$$

When the forcing function $Q(t)$ in (6) is substituted into the expression for the temperature in equation (4), the result (after using integration by parts) is

$$T(t) = e^{-Kt}\left\{ \int e^{Kt}(KB_0 - KB \cos \omega t)\,dt + C \right\}$$

(7) $$T(t) = B_0 - BF(t) + Ce^{-Kt},$$

where

$$F(t) := \frac{\cos \omega t + (\omega/K)\sin \omega t}{1 + (\omega/K)^2}.$$

The constant C is chosen so that at midnight ($t = 0$), the value of the temperature T is equal to some initial temperature T_0. Thus

$$C = T_0 - B_0 + BF(0) = T_0 - B_0 + \frac{B}{1 + (\omega/K)^2}. \quad \blacksquare\ \text{End}$$

Notice that the third term in solution (7) involving the constant C tends to zero exponentially. The constant term B_0 in (7) is equal to $M_0 + H_0/K$ and represents the daily average temperature inside the building (neglecting the exponential term). When there is no additional heating rate inside the building ($H_0 = 0$), this average temperature is equal to the average outside temperature M_0. The term $BF(t)$ in (7) represents the sinusoidal variation of temperature inside the building responding to the outside temperature variation. Since $F(t)$ can be written in the form

(8) $F(t) = [1 + (\omega/K)^2]^{-1/2} \cos(\omega t - \phi),$

where $\tan \phi = \omega/K$ (see Problem 16), the sinusoidal variation inside the building lags behind the outside variation by ϕ hours. Further, the magnitude of the variation inside the building is slightly less, by a factor of $[1 + (\omega/K)^2]^{-1/2}$, than the outside variation. The angular frequency of variation ω is $2\pi/24$ radians/hr (which is about 1/4). Typical values for the dimensionless ratio ω/K lie between 1/2 and 1. For this range, the lag between inside and outside temperature is approximately 1.8 to 3 hr and the magnitude of the inside variation is between 89% and 71% of the variation outside. Figure 3.6 shows the 24-hr sinusoidal variation of the outside temperature for a typical moderate day as well as the temperature variations inside the building for a dimensionless ratio ω/K of unity, which corresponds to a time constant $1/K$ of approximately 4 hr. In sketching the latter curve, we have assumed that the exponential term has died out.

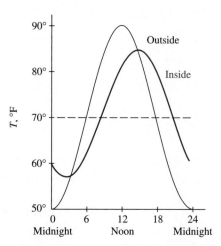

Figure 3.6 Temperature variation inside and outside an unheated building

EXAMPLE 3 Suppose that in the building of Example 2, a simple thermostat is installed that is used to compare the actual temperature inside the building with a desired temperature T_D. If the actual temperature is below the desired temperature, the furnace supplies heating; otherwise it is turned off. If the actual temperature is above the desired temperature, the air conditioner supplies cooling; otherwise it is off. (In practice, there is some dead zone around the desired temperature in which the temperature difference is not sufficient to activate the thermostat, but that is to be ignored here.) Assuming the amount of heating or cooling supplied is proportional to the difference in temperature—that is,

$$U(t) = K_U[T_D - T(t)],$$

where K_U is the (positive) proportionality constant—find $T(t)$.

Solution If the proportional control $U(t)$ is substituted directly into the differential equation (1) for the building temperature, we get

(9) $$\frac{dT(t)}{dt} = K[M(t) - T(t)] + H(t) + K_U[T_D - T(t)].$$

A comparison of equation (9) with the first order linear differential equation (2) shows that for this example, the quantity P is equal to $K + K_U$, while the quantity $Q(t)$ representing the forcing function includes the desired temperature T_D. That is,

$$P = K + K_U,$$

$$Q(t) = KM(t) + H(t) + K_U T_D.$$

When the additional heating rate is a constant H_0 and the outside temperature M varies as a sine wave over a 24-hr period in the same way as it did in Example 2, the forcing function is

$$Q(t) = K(M_0 - B \cos \omega t) + H_0 + K_U T_D.$$

The function $Q(t)$ has a constant term and a cosine term just as in equation (6). This equivalence becomes more apparent after substituting

(10) $$Q(t) = K_1(B_2 - B_1 \cos \omega t),$$

where

$$\omega := \frac{2\pi}{24} = \frac{\pi}{12}, \qquad\qquad K_1 := K + K_U,$$

(11) $$B_2 := \frac{K_U T_D + K M_0 + H_0}{K_1}, \qquad B_1 := \frac{BK}{K_1}.$$

The expressions for the constant P and the forcing function $Q(t)$ of equation (10) are the same as the expressions in Example 2, except that the constants K, B_0, and B are replaced, respectively, by the constants K_1, B_2, and B_1. Hence the solution to the differential equation (9) will be the same as the temperature solution in Example 2, except that the three constant terms are changed. Thus

(12) $$T(t) = B_2 - B_1 F_1(t) + C \exp(-K_1 t),$$

where

$$F_1(t) := \frac{\cos \omega t + (\omega/K_1) \sin \omega t}{1 + (\omega/K_1)^2} .$$

The constant C is chosen so that at time $t = 0$ the value of the temperature T equals T_0. Thus

$$C = T_0 - B_2 + B_1 F_1(0). \quad \blacksquare \text{ End}$$

In the above example, the time constant for equation (9) is $1/P = 1/K_1$, where $K_1 = K + K_U$. Here $1/K_1$ is referred to as the **time constant for the building with heating and air conditioning**. For a typical heating and cooling system, K_U is somewhat less than 2; for a typical building, the constant K is between 1/2 and 1/4. Hence the sum gives a value for K_1 of about 2, and the time constant for the building with heating and air conditioning is about 1/2 hr.

When the heating or cooling is turned on, it takes about 30 min for the exponential term in (12) to die off. If we neglect this exponential term, the average temperature inside the building is B_2. Since K_1 is much larger than K and H_0 is small, it follows from (11) that B_2 is roughly T_D, the desired temperature. In other words, after a certain period of time, the temperature inside the building is roughly T_D, with a small sinusoidal variation. (The outside average M_0 and inside heating rate H_0 have only a small effect.) Thus, to save energy, the heating or cooling system may be left off during the night. When it is turned on in the morning, it will take roughly 30 min for the inside of the building to attain the desired temperature. These observations provide an answer to question (c), regarding the temperature inside the building during summer and winter, that was posed at the beginning of this section.

The assumption made in Example 3 that the amount of heating or cooling is $U(t) = K_U[T_D - T(t)]$ may not always be suitable. We have used it here and in the exercises to illustrate the use of the time constant. More adventuresome readers may want to experiment with other models for $U(t)$, especially if they have available the numerical techniques discussed in Sections 3.5 and 3.6. Project G, page 149, addresses temperature regulation with fixed-rate controllers.

EXERCISES 3.3

1. A cup of hot coffee initially at 95°C cools to 80°C in 5 min while sitting in a room of temperature 21°C. Using just Newton's law of cooling, determine when the temperature of the coffee will be a nice 50°C.

2. A cold beer initially at 35°F warms up to 40°F in 3 min while sitting in a room of temperature 70°F. How warm will the beer be if left out for 20 min?

3. A white wine at room temperature 70°F is chilled in ice (32°F). If it takes 15 min for the wine to chill to 60°F, how long will it take for the wine to reach 56°F?

4. A red wine is brought up from the wine cellar, which is a cool 10°C, and left to breathe in a room of temperature 23°C. If it takes 10 min for the wine to reach 15°C, when will the temperature of the wine reach 18°C?

5. It was noon on a cold December day in Tampa: 16°C. Detective Taylor arrived at the crime scene to find the sergeant leaning over the body. The sergeant said there were several suspects. If they knew the exact time of death, then they could narrow the list. Detective Taylor

took out a thermometer and measured the temperature of the body: 34.5°C. He then left for lunch. Upon returning at 1:00 P.M., he found the body temperature to be 33.7°C. When did the murder occur? [*Hint:* Normal body temperature is 37°C.]

6. On a mild Saturday morning while people are working inside, the furnace keeps the temperature inside the building at 21°C. At noon the furnace is turned off and the people go home. The temperature outside is a constant 12°C the rest of the afternoon. If the time constant for the building is 3 hr, when will the temperature inside the building reach 16°C? If some windows are left open and the time constant drops to 2 hr, when will the temperature inside reach 16°C?

7. On a hot Saturday morning while people are working inside, the air conditioner keeps the temperature inside the building at 24°C. At noon the air conditioner is turned off and the people go home. The temperature outside is a constant 35°C the rest of the afternoon. If the time constant for the building is 4 hr, what will be the temperature inside the building at 2:00 P.M.? At 6:00 P.M.? When will the temperature inside the building reach 27°C?

8. A garage with no heating or cooling has a time constant of 2 hr. If the outside temperature varies as a sine wave with a minimum of 50°F at 2:00 A.M. and a maximum of 80°F at 2:00 P.M., determine the times at which the building reaches its lowest temperature and its highest temperature, assuming the exponential term has died off.

9. A warehouse is being built that will have neither heating nor cooling. Depending on the amount of insulation, the time constant for the building may range from 1 to 5 hr. To illustrate the effect insulation will have on the temperature inside the warehouse, assume the outside temperature varies as a sine wave, with a minimum of 16°C at 2:00 A.M. and a maximum of 32°C at 2:00 P.M. Assuming the exponential term (which involves the initial temperature T_0) has died off, what is the lowest temperature inside the building if the time constant is 1 hr? If the time constant is 5 hr? What is the highest temperature inside the building if the time constant is 1 hr? If it is 5 hr?

10. Early Monday morning, the temperature in the lecture hall has fallen to 40°F, the same as the temperature outside. At 7:00 A.M., the janitor turns on the furnace with the thermostat set at 70°F. The time constant for the building is $1/K = 2$ hr and that for the building along with its heating system is $1/K_1 = 1/2$ hr. Assum-

ing the outside temperature remains constant, what will be the temperature inside the lecture hall at 8:00 A.M.? When will the temperature inside the hall reach 65°F?

11. During the summer the temperature inside a van reaches 55°C while that outside is a constant 35°C. When the driver gets into the van, she turns on the air conditioner with the thermostat set at 16°C. If the time constant for the van is $1/K = 2$ hr and that for the van with its air conditioning system is $1/K_1 = 1/3$ hr, when will the temperature inside the van reach 27°C?

12. Two friends sit down to talk and enjoy a cup of coffee. When the coffee is served, the impatient friend immediately adds a teaspoon of cream to his coffee. The relaxed friend waits 5 min before adding a teaspoon of cream (which has been kept at a constant temperature). The two now begin to drink their coffee. *Who has the hotter coffee?* Assume the cream is cooler than the air and use Newton's law of cooling.

13. A solar hot water heating system consists of a hot water tank and a solar panel. The tank is well insulated and has a time constant of 64 hr. The solar panel generates 2000 Btu/hr during the day, and the tank has a heat capacity of 2°F per thousand Btu. If the water in the tank is initially 110°F and the room temperature outside the tank is 80°F, what will be the temperature in the tank after 12 hr of sunlight?

14. In Problem 13, if a larger tank with a heat capacity of 1°F per thousand Btu and a time constant of 72 hr is used instead (with all other factors being the same), what would be the temperature in the tank after 12 hr?

15. Stefan's law of radiation states that the rate of change of temperature of a body at T degrees Kelvin in a medium at M degrees Kelvin is proportional to $M^4 - T^4$. That is,

$$\frac{dT}{dt} = k(M^4 - T^4),$$

where k is a positive constant. Solve this equation using separation of variables. Explain why Newton's law and Stefan's law are nearly the same when T is close to M and M is constant. [*Hint:* Factor $M^4 - T^4$.]

16. Show that $C_1 \cos \omega t + C_2 \sin \omega t$ can be written in the form $A \cos(\omega t - \phi)$, where $A = \sqrt{C_1^2 + C_2^2}$ and $\tan \phi = C_2/C_1$. [*Hint:* Use a standard trigonometric identity with $C_1 = A \cos \phi$, $C_2 = A \sin \phi$.] Use this fact to verify the alternate representation (8) of $F(t)$ discussed in Example 2.

3.4 NEWTONIAN MECHANICS

Mechanics is the study of the motion of objects and the effect of forces acting on those objects. It is the foundation of several branches of physics and engineering. **Newtonian,** or **classical, mechanics** deals with the motion of **ordinary** objects, that is, objects that are large compared to an atom and slow moving compared with the speed of light. A model for Newtonian mechanics can be based on **Newton's laws of motion:**[†]

1. When a body is subject to no resultant external force, it moves with a constant velocity.
2. When a body is subject to one or more external forces, the time rate of change of the body's momentum is equal to the vector sum of the external forces acting on it.
3. When one body interacts with a second body, the force of the first body on the second is equal in magnitude, but opposite in direction, to the force of the second body on the first.

Experimental results for more than two centuries verify that these laws are extremely useful for studying the motion of ordinary objects in an **inertial reference frame**, that is, a reference frame in which an undisturbed body moves with a constant velocity. It is Newton's second law, which applies only to inertial reference frames, that enables us to formulate the equations of motion for a moving body. We can express Newton's second law by

(1) $$\frac{dp}{dt} = F(t, x, v),$$

where $F(t, x, v)$ is the resultant force on the body at time t, location x, and velocity v, and $p(t)$ is the momentum of the body at time t. The momentum is the product of the mass of the body and its velocity, that is,

$$p(t) = mv(t),$$

so we can express Newton's second law as

(2) $$m\frac{dv}{dt} = ma = F(t, x, v),$$

where $a = dv/dt$ is the acceleration of the body at time t. Observe that equation (2) will be a first order equation in $v(t)$, provided F does not depend on x.

To apply Newton's laws of motion to a problem in mechanics, the reader may find the following general procedure useful.

[†]For a discussion of Newton's laws of motion, see *University Physics*, Seventh Edition, by F. W. Sears, M. W. Zemansky, and H. D. Young (Addison-Wesley Publishing Co., Reading, Mass., 1987).

PROCEDURE FOR NEWTONIAN MODELS

(a) Determine *all* relevant forces acting on the object being studied. It is helpful to draw a simple diagram of the object that depicts these forces.

(b) Choose an appropriate axis or coordinate system in which to represent the motion of the object and the forces acting on it. Keep in mind that this coordinate system must be an inertial reference frame.

(c) Apply Newton's second law as expressed in equation (1) or equation (2), whichever is appropriate, to determine the equations of motion for the object.

In this section, we consider examples in which the motion of the object is one-dimensional and Newton's second law gives rise to a first order differential equation. We use either of two systems of units: the U.S. Customary System or the meter-kilogram-second (MKS) system. The various units in these systems are summarized in Table 3.2, along with approximate values for Earth's gravitational constant.

TABLE 3.2 MECHANICAL UNITS IN THE U.S. CUSTOMARY AND MKS SYSTEMS

Unit	U.S. Customary System	MKS System
Distance	feet (ft)	meters (m)
Mass	slugs	kilograms (kg)
Time	seconds (sec)	seconds (sec)
Force	pounds (lb)	newtons (N)
g (Earth)	32 ft/sec^2	9.81 m/sec^2

EXAMPLE 1 An object of mass m is given an initial downward velocity v_0 and allowed to fall under the influence of gravity. Assuming the gravitational force is constant and the force due to air resistance is proportional to the velocity of the object, determine the equation of motion for this object.

Solution There are two forces acting on the object: a constant force due to the downward pull of gravity and a force due to air resistance that is proportional to the velocity of the object and acts in opposition to the motion of the object. Hence the motion of the object will take place along a vertical axis. On this axis, we choose the origin to be the point where the object was initially dropped and let $x(t)$ denote the distance the object has fallen in time t (see Figure 3.7, page 110).

The forces acting on the object along this axis can be expressed as follows. The force due to gravity is

$$F_1 = mg,$$

Figure 3.7 Forces on a falling object

where g is the acceleration due to gravity at the Earth's surface (see Table 3.2). The force due to air resistance is

$$F_2 = -kv(t),$$

where k (>0) is the proportionality constant and the negative sign is present because air resistance acts in opposition to the motion of the object. Hence the net force acting on the object (see Figure 3.7) is

(3) $F = F_1 + F_2 = mg - kv(t).$

We now apply Newton's second law by substituting (3) into (2) to obtain

$$m\frac{dv}{dt} = mg - kv.$$

Since the initial velocity of the object is v_0, a model for the velocity of the falling body is expressed by the initial value problem

(4) $m\dfrac{dv}{dt} = mg - kv, \qquad v(0) = v_0,$

where g and k are positive constants.

The model (4) is the same as the one we obtained in Section 2.1. Using separation of variables (as we did previously), we get

(5) $v(t) = \dfrac{mg}{k} + \left(v_0 - \dfrac{mg}{k}\right)e^{-kt/m}.$

Since we have taken $x = 0$ when $t = 0$, we can determine the equation of motion of the object by integrating $v = dx/dt$ with respect to t. Thus from (5) we obtain

$$x(t) = \int v(t)\,dt = \frac{mg}{k}t - \frac{m}{k}\left(v_0 - \frac{mg}{k}\right)e^{-kt/m} + c,$$

and setting $x = 0$ when $t = 0$, we find

$$0 = -\frac{m}{k}\left(v_0 - \frac{mg}{k}\right) + c,$$

$$c = \frac{m}{k}\left(v_0 - \frac{mg}{k}\right).$$

Hence the equation of motion is

(6) $$x(t) = \frac{mg}{k}t + \frac{m}{k}\left(v_0 - \frac{mg}{k}\right)(1 - e^{-kt/m}).$$ ■ End

In Figure 3.8, we have sketched the graphs of the velocity and the position as functions of t. Observe that the velocity $v(t)$ approaches the horizontal asymptote $v = mg/k$ as $t \to \infty$ and that the position $x(t)$ asymptotically approaches the line

$$x = \frac{mg}{k}t - \frac{m^2g}{k^2} + \frac{m}{k}v_0$$

as $t \to +\infty$. The value mg/k of the horizontal asymptote for $v(t)$ is called the **limiting**, or **terminal, velocity** of the object.

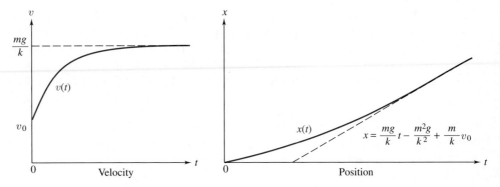

Figure 3.8 Graphs of the position and velocity of a falling object

Now that we have obtained the equation of motion for a falling object with air resistance proportional to v, we can answer a variety of questions.

EXAMPLE 2 An object of mass 3 kg is released from rest 500 m above the ground and allowed to fall under the influence of gravity. Assume the gravitational force is constant, with $g = 9.81$ m/sec^2, and the force due to air resistance is proportional to the velocity of the object[†] with proportionality constant $k = 3$ kg/sec. Determine when the object will strike the ground.

Solution We can use the model discussed in Example 1 with $v_0 = 0$, $m = 3$, $k = 3$, and $g = 9.81$. From (6), the equation of motion in this case is

$$x(t) = \frac{(3)(9.81)}{3}t - \frac{(3)^2(9.81)}{(3)^2}(1 - e^{-3t/3}) = (9.81)t - (9.81)(1 - e^{-t}).$$

[†]The effects of more sophisticated air resistance models (such as a quadratic drag law) are analyzed numerically in Exercises 3.5, Problem 20.

Because the object is released 500 m above the ground, we can determine when the object strikes the ground by setting $x(t) = 500$ and solving for t. Thus we put

$$500 = (9.81)t - 9.81 + (9.81)e^{-t}$$

or

$$t + e^{-t} = \frac{509.81}{9.81} = 51.97,$$

where we have rounded the computations to two decimal places. Unfortunately this equation cannot be solved explicitly for t. We might try to approximate t using Newton's approximation method (see Appendix A), but in this case, it is not necessary. Since e^{-t} will be very small for t near 51.97 ($e^{-51.97} \approx 10^{-22}$), we simply ignore the term e^{-t} and obtain as our approximation $t = 51.97$ sec. ■ End

EXAMPLE 3 A parachutist whose mass is 75 kg drops from a helicopter hovering 4000 m above the ground and falls toward the earth under the influence of gravity. Assume the gravitational force is constant. Assume also that the force due to air resistance is proportional to the velocity of the parachutist, with the proportionality constant $k_1 = 15$ kg/sec when the chute is closed and with constant $k_2 = 105$ kg/sec when the chute is open. If the chute does not open until 1 min after the parachutist leaves the helicopter, after how many seconds will she hit the ground?

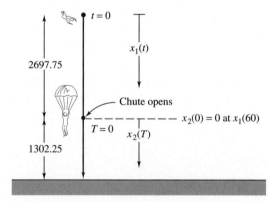

Figure 3.9 The fall of the parachutist

Solution We are interested only in when the parachutist will hit the ground, not where. Thus we consider only the vertical component of her descent. For this, we need to use two equations: one to describe the motion before the chute opens and the other to apply after it opens. Before the chute opens, the model is the same as in Example 1 with $v_0 = 0$, $m = 75$ kg, $k = k_1 = 15$ kg/sec, and $g = 9.81$ m/sec². If we let $x_1(t)$ be the distance the parachutist has

fallen in t seconds and let $v_1 = dx_1/dt$, then substituting into equations (5) and (6), we have

$$v_1(t) = \frac{(75)(9.81)}{15}(1 - e^{-(15/75)t})$$

$$= (49.05)(1 - e^{-0.2t}),$$

and

$$x_1(t) = \frac{(75)(9.81)}{15}t - \frac{(75)^2(9.81)}{(15)^2}(1 - e^{-(15/75)t})$$

$$= 49.05t - 245.25(1 - e^{-0.2t}).$$

Hence, after 1 min, when $t = 60$, the parachutist is falling at the rate

$$v_1(60) = (49.05)(1 - e^{-0.2(60)}) = 49.05 \text{ m/sec},$$

and has fallen

$$x_1(60) = (49.05)(60) - (245.25)(1 - e^{-0.2(60)}) = 2697.75 \text{ m}.$$

(In these and other computations for this problem, we round our answers to two decimal places.)

Now when the chute opens, the parachutist is $4000 - 2697.75$ or 1302.25 m above the ground and traveling at a velocity of 49.05 m/sec. To determine the equation of motion after the chute opens, let $x_2(T)$ denote the position of the parachutist T seconds after the chute opens (so that $T = t - 60$), taking $x_2(0) = 0$ at $x_1(60)$ (see Figure 3.9). Further assume the initial velocity of the parachutist after the chute opens is the same as the final velocity before it opens, that is, $x_2'(0) = x_1'(60) = 49.05$ m/sec. Because the forces acting on the parachutist are the same as those acting on the object in Example 1, we can again use equations (5) and (6). With $v_0 = 49.05$, $m = 75$, $k = k_2 = 105$, and $g = 9.81$, we find from (6) that

$$x_2(T) = \frac{(75)(9.81)}{105}T + \frac{75}{105}\left[49.05 - \frac{(75)(9.81)}{105}\right](1 - e^{-(105/75)T})$$

$$= 7.01T + 30.03(1 - e^{-1.4T}).$$

To determine when the parachutist will hit the ground, we set $x_2(T) = 1302.25$, the height the parachutist was above the ground when her parachute opened. This gives

$$7.01T + 30.03 - 30.03e^{-1.4T} = 1302.25$$

(7) $\qquad T - 4.28e^{-1.4T} - 181.49 = 0.$

Again we cannot solve (7) explicitly for T. However, observe that $e^{-1.4T}$ is very small for T near 181.49, so we ignore the exponential term and obtain $T = 181.49$. Hence the parachutist will strike the earth 181.49 sec after the parachute opens, or 241.49 sec after dropping from the helicoper. ∎ End

In the computation for T in equation (7), we found that the exponential $e^{-1.4T}$ was negligible. Consequently, ignoring the corresponding exponential term in equation (5), we see that the parachutist's velocity at impact is

$$\frac{mg}{k_2} = \frac{(75)(9.81)}{105} = 7.01 \text{ m/sec},$$

which is the limiting velocity for her fall with the chute open.

Unless otherwise stated, in the following problems we assume the gravitational force is constant with g = 9.81 m/sec² in the MKS system and g = 32 ft/sec² in the U.S. Customary System.

1. An object of mass 5 kg is released from rest 1000 m above the ground and allowed to fall under the influence of gravity. Assuming the force due to air resistance is proportional to the velocity of the object with proportionality constant $k = 50$ kg/sec, determine the equation of motion of the object. When will the object strike the ground?

2. A 400-lb object is released from rest 500 ft above the ground and allowed to fall under the influence of gravity. Assuming the force in pounds due to air resistance is $-10v$, where v is the velocity of the object in ft/sec, determine the equation of motion of the object. When will the object hit the ground?

3. If the object in Problem 1 has a mass of 500 kg instead of 5 kg, when will it strike the ground? [*Hint:* Here the exponential term is too large to ignore. Use Newton's method to approximate the time t when the object strikes the ground (see Appendix A).]

4. If the object in Problem 2 is released from rest 30 ft above the ground instead of 500 ft, when will it strike the ground? [*Hint:* Use Newton's method to solve for t.]

5. An object of mass 5 kg is given an initial downward velocity of 50 m/sec and then allowed to fall under the influence of gravity. Assume the force in newtons due to air resistance is $-10v$, where v is the velocity of the object in m/sec. Determine the equation of motion of the object. If the object is initially 500 m above the ground, determine when the object will strike the ground.

6. An object of mass 8 kg is given an upward initial velocity of 20 m/sec and then allowed to fall under the influence of gravity. Assume that the force in newtons due to air resistance is $-16v$, where v is the velocity of the object in m/sec. Determine the equation of motion of the object. If the object is initially 100 m above the ground, determine when the object will strike the ground.

7. A parachutist whose mass is 75 kg drops from a helicopter hovering 2000 m above the ground and falls toward the ground under the influence of gravity. Assume the force due to air resistance is proportional to the velocity of the parachutist, with the proportionality constant $k_1 = 30$ kg/sec when the chute is closed and $k_2 = 90$ kg/sec when the chute is open. If the chute does not open until the velocity of the parachutist reaches 20 m/sec, after how many seconds will she reach the ground?

8. A parachutist whose mass is 100 kg drops from a helicopter hovering 3000 m above the ground and falls under the influence of gravity. Assume the force due to air resistance is proportional to the velocity of the parachutist, with the proportionality constant $k_3 = 20$ kg/sec when the chute is closed and $k_4 = 100$ kg/sec when the chute is open. If the chute does not open until 30 sec after the parachutist leaves the helicopter, after how many seconds will he hit the ground? If the chute does not open until 1 min after he leaves the helicopter, after how many seconds will he hit the ground?

9. An object of mass 100 kg is released from rest from a boat into the water and allowed to sink. While gravity is pulling the object down, a buoyancy force of 1/40 times the weight of the object is pushing the object up (weight = mg). If we assume water resistance exerts a force on the object that is proportional to the velocity of the object, with proportionality constant 10 kg/sec, find the equation of motion of the object. After how many seconds will the velocity of the object be 70 m/sec?

10. An object of mass 2 kg is released from rest from a platform 30 m above the water and allowed to fall under the influence of gravity. After the object strikes the water, it begins to sink with gravity pulling down and a buoyancy force pushing up. Assume the force of gravity is constant, no change in momentum occurs on impact with the water, the buoyancy force is 1/2 the weight (weight = mg), and the force due to air resistance or water resistance is proportional to the velocity, with proportionality constant $k_1 = 10$ kg/sec in the air and $k_2 = 100$ kg/sec in the water. Find the equation of motion of the object. What is the velocity of the object 1 min after it is released?

11. In Example 1, we solved for the velocity of the object as a function of time (equation (5)). In some cases, it

is useful to have an expression, independent of t, that relates v and x. Find this relation for the motion in Example 1. [*Hint:* Letting $v(t) = V(x(t))$, then $dv/dt = (dV/dx)V$.]

12. A shell of mass 2 kg is shot upward with an initial velocity of 200 m/sec. The magnitude of the force on the shell due to air resistance is $|v|/20$. When will the shell reach its maximum height above the ground? What is the maximum height?

13. When the velocity v of an object is very large, the magnitude of the force due to air resistance is proportional to v^2 with the force acting in opposition to the motion of the object. A shell of mass 3 kg is shot upward from the ground with an initial velocity of 500 m/sec. If the magnitude of the force due to air resistance is $(0.1)v^2$, when will the shell reach its maximum height above the ground? What is the maximum height?

14. An object of mass m is released from rest and falls under the influence of gravity. If the magnitude of the force due to air resistance is kv^n, where k and n are positive constants, find the limiting velocity of the object (assuming this limit exists). [*Hint:* Argue that the existence of a (finite) limiting velocity implies that $dv/dt \to 0$ as $t \to \infty$.]

15. A rotating flywheel is being turned by a motor that exerts a constant torque T (see Figure 3.10). A retarding torque due to friction is proportional to the angular velocity ω. If the moment of inertia of the flywheel is I and its initial angular velocity is ω_0, find the equation for the angular velocity ω as a function of time. [*Hint:* Use Newton's second law for rotational motion, that is, moment of inertia × angular acceleration = sum of the torques.]

Figure 3.10 Motor-driven flywheel

16. Find the equation for the angular velocity ω in Problem 15, assuming the retarding torque is proportional to $\sqrt{\omega}$.

17. In Problem 16, let $I = 50$ kg-m^2 and the retarding torque be $5\sqrt{\omega}$ N-m. If the motor is turned off with the angular velocity at 225 rad/sec, determine how long it will take for the flywheel to come to rest.

18. When an object slides on a surface, it encounters a resistance force called **friction**. This force has a magnitude of μN, where μ is the **coefficient of kinetic friction** and N is the magnitude of the normal force that the surface applies to the object. Suppose that an object of mass 30 kg is released from the top of an inclined plane that is inclined 30° to the horizontal (see Figure 3.11). Assume the gravitational force is constant, air resistance is negligible, and the coefficient of kinetic friction $\mu = 0.2$. Determine the equation of motion for the object as it slides down the plane. If the top surface of the plane is 5 m long, what is the velocity of the object when it reaches the bottom?

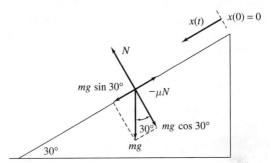

Figure 3.11 Forces on an object on an inclined plane

19. An object of mass 60 kg starts from rest at the top of a 45° inclined plane. Assume that the coefficient of kinetic friction is 0.05 (see Problem 18). If the force due to air resistance is proportional to the velocity of the object, say $-3v$, find the equation of motion of the object. How long will it take the object to reach the bottom of the inclined plane if the incline is 10 m long?

20. An object at rest on an inclined plane will not slide until the component of the gravitational force down the incline is sufficient to overcome the force due to **static friction**. Static friction is governed by an experimental law somewhat like that of kinetic friction (Problem 18); it has a magnitude of *at most* μN, where μ is the coefficient of static friction and N is, again, the magnitude

of the normal force exerted by the surface on the object. If the plane is inclined at an angle α, determine the critical value α_0 for which the object will slide if $\alpha > \alpha_0$ but will not move for $\alpha < \alpha_0$.

21. A sailboat has been running (on a straight course) under a light wind at 1 m/sec. Suddenly the wind picks up, blowing hard enough to apply a constant force of 600 N to the sailboat. The only other force acting on the boat is water resistance that is proportional to the velocity of the boat. If the proportionality constant for water resistance is $k = 100$ kg/sec and the mass of the sailboat is 50 kg, find the equation of motion of the sailboat. What is the limiting velocity of the sailboat under this wind?

22. In Problem 21, it is observed that when the velocity of the sailboat reaches 5 m/sec, the boat begins to rise out of the water and "plane." When this happens, the proportionality constant for the water resistance drops to $k_0 = 60$ kg/sec. Now find the equation of motion of the sailboat. What is the limiting velocity of the sailboat under this wind as it is planing?

23. Sailboats A and B each have a mass of 60 kg and cross the starting line at the same time on the first leg of a race. Each has an initial velocity of 2 m/sec. The wind applies a constant force of 650 N to each boat, and the force due to water resistance is proportional to the velocity of the boat. For sailboat A, the proportionality constants are $k_1 = 80$ kg/sec before planing when the velocity is less than 5 m/sec, and $k_2 = 60$ kg/sec when the velocity is above 5 m/sec. For sailboat B, the proportionality constants are $k_3 = 100$ kg/sec before planing when the velocity is less than 6 m/sec and $k_4 = 50$ kg/sec when the velocity is above 6 m/sec. If the first leg of the race is 500 m long, which sailboat will be leading at the end of the first leg?

24. Rocket Flight. A model rocket having initial mass m_0 kg is launched vertically from the ground. The rocket expels gas at a constant rate of α kg/sec and at a constant velocity of β m/sec relative to the rocket. Assume the magnitude of the gravitational force is proportional to the mass with proportionality constant g. Because the mass is not constant, Newton's second law leads to the equation

$$(m_0 - \alpha t)\frac{dv}{dt} - \alpha\beta = -g(m_0 - \alpha t),$$

where $v = dx/dt$ is the velocity of the rocket, x is its height above the ground, and $m_0 - \alpha t$ is the mass of

the rocket at t seconds after launch. If the initial velocity is zero, solve the above equation to determine the velocity of the rocket and its height above ground for $0 \le t < m_0/\alpha$.

25. Escape Velocity. According to Newton's law of gravitation, the attractive force between two objects varies inversely as the square of the distance between them. That is, $F_g = GM_1M_2/r^2$, where M_1 and M_2 are the masses of the objects, r is the distance between them (center to center), F_g is the attractive force, and G is the constant of proportionality. Consider a projectile of constant mass m being fired vertically from Earth (see Figure 3.12). Let t represent time and v the velocity of the projectile.

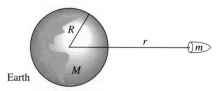

Figure 3.12 Projectile escaping from Earth

(a) Show that the motion of the projectile, under Earth's gravitational force, is governed by the equation

$$\frac{dv}{dt} = -\frac{gR^2}{r^2},$$

where r is the distance between the projectile and the center of Earth, R is the radius of Earth, M is the mass of Earth, and $g = GM/R^2$.

(b) Use the fact that $dr/dt = v$ to obtain

$$v\frac{dv}{dr} = -\frac{gR^2}{r^2}.$$

(c) If the projectile leaves Earth's surface with velocity v_0, show that

$$v^2 = \frac{2gR^2}{r} + v_0^2 - 2gR.$$

(d) Use the result of part (c) to show that the velocity of the projectile remains positive if and only if $v_0^2 - 2gR > 0$. The velocity $v_e = \sqrt{2gR}$ is called the **escape velocity** of Earth.

(e) If $g = 9.81$ m/sec^2 and $R = 6370$ km for Earth, what is Earth's escape velocity?

(f) If the acceleration due to gravity for the moon is $g_m = g/6$ and the radius of the moon is $R_m = 1738$ km, what is the escape velocity of the moon?

3.5 IMPROVED EULER'S METHOD

While the analytical techniques presented in Chapter 2 were useful for the variety of mathematical models presented earlier in this chapter, *the majority of the differential equations encountered in applications cannot be solved either implicitly or explicitly*. This is especially true of higher-order equations and systems of equations, which we study in later chapters. In this section and the next, we discuss methods for obtaining a numerical *approximation* of the solution to an initial value problem for a first order differential equation. Our goal is to develop algorithms that you can use with a calculator or microcomputer. These algorithms also extend naturally to higher-order equations (see Section 5.6). We describe the rationale behind each method but leave the more detailed discussion to texts on numerical analysis.[†]

Consider the initial value problem

(1) $y' = f(x, y), \qquad y(x_0) = y_0.$

To guarantee that (1) has a unique solution, we assume f and $\partial f/\partial y$ are continuous in a rectangle $R := \{(x, y): a < x < b, c < y < d\}$ containing (x_0, y_0). It follows from Theorem 1 in Chapter 1 that the initial value problem (1) has a unique solution $\phi(x)$ in some interval $x_0 - \delta < x < x_0 + \delta$, where δ is a positive number. Because δ is not known a priori, there is no assurance that the solution will exist at a particular point $x(\neq x_0)$, even if x is in the interval (a, b). However, if $\partial f/\partial y$ is continuous and *bounded*[††] on the vertical strip

$$S := \{(x, y): a < x < b, -\infty < y < \infty\},$$

then it turns out that (1) has a unique solution on the whole interval (a, b). In describing numerical methods, we assume this last condition is satisfied and that f possesses as many continuous partial derivatives as needed.

In Section 1.4, we used the concept of direction fields to motivate a scheme for approximating the solution to the initial value problem (1). This scheme, called **Euler's method**, is one of the most basic, so it is worthwhile to discuss its advantages, disadvantages, and possible improvements. We begin with a derivation of Euler's method that is somewhat different from that presented in Section 1.4.

Let $h > 0$ be fixed (h is called the **step size**) and consider the equally spaced points

(2) $x_n := x_0 + nh, \qquad n = 0, 1, 2, \ldots.$

Our goal is to obtain an approximation to the solution $\phi(x)$ of the initial value problem (1) at those points x_n that lie in the interval (a, b). Namely, we will describe a method that generates values $y_0, y_1, y_2, \ldots$ that approximate $\phi(x)$ at the respective points $x_0, x_1, x_2, \ldots$; that is

$$y_n \approx \phi(x_n), \qquad n = 0, 1, 2, \ldots.$$

[†] See, for example, *Elements of Numerical Analysis*, by P. Henrici (John Wiley and Sons, Inc., New York, 1964) or *Numerical Analysis*, Fifth Edition, by R. L. Burden and J. D. Faires (Prindle, Weber & Schmidt, Boston, 1993).

[††] A function $g(x, y)$ is bounded on S if there exists a number M such that $|g(x, y)| \leq M$ for all (x, y) in S.

Of course, the first "approximant" y_0 is exact, since $y_0 = \phi(x_0)$ is given. Thus we must describe how to compute $y_1, y_2, \dots$.

For Euler's method, we begin by integrating both sides of equation (1) from x_n to x_{n+1} to obtain

$$\phi(x_{n+1}) - \phi(x_n) = \int_{x_n}^{x_{n+1}} \phi'(t)\,dt = \int_{x_n}^{x_{n+1}} f(t,\phi(t))\,dt,$$

where we have substituted $\phi(x)$ for y. Solving for $\phi(x_{n+1})$, we have

(3) $$\phi(x_{n+1}) = \phi(x_n) + \int_{x_n}^{x_{n+1}} f(t,\phi(t))\,dt.$$

Without knowing $\phi(t)$, we cannot integrate $f(t,\phi(t))$. Hence, we must approximate the integral in (3). Assuming we have already found $y_n \approx \phi(x_n)$, the simplest approach is to approximate the area under the function $f(t,\phi(t))$ by the rectangle with base $[x_n, x_{n+1}]$ and height $f(x_n,\phi(x_n))$ (see Figure 3.13). This gives

$$\phi(x_{n+1}) \approx \phi(x_n) + (x_{n+1} - x_n)f(x_n,\phi(x_n)).$$

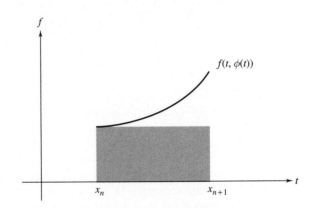

Figure 3.13 Approximation by a rectangle

Substituting h for $x_{n+1} - x_n$ and the approximation y_n for $\phi(x_n)$, we arrive at the numerical scheme

(4) $$y_{n+1} = y_n + hf(x_n, y_n), \qquad n = 0, 1, 2, \dots,$$

which is Euler's method.

Starting with the given value y_0, we use (4) to compute $y_1 = y_0 + hf(x_0, y_0)$ and then use y_1 to compute $y_2 = y_1 + hf(x_1, y_1)$, and so on. Several examples of Euler's method can be found in Section 1.4.

As discussed in Section 1.4, if we wish to use Euler's method to approximate the solution to the initial value problem (1) at a particular value of x, say $x = c$, then we must

first determine a suitable step size h so that $x_0 + Nh = c$ for some integer N. For example, we can take $N = 1$ and $h = c - x_0$ in order to arrive at the approximation after just one step:

$$\phi(c) = \phi(x_0 + h) \approx y_1.$$

If, instead, we wish to take 10 steps in Euler's method, we choose $h = (c - x_0)/10$ and ultimately obtain

$$\phi(c) = \phi(x_0 + 10h) = \phi(x_{10}) \approx y_{10}.$$

In general, depending on the size of h, we will get different approximations to $\phi(c)$. It is reasonable to expect that as h gets smaller (or, equivalently, as N gets larger), the Euler approximations approach the exact value $\phi(c)$. On the other hand, as h gets smaller, the number (and cost) of computations increases and hence so do machine errors that arise from round-off. Thus it is important to analyze how the error in the approximation scheme varies with h.

If Euler's method is used to approximate the solution $\phi(x) = e^x$ to the problem

(5) $y' = y, \qquad y(0) = 1,$

at $x = 1$, then we obtain approximations to the constant $e = \phi(1)$. It turns out that these approximations take a particularly simple form that enables us to compare the error in the approximation with the step size h. Indeed, setting $f(x, y) = y$ in (4) yields

$$y_{n+1} = y_n + hy_n = (1 + h)y_n, \qquad n = 0, 1, 2, \ldots.$$

Since $y_0 = 1$, we get

$$y_1 = (1 + h)y_0 = 1 + h$$
$$y_2 = (1 + h)y_1 = (1 + h)(1 + h) = (1 + h)^2$$
$$y_3 = (1 + h)y_2 = (1 + h)(1 + h)^2 = (1 + h)^3,$$

and, in general,

(6) $y_n = (1 + h)^n, \qquad n = 0, 1, 2, \ldots.$

For the problem in (5) we have $x_0 = 0$, so to obtain approximations at $x = 1$, we must set $nh = 1$. That is, h must be the reciprocal of an integer ($h = 1/n$). Replacing n by $1/h$ in (6), we see that Euler's method gives the (familiar) approximation $(1 + h)^{1/h}$ to the constant e. In Table 3.3 on page 120, we list this approximation for $h = 1, 10^{-1}, 10^{-2}, 10^{-3}, 10^{-4}$, along with the corresponding errors

$$e - (1 + h)^{1/h}.$$

From the second and third columns in Table 3.3, we see that the approximation gains roughly one decimal place in accuracy as h decreases by a factor of 10; that is, the error is roughly proportional to h. This observation is further confirmed by the entries in the last column of Table 3.3. In fact, using methods of calculus (see Exercises 1.4, Problem 13, page

	Euler's Approximation	**Error**	**Error**
h	$(1 + h)^{1/h}$	$e - (1 + h)^{1/h}$	h
1	2.00000	0.71828	0.71828
10^{-1}	2.59374	0.12454	1.24539
10^{-2}	2.70481	0.01347	1.34680
10^{-3}	2.71692	0.00136	1.35790
10^{-4}	2.71815	0.00014	1.35902

TABLE 3.3 EULER'S APPROXIMATIONS TO $e = 2.71828\ldots$

29), it can be shown that

$$(7) \qquad \lim_{h \to 0} \frac{\text{error}}{h} = \lim_{h \to 0} \frac{e - (1 + h)^{1/h}}{h} = \frac{e}{2} \approx 1.35914.$$

The general situation is similar: When Euler's method is used to approximate the solution to the initial value problem (1), the error in the approximation is at worst a constant times the step size h. Moreover, in view of (7), this is the best one can say.

Numerical analysts have a convenient notation for describing the convergence behavior of a numerical scheme. For fixed x, we denote by $y(x; h)$ the approximation to the solution $\phi(x)$ of (1) obtained via the scheme when using a step size of h. We say that the numerical scheme **converges** at x if

$$\lim_{h \to 0} y(x; h) = \phi(x).$$

In other words, as the step size h decreases to zero, the approximations for a convergent scheme approach the exact value $\phi(x)$. The rate at which $y(x; h)$ tends to $\phi(x)$ is often expressed in terms of a suitable power of h. If the error $\phi(x) - y(x; h)$ tends to zero like a constant times h^p, we write

$$\phi(x) - y(x; h) = O(h^p)$$

and say that the **method is of order p**. Of course, the higher the power p, the faster is the rate of convergence as $h \to 0$.

As seen from our earlier discussion, the rate of convergence of Euler's method is $O(h)$; that is, *Euler's method is of order $p = 1$*. In fact, the limit in (7) shows that for equation (5), the error is roughly $1.36h$ for small h. This means that to have an error less than .01 requires $h < .01/1.36$, or $n = 1/h > 136$ computation steps. Thus Euler's method converges too slowly to be of practical use.

How can we improve Euler's method? To answer this, let's return to the derivation expressed in formulas (3) and (4) and analyze the "errors" that were introduced to get the approximation. The crucial step in the process was to approximate the integral

$$\int_{x_n}^{x_{n+1}} f(t, \phi(t))\, dt$$

by using a rectangle (recall Figure 3.13). This step gives rise to what is called the local truncation error in the method. From calculus, we know that a better (more accurate) approach to approximating the integral is to use a trapezoid; that is, to apply the trapezoidal rule (see Figure 3.14). This gives

$$\int_{x_n}^{x_{n+1}} f(t,\phi(t))\, dt \approx \frac{h}{2}\left[f(x_n,\phi(x_n)) + f(x_{n+1},\ \phi(x_{n+1}))\right],$$

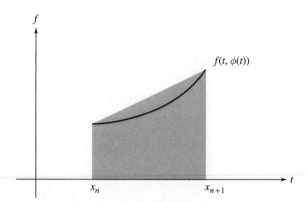

Figure 3.14 Approximation by a trapezoid

which leads to the numerical scheme

$$(8) \qquad y_{n+1} = y_n + \frac{h}{2}\left[f(x_n, y_n) + f(x_{n+1}, y_{n+1})\right], \qquad n = 0, 1, 2, \ldots.$$

We call equation (8) the **trapezoid scheme**. It is an example of an **implicit method**. That is, unlike Euler's method, equation (8) gives only an implicit formula for y_{n+1}, since y_{n+1} appears as an argument of f. Assuming we have already computed y_n, some root-finding technique such as Newton's method (see Appendix A) might be needed to compute y_{n+1}. Despite the inconvenience of working with an implicit method, the trapezoid scheme has two advantages over Euler's method. First, it is a method of order $p = 2$; that is, it converges at a rate that is proportional to h^2 and hence is faster than Euler's method. Second, as described in Project E, the trapezoid scheme has the desirable feature of being **stable**.

Can we somehow modify the trapezoid scheme in order to obtain an explicit method? One idea is first to get an estimate, say, y_{n+1}^*, of the value y_{n+1} using Euler's method and then use formula (8) with y_{n+1} replaced by y_{n+1}^* on the right-hand side. This two-step process is an example of a **predictor-corrector method**. That is, we predict y_{n+1} using Euler's method and then use that value in (8) to obtain a "more correct" approximation. Setting $y_{n+1} = y_n + hf(x_n, y_n)$ in the right-hand side of (8), we obtain

$$(9) \qquad y_{n+1} = y_n + \frac{h}{2}\left[f(x_n, y_n) + f(x_n + h, y_n + hf(x_n, y_n))\right], \qquad n = 0, 1, \ldots,$$

where $x_{n+1} = x_n + h$. This explicit scheme is known as the **improved Euler's method**.

EXAMPLE 1 Compute the improved Euler's method approximation to the solution $\phi(x) = e^x$ of

$$y' = y, \qquad y(0) = 1$$

at $x = 1$ using step sizes of $h = 1,\ 10^{-1},\ 10^{-2},\ 10^{-3},\ 10^{-4}$.

Solution The starting values are $x_0 = 0$ and $y_0 = 1$. Since $f(x, y) = y$, formula (9) becomes

$$y_{n+1} = y_n + \frac{h}{2}[y_n + (y_n + hy_n)] = y_n + hy_n + \frac{h^2}{2}y_n,$$

that is,

(10) $$y_{n+1} = \left(1 + h + \frac{h^2}{2}\right)y_n.$$

Since $y_0 = 1$, we see inductively that

$$y_n = \left(1 + h + \frac{h^2}{2}\right)^n, \qquad n = 0, 1, 2, \ldots .$$

To obtain approximations at $x = 1$, we must have $1 = x_0 + nh = nh$, and so $n = 1/h$. Hence the improved Euler's approximations to $e = \phi(1)$ are just

$$\left(1 + h + \frac{h^2}{2}\right)^{1/h}.$$

In Table 3.4, we have computed this approximation for the specified values of h, along with the corresponding errors

$$e - \left(1 + h + \frac{h^2}{2}\right)^{1/h}.$$

	Approximation $\left(1 + h + \dfrac{h^2}{2}\right)^{1/h}$		
TABLE 3.4	**IMPROVED EULER'S APPROXIMATION TO $e = 2.71828\ldots$**		
h	$\left(1 + h + \dfrac{h^2}{2}\right)^{1/h}$	**Error**	$\dfrac{\textbf{Error}}{h^2}$
1	2.50000	0.21828	21.828
10^{-1}	2.71408	0.00420	0.4201
10^{-2}	2.71824	0.00004	0.4496
10^{-3}	2.71828	0.00000	0.5430
10^{-4}	2.71828	0.00000	0.0000

Comparing the entries of this table with those of Table 3.3, we observe that the improved Euler's method converges much more rapidly than the original Euler's method. In fact, from the first few entries in the second and third columns of Table 3.4, it appears that the

approximation gains two decimal places in accuracy each time h is decreased by a factor of 10. In other words, the error is roughly proportional to h^2 (see the last column of the table and also Problem 4). The entries in the last two rows of the table must be regarded with caution. Indeed, when $h = 10^{-3}$ or $h = 10^{-4}$, the true error is so small that our calculator rounded this error to zero in its display (but internally maintained additional digits). The entries in color in the last column are inaccurate due to the loss of significant figures in the calculator arithmetic. ■ End

As the preceding example suggests, the improved Euler's method converges at the rate $O(h^2)$ and, indeed, it can be proved that in general *this method is of order $p = 2$.*

A step-by-step outline for a subroutine that implements the improved Euler's method over a given interval $[x_0, c]$ is described below. For programming purposes, it is usually more convenient to input the number of steps N in the interval rather than the step size h itself. For an interval starting at $x = x_0$ and ending at $x = c$, the relation between h and N is

(11) $Nh = c - x_0.$

(Note that the subroutine includes an option for printing x and y.)

IMPROVED EULER'S METHOD SUBROUTINE

Purpose To approximate the solution $\phi(x)$ to the initial value problem

$$y' = f(x, y), \qquad y(x_0) = y_0,$$

for $x_0 \leq x \leq c$.

INPUT x_0, y_0, c, N (number of steps), PRNTR (=1 to print table)

Step 1 Set step size $h = (c - x_0)/N$, $x = x_0$, $y = y_0$

Step 2 For $i = 1$ to N, do Steps 3–5

Step 3 Set

$$F = f(x, y)$$
$$G = f(x + h, y + hF)$$

Step 4 Set

$$x = x + h$$
$$y = y + h(F + G)/2$$

Step 5 If PRNTR = 1, print x, y

Now we want to devise a program that will compute $\phi(c)$ *to a desired accuracy.* As we have seen, the accuracy of the approximation depends on the step size h. Our strategy, then, will be to estimate $\phi(c)$ for a given step size and then halve the step size and recompute the estimate; halve again; and so on. When two consecutive estimates of $\phi(c)$ differ by less than some prescribed tolerance ε, we take the final estimate as our approximation to $\phi(c)$. (Admittedly this does not guarantee that $\phi(c)$ is known to within ε, but it provides

a reasonable stopping procedure in practice.) The following procedure also contains a safe-guard to stop if the desired tolerance is not reached after M halvings of h.

IMPROVED EULER'S METHOD WITH TOLERANCE

Purpose To approximate the solution to the initial value problem

$$y' = f(x, y), \qquad y(x_0) = y_0,$$

at $x = c$, to within ε

INPUT $x_0,\ y_0,\ c,\ \varepsilon,$
 M (maximum number of halvings of step size)

Step 1 Set $z = y_0$, PRNTR $= 0$
Step 2 For $m = 0$ to M, do Steps 3–7†
Step 3 Set $N = 2^m$
Step 4 Call IMPROVED EULER'S METHOD SUBROUTINE
Step 5 Print h, y
Step 6 If $|y - z| < \varepsilon$, go to Step 10
Step 7 Set $z = y$
Step 8 Print "$\phi(c)$ is approximately"; y; "but may not be within the toler-ance"; ε
Step 9 Go to Step 11
Step 10 Print "$\phi(c)$ is approximately"; y; "with tolerance"; ε
Step 11 STOP

OUTPUT Approximations of the solution to the initial value problem at $x = c$ using 2^m steps

If one desires a stopping procedure that simulates the **relative error**

$$\left| \frac{\text{approximation} - \text{true value}}{\text{true value}} \right|,$$

then replace Step 6 by

Step 6′ If $\left| \dfrac{z - y}{y} \right| < \varepsilon,$ go to Step 10.

EXAMPLE 2 Use the improved Euler's method with tolerance to approximate the solution to the initial value problem

(12) $y' = x + 2y, \qquad y(0) = 0.25,$

at $x = 2$. For a tolerance of $\varepsilon = 0.001$, use a stopping procedure based on the absolute error.

†To save time, one can start with $m = K < M$ rather than with $m = 0$.

Solution The starting values are $x_0 = 0$, $y_0 = 0.25$. Because we are computing the approximations for $c = 2$, the initial value for h is

$$h = (2 - 0)2^{-0} = 2.$$

For equation (12), we have $f(x, y) = x + 2y$, so the numbers F and G in the subroutine are

$$F = x + 2y$$
$$G = (x + h) + 2(y + hF) = x + 2y + h(1 + 2x + 4y),$$

and we find

$$x = x + h$$
$$y = y + \frac{h}{2}(F + G) = y + \frac{h}{2}(2x + 4y) + \frac{h^2}{2}(1 + 2x + 4y).$$

Thus, with $x_0 = 0$, $y_0 = 0.25$, and $h = 2$, we get for the first approximation

$$y = 0.25 + (0 + 1) + 2(1 + 1) = 5.25.$$

To describe the further outputs of the algorithm, we use the notation $y(2; h)$ for the approximation obtained with step size h. Thus $y(2; 2) = 5.25$, and we find from the algorithm

$y(2; 1) = 11.25000$	$y(2; 2^{-5}) = 25.98132$
$y(2; 2^{-1}) = 18.28125$	$y(2; 2^{-6}) = 26.03172$
$y(2; 2^{-2}) = 23.06067$	$y(2; 2^{-7}) = 26.04468$
$y(2; 2^{-3}) = 25.12012$	$y(2; 2^{-8}) = 26.04797$
$y(2; 2^{-4}) = 25.79127$	$y(2; 2^{-9}) = 26.04880.$

Since $|y(2; 2^{-9}) - y(2; 2^{-8})| = 0.00083$, which is less than $\varepsilon = 0.001$, we stop.

The exact solution to (12) is $\phi(x) = \frac{1}{2}(e^{2x} - x - \frac{1}{2})$, so we have determined that

$$\phi(2) = \frac{1}{2}\left(e^4 - \frac{5}{2}\right) \approx 26.04880.$$ ■ End

In the next section, we discuss methods with higher rates of convergence than either Euler's or the improved Euler's methods.

EXERCISES 3.5

In many of the following problems, it will be helpful to have a calculator or microcomputer available. The reader may also find it convenient to write a program for solving initial value problems using the improved Euler's method algorithms on pages 123 and 124. (Remember, all trigonometric calculations are done in radians.)

1. Show that when Euler's method is used to approximate the solution of the initial value problem

$$y' = 5y, \qquad y(0) = 1,$$

at $x = 1$, then the approximation with step size h is $(1 + 5h)^{1/h}$.

2. Show that when Euler's method is used to approximate the solution of the initial value problem

$$y' = -\frac{1}{2}y, \qquad y(0) = 3,$$

at $x = 2$, then the approximation with step size h is

$$3\left(1 - \frac{h}{2}\right)^{2/h}.$$

3. Show that when the trapezoid scheme given in formula (8) is used to approximate the solution $\phi(x) = e^x$ of

$$y' = y, \qquad y(0) = 1,$$

at $x = 1$, then we get

$$y_{n+1} = \left(\frac{1 + h/2}{1 - h/2}\right) y_n, \qquad n = 0, 1, 2, \dots,$$

which leads to the approximation

$$\left(\frac{1 + h/2}{1 - h/2}\right)^{1/h}$$

for the constant e. Compute this approximation for $h = 1, 10^{-1}, 10^{-2}, 10^{-3},$ and 10^{-4} and compare your results with those in Tables 3.3 and 3.4.

4. In Example 1, the improved Euler's method approximation to e with step size h was shown to be

$$\left(1 + h + \frac{h^2}{2}\right)^{1/h}.$$

First prove that the error $:= e - (1 + h + h^2/2)^{1/h}$ approaches zero as $h \to 0$. Then use L'Hôpital's rule to show that

$$\lim_{h \to 0} \frac{\text{error}}{h^2} = \frac{e}{6} \approx 0.45304.$$

Compare this constant with the entries in the last column of Table 3.4.

5. Show that when the improved Euler's method is used to approximate the solution of the initial value problem

$$y' = 4y, \qquad y(0) = \frac{1}{3},$$

at $x = 1/2$, then the approximation with step size h is

$$\frac{1}{3}(1 + 4h + 8h^2)^{1/(2h)}.$$

6. Since the integral $y(x) := \int_0^x f(t)\,dt$ with variable upper limit satisfies (for continuous f) the initial value problem

$$y' = f(x), \qquad y(0) = 0,$$

any numerical scheme that is used to approximate the solution at $x = 1$ will give an approximation to the definite integral

$$\int_0^1 f(t)\,dt.$$

Derive a formula for this approximation of the integral using

(a) Euler's method.
(b) Trapezoid scheme.
(c) Improved Euler's method.

7. Use the improved Euler's method subroutine with step size $h = 0.1$ to approximate the solution to the initial value problem

$$y' = x - y^2, \qquad y(1) = 0,$$

at the points $x = 1.1, 1.2, 1.3, 1.4,$ and 1.5. (Thus input $N = 5$.) Compare these approximations with those obtained using Euler's method (see Exercises 1.4, Problem 6).

8. Use the improved Euler's method subroutine with step size $h = 0.2$ to approximate the solution to the initial value problem

$$y' = \frac{1}{x}(y^2 + y), \qquad y(1) = 1,$$

at the points $x = 1.2, 1.4, 1.6,$ and 1.8. (Thus input $N = 8$.) Compare these approximations with those obtained using Euler's method (see Exercises 1.4, Problem 5).

9. Use the improved Euler's method subroutine with step size $h = 0.2$ to approximate the solution to

$$y' = x + 3\cos(xy), \qquad y(0) = 0,$$

at the points $x = 0, 0.2, 0.4, \dots, 2.0$. Use your answers to make a rough sketch of the solution on $[0, 2]$.

10. Use the improved Euler's method subroutine with step size $h = 0.1$ to approximate the solution to

$$y' = 4\cos(x + y), \qquad y(0) = 1,$$

at the points $x = 0, 0.1, 0.2, \dots, 1.0$. Use your answers to make a rough sketch of the solution on $[0, 1]$.

11. Use the improved Euler's method with tolerance to approximate the solution to

$$\frac{dx}{dt} = 1 + t\sin(tx), \qquad x(0) = 0,$$

at $t = 1$. For a tolerance of $\varepsilon = 0.01$, use a stopping procedure based on the absolute error.

12. Use the improved Euler's method with tolerance to approximate the solution to

$$y' = 1 - \sin y, \qquad y(0) = 0,$$

at $x = \pi$. For a tolerance of $\varepsilon = 0.01$, use a stopping procedure based on the absolute error.

13. Use the improved Euler's method with tolerance to approximate the solution to

$$y' = 1 - y + y^3, \qquad y(0) = 0,$$

at $x = 1$. For a tolerance of $\varepsilon = 0.003$, use a stopping procedure based on the absolute error.

14. By experimenting with the improved Euler's method subroutine, find the maximum value over the interval [0, 2] of the solution to the initial value problem

$$y' = \sin(x + y), \qquad y(0) = 2.$$

Where does this maximum value occur? Give answers to two decimal places.

15. The solution to the initial value problem

$$\frac{dy}{dx} = (x + y + 2)^2, \qquad y(0) = -2$$

crosses the x-axis at a point in the interval [0, 1.4]. By experimenting with the improved Euler's method subroutine determine this point to two decimal places.

16. The solution to the initial value problem

$$\frac{dy}{dx} + \frac{y}{x} = x^3 y^2, \qquad y(1) = 3$$

has a vertical asymptote ("blows up") at some point in the interval [1, 2]. By experimenting with the improved Euler's method subroutine, determine this point to two decimal places.

17. Use Euler's method (4) with $h = 0.1$ to approximate the solution to the initial value problem

$$y' = -20y, \qquad y(0) = 1,$$

on the interval $0 \le x \le 1$ (that is, at $x = 0, 0.1, \dots,$ 1.0). Compare your answers with the actual solution $y = e^{-20x}$. What went wrong? Next, try the step size $h = 0.025$ and also $h = 0.2$. What conclusions can you draw concerning the choice of step size?

18. Local versus Global Error. In deriving formula (4) for Euler's method, a rectangle was used to approximate the area under a curve (see Figure 3.13). With $g(t) := f(t, \phi(t))$, this approximation can be written as

$$\int_{x_n}^{x_{n+1}} g(t)\,dt \approx hg(x_n), \quad \text{where} \quad h = x_{n+1} - x_n.$$

(a) Show that if g has a continuous derivative that is bounded in absolute value by B, then the rectangle approximation has error $O(h^2)$, that is, for some constant M,

$$\left| \int_{x_n}^{x_{n+1}} g(t)\,dt - hg(x_n) \right| \le Mh^2.$$

This is called the *local truncation error* of the scheme. [*Hint:* Write

$$\int_{x_n}^{x_{n+1}} g(t)\,dt - hg(x_n) = \int_{x_n}^{x_{n+1}} [g(t) - g(x_n)]\,dt.$$

Next, using the mean value theorem, show that $|g(t) - g(x_n)| \le B|t - x_n|$. Then integrate to obtain the error bound $(B/2)h^2$.]

(b) In applying Euler's method, local truncation errors occur in each step of the process and are propagated throughout the further computations. Show that the *sum* of the local truncation errors in part (a) that arise after n steps is $O(h)$. This is the *global error*, which is the same as the convergence rate of Euler's method.

19. Logistic Model. In Section 3.2, we discussed the logistic equation

$$\frac{dp}{dt} = ap - bp^2, \qquad p(0) = p_0,$$

and its use in modeling population growth. A more general model might involve the equation

(13) $$\frac{dp}{dt} = ap - bp^r, \qquad p(0) = p_0,$$

where $r > 1$. To see the effect of changing the parameter r in (13), take $a = 3$, $b = 1$, and $p_0 = 1$. Then use the improved Euler's method subroutine with $h = 0.25$ to approximate the solution to (13) on the interval $0 \le t \le 5$ for $r = 1.5$, 2, and 3. What is the limiting population in each case? For $r > 1$, determine a general formula for the limiting population.

20. Falling Body. In Example 1 of Section 3.4, we modeled the velocity of a falling body by the initial value problem

$$m\frac{dv}{dt} = mg - kv, \qquad v(0) = v_0,$$

under the assumption that the force due to air resistance is $-kv$. However, in certain cases the force due to air

resistance behaves more like $-kv^r$, where r (>1) is some constant. This leads to the model

(14) $m\dfrac{dv}{dt} = mg - kv^r, \qquad v(0) = v_0.$

To study the effect of changing the parameter r in (14), take $m = 1$, $g = 9.81$, $k = 2$, and $v_0 = 0$. Then use the improved Euler's method subroutine with $h = 0.2$ to approximate the solution to (14) on the interval $0 \le t \le 5$ for $r = 1.0$, 1.5, and 2.0. What is the relationship between these solutions and the constant solution $v(t) \equiv (9.81/2)^{1/r}$?

21. Building Temperature. In Section 3.3, we modeled the temperature inside a building by the initial value problem

(15) $\dfrac{dT}{dt} = K[M(t) - T(t)] + H(t) + U(t),$

$T(t_0) = T_0,$

where M is the temperature outside the building, T is the temperature inside the building, H is the additional heating rate, U is the furnace heating or air conditioner cooling rate, K is a positive constant, and T_0 is the initial temperature at time t_0. In a typical model, $t_0 = 0$ (midnight), $T_0 = 65°\text{F}$, $H(t) = 0.1$, $U(t) = 1.5[70 - T(t)]$, and $M(t) = 75 - 20 \cos(\pi t/12)$. The constant K is usually between $1/4$ and $1/2$, depending on such things as insulation. To study the effect of insulating this building, consider the typical building described above and use the improved Euler's method subroutine with $h = 2/3$ to approximate the solution to (15) on the interval $0 \le t \le 24$ (1 day) for $K = 0.2$, 0.4, and 0.6.

3.6 HIGHER-ORDER NUMERICAL METHODS: TAYLOR AND RUNGE-KUTTA

In Sections 1.4 and 3.5, we discussed a simple numerical procedure, Euler's method, for obtaining a numerical approximation of the solution $\phi(x)$ to the initial value problem

(1) $y' = f(x, y), \qquad y(x_0) = y_0.$

Euler's method is easy to implement because it involves only linear approximations to the solution $\phi(x)$. But it suffers from slow convergence, being a method of order 1, that is, the error is O(h). Even the improved Euler's method discussed in Section 3.5 has order of only 2. In this section, we present numerical methods that have faster rates of convergence. These include **Taylor methods,** which are natural extensions of the Euler procedure, and **Runge-Kutta methods,** which are the more popular schemes for solving initial value problems because they have fast rates of convergence and are easy to program.

As in the previous section, we assume that f and $\partial f/\partial y$ are continuous and bounded on the vertical strip $\{(x, y): a < x < b, -\infty < y < \infty\}$ and that f possesses as many continuous partial derivatives as needed.

To derive the Taylor method, let $\phi_n(x)$ be the *exact* solution of the related initial value problem

(2) $\phi_n' = f(x, \phi_n) \qquad \phi_n(x_n) = y_n.$

The Taylor series for $\phi_n(x)$ about the point x_n is

$$\phi_n(x) = \phi_n(x_n) + h\phi_n'(x_n) + \frac{h^2}{2!}\phi_n''(x_n) + \cdots,$$

where $h = x - x_n$. Since ϕ_n satisfies (2), we can write this series in the form

(3) $\qquad \phi_n(x) = y_n + hf(x_n, y_n) + \dfrac{h^2}{2!} \phi_n''(x_n) + \cdots.$

Observe that the recursive formula for y_{n+1} in Euler's method is obtained by truncating the Taylor series after the linear term. For a better approximation, we will use more terms in the Taylor series. This requires that we express the higher-order derivatives of the solution in terms of the function $f(x, y)$.

If y satisfies $y' = f(x, y)$, we can compute y'' by using the chain rule:

(4) $\qquad y'' = \dfrac{\partial f}{\partial x}(x, y) + \dfrac{\partial f}{\partial y}(x, y)y'$

$\qquad\qquad = \dfrac{\partial f}{\partial x}(x, y) + \dfrac{\partial f}{\partial y}(x, y)f(x, y)$

$\qquad\qquad =: f_2(x, y).$

In a similar fashion, define $f_3, f_4, \ldots$, that correspond to the expressions for $y'''(x)$, $y^{(4)}(x)$, etc. If we truncate the expansion in (3) after the h^p term, then, with the above notation, the recursive formulas for the **Taylor method of order p** are

(5) $\qquad x_{n+1} = x_n + h,$

(6) $\qquad y_{n+1} = y_n + hf(x_n, y_n) + \dfrac{h^2}{2!} f_2(x_n, y_n) + \cdots + \dfrac{h^p}{p!} f_p(x_n, y_n).$

As before, $y_n \approx \phi(x_n)$, where $\phi(x)$ is the solution to the initial value problem (1). It can be shown[†] that **the Taylor method of order p has the rate of convergence $O(h^p)$.**

EXAMPLE 1 Determine the recursive formulas for the Taylor method of order 2 for the initial value problem

(7) $\qquad y' = \sin(xy), \qquad y(0) = \pi.$

Solution We must compute $f_2(x, y)$ as defined in (4). Since $f(x, y) = \sin(xy)$,

$$\dfrac{\partial f}{\partial x}(x, y) = y\cos(xy), \qquad \dfrac{\partial f}{\partial y}(x, y) = x\cos(xy).$$

Substituting into (4), we have

$$f_2(x, y) = \dfrac{\partial f}{\partial x}(x, y) + \dfrac{\partial f}{\partial y}(x, y)f(x, y)$$

$$= y\cos(xy) + x\cos(xy)\sin(xy)$$

$$= y\cos(xy) + \dfrac{x}{2}\sin(2xy),$$

[†] See P. Henrici, *Elements of Numerical Analysis* (John Wiley and Sons, Inc., New York, 1964), Chapter 14.

and the recursive formulas (5) and (6) become

$$x_{n+1} = x_n + h,$$

$$y_{n+1} = y_n + h \sin(x_n y_n) + \frac{h^2}{2}\left[y_n \cos(x_n y_n) + \frac{x_n}{2}\sin(2x_n y_n)\right],$$

where $x_0 = 0$, $y_0 = \pi$ are the starting values. ∎ End

The difficulty in employing higher-order Taylor methods is the tedious computation of the partial derivatives needed to determine f_p (typically these computations grow exponentially with p). Fortunately, this difficulty can be circumvented by using one of the **Runge-Kutta methods.**[†]

Observe that the general Taylor method has the form

(8) $y_{n+1} = y_n + hF(x_n, y_n; h),$

where the choice of F depends on p. In particular (cf. (6)), for

$$p = 1, \quad F = T_1(x, y; h) := f(x, y),$$

(9) $p = 2, \quad F = T_2(x, y; h) := f(x, y) + \frac{h}{2}\left[\frac{\partial f}{\partial x}(x, y) + \frac{\partial f}{\partial y}(x, y)f(x, y)\right].$

The idea behind the Runge-Kutta method of order 2 is to choose F in (8) of the form

(10) $F = K_2(x, y; h) := f(x + \alpha h, y + \beta h f(x, y)),$

where the constants α, β are to be selected so that (8) has the rate of convergence $O(h^2)$. The advantage here is that K_2 is computed by two evaluations of the original function $f(x, y)$ and does not involve the derivatives of $f(x, y)$.

To ensure $O(h^2)$ convergence, we compare this new scheme with the Taylor method of order 2 and require

$$T_2(x, y; h) - K_2(x, y; h) = O(h^2), \quad \text{as} \quad h \to 0.$$

That is, we choose α, β so that the Taylor expansions for T_2 and K_2 agree through terms of order h. For (x, y) fixed, when we expand $K_2 = K_2(h)$ as given in (10) about $h = 0$, we find

(11) $K_2(h) = K_2(0) + \frac{dK_2}{dh}(0)h + O(h^2)$

$$= f(x, y) + \left[\alpha\frac{\partial f}{\partial x}(x, y) + \beta\frac{\partial f}{\partial y}(x, y)f(x, y)\right]h + O(h^2),$$

[†] ***Historical Footnote:*** These methods were developed by C. Runge in 1895 and W. Kutta in 1901.

where the expression in brackets for dK_2/dh evaluated at $h = 0$ follows from the chain rule. Comparing (11) with (9), we see that for T_2 and K_2 to agree through terms of order h, we must have $\alpha = \beta = 1/2$. Thus

$$K_2(x, y; h) = f\left(x + \frac{h}{2}, y + \frac{h}{2}f(x, y)\right).$$

The Runge-Kutta method we have derived is called the **midpoint method** and it has the recursive formulas

(12) $x_{n+1} = x_n + h,$

(13) $y_{n+1} = y_n + hf\left(x_n + \frac{h}{2}, y_n + \frac{h}{2}f(x_n, y_n)\right).$

By construction, the midpoint method has the same rate of convergence as the Taylor method of order 2; that is, $O(h^2)$. This is the same rate as the improved Euler's method.

In a similar fashion, one can work with the Taylor method of order 4 and, after some elaborate calculations, obtain the **classical** or **fourth order Runge-Kutta method.** The recursive formulas for this method are

(14)
$$x_{n+1} = x_n + h,$$
$$y_{n+1} = y_n + \frac{1}{6}(k_1 + 2k_2 + 2k_3 + k_4),$$

where

$$k_1 = hf(x_n, y_n),$$
$$k_2 = hf\left(x_n + \frac{h}{2}, y_n + \frac{k_1}{2}\right),$$
$$k_3 = hf\left(x_n + \frac{h}{2}, y_n + \frac{k_2}{2}\right),$$
$$k_4 = hf(x_n + h, y_n + k_3).$$

The fourth order Runge-Kutta method is one of the more popular methods because its rate of convergence is $O(h^4)$ and it is easy to program. Typically, it produces very accurate approximations even when the number of iterations is reasonably small. However, as the number of iterations becomes large, other types of errors may creep in.

Program outlines for the fourth order Runge-Kutta method are given below. Just as with the algorithms for improved Euler's method, the first program (the Runge-Kutta subroutine) is useful for approximating the solution over an interval $[x_0, c]$ and takes the number of steps in the interval as input. As in Section 3.5, the number of steps N is related to the step size h and the interval $[x_0, c]$ by

$$Nh = c - x_0.$$

The subroutine has the option to print out a table of values of x and y. The second algorithm (Runge-Kutta with tolerance) is used to approximate, within a given tolerance, the solution at an inputed value $x = c$. This algorithm[†] automatically halves the step sizes successively until the two approximations $y(c; h)$ and $y(c; h/2)$ differ by less than the prescribed tolerance ε. For a stopping procedure based on the relative error, Step 6 of the algorithm should be replaced by

Step 6′ If $\left| \dfrac{y - z}{y} \right| < \varepsilon$, go to Step 10.

FOURTH ORDER RUNGE-KUTTA SUBROUTINE

Purpose To approximate the solution to the initial value problem

$$y' = f(x, y), \qquad y(x_0) = y_0$$

for $x_0 \leq x \leq c$

INPUT x_0, y_0, c, N (number of steps), PRNTR (=1 to print table)

Step 1 Set step size $h = (c - x_0)/N$, $x = x_0$, $y = y_0$

Step 2 For $i = 1$ to N do Steps 3–5

Step 3 Set

$$k_1 = hf(x, y)$$

$$k_2 = hf\left(x + \frac{h}{2}, y + \frac{k_1}{2} \right)$$

$$k_3 = hf\left(x + \frac{h}{2}, y + \frac{k_2}{2} \right)$$

$$k_4 = hf(x + h, y + k_3)$$

Step 4 Set

$$x = x + h$$

$$y = y + \frac{1}{6}(k_1 + 2k_2 + 2k_3 + k_4)$$

Step 5 If PRNTR = 1, print x, y

[†] Note that the form of this algorithm is the same as that for the improved Euler's method on page 124 except for Step 4 where the Runge-Kutta subroutine is called.

FOURTH ORDER RUNGE-KUTTA ALGORITHM WITH TOLERANCE

Purpose To approximate the solution to the initial value problem

$$y' = f(x, y), \qquad y(x_0) = y_0$$

at $x = c$, to within ε

INPUT x_0, y_0, c, ε, M (maximum number of iterations)

Step 1 Set $z = y_0$, PRNTR = 0

Step 2 For $m = 0$ to M, do Steps 3–7 (or, to save time, start with $m > 0$)

Step 3 Set $N = 2^m$

Step 4 Call FOURTH ORDER RUNGE-KUTTA SUBROUTINE

Step 5 Print h, y

Step 6 If $|z - y| < \varepsilon$, go to Step 10

Step 7 Set $z = y$

Step 8 Print "$\phi(c)$ is approximately,"; y; "but may not be within the tolerance"; ε

Step 9 Go to Step 11

Step 10 Print "$\phi(c)$ is approximately,"; y; "with tolerance"; ε

Step 11 STOP

OUTPUT Approximations of the solution to the initial value problem at $x = c$, using 2^m steps.

EXAMPLE 2 Use the fourth order Runge-Kutta algorithm to approximate the solution $\phi(x)$ of the initial value problem

$$y' = y, \qquad y(0) = 1,$$

at $x = 1$ with a tolerance of 0.001.

Solution The inputs are $x_0 = 0$, $y_0 = 1$, $c = 1$, $\varepsilon = 0.001$, and $M = 25$ (say). Since $f(x, y) = y$, the formulas in Step 3 of the subroutine become

$$k_1 = hy, \qquad k_2 = h\left(y + \frac{k_1}{2}\right), \qquad k_3 = h\left(y + \frac{k_2}{2}\right), \qquad k_4 = h(y + k_3).$$

The initial value for N in the algorithm is $N = 1$, so

$$h = (1 - 0)/1 = 1.$$

Thus, in Step 3, we compute

$$k_1 = (1)(1) = 1, \qquad k_2 = (1)(1 + 0.5) = 1.5,$$

$$k_3 = (1)(1 + 0.75) = 1.75, \qquad k_4 = (1)(1 + 1.75) = 2.75,$$

and, in Step 4, of the subroutine we get for the first approximation

$$y = y_0 + \frac{1}{6}(k_1 + 2k_2 + 2k_3 + k_4)$$

$$= 1 + \frac{1}{6}[1 + 2(1.5) + 2(1.75) + 2.75]$$

$$= 2.70833,$$

where we have rounded to five decimal places. Because

$$|z - y| = |y_0 - y| = |1 - 2.70833| = 1.70833 > \varepsilon,$$

we start over and reset $N = 2$, $h = 0.5$.

Doing Steps 3 and 4 for $i = 1$ and 2, we ultimately obtain (for $i = 2$) the approximation

$$y = 2.71735.$$

Since $|z - y| = |2.70833 - 2.71735| = 0.00902 > \varepsilon$, we again start over and reset $N = 4$, $h = 0.25$. This leads to the approximation

$$y = 2.71821,$$

so that

$$|z - y| = |2.71735 - 2.71821| = 0.00086,$$

which is less than $\varepsilon = 0.001$. Hence $\phi(1) = e \approx 2.71821$. ■ End

In Example 2, we were able to obtain a better approximation for $\phi(1) = e$ with $h = 0.25$ than we obtained in Section 3.5 using Euler's method with $h = 0.001$ (see Table 3.3) and roughly the same accuracy as we obtained in Section 3.5 using the improved Euler's method with $h = 0.01$ (see Table 3.4).

EXAMPLE 3 Use the fourth order Runge-Kutta subroutine to approximate the solution $\phi(x)$ of the initial value problem

(15) $y' = y^2,$ $y(0) = 1,$

on the interval $0 \le x \le 2$ using $N = 8$ steps (i.e., $h = 0.25$).

Solution Here the starting values are $x_0 = 0$ and $y_0 = 1$. Since $f(x, y) = y^2$, the formulas in Step 3 of the subroutine are

$$k_1 = hy^2, \qquad k_2 = h\left(y + \frac{k_1}{2}\right)^2,$$

$$k_3 = h\left(y + \frac{k_2}{2}\right)^2, \qquad k_4 = h(y + k_3)^2.$$

From the output, we find

$$
\begin{array}{ll}
x = 0.25 & y = 1.33322, \\
x = 0.50 & y = 1.99884, \\
x = 0.75 & y = 3.97238, \\
x = 1.00 & y = 32.82820, \\
x = 1.25 & y = 4.09664 \times 10^{11}, \\
x = 1.50 & y = \text{overflow.}
\end{array}
$$

What happened? Fortunately the equation in (15) is separable, and, solving for $\phi(x)$, we obtain $\phi(x) = (1 - x)^{-1}$. It is now obvious where the problem lies: The true solution $\phi(x)$ is not defined at $x = 1$. If we had been more cautious, we would have realized that

$\partial f/\partial y = 2y$ is *not* bounded for all y. Hence the existence of a unique solution is not guaranteed for all x between 0 and 2, and in this case, the method does *not* give meaningful approximations for x near or greater than 1. ∎ End

EXAMPLE 4 Use the fourth order Runge-Kutta algorithm to approximate the solution $\phi(x)$ of the initial value problem

$$y' = x - y^2, \qquad y(0) = 1,$$

at $x = 2$ with a tolerance of 0.0001.

Solution This time we check to see whether $\partial f/\partial y$ is bounded. Here $\partial f/\partial y = -2y$, which is certainly unbounded in any vertical strip. However, let's consider the qualitative behavior of the solution $\phi(x)$. The solution curve starts at $(0, 1)$, where $\phi'(0) = 0 - 1 < 0$, so $\phi(x)$ begins decreasing and continues to decrease until it crosses the curve $y = \sqrt{x}$. After crossing this curve, $\phi(x)$ begins to increase, since $\phi'(x) = x - \phi^2(x) > 0$. As $\phi(x)$ increases, it remains below the curve $y = \sqrt{x}$. This is because if the solution were to get ''close'' to the curve $y = \sqrt{x}$, then the derivative of $\phi(x)$ would approach zero, so that overtaking the function $\sqrt{x}$ is impossible.

Therefore, although the existence-uniqueness theorem does not guarantee a solution, we are inclined to try the algorithm anyway. The above argument shows that $\phi(x)$ probably exists for $x > 0$, so we feel reasonably sure the fourth order Runge-Kutta method will give a good approximation of the true solution $\phi(x)$. Proceeding with the algorithm, we use the starting values $x_0 = 0$ and $y_0 = 1$. Since $f(x, y) = x - y^2$, the formulas in Step 3 of the subroutine become

$$k_1 = h(x - y^2), \qquad k_2 = h\left[\left(x + \frac{h}{2}\right) - \left(y + \frac{k_1}{2}\right)^2\right],$$

$$k_3 = h\left[\left(x + \frac{h}{2}\right) - \left(y + \frac{k_2}{2}\right)^2\right], \qquad k_4 = h[(x + h) - (y + k_3)^2].$$

In Table 3.5, we give the approximations $y(2; 2^{-m+1})$ for $\phi(2)$ for $m = 0, 1, 2, 3$, and 4. The algorithm stops at $m = 4$, since

$$|y(2; 0.125) - y(2; 0.25)| = 0.00000.$$

Hence $\phi(2) \approx 1.25132$ with a tolerance of 0.0001. ∎ End

		TABLE 3.5 FOURTH ORDER RUNGE-KUTTA APPROXIMATION FOR $\phi(2)$	
m	h	Approximation for $\phi(2)$	$\|y(2; h) - y(2; 2h)\|$
0	2.0	-8.33334	
1	1.0	1.27504	9.60838
2	0.5	1.25170	0.02334
3	0.25	1.25132	0.00038
4	0.125	1.25132	0.00000

EXERCISES 3.6

 As in Exercises 3.5, the reader will find it helpful to have a calculator or microcomputer available. For Problems 1–17, note whether or not $\partial f/\partial y$ is bounded.

1. Determine the recursive formulas for the Taylor method of order 2 for the initial value problem
$$y' = \cos(x + y), \qquad y(0) = \pi.$$

2. Determine the recursive formulas for the Taylor method of order 2 for the initial value problem
$$y' = xy - y^2, \qquad y(0) = -1.$$

3. Determine the recursive formulas for the Taylor method of order 4 for the initial value problem
$$y' = x - y, \qquad y(0) = 0.$$

4. Determine the recursive formulas for the Taylor method of order 4 for the initial value problem
$$y' = x^2 + y, \qquad y(0) = 0.$$

5. Use the Taylor methods of orders 2 and 4 with $h = 0.25$ to approximate the solution to the initial value problem
$$y' = x + 1 - y, \qquad y(0) = 1,$$
at $x = 1$. Compare these approximations to the actual solution $y = x + e^{-x}$ evaluated at $x = 1$.

6. Use the Taylor methods of orders 2 and 4 with $h = 0.25$ to approximate the solution to the initial value problem
$$y' = 1 - y, \qquad y(0) = 0,$$
at $x = 1$. Compare these approximations to the actual solution $y = 1 - e^{-x}$ evaluated at $x = 1$.

7. Use the fourth order Runge-Kutta subroutine with $h = 0.25$ to approximate the solution to the initial value problem
$$y' = 2y - 6, \qquad y(0) = 1,$$
at $x = 1$. (Thus input $N = 4$.) Compare this approximation to the actual solution $y = 3 - 2e^{2x}$ evaluated at $x = 1$.

8. Use the fourth order Runge-Kutta subroutine with $h = 0.25$ to approximate the solution to the initial value problem
$$y' = 1 - y, \qquad y(0) = 0,$$
at $x = 1$. Compare this approximation with the one obtained in Problem 6 using the Taylor method of order 4.

9. Use the fourth order Runge-Kutta subroutine with $h = 0.25$ to approximate the solution to the initial value problem
$$y' = x + 1 - y, \qquad y(0) = 1,$$
at $x = 1$. Compare this approximation with the one obtained in Problem 5 using the Taylor method of order 4.

10. Use the fourth order Runge-Kutta algorithm to approximate the solution to the initial value problem
$$y' = 1 - xy, \qquad y(1) = 1,$$
at $x = 2$. For a tolerance of $\varepsilon = 0.001$, use a stopping procedure based on the absolute error.

11. The solution to the initial value problem
$$y' = \frac{2}{x^4} - y^2, \qquad y(1) = -0.414$$
crosses the x-axis at a point in the interval $[1, 2]$. By experimenting with the fourth order Runge-Kutta subroutine determine this point to two decimal places.

12. By experimenting with the fourth order Runge-Kutta subroutine, find the maximum value over the interval $[1, 2]$ of the solution to the initial value problem
$$y' = \frac{1.8}{x^4} - y^2, \qquad y(1) = -1.$$
Where does this maximum occur? Give your answers to two decimal places.

13. The solution to the initial value problem
$$\frac{dy}{dx} = y^2 - 2e^x y + e^{2x} + e^x, \qquad y(0) = 3$$
has a vertical asymptote ("blows up") at some point in the interval $[0, 2]$. By experimenting with the fourth order Runge-Kutta subroutine, determine this point to two decimal places.

14. Use the fourth order Runge-Kutta algorithm to approximate the solution to the initial value problem
$$y' = y \cos x, \qquad y(0) = 1,$$
at $x = \pi$. For a tolerance of $\varepsilon = 0.01$, use a stopping procedure based on the absolute error.

15. Use the fourth order Runge-Kutta subroutine with $h = 0.1$ to approximate the solution to
$$y' = \cos(5y) - x, \qquad y(0) = 0,$$

at the points $x = 0, 0.1, 0.2, \ldots, 3.0$. Use your answers to make a rough sketch of the solution on $[0, 3]$.

16. Use the fourth order Runge-Kutta subroutine with $h = 0.1$ to approximate the solution to

$$y' = 3\cos(y - 5x), \qquad y(0) = 0,$$

at the points $x = 0, 0.1, 0.2, \ldots, 4.0$. Use your answers to make a rough sketch of the solution on $[0, 4]$.

17. The Taylor method of order 2 can be used to approximate the solution to the initial value problem

$$y' = y, \qquad y(0) = 1,$$

at $x = 1$. Show that the approximation y_n obtained by using the Taylor method of order 2 with the step size $1/n$ is given by the formula

$$y_n = \left(1 + \frac{1}{n} + \frac{1}{2n^2}\right)^n, \qquad n = 1, 2, \ldots.$$

The solution to the initial value problem is $y = e^x$, so y_n is an approximation to the constant e.

18. If the Taylor method of order p is used in Problem 17, show that

$$y_n = \left(1 + \frac{1}{n} + \frac{1}{2n^2} + \frac{1}{6n^3} + \cdots + \frac{1}{p!n^p}\right)^n,$$

$$n = 1, 2, \ldots.$$

19. Fluid Flow. In the study of the nonisothermal flow of a Newtonian fluid between parallel plates, the equation

$$\frac{d^2y}{dx^2} + x^2 e^y = 0, \qquad x > 0,$$

was encountered. By a series of substitutions, this equation can be transformed into the first order equation

$$\frac{dv}{du} = u\left(\frac{u}{2} + 1\right)v^3 + \left(u + \frac{5}{2}\right)v^2.$$

Use the fourth order Runge-Kutta algorithm to approximate $v(3)$ if $v(u)$ satisfies $v(2) = 0.1$. For a tolerance of $\varepsilon = 0.0001$, use a stopping procedure based on the relative error.

20. Chemical Reactions. The reaction between nitrous oxide and oxygen to form nitrogen dioxide is given by the balanced chemical equation $2NO + O_2 = 2NO_2$. At high temperatures, the dependence of the rate of this reaction on the concentrations of NO, O_2, and NO_2 is complicated. However, at 25°C the rate at which NO_2 is formed obeys the law of mass action and is given by the rate equation

$$\frac{dx}{dt} = k(\alpha - x)^2\left(\beta - \frac{x}{2}\right),$$

where $x(t)$ denotes the concentration of NO_2 at time t, k is the rate constant, α is the initial concentration of NO, and β is the initial concentration of O_2. At 25°C, the constant k is 7.13×10^3 (liter)2/(mole)2(second). Let $\alpha = 0.0010$ mole/L, $\beta = 0.0041$ mole/L, and $x(0) = 0$ mole/L. Use the fourth order Runge-Kutta algorithm to approximate $x(10)$. For a tolerance of $\varepsilon = 0.000001$, use a stopping procedure based on the relative error.

21. Transmission Lines. In the study of the electric field that is induced by two nearby transmission lines, an equation of the form

$$\frac{dz}{dx} + g(x)z^2 = f(x)$$

arises. Let $f(x) = 5x + 2$ and $g(x) = x^2$. If $z(0) = 1$, use the fourth order Runge-Kutta algorithm to approximate $z(1)$. For a tolerance of $\varepsilon = 0.0001$, use a stopping procedure based on the absolute error.

3.7 SOME AVAILABLE CODES FOR INITIAL VALUE PROBLEMS

Before describing some commercially available codes for solving initial value problems, we must first introduce multistep methods and the concept of step size control.

In the previous two sections, we discussed numerical methods for solving initial value problems for a first order differential equation. These techniques are examples of **one-step methods** because the approximation y_{n+1} of the true value $y(x_{n+1})$ is obtained using the differential equation and the previous approximation y_n of $y(x_n)$. More generally, we can use **multistep methods.** For such schemes, the differential equation and the previous k approximations

$$y_n, y_{n-1}, \ldots, y_{n-(k-2)}, y_{n-(k-1)}$$

are used to obtain the approximation y_{n+1}. (In this case the method is called a **k-step method.**)

A two-step method that is easy to derive is based on Simpson's rule for approximating integrals (see Appendix B). Let y_n be the approximation of the solution $y(x)$ to the initial value problem

(1) $y' = f(x, y), \qquad y(x_0) = y_0,$

at $x_n = x_0 + nh$, where $h > 0$ is the step size. If we integrate both sides of the equation in (1) from x_{n-1} to x_{n+1}, then we have

(2) $y(x_{n+1}) - y(x_{n-1}) = \displaystyle\int_{x_{n-1}}^{x_{n+1}} f(x, y(x))\,dx.$

We now approximate the integral in (2) using Simpson's rule. This gives

$$y(x_{n+1}) - y(x_{n-1}) \approx \frac{h}{3}[f(x_{n-1}, y(x_{n-1})) + 4f(x_n, y(x_n)) + f(x_{n+1}, y(x_{n+1}))].$$

Substituting the approximations y_{n-1}, y_n, and y_{n+1} for $y(x_{n-1})$, $y(x_n)$, and $y(x_{n+1})$, respectively, we obtain the numerical scheme

(3) $y_{n+1} - y_{n-1} = \dfrac{h}{3}[f(x_{n-1}, y_{n-1}) + 4f(x_n, y_n) + f(x_{n+1}, y_{n+1})].$

This method is also called **Simpson's rule.** As you can see, it is an implicit, two-step method. Starting with y_0 and y_1, formula (3) determines y_2; knowing y_1 and y_2, we again use (3) to determine y_3; etc. Formula (3) can also be used in a predictor-corrector procedure, where y_{n+1}^*, a predicted value for y_{n+1}, is used in the right-hand side of (3) to obtain a ''corrected'' estimate y_{n+1}.

Multistep methods were introduced by J. C. Adams prior to the Runge-Kutta methods. In fact, two important classes of multistep methods are the explicit Adams methods called the **Adams-Bashford methods** and the implicit Adams methods called the **Adams-Moulton methods.** Like the Runge-Kutta methods, there exist Adams methods with different orders of convergence.

A disadvantage of the multistep methods is that they are more susceptible to instability. (Stability is discussed in Project E and Section 4.11, Example 4.) From a programming viewpoint, the multistep methods also have the disadvantage of not being self-starting; that is, the first k approximations $y_0, y_1, \ldots, y_{k-1}$ are needed in a k-step method in order to determine y_k. Therefore one must use some other method to generate the starting values $y_1, y_2, \ldots, y_{k-1}$ (recall that y_0 is given). This means the Runge-Kutta formulas are easier to code. However, the Runge-Kutta formulas require several evaluations of the function $f(x, y)$ at each step, whereas the multistep formulas require only one evaluation of $f(x, y)$ at each stage. This makes the multistep methods less expensive (and faster) when the function $f(x, y)$ is complicated. For a discussion of multistep methods, we refer the reader to the texts *Numerical Initial Value Problems in Ordinary Differential Equations*, by C. W. Gear, Prentice-Hall, 1971, and *Discrete Variable Methods in Ordinary Differential Equations*, by P. Henrici, John Wiley and Sons, 1962.

In practice, one is confronted with the problem of approximating the solution to within a desired tolerance and obtaining this approximation as inexpensively or as quickly as possible. This is seldom achieved using a *constant* step size h. To illustrate the problem, consider the periodic function $y(x)$ whose graph is given in Figure 3.15. (The solutions to the "Brusselator" equation, which models a chemical reaction that possesses periodic solutions, have graphs similar to the one in Figure 3.15.) If we are trying to approximate $y(x)$, then, over most of its period, we can use a step size h that is relatively large. However, at the "spike," we must take h to be very small or else we will miss the spike entirely or poorly approximate the values of y on the spike. Therefore it is advantageous to be able to control the step size h.

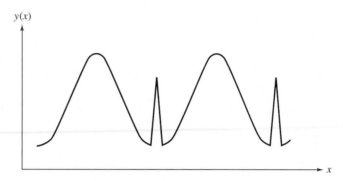

Figure 3.15 Periodic function with spike

Much work has been done on writing codes that vary the step size h. These codes can choose h at each step so that the local error is within the given tolerance; shorten the step size, if necessary, to keep the error within tolerance; and decide what the next step size should be. The codes using multistep methods can vary the order of approximation by shifting from one type of multistep method to another. This allows them to begin with a one-step method, which resolves the difficulty of determining the starting values y_1, y_2, ..., y_{k-1} in a k-step method. Even when using a code with variable step size and order strategies, it is important to use whatever information one might have to assist in approximating the solution on those intervals where it is changing rapidly.

Because of the considerable effort involved in developing and testing codes for numerically solving initial value problems, a scientist or engineer will often use one of the commercially available packages for "real life" applications. Some packages have been specifically written to handle what are called **stiff problems.** Roughly speaking, this refers to problems that involve one or more fast decaying processes in time, with a time constant (cf. Section 3.3) that is short compared to the time span of interest. For example, the problem

$$\frac{dy}{dt} = -1000y + 999e^{-t}, \qquad y(0) = 0,$$

which has the solution

$$y(t) = e^{-t} - e^{-1000t},$$

would be considered stiff on the interval [0, 2], since the time constant for the term e^{-1000t} is 1/1000, which is much less than the length of the interval [0, 2].

In addition to the subroutines available on **MATHEMATICA**, **MAPLE**, and **MATLAB**, several different codes have been specifically designed for handling differential equations. Here we give a brief description of some of the more popular ones.

DEABM was developed by L. F. Shampine and H. A. Watts and is a modification of the code **DE** that is described in the book *Computer Solution of Ordinary Differential Equations, the Initial Value Problem*, by L. F. Shampine and M. K. Gordon, Freeman and Company, San Francisco, 1975. This code uses a predictor-corrector approach involving the Adams methods. In this program, the step size is adjusted to maintain accuracy in the intervals where the solution varies rapidly and the order is changed to maintain accuracy over the intervals where the solution is slowly varying.

DOPRI8 is a one-step method that uses an eighth order Runge-Kutta formula with a seventh order error estimate and step size control. The formulas were obtained by P. J. Prince and J. R. Dormand. The code is given in the appendix to the book *Solving Ordinary Differential Equations I–Nonstiff Problems*, by E. Hairer, S. P. Nørsett, and G. Wanner, which is published by Springer-Verlag, Berlin, 1987.

DVERK is an initial value problem solver written by T. E. Hull, W. H. Enright, and K. R. Jackson. It uses an explicit Runge-Kutta method that is very effective for nonstiff problems. DVERK is available on NETLIB.

ODEPACK is a collection of six initial value problem solvers for stiff and nonstiff problems. The basic member is LSODE, which also uses Adams methods. Other component routines are designed for special types of problems such as those involving implicit equation forms and sparse Jacobians. ODEPACK is available from the National Energy Software Center, Building 221, Argonne National Laboratory, 9700 South Cass Avenue, Argonne, IL 60439.

RKF45 is an initial value problem solver written by L. F. Shampine and H. A. Watts. It uses an explicit Runge-Kutta method. It is also available on NETLIB.

SDRIV is an initial value problem solver that works on a great variety of machines (including PCs) and can solve stiff problems. The algorithm has the "G-stop" capability; that is, it can stop the integration based on a user-supplied subroutine. It is included with the book *Numerical Methods and Software*, by David Kahaner, Cleve Moler, and Stephen Nash, Prentice-Hall, Englewood Cliffs, New Jersey, 1989.

VODE is an initial value problem solver for stiff as well as nonstiff problems. It was developed by P. N. Brown, G. D. Byrne, and A. C. Hindmarsh. It uses variable-coefficient Adams-Moulton and Backward Differentiation Formula methods. VODE supersedes EPISODE and is available from the National Energy Software Center (see ODEPACK).

The following software packages are written for IBM-compatible PCs to be used for both teaching and research:

PHASEPLANE is an interactive software package for studying dynamic systems. It allows the user to choose between Euler's, improved Euler's, Runge-Kutta, Adams, and Gear's methods. It was developed by B. Ermentrout and is available from Brooks/Cole Publishing Company, Pacific Grove, California. The software comes with a user's guide.

PHASER is a simulation program written by H. Koçak. It allows the user to choose between Euler's, improved Euler's, and fourth order Runge-Kutta methods. PHASER comes with the book *Differential and Difference Equations through Computer Experiments*, by H. Koçak, Springer-Verlag, New York, 1986, which describes how to use it.

SIL is a simulation language written by N. Houbak. It uses an implicit Euler method with variable order and variable step size. It can handle discontinuities (switching between several states) and combined discrete/continuous systems. SIL is described in *SIL—A Simulation Language*, by N. Houbak, Lecture Notes in Computer Science No. 426, Springer-Verlag, Berlin, 1990.

The following software packages are written for the Macintosh:

MacMath, developed by John H. Hubbard and Beverly H. West, is available from Springer-Verlag, New York, 1995. It is a scientific toolkit consisting of 12 graphics programs, which is especially useful in experimentation with dynamical systems.

xFunctions was written by David Eck, Department of Mathematics, Hobart and William Smith Colleges, Geneva, New York 14456 and is now in the public domain. It is particularly helpful for plotting functions, making direction field plots, and constructing solutions in the phase plane for 2-dimensional autonomous systems.

GROUP PROJECTS FOR CHAPTER 3

A. Delay Differential Equations

In our discussion of mixing problems in Section 3.2, we encountered the initial value problem

$$(1) \qquad x'(t) = 6 - \frac{3}{500} x(t - t_0), \qquad x(t) = 0 \quad \text{for} \quad t \in [-t_0, 0],$$

where t_0 is a positive constant. The equation in (1) is an example of a **delay differential equation.** These equations differ from the usual differential equations by the presence of the shift $(t - t_0)$ in the argument of the unknown function $x(t)$. In general, these equations are more difficult to work with than are regular differential equations, but quite a bit is known about them.[†]

(a) Show that the simple linear delay differential equation

$$(2) \qquad x'(t) = ax(t - b),$$

where a and b are constants, has a solution of the form $x(t) = Ce^{st}$, for any constant C, provided s satisfies the transcendental equation $s = ae^{-bs}$.

(b) A solution to (2) for $t > 0$ can also be found using the **method of steps.** Assume that $x(t) = f(t)$ for $-b \leq t \leq 0$. For $0 \leq t \leq b$, equation (2) becomes

$$x'(t) = ax(t - b) = af(t - b),$$

and so

$$x(t) = \int_0^t af(v - b)\,dv + x(0).$$

Now that we know $x(t)$ on $[0, b]$, we can repeat this procedure to obtain

$$x(t) = \int_b^t ax(v - b)\,dv + x(b)$$

for $b \leq t \leq 2b$. This process can be continued indefinitely.

Use the method of steps to show that the solution to the initial value problem

$$x'(t) = x(t - 1), \qquad x(t) = 1 \quad \text{on} \quad [-1, 0]$$

is given by

$$x(t) = \sum_{k=0}^{n} \frac{[t - (k - 1)]^k}{k!}, \quad \text{for} \quad n - 1 \leq t \leq n,$$

where n is a nonnegative integer. (This problem can also be solved using the Laplace transform method of Chapter 7.)

[†]See, for example, *Differential-Difference Equations*, by R. Bellman and K. L. Cooke, Academic Press, New York, 1963, or *Ordinary and Delay Differential Equations*, by R. D. Driver, Springer-Verlag, New York, 1977.

(c) Use the method of steps to compute the solution to the initial value problem given in (1) on the interval $0 \le t \le 15$ for $t_0 = 3$.

B. *Aquaculture*

Aquaculture is the art of cultivating the plants and animals indigenous to water. In the example considered here, it is assumed that a batch of catfish are raised in a pond. We are interested in determining the best time for harvesting the fish so that the cost per pound for raising the fish is minimized.

A differential equation describing the growth of fish may be expressed as

(3) $$\frac{dW}{dt} = KW^{\alpha},$$

where $W(t)$ is the weight of the fish at time t, and K and α are empirically determined growth constants. The functional form of this relationship is similar to that of the growth models for other species. Modeling the growth rate or metabolic rate by a term like W^{α} is a common assumption. Biologists often refer to equation (3) as the **allometric equation.** It can be supported by plausibility arguments such as growth rate depending on the surface area of the gut (which varies like $W^{2/3}$) or depending on the volume of the animal (which varies like W).

(a) Solve equation (3) when $\alpha \ne 1$.

(b) The solution obtained in part (a) grows large without bound, but in practice there is some limiting maximum weight W_{MAX} for the fish. This limiting weight may be included in the differential equation describing growth by inserting a dimensionless variable S that can range between 0 and 1 and involves an empirically determined parameter μ. Namely we now assume

(4) $$\frac{dW}{dt} = KW^{\alpha}S,$$

where $S := 1 - (W/W_{\text{MAX}})^{\mu}$.

When $\mu = 1 - \alpha$, equation (4) has a closed form solution. Solve equation (4) when $K = 12$, $\alpha = 3/4$, $\mu = 1/4$, $W_{\text{MAX}} = 81$ (ounces), and $W(0) = 1$ (ounce). The constants are given for t measured in months.

(c) The differential equation describing the total cost in dollars $C(t)$ of raising a fish for t months has one constant term K_1 that specifies the cost per month (due to costs such as interest, depreciation, and labor) and a second constant K_2 that multiplies the growth rate (because the amount of food consumed by the fish is approximately proportional to the growth rate). That is,

(5) $$\frac{dC}{dt} = K_1 + K_2\frac{dW}{dt}.$$

Solve equation (5) when $K_1 = 0.4$, $K_2 = 0.1$, $C(0) = 1.1$ (dollars), and $W(t)$ is as determined in part (b).

(d) Sketch the curve obtained in part (b) that represents the weight of the fish as a function of time. Next, sketch the curve obtained in part (c) that represents the total cost of raising the fish as a function of time.

(e) To determine the optimal time for harvesting the fish, sketch the ratio $C(t)/W(t)$. This

ratio represents the total cost per ounce as a function of time. When this ratio reaches its minimum—that is, when the total cost per ounce is at its lowest—is the optimal time to harvest the fish. Determine this optimal time to the nearest month.

C. Curve of Pursuit

An interesting geometric model arises when one tries to determine the path of a pursuer chasing its prey. This path is called a *curve of pursuit*. These problems were analyzed using methods of calculus circa 1730 (more than two centuries after Leonardo da Vinci had considered them). The simplest problem is to find the curve along which a vessel moves in pursuing another vessel that flees along a straight line, assuming the speeds of the two vessels are constant.

Let's assume vessel A, traveling at a speed α, is pursuing vessel B, which is traveling at a speed β. In addition, assume vessel A begins (at time $t = 0$) at the origin and pursues vessel B, which begins at the point $(b, 0)$, $b > 0$ and travels up the line $x = b$. After t hours, vessel A is located at the point $P = (x, y)$ and vessel B is located at the point $Q = (b, \beta t)$ (see Figure 3.16). The goal is to describe the locus of points P; that is, to find y as a function of x.

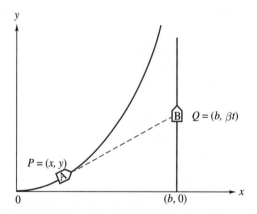

Figure 3.16 The path of vessel A as it pursues vessel B

(a) Vessel A is pursuing vessel B, so at time t, vessel A must be heading right at vessel B. That is, the tangent line to the curve of pursuit at P must pass through the point Q (see Figure 3.16). For this to be true, show that

(6) $$\frac{dy}{dx} = \frac{y - \beta t}{x - b}.$$

(b) We know the speed at which vessel A is traveling, so we know that the distance it travels in time t is αt. This distance is also the length of the pursuit curve from $(0, 0)$ to (x, y). Using the arclength formula from calculus, show that

(7) $$\alpha t = \int_0^x \sqrt{1 + [y'(u)]^2}\, du.$$

Solving for t in equations (6) and (7), conclude that

(8) $$\frac{y - (x - b)(dy/dx)}{\beta} = \frac{1}{\alpha}\int_0^x \sqrt{1 + [y'(u)]^2}\,du.$$

(c) Differentiating both sides of (8) with respect to x, derive the first order equation

$$(x - b)\frac{dw}{dx} = -\frac{\beta}{\alpha}\sqrt{1 + w^2},$$

where $w := dy/dx$.

(d) Using separation of variables and the initial conditions $x = 0$ and $w = dy/dx = 0$ when $t = 0$, show that

(9) $$\frac{dy}{dx} = w = \frac{1}{2}\left[\left(1 - \frac{x}{b}\right)^{-\beta/\alpha} - \left(1 - \frac{x}{b}\right)^{\beta/\alpha}\right].$$

(e) For $\alpha > \beta$, that is, the pursuing vessel A travels faster than the pursued vessel B, use equation (9) and the initial conditions $x = 0$ and $y = 0$, when $t = 0$, to derive the curve of pursuit

(10) $$y = \frac{b}{2}\left[\frac{(1 - x/b)^{1+\beta/\alpha}}{1 + \beta/\alpha} - \frac{(1 - x/b)^{1-\beta/\alpha}}{1 - \beta/\alpha}\right] + \frac{b\alpha\beta}{\alpha^2 - \beta^2}.$$

(f) Find the location where vessel A intercepts vessel B if $\alpha > \beta$.

(g) Show that if $\alpha = \beta$, then the curve of pursuit is given by

(11) $$y = \frac{b}{2}\left\{\frac{1}{2}\left[\left(1 - \frac{x}{b}\right)^2 - 1\right] - \ln\left(1 - \frac{x}{b}\right)\right\}.$$

Will vessel A ever reach vessel B?

D. Aircraft Guidance in a Crosswind

T.L. Pearson, Acadia University, Nova Scotia

An aircraft flying under the guidance of a nondirectional beacon (a fixed radio transmitter, abbreviated NDB) moves so that its longitudinal axis always points toward the beacon (see Figure 3.17). A pilot sets out toward an NDB from a point at which the wind is at right angles to the initial direction of the aircraft; the wind maintains this direction. Assume the windspeed and the speed of the aircraft through the air (its "airspeed") remain constant.

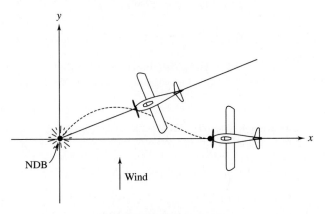

Figure 3.17 Guided aircraft

(a) Locate the flight in the xy-plane, placing the start of the trip at $(1, 0)$ and the destination at $(0, 0)$. Set up the differential equation describing the aircraft's path over the ground.

(b) Make an appropriate substitution and solve this equation. [*Hint:* See Section 2.6.]

(c) Use the fact that $x = 1$ and $y = 0$ at $t = 0$ to determine the appropriate value of the arbitrary constant in the solution set.

(d) Solve to get y explicitly in terms of x. Write your solution in terms of a hyperbolic function.

(e) Let γ be the ratio of windspeed to airspeed. Using a software package, graph the solutions for the cases $\gamma = 0.1, 0.3, 0.5,$ and 0.7 all on the same set of axes. Interpret these graphs.

(f) Discuss the (terrifying!) cases $\gamma = 1$ and $\gamma > 1$.

E. Stability of Numerical Methods

Numerical methods are often tested on simple initial value problems of the form

(12) $y' + \lambda y = 0,$ $y(0) = 1,$ $(\lambda = \text{constant}),$

which has the solution $\phi(x) = e^{-\lambda x}$. Notice that for each $\lambda > 0$, the solution $\phi(x)$ tends to zero as $x \to \infty$. Thus a desirable property for any numerical scheme that generates approximations $y_0, y_1, y_2, y_3, \ldots$ to $\phi(x)$ at the points $0, h, 2h, 3h, \ldots$ is that, for $\lambda > 0$,

(13) $y_n \to 0$ as $n \to \infty.$

For single-step linear methods, property (13) is called **absolute stability**.

(a) Show that for $x_n = nh$ Euler's method, when applied to the initial value problem (12), yields the approximations

$$y_n = (1 - \lambda h)^n, \qquad n = 0, 1, 2, \ldots,$$

and deduce that this method is absolutely stable only when $0 < \lambda h < 2$. (This means that for a given $\lambda > 0$, we must choose the step size h sufficiently small in order for property (13) to hold.) Further show that for $h > 2/\lambda$, the error $y_n - \phi(x_n)$ grows large exponentially!

(b) Show that for $x_n = nh$, the trapezoid scheme of Section 3.5 applied to problem (12) yields the approximations

$$y_n = \left(\frac{1 - \lambda h/2}{1 + \lambda h/2}\right)^n, \qquad n = 0, 1, 2, \ldots,$$

and deduce that this scheme is absolutely stable for all $\lambda > 0, h > 0$.

(c) Show that the improved Euler's method applied to problem (12) is absolutely stable for $0 < \lambda h < 2$.

Multistep Methods. When multistep numerical methods are used, instability problems may arise that cannot be circumvented by simply choosing a sufficiently small step size h. This is because multistep methods yield "extraneous solutions," which may dominate the calculations. To see what can happen, consider the two-step method

(14) $y_{n+1} = y_{n-1} + 2hf(x_n, y_n),$ $n = 1, 2, \ldots,$

for the equation $y' = f(x, y)$.

(d) Show that for the initial value problem

(15) $\qquad y' + 2y = 0, \qquad y(0) = 1,$

the recurrence formula (14), with $x_n = nh$, becomes

(16) $\qquad y_{n+1} + 4hy_n - y_{n-1} = 0.$

Equation (16), which is called a **difference equation**, can be solved by using an approach that is similar to solving linear differential equations with constant coefficients (see Section 4.5). Namely, we try a solution of the form $y_n = r^n$, where r is a constant to be determined.

(e) Show that substituting $y_n = r^n$ in (16) leads to the "characteristic equation"

$$r^2 + 4hr - 1 = 0,$$

which has roots

$$r_1 = -2h + \sqrt{1 + 4h^2} \quad \text{and} \quad r_2 = -2h - \sqrt{1 + 4h^2}.$$

By analogy with the theory for second order differential equations, it can be shown (see Section 4.11) that a general solution of (16) is

$$y_n = c_1 r_1^n + c_2 r_2^n,$$

where c_1 and c_2 are arbitrary constants. Thus the general solution to the difference equation (16) has two independent constants, whereas the differential equation in (15) has only one, namely, $\phi(x) = ce^{-2x}$.

(f) Show that for each $h > 0$,

$$\lim_{n \to \infty} r_1^n = 0, \quad \text{but} \quad \lim_{n \to \infty} |r_2^n| = \infty.$$

Hence the term r_1^n behaves like the solution $\phi(x_n) = e^{-2x_n}$ as $n \to \infty$. However, the extraneous solution r_2^n grows large without bound.

(g) Applying the scheme of (14) to the initial value problem (15) requires two starting values y_0, y_1. The exact values are $y_0 = 1$, $y_1 = e^{-2h}$. However, regardless of the choice of starting values and the size of h, the term $c_2 r_2^n$ will eventually dominate the full solution to the recurrence equation as x_n increases. Illustrate this instability by taking $y_0 = 1$, $y_1 = e^{-2h}$, and using a calculator or microcomputer to compute $y_2, y_3, \ldots, y_{100}$ from the recurrence formula (16) for $h = 0.5$ and $h = 0.05$. (*Note:* Even if initial conditions are chosen so that $c_2 = 0$, round-off error will inevitably "excite" the extraneous dominant solution.)

F. *Period Doubling and Chaos*

In the study of dynamical systems, the phenomena of *period doubling* and *chaos* are observed. These phenomena can be seen when one uses a numerical scheme to approximate the solution to an initial value problem for a nonlinear differential equation such as the following logistic model for population growth:

(17) $\qquad \dfrac{dp}{dt} = 10p(1 - p), \qquad p(0) = 0.1.$

(see Section 3.2).

(a) Solve the initial value problem (17) and show that $p(t)$ approaches 1 as t approaches infinity.

(b) Show that using Euler's method (see Sections 1.4 and 3.5) with step size h to approximate the solution to (17) gives

(18) $p_{n+1} = (1 + 10h)p_n - (10h)p_n^2, \qquad p_0 = 0.1.$

(c) For $h = 0.18, 0.23, 0.25,$ and 0.3, show that the first 40 iterations of (18) appear to (i) converge to 1 when $h = 0.18$, (ii) jump between 1.18 and 0.69 when $h = 0.23$, (iii) jump between 1.23, 0.54, 1.16, and 0.70 when $h = 0.25$, and (iv) display no discernible pattern when $h = 0.3$.

The transitions from convergence to jumping between two numbers, then four numbers, and so on, are called **period doubling**. The phenomenon displayed when $h = 0.3$ is referred to as **chaos**. This transition from period doubling to chaos as h increases is frequently observed in dynamical systems.

The transition to chaos is nicely illustrated in the bifurcation diagram (see Figure 3.18). This diagram is generated for equation (18) as follows. Beginning at $h = 0.18$, compute the sequence $\{p_n\}$ using (18) and, starting at $n = 201$, plot the next 30 values, that is, $p_{201}, p_{202}, \ldots, p_{230}$. Next, increment h by 0.001 to 0.181 and repeat. Continue this process until $h = 0.30$. Notice how the figure splits from one branch to two, then four, and then finally gives way to chaos.

Our concern is with the instabilities of the numerical procedure when h is not chosen small enough. Fortunately, the instability observed for Euler's method—the period doubling and chaos—was immediately recognized because we know that this type of behavior is not expected of a solu-

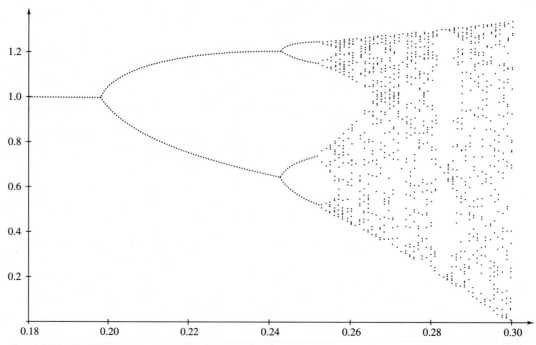

Figure 3.18 Period doubling to chaos

tion to the logistic equation. Consequently, if we had tried Euler's method with $h = 0.23$, 0.25, or 0.3 to numerically solve (17), we would have realized that h was not chosen small enough.

The situation for the fourth order Runge-Kutta method (see Section 3.6) is more troublesome. It may happen that for a certain choice of h period doubling occurs, but it is also possible that for other choices of h the numerical solution actually converges to a limiting value that is *not* the limiting value for any solution to the logistic equation in (17).

(d) Approximate the solution to (17) by computing the first 60 iterations of the fourth order Runge-Kutta method using the step size $h = 0.3$. (Thus, for the subroutine on page 132, input $N = 60$ and $c = 60(0.3) = 18$.) Repeat with $h = 0.325$ and $h = 0.35$. Which values of h (if any) do you feel are giving the ''correct'' approximation to the solution? Why?

A further discussion of chaos appears in Section 5.8.

G. Bang-Bang Controls

In Example 3 of Section 3.3 (page 105), it was assumed that the amount of heating or cooling supplied by a furnace or air conditioner is proportional to the difference between the actual temperature and the desired temperature; recall the equation

$$U(t) = K_U[T_D - T(t)].$$

In many homes, the heating/cooling mechanisms deliver a *constant* rate of heat flow, say,

$$U(t) = \begin{cases} K_1 & \text{if } T(t) > T_D, \\ K_2 & \text{if } T(t) < T_D \end{cases}$$

(with $K_1 < 0$).

(a) Modify the differential equation (9) in Example 3 so that it describes the temperature of a home employing this ''bang-bang'' control law.

(b) Suppose the initial temperature $T(0)$ is greater than T_D. Modify the constants in the solution (12), page 105, so that the formula is valid as long as $T(t) > T_D$.

(c) If the initial temperature $T(0)$ is less than T_D, what values should the constants in (12) take to make the formula valid for $T(t) < T_D$?

(d) How does one piece together the solutions in (b) and (c) to obtain a complete time description of the temperature $T(t)$?

Chapter 4

Linear Second Order Equations

4.1 INTRODUCTION: THE SIMPLE PENDULUM

> A simple pendulum consists of a mass m suspended by a cable of length l and negligible mass (see Fig. 4.1). If the cable is always straight and the mass is free to swing in a vertical plane, find the period of oscillation.

The simplest way to express the equation of motion for the pendulum is to analyze its angular momentum and the torque exerted on it by gravity. The angular momentum of the pendulum mass m, measured about the support O in Figure 4.1, is given by the product of the "lever arm" l and the component of vector momentum $m\mathbf{v}$ perpendicular to the lever arm. The torque produced by gravity equals the product of the lever arm l and the component of gravitational (vector) force $m\mathbf{g}$ perpendicular to the lever arm. Just as Newton's second law says force equals rate of change of momentum, the basic equation for rotational motion is "torque equals rate of change of angular momentum." This is proved in most books on classical mechanics, and a derivation is presented in Problem 1.

Because the length of the cable l of the pendulum is constant, the total vector velocity $\mathbf{v}$ of the mass m is perpendicular to the lever arm and its magnitude equals the rate at

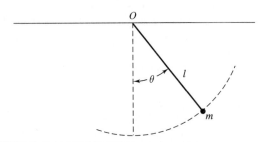

Figure 4.1 The simple pendulum

which the tip of the cable sweeps out arc length along the circular trajectory. The arc length is given by $l\theta$ (θ measured in radians), so this velocity is $l\,d\theta/dt$. The pendulum's angular momentum thus equals

$$\text{angular momentum} = l\left(ml\frac{d\theta}{dt}\right) = ml^2\frac{d\theta}{dt}.$$

In Figure 4.2, the component of gravitational force perpendicular to the lever arm is seen to be $mg \sin \theta$. Consequently the gravitational torque equals

$$\text{torque} = l\,mg \sin \theta.$$

Figure 4.2 Torque on the simple pendulum

The gravitational torque in Figure 4.2 acts *against* the angle θ, so the basic equation becomes

$$- l\,mg \sin \theta = \frac{d}{dt}\left[ml^2\frac{d\theta}{dt}\right] = ml^2\frac{d^2\theta}{dt^2},$$

which reduces to

(1) $$\frac{d^2\theta}{dt^2} + \frac{g}{l}\sin \theta = 0.$$

Equation (1), which governs the motion of the simple pendulum, is an example of a nonlinear equation. In most cases, one cannot find a solution to a nonlinear equation, and even when a solution is found, it often cannot be expressed using elementary functions. (Some exceptions are described in Project D.) In practice, when one encounters a nonlinear equation, it is helpful to study the **linearization** of the equation. This is an approximation of the nonlinear equation by a linear one. In our case, recall from calculus that for θ small, $\sin \theta \approx \theta.^{\dagger}$ Using this approximation in equation (1), we obtain the linear equation

(2) $$\frac{d^2\theta}{dt^2} + \frac{g}{l}\theta = 0.$$

†In fact, the Taylor series for $\sin \theta$, expanded around $\theta = 0$, is $\theta - \theta^3/3! + \theta^5/5! - \dots$.

Because of the difficulty in solving the pendulum equation (1), we focus instead on its linearization (2). For small oscillations of the pendulum, we expect that the period as determined from (2) will be a good approximation to the actual period.

To solve equation (2) we observe that we are looking for a function $\theta(t)$ with the property that its second derivative is just a negative constant times itself. Recalling that this property holds for sine and cosine functions, we are led to guess that our desired function has the form

$$\theta(t) = \cos \omega t \quad \text{or} \quad \theta(t) = \sin \omega t,$$

for some suitable choice of the constant ω. Substituting $\theta(t) = \cos \omega t$ into equation (2) gives

$$-\omega^2 \cos \omega t + \frac{g}{l} \cos \omega t = 0,$$

or

$$\left(\frac{g}{l} - \omega^2 \right) \cos \omega t = 0.$$

Hence, if we choose $\omega = \sqrt{g/l}$, then $\theta(t) = \cos(\sqrt{g/l}\, t)$ is a solution. Similarly, substituting $\theta(t) = \sin \omega t$, we again find $\omega = \sqrt{g/l}$. Therefore both

$$\theta_1(t) := \cos \left(\sqrt{\frac{g}{l}}\, t \right) \quad \text{and} \quad \theta_2(t) := \sin \left(\sqrt{\frac{g}{l}}\, t \right)$$

are solutions to (2). In fact, because (2) is a *linear* differential equation, a short computation shows that any function of the form

(3) $$\theta(t) = c_1 \cos \left(\sqrt{\frac{g}{l}}\, t \right) + c_2 \sin \left(\sqrt{\frac{g}{l}}\, t \right),$$

where c_1 and c_2 are arbitrary constants, is also a solution to (2) (see Section 4.2). Moreover (as we shall see in Section 4.3), every solution to (2) must be of the form (3).

If we are given the initial angular displacement $\theta(0)$ and initial angular velocity $\theta'(0)$, then we can determine c_1 and c_2 in equation (3). However, for us to calculate the period of oscillation, it is not necessary to know c_1 and c_2. It is clear from (3) that because the period of $\cos \omega t$ and $\sin \omega t$ is $2\pi/\omega$, the period of oscillation is

$$P = \frac{2\pi}{\sqrt{g/l}} = 2\pi \sqrt{\frac{l}{g}}.$$

This formula provides an approximation to the period of oscillation for the pendulum.

The oscillation described by equation (3) is called **simple harmonic motion.** We will study this motion in Section 5.1 in connection with mechanical vibrations. The nonlinear equation (1) that governs the motion of the simple pendulum is studied in more detail in Projects E and F. Linearization of nonlinear problems is discussed in Project C.

In this chapter, we concentrate on the theory of linear second order equations and methods for solving them. In many cases the solution steps are easy to execute, but it is tricky to organize them into a cohesive solution procedure. Therefore we devote the next two sections to the theory and defer the construction of explicit solutions to later sections.

EXERCISES 4.1

1. In Figure 4.2, let **r** denote the vector corresponding to the lever arm l. Show that taking the vector product of **r** with both sides of

$$\mathbf{F} = m\mathbf{a} = m\frac{d\mathbf{v}}{dt}$$

results in the equation of motion (1). [*Hint:* First argue that

$$\frac{d}{dt}(\mathbf{r} \times \mathbf{v}) = \mathbf{r} \times \frac{d\mathbf{v}}{dt}.]$$

4.2 LINEAR DIFFERENTIAL OPERATORS

A **linear** second order equation is an equation that can be written in the form

$$(1) \qquad a_2(x)\frac{d^2y}{dx^2} + a_1(x)\frac{dy}{dx} + a_0(x)y = b(x).$$

We will assume that $a_0(x)$, $a_1(x)$, $a_2(x)$, and $b(x)$ are continuous functions of x on an interval I. When a_0, a_1, and a_2 are constants, we say that the equation has **constant coefficients;** otherwise it has **variable coefficients.**

For now, we are interested only in those linear equations for which $a_2(x) \neq 0$ on I. In this case, we can rewrite (1) in the **standard form**

$$(2) \qquad \frac{d^2y}{dx^2} + p(x)\frac{dy}{dx} + q(x)y = g(x),$$

where $p(x) := a_1(x)/a_2(x)$, $q(x) := a_0(x)/a_2(x)$, and $g(x) := b(x)/a_2(x)$ are continuous on I. (Recall the standard form for linear first order equations, Section 2.3, page 47.)

Associated with equation (2) is the equation

$$(3) \qquad \frac{d^2y}{dx^2} + p(x)\frac{dy}{dx} + q(x)y = 0,$$

which is obtained from (2) by replacing $g(x)$ with the zero function. We say that (2) is a **nonhomogeneous** equation and that (3) is the corresponding **homogeneous** equation. Here the meaning of *homogeneous* is not related to the use of the term for first order equations (Section 2.6) but is used more in the sense of a homogeneous system of linear equations as studied in linear algebra.

Let's consider the expression on the left-hand side of equation (3), which, for convenience, we now write with prime notation:

$$(4) \qquad y''(x) + p(x)y'(x) + q(x)y(x).$$

Given any function y with a continuous second derivative on the interval I, then (4) generates a new function that we will denote by $L[y]$. This is,

(5) $L[y] := y'' + py' + qy.$

What we have done is to associate with each function y the function $L[y]$.

The term *function*, in its general sense, means a mapping that associates with each element in its domain a unique element in its range. Thus L can be interpreted as a function that is defined *on a set of functions*. Its domain is the collection of functions with a continuous second derivative; its range consists of continuous functions; and the rule of correspondence is given by (5). To avoid confusion, we call such mappings **operators.** Because L involves differentiation, we refer to L as a **differential operator.**

The image of a function y under the operator L is the function $L[y]$. If we want to evaluate this image function at some point x, we write $L[y](x)$.

For example, let $p(x) = x$ and $q(x) = x - 1$. Then

$$L[y](x) = y''(x) + xy'(x) + (x - 1)y(x).$$

If $y_1(x) = x^3$, we get

$$L[y_1](x) = 6x + x(3x^2) + (x - 1)x^3 = x^4 + 2x^3 + 6x.$$

Thus L maps the function x^3 to the function $x^4 + 2x^3 + 6x$. In particular, at $x = 2$,

$$L[y_1](2) = 2^4 + 2 \cdot 2^3 + 6 \cdot 2 = 44.$$

If $y_2(x) = \sin 2x$, then we similarly find

$$L[y_2](x) = -4 \sin 2x + x(2 \cos 2x) + (x - 1)\sin 2x$$
$$= x \sin 2x - 5 \sin 2x + 2x \cos 2x.$$

For $x = \pi$, we have $L[y_2](\pi) = 2\pi.$

The differential operator L defined by (5) has two very important properties that are extensions of two familiar facts: "The derivative of the sum is the sum of the derivatives," and "the derivative of a constant times a function is the constant times the derivative of the function."

LINEARITY OF THE DIFFERENTIAL OPERATOR L

Lemma 1. Let $L[y] := y'' + py' + qy$. If y, y_1, and y_2 are any twice-differentiable functions on the interval I and if c is any constant, then

(6) $L[y_1 + y_2] = L[y_1] + L[y_2],$

(7) $L[cy] = cL[y].$

In (6) and (7) equality is meant in the sense of equal functions on I.

Proof. On I, we have

$$L[y_1 + y_2] = (y_1 + y_2)'' + p(y_1 + y_2)' + q(y_1 + y_2)$$
$$= (y_1'' + y_2'') + p(y_1' + y_2') + q(y_1 + y_2).$$

On regrouping, this gives

$$L[y_1 + y_2] = (y_1'' + py_1' + qy_1) + (y_2'' + py_2' + qy_2)$$
$$= L[y_1] + L[y_2],$$

which verifies property (6). We leave the proof of property (7) for the exercises (see Problem 28). ▪ End

An operator that satisfies properties (6) and (7) for any constant c and any functions y_1 and y_2 in its domain is called a **linear operator.** Sometimes we abbreviate (6) and (7) by simply saying ''L preserves linear combinations.'' If (6) or (7) fails to hold, the operator is **nonlinear.**

As we showed in Lemma 1, the operator L defined by (5) is linear. We now give an example of a nonlinear operator.

EXAMPLE 1 Show that the operator T defined by

$$T[y](x) := y''(x) + \sin(y(x)),$$

where y is any function whose second derivative is continuous for all real x, is nonlinear.

Solution To demonstrate that T is a nonlinear operator, we attack the $\sin y$ term. It suffices to show that property (7) is *not always* satisfied. Let's try $y_1(x) = x$. Because $y_1''(x) \equiv 0$, we have

$$T[cy_1](x) = \sin(cx), \qquad cT[y_1](x) = c \sin x.$$

But, in general, $\sin(cx) \neq c \sin x$ (take, for example, $c = 2$ and $x = \pi/2$), so property (7) is violated. Hence T is a nonlinear operator. ▪ End

The linearity of the differential operator L in (5) can be used to prove the following theorem. Note that the theorem addresses only *homogeneous* equations.

LINEAR COMBINATIONS OF SOLUTIONS

Theorem 1. Let y_1 and y_2 be solutions to the homogeneous equation

(8) $y'' + py' + qy = 0.$

Then any linear combination $c_1 y_1 + c_2 y_2$ of y_1 and y_2, where c_1 and c_2 are constants, is also a solution to (8).

Proof. If we let $L[y] := y'' + py' + qy$, then $L[y_1] = 0$ and $L[y_2] = 0$, since y_1 and y_2 are solutions to (8). Using the linearity of L as expressed in properties (6) and (7), we have

$$L[c_1 y_1 + c_2 y_2] = L[c_1 y_1] + L[c_2 y_2]$$
$$= c_1 L[y_1] + c_2 L[y_2].$$
$$= 0 + 0 = 0.$$

Thus $c_1 y_1 + c_2 y_2$ is a solution to equation (8). ▪ End

Remark. Theorem 1 applies only to solutions of *homogeneous* equations. Consider, for example, two solutions $y_1(x)$ and $y_2(x)$ of $L[y] = \sin x$. Then the combination $y_1 + y_2$ would satisfy $L[y_1 + y_2] = 2 \sin x$. However, we shall see in Section 4.7 that the "structure" of the solutions to a nonhomogeneous linear differential equation $L[y] = g$ is strongly governed by the associated homogeneous equation $L[y] = 0$.

EXAMPLE 2 Given that $y_1(x) = e^{2x} \cos 3x$ and $y_2(x) = e^{2x} \sin 3x$ are solutions to the homogeneous equation

(9) $y'' - 4y' + 13y = 0,$

find a solution to (9) that satisfies the initial conditions

(10) $y(0) = 5$ and $y'(0) = -2.$

Solution As a consequence of Theorem 1, any linear combination

(11) $y(x) = c_1 e^{2x} \cos 3x + c_2 e^{2x} \sin 3x,$

with c_1 and c_2 arbitrary constants, will be a solution to (9). Thus we try to select c_1 and c_2 so as to satisfy the initial conditions. On differentiating (11), we find

(12) $y'(x) = c_1(2e^{2x} \cos 3x - 3e^{2x} \sin 3x) + c_2(2e^{2x} \sin 3x + 3e^{2x} \cos 3x).$

Substituting (11) and (12) into the initial conditions (10) yields the system of equations

$$c_1 = 5, \qquad 2c_1 + 3c_2 = -2,$$

whose solution is given by $c_1 = 5$, $c_2 = -4$. Hence, the solution to (9) that satisfies (10) is

$$y(x) = 5e^{2x} \cos 3x - 4e^{2x} \sin 3x. \quad \blacksquare \text{ End}$$

Let D denote differentiation with respect to x and D^2 differentiation with respect to x twice; that is, $Dy := dy/dx$, $D^2y := d^2y/dx^2$. In general, let $D^n y := d^n y/dx^n$. Using this nota-tion,[†] we can express L as defined in (5) by

$$L[y] = D^2y + pDy + qy = (D^2 + pD + q)y.$$

For example, if

$$L[y] = y'' + 4y' + 3y,$$

then we can write L in the form

$$L[y] = D^2y + 4Dy + 3y = (D^2 + 4D + 3)y.$$

When p and q are constants, we can even treat $D^2 + pD + q$ as a polynomial in D and factor it. This is illustrated in the next example.

EXAMPLE 3 Show that the operator $D^2 + 4D + 3$ is the same as the product $(D + 1)(D + 3)$, where by *product* we mean the composition of operators.

[†] *Historical Footnote:* The symbol D was introduced by the French mathematician F. J. Servois. The notion of a symbolic operation dates from B. Brisson in 1808 and was extended by Augustin Cauchy in 1827.

Solution For an arbitrary twice-differentiable function y, we have

$$(D + 1)(D + 3)y = (D + 1)[(D + 3)y] = (D + 1)[y' + 3y]$$
$$= D[y' + 3y] + 1[y' + 3y] = (y'' + 3y') + (y' + 3y)$$
$$= y'' + 4y' + 3y = (D^2 + 4D + 3)y.$$

Hence $(D + 1)(D + 3) = D^2 + 4D + 3$. ∎ End

Further discussion of the algebra of differential operators is given in Problems 32–36.
We close this section with a discussion of the existence and uniqueness properties of initial value problems for linear second order differential equations.

EXISTENCE AND UNIQUENESS OF SOLUTION

Theorem 2. Suppose $p(x)$, $q(x)$, and $g(x)$ are continuous on an interval (a, b) that contains the point x_0. Then, for any choice of the initial values y_0 and y_1, there exists a unique solution $y(x)$ on the whole interval (a, b) to the initial value problem

$$\frac{d^2y}{dx^2} + p(x)\frac{dy}{dx} + q(x)y = g(x);$$

$$y(x_0) = y_0, \qquad y'(x_0) = y_1.$$

Remark. Geometrically, the uniqueness part of Theorem 2 asserts that two solution curves passing through the same point and *with the same slope* must be identical throughout the interval (a, b).

The proof of Theorem 2 is more complicated than the proof of Theorem 1 in Section 2.3 for linear first order equations; it is given in Chapter 13.[†]

EXAMPLE 4 Determine the largest interval for which Theorem 2 ensures the existence and uniqueness of a solution to the initial value problem

(13)
$$\frac{d^2y}{dx^2} + \frac{1}{x - 3}\frac{dy}{dx} + \sqrt{x}\,y = \ln x;$$
$$y(1) = 3, \qquad y'(1) = -5.$$

Solution Here $p(x) = 1/(x - 3)$, $q(x) = \sqrt{x}$, $g(x) = \ln x$, and $x_0 = 1$. Now $p(x)$ is continuous for $x \neq 3$, and both $q(x)$ and $g(x)$ are continuous for $x > 0$. Hence the largest open interval containing $x_0 = 1$ for which $p(x)$, $q(x)$, and $g(x)$ are simultaneously continuous is the interval $(0, 3)$. From Theorem 2, we can conclude that the initial value problem (13) has a unique solution on the interval $(0, 3)$. ∎ End

[†] All references to Chapters 11–13 refer to the expanded text *Fundamentals of Differential Equations and Boundary Value Problems*, 2nd edition.

EXERCISES 4.2

In Problems 1 through 8, determine whether the equation is linear or nonlinear. If it is linear, classify it as being homogeneous or nonhomogeneous, with constant coefficients or with variable coefficients.

1. $6\dfrac{d^2y}{dx^2} + \dfrac{dy}{dx} = xy.$

2. $y'' + (1 - x)y' + xy = \sin x.$

3. $xy'' - yy' = \sin x.$

4. $\dfrac{d^2y}{dx^2} - x\dfrac{dy}{dx} + y = 0.$

5. $3t^2\dfrac{d^2x}{dt^2} = t\dfrac{dx}{dt} + 4x - \ln t.$

6. $\dfrac{d^2\theta}{dx^2} = \cos\theta.$

7. $\dfrac{d^2y}{d\theta^2} + y = \tan\theta.$

8. $\dfrac{d^2s}{dx^2} + 3\dfrac{ds}{dx} - s^{1/2} = x^2.$

9. Let $L[y](x) = x^2y''(x) - 3xy'(x) - 5y(x)$. Compute
 (a) $L[\cos x]$. (b) $L[x^{-1}]$.
 (c) $L[x^r]$, r a constant.

10. Let $L[y](x) = y''(x) - 4y'(x) + 3y(x)$. Compute
 (a) $L[x^2]$. (b) $L[e^{3x}]$.
 (c) $L[e^{rx}]$, r a constant.

11. Show that T defined by
 $$T[y](x) := y''(x) - 3y'(x) + y^2(x),$$
 where y is any function whose second derivative is continuous for all real x, is a nonlinear operator.

12. Show that T defined by
 $$T[y](x) := y''(x) + \{y'(x)y^2(x)\}^{1/3}$$
 is a nonlinear operator. Which property, (6) or (7), fails?

13. Given that $y_1(x) = e^{2x}\cos x$ and $y_2(x) = e^{2x}\sin x$ are solutions to the homogeneous equation
 $$y'' - 4y' + 5y = 0,$$
 find solutions to this equation that satisfy the following initial conditions:
 (a) $y(0) = 2$, $y'(0) = 1$.
 (b) $y(\pi) = 4e^{2\pi}$, $y'(\pi) = 5e^{2\pi}$.

14. Given that $y_1(x) = e^{2x}$ and $y_2(x) = e^{-x}$ are solutions to the homogeneous equation
 $$y'' - y' - 2y = 0,$$
 find solutions to this equation that satisfy the following initial conditions:
 (a) $y(0) = -1$, $y'(0) = 4$.
 (b) $y(0) = 3/2$, $y'(0) = 0$.

In Problems 15 through 18, use Theorem 2 to discuss the existence and uniqueness of a solution to the differential equation that satisfies the initial conditions $y(1) = y_0$, $y'(1) = y_1$, where y_0 and y_1 are real constants.

15. $(1 + x^2)y'' + xy' - y = \tan x.$

16. $x(x - 3)y'' + 2xy' - y = x^2.$

17. $e^x y'' - \dfrac{y'}{x - 3} + y = \ln x.$

18. $x^2y'' + y = \cos x.$

In Problems 19 through 22, determine whether Theorem 2 applies. If it does, then discuss what conclusions can be drawn. If it does not, explain why.

19. $y'' + yy' = x^2 - 1;$ $y(0) = 1,$ $y'(0) = -1.$

20. $x^2z'' + xz' + z = \cos x;$ $z(0) = 1,$ $z'(0) = 0.$

21. $(1 - t)x'' + tx' - 2x = \sin t;$ $x(0) = 1,$ $x'(0) = 1.$

22. $y'' + xy' - x^2y = 0;$ $y(0) = 0,$ $y(1) = 0.$

In Problems 23 through 26 express the given operator L using the differential operator D.

23. $L[y] := y'' + 7y' + 12y.$

24. $L[z] := 2z'' + z' - z.$

25. $L[y] := y'' + xy.$

26. $L[y] := x^2y'' - xy' + y.$

27. Let L be defined by $L[y] := (D^2 - xD + 2)y$. Compute
 (a) $L[x^2]$. (b) $L[\cos x]$.

28. Complete the proof of Lemma 1 on page 154 by verifying that the differential operator L defined by (5) satisfies property (7).

29. Show that the integro-differential operator L defined by
 $$L[y](x) := \int_0^1 (x - t)^2 y(t)dt + x^2y'(x)$$

is linear for any function y having a continuous derivative on [0, 1]; that is, L satisfies properties (6) and (7).

30. Boundary Value Problems. When the values of a solution to a differential equation are specified at two different points, these conditions are called **boundary conditions.** (In contrast, initial conditions specify the values of a function and its derivative at the same point.) Given that every solution to

(14) $y'' + y = 0$

is of the form

$$y(x) = c_1 \cos x + c_2 \sin x,$$

where c_1 and c_2 are arbitrary constants, show that

(a) there is a unique solution to (14) that satisfies the boundary conditions $y(0) = 2$ and $y(\pi/2) = 0$;

(b) there is no solution to (14) that satisfies $y(0) = 2$ and $y(\pi) = 0$;

(c) there are infinitely many solutions to (14) that satisfy $y(0) = 2$ and $y(\pi) = -2$.

31. Let $y = \phi(x)$ be the solution to the initial value problem

$$y'' + x^2 y' + e^x y = x - 1;$$

$$y(0) = 1, \qquad y'(0) = -1.$$

Assuming that the differential equation can be repeatedly differentiated, determine $\phi''(0)$, $\phi'''(0)$, and $\phi^{(4)}(0)$.

32. Differential Operators with Constant Coefficients. Let A and B represent two linear differential operators with constant coefficients; that is,

$$A := a_2 D^2 + a_1 D + a_0,$$

$$B := b_2 D^2 + b_1 D + b_0,$$

where a_0, a_1, a_2, b_0, b_1, and b_2 are constants. We define the **sum** $A + B$ by

$$(A + B)[y] := A[y] + B[y]$$

and the **product** AB by

$$(AB)[y] := A[B[y]].$$

Notice that the product AB is the composite operator obtained by applying first B, then A. We say two operators A and B are **equal** if $A[y] = B[y]$ for all functions y with the necessary derivatives. With these definitions, verify the following properties:

(a) Commutative laws: $A + B = B + A$, $AB = BA$.

(b) Associative laws: $(A + B) + C = A + (B + C)$, $(AB)C = A(BC)$.

(c) Distributive law: $A(B + C) = AB + AC$.

33. Let $A = (D - 1)$, $B = (D + 2)$, and $y = x^3$. Compute

(a) $A[y]$. **(b)** $B[A[y]]$.

(c) $B[y]$. **(d)** $A[B[y]]$.

(e) AB. **(f)** $(AB)[y]$.

34. Verify, as in Example 3, each of the following operator equations:

(a) $(D - 1)(D + 6) = D^2 + 5D - 6$.

(b) $(D + 1)(2D^2 - 1) = 2D^3 + 2D^2 - D - 1$.

35. Factor the following differential operators:

(a) $D^2 + 3D - 4$. **(b)** $D^2 + D - 6$.

(c) $2D^2 + 9D - 5$. **(d)** $D^2 - 2$.

36. Let $A = (xD - 1)$ and $B = (D + 1)$. Show for the function $y(x) = x$ that $(AB)[y] \neq (BA)[y]$. **Hence, for operators with *variable* coefficients, the commutative law need not hold.**

4.3 FUNDAMENTAL SOLUTIONS OF HOMOGENEOUS EQUATIONS

In this section, we discuss those properties of homogeneous equations that help us obtain their solutions. First, let's examine the particular equation

(1) $$L[y] = \frac{d^2 y}{dx^2} - y = 0.$$

It is easy to see that the two functions $y_1 = e^x$ and $y_2 = e^{-x}$ are solutions to this homogeneous equation. Further, since L is linear, we know from Theorem 1 in Section 4.2 that $c_1 e^x + c_2 e^{-x}$ is also a solution for any choice of constants c_1 and c_2. The question is, can all solutions of (1) be represented by

(2) $c_1 e^x + c_2 e^{-x},$

with appropriate choices for c_1 and c_2? As we now show, the answer lies in interpreting the existence and uniqueness theorem (Theorem 2, page 157).

Let $\phi(x)$ be a solution to (1) and let x_0 be a fixed real number. If we can choose c_1 and c_2 so that

(3)
$$\begin{cases} c_1 e^{x_0} + c_2 e^{-x_0} = \phi(x_0), \\ c_1 e^{x_0} - c_2 e^{-x_0} = \phi'(x_0), \end{cases}$$

then, because $\phi(x)$ and $c_1 e^x + c_2 e^{-x}$ satisfy the same initial conditions at x_0, the uniqueness conclusion of Theorem 2 guarantees that $\phi(x) = c_1 e^x + c_2 e^{-x}$ for all x. Note that we have reduced the question of the general validity of the representation (2), for all solutions to the differential equation, to the simpler question of the existence of solutions to the algebraic system (3).

To solve the system (3) for c_1, we add the two equations and then divide by $2e^{x_0}$ to obtain

(4) $c_1 = \dfrac{\phi(x_0) + \phi'(x_0)}{2e^{x_0}} .$

Subtracting the equations and dividing by $2e^{-x_0}$ gives

(5) $c_2 = \dfrac{\phi(x_0) - \phi'(x_0)}{2e^{-x_0}} .$

Notice that since e^{x_0} and e^{-x_0} are never zero, we can always carry out the above procedure for finding c_1 and c_2. Thus the algebraic system (3) does indeed possess a solution, and consequently the formula

$$\phi(x) = c_1 e^x + c_2 e^{-x}$$

is valid with suitable c_1, c_2.

In other words, every solution of (1) can be expressed as a linear combination of the two particular solutions y_1, y_2 of (1). If, on the other hand, we had started with the two solutions $y_1 = e^x$ and $y_3 = 5e^x$ of equation (1), the combination $c_1 y_1 + c_2 y_3 = c_1 e^x + c_2 5e^x$ would not be rich enough to meet arbitrary initial conditions like (3) (because this combination is simply a multiple of e^x; see Problem 14). The general situation is characterized by the following theorem.

REPRESENTATION OF SOLUTIONS (HOMOGENEOUS CASE)

Theorem 3. Let y_1 and y_2 denote two solutions on (a, b) of

(6) $y'' + p(x)y' + q(x)y = 0,$

where p and q are continuous on (a, b). Suppose at some point x_0 in (a, b) these solutions satisfy

(7) $y_1(x_0)y_2'(x_0) - y_1'(x_0)y_2(x_0) \neq 0.$

Then every solution of (6) on (a, b) can be expressed in the form

(8) $y(x) = c_1 y_1(x) + c_2 y_2(x),$

where c_1 and c_2 are constants.

In the proof that follows, we will see that statement (7) is precisely the condition we need to guarantee existence of the solution of the *algebraic* equations determining c_1 and c_2. The linear combination of y_1 and y_2 in (8), written with arbitrary constants c_1 and c_2, is referred to as a **general solution** to (6).

Proof of Theorem 3. Let $\phi(x)$ be any solution to (6). As before, we consider the system

(9) $c_1 y_1(x_0) + c_2 y_2(x_0) = \phi(x_0),$

(10) $c_1 y_1'(x_0) + c_2 y_2'(x_0) = \phi'(x_0),$

where x_0 is a point at which (7) holds. To solve for c_1, multiply (9) by $y_2'(x_0)$, multiply (10) by $y_2(x_0)$, and subtract to obtain

$$c_1[y_1(x_0)y_2'(x_0) - y_1'(x_0)y_2(x_0)] = \phi(x_0)y_2'(x_0) - \phi'(x_0)y_2(x_0).$$

Because of condition (7), the expression in brackets is not zero. Hence

$$c_1 = \frac{\phi(x_0)y_2'(x_0) - \phi'(x_0)y_2(x_0)}{y_1(x_0)y_2'(x_0) - y_1'(x_0)y_2(x_0)}.$$

Similarly we find

$$c_2 = \frac{\phi'(x_0)y_1(x_0) - \phi(x_0)y_1'(x_0)}{y_1(x_0)y_2'(x_0) - y_1'(x_0)y_2(x_0)}.$$

With these choices for c_1 and c_2, the functions $c_1 y_1 + c_2 y_2$ and ϕ are solutions to (6) that satisfy the same initial conditions. Hence they must be the same function on (a, b), according to the uniqueness result in Theorem 2. ∎ End

Because the expression on the left-hand side of (7) plays an important role in the theory, we adopt the following terminology.

WRONSKIAN

Definition 1. For any two differentiable functions y_1 and y_2, the function

(11) $W[y_1, y_2](x) := y_1(x)y_2'(x) - y_1'(x)y_2(x)$

is called the **Wronskian** of y_1 and y_2.[†]

A convenient way of writing the Wronskian $W[y_1, y_2]$, which will generalize to higher-order linear equations, is in terms of the determinant

$$W[y_1, y_2](x) = \begin{vmatrix} y_1(x) & y_2(x) \\ y_1'(x) & y_2'(x) \end{vmatrix}.$$

[†] *Historical Footnote:* The Wronskian was named after the Polish mathematician H. Wronski (1778–1853).

> ### FUNDAMENTAL SOLUTION SET
>
> **Definition 2.** A pair of solutions $\{y_1, y_2\}$ of $y'' + py' + qy = 0$ on (a, b) is called a **fundamental solution set** if
>
> $$W[y_1, y_2](x_0) \neq 0$$
>
> at some x_0 in (a, b).[†]

Logically speaking, Definition 2 requires only that the Wronskian be nonzero at *one* point. However, it will soon become clear that if the Wronskian of two solutions is nonzero at one point, it will actually be nonzero at *all* points in the interval (a, b). For example, the solutions $y_1 = e^x$, $y_2 = e^{-x}$ of the differential equation (1) form a fundamental solution set for this equation on $(-\infty, \infty)$, since

$$W[y_1, y_2](x) = e^x(-e^{-x}) - e^x e^{-x} = -2 \neq 0$$

for any x.

Using the terminology of Definition 2, Theorem 3 can be summarized as follows.

> ### PROCEDURE FOR SOLVING HOMOGENEOUS EQUATIONS
>
> To determine all solutions to $y'' + py' + qy = 0$:
>
> **(a)** Find two solutions y_1, y_2 that constitute a fundamental solution set.
> **(b)** Form the linear combination
>
> $$y(x) = c_1 y_1(x) + c_2 y_2(x),$$
>
> where c_1 and c_2 are arbitrary constants, to obtain a general solution.

EXAMPLE 1 Given that $y_1(x) = \cos 3x$ and $y_2(x) = \sin 3x$ are solutions to

(12) $y'' + 9y = 0$

on $(-\infty, \infty)$, find a general solution to (12).

Solution First we verify that $\{\cos 3x, \sin 3x\}$ is a fundamental solution set. We are given that both $y_1(x) = \cos 3x$ and $y_2(x) = \sin 3x$ are solutions to (12) (the reader can quickly verify this), so we need show only that $W[y_1, y_2](x) \neq 0$ for some x in $(-\infty, \infty)$. Substituting y_1 and y_2 into formula (11) yields

$$W[y_1, y_2](x) = (\cos 3x)(3 \cos 3x) - (-3 \sin 3x)(\sin 3x)$$
$$= 3(\cos^2 3x + \sin^2 3x)$$
$$= 3 \neq 0.$$

[†] ***Historical Footnote:*** Lazarus Fuchs introduced the terminology *fundamental system* in 1866.

Thus $\{\cos 3x, \sin 3x\}$ forms a fundamental solution set, and a general solution to (12) is

$$y(x) = c_1 \cos 3x + c_2 \sin 3x. \quad \blacksquare \text{ End}$$

It is easy to see that a fundamental solution set for $y'' + py' + qy = 0$ always exists. Indeed, let $x_0 \in (a, b)$ and take $y_1(x)$ and $y_2(x)$ to be the solutions that satisfy the initial conditions

$$y_1(x_0) = 1, \quad y_2(x_0) = 0,$$
$$y_1'(x_0) = 0, \quad y_2'(x_0) = 1.$$

Then $W[y_1, y_2](x_0) = 1 \neq 0$, and so $\{y_1, y_2\}$ is a fundamental solution set.

There is another method that can be used to determine quickly whether a pair of solutions form a fundamental solution set. To describe this test, we need to recall some basic facts about **vectors** (directed line segments) in the plane.

A planar vector is prescribed by an ordered pair of real numbers, say (u_1, u_2). Recall that vector addition is done componentwise, as is scalar multiplication; that is,

$$(u_1, u_2) + (v_1, v_2) = (u_1 + v_1, u_2 + v_2), \quad c(u_1, u_2) = (cu_1, cu_2).$$

Two vectors are parallel (point in the same or opposite directions) if and only if one vector is a scalar multiple of the other. This is equivalent to saying that there exist two constants c_1, c_2 *not both zero* such that

$$c_1(u_1, u_2) + c_2(v_1, v_2) = (0, 0).$$

Parallel vectors are said to be **linearly dependent,** whereas two nonparallel vectors are **linearly independent.**

These notions also apply to functions.

LINEAR DEPENDENCE OF FUNCTIONS

Definition 3. Two functions y_1 and y_2 are said to be **linearly dependent on an interval I** if there exist constants c_1 and c_2, not both zero, such that

$$c_1 y_1(x) + c_2 y_2(x) = 0$$

for all x in I. If two functions are not linearly dependent, they are said to be **linearly independent.**

The preceding definition extends in a natural way to sets of more than two functions; this is studied in Chapter 6 (see also Problems 24, 25, and 26 in Exercises 4.3). Observe that **in the case of just two functions, linear dependence on I is equivalent to one function being a constant multiple of the other function on I (see Problem 16).**

EXAMPLE 2 Determine whether the following pairs of functions y_1 and y_2 are linearly dependent on $(-5, 5)$:

(a) $y_1(x) = e^{3x}, \qquad y_2(x) = x + 1.$
(b) $y_1(x) = \sin 2x, \qquad y_2(x) = \cos x \sin x.$
(c) $y_1(x) = x, \qquad y_2(x) = |x|.$

Solution (a) A glance at the functions $y_1(x) = e^{3x}$ and $y_2(x) = x + 1$ indicates that neither is a constant multiple of the other. Indeed, if a constant c exists such that

$$e^{3x} = c(x + 1) \quad \text{for all } x \text{ in } (-5, 5),$$

then we arrive at a contradiction by setting $x = 0$ and $x = 1$:

$$e^0 = c(0 + 1) \Rightarrow c = 1,$$
$$e^3 = c(1 + 1) \Rightarrow c = e^3/2 \neq 1.$$

Hence e^{3x} and $x + 1$ are linearly independent.

(b) Because $y_1(x) = \sin 2x = 2 \sin x \cos x$, we see that $y_1(x) = 2y_2(x)$. Hence y_1 and y_2 are linearly dependent on $(-5, 5)$.

(c) Here the two functions $y_1(x) = x$ and $y_2(x) = |x|$ are identical on the subinterval $(0, 5)$. (In particular, they are linearly dependent on $(0, 5)$.) However, neither function is a fixed constant multiple of the other on the *whole* interval $(-5, 5)$. (On $(0, 5)$, $y_1(x) = 1 \cdot y_2(x)$, but on $(-5, 0)$, $y_1(x) = (-1) \cdot y_2(x)$.) Thus y_1 and y_2 are linearly independent on $(-5, 5)$. ■End

Part (c) depicted a situation where $y_1(x)$ and $y_2(x)$ were linearly dependent on a subinterval but not on the whole interval. The next theorem shows (among other things) that this *cannot happen* if y_1 and y_2 are both solutions to the same differential equation (on the whole interval).

CONDITION FOR LINEAR DEPENDENCE OF SOLUTIONS

Theorem 4. Let y_1, y_2 be solutions to $y'' + py' + qy = 0$ on (a, b), and let x_0 be any point in (a, b). Then y_1 and y_2 are linearly dependent on (a, b) if and only if the (constant) vectors[†]

$$\begin{bmatrix} y_1(x_0) \\ y_1'(x_0) \end{bmatrix} \quad \text{and} \quad \begin{bmatrix} y_2(x_0) \\ y_2'(x_0) \end{bmatrix}$$

are linearly dependent.

Remark. Although not explicitly mentioned, we assume in Theorem 4 and in subsequent results that the coefficient functions p and q are continuous.

[†] Here and in later discussions, we represent planar vectors in column rather than row form.

Proof of Theorem 4. ($\Rightarrow$) Assume that the *functions* y_1 and y_2 are linearly dependent. Then one of these functions is a constant multiple of the other, say $y_1 = cy_2$ on (a, b). Thus $y_1' = cy_2'$, and we have

$$\begin{bmatrix} y_1(x_0) \\ y_1'(x_0) \end{bmatrix} = \begin{bmatrix} cy_2(x_0) \\ cy_2'(x_0) \end{bmatrix} = c \begin{bmatrix} y_2(x_0) \\ y_2'(x_0) \end{bmatrix}.$$

Since one initial vector is a constant multiple of the other, they are linearly dependent.
 ($\Leftarrow$) Assume now that the *initial vectors* are linearly dependent, say

$$\begin{bmatrix} y_1(x_0) \\ y_1'(x_0) \end{bmatrix} = c \begin{bmatrix} y_2(x_0) \\ y_2'(x_0) \end{bmatrix}.$$

This implies that the functions y_1 and cy_2 satisfy the same initial conditions; that is,

$$y_1(x_0) = cy_2(x_0) \quad \text{and} \quad y_1'(x_0) = cy_2'(x_0).$$

Because y_1 and cy_2 are solutions to the same initial value problem, by the uniqueness property they must be the same function on (a, b). Thus y_1 is a constant multiple of y_2 on (a, b); that is, y_1 and y_2 are linearly dependent on (a, b). ∎ End

A simple algebraic test for the linear independence of two column vectors

(13) $\begin{bmatrix} u_1 \\ u_2 \end{bmatrix}$ and $\begin{bmatrix} v_1 \\ v_2 \end{bmatrix}$

is to compute the determinant

$$\begin{vmatrix} u_1 & v_1 \\ u_2 & v_2 \end{vmatrix} = u_1 v_2 - v_1 u_2.$$

The vectors (13) *are linearly independent if and only if this determinant is not zero* (see Problem 29). Assembling all these considerations, we obtain the following characterization.

FUNDAMENTAL SETS, LINEAR INDEPENDENCE, AND THE WRONSKIAN

Corollary 1. If y_1 and y_2 are solutions to $y'' + py' + qy = 0$ on (a, b), then the following statements are equivalent:

(i) $\{y_1, y_2\}$ is a fundamental solution set on (a, b).

(ii) y_1 and y_2 are linearly independent on (a, b).

(iii) The Wronskian $W[y_1, y_2](x)$ is never zero on (a, b).

Proof. If $\{y_1, y_2\}$ is a fundamental set, then there is *some* point x_0 in (a, b), where

(14) $W[y_1, y_2](x_0) = \begin{vmatrix} y_1(x_0) & y_2(x_0) \\ y_1'(x_0) & y_2'(x_0) \end{vmatrix} \neq 0,$

by Definition 2. By the determinant test, then, the vectors

(15) $\begin{bmatrix} y_1(x_0) \\ y_1'(x_0) \end{bmatrix}$ and $\begin{bmatrix} y_2(x_0) \\ y_2'(x_0) \end{bmatrix}$

are independent. Theorem 4 thus implies the *functions* y_1 and y_2 are linearly independent. Therefore condition (i) implies condition (ii).

But Theorem 4 also states that linear independence of y_1 and y_2 implies that the vectors in (15) are linearly independent, regardless of the choice of x_0. Thus, because of the determinant test (again), the Wronskian of y_1 and y_2 is nonzero everywhere in the interval, and condition (ii) therefore implies condition (iii).

Finally, condition (iii) implies condition (i) trivially because of our definition of fundamental solution sets (Definition 2). ■ End

In particular, the Wronskian of two solutions y_1 and y_2 is either *identically* zero or *never* zero on (a, b), depending on whether the solutions are linearly dependent or linearly independent. In Example 1, for instance, the Wronskian has the constant value 3.

EXAMPLE 3 Show that $y_1(x) = x^{-1}$ and $y_2(x) = x^3$ are solutions to

(16) $x^2 y'' - xy' - 3y = 0$

on the interval $(0, \infty)$ and give a general solution.

Solution The verification that y_1 and y_2 are solutions to (16) is straightforward; substituting $y = x^{-1}$ and $y = x^3$ in (16) gives, respectively, the identities

$$x^2(2x^{-3}) - x(-x^{-2}) - 3(x^{-1}) = 0 \quad \text{and} \quad x^2(6x) - x(3x^2) - 3(x^3) = 0.$$

Furthermore, the solution functions x^{-1} and x^3 are linearly independent on $(0, \infty)$ (neither is a constant multiple of the other on $(0, \infty)$). Hence, by Corollary 1, $\{x^{-1}, x^3\}$ is a fundamental solution set on $(0, \infty)$, and so a general solution is

$$y(x) = c_1 x^{-1} + c_2 x^3. \quad \text{■ End}$$

EXAMPLE 4 Can the function $w(x) = 3(x - 1)^2$ be the Wronskian on the interval $(0, 2)$ for some homogeneous linear second order equation (with p and q continuous)?

Solution Because $w(x) = 0$ only for $x = 1$, then by Corollary 1, $w(x)$ cannot be such a Wronskian on any open interval containing $x = 1$. ■ End

If two differentiable functions are linearly independent on I, can their Wronskian be identically zero? Surprisingly, the answer is yes. The reader can verify (see Problem 18) that the Wronskian of the functions $y_1(x) = x^3$ and $y_2(x) = |x^3|$ is identically zero on $(-\infty, \infty)$. Yet these functions are certainly linearly independent on $(-\infty, \infty)$ because one is not a constant multiple of the other for all x in $(-\infty, \infty)$. Does this example contradict Corollary 1? No, because Corollary 1 refers only to functions $y_1(x)$ and $y_2(x)$ that are solutions to the **same** homogeneous linear second order differential equation. What we *can*

conclude from Corollary 1 is that $y_1(x) = x^3$ and $y_2(x) = |x^3|$ cannot be solutions to the same equation on $(-\infty, \infty)$.

Another representation of the Wronskian for two solutions y_1, y_2 to the equation $y'' + py' + qy = 0$ on (a, b) is **Abel's identity**[†]:

$$(17) \qquad W[y_1, y_2](x) = C \exp\left(-\int_{x_0}^{x} p(t)\,dt\right),$$

where $x_0 \in (a, b)$ and C is a constant that depends on y_1 and y_2. This formula reaffirms our earlier observation that the Wronskian of two solutions is always zero ($C = 0$) or never zero ($C \neq 0$), since the exponential factor never vanishes. We leave the proof of (17) as an exercise (see Problem 20).

The properties of the Wronskian and linear independence will be extended to sets of n functions $y_1(x), \ldots, y_n(x)$ when we consider higher-order linear equations in Chapter 6. Although the proofs are a bit more complicated, the ideas are essentially the same as in the case of linear second order equations.

To summarize, we have shown in this section that to find *all* the solutions to a homogeneous second order linear differential equation, it suffices to obtain just two linearly independent solutions and then take linear combinations of them. In Sections 4.5 and 4.6, we concentrate on methods for actually finding such a pair of functions.

EXERCISES 4.3

In Problems 1 through 6 use Definition 3 to determine whether the functions y_1 and y_2 are linearly dependent on the interval $(0, 1)$. Also compute the Wronskian $W[y_1, y_2](x)$.

1. $y_1(x) = e^{-x} \cos 2x$, $\quad y_2(x) = e^{-x} \sin 2x$.

2. $y_1(x) = e^{3x}$, $\quad y_2(x) = e^{-4x}$.

3. $y_1(x) = xe^{2x}$, $\quad y_2(x) = e^{2x}$.

4. $y_1(x) = x^2 \cos(\ln x)$, $\quad y_2(x) = x^2 \sin(\ln x)$.

5. $y_1(x) = \tan^2 x - \sec^2 x$, $\quad y_2(x) = 3$.

6. $y_1(x) = 0$, $\quad y_2(x) = e^x$.

In Problems 7 through 12, (a) verify that the functions y_1 and y_2 are linearly independent solutions of the given differential equation, (b) find a general solution to the given differential equation, and (c) find the solution that satisfies the given initial conditions.

7. $x^2 y'' - 2y = 0$; $\quad y_1(x) = x^2$, $\quad y_2(x) = x^{-1}$;
$y(1) = -2$, $\quad y'(1) = -7$.

8. $y'' - 5y' + 6y = 0$; $\quad y_1(x) = e^{2x}$, $\quad y_2(x) = e^{3x}$;
$y(0) = -1$, $\quad y'(0) = -4$.

9. $y'' - 2y' + 5y = 0$; $\quad y_1(x) = e^x \cos 2x$,
$y_2(x) = e^x \sin 2x$; $\quad y(0) = 2$, $\quad y'(0) = 0$.

10. $y'' - 5y' = 0$; $\quad y_1(x) = 2$, $\quad y_2(x) = e^{5x}$;
$y(0) = 2$, $\quad y'(0) = 5$.

11. $ty'' - (t + 2)y' + 2y = 0$; $\quad y_1(t) = e^t$,
$y_2(t) = t^2 + 2t + 2$; $\quad y(1) = 0$, $\quad y'(1) = 1$.

12. $y'' - y = 0$; $\quad y_1(x) = \cosh x$, $\quad y_2(x) = \sinh x$;
$y(0) = 1$, $\quad y'(0) = -1$.

13. Consider the differential equation

$$(18) \qquad y'' + 5y' - 6y = 0.$$

(a) Show that $S_1 := \{e^x, e^x - e^{-6x}\}$ is a fundamental solution set for (18).

(b) Show that $S_2 := \{e^x, 3e^x + e^{-6x}\}$ is another fundamental solution set for (18).

(c) Verify that $\phi(x) = e^{-6x}$ is a solution to (18). Then

[†] **Historical Footnote:** Abel's identity was derived by Niels Abel in 1827.

express ϕ as a linear combination of functions in S_1. Likewise, express ϕ as a linear combination of functions in S_2.

14. **(a)** Verify that $y_1 = e^x$ and $y_3 = 5e^x$ are two solutions to $y'' - y = 0$ (equation (1) of the text), but the solution $\phi(x)$ to this equation that satisfies $\phi(0) = 1$, $\phi'(0) = 0$ cannot be expressed as $c_1 y_1(x) + c_2 y_3(x)$, regardless of the choice of the constants c_1 and c_2.
 (b) Compute the Wronskian of the solutions y_1 and y_3 in part (a).

15. Determine whether the following functions can be Wronskians on $-1 < x < 1$ for some homogeneous linear second order equation (with p and q continuous).

 (a) $w(x) = 6e^{4x}$. **(b)** $w(x) = x^3$.
 (c) $w(x) = (x + 1)^{-1}$. **(d)** $w(x) \equiv 0$.

16. Prove that two functions are linearly dependent on an interval I in the sense of Definition 3 if and only if one is a constant times the other on I.

17. Let y_1 and y_2 be two functions defined on $(-\infty, \infty)$:

 (a) (True or False) If y_1 and y_2 are linearly dependent on the interval $[a, b]$, then y_1 and y_2 are linearly dependent on the smaller interval $[c, d] \subset [a, b]$.
 (b) (True or False) If y_1 and y_2 are linearly dependent on the interval $[a, b]$, then y_1 and y_2 are linearly dependent on the larger interval $[C, D] \supset [a, b]$.

18. Let $y_1(x) = x^3$ and $y_2(x) = |x^3|$. Are y_1 and y_2 linearly independent on the following intervals?

 (a) $[0, \infty)$. **(b)** $(-\infty, 0]$. **(c)** $(-\infty, \infty)$.
 (d) Compute the Wronskian $W[y_1, y_2](x)$ on the interval $(-\infty, \infty)$.

19. Use Abel's identity (17) to determine (up to a constant multiple) the Wronskian of two solutions on $(0, \infty)$ to

$$xy'' + (x - 1)y' + 3y = 0.$$

20. Prove Abel's identity (17) by completing the following steps:

 (a) Show that the Wronskian W satisfies the equation $W' + p(x)W = 0$.
 (b) Solve the separable equation in part (a).

21. Prove that if y_1 and y_2 are linearly independent solutions of $y'' + py' + qy = 0$ on (a, b), then they cannot both be zero at the same point x_0 in (a, b).

22. Show that if y_1 and y_2 are linearly independent solutions of $y'' + py' + qy = 0$ on (a, b), then they cannot both have an extremum at the same point x_0 in (a, b).

23. **Normal Form.** Show that the substitution $y(x) = u(x)v(x)$, where

$$v(x) := \exp\left(-\frac{1}{2}\int p(x)\,dx\right),$$

transforms the differential equation

$$y'' + p(x)y' + q(x)y = 0$$

into an equation of the form

$$u'' + f(x)u = 0.$$

The last equation is called the **normal form** of a homogeneous linear second order equation.

24. **Linear Dependence of Three Functions.** Three functions $y_1(x)$, $y_2(x)$, and $y_3(x)$ are said to be **linearly dependent on an interval I** if there exist constants C_1, C_2, C_3, not all zero, such that

$$C_1 y_1(x) + C_2 y_2(x) + C_3 y_3(x) = 0, \quad \text{all } x \text{ in } I.$$

Otherwise, we say they are **linearly independent.**

 (a) Show that if y_1 and y_2 are two linearly dependent functions on I, then y_1, y_2, and y_3 are linearly dependent on I for any function y_3.
 (b) Show that $y_1(x) = e^x$, $y_2(x) = e^{2x}$, and $y_3(x) = e^{-3x}$ are linearly independent on $(-\infty, \infty)$.
 (c) Show that $y_1(x) = \cos 2x$, $y_2(x) = \sin^2 x$, and $y_3(x) = \cos^2 x$ are linearly dependent on $(-\infty, \infty)$.

25. Using the definitions in Problem 24, determine whether the given three functions are linearly dependent or linearly independent on $(-\infty, \infty)$:

 (a) $y_1(x) = 1$, $y_2(x) = x$, $y_3(x) = x^2$.
 (b) $y_1(x) = -3$, $y_2(x) = 5\sin^2 x$, $y_3(x) = \cos^2 x$.
 (c) $y_1(x) = e^x$, $y_2(x) = xe^x$, $y_3(x) = x^2 e^x$.
 (d) $y_1(x) = e^x$, $y_2(x) = e^{-x}$, $y_3(x) = \cosh x$.

26. Using the definition in Problem 24, prove that $y_1(x) = e^{r_1 x}$, $y_2(x) = e^{r_2 x}$, and $y_3(x) = e^{r_3 x}$ are linearly independent on $(-\infty, \infty)$ if and only if the real numbers r_1, r_2, and r_3 are distinct.

27. **Wronskian of Three Functions.** The Wronskian of three functions is defined in terms of a determinant as follows:

$$W[y_1, y_2, y_3](x) := \begin{vmatrix} y_1(x) & y_2(x) & y_3(x) \\ y_1'(x) & y_2'(x) & y_3'(x) \\ y_1''(x) & y_2''(x) & y_3''(x) \end{vmatrix}.$$

Find the Wronskian of the three functions given in

(a) Part (a) of Problem 25.
(b) Part (b) of Problem 25.
(c) Part (c) of Problem 25.
(d) What do you think is the connection between the Wronskian of three functions and their linear dependence?

28. Let y_1 and y_2 be solutions to $y'' + py' + qy = 0$ on (a, b). Prove that if they are linearly dependent on a subinterval (c, d), they are linearly dependent on all of (a, b). [*Hint:* Choose $x_0 \in (c, d)$ and apply Theorem 4.]

29. Show that the determinant

$$\begin{vmatrix} u_1 & v_1 \\ u_2 & v_2 \end{vmatrix} \quad (u_1, u_2, v_1, v_2 \text{ constants})$$

is not zero

(a) if and only if the two column vectors

$$\begin{bmatrix} u_1 \\ u_2 \end{bmatrix} \quad \text{and} \quad \begin{bmatrix} v_1 \\ v_2 \end{bmatrix}$$

are linearly independent.

(b) if and only if the system with unknowns c_1 and c_2

$$u_1 c_1 + v_1 c_2 = a,$$
$$u_2 c_1 + v_2 c_2 = b,$$

has a unique solution for all a, b.

4.4 REDUCTION OF ORDER

As we found in the previous section, a general solution to a linear second order homogeneous equation is given by a linear combination of two linearly independent solutions. In the next section, we describe a method for finding such solutions when the equation has constant coefficients. However, the method may produce only a *single* solution (and its multiples), rather than the required *pair* of solutions. So although it may seem like "jumping the gun," we devote the present section to a technique that uses a known solution to reduce any homogeneous linear second order equation to a separable first order equation—a type discussed in Section 2.2—yielding a solution which is independent of the given solution. This procedure is referred to as **reduction of order**.[†]

Let f be a nontrivial (i.e., not identically zero) solution to the homogeneous equation

(1) $y'' + p(x)y' + q(x)y = 0.$

For us to find a second solution that is linearly independent of f, the simplest way would be to use function addition or multiplication to modify f. However, trying a solution of the form $y(x) = v(x) + f(x)$ leads us right back to where we started: to the equation $v'' + pv' + qv = 0$. (Check this!) So, let's try to find a solution of the form

(2) $y(x) = v(x)f(x),$

where the nonconstant function v is to be determined. Differentiating the product vf, we have

$$y' = vf' + v'f,$$
$$y'' = vf'' + 2v'f' + v''f.$$

Substituting these expressions into (1) yields

$$(vf'' + 2v'f' + v''f) + p(vf' + v'f) + qvf = 0,$$

[†] ***Historical Footnote:*** The method for the reduction of order is credited to Jean le Rond d'Alembert circa 1760.

or, on regrouping,

(3) $(f'' + pf' + qf)v + fv'' + (2f' + pf)v' = 0.$

Because f is a solution to (1), the factor in front of v is zero. Thus (3) reduces to

(4) $fv'' + (2f' + pf)v' = 0.$

This is a separable first order equation in the variable $w = v'$. Indeed, on substituting w for v' in (4), we have

$$fw' + (2f' + pf)w = 0.$$

Then, separating the variables and integrating gives

$$\int \frac{dw}{w} = -2 \int \frac{f'}{f} \, dx - \int p \, dx,$$

$$\ln w = \ln(f^{-2}) - \int p \, dx,$$

$$w = f^{-2} e^{-\int p \, dx}.$$

If we replace w by v' in the last equation and integrate, we find

(5) $$v = \int \frac{e^{-\int p(x) \, dx}}{[f(x)]^2} \, dx.$$

Hence a second linearly independent solution is $y(x) = v(x)f(x)$, where $v(x)$ is given in equation (5).

The above discussion can be summarized as follows.

REDUCTION OF ORDER PROCEDURES

Given a nontrivial solution $f(x)$ to $y'' + py' + qy = 0$, a second linearly independent solution $y(x)$ can be determined in either of the following ways:

1. Set $y(x) = v(x)f(x)$ and substitute for y, y', and y'' in the given equation. This gives a separable equation for v'. Solve for v' and integrate v' to obtain v. The desired second solution is given by $v(x)f(x)$.

2. The solution $y(x)$ can also be obtained by plugging $p(x)$ and $f(x)$ directly into the **reduction of order formula:**

(6) $$y(x) = f(x) \int \frac{e^{-\int p(x) \, dx}}{[f(x)]^2} \, dx.$$

Caution: If the equation is $a_2(x)y'' + a_1(x)y' + a_0(x)y = 0$, then it must first be put in the form $y'' + p(x)y' + q(x)y = 0$, so $p(x) = a_1(x)/a_2(x)$.

The first procedure is a technique for which we only have to remember the substitution $y = v(x)f(x)$. The second procedure is quicker, but we must remember precisely the reduction of order formula. In the next example, we illustrate both of these methods.

EXAMPLE 1 Given that $f(x) = e^x$ is a solution to

(7) $y'' - 2y' + y = 0$,

determine a second linearly independent solution.

Solution Let $y = v(x)f(x) = ve^x$. Then

$$y' = ve^x + v'e^x,$$
$$y'' = ve^x + 2v'e^x + v''e^x.$$

Substituting these representations into (7) gives

$$(ve^x + 2v'e^x + v''e^x) - 2(ve^x + v'e^x) + ve^x = 0,$$

which simplifies to $v'' = 0$. Hence $v = c_1 x + c_2$. Because we want y to be linearly independent of e^x, we must take $c_1 \neq 0$, so for simplicity, let $c_1 = 1$ and $c_2 = 0$. Then a second linearly independent solution is

$$y = vf = xe^x.$$

The alternative approach is to use the reduction of order formula given by equation (6). In this example, $p(x) = -2$ and $f(x) = e^x$. Hence

$$-\int p(x)\,dx = \int 2\ dx = 2x + c.$$

We will take $c = 0$. Now

$$y(x) = f(x) \int \frac{e^{-\int p(x)\,dx}}{[f(x)]^2}\,dx = e^x \int \frac{e^{2x}}{(e^x)^2}\,dx$$

$$= e^x \int dx = e^x(x + C).$$

Again, we set $C = 0$. This gives us the same linearly independent solution, $y = xe^x$. ∎ End

EXAMPLE 2 Given that $f(x) = x$ is a solution to

(8) $y'' - 2xy' + 2y = 0$,

determine a second linearly independent solution.

Solution Here $p(x) = -2x$, and so

$$-\int p(x)\,dx = \int 2x\ dx = x^2,$$

where we have taken the constant of integration to be zero. The reduction of order formula in (6) gives

(9) $$y(x) = x \int \frac{e^{x^2}}{x^2}\,dx$$

as a second linearly independent solution. Unfortunately, we are not able to evaluate the integral in (9) in terms of elementary functions. One alternative is to write $y(x)$ in terms of a definite integral, say,

$$y(x) = x \int_1^x \frac{e^{t^2}}{t^2} \, dt.$$

Then $y(x)$ can be approximated for various values of x by using a numerical integration method such as Simpson's rule (see Problem 17). Another approach is to obtain a **power series expansion** for the solution in (9). This can be derived as follows.

Recall that the Maclaurin series for e^x is

$$e^x = 1 + x + \frac{x^2}{2!} + \frac{x^3}{3!} + \cdots + \frac{x^n}{n!} + \cdots.$$

Hence the expansion for e^{x^2} is just

$$e^{x^2} = 1 + x^2 + \frac{x^4}{2!} + \frac{x^6}{3!} + \cdots + \frac{x^{2n}}{n!} + \cdots.$$

Substituting this series for e^{x^2} into (9), we integrate term by term to obtain

$$y(x) = x \int x^{-2} \left\{ 1 + x^2 + \frac{x^4}{2} + \cdots + \frac{x^{2n}}{n!} + \cdots \right\} dx$$

$$= x \int \left\{ x^{-2} + 1 + \frac{x^2}{2} + \cdots + \frac{x^{2n-2}}{n!} + \cdots \right\} dx$$

$$= x \left\{ -x^{-1} + x + \frac{x^3}{6} + \cdots + \frac{x^{2n-1}}{n!(2n - 1)} + \cdots \right\},$$

where we have taken the constant of integration to be zero. Hence

(10) $y(x) = -1 + x^2 + \frac{x^4}{6} + \cdots + \frac{x^{2n}}{n!(2n - 1)} + \cdots.$ ■ End

In the last example, we did not bother to justify the use of term-by-term integration to obtain equation (10). Such questions are addressed in Chapter 8.

The reduction of order procedure can also be used to reduce a linear *nonhomogeneous* second order equation to a linear first order equation (see Problem 18 in Exercises 4.7).

EXERCISES 4.4

In Problems 1 through 6, a differential equation and a non-trivial solution f are given. Find a second linearly independent solution.

1. $y'' + 2y' - 15y = 0$; $f(x) = e^{3x}$.

2. $z'' - 3z' + 2z = 0$; $f(x) = e^x$.

3. $x^2 y'' - 2xy' - 4y = 0$, $x > 0$; $f(x) = x^{-1}$.

4. $x^2 y'' + 6xy' + 6y = 0$, $x > 0$; $f(x) = x^{-2}$.

5. $tx'' - (t + 1)x' + x = 0$, $t > 0$; $f(t) = e^t$.

6. $xy'' + (1 - 2x)y' + (x - 1)y = 0$, $x > 0$; $f(x) = e^x$.

7. The reduction of order procedure can be used more generally to reduce a homogeneous linear nth order equation to a homogeneous linear $(n-1)$th order equation. For the equation

$$xy''' - xy'' + y' - y = 0,$$

which has $f(x) = e^x$ as a solution, use the substitution $y(x) = v(x)f(x)$ to reduce this third order equation to a homogeneous linear second order equation in the variable $w = v'$.

8. The equation

$$xy''' + (1 - x)y'' + xy' - y = 0$$

has $f(x) = x$ as a solution. Use the substitution $y(x) = v(x)f(x)$ to reduce this third order equation to a homogeneous linear second order equation in the variable $w = v'$.

9. Given that $f(x) = x$ is a solution to

$$y'' - xy' + y = 0;$$

(a) Obtain an integral representation for a second linearly independent solution.

(b) Using the integral representation from part (a), obtain a Maclaurin series expansion for this solution.

10. Given that $f(x) = x$ is a solution to

$$(1 - x^2)y'' - 2xy' + 2y = 0, \qquad -1 < x < 1;$$

(a) Obtain an integral representation for a second linearly independent solution.

(b) Using the integral representation from part (a), obtain a Maclaurin series expansion for this solution. [*Hint:* $(1 - x)^{-1} = 1 + x + x^2 + \cdots$.]

(c) Use partial fractions to evaluate the integral in part (a) and give a general solution to the differential equation.

11. The reduction of order formula (6) can also be derived from Abel's identity [formula (17) of Section 4.3]. Let $f(x)$ be a nontrivial solution to (1) and $y(x)$ a second linearly independent solution. Show that

$$\left(\frac{y}{f}\right)' = \frac{W[f, y]}{f^2},$$

and then use Abel's identity for the Wronskian $W[f, y]$ to obtain the reduction of order formula.

12. Verify, by direct substitution, that the power series (10) is a solution to the differential equation (8).

13. In quantum mechanics, the study of the Schrödinger equation for the case of a harmonic oscillator leads to a consideration of **Hermite's equation,**

$$y'' - 2xy' + \lambda y = 0,$$

where λ is a parameter. Use the reduction of order formula to obtain an integral representation of a second linearly independent solution to Hermite's equation for the given value of λ and corresponding solution $f(x)$.

(a) $\lambda = 4, \qquad f(x) = 1 - 2x^2$.
(b) $\lambda = 6, \qquad f(x) = 3x - 2x^3$.

14. In mathematical physics, many problems with spherical symmetry involve the study of **Legendre's equation,**

$$(1 - x^2)y'' - 2xy' + \lambda(\lambda + 1)y = 0, \qquad 1 < x < 1,$$

where λ is a parameter. Use the reduction of order formula to obtain an integral representation of a second linearly independent solution to Legendre's equation for the given value of λ and corresponding solution $f(x)$.

(a) $\lambda = 1, \qquad f(x) = x$.
(b) $\lambda = 2, \qquad f(x) = 3x^2 - 1$.
(c) $\lambda = 3, \qquad f(x) = 5x^3 - 3x$.

15. In quantum mechanics, the study of the Schrödinger equation for the hydrogen atom leads to a consideration of **Laguerre's equation,**

$$xy'' + (1 - x)y' + \lambda y = 0,$$

where λ is a parameter. Use the reduction of order formula to obtain an integral representation of a second linearly independent solution to Laguerre's equation for the given value of λ and corresponding solution $f(x)$.

(a) $\lambda = 1, \qquad f(x) = x - 1$.
(b) $\lambda = 2, \qquad f(x) = x^2 - 4x + 2$.

16. For equation (8), we showed in Example 2 that

$$f(x) = x$$

and

$$h(x) = -1 + x^2 + \frac{x^4}{6} + \cdots + \frac{x^{2n}}{n!(2n-1)} + \cdots$$

are two linearly independent solutions. Show that the solution to the initial value problem

(11) $\qquad y'' - 2xy' + 2y = 0;$

$$y(0) = -2,$$

$$y'(0) = -1$$

is given by $y(x) = 2h(x) - f(x)$. Obtain an approximation for the solution to the initial value problem (11) at $x = 1$ by using the first five nonzero terms of the power series for $h(x)$.

 17. In Example 2, we showed that a solution to equation (8) is given by

$$g(x) = x \int_1^x \frac{e^{t^2}}{t^2}\, dt.$$

(a) Show that this solution satisfies $g(1) = 0$ and $g'(1) = e$.

(b) Using some numerical integration algorithm (see Appendix B), approximate the solution to the initial value problem

$$y'' - 2xy' + 2y = 0; \qquad y(1) = 0, \qquad y'(1) = e,$$

at $x = 2$.

18. In Example 2, we showed that

$$f(x) = x, \qquad g(x) = x \int_1^x \frac{e^{t^2}}{t^2}\, dt,$$

and

$$h(x) = -1 + x^2 + \frac{x^4}{6} + \cdots + \frac{x^{2n}}{n!(2n-1)} + \cdots$$

are all solutions to equation (8). Hence, by the representation theorem for homogeneous equations, we know that $h(x)$ can be expressed as a linear combination of $f(x)$ and $g(x)$. This representation can be determined as follows:

(a) Using integration by parts, show that

$$g(x) = ex - e^{x^2} + 2x \int_1^x e^{t^2}\, dt.$$

(b) Prove that

$$h(x) = g(x) + \left(2\int_0^1 e^{t^2}\, dt - e \right) f(x).$$

4.5 HOMOGENEOUS LINEAR EQUATIONS WITH CONSTANT COEFFICIENTS

Let's consider the homogeneous linear second order differential equation with constant coefficients

(1) $ay'' + by' + cy = 0,$

where $a\ (\neq 0)$, b, and c are real constants. Because constant functions are everywhere continuous, Theorem 2 on page 157 guarantees that equation (1) has solutions defined for all x in $(-\infty, \infty)$. If we can find two linearly independent solutions to (1), say y_1 and y_2, then we can express a general solution in the form

$$y = c_1 y_1 + c_2 y_2,$$

where c_1 and c_2 are arbitrary constants.

A look at equation (1) tells us that a solution of (1) must have the property that its second derivative is expressible as a linear combination of its first and zero-th derivatives. This suggests that we try to find a solution of the form $y = e^{rx}$, since derivatives of e^{rx} are just constants times e^{rx}. If we substitute $y = e^{rx}$ into (1), we obtain

$$ar^2 e^{rx} + bre^{rx} + ce^{rx} = 0,$$
$$e^{rx}(ar^2 + br + c) = 0.$$

Because e^{rx} is never zero, we can divide by it to obtain

(2) $ar^2 + br + c = 0.$

Consequently, $y = e^{rx}$ is a solution to (1) if and only if r satisfies equation (2). Equation (2) is called the **auxiliary equation** associated with the homogeneous equation (1).

Now the auxiliary equation is just a quadratic, and its roots are

$$r_1 = \frac{-b + \sqrt{b^2 - 4ac}}{2a} \quad \text{and} \quad r_2 = \frac{-b - \sqrt{b^2 - 4ac}}{2a}.$$

When the discriminant, $b^2 - 4ac$, is positive, the roots r_1 and r_2 are real and distinct. If $b^2 - 4ac = 0$, the roots are real and equal. And when $b^2 - 4ac < 0$, the roots are complex conjugate numbers. We consider the first two cases here; the complex case is deferred to the next section.

DISTINCT REAL ROOTS

If the auxiliary equation has distinct real roots r_1 and r_2, then $e^{r_1 x}$ and $e^{r_2 x}$ are linearly independent solutions to (1). Therefore a general solution of (1) is

$$y(x) = c_1 e^{r_1 x} + c_2 e^{r_2 x},$$

where c_1 and c_2 are arbitrary constants.

Observe that the Wronskian of $e^{r_1 x}$ and $e^{r_2 x}$, evaluated at $x = 0$, is $(r_2 - r_1)$. Thus the linear independence is assured if the roots are distinct.

EXAMPLE 1 Find a general solution to

(3) $y'' + 5y' - 6y = 0.$

Solution The auxiliary equation associated with (3) is

$$r^2 + 5r - 6 = (r - 1)(r + 6) = 0,$$

which has the roots $r_1 = 1$, $r_2 = -6$. Thus $\{e^x, e^{-6x}\}$ is a fundamental solution set for equation (3), and a general solution is

$$y(x) = c_1 e^x + c_2 e^{-6x}. \quad \blacksquare \text{ End}$$

EXAMPLE 2 Solve the initial value problem

(4) $y'' + 2y' - y = 0; \qquad y(0) = 0, \qquad y'(0) = -1.$

Solution Here the auxiliary equation is

$$r^2 + 2r - 1 = 0.$$

Using the quadratic formula, we find that the roots of this equation are

$$r_1 = -1 + \sqrt{2} \quad \text{and} \quad r_2 = -1 - \sqrt{2}.$$

Consequently a general solution to the differential equation in (4) is

(5) $y(x) = c_1 e^{(-1+\sqrt{2})x} + c_2 e^{(-1-\sqrt{2})x},$

where c_1 and c_2 are arbitrary constants. To find the specific solution that satisfies the initial conditions given in (4), we first differentiate y as given in (5), then plug y and y' into the initial conditions of (4). This gives

$$y(0) = c_1e^0 + c_2e^0,$$
$$y'(0) = (-1 + \sqrt{2})c_1e^0 + (-1 - \sqrt{2})c_2e^0,$$

or

$$0 = c_1 + c_2,$$
$$-1 = (-1 + \sqrt{2})c_1 + (-1 - \sqrt{2})c_2.$$

Solving this system yields $c_1 = -\sqrt{2}/4$ and $c_2 = \sqrt{2}/4$. Thus

$$y(x) = -\frac{\sqrt{2}}{4}e^{(-1+\sqrt{2})x} + \frac{\sqrt{2}}{4}e^{(-1-\sqrt{2})x}$$

is the desired solution. ∎ End

When the roots of the auxiliary equation (2) are equal, that is $r_1 = r_2 = r$, then unlike the previous case, we get only one nontrivial solution, namely $y = e^{rx}$. Of course, constant multiples of this function are also solutions, but they are of no help in finding a second **linearly independent** solution. One way to resolve this shortcoming is to use the reduction-of-order method discussed in Section 4.4 (see Problem 50). Another method, based on operators, is discussed in Problems 51 and 52. Both procedures ultimately show that in the case of a repeated root r a second, linearly independent, solution is given simply by xe^{rx}, that is, x times the original solution e^{rx}.

REPEATED ROOT

If the auxiliary equation has a repeated root r, then two linearly independent solutions to (1) are e^{rx} and xe^{rx} and a general solution is

$$y(x) = c_1e^{rx} + c_2xe^{rx},$$

where c_1 and c_2 are arbitrary constants.

EXAMPLE 3 Find a general solution to

(6) $y'' + 4y' + 4y = 0.$

Solution The auxiliary equation for (6) is

$$r^2 + 4r + 4 = (r + 2)^2 = 0.$$

Because $r = -2$ is a double root, two linearly independent solutions are e^{-2x} and xe^{-2x}. These give

$$y(x) = c_1e^{-2x} + c_2xe^{-2x}$$

as a general solution. ∎ End

The method we have described for solving homogeneous linear second order equations with constant coefficients applies to any order (even first order) homogeneous linear equations with constant coefficients. We give a detailed treatment of such higher-order equations in Chapter 6. For now, we shall be content to illustrate the method by means of an example. We remark that, in general, a homogeneous linear nth order equation has a fundamental solution set consisting of n linearly independent solutions.

EXAMPLE 4 Find a general solution to

(7) $y''' + 3y'' - y' - 3y = 0.$

Solution If we try to find solutions of the form $y = e^{rx}$, then, as with second order equations, we are led to finding roots of the auxiliary equation

(8) $r^3 + 3r^2 - r - 3 = 0.$

We observe that $r = 1$ is a root of the above equation, and dividing the polynomial on the left-hand side of (8) by $r - 1$ leads to the factorization

$$(r - 1)(r^2 + 4r + 3) = (r - 1)(r + 1)(r + 3) = 0.$$

Hence the roots of the auxiliary equation are 1, -1, and -3, and so three solutions of (7) are e^x, e^{-x}, and e^{-3x}. A discussion of linear independence for sets containing more than two functions is given in Section 6.1. For now, we assume these form a fundamental solution set. A general solution to (7) is then

(9) $y(x) = c_1 e^x + c_2 e^{-x} + c_3 e^{-3x}.$ ■ End

In the last example, we can actually verify that (9) is a ''general solution'' by showing, as we did in the proof of Theorem 3 on page 160, that every solution to equation (7) can be written in the form (9) for a suitable choice of constants c_1, c_2, and c_3. For this verification, we need a generalization of the existence and uniqueness theorem to linear third order equations (see Section 6.1).

In the general case of a homogeneous linear nth order equation with constant coefficients, the auxiliary equation is a polynomial equation of order n. The fundamental theorem of algebra[†] states that a polynomial equation of order n has n roots, some of which may be complex. Unfortunately a result from Galois theory by Abel and Ruffini[††] also states that for arbitrary polynomial equations of order greater than four, there does *not* exist a formula (like the familiar quadratic formula) involving radicals for determining the roots of the equation. Thus, when $n > 4$, we will usually have to rely on numerical techniques for approximating the roots (see Problem 49).

[†] See, for example, *Elements of Abstract Algebra*, by Richard A. Dean (John Wiley and Sons, Inc., New York, 1966), Section 4.6.
[††] Ibid., Section 10.5.

Cauchy-Euler Equations

There is a special class of equations with variable coefficients that can be transformed into equations with constant coefficients.

CAUCHY-EULER, OR EQUIDIMENSIONAL, EQUATIONS

Definition 4. A linear second order equation that can be expressed in the form

$$(10) \qquad ax^2 \frac{d^2y}{dx^2} + bx\frac{dy}{dx} + cy = h(x),$$

where a, b, and c are constants is called a **Cauchy-Euler, or equidimensional, equation.**[†]

For example, the differential equation

$$3x^2 y'' - 2xy' + 7y = \sin x$$

is a Cauchy-Euler equation, whereas

$$2y'' - 3xy' + 11y = 3x - 1$$

is *not* because the coefficient of y'' is 2, which is not a constant times x^2.

To solve a homogeneous Cauchy-Euler equation (i.e., $h(x) \equiv 0$), we make the substitution $x = e^t$, which transforms equation (10) into an equation with constant coefficients. The technique is illustrated in the next example.

EXAMPLE 5 Find a general solution to

$$(11) \qquad 3x^2 \frac{d^2y}{dx^2} + 11x\frac{dy}{dx} - 3y = 0, \qquad x > 0.$$

Solution We make the substitution $x = e^t$ to transform (11) into an equation in which the new independent variable is t. Because $x = e^t$, it follows by the chain rule that

$$\frac{dy}{dt} = \frac{dy}{dx}\frac{dx}{dt} = \frac{dy}{dx}e^t = x\frac{dy}{dx},$$

and hence

$$(12) \qquad x\frac{dy}{dx} = \frac{dy}{dt}.$$

[†] *Historical Footnote:* Although work on this equation was published by Leonhard Euler in 1769 and later by Augustin Cauchy, its solution was known to John Bernoulli prior to 1700.

Differentiating (12) with respect to t, we find from the product rule that

$$\frac{d^2y}{dt^2} = \frac{d}{dt}\left(x\frac{dy}{dx}\right) = \frac{dx}{dt}\frac{dy}{dx} + x\frac{d}{dt}\left(\frac{dy}{dx}\right)$$

$$= \frac{dy}{dt} + x\frac{d^2y}{dx^2}\frac{dx}{dt} = \frac{dy}{dt} + x\frac{d^2y}{dx^2}e^t$$

$$= \frac{dy}{dt} + x^2\frac{d^2y}{dx^2}.$$

Thus

(13) $$x^2\frac{d^2y}{dx^2} = \frac{d^2y}{dt^2} - \frac{dy}{dt}.$$

Substituting into (11) the expressions for $x\,dy/dx$ and x^2d^2y/dx^2 given in (12) and (13) yields

$$3\left(\frac{d^2y}{dt^2} - \frac{dy}{dt}\right) + 11\frac{dy}{dt} - 3y = 0,$$

(14) $$3\frac{d^2y}{dt^2} + 8\frac{dy}{dt} - 3y = 0.$$

The auxiliary equation associated with (14) is

$$3r^2 + 8r - 3 = (3r - 1)(r + 3) = 0.$$

The roots of this equation are $1/3$ and -3, so the general solution to (14) is

$$y(t) = c_1e^{t/3} + c_2e^{-3t} = c_1(e^t)^{1/3} + c_2(e^t)^{-3}.$$

Expressing y in terms of the original variable x, we find

(15) $$y(x) = c_1x^{1/3} + c_2x^{-3} \quad \text{for } x > 0. \quad \blacksquare \text{ End}$$

It turns out that the solution (15) for the special equation (11) is valid for all $x \neq 0$ even though it was derived only for $x > 0$. In general, it can be shown that if $y(x)$ is a solution to a homogeneous Cauchy-Euler equation for $x > 0$, then $\phi(x) := y(-x)$ is a solution for $x < 0$ (see Problem 43).

Using formulas (12) and (13), the reader can verify that the substitution $x = e^t$ transforms the general Cauchy-Euler equation

(16) $$ax^2\frac{d^2y}{dx^2} + bx\frac{dy}{dx} + cy = h(x), \quad x > 0$$

into the constant coefficient equation

(17) $$a\frac{d^2y}{dt^2} + (b - a)\frac{dy}{dt} + cy = h(e^t).$$

A method similar to the one discussed here for second order Cauchy-Euler equations can be derived for higher-order Cauchy-Euler equations of the form

$$a_n x^n y^{(n)}(x) + a_{n-1} x^{n-1} y^{(n-1)}(x) + \cdots + a_0 y(x) = h(x),$$

where $a_n, a_{n-1}, \ldots, a_0$ are real constants (see Problems 46–48).

Another method for finding solutions to a homogeneous Cauchy-Euler equation involves guessing a solution of the form $y = x^r$. Doing this renders each term xy', $x^2 y''$, ... as a constant times x^r, so the result is an auxiliary equation for r. This method is discussed in Section 8.5. However, there is often an advantage in making the substitution $x = e^t$ when trying to solve a *nonhomogeneous* Cauchy-Euler equation by the methods of undetermined coefficients (see Section 4.8) or Laplace transforms (see Chapter 7).

EXERCISES 4.5

In Problems 1 through 12, find a general solution to the given differential equation.

1. $y'' + 5y' + 6y = 0.$

2. $y'' - y' - 2y = 0.$

3. $y'' + 8y' + 16y = 0.$

4. $y'' + 6y' + 9y = 0.$

5. $z'' + z' - z = 0.$

6. $y'' - 5y' + 6y = 0.$

7. $2u'' + 7u' - 4u = 0.$

8. $6y'' + y' - 2y = 0.$

9. $y'' - y' - 11y = 0.$

10. $4y'' - 4y' + y = 0.$

11. $4w'' + 20w' + 25w = 0.$

12. $3y'' + 11y' - 7y = 0.$

In Problems 13 through 20, solve the given initial value problem.

13. $y'' + 2y' - 8y = 0;$ $y(0) = 3,$ $y'(0) = -12.$

14. $y'' + y' = 0;$ $y(0) = 2,$ $y'(0) = 1.$

15. $y'' + 2y' + y = 0;$ $y(0) = 1,$ $y'(0) = -3.$

16. $y'' - 4y' + 3y = 0;$ $y(0) = 1,$ $y'(0) = 1/3.$

17. $z'' - 2z' - 2z = 0;$ $z(0) = 0,$ $z'(0) = 3.$

18. $y'' - 6y' + 9y = 0;$ $y(0) = 2,$ $y'(0) = 25/3.$

19. $y'' - 4y' - 5y = 0;$ $y(-1) = 3,$ $y'(-1) = 9.$

20. $y'' - 4y' + 4y = 0;$ $y(1) = 1,$ $y'(1) = 1.$

21. First Order Constant Coefficient Equations.

 (a) Substituting $y = e^{rx}$, find the auxiliary equation for the first order linear equation

 $ay' + by = 0,$

 where a, b are constants with $a \neq 0$.

 (b) Use the result of part (a) to find the general solution.

In Problems 22 through 25, use the method described in Problem 21 to find a general solution to the given equation.

22. $3y' - 7y = 0.$

23. $5y' + 4y = 0.$

24. $3z' + 11z = 0.$

25. $6w' - 13w = 0.$

In Problems 26 through 31, find a general solution to the given third order differential equation.

26. $y''' + y'' - 6y' + 4y = 0.$

27. $y''' - 6y'' - y' + 6y = 0.$

28. $z''' + 2z'' - 4z' - 8z = 0.$

29. $y''' - 7y'' + 7y' + 15y = 0.$

30. $y''' + 3y'' - 4y' - 12y = 0.$

31. $w''' + w'' - 4w' + 2w = 0.$

32. Solve the initial value problem

 $y''' - y' = 0;$

 $y(0) = 2,$ $y'(0) = 3,$ $y''(0) = -1.$

In Problems 33 through 38, use the method discussed under "Cauchy-Euler Equations" to find a general solution to the given equation for $x > 0$.

33. $x^2 y''(x) + 7xy'(x) - 7y(x) = 0.$

34. $x^2 \dfrac{d^2 y}{dx^2} + 2x \dfrac{dy}{dx} - 6y = 0.$

35. $\dfrac{d^2 w}{dx^2} + \dfrac{6}{x} \dfrac{dw}{dx} + \dfrac{4}{x^2} w = 0.$

36. $x^2 \dfrac{d^2 z}{dx^2} + 5x \dfrac{dz}{dx} + 4z = 0.$

37. $9x^2 y''(x) + 15xy'(x) + y(x) = 0.$

38. $x^2 y''(x) - 3xy'(x) + 4y(x) = 0.$

In Problems 39 and 40, use the method discussed under "Cauchy-Euler Equations" to solve the given initial value problem.

39. $x^2 y''(x) - 4xy'(x) + 4y(x) = 0;$
 $y(1) = -2, \quad y'(1) = -11.$

40. $x^2 y''(x) + 7xy'(x) + 5y(x) = 0;$
 $y(1) = -1, \quad y'(1) = 13.$

In Problems 41 and 42, devise a modification of the method discussed under "Cauchy-Euler Equations" to find a general solution to the given equation.

41. $(x - 2)^2 y''(x) - 7(x - 2)y'(x) + 7y(x) = 0, \quad x > 2.$

42. $(x + 1)^2 y''(x) + 10(x + 1)y'(x) + 14y(x) = 0, \quad x > -1.$

43. Let $y(x)$ be a solution for $x > 0$ to the *homogeneous* Cauchy-Euler equation $ax^2 y'' + bxy' + cy = 0$. Show that $\phi(x) := y(-x)$ is a solution to the equation for $x < 0$. Is the analogous statement true for the *non-homogeneous* Cauchy-Euler equation?

In Problems 44 and 45, use the result of Problem 43 to find a general solution to the given differential equation for $x < 0$.

44. $x^2 y''(x) + xy'(x) - 2y(x) = 0.$

45. $2x^2 y''(x) + 7xy'(x) + 2y(x) = 0.$

46. Use the substitution $x = e^t$ to show that the third order Cauchy-Euler equation

$$ax^3 y'''(x) + bx^2 y''(x) + cxy'(x) + dy(x) = 0,$$

$$x > 0$$

is equivalent to the constant coefficient equation

$$ay'''(t) + (b - 3a)y''(t)$$
$$+ (2a - b + c)y'(t) + dy(t) = 0.$$

In Problems 47 and 48, use the result of Problem 46 to find a general solution to the given differential equation for $x > 0$.

47. $x^3 y'''(x) - 2x^2 y''(x) + 3xy'(x) - 3y(x) = 0.$

48. $x^3 y'''(x) + x^2 y''(x) - 8xy'(x) - 4y(x) = 0.$

49. By using Newton's method or some other numerical procedure to approximate the roots of the auxiliary equation, find general solutions to the following equations:

 (a) $3y''' + 18y'' + 13y' - 19y = 0.$
 (b) $y^{iv} - 5y'' + 5y = 0.$
 (c) $y^v - 3y^{iv} - 5y''' + 15y'' + 4y' - 12y = 0.$

50. When $b^2 - 4ac = 0$, the roots of the auxiliary equation (2) are equal; that is, $r_1 = r_2 = r = -b/2a$. Hence $f(x) = e^{(-b/2a)x}$ is a solution to (1). Use the reduction of order formula (6) of Section 4.4 to derive $xe^{(-b/2a)x}$ as a second linearly independent solution to (1).

51. Suppose the auxiliary equation (2) has a repeated root r. Then, on dividing by a, equation (1) can be written in the equivalent form

$$(D - r)^2 y = y'' - 2ry' + r^2 y = 0,$$

where D is the differential operator d/dx.

 (a) Show that for any twice differentiable function u,

$$(D - r)^2 [e^{rx} u] = e^{rx} D^2 u,$$

 and hence that the problem of solving $(D - r)^2 y = 0$ reduces to that of solving $D^2 u = 0$, where $y = e^{rx} u$.

 (b) Use the result of part (a) to show that e^{rx} and xe^{rx} are linearly independent solutions of $(D - r)^2 y = 0$.

52. Let $L[y] := ay'' + by' + c$, where $b^2 - 4ac = 0$.

 (a) Show that $L[e^{rx}] = a(r - r_0)^2 e^{rx}$, where $r_0 = -b/2a$.
 (b) Show that $L[(\partial/\partial r)e^{rx}] = (\partial/\partial r)L[e^{rx}]$, and thus

$$L[xe^{rx}] = 2a(r - r_0)e^{rx} + ax(r - r_0)^2 e^{rx}.$$

 (c) Use the results of parts (a) and (b) to show that $e^{r_0 x}$ and $xe^{r_0 x}$ are linearly independent solutions to (1).

53. The suspension in an automobile can be modeled as a vibrating spring with damping due to the shock absorbers (see Figure 4.3). This leads to the equation

$$mx''(t) + bx'(t) + kx(t) = 0,$$

Figure 4.3 Diagram of an automobile suspension

where m is the mass of the automobile, b is the damping constant of the shocks, k is the spring constant, and $x(t)$ is the vertical displacement of the automobile at time t. In the next section, we will find that the solution of a homogeneous linear second order equation with constant coefficients oscillates when the auxiliary equation has complex roots. If the mass of an automobile is 1000 kg and the spring constant is 3000 kg/sec^2, determine the minimum value for the damping constant in kilograms per second that will provide a smooth, oscillation-free (theoretically!) ride. If we replace the springs with heavy duty ones having twice the spring constant, how does this minimum value for b change?

54. Consider the differential equation $y'' - s^2y = 0$, where s is a positive constant.

(a) Show that a general solution can be written as $c_1e^{sx} + c_2e^{-sx}$.

(b) Show that the form $d_1 \cosh sx + d_2 \sinh sx$ is also a general solution.

(c) Use each of these general solution formats to solve the initial value problem $y'' - s^2y = 0$, $y(0) = a$, $y'(0) = b$. Which is more convenient?

4.6 AUXILIARY EQUATIONS WITH COMPLEX ROOTS

The *simple harmonic equation* $y'' + y = 0$, so called because of its relation to the fundamental vibration of a musical tone, has as solutions $y_1(x) = \cos x$ and $y_2(x) = \sin x$. Notice, however, that the auxiliary equation associated with the harmonic equation is $r^2 + 1 = 0$, which has imaginary roots $r = \pm i$, where i denotes $\sqrt{-1}$.[†] In the previous section, we expressed the solutions to a linear second order equation with constant coefficients in terms of exponential functions. It would appear, then, that one might be able to attribute a meaning to the forms e^{ix} and e^{-ix} and that these "functions" should be related to $\cos x$ and $\sin x$. This matchup is accomplished by Euler's formula, which is discussed in this section.

When $b^2 - 4ac < 0$, the roots of the auxiliary equation

(1) $ar^2 + br + c = 0$

associated with the homogeneous equation

(2) $ay'' + by' + cy = 0$

are the complex conjugate numbers

$$r_1 = \alpha + i\beta \quad \text{and} \quad r_2 = \alpha - i\beta \qquad (i = \sqrt{-1}),$$

where α, β are the real numbers

(3) $\alpha = -\dfrac{b}{2a} \quad \text{and} \quad \beta = \dfrac{\sqrt{4ac - b^2}}{2a}.$

As in the previous section, we would like to assert that the functions e^{r_1x} and e^{r_2x} are solutions to the equation (2). This is, in fact, the case, but before we can proceed, we need to address some fundamental questions. For example, if $r_1 = \alpha + i\beta$ is a complex number, what do we mean by the expression $e^{(\alpha+i\beta)x}$? If we assume that the law of exponents applies

[†] Electrical engineers frequently use the symbol j to denote $\sqrt{-1}$.

to complex numbers, then

(4) $e^{(\alpha+i\beta)x} = e^{\alpha x+i\beta x} = e^{\alpha x}e^{i\beta x}.$

We now need only clarify the meaning of $e^{i\beta x}$.

For this purpose, let's assume that the Maclaurin series for e^z is the same for complex numbers z as it is for real numbers. Observing that $i^2 = -1$, then for θ real we have

$$e^{i\theta} = 1 + (i\theta) + \frac{(i\theta)^2}{2!} + \cdots + \frac{(i\theta)^n}{n!} + \cdots$$

$$= 1 + i\theta - \frac{\theta^2}{2!} - \frac{i\theta^3}{3!} + \frac{\theta^4}{4!} + \frac{i\theta^5}{5!} + \cdots$$

$$= \left(1 - \frac{\theta^2}{2!} + \frac{\theta^4}{4!} + \cdots\right) + i\left(\theta - \frac{\theta^3}{3!} + \frac{\theta^5}{5!} + \cdots\right).$$

Now recall the Maclaurin series for $\cos\theta$ and $\sin\theta$:

$$\cos\theta = 1 - \frac{\theta^2}{2!} + \frac{\theta^4}{4!} + \cdots,$$

$$\sin\theta = \theta - \frac{\theta^3}{3!} + \frac{\theta^5}{5!} + \cdots.$$

Recognizing these expansions in the proposed series for $e^{i\theta}$, we make the identification

(5) $e^{i\theta} = \cos\theta + i\sin\theta,$

which is known as **Euler's formula.**[†]

When Euler's formula (with $\theta = \beta x$) is used in equation (4), we find

(6) $e^{(\alpha+i\beta)x} = e^{\alpha x}(\cos\beta x + i\sin\beta x),$

which expresses the complex function $e^{(\alpha+i\beta)x}$ in terms of familiar real functions. Having made sense out of $e^{(\alpha+i\beta)x}$, we can now show (see Problem 30) that

(7) $\dfrac{d}{dx}e^{(\alpha+i\beta)x} = (\alpha + i\beta)e^{(\alpha+i\beta)x},$

and, with the choices of α and β as given in (3), the complex function $e^{(\alpha+i\beta)x}$ is indeed a solution to equation (2). In general, if $z(x)$ is a complex-valued function of the real variable x, we can write

$$z(x) = u(x) + iv(x),$$

where $u(x)$ and $v(x)$ are real-valued functions. The derivatives of $z(x)$ are then given by

$$\frac{dz}{dx} = \frac{du}{dx} + i\frac{dv}{dx}, \qquad \frac{d^2z}{dx^2} = \frac{d^2u}{dx^2} + i\frac{d^2v}{dx^2}.$$

[†] ***Historical Footnote:*** This formula first appeared in Leonhard Euler's monumental two-volume *Introductio in Analysin Infinitorum* (1748).

With the following lemma, we show that the complex-valued solution $e^{(\alpha+i\beta)x}$ gives rise to two linearly independent *real-valued* solutions.

REAL SOLUTIONS DERIVED FROM COMPLEX SOLUTIONS

Lemma 2. Let $z(x) = u(x) + iv(x)$ be a solution to equation (2), where a, b, and c are real numbers. Then, the real part $u(x)$ and the imaginary part $v(x)$ are real-valued solutions of (2).

Proof. By assumption, $az'' + bz' + cz = 0$, and hence

$$a(u'' + iv'') + b(u' + iv') + c(u + iv) = 0,$$
$$(au'' + bu' + cu) + i(av'' + bv' + cv) = 0.$$

But a complex number is zero if and only if its real and imaginary parts are both zero. Thus we must have

$$au'' + bu' + cu = 0 \quad \text{and} \quad av'' + bv' + cv = 0,$$

which means that both $u(x)$ and $v(x)$ are real-valued solutions of (2). ■ End

When we apply Lemma 2 to the solution

$$e^{(\alpha+i\beta)x} = e^{\alpha x} \cos \beta x + ie^{\alpha x} \sin \beta x,$$

we obtain the following.

COMPLEX CONJUGATE ROOTS

If the auxiliary equation has complex conjugate roots $\alpha \pm i\beta$, then two linearly independent solutions to (2) are

$$e^{\alpha x} \cos \beta x \quad \text{and} \quad e^{\alpha x} \sin \beta x,$$

and a general solution is

(8) $y(x) = c_1 e^{\alpha x} \cos \beta x + c_2 e^{\alpha x} \sin \beta x,$

where c_1 and c_2 are arbitrary constants.

In the preceding discussion, we glossed over some important details concerning complex numbers and complex-valued functions. In particular, further analysis is required to justify the use of the law of exponents, Euler's formula, and even the fact that the derivative of e^{rx} is re^{rx} when r is a complex constant.[†] Readers who feel uneasy about our conclusions can readily check the answer by substituting it into (2).

[†] For a detailed treatment of these topics see, for example, *Fundamentals of Complex Analysis*, Second Edition, by E. B. Saff and A. D. Snider (Prentice Hall, Inc., Englewood Cliffs, New Jersey, 1993).

The reader may also be wondering what would have happened if we had worked with the function $e^{(\alpha-i\beta)x}$ instead of $e^{(\alpha+i\beta)x}$. We leave it as an exercise to verify that $e^{(\alpha-i\beta)x}$ gives rise to the same general solution (8).

EXAMPLE 1 Find a general solution to

(9) $y'' + 2y' + 4y = 0.$

Solution The auxiliary equation is

$$r^2 + 2r + 4 = 0,$$

which has roots

$$r = \frac{-2 \pm \sqrt{4-16}}{2} = \frac{-2 \pm \sqrt{-12}}{2} = -1 \pm i\sqrt{3}.$$

Hence with $\alpha = -1$, $\beta = \sqrt{3}$, a general solution for (9) is

$$y(x) = c_1 e^{-x} \cos(\sqrt{3}x) + c_2 e^{-x} \sin(\sqrt{3}x). \quad \blacksquare \text{ End}$$

When the auxiliary equation has complex conjugate roots, the (real) solutions oscillate between positive and negative values. This type of behavior is observed in vibrating springs.

EXAMPLE 2 In the study of a vibrating spring with damping, we are led to an initial value problem of the form

(10) $mx''(t) + bx'(t) + kx(t) = 0; \quad x(0) = x_0, \quad x'(0) = v_0,$

where m is the mass of the spring system, b is the damping constant, k is the spring constant, x_0 is the initial displacement, v_0 is the initial velocity, and $x(t)$ is the displacement from equilibrium of the spring system at time t (see Figure 4.4). Determine the equation of motion for this spring system when $m = 36$ kg, $b = 12$ kg/sec, $k = 37$ kg/sec^2, $x_0 = 70$ cm, and $v_0 = 10$ cm/sec. Also find $x(10)$, the displacement after 10 sec.

System in equilibrium position System in motion

Figure 4.4 Vibrating spring with damping

Solution The equation of motion is given by $x(t)$, the solution of the initial value problem (10) for the specified values of m, b, k, x_0, and v_0. That is, we seek the solution to

(11) $36x'' + 12x' + 37x = 0;$ $x(0) = 70,$ $x'(0) = 10.$

The auxiliary equation for (11) is

$$36r^2 + 12r + 37 = 0,$$

which has roots

$$r = \frac{-12 \pm \sqrt{144 - 4(36)(37)}}{72} = \frac{-12 \pm 12\sqrt{1 - 37}}{72} = -\frac{1}{6} \pm i.$$

Hence with $\alpha = -1/6$, $\beta = 1$, the displacement $x(t)$ can be expressed in the form

(12) $x(t) = c_1 e^{-t/6} \cos t + c_2 e^{-t/6} \sin t.$

We can find c_1 and c_2 by substituting $x(t)$ and $x'(t)$ into the initial conditions given in (11). Differentiating (12), we get a formula for $x'(t)$:

$$x'(t) = \left(-\frac{c_1}{6} + c_2\right) e^{-t/6} \cos t + \left(-c_1 - \frac{c_2}{6}\right) e^{-t/6} \sin t.$$

Substituting into the initial conditions now results in the system

$$c_1 = 70,$$

$$-\frac{c_1}{6} + c_2 = 10.$$

Upon solving, we find $c_1 = 70$ and $c_2 = 65/3$. With these values, the equation of motion becomes

$$x(t) = 70e^{-t/6} \cos t + \frac{65}{3} e^{-t/6} \sin t,$$

and

$$x(10) = 70e^{-5/3} \cos 10 + \frac{65}{3} e^{-5/3} \sin 10$$

$$\approx -13.32 \text{ cm.} \quad \blacksquare \text{ End}$$

If the constants a, b, and c in (2) are complex numbers, then the roots r_1, r_2 of the auxiliary equation (1) are, in general, complex numbers, but not necessarily complex conjugates of each other. If $r_1 \neq r_2$, then a general solution is

$$y(x) = c_1 e^{r_1 x} + c_2 e^{r_2 x},$$

where c_1 and c_2 are now arbitrary *complex-valued* constants.

We remark that a complex differential equation can be regarded as a system of two real differential equations, since we can always work separately with its real and imaginary parts. Systems are discussed in Section 4.10.

EXERCISES 4.6

In Problems 1 through 8, the auxiliary equation for the given differential equation has complex roots. Find a general solution.

1. $y'' + 9y = 0$.
2. $y'' + y = 0$.
3. $z'' - 6z' + 10z = 0$.
4. $y'' - 10y' + 26y = 0$.
5. $w'' + 4w' + 6w = 0$.
6. $y'' - 4y' + 7y = 0$.
7. $4y'' - 4y' + 26y = 0$.
8. $4y'' + 4y' + 6y = 0$.

In Problems 9 through 20, find a general solution.

9. $y'' - 8y' + 7y = 0$.
10. $y'' + 4y' + 8y = 0$.
11. $z'' + 10z' + 25z = 0$.
12. $u'' + 7u = 0$.
13. $y'' + 2y' + 5y = 0$.
14. $y'' - 2y' + 26y = 0$.
15. $y'' + 10y' + 41y = 0$.
16. $y'' - 3y' - 11y = 0$.
17. $y'' - y' + 7y = 0$.
18. $2y'' + 13y' - 7y = 0$.
19. $y''' + y'' + 3y' - 5y = 0$.
20. $y''' - y'' + 2y = 0$.

In Problems 21 through 27, solve the given initial value problem.

21. $y'' + 2y' + 2y = 0$; $y(0) = 2$, $y'(0) = 1$.
22. $y'' + 2y' + 17y = 0$; $y(0) = 1$, $y'(0) = -1$.
23. $w'' - 4w' + 2w = 0$; $w(0) = 0$, $w'(0) = 1$.
24. $y'' + 9y = 0$; $y(0) = 1$, $y'(0) = 1$.
25. $y'' - 2y' + 2y = 0$; $y(\pi) = e^\pi$, $y'(\pi) = 0$.
26. $y'' - 2y' + y = 0$; $y(0) = 1$, $y'(0) = -2$.
27. $y''' - 4y'' + 7y' - 6y = 0$; $y(0) = 1$, $y'(0) = 0$,
 $y''(0) = 0$.

28. To see the effect of changing the parameter b in the initial value problem
 $$y'' + by' + 4y = 0; \quad y(0) = 1, \quad y'(0) = 0,$$
 solve the problem for $b = 5$, 4, and 2 and sketch the solutions.

29. Find a general solution to the following higher-order equations.
 (a) $y''' - y'' + y' + 3y = 0$.
 (b) $y''' + 2y'' + 5y' - 26y = 0$.
 (c) $y^{iv} + 13y'' + 36 = 0$.

30. Using the representation for $e^{(\alpha+i\beta)x}$ in (6), verify the differentiation formula (7).

31. Show that the two real, linearly independent solutions $e^{\alpha x} \cos \beta x$ and $e^{\alpha x} \sin \beta x$ to equation (2), with α and β defined by (3), can be obtained from the complex gen-

eral solution $c_1 e^{(\alpha+i\beta)x} + c_2 e^{(\alpha-i\beta)x}$ by making appropriate choices for the (possibly complex) constants c_1 and c_2.

In Problems 32 through 35, find a general solution to the following Cauchy-Euler equation for $x > 0$ by using the substitution $x = e^t$ as discussed in Section 4.5.

32. $y''(x) - \dfrac{1}{x} y'(x) + \dfrac{5}{x^2} y(x) = 0$.

33. $x^2 y''(x) - 3xy'(x) + 6y(x) = 0$.

34. $x^2 y''(x) + 9xy'(x) + 17y(x) = 0$.

35. $x^2 y''(x) + 3xy'(x) + 5y(x) = 0$.

36. **Vibrating Spring without Damping.** A vibrating spring without damping can be modeled by the initial value problem (10) in Example 2 by taking $b = 0$.
 (a) If $m = 10$ kg, $k = 250$ kg/sec^2, $x_0 = 30$ cm, and $v_0 = -10$ cm/sec, find the equation of motion for this undamped vibrating spring.
 (b) When the equation of motion is of the form displayed in (8), the motion is said to be **oscillatory** with **frequency** $\beta/2\pi$. Find the frequency of oscillation for the spring system of part (a).

37. **Vibrating Spring with Damping.** Using the model for a vibrating spring with damping discussed in Example 2:
 (a) Find the equation of motion for the vibrating spring with damping if $m = 10$ kg, $b = 60$ kg/sec, $k = 250$ kg/sec^2, $x_0 = 30$ cm, and $v_0 = -10$ cm/sec.
 (b) Find the frequency of oscillation for the spring system of part (a). [*Hint:* See the definition of frequency given in Problem 36(b).]
 (c) Compare the results of Problems 36 and 37 and determine what effect the damping has on the frequency of oscillation. What other effects does it have on the solution?

38. **RLC Series Circuit.** In the study of an electrical circuit consisting of a resistor, capacitor, inductor, and an electromotive force (see Figure 4.5 on page 188), we are led to an initial value problem of the form

 (13) $$L\frac{dI}{dt} + RI + \frac{q}{C} = E(t);$$

 $$q(0) = q_0,$$
 $$I(0) = I_0,$$

Figure 4.5 *RLC* series circuit

where L is the inductance in henrys, R is the resistance in ohms, C is the capacitance in farads, $E(t)$ is the electromotive force in volts, $q(t)$ is the charge in coulombs on the capacitor at time t, and $I = dq/dt$ is the current in amperes. Find the current at time t if the charge on the capacitor is initially zero, the initial current is zero, $L = 10$ henrys, $R = 20$ ohms, $C = (6260)^{-1}$ farads, and $E(t) = 100$ volts. [*Hint:* Differentiate both sides of the differential equation in (13) to obtain a homogeneous linear second order equation for $I(t)$. Then use (13) to determine dI/dt at $t = 0$.]

39. Swinging Door. The motion of a swinging door with an adjustment screw that controls the amount of friction on the hinges is governed by the initial value problem

$$I\theta'' + b\theta' + k\theta = 0; \qquad \theta(0) = \theta_0, \qquad \theta'(0) = v_0,$$

where θ is the angle that the door is open, I is the moment of inertia of the door about its hinges, $b > 0$ is a damping constant that varies with the amount of friction on the door, $k > 0$ is the spring constant associated with the swinging door, θ_0 is the initial angle that the door is opened, and v_0 is the initial angular

velocity imparted to the door (see Figure 4.6). If I and k are fixed, determine for which values of b the door will *not* continually swing back and forth when closing.

Figure 4.6 Top view of swinging door

40. Although the real general solution form (8) is convenient, it is possible to also use the form

(14) $\qquad d_1 e^{(\alpha+i\beta)x} + d_2 e^{(\alpha-i\beta)x}$

to solve initial value problems. The coefficients d_1 and d_2 are complex constants.

(a) Use the form (14) to solve Problem 21. Verify that your form is equivalent to the one derived using (8).
(b) Show that, in general, d_1 and d_2 in (14) must be complex conjugates in order that the solution be real.

41. The auxiliary equations for the following differential equations have repeated complex roots. Adapt the "repeated root" procedure of Section 4.5 to find their general solutions:

(a) $y^{iv} + 2y'' + y = 0$.
(b) $y^{iv} + 4y''' + 12y'' + 16y' + 16 = 0$. [*Hint:* The auxiliary equation is $(r^2 + 2r + 4)^2 = 0$.]

4.7 SUPERPOSITION AND NONHOMOGENEOUS EQUATIONS

The differential operator L defined by

(1) $\qquad L[y](x) := y'' + p(x)y' + q(x)y$

can be viewed as a "black box" with input the function $y(x)$ and output the function on the right-hand side of (1) (see Figure 4.7).

Suppose the input functions $y_1(x)$ and $y_2(x)$ yield, respectively, the output functions $g_1(x)$ and $g_2(x)$; that is,

$$L[y_1](x) = g_1(x), \qquad L[y_2](x) = g_2(x).$$

Figure 4.7 Black box diagram for L

Then, since L is a linear operator, an input consisting of the linear combination $c_1 y_1(x) + c_2 y_2(x)$ produces the output $c_1 g_1(x) + c_2 g_2(x)$ in the same combination. Stated in terms of solutions to differential equations, we have the following result.

SUPERPOSITION PRINCIPLE

Theorem 5. Let y_1 be a solution of the differential equation

$$L[y](x) = g_1(x)$$

and let y_2 be a solution of

$$L[y](x) = g_2(x),$$

where L is a linear differential operator. Then for any constants c_1 and c_2, the function $c_1 y_1 + c_2 y_2$ is a solution to the differential equation

$$L[y](x) = c_1 g_1(x) + c_2 g_2(x).$$

EXAMPLE 1 Given that $y_1(x) = -x/3 - 2/9$ is a solution to

$$y'' + 2y' - 3y = x$$

and $y_2(x) = e^{2x}/5$ is a solution to

$$y'' + 2y' - 3y = e^{2x},$$

find a solution to

(2) $y'' + 2y' - 3y = 4x - 5e^{2x}.$

Solution Let $L[y] := y'' + 2y' - 3y$. We are given that

$$L[y_1](x) = g_1(x) = x \quad \text{and} \quad L[y_2](x) = g_2(x) = e^{2x}.$$

Since we can express

$$4x - 5e^{2x} = 4g_1(x) - 5g_2(x),$$

then, by the superposition principle, the function

$$4y_1(x) - 5y_2(x) = 4\left(\frac{-x}{3} - \frac{2}{9}\right) - 5\left(\frac{e^{2x}}{5}\right)$$

$$= \frac{-4x}{3} - \frac{8}{9} - e^{2x}$$

is a solution to equation (2). ■ End

Now we will show how the solution structure for a nonhomogeneous linear differential equation is determined by solutions to the associated homogeneous equation. By combining the superposition principle with the representation theorem for solutions of homogeneous equations, we can prove that *all* solutions of a *non*homogeneous equation can be found if we know just one of its solutions (called a *particular solution*) and a general solution to the corresponding homogeneous equation.

REPRESENTATION OF SOLUTIONS (NONHOMOGENEOUS CASE)

Theorem 6. Let $y_p(x)$ be a particular solution to the nonhomogeneous equation

(3) $y'' + p(x)y' + q(x)y = g(x)$

on the interval (a, b) and let $y_1(x)$, $y_2(x)$ be linearly independent solutions on (a, b) of the corresponding homogeneous equation

(4) $y'' + p(x)y' + q(x)y = 0.$

Then every solution of (3) on the interval (a, b) can be expressed in the form

(5) $y(x) = y_p(x) + c_1y_1(x) + c_2y_2(x).$

The linear combination of y_p, y_1, and y_2 in (5), written with arbitrary constants c_1 and c_2, is referred to as a **general solution** to (3).

Proof. Let $\phi(x)$ be any solution to (3). Because both $\phi(x)$ and $y_p(x)$ are solutions to (3), the superposition principle states that the difference $\phi(x) - y_p(x)$ is a solution to the homogeneous equation (4). It then follows from Theorem 3 on page 160 that

$$\phi(x) - y_p(x) = c_1y_1(x) + c_2y_2(x)$$

for suitable constants c_1 and c_2. The last equation is equivalent to (5) (with $\phi(x)$ in place of $y(x)$), so the theorem is proved. ■ End

The procedure we advocate for analyzing nonhomogeneous equations is described as follows.

PROCEDURE FOR SOLVING NONHOMOGENEOUS EQUATIONS

To solve $y'' + py' + qy = g$:

(a) Determine a general solution, $c_1 y_1 + c_2 y_2$, of the corresponding homogeneous equation.
(b) Find a particular solution, y_p, of the given nonhomogeneous equation.
(c) Form the sum of the particular solution and a general solution to the homogeneous equation; that is,

$$y = y_p + c_1 y_1 + c_2 y_2,$$

to obtain a general solution to the given equation.

EXAMPLE 2 Given that $y_p(x) = x^2$ is a particular solution to

(6) $y'' - y = 2 - x^2,$

find a general solution of (6).

Solution The corresponding homogeneous equation,

$$y'' - y = 0,$$

has the associated auxiliary equation $r^2 - 1 = 0$. Because $r = \pm 1$ are the roots of this equation, a general solution to the homogeneous equation is $c_1 e^x + c_2 e^{-x}$. Combining this with the particular solution $y_p(x) = x^2$ of the nonhomogeneous equation (6), we find that a general solution is

$$y(x) = x^2 + c_1 e^x + c_2 e^{-x}. \quad \blacksquare \text{ End}$$

Obviously our next job is to learn how to find particular solutions to nonhomogeneous equations. This is the objective of Sections 4.8 and 4.9.

EXERCISES 4.7

In Problems 1 through 10, a nonhomogeneous equation and a particular solution are given. Find a general solution for the equation.

1. $y'' - y = x,$ $y_p(x) = -x.$

2. $y'' + y' = 1,$ $y_p(x) = x.$

3. $\theta'' - \theta' - 2\theta = 1 - 2x,$ $\theta_p(x) = x - 1.$

4. $\dfrac{d^2 x}{dt^2} - 4\dfrac{dx}{dt} + 3x = -2e^t,$ $x_p(t) = te^t.$

5. $y'' + 2y' + 4y - 4\cos 2x = 0,$ $y_p(x) = \sin 2x.$

6. $\dfrac{d^2 y}{dx^2} + y = 5e^{2x},$ $y_p(x) = e^{2x}.$

7. $\dfrac{d^2 \theta}{dt^2} - \dfrac{d\theta}{dt} + \theta = \sin t,$ $\theta_p(t) = \cos t.$

8. $y'' + 5y' + 6y = 6x^2 + 10x + 2 + 12e^x,$
 $y_p(x) = e^x + x^2.$

9. $y'' = 2y' - y + 2e^x,$ $y_p(x) = x^2 e^x.$

10. $y'' = 2y + 2\tan^3 x,$ $y_p(x) = \tan x.$

11. Given that $y_1(x) = \cos x$ is a solution to

$$y'' - y' + y = \sin x,$$

and $y_2(x) = e^{2x}/3$ is a solution to

$$y'' - y' + y = e^{2x},$$

find solutions to the following differential equations:

(a) $y'' - y' + y = 5 \sin x$.

(b) $y'' - y' + y = \sin x - 3e^{2x}$.

(c) $y'' - y' + y = 4 \sin x + 18e^{2x}$.

12. Let $L[y] := y'' + 2y' + 4y$, $y_1(x) := (1/4) \sin 2x$, and $y_2(x) := x/4 - 1/8$. Verify that $L[y_1](x) = \cos 2x$ and $L[y_2](x) = x$. Then use the superposition principle to find solutions to the following differential equations:

(a) $L[y](x) = x + \cos 2x$.

(b) $L[y](x) = 2x - 3 \cos 2x$.

(c) $L[y](x) = 11x - 12 \cos 2x$.

(d) $L[y](x) = 0$.

13. Let $L[y] := y'' - 2y$, $y_1(x) := (\tan x)/2$, $y_2(x) := -x/2$, and $y_3(x) := -1/2$. Verify that $L[y_1](x) = \tan^3 x$, $L[y_2](x) = x$, and $L[y_3](x) = 1$. Find a general solution to each of the following differential equations:

(a) $L[y](x) = \tan^3 x - 1$.

(b) $L[y](x) = x - 2$.

(c) $L[y](x) = 2 \tan^3 x + 3x - 4$.

(d) $L[y](x) = (x - \tan^3 x)/4$.

14. Let $L[y] := y'' - 4y' + 3y$, $y_1(x) := x$, and $y_2(x) := e^{-x}$. Verify that $L[y_1](x) = 3x - 4$ and $L[y_2](x) = 8e^{-x}$. Find the solution to the given initial value problem.

(a) $L[y](x) = 8e^{-x} + 8 - 6x$; $y(0) = 2$, $y'(0) = -2$.

(b) $L[y](x) = 3x - 4 - 8e^{-x}$; $y(0) = 1$, $y'(0) = 6$.

15. Equivalence of General Solutions. Let y_p and $\hat{y}_p$ be two particular solutions to $y'' + py' + qy = g$, and let $\{y_1, y_2\}$ be a fundamental solution set for the corresponding homogeneous equation. Show that any function of the form $y_p + c_1 y_1 + c_2 y_2$ can be written in the form $\hat{y}_p + k_1 y_1 + k_2 y_2$ for suitable choice of constants k_1, k_2 (and vice versa). Consequently it makes no difference which particular solution is used in the expression for the general solution of a nonhomogeneous equation.

16. Third Order Superposition Principle. Let $L[y] := y''' + py'' + qy' + ry$. Show that if $L[y_1] = g_1$ and $L[y_2] = g_2$, then for any constants c_1 and c_2, the function $c_1 y_1 + c_2 y_2$ is a solution of the differential equation $L[y] = c_1 g_1 + c_2 g_2$.

17. Let $L[y] := y''' + 2y'' - 13y' + 10y$, $y_1(x) := x^2$, and $y_2(x) := x^3$. Verify that $L[y_1](x) = 10x^2 - 26x + 4$ and $L[y_2](x) = 10x^3 - 39x^2 + 12x + 6$. Using the third order superposition principle (see Problem 16), find a general

solution to each of the following differential equations.

(a) $L[y](x) = -10x^3 + 59x^2 - 64x + 2$.

(b) $L[y](x) = 20x^3 - 68x^2 - 2x + 16$.

18. Reduction of Order. The reduction of order procedure for homogeneous equations (see Section 4.4) can be extended to the nonhomogeneous equation

$$y'' + p(x)y' + q(x)y = g(x).$$

(a) Show that the substitution $y = v(x)f(x)$, where $f(x)$ is a known nontrivial solution to the corresponding *homogeneous* equation, reduces the nonhomogeneous equation to the linear first order equation

$$fw' + (2f' + pf)w = g,$$

where $w = v'$. (Such an equation can be solved using the techniques of Section 2.3)

(b) Use the procedure in part (a) to find a general solution to

$$y'' + x^{-1}y' - 4x^{-2}y = 1 - x^{-3}, \qquad x > 0,$$

given that $f(x) = x^2$ is a solution to the corresponding homogeneous equation.

19. There is another type of reduction-of-order procedure that can be applied to equations of the form $L[y](x) = g(x)$, provided L is a differential operator that can be written as the product of first order factors. To illustrate this method, consider the equation

$$(7) \qquad L[y](x) := y'' - y' - 2y = e^x.$$

Notice that $L[y] = (D^2 - D - 2)y$ can be written as $L[y] = (D + 1)(D - 2)y$. Let $v = (D - 2)y$ and observe that solving (7) is equivalent to solving the two first order equations

$$(8) \qquad (D + 1)v = e^x,$$

$$(9) \qquad (D - 2)y = v,$$

which, in turn, can be solved using the method discussed in Section 2.3. Carry out this procedure by completing the following steps:

(a) Solve equation (8) for v.

(b) Substitute the v obtained in part (a) into equation (9) and solve for y. This gives a general solution to (7).

20. Use the method of Problem 19 to find a general solution to

$$y'' + 6y' + 5y = 10x + 5.$$

4.8 METHOD OF UNDETERMINED COEFFICIENTS

In this section, we give a simple procedure for finding a particular solution to a nonhomogeneous linear equation with constant coefficients when the nonhomogeneous term $g(x)$ is of a special type. To motivate the procedure, let's first look at a few examples.

EXAMPLE 1 Find a particular solution to

$$L[y](x) := y'' + 3y' + 2y = 3x + 1.$$

Solution We want to find a function $y_p(x)$ such that $L[y_p](x)$ is a linear function of x, namely, $3x + 1$. Now what kind of functions y_p "end up" as linear functions after applications of the operator L? Certainly if y_p is linear, so is $L[y_p]$. So let's try the form

$$y_p(x) = Ax + B$$

and attempt to match up $L[y_p]$ with $3x + 1$. Since

$$L[y_p](x) = 0 + 3A + 2(Ax + B) = 2Ax + (3A + 2B),$$

we need to solve the equation

$$2Ax + (3A + 2B) = 3x + 1.$$

Two polynomials are equal when their corresponding coefficients are equal, so we set

$$2A = 3,$$
$$3A + 2B = 1.$$

Solving this system gives $A = 3/2$ and $B = -7/4$. Thus the function

$$y_p(x) = \frac{3}{2}x - \frac{7}{4}$$

is a particular solution since $L[y_p](x) = 3x + 1$. ∎ End

Throughout this section, L denotes a linear differential operator with *constant coefficients*. For such operators, Example 1 suggests the following method for finding a particular solution to the equation

$$L[y](x) = p_n(x),$$

where $p_n(x)$ is a polynomial of degree n. Guess a solution of the indefinite form

$$y_p(x) = A_n x^n + \cdots + A_1 x + A_0$$

and match the coefficients of $L[y_p](x)$ with those of $p_n(x)$:

$$L[y_p](x) = p_n(x).$$

This procedure involves solving $n + 1$ linear equations in the $n + 1$ unknowns $A_0, \ldots, A_n$ and usually has a solution. The technique is called the **method of undetermined coefficients.**[†]

The same strategy can be applied to other forms for the nonhomogeneous term.

EXAMPLE 2 Find a particular solution to

$$L[y](x) := y'' + 3y' + 2y = e^{3x}.$$

Solution Here we seek a function y_p such that $L[y_p](x) = e^{3x}$. If we guess $y_p(x) = Ae^{3x}$, where A is a constant, then

$$L[y_p](x) = 9Ae^{3x} + 3(3Ae^{3x}) + 2(Ae^{3x}) = 20Ae^{3x}.$$

Setting $20Ae^{3x} = e^{3x}$ and solving for A gives $A = 1/20$. Hence

$$y_p(x) = \frac{e^{3x}}{20}$$

is a particular solution, since $L[y_p](x) = e^{3x}$. ∎ End

The method of undetermined coefficients will also work for equations of the form

$$L[y](x) = ae^{\alpha x},$$

where a and α are given constants. Here we would guess

$$y_p(x) = Ae^{\alpha x}$$

and solve $L[y_p](x) = ae^{\alpha x}$ for the unknown coefficient A.

EXAMPLE 3 Find a particular solution to

(1) $L[y](x) := y'' - y' - y = \sin x.$

Solution This time we seek a function y_p that satisfies $L[y_p](x) = \sin x$. Our initial action might be to guess $y_p(x) = A \sin x$, in which case we get

$$L[y_p](x) = -2A \sin x - A \cos x.$$

However, since the right-hand side of (1) involves only $\sin x$, this choice of solution would force A (and hence y_p itself) to be zero. This will not work, since $y_p(x) \equiv 0$ is obviously not a solution.

So, to compensate for the $\cos x$ term, let's try

$$y_p(x) = A \cos x + B \sin x,$$

which the reader can verify gives

$$L[y_p](x) = (-2A - B)\cos x + (A - 2B)\sin x.$$

[†] **Historical Footnote:** This method was developed by Leonhard Euler in his 1753 theory of the motion of the moon.

Then setting

$$(-2A - B)\cos x + (A - 2B)\sin x = \sin x$$

and equating the coefficients of $\cos x$ and of $\sin x$ leads to the system of equations

$$-2A - B = 0,$$
$$A - 2B = 1.$$

Solving for A and B yields $A = 1/5$ and $B = -2/5$. Hence

$$y_p(x) = \frac{1}{5}\cos x - \frac{2}{5}\sin x$$

is a particular solution to equation (1). ■ End

More generally, for an equation of the form

$$L[y](x) = a \cos \beta x + b \sin \beta x,$$

the method of undetermined coefficients suggests that we guess

$$y_p(x) = A \cos \beta x + B \sin \beta x$$

and solve $L[y_p](x) = a \cos \beta x + b \sin \beta x$ for the unknowns A and B.

EXAMPLE 4 Find a particular solution to

$$(2) \qquad L[y](x) := y'' + y' = 5.$$

Solution The right-hand side is the constant polynomial $p_0(x) \equiv 5$, so Example 1 suggests that we guess $y_p(x) = A$ and set $L[y_p](x) = 5$. But then we get

$$L[y_p](x) = 0 \neq 5.$$

This unfortunate circumstance occurs because any constant function is a solution to the corresponding homogeneous equation $L[y](x) = 0$. To circumvent this difficulty, we temporarily abandon the method of undetermined coefficients and observe that we can integrate both sides of equation (2) to obtain the linear first order equation

$$(3) \qquad y' + y = 5x + c_1,$$

where c_1 is an arbitrary constant. *Now* the method of undetermined coefficients suggests that we approach (3) by trying $y = Ax + B$. This guess leads to

$$A + Ax + B = 5x + c_1$$

or

$$(4) \qquad A = 5,$$
$$A + B = c_1,$$

and so

$$y_p(x) = 5x + (c_1 - 5).$$

In fact, if we take $c_1 = 5$, we get

(5) $y_p(x) = 5x$

as a particular solution to equation (2). This suggests that in our original attempt to use undetermined coefficients, we should have guessed

$$y_p(x) = Ax$$

instead of $y_p(x) = A$. Indeed, with $y_p(x) = Ax$, setting $L[y_p](x) = 5$ immediately gives $A = 5$, and we again obtain (5). ■ End

In the preceding example, the trial choice for y_p did not work because it was a solution to the corresponding homogeneous equation. However, when we replaced y_p by the new function $\hat{y}_p = xy_p$, we were able to find a particular solution to the nonhomogeneous equation. This procedure for ''repairing'' the method of undetermined coefficients is generalized as follows.

If any term in the trial expression for y_p is a solution to the corresponding homogeneous equation, then replace y_p by xy_p. If the latter is also a solution, try $x^2 y_p$, then $x^3 y_p$, etc., ultimately employing $x^s y_p$, where s is the smallest nonnegative integer such that no term in $x^s y_p$ is a solution to the corresponding homogeneous equation (see Problems 66, 67, and 68).

EXAMPLE 5 Determine the form of a particular solution to

(6) $y'' - y' - 12y = e^{4x}$.

Solution The auxiliary equation for the corresponding homogeneous equation

(7) $y'' - y' - 12y = 0$

has the roots $r_1 = 4$ and $r_2 = -3$. A general solution for (7) is therefore

$$y_h(x) = c_1 e^{4x} + c_2 e^{-3x}.$$

Now, since the right-hand side of (6) is e^{4x}, we first guess $y_p = Ae^{4x}$. But, because e^{4x} is a solution to equation (7), we replace this choice by $y_p = Axe^{4x}$. Since xe^{4x} is not a solution to equation (7), there exists a particular solution to (6) of the form

$$y_p(x) = Axe^{4x}. \quad ■ \text{End}$$

As our examples illustrate, when the nonhomogeneous term $g(x)$ is a polynomial, an exponential, a sine, or a cosine function, the function $g(x)$ itself suggests the form of the particular solution $y_p(x)$. In fact, we can expand the list of functions $g(x)$ for which the method of undetermined coefficients can be applied to include products of these functions as well. In Table 4.1, we have listed the various forms for $g(x)$ and the corresponding forms for the particular solution $y_p(x)$. Observe that Type VII includes as special cases all the previous types; we have included the first six for easy reference. In using the table, the reader is reminded to choose s as the smallest nonnegative integer such that no term in y_p is a solution to the corresponding homogeneous equation. For linear second order equations,

TABLE 4.1 PARTICULAR SOLUTIONS TO $L[y] = g$ (CONSTANT COEFFICIENTS)		
Type	$g(x)$	$y_p(x)$
(I) $p_n(x) = a_n x^n + \cdots + a_1 x + a_0$		$x^s P_n(x) = x^s\{A_n x^n + \cdots + A_1 x + A_0\}^{\dagger}$
(II) $ae^{\alpha x}$		$x^s A e^{\alpha x}$
(III) $a \cos \beta x + b \sin \beta x$		$x^s\{A \cos \beta x + B \sin \beta x\}$
(IV) $p_n(x)e^{\alpha x}$		$x^s P_n(x)e^{\alpha x}$
(V) $p_n(x)\cos \beta x + q_m(x)\sin \beta x,$ where $q_m(x) = b_m x^m + \cdots + b_1 x + b_0$		$x^s\{P_N(x)\cos \beta x + Q_N(x)\sin \beta x\},$ where $Q_N(x) = B_N x^N + \cdots + B_1 x + B_0$ and $N = \max(n, m)$
(VI) $ae^{\alpha x}\cos \beta x + be^{\alpha x}\sin \beta x$		$x^s\{Ae^{\alpha x}\cos \beta x + Be^{\alpha x}\sin \beta x\}$
(VII) $p_n(x)e^{\alpha x}\cos \beta x + q_m(x)e^{\alpha x}\sin \beta x$		$x^s e^{\alpha x}\{P_N(x)\cos \beta x + Q_N(x)\sin \beta x\},$ where $N = \max(n, m)$

The nonnegative integer s is chosen to be the smallest integer so that no term in the particular solution $y_p(x)$ is a solution to the corresponding homogeneous equation $L[y](x) = 0$.

† $P_n(x)$ must include all its terms even if $p_n(x)$ has some terms that are zero.

s will be either 0, 1, or 2. An incorrect choice for s leads to an inconsistent system of equations for the coefficients.

EXAMPLE 6 Using Table 4.1, find the form for a particular solution y_p to

(8) $y'' + 2y' - 3y = g(x),$

where $g(x)$ equals

 (a) $7 \cos 3x,$ **(b)** $5e^{-3x},$ **(c)** $x^2 \cos \pi x,$ **(d)** $2xe^x \sin x - e^x \cos x,$
 (e) $x^2 e^x + 3xe^x,$ **(f)** $\tan x.$

Solution The auxiliary equation for the corresponding homogeneous equation,

(9) $y'' + 2y' - 3y = 0,$

has roots $r_1 = 1$ and $r_2 = -3$. Thus a general solution for (9) is

$$y_h(x) = c_1 e^x + c_2 e^{-3x}.$$

 (a) From Table 4.1, the function $g(x) = 7 \cos 3x$ is of Type III (with $a = 7$, $b = 0$, and $\beta = 3$). Hence y_p has the form

(10) $y_p(x) = A \cos 3x + B \sin 3x,$

since no term in (10) is a solution to (9).

(b) The function $g(x) = 5e^{-3x}$ is of Type II, so y_p has the form $x^s A e^{-3x}$. Here we take $s = 1$, since e^{-3x} is a solution to equation (9) and xe^{-3x} is not. Thus

$$y_p(x) = Axe^{-3x}.$$

(c) The function $g(x) = x^2 \cos \pi x$ is of Type V, with $p_2(x) = x^2$, $q_0(x) \equiv 0$. Because $N = \max\{2, 0\} = 2$,

(11) $y_p(x) = (A_2 x^2 + A_1 x + A_0) \cos \pi x + (B_2 x^2 + B_1 x + B_0) \sin \pi x,$

where we have taken $s = 0$ because no term in (11) is a solution to the homogeneous equation.

(d) The function $g(x) = 2xe^x \sin x - e^x \cos x$ is of Type VII, with $p_0(x) = -1$ and $q_1(x) = 2x$. For these polynomials, $N = \max\{0, 1\} = 1$, so

$$x^s e^x \{(A_1 x + A_0) \cos x + (B_1 x + B_0) \sin x\}$$

is the form for the particular solution. None of the terms $e^x x \cos x$, $e^x \cos x$, $e^x x \sin x$, and $e^x \sin x$ is a solution of the homogeneous equation (9), so we set $s = 0$ and obtain

$$y_p(x) = e^x \{(A_1 x + A_0) \cos x + (B_1 x + B_0) \sin x\}.$$

(e) Here $g(x) = x^2 e^x + 3xe^x = (x^2 + 3x)e^x$ is of Type IV with $p_2(x) = x^2 + 3x$. Thus

$$y_p(x) = x(A_2 x^2 + A_1 x + A_0)e^x,$$

where we have taken $s = 1$, since the term $A_0 e^x$ in

$$(A_2 x^2 + A_1 x + A_0)e^x$$

is a solution to equation (9).

(f) Unfortunately, $g(x) = \tan x$ is not one of the forms for which the method of undetermined coefficients can be used. In Section 4.10, we discuss a different method that can handle such nonhomogeneous terms (also see Project A, the Convolution Method). ■ End

Let's summarize.

METHOD OF UNDETERMINED COEFFICIENTS

To determine a particular solution to $L[y] = g$:

(a) Check that the linear equation has constant coefficients and that the nonhomogeneous term $g(x)$ is one of the types suitable for the method.

(b) Solve the corresponding homogeneous equation $L[y](x) = 0$.

(c) On the basis of the form of the nonhomogeneous term $g(x)$, determine the appropriate form of the particular solution $y_p(x)$. Remember that no term in the trial expression for $y_p(x)$ may be a solution to the corresponding homogeneous equation.

(d) Since we want $L[y_p](x) = g(x)$, set the corresponding coefficients from both sides of this equation equal to each other to form a system of linear equations.

(e) Solve this system of linear equations for the coefficients of $y_p(x)$.

If we combine the method of undetermined coefficients with the superposition principle, we can solve for $y_p(x)$ when $g(x)$ is a finite *sum* of terms like those listed in Table 4.1.

EXAMPLE 7 Find a general solution to the equation

(12) $y'' - y = 2e^{-x} - 4xe^{-x} + 10 \cos 2x.$

Solution We first solve the associated homogeneous equation $y'' - y = 0$ and obtain as a general solution $y_h(x) = c_1e^x + c_2e^{-x}$.

Notice that the right-hand side of (12) is not one of the forms given in Table 4.1 but is the sum of two such forms. This suggests that we use the superposition principle and consider separately the equations

(13) $y'' - y = 2e^{-x} - 4xe^{-x}$

and

(14) $y'' - y = 10 \cos 2x.$

A particular solution to equation (13) has the form $y_q(x) = x^s(A_1x + A_0)e^{-x}$. Since A_0e^{-x} is a solution to the corresponding homogeneous equation, we take $s = 1$. Then

$$y_q(x) = (A_1x^2 + A_0x)e^{-x}.$$

Substituting y_q for y in (13) gives

(15) $-4A_1xe^{-x} + (2A_1 - 2A_0)e^{-x} = -4xe^{-x} + 2e^{-x}.$

When we equate the coefficients of xe^{-x} and e^{-x} on both sides of equation (15), we obtain the system

$$-4A_1 = -4,$$
$$2A_1 - 2A_0 = 2.$$

Solving, we find $A_1 = 1$ and $A_0 = 0$; thus

$$y_q(x) = x^2e^{-x}$$

is a particular solution to (13).

Next, consider equation (14). A particular solution to this equation has the form

$$y_r(x) = A \cos 2x + B \sin 2x$$

(here we can take $s = 0$). Substituting y_r for y in (14) gives

(16) $-5A \cos 2x - 5B \sin 2x = 10 \cos 2x,$

which yields $A = -2$ and $B = 0$. Thus

$$y_r(x) = -2 \cos 2x.$$

It follows from the superposition principle that a particular solution to the original equation (12) is given by the sum $y_q + y_r$; that is,

$$y_p(x) = x^2e^{-x} - 2 \cos 2x.$$

Hence a general solution to equation (12), given by the sum of the general solution to the corresponding homogeneous equation and a particular solution, is

$$y(x) = c_1 e^x + c_2 e^{-x} + x^2 e^{-x} - 2 \cos 2x. \quad \blacksquare \text{ End}$$

Success in applying the method of undetermined coefficients lies in the ability to guess the form of a particular solution. As we have seen, when L is a second order linear differential operator with constant coefficients, there is a class of functions $g(x)$ for which the choice of the form of a particular solution to the equation $L[y](x) = g(x)$ is quite easy to determine. This class consists of functions of the form

$$p_n(x) e^{\alpha x} \cos \beta x + q_m(x) e^{\alpha x} \sin \beta x,$$

where $p_n(x)$ and $q_m(x)$ are polynomials. (The use of complex numbers can make the manipulation of this form quite efficient; see Project H, page 238.) The method works because when we apply the operator L to a function of this form, we get back a function of the same type. As you might have guessed, these functions are themselves solutions of some (higher-order) homogeneous linear differential equation with constant coefficients. We verify in Chapter 6 that the method of undetermined coefficients applies to all such functions.

EXERCISES 4.8

In Problems 1 through 16, find a particular solution to the differential equation.

1. $y'' + 3y = -9$.

2. $y'' + 2y' - y = 10$.

3. $2z'' + z = 9e^{2x}$.

4. $x'' - x = 3e^{-2t}$.

5. $2x' + x = 3t^2 + 10t$.

6. $y'' - y' - 2y = -2x^3 - 3x^2 + 8x + 1$.

7. $y'' - y' + 9y = 3 \sin 3x$.

8. $y'' + y' + y = 2 \cos 2x - 3 \sin 2x$.

9. $\dfrac{d^2\theta}{dr^2} - 5\dfrac{d\theta}{dr} + 6\theta = re^r$.

10. $\dfrac{d^2x}{dt^2} + 2\dfrac{dx}{dt} - 8x = te^{-t} + e^{-t}$.

11. $y'' + y = 2^x$.

12. $\theta'' - \theta = t \sin t$.

13. $y'' - 2y' + y = 8e^x$.

14. $w'' - 4w' = x^2 - 1$.

15. $z'' - 6z' + 9z = x^2 + e^x$.

16. $y'' - 4y' - 5y = 2e^{-x}$.

In Problems 17 through 30, find a general solution to the differential equation.

17. $y'' - y = -11x + 1$.

18. $\theta'' + \theta = 7x + 5$.

19. $y'' + y' - 2y = x^2 - 2x + 3$.

20. $y'' - 2y' - 3y = 3x^2 - 5$.

21. $y'' - 3y' + 2y = e^t \sin t$.

22. $y'' + 4y = \sin \theta - \cos \theta$.

23. $y'' + 2y' + 2y = e^{-\theta} \cos \theta$.

24. $y'' + 6y' + 10y = 10x^4 + 24x^3 + 2x^2 - 12x + 18$.

25. $x'' - 4x' + 4x = te^{2t}$.

26. $y'' - 7y' = x^2$.

27. $y'' - 3y = x^2 - e^x$.

28. $y'' + 4y' + 5y = e^{-x} - \sin 2x$.

29. $y'' + y' + y = \cos x - x^2 e^x$.

30. $y'' - 2y' - 35y = 13 \sin x - e^{3x} + 1$.

In Problems 31 through 40, find the solution to the initial value problem.

31. $y' - y = 1, \qquad y(0) = 0$.

32. $h'' = 6x; \qquad h(0) = 3, \qquad h'(0) = -1$.

33. $z'' + z = 2e^{-x}; \qquad z(0) = 0, \qquad z'(0) = 0$.

34. $y'' + 9y = 27; \qquad y(0) = 4, \qquad y'(0) = 6$.

35. $y'' - y' - 2y = \cos x - \sin 2x;$
$y(0) = -7/20, \qquad y'(0) = 1/5$.

36. $y'' - y = e^\theta - e^{-\theta} + 2; \qquad y(0) = 0, \qquad y'(0) = 0$.

37. $y'' - 7y' + 10y = x^2 - 4 + e^x;$
$y(0) = 3, \qquad y'(0) = -3$.

38. $y'' + y' - 12y = e^x + e^{2x} - 1$;
 $y(0) = 1, \qquad y'(0) = 3$.

39. $y'' - y = \sin \theta - e^{2\theta}$; $\qquad y(0) = 1, \qquad y'(0) = -1$.

40. $y'' + 2y' + y = x^2 + 1 - e^x$; $\qquad y(0) = 0, \qquad y'(0) = 2$.

In Problems 41 through 48, determine the form of a particular solution for the differential equation.

41. $y'' + y = \sin x + x \cos x + 10^x$.

42. $y'' - y = e^{2x} + xe^{2x} + x^2 e^{2x}$.

43. $x'' - x' - 2x = e^t \cos t - t^2 + t + 1$.

44. $y'' + 5y' + 6y = \sin x - \cos 2x$.

45. $y'' - 4y' + 5y = e^{5x} + x \sin 3x - \cos 3x$.

46. $y'' - 4y' + 4y = x^2 e^{2x} - e^{2x}$.

47. $v'' - v = e^u - 7 + \cos u$.

48. $y'' + 2y' + 2y = x^2 + e^{-x} \cos x + xe^{-x} \sin x$.

In Problems 49 through 54, decide whether the method of undetermined coefficients (possibly together with superposition) can be applied to find a particular solution of the given equation.

49. $y'' + 2y' - y = x^{-1} e^x$.

50. $y'' + 3y' - y = \sec \theta$.

51. $x''(t) - 5x'(t) + 2x(t) = 3^t$.

52. $x''(t) + tx'(t) + x(t) = \sin t$.

53. $xy'' - y' + 2y = \cos x$.

54. $y'' - y' + 2y = x^2 + \sin x$.

In Problems 55 through 58, find a particular solution to the given higher-order equation.

55. $y''' - y'' + y = \sin x$.

56. $2y''' + 3y'' + y' - 4y = e^{-x}$.

57. $y''' + y'' - 2y = xe^x + 1$.

58. $y^{iv} - 3y'' - 8y = \sin x$.

In Problems 59 through 64, use the method discussed under "Cauchy-Euler Equations" in Section 4.5 to find a general solution to the given equation for $x > 0$.

59. $xy''(x) + 3y'(x) - \dfrac{3}{x} y(x) = x^2$.

60. $x^2 y''(x) - y(x) = (\ln x)^2 - 1$.

61. $x^4 y''(x) - 6x^2 y(x) = 1 - 6x^2$.

62. $xy''(x) + y'(x) = 9x^2$.

63. $x^2 y''(x) + xy'(x) + y(x) = (\ln x) \sin(\ln x)$.

64. $x^2 y''(x) + 3xy'(x) + 4y(x) = \cos(4 \ln x)$.

65. Discontinuous Forcing Term. In certain physical models, the nonhomogeneous term, or **forcing term,** $g(x)$ in the equation $L[y](x) = g(x)$ may not be continuous but may have a jump discontinuity. If this occurs, we can still obtain a reasonable solution using the following procedure. Consider the initial value problem

$$y'' + 2y' + 5y = g(x); \qquad y(0) = 0, \qquad y'(0) = 0,$$

where

$$g(x) := \begin{cases} 10, & 0 \le x \le \dfrac{3\pi}{2}, \\ 0, & x > \dfrac{3\pi}{2}. \end{cases}$$

(a) Find a solution to the initial value problem for $0 \le x \le 3\pi/2$.

(b) Find a general solution for $x > 3\pi/2$.

(c) Now choose the constants in the general solution from part (b) so that the solution from part (a) and the solution from part (b) agree, together with their first derivatives, at $x = 3\pi/2$. This gives us a continuously differentiable function that satisfies the differential equation except at $x = 3\pi/2$.

Problems 66 through 68 concern justifications for the method of undetermined coefficients.

66. Consider the nonhomogeneous equation

$$(17) \qquad ay'' + by' + cy = a_n x^n + \cdots + a_1 x + a_0,$$

where $a_n \ne 0$. We want to determine the form of a particular solution y_p to equation (17).

(a) Assume

$$(18) \qquad y_p(x) = A_n x^n + \cdots + A_0$$

and substitute y_p into equation (17) to obtain, after simplifying,

$$cA_n x^n + (cA_{n-1} + nbA_n)x^{n-1} + \cdots + (2aA_2 + bA_1 + cA_0)$$
$$= a_n x^n + \cdots + a_0.$$

(b) Show that if $c \ne 0$, then no (nonzero) term in (18) is a solution to the homogeneous equation

$$(19) \qquad ay'' + by' + cy = 0.$$

(c) Show that if $c \ne 0$, then you can solve for the constants $A_n, A_{n-1}, \ldots, A_0$ in terms of the constants $a_n, a_{n-1}, \ldots, a_0$. Hence there exists a solution to equation (17) of the form (18).

67. Consider the nonhomogeneous equation

$$(20) \qquad ay'' + by' + cy = e^{\alpha x}(a_n x^n + \cdots + a_0).$$

We want to determine the form of a particular solution y_p to equation (20).

(a) Let $y_p(x) = e^{\alpha x} v(x)$. Substitute for y, y', and y'' in equation (20) and divide by the nonzero factor $e^{\alpha x}$ to obtain

$$(21) \qquad av'' + (2a\alpha + b)v' + (a\alpha^2 + b\alpha + c)v$$
$$= a_n x^n + \cdots + a_0.$$

(b) Using the results from Problem 66, show that if $e^{\alpha x}$ is not a solution to the homogeneous equation (19), then equation (20) has a solution of the form

$$y_p(x) = e^{\alpha x}(A_n x^n + \cdots + A_0).$$

68. Annihilator Method. The annihilator method, which is discussed in Chapter 6, can be used to justify the method of undetermined coefficients. To illustrate the annihilator approach, consider the equation

$$(22) \qquad (D - 1)(D + 1)y = y'' - y = e^{3x}.$$

(a) Show that applying $(D - 3)$ to both sides of equation (22) gives

$$(23) \qquad (D - 3)(D - 1)(D + 1)y$$
$$= y''' - 3y'' - y' + 3y = 0.$$

(Since $(D - 3)e^{3x} \equiv 0$, we say that $(D - 3)$ "annihilates" e^{3x}.)

(b) If y_p satisfies equation (22), it will also satisfy equation (23). Now verify that

$$(24) \qquad c_1 e^{3x} + c_2 e^x + c_3 e^{-x}$$

is a general solution to (23) and conclude that there exists a particular solution to (22) of the form (24).

(c) Using the fact that $c_2 e^x + c_3 e^{-x}$ is a solution to the homogeneous equation

$$(D - 1)(D + 1)y = 0,$$

conclude that there must be a particular solution to (22) of the form $y_p(x) = c_1 e^{3x}$.

(d) Use the above procedure to show that there is a particular solution to

$$(D - 1)(D + 1)y = e^x$$

of the form $y_p(x) = Cxe^x$. [*Hint:* Use the annihilator $(D - 1)$.]

69. Forced Vibrations. A vibrating spring with damping that is under external force can be modeled by

$$(25) \qquad mx'' + bx' + kx = g(t),$$

where $m > 0$ is the mass of the spring system, $b > 0$ is the damping constant, $k > 0$ is the spring constant, $g(t)$ is the force on the system at time t, and $x(t)$ is the displacement from the equilibrium of the spring system at time t. Assume $b^2 < 4mk$.

(a) Determine the form of the equation of motion for the spring system when $g(t) = \sin \beta t$ by finding a general solution to equation (25).

(b) Discuss the long-term behavior of this system. [*Hint:* Consider what happens to the general solution obtained in part (a) as $t \to +\infty$.]

4.9 VARIATION OF PARAMETERS

We have seen that the method of undetermined coefficients is a simple procedure for determining a particular solution when the equation has constant coefficients and the nonhomogeneous term is of a special type. Here we present a more general method, called **variation of parameters,**[†] for finding a particular solution. This method applies even when the coefficients of the differential equation are functions of x, *provided* we know a fundamental solution set for the corresponding homogeneous equation.

Consider the nonhomogeneous linear second order equation

$$(1) \qquad L[y](x) := y'' + p(x)y' + q(x)y = g(x),$$

[†] *Historical Footnote:* The method of variation of parameters was invented by Joseph Lagrange in 1774.

where the coefficient of y'' is taken to be 1, and let $\{y_1(x), y_2(x)\}$ be a fundamental solution set for the corresponding homogeneous equation

(2) $L[y](x) = 0.$

Then we know that the solutions to the homogeneous equation (2) are given by

(3) $y_h(x) = c_1 y_1(x) + c_2 y_2(x),$

where c_1 and c_2 are constants. To find a particular solution to the nonhomogeneous equation, the idea behind variation of parameters is to replace the constants in (3) by functions of x. That is, we seek a solution of (1) of the form[†]

(4) $y_p(x) = v_1(x)y_1(x) + v_2(x)y_2(x).$

Because we have introduced two unknown functions, $v_1(x)$ and $v_2(x)$, it is reasonable to expect that we can impose two equations on these functions. Naturally, one of these equations should come from (1). Let's therefore plug $y_p(x)$ given by (4) into (1). To accomplish this, we must first compute $y_p'(x)$ and $y_p''(x)$. From (4), we obtain

(5) $y_p' = (v_1'y_1 + v_2'y_2) + (v_1 y_1' + v_2 y_2').$

To simplify the computation and to avoid second order derivatives for the unknowns v_1, v_2 in the expression for y_p'', we impose the requirement

(6) $v_1'y_1 + v_2'y_2 = 0;$

thus (5) becomes

(7) $y_p' = v_1 y_1' + v_2 y_2'$

and so

(8) $y_p'' = v_1'y_1' + v_1 y_1'' + v_2'y_2' + v_2 y_2''.$

Now, substituting y_p, y_p', and y_p'', as given in (4), (7), and (8), into (1), we find

(9) $g = L[y_p]$

$\qquad = (v_1'y_1' + v_1 y_1'' + v_2'y_2' + v_2 y_2'') + p(v_1 y_1' + v_2 y_2') + q(v_1 y_1 + v_2 y_2)$

$\qquad = (v_1'y_1' + v_2'y_2') + v_1(y_1'' + py_1' + qy_1) + v_2(y_2'' + py_2' + qy_2)$

$\qquad = (v_1'y_1' + v_2'y_2') + v_1 L[y_1] + v_2 L[y_2].$

Since y_1 and y_2 are solutions to the homogeneous equation, we have $L[y_1] = L[y_2] = 0.$ Thus (9) reduces to

(10) $v_1'y_1' + v_2'y_2' = g.$

To summarize, if we can find v_1 and v_2 that satisfy both (6) and (10), that is,

(11) $y_1 v_1' + y_2 v_2' = 0,$

$\qquad\quad y_1' v_1' + y_2' v_2' = g,$

[†] In Exercises 2.3, Problem 36, we developed this approach for first order linear equations. Because of the similarity of equations (3) and (4), this technique is sometimes known as ''variation of constants.''

then y_p given by (4) will be a particular solution to (1). To determine v_1 and v_2, we first solve the linear system (11) for v_1' and v_2'. Algebraic manipulation or Cramer's rule immediately gives

$$v_1'(x) = \frac{-g(x)y_2(x)}{W[y_1,\ y_2](x)}, \qquad v_2'(x) = \frac{g(x)y_1(x)}{W[y_1,\ y_2](x)},$$

where $W[y_1,\ y_2](x)$, which occurs in the denominator, is the Wronskian of $y_1(x)$ and $y_2(x)$. Notice that this Wronskian is never zero because $\{y_1,\ y_2\}$ is a fundamental solution set. Upon integrating these equations, we finally obtain

$$(12) \qquad v_1(x) = \int \frac{-g(x)y_2(x)}{W[y_1,\ y_2](x)}\,dx, \qquad v_2(x) = \int \frac{g(x)y_1(x)}{W[y_1,\ y_2](x)}\,dx.$$

Let's review this procedure.

METHOD OF VARIATION OF PARAMETERS

To determine a particular solution to $y'' + py' + qy = g$:

(a) Find a fundamental solution set $\{y_1(x),\ y_2(x)\}$ for the corresponding homogeneous equation and take

$$y_p(x) = v_1(x)y_1(x) + v_2(x)y_2(x).$$

(b) Determine $v_1(x)$ and $v_2(x)$ by solving the system in (11) for $v_1'(x)$ and $v_2'(x)$ and integrating. (If the given equation is $a_2(x)y'' + a_1(x)y' + a_0(x)y = b(x)$, then it must be put in the form $y'' + p(x)y' + q(x)y = g(x)$ so that $g(x) = b(x)/a_2(x)$.)

(c) Substitute $v_1(x)$ and $v_2(x)$ into the expression for $y_p(x)$ to obtain a particular solution.

Of course, in step (b) one could use the formulas in (12), but $v_1(x)$ and $v_2(x)$ are so easy to derive that the reader is advised not to memorize them.

EXAMPLE 1 Find a general solution on $(-\pi/2,\ \pi/2)$ to

$$(13) \qquad \frac{d^2y}{dx^2} + y = \tan x.$$

Solution Observe that a fundamental solution set for the homogeneous equation $y'' + y = 0$ is $\{\cos x,\ \sin x\}$. We now set

$$(14) \qquad y_p(x) = v_1(x)\cos x + v_2(x)\sin x$$

and, referring to (11), solve the system

$$(\cos x)v_1'(x) + (\sin x)v_2'(x) = 0,$$

$$(-\sin x)v_1'(x) + (\cos x)v_2'(x) = \tan x.$$

for $v_1'(x)$ and $v_2'(x)$. This gives

$$v_1'(x) = -\tan x \sin x,$$

$$v_2'(x) = \tan x \cos x = \sin x.$$

Integrating, we obtain

(15) $\quad v_1(x) = -\int \tan x \sin x \, dx = -\int \frac{\sin^2 x}{\cos x} \, dx$

$$= -\int \frac{1 - \cos^2 x}{\cos x} \, dx = \int (\cos x - \sec x) \, dx$$

$$= \sin x - \ln(\sec x + \tan x) + C_1,$$

(16) $\quad v_2(x) = \int \sin x \, dx = -\cos x + C_2.$

We need only one particular solution, so we take both C_1 and C_2 to be zero for simplicity. Then, substituting $v_1(x)$ and $v_2(x)$ in (14), we obtain

$$y_p(x) = [\sin x - \ln(\sec x + \tan x)]\cos x - \cos x \sin x,$$

which simplifies to

$$y_p(x) = -(\cos x)\ln(\sec x + \tan x).$$

Recall from Theorem 6 in Section 4.7 that a general solution to a nonhomogeneous equation is given by the sum of a general solution to the homogeneous equation and a particular solution. Consequently a general solution to equation (13) on the interval $(-\pi/2, \pi/2)$ is

(17) $\quad y(x) = c_1 \cos x + c_2 \sin x - (\cos x)\ln(\sec x + \tan x).$ ▪ End

Note that in the above example, the constants C_1 and C_2 appearing in (15) and (16) were chosen to be zero. If we had retained these arbitrary constants, the ultimate effect would be just to add $C_1 \cos x + C_2 \sin x$ to (17), which is clearly redundant.

EXAMPLE 2 Find a particular solution on $(-\pi/2, \pi/2)$ to

(18) $\quad \dfrac{d^2 y}{dx^2} + y = \tan x + 3x - 1.$

Solution With $g(x) = \tan x + 3x - 1$, the variation of parameters procedure will lead to a solution of (18). But it is simpler in this case to consider separately the equations

(19) $\quad \dfrac{d^2 y}{dx^2} + y = \tan x,$

(20) $\quad \dfrac{d^2 y}{dx^2} + y = 3x - 1$

and then use the superposition principle (Theorem 5, page 189).

In Example 1, we found that

$$y_q(x) = -(\cos x)\ln(\sec x + \tan x)$$

is a particular solution for equation (19). For equation (20), the method of undetermined coefficients can be applied. On seeking a solution to (20) of the form $y_r(x) = Ax + B$, we quickly obtain

$$y_r(x) = 3x - 1.$$

Finally, we apply the superposition principle to get

$$y_p(x) = y_q(x) + y_r(x)$$
$$= -(\cos x)\ln(\sec x + \tan x) + 3x - 1$$

as a particular solution for equation (18). ∎ End

One important advantage the method of variation of parameters has over the method of undetermined coefficients is its applicability to linear equations whose coefficients are functions of x (see Problems 23–26). The fly in the ointment is that variation of parameters requires that we know a fundamental solution set for the corresponding homogeneous equation. For equations with variable coefficients, this set may be extremely difficult to determine.

EXERCISES 4.9

In Problems 1 through 12, find a general solution to the differential equation using the method of variation of parameters.

1. $y'' + 4y = \tan 2x.$

2. $y'' + y = \sec x.$

3. $2x'' - 2x' - 4x = 2e^{3t}.$

4. $y'' - y = 2x + 4.$

5. $y'' - 2y' + y = x^{-1}e^x.$

6. $y'' + 2y' + y = e^{-x}.$

7. $y'' + 16y = \sec 4\theta.$

8. $y'' + 9y = \sec^2(3x).$

9. $y'' + 4y = \csc^2(2x).$

10. $y'' + 4y' + 4y = e^{-2x}\ln x.$

11. $x^2z'' + xz' + 9z = -\tan(3\ln x).$

12. $x^2y'' + 3xy' + y = x^{-1}.$

In Problems 13 through 21, find a general solution to the differential equation.

13. $y'' + y = \tan^2 x.$

14. $y'' + y = \tan x + e^{3x} - 1.$

15. $v'' + 4v = \sec^4(2x).$

16. $y'' + y = \sec^3\theta.$

17. $y'' + y = 3\sec x - x^2 + 1.$

18. $y'' + 5y' + 6y = 18x^2.$

19. $\frac{1}{2}y'' + 2y = \tan 2x - \frac{1}{2}e^x.$

20. $y'' - 6y' + 9y = x^{-3}e^{3x}.$

21. $x^2z'' - xz' + z = x\left(1 + \dfrac{3}{\ln x}\right).$

22. The **Bessel equation** of order one half,

$$x^2y'' + xy' + \left(x^2 - \frac{1}{4}\right)y = 0, \qquad x > 0,$$

has two linearly independent solutions,

$$y_1(x) = x^{-1/2}\cos x, \qquad y_2(x) = x^{-1/2}\sin x.$$

Find a general solution to the nonhomogeneous equation

$$x^2y'' + xy' + \left(x^2 - \frac{1}{4}\right)y = x^{5/2}, \qquad x > 0.$$

In Problems 23 through 26, find a general solution to the differential equation given that the functions y_1 and y_2 are

linearly independent solutions to the corresponding homogeneous equation for $x > 0$. Remember to put the equation in the same form as equation (1).

23. $xy'' - (x + 1)y' + y = x^2;$ $y_1 = e^x,$ $y_2 = x + 1.$

24. $x^2y'' - 4xy' + 6y = x^3 + 1;$ $y_1 = x^2,$ $y_2 = x^3.$

25. $xy'' + (5x - 1)y' - 5y = x^2e^{-5x};$ $y_1 = 5x - 1,$
$y_2 = e^{-5x}.$

26. $xy'' + (1 - 2x)y' + (x - 1)y = xe^x;$ $y_1 = e^x,$
$y_2 = e^x \ln x.$

27. Express the solution to the initial value problem

$$y'' - y = \frac{1}{x}; \qquad y(1) = 0, \qquad y'(1) = -2$$

using definite integrals. Using numerical integration (Appendix B) to approximate the integrals, find an approximation for $y(2)$ to two decimal places.

28. Use the method of variation of parameters to show that

$$y(x) = c_1 \cos x + c_2 \sin x + \int_0^x f(s)\sin(x - s)\,ds$$

is a general solution to the differential equation

$$y'' + y = f(x),$$

where $f(x)$ is a continuous function on $(-\infty, \infty)$. [*Hint:* Use the trigonometric identity $\sin(x - s) = \sin x \cos s - \sin s \cos x$.]

4.10 INTRODUCTION TO SYSTEMS OF FIRST ORDER EQUATIONS

The algebraic notion of a linear *system* of simultaneous equations, such as

$$3x_1 + 4x_2 = 7$$
$$2x_1 - 3x_2 = 10,$$

has an analog in the theory of differential equations. The general form for a linear system of two first order differential equations is

(1)
$$a_1(t)x_1'(t) + a_2(t)x_1(t) + b_1(t)x_2'(t) + b_2(t)x_2(t) = g_1(t),$$
$$c_1(t)x_1'(t) + c_2(t)x_1(t) + d_1(t)x_2'(t) + d_2(t)x_2(t) = g_2(t),$$

where the coefficients a_1, a_2, b_1, b_2, c_1, c_2, d_1, and d_2, and the nonhomogeneous terms g_1 and g_2 are given functions of t. A solution to system (1) is a pair of functions $x_1(t)$, $x_2(t)$ that satisfy both members of (1) on some interval I.

A linear first order system that is expressed as

$$x_1'(t) = a_1(t)x_1(t) + a_2(t)x_2(t) + g_1(t),$$
$$x_2'(t) = b_1(t)x_1(t) + b_2(t)x_2(t) + g_2(t)$$

is said to be in **normal form.** More generally, the normal form for a first order system of m (not necessarily linear) equations in m unknown functions $x_1(t), x_2(t), \ldots, x_m(t)$ is given by

(2)
$$x_1'(t) = f_1(t, x_1, x_2, \ldots, x_m)$$
$$x_2'(t) = f_2(t, x_1, x_2, \ldots, x_m)$$
$$\vdots$$
$$x_m'(t) = f_m(t, x_1, x_2, \ldots, x_m).$$

It is often useful to rewrite an mth order differential equation as a first order system of m equations. For example, most numerical procedures for approximating the solution to an initial value problem for a higher-order differential equation require that the problem first be restated as a system of first order equations in normal form (see Section 5.6). Further, as we shall see in Chapter 9, the powerful machinery of linear algebra can be readily applied in such a setting if the system is linear.

The trick for converting the mth order differential equation

$$(3) \qquad y^{(m)}(t) = f(t, y, y', \ldots, y^{(m-1)})$$

into a first order system is to introduce, as additional unknowns, the sequence of derivatives of y:

$$x_1(t) := y(t), \quad x_2(t) := y'(t), \quad \ldots, \quad x_m(t) := y^{(m-1)}(t).$$

With this scheme, we obtain $m - 1$ first order equations quite trivially:

$$x_1'(t) = y'(t) = x_2(t),$$
$$(4) \qquad x_2'(t) = y''(t) = x_3(t),$$
$$\vdots$$
$$x_{m-1}'(t) = y^{(m-1)}(t) = x_m(t).$$

The mth and final equation then constitutes a restatement of the original equation (3) in terms of the new unknowns:

$$(5) \qquad x_m'(t) = y^{(m)}(t) = f(t, x_1, x_2, \ldots, x_m).$$

If equation (3) has initial conditions $y(t_0) = a_1$, $y'(t_0) = a_2$, $\ldots$, $y^{(m-1)}(t_0) = a_m$, then the system (4)–(5) has initial conditions $x_1(t_0) = a_1$, $x_2(t_0) = a_2$, $\ldots$, $x_m(t_0) = a_m$.

EXAMPLE 1 Convert the initial value problem

$$(6) \qquad y''(t) + 3y'(t) + 2y(t) = 0; \qquad y(0) = 1, \qquad y'(0) = 3$$

into an initial value problem for a system in normal form.

Solution We first express the differential equation in (6) as

$$y''(t) = -3y'(t) - 2y(t).$$

Setting $x_1(t) := y(t)$ and $x_2(t) := y'(t)$, we obtain

$$x_1'(t) = x_2(t),$$
$$x_2'(t) = -3x_2(t) - 2x_1(t).$$

The initial conditions transform to $x_1(0) = 1$, $x_2(0) = 3$. ■ End

In later chapters, we discuss several analytic and numerical methods for solving first order systems. Here we confine our attention to the **elimination method**, which accomplishes precisely the reverse of what we have just studied: It replaces the system by a single higher-order equation in one unknown. The method works on more general systems than

(1); however, we must assume that the system has constant coefficients. In such a case, the procedure is facilitated if the system is expressed in operator notation, using $D := d/dt$ (see Section 4.2). First, we demonstrate the method for a system of two first order equations.

EXAMPLE 2 Solve the system

$$
(7) \qquad
\begin{aligned}
x'(t) &= 3x(t) - 4y(t), \\
y'(t) &= 4x(t) - 7y(t).
\end{aligned}
$$

Solution The alert reader may observe that we could use the first equation to express y in terms of x and x' and substitute into the second equation to derive an "uncoupled" equation containing only x and its derivatives. However, this simple trick will not work on more general systems (see Problem 22).

To utilize the elimination method, we first write the system using the operator notation:

$$
(8) \qquad
\begin{aligned}
(D - 3)[x] + 4y &= 0, \\
-4x + (D + 7)[y] &= 0.
\end{aligned}
$$

Modeling the elimination procedure for algebraic systems, we can eliminate x from this system by adding 4 times the first equation to $(D - 3)$ applied to the second equation. Doing this gives

$$
(16 + (D - 3)(D + 7))[y] = 4 \cdot 0 + (D - 3)[0] = 0,
$$

which simplifies to

$$
(9) \qquad (D^2 + 4D - 5)[y] = 0.
$$

Now equation (9) is just a second order linear equation in y with constant coefficients that has the general solution

$$
(10) \qquad y(t) = C_1 e^{-5t} + C_2 e^t.
$$

To find $x(t)$, we have two options. One is to return to system (7) and eliminate y. This is accomplished by "multiplying" the first equation in (8) by $(D + 7)$ and the second equation by -4 and then adding to obtain

$$
(D^2 + 4D - 5)[x] = 0.
$$

Coincidentally this equation is the same as equation (9) except that here the unknown function is $x(t)$. Hence

$$
(11) \qquad x(t) = K_1 e^{-5t} + K_2 e^t,
$$

where we have taken K_1 and K_2 to be the arbitrary constants, which are not necessarily the same as C_1 and C_2 used in equation (10).

It is reasonable to expect that system (7) will involve only *two* arbitrary constants, since it consists of two first order equations. Thus the four constants C_1, C_2, K_1, and K_2 are not independent. To determine the relationships, we substitute the expressions for $x(t)$ and $y(t)$ given in (10) and (11) into one of the equations in (7), say, the first one. This yields

$$
-5K_1 e^{-5t} + K_2 e^t = 3K_1 e^{-5t} + 3K_2 e^t - 4C_1 e^{-5t} - 4C_2 e^t,
$$

which simplifies to

$$(4C_1 - 8K_1)e^{-5t} + (4C_2 - 2K_2)e^t = 0.$$

Because e^t and e^{-5t} are linearly independent functions on any interval, this last equation holds for all t only if

$$4C_1 - 8K_1 = 0 \quad \text{and} \quad 4C_2 - 2K_2 = 0.$$

Therefore $K_1 = C_1/2$ and $K_2 = 2C_2$.

A solution to system (7) is then given by the pair

(12) $\qquad x(t) = \dfrac{1}{2}C_1e^{-5t} + 2C_2e^t, \qquad y(t) = C_1e^{-5t} + C_2e^t.$

As you might expect, the pair in (12) is a **general solution** to (7) in the sense that *any* solution to (7) can be expressed in this fashion.

A simpler method for determining $x(t)$ once $y(t)$ is known is to use the system to obtain an equation for $x(t)$ in terms of $y(t)$ and $y'(t)$. In this example, solving the second equation in (7) for $x(t)$ gives

$$x(t) = \frac{1}{4}y'(t) + \frac{7}{4}y(t).$$

Substituting $y(t)$ as given in (10) yields

$$x(t) = \frac{1}{4}[-5C_1e^{-5t} + C_2e^t] + \frac{7}{4}[C_1e^{-5t} + C_2e^t]$$

$$= \frac{1}{2}C_1e^{-5t} + 2C_2e^t,$$

which agrees with (12). ∎ End

The above procedure works, more generally, for any linear system of two equations and two unknowns with *constant coefficients* regardless of the order of the equations. For example, if we let L_1, L_2, L_3, and L_4 denote linear differential operators with constant coefficients (i.e., polynomials in D), then the method can be applied to the linear system

$$L_1[x] + L_2[y] = f_1,$$
$$L_3[x] + L_4[y] = f_2.$$

Because the system has constant coefficients, the operators commute (e.g., $L_2L_4 = L_4L_2$) and we can eliminate variables in the usual algebraic fashion. Eliminating the variable y gives

(13) $\qquad (L_1L_4 - L_2L_3)[x] = g_1,$

where $g_1 := L_4[f_1] - L_2[f_2]$. Similarly, eliminating the variable x yields

(14) $\qquad (L_1L_4 - L_2L_3)[y] = g_2,$

where $g_2 := L_1[f_2] - L_3[f_1]$. Now if $L_1L_4 - L_2L_3$ is a differential operator of order n, then a general solution for (13) contains n arbitrary constants, and a general solution for (14)

also contains n arbitrary constants. Thus a total of $2n$ constants arise. However, as we saw in Example 2, there are only n of these that are independent for the system; the remaining constants can be expressed in terms of these.[†] The pair of general solutions to (13) and (14) written in terms of the n independent constants is called a **general solution for the system.**

If it turns out that $L_1 L_4 - L_2 L_3$ is the zero operator, the system is said to be **degenerate.** As with the anomalous problem of solving for the points of intersection of two parallel or coincident lines, a degenerate system may have no solutions, or if it does possess solutions, they may involve any number of arbitrary constants (see Problems 25 and 26).

ELIMINATION PROCEDURE FOR 2 x 2 SYSTEMS

To find a general solution for the system

$$L_1[x] + L_2[y] = f_1,$$
$$L_3[x] + L_4[y] = f_2,$$

where L_1, L_2, L_3, and L_4 are polynomials in $D = d/dt$:

(a) Make sure the system is written in operator form.

(b) Eliminate one of the variables, say y, and solve the resulting equation for $x(t)$. If the system is degenerate, stop! A separate analysis is required to determine whether there are no solutions or infinitely many linearly independent ones.

(c) (*Short cut*) If possible, use the system to derive an equation that involves $y(t)$ but not its derivatives. (Otherwise, go to step d.) Substitute the expression for $x(t)$ into this equation to get a formula for $y(t)$. The expressions for $x(t)$, $y(t)$ give the desired general solution.

(d) Eliminate x from the system and solve for $y(t)$. (Solving for $y(t)$ gives more constants—in fact, twice as many as needed.)

(e) Remove the extra constants by substituting the expressions for $x(t)$ and $y(t)$ into one or both of the equations in the system. Write the expressions for $x(t)$ and $y(t)$ in terms of the remaining constants.

EXAMPLE 3 Find a general solution for

(15)
$$x''(t) + y'(t) - x(t) + y(t) = -1,$$
$$x'(t) + y'(t) - x(t) = t^2.$$

Solution We begin by expressing the system in operator notation:

(16)
$$(D^2 - 1)[x] + (D + 1)[y] = -1,$$
$$(D - 1)[x] + D[y] = t^2.$$

Here $L_1 := D^2 - 1$, $L_2 := D + 1$, $L_3 := D - 1$, and $L_4 := D$.

[†] For a proof of this fact, see *Ordinary Differential Equations*, by M. Tenebaum and H. Pollard (Dover, New York, 1985), Chapter 7.

Eliminating y gives (cf. (13)):

$$((D^2 - 1)D - (D + 1)(D - 1))[x] = D[-1] - (D + 1)[t^2],$$

which reduces to

$$(D - 1)((D + 1)D - (D + 1))[x] = -2t - t^2,$$

(17) $(D - 1)^2(D + 1)[x] = -2t - t^2.$

Since $(D - 1)^2(D + 1)$ is third order, there will be three arbitrary constants in a general solution to system (15).

Although the methods of this chapter focused on solving second order equations, we have seen several examples of how they extend in a natural way to higher-order equations. Making this leap for the third order equation (17) we observe that the corresponding homogeneous equation has the auxiliary equation $(r - 1)^2(r + 1) = 0$ with roots $r = 1, 1, -1$. Hence a general solution for the homogeneous equation is

$$x_h(t) = C_1e^t + C_2te^t + C_3e^{-t}.$$

To find a particular solution to (17) we use the method of undetermined coefficients with $x_p(t) = At^2 + Bt + C$. Substituting into (17) and solving for A, B, and C yields (after a little algebra)

$$x_p(t) = -t^2 - 4t - 6.$$

Thus a general solution to equation (17) is

(18) $x(t) = x_h(t) + x_p(t) = C_1e^t + C_2te^t + C_3e^{-t} - t^2 - 4t - 6.$

To find $y(t)$, we take the short cut described in step (c) of the elimination procedure box. Subtracting the second equation in (16) from the first, we find

$$(D^2 - D)[x] + y = -1 - t^2,$$

so that

$$y = (D - D^2)[x] - 1 - t^2.$$

Inserting the expression for $x(t)$, given in (18), we obtain

$$y(t) = C_1e^t + C_2(te^t + e^t) - C_3e^{-t} - 2t - 4$$
$$-[C_1e^t + C_2(te^t + 2e^t) + C_3e^{-t} - 2] - 1 - t^2,$$

(19) $y(t) = -C_2e^t - 2C_3e^{-t} - t^2 - 2t - 3.$

The formulas for $x(t)$ in (18) and $y(t)$ in (19) give the desired general solution to (15). ■ End

The next example is a mixing problem involving two interconnected tanks. (See Section 3.2 for a discussion of mixing problems with one tank.)

EXAMPLE 4 Two large tanks, each holding 100 L of liquid, are interconnected by pipes, with the liquid flowing from tank A into tank B at a rate of 3 L/min and from B into A at a rate of

1 L/min (see Figure 4.8). The liquid inside each tank is kept well stirred. A brine solution with a concentration of 2 kg/L of salt flows into tank A at a rate of 6 L/min. The (diluted) solution flows out of the system from tank A at 4 L/min and from tank B at 2 L/min. If, initially, tank A contains only water and tank B contains 200 kg of salt, determine the mass of salt in each tank at time $t \geq 0$.

Figure 4.8 Mixing problem for interconnected tanks

Solution We approach this interconnected tank problem as a compartmental analysis problem involving two compartments.

Let $x(t)$ denote the mass of salt in tank A at time t and $y(t)$ the mass of salt in tank B. Liquid flows into tank A at 7 L/min and flows out at the same rate, so the volume of liquid in tank A remains constant. Similarly the volume of liquid in tank B remains constant. Let's now determine the rate of change in the amount of salt in tank A. Salt enters tank A from the outside at a rate of $(6)(2) = 12$ kg/min and from tank B at a rate of $y/100$ kg/min. Salt leaves tank A to enter tank B at $3x/100$ kg/min and leaves the system at $4x/100$ kg/min. Using the compartmental analysis model,

$$\frac{dx}{dt} = \text{INPUT RATE} - \text{OUTPUT RATE},$$

the rate of change in the mass of salt in tank A is

(20) $$\frac{dx}{dt} = 12 + \frac{1}{100}y - \frac{3}{100}x - \frac{4}{100}x.$$

In a similar fashion, we can show that the rate of change in the mass of salt in tank B satisfies

(21) $$\frac{dy}{dt} = \frac{3}{100}x - \frac{3}{100}y.$$

We can rewrite (20) and (21) as the system

$$(D + 0.07)[x] - 0.01y = 12,$$
$$-0.03x + (D + 0.03)[y] = 0.$$

Eliminating x from this system gives

$$(-0.0003 + (D + 0.07)(D + 0.03))[y] = 0.36,$$

which simplifies to

(22) $(D^2 + 0.1D + 0.0018)[y] = 0.36.$

The auxiliary equation for the homogeneous equation associated with (22) is $r^2 + 0.1r + 0.0018 = 0$, which has roots

(23) $r_1 = \dfrac{-5 - \sqrt{7}}{100} \approx -0.0765, \qquad r_2 = \dfrac{-5 + \sqrt{7}}{100} \approx -0.0235.$

To determine a particular solution to (22), we just substitute $y_p = a$ into (22) and solve for the constant a. This gives $y_p = 200$. Therefore a general solution to (22) is

(24) $y(t) = C_1 e^{r_1 t} + C_2 e^{r_2 t} + 200,$

where r_1 and r_2 are given in (23).

To determine $x(t)$, we solve for x in (21) to obtain

$$x = \frac{100}{3}(D + 0.03)[y],$$

and then insert the expression for $y(t)$ given in (24). This yields

(25) $x(t) = \dfrac{-2 - \sqrt{7}}{3} C_1 e^{r_1 t} + \dfrac{-2 + \sqrt{7}}{3} C_2 e^{r_2 t} + 200.$

Finally, to determine the constants C_1 and C_2, we use the initial conditions $x(0) = 0$ and $y(0) = 200$. Substituting x and y given in (24) and (25) into these initial conditions and solving yields $C_1 = -C_2 = 300/\sqrt{7} \approx 113$. Hence the masses of salt at time t in tanks A and B are, respectively.

(26) $x(t) = -\left(100 + \dfrac{200}{\sqrt{7}}\right) e^{r_1 t} - \left(100 - \dfrac{200}{\sqrt{7}}\right) e^{r_2 t} + 200$ kg,

(27) $y(t) = \dfrac{300}{\sqrt{7}} e^{r_1 t} - \dfrac{300}{\sqrt{7}} e^{r_2 t} + 200$ kg. ∎ End

An important observation we can make here is that since r_1 and r_2 are negative, the amount of salt in each tank approaches 200 kg as $t \to \infty$. Hence the concentration of salt in each tank approaches 2 kg/L, which is the concentration of salt in the brine entering tank A.

EXERCISES 4.10

In Problems 1 through 4, convert the given initial value problem into an initial value problem for a system in normal form.

1. $y''(t) + ty'(t) - 3y(t) = t^2$;
$y(0) = 3, \qquad y'(0) = -6.$

2. $y''(t) = \cos(t - y) + y^2(t); \qquad y(0) = 1, \qquad y'(0) = 0.$

3. $y^{(4)}(t) - y^{(3)}(t) + 7y(t) = \cos t$;
$y(0) = y'(0) = 1, \qquad y''(0) = 0, \qquad y^{(3)}(0) = 2.$

4. $y^{(6)}(t) = [y'(t)]^3 - \sin(y(t)) + e^{2t}$;
$y(0) = y'(0) = \cdots = y^{(5)}(0) = 0.$

In Problems 5 through 22, use the elimination method to find a general solution for the given linear system, where x', y', z' denote differentiation with respect to t.

5. $x' + y' = -2y$,
$x - 2y = y'$.

6. $x' = 3y$,
$y' = 2x - y$.

7. $x' + 2y = 0$,
$x' - y' = 0$.

8. $x' = x - y$,
$y' = y - 4x$.

9. $x' + y' - x = 5$,
$x' + y' + y = 1$.

10. $x' = 3x - 2y + \sin t$,
$y' = 4x - y - \cos t$.

11. $(D + 1)[u] - (D + 1)[v] = e^t$,
$(D - 1)[u] + (2D + 1)[v] = 5$.

12. $(D - 3)[x] + (D - 1)[y] = t$,
$(D + 1)[x] + (D + 4)[y] = 1$.

13. $x' + y' + 2x = 0$,
$x' + y' - x - y = \sin t$.

14. $2x' + y' - x - y = e^{-t}$,
$x' + y' + 2x + y = e^t$.

15. $(D^2 - 1)[u] + 5v = e^t$,
$2u + (D^2 + 2)[v] = 0$.

16. $D^2[u] + D[v] = 2$,
$4u + D[v] = 6$.

17. $\dfrac{dx}{dt} = x - 4y$,

$\dfrac{dy}{dt} = x + y$.

18. $\dfrac{dx}{dt} + y = t^2$,

$-x + \dfrac{dy}{dt} = 1$.

19. $\dfrac{dw}{dt} = 5w + 2z + 5t$,

$\dfrac{dz}{dt} = 3w + 4z + 17t$.

20. $\dfrac{dx}{dt} + x + \dfrac{dy}{dt} = e^{4t}$,

$2x + \dfrac{d^2y}{dt^2} = 0$.

21. $x'' + 5x - 4y = 0$,
$-x + y'' + 2y = 0$.

22. $x'' + y'' - x' = 2t$,
$x'' + y' - x + y = -1$.

In Problems 23 and 24, describe how solutions to the given system behave as $t \to +\infty$ for each (fixed) value of the parameter λ between $-\infty$ and $+\infty$.

23. $\dfrac{dx}{dt} = -x + \lambda y$,

$\dfrac{dy}{dt} = x - y$.

24. $\dfrac{dx}{dt} = x + \lambda y$,

$\dfrac{dy}{dt} = x - y$.

In Problems 25 and 26, show that the given linear system is degenerate. In attempting to solve the system, determine whether it has no solutions or infinitely many linearly independent solutions.

25. $(D - 1)[x] + (D - 1)[y] = -3e^{-2t}$,
$(D + 2)[x] + (D + 2)[y] = 3e^t$.

26. $D[x] + (D + 1)[y] = e^t$,
$D[x] + (D + 1)[y] = 0$.

In Problems 27 and 28, use the elimination method to find a general solution for the given system of 3 equations in the 3 unknown functions $x(t)$, $y(t)$, $z(t)$.

27. $x' = x + 2y - z$
$y' = x + z$
$z' = 4x - 4y + 5z$

28. $x' = 3x + y - z$
$y' = x + 2y - z$
$z' = 3x + 3y - z$.

29. Two large tanks, each holding 50 L of liquid, are interconnected by a pipe with liquid flowing from tank A into tank B at a rate of 5 L/min (see Figure 4.9). The liquid inside each tank is kept well stirred. A salt brine with concentration 3 kg/L of salt flows into tank A at a rate of 5 L/min. The solution flows out of the system from tank B at 5 L/min. If, initially, tank A contains 50 kg of salt and tank B contains 100 kg, determine the mass of salt in each tank at time $t \geq 0$.

Figure 4.9 Mixing problem for the flow through two tanks

30. In Example 4, 3 L/min of liquid flowed from tank A into tank B and 1 L/min from B into A. Determine the mass of salt in each tank at time $t \geq 0$ if, instead, 5 L/min flows from A into B and 3 L/min flows from B into A, with all other data the same.

31. In Example 4, assume no solution flows out of the system from tank B, only 1 L/min flows from A into B, and only 4 L/min of brine flows into the system at tank A, other data being the same. Determine the mass of salt in each tank at time $t \geq 0$.

32. Feedback System with Pooling Delay. Many physical and biological systems involve time delays. A pure time delay has the output the same as the input but is shifted in time. A more common type of delay is *pooling delay*. An example of such a feedback system is shown in Figure 4.10 on page 216. Here the level of fluid in tank B determines the rate at which fluid enters tank A. Suppose this rate is given by $R_1(t) = \alpha[V - V_2(t)]$, where α and V are positive constants and $V_2(t)$ is the volume of fluid in tank B at time t.

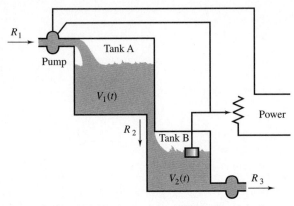

Figure 4.10 Feedback system with pooling delay

(a) If the outflow rate R_3 from tank B is constant and the flow rate R_2 from tank A into B is $R_2(t) = KV_1(t)$, where K is a positive constant and $V_1(t)$ is the volume of fluid in tank A at time t, then show that this feedback system is governed by the system

$$\frac{dV_1}{dt} = \alpha(V - V_2(t)) - KV_1(t),$$

$$\frac{dV_2}{dt} = KV_1(t) - R_3.$$

(b) Find a general solution for the system in part (a) when $\alpha = 5\,(\min)^{-1}$, $V = 20$ L, $K = 2\,(\min)^{-1}$, and $R_3 = 10$ L/min.

(c) Using the general solution obtained in part (b), what can be said about the volume of fluid in each of the tanks as $t \to \infty$?

33. A house, for cooling purposes, consists of two zones: the attic area zone A and the living area zone B (see Figure 4.11). The living area is cooled by a 2-ton air conditioning unit that removes 24,000 Btu/hr. The heat capacity of zone B is 1/2°F per thousand Btu. The time constant for heat transfer between zone A and the outside is 2 hr, between zone B and the outside is 4 hr, and between the two zones is 4 hr. If the outside temperature stays at 100°F, how warm can it eventually get in the attic zone A? (Heating and cooling of buildings was treated in Section 3.3.)

Figure 4.11 Air-conditioned house with attic

34. A building consists of two zones A and B (see Figure 4.12). Only zone A is heated by a furnace, which generates 80,000 Btu/hr. The heat capacity of zone A is 1/4°F per thousand Btu. The time constant for heat transfer between zone A and the outside is 4 hr, between the unheated zone B and the outside is 5 hr, and between the two zones is 2 hr. If the outside temperature stays at 0°F, how cold can it eventually get in the unheated zone B?

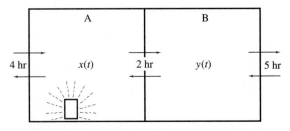

Figure 4.12 Two-zone building with one zone heated

35. In Problem 34, if a small furnace that generates 1000 Btu/hr is placed in zone B, determine the coldest it could eventually get in zone B if zone B has a heat capacity of 2°F per thousand Btu.

36. Arms Race. A mathematical model for an arms race between two countries whose expenditures for defense are expressed by the variables $x(t)$ and $y(t)$ is given by the linear system

$$\frac{dx}{dt} = 2y - x + a, \qquad x(0) = 1,$$

$$\frac{dy}{dt} = 4x - 3y + b, \qquad y(0) = 4,$$

where a and b are constants that measure the trust (or distrust) each country has for the other. Determine whether there is going to be disarmament (x and y approach 0 as t increases), a stabilized arms race (x and y approach a constant as $t \to \infty$), or a runaway arms race (x and y approach ∞ as $t \to \infty$).

4.11 LINEAR DIFFERENCE EQUATIONS

In this section, we give a brief discussion of linear *difference* equations.[†] These equations occur in mathematical models of physical processes and as tools in numerical analysis. For example, a model for population dynamics under immigration involves the equation

$$P_{t+1} - P_t = aP_t + b,$$

where $P_{t+1} - P_t$ is the difference between the population at time $t + 1$ and time t, the constant a is the difference between the average birth rate and average death rate, and b is the rate at which people immigrate to the country. Another example is the equation

$$y_{n+1} - y_{n-1} = -2hy_n,$$

which arises in the study of the stability of a numerical scheme that approximates the solution to a differential equation. Here y_n is the approximation of the solution at x_n and h is the step size.

As these examples suggest, difference equations can be thought of as discrete analogs of differential equations. In fact, the theory of linear difference equations parallels the theory of linear differential equations as discussed earlier in this chapter. Using the results of this chapter as a model both for the statements of the theorems and for their proofs, we can develop a theory and devise methods for solving linear difference equations.

Theory

A ***k*th-order linear difference equation** is an equation of the form

(1) $$a_k(n)y_{n+k} + a_{k-1}(n)y_{n+k-1} + \cdots + a_1(n)y_{n+1} + a_0(n)y_n = g_n,$$

$n = 0, 1, 2, \ldots$, where $a_k(n), \ldots, a_0(n)$, and g_n are defined for all nonnegative integers n. By a **solution** to (1), we mean a sequence of numbers $\{y_n\}_{n=0}^{\infty}$ that satisfies (1) for all integers $n \geq 0$. For example, the sequence $\{2^n\}_{n=0}^{\infty}$ is a solution to the equation

$$y_{n+2} + y_{n+1} - 6y_n = 0,$$

since

$$2^{n+2} + 2^{n+1} - (6)2^n = 2^n(4 + 2 - 6) = 0$$

for $n = 0, 1, 2, \ldots$.

When $a_k(n) \neq 0$ for all integers $n \geq 0$, we can solve equation (1) for y_{n+k}:

(2) $$y_{n+k} = \frac{-1}{a_k(n)} \{a_{k-1}(n)y_{n+k-1} + \cdots + a_0(n)y_n - g_n\}.$$

If we are given the first k values for $\{y_n\}$, that is, $y_0, y_1, \ldots, y_{k-1}$, then we can use equation (2) with $n = 0$ to solve uniquely for y_k. But knowing y_k then allows us to use equation (2)

[†] For a more in-depth discussion of difference equations and their applications, see *Introduction to Difference Equations* by Samuel Goldberg (Dover Publications, Inc., New York, 1986), or *Differential and Difference Equations* by Louis Brand (John Wiley and Sons, Inc., New York, 1966).

with $n = 1$ to determine y_{k+1} uniquely. Since we can continue this process indefinitely, we see that there is a unique solution to (1) that satisfies the given initial conditions. We summarize these observations in the following theorem.

EXISTENCE AND UNIQUENESS OF SOLUTION

Theorem 7. Let $Y_0, Y_1, \ldots, Y_{k-1}$ be given constants and assume that $a_k(n) \neq 0$ for all integers $n \geq 0$. Then there exists a unique solution $\{y_n\}_{n=0}^{\infty}$ to (1) that satisfies the initial conditions $y_0 = Y_0, y_1 = Y_1, \ldots, y_{k-1} = Y_{k-1}$.

When $g_n = 0$ for $n \geq 0$, equation (1) becomes the **homogeneous** (or **reduced**) equation

(3) $$a_k(n)y_{n+k} + \cdots + a_0(n)y_n = 0, \qquad n = 0, 1, 2, \ldots .$$

As is the case for homogeneous differential equations, if two sequences $\{y_n\}_{n=0}^{\infty}$ and $\{z_n\}_{n=0}^{\infty}$ are solutions to the homogeneous equation (3), then so is any linear combination of these sequences (see Problem 28). In this context, by a linear combination of sequences $\{y_n\}_{n=0}^{\infty}$ and $\{z_n\}_{n=0}^{\infty}$, we mean a sequence of the form

$$\{c_1y_n + c_2z_n\}_{n=0}^{\infty},$$

where c_1 and c_2 are constants.

As you would probably guess, if we can find two "different" solutions $\{y_n\}_{n=0}^{\infty}$ and $\{z_n\}_{n=0}^{\infty}$ to a homogeneous linear *second* order equation

(4) $$a_2(n)y_{n+2} + a_1(n)y_{n+1} + a_0(n)y_n = 0, \qquad n = 0, 1, 2, \ldots ,$$

then we can express any solution to (4) as a linear combination of these solutions. To show this (and to decide what we should mean by "different"), let $\{w_n\}_{n=0}^{\infty}$ be a solution to (4). If it is possible to choose constants c_1 and c_2 so that

(5) $$c_1y_0 + c_2z_0 = w_0,$$
$$c_1y_1 + c_2z_1 = w_1,$$

then because $\{w_n\}_{n=0}^{\infty}$ and $\{c_1y_n + c_2z_n\}_{n=0}^{\infty}$ are two solutions to (4) that satisfy the same initial conditions (5), the uniqueness conclusion of Theorem 7 gives

$$w_n = c_1y_n + c_2z_n, \qquad n = 0, 1, 2, \ldots .$$

A necessary and sufficient condition for (5) to have a unique solution for every w_0, w_1 is that

(6) $$\begin{vmatrix} y_0 & z_0 \\ y_1 & z_1 \end{vmatrix} = y_0z_1 - y_1z_0 \neq 0.$$

This is equivalent to saying that the first two terms in the sequences are not multiples of each other, that is, $(z_0, z_1) \neq c(y_0, y_1)$.

We now state the representation theorem for solutions to the homogeneous equation that we have just proved.

REPRESENTATION OF SOLUTIONS (HOMOGENEOUS CASE)

Theorem 8. Assume $a_2(n) \neq 0$ for all integers $n \geq 0$ and let $\{y_n\}_{n=0}^{\infty}$ and $\{z_n\}_{n=0}^{\infty}$ be any two solutions to (4) that satisfy the condition[†]

(7) $y_0 z_1 - y_1 z_0 \neq 0.$

Then, every solution $\{w_n\}_{n=0}^{\infty}$ to (4) can be expressed in the form

(8) $w_n = c_1 y_n + c_2 z_n, \qquad n = 0, 1, 2, \ldots,$

where c_1 and c_2 are constants.

A solution set $\{\{y_n\}_{n=0}^{\infty}, \{z_n\}_{n=0}^{\infty}\}$ that satisfies condition (7) is called a **fundamental solution set** for (4). The linear combination of $\{y_n\}_{n=0}^{\infty}$ and $\{z_n\}_{n=0}^{\infty}$ in (8), written with arbitrary constants c_1 and c_2, is referred to as a **general solution** to (4).

EXAMPLE 1 Given that $\{2^{-n}\}_{n=0}^{\infty}$ and $\{(-3)^n\}_{n=0}^{\infty}$ are solutions to

(9) $2y_{n+2} + 5y_{n+1} - 3y_n = 0, \qquad n = 0, 1, 2, \ldots,$

find the solution to (9) that satisfies $y_0 = 0$ and $y_1 = 1$.

Solution Since $(2^{-0})(-3) - (2^{-1})(-3)^0 = -3 - 1/2 \neq 0$, condition (7) is satisfied. So a general solution to (9) is

(10) $y_n = c_1 2^{-n} + c_2(-3)^n, \qquad n = 0, 1, 2, \ldots,$

where c_1 and c_2 are arbitrary constants. Using the initial conditions to solve for c_1 and c_2 gives the system

$$y_0 = c_1 + c_2 = 0,$$

$$y_1 = (1/2)c_1 - 3c_2 = 1,$$

whose solution is $c_1 = -c_2 = 2/7$. Thus the solution we seek is

(11) $y_n = \left(\frac{2}{7}\right) 2^{-n} - \left(\frac{2}{7}\right)(-3)^n, \qquad n = 0, 1, 2, \ldots.$

The first 12 terms of the solution are displayed in Table 4.2. ■ End

TABLE 4.2 SOME TERMS IN THE SOLUTION TO THE INITIAL VALUE PROBLEM IN EXAMPLE 1

n	y_n	n	y_n	n	y_n
0	0	4	-23.125	8	-1874.57031
1	1	5	69.4375	9	5623.71485
2	-2.5	6	-208.28125	10	-16871.14258
3	7.75	7	624.85938	11	50613.42872

[†] See footnote on page 220.

When $\{g_n\}$ is not the zero sequence, equation (1) is said to be **nonhomogeneous** (or **complete**). Using the superposition principle for linear difference equations (see Problem 29) and the representation theorem for homogeneous equations, we can prove the following representation theorem for nonhomogeneous equations. We leave its proof as an exercise (see Problem 30).

REPRESENTATION OF SOLUTIONS (NONHOMOGENEOUS CASE)

Theorem 9. Assume $a_2(n) \neq 0$ for all integers $n \geq 0$ and let $\{p_n\}_{n=0}^{\infty}$ be a particular solution to the nonhomogeneous equation

(12) $a_2(n)y_{n+2} + a_1(n)y_{n+1} + a_0(n)y_n = g_n, \qquad n = 0, 1, 2, \ldots .$

Then, every solution to (12) can be expressed in the form

(13) $w_n = p_n + c_1 y_n + c_2 z_n, \qquad n = 0, 1, 2, \ldots ,$

where $\{\{y_n\}_{n=0}^{\infty}, \{z_n\}_{n=0}^{\infty}\}$ is a fundamental solution set for the corresponding homogeneous equation and c_1 and c_2 are constants.

The linear combination of $\{p_n\}_{n=0}^{\infty}$, $\{y_n\}_{n=0}^{\infty}$, and $\{z_n\}_{n=0}^{\infty}$ in (13) written with arbitrary constants c_1 and c_2 is referred to as a **general solution** to (12).

Constant Coefficient Equations

When a_0, a_1, and a_2 are constants, the equation

(14) $a_2 y_{n+2} + a_1 y_{n+1} + a_0 y_n = g_n, \qquad n = 0, 1, 2, \ldots$

is said to have **constant coefficients.** To find a general solution to (14), we begin with the homogeneous equation

(15) $a_2 y_{n+2} + a_1 y_{n+1} + a_0 y_n = 0, \qquad n = 0, 1, 2, \ldots .$

We assume both a_0 and a_2 are not zero so that we have a genuine second order difference equation.

Our experience with differential equations suggests we try to find solutions of the form $y_n = r^n$, $n = 0, 1, 2, \ldots$, where $r \neq 0$ is a fixed (real or complex) number. Substituting in (15) gives

$$a_2 r^{n+2} + a_1 r^{n+1} + a_0 r^n = r^n(a_2 r^2 + a_1 r + a_0) = 0.$$

Hence a nontrivial solution to (15) is given by $y_n = r^n$, $n = 0, 1, 2, \ldots$ if and only if r satisfies the **auxiliary equation**

(16) $a_2 r^2 + a_1 r + a_0 = 0.$

[†] Condition (7) is similar to the condition we encountered in Section 4.3 for linear second order differential equations. There we introduced the Wronskian function and discussed its connection with linear independence. The analog of the Wronskian function is the **Casoratian** sequence, which is named for the Italian mathematician Felice Casorati (1835–1890).

Now the form of a fundamental solution set for (15) depends on the type of roots to the auxiliary equation. These results are summarized in the following theorem.

FORM OF GENERAL SOLUTION TO $a_2 y_{n+2} + a_1 y_{n+1} + a_0 y_n = 0$

Theorem 10. Let r_1 and r_2 be the roots of the auxiliary equation (16), where a_2, a_1, a_0 are real constants, $a_2 \neq 0$, $a_0 \neq 0$.

(a) If r_1 and r_2 are distinct real roots, then a general solution to (15) is

$$y_n = c_1 r_1^n + c_2 r_2^n, \qquad n = 0, 1, 2, \ldots .$$

(b) If $r_1 = r_2 = r = -a_1/(2a_2)$ is a repeated root, then a general solution to (15) is

$$y_n = c_1 r^n + c_2 n r^n, \qquad n = 0, 1, 2, \ldots .$$

(c) If r_1 and r_2 are complex conjugate roots and we express r_1 in the polar form $r_1 = \rho e^{i\theta}$, then a general solution to (15) is

$$y_n = c_1 \rho^n \cos n\theta + c_2 \rho^n \sin n\theta, \qquad n = 0, 1, 2, \ldots .$$

Proof. Parts (a) and (b) are analogous to the results for linear second order differential equations with constant coefficients, so we omit their proofs. To prove part (c), we observe that if $r_1 = \rho e^{i\theta}$, then $r_1^n = \rho^n e^{in\theta}$. Using Euler's formula, we find

$$r_1^n = \rho^n \{\cos n\theta + i \sin n\theta\}, \qquad n = 0, 1, 2, \ldots .$$

If $\{r_1^n\}_{n=0}^{\infty}$ is a complex-valued solution, then the two real-valued sequences that make up the real and imaginary parts of $\{r_1^n\}_{n=0}^{\infty}$ must also be solutions. Hence two solutions are

$$\{\rho^n \cos n\theta\}_{n=0}^{\infty} \quad \text{and} \quad \{\rho^n \sin n\theta\}_{n=0}^{\infty}.$$

Here condition (7) is $\rho \sin \theta = \text{Im}(r_1) \neq 0$. But we have assumed that r_1 is complex, and so $\text{Im}(r_1) \neq 0$. Thus these two real-valued sequences form a fundamental solution set. ■ End

EXAMPLE 2 Find a general solution to

$$y_{n+2} + 6y_{n+1} + 9y_n = 0, \qquad n = 0, 1, 2, \ldots .$$

Solution The auxiliary equation is

$$r^2 + 6r + 9 = (r + 3)^2 = 0,$$

which has roots $r_1 = r_2 = -3$. Hence, by Theorem 10, part (b), a general solution is

$$y_n = c_1(-3)^n + c_2 n(-3)^n, \qquad n = 0, 1, 2, \ldots . \quad \blacksquare \text{ End}$$

EXAMPLE 3 Find a general solution to

$$y_{n+2} + 2y_{n+1} + 5y_n = 0, \quad n = 0, 1, 2, \ldots .$$

Solution The auxiliary equation is

$$r^2 + 2r + 5 = (r + 1)^2 + 2^2 = 0,$$

which has roots $r = -1 \pm 2i$. Writing $r_1 = -1 + 2i$ in polar form, we find $\rho = \sqrt{(-1)^2 + 2^2} = \sqrt{5}$ and $\tan \theta = 2/(-1) = -2$. Since θ lies in the second quadrant, we have $\theta = \pi - \arctan 2 \approx 2.03$ radians. Hence, by Theorem 10, part (c), a general solution is

$$y_n = c_1(\sqrt{5})^n \cos n\theta + c_2(\sqrt{5})^n \sin n\theta, \qquad n = 0, 1, 2, \ldots,$$

where $\theta = \pi - \arctan 2$. ∎ End

EXAMPLE 4 To approximate the solution to the initial value problem

(17) $y' = -y, \qquad y(0) = 1,$

a method based on the centered-difference formula, $y'(x) \approx [y(x + h) - y(x - h)]/2h$, leads to the difference equation

(18) $y_{n+1} = y_{n-1} - 2hy_n, \qquad n = 1, 2, 3, \ldots, \qquad y_0 = 1,$

where h (>0) is the step size and y_n is the approximation to y at the point $x_n = nh$, $n = 0$, 1, 2, Discuss the stability of this method; that is, compare the behavior (as $n \to \infty$) of the solutions to (18) with the solution to (17).

Solution The exact solution to problem (17) is just $\phi(x) = e^{-x}$. For fixed h (>0), if $x_n = nh$, then $\phi(x_n) = e^{-x_n} = e^{-nh}$ and $\phi(x_n) \to 0$ as $n \to \infty$.

Equation (18) is a linear second order difference equation with auxiliary equation

$$r^2 + 2hr - 1 = 0,$$

which has roots

$$r_1 = -h + \sqrt{1 + h^2} \quad \text{and} \quad r_2 = -h - \sqrt{1 + h^2}.$$

Hence a general solution to (18) is

$$y_n = c_1 r_1^n + c_2 r_2^n.$$

To solve for c_1 and c_2, we must know y_0 and y_1. The initial condition for the initial value problem (17) provides only $y_0 = 1$. Thus the value for y_1 must be determined by some other numerical method that provides an approximation to the solution of (17) at $x = h$. This, however, is not the problem. To understand the difficulty, let's consider what happens to r_1^n and r_2^n as $n \to \infty$.

For any $h > 0$, it is easy to check that $0 < r_1 < 1$. Hence $r_1^n \to 0$ as $n \to \infty$. This is the same behavior as the exact solution $\phi(x) = e^{-x}$. But, $r_2 = -h - \sqrt{1 + h^2} < -1$, so r_2^n diverges as $n \to \infty$. Consequently, if $c_2 \neq 0$ then $|y_n| \to \infty$.

Thus the centered-difference method for approximating the solution to (17) gives rise to the extraneous "solution" $c_2 r_2^n$, which exhibits larger and larger oscillations. Moreover, since y_1 can only be approximated due to truncation error (or because of round-off), the coefficient c_2, however small, will be nonzero. Hence, at some point the term $c_2 r_2^n$ will dominate the term $c_1 r_1^n$, giving rise to wild oscillations of the form $(-h - \sqrt{1 + h^2})^n$ (see Figure 4.13). Therefore this method is unstable for any choice of the step size h. ∎ End

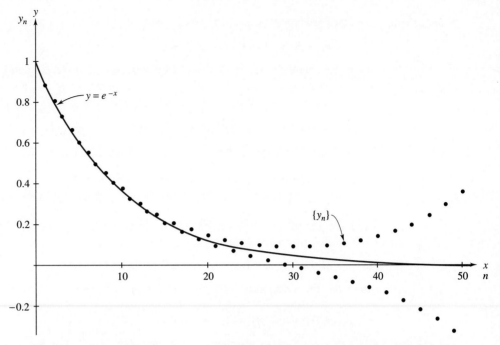

Figure 4.13 Approximation of $y = e^{-x}$ using the solution $\{y_n\}_{n=0}^{\infty}$ to the difference equation (18) with $h = 0.1$ and $y_1 = 0.9$

Undetermined Coefficients

To find a solution to the nonhomogeneous equation (14), we use the analog of the **method of undetermined coefficients** discussed in Section 4.8. We can use Table 4.1 on page 197 with a minor modification. For difference equations, it is more convenient to express an exponential function in the form α^n rather than $e^{\alpha t}$. (Recall $\alpha^n = e^{(\ln \alpha)n}$.) The method of undetermined coefficients is illustrated in the following example.

EXAMPLE 5 Find a general solution to

(19) $y_{n+2} - 2y_{n+1} + 2y_n = 3 \sin n, \qquad n = 0, 1, 2, \ldots .$

Solution We first solve the associated homogeneous equation

$$y_{n+2} - 2y_{n+1} + 2y_n = 0,$$

which has the auxiliary equation $r^2 - 2r + 2 = (r - 1)^2 + 1^2 = 0$. The roots of the auxiliary equation are $r = 1 \pm i$. Writing $1 + i = \rho e^{i\theta}$, we find $\rho = \sqrt{1^2 + 1^2} = \sqrt{2}$ and $\theta = \arctan 1 = \pi/4$. Hence a general solution to the homogeneous equation is

(20) $h_n = c_1(\sqrt{2})^n \cos\left(\dfrac{n\pi}{4}\right) + c_2(\sqrt{2})^n \sin\left(\dfrac{n\pi}{4}\right), \qquad n = 0, 1, 2, \ldots .$

The nonhomogeneous term is $g_n = 3 \sin n$, so we seek a particular solution of the form

(21) $p_n = A \cos n + B \sin n,$

where the constants A and B are to be determined. Substituting this expression for p_n into (19) yields

$$A \cos(n + 2) + B \sin(n + 2) - 2(A \cos(n + 1) + B \sin(n + 1))$$
$$+ 2(A \cos n + B \sin n) = 3 \sin n, \qquad n = 0, 1, 2, \ldots .$$

We use the addition formulas for the trigonometric functions and simplify to obtain

$$[(2 + \cos 2 - 2 \cos 1)A + (\sin 2 - 2 \sin 1)B] \cos n$$
$$+ [(2 \sin 1 - \sin 2)A + (2 + \cos 2 - 2 \cos 1)B] \sin n = 3 \sin n.$$

When we set corresponding coefficients equal, we obtain the system

$$(2 + \cos 2 - 2 \cos 1)A + (\sin 2 - 2 \sin 1)B = 0,$$
$$(2 \sin 1 - \sin 2)A + (2 + \cos 2 - 2 \cos 1)B = 3,$$

which has the solution

(22)
$$A = \frac{-3(\sin 2 - 2 \sin 1)}{(2 + \cos 2 - 2 \cos 1)^2 + (\sin 2 - 2 \sin 1)^2},$$
$$B = \frac{3(2 + \cos 2 - 2 \cos 1)}{(2 + \cos 2 - 2 \cos 1)^2 + (\sin 2 - 2 \sin 1)^2}.$$

Thus a general solution to (26) is

$$y_n = h_n + p_n, \qquad n = 0, 1, 2, \ldots,$$

where h_n is given by equation (20) and p_n is given by equation (21) with A and B given in (22) ■ End

When one of the terms in our initial guess $\{p_n\}_{n=0}^{\infty}$ is a solution to the corresponding homogeneous equation, we must modify our guess by multiplying by n^s, where s is the smallest nonnegative integer such that no term in $\{n^s p_n\}_{n=0}^{\infty}$ is a solution to the corresponding homogeneous equation.

EXERCISES 4.11

In Problems 1 through 10, find a general solution to the given homogeneous equation for $n = 0, 1, 2, \ldots$.

1. $y_{n+2} - 6y_{n+1} + 8y_n = 0.$

2. $y_{n+2} - 4y_{n+1} + 3y_n = 0.$

3. $y_{n+2} + 3y_{n+1} - y_n = 0.$

4. $y_{n+2} + 4y_{n+1} + 4y_n = 0.$

5. $y_{n+2} - 6y_{n+1} + 11y_n = 0.$

6. $y_{n+2} + 4y_{n+1} + 9y_n = 0.$

7. $y_{n+2} + 10y_{n+1} + 25y_n = 0.$

8. $4y_{n+2} + 4y_{n+1} - 3y_n = 0.$

9. $y_{n+2} - y_{n+1} + 4y_n = 0.$

10. $y_{n+2} + y_{n+1} - y_n = 0.$

In Problems 11 through 18, find a general solution to the given nonhomogeneous equation for $n = 0, 1, 2, \ldots$.

11. $y_{n+2} + y_n = n^2$.

12. $y_{n+2} + y_{n+1} - 2y_n = n - 1$.

13. $y_{n+2} - y_n = \cos 3n$.

14. $y_{n+2} - 4y_n = \sin 2n$.

15. $y_{n+2} + 6y_{n+1} + 5y_n = 2^n$.

16. $y_{n+2} + 2y_{n+1} - y_n = (-1)^n$.

17. $y_{n+2} - 4y_{n+1} + 4y_n = 2^n$.

18. $y_{n+2} - y_{n+1} - 2y_n = (-1)^n$.

In Problems 19 through 22, find a general solution to the given third order equation for $n = 0, 1, 2, \ldots$.

19. $y_{n+3} + y_{n+2} - 4y_{n+1} - 4y_n = 0$.

20. $y_{n+3} - 4y_{n+2} + 5y_{n+1} - 2y_n = 0$.

21. $y_{n+3} - 3y_{n+2} + 4y_{n+1} - 2y_n = 0$.

22. $y_{n+3} + 3y_{n+2} + 3y_{n+1} + y_n = 0$.

In Problems 23 through 27, find the solution to the initial value problem for $n = 0, 1, 2, \ldots$.

23. $y_{n+2} + 2y_{n+1} + 5y_n = 0$; $\quad y_0 = 1, \quad y_1 = -1$.

24. $y_{n+2} - 5y_{n+1} + 6y_n = 0$; $\quad y_0 = 3, \quad y_1 = -2$.

25. $y_{n+2} - 6y_{n+1} + 9y_n = 1$; $\quad y_0 = 0, \quad y_1 = 0$.

26. $y_{n+2} - 4y_{n+1} + 3y_n = 3^n$; $\quad y_0 = 1, \quad y_1 = 0$.

27. $y_{n+2} - 2y_{n+1} + y_n = n$; $\quad y_0 = 1, \quad y_1 = 3$.

28. Prove that if the two sequences $\{y_n\}_{n=0}^\infty$ and $\{z_n\}_{n=0}^\infty$ are solutions to the homogeneous equation (3), then so is any linear combination $\{c_1 y_n + c_2 z_n\}_{n=0}^\infty$.

29. Superposition Principle. Let $\{y_n\}_{n=0}^\infty$ be a solution to the nonhomogeneous equation (1) with nonhomogeneous term g_n and $\{z_n\}_{n=0}^\infty$ be a solution to (1) with non-homogeneous term h_n. Show that the sequence $\{c_1 y_n + c_2 z_n\}_{n=0}^\infty$ is a solution to (1) with nonhomogeneous term $c_1 g_n + c_2 h_n$.

30. Using the representation theorem for homogeneous equations and the superposition principle given in Problem 29, prove Theorem 9, the representation theorem for nonhomogeneous equations.

31. The linear multistep numerical method

$$y_{n+2} - y_{n+1} = \frac{h}{2}(3f_{n+1} - f_n), \qquad h > 0,$$

is used to approximate the solution to

$$y' = f(x, y), \qquad y(0) = y_0,$$

where $x_n := nh$, $f_n := f(x_n, y_n)$. As in Example 4, discuss the stability (as $n \to \infty$) of this method when it is applied to

$$y' = -y, \qquad y(0) = 1.$$

32. National Income Model. P. A. Samuelson[†] developed a model for national income Y_t in period t under fixed governmental spending. His model involved the difference equation

$$Y_{t+2} - \alpha(1 + \beta)Y_{t+1} + \alpha\beta Y_t = 1, \quad t = 0, 1, 2, \ldots,$$

where α is a constant called the *marginal propensity to consume* and β is a constant of proportionality called the *relation*, which is a consequence of the so-called acceleration principle of economics.

(a) Solve for the national income in the special case when $\alpha = 1/2$, $\beta = 1$, $Y_0 = 2$, and $Y_1 = 3$.

(b) Show that as $t \to \infty$, the national income Y_t experiences damped oscillations that converge to $1/(1 - \alpha) = 2$. (Samuelson concluded that a fixed level of governmental spending will result in damped oscillations in the national income that will gradually approach the quantity $1/(1 - \alpha)$ times the level of governmental spending.)

CHAPTER SUMMARY

In this chapter, we discussed the theory of linear second order equations and presented methods for solving these equations. We also introduced first order systems of differential equations as well as linear difference equations. The important features and techniques are listed below.

[†] P. A. Samuelson, "Interactions Between the Multiplier Analysis and the Principle of Acceleration," *Review of Economic Statistics*, 21 (1939), 75–78.

Homogeneous Linear Equations

$$y'' + p(x)y' + q(x)y = 0$$

Fundamental Solution Set: $\{y_1, y_2\}$

Two solutions y_1 and y_2 to the homogeneous equation on the interval I form a fundamental solution set, provided their Wronskian,

$$W[y_1, y_2](x) := y_1(x)y_2'(x) - y_1'(x)y_2(x),$$

is different from zero for some (and hence all) x in I. If y_1 and y_2 are linearly independent solutions on I, then $W[y_1, y_2](x) \neq 0$ on I and y_1 and y_2 form a fundamental solution set.

General Solution to Homogeneous Equation: $c_1y_1 + c_2y_2$

If y_1 and y_2 are linearly independent solutions to the homogeneous equation, then a general solution is

$$y(x) = c_1y_1(x) + c_2y_2(x),$$

where c_1 and c_2 are arbitrary constants.

Homogeneous Equation with Constant Coefficients: $ay'' + by' + cy = 0$

The form of a general solution for a homogeneous equation with constant coefficients depends on the roots

$$r_1 = \frac{-b + \sqrt{b^2 - 4ac}}{2a}, \qquad r_2 = \frac{-b - \sqrt{b^2 - 4ac}}{2a}$$

of the auxiliary equation

$$ar^2 + br + c = 0, \qquad a \neq 0.$$

(a) When $b^2 - 4ac > 0$, the auxiliary equation has two distinct real roots r_1 and r_2 and a general solution is

$$y(x) = c_1e^{r_1x} + c_2e^{r_2x}.$$

(b) When $b^2 - 4ac = 0$, the auxiliary equation has a repeated real root $r = r_1 = r_2$ and a general solution is

$$y(x) = c_1e^{rx} + c_2xe^{rx}.$$

(c) When $b^2 - 4ac < 0$, the auxiliary equation has complex conjugate roots $r = \alpha \pm i\beta$ and a general solution is

$$y(x) = c_1e^{\alpha x}\cos \beta x + c_2e^{\alpha x}\sin \beta x.$$

Reduction of Order Formula: $y(x) = v(x)f(x)$

Let $f(x)$ be a nontrivial solution to the homogeneous equation. The substitution $y(x) = v(x)f(x)$ reduces the homogeneous equation to a first order equation in v'. When solved for v' and integrated to obtain v, a second linearly independent solution $y(x)$ is given

by the reduction of order formula

$$y(x) = f(x) \int \frac{e^{-\int p(x)\,dx}}{[f(x)]^2}\,dx.$$

Nonhomogeneous Linear Equations

$$y'' + p(x)y' + q(x)y = g(x)$$

General Solution to Nonhomogeneous Equation: $y_p + c_1 y_1 + c_2 y_2$

If y_p is any particular solution to the nonhomogeneous equation and y_1 and y_2 are linearly independent solutions to the corresponding homogeneous equation, then a general solution is

$$y(x) = y_p(x) + c_1 y_1(x) + c_2 y_2(x),$$

where c_1 and c_2 are arbitrary constants.

Two methods for finding a particular solution y_p are those of undetermined coefficients and variation of parameters.

Undetermined Coefficients: $g(x) = e^{\alpha x} p_n(x) \begin{Bmatrix} \cos \beta x \\ \sin \beta x \end{Bmatrix}$

If the right-hand side $g(x)$ of a nonhomogeneous equation with constant coefficients is a polynomial, an exponential of the form $e^{\alpha x}$, a trigonometric function of the form $\cos \beta x$ or $\sin \beta x$, or any product of these special types of functions, then a particular solution of an appropriate form exists. This special form is given in Table 4.1.

Variation of Parameters: $y(x) = v_1(x)y_1(x) + v_2(x)y_2(x)$

If y_1 and y_2 are two linearly independent solutions to the corresponding homogeneous equation, then a particular solution to the nonhomogeneous equation is

$$y(x) = v_1(x)y_1(x) + v_2(x)y_2(x),$$

where

$$v_1(x) = \int \frac{-g(x)y_2(x)}{W[y_1, y_2](x)}\,dx, \qquad v_2(x) = \int \frac{g(x)y_1(x)}{W[y_1, y_2](x)}\,dx.$$

Superposition Principle: $L[y] = c_1 g_1 + c_2 g_2$

If y_1 and y_2 are solutions to the equations

$$y'' + py' + qy = g_1 \quad \text{and} \quad y'' + py' + qy = g_2,$$

respectively, then $c_1 y_1 + c_2 y_2$ is a solution to the equation

$$y'' + py' + qy = c_1 g_1 + c_2 g_2.$$

Cauchy-Euler Equations: $ax^2 \dfrac{d^2y}{dx^2} + bx \dfrac{dy}{dx} + cy = h(x), \qquad x > 0$

Let $x = e^t$. This substitution transforms the original equation into the following constant coefficient equation in t:

$$a \frac{d^2y}{dt^2} + (b - a)\frac{dy}{dt} + cy = h(e^t).$$

First Order Systems

In certain instances, it is helpful to express a higher-order differential equation as a system of first order equations in **normal form.** To reduce the mth order equation $y^{(m)}(t) = f(t, y, y', \ldots, y^{(m-1)})$ to a normal system, set $x_1 = y$, $x_2 = y'$, $\ldots$, and $x_m = y^{(m-1)}$. This leads to

$$x_1' = x_2, \quad x_2' = x_3, \ldots, \quad x_{m-1}' = x_m, \quad x_m' = f(t, x_1, \ldots, x_m).$$

Elimination Method for Systems with Constant Coefficients: For this method, we begin by expressing the system in operator form. For example, a linear system of two equations and two unknowns has the form

(1) $L_1[x] + L_2[y] = f_1,$

(2) $L_3[x] + L_4[y] = f_2.$

If we formally eliminate y from the above system, we obtain the single equation

(3) $(L_1L_4 - L_2L_3)[x] = g_1,$

where $g_1 := L_4[f_1] - L_2[f_2]$. We can solve this linear equation for x by using the techniques described above. This solution will involve n arbitrary constants, where n is the order of the operator $L_1L_4 - L_2L_3$. We then solve for y in terms of these same constants.

Difference Equations

A linear second order difference equation is often a discretized version of a differential equation and has the form

$$a_2(n)y_{n+2} + a_1(n)y_{n+1} + a_0(n)y_n = g_n, \quad n = 0, 1, 2, 3 \ldots.$$

Solution sequences $\{y_n\}_{n=0}^{\infty}$ to this equation can be found using analogs of the techniques for differential equations. For example, if a_0, a_1, and a_2 are constants (independent of n), then the homogeneous difference equation

$$a_2 y_{n+2} + a_1 y_{n+1} + a_0 y_n = 0$$

has a solution of the form $y_n = r^n$, $n = 0, 1, 2, \ldots$, where r satisfies the auxiliary equation

$$a_2 r^2 + a_1 r + a_0 = 0.$$

A general solution to a nonhomogeneous linear difference equation is the sum of a particular solution and a general solution to the corresponding homogeneous equation.

REVIEW PROBLEMS

In Problems 1 through 28, find a general solution to the given differential equation.

1. $y'' + 8y' - 9y = 0.$ **2.** $49y'' + 14y' + y = 0.$

3. $4y'' - 4y' + 10y = 0.$ **4.** $9y'' - 30y' + 25y = 0.$

5. $6y'' - 11y' + 3y = 0.$ **6.** $y'' + 8y' - 14y = 0.$

7. $36y'' + 24y' + 5y = 0.$ **8.** $25y'' + 20y' + 4y = 0.$

9. $16z'' - 56z' + 49z = 0.$ **10.** $u'' + 11u = 0.$

11. $t^2 x''(t) + 5x(t) = 0,$ $t > 0.$

12. $2y''' - 3y'' - 12y' + 20y = 0.$

13. $y'' + 16y = xe^x.$

14. $v'' - 4v' + 7v = 0.$

15. $3y''' + 10y'' + 9y' + 2y = 0.$

16. $y''' + 3y'' + 5y' + 3y = 0.$

17. $y''' + 10y' - 11y = 0.$

18. $h^{(4)} = 120x.$

19. $4y''' + 8y'' - 11y' + 3y = 0.$

20. $2y'' - y = x \sin x.$

21. $y'' - 3y' + 7y = 7x^2 - e^x.$

22. $y'' - 8y' - 33y = 546 \sin x.$

23. $y'' + 16y = \tan 4\theta.$

24. $10y'' + y' - 3y = x - e^{x/2}.$

25. $4y'' - 12y' + 9y = e^{5x} + e^{3x}.$

26. $y'' + 6y' + 15y = e^{2x} + 75.$

27. $x^2 y'' + 2xy' - 2y = 6x^{-2} + 3x,$ $x > 0.$

28. $y'' = 5x^{-1}y' - 13x^{-2}y,$ $x > 0.$

In Problems 29 through 36, find the solution to the initial value problem.

29. $y'' + 4y' + 7y = 0;$ $y(0) = 1,$ $y'(0) = -2.$

30. $y'' + 2y' + y = 2 \cos \theta;$ $y(0) = 3,$ $y'(0) = 0.$

31. $y'' - 2y' + 10y = 6 \cos 3x - \sin 3x;$
$y(0) = 2,$ $y'(0) = -8.$

32. $4y'' - 4y' + 5y = 0;$ $y(0) = 1,$ $y'(0) = -11/2.$

33. $y''' - 12y'' + 27y' + 40y = 0;$
$y(0) = -3,$ $y'(0) = -6,$ $y''(0) = -12.$

34. $y'' + 5y' - 14y = 0;$ $y(0) = 5,$ $y'(0) = 1.$

35. $y'' + y = \sec \theta;$ $y(0) = 1,$ $y'(0) = 2.$

36. $9y'' + 12y' + 4y = 0;$ $y(0) = -3,$ $y'(0) = 3.$

In Problems 37 through 39, a differential equation and a nontrivial solution $f(x)$ are given. Find a general solution for the equation.

37. $xy'' + (x - 1)y' - y = 0;$ $x > 0;$ $f(x) = e^x.$

38. $(x - 1)y'' - xy' + y = 0,$ $x > 1;$ $f(x) = e^x.$

39. $(1 - x^2)w'' - 2xw' + 2w = 0,$ $-1 < x < 1;$
$f(x) = x.$

40. Determine whether the statement made is *always true* or *sometimes false*.

 (a) Two differentiable functions are linearly independent on $[a, b]$ if and only if their Wronskian is not zero on $[a, b]$.

 (b) The differential operator

$$L[y](x) := 2y''(x) + x^2 y'(x) + \frac{1}{x} y(x)$$

 defined for twice differentiable functions on $(0, \infty)$ is linear.

 (c) The initial value problem

$$xy'' + \frac{5}{x - 3} y' + 2y = \cos x;$$

$$y(1) = 6, \quad y'(1) = 23$$

 has a unique solution on the interval $(0, 3)$.

 (d) A general solution of $y'' + y = x$ is given by

$$y = c_1 \cos x + c_2 \sin x + c_3 x,$$

 where c_1, c_2, and c_3 are arbitrary constants.

In Problems 41 through 46, use the elimination method to find a general solution for the given linear system, where x', y', and D denote differentiation with respect to t.

41. $x' = y,$
$y' = 3x.$

42. $x' + y' = 2y,$
$y' = x - y.$

43. $(D + 2)[x] + (D - 1)[y] = t.$
$(D - 1)[x] + (D + 1)[y] = 0.$

44. $(D - 3)[x] + D[y] = 0,$
$(D + 1)[x] - y = 0.$

45. $x'' + y' + y = 0,$
$x' - x + y = 0.$

46. $x'' - x + 5y = 2,$
$2x + y'' + 2y = 0.$

TECHNICAL WRITING EXERCISES

1. Compare the two methods—undetermined coefficients and variation of parameters—for determining a particular solution to a nonhomogeneous equation. What are the advantages and disadvantages of each?

2. Consider the differential equation

$$\frac{d^2y}{dx^2} + 2b\frac{dy}{dx} + y = 0,$$

where b is a constant. Describe how the behavior of solutions to this equation changes as b varies.

3. Consider the differential equation

$$\frac{d^2y}{dx^2} + cy = 0,$$

where c is a constant. Describe how the behavior of solutions to this equation changes as c varies.

4. For students with a background in linear algebra: Compare the theory for linear second order equations with that for systems of n linear equations in n unknowns whose coefficient matrix has rank $n - 2$. Use the terminology from linear algebra; for example, subspace, basis, dimension, linear transformation, and kernel. Discuss both homogeneous and nonhomogeneous equations.

GROUP PROJECTS FOR CHAPTER 4

A. Convolution Method

The **convolution** of two functions g and f is the function $g * f$ defined by

$$(g * f)(x) := \int_0^x g(x - v)f(v)\,dv.$$

The aim of this project is to show how convolutions can be used to obtain a particular solution to a nonhomogeneous equation of the form

$$(1) \qquad ay'' + by' + cy = f(x),$$

where a, b, and c are constants, $a \neq 0$.

(a) Use Leibniz's rule,

$$\frac{d}{dx}\int_a^x h(x, v)\,dv = \int_a^x \frac{\partial h}{\partial x}(x, v)\,dv + h(x, x),$$

to show that

$$(y * f)'(x) = (y' * f)(x) + y(0)f(x)$$

and

$$(y * f)''(x) = (y'' * f)(x) + y'(0)f(x) + y(0)f'(x),$$

assuming y and f are sufficiently differentiable.

(b) Let $y_s(x)$ be the solution to the homogeneous equation $ay'' + by' + cy = 0$ that satisfies $y_s(0) = 0$, $y_s'(0) = 1/a$. Show that $y_s * f$ is the particular solution to equation (1) satisfying $y(0) = y'(0) = 0$.

(c) Let $y_k(x)$ be the solution to the homogeneous equation $ay'' + by' + cy = 0$ that satisfies $y(0) = y_0$, $y'(0) = y_1$, and let y_s be as defined in part (b). Show that

$$(y_s * f)(x) + y_k(x)$$

is the unique solution to the initial value problem

$$(2) \qquad ay'' + by' + cy = f(x); \qquad y(0) = y_0, \qquad y'(0) = y_1.$$

(d) Use the result of part (c) to determine the solution to each of the following initial value problems. Carry out all integrations and express your answers in terms of elementary functions.
 (i) $y'' + y = \tan x;$ $\qquad y(0) = 1,$ $\qquad y'(0) = -1.$
 (ii) $2y'' + y' - y = e^{-x}\sin x;$ $\qquad y(0) = 1,$ $\qquad y'(0) = 1.$
 (iii) $y'' - 2y' + y = \sqrt{x}e^x;$ $\qquad y(0) = 2,$ $\qquad y'(0) = 0.$

B. *Asymptotic Behavior of Solutions*

In the application of linear systems theory to mechanical and electrical problems, one often encounters the equation

(3) $y'' + py' + qy = f(x),$

where p and q are positive constants with $p^2 < 4q$ and $f(x)$ is a forcing function for the system. In many cases, it is important for the design engineer to know that a bounded forcing function gives rise only to bounded solutions. More specifically, how does the behavior of $f(x)$ for large values of x affect the asymptotic behavior of the solution? To answer this question, do the following:

(a) Show that the homogeneous equation associated with equation (3) has two linearly independent solutions given by

$$e^{\alpha x} \cos \beta x, \qquad e^{\alpha x} \sin \beta x,$$

where $\alpha = -p/2 < 0$ and $\beta = \,.$

(b) Let $f(x)$ be a continuous function defined on the interval $[0, \infty)$. Use the variation of parameters formula to show that any solution to (3) on $[0, \infty)$ can be expressed in the form

(4) $y(x) = c_1 e^{\alpha x} \cos \beta x + c_2 e^{\alpha x} \sin \beta x$

$$- \frac{1}{\beta} e^{\alpha x} \cos \beta x \int_0^x f(t) e^{-\alpha t} \sin \beta t \, dt$$

$$+ \frac{1}{\beta} e^{\alpha x} \sin \beta x \int_0^x f(t) e^{-\alpha t} \cos \beta t \, dt.$$

(c) Assuming f is bounded on $[0, \infty)$, that is, there exists a constant K such that $|f(t)| \le K$ for all $t \ge 0$, use the triangle inequality and other properties of the absolute value to show that $y(x)$ given in (4) satisfies

$$|y(x)| \le \{|c_1| + |c_2|\} e^{\alpha x} + \frac{2K}{|\alpha|\beta} (1 - e^{\alpha x})$$

for all $x > 0$.

(d) In a similar fashion, show that if $f_1(x)$ and $f_2(x)$ are two bounded continuous functions on $[0, \infty)$ such that $|f_1(x) - f_2(x)| \le \varepsilon$ for all $x > x_0$, and if ϕ_1 is a solution to (3) with $f = f_1$ and ϕ_2 is a solution to (3) with $f = f_2$, then

$$|\phi_1(x) - \phi_2(x)| \le M e^{\alpha x} + \frac{2\varepsilon}{|\alpha|\beta} (1 - e^{\alpha(x-x_0)})$$

for all $x > x_0$, where M is a constant that depends on ϕ_1 and ϕ_2 but not on x.

(e) Now assume $f(x) \to F_0$ as $x \to \infty$, where F_0 is a constant. Use the result of part (d) to prove that any solution ϕ to (3) must satisfy $\phi(x) \to F_0/q$ as $x \to \infty$. [*Hint:* Choose $f_1 = f, f_2 \equiv F_0, \phi_1 = \phi, \phi_2 \equiv F_0/q$.]

C. *Linearization of Nonlinear Problems*

A useful approach to analyzing a nonlinear equation is to study its **linearized equation,** which is obtained by replacing the nonlinear terms by linear approximations. For example, the nonlinear

equation

$$(5) \qquad \frac{d^2\theta}{dt^2} + \sin\theta = 0,$$

which governs the motion of a simple pendulum, has

$$(6) \qquad \frac{d^2\theta}{dt^2} + \theta = 0$$

as a linearization for small θ. (The nonlinear term $\sin\theta$ has been replaced by the linear approximation θ.)

A general solution to equation (5) involves Jacobi elliptic functions (see Project E), which are rather complicated, so let's try to approximate the solutions. For this purpose, we consider two methods: Taylor series and linearization.

(a) Derive the first six terms of the Taylor series about $t = 0$ of the solution to equation (5) with initial conditions $\theta(0) = \pi/12$, $\theta'(0) = 0$. (The Taylor series method is discussed in Project A of Chapter 1 and Section 8.2.)

(b) Solve equation (6) subject to the same initial conditions $\theta(0) = \pi/12$, $\theta'(0) = 0$.

(c) On the same coordinate axes, graph the two approximations found in parts (a) and (b).

(d) Discuss the advantages and disadvantages of the Taylor series method and the linearization method.

(e) Give a linearization for the initial value problem

$$x''(t) + 0.1[1 - x^2(t)]x'(t) + x(t) = 0; \qquad x(0) = 0.4, \qquad x'(0) = 0$$

for x small. Solve this linearized problem to obtain an approximation for the nonlinear problem.

D. Nonlinear Equations Solvable by First Order Techniques

Certain *nonlinear* second order equations, namely those with dependent or independent variables missing, can be solved by reducing them to a pair of *first order equations*. This is accomplished by making the substitution $w = dy/dx$.

(a) To solve an equation of the form $y'' = F(x, y')$ in which the dependent variable y is missing, setting $w = y'$ (so that $w' = y''$) yields the pair of equations

$$w' = F(x, w),$$

$$y' = w.$$

Because $w' = F(x, w)$ is a first order equation, we have available the techniques of Chapter 2 to solve it for $w(x)$. Once $w(x)$ is determined, we integrate it to obtain $y(x)$.

Using this method, solve

$$2xy'' - y' + \frac{1}{y'} = 0, \qquad x > 0.$$

(b) To solve an equation of the form $y'' = F(y, y')$ in which the independent variable x is missing, setting $w = dy/dx$ yields, via the chain rule,

$$\frac{d^2y}{dx^2} = \frac{dw}{dx} = \frac{dw}{dy}\frac{dy}{dx} = w\frac{dw}{dy}.$$

Thus $y'' = F(y, y')$ is equivalent to the pair of equations

(7) $\qquad w \dfrac{dw}{dy} = F(y, w),$

(8) $\qquad \dfrac{dy}{dx} = w.$

In equation (7), notice that y plays the role of the *independent* variable; hence solving it yields $w(y)$. Then substituting $w(y)$ into (8), we obtain a separable equation that determines $y(x)$.

Using this method, solve the following equations.

(i) $\quad 2y\dfrac{d^2y}{dx^2} = 1 + \left(\dfrac{dy}{dx}\right)^2.$

(ii) $\quad \dfrac{d^2y}{dx^2} + y\dfrac{dy}{dx} = 0.$

(c) **Suspended Cable.** In the study of a cable suspended between two fixed points (see Figure 4.14), one encounters the initial value problem

$$\dfrac{d^2y}{dx^2} = \dfrac{1}{a}\sqrt{1 + \left(\dfrac{dy}{dx}\right)^2}; \qquad y(0) = a, \qquad y'(0) = 0,$$

where $a\ (\neq 0)$ is a constant. Solve this initial value problem for y. The resulting curve is called a **catenary**.

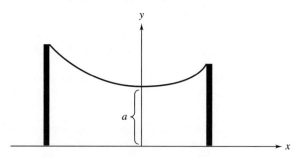

Figure 4.14 Suspended cable

(d) **Newton Equation.** A mechanical system with one degree of freedom is a conservative system if the total energy remains constant during motion. Such a system can always be expressed (using generalized coordinates) in the form known as **Newton's equation,**

$$\dfrac{d^2y}{dx^2} = f(y),$$

where $f(y)$ is a continuous function for all y. Derive the following implicit solution to the Newton equation:

$$\int \left(2\int f(y)\,dy + c_1\right)^{-1/2} dy = x + c_2.$$

E. Simple Pendulum

In Section 4.1, we discussed the simple pendulum consisting of a mass m suspended by a rod of length l having negligible mass and derived the nonlinear initial value problem

(9) $$\frac{d^2\theta}{dt^2} + \frac{g}{l}\sin\theta = 0; \qquad \theta(0) = \alpha, \qquad \theta'(0) = 0,$$

where g is the acceleration due to gravity and $\theta(t)$ is the angle the rod makes with the vertical at time t (see Figure 4.1). Here it is assumed that the mass is released with zero velocity at an initial angle α, $0 < \alpha < \pi$. We would like to determine the equation of motion for the pendulum and its period of oscillation.

(a) Use equation (9) and the technique discussed in Project D, part (b), to show that

$$\left(\frac{d\theta}{dt}\right)^2 = \frac{2g}{l}(\cos\theta - \cos\alpha)$$

and hence

$$dt = -\sqrt{\frac{l}{2g}}\frac{d\theta}{\sqrt{\cos\theta - \cos\alpha}}.$$

(b) Use the trigonometric identity $\cos x = 1 - 2\sin^2(x/2)$ to express dt by

$$dt = -\frac{1}{2}\sqrt{\frac{l}{g}}\frac{d\theta}{\sqrt{\sin^2(\alpha/2) - \sin^2(\theta/2)}}.$$

(c) Make the change of variables $\sin(\theta/2) = \sin(\alpha/2)\sin\phi$ to rewrite dt in the form

(10) $$dt = -\sqrt{\frac{l}{g}}\frac{d\phi}{\sqrt{1 - k^2\sin^2\phi}}, \qquad \text{where } k := \sin(\alpha/2).$$

(d) The period $P(\alpha)$ of the pendulum is defined to be the time required for the pendulum to swing from one extreme to the other and back, that is, from α to $-\alpha$ back to α. Show that the period of oscillation is given by

(11) $$P(\alpha) = 4\sqrt{\frac{l}{g}}\int_0^{\pi/2}\frac{d\phi}{\sqrt{1 - k^2\sin^2\phi}},$$

where $k := \sin(\alpha/2)$. [*Hint:* The period is just four times the time it takes the pendulum to go from $\theta = 0$ to $\theta = \alpha$.]

The integral in (11) is called an **elliptic integral of the first kind** and is denoted by $F(k, \pi/2)$. As you might expect, the period of the simple pendulum depends on the length l of the rod and the initial displacement α. In fact, a check of an elliptic integral

table will show that the period nearly doubles as the initial displacement increases from $\pi/8$ to $15\pi/16$ (for fixed l). What happens as α approaches π?

(e) To determine the equation of motion of the pendulum, we observe that as t varies from 0 to t, then θ varies from α to $\theta(t)$ and ϕ varies from $\pi/2$ to $\phi(t)$. Integrating equation (10), show that

$$(12) \qquad t + \frac{P(\alpha)}{4} = \sqrt{\frac{l}{g}} \int_0^\phi \frac{ds}{\sqrt{1 - k^2 \sin^2 s}} = \sqrt{\frac{l}{g}} F(k, \phi).$$

For fixed k, $F(k, \phi)$ has an "inverse," denoted by $\operatorname{sn}(k, u)$, that satisfies $u = F(k, \phi)$ if and only if $\operatorname{sn}(k, u) = \sin \phi$. The function $\operatorname{sn}(k, u)$ is called a **Jacobi elliptic function** and has many properties that resemble those of the sine function. Using the Jacobi elliptic function $\operatorname{sn}(k, u)$ express the equation of motion for the pendulum in the form

$$(13) \qquad \theta = 2 \arcsin\left\{ k \, \operatorname{sn}\left(k, \sqrt{\frac{g}{l}}\left(t + \frac{P(\alpha)}{4} \right) \right) \right\}, \quad 0 < t \le P(\alpha)/4,$$

where $k := \sin(\alpha/2)$.

F. Phase Plane Diagrams and Periodic Solutions

In Project E, a procedure is given for computing the period of a simple pendulum in terms of elliptic integrals. That computation was based on the premise that the simple pendulum problem *has* a periodic solution.

The differential equation that governs the motion of a simple pendulum is

$$(14) \qquad \frac{d^2\theta}{dt^2} + \frac{g}{l} \sin \theta = 0,$$

where θ is the angular displacement from the vertical, l is the length of the pendulum, and g is the acceleration due to gravity (see Figure 4.1 on page 150). To simplify the computations, let's assume $l = g$ (in some units) so that equation (14) becomes

$$(15) \qquad \frac{d^2\theta}{dt^2} + \sin \theta = 0.$$

To show that equation (15) has a periodic solution, proceed as follows:

(a) Show that solving (15) is equivalent to solving the system

$$(16) \qquad \begin{aligned} \frac{d\theta}{dt} &= v, \\ \frac{dv}{dt} &= -\sin \theta. \end{aligned}$$

(b) The **phase plane** for system (16) is the plane of the variables $v = d\theta/dt$ and θ. A solution to system (16) traces out a **path** in the θv-phase plane. Show that this path is a

solution to the differential equation

(17) $\dfrac{dv}{d\theta} = -\dfrac{\sin\theta}{v}.$

(c) Solve equation (17).

(d) Show that there are closed paths in the phase plane associated with the simple pendulum equation and hence that the simple pendulum problem has periodic solutions (see Figure 4.15).

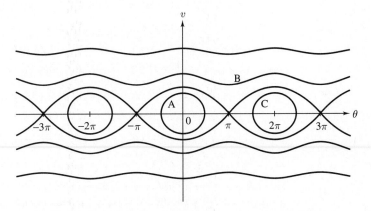

Figure 4.15 Phase plane trajectories for a simple pendulum

(e) Give a physical interpretation of the motion of a pendulum that corresponds to the three trajectories labeled A, B, and C in Figure 4.15.

(f) The procedure outlined in parts (a)–(d) can be used to show the existence of periodic solutions for other nonlinear equations. In particular, show that the **nonlinear spring equation**

$$\frac{d^2x}{dt^2} + x + \beta x^3 = 0,$$

where $\beta > 0$ is a fixed parameter, has a nontrivial periodic solution.

G. *Critical Damping of a Vibrating Spring*

In Example 2 of Section 4.6, the motion of a damped vibrating spring was analyzed. Use a computer to graph the solution (12) on page 185 to the problem, with the data as given. Now use the computer graphics to study the effect of increasing the damping factor b from 12 kg/sec to 80 kg/sec. Note that the qualitative nature of the solution has changed from an oscillatory motion to an exponential decay of the solution to zero.

Experiment with intermediate values of b until you find where this "changeover" occurs. How is this phenomenon related to the auxiliary equation?

H. Undetermined Coefficients Using Complex Arithmetic

The technique of undetermined coefficients described in Section 4.8 can be streamlined if the reader is familiar with complex arithmetic and the properties of the complex exponential function. The essential formulas are

$$e^{(\alpha+i\beta)x} = e^{\alpha x}(\cos \beta x + i \sin \beta x),$$

$$\frac{d}{dx} e^{(\alpha+i\beta)x} = (\alpha + i\beta)e^{(\alpha+i\beta)x},$$

$$\text{Re } e^{(\alpha+i\beta)x} = e^{\alpha x} \cos \beta x, \qquad \text{Im } e^{(\alpha+i\beta)x} = e^{\alpha x} \sin \beta x,$$

and, more generally,

(18) $\text{Re}\{(a + ib)e^{(\alpha+i\beta)x}\} = e^{\alpha x}(a \cos \beta x - b \sin \beta x),$

(19) $\text{Im}\{(a + ib)e^{(\alpha+i\beta)x}\} = e^{\alpha x}(b \cos \beta x + a \sin \beta x).$

Now consider a second order equation of the form

(20) $L[y] := ay'' + by' + cy = g,$

where a, b, and c are real numbers and g is of the special form

(21) $g(x) = e^{\alpha x}\{(a_n x^n + \cdots + a_1 x + a_0)\cos \beta x + (b_n x^n + \cdots + b_1 x + b_0)\sin \beta x\},$

with the a_j's, b_j's, α, and β real numbers. Such a function can always be expressed as the real or imaginary part of a function involving the complex exponential. For example, using equation (18), one can quickly check that

(22) $g(x) = \text{Re}\{G(x)\},$

where

(23) $G(x) = e^{(\alpha-i\beta)x}[(a_n + ib_n)x^n + \cdots + (a_1 + ib_1)x + (a_0 + ib_0)].$

Now suppose for the moment that we can find a complex-valued solution Y to the equation

(24) $L[Y] = aY'' + bY' + cY = G.$

Then, since a, b, and c are real numbers, we get a real-valued solution y to (20) by simply taking the real part of Y; that is, $y = \text{Re } Y$ solves (20). (Recall that in Lemma 2, page 184, we proved this fact for homogeneous equations!) Thus we need focus only on finding a solution to (24).

The method of undetermined coefficients asserts that any differential equation of the form

(25) $L[Y] = e^{(\alpha\pm i\beta)x}[(a_n + ib_n)x^n + \cdots + (a_1 + ib_1)x + (a_0 + ib_0)]$

has a solution of the form

(26) $Y_p(x) = x^s e^{(\alpha\pm i\beta)x}[A_n x^n + \cdots + A_1 x + A_0],$

where $A_n, \ldots, A_0$ are complex constants and s is the smallest nonnegative integer such that no term in (26) is a (complex) solution to the corresponding homogeneous equation $L[Y] = 0$. We can solve for the unknown constants A_j by substituting (26) into (25) and equating coefficients of like terms. With these facts in mind, we can (for the small price of using complex arithmetic) dispense with Table 4.1 on page 197 and avoid the unpleasant task of computing derivatives of a function like $e^{3x}(2 + 3x + x^2)\sin(2x)$, which involves both exponential and trigonometric factors.

Carry out this procedure to determine particular solutions to the following equations:

(a) $y'' + y = e^{-x}[\cos 2x - 3 \sin 2x]$.

(b) $y'' - 2y' + 10y = xe^x \sin 3x$.

The use of complex arithmetic not only streamlines the computations but proves very useful in analyzing the response of a linear system to a sinusoidal input. Electrical engineers make good use of this in their study of RLC circuits by introducing the concept of **impedance.** This will be illustrated in Chapter 5.

Chapter 5

Applications and Numerical Methods for Second Order Equations and Systems

5.1 MECHANICAL VIBRATIONS AND SIMPLE HARMONIC MOTION

Each day, we encounter many types of mechanical vibrations. Some common examples are the bouncing motion of an automobile due to the bumps and cracks in the pavement, the vibrations of a bridge caused by traffic and wind, and the normal flutter of an airplane wing due to the vibration of the engine and the air rushing past. In our study of mechanical vibrations, we start with the simple mechanical system consisting of a coil spring suspended from a rigid support with a mass attached to the end of the spring. See Figure 5.1.

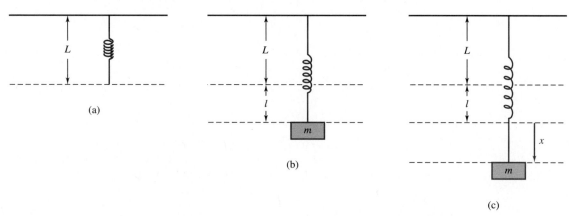

Figure 5.1 Spring (a) in natural position, (b) in equilibrium, and (c) in motion

To analyze this spring-mass system, we need to recall two laws of physics: Hooke's law and Newton's second law of motion. Hooke's law states that the spring exerts a restoring force opposite to the direction of elongation of the spring and with a magnitude directly

proportional to the amount of elongation.[†] That is, the spring exerts a restoring force $\mathscr{F}$ whose magnitude is ks, where s is the amount of elongation and $k(>0)$ is the **spring constant.**

For example, if a 20-lb weight stretches a spring 6 in., then (in the ft-lb system) Hooke's law gives

(1) $$20 = |\mathscr{F}| = ks = k\left(\frac{1}{2}\right).$$

Hence the spring constant is $k = 40$ lb/ft.

In Section 3.4, we discussed Newton's second law and how to apply it to problems in mechanics. Recall that this law can be expressed as

(2) $$m\frac{d^2x}{dt^2} = ma = F(t, x, dx/dt),$$

where x is the position, dx/dt is the velocity, and d^2x/dt^2 is the acceleration of the mass at time t. Here F, the total force acting on the mass, is assumed to depend only on time, position, and velocity.

A first step in the analysis of the spring-mass system is to choose a convenient coordinate axis in which to represent the motion of the mass. For this purpose, we observe that the spring has a certain natural length L when hanging from its support (see Figure 5.1(a)). Attaching a mass m elongates the spring, and when the spring comes to rest (equilibrium), it has been stretched an amount, say l, beyond its natural length (Figure 5.1(b)). Therefore let's choose a vertical coordinate axis passing through the spring, with the origin at the equilibrium position of the mass. We let x denote the displacement of the mass from its equilibrium position, taking x positive when the mass is below its equilibrium position, as shown in Figure 5.1(c).

We now consider the various forces acting on the mass m.

Gravity. The force of gravity F_1 is a downward force with magnitude mg, where g is the acceleration due to gravity. Hence

$$F_1 = mg.$$

Restoring Force. The spring exerts a restoring force F_2 whose magnitude is proportional to the elongation of the spring. From Figure 5.1(c), we see that the spring is stretched $x + l$ units beyond its natural length. Hence the magnitude of F_2 is $k(x + l)$, where k is the spring constant. Since the spring pulls upward (in the negative x direction), we have

$$F_2 = -k(x + l).$$

Note that when $x = 0$, that is, when the system is at equilibrium, the force of gravity and the force due to the spring balance each other. Thus $mg = kl$. We can now express the

[†] Hooke's law is a reasonable approximation until the spring is stretched to near its elastic limit.

restoring force as

$$F_2 = -kx - mg.$$

Damping Force. There is a damping or frictional force F_3 acting on the mass. For example, this force may be air resistance or friction due to a dashpot or a shock absorber. In either case, we *assume* the damping force is proportional to the magnitude of the velocity of the mass, but opposite in direction. That is,

$$F_3 = -b\frac{dx}{dt}, \qquad b > 0,$$

where b is the **damping constant** given in units of mass/time (or force-time/length). In certain circumstances, a more elaborate formula for the damping force may be required, but empirical studies show that when the velocity is small, the above expression for F_3 is a reasonable one.

External Forces. Any external forces acting on the mass (for example, a magnetic force or the forces exerted on a car by bumps in the pavement) will be denoted by $F_4 = f(t)$. For simplicity, we assume these forces depend only on time and *not* on the location of the mass or its velocity. (We can model the vertical displacements of a car in this way if its horizontal position is known as a function of time.)

The total force F acting on the mass m is the sum of the four forces F_1, F_2, F_3, and F_4:

(3) $$F(t, x, dx/dt) = mg - kx - mg - b\frac{dx}{dt} + f(t).$$

Applying Newton's second law to the system gives

$$m\frac{d^2x}{dt^2} = mg - kx - mg - b\frac{dx}{dt} + f(t),$$

which simplifies to

(4) $$m\frac{d^2x}{dt^2} + b\frac{dx}{dt} + kx = f(t).$$

When $b = 0$, the system is said to be **undamped;** otherwise it is **damped.** When $f(t) \equiv 0$, the motion is said to be **free;** otherwise the motion is **forced.** A convenient schematic representation for a simple spring-mass system with damping is given in Figure 5.2.

Let's begin with the simple system in which $b = 0$ and $f(t) \equiv 0$, the so-called **undamped, free** case. In this case, equation (4) reduces to

(5) $$m\frac{d^2x}{dt^2} + kx = 0,$$

Figure 5.2 Spring-mass system with spring constant k, damping constant b, and mass m

and, when divided by m, becomes

(6) $$\frac{d^2x}{dt^2} + \omega^2 x = 0,$$

where $\omega = \sqrt{k/m}$. The auxiliary equation associated with (6) is $r^2 + \omega^2 = 0$, which has complex conjugate roots $\pm\omega i$. Hence a general solution to (6) is

(7) $x(t) = C_1 \cos \omega t + C_2 \sin \omega t.$

We can express $x(t)$ in the more convenient form

(8) $x(t) = A \sin(\omega t + \phi),$

with $A \geq 0$, by letting $C_1 = A \sin \phi$ and $C_2 = A \cos \phi$. That is,

$$A \sin(\omega t + \phi) = A \cos \omega t \sin \phi + A \sin \omega t \cos \phi$$
$$= C_1 \cos \omega t + C_2 \sin \omega t.$$

Solving for A and ϕ in terms of C_1 and C_2, we find

(9) $$A = \sqrt{C_1^2 + C_2^2} \quad \text{and} \quad \tan \phi = \frac{C_1}{C_2},$$

where the quadrant in which ϕ lies is determined by the signs of C_1 and C_2. This is because $\sin \phi$ has the same sign as C_1 ($\sin \phi = C_1/A$) and $\cos \phi$ has the same sign as C_2 ($\cos \phi = C_2/A$). For example, if $C_1 > 0$ and $C_2 < 0$, then ϕ is in Quadrant II. (Note, in particular, that ϕ is not simply the arctangent of C_1/C_2, which would lie in Quadrant IV.)

It is evident from (8) that the motion of a mass in an *undamped*, *free* system is simply a sine wave, or what is called **simple harmonic motion.** The constant A is the amplitude of the motion and ϕ is the phase angle. The motion is periodic with **period** $2\pi/\omega$ and **natural frequency** $\omega/2\pi$, where $\omega = \sqrt{k/m}$. The period is measured in units of time, and the natural frequency has the dimensions of periods (or cycles) per unit time. The constant ω is the **angular frequency** for the sine function in (8) and has dimensions of radians per unit time.

Observe that the amplitude and phase angle depend on the constants C_1 and C_2, which, in turn, are determined by the initial position and initial velocity of the mass. However, the period and frequency depend only on k and m and not on the initial conditions (see Figure 5.3 on page 244).

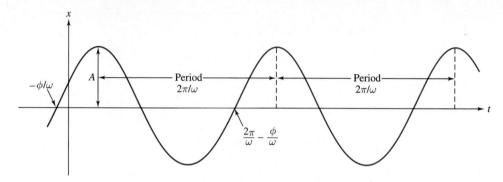

Figure 5.3 Simple harmonic motion of undamped, free vibrations

EXAMPLE 1 A mass weighing 4 lb stretches a spring 3 in. after coming to rest at equilibrium. The mass is then pulled down 6 in. below the equilibrium point and given a downward velocity of $\sqrt{2}$ ft/sec. Neglecting any damping or external forces that may be present, determine the equation of motion of the mass along with its amplitude, period, and natural frequency. How long after release does the mass pass through the equilibrium position?

Solution Because we have a case of undamped, free vibration, the equation of motion is given by (7). To find k, we observe that the 4-lb mass stretches the spring 3 in., or 1/4 ft. Using Hooke's law, we have

$$4 = mg = k\left(\frac{1}{4}\right),$$

or $k = 16$ lb/ft. Since $g \approx 32$ ft/sec^2, we find $m = 4/32 = 1/8$ slugs, and so

$$\omega = \sqrt{\frac{k}{m}} = \sqrt{\frac{16}{1/8}} = 8\sqrt{2}.$$

Substituting this value for ω into (7) gives

(10) $x(t) = C_1 \cos(8\sqrt{2}t) + C_2 \sin(8\sqrt{2}t).$

Now we use the initial conditions, $x(0) = 6$ in. $= 1/2$ ft and $x'(0) = \sqrt{2}$ ft/sec, to solve for C_1 and C_2 in (10). That is,

$$1/2 = x(0) = C_1,$$
$$\sqrt{2} = x'(0) = 8\sqrt{2}C_2,$$

and so $C_1 = 1/2$ and $C_2 = 1/8$. Hence the equation of motion of the mass is

(11) $x(t) = \dfrac{1}{2}\cos(8\sqrt{2}t) + \dfrac{1}{8}\sin(8\sqrt{2}t).$

To express $x(t)$ in the alternative form (8), we set

$$A = \sqrt{C_1^2 + C_2^2} = \sqrt{(1/2)^2 + (1/8)^2} = \frac{\sqrt{17}}{8},$$

$$\tan \phi = \frac{C_1}{C_2} = \frac{1/2}{1/8} = 4.$$

Since both C_1 and C_2 are positive, ϕ is in the first quadrant, so $\phi = \arctan 4 \approx 1.326$. Hence

(12) $x(t) = \dfrac{\sqrt{17}}{8} \sin(8\sqrt{2}t + \phi).$

Thus the amplitude A is $\sqrt{17}/8$ ft and the phase angle ϕ is approximately 1.326 radians. The period is $P = 2\pi/\omega = 2\pi/(8\sqrt{2}) = \sqrt{2}\pi/8$ sec, and the natural frequency is $1/P = 8/(\sqrt{2}\pi)$ cycles per sec.

Finally, to determine when the mass will pass through the equilibrium position, $x = 0$, we must solve the trigonometric equation

(13) $x(t) = \dfrac{\sqrt{17}}{8} \sin(8\sqrt{2}t + \phi) = 0$

for t. Equation (13) will be satisfied whenever

(14) $8\sqrt{2}t + \phi = n\pi$ or $t = \dfrac{n\pi - \phi}{8\sqrt{2}} \approx \dfrac{n\pi - 1.326}{8\sqrt{2}},$

n an integer. Putting $n = 1$ in (14) determines the first time t when the mass crosses its equilibrium position:

$$t = \frac{\pi - \phi}{8\sqrt{2}} \approx 0.16 \text{ sec.} \quad \blacksquare \text{ End}$$

EXERCISES 5.1

In the following problems, take $g = 32$ ft/sec^2 for the U.S. Customary System and $g = 9.8$ m/sec^2 for the MKS system.

1. A mass of 100 kg is attached to a spring suspended from the ceiling, thereby causing the spring to stretch 20 cm on coming to rest at equilibrium. The mass is then pulled down 5 cm below the equilibrium point and released (zero initial velocity). Neglecting any damping or external forces that may be present, determine the equation of motion of the mass, along with its amplitude, period, and natural frequency. Sketch the graph of this simple harmonic motion.

2. A mass weighing 8 lb is attached to a spring suspended from the ceiling. When the mass comes to rest at equilibrium, the spring has been stretched 6 in. The mass is then pulled down 3 in. below the equilibrium point and given an upward velocity of 0.5 ft/sec. Neglecting any damping or external forces that may be present, determine the equation of motion of the mass, along with its amplitude, period, and natural frequency. Sketch the graph of this simple harmonic motion.

3. A mass weighing 8 lb stretches a spring 2 ft on coming to rest at equilibrium. The mass is then lifted up 6 in.

above the equilibrium point and given a downward velocity of 1 ft/sec. Determine the simple harmonic motion of the mass. How fast and in what direction will the mass be moving 10 sec after being released?

4. Show that the period of the simple harmonic motion of a mass hanging from a spring is $2\pi\sqrt{l/g}$, where l denotes the amount (beyond its natural length) that the spring is stretched when the mass is at equilibrium.

5. A mass of 5 kg is attached to a spring hanging from a ceiling, thereby stretching the spring 0.5 m on coming to rest at equilibrium. The mass is then pulled down 0.1 m below the equilibrium point and given an upward velocity of 0.1 m/sec. Determine the equation for the simple harmonic motion of the mass. When will the mass first reach its minimum height after being set in motion?

6. A mass of 5 kg is attached to a spring suspended from a high ceiling, thereby stretching the spring 2 m on

coming to rest at equilibrium. The mass is then lifted up 1 m above the equilibrium point and given an upward velocity of 1/3 m/sec. Determine the equation for the simple harmonic motion of the mass. When will the mass first reach its maximum height after being set in motion?

7. A mass of 1/2 kg is attached to the lower end of a spring, thereby stretching the spring 2 m on coming to rest at equilibrium. The mass is then pulled down 1/2 m and given an upward velocity of 1/2 m/sec. Determine the equation for the simple harmonic motion of the mass. How long until the mass first returns to the equilibrium position?

8. A weight hanging on a spring oscillates with a period of 3 sec. After 2 lb are added, the period becomes 4 sec. Assuming we can neglect any damping or external forces, determine how much weight was originally attached to the spring.

5.2 DAMPED FREE VIBRATIONS

In the previous section we considered vibrations in a rather idyllic setting—no external frictional forces were present—and the result was simple harmonic motion. In most applications, however, there is at least some type of frictional or damping force that plays a significant role. This force may be due to a component in the system, such as a shock absorber in a car, or to the medium that surrounds the system, such as air or some liquid.

In this section we study the effect of a damping force on free vibrations. In particular, setting $f(t) \equiv 0$ in equation (4) of Section 5.1, we consider the motion of a system that is governed by

$$(1)\qquad m\frac{d^2x}{dt^2} + b\frac{dx}{dt} + kx = 0.$$

The auxiliary equation associated with (1) is

$$(2)\qquad mr^2 + br + k = 0,$$

and its roots are

$$(3)\qquad \frac{-b \pm \sqrt{b^2 - 4mk}}{2m} = -\frac{b}{2m} \pm \frac{1}{2m}\sqrt{b^2 - 4mk}.$$

As we found in Chapter 4, the form of the solution to (1) depends on the nature of these roots and, in particular, on the discriminant $b^2 - 4mk$.

Underdamped or Oscillatory Motion ($b^2 < 4mk$)

When $b^2 < 4mk$, the discriminant $b^2 - 4mk$ is negative and there are two complex conjugate roots to the auxiliary equation (2). These roots are $\alpha \pm i\beta$, where

(4) $\alpha := -\dfrac{b}{2m}, \qquad \beta := \dfrac{1}{2m}\sqrt{4mk - b^2}.$

Hence a general solution to (1) is

(5) $x(t) = e^{\alpha t}(C_1 \cos \beta t + C_2 \sin \beta t).$

As we did with simple harmonic motion, we can express $x(t)$ in the alternate form

(6) $x(t) = A e^{\alpha t} \sin(\beta t + \phi),$

where $A = \sqrt{C_1^2 + C_2^2}$ and $\tan \phi = C_1/C_2$. It is now evident that $x(t)$ is the product of an exponential factor,

$$A e^{\alpha t} = A e^{-(b/2m)t},$$

and a sine factor, $\sin(\beta t + \phi)$, which accounts for the oscillatory motion. Because the sine factor varies between -1 and 1 with period $2\pi/\beta$, the solution $x(t)$ varies between $-Ae^{\alpha t}$ and $Ae^{\alpha t}$ with **quasiperiod** $P = 2\pi/\beta = 4m\pi/\sqrt{4mk - b^2}$ and **quasifrequency** $1/P$. Moreover, since b and m are positive, $\alpha = -b/2m$ is negative, and thus the exponential factor tends to zero at $t \to \infty$. A graph of a typical solution $x(t)$ is given in Figure 5.4. The system is called **underdamped** because there is not enough damping present (b is too small) to prevent the system from oscillating.

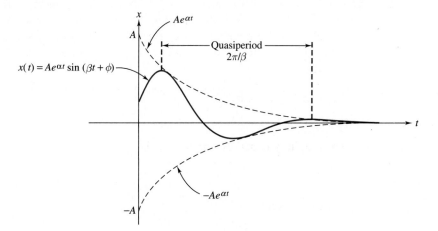

Figure 5.4 Oscillatory motion

It is easily seen that as $b \to 0$, the exponential factor approaches the constant A and the quasifrequency approaches the natural frequency of the simple harmonic motion cor-

responding to $b = 0$. Figure 5.4 demonstrates that the values of t where the graph of $x(t)$ touches the exponential curves $\pm Ae^{\alpha t}$ are *not* the same values of t at which $x(t)$ attains its relative maximum and minimum values (see Problem 11).

Critically Damped Motion ($b^2 = 4mk$)

When $b^2 = 4mk$, the discriminant $b^2 - 4mk$ is zero and the auxiliary equation has the repeated root $-b/2m$. Hence a general solution to (1) is now

(7) $\qquad x(t) = (C_1 + C_2 t)e^{-(b/2m)t}$.

To understand the motion described by $x(t)$ in (7), we first consider the behavior of $x(t)$ as $t \to \infty$. Using L'Hôpital's rule,

(8) $\qquad \lim_{t\to\infty} x(t) = \lim_{t\to\infty} \dfrac{C_1 + C_2 t}{e^{(b/2m)t}} = \lim_{t\to\infty} \dfrac{C_2}{(b/2m)e^{(b/2m)t}} = 0,$

(recall that $b/2m > 0$). Hence $x(t)$ dies off to zero as $t \to \infty$. Next, since

$$x'(t) = \left(C_2 - \frac{b}{2m}C_1 - \frac{b}{2m}C_2 t \right) e^{-(b/2m)t},$$

we see that the derivative is either identically zero (when $C_1 = C_2 = 0$) or vanishes for at most one point (when the linear factor in parentheses is zero). If the trivial solution $x(t) \equiv 0$ is ignored, it follows that $x(t)$ has at most one local maximum or minimum for $t > 0$. Therefore $x(t)$ *does not oscillate*. This leaves, qualitatively, only three possibilities for the motion of $x(t)$, depending on the initial conditions. These are illustrated in Figure 5.5. This special case when $b^2 = 4mk$ is called **critically damped** motion, since if b were any smaller, oscillation would occur.

Overdamped Motion ($b^2 > 4mk$)

When $b^2 > 4mk$, the discriminant $b^2 - 4mk$ is positive and there are two distinct real roots to the auxiliary equation (2):

(9) $\qquad r_1 = -\dfrac{b}{2m} + \dfrac{1}{2m}\sqrt{b^2 - 4mk}, \qquad r_2 = -\dfrac{b}{2m} - \dfrac{1}{2m}\sqrt{b^2 - 4mk}.$

Hence a general solution to (1) is

(10) $\qquad x(t) = C_1 e^{r_1 t} + C_2 e^{r_2 t}$.

Obviously r_2 is negative. And since $b^2 > b^2 - 4mk$ (that is, $b > \sqrt{b^2 - 4mk}$), it follows that r_1 is also negative. Therefore, as $t \to \infty$, both of the exponentials in (10) decay and $x(t) \to 0$. Moreover, since

$$x'(t) = C_1 r_1 e^{r_1 t} + C_2 r_2 e^{r_2 t} = e^{r_1 t}(C_1 r_1 + C_2 r_2 e^{(r_2 - r_1)t}),$$

it follows that $x'(t) = 0$ only when $C_1 r_1 + C_2 r_2 e^{(r_2 - r_1)t} = 0$. Thus a nontrivial solution $x(t)$ can have at most one local maximum or minimum for $t > 0$. As in the case of critical damping, the motion is *nonoscillatory* and must be qualitatively like one of the three sketches in Figure 5.5. This case where $b^2 > 4mk$ is called **overdamped** motion.

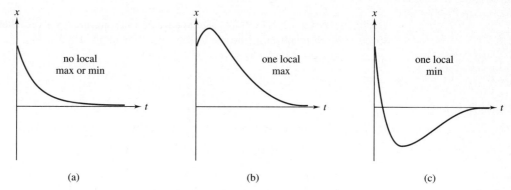

no local
max or min

one local
max

one local
min

(a) (b) (c)

Figure 5.5 Critically damped vibrations

EXAMPLE 1 Assume the motion of a spring-mass system with damping is governed by

(11) $\dfrac{d^2x}{dt^2} + b\dfrac{dx}{dt} + 25x = 0;$ $x(0) = 1,$ $x'(0) = 0.$

Find the equation of motion and sketch its graph for the three cases where $b = 8$, 10, and 12.

Solution The auxiliary equation for (11) is

(12) $r^2 + br + 25 = 0,$

whose roots are

(13) $r = -\dfrac{b}{2} \pm \dfrac{1}{2}\sqrt{b^2 - 100}.$

Case 1. When $b = 8$, the roots (13) are $-4 \pm 3i$. This is thus a case of underdamping, and the equation of motion has the form

(14) $x(t) = C_1 e^{-4t} \cos 3t + C_2 e^{-4t} \sin 3t.$

Setting $x(0) = 1$ and $x'(0) = 0$ gives the system

$C_1 = 1,$ $-4C_1 + 3C_2 = 0,$

whose solution is $C_1 = 1$, $C_2 = 4/3$. To express $x(t)$ as the product of a damping factor and a sine factor (recall equation (6)), we set

$$A = \sqrt{C_1^2 + C_2^2} = \frac{5}{3}, \qquad \tan\phi = \frac{C_1}{C_2} = \frac{3}{4},$$

where ϕ is a first-quadrant angle, since C_1 and C_2 are both positive. Then

(15) $x(t) = \dfrac{5}{3} e^{-4t} \sin(3t + \phi),$

where $\phi = \arctan(3/4) \approx 0.64$. The spring is underdamped, but the damping is nonetheless strong enough to obscure the oscillations shown in Figure 5.6(a).

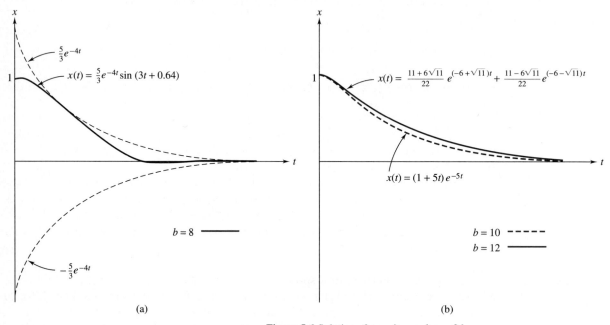

Figure 5.6 Solutions for various values of b

Case 2. When $b = 10$, there is only one (repeated) root to the auxiliary equation (12), namely $r = -5$. This is a case of critical damping, and the equation of motion has the form

(16) $x(t) = (C_1 + C_2 t)e^{-5t}$.

Setting $x(0) = 1$ and $x'(0) = 0$ now gives

$$C_1 = 1, \qquad C_2 - 5C_1 = 0,$$

and so $C_1 = 1$, $C_2 = 5$. Thus

(17) $x(t) = (1 + 5t)e^{-5t}$.

The graph of $x(t)$ given in (17) is represented by the dashed line in Figure 5.6(b). Notice that $x(t)$ is zero only for $t = -1/5$ and hence doesn't cross the t-axis for $t > 0$.

Case 3. When $b = 12$, the roots to the auxiliary equation are $-6 \pm \sqrt{11}$. This is a case of overdamping, and the equation of motion has the form

(18) $x(t) = C_1 e^{(-6+\sqrt{11})t} + C_2 e^{(-6-\sqrt{11})t}$.

Setting $x(0) = 1$ and $x'(0) = 0$ gives

$$C_1 + C_2 = 1, \qquad (-6 + \sqrt{11})C_1 + (-6 - \sqrt{11})C_2 = 0,$$

from which we find $C_1 = (11 + 6\sqrt{11})/22$ and $C_2 = (11 - 6\sqrt{11})/22$. Hence

(19)
$$x(t) = \frac{11 + 6\sqrt{11}}{22} e^{(-6+\sqrt{11})t} + \frac{11 - 6\sqrt{11}}{22} e^{(-6-\sqrt{11})t}$$

$$= \frac{e^{(-6+\sqrt{11})t}}{22} \{11 + 6\sqrt{11} + (11 - 6\sqrt{11})e^{-2\sqrt{11}t}\}.$$

The graph of this overdamped motion is represented by the solid line in Figure 5.6(b).

■ End

It is interesting to observe in Example 1 that when the system is underdamped ($b = 8$), the solution goes to zero like e^{-4t}; when the system is critically damped ($b = 10$), the solution tends to zero roughly like e^{-5t}; and when the system is overdamped ($b = 12$), the solution goes to zero like $e^{(-6+\sqrt{11})t} \approx e^{-2.68t}$. This means that if the system is underdamped, it not only oscillates, but also dies off slower than if it were critically damped. Moreover, if the system is overdamped, it again dies off more slowly than if it were critically damped. (This agrees with our physical intuition that the damping forces hinder the return to equilibrium.)

EXAMPLE 2 An 8-lb weight is attached to a spring hanging from the ceiling. When the weight comes to rest at equilibrium, the spring has been stretched 2 ft. The damping constant b for the system is 1 lb-sec/ft. If the weight is raised 6 in. above equilibrium and given an initial upward velocity of 1 ft/sec, find the equation of motion for the weight. What is the maximum displacement above equilibrium that the weight will attain?

Solution An 8-lb weight stretches the spring 2 ft, so Hooke's law gives $8 = k2$ and the spring constant is $k = 4$ lb/ft. The 8-lb weight has mass $m = W/g$, where g, the acceleration due to gravity, is 32 ft/sec^2. Thus $m = 8/32 = 1/4$ slugs. Substituting these values for m, k, and the given value $b = 1$ into equation (1) and using the initial conditions, we obtain the initial value problem

(20)
$$\frac{1}{4}\frac{d^2x}{dt^2} + \frac{dx}{dt} + 4x = 0; \qquad x(0) = -\frac{1}{2}, \qquad x'(0) = -1.$$

The negative signs for the initial conditions reflect the facts that the initial displacement and push are *upward* (see Figure 5.7(a) on page 252).

The reader can verify that the solution to Problem (20) is

(21)
$$x(t) = -\frac{1}{2}e^{-2t}\cos(2\sqrt{3}t) - \frac{1}{\sqrt{3}}e^{-2t}\sin(2\sqrt{3}t),$$

or

(22)
$$x(t) = \sqrt{\frac{7}{12}}e^{-2t}\sin(2\sqrt{3}t + \phi),$$

Figure 5.7 Spring-mass system and graph of motion for Example 2

where $\tan \phi = \sqrt{3}/2$ and ϕ lies in the third quadrant because $C_1 = -1/2$ and $C_2 = -1/\sqrt{3}$ are both negative. (See Figure 5.7(b) for a sketch of $x(t)$ where we have inverted the x-axis so that the graph corresponds to the motion of the system in Figure 5.7(a).)

To determine the maximum displacement above equilibrium, we must determine the *highest* point on the graph of $x(t)$ in Figure 5.7(b). Because $x(t)$ dies off exponentially, this will occur at the first positive critical point of $x(t)$. Computing $x'(t)$ from (21), setting it equal to zero, and solving gives

$$x'(t) = e^{-2t} \left\{ \frac{5}{\sqrt{3}} \sin(2\sqrt{3}t) - \cos(2\sqrt{3}t) \right\} = 0,$$

$$\frac{5}{\sqrt{3}} \sin(2\sqrt{3}t) = \cos(2\sqrt{3}t),$$

$$\tan(2\sqrt{3}t) = \frac{\sqrt{3}}{5}.$$

Thus the first positive root is

$$t = \frac{1}{2\sqrt{3}} \arctan \frac{\sqrt{3}}{5} \approx 0.096.$$

Substituting this value for t back into equation (21) or (22) gives $x(0.096) \approx -0.55$. Hence the maximum displacement above equilibrium is approximately 0.55 ft. ■ End

EXERCISES 5.2

In the following problems, take g = 32 ft/sec² for the U.S. Customary System and g = 9.8 m/sec² for the MKS system.

1. The motion of a spring-mass system with damping is governed by

 $$x''(t) + bx'(t) + 16x(t) = 0; \quad x(0) = 1, \quad x'(0) = 0.$$

 Find the equation of motion and sketch its graph for $b = 6$, 8, and 10.

2. The motion of a spring-mass system with damping is governed by

 $$x''(t) + bx'(t) + 64x(t) = 0; \quad x(0) = 1, \quad x'(0) = 0.$$

 Find the equation of motion and sketch its graph for $b = 10$, 16, and 20.

3. The motion of a spring-mass system with damping is governed by

 $$x''(t) + 10x'(t) + kx(t) = 0; \quad x(0) = 1, \quad x'(0) = 0.$$

 Find the equation of motion and sketch its graph for $k = 20$, 25, and 30.

4. The motion of a spring-mass system with damping is governed by

 $$x''(t) + 4x'(t) + kx(t) = 0; \quad x(0) = 1, \quad x'(0) = 0.$$

 Find the equation of motion and sketch its graph for $k = 2$, 4, and 6.

5. A 4-lb weight is attached to a spring suspended from the ceiling. When the weight comes to rest at equilibrium, the spring has been stretched 3 in. The damping constant for the system is 2 lb-sec/ft. If the weight is raised 9 in. above equilibrium and given an initial upward velocity of 2 ft/sec, determine the equation of motion of the weight and give its damping factor, quasiperiod, and quasifrequency.

6. A mass of 20 kg stretches a spring 98 cm on coming to rest at equilibrium. The damping constant for the system is 140 N-sec/m. If the mass is pulled 25 cm below equilibrium and given an upward velocity of 1 m/sec, when will it return to its equilibrium position?

7. A mass of 2 kg stretches a spring 49 cm on coming to rest at equilibrium. The damping constant for the system is $8\sqrt{5}$ kg/sec. If the mass is pulled 10 cm below equilibrium and given a downward velocity of 2 m/sec, what is the maximum displacement from equilibrium that it will attain?

8. An 8-lb weight stretches a spring 1 ft on coming to rest at equilibrium. The damping constant for the system is 1/4 lb-sec/ft. If the weight is raised 1 ft above equilibrium and released, what is the maximum displacement below equilibrium that it will attain?

9. A mass of 1 kg stretches a spring 9.8 cm on coming to rest at equilibrium. The damping constant for the system is 0.2 N-sec/m. If the mass is pushed downward from the equilibrium position with a velocity of 1 m/sec, when will it attain its maximum displacement below equilibrium?

10. An 8-lb weight stretches a spring 1 ft on coming to rest at equilibrium. The damping constant for the system is 2 lb-sec/ft. If the weight is raised 6 in. above equilibrium and given an upward velocity of 2 ft/sec, when will the mass attain its maximum displacement above equilibrium?

11. Show that for the underdamped system of Example 2, the times when the solution curve $x(t)$ in (22) touches the exponential curves $\pm\sqrt{7/12}\,e^{-2t}$ are not the same values of t for which the function $x(t)$ attains its relative extrema.

12. For an underdamped system, verify that as $b \to 0$ the damping factor approaches the constant A and the quasifrequency approaches the natural frequency $\sqrt{k/m}/(2\pi)$.

13. How can one deduce the value of the damping constant b by observing the motion of an underdamped system? Assume that the mass m is known.

5.3 FORCED VIBRATIONS

We now consider the vibrations of a spring-mass system when an external force is applied. Of particular interest is the response of the system to a *periodic* forcing term. As a paradigm, let's investigate the effect of a cosine forcing function on the system governed by

the differential equation

$$(1) \qquad m\frac{d^2x}{dt^2} + b\frac{dx}{dt} + kx = F_0 \cos \gamma t,$$

where F_0, γ are nonnegative constants and $0 < b^2 < 4mk$ (so the system is *underdamped*).

A solution to (1) has the form $x = x_h + x_p$, where x_p is a particular solution and x_h is a general solution to the corresponding homogeneous equation. We found in equation (6) of Section 5.2 that

$$(2) \qquad x_h(t) = Ae^{-(b/2m)t} \sin\left(\frac{\sqrt{4mk - b^2}}{2m}t + \phi\right),$$

where A and ϕ are constants.

To determine x_p, we can use the method of undetermined coefficients (Section 4.8). From the form of the nonhomogeneous term, we know that

$$(3) \qquad x_p(t) = A_1 \cos \gamma t + A_2 \sin \gamma t,$$

where A_1, A_2 are constants to be determined. Substituting this expression into equation (1) and simplifying gives

$$(4) \qquad [(k - m\gamma^2)A_1 + b\gamma A_2]\cos \gamma t + [(k - m\gamma^2)A_2 - b\gamma A_1]\sin \gamma t = F_0 \cos \gamma t.$$

Setting the corresponding coefficients on both sides equal, we have

$$(k - m\gamma^2)A_1 + b\gamma A_2 = F_0,$$
$$-b\gamma A_1 + (k - m\gamma^2)A_2 = 0.$$

Solving, we obtain

$$(5) \qquad A_1 = \frac{F_0(k - m\gamma^2)}{(k - m\gamma^2)^2 + b^2\gamma^2}, \qquad A_2 = \frac{F_0 b\gamma}{(k - m\gamma^2)^2 + b^2\gamma^2}.$$

Hence a particular solution to (1) is

$$(6) \qquad x_p(t) = \frac{F_0}{(k - m\gamma^2)^2 + b^2\gamma^2}\{(k - m\gamma^2)\cos \gamma t + b\gamma \sin \gamma t\}.$$

The expression in braces can also be written as

$$\sqrt{(k - m\gamma^2)^2 + b^2\gamma^2}\,\sin(\gamma t + \theta),$$

so we can express x_p in the alternative form

$$(7) \qquad x_p(t) = \frac{F_0}{\sqrt{(k - m\gamma^2)^2 + b^2\gamma^2}}\,\sin(\gamma t + \theta),$$

where $\tan \theta = A_1/A_2 = (k - m\gamma^2)/(b\gamma)$ and the quadrant in which θ lies is determined by the signs of A_1 and A_2.

Combining equations (2) and (7), we have the following representation of a general solution to (1) in the case $0 < b^2 < 4mk$:

(8) $$x(t) = Ae^{-(b/2m)t}\sin\left(\frac{\sqrt{4mk - b^2}}{2m}t + \phi\right) + \frac{F_0}{\sqrt{(k - m\gamma^2)^2 + b^2\gamma^2}}\sin(\gamma t + \theta).$$

The solution (8) is the sum of two terms. The first term, x_h, represents damped oscillation and depends only on the parameters of the system and the initial conditions. Because of the damping factor $e^{-(b/2m)t}$, this term tends to zero as $t \to \infty$. Consequently, it is referred to as the **transient** part of the solution. The second term, x_p, in (8) is the offspring of the external forcing function $f(t) = F_0 \cos \gamma t$. Like the forcing function, x_p is a sinusoid with angular frequency γ. However, x_p is out of phase with $f(t)$ (by the angle $\theta - \pi/2$), and its magnitude is different by the factor

(9) $$\frac{1}{\sqrt{(k - m\gamma^2)^2 + b^2\gamma^2}}.$$

As the transient term dies off, the motion of the spring-mass system becomes essentially that of the second term x_p (see Figure 5.8). Hence this term is called the **steady-state** solution. The factor appearing in (9) is referred to as the **frequency gain** or **gain factor** since it represents the ratio of the magnitude of the sinusoidal response to that of the input force. Note that this factor depends on the frequency γ and has units of length/force.

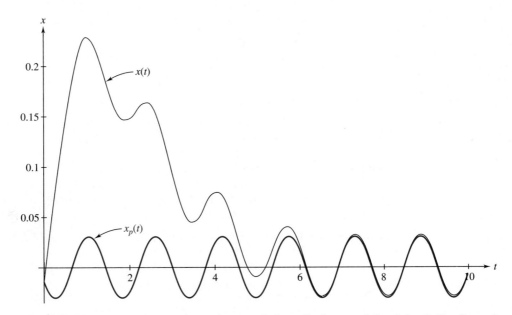

Figure 5.8 Convergence of $x(t)$ to the steady-state solution $x_p(t)$ when $m = 4$, $b = 6$, $k = 3$, $F_0 = 2$, $\gamma = 4$

EXAMPLE 1 A 10-kg mass is attached to a spring hanging from the ceiling, thereby causing the spring to stretch 2 m on coming to rest at equilibrium. At time $t = 0$, an external force

$f(t) = 20 \cos 4t$ N is applied to the system. The damping constant for the system is 3 N-sec/m. Determine the steady-state solution for the system.

Solution The mass m, damping constant b, and forcing term are given, so we need to determine only the spring constant k to obtain the differential equation that governs the system. The constant acceleration g due to gravity is approximately 9.8 m/sec^2. Thus the force of gravity on the system is $mg = (10)(9.8) = 98$ N. From Hooke's law, we have $98 = k2$, and so $k = 49$ N/m. Now, substituting into equation (1), we obtain

$$(10) \qquad 10\frac{d^2x}{dt^2} + 3\frac{dx}{dt} + 49x = 20 \cos 4t,$$

where $x(t)$ is the displacement (from equilibrium) of the mass at time t.

To find the steady-state response we must produce a particular solution to (10) that is a sinusoid. We can do this using the method of undetermined coefficients, guessing a solution of the form $A_1 \cos 4t + A_2 \sin 4t$. But this is precisely how we derived equation (7). Thus we substitute directly into (7) and find

$$(11) \qquad x_p(t) = \frac{20}{\sqrt{(49 - 160)^2 + (9)(16)}} \sin(4t + \theta) \approx (0.18)\sin(4t + \theta),$$

where $\tan \theta = (49 - 160)/12 \approx -9.25$. Since the numerator, $(49 - 160)$, is negative and the denominator, 12, positive, θ is a fourth-quadrant angle. Thus

$$\theta \approx \arctan(-9.25) \approx -1.46,$$

and the steady-state solution is given (approximately) by

$$(12) \qquad x_p(t) = (0.18)\sin(4t - 1.46). \quad \blacksquare \text{ End}$$

The above example illustrates an important point made earlier: The steady-state response (12) to the sinusoidal forcing function $20 \cos 4t$ is a sinusoid of the same frequency but different amplitude. The gain factor (see (9)) in this case is $(0.18)/20 = 0.009$ m/N.

In general, the amplitude of the steady-state solution to equation (1) depends on the angular frequency γ of the forcing function and is given by $A(\gamma) = F_0 M(\gamma)$, where

$$(13) \qquad M(\gamma) := \frac{1}{\sqrt{(k - m\gamma^2)^2 + b^2\gamma^2}} = \frac{1/m}{\sqrt{\left(\frac{k}{m} - \gamma^2\right)^2 + \left(\frac{b}{m}\right)^2 \gamma^2}}$$

is the frequency gain (see (9)). This formula is valid even when $b^2 \geq 4mk$. For a given system (m, b, and k fixed), it is often of interest to know how this system reacts to sinusoidal inputs of various frequencies (γ is a variable). For this purpose, the graph of the gain $M(\gamma)$, called the **frequency response curve** or **resonance curve** for the system, is enlightening.

To sketch the frequency response curve for $\gamma \geq 0$, we observe that when $\gamma = 0$, we have $M(0) = 1/k$. Also note that as $\gamma \to \infty$, the gain $M(\gamma) \to 0$. As a further aid in describing

the graph, we compute from (13)

$$(14) \qquad M'(\gamma) = \frac{-\left(\dfrac{2\gamma}{m}\right)\left[\gamma^2 - \left(\dfrac{k}{m} - \dfrac{b^2}{2m^2}\right)\right]}{\left[\left(\dfrac{k}{m} - \gamma^2\right)^2 + \left(\dfrac{b}{m}\right)^2 \gamma^2\right]^{3/2}}.$$

It follows from (14) that $M'(\gamma) = 0$ if and only if

$$(15) \qquad \gamma = 0 \quad \text{or} \quad \gamma = \gamma_r := \sqrt{\frac{k}{m} - \frac{b^2}{2m^2}}.$$

Now when $b^2 > 2mk$, the term inside the radical in (15) is negative and hence $M'(\gamma) = 0$ only when $\gamma = 0$. In this case, as γ increases from zero to infinity, $M(\gamma)$ decreases from $M(0) = 1/k$ to a limit value of zero.

When $b^2 < 2mk$, then γ_r is real and positive, and it is easy to verify that $M(\gamma)$ has a *maximum* at γ_r. Substituting γ_r into (13) gives

$$(16) \qquad M(\gamma_r) = \frac{1}{b\sqrt{\dfrac{k}{m} - \dfrac{b^2}{4m^2}}}.$$

The value $\gamma_r/2\pi$ is called the **resonance frequency** for the system. When the system is stimulated by an external force at this frequency, it is said to be **at resonance.**

Observe that since we must have $b^2 < 2mk$ for resonance to occur, a system cannot be at resonance unless it is underdamped ($b^2 < 4mk$).

To illustrate the effect of the damping constant b on the resonance curve, we consider a system in which $m = k = 1$. In this case, the frequency response curves are given by

$$(17) \qquad M(\gamma) = \frac{1}{\sqrt{(1 - \gamma^2)^2 + b^2\gamma^2}},$$

and, for $b < \sqrt{2}$, the resonance frequency is $\gamma_r/2\pi = (1/2\pi)\sqrt{1 - b^2/2}$. Figure 5.9 on page 258 displays the graphs of these frequency response curves for $b = 1/4, 1/2, 1, 3/2$, and 2. Observe that as $b \to 0$, the maximum magnitude of the frequency gain increases and the resonance frequency $\gamma_r/2\pi$ for the damped system approaches $1/2\pi = \sqrt{k/m}/2\pi$, the natural frequency for the undamped system.

To understand what is occurring, consider the undamped system ($b = 0$) with forcing term $F_0 \cos \gamma t$. This system is governed by

$$(18) \qquad m\frac{d^2x}{dt^2} + kx = F_0 \cos \gamma t.$$

A general solution to (18) is the sum of a particular solution and a general solution to the homogeneous equation. In Section 5.1, we showed that the latter describes simple harmonic motion:

$$(19) \qquad x_h(t) = A \sin(\omega t + \phi), \qquad \omega := \sqrt{k/m}.$$

The formula for the particular solution given in (7) is valid for $b = 0$, provided $\gamma \neq \omega = \sqrt{k/m}$. However, when $b = 0$ and $\gamma = \omega$, then the form we used with undetermined coefficients to derive (7) does not work because $\cos \omega t$ and $\sin \omega t$ are solutions to the corresponding homogeneous equation. The correct form is

$$\textbf{(20)} \qquad x_p(t) = A_1 t \cos \omega t + A_2 t \sin \omega t,$$

which leads to the solution

$$\textbf{(21)} \qquad x_p(t) = \frac{F_0}{2m\omega} t \sin \omega t.$$

(The verification of (21) is left to the reader.) Hence, in the *undamped resonant case* (when $\gamma = \omega$), a general solution to (18) is

$$\textbf{(22)} \qquad x(t) = A \sin(\omega t + \phi) + \frac{F_0}{2m\omega} t \sin \omega t.$$

Returning to the question of resonance, observe that the particular solution in (21) oscillates between $\pm(F_0 t)/(2m\omega)$. Hence as $t \to \infty$, the maximum magnitude of (21) approaches ∞ (see Figure 5.10).

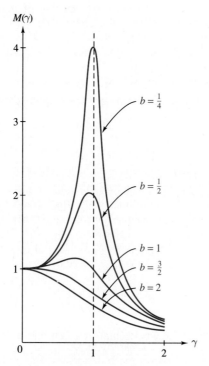

Figure 5.9 Frequency response curves for various values of b

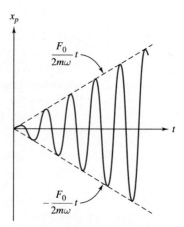

Figure 5.10 Undamped oscillation of the particular solution in (21)

It is obvious from the above discussion that **if the damping constant b is very small, the system is subject to large oscillations when the forcing function has a frequency near the resonance frequency for the system.** It is these large vibrations at resonance that concern engineers. Indeed, resonance vibrations have been known to cause airplane wings to snap, bridges to collapse,[†] and (less catastrophically) wine glasses to shatter.

EXERCISES 5.3

In the following problems, take $g = 32$ ft/sec^2 for the U.S. Customary System and $g = 9.8$ m/sec^2 for the MKS system.

1. An 8-kg mass is attached to a spring hanging from the ceiling, thereby causing the spring to stretch 1.96 m upon coming to rest at equilibrium. At time $t = 0$, an external force $f(t) = \cos 2t$ N is applied to the system. The damping constant for the system is 3 N-sec/m. Determine the steady-state solution for the system.

2. A mass weighing 32 lb is attached to a spring hanging from the ceiling and comes to rest at its equilibrium position. At time $t = 0$, an external force $f(t) = 3 \cos 4t$ lb is applied to the system. If the spring constant is 5 lb/ft and the damping constant is 2 lb-sec/ft, find the steady-state solution for the system.

3. A mass weighing 8 lb is attached to a spring hanging from the ceiling and comes to rest at its equilibrium position. At time $t = 0$, an external force $f(t) = 2 \cos 2t$ lb is applied to the system. If the spring constant is 10 lb/ft and the damping constant is 1 lb-sec/ft, find the equation of motion of the mass. What is the resonance frequency for the system?

4. A 2-kg mass is attached to a spring hanging from the ceiling, thereby causing the spring to stretch 20 cm upon coming to rest at equilibrium. At time $t = 0$, the mass is displaced 5 cm below the equilibrium position and released. At this same instant, an external force $f(t) = 0.3 \cos t$ N is applied to the system. If the damping constant for the system is 5 N-sec/m, determine the equation of motion for the mass. What is the resonance frequency for the system?

5. Sketch the frequency response curve (13) for the system in which $m = 4$, $k = 1$, $b = 2$.

6. Sketch the frequency response curve (13) for the system in which $m = 2$, $k = 3$, $b = 3$.

7. Determine the equation of motion for an undamped system at resonance governed by

$$\frac{d^2x}{dt^2} + 9x = 2 \cos 3t; \qquad x(0) = 1, \qquad x'(0) = 0.$$

Sketch the solution.

8. Determine the equation of motion for an undamped system at resonance governed by

$$\frac{d^2x}{dt^2} + x = 5 \cos t; \qquad x(0) = 0, \qquad x'(0) = 1.$$

Sketch the solution.

9. An undamped system is governed by

$$m\frac{d^2x}{dt^2} + kx = F_0 \cos \gamma t, \qquad x(0) = x'(0) = 0,$$

where $\gamma \neq \omega := \sqrt{k/m}$.

(a) Find the equation of motion of the system.
(b) Use trigonometric identities to show that the solution can be written in the form

$$x(t) = \frac{2F_0}{m(\omega^2 - \gamma^2)} \sin\left(\frac{\omega + \gamma}{2}t\right) \sin\left(\frac{\omega - \gamma}{2}t\right).$$

(c) When γ is near ω, then $\omega - \gamma$ is small, while $\omega + \gamma$ is relatively large compared with $\omega - \gamma$. Hence $x(t)$ can be viewed as the product of a slowly vary-

[†] An interesting discussion of one such disaster, involving the Tacoma Narrows bridge in Washington State, can be found in *Differential Equations and Their Applications*, by M. Braun (Springer-Verlag, New York, 1978). Researchers are still trying to understand this disaster; see the recent articles "Large-amplitude periodic oscillations in suspension bridges: some new connections with nonlinear analysis," by A. C. Lazer and P. J. McKenna, *SIAM Review*, Vol. 32 (1990), pp. 537–578 or "Still twisting," by Henry Petroski, *American Scientist*, Vol. 19 (1991), pp. 398–401.

ing sine function, $\sin[(\omega - \gamma)t/2]$, and a rapidly varying sine function, $\sin[(\omega + \gamma)t/2]$. The net effect is a sine function $x(t)$ with frequency $(\omega - \gamma)/4\pi$, which serves as the time-varying amplitude of a sine function with frequency $(\omega + \gamma)/4\pi$. This vibration phenomenon is referred to as **beats** and is used in tuning stringed instruments. This same phenomenon in electronics is called **amplitude modulation.** To illustrate this phenomenon, sketch the curve $x(t)$ for $F_0 = 32$, $m = 2$, $\omega = 9$, and $\gamma = 7$.

10. Derive the formula for $x_p(t)$ given in (21).

11. Shock absorbers in automobiles, aircraft, etc. can be described as forced *overdamped* spring-mass systems. Derive an expression analogous to equation (8) for the general solution to the differential equation (1) when $b^2 > 4mk$.

12. The response of an overdamped system to a constant force is governed by equation (1) with $m = 2$, $b = 8$, $k = 6$, $F_0 = 18$, and $\gamma = 0$. If the system starts from rest $(x(0) = x'(0) = 0)$, compute and sketch the displacement $x(t)$. What is the limiting value of $x(t)$ as $t \to \infty$? Interpret this physically.

5.4 COUPLED SPRING-MASS SYSTEMS

In this section, we extend our spring-mass model to include situations where coupled springs connect two masses, both of which are free to move. The resulting motions can be very intriguing. For simplicity, we shall neglect the effects of friction, gravity, and external forces. Consider the following experiment.

EXAMPLE 1 On a smooth horizontal surface, a mass $m_1 = 2$ kg is attached to a fixed wall by a spring with spring constant $k_1 = 4$ N/m. Another mass $m_2 = 1$ kg is attached to the first object by a spring with spring constant $k_2 = 2$ N/m. The objects are aligned horizontally so that the springs are their natural lengths (Figure 5.11). If both objects are displaced 3 m to the right of their equilibrium positions (Figure 5.12) and then released, what are the equations of motion for the two objects?

Figure 5.11 Coupled system at equilibrium

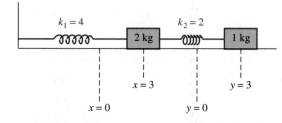

Figure 5.12 Coupled system at initial displacement

Solution From our assumptions, the only forces we need to consider are those due to the springs themselves. Recall that Hooke's law asserts that the force acting on an object due to a spring has magnitude proportional to the displacement of the spring from its natural length and has direction opposite to its displacement. That is, if the spring is either stretched or compressed, then it tries to return to its natural length.

Because each mass is free to move, we apply Newton's second law to each object. Let $x(t)$ denote the displacement (to the right) of the 2-kg mass from its equilibrium position, and, similarly, let $y(t)$ denote the corresponding displacement for the 1-kg mass. The 2-kg

mass has a force F_1 acting on its left side due to one spring and a force F_2 acting on its right side due to the second spring. Referring to Figure 5.11 and applying Hooke's law, we see that

$$F_1 = -k_1 x, \qquad F_2 = +k_2(y - x),$$

since $(y - x)$ is the net displacement of the second spring from its natural length. There is only one force acting on the 1-kg mass: the force due to the second spring, which is

$$F_3 = -k_2(y - x).$$

Applying Newton's second law to these objects, we obtain

(1)
$$\begin{cases} m_1 \dfrac{d^2x}{dt^2} = F_1 + F_2 = -k_1 x + k_2(y - x), \\[2mm] m_2 \dfrac{d^2y}{dt^2} = F_3 = -k_2(y - x), \end{cases}$$

or

(2)
$$\begin{cases} m_1 \dfrac{d^2x}{dt^2} + (k_1 + k_2)x - k_2 y = 0 \\[2mm] m_2 \dfrac{d^2y}{dt^2} + k_2 y - k_2 x = 0. \end{cases}$$

In this problem, we know that $m_1 = 2$, $m_2 = 1$, $k_1 = 4$, and $k_2 = 2$. Substituting these values into system (2) yields

(3) $\qquad 2\dfrac{d^2x}{dt^2} + 6x - 2y = 0,$

(4) $\qquad \dfrac{d^2y}{dt^2} + 2y - 2x = 0.$

We shall use the elimination method of Section 4.10 to solve (3)–(4). With $D := d/dt$, we rewrite the latter as

(5) $\qquad (2D^2 + 6)[x] - 2y = 0,$

(6) $\qquad -2x + (D^2 + 2)[y] = 0.$

Adding $(D^2 + 2)$ applied to equation (5) to 2 times equation (6) eliminates y:

$$[(D^2 + 2)(2D^2 + 6) - 4][x] = 0,$$

which simplifies to

(7) $\qquad 2\dfrac{d^4x}{dt^4} + 10\dfrac{d^2x}{dt^2} + 8x = 0.$

Notice that equation (7) is linear with constant coefficients. To solve it, let's proceed as we did with linear second order equations and try to find solutions of the form $x = e^{rt}$. Substituting e^{rt} in equation (7) gives

$$2(r^4 + 5r^2 + 4)e^{rt} = 0.$$

Thus we get a solution to (7) when r satisfies the auxiliary equation

$$r^4 + 5r^2 + 4 = 0.$$

From the factorization $r^4 + 5r^2 + 4 = (r^2 + 1)(r^2 + 4)$, we see that the roots of the auxiliary equation are the complex numbers $i, -i, 2i, -2i$. Using Euler's formula, it follows that

$$z_1(t) = e^{it} = \cos t + i \sin t$$

and

$$z_2(t) = e^{2it} = \cos 2t + i \sin 2t$$

are complex-valued solutions to equation (7). To obtain real-valued solutions, we take the real and imaginary parts of $z_1(t)$ and $z_2(t)$. Thus four real-valued solutions are

$$x_1(t) = \cos t, \qquad x_2(t) = \sin t,$$
$$x_3(t) = \cos 2t, \qquad x_4(t) = \sin 2t,$$

and a general solution is

(8) $x(t) = a_1 \cos t + a_2 \sin t + a_3 \cos 2t + a_4 \sin 2t,$

where a_1, a_2, a_3, and a_4 are arbitrary constants.[†]
 To obtain a formula for $y(t)$, we use equation (3) to express y in terms of x,

$$y(t) = \frac{d^2x}{dt^2} + 3x$$

$$= -a_1 \cos t - a_2 \sin t - 4a_3 \cos 2t - 4a_4 \sin 2t$$

$$+ 3a_1 \cos t + 3a_2 \sin t + 3a_3 \cos 2t + 3a_4 \sin 2t,$$

and so

(9) $y(t) = 2a_1 \cos t + 2a_2 \sin t - a_3 \cos 2t - a_4 \sin 2t.$

To determine the constants a_1, a_2, a_3, and a_4, let's return to the original problem. We were told that the objects were originally displaced 3 m to the right and then released. Hence

(10) $x(0) = 3, \qquad \dfrac{dx}{dt}(0) = 0; \qquad y(0) = 3, \qquad \dfrac{dy}{dt}(0) = 0.$

[†] A more detailed discussion of general solutions is given in Chapter 6.

On differentiating equations (8) and (9), we find

$$\frac{dx}{dt} = -a_1 \sin t + a_2 \cos t - 2a_3 \sin 2t + 2a_4 \cos 2t,$$

$$\frac{dy}{dt} = -2a_1 \sin t + 2a_2 \cos t + 2a_3 \sin 2t - 2a_4 \cos 2t.$$

Now, if we put $t = 0$ in the formulas for x, dx/dt, y, and dy/dt, the initial conditions (10) give the four equations

$$x(0) = a_1 + a_3 = 3; \qquad \frac{dx}{dt}(0) = a_2 + 2a_4 = 0,$$

$$y(0) = 2a_1 - a_3 = 3, \qquad \frac{dy}{dt}(0) = 2a_2 - 2a_4 = 0.$$

From this system, we find $a_1 = 2$, $a_2 = 0$, $a_3 = 1$, and $a_4 = 0$. Hence the equations of motion for the two objects are

$$x(t) = 2 \cos t + \cos 2t,$$
$$y(t) = 4 \cos t - \cos 2t,$$

which are depicted in Figure 5.13. ∎ End

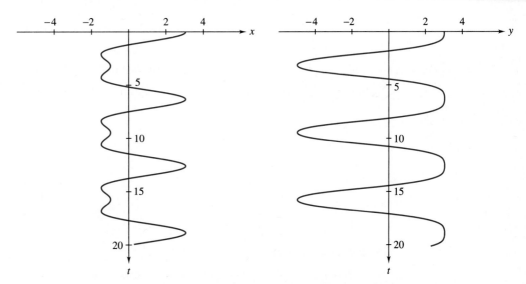

Figure 5.13 Graphs of the motion of the two masses in the coupled spring-mass system

The general solutions (8) and (9) that we have obtained are combinations of sinusoids oscillating at two different angular frequencies: 1 rad/sec and 2 rad/sec. These frequencies extend the notion of the *natural frequency* of the single (free, undamped) spring-mass oscillator (page 243, Section 5.1) and are called the **natural** (or **normal**) **angular fre-**

quencies of the system. A complex system consisting of more masses and springs would have many normal frequencies.

Notice that if the initial conditions were altered so that the constants a_3 and a_4 in (8) and (9) were zero, the motion would be a pure sinusoid oscillating at the single frequency 1 rad/sec. Similarly, if a_1 and a_2 were zero, only the 2 rad/sec oscillation would be "excited." Such solutions, wherein the entire motion is described by a single sinusoid, are called the **normal modes** of the system.[†] The normal modes in the following example are particularly easy to visualize because we shall take all the masses and all the spring constants to be equal.

EXAMPLE 2 Three identical springs with spring constant k and two identical masses m are attached in a straight line with the ends of the outside springs fixed (see Figure 5.14). Determine and interpret the normal modes of the system.

Figure 5.14 Coupled spring-mass system with fixed ends

Solution We define the displacements from equilibrium, x and y, as in Example 1. The equations expressing Newton's second law for the masses are quite analogous to (1), except for the effect of the third spring on the second mass:

(11) $mx'' = -kx + k(y - x),$

(12) $my'' = -k(y - x) - ky$

or

$$(mD^2 + 2k)[x] - ky = 0,$$
$$-kx + (mD^2 + 2k)[y] = 0.$$

Eliminating y in the usual manner results in

(13) $((mD^2 + 2k)^2 - k^2)[x] = 0.$

This has the auxiliary equation

$$(mr^2 + 2k)^2 - k^2 = (mr^2 + k)(mr^2 + 3k) = 0,$$

[†] The normal modes are more naturally characterized in terms of *eigenvalues* (see Section 9.5).

with roots $\pm i\sqrt{k/m}$, $\pm i\sqrt{3k/m}$. Setting $\omega := \sqrt{k/m}$, we get the following general solution to (13):

(14) $x(t) = C_1 \cos \omega t + C_2 \sin \omega t + C_3 \cos(\sqrt{3}\omega t) + C_4 \sin(\sqrt{3}\omega t)$.

To obtain $y(t)$, we solve for $y(t)$ in (11) and substitute $x(t)$ as given in (14). Upon simplifying, we get

(15) $y(t) = C_1 \cos \omega t + C_2 \sin \omega t - C_3 \cos(\sqrt{3}\omega t) - C_4 \sin(\sqrt{3}\omega t)$.

From equations (14) and (15), we see that the normal angular frequencies are ω and $\sqrt{3}\omega$. The system oscillates at the first frequency ω if $C_3 = C_4 = 0$; but then in fact the displacements x and y are equal. Therefore in this normal mode, the middle spring is never compressed and the pair of masses moves as a rigid body of mass $2m$, forced by a "double" spring of effective spring constant $2k$. The motion is depicted in Figure 5.15. Recall that in Section 5.1, we predicted that the undamped, free motion of a body of mass $2m$ with a spring constant $2k$ would be a sinusoid of angular frequency $\sqrt{(2k)/(2m)} = \omega$ (see equations (5), (6), and (7), pages 242–243).

The second normal mode, oscillating at frequency $\sqrt{3}\omega$, arises if $C_1 = C_2 = 0$ in equations (14) and (15), and the displacements x and y are equal but opposite. This symmetrical *mirror image* mode is depicted in Figure 5.16; the center of the middle spring is stationary and each mass feels the force of a "a spring and a half." ∎ End

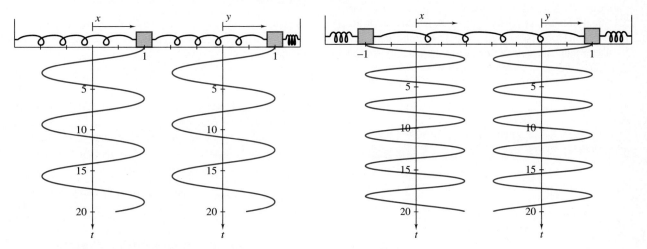

Figure 5.15 First normal mode for Example 2 **Figure 5.16** Second normal mode for Example 2

The study of the natural frequencies of complex systems is known as **modal analysis.** It is an important tool in mechanical and civil engineering, particularly in the stress analysis of such structures as buildings and bridges.

EXERCISES 5.4

1. Two springs and two masses are attached in a straight line on a horizontal frictionless surface as illustrated in Figure 5.17. The system is set in motion by holding the mass m_2 at its equilibrium position and pulling the mass m_1 to the left of its equilibrium position a distance 1 m and then releasing both masses. Express Newton's law for the system and determine the equations of motion for the two masses if $m_1 = 1$ kg, $m_2 = 2$ kg, $k_1 = 4$ N/m, and $k_2 = 10/3$ N/m.

Figure 5.17 Coupled spring-mass system with one end free

2. Determine the equations of motion for the two masses described in Problem 1 if $m_1 = 1$ kg, $m_2 = 1$ kg, $k_1 = 3$ N/m, and $k_2 = 2$ N/m.

3. Four springs with the same spring constant and three equal masses are attached in a straight line on a horizontal frictionless surface as illustrated in Figure 5.18. Determine the normal frequencies for the system and describe the three normal modes of vibration.

Figure 5.18 Coupled spring-mass system with three degrees of freedom

4. Two springs, two masses, and a dashpot are attached in a straight line on a horizontal frictionless surface as shown in Figure 5.19. Derive the system of differential equations for the displacements x and y. [*Hint:* For the force due to the dashpot, see page 242, Section 5.1.]

Figure 5.19 Coupled spring-mass system with one end damped

5. Two springs, two masses, and a dashpot are attached in a straight line on a horizontal frictionless surface as shown in Figure 5.20. The system is set in motion by holding the mass m_2 at its equilibrium position and pushing the mass m_1 to the left of its equilibrium position a distance 2 m and then releasing both masses. Determine the equations of motion for the two masses if $m_1 = m_2 = 1$ kg, $k_1 = k_2 = 1$ N/m, and $b = 1$ N-sec/m. [*Hint:* For the force due to the dashpot, see page 242, Section 5.1.]

Figure 5.20 Coupled spring-mass system with damping between the masses

6. Referring to the coupled spring-mass system discussed in Example 1, suppose an external force $E(t) = 37 \cos 3t$ is applied to the second object of mass 1 kg. The displacement functions $x(t)$ and $y(t)$ now satisfy the system

(16) $2x''(t) + 6x(t) - 2y(t) = 0,$

(17) $y''(t) + 2y(t) - 2x(t) = 37 \cos 3t.$

(a) Show that $x(t)$ satisfies the equation

(18) $x^{(4)}(t) + 5x''(t) + 4x(t) = 37 \cos 3t.$

(b) Find a general solution $x(t)$ to equation (18). [*Hint:* Use undetermined coefficients with $x_p = A \cos 3t + B \sin 3t$.]

(c) Substitute $x(t)$ back into (16) to obtain a general solution for $y(t)$.

(d) If both masses are displaced 2 m to the right of their equilibrium positions and then released, find the displacement functions $x(t)$ and $y(t)$.

7. Suppose the displacement functions $x(t)$ and $y(t)$ for a coupled spring-mass system (similar to the one discussed in Problem 6) satisfy the initial value problem

$$x''(t) + 5x(t) - 2y(t) = 0,$$

$$y''(t) + 2y(t) - 2x(t) = 3 \sin 2t;$$

$$x(0) = x'(0) = 0,$$

$$y(0) = 1, \quad y'(0) = 0.$$

Solve for $x(t)$ and $y(t)$.

8. A double pendulum swinging in a vertical plane under the influence of gravity (see Figure 5.21) satisfies the system

$$(m_1 + m_2)l_1^2\theta_1'' + m_2l_1l_2\theta_2'' + (m_1 + m_2)l_1 g\theta_1 = 0,$$

$$m_2l_2^2\theta_2'' + m_2l_1l_2\theta_1'' + m_2l_2 g\theta_2 = 0,$$

when θ_1 and θ_2 are small angles. Solve the system when $m_1 = 3$ kg, $m_2 = 2$ kg, $l_1 = l_2 = 5$ m, $\theta_1(0) = \pi/6$, $\theta_2(0) = \theta_1'(0) = \theta_2'(0) = 0$.

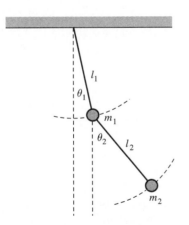

Figure 5.21 Double pendulum

9. The motion of a pair of identical pendulums coupled by a spring is modeled by the system

$$mx_1'' = -\frac{mg}{l}x_1 - k(x_1 - x_2),$$

$$mx_2'' = -\frac{mg}{l}x_2 + k(x_1 - x_2)$$

for small displacements (see Figure 5.22). Determine the two normal frequencies for the system.

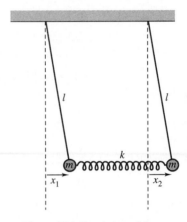

Figure 5.22 Coupled pendulums

10. Suppose the coupled spring-mass system of Problem 1 (Figure 5.17) is hung *vertically* from a support (with mass m_2 above m_1), as in Section 5.1.

(a) Argue that at equilibrium, the lower spring is stretched a distance l_1 from its natural length L_1, given by $l_1 = m_1g/k_1$.

(b) Argue that at equilibrium, the upper spring is stretched a distance $l_2 = (m_1 + m_2)g/k_2$.

(c) Show that if x_1 and x_2 are redefined to be displacements from the equilibrium positions of the masses m_1 and m_2, then the equations of motion are identical with those derived in Problem 1.

[*Hint*: Reread Section 5.1.]

5.5 ELECTRIC CIRCUITS

In this section, we consider the application of differential equations to simple electric circuits consisting of voltage sources (e.g., batteries or generators), resistors, inductors, and capacitors. The so-called series *RLC* circuit is shown schematically in Figure 5.23 on page 268.

Figure 5.23 Schematic representation of an *RLC* series circuit

The physical principles governing electric circuits were formulated by G. R. Kirchhoff in 1859. They are the following:

1. **Kirchhoff's current law:** The algebraic sum of the currents flowing into any junction point must be zero.
2. **Kirchhoff's voltage law**: The algebraic sum of the instantaneous changes in potential (voltage drops) around any closed loop must be zero.

Kirchhoff's current law implies that the same current, I, passes through each element in the series circuit of Figure 5.23.

To apply Kirchhoff's voltage law, we need to know the voltage drop across each element of the circuit. These voltage formulas are stated below (the reader can consult an introductory physics text for further details).

(a) According to Ohm's law, the voltage drop E_R across a resistor is proportional to the current I passing through the resistor:

$$E_R = RI.$$

The proportionality constant R is called the **resistance.**

(b) It can be shown using Faraday's law and Lenz's law that the voltage drop E_L across an inductor is proportional to the instantaneous rate of change of the current I:

$$E_L = L\frac{dI}{dt}.$$

The proportionality constant L is called the **inductance.**

(c) The voltage drop E_C across a capacitor is proportional to the electric charge q on the capacitor:

$$E_C = \frac{1}{C}q.$$

The constant C is called the **capacitance.**

A voltage source is assumed to *add* voltage or potential energy to the circuit. If we let $E(t)$ denote the voltage supplied to the circuit at time t, then Kirchhoff's voltage law gives

(1) $E_L + E_R + E_C = E(t).$

Substituting into (1) the expressions for E_L, E_R, and E_C gives

(2) $$L\frac{dI}{dt} + RI + \frac{1}{C}q = E(t).$$

Now, the current is just the instantaneous rate of change in charge; that is, $I = dq/dt$. Therefore we can express (2) in terms of the single variable q as follows:

(3) $$L\frac{d^2q}{dt^2} + R\frac{dq}{dt} + \frac{1}{C}q = E(t).$$

In most applications, we will be interested in determining the current $I(t)$. If we differentiate (3) with respect to t and substitute I for dq/dt, we obtain

(4) $$L\frac{d^2I}{dt^2} + R\frac{dI}{dt} + \frac{1}{C}I = \frac{dE}{dt}.$$

For an *RLC* series circuit, one is usually given the initial charge on the capacitor $q(0)$ and the initial current $I(0) = q'(0)$. These constants provide the initial conditions for equation (3). However, for equation (4), we also need to know the initial rate of change in the current, $I'(0)$. This can be obtained from equation (2) by substituting in the initial values $q(0)$, $I(0)$, and $E(0)$.

The common units and symbols used for electric circuits are listed in Table 5.1.

TABLE 5.1 COMMON UNITS AND SYMBOLS USED WITH ELECTRIC CIRCUITS

Quantity	Letter Representation	Units	Symbol Representation
Voltage source	E	volt (V)	—O— Generator —╢├— Battery
Resistance	R	ohm (Ω)	—W—
Inductance	L	henry (H)	—0000—
Capacitance	C	farad (F)	—╢├—
Charge	q	coulomb	
Current	I	ampere	

EXAMPLE 1 An *RLC* series circuit has a voltage source given by $E(t) = \sin 100t$ volts, a resistor of 0.02 ohms, an inductor of 0.001 henrys, and a capacitor of 2 farads. (These values are selected for convenience; typical values for capacitance are much smaller.) If the initial current and the initial charge on the capacitor are both zero, determine the current in the circuit for $t > 0$.

Solution Here $L = 0.001$, $R = 0.02$, $C = 2$, and $E(t) = \sin 100t$. Substituting these into equation (4) for the current gives

$$(0.001)\frac{d^2I}{dt^2} + (0.02)\frac{dI}{dt} + (0.5)I = 100 \cos 100t,$$

or, equivalently,

$$(5) \qquad \frac{d^2I}{dt^2} + 20\frac{dI}{dt} + 500I = 100{,}000 \cos 100t.$$

The homogeneous equation associated with (5) has the auxiliary equation

$$r^2 + 20r + 500 = (r + 10)^2 + (20)^2 = 0,$$

whose roots are $-10 \pm 20i$. Hence the solution to the homogeneous equation is

$$(6) \qquad I_h(t) = C_1 e^{-10t} \cos 20t + C_2 e^{-10t} \sin 20t.$$

To find a particular solution for (5), we can use the method of undetermined coefficients. Setting

$$I_p(t) = A \cos 100t + B \sin 100t$$

and carrying out the procedure in Section 4.8, we ultimately find

$$A = -\frac{95}{9.425}, \qquad B = \frac{20}{9.425}.$$

Hence a particular solution to (5) is given by

$$(7) \qquad I_p(t) = -\frac{95}{9.425} \cos 100t + \frac{20}{9.425} \sin 100t.$$

Since $I = I_h + I_p$, we find from (6) and (7) that

$$(8) \qquad I(t) = e^{-10t}(C_1 \cos 20t + C_2 \sin 20t) - \frac{95}{9.425} \cos 100t + \frac{20}{9.425} \sin 100t.$$

To determine the constants C_1 and C_2, we need the values $I(0)$ and $I'(0)$. We were given $I(0) = q(0) = 0$. To find $I'(0)$, we substitute the values for L, R, and C into equation (2) and equate the two sides at $t = 0$. This gives

$$(0.001)I'(0) + (0.02)I(0) + (0.5)q(0) = \sin 0.$$

Because $I(0) = q(0) = 0$, we find $I'(0) = 0$. Finally, using $I(t)$ in (8) and the initial conditions $I(0) = I'(0) = 0$, we obtain the system

$$I(0) = C_1 - \frac{95}{9.425} = 0,$$

$$I'(0) = -10C_1 + 20C_2 + \frac{2000}{9.425} = 0.$$

Solving this system yields $C_1 = 95/9.425$ and $C_2 = -105/18.85$. Hence the current in the *RLC* series circuit is

(9) $$I(t) = e^{-10t}\left(\frac{95}{9.425}\cos 20t - \frac{105}{18.85}\sin 20t\right)$$
$$- \frac{95}{9.425}\cos 100t + \frac{20}{9.425}\sin 100t. \quad \blacksquare \text{ End}$$

Observe that, as was the case with forced mechanical vibrations, the current in (9) is made up of two components. The first, I_h, is a **transient current** that tends to zero as $t \to \infty$. The second,

$$I_p(t) = -\frac{95}{9.425}\cos 100t + \frac{20}{9.425}\sin 100t,$$

is a sinusoidal **steady-state current** that remains.

We leave it for the reader to verify that the steady-state solution $I_p(t)$ that arises from the more general voltage source $E(t) = E_0 \sin \gamma t$ is

(10) $$I_p(t) = \frac{E_0 \sin(\gamma t + \theta)}{\sqrt{R^2 + [\gamma L - 1/(\gamma C)]^2}},$$

where $\tan \theta = (1/C - L\gamma^2)/(\gamma R)$.

EXAMPLE 2 At time $t = 0$, the charge on the capacitor in the electric network shown in Figure 5.24 is 2 coulombs, while the current through the capacitor is zero. Determine the charge on the capacitor and the currents in the various branches of the network at any time $t > 0$.

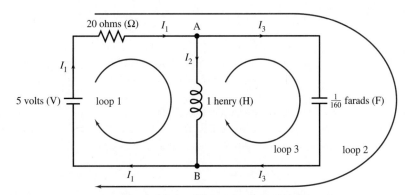

Figure 5.24 Schematic of an electric network

Solution To determine the charge and currents in the electric network, we begin by observing that the network consists of three closed circuits: loop 1 through the battery, resistor, and inductor; loop 2 through the battery, resistor, and capacitor; and loop 3 containing the capacitor and inductor. Taking advantage of Kirchhoff's current law, we denote the current passing

through the battery and the resistor by I_1, the current through the inductor by I_2, and the current through the capacitor by I_3. For consistency of notation, we denote the charge on the capacitor by q_3. Hence $I_3 = dq_3/dt$.

As discussed at the beginning of this section, the voltage drop at a resistor is RI, at an inductor $L\,dI/dt$, and at a capacitor q/C. So, applying Kirchhoff's voltage law to the electric network in Figure 5.24, we find for loop 1,

(11) $$\frac{dI_2}{dt} + 20I_1 = 5;$$

<center>(inductor) (resistor) (battery)</center>

for loop 2,

(12) $$20I_1 + 160q_3 = 5;$$

<center>(resistor) (capacitor) (battery)</center>

and for loop 3,

(13) $$-\frac{dI_2}{dt} + 160q_3 = 0.$$

<center>**(inductor)** **(capacitor)**</center>

(The minus sign in (13) arises from taking a clockwise path around loop 3 so that the current passing through the inductor is $-I_2$.) Notice that these three equations are not independent: We can obtain equation (13) by subtracting (11) from (12). Hence we have only two equations from which to determine the three unknowns I_1, I_2, and q_3. If we now apply Kirchhoff's current law to the two junction points in the network, we find at point A that $I_1 - I_2 - I_3 = 0$ and at point B that $I_2 + I_3 - I_1 = 0$. In both cases, we get

(14) $$I_1 - I_2 - \frac{dq_3}{dt} = 0,$$

since $I_3 = dq_3/dt$. Assembling equations (11), (12), and (14) into a system, we have (with $D = d/dt$)

(15) $$DI_2 + 20I_1 \qquad\qquad = 5,$$
(16) $$\qquad 20I_1 + 160q_3 = 5,$$
(17) $$-I_2 \quad + I_1 \quad - Dq_3 = 0.$$

We solve these by the elimination method of Section 4.10. Using equation (16) to eliminate I_1 from the others, we are left with

(18) $$DI_2 - 160q_3 \qquad\qquad = 0,$$
(19) $$20I_2 + (20D + 160)q_3 = 5.$$

Elimination of I_2 then leads to

(20) $$20D^2q_3 + 160Dq_3 + 3200q_3 = 0.$$

To obtain the initial conditions for (20), recall that at time $t = 0$, the charge on the capacitor is 2 coulombs and the current is zero. Hence

(21) $q_3(0) = 2, \quad \dfrac{dq_3}{dt}(0) = 0.$

We can now solve the initial value problem (20)–(21) using the techniques of Chapter 4. Ultimately we find

$$q_3(t) = 2e^{-4t} \cos 12t + \frac{2}{3} e^{-4t} \sin 12t,$$

$$I_3(t) = \frac{dq_3}{dt}(t) = -\frac{80}{3} e^{-4t} \sin 12t.$$

Next, to determine I_2, we substitute these expressions into (19) and obtain

$$I_2(t) = \frac{1}{4} - \frac{dq_3}{dt}(t) - 8q_3(t)$$

$$= \frac{1}{4} - 16e^{-4t} \cos 12t + \frac{64}{3} e^{-4t} \sin 12t.$$

Finally, from $I_1 = I_2 + I_3$, we get

$$I_1(t) = \frac{1}{4} - 16e^{-4t} \cos 12t - \frac{16}{3} e^{-4t} \sin 12t. \quad \blacksquare \text{ End}$$

Note that the differential equations that describe mechanical vibrations and *RLC* series circuits are essentially the same. And, in fact, there is a natural identification of the parameters m, b, and k for a spring-mass system with the parameters L, R, and C that describe circuits. This is illustrated in Table 5.2. Moreover, the terms *transient, steady-state, overdamped, critically damped, underdamped*, and *resonance frequency* described in the preceding sections apply to electric circuits as well.

TABLE 5.2 ANALOGY BETWEEN MECHANICAL AND ELECTRIC SYSTEMS			
Mechanical Spring-Mass System with Damping		**Electric *RLC* Series Circuit**	
$mx'' + bx' + kx = f(t)$		$Lq'' + Rq' + (1/C)q = E(t)$	
Displacement	x	Charge	q
Velocity	x'	Current	$q' = I$
Mass	m	Inductance	L
Damping constant	b	Resistance	R
Spring constant	k	(Capacitance)$^{-1}$	$1/C$
External force	$f(t)$	Voltage source	$E(t)$

This analogy between a mechanical system and an electric circuit extends to large-scale systems and circuits. An interesting consequence of this is the use of **analog simulation** and, in particular, analog computers to analyze mechanical systems. Large-scale mechanical systems are modeled by building a corresponding electric system and then measuring the charges $q(t)$ and currents $I(t)$.

While such analog simulations are important, *both* large-scale mechanical and electric systems are currently modeled using digital computer simulation. This involves the numerical solution of the initial value problem governing the system. Still, the analogy between mechanical and electric systems means that basically the same computer software can be used to analyze both systems.

EXERCISES 5.5

1. An *RLC* series circuit has a voltage source given by $E(t) = 20$ volts, a resistor of 100 ohms, an inductor of 4 henrys, and a capacitor of 0.01 farads. If the initial current is zero and the initial charge on the capacitor is 4 coulombs, determine the current in the circuit for $t > 0$.

2. An *RLC* series circuit has a voltage source given by $E(t) = 40 \cos 2t$ volts, a resistor of 2 ohms, an inductor of 1/4 henrys, and a capacitor of 1/13 farads. If the initial current is zero and the initial charge on the capacitor is 3.5 coulombs, determine the charge on the capacitor for $t > 0$.

3. An *RLC* series circuit has a voltage source given by $E(t) = 10 \cos 20t$ volts, a resistor of 120 ohms, an inductor of 4 henrys, and a capacitor of $(2200)^{-1}$ farads. Find the steady-state current (solution) for this circuit. What is the resonance frequency of the circuit?

4. An *LC* series circuit has a voltage source given by $E(t) = 30 \sin 50t$ volts, an inductor of 2 henrys, and a capacitor of 0.02 farads (but no resistor). What is the current in this circuit for $t > 0$ if at $t = 0$, $I(0) = q(0) = 0$?

5. An *RLC* series circuit has a voltage source of the form $E(t) = E_0 \cos \gamma t$ volts, a resistor of 10 ohms, an inductor of 4 henrys, and a capacitor of 0.01 farads. Sketch the frequency response curve for this circuit.

6. Show that when the voltage source in (3) is of the form $E(t) = E_0 \sin \gamma t$, then the steady-state solution I_p is as given in equation (10).

7. A spring-mass system with damping consists of a 7-kg mass, a spring with spring constant 3 N/m, a frictional component with damping constant 2 N-sec/m, and an external force given by $f(t) = 10 \cos 10t$ N. Using a 10-ohm resistor, construct an *RLC* series circuit that is the analog of this mechanical system in the sense that the two systems are governed by the same differential equation.

8. A spring-mass system with damping consists of a 16-lb weight, a spring with spring constant 64 lb/ft, a frictional component with damping constant 10 lb-sec/ft, and an external force given by $f(t) = 20 \cos 8t$ lb. Using an inductor of 0.01 henrys, construct an *RLC* series circuit that is the analog of this mechanical system.

9. Because of Euler's formula, $e^{i\theta} = \cos \theta + i \sin \theta$, it is often convenient to treat the voltage sources $E_0 \cos \gamma t$ and $E_0 \sin \gamma t$ simultaneously, using $E(t) = E_0 e^{i\gamma t}$. In this case, equation (3) becomes

(22) $$L\frac{d^2q}{dt^2} + R\frac{dq}{dt} + \frac{1}{C}q = E_0 e^{i\gamma t}.$$

(a) Show that the steady-state solution to (22) is

$$q_p(t) = \frac{E_0}{1/C - \gamma^2 L + i\gamma R} e^{i\gamma t}.$$

[*Hint:* Use the method of undetermined coefficients with the guess $q_p = Ae^{i\gamma t}$, where A denotes a complex constant. The technique is discussed in detail in Project H at the end of Chapter 4.]

(b) Now show that the steady-state current is

$$I_p(t) = \frac{E_0}{R + i[\gamma L + 1/(\gamma C)]} e^{i\gamma t}.$$

(c) Use the relation $\alpha + i\beta = \sqrt{\alpha^2 + \beta^2}\, e^{i\theta}$, where $\tan\theta = \beta/\alpha$, to show that I_p can be expressed in the form

$$I_p(t) = \frac{E_0}{\sqrt{R^2 + [\gamma L - 1/(\gamma C)]^2}}\, e^{i(\gamma t + \theta)},$$

where $\tan\theta = (1/C - L\gamma^2)/(\gamma R)$.

(d) The imaginary part of $e^{i\gamma t}$ is $\sin\gamma t$, so the imaginary part of the solution to (22) must be the solution to equation (3) for $E(t) = E_0 \sin\gamma t$. Verify that this is also the case for the current by showing that the imaginary part of I_p in part (c) is the same as that given in equation (10).

In Problems 10 through 13, find a system of differential equations and initial conditions for the currents in the networks given in the schematic diagram. Assume all initial currents are zero. Solve for the currents in each branch of the network.

10.

Figure 5.25 *RL* network for Problem 10

11.

Figure 5.26 *RLC* network for Problem 11

12.

Figure 5.27 *RL* network for Problem 12

13.

Figure 5.28 *RLC* network for Problem 13

5.6 NUMERICAL METHODS FOR HIGHER-ORDER EQUATIONS AND SYSTEMS

In Section 3.6, we discussed the Runge-Kutta algorithm for obtaining numerical approximations to solutions of first order equations. We now extend the algorithm so that it provides a technique for solving initial value problems for a system of equations in normal form. Since we showed in Section 4.10 (page 208) how to convert any mth order equation into such a first order system, the extended method will solve higher-order equations as well.

To see how these extensions are done let's recall, from Section 3.6, the classical Runge-Kutta method of order four. For the initial value problem

$$x' = f(t, x), \qquad x(t_0) = x_0,$$

the recursive formulas for the fourth order Runge-Kutta method are

(1)
$$t_{n+1} = t_n + h, \qquad n = 0, 1, \ldots,$$
$$x_{n+1} = x_n + \frac{1}{6}(k_1 + 2k_2 + 2k_3 + k_4),$$

where h is the step size and

(2)
$$k_1 = hf(t_n, x_n), \qquad k_2 = hf\left(t_n + \frac{h}{2}, x_n + \frac{1}{2}k_1\right),$$
$$k_3 = hf\left(t_n + \frac{h}{2}, x_n + \frac{1}{2}k_2\right), \qquad k_4 = hf(t_n + h, x_n + k_3).$$

Now suppose we want to approximate the solution $x_1(t)$, $x_2(t)$ to the *system*

(3)
$$x_1'(t) = f_1(t, x_1, x_2),$$
$$x_2'(t) = f_2(t, x_1, x_2)$$

that satisfies the initial conditions

(4) $x_1(t_0) = a_1, \qquad x_2(t_0) = a_2.$

Let $x_{n,1}$ and $x_{n,2}$ denote approximations to $x_1(t_n)$ and $x_2(t_n)$, respectively, where $t_n = t_0 + nh$ for $n = 0, 1, 2, \ldots$. The recursive formulas for the fourth order Runge-Kutta method for the system (3)–(4) are obtained from the formulas for a single equation given in (1)–(2) by treating each of the quantities x_n, k_1, k_2, k_3, and k_4 as *vectors*. Namely, we set $\mathbf{x}_n :=$ $(x_{n,1}, x_{n,2})$, $\mathbf{k}_1 := (k_{1,1}, k_{1,2})$, $\mathbf{k}_2 := (k_{2,1}, k_{2,2})$, etc. This approach gives the following recursive formulas for the fourth order Runge-Kutta method for system (3)–(4):

(5) $t_{n+1} := t_n + h, \qquad n = 0, 1, 2, \ldots,$

(6) $x_{n+1,1} := x_{n,1} + \dfrac{1}{6}(k_{1,1} + 2k_{2,1} + 2k_{3,1} + k_{4,1}),$

(7) $x_{n+1,2} := x_{n,2} + \dfrac{1}{6}(k_{1,2} + 2k_{2,2} + 2k_{3,2} + k_{4,2}),$

where h is the step size and, for $i = 1$ and 2,

$$k_{1,i} := hf_i(t_n, x_{n,1}, x_{n,2}),$$
$$k_{2,i} := hf_i\left(t_n + \frac{h}{2}, x_{n,1} + \frac{1}{2}k_{1,1}, x_{n,2} + \frac{1}{2}k_{1,2}\right),$$
$$k_{3,i} := hf_i\left(t_n + \frac{h}{2}, x_{n,1} + \frac{1}{2}k_{2,1}, x_{n,2} + \frac{1}{2}k_{2,2}\right),$$
$$k_{4,i} := hf_i(t_n + h, x_{n,1} + k_{3,1}, x_{n,2} + k_{3,2}).$$

It is important to note that both $k_{1,1}$ and $k_{1,2}$ must be computed before either $k_{2,1}$ or $k_{2,2}$. Similarly, both $k_{2,1}$ and $k_{2,2}$ are needed to compute $k_{3,1}$ and $k_{3,2}$, etc.

To adapt this algorithm to a single second order equation, we recall the procedure from Section 4.10. The initial value problem

(8) $y''(t) = F(t, y, y'); \qquad y(t_0) = y_0, \qquad y'(t_0) = y_1$

is converted by the identifications

(9) $x_1(t) := y(t),$ $x_2(t) := y'(t)$

into the system

(10)
$$x_1'(t) = y'(t) = x_2(t),$$
$$x_2'(t) = y''(t) = F(t, x_1, x_2),$$

and the initial conditions become

(11) $x_1(t_0) = y(t_0) = y_0,$ $x_2(t_0) = y'(t_0) = y_1.$

Thus we identify $f_1(t, x_1, x_2)$ in (3) with x_2, and $f_2(t, x_1, x_2)$ with $F(t, x_1, x_2)$.

EXAMPLE 1 Use the Runge-Kutta method with step size $h = 0.125$ to find an approximation for the solution to the initial value problem

(12) $y''(t) + 3y'(t) + 2y(t) = 0;$ $y(0) = 1,$ $y'(0) = 1,$

on the interval $[0, 1]$.

Solution For the given step size, the method will yield approximations for $y(0.125)$, $y(0.250)$, ..., $y(1)$. The substitutions (9)–(11) convert the initial value problem in (12) to

(13)
$$x_1' = x_2; \qquad x_1(0) = 1,$$
$$x_2' = -3x_2 - 2x_1; \qquad x_2(0) = 1.$$

Comparing this system with (3), we see that $f_1(t, x_1, x_2) = x_2$ and $f_2(t, x_1, x_2) = -3x_2 - 2x_1$. With the starting values of $t_0 = 0$, $x_{0,1} = 1$, and $x_{0,2} = 1$, we compute

$$k_{1,1} = hf_1(t_0, x_{0,1}, x_{0,2}) = hx_{0,2} = (0.125)(1) = 0.125,$$

$$k_{1,2} = hf_2(t_0, x_{0,1}, x_{0,2}) = h(-3x_{0,2} - 2x_{0,1})$$

$$= (0.125)[(-3)(1) - 2(1)] = -0.625,$$

$$k_{2,1} = hf_1\left(t_0 + \frac{h}{2}, x_{0,1} + \frac{1}{2}k_{1,1}, x_{0,2} + \frac{1}{2}k_{1,2}\right)$$

$$= h\left(x_{0,2} + \frac{1}{2}k_{1,2}\right) = (0.125)(1 - 0.3125) = 0.08594,$$

$$k_{2,2} = hf_2\left(t_0 + \frac{h}{2}, x_{0,1} + \frac{1}{2}k_{1,1}, x_{0,2} + \frac{1}{2}k_{1,2}\right)$$

$$= -3h\left(x_{0,2} + \frac{1}{2}k_{1,2}\right) - 2h\left(x_{0,1} + \frac{1}{2}k_{1,1}\right)$$

$$= -3(0.125)(1 - 0.3125) - 2(0.125)(1 + 0.0625)$$

$$= -0.52344.$$

Similarly, we find

$$k_{3,1} = hf_1\left(t_0 + \frac{h}{2}, x_{0,1} + \frac{1}{2}k_{2,1}, x_{0,2} + \frac{1}{2}k_{2,2}\right) = 0.09229,$$

$$k_{3,2} = hf_2\left(t_0 + \frac{h}{2}, x_{0,1} + \frac{1}{2}k_{2,1}, x_{0,2} + \frac{1}{2}k_{2,2}\right) = -0.53760,$$

$$k_{4,1} = hf_1(t_0 + h, x_{0,1} + k_{3,1}, x_{0,2} + k_{3,2}) = 0.05780,$$

$$k_{4,2} = hf_2(t_0 + h, x_{0,1} + k_{3,1}, x_{0,2} + k_{3,2}) = -0.44647.$$

Hence from (6) and (7), we compute

$$x_{1,1} = x_{0,1} + \frac{1}{6}(k_{1,1} + 2k_{2,1} + 2k_{3,1} + k_{4,1})$$

$$= 1 + \frac{1}{6}(0.125 + 2(0.08594) + 2(0.09229) + 0.05780)$$

$$= 1.08987,$$

$$x_{1,2} = x_{0,2} + \frac{1}{6}(k_{1,2} + 2k_{2,2} + 2k_{3,2} + k_{4,2})$$

$$= 1 - \frac{1}{6}(0.625 + 2(0.52344) + 2(0.53760) + 0.44647)$$

$$= 0.46774.$$

Continuing the algorithm, we compute $x_{i,1}$ and $x_{i,2}$ for $i = 2, 3, \ldots, 8$. These values, along with the values of the actual solution to (12), $y(t) = x_1(t) = 3e^{-t} - 2e^{-2t}$, are given in Table 5.3 (to five decimal places). Note that the $x_{i,2}$ column gives approximations for $y'(t)$, since $x_2(t) = y'(t)$. ■ End

t_i	$x_{i,1}$	$x_{i,2}$	$y(t_i) = x_1(t_i)$	$\lvert y(t_i) - x_{i,1}\rvert$
0.000	1.00000	1.00000	1.00000	0
0.125	1.08987	0.46774	1.08989	2×10^{-5}
0.250	1.12332	0.08977	1.12334	2×10^{-5}
0.375	1.11711	-0.17235	1.11713	2×10^{-5}
0.500	1.08381	-0.34802	1.08383	2×10^{-5}
0.625	1.03275	-0.45971	1.03277	2×10^{-5}
0.750	0.97981	-0.52453	0.97984	3×10^{-5}
0.875	0.90302	-0.55544	0.90304	2×10^{-5}
1.000	0.83295	-0.56226	0.83297	2×10^{-5}

TABLE 5.3 FOURTH ORDER RUNGE-KUTTA APPROXIMATION OF THE SOLUTION TO (12) USING $h = 0.125$

The Runge-Kutta program outlines of Section 3.6 for a single first order equation can be easily modified to handle second order systems (see pages 132–133). For systems, the algorithm "with tolerance" stops when *each* component of two successive approximations differs by less than a prescribed tolerance ε or after a prescribed maximum number of iterations.

FOURTH ORDER RUNGE-KUTTA SUBROUTINE (SYSTEMS)

Purpose To approximate the solution to the initial value problem

$$x_1' = f_1(t, x_1, x_2), \qquad x_2' = f_2(t, x_1, x_2);$$
$$x_1(t_0) = a_1, \qquad x_2(t_0) = a_2,$$

for $t_0 \leq t \leq c$

INPUT t_0, a_1, a_2, c, N (number of steps), PRNTR (=1 to print table)

Step 1 Set step size $h = (c - t_0)/N$, $x_1 = a_1$, $x_2 = a_2$

Step 2 For $i = 1$ to N, do Steps 3 through 5

Step 3 Set

$$k_{1,1} = hf_1(t, x_1, x_2)$$
$$k_{1,2} = hf_2(t, x_1, x_2)$$
$$k_{2,1} = hf_1\left(t + \frac{h}{2}, x_1 + \frac{1}{2}k_{1,1}, x_2 + \frac{1}{2}k_{1,2}\right)$$
$$k_{2,2} = hf_2\left(t + \frac{h}{2}, x_1 + \frac{1}{2}k_{1,1}, x_2 + \frac{1}{2}k_{1,2}\right)$$
$$k_{3,1} = hf_1\left(t + \frac{h}{2}, x_1 + \frac{1}{2}k_{2,1}, x_2 + \frac{1}{2}k_{2,2}\right)$$
$$k_{3,2} = hf_2\left(t + \frac{h}{2}, x_1 + \frac{1}{2}k_{2,1}, x_2 + \frac{1}{2}k_{2,2}\right)$$
$$k_{4,1} = hf_1(t + h, x_1 + k_{3,1}, x_2 + k_{3,2})$$
$$k_{4,2} = hf_2(t + h, x_1 + k_{3,1}, x_2 + k_{3,2})$$

Step 4 Set

$$t = t + h$$
$$x_1 = x_1 + \frac{1}{6}(k_{1,1} + 2k_{2,1} + 2k_{3,1} + k_{4,1})$$
$$x_2 = x_2 + \frac{1}{6}(k_{1,2} + 2k_{2,2} + 2k_{3,2} + k_{4,2})$$

Step 5 If PRNTR = 1, print t, x_1, x_2

FOURTH ORDER RUNGE-KUTTA ALGORITHM WITH TOLERANCE (SYSTEMS)

Purpose To approximate the solution to the initial value problem

$$x_1' = f_1(t, x_1, x_2), \qquad x_2' = f_2(t, x_1, x_2);$$

$$x_1(t_0) = a_1, \qquad x_2(t_0) = a_2,$$

at $t = c$, with tolerance ε

INPUT $t_0, a_1, a_2, c, \varepsilon, M$ (maximum number of iterations)

Step 1 Set $z_1 = a_1, z_2 = a_2$, PRNTR $= 0$

Step 2 For $m = 0$ to M, do Steps 3 through 7 (Or, to save time, start with $m > 0$.)

Step 3 Set $N = 2^m$

Step 4 Call FOURTH ORDER RUNGE-KUTTA SUBROUTINE (SYSTEMS)

Step 5 Print h, x_1, x_2

Step 6 If $|z_1 - x_1| < \varepsilon$ and $|z_2 - x_2| < \varepsilon$, go to Step 10

Step 7 Set $z_1 = x_1, z_2 = x_2$

Step 8 Print ''$x_1(c)$ is approximately''; x_1; ''and $x_2(c)$ is approximately''; x_2; ''but may not be within the tolerance''; ε

Step 9 Go to Step 11

Step 10 Print ''$x_1(c)$ is approximately''; x_1; ''and $x_2(c)$ is approximately''; x_2; ''with tolerance''; ε

Step 11 STOP

OUTPUT Approximations of the solution to the initial value problem at $t = c$, using 2^m steps

For a stopping procedure based on the *relative* error, replace Step 6 by

Step 6' If $\left| \dfrac{z_1 - x_1}{x_1} \right| < \varepsilon$ and $\left| \dfrac{z_2 - x_2}{x_2} \right| < \varepsilon$, go to Step 10.

EXAMPLE 2 Use the preceding Runge-Kutta algorithm with a stopping procedure based on the absolute error and a tolerance of $\varepsilon = 0.001$ to approximate the solution $y(t)$ of the initial value problem

(14) $y'' + ty' + y^3 = 0; \qquad y(0) = 1, \qquad y'(0) = 0,$

at $t = 2$.

Solution On converting (14) to the equivalent system

$$x_1' = x_2, \qquad x_1(0) = 1,$$

$$x_2' = -tx_2 - x_1^3, \qquad x_2(0) = 0,$$

where $x_1 = y$, $x_2 = y'$, we use as inputs for the algorithm $t_0 = 0$, $a_1 = y(0) = 1$, $a_2 = y'(0) = 0$, $c = 2$, $\varepsilon = 0.001$, and $M = 10$ (say). Then, the formulas in Step 3 of the

SUBROUTINE are

$$k_{1,1} = hx_2, \qquad\qquad k_{1,2} = h(-tx_2 - x_1^3),$$

$$k_{2,1} = h\left(x_2 + \frac{1}{2}k_{1,2}\right), \qquad k_{2,2} = h\left[-\left(t + \frac{h}{2}\right)\left(x_2 + \frac{1}{2}k_{1,2}\right) - \left(x_1 + \frac{1}{2}k_{1,1}\right)^3\right],$$

$$k_{3,1} = h\left(x_2 + \frac{1}{2}k_{2,2}\right), \qquad k_{3,2} = h\left[-\left(t + \frac{h}{2}\right)\left(x_2 + \frac{1}{2}k_{2,2}\right) - \left(x_1 + \frac{1}{2}k_{2,1}\right)^3\right],$$

$$k_{4,1} = h(x_2 + k_{3,2}), \qquad k_{4,2} = h[-(t + h)(x_2 + k_{3,2}) - (x_1 + k_{3,1})^3].$$

In Table 5.4, we give the approximations $x_1(2; 2^{-m+1})$ for $y(2)$ and $x_2(2; 2^{-m+1})$ for $y'(2)$ for $m = 0$, 1, 2, and 3. (Recall that $x(c; h)$ denotes the approximation of $x(c)$ with step size h.) The algorithm stops at $m = 3$ because

$$|x_1(2; 0.5) - x_1(2; 0.25)| = 0.00046 < \varepsilon = 0.001,$$

and

$$|x_2(2; 0.5) - x_2(2; 0.25)| = 0.00040 < \varepsilon = 0.001.$$

Hence $y(2) \approx x_1(2) = 0.34400$ with a tolerance of 0.001. ■ End

TABLE 5.4 RUNGE-KUTTA APPROXIMATION FOR $y(2)$ FROM EXAMPLE 2

		Approximations	
m	h	$y(2) \approx x_1(2; h)$	$y'(2) \approx x_2(2; h)$
0	2	0.33333	−0.66667
1	1	0.33417	−0.13764
2	0.5	0.34446	−0.14822
3	0.25	0.34400	−0.14782

$y(2)$ is approximately 0.34400 with a tolerance of $\varepsilon = 0.001$.

An Application to Population Dynamics

A mathematical model for the population dynamics of competing species, one a predator with population $x_2(t)$ and the other its prey with population $x_1(t)$, was developed independently in the early 1900s by A. J. Lotka and V. Volterra. It assumes there is plenty of food available for the prey to eat, so the birth rate of the prey should follow the Malthusian or exponential law (see Section 3.2); that is, the birth rate of the prey is Ax_1, where A is a positive constant. The death rate of the prey depends on the number of interactions between the predators and the prey. This is modeled by the expression Bx_1x_2, where B is a positive constant. Therefore the rate of change in the population of the prey per unit time is $dx_1/dt = Ax_1 - Bx_1x_2$. Assuming the predators depend entirely on the prey for their food, it is argued that the birth rate of the predators depends on the number of interactions with

the prey; that is, the birth rate of predators is Dx_1x_2, where D is a positive constant. The death rate of the predators is assumed to be Cx_2 because without food the population would die off at a rate proportional to the population present. Hence the rate of change in the population of predators per unit time is $dx_2/dt = -Cx_2 + Dx_1x_2$. Combining these two equations, we obtain the Volterra-Lotka system for the population dynamics of two competing species:

$$x_1' = Ax_1 - Bx_1x_2,$$

$$x_2' = -Cx_2 + Dx_1x_2.$$

Such systems are in general not explicitly solvable. In the following example, we obtain an approximate solution for such a system.

EXAMPLE 3 Use the fourth order Runge-Kutta algorithm for systems to approximate the solution of the initial value problem

(15)
$$x_1' = 2x_1 - 2x_1x_2; \qquad x_1(0) = 1,$$
$$x_2' = x_1x_2 - x_2; \qquad x_2(0) = 3$$

at $t = 1$ with a tolerance of 0.0001.

Solution Here $f_1(t, x_1, x_2) = 2x_1 - 2x_1x_2$ and $f_2(t, x_1, x_2) = x_1x_2 - x_2$. With the inputs $t_0 = 0$, $a_1 = x_1(0) = 1$, $a_2 = x_2(0) = 3$, and $M = 10$ (say) we proceed with the algorithm to compute $x_1(1; 1)$ and $x_2(1; 1)$, the approximations to $x_1(1)$, $x_2(1)$ using $h = 1$. We find from the formulas in Step 3 of the SUBROUTINE (with $x_1 = 1$, $x_2 = 3$, and $h = 1$)

$$k_{1,1} = hf_1(t, x_1, x_2) = h(2x_1 - 2x_1x_2) = 2(1) - 2(1)(3) = -4,$$

$$k_{1,2} = hf_2(t, x_1, x_2) = h(x_1x_2 - x_2) = (1)(3) - 3 = 0,$$

$$k_{2,1} = hf_1\left(t + \frac{h}{2}, x_1 + \frac{1}{2}k_{1,1}, x_2 + \frac{1}{2}k_{1,2}\right)$$

$$= h\left[2\left(x_1 + \frac{1}{2}k_{1,1}\right) - 2\left(x_1 + \frac{1}{2}k_{1,1}\right)\left(x_2 + \frac{1}{2}k_{1,2}\right)\right]$$

$$= 2\left(1 + \frac{1}{2}(-4)\right) - 2\left(1 + \frac{1}{2}(-4)\right)\left(3 + \frac{1}{2}(0)\right)$$

$$= -2 + 2(3) = 4,$$

$$k_{2,2} = hf_2\left(t + \frac{h}{2}, x_1 + \frac{1}{2}k_{1,1}, x_2 + \frac{1}{2}k_{1,2}\right)$$

$$= h\left[\left(x_1 + \frac{1}{2}k_{1,1}\right)\left(x_2 + \frac{1}{2}k_{1,2}\right) - \left(x_2 + \frac{1}{2}k_{1,2}\right)\right]$$

$$= \left[1 + \frac{1}{2}(-4)\right]\left[3 + \frac{1}{2}(0)\right] - \left[3 + \frac{1}{2}(0)\right]$$

$$= (-1)(3) - 3 = -6.$$

Similarly, we find

$$k_{3,1} = h\left[2\left(x_1 + \frac{1}{2}k_{2,1}\right) - 2\left(x_1 + \frac{1}{2}k_{2,1}\right)\left(x_2 + \frac{1}{2}k_{2,2}\right)\right] = 6,$$

$$k_{3,2} = h\left[\left(x_1 + \frac{1}{2}k_{2,1}\right)\left(x_2 + \frac{1}{2}k_{2,2}\right) - \left(x_2 + \frac{1}{2}k_{2,2}\right)\right] = 0,$$

$$k_{4,1} = h[2(x_1 + k_{3,1}) - 2(x_1 + k_{3,1})(x_2 + k_{3,2})] = -28,$$

$$k_{4,2} = h[(x_1 + k_{3,1})(x_2 + k_{3,2}) - (x_2 + k_{3,2})] = 18.$$

Hence in Step 4 of the SUBROUTINE we compute

$$x_1 = x_1 + \frac{1}{6}(k_{1,1} + 2k_{2,1} + 2k_{3,1} + k_{4,1})$$

$$= 1 + \frac{1}{6}[-4 + 8 + 12 - 28] = -1,$$

$$x_2 = x_2 + \frac{1}{6}(k_{1,2} + 2k_{2,2} + 2k_{3,2} + k_{4,2})$$

$$= 3 + \frac{1}{6}(0 - 12 + 0 + 18) = 4.$$

Repeating the algorithm with $h = 1/2$ ($N = 2$) we obtain the approximations $x_1(1; 2^{-1})$ and $x_2(1; 2^{-1})$ for $x_1(1)$ and $x_2(1)$. In Table 5.5, we list the approximations $x_1(1; 2^{-m})$ and $x_2(1; 2^{-m})$ for $x_1(1)$ and $x_2(1)$ using step size $h = 2^{-m}$ for $m = 0, 1, 2, 3,$ and 4. We stopped at $m = 4$, since both

$$|x_1(1; 2^{-3}) - x_1(1; 2^{-4})| = 0.00006 < 0.0001$$

and

$$|x_2(1; 2^{-3}) - x_2(1; 2^{-4})| = 0.00001 < 0.0001.$$

Hence $x_1(1) \approx 0.07735$ and $x_2(1) \approx 1.46445$, with tolerance 0.0001. ∎ End

TABLE 5.5	APPROXIMATIONS OF THE SOLUTION TO SYSTEM (15) IN EXAMPLE 3		
m	h	$x_1(1; h)$	$x_2(1; h)$
0	1.0	-1.0	4.0
1	0.5	0.14662	1.47356
2	0.25	0.07885	1.46469
3	0.125	0.07741	1.46446
4	0.0625	0.07735	1.46445

To get a better feel for the solution to system (15), we have graphed in Figure 5.29 an approximation of the solution for $0 \leq t \leq 12$, using linear interpolation to connect the points. This approximation was obtained using step size $h = 0.125$ ($N = 96$) and the fourth order Runge-Kutta **subroutine** for systems. From the graph it appears that the components x_1 and x_2 are periodic in the variable t. In the next section we study the qualitative behavior of a system of two first order equations using a technique called **phase plane analysis.** This technique is also used in Project D to study general Volterra-Lotka equations and, in particular, to show that these equations have periodic solutions.

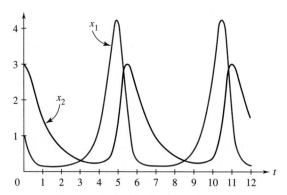

Figure 5.29 Graphs of the components of an approximate solution to the Volterra-Lotka system (5)

The Runge-Kutta algorithm on page 280 is readily generalized to systems of n equations. Appendix E lists the corresponding instructions.

EXERCISES 5.6

 The reader will find it helpful to have access to either a microcomputer or a mainframe. Software such as MAC-MATH, PHASEPLANE, or PHASER may be used.

In Problems 1 through 4, use the Runge-Kutta subroutine with $h = 0.25$ to find an approximation for the solution to the given initial value problem on the specified interval. (For an interval $t_0 \leq t \leq c$, recall that the input number of steps N and the step size h are related by $Nh = c - t_0$.)

1. $y'' + ty' + y = 0;$ $y(0) = 1,$ $y'(0) = 0$ on $[0, 1]$.

2. $(1 + t^2)y'' + y' - y = 0;$
 $y(0) = 1,$ $y'(0) = -1$ on $[0, 1]$.

3. $t^2y'' + y = t + 2;$
 $y(1) = 1,$ $y'(1) = -1$ on $[1, 2]$.

4. $y'' = t^2 - y^2;$ $y(0) = 0,$ $y'(0) = 1$ on $[0, 1]$.
 (Can you guess the solution?)

5. Using the Runge-Kutta subroutine with $h = 0.5$, approximate the solution to the initial value problem

$$3t^2y'' - 5ty' + 5y = 0; \quad y(1) = 0, \quad y'(1) = \frac{2}{3}$$

at $t = 8$. Compare this approximation to the actual solution $y(t) = t^{5/3} - t$.

6. Using the Runge-Kutta algorithm with the stopping procedure based on the absolute error and a tolerance of $\varepsilon = 0.01$, approximate the solution to the initial value problem

$$y'' = t^2 + y^2; \quad y(0) = 1, \quad y'(0) = 0$$

at $t = 1$.

7. Using the Runge-Kutta algorithm with the stopping procedure based on the relative error and a tolerance

of $\varepsilon = 0.02$, approximate the solution to the initial value problem

$$y'' = ty^3; \qquad y(0) = 1, \qquad y'(0) = 1$$

at $t = 1$.

8. In Section 1.4, we discussed Euler's method for approximating the solution to a first order equation. Extend Euler's method to systems and give the recursive formulas in component form.

9. In Section 3.5, we discussed the improved Euler's method for approximating the solution to a first order equation. Extend this method to systems and give the recursive formulas for solving the initial value problem.

10. In Project F of Chapter 4, it was shown that the simple pendulum equation

$$\theta''(t) + \sin \theta(t) = 0$$

has periodic solutions when the initial displacement and velocity are small. Show that the period of the solution may depend on the initial conditions by using the Runge-Kutta subroutine with $h = 0.02$ to approximate the solutions to the simple pendulum problem on $[0, 4]$ for the initial conditions:

(a) $\theta(0) = 0.1;$ $\theta'(0) = 0.$
(b) $\theta(0) = 0.5;$ $\theta'(0) = 0.$
(c) $\theta(0) = 1.0;$ $\theta'(0) = 0.$

[*Hint:* Approximate the length of time it takes to reach $-\theta(0)$.]

11. Fluid Ejection. In the design of a sewage treatment plant, the following equation arose:[†]

$$60 - H = (77.7)H'' + (19.42)(H')^2;$$

$$H(0) = H'(0) = 0,$$

where H is the level of the fluid in an ejection chamber and t is the time in seconds. Use the Runge-Kutta subroutine with $h = 0.5$ to approximate $H(t)$ over the interval $[0, 5]$.

12. Oscillations and Nonlinear Equations. For the initial value problem

$$x'' + (0.1)(1 - x^2)x' + x = 0;$$

$$x(0) = x_0, \qquad x'(0) = 0,$$

use the Runge-Kutta subroutine with $h = 0.02$ to illustrate that as t increases from 0 to 20, the solution x

exhibits damped oscillations with $x_0 = 1$, whereas x exhibits expanding oscillations when $x_0 = 3$.

13. Nonlinear Spring. The **Duffing equation**

$$y'' + y + ry^3 = 0,$$

where r is a constant, is a model for the vibrations of a mass attached to a *nonlinear* spring. For this model, does the period of vibration vary as the parameter r is varied? Does the period vary as the initial conditions are varied? [*Hint:* Use the Runge-Kutta subroutine with $h = 0.1$ to approximate the solutions for $r = 1$ and 2, with initial conditions $y(0) = a$, $y'(0) = 0$ for $a = 1, 2,$ and 3.]

14. Pendulum with Varying Length. A pendulum is formed by a mass m attached to the end of a wire that is attached to the ceiling. Assume the length $l(t)$ of the wire varies with time in some predetermined fashion. If $\theta(t)$ is the angle between the pendulum and the vertical, then the motion of the pendulum is governed by the initial value problem

$$l^2(t)\theta''(t) + 2l(t)l'(t)\theta'(t) + gl(t)\theta(t) = 0;$$

$$\theta(0) = \theta_0, \qquad \theta'(0) = \theta_1,$$

where g is the acceleration due to gravity. Assume

$$l(t) = l_0 + l_1\cos(\omega t - \phi),$$

where l_1 is much smaller than l_0. (This might be a model for a person on a swing, where the *pumping* action changes the distance from the center of mass of the swing to the point where the swing is attached.) To simplify the computations, take $g = 1$. Using the Runge-Kutta subroutine with $h = 0.1$, study the motion of the pendulum when $\theta_0 = 0.5$, $\theta_1 = 0$, $l_0 = 1$, $l_1 = 0.1$, $\omega = 1$, and $\phi = 0.02$. In particular, does the pendulum ever attain an angle greater in absolute value than the initial angle θ_0? Does the total arc traversed during one half of a swing ever exceed 1?

15. Using the Runge-Kutta subroutine for systems with $h = 0.125$, approximate the solution to the initial value problem

$$x' = 2x - y, \qquad x(0) = 0,$$

$$y' = 3x + 6y, \qquad y(0) = -2$$

at $t = 1$. Compare this approximation to the actual solution

$$x(t) = e^{5t} - e^{3t}, \qquad y(t) = e^{3t} - 3e^{5t}.$$

[†] See *Numerical Solution of Differential Equations*, by William Milne (Dover Publications, Inc., New York, 1970), p. 82.

16. Using the Runge-Kutta algorithm for systems with the stopping procedure based on the relative error and a tolerance of $\varepsilon = 0.02$, approximate the solution to the initial value problem

$$x' = 6x - 3y, \qquad x(0) = -1,$$
$$y' = 2x + y, \qquad y(0) = 0$$

at $t = 1$.

17. Using the Runge-Kutta algorithm for systems with the stopping procedure based on the absolute error and a tolerance of $\varepsilon = 0.001$, approximate the solution to the initial value problem

$$\frac{du}{dx} = 3u - 4v, \qquad u(0) = 1,$$

$$\frac{dv}{dx} = 2u - 3v, \qquad v(0) = 1$$

at $x = 1$.

18. **Combat Model.** A mathematical model for conventional versus guerrilla combat is given by the system

$$x_1' = -(0.1)x_1x_2, \qquad x_1(0) = 10,$$
$$x_2' = -x_1, \qquad x_2(0) = 15,$$

where x_1 and x_2 are the strengths of guerrilla and conventional troops, respectively, and 0.1 and 1 are the *combat effectiveness coefficients*. Who will win the conflict: the conventional troops or the guerrillas? [*Hint:* Use the Runge-Kutta subroutine for systems with $h = 0.1$ to approximate the solutions.]

19. **Predator-Prey Model.** The Volterra-Lotka predator-prey model predicts some rather interesting behavior that is evident in certain biological systems. For example, suppose you fix the initial population of prey but increase the initial population of predators. Then the population cycle for the prey becomes more severe in the sense that there is a long period of time with a reduced population of prey followed by a short period when the population of prey is very large. To demonstrate this behavior, use the fourth order Runge-Kutta subroutine for systems with $h = 0.5$ to approximate the populations of prey x and of predators y over the period $[0, 5]$ that satisfy the Volterra-Lotka system

$$x' = x(3 - y),$$
$$y' = y(x - 3)$$

under each of the following initial conditions:

(a) $x(0) = 2$, $y(0) = 4$.
(b) $x(0) = 2$, $y(0) = 5$.
(c) $x(0) = 2$, $y(0) = 7$.

In Problems 20 through 27, use the Runge-Kutta subroutine for systems with $h = 0.25$ to find an approximation for the solution to the given initial value problem at the times $t = 0.25$, $t = 0.5$, $t = 0.75$, and $t = 1.0$. For Problems 24 through 27, you may use the subroutine in Appendix E.

20. $x' = -x^2 + y^2 + 1$, $x(0) = 0$,
$\quad y' = -y$, $\qquad\qquad\quad y(0) = 1$.

21. $x' = 3x - y$, $\qquad x(0) = 0$,
$\quad y' = x - 2y$, $\qquad y(0) = 2$.

22. $x' = e^{-x^2} + y$, $\qquad x(0) = 1$,
$\quad y' = x^2 - y + t$, $\quad y(0) = 1$.

23. $x' = \cos(x + y)$, $\qquad x(0) = 1$,
$\quad y' = (x^2 + t)^{-1} + t$, $\quad y(0) = 0.5$.

24. $y^{(4)} + 3y''' + 2y' + (\sin t)y = 0$;
$\quad y(0) = 0$, $y'(0) = 1$, $y''(0) = 0$, $y'''(0) = 0$.

25. $y''' - ty'' + ty' - 3y = 0$;
$\quad y(0) = 1$, $\quad y'(0) = -1$, $\quad y''(0) = 0$.

26. $y''' - \sin y = t$;
$\quad y(0) = 1$, $\quad y'(0) = 0$, $\quad y''(0) = 0$.

27. $y^{(4)} - t^2y' + e^ty = t + 1$;
$\quad y(0) = -1$, $y'(0) = 1$, $y''(0) = 1$, $y'''(0) = -1$.

For Problems 28 through 34 use a commercial software package or, if none is available, the program in Appendix E.

28. Using the Runge-Kutta algorithm for systems with the stopping procedure based on the absolute error and a tolerance of $\varepsilon = 0.0001$, approximate the solution to the initial value problem

$$y''' - y'' - y = t^2;$$
$$y(0) = 0, \qquad y'(0) = 1, \qquad y''(0) = 0$$

at $t = 1$.

29. Using the Runge-Kutta algorithm for systems with the stopping procedure based on the relative error and a tolerance of $\varepsilon = 0.01$, approximate the solution to the initial value problem

$$y''' + y'' + y^2 = t;$$
$$y(0) = 1, \qquad y'(0) = 0, \qquad y''(0) = 1$$

at $t = 1$.

30. Use the Runge-Kutta subroutine for systems with $h = 0.1$ to approximate the solution to the initial value problem

$$x' = yz, \qquad\qquad x(0) = 0,$$
$$y' = -xz, \qquad\quad y(0) = 1,$$
$$z' = -xy/2, \qquad z(0) = 1$$

at $t = 1$.

31. Generalized Blasius Equation. H. Blasius, in his study of laminar flow of a fluid, encountered an equation of the form

$$y''' + yy'' = (y')^2 - 1.$$

Use the Runge-Kutta subroutine for systems with $h = 0.1$ to approximate the solution that satisfies the initial conditions $y(0) = 0$, $y'(0) = 0$, and $y''(0) = 1.32824$. Sketch this solution on the interval [0, 2].

32. Lunar Orbit. The motion of a moon moving in a planar orbit about a planet is governed by the equations

$$\frac{d^2x}{dt^2} = -G\frac{mx}{r^3}, \qquad \frac{d^2y}{dt^2} = -G\frac{my}{r^3},$$

where $r := (x^2 + y^2)^{1/2}$, G is the gravitational constant, and m is the mass of the moon. Assume $Gm = 1$. When $x(0) = 1$, $x'(0) = y(0) = 0$, and $y'(0) = 1$, the motion is a circular orbit of radius 1 and period 2π.

(a) Setting $x_1 = x$, $x_2 = x'$, $x_3 = y$, $x_4 = y'$, express the governing equations as a first order system in normal form.
(b) Using $h = 2\pi/100 \approx 0.0628318$, compute one orbit of this moon (i.e., let $N = 100$). Do your approximations agree with the fact that the orbit is a circle of radius 1?

33. Competing Species. Let $p_i(t)$ denote, respectively, the populations of three competing species S_i, $i = 1, 2, 3$. Suppose these species have the same growth rates and the maximum population that the habitat can support is the same for each species. (We assume it to be one unit.) Also suppose the competitive advantage that S_1 has over S_2 is the same as that of S_2 over S_3 and S_3 over S_1. This situation is modeled by the system

$$p_1' = p_1(1 - p_1 - ap_2 - bp_3),$$
$$p_2' = p_2(1 - bp_1 - p_2 - ap_3).$$
$$p_3' = p_3(1 - ap_1 - bp_2 - p_3),$$

where a and b are positive constants. To demonstrate the population dynamics of this system when $a = b = 0.5$, use the Runge-Kutta subroutine for systems with $h = 0.1$ to approximate the populations p_i over the time interval [0, 10] under each of the following initial conditions:

(a) $p_1(0) = 1.0$, $p_2(0) = 0.1$, $p_3(0) = 0.1$.
(b) $p_1(0) = 0.1$, $p_2(0) = 1.0$, $p_3(0) = 0.1$.
(c) $p_1(0) = 0.1$, $p_2(0) = 0.1$, $p_3(0) = 1.0$.

On the basis of the results of parts (a)–(c), decide what you think will happen to these populations as $t \to \infty$.

When $a = 0.5$ and $b = 2.0$, the behavior of these populations is entirely different (see Project F).

34. Spring Pendulum. Let a mass be attached to one end of a spring with spring constant k and the other end attached to the ceiling. Let l_0 be the natural length of the spring and let $l(t)$ be its length at time t. If $\theta(t)$ is the angle between the pendulum and the vertical, then the motion of the spring pendulum is governed by the system

$$l''(t) - l(t)\theta'(t) - g\cos\theta(t) + \frac{k}{m}(l - l_0) = 0,$$

$$l^2(t)\theta''(t) + 2l(t)l'(t)\theta'(t) + gl(t)\sin\theta(t) = 0.$$

Assume $g = 1$, $k = m = 1$, and $l_0 = 4$. When the system is at rest, $l = l_0 + mg/k = 5$.

(a) Describe the motion of the pendulum when $l(0) = 5.5$, $l'(0) = 0$, $\theta(0) = 0$, and $\theta'(0) = 0$. (*Hint:* Use the Runge-Kutta subroutine for systems with $h = 0.1$ to approximate the solution on the interval [0, 10].)
(b) When the pendulum is both stretched and given an angular displacement, the motion of the pendulum is more complicated. Using the Runge-Kutta subroutine for systems with $h = 0.1$ to approximate the solution, sketch the graph of the length l and the angular displacement θ on the interval [0, 10] if $l(0) = 5.5$, $l'(0) = 0$, $\theta(0) = 0.5$, and $\theta'(0) = 0$.

5.7 INTRODUCTION TO PHASE PLANE ANALYSIS

In this section, we study systems of two first order equations of the form

(1)
$$\frac{dx}{dt} = f(x, y),$$
$$\frac{dy}{dt} = g(x, y),$$

where f and g are real-valued functions that do not depend explicitly on t. Such systems are called **autonomous.** An example of an autonomous system is the Volterra-Lotka system

$$\frac{dx}{dt} = Ax - Bxy,$$

$$\frac{dy}{dt} = -Cy + Dxy$$

discussed in Section 5.6. As with all autonomous systems, the parameters involved $(A, B, C,$ and $D)$ are *time independent.*

A **solution** to (1) on an interval I is a pair of functions $x(t)$, $y(t)$ that satisfy (1) for all t in I. The set of points $\{(x(t), y(t)); t \in I\}$ in the xy-plane is called a **trajectory,**[†] and the xy-plane is referred to as the **phase plane.**

If the pair $x(t)$, $y(t)$ is a solution to (1), then so is the pair $x(t + c)$, $y(t + c)$ for any constant c because the system (1) is not time dependent. To be more precise, let $X(t) := x(t + c)$ and $Y(t) := y(t + c)$. Then, by the chain rule,

$$\frac{dX}{dt}(t) = \frac{dx}{dt}(t + c) = f(x(t + c), y(t + c)) = f(X(t), Y(t)),$$

$$\frac{dY}{dt}(t) = \frac{dy}{dt}(t + c) = g(x(t + c), y(t + c)) = g(X(t), Y(t)),$$

which proves that $X(t)$, $Y(t)$ is also a solution to (1).

For nonlinear autonomous systems, obtaining explicit solutions $x(t)$, $y(t)$ may be an extraordinarily difficult task. However, for certain systems, it is possible to obtain some trajectories that give us valuable qualitative information about the solution. Here we give a brief introduction to the study of the qualitative behavior of autonomous systems.[††] More detailed discussion of stability issues for linear and nonlinear autonomous systems appears in Chapter 12.[†††]

To illustrate the idea of trajectories, we start with a simple example of two uncoupled linear equations.

EXAMPLE 1 Find solutions to the autonomous system

(2) $\qquad \dfrac{dx}{dt} = x, \qquad \dfrac{dy}{dt} = 2y$

and sketch their trajectories.

[†] Trajectories are also called **solution curves, paths,** or **orbits.**
[††] *Historical Footnote:* The study of the qualitative behavior of autonomous systems was begun by the great French mathematician **J. Henri Poincaré** (1854–1912) in 1881 in his work in celestial mechanics. His work was continued by **I. Bendixson** in 1901 in his study of closed trajectories.
[†††] All references to Chapters 11–13 refer to the expanded text *Fundamentals of Differential Equations and Boundary Value Problems*, Second Edition. See also Chapter 5 of *Ordinary Differential Equations* by G. Birkhoff and G. C. Rota (John Wiley and Sons, Inc., New York, 1978).

Solution Solving separately each equation in (2), we quickly find that $x(t) = c_1 e^t$, $y(t) = c_2 e^{2t}$ are solutions. Expressing y in terms of x, we obtain the trajectories

(3) $y = c_2(e^t)^2 = c_2(x/c_1)^2 = kx^2,$

where $k = c_2/c_1^2$. These parabolic trajectories are sketched in the phase plane diagram, Figure 5.30.

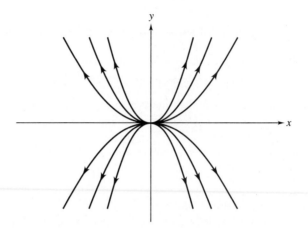

Figure 5.30 Phase diagram for $x' = x$, $y' = 2y$

The arrows on the trajectories in Figure 5.30 indicate the direction, or **flow,** of a particle as it travels along the trajectory. In equation (3), we see that as x moves away from zero, y increases if k is positive (this is the case for the trajectories above the x-axis) and decreases if k is negative (the trajectories below the x-axis). Hence the flow is always away from the origin. ■ End

In the preceding example, the origin is the trajectory of the constant solution $x \equiv 0$, $y \equiv 0$. Such points and solutions play an important role in the study of autonomous systems.

CRITICAL POINTS AND EQUILIBRIUM SOLUTIONS

Definition 1. A point (x_0, y_0), where $f(x_0, y_0) = 0$ and $g(x_0, y_0) = 0$, is called a **critical point** of the system $dx/dt = f(x, y)$, $dy/dt = g(x, y)$, and the corresponding constant solution $x(t) \equiv x_0$, $y(t) \equiv y_0$ is called an **equilibrium solution.** The set of all critical points is called the **critical point set.**

Roughly speaking, an equilibrium solution is said to be *stable* if all solutions passing near the critical point stay close to it.[†] The point $(0, 0)$ in Figure 5.30 is an example of an *unstable* equilibrium solution.

[†] See Section 12.3 for a rigorous definition of stability.

EXAMPLE 2 Find the critical points and corresponding equilibrium solutions of

$$\frac{dx}{dy} = -y(y - 2),$$

(4)

$$\frac{dy}{dt} = (x - 2)(y - 2).$$

Solution To find the critical points, we solve the system

$$-y(y - 2) = 0, \qquad (x - 2)(y - 2) = 0.$$

One family of solutions to this system is given by $y = 2$ with x arbitrary; that is, the line $y = 2$. If $y \neq 2$, then the system simplifies to $-y = 0$, and $x - 2 = 0$, which has the solution $x = 2$, $y = 0$. Hence the critical point set consists of the isolated point $(2, 0)$ and the horizontal line $y = 2$. The corresponding equilibrium solutions are $x(t) \equiv 2$, $y(t) \equiv 0$, and the family $x(t) \equiv c$, $y(t) \equiv 2$, where c is an arbitrary constant. ■ End

Although we succeeded in finding the equilibrium solutions to the nonlinear system (4) of Example 2, determining other explicit solutions to this system is beyond our present capabilities. However, we can determine the trajectories for the system using techniques for solving first order equations discussed in Chapter 2.

EXAMPLE 3 Determine the trajectories of system (4) and discuss the stability of its critical points.

Solution Unlike in Example 1, we are not able to determine the trajectories by first computing the solutions. But we do know that when $dx/dt \neq 0$, the chain rule yields

(5) $$\frac{dy}{dx} = \frac{dy/dt}{dx/dt} = \frac{(x - 2)(y - 2)}{-y(y - 2)} = -\frac{x - 2}{y}.$$

In (5) we have a first order differential equation whose solutions in x and y contain the trajectories of system (4). Solving (5) by separation of variables, we obtain

$$\frac{y^2}{2} = \int y\,dy = -\int (x - 2)\,dx = -\frac{(x - 2)^2}{2} + c,$$

which is just the family of concentric circles

(6) $$y^2 + (x - 2)^2 = k,$$

with centers located at $(2, 0)$. Thus each trajectory of (4) lies on a circle given in (6). In Figure 5.31, we have sketched these circles and the critical points of (4).

Now that we know that the trajectories lie on the circles in (6), what remains is to analyze the flow along each trajectory. For this purpose, we use the given equation $dx/dt = -y(y - 2)$. If $y > 2$, we see that $dx/dt < 0$. Hence x is decreasing when $y > 2$.

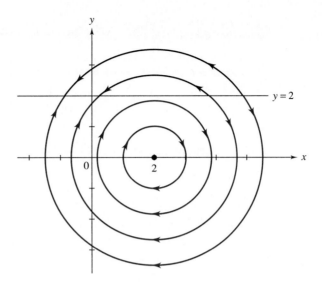

Figure 5.31 Phase plane diagram for $x' = -y(y - 2)$, $y' = (x - 2)(y - 2)$

This means the flow is from right to left along the arc of a circle that lies above the line $y = 2$. For $0 < y < 2$, we have $dx/dt > 0$, so in this region the flow is from left to right. Furthermore, for $y < 0$, we have $dx/dt < 0$, and again the flow is from right to left.

We now observe in Figure 5.31 that there are depicted four types of trajectories associated with system (4): (a) those that begin above the line $y = 2$ and follow the arc of a circle counterclockwise back to that line; (b) those that begin below the line $y = 2$ and follow the arc of a circle clockwise back to that line; (c) those that continually move clockwise around a circle centered at $(2, 0)$ with radius less than 2 (i.e., they do not intersect the line $y = 2$); and finally, (d) the critical points $(2, 0)$ and $y = 2$, x arbitrary.

From Figure 5.31, one can see that solutions starting close to a critical point that lies on the half-line $y = 2$, $x < 2$ remain close to that critical point. The same is true of solutions starting near $(2, 0)$. These critical points are classified as stable. The critical points on the half-line $y = 2$, $x \geq 2$ have nearby solutions that diverge from them, so they are unstable. ∎ End

The curves given in equation (6) are called the **integral curves** of (4). They are the level curves of the function $V(x, y) := (x - 2)^2 + y^2$, which is called a **first integral of** (4).[†] As we observed earlier, a trajectory must lie on an integral curve.

In Example 3, we noted that the nonperiodic trajectories (a) and (b) begin and end at a critical point, namely at a point on the line $y = 2$. This fact is a consequence of the following theorem.

[†] In classical mechanics, the first integral V is associated with the total energy of the system and hence a level curve is a curve along which the energy is constant. In fluid mechanics, V is called the **stream function.**

LIMIT POINTS ARE CRITICAL POINTS

Theorem 1. Let the pair $x(t)$, $y(t)$ be a solution on $[0, \infty)$ to the autonomous system $dx/dt = f(x, y)$, $dy/dt = g(x, y)$, where f and g are continuous in the plane. If the limits

$$x_\omega := \lim_{t \to \infty} x(t) \quad \text{and} \quad y_\omega := \lim_{t \to \infty} y(t)$$

exist and are finite, then the limit point (x_ω, y_ω) is a critical point for the system.

Proof. To show that (x_ω, y_ω) is a critical point, we must prove that $f(x_\omega, y_\omega) = g(x_\omega, y_\omega) = 0$. First observe that, by the continuity of f,

$$\lim_{t \to \infty} x'(t) = \lim_{t \to \infty} f(x(t), y(t)) = f(x_\omega, y_\omega).$$

Let's suppose to the contrary that $f(x_\omega, y_\omega) \neq 0$; say, $w := f(x_\omega, y_\omega) > 0$ (a similar argument will handle the case when $w < 0$). For any $\varepsilon > 0$, we can choose N sufficiently large so that

$$\left| f(x_\omega, y_\omega) - f(x(t), y(t)) \right| < \varepsilon \quad \text{for all } t \geq N.$$

Taking $\varepsilon = w/2$, we have $\left| w - f(x(t), y(t)) \right| < w/2$, so $x'(t) = f(x(t), y(t)) > w/2$. But if $x'(t) > w/2$ for all $t \geq N$, then $x(t)$ must lie above some line with positive slope $1/2$. Hence $\lim_{t \to \infty} x(t) = +\infty$, which contradicts the assumption that this limit is finite. Therefore $f(x_\omega, y_\omega) = 0$. A similar argument also shows that $g(x_\omega, y_\omega) = 0$. Hence the limit point (x_ω, y_ω) is a critical point. ∎ End

The proof of Theorem 1 can be easily modified to show that a limit point (x_α, y_α), where

$$x_\alpha := \lim_{t \to -\infty} x(t), \qquad y_\alpha := \lim_{t \to -\infty} y(t)$$

is also a critical point. As a consequence of Theorem 1, we observe that *the only points in the phase plane where a nonperiodic trajectory can begin $(t \to -\infty)$ or end $(t \to +\infty)$ are the critical points.* For example, in Figure 5.31, the trajectories that lie above the line $y = 2$ begin at a critical point on the line $y = 2$ and end at a critical point on the line.

A sketch of the critical point set for a system, along with representative integral curves and their trajectories with arrows indicating the flow, is called a **phase plane diagram.** The use of these diagrams to obtain qualitative information about the solutions of the system is referred to as **phase plane analysis.** In many cases, the integral curves or the trajectories must be obtained using numerical techniques; several software packages are available for this purpose. Two such packages that run on the IBM-PC or its clones are PHASEPLANE by Brad Ermentrout available from Brooks/Cole Publishing Company, Pacific Groves, California, and PHASER by Hüseyin Koçak, available from Springer-Verlag, New York. Both the DOS-based software and the MAPLE-based supplement designed for this text include the plotting of phase diagrams.

PHASE PLANE ANALYSIS

To sketch a phase plane diagram for the system $dx/dt = f(x, y)$, $dy/dt = g(x, y)$:

(a) Find the critical point set consisting of the solutions to the system $f(x, y) = 0$, $g(x, y) = 0$. Sketch this set in the xy-plane.

(b) Solve the first order equation $dy/dx = g(x, y)/f(x, y)$. Use the solution to sketch some representative integral curves for the system. (Numerical techniques may be needed for this purpose.)

(c) Determine the direction of the flow along the integral curves using either of the equations $dx/dt = f(x, y)$ or $dy/dt = g(x, y)$. This gives the trajectories.

It can be shown that system (1) has a unique solution satisfying $x(t_0) = x_0$, $y(t_0) = y_0$, provided f and g have continuous partial derivatives in the xy-plane. This is helpful in sketching the trajectories of a system because it means that two different trajectories cannot intersect (in finite time).

In the next example, we use the software PHASEPLANE to construct a phase plane diagram.

EXAMPLE 4 Use a software package to construct a phase plane diagram about the critical point $(0, 0)$ for the linear system

$$\frac{dx}{dt} = -2x + y,$$

(7)

$$\frac{dy}{dt} = -5x - 4y.$$

Solution With the aid of the software package PHASEPLANE, we graphed in Figure 5.32 several solutions to system (7) starting at various initial locations within the window $-4 \le x \le 4$,

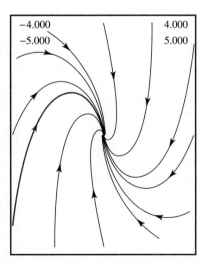

Figure 5.32 Phase plane diagram for system (7) constructed using the software PHASEPLANE

$-5 \leq y \leq 5$. The trajectories all appear to converge to the origin. In Chapter 12[†], we shall see that the origin is an example of an *asymptotically stable spiral point*. Here we use the zoom feature of PHASEPLANE to observe the spiraling of a trajectory about the origin. In the series of graphs in Figure 5.33, we have zoomed in on the trajectory (in color in Figure 5.32) that passes through the point $(-3.875000, -3.111111)$. Observe how the trajectory spirals about the origin as the window about the origin gets smaller $\{-0.089 \leq x \leq 0.125, -0.111 \leq y \leq 0.222\}$ and smaller $\{-0.002 \leq x \leq 0.002, -0.004 \leq y \leq 0.004\}$ and smaller (the software quotes ''$-0.000 \leq x \leq 0.000, -0.000 \leq y \leq 0.000$'' [to three decimals, of course]). ∎ End

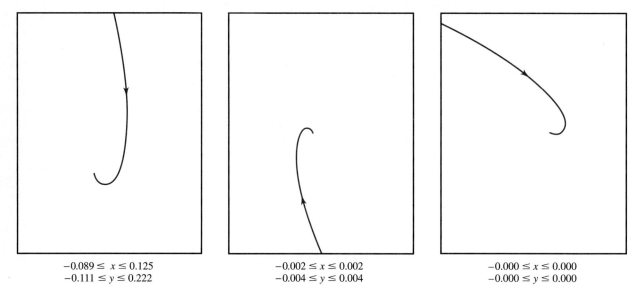

$-0.089 \leq x \leq 0.125$
$-0.111 \leq y \leq 0.222$

$-0.002 \leq x \leq 0.002$
$-0.004 \leq y \leq 0.004$

$-0.000 \leq x \leq 0.000$
$-0.000 \leq y \leq 0.000$

Figure 5.33 Spiraling of a trajectory about the origin

Autonomous systems arise in the study of the motion of a particle governed by Newton's second law, $m\,d^2x/dt^2 = F(x, dx/dt)$, when the forcing function F does not depend explicitly on t. In the next example, we use a phase plane analysis to study a nonlinear spring-mass system.

EXAMPLE 5 When we considered mass-spring oscillators in Sections 5.1 through 5.3, we modeled the friction force with a term proportional to the velocity $-b\,dx/dt = -bx'$, where b was the damping constant. However, for a mass m sliding on a surface as in Figure 5.34, the friction effect may be more complicated. We observe, for example, that even if the mass is displaced slightly off of the equilibrium location $x = 0$, it may nonetheless remain stationary due to *static* friction; the spring force $-kx$ is insufficient to break the static friction's grip. If the

[†] All references to Chapters 11–13 refer to the expanded text *Fundamentals of Differential Equations and Boundary Value Problems*, Second Edition.

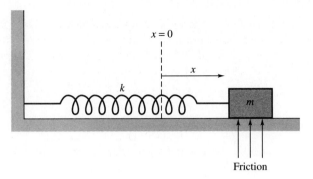

Figure 5.34 Spring-mass system with friction

maximum force that this ''sticky friction'' can exert is denoted by μ, then a feasible model is given by

$$F_{\text{friction}} = \begin{cases} kx, & \text{if } |kx| < \mu \text{ and } x' = 0, \\ \mu \operatorname{sign}(x), & \text{if } |kx| \geq \mu \text{ and } x' = 0, \\ -\mu \operatorname{sign}(x'), & \text{if } x' \neq 0. \end{cases}$$

(The function $\operatorname{sign}(s)$ is $+1$ when $s > 0$ and -1 when $s < 0$.) Thus if the mass is at rest, friction *balances* the spring force if $|x| < \mu/k$ but simply *opposes* it with intensity μ if $|x| \geq \mu/k$. If the mass is moving, friction opposes the velocity with the same intensity μ, and the motion is governed by

$$m\frac{d^2x}{dt^2} = -kx + F_{\text{friction}}$$

or

(8) $\qquad m\dfrac{d^2x}{dt^2} + \mu \operatorname{sign}\left(\dfrac{dx}{dt}\right) + kx = 0.$

Describe the motion for the case where $m = \mu = k = 1$.

Solution To study this motion, we convert (8) to a system by introducing (as usual) the velocity as a second dependent variable.[†] Thus we get (on substituting the prescribed values for m, μ, and k)

(9) $\qquad \begin{aligned} x' &= v, \\ v' &= x'' = -x \mp 1, \qquad x' \gtrless 0, \end{aligned}$

where we take $\mp$ to be *minus* for $x' > 0$ and plus for $x' < 0$.

[†]***Historical Footnote:*** The resulting xv-plane is referred to as the **Poincaré phase plane** in honor of **J. Henri Poincaré.**

To find the integral curves of (9), we solve

$$\frac{dv}{dx} = \frac{dv/dt}{dx/dt} = \frac{-x \mp 1}{v}, \qquad v \gtrless 0.$$

Separating variables and integrating gives

$$v^2 + (x \pm 1)^2 = c, \qquad v \gtrless 0.$$

Therefore the integral curves in the xv-plane are concentric semicircles in the upper half plane with centers at $(-1, 0)$ and concentric semicircles in the lower half plane with centers at $(1, 0)$ (see Figure 5.35). Since $x' = v$, the flow is left to right for $v > 0$ and right to left for $v < 0$.

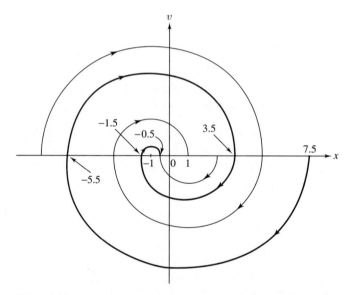

Figure 5.35 Phase plane diagram for a spring-mass system with friction

To understand the motion of the mass, first observe that $\mu/k = 1$. Thus if $|x| \leq 1$ and $x' = 0$, the system remains stationary and we can interpret such points as the critical points in the phase plane.

Next, consider the case when the mass is released (with zero velocity) at $x = 7.5$. In the phase plane, the mass begins at the point $(7.5, 0)$ and follows the flow along the semicircular trajectory in the lower half plane until it reaches the point $(-5.5, 0)$ (see the colored curve in Figure 5.35). (This corresponds to the mass moving from $x = 7.5$ to $x = -5.5$.) The flow now takes the mass from $(-5.5, 0)$ along the semicircular trajectory in the upper half plane to the point $(3.5, 0)$. (The mass has moved from $x = -5.5$ to $x = 3.5$.) The mass now follows the trajectory in the lower half plane to $(-1.5, 0)$ and then the trajectory in the upper half plane to $(-0.5, 0)$. But $(-0.5, 0)$ is a critical point. When

$x = -0.5$, the force on the mass due to the spring is just $kx = -0.5$, and friction can apply a force up to 1. Therefore friction cancels out the force due to the spring and prevents the mass from moving away from the spot $x = -0.5$. Consequently, the motion is stopped at $x = -0.5$. ▪ End

EXERCISES 5.7

In Problems 1 and 2, verify that the pair $x(t)$, $y(t)$ is a solution to the given system. Find the trajectory of the given solution and sketch its graph in the phase plane.

1. $\dfrac{dx}{dt} = 3y^3$, $\dfrac{dy}{dt} = y$; $x(t) = e^{3t}$, $y(t) = e^t$.

2. $\dfrac{dx}{dt} = 1$, $\dfrac{dy}{dt} = 3x^2$;

$x(t) = t + 1$, $y(t) = t^3 + 3t^2 + 3t$.

In Problems 3 through 8, find the critical point set for the given system.

3. $\dfrac{dx}{dt} = x - y$,

$\dfrac{dy}{dt} = x^2 + y^2 - 1$.

4. $\dfrac{dx}{dt} = y - 1$,

$\dfrac{dy}{dt} = x + y + 5$.

5. $\dfrac{dx}{dt} = x^2 - 2xy$,

$\dfrac{dy}{dt} = 3xy - y^2$.

6. $\dfrac{dx}{dt} = 3x^2 - xy$,

$\dfrac{dy}{dt} = 4xy - 3y^2$.

7. $\dfrac{dx}{dt} = y^2 - 3y + 2$,

$\dfrac{dy}{dt} = (x - 1)(y - 2)$.

8. $\dfrac{dx}{dt} = (x + 1)(y - 2)$,

$\dfrac{dy}{dt} = x^2 - x - 2$.

In Problems 9 through 14, determine the integral curves for the given system.

9. $\dfrac{dx}{dt} = y - 1$,

$\dfrac{dy}{dt} = e^{x+y}$,

10. $\dfrac{dx}{dt} = x^2$,

$\dfrac{dy}{dt} = x^2 + y^2 + xy.^\dagger$

11. $\dfrac{dx}{dt} = \dfrac{x^2}{y}$,

$\dfrac{dy}{dt} = x - y.^\dagger$

12. $\dfrac{dx}{dt} = x^2 - 2y^{-3}$,

$\dfrac{dy}{dt} = 3x^2 - 2xy$.

13. $\dfrac{dx}{dt} = 2y - x$,

$\dfrac{dy}{dt} = e^x + y$.

14. $\dfrac{dx}{dt} = x + y + 6$,

$\dfrac{dy}{dt} = 3x - y - 6.^\dagger$

In Problems 15 through 20, sketch (by hand or using software) a phase plane diagram for the given system.

15. $\dfrac{dx}{dt} = 2y$,

$\dfrac{dy}{dt} = 2x$.

16. $\dfrac{dx}{dt} = -8y$,

$\dfrac{dy}{dt} = 18x$.

17. $\dfrac{dx}{dt} = (y - x)(y - 1)$,

$\dfrac{dy}{dt} = (x - y)(x - 1)$.

18. $\dfrac{dx}{dt} = (x - 4)(1 - y)$,

$\dfrac{dy}{dt} = (x + 1)(x - 4)$.

19. $\dfrac{dx}{dt} = \dfrac{3}{y}$,

$\dfrac{dy}{dt} = \dfrac{2}{x}$.

20. $\dfrac{dx}{dt} = xy^2 - x$,

$\dfrac{dy}{dt} = 2y^3 - 4y$.

In Problems 21 through 26, use a software package such as PHASEPLANE to construct a phase plane diagram containing all the critical points for the given system. Discuss the stability of the critical points.

21. $\dfrac{dx}{dt} = 2x + y + 3$,

$\dfrac{dy}{dt} = -3x - 2y - 4$.

22. $\dfrac{dx}{dt} = -5x + 2y$,

$\dfrac{dy}{dt} = x - 4y$.

† Use techniques from Section 2.6.

23. $\dfrac{dx}{dt} = 2x + 13y,$

$\dfrac{dy}{dt} = -x - 2y.$

24. $\dfrac{dx}{dt} = -2x + y,$

$\dfrac{dy}{dt} = -5x - 4y.$

25. $\dfrac{dx}{dt} = x(7 - x - 2y),$

$\dfrac{dy}{dt} = y(5 - x - y).$

26. $\dfrac{dx}{dt} = 5x - 3y,$

$\dfrac{dy}{dt} = 4x - 3y.$

In Problems 27 through 30, sketch (by hand or using soft-ware) an xv-phase plane diagram for the given equation.

27. $\dfrac{d^2x}{dt^2} - x = 0.$

28. $\dfrac{d^2x}{dt^2} + x = 0.$

29. $\dfrac{d^2x}{dt^2} + x + x^5 = 0.$

30. $\dfrac{d^2x}{dt^2} + x^3 = 0.$

In Problems 31 and 32, use a software package such as PHASEPLANE to construct an xv-phase plane diagram con-taining all the critical points for the given equation. Discuss the stability of the critical points.

31. $x''(t) + x(t) - x^4(t) = 0.$

32. $x''(t) + x(t) - x^3(t) = 0.$

33. Epidemic Models. A model for the spread of a dis-ease through a population (an epidemic model) is given by the system

$$\frac{dS}{dt} = -aSI,$$

$$\frac{dI}{dt} = aSI - bI,$$

where a and b are positive constants. Here $S(t)$ repre-sents the susceptible population and $I(t)$ the infected population after t days. The constant a is a measure of how fast the disease is transmitted from an infected person to the susceptible population. The constant b represents the rate at which infected people become well, i.e., are removed from the infected population (and are thus presumably immune).

(a) Suppose that for a certain disease it is determined $a = 0.003$ and $b = 0.5$. In the SI-phase plane, sketch the trajectory corresponding to the initial condition that one person is infected and 700 persons are susceptible.

(b) From your graph in part (a), estimate the peak (largest possible) number of infected persons.

(c) Using the differential equations that model the spread of disease, argue that the peak number of infected persons occurs when $S = b/a = 1000/6 \approx$ 167 persons.

34. (a) For the epidemic model in Problem 33, show that the peak value for I will always occur when $S = b/a = 1000/6$, provided $S(0) > b/a$. (The num-ber b/a is called the **threshold value.**)

(b) The SI-phase plane analysis of Problem 33 indi-cates that an epidemic occurs (the number of infected persons first increases and then decreases to zero) when $S(0) > b/a$. Verify this fact directly from the expression for dI/dt.

(c) As time increases, what happens to the number of infected persons if $S(0) < b/a$?

35. (a) Use a numerical algorithm (such as Runge-Kutta for systems) to graph the solution functions $S(t)$ and $I(t)$ in Problem 33 with $S(0) = 700$ and $I(0) = 1$.

(b) After how many days does the number of infected persons peak?

(c) After approximately how many days does the epi-demic run its course (that is, $I(t) < 1$)?

36. Simple Pendulum. The motion of a simple pendulum is governed by the equation

$$\frac{d^2\theta}{dt^2} + \sin\theta = 0,$$

where θ represents the angular displacement from the vertical of the pendulum. Give a phase plane analysis for the equation and interpret the results physically.

37. Hard Spring. The motion of a spring-mass system in which the spring is "hard" and does not satisfy Hooke's law is governed by the equation

$$\frac{d^2x}{dt^2} + x + x^3 = 0,$$

where x represents the displacement from equilibrium. Give a phase plane analysis for the equation and inter-pret the results physically.

38. Falling Object. The motion of an object falling through the air is governed by the equation

$$\frac{d^2x}{dt^2} = g - \frac{g}{V^2}\frac{dx}{dt}\left|\frac{dx}{dt}\right|,$$

where x is the distance fallen and V is a constant called the terminal velocity. Give a phase plane analysis for the equation and interpret the results physically.

39. Trace the trajectory in the phase plane in Figure 5.35, page 296, that begins at $(6.5, 0)$; $(-6.5, 0)$.

40. Trace the trajectories in the phase plane in Figure 5.35, page 296, that terminate at the origin $(0, 0)$.

41. Restrained Bar. The motion of a bar restrained by springs and attracted by a parallel current passing through a wire (see Figure 5.36) is governed by the equation

$$\frac{d^2x}{dt^2} + a\left[x - \frac{b}{C - x}\right] = 0,$$

where a, b, and C are positive constants. Give a phase plane analysis for the equation for $-x_0 < x < C$, where x_0 is the distance from the wall to the equilibrium position. Interpret the results physically.

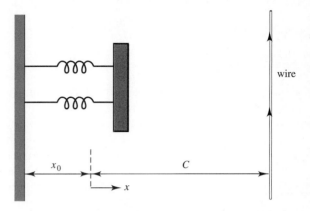

Figure 5.36 Bar restrained by springs and attracted by a parallel current

5.8 DYNAMICAL SYSTEMS, POINCARÉ MAPS, AND CHAOS

In this section, we take an excursion through an area of mathematics that is currently receiving a lot of attention both for the interesting mathematical phenomena being observed and for its application to fields such as meteorology, heat conduction, fluid mechanics, lasers, chemical reactions, and nonlinear circuits, among others. The area is that of dynamical systems and, in particular, nonlinear dynamical systems.[†]

A **dynamical system** is any system that allows one to determine (at least theoretically) the future states of the system given its present or past state. For example, the map (recurrence relation)

$$x_{n+1} = (1.05)x_n, \qquad n = 0, 1, 2, \ldots$$

is a dynamical system, since we can determine the next state, x_{n+1}, given the previous state, x_n. If we know x_0, then we can compute any future state (indeed $x_{n+1} = x_0(1.05)^{n+1}$). As this example illustrates, a dynamical system may be defined by a difference equation, such as those studied in Section 4.11.

Another example of a dynamical system is provided by the differential equation

$$\frac{dx}{dt} = -2x,$$

where the solution $x(t)$ specifies the state of the system at "time" t. If we know $x(t_0) = x_0$, then we can determine the state of the system at any future time $t > t_0$ by solving the initial

[†] For a more detailed study of dynamical systems, we refer the reader to *An Introduction to Chaotic Dynamical Systems*, Second Edition, by R. L. Devaney (Addison-Wesley Publishing Co., Reading, Mass., 1989) and *Nonlinear Oscillations, Dynamical Systems and Bifurcations of Vector Fields* by J. Guckenheimer and P. J. Holmes (Springer-Verlag, New York, 1983).

value problem

$$\frac{dx}{dt} = -2x, \qquad x(t_0) = x_0.$$

Indeed, a simple calculation yields $x(t) = x_0 e^{-2(t-t_0)}$ for $t \geq t_0$.

For a dynamical system defined by a differential equation, it is often helpful to work with a related dynamical system defined by a difference equation. For example, when we cannot express the solution to a differential equation using elementary functions, we can use a numerical technique such as improved Euler's method or Runge-Kutta to approximate the solution to an initial value problem. This numerical scheme defines a new (but related) dynamical system that is often easier to study.

In the previous section, we used phase plane diagrams to study autonomous systems in the plane. Many important features of the system can be detected just by looking at these diagrams. For example, a closed trajectory corresponds to a periodic solution. The phase plane diagrams for *non*autonomous systems are much more complicated to decipher. One technique that is helpful in this regard is the so-called **Poincaré map.** As we shall see, these maps replace the study of a nonautonomous system with the study of a dynamical system defined by the location in the *xv*-plane of the solution at regularly spaced moments in time such as $t = 2\pi n$, where $n = 0, 1, 2, \ldots$. The advantage in using the Poincaré map will become clear when the method is applied to a nonlinear problem for which no explicit solution is known. In such a case, the trajectories are computed using a numerical scheme such as Runge-Kutta. Software packages such as PHASEPLANE and PHASER have options that will construct Poincaré maps for a given system.

To illustrate the Poincaré map, consider the equation

(1) $x''(t) + \omega^2 x(t) = F \cos t,$

where F and ω are positive constants. We studied similar equations in Section 5.3 and found that a general solution for $\omega \neq 1$ is given by

(2) $x(t) = A \sin(\omega t + \phi) + \dfrac{F}{\omega^2 - 1} \cos t,$

where the amplitude A and the phase angle ϕ are arbitrary constants. If we let $v = x'$, then

$$v(t) = \omega A \cos(\omega t + \phi) - \frac{F}{\omega^2 - 1} \sin t.$$

Because the forcing function $F \cos t$ is 2π-periodic, it is natural to seek 2π-periodic solutions to (1). For this purpose, we define the Poincaré map

(3)
$$x_n := x(2\pi n) = A \sin(2\pi\omega n + \phi) + F/(\omega^2 - 1),$$
$$v_n := v(2\pi n) = \omega A \cos(2\pi\omega n + \phi),$$

for $n = 0, 1, 2, \ldots$. In Figure 5.37, we plotted the first 100 values of (x_n, v_n) in the *xv*-plane for different choices of ω. For simplicity, we have taken $A = F = 1$ and $\phi = 0$. These graphs are called **Poincaré sections.** We will interpret them shortly.

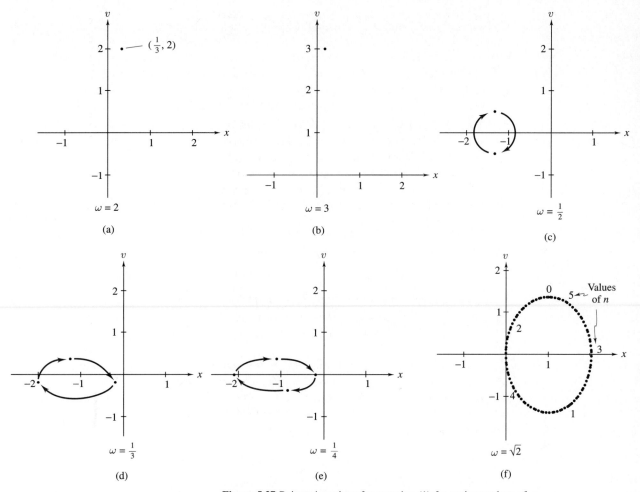

Figure 5.37 Poincaré sections for equation (1) for various values of ω

Now let's play the following game. We agree to ignore the fact that we already know the formula for $x(t)$ for all $t \geq 0$. We want to see what information about the solution we can glean just from the Poincaré section and the form of the differential equation.

Notice that the first two Poincaré sections in Figure 5.37, corresponding to $\omega = 2$ and 3, consist of a single point. This tells us that, starting with $t = 0$, every increment 2π of t returns us to the same point in the phase plane. This in turn implies that equation (1) has a 2π-periodic solution, which can be seen as follows: For $\omega = 2$, let $x(t)$ be the solution to (1) with $(x(0), v(0)) = (1/3, 2)$ and let $X(t) := x(t + 2\pi)$. Since the Poincaré section is just the point $(1/3, 2)$, we have $X(0) = x(2\pi) = 1/3$ and $X'(0) = x'(2\pi) = 2$. Thus $x(t)$ and $X(t)$ have the same initial values at $t = 0$. Further, because $\cos t$ is 2π-periodic, we also have

$$X''(t) + \omega^2 X(t) = x''(t + 2\pi) + \omega^2 x(t + 2\pi) = \cos(t + 2\pi) = \cos t.$$

Consequently, $x(t)$ and $X(t)$ satisfy the same initial value problem. By the existence and uniqueness theorem (see Theorem 2 on page 157), these functions must agree on the interval $[0, \infty)$. Hence $x(t) = X(t) = x(t + 2\pi)$ for all $t \geq 0$; that is, $x(t)$ is 2π-periodic. (The same reasoning works for $\omega = 3$.) With a similar argument, it follows from the Poincaré section for $\omega = 1/2$ that there is a solution of period 4π that alternates between the two points displayed in Figure 5.37(c) as t is incremented by 2π. For the case $\omega = 1/3$, we deduce that there is a solution of period 6π rotating among three points, and for $\omega = 1/4$, there is an 8π-periodic solution rotating among four points. We call these last three solutions **subharmonics.**

The case $\omega = \sqrt{2}$ is different. So far, in Figure 5.37(f), none of the points has repeated. Did we stop too soon? Will the points ever repeat? Here, the fact that $\sqrt{2}$ is irrational plays a crucial role. It turns out that every integer n yields a distinct point in the Poincaré section (see Problem 8). However, there is a pattern developing. The points all appear to lie on a simple curve, possibly an ellipse. To see that this is indeed the case, notice that when $\omega = \sqrt{2}$, $A = F = 1$, and $\phi = 0$, we have

$$x_n = \sin(2\sqrt{2}\pi n) + 1, \qquad v_n = \sqrt{2}\cos(2\sqrt{2}\pi n), \qquad n = 0, 1, 2, \ldots .$$

It is then an easy computation to show that each (x_n, ω_n) lies on the ellipse

$$(x - 1)^2 + \frac{v^2}{2} = 1.$$

In our investigation of equation (1), we concentrated on 2π-periodic solutions because the forcing term $F\cos t$ has period 2π. (We observed subharmonics when $\omega = 1/2$, $1/3$, and $1/4$, that is, solutions with periods $2(2\pi)$, $3(2\pi)$, and $4(2\pi)$.) When a damping term is introduced into the differential equation, the Poincaré map displays a different behavior. Recall that the solution will now be the sum of a transient and a steady-state term. For example, let's consider the equation

(4) $$x''(t) + bx'(t) + \omega^2 x(t) = F\cos t,$$

where b, F, and ω are positive constants.

When $b^2 < 4\omega^2$, the solution to (4) can be expressed as

(5) $$x(t) = Ae^{-(b/2)t}\sin\left(\frac{\sqrt{4\omega^2 - b^2}}{2}t + \phi\right) + \frac{F}{\sqrt{(\omega^2 - 1)^2 + b^2}}\sin(t + \theta),$$

where $\tan\theta = (\omega^2 - 1)/b$ and A and ϕ are arbitrary constants (cf. equations (7) and (8) in Section 5.3). The first term on the right-hand side of (5) is the transient and the second term, the steady-state solution. Let's construct the Poincaré map using $t = 2\pi n$, $n = 0, 1, 2, \ldots$. We will take $b = 0.22$, $\omega = A = F = 1$, and $\phi = 0$ to simplify the computations. Because $\tan\theta = (\omega^2 - 1)/b = 0$, we will take $\theta = 0$ as well. Then we have

$$x(2\pi n) = x_n = e^{-0.22\pi n}\sin(\sqrt{0.9879}\,2\pi n),$$
$$x'(2\pi n) = v_n = -0.11e^{-0.22\pi n}\sin(\sqrt{0.9879}\,2\pi n)$$
$$+ \sqrt{0.9879}\,e^{-0.22\pi n}\cos(\sqrt{0.9879}\,2\pi n) + \frac{1}{(0.22)}.$$

The Poincaré maps for $n = 0, 1, 2, \ldots, 10$ are shown in Figure 5.38 (black points). After just a few iterations, we observe that $x_n \approx 0$ and $v_n \approx 1/(0.22) \approx 4.545$; that is, the points of the Poincaré section are approaching a single point in the xv-plane (colored point). Thus we might expect that there is a 2π-periodic solution corresponding to a particular choice of A and ϕ. (In this example, where we can explicitly represent the solution, we see that indeed a 2π-periodic solution arises when we take $A = 0$ in (5).)

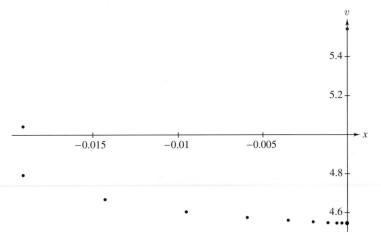

Figure 5.38 Poincaré section for equation (4) with $F = 1$, $b = 0.22$, and $\omega = 1$

There is an important difference between the Poincaré sections for equation (2) and those for equation (4). In Figure 5.37, the location of all of the points in (a)–(e) depends on the initial value selected (here $A = 1$ and $\phi = 0$). (See Problem 10.) However, in Figure 5.38, the first few points (black points) depend on the initial conditions, while the limit point (colored point) does *not* (see Problem 6). The latter behavior is typical for equations that have a "damping" term (i.e., $b > 0$); namely, the Poincaré section has a limit set[†] that is essentially independent of the initial conditions.

For equations with damping, the limit set may be more complicated than just a point. For example, the Poincaré map for the equation

(6) $x''(t) + (0.22)x'(t) + x(t) = \cos t + \cos(\sqrt{2}t)$

has a limit set consisting of an ellipse (see Problem 11). This is illustrated in Figure 5.39 on page 304 for the initial values $x_0 = 2$, $v_0 = 4$ and $x_0 = 2$, $v_0 = 6$.

So far we have seen limit sets for the Poincaré map that were either a single point or an ellipse—independent of the initial values. These particular limit sets are **attractors.** In general, an attractor is a set A with the property that there exists an open set[††] B containing

[†] The **limit set** for a map (x_n, v_n), $n = 1, 2, 3, \ldots$, is the set of points (p, q) such that $\lim_{k \to \infty}(x_{n_k}, v_{n_k}) = (p, q)$, where $n_1 < n_2 < n_3 < \cdots$ is some subsequence of the positive integers.

[††] A set $B \subset \mathbf{R}^2$ is an **open set** if for each point $p \in B$ there is an open disk V containing p such that $V \subset B$.

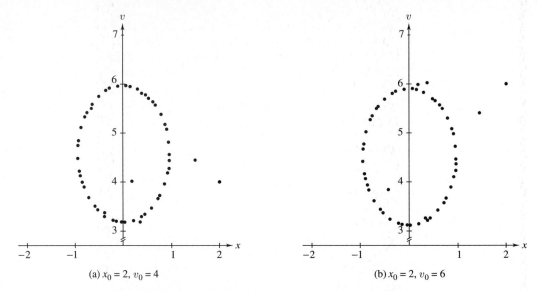

<div align="center">(a) $x_0 = 2$, $v_0 = 4$ (b) $x_0 = 2$, $v_0 = 6$</div>

<div align="center">**Figure 5.39** Poincaré section for equation (6) with initial values x_0, v_0</div>

A such that whenever the Poincaré map enters B, its points remain in B *and* the limit set of the Poincaré map is a subset of A. Further, we assume A has the *invariance property*: Whenever the Poincaré map starts at a point in A, it remains in A.

In the previous examples, the attractors of the dynamical system (Poincaré map) were easy to describe. In recent years, however, many investigators, working on a variety of applications, have encountered dynamical systems that do *not* behave orderly—their attractor sets are very complicated (not just isolated points or familiar geometric objects such as ellipses). The behavior of such systems is called **chaotic**, and the corresponding limit sets are referred to as **strange attractors.**

To illustrate chaotic behavior and what is meant by a strange attractor, we discuss two nonlinear differential equations and a simple difference equation. First, let's consider the **Duffing equation**

$$(7) \qquad x''(t) + bx'(t) - x(t) + x^3(t) = F \sin \gamma t.$$

We cannot express the solution to (7) in any explicit form, so we must obtain the Poincaré map by numerically approximating the solution to (7) for fixed initial values and then plot the approximations for $x(2\pi n/\gamma)$ and $v(2\pi n/\gamma) = x'(2\pi n/\gamma)$. (Because the forcing term $F \sin \gamma t$ has period $2\pi/\gamma$, we seek $2\pi/\gamma$-periodic solutions and subharmonics.) In Figure 5.40, we display the limit sets (attractors) when $b = 0.3$ and $\gamma = 1.2$ in the cases (a) $F = 0.2$, (b) $F = 0.28$, (c) $F = 0.29$, and (d) $F = 0.37$.

Notice that as the constant F increases, the Poincaré map changes character. When $F = 0.2$, the Poincaré section tells us that there is a $2\pi/\gamma$-periodic solution. For $F = 0.28$, there is a subharmonic of period $4\pi/\gamma$, and for $F = 0.29$ and 0.37, there are subharmonics with periods $8\pi/\gamma$ and $10\pi/\gamma$, respectively.

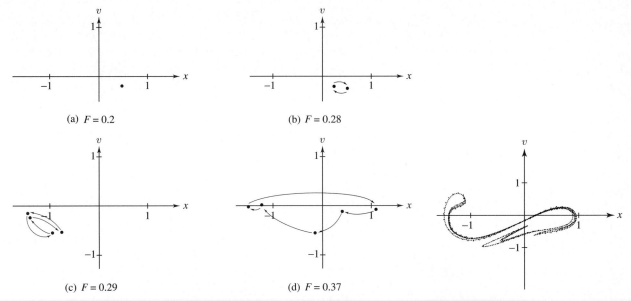

(a) $F = 0.2$

(b) $F = 0.28$

(c) $F = 0.29$

(d) $F = 0.37$

Figure 5.40 Poincaré sections for the Duffing equation (7) with $b = 0.3$ and $\gamma = 1.2$

Figure 5.41 Poincaré section for the Duffing equation (7) with $b = 0.3$, $\gamma = 1.2$, and $F = 0.5$

Things are dramatically different when $F = 0.5$: The solution is neither $2\pi/\gamma$-periodic nor subharmonic. The Poincaré section for $F = 0.5$ is illustrated in Figure 5.41. This section was generated by numerically approximating the solution to (7) when $\gamma = 1.2$, $b = 0.3$, and $F = 0.5$, for fixed initial values[†]. Not all of the approximations $x(2\pi n/\gamma)$ and $v(2\pi n/\gamma)$ that were calculated are graphed; because of the presence of a transient solution, the first few points were omitted. It turns out that the plotted set is essentially independent of the initial values and has the property that once a point is in the set, all subsequent points lie in the set. Because of the complicated shape of the set, it is indeed a strange attractor. While the shape of the strange attractor does not depend on the initial values, the picture does change if we consider different sections; for example, $t = (2\pi n + \pi/2)/\gamma$, $n = 0, 1, 2, \ldots$ yields a different configuration.

Another example of a strange attractor occurs when we consider the **forced pendulum equation**

$$(8) \qquad x''(t) + bx'(t) + \sin(x(t)) = F \cos t,$$

where the $x(t)$ term in (4) has been replaced by $\sin(x(t))$. Here $x(t)$ is the angle between the pendulum and the vertical rest position, b is related to friction, and F represents the

[†]**Historical Footnote:** When researchers first encountered these strange-looking Poincaré sections, they would check their computations using different computers and different numerical schemes (see Hénon and Heiles, *The applicability of the third integral of motion: some numerical experiments*, Astronomical Journal, Vol. 69 (1964), p. 75). For special types of dynamical systems, such as the Hénon map, it can be shown that there exists a true trajectory that *shadows* the numerical trajectory (see Hammel, Yorke, and Grebogi, *Numerical orbits of chaotic processes represent true orbits*, Bulletin American Mathematical Society, Vol. 19 (1988), pp. 466–469).

strength of the forcing function (see Figure 5.42). For $F = 2.7$ and $b = 0.22$, we have graphed in Figure 5.43 approximately 90,000 points in the Poincaré map. Since we cannot express the solution to (8) in any explicit form, the Poincaré map was obtained by numerically approximating the solution to (8) for fixed initial values and plotting the approximations for $x(2\pi n)$ and $v(2\pi n) = x'(2\pi n)$.

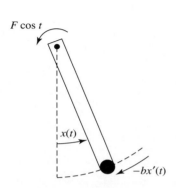

Figure 5.42 Forced damped pendulum

Figure 5.43 Poincaré section for the forced pendulum equation (8) with $b = 0.22$ and $F = 2.7$

The Poincaré maps for the Duffing equation and for the forced pendulum equation not only illustrate the idea of a strange attractor; they also exhibit another peculiar behavior called **chaos**. Chaos occurs when small changes in the initial conditions lead to major changes in the behavior of the solution. Henri Poincaré described the situation as follows:

> "... it may happen that small differences in the initial conditions will produce very large ones in the final phenomena. A small error in the former produces an enormous error in the latter. Prediction becomes impossible"

In a physical experiment, we can never *exactly* (with infinite accuracy) reproduce the same initial conditions. Consequently, if the behavior is chaotic, even a slight difference in the initial conditions will lead to quite different values for the corresponding Poincaré map when n is large. Such behavior does not occur for solutions to either equation (4) or equation (1) (see Problems 6 and 7). However, two solutions to the Duffing equation (7) with $F = 0.5$ that correspond to two different, but close, initial values have Poincaré maps that do *not* remain close together. While they both are attracted to the same set, their locations with respect to this set may be relatively far apart.

The phenomenon of chaos can also be illustrated by the following simple map. Let x_0 lie in $[0, 1)$ and define

(9) $x_{n+1} = 2x_n \pmod 1,$

where by (mod 1) we mean the decimal part of the number if it is greater than or equal to 1; that is

$$x_{n+1} = \begin{cases} 2x_n, & \text{for } 0 \le x_n < 1/2, \\ 2x_n - 1, & \text{for } 1/2 \le x_n < 1. \end{cases}$$

When $x_0 = 1/3$, we find

$$x_1 = 2 \,(1/3) \,(\text{mod } 1) = 2/3,$$

$$x_2 = 2 \,(2/3) \,(\text{mod } 1) = 4/3 \,(\text{mod } 1) = 1/3,$$

$$x_3 = 2 \,(1/3) \,(\text{mod } 1) = 2/3,$$

$$x_4 = 2 \,(2/3) \,(\text{mod } 1) = 1/3,$$

$$\text{etc.}$$

Written as a sequence, we get $\{1/3, \, 2/3, \, \overline{1/3, \, 2/3}, \ldots\}$.

What happens when we pick a starting value x_0 near 1/3? Does the sequence cluster about 1/3 and 2/3 as does the mapping when $x_0 = 1/3$? For example, when $x_0 = 0.3$, we get the sequence

$$\{0.3, \, 0.6, \, 0.2, \, 0.4, \, 0.8, \, \overline{0.6, \, 0.2, \, 0.4, \, 0.8}, \ldots\}.$$

In Figure 5.44, we have plotted the values of x_n for $x_0 = 0.3$, 0.33, and 0.333. We have not plotted the first few terms, but only those that repeat. (This omission of the first few terms parallels the situation depicted in Figure 5.41 where transient solutions arise.)

Figure 5.44 Plots of the map $x_{n+1} = 2x_n \,(\text{mod } 1)$ for $x_0 = 0.3$, 0.33, and 0.333

It is clear from Figure 5.44 that while the values for x_0 are getting closer to 1/3, the corresponding maps are spreading out over the whole interval [0, 1] and *not* clustering near 1/3 and 2/3. This behavior is chaotic, since the Poincaré maps for initial values near 1/3 behave quite differently from the map for $x_0 = 1/3$. If we had selected x_0 to be irrational (which we can't do with a calculator), the sequence would *not* repeat and would be dense in [0, 1].

Systems that exhibit chaotic behavior arise in many applications. The challenge to engineers is to design systems that avoid this chaos and, instead, enjoy the property of **stability.** The topic of stable systems is discussed in Chapter 12.[†]

[†] All references to Chapters 11–13 refer to the expanded text *Fundamentals of Differential Equations and Boundary Value Problems*, Second Edition.

EXERCISES 5.8

 The reader will find it helpful to have a microcomputer or mainframe available. Software such as MACMATH, PHASEPLANE, or PHASER may be used.

1. Compute and graph the points of the Poincaré map with $t = 2\pi n$, $n = 0, 1, \ldots, 20$ for equation (1) taking $A = F = 1$, $\phi = 0$, and $\omega = 3/2$. Repeat, taking $\omega = 3/5$. Do you think the equation has a 2π-periodic solution for either choice of ω? A subharmonic solution?

2. Compute and graph the points of the Poincaré map with $t = 2\pi n$, $n = 0, 1, \ldots, 20$ for equation (1) taking $A = F = 1$, $\phi = 0$, and $\omega = 1/\sqrt{3}$. Describe the limit set for this system.

3. Compute and graph the points of the Poincaré map with $t = 2\pi n$, $n = 0, 1, \ldots, 20$ for the equation (5) taking $A = F = 1$, $\phi = 0$, $\omega = 1$, and $b = -0.1$. What is happening to these points as $n \to \infty$?

4. Compute and graph the Poincaré map with $t = 2\pi n$, $n = 0, 1, \ldots, 20$ for equation (5) taking $A = F = 1$, $\phi = 0$, $\omega = 1$, and $b = 0.1$. Describe the attractor for this system.

5. Compute and graph the Poincaré map with $t = 2\pi n$, $n = 0, 1, \ldots, 20$ for equation (5) taking $A = F = 1$, $\phi = 0$, $\omega = 1/3$, and $b = 0.22$. Describe the attractor for this system.

6. Show that the Poincaré map for equation (4) is not chaotic by showing that as t gets large

$$x_n = x(2\pi n) \approx \frac{F}{\sqrt{(\omega^2 - 1)^2 + b^2}} \sin(2\pi n + \theta),$$

$$v_n = x'(2\pi n) \approx \frac{F}{\sqrt{(\omega^2 - 1)^2 + b^2}} \cos(2\pi n + \theta)$$

independent of the initial values $x_0 = x(0)$ and $v_0 = x'(0)$.

7. Show that the Poincaré map for equation (1) is not chaotic by showing that if (x_0, v_0) and (x_0^*, v_0^*) are two initial values that define the Poincaré maps $\{(x_n, v_n)\}$ and $\{(x_n^*, v_n^*)\}$, respectively, using the recursive formulas in (3), then one can make the distance between (x_n, v_n) and (x_n^*, v_n^*) small by making the distance between (x_0, v_0) and (x_0^*, v_0^*) small. [*Hint:* Let (A, ϕ) and (A^*, ϕ^*) be the polar coordinates of two points in the plane. From the law of cosines, it follows that the distance d between them is given by $d^2 = (A - A^*)^2 + 2AA^*[1 - \cos(\phi - \phi^*)]$.]

8. Consider the Poincaré maps defined in (3) with $\omega = \sqrt{2}$, $A = F = 1$, and $\phi = 0$. If this map were ever to repeat, then, for two distinct positive integers n and m, $\sin(2\sqrt{2}\pi n) = \sin(2\sqrt{2}\pi m)$. Using basic properties of the sine function, show that this would imply that $\sqrt{2}$ is rational. It follows from this contradiction that the points of the Poincaré map do not repeat.

9. The doubling modulo 1 map defined by equation (9) exhibits some fascinating behavior. Compute the sequence obtained when

 (a) $x_0 = k/7$ for $k = 1, 2, \ldots, 6$.
 (b) $x_0 = k/15$ for $k = 1, 2, \ldots, 14$.
 (c) $x_0 = k/2^j$, where j is a positive integer and $k = 1, 2, \ldots, 2^j - 1$.

 Numbers of the form $k/2^j$ are called **dyadic numbers** and are dense in [0, 1]. That is, there is a dyadic number arbitrarily close to any real number (rational or irrational).

10. To show that the limit set of the Poincaré map given in (3) depends upon the initial values, do the following:

 (a) Show that when $\omega = 2$ or 3, the Poincaré map consists of the single point
 $$\left(A \sin \phi + \frac{F}{\omega^2 - 1}, \quad \omega A \cos \phi \right).$$

 (b) Show that when $\omega = 1/2$, the Poincaré map alternates between the two points
 $$\left(\frac{F}{\omega^2 - 1} \pm A \sin \phi, \quad \pm \omega A \cos \phi \right).$$

 (c) Use the results of parts (a) and (b) to show that when $\omega = 2$, 3, or 1/2, the Poincaré map (3) depends on the initial values (x_0, v_0).

11. To show that the limit set for the Poincaré map $x_n := x(2\pi n)$, $v_n := x'(2\pi n)$, where $x(t)$ is a solution to equation (6), is an ellipse and that this ellipse is the same for any initial values x_0, v_0, do the following:

 (a) Argue that since the initial values affect only the transient solution to (6), the limit set for the Poincaré map is independent of the initial values.

 (b) Now show that for n large
 $$x_n \approx a \sin(2\sqrt{2}\pi n + \psi),$$
 $$v_n \approx c + \sqrt{2}a \cos(2\sqrt{2}\pi n + \psi),$$
 where $a = (1 + 2(0.22)^2)^{-1/2}$, $c = (0.22)^{-1}$, and $\psi = \arctan\{-[(0.22)\sqrt{2}]^{-1}\}$.

(c) Use the result of part (b) to conclude that the ellipse

$$x^2 + \frac{(v - c)^2}{2} = a^2$$

contains the limit set of the Poincaré map.

12. Using a numerical scheme such as Runge-Kutta or software such as PHASEPLANE, calculate the Poincaré map for equation (7) when $b = 0.3$, $\gamma = 1.2$, and $F = 0.2$. (Notice that the closer you start to the limiting point, the sooner the transient part will die out.) Compare your map with Figure 5.40(a) on page 305. Redo for $F = 0.28$.

13. Redo Problem 12 with $F = 0.31$. What kind of behavior does the solution exhibit?

14. Redo Problem 12 with $F = 0.65$. What kind of behavior does the solution exhibit?

15. **Chaos Machine.** Chaos can be illustrated using a long ruler, a short ruler, a pin, and a tie tack (pivot). Construct the double pendulum as shown in Figure 5.45(a). The pendulum is set in motion by releasing it from a position such as the one shown in Figure 5.45(b). Repeatedly set the pendulum in motion, each time trying to release it from the same position. Record the number of times the short ruler flips over and the direction in which is was moving. If the pendulum was released in *exactly* the same position each time, then the motion would be the same. However, from your experiments you will observe that even beginning close

to the same position leads to very different motions. This double pendulum exhibits chaotic behavior.

(a) double pendulum

(b) release position

Figure 5.45 Double pendulum as a chaos machine

CHAPTER SUMMARY

The various motions of a mass-spring oscillator illustrate many of the aspects of the theory of linear second order ordinary differential equations. The free, undamped oscillator exhibits simple harmonic motion. When friction is present, the free motion may be underdamped, overdamped, or critically damped. If the oscillator is externally forced, the response contains a transient motion corresponding to a solution for the unforced oscillator, plus a synchronous motion corresponding to a particular solution. When two oscillators are coupled, they are governed by a linear system of differential equations.

Entirely analogous responses are exhibited by electric circuits consisting of voltage sources, resistors, inductors, and capacitors. Circuits containing multiple loops are analyzed by applying Kirchhoff's laws, which result in a system of differential equations.

The numerical techniques for first order equations discussed in Sections 3.5 and 3.6 each have extensions to first order systems in normal form. These techniques also apply to

higher-order equations once they have been reduced to a system in normal form. In Section 5.6, we discussed the extension of the fourth order Runge-Kutta algorithm to systems.

A technique for studying the qualitative behavior of solutions to the nonlinear **autonomous system**

(1) $$\frac{dx}{dt} = f(x, y), \qquad \frac{dy}{dt} = g(x, y)$$

is the method of **phase plane analysis.** To sketch a phase plane diagram for (1), we begin by finding the **critical points,** which are solutions to the system

$$f(x, y) = 0, \qquad g(x, y) = 0.$$

Next we solve (either analytically or numerically) the first order equation $dy/dx = g(x, y)/f(x, y)$ and use the solutions to sketch some integral curves for (1). (Software such as MACMATH, PHASEPLANE, or PHASER is helpful in constructing a phase plane diagram.) Returning to the original equations in (1), we can determine the **flow** along each integral curve and hence sketch the **trajectories** of the system. This information can be used to describe the behavior of the solutions to the system even though the solutions themselves cannot be found.

Nonautonomous systems can be studied by considering a Poincaré map for the system. In Section 5.8, we used a Poincaré map to detect periodic and subharmonic solutions. We also used it to study systems whose solutions exhibited chaotic behavior.

GROUP PROJECTS FOR CHAPTER 5

A. Designing a Landing System for Interplanetary Travel

Alfred Clark, Jr., University of Rochester

You are a second-year Starfleet Academy Cadet aboard the U.S.S. Enterprise on a continuing study of the starsystem Glia. The object of study on the present expedition is the large airless planet Glia-4. A Class 1 Sensor Probe of mass m is to be sent to the planet's surface to collect data. The probe has a modifiable landing system so that it can be used on planets of different gravity. The system consists of a linear spring (force = $-kx$, where x is displacement), a nonlinear spring (force = $-ax^3$), and a shock-damper (force = $-b\dot{x}$),[†] all in parallel. Figure 5.46 shows a schematic of the system. During the landing process, the probe's thrusters are used to create a constant rate of descent. The velocity at impact varies; the symbol V_L is used to denote the largest velocity likely to happen in practice. At the instant of impact, (1) the thrust is turned off, and (2) the suspension springs are at their unstretched natural length.

(a) Let the displacement x be measured from the unstretched length of the springs and be taken negative downward (i.e., compression gives a negative x). Show that the equation governing the oscillations after impact is

$$m\ddot{x} + b\dot{x} + kx + ax^3 = -mg.$$

(b) The probe has a mass $m = 1220$ kg. The linear spring is permanently installed and has a stiffness $k = 35{,}600$ N/m. The gravity on the surface of Glia-4 is $g = 17.5$ m/sec². The nonlinear spring is removable; an appropriate spring must be chosen for each mission. These nonlinear springs are made of coralidium, a rare and difficult-to-fabricate alloy.

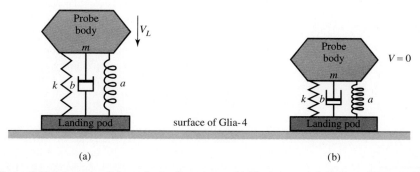

(a) (b)

Figure 5.46 Schematic of the probe landing system. (a) The system at the instant of impact. The springs are not stretched or compressed, the thrusters have turned off, and the velocity is V_L downward. (b) The probe has reached a state of rest on the surface, and the springs are compressed enough to support the weight. Between states (a) and (b), the probe oscillates relative to the landing pod.

[†] The symbol $\dot{x}$ denotes dx/dt.

Therefore the Enterprise stocks only four different strengths: $a = 150,000$; $300,000$; $450,000$; and $600,000$ N/m³. Determine which springs give a compression as close as possible to 0.3 m without exceeding 0.3 m, when the ship is resting on the surface of Glia-4. (The limit of 0.3 m is imposed by unloading clearance requirements.)

(c) The other adjustable component on the landing system is the linear shock-damper, which may be adjusted in increments of $\Delta b = 500$ N-sec/m, from a low value of 1000 N-sec/m to a high value of 10,000 N-sec/m. It is desirable to make b as small as possible because a large b produces large forces at impact. However, if b is too small, there is some danger that the probe will rebound after impact. To minimize the chance of this, find the smallest value of b such that the springs are always in compression during the oscillations after impact. Use a maximum impact velocity $V_L = 5$ m/sec downward. To find this value of b, you will need to use a software package such as PHASER, PHASEPLANE, MACMATH, or XFUNCTIONS to integrate the differential equation.

B. Things That Bob

Richard Bernatz, Luther College

The motion of various shaped objects that bob in a pool of water can be modeled by a second order differential equation derived from Newton's second law of motion, $F = ma$. The forces acting on the object include the force due to gravity, a frictional force due to the motion of the object in the water, and a buoyant force based on Archimedes' principle: An object that is completely or partially submerged in a fluid is acted on by an upward (buoyant) force equal to the weight of the water it displaces.

(a) The first step is to write down the governing differential equation. The dependent variable is the depth z of the object's lowest point in the water. Take z to be negative downward so that $z = -1$ means 1 ft of the object has submerged. Let $V(z)$ be the submerged volume of the object, m be the mass of the object, ρ be the density of water (in pounds per cubic foot), g be the acceleration due to gravity, and γ_w be the coefficient of friction for water. Assuming the frictional force is proportional to the vertical velocity of the object, write down the governing second order ODE.

(b) For the time being, neglect the effect of friction and assume the object is a cube measuring L feet on a side. Write down the governing differential equation for this case. Next, designate $z = l$ to be the depth of submersion such that the buoyant force is equal and opposite the gravitational force. Introduce a new variable, ζ, that gives the displacement of the object from its equilibrium position l (that is, $z = \zeta + l$). You can now write the ODE in a more familiar form. [*Hint:* Recall the spring-mass system and the equilibrium case.] Now you should recognize the type of solution for this problem. What is the natural frequency?

(c) In this task, you will consider the effect of friction. The bobbing object is a cube, one foot on a side, that weighs 32 lb. Let $\gamma_w = 3$ lb-sec/ft and suppose the object is initially placed on the surface of the water. Solve the governing ODE by hand to find the general solution. Next, find the particular solution for the case in which the cube is initially placed on the surface of the water and is given no initial velocity. Provide a plot of the position of the object as a function of time t.

(d) In this step of the project, you will develop a numerical solution to the same problem presented in part (c). The numerical solution will be useful (indeed necessary) for subsequent parts of the project. This case will provide a trial to verify that your numerical solution is correct. Go back to the initial ODE you developed in part (a). Using param-

eter values given in part (c), solve the initial value problem for the cube starting on the surface with no initial velocity. To solve this problem numerically, you will have to write the second order ODE as a system of two first order ODEs, one for vertical position z and one for vertical velocity w. Plot your results for vertical position as a function of time t for the first 3 or 4 sec and compare with the analytical solution you found in part (c). Are they in close agreement? What might you have to do in order to compare these solutions? Provide a plot of both your analytical and numerical solutions on the same graph.

(e) Suppose a sphere of radius R is allowed to bob in the water. Derive the governing second order equation for the sphere using Archimedes' principles and allowing for friction due to its motion in the water. Suppose a sphere weighs 32 lb, has a radius of 1/2 ft, and $\gamma_w = 3.0$ lb-sec/ft. Determine the limiting value of the position of the sphere without solving the ODE. Next, solve the governing ODE for the velocity and position of the sphere as a function of time for a sphere placed on the surface of the water. You will need to write the governing second order ODE as a system of two ODEs, one for velocity and one for position. What is the limiting position of the sphere for your solution? Does it agree with the equilibrium solution you found above? How does it compare with the equilibrium position of the cube? If it is different, explain why.

(f) Suppose the sphere in part (d) is a volleyball. Calculate the position of the sphere as a function of time t for the first 3 sec if the ball is submerged so that its lowest point is 5 ft under water. Will the ball leave the water? How high will it go? Next, calculate the ball's trajectory for initial depths lower than 5 ft, and higher than 5 ft. Provide plots of velocity and position for each case and comment on what you see. Specifically, comment on the relationship between the initial depth of the ball and the maximum height the ball eventually attains.

You might consider taking a volleyball into a swimming pool to gather real data in order to verify and improve on your model. If you do so, report the data you found and explain how you used it for verification and improvement of your model.

C. Effects of Hunting on Predator-Prey Systems

Cyclic variations in the populations of predators and their prey have been studied using the Volterra-Lotka predator-prey model given by the system

(1) $$\frac{dx}{dt} = Ax - Bxy,$$

(2) $$\frac{dy}{dt} = -Cy + Dxy,$$

where A, B, C, and D are positive constants, $x(t)$ is the population of prey at time t, and $y(t)$ is the population of predators. It can be shown that such a system has a periodic solution (see Project D). That is, there exists some constant T such that $x(t) = x(t + T)$ and $y(t) = y(t + T)$ for all t. This periodic or cyclic variation in the populations has been observed in various systems such as sharks-food fish, lynx-rabbits, and ladybird beetles-cottony cushion scale. Because of this periodic behavior, it is useful to consider the average populations $\bar{x}$ and $\bar{y}$ defined by

$$\bar{x} := \frac{1}{T}\int_0^T x(t)\,dt, \qquad \bar{y} := \frac{1}{T}\int_0^T y(t)\,dt.$$

(a) Show that $\bar{x} = C/D$ and $\bar{y} = A/B$. [*Hint:* Use equation (1) and the fact that $x(0) = x(T)$ to show that

$$\int_0^T (A - By(t))\,dt = \int_0^T \frac{x'(t)}{x(t)}\,dt = 0.]$$

(b) To determine the effect of indiscriminate hunting on the populations, assume hunting reduces the rate of change in a population by a constant times the population. Then, the predator-prey system satisfies the new set of equations

(3) $\dfrac{dx}{dt} = Ax - Bxy - \varepsilon x = (A - \varepsilon)x - Bxy,$

(4) $\dfrac{dy}{dt} = -Cy + Dxy - \delta y = -(C + \delta)y + Dxy,$

where ε and δ are positive constants with $\varepsilon < A$. What effect does this have on the average population of prey? On the average population of predators?

(c) Assume the hunting is done selectively, as in shooting only rabbits (or shooting only lynx). Then we have $\varepsilon > 0$ and $\delta = 0$ (or $\varepsilon = 0$ and $\delta > 0$) in (3)–(4). What effect does this have on the average populations of predator and prey?

(d) In a rural county, foxes prey mainly on rabbits but occasionally include a chicken in their diet. The farmers decide to put a stop to the chicken killing by hunting the foxes. What do you predict will happen? What happens to the farmers' gardens?

For a discussion of other differential equation models in population biology, see *Differential Equation Models*, by Martin Braun, Courtney S. Coleman, and Donald A. Drew (eds.) (Springer-Verlag, New York, 1983), Part V.

D. Periodic Solutions to Volterra-Lotka Systems

As stated in Project C, the Volterra-Lotka predator-prey model is given by the system (1)–(2), where $x(t)$ and $y(t)$ are the populations of the prey and the predators, respectively, and A, B, C, and D are positive constants. For autonomous systems such as (1)–(2), the existence of a closed integral curve that does not contain a critical point implies the existence of a periodic solution (see Section 5.7). To show that (1)–(2) has such an integral curve, proceed as follows:

(a) Determine the critical points of system (1)–(2).
(b) Show that the integral curves of (1)–(2) can be expressed in the form $w(x)z(y) = k$, where $w(x) := x^C e^{-Dx}$ and $z(y) := y^A e^{-By}$.
(c) Show that $w(x) = x^C e^{-Dx}$ has the properties $w(0) = 0$, $\lim_{x \to \infty} w(x) = 0$, and $w(x)$ attains a maximum value of $M_w := (C/D)^C e^{-C}$ for $x > 0$ at $x = C/D$. Sketch the graph of $w(x)$. Similarly, show that $z(y) = y^A e^{-By}$ has the properties $z(0) = 0$, $\lim_{y \to \infty} z(y) = 0$, and $z(y)$ attains a maximum value of $M_z := (A/B)^A e^{-A}$ for $y > 0$ at $y = A/B$. Sketch the graph of $z(y)$.
(d) Show that the product $w(x)z(y)$ is bounded by $M_w M_z$ for $x, y > 0$, and hence there is no solution with $x, y > 0$ for $w(x)z(y) = k$ when $k > M_w M_z$. Also show that the only solution when $k = M_w M_z$ is $x = C/D$, $y = A/B$.
(e) Assume $0 < k < M_w M_z$. Show that when $y = A/B$, there are two solutions to $w(x)z(y) = k$: one being $x_{\min} < C/D$ and the other $x_{\max} > C/D$.
(f) Again assume $0 < k < M_w M_z$. Show that for x fixed, the equation $w(x)z(y) = k$ has (i) no

solution when $x < x_{min}$ or $x > x_{max}$, (ii) one solution $y = A/B$ when $x = x_{min}$ or x_{max}, and (iii) two solutions $y_1(x) < A/B < y_2(x)$ when $x_{min} < x < x_{max}$ (see Figure 5.47).

(g) Using the results of parts (e) and (f), argue that system (1)–(2) has a family of closed integral curves in the first quadrant that surround the critical point $(C/D, A/B)$. Hence the system has a periodic solution.

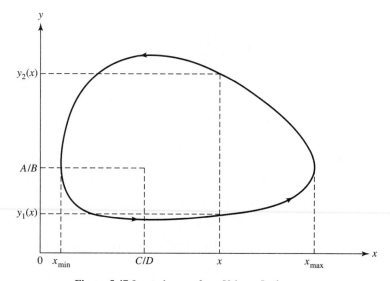

Figure 5.47 Integral curve for a Volterra-Lotka system

E. Limit Cycles and the van der Pol Equation

In the study of triode vacuum tubes, one encounters the **van der Pol equation:**[†]

(5) $x''(t) - \mu(1 - x^2(t))x'(t) + x(t) = 0, \qquad \mu > 0.$

We will study this equation by constructing a phase plane diagram using available software or by using a numerical scheme to generate the trajectories. To obtain better numerical approximations, it is advantageous to work with an equivalent equation called a **Rayleigh equation:**

(6) $y''(t) - \mu[1 - (y'(t))^2]y'(t) + y(t) = 0, \qquad \mu > 0.$

(a) Show that the change of variables $x(t) = \sqrt{3}\, y'(t)$ transforms equation (6) into equation (5).

[†]**Historical Footnote:** Experimental research by **E. V. Appleton** and **B. van der Pol** in 1921 on the oscillations of an electric circuit containing a triode generator (vacuum tube) led to the nonlinear equation now called **van der Pol's equation.** Methods of solution were developed by van der Pol in 1926–1927. **Mary L. Cartwright** continued research into nonlinear oscillation theory and together with **J. E. Littlewood** obtained existence results for forced oscillations in nonlinear systems in 1945.

(b) Show that the Rayleigh equation (6) is equivalent to the system

(7)

$$\frac{dv}{dt} = -w + \mu v(1 - v^2),$$

$$\frac{dw}{dt} = v,$$

where $w(t) = y(t)$. (When you call up the file vdp.ode in PHASEPLANE, you get system (7) with the parameters AL = μ and EPS = 1. When you call up the file vanderpol in PHASER, you get a related system, but with a parameter a unrelated to μ. To do this project in PHASER, you should use the file forcevdp.)

(c) Construct a phase plane diagram for the system (7) with $\mu = 0.1$ about the origin and large enough to include the rectangle $-4 < x < 4$, $-5 < y < 5$. Include trajectories beginning near, but not at, the origin and some beginning outside the circle of radius 3.

Notice that some of the trajectories spiral in while others spiral out. If you continued the approximation long enough in time, then you would find that the curves approach a closed curve that is approximately a circle centered at the origin. This closed curve is called a **limit cycle** for the van der Pol equation and is the trajectory of a periodic solution to system (7) and hence to van der Pol's equation.

(d) To investigate how the limit cycle for van der Pol's equation depends upon the parameter μ, construct a set of phase plane diagrams corresponding to $\mu = 0.1$, 0.2, 1.0, 2.0, 5.0, 12.0, 0, -0.5, and -2.0. How many limit cycles do you find for each μ? How does the shape change? Do the trajectories spiral toward the limit cycle or away from it?

F. Strange Behavior of Competing Species—Part I

Let $p_i(t)$ denote, respectively, the populations of three competing species S_i, $i = 1, 2, 3$. Suppose that these species have the same growth rates and that the maximum population of the species the habitat can support is the same for each. (We assume it to be one unit.) Moreover, suppose the competitive advantage that S_1 has over S_2 is the same as that of S_2 over S_3 and S_3 over S_1. This situation is modeled by the system

(8)

$$p_1' = p_1(1 - p_1 - ap_2 - bp_3),$$

$$p_2' = p_2(1 - bp_1 - p_2 - ap_3),$$

$$p_3' = p_3(1 - ap_1 - bp_2 - p_3),$$

where a and b are positive constants.

(a) Show that system (8) has five equilibrium populations (solutions) corresponding to the critical points $(0, 0, 0)$, $(1, 0, 0)$, $(0, 1, 0)$, $(0, 0, 1)$, and (δ, δ, δ), where $\delta := (1 + a + b)^{-1}$.

(b) Study the system when $a = b = 0.5$. Since the system is *not* linear, we cannot apply the technique discussed in Section 4.10. The system has three unknowns, so the phase plane analysis is much more complex than that presented in Section 5.7. Instead, we perform numerical experiments on the system to see if we can determine the behavior of its solutions. Use a scheme such as fourth order Runge-Kutta method for systems (see Appendix E) to perform these experiments. Use various initial values, in particular, values near the four equilibrium points. Describe the behavior of the populations as $t \to \infty$.

(c) The system behaves quite differently when $a = 0.5$ and $b = 2.0$. Perform numerical

experiments to show that the solution for this system approaches one of the equilibrium solutions—(1, 0, 0), (0, 1, 0), (0, 0, 1)—and then moves toward another, then toward the third, then back toward the first, and so on. Furthermore, the solutions linger longer and longer about these points before rapidly moving to the next one.

(d) Use the results of part (c) to argue that "most of the time,"

$$p_1 p_2 \approx 0, \qquad p_2 p_3 \approx 0, \qquad p_1 p_3 \approx 0.$$

Moreover, $T := p_1 + p_2 + p_3 \approx 1$.

(e) From system (8), derive the equations

$$\frac{dT}{dt} = T(1 - T) + \eta(p_1 p_2 + p_2 p_3 + p_1 p_3),$$

$$\frac{d \ln P}{dt} = -\eta + (3 + \eta)(1 - T),$$

where $T := p_1 + p_2 + p_3$, $P := p_1 p_2 p_3$, and $\eta := a + b - 2 = 0.5$.

(f) Using the results of parts (d) and (e), show that

$$\frac{dT}{dt} = T(1 - T), \qquad \frac{d \ln P}{dt} = -\eta$$

are reasonable simplifications.

(g) Solve the equations in part (f) and verify that as $t \to \infty$, the sum T approaches 1 and the product P approaches zero.

This system is studied in more detail in Project D in Chapter 9.

G. Cleaning Up the Great Lakes

A simple mathematical model that can be used to determine the time it would take to clean up the Great Lakes can be developed using a multiple compartmental analysis approach. In particular, we can view each lake as a tank that contains a liquid in which is dissolved a particular pollutant (DDT, phosphorus, mercury). Schematically, we view the lakes as consisting of five tanks connected as indicated in Figure 5.48.

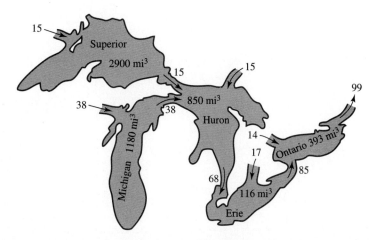

Figure 5.48 Compartmental model of the Great Lakes with flow rates (mi³/year) and volumes (mi³)

For our model, we make the following assumptions:

1. The volume of each lake remains constant.
2. The flow rates are constant throughout the year.
3. When a liquid enters the lake, perfect mixing occurs and the pollutants are uniformly distributed.
4. Pollutants are dissolved in the water and enter or leave by inflow or outflow of solution.

Before using this model to obtain estimates on the cleanup times for the lakes, we consider some simpler models:

(a) Use the outflow rates given in Figure 5.48 to determine the time it would take to ''drain'' each lake. This gives a lower bound on how long it would take to remove all the pollutants.
(b) A better estimate is obtained by assuming each lake is a separate tank with *only* clean water flowing in. Use this approach to determine how long it would take the pollution level in each lake to be reduced to 50% of its original level. How long would it take to reduce the pollution to 5% of its original level?
(c) Finally, to take into account the fact that pollution from one lake flows into the next lake in the chain, use the entire multiple compartment model given in Figure 5.48 to determine when the pollution level in each lake has been reduced to 50% of its original level, assuming pollution has ceased (that is, inflows not from a lake are clean water). Assume all the lakes initially have the same pollution concentration p. How long would it take for the pollution to be reduced to 5% of its original level?

For a detailed discussion of this model, see *An Introduction to Mathematical Modeling* by Edward A. Bender (Krieger, New York, 1991), Chapter 8.

H. A Growth Model for Phytoplankton—Part I

Olivier Bernard, Station Zoologique de Villefranche-sur-Mer, France, and Jean-Luc Gouzé, INRIA Sophia-Antipolis, France

A chemostat is a stirred tank in which phytoplankton grow by consuming a nutrient (e.g., nitrate). The nutrient is supplied to the tank at a given rate, and a solution containing the phytoplankton and remaining nutrient is removed at an equal rate (cf. Figure 5.49). The chemostat reproduces *in vitro*

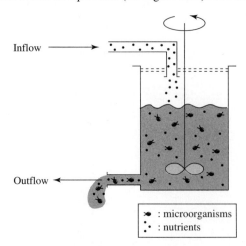

Inflow

Outflow

⚫ : microorganisms
: nutrients

Figure 5.49 Chemostat

the conditions of the growth of phytoplankton in the ocean; the phytoplankton is the first element of the marine food chain.

Let S denote the concentration (in μmol/liter) of the nutrient and X the biovolume (which is to be taken as an estimation of the biomass) of phytoplankton (in mm^3 of cells per liter of solution). A classical model (J. Monod, *La technique de culture continue: théorie et applications.* Annales de l'Institut Pasteur, 79, 1950) of the behavior of the chemostat is the following:

(9)
$$\begin{cases} \dfrac{dX}{dt} = \rho\,\dfrac{SX}{S+k} - \alpha X \\[2mm] \dfrac{dS}{dt} = \alpha(s_i - S) - \dfrac{\rho}{y}\,\dfrac{SX}{S+k}. \end{cases}$$

The unit for time is the day; the dilution rate α and the growth rate ρ are in day^{-1}; and the input concentration s_i and the constant k have the same units as S. Experimental (smoothed) data for S and X are given in Table 5.6 and illustrated in Figure 5.50 on page 320 (data obtained from the Station Zoologique of Villefranche-sur-Mer, France).

TABLE 5.6	NUTRIENT/BIOVOLUME DATA				
Time	**S**	**X**	**Time**	**S**	**X**
0.00	45.84	0.15	3.90	22.84	5.91
0.15	45.29	0.19	4.05	21.68	6.52
0.30	44.66	0.25	4.20	20.07	7.54
0.45	43.95	0.33	4.35	18.17	8.80
0.60	43.18	0.42	4.50	15.98	10.13
0.75	42.34	0.52	4.65	13.50	11.53
0.90	41.43	0.65	4.80	10.73	12.83
1.05	40.45	0.78	4.95	7.67	13.85
1.20	39.37	0.94	5.10	4.32	14.59
1.35	38.03	1.14	5.25	0.99	15.12
1.50	36.36	1.39	5.40	0.24	15.71
1.65	34.39	1.69	5.55	0.13	16.37
1.80	32.05	2.03	5.70	0.08	17.08
1.95	29.13	2.43	5.85	0.07	17.69
2.10	25.57	2.86	6.00	0.05	18.15
2.25	21.39	3.34	6.15	0.03	18.47
2.40	16.87	3.83	6.30	0.01	18.66
2.55	12.20	4.34	6.45	0.00	18.78
2.70	13.50	4.86	6.60	0.00	18.84
2.85	15.39	5.21	6.75	0.02	18.86
3.00	17.60	5.31	6.90	0.02	18.82
3.15	20.00	5.42	7.05	0.02	18.73
3.30	21.89	5.60	7.20	0.02	18.61
3.45	23.07	5.73	7.35	0.01	18.52
3.60	23.56	5.75	7.50	0.01	18.48
3.75	23.47	5.70			

Figure 5.50 Nutrient/biovolume data

(i) When $t < 2.5$, the dilution rate α is zero ("batch culture"). It is known that k is in the range $0.1 \leq k \leq 1$.

(a) What are the units for the yield factor y?

(b) Write a linear approximation of the system (9) for $S \gg k$ (i.e., S is much larger than k).

(c) Solve the approximated system in part (b) and use the solution for X and the experimental data in Table 5.6 to obtain a numerical value for ρ [*Hint:* Plot the logarithm of X against the time, and estimate the slope.] Use the equation for S and the data to obtain an estimation of y.

(d) Take $k = 0.5$ and the values for ρ and y obtained in (c). Using a computer software package such as MACMATH, MATLAB, or PHASEPLANE and the initial conditions given in Table 5.6, draw the numerical solutions $X(t)$, $S(t)$ for the system (9) and for the approximated system of part (b). Is the approximation of part (b) a reasonable one?

(ii) For $t > 2.5$, the dilution rate is $\alpha = 1.06$ day^{-1}. After a delay, the growth rate ρ of the phytoplankton changes because the cells adapt themselves to their new environment.

(e) Estimate the time T when the growth rate changes and obtain the new value for ρ. (As above, take $k = 0.5$.)

Further discussion of this model appears in Project E of Chapter 12.

Chapter 6

Theory of Higher-Order Linear Differential Equations

In this chapter, we discuss the basic theory of linear higher-order differential equations. The material is a generalization of the results we obtained in Chapter 4 for second order equations. In the statements and proofs of these results, we use concepts usually covered in an elementary linear algebra course; namely, linear dependence, determinants, and solving systems of linear equations. These concepts also arise in the matrix approach for solving systems of differential equations and are discussed in Section 9.2.

6.1 BASIC THEORY OF LINEAR DIFFERENTIAL EQUATIONS

Recall that a *linear* differential equation is an equation that can be written in the form

(1) $$a_n(x)y^{(n)}(x) + a_{n-1}(x)y^{(n-1)}(x) + \cdots + a_0(x)y(x) = b(x),$$

where $a_0(x)$, $a_1(x)$, ..., $a_n(x)$ and $b(x)$ depend only on x, not y. When a_0, a_1, ..., a_n are constants, we say equation (1) has **constant coefficients;** otherwise it has **variable coefficients.** If $b(x) \equiv 0$, equation (1) is called **homogeneous;** otherwise it is **nonhomogeneous.**

In developing a basic theory, we assume $a_0(x)$, $a_1(x)$, ..., $a_n(x)$ and $b(x)$ are continuous on an interval I and $a_n(x) \neq 0$ on I. Then, on dividing by $a_n(x)$, we can rewrite (1) in the **standard form**

(2) $$y^{(n)}(x) + p_1(x)y^{(n-1)}(x) + \cdots + p_n(x)y(x) = g(x),$$

where the functions $p_1(x)$, ..., $p_n(x)$, and $g(x)$ are continuous on I.

For a linear higher-order differential equation, the initial value problem always has a unique solution.

EXISTENCE AND UNIQUENESS

Theorem 1. Suppose that $p_1(x), \ldots, p_n(x)$ and $g(x)$ are each continuous on an interval (a, b) that contains the point x_0. Then, for any choice of the initial values $y_0, y_1, \ldots, y_{n-1}$, there exists a unique solution $y(x)$ on the whole interval (a, b) to the initial value problem

(3) $y^{(n)}(x) + p_1(x)y^{(n-1)}(x) + \cdots + p_n(x)y(x) = g(x),$

(4) $y(x_0) = y_0, \; y'(x_0) = y_1, \; \ldots, \; y^{(n-1)}(x_0) = y_{n-1}.$

The proof of Theorem 1 can be found in Chapter 13.[†]

EXAMPLE 1 For the initial value problem

(5) $x(x-1)y''' - 3xy'' + 6x^2y' - (\cos x)y = \sqrt{x+5};$

(6) $y(x_0) = 1, \qquad y'(x_0) = 0, \qquad y''(x_0) = 7,$

determine the values of x_0 and the intervals (a, b) containing x_0 for which Theorem 1 guarantees the existence of a unique solution on (a, b).

Solution Putting equation (5) in standard form, we find that $p_1(x) = -3/(x-1)$, $p_2(x) = 6x/(x-1)$, $p_3(x) = -(\cos x)/[x(x-1)]$, and $g(x) = \sqrt{x+5}/[x(x-1)]$. Now $p_1(x)$ and $p_2(x)$ are continuous on every interval not containing $x = 1$, while $p_3(x)$ is continuous on every interval not containing $x = 0$ or $x = 1$. The function $g(x)$ is not defined for $x < -5$, $x = 0$, and $x = 1$, but is continuous on $(-5, 0)$, $(0, 1)$, and $(1, \infty)$. Hence the functions p_1, p_2, p_3, and g are *simultaneously* continuous on the intervals $(-5, 0)$, $(0, 1)$, and $(1, \infty)$. From Theorem 1, it follows that if we choose $x_0 \in (-5, 0)$, then there exists a unique solution to the initial value problem (5)–(6) on the whole interval $(-5, 0)$. Similarly, for $x_0 \in (0, 1)$, there is a unique solution on $(0, 1)$ and for $x_0 \in (1, \infty)$, a unique solution on $(1, \infty)$. ■ End

If we let the left-hand side of equation (3) define the differential operator L,

(7) $L[y] := \dfrac{d^n y}{dx^n} + p_1 \dfrac{d^{n-1}y}{dx^{n-1}} + \cdots + p_n y,$

then we can express equation (3) in the operator form

(8) $L[y](x) = g(x).$

It is essential to keep in mind that L is a *linear* operator; it satisfies

(9) $L[y_1 + y_2] = L[y_1] + L[y_2],$

(10) $L[cy] = cL[y], \qquad c \text{ any constant}$

(see Problem 25).

[†] All references to Chapters 11–13 refer to the expanded text *Fundamentals of Differential Equations and Boundary Value Problems*, Second Edition.

As a consequence of this linearity, if $y_1, \ldots, y_m$ are solutions to the *homogeneous* equation

(11) $L[y](x) = 0,$

then any linear combination of these functions, $C_1 y_1 + \cdots + C_m y_m$, is also a solution. Imagine now that we have found n solutions $y_1, \ldots, y_n$ to the nth order linear equation (11). Is it true that *every* solution to (11) can be represented by

(12) $C_1 y_1 + \cdots + C_n y_n$

for appropriate choices of the constants $C_1, \ldots, C_n$? The answer is yes, provided the solutions $y_1, \ldots, y_n$ satisfy a certain property of independence that we shall now derive.

Let $\phi(x)$ be a solution to (11) on the interval (a, b) and let x_0 be a fixed number in (a, b). If it is possible to choose the constants $C_1, \ldots, C_n$ so that

(13)
$$
\begin{aligned}
C_1 y_1(x_0) \quad &+ \cdots + C_n y_n(x_0) \quad = \phi(x_0), \\
C_1 y_1'(x_0) \quad &+ \cdots + C_n y_n'(x_0) \quad = \phi'(x_0), \\
&\vdots \\
C_1 y_1^{(n-1)}(x_0) &+ \cdots + C_n y_n^{(n-1)}(x_0) = \phi^{(n-1)}(x_0),
\end{aligned}
$$

then, since $\phi(x)$ and $C_1 y_1(x) + \cdots + C_n y_n(x)$ are two solutions satisfying the same initial conditions at x_0, the uniqueness conclusion of Theorem 1 gives

(14) $\phi(x) = C_1 y_1(x) + \cdots + C_n y_n(x)$

for all x in (a, b).

The system (13) consists of n linear equations in the n unknowns $C_1, \ldots, C_n$. It has a unique solution for all possible values of $\phi(x_0), \ldots, \phi^{(n-1)}(x_0)$ if and only if the determinant[†] of the coefficients is different from zero; that is, if and only if

(15)
$$
\begin{vmatrix}
y_1(x_0) & y_2(x_0) & \cdots & y_n(x_0) \\
y_1'(x_0) & y_2'(x_0) & \cdots & y_n'(x_0) \\
\vdots & \vdots & & \vdots \\
y_1^{(n-1)}(x_0) & y_2^{(n-1)}(x_0) & \cdots & y_n^{(n-1)}(x_0)
\end{vmatrix} \neq 0.
$$

Hence, if $y_1, \ldots, y_n$ are solutions to equation (11) and there is some point x_0 in (a, b) such that (15) holds, then every solution $\phi(x)$ to (11) is a linear combination of $y_1, \ldots, y_n$.

Before putting this observation in the form of a theorem, it is convenient for us to extend the Wronskian concept to a set of n functions.

[†] Determinants are discussed in Section 9.2.

WRONSKIAN

Definition 1. Let $f_1, \ldots, f_n$ be any n functions that are $(n-1)$ times differentiable. The function

$$(16) \qquad W[f_1, \ldots, f_n](x) := \begin{vmatrix} f_1(x) & f_2(x) & \cdots & f_n(x) \\ f_1'(x) & f_2'(x) & \cdots & f_n'(x) \\ \vdots & \vdots & & \vdots \\ f_1^{(n-1)}(x) & f_2^{(n-1)}(x) & \cdots & f_n^{(n-1)}(x) \end{vmatrix}$$

is called the **Wronskian** of $f_1, \ldots, f_n$.

We now state the representation theorem that we proved above for solutions to homogeneous linear differential equations.

REPRESENTATION OF SOLUTIONS (HOMOGENEOUS CASE)

Theorem 2. Let $y_1, \ldots, y_n$ be n solutions on (a, b) of

$$(17) \qquad y^{(n)}(x) + p_1(x)y^{(n-1)}(x) + \cdots + p_n(x)y(x) = 0,$$

where $p_1, \ldots, p_n$ are continuous on (a, b). If at some point x_0 in (a, b) these solutions satisfy

$$(18) \qquad W[y_1, \ldots, y_n](x_0) \neq 0,$$

then every solution of (17) on (a, b) can be expressed in the form

$$(19) \qquad y(x) = C_1 y_1(x) + \cdots + C_n y_n(x),$$

where $C_1, \ldots, C_n$ are constants.

A set of solutions $\{y_1, \ldots, y_n\}$ satisfying (18) for some x_0 in (a, b) is called a **fundamental solution set** for (17) on (a, b). The linear combination of $y_1, \ldots, y_n$ in (19), written with arbitrary constants $C_1, \ldots, C_n$, is referred to as a **general solution** to (17).

EXAMPLE 2 Given that $y_1(x) = x$, $y_2(x) = x^2$, and $y_3(x) = x^{-1}$ are solutions to

$$(20) \qquad x^3 y''' + x^2 y'' - 2xy' + 2y = 0, \qquad x > 0,$$

find a general solution.

Solution We first show that $\{y_1, y_2, y_3\}$ is a fundamental solution set for equation (20) on $(0, \infty)$. Since we are told that $y_1, y_2,$ and y_3 satisfy (20) (the reader can easily verify this), we need only consider

$$W[y_1, y_2, y_3](x) = \begin{vmatrix} y_1(x) & y_2(x) & y_3(x) \\ y_1'(x) & y_2'(x) & y_3'(x) \\ y_1''(x) & y_2''(x) & y_3''(x) \end{vmatrix} = \begin{vmatrix} x & x^2 & x^{-1} \\ 1 & 2x & -x^{-2} \\ 0 & 2 & 2x^{-3} \end{vmatrix}.$$

Evaluating the determinant, we find after a little algebra that

$$W[y_1, y_2, y_3](x) = 6x^{-1},$$

which is not zero for $x > 0$. Thus $\{y_1, y_2, y_3\}$ is a fundamental solution set, and hence a general solution is

(21) $y(x) = C_1 x + C_2 x^2 + C_3 x^{-1}, \qquad x > 0.$ ■ End

The important condition (18) concerning the nonvanishing of the Wronskian can also be described in terms of the linear independence of the solutions $y_1, \ldots, y_n$ (recall Definition 3, Section 4.3, page 163).

LINEAR DEPENDENCE OF FUNCTIONS

Definition 2. The m functions $f_1, \ldots, f_m$ are said to be **linearly dependent on an interval I** if there exist constants $c_1, \ldots, c_m$, not all zero, such that

(22) $c_1 f_1(x) + \cdots + c_m f_m(x) = 0$

for all x in I. If the functions $f_1, \ldots, f_m$ are not linearly dependent on I, they are said to be **linearly independent on I**.

Remark. Linear dependence of $f_1, \ldots, f_m$ on I means that at least one of these functions is expressible as a linear combination of the remaining functions (on I).

EXAMPLE 3 Show that the functions $f_1(x) = e^x$, $f_2(x) = e^{-2x}$, and $f_3(x) = 3e^x - 2e^{-2x}$ are linearly dependent on $(-\infty, \infty)$.

Solution Notice that f_3 is a linear combination of f_1 and f_2:

$$f_3(x) = 3e^x - 2e^{-2x} = 3f_1(x) - 2f_2(x).$$

Therefore we have

(23) $3f_1(x) - 2f_2(x) - f_3(x) = 0$

for all x in $(-\infty, \infty)$. Consequently, f_1, f_2, and f_3 are linearly dependent on $(-\infty, \infty)$. ■ End

To prove that functions $f_1, f_2, \ldots, f_m$ are linearly *independent* on (a, b), a convenient approach is the following: *Assume* equation (22) holds on (a, b) and show that this forces $c_1 = c_2 = \cdots = c_m = 0$.

EXAMPLE 4 Show that the functions $f_1(x) = x$, $f_2(x) = x^2$, and $f_3(x) = 1 - 2x^2$ are linearly independent on $(-\infty, \infty)$.

Solution Assume c_1, c_2, and c_3 are constants for which

(24) $c_1 x + c_2 x^2 + c_3(1 - 2x^2) = 0$

holds at every x. If we can prove that (24) implies $c_1 = c_2 = c_3 = 0$, then linear independence follows. Let's set $x = 0$, 1, and -1 in equation (24). These x values are, essentially, ''picked out of a hat'' but will get the job done. Substituting in (24) gives

$$c_3 = 0 \qquad (x = 0),$$

(25) $$c_1 + c_2 - c_3 = 0 \qquad (x = 1),$$

$$-c_1 + c_2 - c_3 = 0 \qquad (x = -1).$$

When we solve this system (or compute the determinant of the coefficients), we find that the only possible solution is $c_1 = c_2 = c_3 = 0$. Consequently, the functions f_1, f_2, and f_3 are linearly independent on $(-\infty, \infty)$. ■ End

The connection between linear independence and fundamental solution sets is stated in the next theorem.

FUNDAMENTAL SETS, LINEAR INDEPENDENCE, AND THE WRONSKIAN

Theorem 3. If $y_1, \ldots, y_n$ are n solutions to $y^{(n)} + p_1 y^{(n-1)} + \cdots + p_n y = 0$, then the following statements are equivalent:

(i) $\{y_1, \ldots, y_n\}$ is a fundamental solution set on (a, b).
(ii) $y_1, \ldots, y_n$ are linearly independent on (a, b).
(iii) The Wronskian $W[y_1, \ldots, y_n](x)$ is never zero on (a, b).

Theorem 3 is a generalization of Corollary 1 in Section 4.3. Its proof is similar to that corollary's, so we leave it as an exercise (see Problem 30).

As with linear second order equations, *the Wronskian of solutions is either identically zero or never zero on* (a, b). This fact follows from Theorem 3 or from **Abel's identity**

$$\textbf{(26)} \qquad W[y_1, \ldots, y_n](x) = W[y_1, \ldots, y_n](x_0) \exp\left(-\int_{x_0}^{x} p_1(t)\, dt\right),$$

which holds for any n solutions $y_1, \ldots, y_n$ to $y^{(n)} + p_1 y^{(n-1)} + \cdots + p_n y = 0$ on (a, b). In (26), the point x_0 can be anywhere in (a, b). For a proof of Abel's identity when $n = 3$, see Problem 31.

Notice that Theorem 3 provides a simpler approach to proving that the solutions $y_1(x) = x$, $y_2(x) = x^2$, $y_3(x) = x^{-1}$ in Example 2 form a fundamental solution set. For, without having to compute a determinant, we can see (practically at a glance) that no one of these functions is a linear combination of the remaining two, so these functions are linearly independent on $(0, \infty)$.

It is useful to keep in mind that the following sets consist of functions that are linearly independent on every open interval (a, b):

$$\{1, x, x^2, \ldots, x^n\},$$

$$\{1, \cos x, \sin x, \cos 2x, \sin 2x, \ldots, \cos nx, \sin nx\},$$

$$\{e^{\alpha_1 x}, e^{\alpha_2 x}, \ldots, e^{\alpha_n x}\}, \qquad \alpha_i\text{'s distinct constants.}$$

(See Problems 27 and 28.)

Recall that the **superposition principle** (Section 4.7, page 189) for linear differential operators states that if $L[y_1] = g_1$ and $L[y_2] = g_2$, then the linear combination $C_1 y_1 + C_2 y_2$ satisfies

$$L[C_1 y_1 + C_2 y_2] = C_1 g_1 + C_2 g_2.$$

If we combine this principle with the representation theorem for solutions of the homogeneous equation, we obtain the following representation theorem for nonhomogeneous equations.

REPRESENTATION OF SOLUTIONS (NONHOMOGENEOUS CASE)

Theorem 4. Let $y_p(x)$ be a particular solution to the nonhomogeneous equation

(27) $y^{(n)}(x) + p_1(x)y^{(n-1)}(x) + \cdots + p_n(x)y(x) = g(x)$

on the interval (a, b) and let $\{y_1, \ldots, y_n\}$ be a fundamental solution set on (a, b) for the corresponding homogeneous equation

(28) $y^{(n)}(x) + p_1(x)y^{(n-1)}(x) + \cdots + p_n(x)y(x) = 0.$

Then every solution of (27) on the interval (a, b) can be expressed in the form

(29) $y(x) = y_p(x) + C_1 y_1(x) + \cdots + C_n y_n(x).$

Proof. Let $\phi(x)$ be any solution to (27). Because both $\phi(x)$ and $y_p(x)$ are solutions to (27), the superposition principle states that the difference $\phi(x) - y_p(x)$ is a solution to the homogeneous equation (28). It then follows from Theorem 2 that

$$\phi(x) - y_p(x) = C_1 y_1(x) + \cdots + C_n y_n(x)$$

for suitable constants $C_1, \ldots, C_n$. The last equation is equivalent to (29) (with $\phi(x)$ in place of $y(x)$), so the theorem is proved. ■ End

The linear combination of $y_p, y_1, \ldots, y_n$ in (29) written with arbitrary constants $C_1, \ldots, C_n$ is, for obvious reasons, referred to as a **general solution** to (27).

EXAMPLE 5 Given that $y_p(x) = x^2$ is a particular solution to

(30) $y''' - 2y'' - y' + 2y = 2x^2 - 2x - 4$

on the interval $(-\infty, \infty)$ and that $y_1(x) = e^{-x}$, $y_2(x) = e^x$, $y_3(x) = e^{2x}$ are solutions to the corresponding homogeneous equation, find a general solution to (30).

Solution We previously remarked that the functions e^{-x}, e^x, e^{2x} are linearly independent because the exponents -1, 1, and 2 are distinct. Because each of these functions is a solution to the corresponding homogeneous equation, then $\{e^{-x}, e^x, e^{2x}\}$ is a fundamental solution set. It now follows from the representation theorem for nonhomogeneous equations that a general solution is

(31) $y(x) = x^2 + C_1 e^{-x} + C_2 e^x + C_3 e^{2x}.$ ■ End

EXERCISES 6.1

In Problems 1 through 6, determine the largest interval (a, b) for which Theorem 1 guarantees the existence of a unique solution on (a, b) to the given initial value problem.

1. $xy''' - 3y' + e^x y = x^2 - 1$;
 $y(-2) = 1$, $y'(-2) = 0$, $y''(-2) = 2$.

2. $y''' - \sqrt{x}y = \sin x$;
 $y(\pi) = 0$, $y'(\pi) = 11$, $y''(\pi) = 3$.

3. $y''' - y'' + \sqrt{x-1}\,y = \tan x$;
 $y(5) = y'(5) = y''(5) = 1$.

4. $x(x + 1)y''' - 3xy' + y = 0$;
 $y(-1/2) = 1$, $y'(-1/2) = y''(-1/2) = 0$.

5. $x\sqrt{x+1}\,y''' - y' + xy = 0$;
 $y(1/2) = y'(1/2) = -1$, $y''(1/2) = 1$.

6. $(x^2 - 1)y''' + e^x y = \ln x$;
 $y(3/4) = 1$, $y'(3/4) = y''(3/4) = 0$.

In Problems 7 through 14, determine whether the given functions are linearly dependent or linearly independent on the specified interval. Also compute their Wronskian.

7. $\{e^{3x}, e^{5x}, e^{-x}\}$ on $(-\infty, \infty)$.

8. $\{x^2, x^2 - 1, 5\}$ on $(-\infty, \infty)$.

9. $\{\sin^2 x, \cos^2 x, 1\}$ on $(-\infty, \infty)$.

10. $\{\sin x, \cos x, \tan x\}$ on $(-\pi/2, \pi/2)$.

11. $\{x^{-1}, x^{1/2}, x\}$ on $(0, \infty)$.

12. $\{\cos 2x, \cos^2 x, \sin^2 x\}$ on $(-\infty, \infty)$.

13. $\{x, x^2, x^3, x^4\}$ on $(-\infty, \infty)$.

14. $\{x, xe^x, 1\}$ on $(-\infty, \infty)$.

In Problems 15 through 18, verify that the given functions form a fundamental solution set for the given differential equation and find a general solution.

15. $y''' + 2y'' - 11y' - 12y = 0$; $\{e^{3x}, e^{-x}, e^{-4x}\}$.

16. $y''' - y'' + 4y' - 4y = 0$; $\{e^x, \cos 2x, \sin 2x\}$.

17. $x^3y''' - 3x^2y'' + 6xy' - 6y = 0$, $x > 0$;
 $\{x, x^2, x^3\}$.

18. $y^{(4)} - y = 0$; $\{e^x, e^{-x}, \cos x, \sin x\}$.

In Problems 19 through 22, a particular solution and a fundamental solution set are given for a nonhomogeneous

equation and its corresponding homogeneous equation. (a) Find a general solution to the nonhomogeneous equation. (b) Find the solution that satisfies the specified initial conditions.

19. $y''' + y'' + 3y' - 5y = 2 + 6x - 5x^2$;
 $y(0) = -1$, $y'(0) = 1$, $y''(0) = -3$;
 $y_p = x^2$; $\{e^x, e^{-x}\cos 2x, e^{-x}\sin 2x\}$.

20. $xy''' - y'' = -2$; $y(1) = 2$, $y'(1) = -1$, $y''(1) = -4$;
 $y_p = x^2$; $\{1, x, x^3\}$.

21. $x^3y''' + xy' - y = 3 - \ln x$;
 $y(1) = 3$, $y'(1) = 3$, $y''(1) = 0$; $y_p = \ln x$;
 $\{x, x\ln x, x(\ln x)^2\}$.

22. $y^{(4)} + 4y = 5\cos x$;
 $y(0) = 2$, $y'(0) = 1$, $y''(0) = -1$, $y'''(0) = -2$;
 $y_p = \cos x$; $\{e^x\cos x, e^x\sin x, e^{-x}\cos x, e^{-x}\sin x\}$.

23. Let $L[y] := y''' + y' + xy$, $y_1(x) := \sin x$, and $y_2(x) := x$. Verify that $L[y_1](x) = x\sin x$ and $L[y_2](x) = x^2 + 1$. Then use the superposition principle to find a solution to the differential equation:

 (a) $L[y] = 2x\sin x - x^2 - 1$.
 (b) $L[y] = 4x^2 + 4 - 6x\sin x$.

24. Let $L[y] := y''' - xy'' + 4y' - 3xy$, $y_1(x) := \cos 2x$, and $y_2(x) := -1/3$. Verify that $L[y_1](x) = x\cos 2x$ and $L[y_2](x) = x$. Then use the superposition principle to find a solution to the differential equation:

 (a) $L[y] = 7x\cos 2x - 3x$.
 (b) $L[y] = -6x\cos 2x + 11x$.

25. Prove that L defined in (7) is a linear operator by verifying that properties (9) and (10) hold for any n-times differentiable functions y, y_1, y_2 on (a, b).

26. Show that a fundamental solution set always exists for equation (17). [*Hint:* By Theorem 1, we know that there is a unique solution for each choice of initial conditions. Choose n sets of initial conditions wisely!]

27. Show that the set of functions $\{1, x, x^2, \ldots, x^n\}$, where n is a positive integer, is linearly independent on every open interval (a, b). [*Hint:* Use the fact that a polynomial of degree at most n has no more than n zeros unless it is identically zero.]

28. The set of functions

$$\{1, \cos x, \sin x, \ldots, \cos nx, \sin nx\},$$

where n is a positive integer, is linearly independent on every interval (a, b). Prove this in the special case $n = 2$ and $(a, b) = (-\infty, \infty)$.

29. (a) Show that if $f_1, \ldots, f_m$ are linearly independent on $(-1, 1)$, then they are linearly independent on $(-\infty, \infty)$.

(b) Give an example to show that if $f_1, \ldots, f_m$ are linearly independent on $(-\infty, \infty)$, then they need not be linearly independent on $(-1, 1)$.

30. To prove Theorem 3, proceed as follows:

(a) Let $x_0 \in (a, b)$. Prove that the solutions $y_1, \ldots, y_n$ are linearly independent on (a, b) if and only if the initial vectors $(y_k(x_0), y_k'(x_0), \ldots, y_k^{(n-1)}(x_0))$, $k = 1, 2, \ldots, n$, are linearly independent.

(b) To complete the proof, use the result of part (a) and the fact that *the column vectors of a matrix are linearly independent if and only if the determinant of the matrix is not zero.*

31. To prove Abel's identity for $n = 3$, proceed as follows:

(a) Let $W(x) := W[y_1, y_2, y_3](x)$. Use the product rule for differentiation to show

$$W'(x) = \begin{vmatrix} y_1' & y_2' & y_3' \\ y_1' & y_2' & y_3' \\ y_1'' & y_2'' & y_3'' \end{vmatrix} + \begin{vmatrix} y_1 & y_2 & y_3 \\ y_1'' & y_2'' & y_3'' \\ y_1'' & y_2'' & y_3'' \end{vmatrix}$$

$$+ \begin{vmatrix} y_1 & y_2 & y_3 \\ y_1' & y_2' & y_3' \\ y_1''' & y_2''' & y_3''' \end{vmatrix}.$$

(b) Show that the above expression reduces to

$$(32) \qquad W'(x) = \begin{vmatrix} y_1 & y_2 & y_3 \\ y_1' & y_2' & y_3' \\ y_1''' & y_2''' & y_3''' \end{vmatrix}.$$

(c) Since each y_i satisfies (17), show that

$$(33) \qquad y_i^{(3)}(x) = -\sum_{k=1}^{3} p_k(x) y_i^{(3-k)}(x), \quad i = 1, 2, 3.$$

(d) Substituting the expressions in (33) into (32), show that

$$(34) \qquad W'(x) = -p_1(x) W(x).$$

(e) Deduce Abel's identity by solving equation (34).

6.2 HOMOGENEOUS LINEAR EQUATIONS WITH CONSTANT COEFFICIENTS

Our goal in this section is to obtain a general solution to an nth order linear differential equation with constant coefficients. Based on the experience gained with second order equations in Section 4.5, the reader should have little trouble guessing the form of such a solution. However, our interest here is to help the reader understand *why* these techniques work. This is done using an operator approach—a technique that is useful in tackling many other problems in analysis such as solving partial differential equations.

Let's consider the homogeneous linear nth order differential equation

$$(1) \qquad a_n y^{(n)}(x) + a_{n-1} y^{(n-1)}(x) + \cdots + a_1 y'(x) + a_0 y(x) = 0,$$

where $a_n \, (\neq 0), a_{n-1}, \ldots, a_0$ are real constants.[†] Since constant functions are everywhere continuous, equation (1) has solutions defined for all x in $(-\infty, \infty)$ (recall Theorem 1 in Section 6.1). If we can find n linearly independent solutions to (1) on $(-\infty, \infty)$, say $y_1, \ldots, y_n$, then we can express a general solution to (1) in the form

$$(2) \qquad y(x) = C_1 y_1(x) + \cdots + C_n y_n(x),$$

with $C_1, \ldots, C_n$ as arbitrary constants.

[†] ***Historical Footnote:*** In a letter to John Bernoulli dated September 15, 1739, Leonhard Euler claimed to have solved the general case of the homogeneous linear nth order equation with constant coefficients.

To find these n linearly independent solutions, we shall capitalize on our previous success with second order equations. Namely, experience suggests that we begin by trying a function of the form $y = e^{rx}$.

If we let L be the differential operator defined by the left-hand side of (1), that is,

(3) $L[y] := a_n y^{(n)} + a_{n-1} y^{(n-1)} + \cdots + a_1 y' + a_0 y,$

then we can write (1) in the operator form

(4) $L[y](x) = 0.$

For $y = e^{rx}$, we find

(5) $L[e^{rx}](x) = a_n r^n e^{rx} + a_{n-1} r^{n-1} e^{rx} + \cdots + a_0 e^{rx}$

$$= e^{rx}(a_n r^n + a_{n-1} r^{n-1} + \cdots + a_0) = e^{rx} P(r),$$

where $P(r)$ is the polynomial $a_n r^n + a_{n-1} r^{n-1} + \cdots + a_0$. Thus e^{rx} is a solution to equation (4), provided r is a root of the **auxiliary equation**

(6) $P(r) = a_n r^n + a_{n-1} r^{n-1} + \cdots + a_0 = 0.$

According to the fundamental theorem of algebra, the auxiliary equation has n roots (counting multiplicities), which may be either real or complex. As we mentioned in Chapter 4, there are no formulas for determining the zeros of an arbitrary polynomial of degree greater than four. However, if we can determine one zero r_1, then we can divide out the factor $r - r_1$ and be left with a polynomial of lower degree. (For convenience, we have chosen most of our examples and exercises so that 0, ± 1, or ± 2 are zeros of any polynomial of degree greater than two that we must factor.) When a zero cannot be exactly determined, numerical algorithms such as Newton's method or the quotient-difference algorithm can be used to compute approximate roots of the polynomial equation.[†] Some pocket calculators even have these algorithms built in.

We proceed to discuss the various possibilities.

Distinct Real Roots

If the roots $r_1, \ldots, r_n$ of the auxiliary equation (6) are real and distinct, then n solutions to equation (1) are

(7) $y_1(x) = e^{r_1 x}, \ y_2(x) = e^{r_2 x}, \ \ldots, \ y_n(x) = e^{r_n x}.$

As stated in the previous section, these functions are linearly independent on $(-\infty, \infty)$, a fact that we shall now officially verify. Let's assume that $c_1, \ldots, c_n$ are constants such that

(8) $c_1 e^{r_1 x} + \cdots + c_n e^{r_n x} = 0$

for all x in $(-\infty, \infty)$. Our goal is to prove that $c_1 = c_2 = \cdots = c_n = 0$.

[†] See, for example, *Applied and Computational Complex Analysis*, by P. Henrici (Wiley-Interscience, New York, 1974), Volume 1, or *Numerical Analysis*, Fifth Edition, by R. L. Burden, J. D. Faires, and A. C. Reynolds (Prindle, Weber & Schmidt, Boston, 1993).

One way to show this is to construct a linear operator L_k that annihilates (maps to zero) everything on the left-hand side of (8) except the kth term. For this purpose, we note that since $r_1, \ldots, r_n$ are the zeros of the auxiliary polynomial $P(r)$, then $P(r)$ can be factored as

(9) $P(r) = a_n(r - r_1) \cdots (r - r_n).$

Consequently, the operator $L[y] = a_n y^{(n)} + a_{n-1} y^{(n-1)} + \cdots + a_0 y$ can be expressed as the following composition:[†]

(10) $L = P(D) = a_n(D - r_1) \cdots (D - r_n).$

We now construct the polynomial $P_k(r)$ by deleting the factor $(r - r_k)$ from $P(r)$. Then we set $L_k := P_k(D)$; that is,

(11) $L_k := P_k(D) = a_n(D - r_1) \cdots (D - r_{k-1})(D - r_{k+1}) \cdots (D - r_n).$

Applying L_k to both sides of (8), we get, via linearity,

(12) $c_1 L_k[e^{r_1 x}] + \cdots + c_n L_k[e^{r_n x}] = 0.$

Also, since $L_k = P_k(D)$, we find (just as in equation (5)) that $L_k[e^{rx}](x) = e^{rx} P_k(r)$ for all r. Thus (12) can be written as

$$c_1 e^{r_1 x} P_k(r_1) + \cdots + c_n e^{r_n x} P_k(r_n) = 0,$$

which simplifies to

(13) $c_k e^{r_k x} P_k(r_k) = 0,$

because $P_k(r_i) = 0$ for $i \neq k$. Since r_k is not a root of $P_k(r)$, then $P_k(r_k) \neq 0$. It now follows from (13) that $c_k = 0$. But as k is arbitrary, all the constants $c_1, \ldots, c_n$ must be zero. Thus $y_1(x), \ldots, y_n(x)$ as given in (7) are linearly independent. (See Problem 26 for an alternative proof.)

We have proved that, in the case of n distinct real roots, a general solution to (1) is

(14) $\mathbf{y(x) = C_1 e^{r_1 x} + \cdots + C_n e^{r_n x}},$

where $C_1, \ldots, C_n$ are arbitrary constants.

EXAMPLE 1 Find a general solution to

(15) $y''' - 2y'' - 5y' + 6y = 0.$

Solution The auxiliary equation is

(16) $r^3 - 2r^2 - 5r + 6 = 0.$

By inspection, we find that $r = 1$ is a root. Then, using polynomial division, we get

$$r^3 - 2r^2 - 5r + 6 = (r - 1)(r^2 - r - 6),$$

[†] ***Historical Footnote:*** The symbolic notation $P(D)$ was introduced by Augustin Cauchy in 1827.

which further factors into $(r - 1)(r + 2)(r - 3)$. Hence the roots of equation (16) are $r_1 = 1$, $r_2 = -2$, $r_3 = 3$. Since these roots are real and distinct, a general solution to (15) is

$$y(x) = C_1 e^x + C_2 e^{-2x} + C_3 e^{3x}. \quad \blacksquare \text{ End}$$

Complex Roots

If $\alpha + i\beta$ (α, β real) is a complex root of the auxiliary equation (6), then so is its complex conjugate $\alpha - i\beta$, since the coefficients of $P(r)$ are real valued (see Problem 24). If we accept complex-valued functions as solutions, then both $e^{(\alpha+i\beta)x}$ and $e^{(\alpha-i\beta)x}$ are solutions to (1). Moreover, if there are no repeated roots, then a general solution to (1) is again given by (14). To find two real-valued solutions corresponding to the roots $\alpha \pm i\beta$, we can just take the real and imaginary parts of $e^{(\alpha+i\beta)x}$. That is, since

(17) $e^{(\alpha+i\beta)x} = e^{\alpha x} \cos \beta x + i e^{\alpha x} \sin \beta x,$

then two linearly independent solutions to (1) are

(18) $e^{\alpha x} \cos \beta x,$ $e^{\alpha x} \sin \beta x.$

In fact, using these solutions in place of $e^{(\alpha+i\beta)x}$ and $e^{(\alpha-i\beta)x}$ in (14) preserves the linear independence of the set of n solutions. Thus, treating each of the conjugate pairs of roots in this manner, we obtain a real-valued general solution to (1).

EXAMPLE 2 Find a general solution to

(19) $y''' + y'' + 3y' - 5y = 0.$

Solution The auxiliary equation is

(20) $r^3 + r^2 + 3r - 5 = (r - 1)(r^2 + 2r + 5) = 0,$

which has distinct roots $r_1 = 1$, $r_2 = -1 + 2i$, $r_3 = -1 - 2i$. Thus a general solution is

(21) $y(x) = C_1 e^x + C_2 e^{-x} \cos 2x + C_3 e^{-x} \sin 2x. \quad \blacksquare \text{ End}$

Repeated Roots

If r_1 is a root of multiplicity m, then the n solutions given in (7) are not even distinct, let alone linearly independent. Recall that for a second order equation, when we had a repeated root r_1 to the auxiliary equation, we obtained two linearly independent solutions by taking $e^{r_1 x}$ and $x e^{r_1 x}$. So if r_1 is a root of (6) of multiplicity m, we might expect that m linearly independent solutions are

(22) $e^{r_1 x}, \quad x e^{r_1 x}, \quad x^2 e^{r_1 x}, \quad \dots, \quad x^{m-1} e^{r_1 x}.$

To see that this is the case, observe that if r_1 is a root of multiplicity m, then the auxiliary equation can be written in the form

(23) $a_n (r - r_1)^m (r - r_{m+1}) \cdots (r - r_n) = 0,$

$$(r - r_1)^m \tilde{P}(r) = 0,$$

where $\tilde{P}(r) := a_n(r - r_{m+1}) \cdots (r - r_n)$ and $\tilde{P}(r_1) \neq 0$. With this notation, we have the identity

(24) $L[e^{rx}](x) = e^{rx}(r - r_1)^m \tilde{P}(r)$

(see (5)). Setting $r = r_1$ in (24), we again see that $e^{r_1 x}$ is a solution.

To find other solutions, we take the kth partial derivative with respect to r of both sides of (24):

(25) $\dfrac{\partial^k}{\partial r^k} L[e^{rx}](x) = \dfrac{\partial^k}{\partial r^k}[e^{rx}(r - r_1)^m \tilde{P}(r)]$.

Carrying out the differentiation on the right-hand side of (25), we find that the resulting expression will still have $(r - r_1)$ as a factor, provided $k \leq m - 1$. Thus setting $r = r_1$ in (25) gives

(26) $\dfrac{\partial^k}{\partial r^k} L[e^{rx}](x)\big|_{r=r_1} = 0$ if $k \leq m - 1$.

Now notice that the function e^{rx} has continuous partial derivatives of all orders with respect to r and x. Hence for mixed partial derivatives of e^{rx}, it makes no difference whether the differentiation is done first with respect to x, then with respect to r, or vice versa. Since L involves derivatives with respect to x, this means we can interchange the order of differentiation in (26) to obtain

$$L\left[\dfrac{\partial^k}{\partial r^k}(e^{rx})\big|_{r=r_1}\right](x) = 0.$$

Thus

(27) $\dfrac{\partial^k}{\partial r^k}(e^{rx})\big|_{r=r_1} = x^k e^{r_1 x}$

will be a solution to (1) for $k = 0, 1, \ldots, m - 1$. So m distinct solutions to (1), due to the root $r = r_1$ of multiplicity m, are indeed given by (22). We leave it as an exercise to show that the m functions in (22) are linearly independent on $(-\infty, \infty)$ (see Problem 25).

If $\alpha + i\beta$ is a repeated complex root of multiplicity m, then we can replace the $2m$ complex-valued functions

$$e^{(\alpha+i\beta)x}, \quad xe^{(\alpha+i\beta)x}, \quad \ldots, \quad x^{m-1}e^{(\alpha+i\beta)x},$$
$$e^{(\alpha-i\beta)x}, \quad xe^{(\alpha-i\beta)x}, \quad \ldots, \quad x^{m-1}e^{(\alpha-i\beta)x}$$

by the $2m$ linearly independent real-valued functions

(28) $\begin{aligned} &e^{\alpha x}\cos\beta x, \quad xe^{\alpha x}\cos\beta x, \quad \ldots, \quad x^{m-1}e^{\alpha x}\cos\beta x, \\ &e^{\alpha x}\sin\beta x, \quad xe^{\alpha x}\sin\beta x, \quad \ldots, \quad x^{m-1}e^{\alpha x}\sin\beta x. \end{aligned}$

Using the results of the three cases discussed above, we can obtain a set of n linearly independent solutions that yield a real-valued general solution for (1).

EXAMPLE 3 Find a general solution to

(29) $y^{(4)} - y^{(3)} - 3y'' + 5y' - 2y = 0.$

Solution The auxiliary equation is

$$r^4 - r^3 - 3r^2 + 5r - 2 = (r - 1)^3(r + 2) = 0,$$

which has roots $r_1 = 1$, $r_2 = 1$, $r_3 = 1$, $r_4 = -2$. Because the root at 1 has multiplicity 3, a general solution is

(30) $y(x) = C_1 e^x + C_2 x e^x + C_3 x^2 e^x + C_4 e^{-2x}.$ ■ End

EXAMPLE 4 Find a general solution to

(31) $y^{(4)} - 8y^{(3)} + 26y'' - 40y' + 25y = 0,$

whose auxiliary equation can be factored as

(32) $r^4 - 8r^3 + 26r^2 - 40r + 25 = (r^2 - 4r + 5)^2 = 0.$

Solution The auxiliary equation (32) has repeated complex roots: $r_1 = 2 + i$, $r_2 = 2 + i$, $r_3 = 2 - i$, and $r_4 = 2 - i$. Hence a general solution is

$$y(x) = C_1 e^{2x} \cos x + C_2 x e^{2x} \cos x + C_3 e^{2x} \sin x + C_4 x e^{2x} \sin x.$$ ■ End

When the auxiliary equation is a polynomial of order greater than four, there are no formulas involving radicals for determining the roots of the equation (see discussion at the end of Section 4.5 on page 177). Numerical routines such as Newton's method can be used to approximate both the real *and* complex roots (see Problems 27–29).

EXERCISES 6.2

In Problems 1 through 14, find a general solution for the differential equation.

1. $y''' + 2y'' - 8y' = 0.$

2. $y''' - 3y'' - y' + 3y = 0.$

3. $6z''' + 7z'' - z' - 2z = 0.$

4. $y''' + 2y'' - 19y' - 20y = 0.$

5. $y''' + 3y'' + 28y' + 26y = 0.$

6. $y''' - y'' + 2y = 0.$

7. $2y''' - y'' - 10y' - 7y = 0.$

8. $y''' + 5y'' - 13y' + 7y = 0.$

9. $u''' - 9u'' + 27u' - 27u = 0.$

10. $y''' + 3y'' - 4y' - 6y = 0.$

11. $y^{(4)} + 4y''' + 6y'' + 4y' + y = 0.$

12. $y''' + 5y'' + 3y' - 9y = 0.$

13. $y^{(4)} + 4y'' + 4y = 0.$

14. $\theta''' + 5\theta'' + 5\theta' - 11\theta = 0.$

In Problems 15 through 18, find a general solution to the linear homogeneous differential equation with constant coefficients whose auxiliary equation is given.

15. $(r - 1)^2(r + 3)(r^2 + 2r + 5)^2 = 0.$

16. $(r + 1)^2(r - 6)^3(r + 5)(r^2 + 1)(r^2 + 4) = 0.$

17. $(r + 4)(r - 3)(r + 2)^3(r^2 + 4r + 5)^2 r^5 = 0.$

18. $(r - 1)^3(r - 2)(r^2 + r + 1)(r^2 + 6r + 10)^3 = 0.$

In Problems 19 through 21, solve the given initial value problem.

19. $y''' - y'' - 4y' + 4y = 0;$
 $y(0) = -4,$ $y'(0) = -1,$ $y''(0) = -19.$

20. $y''' + 7y'' + 14y' + 8y = 0$;
$\qquad y(0) = 1, \qquad y'(0) = -3, \qquad y''(0) = 13.$

21. $y''' - 4y'' + 7y' - 6y = 0$;
$\qquad y(0) = 1, \qquad y'(0) = 0, \qquad y''(0) = 0.$

In Problems 22 and 23, find a general solution for the given linear system using the elimination method of Section 4.10.

22. $\dfrac{d^2x}{dt^2} - x + 5y = 0$

$\qquad 2x + \dfrac{d^2y}{dt^2} + 2y = 0.$

23. $\dfrac{d^3x}{dt^3} - x + \dfrac{dy}{dt} + y = 0$

$\qquad \dfrac{dx}{dt} - x + y = 0.$

24. Let $P(r) = a_n r^n + \cdots + a_1 r + a_0$ be a polynomial with real coefficients $a_n, \ldots, a_0$. Prove that if r_1 is a zero of $P(r)$, then so is its complex conjugate $\bar{r}_1$. [*Hint:* Show that $\overline{P(r)} = P(\bar{r})$, where the bar denotes complex conjugation.]

25. Show that the m functions $e^{rx}, xe^{rx}, \ldots, x^{m-1}e^{rx}$ are linearly independent on $(-\infty, \infty)$. [*Hint:* Show that these functions are linearly independent if and only if $1, x, \ldots, x^{m-1}$ are linearly independent.]

26. As an alternative proof that the functions $e^{r_1 x}, e^{r_2 x}, \ldots, e^{r_n x}$ are linearly independent on $(-\infty, \infty)$ when $r_1, r_2, \ldots, r_n$ are distinct, assume

(33) $\qquad C_1 e^{r_1 x} + C_2 e^{r_2 x} + \cdots + C_n e^{r_n x} = 0$

holds for all x in $(-\infty, \infty)$ and proceed as follows:

(a) Because the r_i's are distinct we can (if necessary) relabel them so that

$\qquad r_1 > r_2 > \cdots > r_n.$

Divide equation (33) by $e^{r_1 x}$ to obtain

$\qquad C_1 + C_2 \dfrac{e^{r_2 x}}{e^{r_1 x}} + \cdots + C_n \dfrac{e^{r_n x}}{e^{r_1 x}} = 0.$

Now let $x \to \infty$ on the left-hand side to obtain $C_1 = 0$.

(b) Since $C_1 = 0$, equation (33) becomes

$\qquad C_2 e^{r_2 x} + C_3 e^{r_3 x} + \cdots + C_n e^{r_n x} = 0$

for all x in $(-\infty, \infty)$. Divide this equation by $e^{r_2 x}$ and let $x \to \infty$ to conclude that $C_2 = 0$.

(c) Continuing in the manner of (b), argue that all the coefficients $C_1, C_2, \ldots, C_n$ are zero and hence $e^{r_1 x}, e^{r_2 x}, \ldots, e^{r_n x}$ are linearly independent on $(-\infty, \infty)$.

27. Find a general solution to

$$y^{(4)} + 2y''' - 3y'' - y' + \frac{1}{2}y = 0$$

by using Newton's method (Appendix A) or some other numerical procedure to approximate the roots of the auxiliary equation.

28. Find a general solution to $y''' - 3y' - y = 0$ by using Newton's method or some other numerical procedure to approximate the roots of the auxiliary equation.

29. Find a general solution to

$$y^{(4)} + 2y^{(3)} + 4y'' + 3y' + 2y = 0$$

by using Newton's method to approximate numerically the roots of the auxiliary equation. [*Hint:* To find complex roots, use the Newton recursion formula $z_{n+1} = z_n - f(z_n)/f'(z_n)$ and start with a nearby *complex* initial guess z_0.]

30. (a) Show that a general solution for the equation $y^{(n)} - y = 0$ is given by

$$y(x) = A_0 e^x + \sum_{k=1}^{(n-1)/2} \left\{ A_k \cos\frac{2\pi kx}{n} + B_k \sin\frac{2\pi kx}{n} \right\}$$

if n is odd and by

$$y(x) = A_0 e^x + B_0 e^{-x}$$
$$+ \sum_{k=1}^{(n/2)-1} \left\{ A_k \cos\frac{2\pi kx}{n} + B_k \sin\frac{2\pi kx}{n} \right\}$$

if n is even.

[*Hint:* Recall that the nth roots of unity are given by $e^{2\pi ki/n}$, $k = 0, 1, \ldots, n - 1$.]

(b) Find a general solution for the equation

$$y^{(n)} - 2y = 0.$$

31. Higher-Order Cauchy-Euler Equations. A differential equation that can be expressed in the form

$$a_n x^n y^{(n)}(x) + a_{n-1} x^{n-1} y^{(n-1)}(x) + \cdots + a_0 y(x) = 0,$$

where $a_n, a_{n-1}, \ldots, a_0$ are constants, is called a homogeneous **Cauchy-Euler** equation. (The second order case is discussed in Sections 4.5 and 8.5.) Use the substitution $y = x^r$ to help determine a fundamental solution set for the following Cauchy-Euler equations:

(a) $x^3 y''' + x^2 y'' - 2xy' + 2y = 0, \qquad x > 0.$

(b) $x^4 y^{(4)} + 6x^3 y''' + 2x^2 y'' - 4xy' + 4y = 0$, $x > 0$.

(c) $x^3 y''' - 2x^2 y'' + 13xy' - 13y = 0$, $\quad x > 0$.
[*Hint:* $x^{\alpha + i\beta} = e^{(\alpha + i\beta)\ln x}$
$\qquad = x^{\alpha}\{\cos(\beta \ln x) + i \sin(\beta \ln x)\}$.]

32. Let $y(x) = Ce^{rx}$, where $C \ (\neq 0)$ and r are real numbers, be a solution to a differential equation. Suppose we can not determine r exactly but can only approximate it by $\tilde{r}$. Let $\tilde{y}(x) := Ce^{\tilde{r}x}$ and consider the error $y(x) - \tilde{y}(x)$.

 (a) If r and $\tilde{r}$ are positive, $r \neq \tilde{r}$, show that the error grows exponentially large as x approaches ∞.

 (b) If r and $\tilde{r}$ are negative, $r \neq \tilde{r}$, show that the error goes to zero exponentially as x approaches ∞.

33. On a smooth horizontal surface, a mass of m_1 kg is attached to a vertical surface by a spring with spring constant k_1 N/m. Another mass of m_2 kg is attached to the first object by a spring with spring constant k_2 N/m. The objects are aligned horizontally so that the springs are their natural lengths. As we showed in Section 5.4, this coupled spring-mass system is governed by the system of differential equations

(34) $\quad m_1 \dfrac{d^2 x}{dt^2} + (k_1 + k_2)x - k_2 y = 0,$

(35) $\quad m_2 \dfrac{d^2 y}{dt^2} - k_2 x + k_2 y = 0.$

 Let's assume that $m_1 = m_2 = 1$, $k_1 = 3$, and $k_2 = 2$. If both objects are displaced 1 m to the right of their equilibrium positions and then released, determine the equations of motion for the objects as follows:

 (a) Show that $x(t)$ satisfies the equation

(36) $\qquad x^{(4)}(t) + 7x''(t) + 6x(t) = 0.$

 (b) Find a general solution $x(t)$ to (36).

 (c) Substitute $x(t)$ back into (34) to obtain a general solution for $y(t)$.

 (d) Use the initial conditions to determine the solutions, $x(t)$ and $y(t)$, which are the equations of motion.

34. Suppose the two springs in the coupled spring-mass system discussed in Problem 33 are switched, giving the new data $m_1 = m_2 = 1$, $k_1 = 2$, and $k_2 = 3$. If both objects are now displaced 1 m to the right of their equilibrium positions and then released, determine the equations of motion of the two objects.

35. **Vibrating Beam.** In studying the transverse vibrations of a beam, one encounters the homogeneous equation

$$EI \dfrac{d^4 y}{dx^4} - ky = 0,$$

where $y(x)$ is related to the displacement of the beam at position x, the constant E is Young's modulus, I is the area moment of inertia, and k is a parameter. Assuming E, I, and k are positive constants, find a general solution in terms of sines, cosines, hyperbolic sines, and hyperbolic cosines.

6.3 UNDETERMINED COEFFICIENTS AND THE ANNIHILATOR METHOD

When a higher-order linear equation has constant coefficients, the method of undetermined coefficients discussed in Section 4.8 can still be used to obtain a particular solution y_p, provided the forcing term is a sum of functions of the type given in Table 4.1 on page 197 (or inside back cover).

EXAMPLE 1 Use the method of undetermined coefficients to find a general solution to

(1) $\qquad y'''(x) - 3y''(x) + 4y(x) = xe^{2x}.$

Solution We first solve the corresponding homogeneous equation

(2) $\qquad y'''(x) - 3y''(x) + 4y(x) = 0.$

The auxiliary equation

$$r^3 - 3r^2 + 4 = (r + 1)(r - 2)^2 = 0$$

has a single root at -1 and a double root at 2, and hence a general solution to (2) is

(3) $y_h(x) = C_1 e^{-x} + C_2 e^{2x} + C_3 x e^{2x}.$

We can find the form for a particular solution y_p to equation (1) by referring to Table 4.1 on page 197. Since $g(x) = x e^{2x}$ is of Type IV, y_p has the form

$$y_p(x) = x^s (A_1 x + A_0) e^{2x}.$$

Because e^{2x} and $x e^{2x}$ are solutions to the corresponding homogeneous equation (2), we take $s = 2$. That is, y_p has the form

(4) $y_p(x) = (A_1 x^3 + A_0 x^2) e^{2x}.$

Substituting y_p for y in (1) gives

$$[8A_1 x^3 + (36A_1 + 8A_0)x^2 + (36A_1 + 24A_0)x + (6A_1 + 12A_0)]e^{2x}$$
$$- 3[4A_1 x^3 + (12A_1 + 4A_0)x^2 + (6A_1 + 8A_0)x + 2A_0]e^{2x}$$
$$+ 4[A_1 x^3 + A_0 x^2]e^{2x} = x e^{2x},$$

which reduces to

$$[18A_1 x + (6A_1 + 6A_0)]e^{2x} = x e^{2x}.$$

This leads to the system of equations

$$18A_1 = 1, \qquad 6A_1 + 6A_0 = 0,$$

which yields $A_1 = 1/18$ and $A_0 = -1/18$. Thus

(5) $y_p(x) = \dfrac{1}{18}(x^3 - x^2)e^{2x}.$

A general solution to (1) is $y_h + y_p$, that is,

(6) $y(x) = C_1 e^{-x} + C_2 e^{2x} + C_3 x e^{2x} + \dfrac{1}{18}(x^3 - x^2)e^{2x}.$ ■ End

For a better understanding of why the method of undetermined coefficients works for linear equations with constant coefficients, we will study a procedure for solving nonhomogeneous equations that begins by transforming the nonhomogeneous equation into a homogeneous one. For this purpose, we introduce the concept of an annihilator.

ANNIHILATOR

Definition 3. A linear differential operator A is said to **annihilate** a function f if

(7) $A[f](x) = 0, \quad$ for all $x.$

That is, A annihilates f if f is a solution to the homogeneous linear differential equation (7) on $(-\infty, \infty)$.

For example, $A = D - 3$ annihilates $f(x) = e^{3x}$, since

$$(D - 3)[e^{3x}] = 3e^{3x} - 3e^{3x} = 0.$$

Also, $A = D^2 - 4D + 20$ is an annihilator of $e^{2x} \sin 4x$, since this function satisfies the equation

$$y'' - 4y' + 20y = 0.$$

From what we learned in the previous section about auxiliary equations with repeated roots, it follows that the differential operator $(D - r)^m$, m a positive integer, annihilates each of the functions

$$(8) \qquad e^{rx}, xe^{rx}, \ldots, x^{m-1}e^{rx}.$$

Moreover, the differential operator $[(D - \alpha)^2 + \beta^2]^m$ annihilates each of the functions

$$(9) \qquad \begin{aligned} &e^{\alpha x} \cos \beta x, \; xe^{\alpha x} \cos \beta x, \; \ldots, \; x^{m-1}e^{\alpha x} \cos \beta x, \\ &e^{\alpha x} \sin \beta x, \; xe^{\alpha x} \sin \beta x, \; \ldots, \; x^{m-1}e^{\alpha x} \sin \beta x, \end{aligned}$$

since these are the $2m$ linearly independent solutions to $[(D - \alpha)^2 + \beta^2]^m y = 0$.

In the next example, we make use of the fact that linear differential operators with *constant* coefficients commute (see Exercises 4.2, Problem 32).

EXAMPLE 2 Find a differential operator that annihilates

$$(10) \qquad 6xe^{-4x} + 5e^x \sin 2x.$$

Solution Consider the two functions whose sum appears in (10). Observe that $(D + 4)^2$ annihilates the function $f_1(x) := 6xe^{-4x}$. Further, $f_2(x) := 5e^x \sin 2x$ is annihilated by the operator $(D - 1)^2 + 4$. Hence the composite operator

$$A := (D + 4)^2[(D - 1)^2 + 4],$$

which is the same as the operator

$$[(D - 1)^2 + 4](D + 4)^2,$$

annihilates both f_1 and f_2. But then, by linearity, A also annihilates the sum $f_1 + f_2$. ■ End

We now show how annihilators can be used to determine particular solutions to certain *non*homogeneous equations. Consider the nth order differential equation with constant coefficients

$$(11) \qquad a_n y^{(n)}(x) + a_{n-1} y^{(n-1)}(x) + \cdots + a_0 y(x) = g(x),$$

which can be written in the operator form

$$(12) \qquad L[y](x) = g(x),$$

where

$$L[y] := a_n y^{(n)} + a_{n-1} y^{(n-1)} + \cdots + a_0 y.$$

In this section, we restrict our attention to nonhomogeneous terms that can be annihilated by a linear differential operator with *constant* coefficients. Assuming A is such an annihilator of g, then applying A to both sides of (12) yields

$$A[L[y]](x) = A[g](x) = 0.$$

Therefore, if y is a solution to (12), then y is also a solution to the *homogeneous* linear differential equation

(13) $AL[y](x) = 0$

involving the composition of the operators A and L. Because equation (13) has constant coefficients (why?), we can apply the methods of Section 6.2 to obtain its general solution. Comparing this with a general solution of $L[y] = 0$, it is then possible to determine the *form* of a particular solution to equation (12). This procedure, known as the **annihilator method,** is illustrated in the following examples.

EXAMPLE 3 Find a general solution to

(14) $y''(x) + y(x) = e^{2x} + 1.$

Solution Since $D - 2$ annihilates e^{2x} and D annihilates 1, then $A := D(D - 2)$ annihilates $e^{2x} + 1$. Therefore applying A to both sides of (14) yields

$$A[y'' + y] = A[e^{2x} + 1]$$
(15) $D(D - 2)(D^2 + 1)[y] = 0.$

The auxiliary equation associated with (15) is

$$r(r - 2)(r^2 + 1) = 0,$$

which has the roots i, $-i$, 2, 0. Hence a general solution to (15) is

(16) $y(x) = C_1 \cos x + C_2 \sin x + C_3 e^{2x} + C_4.$

Now recall that a general solution to (14) is of the form $y_h + y_p$, where y_p is a particular solution to (14) and y_h is a general solution to

$$y''(x) + y(x) = 0.$$

Since every solution to (14) is also a solution to (15), then $y_h + y_p$ must have the form displayed on the right-hand side of (16). But we recognize that $y_h(x) = C_1 \cos x + C_2 \sin x$, so there exists a particular solution of the form

(17) $y_p(x) = C_3 e^{2x} + C_4.$

To determine the constants C_3 and C_4, we substitute y_p given by (17) into (14):

$$4C_3 e^{2x} + C_3 e^{2x} + C_4 = e^{2x} + 1,$$
$$5C_3 e^{2x} + C_4 = e^{2x} + 1.$$

Equating the coefficients of e^{2x} and setting the constant terms equal yields $C_3 = 1/5$ and $C_4 = 1$. Thus

$$y_p(x) = \frac{1}{5}e^{2x} + 1,$$

so a general solution to (14) is

$$y(x) = C_1 \cos x + C_2 \sin x + \frac{1}{5}e^{2x} + 1. \quad \blacksquare \text{ End}$$

EXAMPLE 4 Use the annihilator method to determine the form of a particular solution to

(18) $y'' - y = e^{-2x} \sin x.$

Solution The function $g(x) = e^{-2x} \sin x$ is annihilated by the operator

$$A := (D + 2)^2 + 1^2 = D^2 + 4D + 5.$$

If we apply A to both sides of (18), we obtain

$$A[y'' - y] = A[e^{-2x} \sin x]$$

(19) $(D^2 + 4D + 5)(D^2 - 1)[y] = 0.$

Now the auxiliary equation associated with (19) is

$$(r^2 + 4r + 5)(r - 1)(r + 1) = 0,$$

which has roots $1, -1, -2 + i, -2 - i$. Hence a general solution to (19) is

(20) $y(x) = C_1 e^x + C_2 e^{-x} + C_3 e^{-2x} \cos x + C_4 e^{-2x} \sin x.$

A general solution to the corresponding homogeneous equation $y'' - y = 0$ is $y_h(x) = C_1 e^x + C_2 e^{-x}$, so we see that a particular solution for (18) has the form

(21) $y_p(x) = C_3 e^{-2x} \cos x + C_4 e^{-2x} \sin x. \quad \blacksquare \text{ End}$

Of course the forms of the particular solutions derived in Examples 3 and 4 using the annihilator method can be quickly obtained from Table 4.1, page 197, for the method of undetermined coefficients. The essential point is that the annihilator method can be used to *justify* the method of undetermined coefficients; that is, to derive the entries that appear in Table 4.1. This applies not only to second order, but also to higher-order linear differential equations with constant coefficients.

The connection between the two methods is even more evident once one realizes that those functions for which the method of undetermined coefficients works are *exactly* those functions that are solutions to higher-order linear differential equations with constant coefficients, that is, those functions $g(x)$ that are annihilated by a linear differential operator with constant coefficients. In particular, these methods work for polynomials, exponentials, sines, and cosines, and for any product or sum of these functions. (Why?)

We leave it for the reader to show how the annihilator method yields various entries in Table 4.1 (see Problems 34–37). The derivation for the general case, where

$$g(x) = p_n(x)e^{\alpha x} \cos \beta x + q_m(x)e^{\alpha x} \sin \beta x,$$

p_n, q_m polynomials, is discussed in Project A at the end of this chapter.

EXERCISES 6.3

In Problems 1 through 4, use the method of undetermined coefficients (Table 4.1, inside back cover) to determine the form of a particular solution for the given equation.

1. $y''' - 2y'' - 5y' + 6y = e^x + x^2$.

2. $y''' + y'' - 5y' + 3y = e^{-x} + \sin x$.

3. $y''' + 3y'' - 4y = e^{-2x}$.

4. $y''' + y'' - 2y = xe^x + 1$.

In Problems 5 through 10, find a general solution to the given equation.

5. $y''' - 2y'' - 5y' + 6y = e^x + x^2$.

6. $y''' + y'' - 5y' + 3y = e^{-x} + \sin x$.

7. $y''' + 3y'' - 4y = e^{-2x}$.

8. $y''' + y'' - 2y = xe^x + 1$.

9. $y''' - 3y'' + 3y' - y = e^x$.

10. $y''' + 4y'' + y' - 26y = e^{-3x} \sin 2x + x$.

In Problems 11 through 20, find a differential operator that annihilates the given function.

11. $x^4 - x^2 + 11$.

12. $3x^2 - 6x + 1$.

13. e^{-7x}.

14. e^{5x}.

15. $e^{2x} - 6e^x$.

16. $x^2 - e^x$.

17. $x^2 e^{-x} \sin 2x$.

18. $xe^{3x} \cos 5x$.

19. $xe^{-2x} + xe^{-5x} \sin 3x$.

20. $x^2 e^x - x \sin 4x + x^3$.

In Problems 21 through 30, use the annihilator method to determine the form of a particular solution for the given equation.

21. $u'' - 5u' + 6u = \cos 2x + 1$.

22. $y'' + 6y' + 8y = e^{3x} - \sin x$.

23. $y'' - 5y' + 6y = e^{3x} - x^2$.

24. $\theta'' - \theta = xe^x$.

25. $y'' - 6y' + 9y = \sin 2x + x$.

26. $y'' + 2y' + y = x^2 - x + 1$.

27. $y'' + 2y' + 2y = e^{-x} \cos x + x^2$.

28. $y'' - 6y' + 10y = e^{3x} - x$.

29. $z''' - 2z'' + z' = x - e^x$.

30. $y''' + 2y'' - y' - 2y = e^x - 1$.

In Problems 31 through 33, solve the given initial value problem.

31. $y''' + 2y'' - 9y' - 18y = -18x^2 - 18x + 22$;
$y(0) = -2$, $y'(0) = -8$, $y''(0) = -12$.

32. $y''' - 2y'' + 5y' = -24e^{3x}$;
$y(0) = 4$, $y'(0) = -1$, $y''(0) = -5$.

33. $y''' - 2y'' - 3y' + 10y$
$= 34xe^{-2x} - 16e^{-2x} - 10x^2 + 6x + 34$;
$y(0) = 3$, $y'(0) = 0$, $y''(0) = 0$.

34. Use the annihilator method to show that if $a_0 \neq 0$ in equation (11) and $g(x)$ has the form

(22) $g(x) = b_m x^m + b_{m-1} x^{m-1} + \cdots + b_1 x + b_0$,

then

$$y_p(x) = B_m x^m + B_{m-1} x^{m-1} + \cdots + B_1 x + B_0$$

is the form of a particular solution to equation (11).

35. Use the annihilator method to show that if $a_0 = 0$ and $a_1 \neq 0$ in (11) and $g(x)$ has the form given in (22), then equation (11) has a particular solution of the form

$$y_p(x) = x\{B_m x^m + B_{m-1} x^{m-1} + \cdots + B_1 x + B_0\}.$$

36. Use the annihilator method to show that if $g(x)$ in (11) has the form $g(x) = Be^{\alpha x}$, then equation (11) has a particular solution of the form $y_p(x) = x^s Be^{\alpha x}$, where s is chosen to be the smallest nonnegative integer such that $x^s e^{\alpha x}$ is not a solution to the corresponding homogeneous equation.

37. Use the annihilator method to show that if $g(x)$ in (11) has the form

$$g(x) = a \cos \beta x + b \sin \beta x,$$

then equation (11) has a particular solution of the form

(23) $y_p(x) = x^s\{A \cos \beta x + B \sin \beta x\}$,

where s is chosen to be the smallest nonnegative integer such that $x^s \cos \beta x$ and $x^s \sin \beta x$ are not solutions to the corresponding homogeneous equation.

In Problems 38 and 39, use the elimination method of Section 4.10 to find a general solution to the given system.

38. $x - \dfrac{d^2 y}{dt^2} = t + 1$

$\dfrac{dx}{dt} + \dfrac{dy}{dt} - 2y = e^t$.

39. $\dfrac{d^2x}{dt^2} - x + y = 0$

$x + \dfrac{d^2y}{dt^2} - y = e^{3t}.$

40. The currents in the electric network in Figure 6.1 satisfy the system

$$\frac{1}{9}I_1 + 64I_2'' = -2\sin\frac{t}{24},$$

$$\frac{1}{64}I_3 + 9I_3'' - 64I_2'' = 0,$$

$$I_1 = I_2 + I_3,$$

where I_1, I_2, and I_3 are the currents through the different branches of the network. Using the elimination method of Section 4.10, determine the currents if initially $I_1(0) = I_2(0) = I_3(0) = 0$, $I_1'(0) = 73/12$, $I_2'(0) = 3/4$, and $I_3'(0) = 16/3$.

Figure 6.1 An electric network

6.4 METHOD OF VARIATION OF PARAMETERS

In the previous section, we discussed the method of undetermined coefficients and the annihilator method. These methods work only for linear equations with constant coefficients *and* when the nonhomogeneous term is a solution to some homogeneous linear equation with constant coefficients. In this section, we show how the method of **variation of parameters** discussed in Section 4.9 generalizes to higher-order linear equations with variable coefficients.

Our goal is to find a particular solution to

(1) $L[y](x) = g(x),$

where $L[y] := y^{(n)} + p_1 y^{(n-1)} + \cdots + p_n y$ and the coefficient functions $p_1, \ldots, p_n$, as well as g, are continuous on (a, b). The method to be described requires that we already know a fundamental solution set $\{y_1, \ldots, y_n\}$ for the corresponding homogeneous equation

(2) $L[y](x) = 0.$

A general solution to (2) is then

(3) $y_h(x) = C_1 y_1(x) + \cdots + C_n y_n(x),$

where $C_1, \ldots, C_n$ are arbitrary constants. In the method of variation of parameters, we assume there exists a particular solution to (1) of the form

(4) $y_p(x) = v_1(x)y_1(x) + \cdots + v_n(x)y_n(x)$

and try to determine the functions $v_1, \ldots, v_n$.

There are n unknown functions, so we will need n conditions (equations) to determine them. These conditions are obtained as follows. Differentiating y_p in (4) gives

(5) $y_p' = (v_1 y_1' + \cdots + v_n y_n') + (v_1' y_1 + \cdots + v_n' y_n).$

To prevent second order and higher-order derivatives of the unknowns $v_1, \ldots, v_n$ from entering our later computations, we impose the condition

$$v_1' y_1 + \cdots + v_n' y_n = 0.$$

In a like manner, on computing $y_p'', y_p''', \ldots, y_p^{(n-1)}$, we impose $(n-2)$ additional conditions involving $v_1', \ldots, v_n'$; namely,

$$v_1' y_1' + \cdots + v_n' y_n' = 0, \quad \ldots, \quad v_1' y_1^{(n-2)} + \cdots + v_n' y_n^{(n-2)} = 0.$$

Finally, the nth condition that we impose is that y_p satisfy the given equation (1). Using the previous conditions and the fact that $y_1, \ldots, y_n$ are solutions to the homogeneous equation, then $L[y_p] = g$ reduces to

(6) $\qquad v_1' y_1^{(n-1)} + \cdots + v_n' y_n^{(n-1)} = g,$

(see Problem 12). We therefore seek n functions $v_1', \ldots, v_n'$ that satisfy the system

(7)
$$
\begin{aligned}
y_1 v_1' + \cdots + \; & y_n v_n' = 0, \\
\vdots \qquad\qquad & \;\; \vdots \quad \vdots \\
y_1^{(n-2)} v_1' + \cdots + & y_n^{(n-2)} v_n' = 0, \\
y_1^{(n-1)} v_1' + \cdots + & y_n^{(n-1)} v_n' = g.
\end{aligned}
$$

A sufficient condition for the existence of a solution to system (7) for x in (a, b) is that the determinant of the matrix made up of the coefficients of $v_1', \ldots, v_n'$ be different from zero for all x in (a, b). But this determinant is just the Wronskian:

(8)
$$
\begin{vmatrix}
y_1 & \cdots & y_n \\
\vdots & & \vdots \\
y_1^{(n-2)} & \cdots & y_n^{(n-2)} \\
y_1^{(n-1)} & \cdots & y_n^{(n-1)}
\end{vmatrix}
= W[y_1, \ldots, y_n](x),
$$

which is never zero on (a, b) because $\{y_1, \ldots, y_n\}$ is a fundamental solution set. Solving (7) via Cramer's rule, we find

(9) $\qquad v_k'(x) = \dfrac{g(x) W_k(x)}{W[y_1, \ldots, y_n](x)}, \qquad k = 1, \ldots, n,$

where $W_k(x)$ is the determinant of the matrix obtained from the Wronskian $W[y_1, \ldots, y_n](x)$ by replacing the kth column by $\mathrm{col}[0, \ldots, 0, 1]$. Using a cofactor expansion about this column, we can express $W_k(x)$ in terms of an $(n-1)$st order Wronskian:

(10) $\qquad W_k(x) = (-1)^{n-k} W[y_1, \ldots, y_{k-1}, y_{k+1}, \ldots, y_n](x), \qquad k = 1, \ldots, n.$

Integrating $v_k'(x)$ in (9) gives

(11) $\qquad v_k(x) = \displaystyle\int \dfrac{g(x) W_k(x)}{W[y_1, \ldots, y_n](x)}\, dx, \qquad k = 1, \ldots, n.$

Finally, substituting the v_k's back into (4), we obtain the particular solution

(12) $$y_p(x) = \sum_{k=1}^{n} y_k(x) \int \frac{g(x) W_k(x)}{W[y_1, \ldots, y_n](x)} \, dx.$$

Although equation (12) gives a neat formula for a particular solution to (1), its implementation requires one to evaluate $n + 1$ determinants and then perform n integrations. This may entail several tedious computations. However, the method does work in cases when the technique of undetermined coefficients does not apply (provided, of course, we know a fundamental solution set).

EXAMPLE 1 Find a general solution to the Cauchy-Euler equation

(13) $$x^3 y''' + x^2 y'' - 2xy' + 2y = x^3 \sin x, \qquad x > 0,$$

given that $\{x, x^{-1}, x^2\}$ is a fundamental solution set to the corresponding homogeneous equation.

Solution An important first step is to divide (13) by x^3 to obtain the standard form

(14) $$y''' + \frac{1}{x} y'' - \frac{2}{x^2} y' + \frac{2}{x^3} y = \sin x, \qquad x > 0,$$

from which we see that $g(x) = \sin x$. Since $\{x, x^{-1}, x^2\}$ is a fundamental solution set, we can obtain a particular solution of the form

(15) $$y_p(x) = v_1(x)x + v_2(x)x^{-1} + v_3(x)x^2.$$

To use formula (12), we must first evaluate the four determinants:

$$W[x, x^{-1}, x^2](x) = \begin{vmatrix} x & x^{-1} & x^2 \\ 1 & -x^{-2} & 2x \\ 0 & 2x^{-3} & 2 \end{vmatrix} = -6x^{-1},$$

$$W_1(x) = (-1)^{(3-1)} W[x^{-1}, x^2](x) = (-1)^2 \begin{vmatrix} x^{-1} & x^2 \\ -x^{-2} & 2x \end{vmatrix} = 3,$$

$$W_2(x) = (-1)^{(3-2)} \begin{vmatrix} x & x^2 \\ 1 & 2x \end{vmatrix} = -x^2,$$

$$W_3(x) = (-1)^{(3-3)} \begin{vmatrix} x & x^{-1} \\ 1 & -x^{-2} \end{vmatrix} = -2x^{-1}.$$

Substituting the above expressions into (12), we find

$$y_p(x) = x \int \frac{(\sin x)3}{-6x^{-1}} \, dx + x^{-1} \int \frac{(\sin x)(-x^2)}{-6x^{-1}} \, dx + x^2 \int \frac{(\sin x)(-2x^{-1})}{-6x^{-1}} \, dx$$

$$= x \int \left(-\frac{1}{2} x \sin x \right) dx + x^{-1} \int \frac{1}{6} x^3 \sin x \, dx + x^2 \int \frac{1}{3} \sin x \, dx,$$

which simplifies to

(16) $y_p(x) = \cos x - x^{-1} \sin x + C_1 x + C_2 x^{-1} + C_3 x^2,$

where C_1, C_2, and C_3 denote the constants of integration. Since $\{x, x^{-1}, x^2\}$ is a fundamental solution set for the homogeneous equation, we can take C_1, C_2, and C_3 to be arbitrary constants. The right-hand side of (16) then gives the desired general solution. ■ End

In the preceding example, the fundamental solution set $\{x, x^{-1}, x^2\}$ can be derived by substituting $y = x^r$ into the homogeneous equation corresponding to (13) (see Problem 31, Exercises 6.2). However, in dealing with other equations that have variable coefficients, the determination of a fundamental set may be extremely difficult. In Chapter 8 we tackle this problem using power series methods.

EXERCISES 6.4

In Problems 1 through 6, use the method of variation of parameters to determine a particular solution to the given equation.

1. $y''' - 3y'' + 4y = e^{2x}$.

2. $y''' - 2y'' + y' = x$.

3. $z''' + 3z'' - 4z = e^{2x}$.

4. $y''' - 3y'' + 3y' - y = e^x$.

5. $y''' + y' = \tan x$, $0 < x < \pi/2$.

6. $y''' + y' = \sec \theta \tan \theta$, $0 < \theta < \pi/2$.

7. Find a general solution to the Cauchy-Euler equation
$$x^3 y''' - 3x^2 y'' + 6xy' - 6y = x^{-1}, \qquad x > 0,$$
given that $\{x, x^2, x^3\}$ is a fundamental solution set for the corresponding homogeneous equation.

8. Find a general solution to the Cauchy-Euler equation
$$x^3 y''' - 2x^2 y'' + 3xy' - 3y = x^2, \qquad x > 0,$$
given that $\{x, x \ln x, x^3\}$ is a fundamental solution set for the corresponding homogeneous equation.

9. Given that $\{e^x, e^{-x}, e^{2x}\}$ is a fundamental solution set for the homogeneous equation corresponding to the equation
$$y''' - 2y'' - y' + 2y = g(x),$$
determine a formula involving integrals for a particular solution.

10. Given that $\{x, x^{-1}, x^4\}$ is a fundamental solution set for the homogeneous equation corresponding to the equation
$$x^3 y''' - x^2 y'' - 4xy' + 4y = g(x), \qquad x > 0,$$
determine a formula involving integrals for a particular solution.

11. Find a general solution to the Cauchy-Euler equation
$$x^3 y''' - 3xy' + 3y = x^4 \cos x, \qquad x > 0.$$

12. Derive the system (7) in the special case when $n = 3$. [*Hint:* To determine the last equation, require that $L[y_p] = g$ and use the fact that y_1, y_2, and y_3 satisfy the corresponding homogeneous equation.]

13. Show that
$$W_k(x) = (-1)^{(n-k)} W[y_1, \ldots, y_{k-1}, y_{k+1}, \ldots, y_n](x).$$

14. Deflection of a Beam under Axial Force. A uniform beam under a load and subject to a constant axial force is governed by the differential equation
$$y^{(4)}(x) - k^2 y''(x) = q(x), \qquad 0 < x < L,$$
where $y(x)$ is the deflection of the beam, L is the length of the beam, k^2 is proportional to the axial force, and $q(x)$ is proportional to the load (see Figure 6.2).

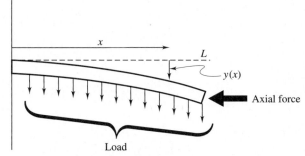

Figure 6.2 Deformation of a beam under axial force and load

(a) Show that a general solution can be written in the form

$$y(x) = C_1 + C_2 x + C_3 e^{kx} + C_4 e^{-kx}$$

$$+ \frac{1}{k^3} \int q(x)(kx - 1)\, dx - \frac{x}{k^2} \int q(x)\, dx$$

$$+ \frac{e^{kx}}{2k^3} \int q(x) e^{-kx}\, dx - \frac{e^{-kx}}{2k^3} \int q(x) e^{kx}\, dx.$$

(b) Show that the general solution in part (a) can be rewritten in the form

$$y(x) = c_1 + c_2 x + c_3 e^{kx} + c_4 e^{-kx}$$

$$+ \int_0^x q(s) G(s, x)\, ds,$$

where

$$G(s, x) := \frac{s}{k^2} - \frac{1}{k^3} - \frac{x}{k^2} + \frac{e^{kx} e^{-ks}}{2k^3} - \frac{e^{-kx} e^{ks}}{2k^3}.$$

(c) Let $q(x) \equiv 1$. First compute the general solution using the formula in part (a) and then using the formula in part (b). Compare these two general solutions with the general solution

$$y(x) = B_1 + B_2 x + B_3 e^{kx} + B_4 e^{-kx} - \frac{1}{2k^2} x^2,$$

which one would obtain using the method of undetermined coefficients.

(d) What are some advantages of the formula in part (b)?

CHAPTER SUMMARY

The theory and techniques for solving an nth order linear differential equation

(1) $$y^{(n)} + p_1(x) y^{(n-1)} + \cdots + p_n(x) y = g(x)$$

are natural extensions of the development for second order equations given in Chapter 4. Assuming $p_1, \ldots, p_n$ and g are continuous functions on an open interval I, there is a unique solution to (1) on I that satisfies the n initial conditions: $y(x_0) = y_0$, $y'(x_0) = y_1, \ldots,$ $y^{(n-1)}(x_0) = y_{n-1}$, where $x_0 \in I$.

For the corresponding homogeneous equation

(2) $$y^{(n)} + p_1(x) y^{(n-1)} + \cdots + p_n(x) y = 0,$$

there exists a set of n **linearly independent** solutions $\{y_1, \ldots, y_n\}$ on I. Such functions are said to form a **fundamental solution set,** and every solution to (2) can be written as a linear combination of these functions:

$$y(x) = C_1 y_1(x) + C_2 y_2(x) + \cdots + C_n y_n(x).$$

The linear independence of solutions to (2) is equivalent to the nonvanishing on I of the **Wronskian**

$$W[y_1, \ldots, y_n](x) := \det \begin{bmatrix} y_1(x) & \cdots & y_n(x) \\ y_1'(x) & \cdots & y_n'(x) \\ \vdots & & \vdots \\ y_1^{(n-1)}(x) & \cdots & y_n^{(n-1)}(x) \end{bmatrix}.$$

When equation (2) has (real) constant coefficients so that it is of the form

(3) $$a_n y^{(n)} + a_{n-1} y^{(n-1)} + \cdots + a_0 y = 0, \qquad a_n \neq 0,$$

then the problem of determining a fundamental solution set can be reduced to the algebraic problem of solving the **auxiliary equation**

(4) $a_n r^n + a_{n-1} r^{n-1} + \cdots + a_0 = 0.$

If the n roots of (4), say $r_1, r_2, \ldots, r_n$, are all distinct, then

(5) $\{e^{r_1 x}, e^{r_2 x}, \ldots, e^{r_n x}\}$

is a fundamental solution set for (3). If some real root, say r_1, occurs with multiplicity m (e.g., $r_1 = r_2 = \cdots = r_m$), then m of the functions in (5) are replaced by

$\qquad e^{r_1 x}, xe^{r_1 x}, \ldots, x^{m-1} e^{r_1 x}.$

When a complex root $\alpha + i\beta$ to (4) occurs with multiplicity m, then so does its conjugate, and $2m$ members of the set (5) are replaced by the real-valued functions

$$e^{\alpha x} \sin \beta x, \; xe^{\alpha x} \sin \beta x, \; \ldots, \; x^{m-1} e^{\alpha x} \sin \beta x,$$
$$e^{\alpha x} \cos \beta x, \; xe^{\alpha x} \cos \beta x, \; \ldots, \; x^{m-1} e^{\alpha x} \cos \beta x.$$

A general solution to the nonhomogeneous equation (1) can be written as

$$y(x) = y_p(x) + y_h(x),$$

where y_p is some particular solution to (1) and y_h is a general solution to the corresponding homogeneous equation. Two useful techniques for finding particular solutions are the **annihilator method** (undetermined coefficients) and the method of **variation of parameters.**

The annihilator method applies to equations of the form

(6) $L[y] = g(x),$

where L is a linear differential operator with constant coefficients and the forcing term $g(x)$ is a polynomial, exponential, sine, or cosine, or a linear combination of products of these. Such a function $g(x)$ is annihilated (mapped to zero) by a linear differential operator A that also has constant coefficients. Every solution to the nonhomogeneous equation (6) is then a solution to the homogeneous equation $AL[y] = 0$, and, by comparing the solutions of the latter equation with a general solution to $L[y] = 0$, we can obtain the *form* of a particular solution to (6). These forms have previously been compiled in Table 4.1, page 197, for the method of undetermined coefficients.

The method of variation of parameters is more general in that it applies to arbitrary equations of the form (1). The idea is, starting with a fundamental solution set $\{y_1, \ldots, y_n\}$ for (2), to determine functions $v_1, \ldots, v_n$ such that

(7) $y_p(x) = v_1(x) y_1(x) + \cdots + v_n(x) y_n(x)$

satisfies (1). This method leads to the formula

(8) $y_p(x) = \displaystyle\sum_{k=1}^{n} y_k(x) \int \frac{g(x) W_k(x)}{W[y_1, \ldots, y_n](x)} \, dx,$

where

$$W_k(x) = (-1)^{n-k} W[y_1, \ldots, y_{k-1}, y_{k+1}, \ldots, y_n](x), \qquad k = 1, \ldots, n.$$

REVIEW PROBLEMS

1. Determine the intervals for which Theorem 1 on page 322 guarantees the existence of a solution in that interval.

 (a) $y^{(4)} - (\ln x)y'' + xy' + 2y = \cos 3x$.
 (b) $(x^2 - 1)y''' + (\sin x)y'' + \sqrt{x + 4}\, y' + e^x y = x^2 + 3$.

2. Determine whether the given functions are linearly dependent or linearly independent on the interval $(0, \infty)$.

 (a) $\{e^{2x}, x^2 e^{2x}, e^{-x}\}$.
 (b) $\{e^x \sin 2x, xe^x \sin 2x, e^x, xe^x\}$.
 (c) $\{2e^{2x} - e^x, e^{2x} + 1, e^{2x} - 3, e^x + 1\}$.

3. Show that the set of functions $\{\sin x, x \sin x, x^2 \sin x, x^3 \sin x\}$ is linearly independent on $(-\infty, \infty)$.

4. Find a general solution for the given differential equation.

 (a) $y^{(4)} + 2y''' - 4y'' - 2y' + 3y = 0$.
 (b) $y''' + 3y'' - 5y' + y = 0$.
 (c) $y^{(5)} - y^{(4)} + 2y''' - 2y'' + y' - y = 0$.
 (d) $y''' - 2y'' - y' + 2y = e^x + x$.

5. Find a general solution for the homogeneous linear differential equation with constant coefficients whose auxiliary equation is

 (a) $(r + 5)^2(r - 2)^3(r^2 + 1)^2 = 0$.

6. Given that $y_p = \sin(x^2)$ is a particular solution to
$$y^{(4)} + y = (16x^4 - 11)\sin(x^2) - 48x^2 \cos(x^2)$$
 on $(0, \infty)$, find a general solution.

7. Find a differential operator that annihilates the given function.

 (a) $x^2 - 2x + 5$. (b) $e^{3x} + x - 1$.
 (c) $x \sin 2x$. (d) $x^2 e^{-2x} \cos 3x$.
 (e) $x^2 - 2x + xe^{-x} + \sin 2x - \cos 3x$.

8. Use the annihilator method to determine the form of a particular solution for the given equation.

 (a) $y'' + 6y' + 5y = e^{-x} + x^2 - 1$.
 (b) $y''' + 2y'' - 19y' - 20y = xe^{-x}$.
 (c) $y^{(4)} + 6y'' + 9y = x^2 - \sin 3x$.
 (d) $y''' - y'' + 2y = x \sin x$.

9. Find a general solution to the Cauchy-Euler equation
$$x^3 y''' - 2x^2 y'' - 5xy' + 5y = x^{-2}, \qquad x > 0,$$
 given that $\{x, x^5, x^{-1}\}$ is a fundamental solution set to the corresponding homogeneous equation.

10. Find a general solution to the given Cauchy-Euler equation.

 (a) $4x^3 y''' + 8x^2 y'' - xy' + y = 0$, $x > 0$.
 (b) $x^3 y''' + 2x^2 y'' + 2xy' + 4y = 0$, $x > 0$.

TECHNICAL WRITING EXERCISES

1. Describe the differences and similarities between second order and higher-order linear differential equations. Include in your comparisons both theoretical results and the methods of solution. For example, what complications arise in solving higher-order equations that are not present for the second order case?

2. Explain the relationship between the method of undetermined coefficients and the annihilator method. What difficulties would you encounter in applying the annihilator method if the linear equation did not have constant coefficients?

3. Section 4.11 contains a discussion of second order linear difference equations. Extend that discussion to nth order linear difference equations using the results for nth order differential equations as a guide.

4. For students with a background in linear algebra: Compare the theory for kth order linear differential equations with that for systems of n linear equations in n unknowns whose coefficient matrix has rank $n - k$. Use the terminology from linear algebra; for example, subspaces, basis, dimension, linear transformation, and kernel. Discuss both homogeneous and nonhomogeneous equations.

GROUP PROJECTS FOR CHAPTER 6

A. Justifying the Method of Undetermined Coefficients

The annihilator method discussed in Section 6.3 can be used to derive the entries in Table 4.1, page 197, for the method of undetermined coefficients. To show this, it suffices to work with Type VII functions—that is, functions of the form

(1) $\qquad g(x) = p_n(x)e^{\alpha x} \cos \beta x + q_m(x)e^{\alpha x} \sin \beta x,$

where p_n and q_m are polynomials of degrees n and m, respectively—since the other types listed in Table 4.1 are just special cases of (1).

Consider the nonhomogeneous equation

(2) $\qquad L[y](x) = g(x),$

where L is the linear operator

(3) $\qquad L[y] := a_n y^{(n)} + a_{n-1} y^{(n-1)} + \cdots + a_0 y,$

with $a_n, a_{n-1}, \ldots, a_0$ constants and $g(x)$ as given in equation (1). Let $N := \max(n, m)$.

(a) Show that

$$A := [(D - \alpha)^2 + \beta^2]^{N+1}$$

is an annihilator for g.

(b) Show that the auxiliary equation associated with $AL[y] = 0$ is of the form

(4) $\qquad a_n[(r - \alpha)^2 + \beta^2]^{s+N+1}(r - r_{2s+1}) \cdots (r - r_n) = 0,$

where $s(\geq 0)$ is the multiplicity of $\alpha \pm i\beta$ as roots of the auxiliary equation associated with $L[y] = 0$, and $r_{2s+1}, \ldots, r_n$ are the remaining roots of this equation.

(c) Find a general solution for $AL[y] = 0$ and compare it with a general solution for $L[y] = 0$ to verify that equation (2) has a particular solution of the form

$$y_p(x) = x^s e^{\alpha x}\{P_N(x) \cos \beta x + Q_N(x) \sin \beta x\},$$

where P_N and Q_N are polynomials of degree N.

B. Transverse Vibrations of a Beam

In applying elasticity theory to study the transverse vibrations of a beam, one encounters the equation

$$EIy^{(4)}(x) - \gamma\lambda y(x) = 0,$$

where $y(x)$ is related to the displacement of the beam at position x, the constant E is Young's modulus, I is the area moment of inertia, which we assume is constant, γ is the constant mass per unit length of the beam, and λ is a positive parameter to be determined. We can simplify the equation

349

by letting $r^4 := \gamma\lambda/EI$, that is, we consider

(5) $y^{(4)}(x) - r^4 y(x) = 0$.

When the beam is clamped at each end, we seek a solution to (5) that satisfies the boundary conditions

(6) $y(0) = y'(0) = 0$ and $y(L) = y'(L) = 0$,

where L is the length of the beam. The problem is to determine those values of r for which equation (5) has a nontrivial solution ($y(x) \not\equiv 0$) that satisfies (6). To do this, proceed as follows:

(a) Represent the general solution to (5) in terms of sines, cosines, hyperbolic sines, and hyperbolic cosines.

(b) Substitute the general solution obtained in part (a) into the equations in (6) to obtain four equations in four unknowns.

(c) Determine those values of r for which the system of equations in part (b) has nontrivial solutions.

(d) For each value of r found in part (c), find the corresponding nontrivial solutions.

Chapter 7

Laplace Transforms

7.1 INTRODUCTION: A MIXING PROBLEM

Figure 7.1 depicts a mixing problem with valved input feeders. At time $t = 0$, valve A is opened, delivering 6 L/min of a brine solution containing 4 kg of salt per liter. At $t = 10$ minutes, valve A is closed and valve B is opened, delivering 6 L/min of brine at a concentration of 2 kg/L. Initially there are 30 kg of salt dissolved in 1000 L of water in the tank. The exit valve C, which empties the tank at 6 L/min, maintains the contents of the tank at constant volume. Assuming the solution is kept well stirred, determine the amount of salt in the tank at all times $t > 0$.

We analyzed a simpler version of this problem in Example 1 of Section 3.2. Let $x(t)$ be the amount of salt (in kilograms) in the tank at time t. Then of course $x(t)/1000$ is the concentration, in kg/L. The salt content is depleted at the rate (6 L/min) $\times$ $(x(t)/1000$ kg/L) $= 3x(t)/500$ kg/min through the exit valve. Simultaneously it is enriched through

Figure 7.1 Mixing tank with valve A open

valves A and B at the rate $g(t)$, given by

(1) $g(t) = \begin{cases} 4 \text{ kg/L} \times 6 \text{ L/min} = 24 \text{ kg/min}, & 0 < t < 10 \text{ (valve } A); \\ 2 \text{ kg/L} \times 6 \text{ L/min} = 12 \text{ kg/min}, & t > 10 \text{ (valve } B). \end{cases}$

Thus $x(t)$ changes at a rate

$$\frac{d}{dt}x(t) = g(t) - \frac{3x(t)}{500},$$

or

(2) $\dfrac{dx}{dt} + \dfrac{3}{500}x = g(t),$

with initial condition

(3) $x(0) = 30.$

To solve the initial value problem (2)–(3) using the techniques of Chapter 4, we would have to break up the time interval $(0, \infty)$ into two subintervals $(0, 10)$ and $(10, \infty)$. On these subintervals, the nonhomogeneous term $g(t)$ is constant, and the method of undetermined coefficients could be applied to equation (2) to determine general solutions for each subinterval, each containing one arbitrary constant (in the associated homogeneous solutions). The initial condition (3) would fix this constant for $0 < t < 10$, but then we would need to evaluate $x(10)$ and use it to reset the constant in the general solution for $t > 10$.

Our purpose here is to illustrate a new approach using Laplace transforms. As we shall see, this method offers several advantages over the previous techniques. For one thing, it is much more convenient in solving initial value problems for linear constant-coefficient equations when the forcing term is discontinuous.

The **Laplace transform** of a function $f(t)$, defined on $[0, \infty)$, is given by[†]

(4) $\mathcal{L}\{f\}(s) := \displaystyle\int_0^\infty e^{-st}f(t)\,dt.$

Thus we multiply $f(t)$ by e^{-st} and integrate with respect to t from 0 to ∞. This takes a function of t and produces a function of s. A listing of the Laplace transforms of some commonly occurring functions appears in the table on the inside of the back cover of this text. In Section 7.2, we'll exhibit conditions on f that ensure the existence of the Laplace transform. For now our goal is to demonstrate its utility, so we shall simply state some of its properties without proof.

The most important property relates the transform of $f(t)$, the transform of its derivative $f'(t)$, and the initial value $f(0)$:

(5) $\mathcal{L}\{f'\}(s) = sF(s) - f'(0), \quad \text{where} \quad F(s) := \mathcal{L}\{f\}(s).$

[†] ***Historical Footnote:*** The Laplace transform was first introduced by Pierre Laplace in 1779 in his research on probability. G. Doetsch helped develop the use of Laplace transforms to solve differential equations. His work in the 1930s served to justify the operational calculus procedures earlier used by Oliver Heaviside.

Formula (5) will be verified in Section 7.3 using integration by parts. Essentially the formula says that we replace the operation "differentiate with respect to t" by "multiply by s" when we move from the *t-domain* to the *s-domain*. No wonder the Laplace transform method is so appealing for solving differential equations! It reduces them to algebraic equations. This is the first advantage.

Let's take the transform of the left-hand side of the differential equation (2). Clearly the Laplace transform (4) is a linear operation; that is,

$$\mathscr{L}\left\{\frac{dx}{dt} + \frac{3}{500}x\right\} = \mathscr{L}\left\{\frac{dx}{dt}\right\} + \frac{3}{500}\mathscr{L}\{x\}.$$

For convenience, we set $X(s) := \mathscr{L}\{x\}(s)$. Then, applying formula (5), we find

$$\mathscr{L}\left\{\frac{dx}{dt}\right\}(s) = sX(s) - x(0) = sX(s) - 30.$$

Thus the transform of $dx/dt + 3x/500$ equals

(6) $$\mathscr{L}\left\{\frac{dx}{dt} + \frac{3}{500}x\right\}(s) = sX(s) - 30 + \frac{3}{500}X(s) = \left(s + \frac{3}{500}\right)X(s) - 30.$$

Next we have to take the Laplace transform of the discontinuous function $g(t)$ in (1). From the definition (4), we have

$$\mathscr{L}\{g\}(s) = \int_{t=0}^{10} e^{-st}24\, dt + \int_{t=10}^{\infty} e^{-st}12\, dt$$

$$= \frac{24}{s}(1 - e^{-10s}) + \left.\frac{12e^{-st}}{-s}\right|_{t=10}^{t=\infty}.$$

Now for $s > 0$, the function e^{-st} tends to zero as $t \to \infty$, so

(7) $$\mathscr{L}\{g\}(s) = \frac{24}{s}(1 - e^{-10s}) + \frac{12}{s}e^{-10s} = \frac{24}{s} - \frac{12}{s}e^{-10s}.$$

Note that although the discontinuous function g was specified in "pieces" (equation (1)), its transform is continuous and is represented by a single formula. This is the second advantage of the method: It handles many discontinuous functions conveniently. (We explore this property in Section 7.6.)

Matching the transforms (6)–(7) of the left- and right-hand sides of equation (2), we solve for $X(s)$ and find

(8) $$X(s) = \frac{30}{s + 3/500} + \frac{24}{s(s + 3/500)} - \frac{12e^{-10s}}{s(s + 3/500)}.$$

Notice that there are no undetermined constants to deal with; the initial condition $x(0) = 30$ was incorporated in deriving the transform of dx/dt for equation (6). This automatic fulfilling of initial conditions is the third advantage of the Laplace transform method.

Now we have the task of recovering $x(t)$ from its transform $X(s)$—that is, taking the *inverse transform*. This isn't always easy, but in Sections 7.4 and 7.6 we'll see how to use

the table on this book's inside back cover to derive from (8) the formula for the salt content $x(t)$:

(9) $$x(t) = 4000 - 3970e^{-3t/500} - 2000 \cdot \begin{cases} 0, & t < 10; \\ [1 - e^{-3(t-10)/500}], & t \geq 10. \end{cases}$$

See Figure 7.2.

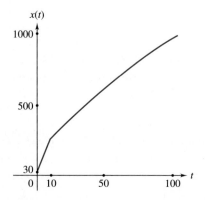

Figure 7.2 Solution to mixing tank example

Note that to arrive at (9), we did not have to take derivatives of trial solutions, break up intervals, or evaluate constants through initial data. The Laplace transform machinery replaces all of these operations by basic algebra: addition, subtraction, multiplication, division—and, of course, the judicious use of the table. Figure 7.3 depicts the advantages of the transform method.

Unfortunately the Laplace transform method is less helpful with equations containing variable coefficients or nonlinear equations (and sometimes determining inverse transforms can be a Herculean task!). But it is ideally suited for many problems arising in applications. Thus we devote Chapter 7 to this important topic.

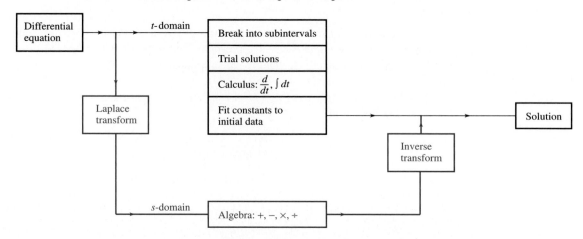

Figure 7.3 Comparison of solution methods

1. Verify that the function defined by (9) is continuous, satisfies the initial condition (3), and satisfies the differential equation (2).

7.2 DEFINITION OF THE LAPLACE TRANSFORM

In earlier chapters we studied differential operators. These operators took a function and mapped or transformed it (via differentiation) into another function. The Laplace transform, which is an integral operator, is another such transformation.

LAPLACE TRANSFORM

Definition 1. Let $f(t)$ be a function on $[0, \infty)$. The **Laplace transform** of f is the function F defined by the integral

$$(1) \qquad F(s) := \int_0^\infty e^{-st} f(t)\, dt.$$

The domain of $F(s)$ is all the values of s for which the integral in (1) exists.[†] The Laplace transform of f is denoted by both F and $\mathscr{L}\{f\}$.

Notice that the integral in (1) is an **improper** integral. More precisely,

$$\int_0^\infty e^{-st} f(t)\, dt := \lim_{N \to \infty} \int_0^N e^{-st} f(t)\, dt$$

whenever the limit exists.

EXAMPLE 1 Determine the Laplace transform of the constant function $f(t) = 1$, $t \geq 0$.

Solution Using the definition of the transform, we compute

$$F(s) = \int_0^\infty e^{-st} \cdot 1 \ dt = \lim_{N \to \infty} \int_0^N e^{-st} \ dt$$

$$= \lim_{N \to \infty} \frac{-e^{-st}}{s} \Bigg|_{t=0}^{t=N} = \lim_{N \to \infty} \left[\frac{1}{s} - \frac{e^{-sN}}{s} \right].$$

[†] We treat s as real valued, but in certain applications s may be a complex variable. For a detailed treatment of complex-valued Laplace transforms, see *Complex Variables and the Laplace Transform for Engineers*, by Wilbur R. LePage (Dover Publications, New York, 1980), or *Fundamentals of Complex Analysis for Mathematics, Science, and Engineering* (Second Edition), by E. B. Saff and A. D. Snider (Prentice Hall, 1993).

Since $e^{-sN} \to 0$ when $s > 0$ is fixed and $N \to \infty$, we get

$$F(s) = \frac{1}{s} \quad \text{for} \quad s > 0.$$

When $s \leq 0$, the integral $\int_0^\infty e^{-st}\, dt$ diverges (Why?). Hence $F(s) = 1/s$, with the domain of $F(s)$ being all $s > 0$. ■ End

EXAMPLE 2 Determine the Laplace transform of $f(t) = e^{at}$, where a is a constant.

Solution Using the definition of the transform,

$$F(s) = \int_0^\infty e^{-st} e^{at}\, dt = \int_0^\infty e^{-(s-a)t}\, dt$$

$$= \lim_{N \to \infty} \int_0^N e^{-(s-a)t}\, dt = \lim_{N \to \infty} \frac{-e^{-(s-a)t}}{s-a} \Big|_0^N$$

$$= \lim_{N \to \infty} \left[\frac{1}{s-a} - \frac{e^{-(s-a)N}}{s-a} \right]$$

$$= \frac{1}{s-a} \quad \text{for} \quad s > a.$$

Again, if $s \leq a$ the integral diverges, and hence the domain of $F(s)$ is all $s > a$. ■ End

It is comforting to note from Example 2 that the transform of the constant function $f(t) = 1 = e^{0t}$ is $1/(s - 0) = 1/s$, which agrees with the solution in Example 1.

EXAMPLE 3 Find $\mathcal{L}\{\sin bt\}$, where b is a nonzero constant.

Solution We need to compute

$$\mathcal{L}\{\sin bt\}(s) = \int_0^\infty e^{-st} \sin bt\, dt = \lim_{N \to \infty} \int_0^N e^{-st} \sin bt\, dt.$$

Using integration by parts twice, we ultimately find

$$\mathcal{L}\{\sin bt\}(s) = \lim_{N \to \infty} \left[-\frac{e^{-st}}{s^2 + b^2} (s \sin bt + b \cos bt) \Big|_0^N \right]$$

$$= \lim_{N \to \infty} \left[\frac{b}{s^2 + b^2} - \frac{e^{-sN}}{s^2 + b^2} (s \sin bN + b \cos bN) \right]$$

$$= \frac{b}{s^2 + b^2} \quad \text{for} \quad s > 0$$

(since for such s we have $\lim_{N \to \infty} e^{-sN}(s \sin bN + b \cos bN) = 0$; see Problem 32). ■ End

EXAMPLE 4 Determine the Laplace transform of

$$f(t) = \begin{cases} 2, & 0 < t < 5, \\ 0, & 5 < t < 10, \\ e^{4t}, & 10 < t. \end{cases}$$

Solution Since $f(t)$ is defined by a different formula on different intervals, we begin by breaking up the integral in (1) into three separate parts.[†] Thus

$$F(s) = \int_0^\infty e^{-st} f(t)\, dt$$

$$= \int_0^5 e^{-st} \cdot 2 \, dt + \int_5^{10} e^{-st} \cdot 0 \, dt + \int_{10}^\infty e^{-st} e^{4t} \, dt$$

$$= 2 \int_0^5 e^{-st} \, dt + \lim_{N \to \infty} \int_{10}^N e^{-(s-4)t} \, dt$$

$$= \frac{2}{s} - \frac{2e^{-5s}}{s} + \lim_{N \to \infty} \left[\frac{e^{-10(s-4)}}{s-4} - \frac{e^{-(s-4)N}}{s-4} \right]$$

$$= \frac{2}{s} - \frac{2e^{-5s}}{s} + \frac{e^{-10(s-4)}}{s-4} \quad \text{for} \quad s > 4. \quad \blacksquare \text{ End}$$

Notice that the function $f(t)$ of Example 4 has jump discontinuities at $t = 5$ and $t = 10$. These values are reflected in the exponential terms e^{-5s} and e^{-10s} that appear in the formula for $F(s)$. We shall make this connection more precise when we discuss the unit step function in Section 7.6.

An important property of the Laplace transform is its **linearity.** That is, the Laplace transform $\mathscr{L}$ is a linear operator.

LINEARITY OF THE TRANSFORM

Theorem 1. Let f, f_1, and f_2 be functions whose Laplace transforms exist for $s > \alpha$ and let c be a constant. Then

(2) $\mathscr{L}\{f_1 + f_2\} = \mathscr{L}\{f_1\} + \mathscr{L}\{f_2\},$

(3) $\mathscr{L}\{cf\} = c\mathscr{L}\{f\}.$

[†] Notice that $f(t)$ is not defined at the points $t = 0$, 5, and 10. Nevertheless, the integral in (1) is still meaningful and unaffected by function values at finitely many points.

Proof. Using the linearity properties of integration, we have for $s > \alpha$

$$\mathcal{L}\{f_1 + f_2\}(s) = \int_0^\infty e^{-st}[f_1(t) + f_2(t)]\,dt$$

$$= \int_0^\infty e^{-st}f_1(t)\,dt + \int_0^\infty e^{-st}f_2(t)\,dt$$

$$= \mathcal{L}\{f_1\}(s) + \mathcal{L}\{f_2\}(s).$$

Hence equation (2) is satisfied. In a similar fashion, we see that

$$\mathcal{L}\{cf\}(s) = \int_0^\infty e^{-st}[cf(t)]\,dt = c\int_0^\infty e^{-st}f(t)\,dt$$

$$= c\mathcal{L}\{f\}(s). \quad \blacksquare\,\text{End}$$

EXAMPLE 5 Determine $\mathcal{L}\{11 + 5e^{4t} - 6\sin 2t\}$.

Solution From the linearity property, we know that the Laplace transform of the sum of any finite number of functions is the sum of their Laplace transforms. Thus

$$\mathcal{L}\{11 + 5e^{4t} - 6\sin 2t\} = \mathcal{L}\{11\} + \mathcal{L}\{5e^{4t}\} + \mathcal{L}\{-6\sin 2t\}$$

$$= 11\mathcal{L}\{1\} + 5\mathcal{L}\{e^{4t}\} - 6\mathcal{L}\{\sin 2t\}.$$

In Examples 1, 2, and 3, we determined that

$$\mathcal{L}\{1\}(s) = \frac{1}{s}, \qquad \mathcal{L}\{e^{4t}\}(s) = \frac{1}{s-4}, \qquad \mathcal{L}\{\sin 2t\}(s) = \frac{2}{s^2 + 2^2}.$$

Using these results, we find

$$\mathcal{L}\{11 + 5e^{4t} - 6\sin 2t\}(s) = 11\left(\frac{1}{s}\right) + 5\left(\frac{1}{s-4}\right) - 6\left(\frac{2}{s^2+4}\right)$$

$$= \frac{11}{s} + \frac{5}{s-4} - \frac{12}{s^2+4}.$$

Since $\mathcal{L}\{1\}$, $\mathcal{L}\{e^{4t}\}$, and $\mathcal{L}\{\sin 2t\}$ are all defined for $s > 4$, so is the transform $\mathcal{L}\{11 + 5e^{4t} - 6\sin 2t\}$. $\blacksquare\,\text{End}$

EXISTENCE OF THE TRANSFORM

There are functions for which the improper integral in (1) fails to converge for any value of s. For example, this is the case for the function $f(t) = 1/t$, which grows too fast near zero. Likewise, no Laplace transform exists for the function $f(t) = e^{t^2}$, which increases too rapidly as $t \to \infty$. Fortunately, the set of functions for which the Laplace transform is defined includes many of the functions that arise in applications involving linear differential equations. We now discuss some properties that will (collectively) ensure the existence of the Laplace transform.

A function $f(t)$ on $[a, b]$ is said to have a **jump discontinuity** at $t_0 \in (a, b)$ if $f(t)$ is discontinuous at t_0, but the one-sided limits

$$\lim_{t \to t_0^-} f(t) \quad \text{and} \quad \lim_{t \to t_0^+} f(t)$$

exist as finite numbers. If the discontinuity occurs at an endpoint, $t_0 = a$ (or b), a jump discontinuity occurs if the one-sided limit of $f(t)$ as $t \to a^+$ ($t \to b^-$) exists as a finite number. We can now define piecewise continuity.

PIECEWISE CONTINUITY

Definition 2. A function $f(t)$ is said to be **piecewise continuous on a finite interval** $[a, b]$ if $f(t)$ is continuous at every point in $[a, b]$, except possibly for a finite number of points at which $f(t)$ has a jump discontinuity.

A function $f(t)$ is said to be **piecewise continuous on $[0, \infty)$** if $f(t)$ is piecewise continuous on $[0, N]$ for all $N > 0$.

EXAMPLE 6 Show that

$$f(t) = \begin{cases} t, & 0 < t < 1, \\ 2, & 1 < t < 2, \\ (t - 2)^2, & 2 \le t \le 3 \end{cases}$$

is piecewise continuous on $[0, 3]$.

Solution From the graph of $f(t)$ sketched in Figure 7.4, we see that $f(t)$ is continuous on the intervals $(0, 1)$, $(1, 2)$, and $(2, 3]$. Moreover, at the points of discontinuity, $t = 0$, 1, and 2, the function has jump discontinuities, since the one-sided limits exist as finite numbers. In particular, at $t = 1$, the left-hand limit is 1 and the right-hand limit is 2. Therefore $f(t)$ is piecewise continuous on $[0, 3]$. ■ End

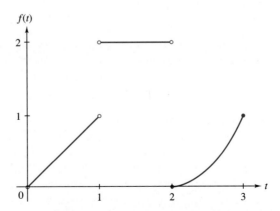

Figure 7.4 Graph of $f(t)$ in Example 6

Observe that the function $f(t)$ of Example 4 is piecewise continuous on $[0, \infty)$ because it is piecewise continuous on every finite interval of the form $[0, N]$, with $N > 0$. In contrast, the function $f(t) = 1/t$ is not piecewise continuous on any interval containing the origin, since it has an "infinite jump" at the origin (see Figure 7.5).

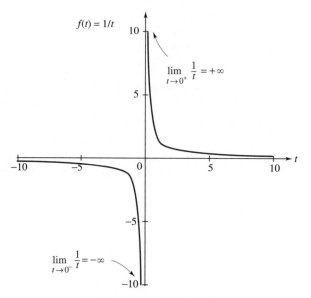

Figure 7.5 Infinite jump at origin

A function that is piecewise continuous on a *finite* interval is necessarily integrable over that interval. However, piecewise continuity on $[0, \infty)$ is not enough to guarantee the existence (as a finite number) of the improper integral over $[0, \infty)$; we also need to consider the growth of the integrand for large t. Roughly speaking, we shall show that the Laplace transform of a piecewise continuous function exists, provided the function does not grow "faster than an exponential."

EXPONENTIAL ORDER α

Definition 3. A function $f(t)$ is said to be of **exponential order** α if there exist positive constants T and M such that

(4) $\qquad |f(t)| \leq Me^{\alpha t}, \quad \text{for all } t \geq T.$

For example, $f(t) = e^{5t} \sin 2t$ is of exponential order $\alpha = 5$ since

$$|e^{5t} \sin 2t| \leq e^{5t},$$

and hence (4) holds with $M = 1$ and T any positive constant.

We use the phrase $f(t)$ *is of exponential order* to mean that for *some* value of α, the function $f(t)$ satisfies the conditions of Definition 3; that is, $f(t)$ grows no faster than a function of the form $Me^{\alpha t}$. The function e^{t^2} is *not* of exponential order. To see this, observe that

$$\lim_{t \to \infty} \frac{e^{t^2}}{e^{\alpha t}} = \lim_{t \to \infty} e^{t(t-\alpha)} = +\infty$$

for any α. Consequently, e^{t^2} grows faster than $e^{\alpha t}$ for every choice of α.

The usual functions encountered in solving linear differential equations with constant coefficients (polynomials, exponentials, sines, and cosines) are both piecewise continuous and of exponential order. As we now show, the Laplace transforms of such functions exist for large enough values of s.

CONDITIONS FOR EXISTENCE OF THE TRANSFORM

Theorem 2. If $f(t)$ is piecewise continuous on $[0, \infty)$ and of exponential order α, then $\mathcal{L}\{f\}(s)$ exists for $s > \alpha$.

Proof. We need to show that the integral

$$\int_0^\infty e^{-st}f(t)\,dt$$

converges for $s > \alpha$. We begin by breaking up this integral into two separate integrals:

(5) $$\int_0^T e^{-st}f(t)\,dt + \int_T^\infty e^{-st}f(t)\,dt,$$

where T is chosen so that inequality (4) holds. The first integral in (5) exists because $f(t)$ and hence $e^{-st}f(t)$ are piecewise continuous on the interval $[0, T]$ for any fixed s. To see that the second integral in (5) converges, we use the **comparison test for improper integrals.**

Since $f(t)$ is of exponential order α, we have for $t \geq T$

$$|f(t)| \leq Me^{\alpha t},$$

and hence

$$\left|e^{-st}f(t)\right| = e^{-st}|f(t)| \leq Me^{-(s-\alpha)t},$$

for all $t \geq T$. Now for $s > \alpha$,

$$\int_T^\infty Me^{-(s-\alpha)t}\,dt = M\int_T^\infty e^{-(s-\alpha)t}\,dt = \frac{Me^{-(s-\alpha)T}}{s-\alpha} < \infty.$$

Since $\left|e^{-st}f(t)\right| \le Me^{-(s-\alpha)t}$ for $t \ge T$ and the improper integral of the larger function converges for $s > \alpha$, then, by the comparison test, the integral

$$\int_T^\infty e^{-st}f(t)\,dt$$

converges for $s > \alpha$. Finally, because the two integrals in (5) exist, the Laplace transform $\mathcal{L}\{f\}(s)$ exists for $s > \alpha$. ■ End

In Table 7.1 we have listed the Laplace transforms of some of the elementary functions. The reader should become familiar with these, since they are frequently encountered in solving linear differential equations with constant coefficients. The entries in the table can be derived from the definition of the Laplace transform. A more elaborate table of transforms is given on the inside back cover of this book.

TABLE 7.1 BRIEF TABLE OF LAPLACE TRANSFORMS

$f(t)$	$F(s) = \mathcal{L}\{f\}(s)$
1	$\dfrac{1}{s}, \quad s > 0$
e^{at}	$\dfrac{1}{s-a}, \quad s > a$
$t^n, \quad n = 1, 2, \ldots$	$\dfrac{n!}{s^{n+1}}, \quad s > 0$
$e^{at}t^n, \quad n = 1, 2, \ldots$	$\dfrac{n!}{(s-a)^{n+1}}, \quad s > a$
$\sin bt$	$\dfrac{b}{s^2 + b^2}, \quad s > 0$
$\cos bt$	$\dfrac{s}{s^2 + b^2}, \quad s > 0$
$e^{at}\sin bt$	$\dfrac{b}{(s-a)^2 + b^2}, \quad s > a$
$e^{at}\cos bt$	$\dfrac{s-a}{(s-a)^2 + b^2}, \quad s > a$

EXERCISES 7.2

In Problems 1 through 12, use Definition 1 to determine the Laplace transform of the given function.

1. t.

2. t^2.

3. e^{6t}.

4. te^{3t}.

5. $\cos 2t$.

6. $\cos bt$, b a constant.

7. $e^{2t} \cos 3t$.

8. $e^{-t} \sin 2t$.

9. $f(t) = \begin{cases} 0, & 0 < t < 2, \\ t, & 2 < t. \end{cases}$

10. $f(t) = \begin{cases} 1 - t, & 0 < t < 1, \\ 0, & 1 < t. \end{cases}$

11. $f(t) = \begin{cases} \sin t, & 0 < t < \pi, \\ 0, & \pi < t. \end{cases}$

12. $f(t) = \begin{cases} e^{2t}, & 0 < t < 3, \\ 1, & 3 < t. \end{cases}$

In Problems 13 through 20, use the Laplace transform table and the linearity of the Laplace transform to determine the following transforms.

13. $\mathcal{L}\{6e^{-3t} - t^2 + 2t - 8\}$.

14. $\mathcal{L}\{5 - e^{2t} + 6t^2\}$.

15. $\mathcal{L}\{t^3 - te^t + e^{4t} \cos t\}$.

16. $\mathcal{L}\{t^2 - 3t - 2e^{-t} \sin 3t\}$.

17. $\mathcal{L}\{e^{3t} \sin 6t - t^3 + e^t\}$.

18. $\mathcal{L}\{t^4 - t^2 - t + \sin \sqrt{2}t\}$.

19. $\mathcal{L}\{t^4 e^{5t} - e^t \cos \sqrt{7}t\}$.

20. $\mathcal{L}\{e^{-2t} \cos \sqrt{3}t - t^2 e^{-2t}\}$.

In Problems 21 through 28, determine whether $f(t)$ is continuous, piecewise continuous, or neither on $[0, 10]$ and sketch the graph of $f(t)$.

21. $f(t) = \begin{cases} 1, & 0 \le t \le 1, \\ (t - 2)^2, & 1 < t \le 10. \end{cases}$

22. $f(t) = \begin{cases} 0, & 0 \le t < 2, \\ t, & 2 \le t \le 10. \end{cases}$

23. $f(t) = \begin{cases} 1, & 0 \le t < 1, \\ t - 1, & 1 < t < 3, \\ t^2 - 4, & 3 < t \le 10. \end{cases}$

24. $f(t) = \dfrac{t^2 - 3t + 2}{t^2 - 4}$.

25. $f(t) = \dfrac{t^2 - t - 20}{t^2 + 7t + 10}$.

26. $f(t) = \dfrac{t}{t^2 - 1}$.

27. $f(t) = \begin{cases} 1/t, & 0 < t < 1, \\ 1, & 1 \le t \le 2, \\ 1 - t, & 2 < t \le 10. \end{cases}$

28. $f(t) = \begin{cases} \dfrac{\sin t}{t}, & t \ne 0, \\ 1, & t = 0. \end{cases}$

29. Which of the following functions are of exponential order?

(a) $t^3 \sin t$.

(b) $100e^{49t}$.

(c) e^{t^3}.

(d) $t \ln t$.

(e) $\cosh(t^2)$.

(f) $\dfrac{1}{t^2 + 1}$.

(g) $\sin(t^2) + t^4 e^{6t}$.

(h) $3 - e^{t^2} + \cos 4t$.

(i) $\exp\{t^2/(t + 1)\}$.

(j) $\sin(e^{t^2}) + e^{\sin t}$.

30. For the transforms $F(s)$ in Table 7.1, what can be said about $\lim_{s \to \infty} F(s)$?

31. Thanks to Euler's formula (page 183) and the algebraic properties of complex numbers, several of the entries of Table 7.1 can be derived from a single formula; namely,

(6) $\quad \mathcal{L}\{e^{(a+ib)t}\}(s) = \dfrac{s - a + ib}{(s - a)^2 + b^2}, \quad s > a.$

(a) Use the definition of Laplace transform to show that

$$\mathcal{L}\{e^{(a+ib)t}\}(s) = \dfrac{1}{s - (a + ib)}, \quad s > a.$$

(b) Deduce (6) from part (a) by showing that

$$\dfrac{1}{s - (a + ib)} = \dfrac{s - a + ib}{(s - a)^2 + b^2}.$$

(c) By equating the real and imaginary parts in formula (6), deduce the last two entries in Table 7.1.

32. Prove that for fixed $s > 0$, we have

$$\lim_{N \to \infty} e^{-sN}(s \sin bN + b \cos bN) = 0.$$

33. Prove that if f is piecewise continuous on $[a, b]$ and g is continuous on $[a, b]$, then the product fg is piecewise continuous on $[a, b]$.

7.3 PROPERTIES OF THE LAPLACE TRANSFORM

In the previous section we defined the Laplace transform of a function $f(t)$ as

$$\mathscr{L}\{f\}(s) := \int_0^\infty e^{-st} f(t)\, dt.$$

Using this definition to get an explicit expression for $\mathscr{L}\{f\}$ requires the evaluation of the improper integral—frequently a tedious task! We have already seen how the linearity property of the transform can help relieve this burden. In this section, we discuss some further properties of the Laplace transform that simplify its computation. These new properties will also enable us to use the Laplace transform to solve initial value problems.

TRANSLATION IN s

Theorem 3. If the Laplace transform $\mathscr{L}\{f\}(s) = F(s)$ exists for $s > \alpha$, then

$$(1) \qquad \mathscr{L}\{e^{at}f(t)\}(s) = F(s - a)$$

for $s > \alpha + a$.

Proof. We simply compute

$$\mathscr{L}\{e^{at}f(t)\}(s) = \int_0^\infty e^{-st} e^{at} f(t)\, dt$$

$$= \int_0^\infty e^{-(s-a)t} f(t)\, dt$$

$$= F(s - a). \quad \blacksquare \text{ End}$$

Theorem 3 illustrates the effect on the Laplace transform of multiplication of a function $f(t)$ by e^{at}.

EXAMPLE 1 Determine the Laplace transform of $e^{at} \sin bt$.

Solution In Example 3 in Section 7.2, we found that

$$\mathscr{L}\{\sin bt\}(s) = F(s) = \frac{b}{s^2 + b^2}.$$

Thus, by the translation property of $F(s)$, we have

$$\mathscr{L}\{e^{at} \sin bt\}(s) = F(s - a) = \frac{b}{(s - a)^2 + b^2} \quad \blacksquare \text{ End}$$

> ### LAPLACE TRANSFORM OF THE DERIVATIVE
>
> **Theorem 4.** Let $f(t)$ be continuous on $[0, \infty)$ and $f'(t)$ be piecewise continuous on $[0, \infty)$, with both of exponential order α. Then, for $s > \alpha$,
>
> **(2)** $\mathscr{L}\{f'\}(s) = s\mathscr{L}\{f\}(s) - f(0).$

Proof. Since $\mathscr{L}\{f'\}$ exists, we can use integration by parts (with $u = e^{-st}$ and $dv = f'(t)\,dt$) to obtain

(3) $$\mathscr{L}\{f'\}(s) = \int_0^\infty e^{-st}f'(t)\,dt = \lim_{N \to \infty} \int_0^N e^{-st}f'(t)\,dt$$

$$= \lim_{N \to \infty}\left[e^{-st}f(t)\Big|_0^N + s\int_0^N e^{-st}f(t)\,dt \right]$$

$$= \lim_{N \to \infty} e^{-sN}f(N) - f(0) + s\lim_{N \to \infty}\int_0^N e^{-st}f(t)\,dt$$

$$= \lim_{N \to \infty} e^{-sN}f(N) - f(0) + s\mathscr{L}\{f\}(s).$$

To evaluate $\lim_{N \to \infty} e^{-sN}f(N)$, we observe that since $f(t)$ is of exponential order α, there exists a constant M such that for N large,

$$\left| e^{-sN}f(N) \right| \le e^{-sN}Me^{\alpha N} = Me^{-(s-\alpha)N}.$$

Hence, for $s > \alpha$,

$$\lim_{N \to \infty}\left| e^{-sN}f(N) \right| \le \lim_{N \to \infty} Me^{-(s-\alpha)N} = 0,$$

so

$$\lim_{N \to \infty} e^{-sN}f(N) = 0$$

for $s > \alpha$. Equation (3) now reduces to

$$\mathscr{L}\{f'\}(s) = s\mathscr{L}\{f\}(s) - f(0). \quad \blacksquare \text{ End}$$

Using induction, we can extend the last theorem to higher-order derivatives of $f(t)$. For example,

$$\mathscr{L}\{f''\}(s) = s\mathscr{L}\{f'\}(s) - f'(0)$$
$$= s[s\mathscr{L}\{f\}(s) - f(0)] - f'(0),$$

which simplifies to

$$\mathscr{L}\{f''\}(s) = s^2\mathscr{L}\{f\}(s) - sf(0) - f'(0).$$

In general we obtain the following result.

> ### LAPLACE TRANSFORM OF HIGHER-ORDER DERIVATIVES
>
> **Theorem 5.** Let $f(t), f'(t), \ldots, f^{(n-1)}(t)$ be continuous on $[0, \infty)$ and let $f^{(n)}(t)$ be piecewise continuous on $[0, \infty)$, with all these functions of exponential order α. Then, for $s > \alpha$,
>
> (4) $\mathcal{L}\{f^{(n)}\}(s) = s^n \mathcal{L}\{f\}(s) - s^{n-1}f(0) - s^{n-2}f'(0) - \cdots - f^{(n-1)}(0).$

The last two theorems shed light on the reason why the Laplace transform is such a useful tool in solving initial value problems. Roughly speaking, they tell us that by using the Laplace transform we can replace "differentiation with respect to t" with "multiplication by s," thereby converting a differential equation into an algebraic one. This idea is explored in Section 7.5. For now, we show how Theorem 4 can be helpful in computing a Laplace transform.

EXAMPLE 2 Using the fact that

$$\mathcal{L}\{\sin bt\}(s) = \frac{b}{s^2 + b^2},$$

determine $\mathcal{L}\{\cos bt\}$.

Solution Let $f(t) = \sin bt$. Then $f(0) = 0$ and $f'(t) = b \cos bt$. Substituting into equation (2), we have

$$\mathcal{L}\{f'\}(s) = s\mathcal{L}\{f\}(s) - f(0),$$
$$\mathcal{L}\{b \cos bt\}(s) = s\mathcal{L}\{\sin bt\}(s) - 0,$$
$$b\mathcal{L}\{\cos bt\}(s) = \frac{sb}{s^2 + b^2}.$$

Dividing by b gives

$$\mathcal{L}\{\cos bt\}(s) = \frac{s}{s^2 + b^2}. \quad \blacksquare \text{ End}$$

EXAMPLE 3 Prove the following identity for continuous functions $f(t)$ (assuming the transforms exist):

(5) $$\mathcal{L}\left\{\int_0^t f(\tau)\,d\tau\right\}(s) = \frac{1}{s}\mathcal{L}\{f(t)\}(s).$$

Use it to verify the solution to Example 2.

Solution Define the function $g(t)$ by the integral

$$g(t) := \int_0^t f(\tau)\,d\tau.$$

Observe that $g(0) = 0$ and $g'(t) = f(t)$. Thus if we apply Theorem 4 to $g(t)$ (instead of $f(t)$), equation (2) on page 365 reads

$$\mathscr{L}\{f(t)\}(s) = s\mathscr{L}\left\{\int_0^t f(\tau)d\tau\right\}(s) - 0,$$

which is equivalent to equation (5).

Now since

$$\sin t = \int_0^t \cos \tau \, d\tau,$$

equation (5) predicts

$$\mathscr{L}\{\sin t\}(s) = \frac{1}{s}\mathscr{L}\{\cos t\}(s).$$

This identity is valid for the transforms in Example 2 with $b = 1$. ▪ End

Another question arises concerning the Laplace transform. If $F(s)$ is the Laplace transform of $f(t)$, is $F'(s)$ also a Laplace transform of some function of t? The answer is yes:

$$F'(s) = \mathscr{L}\{-tf(t)\}(s).$$

In fact, the following more general assertion holds.

DERIVATIVES OF THE LAPLACE TRANSFORM

Theorem 6. Let $F(s) = \mathscr{L}\{f\}(s)$ and assume $f(t)$ is piecewise continuous on $[0, \infty)$ and of exponential order α. Then, for $s > \alpha$,

(6) $$\mathscr{L}\{t^n f(t)\}(s) = (-1)^n \frac{d^n F}{ds^n}(s).$$

Proof. Consider the identity

$$\frac{dF}{ds}(s) = \frac{d}{ds}\int_0^\infty e^{-st}f(t)\,dt.$$

Because of the assumptions on $f(t)$, we can apply a theorem from advanced calculus (sometimes called **Leibniz's rule**) to interchange the order of integration and differentiation:

$$\frac{dF}{ds}(s) = \int_0^\infty \frac{d}{ds}(e^{-st})f(t)\,dt$$

$$= -\int_0^\infty e^{-st}tf(t)\,dt = -\mathscr{L}\{tf(t)\}(s).$$

Thus

$$\mathcal{L}\{tf(t)\}(s) = (-1)\frac{dF}{ds}(s).$$

The general result (6) now follows by induction on n. ∎ End

A consequence of the above theorem is that if $f(t)$ is piecewise continuous and of exponential order, then its transform $F(s)$ has derivatives of all orders.

EXAMPLE 4 Determine $\mathcal{L}\{t\sin bt\}$.

Solution We already know that

$$\mathcal{L}\{\sin bt\}(s) = F(s) = \frac{b}{s^2 + b^2}.$$

Differentiating $F(s)$, we obtain

$$\frac{dF}{ds}(s) = \frac{-2bs}{(s^2 + b^2)^2}.$$

Hence, using formula (6), we have

$$\mathcal{L}\{t\sin bt\}(s) = -\frac{dF}{ds}(s) = \frac{2bs}{(s^2 + b^2)^2}.$$ ∎ End

For easy reference, we have listed in Table 7.2 some of the basic properties of the Laplace transform derived in this section.

TABLE 7.2 PROPERTIES OF LAPLACE TRANSFORMS

1. $\mathcal{L}\{f + g\} = \mathcal{L}\{f\} + \mathcal{L}\{g\}$.

2. $\mathcal{L}\{cf\} = c\mathcal{L}\{f\}$ for any constant c.

3. $\mathcal{L}\{e^{at}f(t)\}(s) = \mathcal{L}\{f\}(s - a)$.

4. $\mathcal{L}\{f'\}(s) = s\mathcal{L}\{f\}(s) - f(0)$.

5. $\mathcal{L}\{f''\}(s) = s^2\mathcal{L}\{f\}(s) - sf(0) - f'(0)$.

6. $\mathcal{L}\{f^{(n)}\}(s) = s^n\mathcal{L}\{f\}(s) - s^{n-1}f(0) - s^{n-2}f'(0) - \cdots - f^{(n-1)}(0)$.

7. $\mathcal{L}\{t^n f(t)\}(s) = (-1)^n\dfrac{d^n}{ds^n}(\mathcal{L}\{f\}(s))$.

EXERCISES 7.3

In Problems 1 through 20, determine the Laplace transform of the given function using Table 7.1, page 362, and the properties of the transform given in Table 7.2. [Hint: In Problems 12 through 20, use an appropriate trigonometric identity.]

1. $t^2 + e^t \sin 2t$.

2. $3t^2 - e^{2t}$.

3. $e^{-t} \cos 3t + e^{6t} - 1$.

4. $3t^4 - 2t^2 + 1$.

5. $2t^2 e^{-t} - t + \cos 4t$.

6. $e^{-2t} \sin 2t + e^{3t} t^2$.

7. $(t - 1)^4$.

8. $(1 + e^{-t})^2$.

9. $e^{-t} t \sin 2t$.

10. $t e^{2t} \cos 5t$.

11. $\cosh bt$.

12. $\sin 3t \cos 3t$.

13. $\sin^2 t$.

14. $e^{7t} \sin^2 t$.

15. $\cos^3 t$.

16. $t \sin^2 t$.

17. $\sin 2t \sin 5t$.

18. $\cos nt \cos mt, \quad m \neq n$.

19. $\cos nt \sin mt, \quad m \neq n$.

20. $t \sin 2t \sin 5t$.

21. Given that $\mathcal{L}\{\cos bt\}(s) = s/(s^2 + b^2)$, use the translation property to compute $\mathcal{L}\{e^{at} \cos bt\}$.

22. Starting with the transform $\mathcal{L}\{1\}(s) = 1/s$, use formula (6) for the derivatives of the Laplace transform to show that $\mathcal{L}\{t^n\}(s) = n!/s^{n+1}$.

23. Use Theorem 4 on the Laplace transform of the derivative of $f(t)$ to determine $\mathcal{L}\{\cos^2 t\}$. Verify your answer by using the half-angle formula to derive $\mathcal{L}\{\cos^2 t\}$.

24. Show that $\mathcal{L}\{e^{at} t^n\}(s) = n!/(s - a)^{n+1}$ in two ways:

(a) Use the translation property for $F(s)$.

(b) Use formula (6) for the derivatives of the Laplace transform.

25. Use formula (6) to help determine

(a) $\mathcal{L}\{t \cos bt\}$.

(b) $\mathcal{L}\{t^2 \cos bt\}$.

26. Let $f(t)$ be piecewise continuous on $[0, \infty)$ and of exponential order.

(a) Show that there exist constants K and α such that

$$|f(t)| \leq K e^{\alpha t} \quad \text{for all } t \geq 0.$$

(b) By using the definition of the transform and estimating the integral with the help of part (a), prove that

$$\lim_{s \to \infty} \mathcal{L}\{f\}(s) = 0.$$

27. Let $f(t)$ be piecewise continuous on $[0, \infty)$ and of exponential order α and assume $\lim_{t \to 0^+}[f(t)/t]$ exists. Show that

$$\mathcal{L}\left\{\frac{f(t)}{t}\right\}(s) = \int_s^\infty F(u)\,du,$$

where $F(s) = \mathcal{L}\{f\}(s)$. [Hint: Use formula (6) for the derivative of the Laplace transform and the result of Problem 26.]

28. Verify the identity in Problem 27 for the following functions. (Use the table of Laplace transforms on the inside back cover.)

(a) t^4. (b) $\sqrt{t}$.

29. The **transfer function** of a linear system is defined as the ratio of the Laplace transform of the output function $y(t)$ to the Laplace transform of the input function $g(t)$, when all initial conditions are zero. If a linear system is governed by the differential equation

$$y''(t) + 6y'(t) + 10y(t) = g(t), \qquad t > 0,$$

use the linearity property of the Laplace transform and Theorem 5 on the Laplace transform of higher-order derivatives to determine the transfer function $H(s) = Y(s)/G(s)$ for this system.

30. Find the transfer function, as defined in Problem 29, for the linear system governed by

$$y''(t) + 5y'(t) + 6y(t) = g(t), \qquad t > 0.$$

31. Translation in t. Show that for $c > 0$, the translated function

$$g(t) = \begin{cases} 0, & 0 < t < c \\ f(t - c), & c < t \end{cases}$$

has Laplace transform

$$\mathcal{L}\{g\}(s) = e^{-cs}\mathcal{L}\{f\}(s).$$

In Problems 32 through 35, let $g(t)$ be the given function $f(t)$ translated to the right by c units. Sketch $f(t)$ and $g(t)$ and find $\mathcal{L}\{g(t)\}(s)$. (See Problem 31.)

32. $f(t) \equiv 1, \qquad c = 2$.

33. $f(t) = t, \qquad c = 1$.

34. $f(t) = \sin t, \qquad c = \pi$.

35. $f(t) = \sin t, \qquad c = \pi/2$.

36. Use equation (5) to derive the formula $\mathcal{L}\{t^n\}(s) = n!/s^{n+1}$. [Hint: Start with $\mathcal{L}\{1\}(s)$ and use induction.]

37. Initial Value Theorem. Apply the relation

(7) $\mathcal{L}\{f'\}(s) = \displaystyle\int_0^\infty e^{-st}f'(t)\,dt = s\mathcal{L}\{f\}(s) - f(0)$

to argue that for any function $f(t)$ whose derivative is piecewise continuous and of exponential order on $[0, \infty)$,

 $f(0) = \displaystyle\lim_{s\to\infty} s\mathcal{L}\{f\}(s).$

38. Verify the Initial Value Theorem (Problem 37) for the following functions. (Use the table of Laplace transforms on the inside back cover.)

(a) 1. **(b)** e^t. **(c)** e^{-t}. **(d)** $\cos t$.
(e) $\sin t$. **(f)** t^2. **(g)** $t \cos t$.

7.4 INVERSE LAPLACE TRANSFORM

In Section 7.2, we defined the Laplace transform as an integral operator that maps a function $f(t)$ into a function $F(s)$. In this section we consider the problem of finding the function $f(t)$ when we are given the transform $F(s)$. That is, we seek an **inverse mapping** for the Laplace transform.

To see the usefulness of such an inverse, let's consider the simple initial value problem

(1) $y'' - y = -t; \qquad y(0) = 0, \qquad y'(0) = 1.$

If we take the transform of both sides of equation (1) and use the linearity property of the transform, we find

$$\mathcal{L}\{y''\}(s) - Y(s) = -\frac{1}{s^2},$$

where $Y(s):= \mathcal{L}\{y\}(s)$. We know the initial values of the solution $y(t)$, so we can use Theorem 5, page 366, on the Laplace transform of higher-order derivatives to express

$$\mathcal{L}\{y''\}(s) = s^2 Y(s) - sy(0) - y'(0) = s^2 Y(s) - 1.$$

Substituting for $\mathcal{L}\{y''\}(s)$ yields

$$s^2 Y(s) - 1 - Y(s) = -\frac{1}{s^2}.$$

Solving this algebraic equation for $Y(s)$ gives

$$Y(s) = \frac{1 - \left(\dfrac{1}{s^2}\right)}{s^2 - 1} = \frac{s^2 - 1}{s^2(s^2 - 1)} = \frac{1}{s^2}.$$

We now recall that $\mathcal{L}\{t\}(s) = 1/s^2$, and since $Y(s) = \mathcal{L}\{y\}(s)$, we have

 $\mathcal{L}\{y\}(s) = 1/s^2 = \mathcal{L}\{t\}(s).$

It therefore seems reasonable to conclude that $y(t) = t$ is the solution to the initial value problem (1). A quick check confirms this!

Notice that in the above procedure, a crucial step is to determine $y(t)$ from its Laplace transform $Y(s) = 1/s^2$. As we noted, $y(t) = t$ is such a function, but it is *not* the only function

whose Laplace function is $1/s^2$. For example, the transform of

$$g(t) := \begin{cases} t, & t \neq 6, \\ 0, & t = 6 \end{cases}$$

is also $1/s^2$. This is because the transform is an integral, and integrals are not affected by changing a function's values at isolated points. The significant difference between $y(t)$ and $g(t)$ as far as we are concerned is that $y(t)$ is continuous on $[0, \infty)$, while $g(t)$ is not. Naturally, we prefer to work with continuous functions, since solutions to differential equations are continuous. Fortunately, it can be shown that if two different functions have the same Laplace transform, at most one of them can be continuous.[†] With this in mind we give the following definition.

INVERSE LAPLACE TRANSFORM

Definition 4. By the **inverse Laplace transform** of the function $F(s)$, we mean the unique function $f(t)$ that is continuous on $[0, \infty)$ and satisfies

(2) $\mathcal{L}\{f\}(s) = F(s).$

We denote f by $\mathcal{L}^{-1}\{F\}$. In case all functions that satisfy (2) are discontinuous on $[0, \infty)$, we select a piecewise continuous function that satisfies (2) to be $\mathcal{L}^{-1}\{F\}$.

Naturally the Laplace transform tables will be a great help in determining the inverse Laplace transform of a given function $F(s)$.

EXAMPLE 1 Determine $\mathcal{L}^{-1}\{F\}$, where

$$\textbf{(a)}\quad F(s) = \frac{2}{s^3}. \qquad \textbf{(b)}\quad F(s) = \frac{3}{s^2 + 9}. \qquad \textbf{(c)}\quad F(s) = \frac{s-1}{s^2 - 2s + 5}.$$

Solution To compute $\mathcal{L}^{-1}\{F\}$, we refer to the Laplace transform table on page 362.

(a) $\mathcal{L}^{-1}\left\{\dfrac{2}{s^3}\right\}(t) = \mathcal{L}^{-1}\left\{\dfrac{2!}{s^3}\right\}(t) = t^2.$

(b) $\mathcal{L}^{-1}\left\{\dfrac{3}{s^2 + 9}\right\}(t) = \mathcal{L}^{-1}\left\{\dfrac{3}{s^2 + 3^2}\right\}(t) = \sin 3t.$

(c) $\mathcal{L}^{-1}\left\{\dfrac{s-1}{s^2 - 2s + 5}\right\}(t) = \mathcal{L}^{-1}\left\{\dfrac{s-1}{(s-1)^2 + 2^2}\right\}(t) = e^t \cos 2t.$

In part (c), we used the technique of completing the square to rewrite the denominator in a form that we could find in the table. ■ End

[†] For this result and further properties of the Laplace transform and its inverse, we refer the reader to *Operational Mathematics*, Third Edition, by R. V. Churchill (McGraw-Hill, New York, 1972).

In practice, we do not always encounter a transform $F(s)$ that exactly corresponds to an entry in the second column of the Laplace transform table. To handle more complicated functions $F(s)$, we use properties of $\mathcal{L}^{-1}$, just as we used properties of $\mathcal{L}$. One such tool is the linearity of the inverse Laplace transform, a property that is inherited from the linearity of the operator $\mathcal{L}$.

LINEARITY OF THE INVERSE TRANSFORM

Theorem 7. Assume $\mathcal{L}^{-1}\{F\}$, $\mathcal{L}^{-1}\{F_1\}$, and $\mathcal{L}^{-1}\{F_2\}$ exist and are continuous on $[0, \infty)$ and let c be any constant. Then

(3) $\qquad \mathcal{L}^{-1}\{F_1 + F_2\} = \mathcal{L}^{-1}\{F_1\} + \mathcal{L}^{-1}\{F_2\}$,

(4) $\qquad\qquad \mathcal{L}^{-1}\{cF\} = c\mathcal{L}^{-1}\{F\}$.

The proof of Theorem 7 is outlined in Problem 37. We illustrate the usefulness of this theorem in the next example.

EXAMPLE 2 Determine $\displaystyle \mathcal{L}^{-1}\left\{ \frac{5}{s-6} - \frac{6s}{s^2+9} + \frac{3}{2s^2+8s+10} \right\}$.

Solution We begin by using the linearity property. Thus

$$\mathcal{L}^{-1}\left\{ \frac{5}{s-6} - \frac{6s}{s^2+9} + \frac{3}{2(s^2+4s+5)} \right\}$$

$$= 5\mathcal{L}^{-1}\left\{ \frac{1}{s-6} \right\} - 6\mathcal{L}^{-1}\left\{ \frac{s}{s^2+9} \right\} + \frac{3}{2}\mathcal{L}^{-1}\left\{ \frac{1}{s^2+4s+5} \right\}.$$

Referring to the Laplace transform tables, we see that

$$\mathcal{L}^{-1}\left\{ \frac{1}{s-6} \right\}(t) = e^{6t} \quad \text{and} \quad \mathcal{L}^{-1}\left\{ \frac{s}{s^2+3^2} \right\}(t) = \cos 3t.$$

This gives us the first two terms. To determine $\mathcal{L}^{-1}\{1/(s^2+4s+5)\}$, we complete the square of the denominator to obtain $s^2+4s+5 = (s+2)^2+1$. We now recognize from the tables that

$$\mathcal{L}^{-1}\left\{ \frac{1}{(s+2)^2+1^2} \right\}(t) = e^{-2t}\sin t.$$

Hence

$$\mathcal{L}^{-1}\left\{ \frac{5}{s-6} - \frac{6s}{s^2+9} + \frac{3}{2s^2+8s+10} \right\}(t) = 5e^{6t} - 6\cos 3t + \frac{3e^{-2t}}{2}\sin t. \quad \blacksquare \text{ End}$$

EXAMPLE 3 Determine $\mathcal{L}^{-1}\left\{\dfrac{5}{(s+2)^4}\right\}$.

Solution The $(s+2)^4$ in the denominator suggests that we work with the formula

$$\mathcal{L}^{-1}\left\{\frac{n!}{(s-a)^{n+1}}\right\}(t) = e^{at}t^n.$$

Here we have $a = -2$ and $n = 3$, so $\mathcal{L}^{-1}\{6/(s+2)^4\}(t) = e^{-2t}t^3$. Using the linearity property, we find

$$\mathcal{L}^{-1}\left\{\frac{5}{(s+2)^4}\right\}(t) = \frac{5}{6}\mathcal{L}^{-1}\left\{\frac{3!}{(s+2)^4}\right\}(t) = \frac{5}{6}e^{-2t}t^3. \quad \blacksquare \text{ End}$$

EXAMPLE 4 Determine $\mathcal{L}^{-1}\left\{\dfrac{3s+2}{s^2+2s+10}\right\}$.

Solution By completing the square, the quadratic term in the denominator can be written as

$$s^2 + 2s + 10 = s^2 + 2s + 1 + 9 = (s+1)^2 + 3^2.$$

The form of $F(s)$ now suggests that we use one or both of the formulas

$$\mathcal{L}^{-1}\left\{\frac{s-a}{(s-a)^2+b^2}\right\}(t) = e^{at}\cos bt,$$

$$\mathcal{L}^{-1}\left\{\frac{b}{(s-a)^2+b^2}\right\}(t) = e^{at}\sin bt.$$

In this case, $a = -1$ and $b = 3$. The next step is to express

(5) $$\frac{3s+2}{s^2+2s+10} = A\frac{s+1}{(s+1)^2+3^2} + B\frac{3}{(s+1)^2+3^2},$$

where A, B are constants to be determined. Multiplying both sides of (5) by $s^2 + 2s + 10$ leaves

$$3s + 2 = A(s+1) + 3B = As + (A + 3B),$$

which is an identity between two polynomials in s. Equating the coefficients of like terms gives

$$A = 3, \qquad A + 3B = 2,$$

so $A = 3$ and $B = -1/3$. Finally, from (5) and the linearity property, we find

$$\mathcal{L}^{-1}\left\{\frac{3s+2}{s^2+2s+10}\right\}(t) = 3\mathcal{L}^{-1}\left\{\frac{s+1}{(s+1)^2+3^2}\right\}(t) - \frac{1}{3}\mathcal{L}^{-1}\left\{\frac{3}{(s+1)^2+3^2}\right\}(t)$$

$$= 3e^{-t}\cos 3t - \frac{1}{3}e^{-t}\sin 3t. \quad \blacksquare \text{ End}$$

Given the choice of finding the inverse Laplace transform of

$$F_1(s) = \frac{7s^2 + 10s - 1}{s^3 + 3s^2 - s - 3}$$

or of

$$F_2(s) = \frac{2}{s - 1} + \frac{1}{s + 1} + \frac{4}{s + 3},$$

which would you select? No doubt $F_2(s)$ is the easier one. Actually, the two functions $F_1(s)$ and $F_2(s)$ are identical. This can be checked by combining the simple fractions that form $F_2(s)$. Thus, if we are faced with the problem of computing $\mathcal{L}^{-1}$ of a rational function such as $F_1(s)$, we will first express it, as we did $F_2(s)$, as a sum of simple rational functions. This is accomplished by the **method of partial fractions.**

We briefly review this method. Recall from calculus that a rational function of the form $P(s)/Q(s)$, where $P(s)$ and $Q(s)$ are polynomials with the degree of P less than the degree of Q, has a partial fraction expansion whose form is based on the linear and quadratic factors of $Q(s)$. (We assume the coefficients of the polynomials to be real numbers.) There are three cases to consider:

1. Nonrepeated linear factors.
2. Repeated linear factors.
3. Quadratic factors.

1. Nonrepeated Linear Factors

If $Q(s)$ can be factored into a product of distinct linear factors,

$$Q(s) = (s - r_1)(s - r_2) \cdots (s - r_n),$$

Where the r_i's are all distinct real numbers, then the partial fraction expansion has the form

$$\frac{P(s)}{Q(s)} = \frac{A_1}{s - r_1} + \frac{A_2}{s - r_2} + \cdots + \frac{A_n}{s - r_n},$$

where the A_i's are real numbers. There are various ways of determining the constants $A_1, \ldots, A_n$. In the next example, we demonstrate two such methods.

EXAMPLE 5 Determine $\mathcal{L}^{-1}\{F\}$, where

$$F(s) = \frac{7s - 1}{(s + 1)(s + 2)(s - 3)}.$$

Solution We begin by finding the partial fraction expansion for $F(s)$. The denominator consists of three distinct linear factors, so the expansion has the form

(6) $$\frac{7s - 1}{(s + 1)(s + 2)(s - 3)} = \frac{A}{s + 1} + \frac{B}{s + 2} + \frac{C}{s - 3},$$

where A, B, and C are real numbers to be determined.

One procedure that works for all partial fraction expansions is first to multiply the expansion equation by the denominator of the given rational function. This leaves us with two identical polynomials. Equating the coefficients of s^k leads to a system of linear equations that we can solve to determine the unknown constants. In this example we multiply (6) by $(s + 1)(s + 2)(s - 3)$ and find

(7) $7s - 1 = A(s + 2)(s - 3) + B(s + 1)(s - 3) + C(s + 1)(s + 2)$,

which reduces to

$$7s - 1 = (A + B + C)s^2 + (-A - 2B + 3C)s + (-6A - 3B + 2C).$$

Equating the coefficients of s^2, s, and 1 gives the system of linear equations

$$A + B + C = 0,$$
$$-A - 2B + 3C = 7,$$
$$-6A - 3B + 2C = -1.$$

Solving this system yields $A = 2$, $B = -3$, and $C = 1$. Hence

(8) $$\frac{7s - 1}{(s + 1)(s + 2)(s - 3)} = \frac{2}{s + 1} - \frac{3}{s + 2} + \frac{1}{s - 3}.$$

An alternative method for finding the constants A, B, and C from (7) is to choose three values for s and substitute them into (7) to obtain three linear equations in the three unknowns. If we are careful in our choice of the values for s, the system is easy to solve. In this case, equation (7) obviously simplifies if $s = -1$, -2, or 3. Putting $s = -1$ gives

$$-7 - 1 = A(1)(-4) + B(0) + C(0),$$
$$-8 = -4A.$$

Hence $A = 2$. Next, setting $s = -2$ gives

$$-14 - 1 = A(0) + B(-1)(-5) + C(0),$$
$$-15 = 5B,$$

and so $B = -3$. Finally, letting $s = 3$, we similarly find that $C = 1$. In the case of nonrepeated linear factors, the alternative method is easier to use.

Now that we have obtained the partial fraction expansion (8), we use linearity to compute

$$\mathscr{L}^{-1}\left\{\frac{7s - 1}{(s + 1)(s + 2)(s - 3)}\right\}(t) = \mathscr{L}^{-1}\left\{\frac{2}{s + 1} - \frac{3}{s + 2} + \frac{1}{s - 3}\right\}(t)$$

$$= 2\mathscr{L}^{-1}\left\{\frac{1}{s + 1}\right\}(t) - 3\mathscr{L}^{-1}\left\{\frac{1}{s + 2}\right\}(t)$$

$$+ \mathscr{L}^{-1}\left\{\frac{1}{s - 3}\right\}(t)$$

$$= 2e^{-t} - 3e^{-2t} + e^{3t}. \quad \blacksquare \text{ End}$$

2. Repeated Linear Factors

Let $s - r$ be a factor of $Q(s)$ and suppose $(s - r)^m$ is the highest power of $s - r$ that divides $Q(s)$. Then the portion of the partial fraction expansion of $P(s)/Q(s)$ that corresponds to the term $(s - r)^m$ is

$$\frac{A_1}{s - r} + \frac{A_2}{(s - r)^2} + \cdots + \frac{A_m}{(s - r)^m},$$

where the A_i's are real numbers.

EXAMPLE 6 Determine $\mathscr{L}^{-1}\left\{\dfrac{s^2 + 9s + 2}{(s - 1)^2(s + 3)}\right\}$.

Solution Since $s - 1$ is a repeated linear factor with multiplicity two and $s + 3$ is a nonrepeated linear factor, the partial fraction expansion has the form

$$\frac{s^2 + 9s + 2}{(s - 1)^2(s + 3)} = \frac{A}{s - 1} + \frac{B}{(s - 1)^2} + \frac{C}{s + 3}.$$

We begin by multiplying both sides by $(s - 1)^2(s + 3)$ to obtain

(9) $s^2 + 9s + 2 = A(s - 1)(s + 3) + B(s + 3) + C(s - 1)^2.$

Now observe that when we set $s = 1$ (or $s = -3$), two terms on the right-hand side of (9) vanish, leaving a linear equation that we can solve for B (or C). Setting $s = 1$ in (9) gives

$$1 + 9 + 2 = A(0) + 4B + C(0),$$
$$12 = 4B,$$

and hence $B = 3$. Similarly, setting $s = -3$ in (9) gives

$$9 - 27 + 2 = A(0) + B(0) + 16C$$
$$-16 = 16C.$$

Thus $C = -1$. Finally, to find A, we pick a different value for s, say $s = 0$. Then, since $B = 3$ and $C = -1$, plugging $s = 0$ into (9) yields

$$2 = -3A + 3B + C = -3A + 9 - 1$$

so that $A = 2$. Hence

(10) $\dfrac{s^2 + 9s + 2}{(s - 1)^2(s + 3)} = \dfrac{2}{s - 1} + \dfrac{3}{(s - 1)^2} - \dfrac{1}{s + 3}.$

We could also have determined the constants A, B, and C by first rewriting equation (9) in the form

$$s^2 + 9s + 2 = (A + C)s^2 + (2A + B - 2C)s + (-3A + 3B + C).$$

Then, equating the corresponding coefficients of s^2, s, and 1 and solving the resulting system, we again find $A = 2$, $B = 3$, and $C = -1$.

Now that we have derived the partial fraction expansion (10) for the given rational function, we can determine its inverse Laplace transform:

$$\mathcal{L}^{-1}\left\{\frac{s^2 + 9s + 2}{(s - 1)^2(s + 3)}\right\}(t) = \mathcal{L}^{-1}\left\{\frac{2}{s - 1} + \frac{3}{(s - 1)^2} - \frac{1}{s + 3}\right\}(t)$$

$$= 2\mathcal{L}^{-1}\left\{\frac{1}{s - 1}\right\}(t) + 3\mathcal{L}^{-1}\left\{\frac{1}{(s - 1)^2}\right\}(t)$$

$$- \mathcal{L}^{-1}\left\{\frac{1}{s + 3}\right\}(t)$$

$$= 2e^t + 3te^t - e^{-3t}. \quad \blacksquare \text{ End}$$

3. Quadratic Factors

Let $(s - \alpha)^2 + \beta^2$ be a quadratic factor of $Q(s)$ that cannot be reduced to linear factors with real coefficients. Suppose m is the highest power of $(s - \alpha)^2 + \beta^2$ that divides $Q(s)$. Then the portion of the partial fraction expansion that corresponds to $(s - \alpha)^2 + \beta^2$ is

$$\frac{C_1 s + D_1}{(s - \alpha)^2 + \beta^2} + \frac{C_2 s + D_2}{[(s - \alpha)^2 + \beta^2]^2} + \cdots + \frac{C_m s + D_m}{[(s - \alpha)^2 + \beta^2]^m}.$$

As we saw in Example 4, it is more convenient to express $C_i s + D_i$ in the form $A_i(s - \alpha) + \beta B_i$ when we look up the inverse Laplace transforms. So let's agree to write this portion of the partial fraction expansion in the equivalent form

$$\frac{A_1(s - \alpha) + \beta B_1}{(s - \alpha)^2 + \beta^2} + \frac{A_2(s - \alpha) + \beta B_2}{[(s - \alpha)^2 + \beta^2]^2} + \cdots + \frac{A_m(s - \alpha) + \beta B_m}{[(s - \alpha)^2 + \beta^2]^m}.$$

EXAMPLE 7 Determine $\mathcal{L}^{-1}\left\{\dfrac{2s^2 + 10s}{(s^2 - 2s + 5)(s + 1)}\right\}$.

Solution We first observe that the quadratic factor $s^2 - 2s + 5$ is irreducible (check the sign of the discriminant in the quadratic formula). Next we write the quadratic in the form $(s - \alpha)^2 + \beta^2$ by completing the square:

$$s^2 - 2s + 5 = (s - 1)^2 + 2^2.$$

Since $s^2 - 2s + 5$ and $s + 1$ are nonrepeated factors, the partial fraction expansion has the form

$$\frac{2s^2 + 10s}{(s^2 - 2s + 5)(s + 1)} = \frac{A(s - 1) + 2B}{(s - 1)^2 + 2^2} + \frac{C}{s + 1}.$$

When we multiply both sides by the common denominator, we obtain

(11) $2s^2 + 10s = [A(s - 1) + 2B](s + 1) + C(s^2 - 2s + 5).$

In equation (11), let's put $s = -1$, 1, and 0. With $s = -1$, we find

$$2 - 10 = [A(-2) + 2B](0) + C(8),$$
$$-8 = 8C,$$

and hence $C = -1$. With $s = 1$ in (11), we obtain

$$2 + 10 = [A(0) + 2B](2) + C(4),$$

and since $C = -1$, the last equation becomes $12 = 4B - 4$. Thus $B = 4$. Finally, setting $s = 0$ in (11) and using $C = -1$ and $B = 4$ gives

$$0 = [A(-1) + 2B](1) + C(5),$$
$$0 = -A + 8 - 5,$$
$$A = 3.$$

Hence $A = 3$, $B = 4$, and $C = -1$ so that

$$\frac{2s^2 + 10s}{(s^2 - 2s + 5)(s + 1)} = \frac{3(s - 1) + 2(4)}{(s - 1)^2 + 2^2} - \frac{1}{s + 1}.$$

With this partial fraction expansion in hand, we can immediately determine the inverse Laplace transform:

$$\mathscr{L}^{-1}\left\{\frac{2s^2 + 10s}{(s^2 - 2s + 5)(s + 1)}\right\}(t) = \mathscr{L}^{-1}\left\{\frac{3(s - 1) + 2(4)}{(s - 1)^2 + 2^2} - \frac{1}{s + 1}\right\}(t)$$

$$= 3\mathscr{L}^{-1}\left\{\frac{s - 1}{(s - 1)^2 + 2^2}\right\}(t)$$

$$+ 4\mathscr{L}^{-1}\left\{\frac{2}{(s - 1)^2 + 2^2}\right\}(t) - \mathscr{L}^{-1}\left\{\frac{1}{s + 1}\right\}(t)$$

$$= 3e^t \cos 2t + 4e^t \sin 2t - e^{-t}. \quad \blacksquare \text{ End}$$

In Section 7.7, we discuss a different method (involving convolutions) for computing inverse transforms that does not require partial fraction decompositions. Moreover, the convolution method is convenient in the case of a rational function with a repeated quadratic factor in the denominator. Other helpful tools are described in Problems 33–36 and 38–43.

EXERCISES 7.4

In Problems 1 through 10, determine the inverse Laplace transform of the given function.

1. $\dfrac{6}{(s-1)^4}$.

2. $\dfrac{2}{s^2+4}$.

3. $\dfrac{s+1}{s^2+2s+10}$.

4. $\dfrac{4}{s^2+9}$.

5. $\dfrac{1}{s^2+4s+8}$.

6. $\dfrac{3}{(2s+5)^3}$.

7. $\dfrac{2s+16}{s^2+4s+13}$.

8. $\dfrac{1}{s^5}$.

9. $\dfrac{3s-15}{2s^2-4s+10}$.

10. $\dfrac{s-1}{2s^2+s+6}$.

In Problems 11 through 20, determine the partial fraction expansion for the given rational function.

11. $\dfrac{s^2-26s-47}{(s-1)(s+2)(s+5)}$.

12. $\dfrac{-s-7}{(s+1)(s-2)}$.

13. $\dfrac{-2s^2-3s-2}{s(s+1)^2}$.

14. $\dfrac{-8s^2-5s+9}{(s+1)(s^2-3s+2)}$.

15. $\dfrac{8s-2s^2-14}{(s+1)(s^2-2s+5)}$.

16. $\dfrac{-5s-36}{(s+2)(s^2+9)}$.

17. $\dfrac{3s+5}{s(s^2+s-6)}$.

18. $\dfrac{3s^2+5s+3}{s^4+s^3}$.

19. $\dfrac{1}{(s-3)(s^2+2s+2)}$.

20. $\dfrac{s}{(s-1)(s^2-1)}$.

In Problems 21 through 30, determine $\mathscr{L}^{-1}\{F\}$.

21. $F(s)=\dfrac{6s^2-13s+2}{s(s-1)(s-6)}$.

22. $F(s)=\dfrac{s+11}{(s-1)(s+3)}$.

23. $F(s)=\dfrac{5s^2+34s+53}{(s+3)^2(s+1)}$.

24. $F(s)=\dfrac{7s^2-41s+84}{(s-1)(s^2-4s+13)}$.

25. $F(s)=\dfrac{7s^2+23s+30}{(s-2)(s^2+2s+5)}$.

26. $F(s)=\dfrac{7s^3-2s^2-3s+6}{s^3(s-2)}$.

27. $s^2F(s)-4F(s)=\dfrac{5}{s+1}$.

28. $s^2F(s)+sF(s)-6F(s)=\dfrac{s^4+4}{s^2+s}$.

29. $sF(s)+2F(s)=\dfrac{10s^2+12s+14}{s^2-2s+2}$.

30. $sF(s)-F(s)=\dfrac{2s+5}{s^2+2s+1}$.

31. Determine the Laplace transform of each of the following functions:

(a) $f_1(t)=\begin{cases}0, & t=2,\\ t, & t\neq 2.\end{cases}$

(b) $f_2(t)=\begin{cases}5, & t=1,\\ 2, & t=6,\\ t, & t\neq 1,\,6.\end{cases}$

(c) $f_3(t)=t$.

Which of the preceding functions is the inverse Laplace transform of $1/s^2$?

32. Determine the Laplace transform of each of the following functions:

(a) $f_1(t)=\begin{cases}t, & t=1,2,3,\dots,\\ e^t, & t\neq 1,2,3,\dots.\end{cases}$

(b) $f_2(t)=\begin{cases}e^t, & t\neq 5,8,\\ 6, & t=5,\\ 0, & t=8.\end{cases}$

(c) $f_3(t)=e^t$.

Which of the preceding functions is the inverse Laplace transform of $1/(s-1)$?

Theorem 6 in Section 7.3 can be expressed in terms of the inverse Laplace transform as

$$\mathscr{L}^{-1}\left\{\dfrac{d^nF}{ds^n}\right\}(t)=(-t)^nf(t),$$

where $f=\mathscr{L}^{-1}\{F\}$. Use this equation in Problems 33 through 36 to compute $\mathscr{L}^{-1}\{F\}$.

33. $F(s)=\ln\left(\dfrac{s+2}{s-5}\right)$.

34. $F(s)=\ln\left(\dfrac{s-4}{s-3}\right)$.

35. $F(s)=\ln\left(\dfrac{s^2+9}{s^2+1}\right)$.

36. $F(s)=\arctan(1/s)$.

37. Prove Theorem 7 on the linearity of the inverse transform. [*Hint:* Show that the right-hand side of equation (3) is a continuous function on $[0, \infty)$ whose Laplace transform is $F_1(s) + F_2(s)$.]

38. Residue Computation. Let $P(s)/Q(s)$ be a rational function with deg $P <$ deg Q and suppose $s - r$ is a nonrepeated linear factor of $Q(s)$. Prove that the portion of the partial fraction expansion of $P(s)/Q(s)$ corresponding to $s - r$ is

$$\frac{A}{s - r},$$

where A (called the **residue**) is given by the formula

$$A = \lim_{s \to r} \frac{(s - r)P(s)}{Q(s)}.$$

39. Use the residue computation formula derived in Problem 38 to determine quickly the partial fraction expansion for

$$F(s) = \frac{2s + 1}{s(s - 1)(s + 2)}.$$

40. Heaviside's Expansion Formula.[†] Let $P(s)$ and $Q(s)$ be polynomials with the degree of $P(s)$ less than the degree of $Q(s)$. Let

$$Q(s) = (s - r_1)(s - r_2) \cdots (s - r_n),$$

where the r_i's are distinct real numbers. Show that

$$\mathcal{L}^{-1}\left\{\frac{P}{Q}\right\}(t) = \sum_{i=1}^{n} \frac{P(r_i)}{Q'(r_i)} e^{r_i t}.$$

41. Use Heaviside's expansion formula derived in Problem 40 to determine the inverse Laplace transform of

$$F(s) = \frac{3s^2 - 16s + 5}{(s + 1)(s - 3)(s - 2)}.$$

42. Complex Residues. Let $P(s)/Q(s)$ be a rational function with deg $P <$ deg Q and suppose $(s - \alpha)^2 + \beta^2$ is a nonrepeated quadratic factor of Q. (That is, $\alpha \pm i\beta$ are complex conjugate zeros of Q.) Prove that the portion of the partial fraction expansion of $P(s)/Q(s)$ corresponding to $(s - \alpha)^2 + \beta^2$ is

$$\frac{A(s - \alpha) + \beta B}{(s - \alpha)^2 + \beta^2},$$

where the **complex residue** $\beta B + i\beta A$ is given by the formula

$$\beta B + i\beta A = \lim_{s \to \alpha + i\beta} \frac{[(s - \alpha)^2 + \beta^2]P(s)}{Q(s)}.$$

(Thus we can determine B and A by taking the real and imaginary parts of the limit and dividing them by β.)

43. Use the residue formulas derived in Problems 38 and 42 to determine the partial fraction expansion for

$$F(s) = \frac{6s^2 + 28}{(s^2 - 2s + 5)(s + 2)}.$$

7.5 SOLVING INITIAL VALUE PROBLEMS

Our goal is to show how Laplace transforms can be used to solve initial value problems for linear differential equations. Recall that we have already studied ways of solving such initial value problems in Chapter 4. These previous methods required that we first find a *general solution* of the differential equation and then use the initial conditions to determine the desired solution. As we shall see, the method of Laplace transforms leads to the solution of the initial value problem *without* first finding a general solution.

Other advantages to the transform method are worth noting. For example, the technique can easily handle equations involving discontinuous forcing functions as illustrated in Section 7.1. Further, the method can be used for certain linear differential equations with variable coefficients, a special class of integral equations, systems of differential equations, and partial differential equations.

[†] *Historical Footnote:* This formula played an important role in the "operational solution" to ordinary differential equations developed by Oliver Heaviside in the 1890s.

METHOD OF LAPLACE TRANSFORMS

To solve an initial value problem:

(a) Take the Laplace transform of both sides of the equation.
(b) Use the properties of the Laplace transform and the initial conditions to obtain an equation for the Laplace transform of the solution and then solve this equation for the transform.
(c) Determine the inverse Laplace transform of the solution by looking it up in a table or by using a suitable method (such as partial fractions) in combination with the table.

In step (a), we are tacitly assuming the solution is piecewise continuous on $[0, \infty)$ and of exponential order. Once we have obtained the inverse Laplace transform in step (c), we can verify that these tacit assumptions are satisfied.

EXAMPLE 1 Solve the initial value problem

$$(1) \qquad y'' - 2y' + 5y = -8e^{-t}; \qquad y(0) = 2, \qquad y'(0) = 12.$$

Solution The differential equation in (1) is an identity between two functions of t. Hence equality holds for the Laplace transforms of these functions:

$$\mathscr{L}\{y'' - 2y' + 5y\} = \mathscr{L}\{-8e^{-t}\}.$$

Using the linearity property of $\mathscr{L}$ and the previously computed transform of the exponential function, we can write

$$(2) \qquad \mathscr{L}\{y''\}(s) - 2\mathscr{L}\{y'\}(s) + 5\mathscr{L}\{y\}(s) = \frac{-8}{s+1}.$$

Now let $Y(s) := \mathscr{L}\{y\}(s)$. From the formulas for the Laplace transform of higher-order derivatives (see Section 7.3) and the initial conditions in (1), we find

$$\mathscr{L}\{y'\}(s) = sY(s) - y(0) = sY(s) - 2,$$
$$\mathscr{L}\{y''\}(s) = s^2Y(s) - sy(0) - y'(0) = s^2Y(s) - 2s - 12.$$

Substituting these expressions into (2) and solving for $Y(s)$ yields

$$[s^2Y(s) - 2s - 12] - 2[sY(s) - 2] + 5Y(s) = \frac{-8}{s+1},$$

$$(s^2 - 2s + 5)Y(s) = 2s + 8 - \frac{8}{s+1},$$

$$(s^2 - 2s + 5)Y(s) = \frac{2s^2 + 10s}{s+1},$$

$$Y(s) = \frac{2s^2 + 10s}{(s^2 - 2s + 5)(s+1)}.$$

Our remaining task is to compute the inverse transform of the rational function $Y(s)$. This was done in Example 7 of Section 7.4, where, using a partial fraction expansion, we found

(3) $y(t) = 3e^t \cos 2t + 4e^t \sin 2t - e^{-t}$,

which is the solution to the initial value problem (1). ∎ End

As a quick check on the accuracy of our computations, the reader is advised to verify that the computed solution satisfies the given initial conditions.

The reader is probably questioning the wisdom of using the Laplace transform method to solve an initial value problem that can be easily handled by the methods discussed in Chapter 4. The objective of the first few examples in this section is simply to make the reader familiar with the Laplace transform procedure. We will see in Example 4 and in later sections that the method is applicable to problems that cannot be readily handled by the techniques discussed in the previous chapters.

EXAMPLE 2 Solve the initial value problem

(4) $y'' + 4y' - 5y = te^t$; $y(0) = 1$, $y'(0) = 0$.

Solution Let $Y(s) := \mathcal{L}\{y\}(s)$. Taking the Laplace transform of both sides of the differential equation in (4) gives

(5) $\mathcal{L}\{y''\}(s) + 4\mathcal{L}\{y'\}(s) - 5Y(s) = \dfrac{1}{(s-1)^2}$.

Using the initial conditions, we can express $\mathcal{L}\{y'\}(s)$ and $\mathcal{L}\{y''\}(s)$ in terms of $Y(s)$. That is,

$$\mathcal{L}\{y'\}(s) = sY(s) - y(0) = sY(s) - 1,$$
$$\mathcal{L}\{y''\}(s) = s^2 Y(s) - sy(0) - y'(0) = s^2 Y(s) - s.$$

Substituting back into (5) and solving for $Y(s)$ gives

$$[s^2 Y(s) - s] + 4[sY(s) - 1] - 5Y(s) = \frac{1}{(s-1)^2},$$

$$(s^2 + 4s - 5)Y(s) = s + 4 + \frac{1}{(s-1)^2},$$

$$(s+5)(s-1)Y(s) = \frac{s^3 + 2s^2 - 7s + 5}{(s-1)^2},$$

$$Y(s) = \frac{s^3 + 2s^2 - 7s + 5}{(s+5)(s-1)^3}.$$

The partial fraction expansion for $Y(s)$ has the form

(6) $\dfrac{s^3 + 2s^2 - 7s + 5}{(s+5)(s-1)^3} = \dfrac{A}{s+5} + \dfrac{B}{s-1} + \dfrac{C}{(s-1)^2} + \dfrac{D}{(s-1)^3}$.

Solving for the numerators, we ultimately obtain $A = 35/216$, $B = 181/216$, $C = -1/36$, and $D = 1/6$. Substituting these values into (6) gives

$$Y(s) = \frac{35}{216}\left(\frac{1}{s+5}\right) + \frac{181}{216}\left(\frac{1}{s-1}\right) - \frac{1}{36}\left(\frac{1}{(s-1)^2}\right) + \frac{1}{12}\left(\frac{2}{(s-1)^3}\right),$$

where we have written $D = 1/6 = (1/12)2$ to facilitate the final step of taking the inverse transform. From the tables, we now obtain

$$(7) \qquad y(t) = \frac{35}{216}e^{-5t} + \frac{181}{216}e^t - \frac{1}{36}te^t + \frac{1}{12}t^2e^t$$

as the solution to the initial value problem (4). ■ End

EXAMPLE 3 Solve the initial value problem

$$(8) \qquad w''(t) - 2w'(t) + 5w(t) = -8e^{\pi-t}; \qquad w(\pi) = 2, \qquad w'(\pi) = 12.$$

Solution To use the method of Laplace transforms, we first move the initial conditions to $t = 0$. This can be done by setting $y(t) := w(t + \pi)$. Then

$$y'(t) = w'(t + \pi), \qquad y''(t) = w''(t + \pi).$$

Replacing t by $t + \pi$ in the differential equation in (8), we have

$$(9) \qquad w''(t + \pi) - 2w'(t + \pi) + 5w(t + \pi) = -8e^{\pi-(t+\pi)} = -8e^{-t}.$$

Substituting $y(t) = w(t + \pi)$ in (9), the initial value problem in (8) becomes

$$y''(t) - 2y'(t) + 5y(t) = -8e^{-t}; \qquad y(0) = 2, \qquad y'(0) = 12.$$

Because the initial conditions are now given at the origin, the Laplace transform method is applicable. In fact, we carried out the procedure in Example 1, where we found

$$(10) \qquad y(t) = 3e^t \cos 2t + 4e^t \sin 2t - e^{-t}.$$

Since $w(t + \pi) = y(t)$, then $w(t) = y(t - \pi)$. Hence, replacing t by $t - \pi$ in (10) gives

$$w(t) = y(t - \pi) = 3e^{t-\pi}\cos 2(t - \pi) + 4e^{t-\pi}\sin 2(t - \pi) - e^{-(t-\pi)}$$
$$= 3e^{t-\pi}\cos 2t + 4e^{t-\pi}\sin 2t - e^{\pi-t}. ■ End$$

Thus far we have applied the Laplace transform method only to linear equations with constant coefficients. Yet several important equations in mathematical physics involve linear equations whose coefficients are polynomials in t. To solve such equations using Laplace transforms, we shall apply Theorem 6, page 367, where we proved that

$$(11) \qquad \mathcal{L}\{t^n f(t)\}(s) = (-1)^n \frac{d^n F}{ds^n}(s).$$

If we let $n = 1$ and $f(t) = y'(t)$, we find

$$\mathcal{L}\{ty'(t)\}(s) = -\frac{d}{ds}\mathcal{L}\{y'\}(s)$$

$$= -\frac{d}{ds}[sY(s) - y(0)] = -sY'(s) - Y(s).$$

Similarly, with $n = 1$ and $f(t) = y''(t)$, we obtain from (11)

$$\mathcal{L}\{ty''(t)\}(s) = -\frac{d}{ds}\mathcal{L}\{y''\}(s)$$

$$= -\frac{d}{ds}[s^2Y(s) - sy(0) - y'(0)]$$

$$= -s^2Y'(s) - 2sY(s) + y(0).$$

Thus we see that for a linear differential equation in $y(t)$ whose coefficients are polynomials in t, the method of Laplace transforms will convert the given equation into a linear differential equation in $Y(s)$ whose coefficients are polynomials in s. Moreover, if the coefficients of the given equation are polynomials of degree one in t, then (regardless of the order of the given equation) the differential equation for $Y(s)$ is just a linear *first order* equation. Since we know how to solve this first order equation, the only serious obstacle we may encounter is obtaining the inverse Laplace transform of $Y(s)$. (This problem may be insurmountable, since the solution $y(t)$ may *not* have a Laplace transform.)

In illustrating the technique, we make use of the following fact. *If $f(t)$ is piecewise continuous on $[0, \infty)$ and of exponential order, then*

(12) $$\lim_{s\to\infty} \mathcal{L}\{f\}(s) = 0.$$

(The reader may have already guessed this from the entries in Table 7.1, page 362.) An outline of the proof of (12) is given in Exercises 7.3, Problem 26.

EXAMPLE 4 Solve the initial value problem

(13) $y'' + 2ty' - 4y = 1, \qquad y(0) = y'(0) = 0.$

Solution Let $Y(s) = \mathcal{L}\{y\}(s)$ and take the Laplace transform of both sides of the equation in (13):

(14) $$\mathcal{L}\{y''\}(s) + 2\mathcal{L}\{ty'(t)\}(s) - 4Y(s) = \frac{1}{s}.$$

Using the initial conditions, we find

$$\mathcal{L}\{y''\}(s) = s^2Y(s) - sy(0) - y'(0) = s^2Y(s)$$

and

$$\mathcal{L}\{ty'(t)\}(s) = -\frac{d}{ds}\mathcal{L}\{y'\}(s)$$

$$= -\frac{d}{ds}[sY(s) - y(0)] = -sY'(s) - Y(s).$$

Substituting these expressions into (14) gives

$$s^2 Y(s) + 2[-sY'(s) - Y(s)] - 4Y(s) = \frac{1}{s},$$

$$-2sY'(s) + (s^2 - 6)Y(s) = \frac{1}{s},$$

(15)
$$Y'(s) + \left(\frac{3}{s} - \frac{s}{2}\right)Y(s) = \frac{-1}{2s^2}.$$

Equation (15) is a linear first order equation and has the integrating factor

$$\mu(s) = e^{\int (3/s - s/2)\, ds} = e^{\ln s^3 - s^2/4} = s^3 e^{-s^2/4}$$

(see Section 2.3). Multiplying (15) by $\mu(s)$, we obtain

$$\frac{d}{ds}\{\mu(s)Y(s)\} = \frac{d}{ds}\{s^3 e^{-s^2/4} Y(s)\} = -\frac{s}{2} e^{-s^2/4}.$$

Integrating and solving for $Y(s)$ yields

$$s^3 e^{-s^2/4} Y(s) = -\int \frac{s}{2} e^{-s^2/4}\, ds = e^{-s^2/4} + C,$$

(16)
$$Y(s) = \frac{1}{s^3} + C \frac{e^{s^2/4}}{s^3}.$$

Now if $Y(s)$ is the Laplace transform of a piecewise continuous function of exponential order, then it follows from equation (12) that

$$\lim_{s \to \infty} Y(s) = 0.$$

For this to occur, the constant C in equation (16) must be zero. Hence $Y(s) = 1/s^3$, and taking the inverse transform gives $y(t) = t^2/2$. We can easily verify that $y(t) = t^2/2$ is the solution to the given initial value problem by substituting it into (13). ∎ End

We end this section with an application from **control theory.** Let's consider a servomechanism that models an automatic pilot. Such a mechanism applies a torque to the steering control shaft so that a plane or boat will follow a prescribed course. If we let $y(t)$ be the true direction (angle) of the craft at time t and $g(t)$ be the desired direction at time t, then

$$e(t) := y(t) - g(t)$$

denotes the **error** or **deviation** between the desired direction and the true direction.

Let's assume the servomechanism can measure the error $e(t)$ and feed back to the steering shaft a component of torque that is proportional to $e(t)$ but opposite in sign (see Figure 7.6 on page 386.) Newton's second law, expressed in terms of torques, states that

(moment of inertia) × (angular acceleration) = total torque.

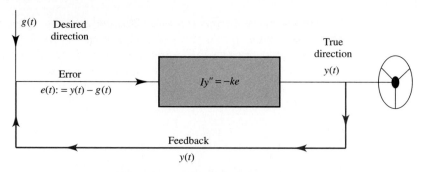

Figure 7.6 Servomechanism with feedback

For the servomechanism described, this becomes

(17) $Iy''(t) = -ke(t),$

where I is the moment of inertia of the steering shaft and k is a positive proportionality constant.

EXAMPLE 5 Determine the error $e(t)$ for the automatic pilot if the steering shaft is initially at rest in the zero direction and the desired direction is given by $g(t) = at$, where a is a constant.

Solution Based on the discussion leading to equation (17), a model for the mechanism is given by the initial value problem

(18) $Iy''(t) = -ke(t);$ $y(0) = 0,$ $y'(0) = 0,$

where $e(t) = y(t) - g(t) = y(t) - at$. We begin by taking the Laplace transform of both sides of (18):

$$I\mathscr{L}\{y''\}(s) = -k\mathscr{L}\{e\}(s),$$
$$I[s^2Y(s) - sy(0) - y'(0)] = -kE(s),$$
(19) $$s^2IY(s) = -kE(s),$$

where $Y(s) = \mathscr{L}\{y\}(s)$ and $E(s) = \mathscr{L}\{e\}(s)$. Since

$$E(s) = \mathscr{L}\{y(t) - at\}(s) = Y(s) - \mathscr{L}\{at\}(s) = Y(s) - as^{-2},$$

we find from (19) that

$$s^2IE(s) + aI = -kE(s).$$

Solving this equation for $E(s)$ gives

$$E(s) = -\frac{aI}{s^2I + k} = \frac{-a}{\sqrt{k/I}}\frac{\sqrt{k/I}}{s^2 + k/I}.$$

Hence, on taking the inverse Laplace transform, we obtain the error

(20) $$e(t) = -\frac{a}{\sqrt{k/I}}\sin(\sqrt{k/I}t).$$ ■ End

As we can see from equation (20), the automatic pilot will oscillate back and forth about the desired course, always "oversteering" by the factor $a\sqrt{k/I}$. Clearly we can make the error small by making k large relative to I, but then the term $\sqrt{k/I}$ becomes large, causing the error to oscillate more rapidly. (See Figure 7.7.) As with vibrations, the oscillations or oversteering can be controlled by introducing a damping torque proportional to $e'(t)$ but opposite in sign (see Problem 40).

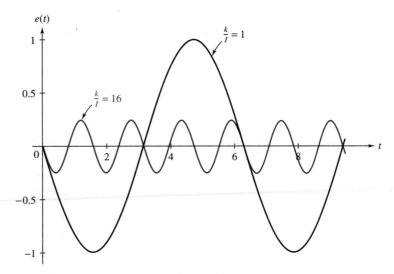

Figure 7.7 Error for automatic pilot when $k/I = 1$ and when $k/I = 16$

EXERCISES 7.5

In Problems 1 through 14, solve the given initial value problem using the method of Laplace transforms.

1. $y'' - 2y' + 5y = 0$; $y(0) = 2$, $y'(0) = 4$.

2. $y'' - y' - 2y = 0$; $y(0) = -2$, $y'(0) = 5$.

3. $y'' + 6y' + 9y = 0$; $y(0) = -1$, $y'(0) = 6$.

4. $y'' + 6y' + 5y = 12e^t$; $y(0) = -1$, $y'(0) = 7$.

5. $w'' + w = t^2 + 2$; $w(0) = 1$, $w'(0) = -1$.

6. $y'' - 4y' + 5y = 4e^{3t}$; $y(0) = 2$, $y'(0) = 7$.

7. $y'' - 7y' + 10y = 9 \cos t + 7 \sin t$;
$y(0) = 5$, $y'(0) = -4$.

8. $y'' + 4y = 4t^2 - 4t + 10$; $y(0) = 0$, $y'(0) = 3$.

9. $z'' + 5z' - 6z = 21e^t$; $z(0) = -1$, $z'(0) = 9$.

10. $y'' - 4y = 4t - 8e^{-2t}$; $y(0) = 0$, $y'(0) = 5$.

11. $y'' - y = t - 2$; $y(2) = 3$, $y'(2) = 0$.

12. $w'' - 2w' + w = 6t - 2$; $w(-1) = 3$; $w'(-1) = 7$.

13. $y'' - y' - 2y = -8 \cos t - 2 \sin t$;
$y(\pi/2) = 1$, $y'(\pi/2) = 0$.

14. $y'' + y = t$; $y(\pi) = 0$, $y'(\pi) = 0$.

In Problems 15 through 24, solve for $Y(s)$, the Laplace transform of the solution $y(t)$ to the given initial value problem.

15. $y'' - 3y' + 2y = \cos t$; $y(0) = 0$, $y'(0) = -1$.

16. $y'' + 6y = t^2 - 1$; $y(0) = 0$, $y'(0) = -1$.

17. $y'' + y' - y = t^3$; $y(0) = 1$, $y'(0) = 0$.

18. $y'' - 2y' - y = e^{2t} - e^t$; $y(0) = 1$, $y'(0) = 3$.

19. $y'' + 5y' - y = e^t - 1$; $y(0) = 1$, $y'(0) = 1$.

20. $y'' + 3y = t^3$; $y(0) = 0$, $y'(0) = 0$.

21. $y'' - 2y' + y = \cos t - \sin t$; $y(0) = 1$, $y'(0) = 3$.

22. $y'' - 6y' + 5y = te^t$; $y(0) = 2$, $y'(0) = -1$.

23. $y'' + 4y = g(t)$; $y(0) = -1$, $y'(0) = 0$, where

$$g(t) = \begin{cases} t, & t < 2, \\ 5, & t > 2. \end{cases}$$

24. $y'' - y = g(t)$; $y(0) = 1$, $y'(0) = 2$, where

$$g(t) = \begin{cases} 1, & t < 3, \\ t, & t > 3. \end{cases}$$

In Problems 25 through 28, solve the given third order initial value problem for y(t) using the method of Laplace transforms.

25. $y''' - y'' + y' - y = 0$;
$y(0) = 1$, $y'(0) = 1$, $y''(0) = 3$.

26. $y''' + 4y'' + y' - 6y = -12$;
$y(0) = 1$, $y'(0) = 4$, $y''(0) = -2$.

27. $y''' + 3y'' + 3y' + y = 0$;
$y(0) = -4$, $y'(0) = 4$, $y''(0) = -2$.

28. $y''' + y'' + 3y' - 5y = 16e^{-t}$;
$y(0) = 0$, $y'(0) = 2$, $y''(0) = -4$.

In Problems 29 through 32, use the method of Laplace transforms to find a general solution to the given differential equation by assuming y(0) = a and y'(0) = b, where a and b are arbitrary constants.

29. $y'' - 4y' + 3y = 0$. **30.** $y'' + 6y' + 5y = t$.

31. $y'' + 2y' + 2y = 5$. **32.** $y'' - 5y' + 6y = -6te^{2t}$.

33. Use Theorem 6 in Section 7.3 to show that
$$\mathcal{L}\{t^2 y'(t)\}(s) = sY''(s) + 2Y'(s),$$
where $Y(s) = \mathcal{L}\{y\}(s)$.

34. Use Theorem 6 in Section 7.3 to show that
$$\mathcal{L}\{t^2 y''(t)\}(s) = s^2 Y''(s) + 4sY'(s) + 2Y(s),$$
where $Y(s) = \mathcal{L}\{y\}(s)$.

In Problems 35 through 38, find solutions to the given initial value problem.

35. $y'' + 3ty' - 6y = 1$; $y(0) = 0$, $y'(0) = 0$.

36. $ty'' - ty' + y = 2$; $y(0) = 2$, $y'(0) = -1$.

37. $ty'' - 2y' + ty = 0$; $y(0) = 1$, $y'(0) = 0$.
[*Hint:* $\mathcal{L}^{-1}\{1/(s^2 + 1)^2\}(t) = (\sin t - t \cos t)/2$.]

38. $y'' + ty' - y = 0$; $y(0) = 0$, $y'(0) = 3$.

39. Determine the error $e(t)$ for the automatic pilot in Example 5 if the shaft is initially at rest in the zero direction and the desired direction is $g(t) = a$, where a is a constant.

40. In Example 5, assume that in order to control oscillations, a component of torque proportional to $e'(t)$, but opposite in sign, is also fed back to the steering shaft. Show that equation (17) is now replaced by
$$Iy''(t) = -ke(t) - \mu e'(t),$$
where μ is a positive constant. Determine the error $e(t)$ for the automatic pilot with mild damping (i.e., $\mu < 2\sqrt{Ik}$) if the steering shaft is initially at rest in the zero direction and the desired direction is given by $g(t) = a$, where a is a constant.

41. In Problem 40, determine the error $e(t)$ when the desired direction is given by $g(t) = at$, where a is a constant.

7.6 TRANSFORMS OF DISCONTINUOUS AND PERIODIC FUNCTIONS

In this section we study special functions that often arise when the method of Laplace transforms is applied to physical problems. Of particular interest are methods for handling functions with jump discontinuities. Jump discontinuities occur naturally in physical problems such as electric circuits with on/off switches. To handle such behavior, Oliver Heaviside introduced the following step function.

UNIT STEP FUNCTION

Definition 5. The **unit step function** $u(t)$ is defined by

$$(1) \qquad u(t) := \begin{cases} 0, & t < 0, \\ 1, & 0 < t. \end{cases}$$

By shifting the argument of $u(t)$, the jump can be moved to a different location. That is,

(2) $u(t - a) = \begin{cases} 0, & t - a < 0, \\ 1, & 0 < t - a \end{cases} = \begin{cases} 0, & t < a, \\ 1, & a < t \end{cases}$

has its jump at $t = a$. By multiplying by a constant M, the height of the jump can also be modified (see Figure 7.8):

$$Mu(t - a) = \begin{cases} 0, & t < a, \\ M, & a < t. \end{cases}$$

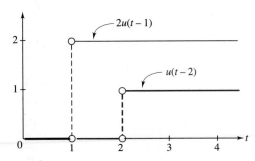

Figure 7.8 Two-step functions expressed using the unit step function

Many discontinuous functions can be expressed in terms of unit step functions. For example,

$$f(t) := \begin{cases} \sin t, & t < \pi, \\ t, & \pi < t \end{cases}$$

can be expressed as

$$f(t) = \sin t + \begin{cases} 0, & t < \pi, \\ t - \sin t, & \pi < t \end{cases}$$

$$= \sin t + (t - \sin t)u(t - \pi).$$

The Laplace transform of $u(t - a)$ with $a \geq 0$ is

(3) $\mathcal{L}\{u(t - a)\}(s) = \dfrac{e^{-as}}{s},$

since, for $s > 0$,

$$\mathcal{L}\{u(t - a)\}(s) = \int_0^\infty e^{-st}u(t - a)\,dt = \int_a^\infty e^{-st}\,dt$$

$$= \lim_{N \to \infty} \left. \frac{-e^{-st}}{s} \right|_a^N = \frac{e^{-as}}{s}.$$

Observe that $\mathcal{L}\{u(t)\} = \mathcal{L}\{1\}$ since $u(t) = 1$ for $t > 0$.

EXAMPLE 1 Find the Laplace transform of

(4) $$f(t) := \begin{cases} 3, & t < 2, \\ -1, & 2 < t < 5, \\ 7, & 5 < t. \end{cases}$$

(See Figure 7.9.)

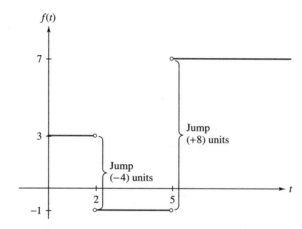

Figure 7.9 Graph of $f(t)$ in equation (4) showing jumps

Solution We could use the definition of the Laplace transform to compute $\mathscr{L}\{f\}$, but instead let's express $f(t)$ in terms of unit step functions and then use formula (3).

The function $f(t)$ equals 3 until t reaches 2, at which point $f(t)$ jumps to -1. We can express this jump by $3 - 4u(t - 2)$, since $u(t - 2)$ is zero until t reaches 2, after which it has the value 1. At $t = 5$, the function $f(t)$ jumps from -1 to 7. This can be expressed by adding on the term $8u(t - 5)$. Hence

$$f(t) = 3 - 4u(t - 2) + 8u(t - 5).$$

Finally, taking the Laplace transform and using formula (3), we find

$$\mathscr{L}\{f\}(s) = 3\,\mathscr{L}\{1\}(s) - 4\mathscr{L}\{u(t - 2)\}(s) + 8\mathscr{L}\{u(t - 5)\}(s)$$
$$= \frac{3}{s} - \frac{4e^{-2s}}{s} + \frac{8e^{-5s}}{s}. \quad \blacksquare \text{ End}$$

The translation property of $F(s)$ discussed in Section 7.3 described the effect on the Laplace transform of multiplying a function by e^{at}. The next theorem illustrates an analogous effect of multiplying the Laplace transform of a function by e^{-as}.

> **TRANSLATION IN *t***
>
> **Theorem 8.** Let $F(s) = \mathcal{L}\{f\}(s)$ exist for $s > \alpha \ge 0$. If a is a positive constant, then
>
> (5) $\mathcal{L}\{f(t - a)u(t - a)\}(s) = e^{-as}F(s),$
>
> and, if $f(t)$ is continuous on $[0, \infty)$, then[†]
>
> (6) $\mathcal{L}^{-1}\{e^{-as}F(s)\}(t) = f(t - a)u(t - a).$

Proof. By the definition of the Laplace transform, we have

$$(7) \qquad \mathcal{L}\{f(t - a)u(t - a)\}(s) = \int_0^\infty e^{-st}f(t - a)u(t - a)\,dt$$

$$= \int_a^\infty e^{-st}f(t - a)\,dt,$$

where, in the last equation, we used the fact that $u(t - a)$ is zero for $t < a$ and equals 1 for $t > a$. Now let $v = t - a$. Then we have $dv = dt$, and equation (7) becomes

$$\mathcal{L}\{f(t - a)u(t - a)\}(s) = \int_0^\infty e^{-as}e^{-sv}f(v)\,dv$$

$$= e^{-as}\int_0^\infty e^{-sv}f(v)\,dv = e^{-as}F(s). \quad \blacksquare \text{ End}$$

Notice that formula (5) includes as a special case the formula for $\mathcal{L}\{u(t - a)\}$; indeed, if we take $f(t) \equiv 1$, then $F(s) = 1/s$ and (5) becomes $\mathcal{L}\{u(t - a)\}(s) = e^{-as}/s$.

In practice it is more common to be faced with the problem of computing the transform of a function expressed as $g(t)u(t - a)$ rather than $f(t - a)u(t - a)$. To compute $\mathcal{L}\{g(t)u(t - a)\}$, we simply identify $g(t)$ with $f(t - a)$ so that $f(t) = g(t + a)$. Equation (5) then gives

$$(8) \qquad \mathcal{L}\{g(t)u(t - a)\}(s) = e^{-as}\mathcal{L}\{g(t + a)\}(s).$$

EXAMPLE 2 Determine the Laplace transform of $t^2 u(t - 1)$.

Solution To apply equation (8), we take $g(t) = t^2$ and $a = 1$. Then

$$g(t + a) = g(t + 1) = (t + 1)^2 = t^2 + 2t + 1.$$

Now the Laplace transform of $g(t + a)$ is

$$\mathcal{L}\{g(t + a)\}(s) = \mathcal{L}\{t^2 + 2t + 1\}(s) = \frac{2}{s^3} + \frac{2}{s^2} + \frac{1}{s}.$$

[†] Although $f(t)$ is continuous on $[0, \infty)$, the function $f(t - a)u(t - a)$ may have a discontinuity at $t = a$. In such cases, it can be shown that there is no continuous function on $[0, \infty)$ whose Laplace transform is $e^{-as}F(s)$.

So, by formula (8), we have

$$\mathcal{L}\{t^2 u(t-1)\}(s) = e^{-s}\left\{\frac{2}{s^3} + \frac{2}{s^2} + \frac{1}{s}\right\}. \quad \blacksquare \text{ End}$$

EXAMPLE 3 Determine $\mathcal{L}\{(\cos t)u(t-\pi)\}$.

Solution Here $g(t) = \cos t$ and $a = \pi$. Hence

$$g(t+a) = g(t+\pi) = \cos(t+\pi) = -\cos t,$$

and so the Laplace transform of $g(t+a)$ is

$$\mathcal{L}\{g(t+a)\}(s) = -\mathcal{L}\{\cos t\}(s) = -\frac{s}{s^2+1}.$$

Thus, from formula (8), we get

$$\mathcal{L}\{(\cos t)u(t-\pi)\}(s) = -e^{-\pi s}\frac{s}{s^2+1}. \quad \blacksquare \text{ End}$$

In Examples 2 and 3, we could also have computed the Laplace transform directly from the definition. In dealing with inverse transforms, however, we do not have a simple alternative formula[†] upon which to rely, and so formula (6) is especially useful whenever the transform has e^{-as} as a factor.

EXAMPLE 4 Determine $\mathcal{L}^{-1}\left\{\dfrac{e^{-2s}}{s^2}\right\}$ and sketch its graph.

Solution To use the translation property (6), we first express e^{-2s}/s^2 as the product $e^{-as}F(s)$. For this purpose, we put $e^{-as} = e^{-2s}$ and $F(s) = 1/s^2$. Thus $a = 2$ and

$$f(t) = \mathcal{L}^{-1}\left\{\frac{1}{s^2}\right\}(t) = t.$$

It now follows from the translation property that

$$\mathcal{L}^{-1}\left\{\frac{e^{-2s}}{s^2}\right\}(t) = f(t-2)u(t-2) = (t-2)u(t-2).$$

See Figure 7.10. $\blacksquare$ End

[†] Under certain conditions, the inverse transform is given by the contour integral

$$\mathcal{L}^{-1}\{F\}(t) = \frac{1}{2\pi i}\int_{\alpha-i\infty}^{\alpha+i\infty} e^{st}F(s)\,ds.$$

See, for example, *Complex Variables and the Laplace Transform for Engineers*, by Wilbur R. LePage (Dover Publications, New York, 1980) or *Fundamentals of Complex Analysis for Mathematics, Science, and Engineering* (Second Edition), by E. B. Saff and A. D. Snider (Prentice Hall, Englewood Cliffs, 1993).

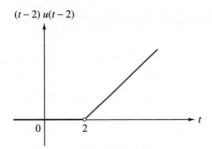

Figure 7.10 Graph of solution to Example 4

As illustrated by the next example, step functions arise in the modeling of on/off switches, changes in polarity, etc.

EXAMPLE 5 The current I in an LC series circuit is governed by the initial value problem

(9) $I''(t) + 4I(t) = g(t);$ $I(0) = 0,$ $I'(0) = 0,$

where

$$g(t) := \begin{cases} 1, & 0 < t < 1, \\ -1, & 1 < t < 2, \\ 0, & 2 < t. \end{cases}$$

Determine the current as a function of time t.

Solution Let $J(s) := \mathcal{L}\{I\}(s)$. Then we have $\mathcal{L}\{I''\}(s) = s^2 J(s)$. Expressing

$$g(t) = u(t) - 2u(t - 1) + u(t - 2),$$

we find that

$$\mathcal{L}\{g\}(s) = \frac{1}{s} - \frac{2e^{-s}}{s} + \frac{e^{-2s}}{s}.$$

Thus, when we take the Laplace transform of both sides of (9), we obtain

$$\mathcal{L}\{I''\}(s) + 4\mathcal{L}\{I\}(s) = \mathcal{L}\{g\}(s),$$

$$s^2 J(s) + 4J(s) = \frac{1}{s} - \frac{2e^{-s}}{s} + \frac{e^{-2s}}{s},$$

$$J(s) = \frac{1}{s(s^2 + 4)} - \frac{2e^{-s}}{s(s^2 + 4)} + \frac{e^{-2s}}{s(s^2 + 4)}.$$

To find $I = \mathcal{L}^{-1}\{J\}$, we first observe that

$$J(s) = F(s) - 2e^{-s}F(s) + e^{-2s}F(s),$$

where

$$F(s) := \frac{1}{s(s^2 + 4)} = \frac{1}{4}\left(\frac{1}{s}\right) - \frac{1}{4}\left(\frac{s}{s^2 + 4}\right).$$

Computing the inverse transform of $F(s)$ gives

$$f(t) := \mathcal{L}^{-1}\{F\}(t) = \frac{1}{4} - \frac{1}{4}\cos 2t.$$

Hence, via the translation property (6), we find

$$\begin{aligned}
I(t) &= \mathcal{L}^{-1}\{F(s) - 2e^{-s}F(s) + e^{-2s}F(s)\}(t) \\
&= f(t) - 2f(t-1)u(t-1) + f(t-2)u(t-2) \\
&= \left(\frac{1}{4} - \frac{1}{4}\cos 2t\right) - \left[\frac{1}{2} - \frac{1}{2}\cos 2(t-1)\right]u(t-1) \\
&\quad + \left[\frac{1}{4} - \frac{1}{4}\cos 2(t-2)\right]u(t-2).
\end{aligned}$$

The current is graphed in Figure 7.11. Note that $I(t)$ is smoother than $g(t)$; the former undergoes discontinuities in its second derivative at the points where the latter has jumps. ■ End

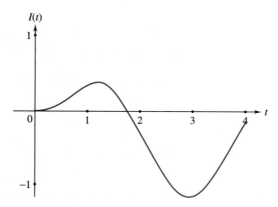

Figure 7.11 Solution to Example 5

Periodic functions are another class of functions that occur frequently in applications.

PERIODIC FUNCTION

Definition 6. A function $f(t)$ is said to be **periodic of period T** if

$$f(t + T) = f(t)$$

for all t in the domain of f.

As we know, the sine and cosine functions are periodic with period 2π and the tangent function is periodic with period π.[†] To specify a periodic function, it is sufficient to give its values over one period. For example, the square wave function in Figure 7.12 can be expressed as

(10) $\quad f(t) := \begin{cases} 1, & 0 < t < 1, \\ 0, & 1 < t < 2 \end{cases}$ and $f(t)$ has period 2.

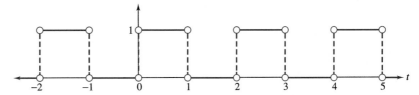

Figure 7.12 Graph of square wave function $f(t)$

It is convenient to introduce a notation for the "windowed" version of a periodic function $f(t)$:

(11) $\quad f_T(t) := \begin{cases} f(t), & 0 < t < T, \\ 0, & \text{otherwise.} \end{cases}$

Figure 7.13 Windowed version of periodic function

(See Figure 7.13.) The Laplace transform of $f_T(t)$ is given by

$$F_T(s) = \int_0^\infty e^{-st} f_T(t)\, dt = \int_0^T e^{-st} f(t)\, dt.$$

It is related to the Laplace transform of $f(t)$ as follows.

[†] A function that has period T will also have period $2T$, $3T$, etc. For example, the sine function has periods 2π, 4π, 6π, etc. Some authors refer to the smallest period as the **fundamental period** or just the period of the function.

TRANSFORM OF PERIODIC FUNCTION

Theorem 9. If f has period T and is piecewise continuous on $[0, T]$, then

$$(12) \qquad \mathscr{L}\{f\}(s) = \frac{F_T(s)}{1 - e^{-sT}} = \frac{\displaystyle\int_0^T e^{-st} f(t)\, dt}{1 - e^{-sT}}.$$

Proof. Observe that $f(t)$ can be written as a sum of translated copies of its windowed version:

$$(13) \qquad f(t) = f_T(t) + f_T(t - T)u(t - T) + f_T(t - 2T)u(t - 2T) + \cdots.$$

Therefore

$$\mathscr{L}\{f\} = \mathscr{L}\{f_T\} + \mathscr{L}\{f_T(t - T)u(t - T)\} + \mathscr{L}\{f_T(t - 2T)u(t - 2T)\} + \cdots,$$

so, by the translation property expressed in Theorem 8 (page 391),

$$\mathscr{L}\{f\} = \mathscr{L}\{f_T\} + \mathscr{L}\{f_T\}e^{-sT} + \mathscr{L}\{f_T\}e^{-2sT} + \cdots$$

$$= F_T(s)[1 + e^{-sT} + e^{-2sT} + \cdots].$$

Now the series in brackets is a geometric series with ratio $r = e^{-sT}$, which sums to $1/(1 - r)$ or $1/(1 - e^{-sT})$ for $s > 0$. Hence

$$\mathscr{L}\{f\} = F_T(s)/(1 - e^{-sT}). \qquad \blacksquare \text{ End}$$

EXAMPLE 6 Determine $\mathscr{L}\{f\}$, where f is the square wave function in (10).

Solution Here $T = 2$. From the formula (11), we express the windowed version of f as

$$f_T(t) = 1 - u(t - 1),$$

so

$$F_T(s) = \mathscr{L}\{1 - u(t - 1)\} = \frac{1}{s} - \frac{1}{s}e^{-s} = \frac{1}{s}(1 - e^{-s}).$$

Therefore, from formula (12), we obtain

$$\mathscr{L}\{f\} = \frac{1}{s}(1 - e^{-s})\frac{1}{1 - e^{-2s}} = \frac{1}{s(1 + e^{-s})}. \qquad \blacksquare \text{ End}$$

In the proof of Theorem 9, we determined the Laplace transform of $f(t)$ by writing $f(t)$ as a series of functions with known Laplace transforms. This approach can also be used for functions that have a power series expansion, since we know that

$$\mathscr{L}\{t^n\}(s) = \frac{n!}{s^{n+1}}, \qquad n = 0, 1, 2, \ldots.$$

EXAMPLE 7 Determine $\mathcal{L}\{f\}$, where

$$f(t) := \begin{cases} \dfrac{\sin t}{t}, & t \neq 0, \\ 1, & t = 0. \end{cases}$$

Solution We begin by expressing $f(t)$ in a Taylor series about $t = 0$. Since

$$\sin t = t - \frac{t^3}{3!} + \frac{t^5}{5!} - \frac{t^7}{7!} + \cdots,$$

then dividing by t, we obtain

$$f(t) = \frac{\sin t}{t} = 1 - \frac{t^2}{3!} + \frac{t^4}{5!} - \frac{t^6}{7!} + \cdots$$

for $t > 0$. This representation also holds at $t = 0$ since

$$\lim_{t \to 0} f(t) = \lim_{t \to 0} \frac{\sin t}{t} = 1.$$

Observe that $f(t)$ is continuous on $[0, \infty)$ and of exponential order. Hence its Laplace transform exists for all s large. Because of the linearity of the Laplace transform, we would expect that

$$\mathcal{L}\{f\}(s) = \mathcal{L}\{1\}(s) - \frac{1}{3!}\mathcal{L}\{t^2\}(s) + \frac{1}{5!}\mathcal{L}\{t^4\}(s) + \cdots$$

$$= \frac{1}{s} - \frac{2!}{3!s^3} + \frac{4!}{5!s^5} - \frac{6!}{7!s^7} + \cdots$$

$$= \frac{1}{s} - \frac{1}{3s^3} + \frac{1}{5s^5} - \frac{1}{7s^7} + \cdots.$$

Indeed, using tools from analysis, it can be verified that this series representation is valid for all $s > 1$. Moreover, one can show that the series converges to the function arctan$(1/s)$ (see Problem 54). Thus

(14) $$\mathcal{L}\left\{\frac{\sin t}{t}\right\}(s) = \arctan \frac{1}{s}. \quad \blacksquare \text{ End}$$

A similar procedure involving the series expansion for $F(s)$ in powers of $1/s$ can be used to compute $f(t) = \mathcal{L}^{-1}\{F\}(t)$ (see Problems 55–57).

We have previously shown, for every nonnegative integer n, that $\mathcal{L}\{t^n\}(s) = n!/s^{n+1}$. But what if the power of t is *not* an integer. Is this formula still valid? To answer this question, we need to extend the idea of "factorial." This is accomplished by the gamma function.[†]

[†] *Historical Footnote:* The gamma function was introduced by Leonhard Euler.

GAMMA FUNCTION

Definition 7. The **gamma function** $\Gamma(t)$ is defined by

$$(15) \qquad \Gamma(t) := \int_0^\infty e^{-u} u^{t-1}\, du, \qquad t > 0.$$

It can be shown that the integral in (15) converges for $t > 0$. A useful property of the gamma function is the recursive relation

$$(16) \qquad \Gamma(t + 1) = t\Gamma(t).$$

This identity follows from the definition (15) after performing an integration by parts:

$$\Gamma(t + 1) = \int_0^\infty e^{-u} u^t\, du = \lim_{N \to \infty} \int_0^N e^{-u} u^t\, du$$

$$= \lim_{N \to \infty} \left\{ -e^{-u} u^t \Big|_0^N + \int_0^N te^{-u} u^{t-1}\, du \right\}$$

$$= \lim_{N \to \infty} (-e^{-N} N^t) + t \lim_{N \to \infty} \int_0^N e^{-u} u^{t-1}\, du$$

$$= 0 + t\Gamma(t) = t\Gamma(t).$$

When t is a positive integer, say $t = n$, then the recursive relation (16) can be repeatedly applied to obtain

$$\Gamma(n + 1) = n\Gamma(n) = n(n - 1)\Gamma(n - 1) = \cdots$$
$$= n(n - 1)(n - 2) \cdots 2\Gamma(1).$$

It follows from the definition (15) that $\Gamma(1) = 1$, so we find

$$\Gamma(n + 1) = n!.$$

Thus the gamma function extends the notion of factorial!

As an application of the gamma function, let's return to the problem of determining the Laplace transform of an arbitrary power of t. We shall verify that the formula

$$(17) \qquad \mathcal{L}\{t^r\}(s) = \frac{\Gamma(r + 1)}{s^{r+1}}$$

holds for every constant $r > -1$.

By definition,

$$\mathcal{L}\{t^r\}(s) = \int_0^\infty e^{-st} t^r\, dt.$$

Let's make the substitution $u = st$. Then $du = s\,dt$, and we find

$$\mathcal{L}\{t^r\}(s) = \int_0^\infty e^{-u}\left(\frac{u}{s}\right)^r\left(\frac{1}{s}\right)du$$

$$= \frac{1}{s^{r+1}}\int_0^\infty e^{-u}u^r\,du = \frac{\Gamma(r+1)}{s^{r+1}}.$$

Notice that when $r = n$ is a nonnegative integer, then $\Gamma(n+1) = n!$, and so formula (17) reduces to the familiar formula for $\mathcal{L}\{t^n\}$.

EXERCISES 7.6

In Problems 1 through 4, sketch the graph of the given function and determine its Laplace transform.

1. $(t-1)^2 u(t-1)$.

2. $u(t-1) - u(t-4)$.

3. $t^2 u(t-2)$.

4. $tu(t-1)$.

In Problems 5 through 10, express the given function using unit step functions and compute its Laplace transform.

5. $g(t) = \begin{cases} 0, & 0 < t < 1, \\ 2, & 1 < t < 2, \\ 1, & 2 < t < 3, \\ 3, & 3 < t. \end{cases}$

6. $g(t) = \begin{cases} 0, & 0 < t < 2, \\ t+1, & 2 < t. \end{cases}$

7.

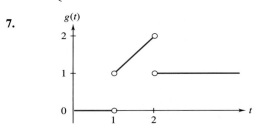

Figure 7.14 Function in Problem 7

8.

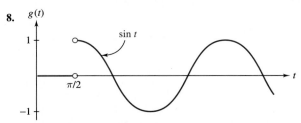

Figure 7.15 Function in Problem 8

9.

Figure 7.16 Function in Problem 9

10.

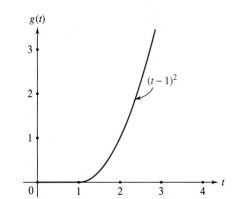

Figure 7.17 Function in Problem 10

In Problems 11 through 18, determine the inverse Laplace transform of the given function.

11. $\dfrac{e^{-2s}}{s-1}$.

12. $\dfrac{e^{-3s}}{s^2}$.

13. $\dfrac{e^{-2s} - 3e^{-4s}}{s+2}$.

14. $\dfrac{e^{-3s}}{s^2+9}$.

15. $\dfrac{se^{-3s}}{s^2 + 4s + 5}$.

16. $\dfrac{e^{-s}}{s^2 + 4}$.

17. $\dfrac{e^{-3s}(s - 5)}{(s + 1)(s + 2)}$.

18. $\dfrac{e^{-s}(3s^2 - s + 2)}{(s - 1)(s^2 + 1)}$.

19. The current $I(t)$ in an *RLC* series circuit is governed by the initial value problem

$$I''(t) + 2I'(t) + 2I(t) = g(t);$$

$$I(0) = 10, \qquad I'(0) = 0,$$

where

$$g(t) := \begin{cases} 20, & 0 < t < 3\pi, \\ 0, & 3\pi < t < 4\pi, \\ 20, & 4\pi < t. \end{cases}$$

Determine the current as a function of time t. Sketch $I(t)$ for $0 < t < 8\pi$.

20. The current $I(t)$ in an *LC* series circuit is governed by the initial value problem

$$I''(t) + 4I(t) = g(t); \qquad I(0) = 1, \qquad I'(0) = 3,$$

where

$$g(t) := \begin{cases} 3 \sin t, & 0 \le t \le 2\pi, \\ 0, & 2\pi < t. \end{cases}$$

Determine the current as a function of time t.

In Problems 21 through 24, determine $\mathcal{L}\{f\}$, where $f(t)$ is periodic with given period. Also graph $f(t)$.

21. $f(t) = t$, $\quad 0 < t < 2$, $\quad$ and $f(t)$ has period 2.

22. $f(t) = e^t$, $\quad 0 < t < 1$, $\quad$ and $f(t)$ has period 1.

23. $f(t) = \begin{cases} e^{-t}, & 0 < t < 1, \\ 1, & 1 < t < 2, \end{cases}$ and $f(t)$ has period 2.

24. $f(t) = \begin{cases} t, & 0 < t < 1, \\ 1 - t, & 1 < t < 2, \end{cases}$ and $f(t)$ has period 2.

In Problems 25 through 28, determine $\mathcal{L}\{f\}$, where the periodic function is described by its graph.

25.

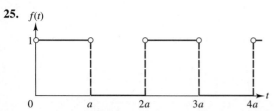

Figure 7.18 Square wave

26. $f(t)$

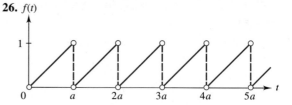

Figure 7.19 Sawtooth wave

27. $f(t)$

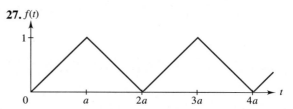

Figure 7.20 Triangular wave

28. $f(t)$

Figure 7.21 Half-rectified sine wave

In Problems 29 through 32, solve the given initial value problem using the method of Laplace transforms. Sketch the graph of the solution.

29. $y'' + y = u(t - 3)$; $\qquad y(0) = 0$, $\qquad y'(0) = 1$.

30. $w'' + w = u(t - 2) - u(t - 4)$;
$\qquad w(0) = 1$, $\qquad w'(0) = 0$.

31. $y'' + y = t - (t - 4)u(t - 2)$;
$\qquad y(0) = 0$, $\qquad y'(0) = 1$.

32. $y'' + y = 3 \sin 2t - 3(\sin 2t)u(t - 2\pi)$;
$\qquad y(0) = 1$, $\qquad y'(0) = -2$.

In Problems 33 through 42, solve the given initial value problem using the method of Laplace transforms.

33. $y'' + 2y' + 2y = u(t - 2\pi) - u(t - 4\pi)$;
$\qquad y(0) = 1$, $\qquad y'(0) = 1$.

34. $y'' + 4y' + 4y = u(t - \pi) - u(t - 2\pi)$;
$\qquad y(0) = 0$, $\qquad y'(0) = 0$.

35. $z'' + 3z' + 2z = e^{-3t}u(t - 2)$;
$\qquad z(0) = 2$, $\qquad z'(0) = -3$.

36. $y'' + 5y' + 6y = tu(t - 2)$;
$y(0) = 0$, $y'(0) = 1$.

37. $y'' + 4y = g(t)$; $y(0) = 1$, $y'(0) = 3$,

where $g(t) = \begin{cases} \sin t, & 0 \le t \le 2\pi, \\ 0, & 2\pi < t. \end{cases}$

38. $y'' + 2y' + 10y = g(t)$; $y(0) = -1$, $y'(0) = 0$,

where $g(t) = \begin{cases} 10, & 0 \le t \le 10, \\ 20, & 10 < t < 20, \\ 0, & 20 < t. \end{cases}$

39. $y'' + 5y' + 6y = g(t)$; $y(0) = 0$, $y'(0) = 2$,

where $g(t) = \begin{cases} 0, & 0 \le t < 1, \\ t, & 1 < t < 5, \\ 1, & 5 < t. \end{cases}$

40. $y'' + 3y' + 2y = g(t)$; $y(0) = 2$, $y'(0) = -1$,

where $g(t) = \begin{cases} e^{-t}, & 0 \le t < 3, \\ 1, & 3 < t. \end{cases}$

41. Show that if $\mathcal{L}\{g\}(s) = [(s + \alpha)(1 - e^{-Ts})]^{-1}$, where $T > 0$ is fixed, then

(18) $g(t) = e^{-\alpha t} + e^{-\alpha(t - T)}u(t - T)$
$+ e^{-\alpha(t - 2T)}u(t - 2T)$
$+ e^{-\alpha(t - 3T)}u(t - 3T) + \cdots$.

[*Hint:* Use the fact that
$1 + x + x^2 + \cdots = 1/(1 - x)$.]

42. The function $g(t)$ in (18) can be expressed in a more convenient form as follows:

(a) Show that for each $n = 0, 1, 2, \ldots$,

$g(t) = e^{-\alpha t}\left[\dfrac{e^{(n+1)\alpha T} - 1}{e^{\alpha T} - 1}\right]$

for $nT < t < (n + 1)T$.

[*Hint:* Use the fact that $1 + x + x^2 + \cdots + x^n = (x^{n+1} - 1)/(x - 1)$.]

(b) Let $v = t - (n + 1)T$. Show that when $nT < t < (n + 1)T$, then $-T < v < 0$ and

(19) $g(t) = \dfrac{e^{-\alpha v}}{e^{\alpha T} - 1} - \dfrac{e^{-\alpha t}}{e^{\alpha T} - 1}$.

(c) Use the facts that the first term in (19) is periodic with period T and the second term is independent of n to sketch the graph of $g(t)$ in (19) for $\alpha = 1$ and $T = 2$.

43. Show that if $\mathcal{L}\{g\}(s) = \beta[(s^2 + \beta^2)(1 - e^{-Ts})]^{-1}$, then
$g(t) = \sin \beta t + [\sin \beta(t - T)]u(t - T)$
$+ [\sin \beta(t - 2T)]u(t - 2T)$
$+ [\sin \beta(t - 3T)]u(t - 3T) + \cdots$.

44. Use the result of Problem 43 to show that

$\mathcal{L}^{-1}\left\{\dfrac{1}{(s^2 + 1)(1 - e^{-\pi s})}\right\}(t) = g(t)$,

where $g(t)$ is periodic with period 2π and

$g(t) := \begin{cases} \sin t, & 0 \le t \le \pi, \\ 0, & \pi \le t \le 2\pi. \end{cases}$

In Problems 45 and 46, use the method of Laplace transforms and the results of Problems 41 and 42 to solve the initial value problem

$y'' + 3y' + 2y = f(t)$; $y(0) = 0$, $y'(0) = 0$,

where $f(t)$ is the periodic function defined in the stated problem.

45. Problem 22 **46.** Problem 25 with $a = 1$.

In Problems 47 through 50, find a Taylor series for $f(t)$ about $t = 0$. Assuming the Laplace transform of $f(t)$ can be computed term by term, find an expansion for $\mathcal{L}\{f\}(s)$ in powers of $1/s$. If possible, sum the series.

47. $f(t) = e^t$. **48.** $f(t) = \sin t$.

49. $f(t) = \dfrac{1 - \cos t}{t}$. **50.** $f(t) = e^{-t^2}$.

51. Using the recursive relation (16) and the fact that $\Gamma(1/2) = \sqrt{\pi}$, determine

(a) $\mathcal{L}\{t^{-1/2}\}$. **(b)** $\mathcal{L}\{t^{7/2}\}$.

52. Use the recursive relation (16) and the fact that $\Gamma(1/2) = \sqrt{\pi}$ to show that

$\mathcal{L}^{-1}\{s^{-(n+1/2)}\}(t) = \dfrac{2^n t^{n-1/2}}{1 \cdot 3 \cdot 5 \cdots (2n - 1)\sqrt{\pi}}$,

where n is a positive integer.

53. Explain why, for each fixed t, the series (13) in the proof of Theorem 9 has only a *finite* number of nonzero terms.

54. By replacing s by $1/s$ in the Maclaurin series expansion for arctan s, show that

$\arctan \dfrac{1}{s} = \dfrac{1}{s} - \dfrac{1}{3s^3} + \dfrac{1}{5s^5} - \dfrac{1}{7s^7} + \cdots$.

55. Find an expansion for $e^{-1/s}$ in powers of $1/s$. Use the expansion for $e^{-1/s}$ to obtain an expansion for $s^{-1/2}e^{-1/s}$

in terms of $1/s^{n+1/2}$. Assuming the inverse Laplace transform can be computed term by term, show that

$$\mathcal{L}^{-1}\{s^{-1/2}e^{-1/s}\}(t) = \frac{1}{\sqrt{\pi t}}\cos 2\sqrt{t}.$$

[*Hint:* Use the result of Problem 52.]

56. Use the procedure discussed in Problem 55 to show that

$$\mathcal{L}^{-1}\{s^{-3/2}e^{-1/s}\}(t) = \frac{1}{\sqrt{\pi}}\sin 2\sqrt{t}.$$

57. Find an expansion for $\ln[1 + (1/s^2)]$ in powers of $1/s$. Assuming the inverse Laplace transform can be computed term by term, show that

$$\mathcal{L}^{-1}\left\{\ln\left(1 + \frac{1}{s^2}\right)\right\}(t) = \frac{2}{t}(1 - \cos t).$$

58. The **unit gate function** $G_a(t)$ is defined by

$$G_a(t) := \begin{cases} 0, & t < 0, \\ 1, & 0 < t < a, \\ 0, & a < t. \end{cases}$$

 (a) Show that $G_a(t) = u(t) - u(t - a)$.
 (b) Verify that $\mathcal{L}\{G_a\}(s) = (1/s)(1 - e^{-as})$.
 (c) Show that
$$\mathcal{L}\{G_a(t - b)\}(s) = (1/s)(e^{-bs} - e^{-(a+b)s}).$$

In Problems 59 and 60, use the results of Problem 58 and the method of Laplace transforms to solve the given initial value problem.

59. $y'' - y = G_3(t - 1); \qquad y(0) = 0, \qquad y'(0) = 2.$

60. $y'' - y = G_4(t - 3); \qquad y(0) = 1, \qquad y'(0) = -1.$

61. The mixing tank in Figure 7.22 initially holds 500 L of a brine solution with a salt concentration of 2 kg/L. For the first 10 min of operation, valve A is open, adding 12 L/min of brine containing a 4 kg/L salt concentration. After 10 min, valve B is switched in, adding a 6 kg/L concentration at 12 L/min. The exit valve C removes 12 L/min, thereby keeping the volume constant. Find the concentration of salt in the tank as a function of time.

12 L/min
4 kg/L

A

B

12 L/min
6 kg/L

C

Figure 7.22 Mixing tank

62. Suppose that in Problem 61 valve B is initially opened for 10 min and then valve A is switched in for 10 min. Finally valve B is switched back in. Find the concentration of salt in the tank as a function of time.

63. Suppose valve C removes only 6 L/min in Problem 61. Can Laplace transforms be used to solve the problem? Discuss.

7.7 CONVOLUTION

Consider the initial value problem

(1) $y'' + y = g(t); \qquad y(0) = 0, \qquad y'(0) = 0.$

If we let $Y(s) = \mathcal{L}\{y\}(s)$ and $G(s) = \mathcal{L}\{g\}(s)$, then taking the Laplace transform of both sides of (1) yields

$$s^2 Y(s) + Y(s) = G(s),$$

and hence

(2) $Y(s) = \left(\dfrac{1}{s^2 + 1}\right) G(s).$

That is, the Laplace transform of the solution to (1) is the product of the Laplace transform of $\sin t$ and the Laplace transform of the forcing term $g(t)$. What we would now like to have is a simple formula for $y(t)$ in terms of $\sin t$ and $g(t)$. Just as the integral of a product is not the product of the integrals, $y(t)$ is not the product of $\sin t$ and $g(t)$. However, we can express $y(t)$ as the "convolution" of $\sin t$ and $g(t)$.

CONVOLUTION

Definition 8. Let $f(t)$ and $g(t)$ be piecewise continuous on $[0, \infty)$. The **convolution** of $f(t)$ and $g(t)$, denoted $f * g$, is defined by

$$(3) \qquad (f * g)(t) := \int_0^t f(t - v)g(v)\,dv.$$

For example, the convolution of t and t^2 is

$$t * t^2 = \int_0^t (t - v)v^2\,dv = \int_0^t (tv^2 - v^3)\,dv$$

$$= \left(\frac{tv^3}{3} - \frac{v^4}{4}\right)\Bigg|_0^t = \frac{t^4}{3} - \frac{t^4}{4} = \frac{t^4}{12}.$$

Convolution is certainly different from ordinary multiplication. For example, $1 * 1 = t \neq 1$ and in general $1 * f \neq f$. However, convolution does satisfy some of the same properties as multiplication.

PROPERTIES OF CONVOLUTION

Theorem 10. Let $f(t)$, $g(t)$, and $h(t)$ be piecewise continuous on $[0, \infty)$. Then

$$(4) \qquad\qquad f * g = g * f,$$
$$(5) \qquad f * (g + h) = (f * g) + (f * h),$$
$$(6) \qquad (f * g) * h = f * (g * h),$$
$$(7) \qquad\qquad f * 0 = 0.$$

Proof. To prove equation (4), we begin with the definition

$$(f * g)(t) := \int_0^t f(t - v)g(v)\,dv.$$

Using the change of variables $w = t - v$, we have

$$(f * g)(t) = \int_t^0 f(w)g(t - w)(-dw) = \int_0^t g(t - w)f(w)\,dw = (g * f)(t),$$

which proves (4). The proofs of equations (5) and (6) are left to the exercises (see Problems 33 and 34). Equation (7) is obvious, since $f(t - v) \cdot 0 \equiv 0$. ∎ End

Returning to our original goal, we now prove that if $Y(s)$ is the product of the Laplace transforms $F(s)$ and $G(s)$, then $y(t)$ is the convolution $(f * g)(t)$.

CONVOLUTION THEOREM

Theorem 11. Let $f(t)$ and $g(t)$ be piecewise continuous on $[0, \infty)$ and of exponential order α and set $F(s) = \mathcal{L}\{f\}(s)$ and $G(s) = \mathcal{L}\{g\}(s)$. Then

(8) $\mathcal{L}\{f * g\}(s) = F(s)G(s),$

or, equivalently,

(9) $\mathcal{L}^{-1}\{F(s)G(s)\}(t) = (f * g)(t).$

Proof. Starting with the left-hand side of (8), we use the definition of convolution to write for $s > \alpha$

$$\mathcal{L}\{f * g\}(s) = \int_0^\infty e^{-st}\left(\int_0^t f(t - v)g(v)\,dv\right)dt.$$

To simplify the evaluation of this iterated integral, we introduce the unit step function $u(t - v)$ and write

$$\mathcal{L}\{f * g\}(s) = \int_0^\infty e^{-st}\left(\int_0^\infty u(t - v)f(t - v)g(v)\,dv\right)dt,$$

where we have used the fact that $u(t - v) = 0$ if $v > t$. Reversing the order of integration[†] gives

(10) $$\mathcal{L}\{f * g\}(s) = \int_0^\infty g(v)\left(\int_0^\infty e^{-st}u(t - v)f(t - v)\,dt\right)dv.$$

Recall from the translation property in Section 7.6 that the integral in parentheses in equation (10) equals $e^{-sv}F(s)$. Hence

$$\mathcal{L}\{f * g\}(s) = \int_0^\infty g(v)e^{-sv}F(s)\,dv = F(s)\int_0^\infty e^{-sv}g(v)\,dv = F(s)G(s).$$

This proves formula (8). ∎ End

[†] This is permitted since, for each $s > \alpha$, the absolute value of the integrand is integrable on $(0, \infty) \times (0, \infty)$.

For the initial value problem (1), recall that we found

$$Y(s) = \left(\frac{1}{s^2 + 1}\right)G(s) = \mathscr{L}\{\sin t\}(s)\,\mathscr{L}\{g\}(s).$$

It now follows from the convolution theorem that

$$y(t) = \sin t * g(t) = \int_0^t \sin(t - v)g(v)\,dv.$$

Thus we have obtained an integral representation for the solution to the initial value problem (1) for any forcing function $g(t)$ that is piecewise continuous on $[0, \infty)$ and of exponential order.

EXAMPLE 1 Use the convolution theorem to solve the initial value problem

(11) $y'' - y = g(t);$ $y(0) = 1,$ $y'(0) = 1,$

where $g(t)$ is piecewise continuous on $[0, \infty)$ and of exponential order.

Solution Let $Y(s) = \mathscr{L}\{y\}(s)$ and $G(s) = \mathscr{L}\{g\}(s)$. Taking the Laplace transform of both sides of the differential equation in (11) and using the initial conditions gives

$$s^2 Y(s) - s - 1 - Y(s) = G(s).$$

Solving for $Y(s)$, we have

$$Y(s) = \frac{s + 1}{s^2 - 1} + \left(\frac{1}{s^2 - 1}\right)G(s) = \frac{1}{s - 1} + \left(\frac{1}{s^2 - 1}\right)G(s).$$

Hence

$$y(t) = \mathscr{L}^{-1}\left\{\frac{1}{s - 1}\right\}(t) + \mathscr{L}^{-1}\left\{\frac{1}{s^2 - 1}\,G(s)\right\}(t)$$

$$= e^t + \mathscr{L}^{-1}\left\{\frac{1}{s^2 - 1}\,G(s)\right\}(t).$$

Referring to the table of Laplace transforms on the inside back cover, we find

$$\mathscr{L}\{\sinh t\}(s) = \frac{1}{s^2 - 1},$$

so we can now express

$$\mathscr{L}^{-1}\left\{\frac{1}{s^2 - 1}\,G(s)\right\}(t) = \sinh t * g(t).$$

Thus

$$y(t) = e^t + \int_0^t \sinh(t - v)g(v)\,dv$$

is the solution to the initial value problem (11). ■ End

EXAMPLE 2 Use the convolution theorem to find $\mathcal{L}^{-1}\{1/(s^2 + 1)^2\}$.

Solution Write

$$\frac{1}{(s^2 + 1)^2} = \left(\frac{1}{s^2 + 1}\right)\left(\frac{1}{s^2 + 1}\right).$$

Since $\mathcal{L}\{\sin t\}(s) = 1/(s^2 + 1)$, it follows from the convolution theorem that

$$\mathcal{L}^{-1}\left\{\frac{1}{(s^2 + 1)^2}\right\}(t) = \sin t * \sin t = \int_0^t \sin(t - v)\sin v \, dv$$

$$= \frac{1}{2}\int_0^t [\cos(2v - t) - \cos t]\, dv^\dagger$$

$$= \frac{1}{2}\left[\frac{\sin(2v - t)}{2}\right]_0^t - \frac{1}{2}t\cos t$$

$$= \frac{1}{2}\left[\frac{\sin t}{2} - \frac{\sin(-t)}{2}\right] - \frac{1}{2}t\cos t$$

$$= \frac{\sin t - t\cos t}{2}. \quad \blacksquare \text{ End}$$

As the preceding example attests, the convolution theorem is useful in determining the inverse transforms of rational functions of s. In fact, it provides an alternative to the method of partial fractions. For example,

$$\mathcal{L}^{-1}\left\{\frac{1}{(s - a)(s - b)}\right\}(t) = \mathcal{L}^{-1}\left\{\left(\frac{1}{s - a}\right)\left(\frac{1}{s - b}\right)\right\}(t) = e^{at} * e^{bt},$$

and all that remains in finding the inverse is to compute the convolution $e^{at} * e^{bt}$.

In the early 1900s, V. Volterra introduced **integro-differential** equations in his study of population growth. These equations enabled him to take into account "hereditary influences." In certain cases, these equations involved a convolution. As the next example shows, the convolution theorem helps to solve such integro-differential equations.

EXAMPLE 3 Solve the integro-differential equation

$$(12) \quad y'(t) = 1 - \int_0^t y(t - v)e^{-2v}\, dv, \qquad y(0) = 1.$$

Solution Equation (12) can be written as

$$(13) \quad y'(t) = 1 - y(t) * e^{-2t}.$$

† Here we used the identity $\sin \alpha \sin \beta = \frac{1}{2}[\cos(\beta - \alpha) - \cos(\beta + \alpha)]$.

Let $Y(s) = \mathcal{L}\{y\}(s)$. Taking the Laplace transform of (13) (with the help of the convolution theorem) and solving for $Y(s)$, we obtain

$$sY(s) - 1 = \frac{1}{s} - Y(s)\left(\frac{1}{s+2}\right)$$

$$sY(s) + \left(\frac{1}{s+2}\right)Y(s) = 1 + \frac{1}{s}$$

$$\left(\frac{s^2 + 2s + 1}{s+2}\right)Y(s) = \frac{s+1}{s}$$

$$Y(s) = \frac{(s+1)(s+2)}{s(s+1)^2} = \frac{s+2}{s(s+1)}$$

$$Y(s) = \frac{2}{s} - \frac{1}{s+1}.$$

Hence $y(t) = 2 - e^{-t}$. ∎ End

The **transfer function** $H(s)$ of a linear system is defined as the ratio of the Laplace transform of the output function $y(t)$ to the Laplace transform of the input function $g(t)$, under the assumption that all initial conditions are zero. That is, $H(s) = Y(s)/G(s)$. If the linear system is governed by the differential equation

(14) $ay'' + by' + cy = g(t), \qquad t > 0,$

where a, b, and c are constants, we can compute the transfer function as follows. Take the Laplace transform of both sides of (14) to get

$$as^2Y(s) - asy(0) - ay'(0) + bsY(s) - by(0) + cY(s) = G(s).$$

Because the initial conditions are assumed to be zero, the equation reduces to

$$(as^2 + bs + c)Y(s) = G(s).$$

Thus the transfer function for equation (14) is

(15) $H(s) = \dfrac{Y(s)}{G(s)} = \dfrac{1}{as^2 + bs + c}.$

The function $h(t) := \mathcal{L}^{-1}\{H\}(t)$ is called the **impulse response function** for the system because, physically speaking, it describes the solution when a spring-mass system is struck by a hammer (see Section 7.8). We can also characterize $h(t)$ as the unique solution to the homogeneous problem

(16) $ah'' + bh' + ch = 0; \qquad h(0) = 0, \qquad h'(0) = 1/a.$

Indeed, observe that taking the Laplace transform of the equation in (16) gives

(17) $a[s^2H(s) - sh(0) - h'(0)] + b[sH(s) - h(0)] + cH(s) = 0.$

Substituting in $h(0) = 0$ and $h'(0) = 1/a$ and solving for $H(s)$ yields

$$H(s) = \frac{1}{as^2 + bs + c},$$

which is the same as the formula for the transfer function given in equation (15).

One nice feature of the impulse response function h is that it can help us describe the solution to the *general* initial value problem

(18) $ay'' + by' + cy = g(t);$ $y(0) = y_0,$ $y'(0) = y_1.$

From the discussion of equation (14), we can see that the convolution $h * g$ is the solution to (18) in the special case when the initial conditions are zero (i.e., $y_0 = y_1 = 0$). To deal with nonzero initial conditions, let y_k denote the solution to the corresponding *homogeneous* initial value problem; that is, y_k solves

(19) $ay'' + by' + cy = 0;$ $y(0) = y_0,$ $y'(0) = y_1.$

Then, the desired solution to the general initial value problem (18) must be $h * g + y_k$. Indeed, it follows from the superposition principle (see Theorem 6 in Section 4.7) that since $h * g$ is a solution to equation (14) and y_k is a solution to the corresponding homogeneous equation, then $h * g + y_k$ is a solution to equation (14). Moreover, since $h * g$ has initial conditions zero,

$$(h * g)(0) + y_k(0) = 0 + y_0 = y_0,$$
$$(h * g)'(0) + y_k'(0) = 0 + y_1 = y_1.$$

We summarize these observations in the following theorem.

IMPULSE RESPONSE FUNCTION

Theorem 12. Let I be an interval containing the origin. The unique solution to the initial value problem

$$ay'' + by' + cy = g;$$ $y(0) = y_0,$ $y'(0) = y_1,$

where a, b, and c are constants and g is continuous on I, is given by

(20) $(h * g)(t) + y_k(t),$

where h is the impulse response function for the system and y_k is the unique solution to (19).

A proof of Theorem 12 that does not involve Laplace transforms is outlined in Project A in Chapter 4.

In the next example we use Theorem 12 to find a formula for the solution to an initial value problem.

EXAMPLE 4 A linear system is governed by the differential equation

(21) $y'' + 2y' + 5y = g(t);$ $y(0) = 2,$ $y'(0) = -2.$

Find the transfer function for the system, the impulse response function, and give a formula for the solution.

Solution According to formula (15), the transfer function for (21) is

$$H(s) = \frac{1}{as^2 + bs + c} = \frac{1}{s^2 + 2s + 5} = \frac{1}{(s + 1)^2 + 2^2}.$$

The inverse Laplace transform of $H(s)$ is the impulse response function

$$h(t) = \mathcal{L}^{-1}\{H\}(t) = \frac{1}{2}\mathcal{L}^{-1}\left\{\frac{2}{(s + 1)^2 + 2^2}\right\}(t)$$

$$= \frac{1}{2}e^{-t}\sin 2t.$$

To solve the initial value problem, we need the solution to the corresponding homogeneous problem. The auxiliary equation for the homogeneous equation is $r^2 + 2r + 5 = 0$, which has roots $r = -1 \pm 2i$. Thus a general solution is $C_1 e^{-t}\cos 2t + C_2 e^{-t}\sin 2t$. Choosing C_1 and C_2 so that the initial conditions in (21) are satisfied, we obtain $y_k(t) = 2e^{-t}\cos 2t$.

Hence, a formula for the solution to the initial value problem (21) is

$$(h * g)(t) + y_k(t) = \frac{1}{2}\int_0^t e^{-(t-v)}\sin[2(t - v)]g(v)\,dv + 2e^{-t}\cos 2t. \quad \blacksquare \text{ End}$$

EXERCISES 7.7

In Problems 1 through 4, use the convolution theorem to obtain a formula for the solution to the given initial value problem, where $g(t)$ is piecewise continuous on $[0, \infty)$ and of exponential order.

1. $y'' - 2y' + y = g(t);$ $y(0) = -1,$ $y'(0) = 1.$

2. $y'' + 9y = g(t);$ $y(0) = 1,$ $y'(0) = 0.$

3. $y'' + 4y' + 5y = g(t);$ $y(0) = 1,$ $y'(0) = 1.$

4. $y'' + y = g(t);$ $y(0) = 0,$ $y'(0) = 1.$

In Problems 5 through 12, use the convolution theorem to find the inverse Laplace transform of the given function.

5. $\dfrac{1}{s(s^2 + 1)}.$

6. $\dfrac{1}{(s + 1)(s + 2)}.$

7. $\dfrac{14}{(s + 2)(s - 5)}.$

8. $\dfrac{1}{(s^2 + 4)^2}.$

9. $\dfrac{s}{(s^2 + 1)^2}.$

10. $\dfrac{1}{s^3(s^2 + 1)}.$

11. $\dfrac{s}{(s - 1)(s + 2)}.$ $\left[\text{Hint: } \dfrac{s}{s - 1} = 1 + \dfrac{1}{s - 1}.\right]$

12. $\dfrac{s + 1}{(s^2 + 1)^2}.$

13. Find the Laplace transform of

$$f(t) := \int_0^t (t - v)e^{3v}\,dv.$$

14. Find the Laplace transform of

$$f(t) := \int_0^t \sin(t - v)e^v\,dv.$$

In Problems 15 through 22, solve the given integral equation or integro-differential equation for $y(t)$.

15. $y(t) + 3\displaystyle\int_0^t y(v)\sin(t - v)\,dv = t.$

16. $y(t) + \int_0^t e^{t-v} y(v)\, dv = \sin t.$

17. $y(t) + \int_0^t (t-v) y(v)\, dv = 1.$

18. $y(t) + \int_0^t (t-v) y(v)\, dv = t^2.$

19. $y(t) + \int_0^t (t-v)^2 y(v)\, dv = t^3 + 3.$

20. $y'(t) + \int_0^t (t-v) y(v)\, dv = t, \qquad y(0) = 0.$

21. $y'(t) + y(t) - \int_0^t y(v) \sin(t-v)\, dv = -\sin t,$

$\quad y(0) = 1.$

22. $y'(t) - 2 \int_0^t e^{t-v} y(v)\, dv = t, \qquad y(0) = 2.$

In Problems 23 through 28, a linear system is governed by the given initial value problem. Find the transfer function H(s) for the system and the impulse response function h(t) and give a formula for the solution to the initial value problem.

23. $y'' + 9y = g(t); \qquad y(0) = 2, \qquad y'(0) = -3.$
24. $y'' - 9y = g(t); \qquad y(0) = 2, \qquad y'(0) = 0.$
25. $y'' - y' - 6y = g(t); \qquad y(0) = 1, \qquad y'(0) = 8.$
26. $y'' + 2y' - 15y = g(t); \qquad y(0) = 0, \qquad y'(0) = 8.$
27. $y'' - 2y' + 5y = g(t); \qquad y(0) = 0, \qquad y'(0) = 2.$
28. $y'' - 4y' + 5y = g(t); \qquad y(0) = 0, \qquad y'(0) = 1.$

In Problems 29 and 30, the current I(t) in an RLC circuit with voltage source E(t) is governed by the initial value problem

$$LI''(t) + RI'(t) + \frac{1}{C} I(t) = e(t), \qquad I(0) = a, \qquad I'(0) = b,$$

where e(t) = E'(t) (see Figure 7.23). Given the constants R, L, C, a, and b find a formula for the solution I(t) in terms of e(t).

Figure 7.23 Schematic representation of an *RLC* series circuit

29. $R = 20$ ohms, $L = 5$ henrys, $C = 0.005$ farads, $a = -1$ amps, $b = 8$ amps/sec.

30. $R = 80$ ohms, $L = 10$ henrys, $C = 1/410$ farads, $a = 2$ amps, $b = -8$ amps/sec.

31. Use the convolution theorem and Laplace transforms to compute $1 * 1 * 1$.

32. Use the convolution theorem and Laplace transforms to compute $1 * t * t^2$.

33. Prove property (5) in Theorem 10.

34. Prove property (6) in Theorem 10.

35. Use the convolution theorem to show that

$$\mathcal{L}^{-1}\left\{\frac{F(s)}{s}\right\}(t) = \int_0^t f(v)\, dv,$$

where $F(s) = \mathcal{L}\{f\}(s)$.

36. Using Theorem 5 in Section 7.3 and the convolution theorem, show that

$$\int_0^t \int_0^v f(z)\, dz\, dv = \mathcal{L}^{-1}\left\{\frac{F(s)}{s^2}\right\}(t)$$

$$= t \int_0^t f(v)\, dv - \int_0^t v f(v)\, dv,$$

where $F(s) = \mathcal{L}\{f\}(s)$.

37. Prove directly that if $h(t)$ is the impulse response function characterized by equation (16), then for any continuous $g(t)$, we have $(h * g)(0) = (h * g)'(0) = 0.$ [*Hint:* Use Leibniz's rule, described in Project A of Chapter 4, page 231.]

38. **Determining System Parameters.** In mechanical design, one sometimes must determine system parameters even though information on the system forces is incomplete. For example, the differential equation governing the motion $x(t)$ of an externally forced damped mass-spring oscillator was shown in Section 5.1 (equation (4), page 242) to be

$$m\frac{d^2x}{dt^2} + b\frac{dx}{dt} + kx = f(t),$$

where $f(t)$ is the external force; the other parameters are defined in Section 5.1. Assume the system is underdamped ($b^2 < 4mk$; see Section 5.2) and starts from rest ($x(0) = 0, x'(0) = 0$) and that the force is bounded: $|f(t)| \le A.$

(a) Show that the transforms $X(s)$ and $F(s)$ of $x(t)$ and $f(t)$ are related by

$$X(s) = \frac{1}{ms^2 + bs + k} F(s).$$

(b) Use the convolution theorem to derive the formula

$$x(t) = \frac{1}{\beta m} \int_0^t f(t - v) e^{-bv/2m} \sin \beta v \, dv,$$

where (as in Section 5.2)

$$\beta := \frac{1}{2m} \sqrt{4mk - b^2}.$$

(c) Show that the motion $x(t)$ is bounded under these circumstances by

$$|x(t)| \le At/\beta m$$

and also by

$$|x(t)| \le 2A/\beta b.$$

(d) Suppose the mass $m = 5$ kg, the spring constant $k = 3000$ N/m, and the force is bounded by $A = 10$ N. What range of values for the damping constant b will ensure a displacement of 1 cm (.01 m) or less?

7.8 IMPULSES AND THE DIRAC DELTA FUNCTION

In mechanical systems, electric circuits, bending of beams, and other applications, one encounters functions that have a very large value over a very short interval. For example, the strike of a hammer exerts a relatively large force over a relatively short time, and a heavy weight concentrated at a spot on a suspended beam exerts a large force over a very small section of the beam. To deal with violent forces of short duration, physicists and engineers use the delta function introduced by Paul A. M. Dirac.

DIRAC DELTA FUNCTION

Definition 9. The **Dirac delta function** $\delta(t)$ is characterized by the following two properties:

(1) $\qquad \delta(t) = \begin{cases} 0, & t \ne 0, \\ \text{``infinite,''} & t = 0 \end{cases}$

and

(2) $\qquad \displaystyle\int_{-\infty}^{\infty} f(t)\delta(t)\, dt = f(0)$

for any function $f(t)$ that is continuous on an open interval containing $t = 0$.

By shifting the argument of $\delta(t)$, we have $\delta(t - a) = 0$, $t \ne a$, and

(3) $\qquad \displaystyle\int_{-\infty}^{\infty} f(t)\delta(t - a)\, dt = f(a)$

for any function $f(t)$ that is continuous on an interval containing $t = a$.

It is obvious that $\delta(t - a)$ is not a function in the usual sense; instead it is an example of what is called a **generalized function** or a **distribution.** In spite of this shortcoming,

the Dirac delta function was successfully used for several years to solve various physics and engineering problems before Laurent Schwartz mathematically justified its use!

A heuristic argument for the existence of the Dirac delta function can be made by considering the impulse of a force over a short interval. If a force $\mathcal{F}(t)$ is applied from time t_0 to time t_1, then the **impulse** due to $\mathcal{F}$ is the integral

$$\text{Impulse} := \int_{t_0}^{t_1} \mathcal{F}(t)\, dt.$$

By Newton's second law, we see that

(4) $$\int_{t_0}^{t_1} \mathcal{F}(t)\, dt = \int_{t_0}^{t_1} m\, \frac{dv}{dt}\, dt = mv(t_1) - mv(t_0),$$

where m denotes mass and v denotes velocity. Since mv represents the momentum, we can interpret equation (4) as saying: **The impulse equals the change in momentum.**

When a hammer strikes an object, it transfers momentum to the object. This change in momentum takes place over a very short period of time, say $[t_0, t_1]$. If we let $\mathcal{F}_1(t)$ represent the force due to the hammer, then the *area* under the curve $\mathcal{F}_1(t)$ is the impulse or change in momentum (see Figure 7.24). If, as is illustrated in Figure 7.25, the same change in momentum takes place over shorter and shorter time intervals, say $[t_0, t_2]$ or $[t_0, t_3]$, then the average force must get greater and greater in order for the impulses (the areas under the curves $\mathcal{F}_n$) to remain the same. In fact, if the forces $\mathcal{F}_n$ having the same impulse act, respectively, over the intervals $[t_0, t_n]$, where $t_n \to t_0$ as $n \to \infty$, then $\mathcal{F}_n$ approaches a function that is zero for $t \neq t_0$ but has an infinite value for $t = t_0$. Moreover, the areas under the $\mathcal{F}_n$'s have a common value. Normalizing this value to be 1 gives

$$\int_{-\infty}^{\infty} \mathcal{F}_n(t)\, dt = 1 \qquad \text{for all } n.$$

When $t_0 = 0$, we derive from the limiting properties of the $\mathcal{F}_n$'s a "function" δ that satisfies property (1) and the integral condition

(5) $$\int_{-\infty}^{\infty} \delta(t)\, dt = 1.$$

Notice that (5) is a special case of property (2) that is obtained by taking $f(t) \equiv 1$. It is interesting to note that (1) and (5) actually imply the general property (2) (see Problem 33).

The Laplace transform of the Dirac delta function can be quickly derived from property (3). Since $\delta(t - a) = 0$ for $t \neq a$, then setting $f(t) = e^{-st}$ in (3), we find for $a \geq 0$

$$\int_0^{\infty} e^{-st} \delta(t - a)\, dt = \int_{-\infty}^{\infty} e^{-st} \delta(t - a)\, dt = e^{-as}.$$

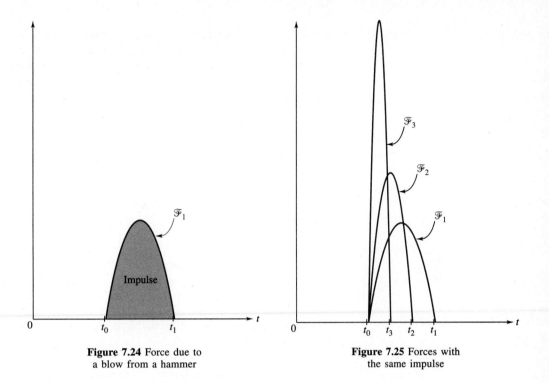

Figure 7.24 Force due to
a blow from a hammer

Figure 7.25 Forces with
the same impulse

Thus, for $a \geq 0$,

(6) $\mathcal{L}\{\delta(t - a)\}(s) = e^{-as}.$

An interesting connection exists between the unit step function and the Dirac delta function. Observe that as a consequence of equation (5) and the fact that $\delta(x - a)$ is zero for $x < a$ and for $x > a$, we have

(7) $$\int_{-\infty}^{t} \delta(x - a)\,dx = \begin{cases} 0, & t < a, \\ 1, & t > a \end{cases}$$

$$= u(t - a).$$

If we formally differentiate both sides of (7) with respect to t (in the spirit of the fundamental theorem of calculus), we find

$$\delta(t - a) = u'(t - a).$$

Thus it appears that the Dirac delta function is the derivative of the unit step function. That is, in fact, the case if we consider "differentiation" in a more general sense.[†]

[†] See *Distributions, Complex Variables, and Fourier Transforms*, by H. J. Bremermann (Addison-Wesley, Reading, Mass., 1966).

The Dirac delta function is used in modeling mechanical vibration problems involving an impulse. For example, a spring-mass system at rest that is struck by a hammer exerting an impulse on the mass might be governed by the *symbolic* initial value problem

$$(8) \qquad x'' + x = \delta(t); \qquad x(0) = 0, \qquad x'(0) = 0,$$

where $x(t)$ denotes the displacement from equilibrium at time t. We refer to this as a symbolic problem because while the left-hand side of equation (8) represents an ordinary function, the right-hand side does *not*. Consequently, it is not clear what we mean by a solution to problem (8). Because $\delta(t)$ is zero everywhere except at $t = 0$, one might be tempted to treat (8) as a homogeneous equation with zero initial conditions. But the solution to the latter is zero everywhere, which certainly does not describe the motion of the spring after being struck by the hammer.

To define what is meant by a solution to (8), recall that $\delta(t)$ is depicted as the limit of forces $\mathscr{F}_n(t)$ having unit impulse and acting over shorter and shorter intervals. If we let $y_n(t)$ be the solution to the initial value problem

$$(9) \qquad y_n'' + y_n = \mathscr{F}_n(t); \qquad y_n(0) = 0, \qquad y_n'(0) = 0,$$

where δ is replaced by $\mathscr{F}_n$, then we can think of the solution $x(t)$ to (8) as the limit (as $n \to \infty$) of the solutions $y_n(t)$.

For example, let

$$\mathscr{F}_n(t) := n - nu(t - 1/n) = \begin{cases} n, & 0 < t < 1/n, \\ 0, & \text{otherwise.} \end{cases}$$

Taking the Laplace transform of equation (9), we find

$$(s^2 + 1)Y_n(s) = \frac{n}{s}(1 - e^{-s/n}),$$

and so

$$Y_n(s) = \frac{n}{s(s^2 + 1)} - e^{-s/n} \frac{n}{s(s^2 + 1)}.$$

Now

$$\frac{n}{s(s^2 + 1)} = \frac{n}{s} - \frac{ns}{s^2 + 1} = \mathscr{L}\{n - n\cos t\}(s).$$

Hence

$$(10) \qquad y_n(t) = n - n\cos t - (n - n\cos(t - 1/n))u(t - 1/n)$$

$$= \begin{cases} n - n\cos t, & 0 < t < 1/n, \\ n\cos(t - 1/n) - n\cos t, & 1/n < t. \end{cases}$$

Fix $t > 0$. Then for n large enough, we have $1/n < t$. Thus

$$\lim_{n \to \infty} y_n(t) = \lim_{n \to \infty} [n \cos(t - 1/n) - n \cos t]$$

$$= -\lim_{n \to \infty} \frac{\cos(t - 1/n) - \cos t}{-1/n}$$

$$= -\lim_{h \to 0} \frac{\cos(t + h) - \cos t}{h}, \quad (\text{where } h = -1/n),$$

$$= -\frac{d}{dt}(\cos t) = \sin t.$$

Also, for $t = 0$, $\lim_{n \to \infty} y_n(0) = 0 = \sin 0$. Therefore

$$\lim_{n \to \infty} y_n(t) = \sin t.$$

Hence the solution to the symbolic initial value problem (8) is $x(t) = \sin t$.

Fortunately, we do not have to go through the tedious process of solving for each y_n in order to find the solution x of the symbolic problem. It turns out that the Laplace transform method when applied directly to (8) yields the derived solution $x(t)$. Indeed, simply taking the Laplace transform of both sides of (8), we obtain from (6) (with $a = 0$)

$$(s^2 + 1)X(s) = 1,$$

$$X(s) = \frac{1}{s^2 + 1},$$

which gives

$$x(t) = \mathcal{L}^{-1}\left\{\frac{1}{s^2 + 1}\right\}(t) = \sin t.$$

A peculiarity of using the Dirac delta function is that the solution $x(t) = \sin t$ of the symbolic initial value problem (8) does not satisfy both initial conditions, that is, $x'(0) = 1 \neq 0$. This reflects the fact that the impulse $\delta(t)$ is applied at $t = 0$. Thus the momentum x' (observe that in equation (8) the mass equals one) jumps abruptly from $x'(0) = 0$ to $x'(0^+) = 1$.

In the next example, the Dirac delta function is used in modeling a mechanical vibration problem.

EXAMPLE 1 A mass attached to a spring is released from rest 1 m below the equilibrium position for the spring-mass system and begins to vibrate. After π seconds, the mass is struck by a hammer exerting an impulse on the mass. The system is governed by the symbolic initial value problem

(11) $\dfrac{d^2x}{dt^2} + 9x = 3\delta(t - \pi); \quad x(0) = 1, \quad \dfrac{dx}{dt}(0) = 0,$

where $x(t)$ denotes the displacement from equilibrium at time t. Determine $x(t)$.

Solution Let $X(s) = \mathcal{L}\{x\}(s)$. Since

$$\mathcal{L}\{x''\}(s) = s^2 X(s) - s \quad \text{and} \quad \mathcal{L}\{\delta(t - \pi)\}(s) = e^{-\pi s},$$

taking the Laplace transform of both sides of (11) and solving for $X(s)$ yields

$$s^2 X(s) - s + 9X(s) = 3e^{-\pi s}$$

$$X(s) = \frac{s}{s^2 + 9} + e^{-\pi s}\frac{3}{s^2 + 9}$$

$$= \mathcal{L}\{\cos 3t\}(s) + e^{-\pi s}\mathcal{L}\{\sin 3t\}(s).$$

Using the translation property (cf. page 391) to determine the inverse Laplace transform of $X(s)$, we find

$$x(t) = \cos 3t + \sin 3(t - \pi)u(t - \pi)$$

$$= \begin{cases} \cos 3t, & t < \pi, \\ \cos 3t - \sin 3t, & \pi < t \end{cases}$$

$$= \begin{cases} \cos 3t, & t < \pi, \\ \sqrt{2} \cos\left(3t + \dfrac{\pi}{4}\right), & \pi < t. \end{cases}$$

The graph of $x(t)$ is given in color in Figure 7.26. For comparison, the dashed curve depicts the displacement of an undisturbed vibrating spring. Note that the impulse effectively adds 3 units to the momentum at time $t = \pi$. ■ End

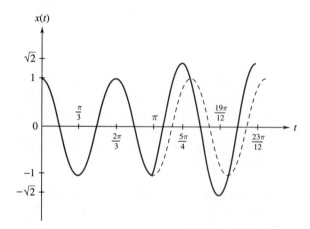

Figure 7.26 Displacement of a vibrating spring that is struck by a hammer at $t = \pi$

In Section 7.7, we defined the **impulse response function** for

(12) $ay'' + by' + cy = g(t)$

as the function $h(t) := \mathcal{L}^{-1}\{H\}(t)$, where $H(s)$ is the **transfer function.** Recall that $H(s)$ is the ratio

$$H(s) := \frac{Y(s)}{G(s)},$$

where $Y(s)$ is the Laplace transform of the solution to (12) with zero initial conditions and $G(s)$ is the Laplace transform of $g(t)$. It is important to note that $H(s)$, and hence $h(t)$, does not depend on the choice of the function $g(t)$ in (12) (see equation (15) in Section 7.7. However, it is useful to think of the impulse response function as the solution of the symbolic initial value problem

(13) $ay'' + by' + cy = \delta(t);$ $y(0) = 0,$ $y'(0) = 0.$

Indeed, with $g(t) = \delta(t)$, we have $G(s) = 1$, and hence $H(s) = Y(s)$. Consequently $h(t) = y(t)$. So we see that the function $h(t)$ is the response to the impulse $\delta(t)$ for a mechanical system governed by the symbolic initial value problem (13).

EXERCISES 7.8

In Problems 1 through 6, evaluate the given integral.

1. $\displaystyle\int_{-\infty}^{\infty} (t^2 - 1)\delta(t)\, dt.$

2. $\displaystyle\int_{-\infty}^{\infty} e^{3t}\delta(t)\, dt.$

3. $\displaystyle\int_{-\infty}^{\infty} (\sin 3t)\delta\left(t - \frac{\pi}{2}\right) dt.$

4. $\displaystyle\int_{-\infty}^{\infty} e^{-2t}\delta(t + 1)\, dt.$

5. $\displaystyle\int_{0}^{\infty} e^{-2t}\delta(t - 1)\, dt.$

6. $\displaystyle\int_{-1}^{1} (\cos 2t)\delta(t)\, dt.$

In Problems 7 through 12, determine the Laplace transform of the given generalized function.

7. $\delta(t - 1) - \delta(t - 3).$ **8.** $3\delta(t - 1).$

9. $t\delta(t - 1).$ **10.** $t^3\delta(t - 3).$

11. $\delta(t - \pi)\sin t.$ **12.** $e^t\delta(t - 3).$

In Problems 13 through 20, solve the given symbolic initial value problem.

13. $w'' + w = \delta(t - \pi);$ $w(0) = 0,$ $w'(0) = 0.$

14. $y'' + 2y' + 2y = \delta(t - \pi);$ $y(0) = 1,$ $y'(0) = 1.$

15. $y'' + 2y' - 3y = \delta(t - 1) - \delta(t - 2);$
 $y(0) = 2,$ $y'(0) = -2.$

16. $y'' - 2y' - 3y = 2\delta(t - 1) - \delta(t - 3);$
 $y(0) = 2,$ $y'(0) = 2.$

17. $y'' - y = 4\delta(t - 2) + t^2;$ $y(0) = 0,$ $y'(0) = 2.$

18. $y'' - y' - 2y = 3\delta(t - 1) + e^t;$ $y(0) = 0,$ $y'(0) = 3.$

19. $w'' + 6w' + 5w = e^t\delta(t - 1);$
 $w(0) = 0,$ $w'(0) = 4.$

20. $y'' + 5y' + 6y = e^{-t}\delta(t - 2);$
 $y(0) = 2,$ $y'(0) = -5.$

In Problems 21 through 24, solve the given symbolic initial value problem and sketch a graph of the solution.

21. $y'' + y = \delta(t - 2\pi);$ $y(0) = 0,$ $y'(0) = 1.$

22. $y'' + y = \delta(t - \pi/2);$ $y(0) = 0,$ $y'(0) = 1.$

23. $y'' + y = -\delta(t - \pi) + \delta(t - 2\pi);$
 $y(0) = 0,$ $y'(0) = 1.$

24. $y'' + y = \delta(t - \pi) - \delta(t - 2\pi);$
 $y(0) = 0,$ $y'(0) = 1.$

In Problems 25 through 28, find the impulse response function $h(t)$ by using the fact that $h(t)$ is the solution to the

symbolic initial value problem with $g(t) = \delta(t)$ *and zero initial conditions.*

25. $y'' + 4y' + 8y = g(t)$. **26.** $y'' - 6y' + 13y = g(t)$.

27. $y'' - 2y' + 5y = g(t)$. **28.** $y'' - y = g(t)$.

29. A mass attached to a spring is released from rest 1 m below the equilibrium position for the spring-mass system and begins to vibrate. After $\pi/2$ sec, the mass is struck by a hammer exerting an impulse on the mass. The system is governed by the symbolic initial value problem

$$\frac{d^2x}{dt^2} + 9x = -3\delta\left(t - \frac{\pi}{2}\right);$$

$$x(0) = 1,$$

$$\frac{dx}{dt}(0) = 0,$$

where $x(t)$ denotes the displacement from equilibrium at time t. What happens to the mass after it is struck?

30. You have probably heard that soldiers are told not to march in cadence when crossing a bridge. By solving the symbolic initial value problem

$$y'' + y = \sum_{k=1}^{\infty} \delta(t - 2k\pi); \quad y(0) = 0, \quad y'(0) = 0,$$

explain why soldiers are so instructed. [*Hint:* See Section 5.3.]

31. A linear system is said to be **stable** if its impulse response function $h(t)$ remains bounded as $t \to \infty$. If the linear system is governed by

$$ay'' + by' + cy = g(t),$$

where b and c are not both zero, show that the system is stable if and only if the real parts of the roots to

$$ar^2 + br + c = 0$$

are less than or equal to zero.

32. A linear system is said to be **asymptotically stable** if its impulse response function satisfies $h(t) \to 0$ as $t \to \infty$. If the linear system is governed by

$$ay'' + by' + cy = g(t),$$

show that the system is asymptotically stable if and only if the real parts of the roots to

$$ar^2 + br + c = 0$$

are strictly less than zero.

33. The Dirac delta function may also be characterized by the properties

$$\delta(t) = \begin{cases} 0, & t \neq 0, \\ \text{``infinite,''} & t = 0, \end{cases} \quad \text{and}$$

$$\int_{-\infty}^{\infty} \delta(t)\, dt = 1.$$

Formally using the mean value theorem for definite integrals, verify that if $f(t)$ is continuous, then the above properties imply

$$\int_{-\infty}^{\infty} f(t)\delta(t)\, dt = f(0).$$

34. Formally using integration by parts, show that

$$\int_{-\infty}^{\infty} f(t)\delta'(t)\, dt = -f'(0).$$

Also show that, in general,

$$\int_{-\infty}^{\infty} f(t)\delta^{(n)}(t)\, dt = (-1)^n f^{(n)}(0).$$

35. Figure 7.27 shows a beam of length 2λ that is imbedded in a support on the left and free on the right. The vertical deflection of the beam a distance x from the support is denoted by $y(x)$. If the beam has a concentrated load L acting on it in the center of the beam, then the deflection must satisfy the symbolic boundary value problem

$$EIy^{(4)}(x) = L\delta(x - \lambda);$$

$$y(0) = y'(0) = y''(2\lambda) = y'''(2\lambda) = 0,$$

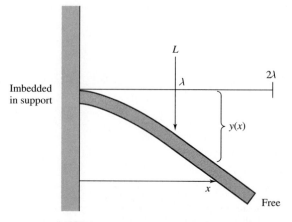

Figure 7.27 Beam imbedded in a support under a concentrated load at $x = \lambda$

where E, the modulus of elasticity, and I, the moment of inertia, are constants. Find a formula for the displacement $y(x)$ in terms of the constants λ, L, E, and I. [*Hint:* Let $y''(0) = A$ and $y'''(0) = B$. First solve the fourth order symbolic initial value problem and then use the conditions $y''(2\lambda) = y'''(2\lambda) = 0$ to determine A and B.]

7.9 SOLVING LINEAR SYSTEMS WITH LAPLACE TRANSFORMS

We can use the Laplace transform to reduce certain *systems* of linear differential equations with initial conditions to a system of linear algebraic equations, where again the unknowns are the transforms of the functions that make up the solution. Solving for these unknowns and taking their inverse Laplace transforms, we can then obtain the solution to the initial value problem for the system.

EXAMPLE 1 Solve the initial value problem

$$
\text{(1)} \qquad
\begin{aligned}
x'(t) - 2y(t) &= 4t, & x(0) &= 4, \\
y'(t) + 2y(t) - 4x(t) &= -4t - 2, & y(0) &= -5.
\end{aligned}
$$

Solution Taking the Laplace transform of both sides of the differential equations gives

$$
\text{(2)} \qquad
\begin{aligned}
\mathscr{L}\{x'\}(s) - 2\mathscr{L}\{y\}(s) &= \frac{4}{s^2}, \\
\mathscr{L}\{y'\}(s) + 2\mathscr{L}\{y\}(s) - 4\mathscr{L}\{x\}(s) &= -\frac{4}{s^2} - \frac{2}{s}.
\end{aligned}
$$

Let $X(s) := \mathscr{L}\{x\}(s)$ and $Y(s) := \mathscr{L}\{y\}(s)$. Then, we have, by Theorem 4 on page 365,

$$
\begin{aligned}
\mathscr{L}\{x'\}(s) &= sX(s) - x(0) = sX(s) - 4, \\
\mathscr{L}\{y'\}(s) &= sY(s) - y(0) = sY(s) + 5.
\end{aligned}
$$

Substituting these expressions into system (2) and simplifying, we find

$$
\text{(3)} \qquad
\begin{aligned}
sX(s) - 2Y(s) &= \frac{4s^2 + 4}{s^2}, \\
-4X(s) + (s + 2)Y(s) &= -\frac{5s^2 + 2s + 4}{s^2}.
\end{aligned}
$$

To eliminate $Y(s)$ from the system, we multiply the first equation by $(s + 2)$ and the second by 2 and then add to obtain

$$
[s(s + 2) - 8]X(s) = \frac{(s + 2)(4s^2 + 4)}{s^2} - \frac{10s^2 + 4s + 8}{s^2}.
$$

This simplifies to

$$
X(s) = \frac{4s - 2}{(s + 4)(s - 2)}.
$$

To compute the inverse transform, we first write $X(s)$ in the partial fraction form

$$X(s) = \frac{3}{s+4} + \frac{1}{s-2}.$$

Hence, from the Laplace transform table on the inside back cover, we find that

(4) $x(t) = 3e^{-4t} + e^{2t}.$

To determine $y(t)$, we could solve system (3) for $Y(s)$ and then compute its inverse Laplace transform. However, it is easier just to solve the first equation in system (1) for $y(t)$ in terms of $x(t)$. Thus

$$y(t) = \frac{1}{2}x'(t) - 2t.$$

Substituting $x(t)$ from equation (4), we find that

(5) $y(t) = -6e^{-4t} + e^{2t} - 2t.$

The solution to the initial value problem (1) consists of the pair of functions $x(t)$, $y(t)$ given by equations (4) and (5). ∎ End

EXERCISES 7.9

In Problems 1 through 19, use the method of Laplace transforms to solve the given initial value problem. Here x', y', etc., denotes differentiation with respect to t; so does the symbol D.

1. $x' = 3x - 2y;$ $x(0) = 1,$
 $y' = 3y - 2x;$ $y(0) = 1.$

2. $x' = x - y;$ $x(0) = -1,$
 $y' = 2x + 4y;$ $y(0) = 0.$

3. $z' + w' = z - w,$ $z(0) = 1,$
 $z' - w' = z - w,$ $w(0) = 0.$

4. $x' - 3x + 2y = \sin t;$ $x(0) = 0,$
 $4x - y' - y = \cos t;$ $y(0) = 0.$

5. $x' = y + \sin t;$ $x(0) = 2,$
 $y' = x + 2\cos t;$ $y(0) = 0.$

6. $x' - x - y = 1;$ $x(0) = 0,$
 $-x + y' - y = 0;$ $y(0) = -5/2.$

7. $(D - 4)[x] + 6y = 9e^{-3t};$ $x(0) = -9,$
 $x - (D - 1)[y] = 5e^{-3t};$ $y(0) = 4.$

8. $D[x] + y = 0;$ $x(0) = 7/4,$
 $4x + D[y] = 3;$ $y(0) = 4.$

9. $x'' + 2y' = -x;$ $x(0) = 2,$ $x'(0) = -7,$
 $-3x'' + 2y'' = 3x - 4y;$ $y(0) = 4,$ $y'(0) = -9.$

10. $x'' + y = 1;$ $x(0) = 1,$ $x'(0) = 1,$
 $x + y'' = -1;$ $y(0) = 1,$ $y'(0) = -1.$

11. $x' + y = 1 - u(t - 2);$ $x(0) = 0,$
 $x + y' = 0;$ $y(0) = 0.$

12. $x' + y = x;$ $x(0) = 0,$
 $2x' + y'' = u(t - 3);$ $y(0) = 1,$ $y'(0) = -1.$

13. $x' - y' = (\sin t)u(t - \pi);$ $x(0) = 1,$
 $x + y' = 0;$ $y(0) = 1.$

14. $x'' = y + u(t - 1);$ $x(0) = 1,$ $x'(0) = 0,$
 $y'' = x + 1 - u(t - 1);$ $y(0) = 0,$ $y'(0) = 0.$

15. $x' - 2y = 2;$ $x(1) = 1,$
 $x' + x - y' = t^2 + 2t - 1;$ $y(1) = 0.$

16. $x' - 2x + y' = -(\cos t + 4 \sin t);$ $x(\pi) = 0,$
$\quad 2x + y' + y = \sin t + 3 \cos t;$ $y(\pi) = 3.$

17. $x' + x - y' = 2(t - 2)e^{t-2};$ $x(2) = 0,$ $x'(2) = 1,$
$\quad x'' - x' - 2y = -e^{t-2};$ $y(2) = 1.$

18. $\quad x' - 2y = 0;$ $x(0) = 0,$
$\quad\quad x' - z' = 0;$ $y(0) = 0,$
$\quad x + y' - z = 3;$ $z(0) = -2.$

19. $x' = 3x + y - 2z;$ $x(0) = -6,$
$\quad y' = -x + 2y + z;$ $y(0) = 2,$
$\quad z' = 4x + y - 3z;$ $z(0) = -12.$

20. Use the method of Laplace transforms to solve

$$x'' + y' = 2; \quad x(0) = 3, \quad x'(0) = 0,$$
$$4x + y' = 6; \quad y(1) = 4.$$

[*Hint:* Let $y(0) = c$ and then solve for c.]

In Problems 21 and 22, find a system of differential equations and initial conditions for the currents in the networks given by the schematic diagrams; the initial currents are all assumed to be zero. Solve for the currents in each branch of the network (see Section 5.5 for a discussion of electric networks).

21.

Figure 7.28 *RL* network for Problem 21

22.

Figure 7.29 *RL* network for Problem 22

CHAPTER SUMMARY

The use of the Laplace transform helps to simplify the process of solving initial value problems for certain differential and integral equations, especially when a discontinuous forcing function is involved. The Laplace transform $\mathscr{L}\{f\}$ of a function $f(t)$ is defined by

$$\mathscr{L}\{f\}(s) := \int_0^\infty e^{-st} f(t)\, dt$$

for all values of s for which the improper integral exists. If $f(t)$ is piecewise continuous on $[0, \infty)$ and of exponential order α (that is, $|f(t)|$ grows no faster than a constant times $e^{\alpha t}$ as $t \to \infty$), then $\mathscr{L}\{f\}(s)$ exists for all $s > \alpha$.

The Laplace transform can be interpreted as an integral operator that maps a function $f(t)$ to a function $F(s)$. The transforms of commonly occurring functions appear in Table 7.1, page 362, and on the inside back cover of this book. The use of these tables is enhanced by several important properties of the operator $\mathscr{L}$.

Linearity: $\mathscr{L}\{af + bg\} = a\mathscr{L}\{f\} + b\mathscr{L}\{g\}.$

Translation in s: $\mathscr{L}\{e^{at}f(t)\}(s) = F(s - a)$, where $F = \mathscr{L}\{f\}$.

Translation in t: $\mathscr{L}\{g(t)u(t - a)\}(s) = e^{-as}\mathscr{L}\{g(t + a)\}(s)$, where $u(t - a)$ is the step function that equals 1 for $t > a$ and 0 for $t < a$. If $f(t)$ is continuous, then

$$\mathscr{L}^{-1}\{e^{-as}F(s)\}(t) = f(t - a)u(t - a),$$

where $f = \mathscr{L}^{-1}\{F\}$.

Convolution Property: $\mathscr{L}\{f * g\} = \mathscr{L}\{f\}\mathscr{L}\{g\}$, where $f * g$ denotes the convolution function

$$(f * g)(t) := \int_0^t f(t - v)g(v)\,dv.$$

One reason for the usefulness of the Laplace transform lies in the simple formula for the transform of the derivative f':

(1) $\qquad \mathscr{L}\{f'\}(s) = sF(s) - f(0)$, where $F = \mathscr{L}\{f\}$.

This formula shows that by using the Laplace transform, "differentiation with respect to t" can essentially be replaced by the simple operation of "multiplication by s." The extension of (1) to higher-order derivatives is

(2) $\qquad \mathscr{L}\{f^{(n)}\}(s) = s^n F(s) - s^{n-1}f(0) - s^{n-2}f'(0) - \cdots - f^{(n-1)}(0).$

To solve an initial value problem of the form

(3) $\qquad ay'' + by' + cy = f(t); \qquad y(0) = \alpha, \qquad y'(0) = \beta$

via the Laplace transform method, one takes the transform of both sides of the differential equation in (3). Using the linearity of $\mathscr{L}$ and formula (2) leads to an equation involving the Laplace transform $Y(s)$ of the (unknown) solution $y(t)$. The next step is to solve this simpler equation for $Y(s)$. Finally, one computes the *inverse* Laplace transform of $Y(s)$ to obtain the desired solution. This last step of finding $\mathscr{L}^{-1}\{Y\}$ is often the most difficult. Sometimes it requires a partial fractions decomposition, a judicious use of the properties of the transform, an excursion through the Laplace transform tables, or the evaluation of a contour integral in the complex plane.

For the special problem in (3), where a, b, and c are constants, the differential equation is transformed to a simple *algebraic* equation for $Y(s)$. Another nice feature of this latter equation is that it incorporates the initial conditions. When the coefficients of the equation in (3) depend on t, the following formula may be helpful in taking the transform:

$$\mathscr{L}\{t^n f(t)\}(s) = (-1)^n \frac{d^n F}{ds^n}(s), \qquad \text{where} \quad F = \mathscr{L}\{f\}.$$

If the forcing function $f(t)$ in equation (3) has jump discontinuities, it is often convenient to write $f(t)$ in terms of unit step functions $u(t - a)$ before proceeding with the Laplace

transform method. The transform of a periodic forcing function $f(t)$ with period T is given by

$$\mathcal{L}\{f\}(s) = \frac{\int_0^T e^{-st} f(t)\, dt}{1 - e^{-sT}}.$$

The Dirac delta function $\delta(t)$ is useful in modeling a system that is excited by a large force applied over a short time interval. It is not a function in the usual sense but can be roughly interpreted as the derivative of a unit step function. The transform of $\delta(t - a)$ is

$$\mathcal{L}\{\delta(t - a)\}(s) = e^{-as}, \qquad a \geq 0.$$

REVIEW PROBLEMS

In Problems 1 and 2, use the definition of the Laplace transform to determine $\mathcal{L}\{f\}$.

1. $f(t) = \begin{cases} 3, & 0 \leq t \leq 2, \\ 6 - t, & 2 < t. \end{cases}$

2. $f(t) = \begin{cases} e^{-t}, & 0 \leq t \leq 5, \\ -1, & 5 < t. \end{cases}$

In Problems 3 through 10, determine the Laplace transform of the given function.

3. $t^2 e^{-9t}$.

4. $e^{3t} \sin 4t$.

5. $e^{2t} - t^3 + t^2 - \sin 5t$.

6. $7e^{2t} \cos 3t - 2e^{7t} \sin 5t$.

7. $t \cos 6t$.

8. $(t + 3)^2 - (e^t + 3)^2$.

9. $t^2 u(t - 4)$.

10. $f(t) = \cos t$, $-\pi/2 \leq t \leq \pi/2$, and $f(t)$ has period π.

In Problems 11 through 17, determine the inverse Laplace transform of the given function.

11. $\dfrac{7}{(s + 3)^3}$.

12. $\dfrac{2s - 1}{s^2 - 4s + 6}$.

13. $\dfrac{4s^2 + 13s + 19}{(s - 1)(s^2 + 4s + 13)}$.

14. $\dfrac{s^2 + 16s + 9}{(s + 1)(s + 3)(s - 2)}$.

15. $\dfrac{2s^2 + 3s - 1}{(s + 1)^2(s + 2)}$.

16. $\dfrac{1}{(s^2 + 9)^2}$.

17. $\dfrac{e^{-2s}(4s + 2)}{(s - 1)(s + 2)}$.

18. Find the Taylor series for $f(t) = e^{-t^2}$ about $t = 0$. Then, assuming the Laplace transform of $f(t)$ can be computed term by term, find an expansion for $\mathcal{L}\{f\}(s)$ in powers of $1/s$.

In Problems 19 through 24, solve the given initial value problem for $y(t)$ using the method of Laplace transforms.

19. $y'' - 7y' + 10y = 0$; $y(0) = 0$, $y'(0) = -3$.

20. $y'' + 6y' + 9y = 0$; $y(0) = -3$, $y'(0) = 10$.

21. $y'' + 2y' + 2y = t^2 + 4t$;
$y(0) = 0$, $y'(0) = -1$.

22. $y'' + 9y = 10e^{2t}$; $y(0) = -1$, $y'(0) = 5$.

23. $y'' + 3y' + 4y = u(t - 1)$; $y(0) = 0$, $y'(0) = 1$.

24. $y'' - 4y' + 4y = t^2 e^t$; $y(0) = 0$, $y'(0) = 0$.

In Problems 25 and 26, find solutions to the given initial value problem.

25. $ty'' + 2(t - 1)y' - 2y = 0$; $y(0) = 0$, $y'(0) = 0$.

26. $ty'' + 2(t - 1)y' + (t - 2)y = 0$;
$y(0) = 1$, $y'(0) = -1$.

In Problems 27 and 28, solve the given equation for $y(t)$.

27. $y(t) + \displaystyle\int_0^t (t - v)y(v)\, dv = e^{-3t}$.

28. $y'(t) - 2\displaystyle\int_0^t y(v) \sin(t - v)\, dv = 1$; $y(0) = -1$.

29. A linear system is governed by

$$y'' - 5y' + 6y = g(t).$$

Find the transfer function and the impulse response function.

30. Solve the symbolic initial value problem

$$y'' + 4y = \delta\left(t - \frac{\pi}{2}\right); \qquad y(0) = 0, \qquad y'(0) = 1.$$

TECHNICAL WRITING EXERCISES

1. Compare the use of Laplace transforms in solving linear differential equations with constant coefficients with the use of logarithms in solving algebraic equations of the form $x^r = a$.

2. Explain why the method of Laplace transforms works so well for linear differential equations with constant coefficients *and* integro-differential equations involving a convolution.

3. Discuss several examples of initial value problems in which the method of Laplace transforms *cannot* be applied.

4. A linear system is said to be **asymptotically stable** if its impulse response function $h(t) \to 0$ as $t \to \infty$.

Assume $H(s)$, the Laplace transform of $h(t)$, is a rational function in reduced form with the degree of its numerator less than the degree of its denominator. Explain in detail how the asymptotic stability of the linear system can be characterized in terms of the zeros of the denominator of $H(s)$. Give examples.

5. Can the method of Laplace transforms be modified to solve *difference* equations (see Section 4.11)? If so, give the precise definition of the transform and describe some of its properties.

6. Compare and contrast the solution of initial value problems by Laplace transforms versus the methods of Chapter 4.

GROUP PROJECTS FOR CHAPTER 7

A. Duhamel's Formulas

For a linear system governed by the equation

(1) $\qquad ay'' + by' + cy = g(t),$

where a, b, c are real constants, the function

(2) $\qquad H(s) := \dfrac{\mathcal{L}\{y\}(s)}{\mathcal{L}\{g\}(s)} = \dfrac{\mathcal{L}\{\text{output}\}}{\mathcal{L}\{\text{input}\}},$

where all initial conditions are taken to be zero, is called the **transfer function** for the system. (As mentioned in Section 7.7, the transfer function $H(s)$ depends only on the constants a, b, c of the system; it is not affected by the choice of g.) If the input function $g(t)$ is the unit step function $u(t)$, then equation (2) yields

$$\mathcal{L}\{y\}(s) = \mathcal{L}\{u\}(s)H(s) = \frac{H(s)}{s}.$$

The solution (output function) in this special case is called the **indicial admittance** and is denoted by $A(t)$. Hence $\mathcal{L}\{A\}(s) = H(s)/s$.

 It is possible to express the response $y(t)$ of the system to a general input function $g(t)$ in terms of $g(t)$ and the indicial admittance $A(t)$. To derive these relations, proceed as follows:

(a) Show that

(3) $\qquad \mathcal{L}\{y\}(s) = s\mathcal{L}\{A\}(s)\mathcal{L}\{g\}(s).$

(b) Now apply the convolution theorem to (3) and show that

(4) $\qquad y(t) = \dfrac{d}{dt}\left[\displaystyle\int_0^t A(t - v)g(v)\,dv\right] = \dfrac{d}{dt}\left[\displaystyle\int_0^t A(v)g(t - v)\,dv\right].$

(c) To perform the differentiation indicated in (4), one can use **Leibniz's rule:**

$$\frac{d}{dt}\left[\int_{a(t)}^{b(t)} f(v,t)\,dv\right] = \int_{a(t)}^{b(t)} \frac{\partial f}{\partial t}(v,t)\,dv + f(b(t),t)\frac{db}{dt}(t) - f(a(t),t)\frac{da}{dt}(t),$$

where f and $\partial f/\partial t$ are assumed continuous in v and t, and $a(t)$, $b(t)$ are differentiable functions of t. Applying this rule to (4), derive the formulas

(5) $\qquad y(t) = \displaystyle\int_0^t A'(t - v)g(v)\,dv,$

(6) $\qquad y(t) = \displaystyle\int_0^t A(v)g'(t - v)\,dv + A(t)g(0).$

[*Hint:* Recall that the initial conditions on A are $A(0) = A'(0) = 0$.]

(d) In equations (5) and (6), make the change of variables $w = t - v$ and show that

(7) $$y(t) = \int_0^t A'(w)g(t - w)\,dw,$$

(8) $$y(t) = \int_0^t A(t - w)g'(w)\,dw + A(t)g(0).$$

Equations (5)–(8) are referred to as **Duhamel's formulas,** in honor of the French mathematician J. M. C. Duhamel. These formulas are helpful in determining the response of the system to a general input $g(t)$, since the indicial admittance of the system can be determined experimentally by measuring the response of the system to a unit step function.

(e) The impulse response function $h(t)$ is defined as $h(t) := \mathcal{L}^{-1}\{H\}(t)$, where $H(s)$ is the transfer function. Show that $h(t) = A'(t)$, so that equations (5) and (7) can be written in the form

(9) $$y(t) = \int_0^t h(t - v)g(v)\,dv = \int_0^t h(v)g(t - v)\,dv.$$

We remark that the indicial admittance is the response of the system to a unit step function, and the impulse response function is the response to the unit impulse or delta function (see Section 7.8). But the delta function is the derivative (in a generalized sense) of the unit step function. Therefore the fact that $h(t) = A'(t)$ is not really surprising.

B. Frequency Response Modeling

Frequency response modeling of a linear system is based on the premise that the dynamics of a linear system can be recovered from a knowledge of how the system responds to sinusoidal inputs. (This will be made mathematically precise in Theorem 13.) In other words, to determine (or identify) a linear system, all one has to do is observe how the system reacts to sinusoidal inputs.

Let's assume that we have a linear system governed by

(10) $$y'' + py' + qy = g(t),$$

where p and q are real constants. The function $g(t)$ is called the **forcing function** or **input function.** When $g(t)$ is a sinusoid, the particular solution to (10) obtained by the method of undetermined coefficients is the **steady-state solution** or **output function** $y_{ss}(t)$ corresponding to $g(t)$. We can think of a linear system as a compartment or block into which goes an input function g and out of which comes the output function y_{ss} (see Figure 7.30). To **identify** a linear system means to determine the coefficients p and q in equation (10).

Input
$g(t)$

Linear system
$y'' + py' + qy = g(t)$

Output
$y_{ss}(t)$

Figure 7.30 Block diagram depicting a linear system

It will be convenient for us to work with complex variables. A complex number z is usually expressed in the form $z = \alpha + i\beta$, with α, β real numbers and i denoting $\sqrt{-1}$. We can also express z in polar form, $z = re^{i\theta}$, where $r^2 = \alpha^2 + \beta^2$ and $\tan \theta = \beta/\alpha$. Here r (≥ 0) is called the **magnitude** and θ the **phase angle** of z.

The following theorem gives the relationship between the linear system and its response to sinusoidal inputs in terms of the **transfer function** $H(s)$ (see Project A, equation (2)).

STEADY-STATE SOLUTIONS TO SINUSOIDAL INPUTS

Theorem 13. Let $H(s)$ be the transfer function for equation (10). If $H(s)$ is finite at $s = i\omega$, with ω real, then the steady-state solution to (10) for $g(t) = e^{i\omega t}$ is

(11) $y_{ss}(t) = H(i\omega)e^{i\omega t} = H(i\omega)\{\cos \omega t + i \sin \omega t\}.$

(a) Prove Theorem 13. [*Hint:* Guess $y_{ss}(t) = Ae^{i\omega t}$ and show that $A = H(i\omega)$.]

(b) Use Theorem 13 to show that if $g(t) = \sin \omega t$, then the steady-state solution to (10) is $M(\omega) \sin(\omega t + N(\omega))$, where $H(i\omega) = M(\omega)e^{iN(\omega)}$ is the polar form for $H(i\omega)$.

(c) Solve for $M(\omega)$ and $N(\omega)$ in terms of p and q.

(d) Experimental results for modeling done by frequency response methods are usually presented in **frequency response**,[†] or **Bode plots.** There are two types of Bode plots. The first is of the log of the magnitude $M(\omega)$ of $H(i\omega)$ versus the angular frequency ω using a log scale for ω. The second is a plot of the phase angle or argument $N(\omega)$ of $H(i\omega)$ versus the angular frequency using a log scale for ω. The Bode plots for the transfer function $H(s) = (1 + 0.2s + s^2)^{-1}$ are given in Figure 7.31.

Sketch the Bode plots of the linear system governed by equation (10) with $p = 0.4$ and $q = 1.0$. Use $\omega = 0.3$, 0.6, 0.9, 1.2, and 1.5 for the plot of $M(\omega)$ and $\omega = 0.5$, 0.8, 1, 2, and 5 for the plot of $N(\omega)$.

 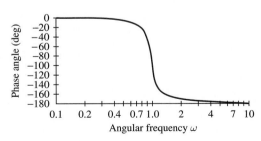

Figure 7.31 Bode plots for $H(i\omega) = [1 + 0.2(i\omega) + (i\omega)^2]^{-1}$

[†]Frequency response curves are also discussed in Section 5.3.

(e) Assume we know that $q = 1$. When we input a sine wave with $\omega = 2$, the system settles into a steady-state sinusoidal output with magnitude $M(2) = 0.333$. Find p and thus identify the linear system.

(f) Suppose a sine wave input with $\omega = 2$ produces a steady-state sinusoidal output with magnitude $M(2) = 0.5$ and that when $\omega = 4$, then $M(4) = 0.01$. Find p and q and thus identify the system.

In most applications, there are some inaccuracies in the measurement of the magnitudes and frequencies. To compensate for these errors, sinusoids with several different frequencies are used as input. A least squares approximation for p and q is then found. For a discussion of frequency response modeling as a mathematical modeling tool, see the chapter by W. F. Powers, "Modeling Linear Systems by Frequency Response Methods," in *Differential Equations Models*, by M. Braun, C. Coleman, and D. Drew (eds.) (Springer-Verlag, New York, 1983), Chapter 9. Additional examples may be found in *Schaum's Outline on Feedback and Control Systems*, by J. J. DiStefano, A. R. Stubberud, and I. J. Williams (McGraw-Hill, New York, 1967), Chapter 15.

Chapter **8**

Series Solution of Differential Equations

8.1 INTRODUCTION: A PROBLEM IN ASTROPHYSICS

In the early 1900s a German astrophysicist, Robert Emden, encountered the following problem in his study of the thermal behavior of a spherical gaseous cloud.

> **Determine the first point on the positive x-axis where the solution to the initial value problem**
>
> **(1)** $\quad xy''(x) + 2y'(x) + xy(x) = 0; \qquad y(0) = 1, \qquad y'(0) = 0$
>
> **is zero.**

The variables x and y in (1) are proportional to, respectively, the distance to the center of the cloud and the gravitational potential of the gas. The initial conditions correspond to the fact that $y(x)$, the potential, has been normalized to be 1 at the center of the cloud and that, at the center of the cloud, the gravitational forces cancel and the net force is zero. Astrophysicists and astronomers use equation (1) to approximate the density and internal temperatures of certain stars and nebula.[†]

The differential equation in (1) appears nice enough, since it is a linear second order equation; however, there are two difficulties. First, while (1) is linear, it has variable coefficients, and hence the methods discussed in Chapter 4 do not apply. Second, we are given initial conditions at $x = 0$, which is precisely where the coefficient of $y''(x)$ is zero. Consequently, even the theoretical results of Chapter 4 do not apply to this problem. While the prospects for finding an explicit solution to the initial value problem appear dim, the physical situation that is modeled suggests that the problem *has* a solution and the solution

[†] Equation (1) is called the **Lane-Emden equation of index $n = 1$** and is discussed in detail in *An Introduction to the Study of Stellar Structure*, by S. Chandrasekhar (Dover Publications, Inc., New York, 1958), Chapter IV.

should be "nice" (analytic). Having faith in the model, let's assume the solution has a power series expansion denoted by[†]

$$(2) \qquad y(x) = a_0 + a_1 x + a_2 x^2 + \cdots = \sum_{n=0}^{\infty} a_n x^n.$$

If we differentiate $y(x)$ as given in (2), term by term, we obtain

$$y'(x) = a_1 + 2a_2 x + 3a_3 x^2 + \cdots = \sum_{n=1}^{\infty} n a_n x^{n-1}.$$

$$y''(x) = 2a_2 + 6a_3 x + 12a_4 x^2 + \cdots = \sum_{n=2}^{\infty} n(n-1) a_n x^{n-2}.$$

Substituting these series representations for y, y', and y'' in equation (1), we get

$$x(2a_2 + 6a_3 x + 12a_4 x^2 + \cdots + (n+1)n a_{n+1} x^{n-1} + \cdots)$$
$$+ 2(a_1 + 2a_2 x + 3a_3 x^2 + 4a_4 x^3 + \cdots + (n+1)a_{n+1} x^n + \cdots)$$
$$+ x(a_0 + a_1 x + a_2 x^2 + \cdots + a_{n-1} x^{n-1} + \cdots) = 0.$$

Grouping like terms gives

$$2a_1 + (6a_2 + a_0)x + (12a_3 + a_1)x^2 + (20a_4 + a_2)x^3$$
$$+ \cdots + ([(n+1)n + 2(n+1)]a_{n+1} + a_{n-1})x^n + \cdots = 0.$$

Now, in order for the power series to sum identically to zero, each of the coefficients must be zero. This means

$$2a_1 = 0, \qquad 6a_2 + a_0 = 0, \qquad 12a_3 + a_1 = 0,$$

and, in general,

$$(n+2)(n+1)a_{n+1} + a_{n-1} = 0.$$

Solving, we find

$$a_1 = 0, \qquad a_2 = -\frac{1}{6}a_0, \qquad a_3 = -\frac{1}{12}a_1 = 0,$$

and, in general,

$$(3) \qquad a_{n+1} = \frac{-a_{n-1}}{(n+2)(n+1)}.$$

[†] As discussed in Section 8.6, the point $x = 0$ is a *regular singular point* for equation (1), and according to the method of Frobenius, one would typically try a solution of the more general form $y(x) = x^r(a_0 + a_1 x + a_2 x^2 + \cdots)$.

Since $a_1 = 0$, it follows from (3) that a_3, a_5, a_7, etc., are zero. For the even coefficients, a_2, a_4, etc., we see from repeated applications of (3) that

$$a_{2k} = \frac{-a_{2k-2}}{(2k+1)(2k)} = \frac{-1}{(2k+1)(2k)} \cdot \frac{-a_{2k-4}}{(2k-1)(2k-2)} = \cdots$$

$$= \frac{(-1)^k a_0}{(2k+1)(2k)\cdots(3)(2)} = \frac{(-1)^k a_0}{(2k+1)!}.$$

Knowing the coefficients of the power series for $y(x)$, we obtain

(4) $$y(x) = a_0\left(1 - \frac{1}{6}x^2 + \frac{1}{120}x^4 + \cdots + \frac{(-1)^k x^{2k}}{(2k+1)!} + \cdots\right)$$

as a solution to the differential equation.

Next, we substitute in the initial conditions $y(0) = 1$, $y'(0) = 0$. From (4) we see that $y(0) = a_0$ and $y'(0) = 0$. Hence $a_0 = 1$. (Notice our good fortune: Equation (4) has only one parameter, a_0, yet we are able to satisfy *both* initial conditions.) Letting $a_0 = 1$ in (4), we find that the solution to the initial value problem (1) is

(5) $$y(x) = 1 - \frac{1}{6}x^2 + \frac{1}{120}x^4 - \frac{1}{5040}x^6 + \cdots + \frac{(-1)^k x^{2k}}{(2k+1)!} + \cdots.$$

Now recall that we are interested in the first (positive) zero of $y(x)$. To locate this zero, it would be desirable to have a closed-form expression for $y(x)$. Again good fortune is with us: The expansion in (5) is a recognizable one. It is the Taylor series for $(\sin x)/x$. Hence $y(x) = (\sin x)/x$, and the first positive zero of $y(x)$ occurs when $\sin x = 0$; that is, when $x = \pi$.

In this chapter, we study methods for determining series expansions for solutions to differential equations with variable coefficients. In particular, we determine the form of the series expansion, a recurrence relation for determining the coefficients a_n, and the interval of convergence of the expansion.

8.2 POWER SERIES, ANALYTIC FUNCTIONS, AND THE TAYLOR SERIES METHOD[†]

The differential equations studied in earlier chapters often possessed solutions expressible in terms of elementary functions such as polynomials, exponentials, sines, and cosines. However, many important equations arise that do not have solutions that can be so conveniently expressed. In previous chapters, when we encountered such an equation we either expressed the solution as an integral (see Exercises 2.2, Problem 27) or tried to approximate it numerically (see Sections 3.5–3.7 and 5.6). In this chapter, our goal is to obtain representations for solutions as power series. We begin with a brief review of the basic definitions and properties of power series and real analytic functions.

[†] This section contains review material from calculus and a discussion of the Taylor series method briefly introduced in Project A in Chapter 1.

Power Series

A **power series** about the point x_0 is an expression of the form

(1) $$\sum_{n=0}^{\infty} a_n(x - x_0)^n = a_0 + a_1(x - x_0) + a_2(x - x_0)^2 + \cdots,$$

where x is a variable and the a_n's are constants. We say that (1) **converges** at the point $x = c$ if the infinite series (of real numbers) $\sum_{n=0}^{\infty} a_n(c - x_0)^n$ converges; that is, the limit of the partial sums,

$$\lim_{N \to \infty} \sum_{n=0}^{N} a_n(c - x_0)^n,$$

exists (as a finite number). If this limit does not exist, the power series is said to **diverge** at $x = c$. Observe that (1) converges at $x = x_0$, since

$$\sum_{n=0}^{\infty} a_n(x_0 - x_0)^n = a_0 + 0 + 0 + \cdots = a_0.$$

But what about convergence for other values of x? As stated in the following Theorem 1, a power series of the form (1) converges for all values of x in some ''interval'' centered at x_0 and diverges for x outside this interval. Moreover, at the interior points of this interval, the power series converges **absolutely** in the sense that $\sum_{n=0}^{\infty} |a_n(x - x_0)^n|$ converges. (Recall that absolute convergence of a series implies (ordinary) convergence of the series.)

RADIUS OF CONVERGENCE

Theorem 1. For each power series of the form (1), there is a number ρ ($0 \le \rho \le \infty$), called the **radius of convergence** of the power series, such that (1) converges absolutely for $|x - x_0| < \rho$ and diverges for $|x - x_0| > \rho$. (See Figure 8.1.)

If the series (1) converges for all values of x, then $\rho = \infty$. When the series (1) converges only at x_0, then $\rho = 0$.

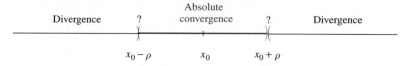

Figure 8.1 Interval of convergence

Notice that Theorem 1 settles the question of convergence except at the endpoints $x_0 \pm \rho$. Thus these two points require separate analysis. To determine the radius of convergence ρ, one method that is often easy to apply is the ratio test.

RATIO TEST

Theorem 2. If

$$\lim_{n \to \infty} \left| \frac{a_{n+1}}{a_n} \right| = L,$$

where $0 \le L \le \infty$, then the radius of convergence of the power series $\sum_{n=0}^{\infty} a_n(x - x_0)^n$ is $\rho = 1/L$, with $\rho = \infty$ if $L = 0$ and $\rho = 0$ if $L = \infty$.

Remark. We caution the reader that if the ratio $|a_{n+1}/a_n|$ does not have a limit, then methods other than the ratio test (e.g., root test) must be used to determine ρ. In particular, if infinitely many of the a_n's are zero, then the ratio test cannot be directly applied. (However, see Problem 7.)

EXAMPLE 1 Determine the convergence set of

(2) $$\sum_{n=0}^{\infty} \frac{(-2)^n}{n + 1} (x - 3)^n.$$

Solution Since $a_n = (-2)^n/(n + 1)$, we have

$$\lim_{n \to \infty} \left| \frac{a_{n+1}}{a_n} \right| = \lim_{n \to \infty} \left| \frac{(-2)^{n+1}(n + 1)}{(-2)^n(n + 2)} \right|$$

$$= \lim_{n \to \infty} \frac{2(n + 1)}{(n + 2)} = 2 = L.$$

By the ratio test, the radius of convergence is $\rho = 1/2$. Hence the series (2) converges absolutely for $|x - 3| < 1/2$ and diverges when $|x - 3| > 1/2$. It remains only to determine what happens when $|x - 3| = 1/2$, that is, when $x = 5/2$ and $x = 7/2$.

Set $x = 5/2$, and the series (2) becomes the harmonic series $\sum_{n=0}^{\infty} (n + 1)^{-1}$, which is known to diverge. When $x = 7/2$, the series (2) becomes an **alternating** harmonic series, which is known to converge. Thus the power series converges for each x in the half-open interval $(5/2, 7/2]$; outside this interval it diverges. ∎ End

For each value of x for which the power series $\sum_{n=0}^{\infty} a_n(x - x_0)^n$ converges, we get a number that is the sum of the series. It is appropriate to denote this sum by $f(x)$, since its value depends on the choice of x. Thus we write

$$f(x) = \sum_{n=0}^{\infty} a_n(x - x_0)^n$$

for all numbers x in the convergence interval. For example, the **geometric series** $\sum_{n=0}^{\infty} x^n$ has the radius of convergence $\rho = 1$ and the sum function $f(x) = 1/(1 - x)$; that is,

$$(3) \qquad \frac{1}{1 - x} = 1 + x + x^2 + \cdots = \sum_{n=0}^{\infty} x^n \quad \text{for} \quad -1 < x < 1.$$

Given two power series

$$(4) \qquad f(x) = \sum_{n=0}^{\infty} a_n(x - x_0)^n, \qquad g(x) = \sum_{n=0}^{\infty} b_n(x - x_0)^n,$$

with nonzero radii of convergence, we want to find power series representations for the sum, product, and quotient of the functions $f(x)$ and $g(x)$. The sum is simply obtained by termwise addition:

$$f(x) + g(x) = \sum_{n=0}^{\infty} (a_n + b_n)(x - x_0)^n$$

for all x in the common interval of convergence of the power series in (4). The power series representation for the product $f(x)g(x)$ is a bit more complicated. To provide motivation for the formula, we treat the power series for $f(x)$ and $g(x)$ as "long polynomials," apply the distributive law, and group the terms in powers of $(x - x_0)$:

$$[a_0 + a_1(x - x_0) + a_2(x - x_0)^2 + \cdots] \cdot [b_0 + b_1(x - x_0) + b_2(x - x_0)^2 + \cdots]$$

$$= a_0 b_0 + (a_0 b_1 + a_1 b_0)(x - x_0) + (a_0 b_2 + a_1 b_1 + a_2 b_0)(x - x_0)^2 + \cdots.$$

The general formula for the product is

$$(5) \qquad f(x)g(x) = \sum_{n=0}^{\infty} c_n(x - x_0)^n,$$

where

$$(6) \qquad c_n := \sum_{k=0}^{n} a_k b_{n-k}.$$

The power series in (5) is called the **Cauchy product,** and it will converge for all x in the common *open* interval of convergence for the power series of f and g.[†]

The quotient $f(x)/g(x)$ will also have a power series expansion about x_0, provided $g(x_0) \neq 0$. However, the radius of convergence for this quotient series may be smaller than that for $f(x)$ or $g(x)$. Unfortunately, there is no nice formula for obtaining the coefficients in the power series for $f(x)/g(x)$. However, we can use the Cauchy product to divide power series indirectly (see Problem 15). The quotient series can also be obtained by formally carrying out polynomial long division (see Problem 16).

The next theorem explains, in part, why power series are so useful.

[†] Actually, it may happen that the radius of convergence of the power series for $f(x)g(x)$ or $f(x) + g(x)$ is larger than that for the power series of f or g.

DIFFERENTIATION AND INTEGRATION OF POWER SERIES

Theorem 3. If the series $f(x) = \sum_{n=0}^{\infty} a_n (x - x_0)^n$ has a positive radius of convergence ρ, then f is differentiable in the interval $|x - x_0| < \rho$ and termwise differentiation gives the power series for the derivative:

$$f'(x) = \sum_{n=1}^{\infty} n a_n (x - x_0)^{n-1} \quad \text{for} \quad |x - x_0| < \rho.$$

Furthermore, termwise integration gives the power series for the integral of f:

$$\int f(x)\, dx = \sum_{n=0}^{\infty} \frac{a_n}{n + 1} (x - x_0)^{n+1} + C \quad \text{for} \quad |x - x_0| < \rho.$$

EXAMPLE 2 Starting with the geometric series (3) for $1/(1 - x)$, find a power series for each of the following functions:

(a) $\dfrac{1}{1 + x^2}$. (b) $\dfrac{1}{(1 - x)^2}$. (c) arctan x.

Solution (a) Replacing x by $-x^2$ in (3) immediately gives

(7) $$\frac{1}{1 + x^2} = 1 - x^2 + x^4 - x^6 + \cdots + (-1)^n x^{2n} + \cdots.$$

(b) Notice that $1/(1 - x)^2$ is the derivative of the function $f(x) = 1/(1 - x)$. Hence on differentiating (3) term by term, we obtain

(8) $$f'(x) = \frac{1}{(1 - x)^2} = 1 + 2x + 3x^2 + 4x^3 + \cdots + nx^{n-1} + \cdots.$$

(c) Since

$$\text{arctan } x = \int_0^x \frac{1}{1 + t^2}\, dt,$$

we can integrate the series in (7) termwise to obtain the series for arctan x. Thus

$$\int_0^x \frac{1}{1 + t^2}\, dt = \int_0^x \{1 - t^2 + t^4 - t^6 + \cdots + (-1)^n t^{2n} + \cdots\}\, dt$$

(9) $$\text{arctan } x = x - \frac{1}{3} x^3 + \frac{1}{5} x^5 - \frac{1}{7} x^7 + \cdots + \frac{(-1)^n x^{2n+1}}{2n + 1} + \cdots. \quad \blacksquare \text{ End}$$

It is important to keep in mind that since the geometric series (3) has the (open) interval of convergence $(-1, 1)$, the representations (7), (8), and (9) are at least valid in this interval. (Actually, the series (9) for arctan x converges for all $|x| \le 1$.)

Shifting the Summation Index

The index of summation in a power series is a dummy index just like the variable of integration in a definite integral. Hence

$$\sum_{n=0}^{\infty} a_n(x - x_0)^n = \sum_{k=0}^{\infty} a_k(x - x_0)^k = \sum_{i=0}^{\infty} a_i(x - x_0)^i.$$

Just as there are times when we want to change the variable of integration, there are situations (and we will encounter many in this chapter) when it is desirable to change or shift the index of summation. This is particularly important when one has to combine two different power series.

EXAMPLE 3 Express the series

$$\sum_{n=2}^{\infty} n(n - 1)a_n x^{n-2}$$

as a series where the generic term is x^k instead of x^{n-2}.

Solution Setting $k = n - 2$, we have $n = k + 2$ and $n - 1 = k + 1$. Note that when $n = 2$, then $k = 0$. Hence, substituting into the given series, we find

$$\sum_{n=2}^{\infty} n(n - 1)a_n x^{n-2} = \sum_{k=0}^{\infty} (k + 2)(k + 1)a_{k+2} x^k. \quad \blacksquare \text{ End}$$

EXAMPLE 4 Show that

$$x^3 \sum_{n=0}^{\infty} n^2(n - 2)a_n x^n = \sum_{n=3}^{\infty} (n - 3)^2(n - 5)a_{n-3} x^n.$$

Solution We start by taking the x^3 inside the summation on the left-hand side:

$$x^3 \sum_{n=0}^{\infty} n^2(n - 2)a_n x^n = \sum_{n=0}^{\infty} n^2(n - 2)a_n x^{n+3}.$$

To rewrite this with generic term x^k, we set $k = n + 3$. Thus $n = k - 3$, and $n = 0$ corresponds to $k = 3$. Straightforward substitution thus yields

$$\sum_{n=0}^{\infty} n^2(n - 2)a_n x^{n+3} = \sum_{k=3}^{\infty} (k - 3)^2(k - 5)a_{k-3} x^k.$$

By replacing k by n, we obtain the desired form. $\blacksquare$ End

EXAMPLE 5 Show that the identity

$$\sum_{n=1}^{\infty} n a_{n-1} x^{n-1} + \sum_{n=2}^{\infty} b_n x^{n+1} = 0$$

implies that $a_0 = a_1 = a_2 = 0$ and $a_n = -b_{n-1}/(n + 1)$ for $n \geq 3$.

Solution First we rewrite both series in terms of x^k. For the first series, we set $k = n - 1$, and hence $n = k + 1$, to write

$$\sum_{n=1}^{\infty} n a_{n-1} x^{n-1} = \sum_{k=0}^{\infty} (k + 1) a_k x^k.$$

Then with $k = n + 1$, $n = k - 1$, the second series becomes

$$\sum_{n=2}^{\infty} b_n x^{n+1} = \sum_{k=3}^{\infty} b_{k-1} x^k.$$

The identity thus states

$$\sum_{k=0}^{\infty} (k + 1) a_k x^k + \sum_{k=3}^{\infty} b_{k-1} x^k = 0.$$

Now the only way a power series can sum to the zero function is if each coefficient, individually, equals zero. For $k = 3, 4, \ldots$, both series contribute to the coefficient of x^k, and thus we confirm that

$$(k + 1) a_k + b_{k-1} = 0$$

or $a_k = -b_{k-1}/(k + 1)$ for $k \geq 3$. For $k = 0$, 1, or 2, only the first series contributes, and we find, in turn,

$$(0 + 1) a_0 = 0$$
$$(1 + 1) a_1 = 0$$
$$(2 + 1) a_2 = 0.$$

Hence $a_0 = a_1 = a_2 = 0$. ■ End

Analytic Functions

Not all functions are expressible as power series. Those distinguished functions that can be so represented are called **analytic**.

ANALYTIC FUNCTION

Definition 1. A function f is said to be **analytic at** x_0 if, in an open interval about x_0, this function is the sum of a power series $\sum_{n=0}^{\infty} a_n (x - x_0)^n$ that has a positive radius of convergence.

For example, a polynomial function $b_0 + b_1 x + \cdots + b_n x^n$ is analytic at every x_0, since we can always rewrite it in the form $a_0 + a_1 (x - x_0) + \cdots + a_n (x - x_0)^n$. A rational function $P(x)/Q(x)$, where $P(x)$ and $Q(x)$ are polynomials without a common factor, is an analytic function except at those x_0 for which $Q(x_0) = 0$. As the reader may recall from calculus, the elementary functions e^x, $\sin x$, and $\cos x$ are analytic for all x,

while $\ln x$ is analytic for $x > 0$. Indeed, we have the familiar representations

(10) $\displaystyle e^x = 1 + x + \frac{x^2}{2!} + \frac{x^3}{3!} + \cdots \qquad\qquad = \sum_{n=0}^{\infty} \frac{x^n}{n!},$

(11) $\displaystyle \sin x = x - \frac{x^3}{3!} + \frac{x^5}{5!} - \cdots \qquad\qquad = \sum_{n=0}^{\infty} \frac{(-1)^n}{(2n+1)!} x^{2n+1},$

(12) $\displaystyle \cos x = 1 - \frac{x^2}{2!} + \frac{x^4}{4!} - \cdots \qquad\qquad = \sum_{n=0}^{\infty} \frac{(-1)^n}{(2n)!} x^{2n},$

(13) $\displaystyle \ln x = (x - 1) - \frac{1}{2}(x - 1)^2 + \frac{1}{3}(x - 1)^3 - \cdots = \sum_{n=1}^{\infty} \frac{(-1)^{n-1}}{n} (x - 1)^n,$

where (10), (11), and (12) are valid for all x, whereas (13) is valid for x in the half-open interval $(0, 2]$. In (13) the expansion is about $x_0 = 1$. However, a power series representation for $\ln x$ can be derived about any $x_0 > 0$.

From Theorem 3 on the differentiation of power series, we see that a function f analytic at x_0 is differentiable in a neighborhood of x_0. Moreover, because f' has a power series representation in this neighborhood, it too is analytic at x_0. Repeating this argument, we see that f'', $f^{(3)}$, etc., exist and are analytic at x_0. Consequently, if a function does not have derivatives of all orders at x_0, then it cannot be analytic at x_0. The function $f(x) = |x - 1|$ is not analytic at $x_0 = 1$ because $f'(1)$ does not exist, and $f(x) = x^{8/3}$ is not analytic at $x_0 = 0$ because $f'''(0)$ does not exist.

A formula for the coefficients in the power series of an analytic function is given in the next theorem.

TAYLOR AND MACLAURIN SERIES

Theorem 4. If f is analytic at x_0, then the representation

$$f(x) = \sum_{n=0}^{\infty} \frac{f^{(n)}(x_0)}{n!} (x - x_0)^n$$

$$= f(x_0) + f'(x_0)(x - x_0) + \frac{f''(x_0)}{2!}(x - x_0)^2 + \cdots$$

holds in some open interval centered at x_0.

This series is called the **Taylor series** for f about x_0. When $x_0 = 0$, it is also referred to as the **Maclaurin series** for f.

To determine the Taylor series for an analytic function f, a direct but sometimes tedious approach is to compute the successive derivatives of f and evaluate them at x_0. For example, the series in (10), (11), (12), and (13) can be derived in this manner.

Power series expansions also have a uniqueness property; namely, if the equation

$$\sum_{n=0}^{\infty} a_n(x - x_0)^n = \sum_{n=0}^{\infty} b_n(x - x_0)^n$$

holds in some *open* interval about x_0, then $a_n = b_n$ for $n = 0, 1, 2, \ldots$. Hence, if "by hook or by crook" we can produce a power series expansion for an analytic function, then this power series must be its Taylor series. For example, the expansion for arctan x given in (9) of Example 2 must be its Maclaurin expansion.

The reader will also find it helpful to keep in mind that if f and g are analytic at x_0, then so are $f + g$, cf, fg, and f/g, provided $g(x_0) \neq 0$. These facts follow from the algebraic properties of power series discussed earlier.

Taylor Series Method

We can use the Taylor series representation to obtain a series solution to an initial value problem. Assuming we know the problem has an analytic solution, say by some existence theorem, then we can use the initial conditions and the differential equation itself to compute the successive derivatives of the solution at the initial point. Substituting these values into the formula for the Taylor series, we then obtain a representation of the solution as a power series about the initial point x_0. This procedure, called the **Taylor series method,** is illustrated in the next example.

EXAMPLE 6 Determine the first few terms of a power series solution for the initial value problem

$$\textbf{(14)} \qquad y' = \frac{1}{x + y + 1}, \qquad y(0) = 0.$$

Solution Let's assume the solution to this initial value problem is analytic at $x = 0$. Then we can write

$$\textbf{(15)} \qquad y(x) = \sum_{n=0}^{\infty} \frac{y^{(n)}(0)}{n!} x^n.$$

We already know that $y(0) = 0$. Substituting $x = 0$ and $y = 0$ into equation (14), we find that $y'(0) = 1$. To determine $y''(0)$, we differentiate both sides of the equation in (14) with respect to x, thereby getting an expression for $y''(x)$ in terms of x, $y(x)$, and $y'(x)$. That is,

$$\textbf{(16)} \qquad y''(x) = (-1)[x + y(x) + 1]^{-2}[1 + y'(x)].$$

Substituting $x = 0$, $y(0) = 0$, and $y'(0) = 1$ in (16), we obtain

$$y''(0) = (-1)(1)^{-2}(1 + 1) = -2.$$

Similarly, differentiating (16) and substituting, we obtain

$$y^{(3)}(x) = 2[x + y(x) + 1]^{-3}[1 + y'(x)]^2 - [x + y(x) + 1]^{-2}y''(x),$$
$$y^{(3)}(0) = 2(1)^{-3}(1 + 1)^2 - (1)^{-2}(-2) = 10.$$

Repeating this procedure, we can determine $y^{(n)}(0)$ for any value of n, although the amount of work needed to compute $y^{(n)}(0)$ increases as n increases. Finally, substituting into (15), we find that the first few terms of the Maclaurin series of the solution are

$$y(x) = x - x^2 + \frac{5}{3}x^3 + \cdots . \quad \blacksquare \, \text{End}$$

The Taylor series method applies to nonlinear as well as linear equations. This is in contrast to the methods in the remainder of this chapter, which are restricted to linear equations.

One disadvantage of the Taylor series method is that by computing finitely many terms of the Taylor expansion, one still has no way of knowing the radius of convergence of the series. Fortunately, when the differential equation is linear, there are existence theorems that give a minimum value for this radius (see Theorem 5 in Section 8.4).

EXERCISES 8.2

In Problems 1 through 6, determine the convergence set of the given power series.

1. $\displaystyle\sum_{n=0}^{\infty} \frac{2^{-n}}{n+1}(x-1)^n$.

2. $\displaystyle\sum_{n=0}^{\infty} \frac{3^n}{n!}x^n$.

3. $\displaystyle\sum_{n=0}^{\infty} \frac{n^2}{2^n}(x+2)^n$.

4. $\displaystyle\sum_{n=1}^{\infty} \frac{4}{n^2+2n}(x-3)^n$.

5. $\displaystyle\sum_{n=1}^{\infty} \frac{3}{n^3}(x-2)^n$.

6. $\displaystyle\sum_{n=0}^{\infty} \frac{(n+2)!}{n!}(x+2)^n$.

7. Sometimes the ratio test (Theorem 2) can be applied to a power series containing an infinite number of zero coefficients, provided the zero pattern is regular. Use Theorem 2 to show, for example, that the series

$$a_0 + a_2 x^2 + a_4 x^4 + a_6 x^6 + \cdots = \sum_{k=0}^{\infty} a_{2k} x^{2k}$$

has a radius of convergence $\rho = 1/\sqrt{L}$, if

$$\lim_{k\to\infty} \left| \frac{a_{2k+2}}{a_{2k}} \right| = L,$$

and that

$$a_1 x + a_3 x^3 + a_5 x^5 + a_7 x^7 + \cdots$$

$$= \sum_{k=0}^{\infty} a_{2k+1} x^{2k+1}$$

has a radius of convergence $\rho = 1/\sqrt{M}$, if

$$\lim_{k\to\infty} \left| \frac{a_{2k+3}}{a_{2k+1}} \right| = M.$$

[*Hint:* Let $z = x^2$.]

8. Determine the convergence set of the given power series.

(a) $\displaystyle\sum_{k=0}^{\infty} 2^{2k} x^{2k}$ **(b)** $\displaystyle\sum_{k=0}^{\infty} 2^{2k+1} x^{2k+1}$

(c) $\sin x$ (equation (11)) **(d)** $\cos x$ (equation (12))

(e) $(\sin x)/x$ (equation (5), Section 8.1)

(f) $\displaystyle\sum_{k=0}^{\infty} 2^{2k} x^{4k}$.

In Problems 9 and 10, find the power series expansion for $f(x) + g(x)$, given the expansions for $f(x)$ and $g(x)$.

9. $f(x) = \displaystyle\sum_{n=0}^{\infty} \frac{1}{n+1} x^n$, $g(x) = \displaystyle\sum_{n=1}^{\infty} 2^{-n} x^{n-1}$.

10. $f(x) = \displaystyle\sum_{n=3}^{\infty} \frac{2^n}{n!}(x-1)^{n-2}$, $g(x) = \displaystyle\sum_{n=1}^{\infty} \frac{n^2}{2^n}(x-1)^{n-1}$.

In Problems 11 through 14, find the first three nonzero terms in the power series expansion for the product $f(x)g(x)$.

11. $f(x) = e^x = \displaystyle\sum_{n=0}^{\infty} \frac{1}{n!} x^n$,

$g(x) = \sin x = \displaystyle\sum_{k=0}^{\infty} \frac{(-1)^k}{(2k+1)!} x^{2k+1}$.

12. $f(x) = \sin x = \displaystyle\sum_{k=0}^{\infty} \frac{(-1)^k}{(2k+1)!} x^{2k+1}$,

$g(x) = \cos x = \displaystyle\sum_{k=0}^{\infty} \frac{(-1)^k}{(2k)!} x^{2k}$.

13. $f(x) = e^{-x} = \displaystyle\sum_{n=0}^{\infty} \frac{(-1)^n}{n!} x^n$,

$g(x) = (1+x)^{-1} = \displaystyle\sum_{n=0}^{\infty} (-1)^n x^n$.

14. $f(x) = e^x = \displaystyle\sum_{n=0}^{\infty} \frac{1}{n!} x^n$, $g(x) = e^{-x} = \displaystyle\sum_{n=0}^{\infty} \frac{(-1)^n}{n!} x^n$.

15. Find the first few terms of the power series for the quotient

$$q(x) = \left(\sum_{n=0}^{\infty} \frac{1}{2^n} x^n \right) \Big/ \left(\sum_{n=0}^{\infty} \frac{1}{n!} x^n \right)$$

by completing the following:

(a) Let $q(x) = \sum_{n=0}^{\infty} a_n x^n$, where the coefficients a_n are to be determined. Argue that $\sum_{n=0}^{\infty} x^n/2^n$ is the Cauchy product of $q(x)$ and $\sum_{n=0}^{\infty} x^n/n!$.

(b) Use formula (6) of the Cauchy product to deduce the equations

$$\frac{1}{2^0} = a_0, \qquad \frac{1}{2} = a_0 + a_1, \qquad \frac{1}{2^2} = \frac{a_0}{2} + a_1 + a_2,$$

$$\frac{1}{2^3} = \frac{a_0}{6} + \frac{a_1}{2} + a_2 + a_3, \ \dots.$$

(c) Solve the equations in part (b) to determine the constants a_0, a_1, a_2, a_3.

16. To find the first few terms in the power series for the quotient $q(x)$ in Problem 15 treat the power series in the numerator and denominator as "long polynomials" and carry out long division. That is, perform

$$1 + x + \frac{1}{2}x^2 + \cdots \,\bigg|\, 1 + \frac{1}{2}x + \frac{1}{4}x^2 + \cdots.$$

In Problems 17 through 20, find a power series expansion for $f'(x)$, given the expansion for $f(x)$.

17. $f(x) = (1 + x)^{-1} = \sum_{n=0}^{\infty} (-1)^n x^n.$

18. $f(x) = \sin x = \sum_{k=0}^{\infty} \frac{(-1)^k}{(2k + 1)!} x^{2k+1}.$

19. $f(x) = \sum_{n=0}^{\infty} a_n x^n.$

20. $f(x) = \sum_{n=1}^{\infty} n a_n x^{n-1}.$

In Problems 21 and 22, find a power series expansion for $g(x) := \int_0^x f(t)\,dt$, given the expansion for $f(x)$.

21. $f(x) = (1 + x)^{-1} = \sum_{n=0}^{\infty} (-1)^n x^n.$

22. $f(x) = \frac{\sin x}{x} = \sum_{k=0}^{\infty} \frac{(-1)^k}{(2k + 1)!} x^{2k}.$

In Problems 23 through 26, express the given power series as a series with generic term x^k.

23. $\sum_{n=1}^{\infty} n a_n x^{n-1}.$

24. $\sum_{n=2}^{\infty} n(n - 1) a_n x^{n+2}.$

25. $\sum_{n=0}^{\infty} a_n x^{n+1}.$

26. $\sum_{n=1}^{\infty} \frac{a_n}{n + 3} x^{n+3}.$

27. Show that

$$x^2 \sum_{n=0}^{\infty} n(n + 1) a_n x^n = \sum_{n=2}^{\infty} (n - 2)(n - 1) a_{n-2} x^n.$$

28. Show that

$$2 \sum_{n=0}^{\infty} a_n x^{n+1} + \sum_{n=1}^{\infty} n b_n x^{n-1}$$

$$= b_1 + \sum_{n=1}^{\infty} [2a_{n-1} + (n + 1)b_{n+1}] x^n.$$

In Problems 29 through 34, determine the Taylor series about the point x_0 for the given functions and values of x_0.

29. $f(x) = \cos x, \qquad x_0 = \pi.$

30. $f(x) = x^{-1}, \qquad x_0 = 1.$

31. $f(x) = \dfrac{1 + x}{1 - x}, \qquad x_0 = 0.$

32. $f(x) = \ln(1 + x), \qquad x_0 = 0.$

33. $f(x) = x^3 + 3x - 4, \qquad x_0 = 1.$

34. $f(x) = \sqrt{x}, \qquad x_0 = 1.$

35. The Taylor series for $f(x) = \ln x$ about $x_0 = 1$ given in equation (13) can also be obtained as follows:

(a) Starting with the expansion $1/(1 - s) = \sum_{n=0}^{\infty} s^n$ and observing that

$$\frac{1}{x} = \frac{1}{1 + (x - 1)},$$

obtain the Taylor series for $1/x$ about $x_0 = 1$.

(b) Since $\ln x = \int_1^x 1/t\,dt$, use the result of part (a) and termwise integration to obtain the Taylor series for $f(x) = \ln x$ about $x_0 = 1$.

36. Let $f(x)$ and $g(x)$ be analytic at x_0. Determine whether the following statements are always true or sometimes false:

(a) $3f(x) + g(x)$ is analytic at x_0.

(b) $f(x)/g(x)$ is analytic at x_0.

(c) $f'(x)$ is analytic at x_0.

(d) $[f(x)]^3 - \int_{x_0}^x g(t)\,dt$ is analytic at x_0.

(e) $\sin[f(x)]$ is analytic at x_0.

In Problems 37 through 44, use the Taylor series method to determine the first three nonzero terms of a series solution for the given initial value problem.

37. $y' = x^2 + y^2$; $y(0) = 1$.

38. $y' = y^2$; $y(0) = 1$.

39. $y' = \sin y + e^x$; $y(0) = 0$.

40. $y' = \sin(x + y)$; $y(0) = 0$.

41. $x'' + tx = 0$; $x(0) = 1$, $x'(0) = 0$.

42. $y'' + y = 0$; $y(0) = 0$, $y'(0) = 1$.

43. $y''(\theta) + y(\theta)^3 = \sin \theta$; $y(0) = 0$, $y'(0) = 0$.

44. $y'' + \sin y = 0$; $y(0) = 1$, $y'(0) = 0$.

45. Let

$$f(x) = \begin{cases} e^{-1/x^2}, & x \neq 0, \\ 0, & x = 0. \end{cases}$$

Show that $f^{(n)}(0) = 0$ for $n = 0, 1, 2, \ldots$ and hence that the Maclaurin series for $f(x)$ is $0 + 0 + 0 + \cdots$, which converges for all x but is equal to $f(x)$ only when $x = 0$.

46. van der Pol Equation. In the study of the vacuum tube, the following equation is encountered.

$$y'' + (0.1)(y^2 - 1)y' + y = 0.$$

Use the Taylor series method to find the first three non-zero terms of the series solution for the initial values $y(0) = 1$, $y'(0) = 0$.

47. Duffing's Equation. In the study of a nonlinear spring with periodic forcing, the following equation arises:

$$y'' + ky + ry^3 = A \cos \omega t.$$

Let $k = r = A = 1$ and $\omega = 10$. Use the Taylor series method to find the first three nonzero terms of the series solution for the initial values $y(0) = 0$, $y'(0) = 1$.

48. Soft versus Hard Springs. For Duffing's equation given in Problem 47, the behavior of the solutions changes as r changes sign. When $r > 0$, the restoring force $ky + ry^3$ becomes stronger than for the linear spring ($r = 0$). Such a spring is called **hard.** When $r < 0$, the restoring force becomes weaker than the linear spring and the spring is called **soft.** Pendulums act like soft springs.

(a) Redo Problem 47 with $r = -1$. Notice that for the initial conditions $y(0) = 0$, $y'(0) = 1$, the soft and hard springs appear to respond in the same way for t small.

(b) Keeping $k = A = 1$ and $\omega = 10$, change the initial conditions to $y(0) = 1$ and $y'(0) = 0$. Now redo Problem 47 with $r = \pm 1$.

(c) Based on the results of part (b), is there a difference between the behavior of soft and hard springs for t small? Describe.

8.3 POWER SERIES SOLUTIONS TO LINEAR DIFFERENTIAL EQUATIONS

In this section we demonstrate a method for obtaining a power series solution to a linear differential equation with polynomial coefficients. This method is easier to use than the Taylor series method discussed in Section 8.2 and sometimes gives a nice expression for the general term in the power series expansion. Knowing the form of the general term also allows us to test for the radius of convergence of the power series.

We begin by writing the linear differential equation

(1) $a_2(x)y'' + a_1(x)y' + a_0(x)y = 0$

in the standard form

(2) $y'' + p(x)y' + q(x)y = 0$,

where $p(x) := a_1(x)/a_2(x)$ and $q(x) := a_0(x)/a_2(x)$.

ORDINARY AND SINGULAR POINTS

Definition 2. A point x_0 is called an **ordinary point** of equation (1) if both $p = a_1/a_2$ and $q = a_0/a_2$ are analytic at x_0. If x_0 is not an ordinary point, it is called a **singular point** of the equation.

EXAMPLE 1 Determine all the singular points of

$$xy'' + x(1 - x)^{-1}y' + (\sin x)y = 0.$$

Solution Dividing the equation by x, we find that

$$p(x) = \frac{x}{x(1 - x)}, \qquad q(x) = \frac{\sin x}{x}.$$

The singular points are those points where $p(x)$ *or* $q(x)$ fails to be analytic. Observe that $p(x)$ and $q(x)$ are the ratios of functions that are everywhere analytic. Hence $p(x)$ and $q(x)$ are analytic except, *perhaps*, when their denominators are zero. For $p(x)$ this occurs at $x = 0$ and $x = 1$. But since we can cancel an x in the numerator and denominator of $p(x)$, that is,

$$p(x) = \frac{x}{x(1 - x)} = \frac{1}{1 - x},$$

we see that $p(x)$ is actually analytic at $x = 0$.[†] Therefore $p(x)$ is analytic except at $x = 1$. For $q(x)$, the denominator is zero at $x = 0$. Just as with $p(x)$, this zero is removable, since $q(x)$ has the power series expansion

$$q(x) = \frac{\sin x}{x} = \frac{x - \dfrac{x^3}{3!} + \dfrac{x^5}{5!} - \cdots}{x} = 1 - \frac{x^2}{3!} + \frac{x^4}{5!} - \cdots.$$

Thus $q(x)$ is everywhere analytic. Consequently the only singular point of the given equation is $x = 1$. ∎ End

At an ordinary point x_0 of equation (1) (or (2)), the coefficient functions $p(x)$ and $q(x)$ are analytic. Hence we might expect that the solutions to these equations inherit this property. From the discussion in Section 4.3 on linear second order equations, the continuity of p and q in a neighborhood of x_0 is sufficient to imply that equation (2) has two linearly independent solutions defined in that neighborhood. But analytic functions are not merely continuous—they possess derivatives of all orders in a neighborhood of x_0. Thus we can differentiate equation (2) to show that $y^{(3)}$ exists and, by a ''bootstrap'' argument, prove that solutions to (2) must likewise possess derivatives of all orders. Although we cannot conclude by this reasoning that the solutions enjoy the stronger property of analyticity, this is

[†] Such points are called **removable singularities**. In this chapter we assume in such cases that the function has been defined (or redefined) so that it is analytic at the point.

nonetheless the case (see Theorem 5 in Section 8.4). Hence, in a neighborhood of an ordinary point x_0, the solutions to (1) (or (2)) can be expressed as a power series about x_0.

To illustrate the power series method about an ordinary point, let's look at a simple *first order* linear differential equation.

EXAMPLE 2 Find a power series solution about $x = 0$ to

$$(3) \qquad y' + 2xy = 0.$$

Solution The coefficient of y is the polynomial $2x$, which is analytic everywhere, so $x = 0$ is an ordinary point[†] of equation (3). Thus we expect to find a power series solution of the form

$$(4) \qquad y(x) = a_0 + a_1 x + a_2 x^2 + \cdots = \sum_{n=0}^{\infty} a_n x^n.$$

Our task is to determine the coefficients a_n.

For this purpose we need the expansion for $y'(x)$ that is given by termwise differentiation of (4):

$$y'(x) = 0 + a_1 + 2a_2 x + 3a_3 x^2 + \cdots = \sum_{n=1}^{\infty} n a_n x^{n-1}.$$

We now substitute the series expansions for y and y' into (3) and obtain

$$\sum_{n=1}^{\infty} n a_n x^{n-1} + 2x \sum_{n=0}^{\infty} a_n x^n = 0,$$

which simplifies to

$$(5) \qquad \sum_{n=1}^{\infty} n a_n x^{n-1} + \sum_{n=0}^{\infty} 2 a_n x^{n+1} = 0.$$

To add the two power series in (5), we add the coefficients of like powers of x. If we write out the first few terms of these summations and add, we get

$$\left(a_1 + 2a_2 x + 3a_3 x^2 + 4a_4 x^3 + \cdots \right) + \left(2a_0 x + 2a_1 x^2 + 2a_2 x^3 + \cdots \right) = 0,$$

$$(6) \qquad a_1 + (2a_2 + 2a_0)x + (3a_3 + 2a_1)x^2 + (4a_4 + 2a_2)x^3 + \cdots = 0.$$

For the power series on the left-hand side of equation (6) to be identically zero, we must have all the coefficients equal to zero. Thus

$$a_1 = 0, \qquad 2a_2 + 2a_0 = 0,$$

$$3a_3 + 2a_1 = 0, \qquad 4a_4 + 2a_2 = 0, \qquad \text{etc.}$$

[†] By an ordinary point of a first order equation $y' + q(x)y = 0$, we mean a point where $q(x)$ is analytic.

Solving the preceding system, we find

$$a_1 = 0, \qquad a_2 = -a_0, \qquad a_3 = -\frac{2}{3}a_1 = 0,$$

$$a_4 = -\frac{1}{2}a_2 = -\frac{1}{2}(-a_0) = \frac{1}{2}a_0.$$

Hence the power series for the solution takes the form

(7) $$y(x) = a_0 - a_0 x^2 + \frac{1}{2}a_0 x^4 + \cdots.$$

While the first few terms displayed in (7) are useful, we would much prefer to have a formula for the *general term* in the power series expansion for the solution. To achieve this goal, let's return to equation (5). This time, instead of just writing out a few terms, let's shift the indices in the two power series so that they sum over the same powers of x, say x^k. To do this, we shift the index in the first summation in (5) by setting $k = n - 1$. Then $n = k + 1$ and $k = 0$ when $n = 1$. Hence the first summation in (5) becomes

(8) $$\sum_{n=1}^{\infty} n a_n x^{n-1} = \sum_{k=0}^{\infty} (k + 1)a_{k+1} x^k.$$

In the second summation of (5), we put $k = n + 1$ so that $n = k - 1$ and $k = 1$ when $n = 0$. This gives

(9) $$\sum_{n=0}^{\infty} 2 a_n x^{n+1} = \sum_{k=1}^{\infty} 2 a_{k-1} x^k.$$

Substituting (8) and (9) into (5) yields

(10) $$\sum_{k=0}^{\infty} (k + 1)a_{k+1} x^k + \sum_{k=1}^{\infty} 2 a_{k-1} x^k = 0.$$

Since the first summation in (10) begins at $k = 0$ and the second at $k = 1$, we break up the first into

$$\sum_{k=0}^{\infty} (k + 1)a_{k+1} x^k = a_1 + \sum_{k=1}^{\infty} (k + 1)a_{k+1} x^k.$$

Then (10) becomes

(11) $$a_1 + \sum_{k=1}^{\infty} [(k + 1)a_{k+1} + 2a_{k-1}]x^k = 0.$$

When we set all the coefficients in (11) equal to zero, we find

$$a_1 = 0,$$

and, for all $k \geq 1$,

(12) $$(k + 1)a_{k+1} + 2a_{k-1} = 0.$$

Equation (12) is a **recurrence relation** that we can use to determine the coefficient a_{k+1} in terms of a_{k-1}, that is,

$$a_{k+1} = -\frac{2}{k+1} a_{k-1}.$$

Setting $k = 1, 2, \ldots, 8$ and using the fact that $a_1 = 0$, we find

$$a_2 = -\frac{2}{2} a_0 = -a_0 \quad (k=1), \qquad a_3 = -\frac{2}{3} a_1 = 0 \quad (k=2),$$

$$a_4 = -\frac{2}{4} a_2 = \frac{1}{2} a_0 \quad (k=3), \qquad a_5 = -\frac{2}{5} a_3 = 0 \quad (k=4),$$

$$a_6 = -\frac{2}{6} a_4 = -\frac{1}{3!} a_0 \quad (k=5), \qquad a_7 = -\frac{2}{7} a_5 = 0 \quad (k=6),$$

$$a_8 = -\frac{2}{8} a_6 = \frac{1}{4!} a_0 \quad (k=7), \qquad a_9 = -\frac{2}{9} a_7 = 0 \quad (k=8).$$

After a moment's reflection, we realize that

$$a_{2n} = \frac{(-1)^n}{n!} a_0, \qquad n = 1, 2, \ldots,$$

$$a_{2n+1} = 0, \qquad n = 0, 1, 2, \ldots.$$

Substituting back into the expression (4), we obtain the power series solution

(13) $$y(x) = a_0 - a_0 x^2 + \frac{1}{2!} a_0 x^4 + \cdots = a_0 \sum_{n=0}^{\infty} \frac{(-1)^n}{n!} x^{2n}.$$

Since a_0 is left undetermined, it serves as an arbitrary constant, and hence (13) gives a general solution to equation (3). ■ End

Using the ratio test, it can be verified that the power series in (13) has radius of convergence $\rho = \infty$. Moreover, (13) is reminiscent of the expansion for the exponential function; the reader should check that it converges to

$$y(x) = a_0 e^{-x^2}.$$

This general solution to the simple equation (3) can also be obtained by the method of separation of variables.

The convergence of the partial sums of (13), with $a_0 = 1$, to the actual solution e^{-x^2} is depicted in Figure 8.2. Notice that taking more terms results in better approximations around $x = 0$. Note also, however, that every partial sum is a *polynomial*, and hence must diverge at $x = \pm\infty$, while e^{-x^2} converges to zero, of course. Thus the approximation eventually deteriorates no matter how many terms we keep. This is a typical feature of power series approximations.

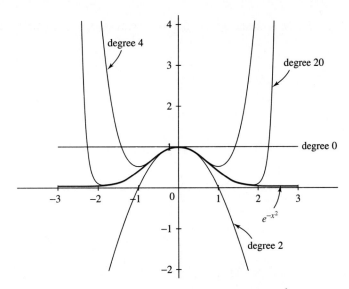

Figure 8.2 Partial sum approximations to e^{-x^2}

In the next example we use the power series method to obtain a general solution to a linear second order differential equation.

EXAMPLE 3 Find a general solution to

(14) $2y'' + xy' + y = 0$

in the form of a power series about the ordinary point $x = 0$.

Solution Writing

(15) $y(x) = a_0 + a_1 x + a_2 x^2 + \cdots = \displaystyle\sum_{n=0}^{\infty} a_n x^n,$

we differentiate termwise to obtain

$$y'(x) = a_1 + 2a_2 x + 3a_3 x^2 + \cdots = \sum_{n=1}^{\infty} n a_n x^{n-1},$$

$$y''(x) = 2a_2 + 6a_3 x + 12a_4 x^2 + \cdots = \sum_{n=2}^{\infty} n(n-1) a_n x^{n-2}.$$

Substituting these power series into equation (14), we find

(16) $\displaystyle\sum_{n=2}^{\infty} 2n(n-1) a_n x^{n-2} + \sum_{n=1}^{\infty} n a_n x^n + \sum_{n=0}^{\infty} a_n x^n = 0.$

To simplify the addition of the three summations in (16), let's shift the indices so that the general term in each is a constant times x^k. For the first summation, we substitute

$k = n - 2$ and get

$$\sum_{n=2}^{\infty} 2n(n - 1)a_n x^{n-2} = \sum_{k=0}^{\infty} 2(k + 2)(k + 1)a_{k+2} x^k.$$

In the second and third summations, we simply substitute k for n. With these changes of indices equation (16) becomes

$$\sum_{k=0}^{\infty} 2(k + 2)(k + 1)a_{k+2} x^k + \sum_{k=1}^{\infty} ka_k x^k + \sum_{k=0}^{\infty} a_k x^k = 0.$$

Next, we separate the x^0 terms from the others and then combine the like powers of x in the three summations to get

$$4a_2 + a_0 + \sum_{k=1}^{\infty} [2(k + 2)(k + 1)a_{k+2} + ka_k + a_k]x^k = 0.$$

Setting the coefficients of this power series equal to zero yields

(17) $4a_2 + a_0 = 0$

and the recurrence relation

(18) $2(k + 2)(k + 1)a_{k+2} + (k + 1)a_k = 0, \qquad k \geq 1.$

We can now use (17) and (18) to determine all the coefficients a_k of the solution in terms of a_0 and a_1. Solving (18) for a_{k+2} gives

(19) $a_{k+2} = \dfrac{-1}{2(k + 2)} a_k, \qquad k \geq 1.$

Thus

$$a_2 = \frac{-1}{2^2} a_0,$$

$$a_3 = \frac{-1}{2 \cdot 3} a_1 \qquad\qquad (k = 1),$$

$$a_4 = \frac{-1}{2 \cdot 4} a_2 = \frac{1}{2^2 \cdot 2 \cdot 4} a_0 \qquad (k = 2),$$

$$a_5 = \frac{-1}{2 \cdot 5} a_3 = \frac{1}{2^2 \cdot 3 \cdot 5} a_1 \qquad (k = 3),$$

$$a_6 = \frac{-1}{2 \cdot 6} a_4 = \frac{-1}{2^3 \cdot 2 \cdot 4 \cdot 6} a_0 = \frac{-1}{2^6 \cdot 3!} a_0 \qquad (k = 4),$$

$$a_7 = \frac{-1}{2 \cdot 7} a_5 = \frac{-1}{2^3 \cdot 3 \cdot 5 \cdot 7} a_1 \qquad (k = 5),$$

$$a_8 = \frac{-1}{2 \cdot 8} a_6 = \frac{1}{2^4 \cdot 2 \cdot 4 \cdot 6 \cdot 8} a_0 = \frac{1}{2^8 \cdot 4!} a_0 \qquad (k = 6).$$

The pattern for the coefficients is now apparent. With a_0 and a_1 taken as arbitrary constants, we find

$$a_{2n} = \frac{(-1)^n}{2^{2n}n!}a_0, \qquad n \geq 1,$$

and

$$a_{2n+1} = \frac{(-1)^n}{2^n[1 \cdot 3 \cdot 5 \cdots (2n + 1)]}a_1, \qquad n \geq 1.$$

From this, two linearly independent solutions emerge; namely,

(20) $$y_1(x) = \sum_{n=0}^{\infty} \frac{(-1)^n}{2^{2n}n!} x^{2n},$$

(21) $$y_2(x) = \sum_{n=0}^{\infty} \frac{(-1)^n}{2^n[1 \cdot 3 \cdot 5 \cdots (2n + 1)]} x^{2n+1}.$$

Hence a general solution to (14) is $a_0 y_1(x) + a_1 y_2(x)$. Approximations to the solutions $y_1(x)$, $y_2(x)$ are depicted in Figure 8.3. ■ End

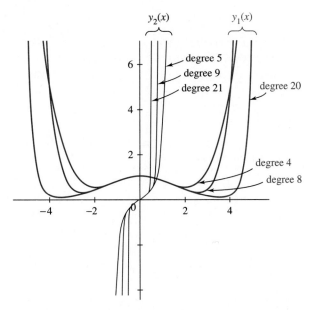

Figure 8.3 Partial sum approximations to solutions for Example 3

The method illustrated in Example 3 can also be used to solve initial value problems. Suppose we are given the value of $y(0)$ and $y'(0)$; then, from equation (15), we see that $a_0 = y(0)$ and $a_1 = y'(0)$. Knowing these two coefficients leads to a unique power series solution for the initial value problem.

The recurrence relation (18) in Example 3 involved just two of the coefficients, a_{k+2} and a_k, and we were fortunate in being able to deduce from this relation the general form for the coefficient a_n. However, many cases arise that lead to more complicated two-term or even to many-term recurrence relations. When this occurs, it may be impossible to determine the general form for the coefficients a_n. In the next example we consider an equation that gives rise to a three-term recurrence relation.

EXAMPLE 4 Find the first few terms in a power series expansion about $x = 0$ for a general solution to

(22) $(1 + x^2)y'' - y' + y = 0.$

Solution Since $p(x) = -(1 + x^2)^{-1}$ and $q(x) = (1 + x^2)^{-1}$ are analytic at $x = 0$, then $x = 0$ is an ordinary point for equation (22). Hence we can express its general solution in the form

$$y(x) = \sum_{n=0}^{\infty} a_n x^n.$$

Substituting this expansion into (22) yields

$$(1 + x^2)\sum_{n=2}^{\infty} n(n - 1)a_n x^{n-2} - \sum_{n=1}^{\infty} na_n x^{n-1} + \sum_{n=0}^{\infty} a_n x^n = 0,$$

(23) $\displaystyle\sum_{n=2}^{\infty} n(n - 1)a_n x^{n-2} + \sum_{n=2}^{\infty} n(n - 1)a_n x^n - \sum_{n=1}^{\infty} na_n x^{n-1} + \sum_{n=0}^{\infty} a_n x^n = 0.$

To sum over like powers x^k, we put $k = n - 2$ in the first summation of (23), $k = n - 1$ in the third, and $k = n$ in the second and fourth. This gives

$$\sum_{k=0}^{\infty} (k + 2)(k + 1)a_{k+2}x^k + \sum_{k=2}^{\infty} k(k - 1)a_k x^k - \sum_{k=0}^{\infty} (k + 1)a_{k+1}x^k + \sum_{k=0}^{\infty} a_k x^k = 0.$$

Separating the terms corresponding to $k = 0$ and $k = 1$ and combining the rest under one summation, we have

$$(2a_2 - a_1 + a_0) + (6a_3 - 2a_2 + a_1)x$$

$$+ \sum_{k=2}^{\infty} [(k + 2)(k + 1)a_{k+2} - (k + 1)a_{k+1} + (k(k - 1) + 1)a_k]x^k = 0.$$

Setting the coefficients equal to zero gives

(24) $2a_2 - a_1 + a_0 = 0,$

(25) $6a_3 - 2a_2 + a_1 = 0$

and the recurrence relation

(26) $(k + 2)(k + 1)a_{k+2} - (k + 1)a_{k+1} + (k^2 - k + 1)a_k = 0, \qquad k \geq 2.$

We can solve (24) for a_2 in terms of a_0 and a_1:

$$a_2 = \frac{a_1 - a_0}{2}.$$

Now that we have a_2, we can use (25) to express a_3 in terms of a_0 and a_1:

$$a_3 = \frac{2a_2 - a_1}{6} = \frac{(a_1 - a_0) - a_1}{6} = \frac{-a_0}{6}.$$

Solving the recurrence relation (26) for a_{k+2}, we obtain

(27) $a_{k+2} = \dfrac{(k+1)a_{k+1} - (k^2 - k + 1)a_k}{(k+2)(k+1)}, \qquad k \geq 2.$

For $k = 2$, 3, and 4, this gives

$$a_4 = \frac{3a_3 - 3a_2}{4 \cdot 3} = \frac{a_3 - a_2}{4}$$

$$= \frac{\dfrac{-a_0}{6} - \left(\dfrac{a_1 - a_0}{2}\right)}{4} = \frac{2a_0 - 3a_1}{24} \qquad (k = 2),$$

$$a_5 = \frac{4a_4 - 7a_3}{5 \cdot 4} = \frac{3a_0 - a_1}{40} \qquad (k = 3),$$

$$a_6 = \frac{5a_5 - 13a_4}{6 \cdot 5} = \frac{36a_1 - 17a_0}{720} \qquad (k = 4).$$

We can now express a general solution in terms up to order 6, using a_0 and a_1 as the arbitrary constants. Thus

(28) $y(x) = a_0 + a_1 x + \left(\dfrac{a_1 - a_0}{2}\right)x^2 - \dfrac{a_0}{6}x^3$

$$+ \left(\frac{2a_0 - 3a_1}{24}\right)x^4 + \left(\frac{3a_0 - a_1}{40}\right)x^5 + \left(\frac{36a_1 - 17a_0}{720}\right)x^6 + \cdots$$

$$= a_0\left(1 - \frac{1}{2}x^2 - \frac{1}{6}x^3 + \frac{1}{12}x^4 + \frac{3}{40}x^5 - \frac{17}{720}x^6 + \cdots\right)$$

$$+ a_1\left(x + \frac{1}{2}x^2 - \frac{1}{8}x^4 - \frac{1}{40}x^5 + \frac{1}{20}x^6 + \cdots\right). \quad \blacksquare \text{ End}$$

Given specific values for a_0 and a_1, will the partial sums of the power series representation (28) yield useful approximations to the solution when $x = 0.5$? What about when $x = 2.3$ or $x = 7.8$? The answers to these questions certainly depend on the radius of convergence of the power series in (28). But since we were not able to determine a general form for the coefficients a_n in this example, we cannot use the ratio test (or other methods such as the root test, integral test, or comparison test) to compute the radius ρ. In the next section, we remedy this situation by giving a simple procedure that determines a lower bound for the radius of convergence of power series solutions.

EXERCISES 8.3

In Problems 1 through 10, determine all the singular points of the given differential equation.

1. $(x + 1)y'' - x^2y' + 3y = 0$.

2. $x^2y'' + 3y' - xy = 0$.

3. $(\theta^2 - 2)y'' + 2y' + (\sin \theta)y = 0$.

4. $(x^2 + x)y'' + 3y' - 6xy = 0$.

5. $(t^2 - t - 2)x'' + (t + 1)x' - (t - 2)x = 0$.

6. $(x^2 - 1)y'' + (1 - x)y' + (x^2 - 2x + 1)y = 0$.

7. $(\sin x)y'' + (\cos x)y = 0$.

8. $e^xy'' - (x^2 - 1)y' + 2xy = 0$.

9. $(\sin \theta)y'' - (\ln \theta)y = 0$.

10. $[\ln(x - 1)]y'' + (\sin 2x)y' - e^xy = 0$.

In Problems 11 through 18, find at least the first four non-zero terms in a power series expansion about $x = 0$ for a general solution to the given differential equation.

11. $y' + (x + 2)y = 0$. **12.** $y' - y = 0$.

13. $z'' - x^2z = 0$. **14.** $(x^2 + 1)y'' + y = 0$.

15. $y'' + (x - 1)y' + y = 0$.

16. $y'' - 2y' + y = 0$.

17. $w'' - x^2w' + w = 0$.

18. $(2x - 3)y'' - xy' + y = 0$.

In Problems 19 through 24, find a power series expansion about $x = 0$ for a general solution to the given differential equation. Your answer should include a general formula for the coefficients.

19. $y' - 2xy = 0$. **20.** $y'' + y = 0$.

21. $y'' - xy' + 4y = 0$. **22.** $y'' - xy = 0$.

23. $z'' - x^2z' - xz = 0$.

24. $(x^2 + 1)y'' - xy' + y = 0$.

In Problems 25 through 28, find at least the first four non-zero terms in a power series expansion about $x = 0$ for the solution to the given initial value problem.

25. $w'' + 3xw' - w = 0$; $w(0) = 2$, $w'(0) = 0$.

26. $(x^2 - x + 1)y'' - y' - y = 0$; $y(0) = 0$, $y'(0) = 1$.

27. $(x + 1)y'' - y = 0$; $y(0) = 0$, $y'(0) = 1$.

28. $y'' + (x - 2)y' - y = 0$; $y(0) = -1$, $y'(0) = 0$.

In Problems 29 through 31, use the first few terms of the power series expansion to find a cubic polynomial approximation for the solution to the given initial value problem. Graph the linear, quadratic, and cubic polynomial approximations for $-5 \leq x \leq 5$.

29. $y'' + y' - xy = 0$; $y(0) = 1$, $y'(0) = -2$.

30. $y'' - 4xy' + 5y = 0$; $y(0) = -1$, $y'(0) = 1$.

31. $(x^2 + 2)y'' + 2xy' + 3y = 0$; $y(0) = 1$, $y'(0) = 2$.

32. Consider the initial value problem
$$y'' - 2xy' - 2y = 0; \quad y(0) = a_0, \quad y'(0) = a_1,$$
where a_0 and a_1 are constants.

(a) Show that if $a_0 = 0$, then the solution will be an odd function (that is, $y(-x) = -y(x)$ for all x. What happens when $a_1 = 0$?

(b) Show that if a_0 and a_1 are positive, then the solution is increasing on $(0, \infty)$.

(c) Show that if a_0 is negative and a_1 is positive, then the solution is increasing on $(-\infty, 0)$.

(d) What conditions on a_0 and a_1 would guarantee that the solution is increasing on $(-\infty, \infty)$?

33. Use the ratio test to show that the radius of convergence of the series in equation (13) is infinite. [*Hint:* First consider the series $\sum_{n=0}^{\infty}(-1)^n x^n/n!$.]

34. Emden's Equation. A classical nonlinear equation that occurs in the study of the thermal behavior of a spherical cloud is **Emden's equation**
$$y'' + \frac{2}{x}y' + y^n = 0,$$
with initial conditions $y(0) = 1$, $y'(0) = 0$. (See Section 8.1 for the case when $n = 1$.) Even though $x = 0$ is *not* an ordinary point for this equation (which is nonlinear for $n \neq 1$), it turns out that there does exist a solution analytic at $x = 0$. Assuming n is a positive integer, show that the first few terms in a power series solution are
$$y = 1 - \frac{x^2}{3!} + n\frac{x^4}{5!} + \cdots.$$

[*Hint:* Substitute $y = 1 + c_2x^2 + c_3x^3 + c_4x^4 + c_5x^5 + \cdots$ into the equation and carefully compute the first few terms in y^n.]

35. Variable Resistor. In Section 5.5, we showed that the charge q on the capacitor in a simple *RLC* circuit is

governed by the equation

$$Lq''(t) + Rq'(t) + \frac{1}{C} q(t) = E(t),$$

where L is the inductance, R the resistance, C the capacitance, and E the voltage source. Since the resistance of a resistor increases with temperature, let's assume the resistor is heated so that the resistance at time t is $R(t) = 1 + t/10$ ohms (see Figure 8.4). If $L = 0.1$ henrys, $C = 2$ farads, $E(t) \equiv 0$, $q(0) = 10$ coulombs, and $q'(0) = 0$ amps, find at least the first four nonzero terms in a power series expansion about $t = 0$ for the charge on the capacitor.

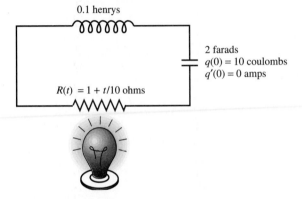

Figure 8.4 An *RLC* circuit whose resistor is being heated

36. Variable Spring Constant. As a spring is heated, its spring "constant" decreases. Suppose the spring is heated so that the spring "constant" at time t is $k(t) = 6 - t$ N/m (see Figure 8.5). If the unforced spring-mass system has mass $m = 2$ kg and a damping constant $b = 1$ N-sec/m with initial conditions $x(0) = 3$ m and $x'(0) = 0$ m/sec, then the displacement $x(t)$ is governed by the initial value problem

$$2x''(t) + x'(t) + (6 - t)x(t) = 0;$$
$$x(0) = 3, \qquad x'(0) = 0.$$

Find at least the first four nonzero terms in a power series expansion about $t = 0$ for the displacement.

Figure 8.5 A spring-mass system whose spring is being heated

8.4 EQUATIONS WITH ANALYTIC COEFFICIENTS

In Section 8.3 we introduced a method for obtaining a power series solution about an ordinary point. In this section we continue the discussion of this procedure. We begin by stating a basic existence theorem for the equation

(1) $y''(x) + p(x)y'(x) + q(x)y(x) = 0,$

which justifies the power series method.

EXISTENCE OF ANALYTIC SOLUTIONS

Theorem 5. Suppose x_0 is an ordinary point for equation (1). Then (1) has two linearly independent analytic solutions of the form

(2) $y(x) = \displaystyle\sum_{n=0}^{\infty} a_n(x - x_0)^n.$

Moreover, the radius of convergence of any power series solution of the form given by (2) is at least as large as the distance from x_0 to the nearest singular point (real or complex-valued) of equation (1).

The key element in the proof of Theorem 5 is the construction of a convergent geometric series that dominates the series expansion (2) of a solution to equation (1). The convergence of the series in (2) then follows by the comparison test. The details of the proof can be found in more advanced books on differential equations.[†]

As we saw in Section 8.3, the power series method gives us a general solution in the same form as (2), with a_0 and a_1 as arbitrary constants. The two linearly independent solutions referred to in Theorem 5 can be obtained by taking $a_0 = 1$, $a_1 = 0$ for the first and $a_0 = 0$, $a_1 = 1$ for the second. Thus we can extend Theorem 5 by saying that *equation (1) has a general solution of the form (2) with a_0 and a_1 as the arbitrary constants.*

The second part of Theorem 5 gives a simple way of determining a minimum value for the radius of convergence of the power series. We need only find the singular points of equation (1) and then determine the distance between the ordinary point x_0 and the nearest singular point.

EXAMPLE 1 Find a minimum value for the radius of convergence of a power series solution about $x = 0$ to

(3) $2y'' + xy' + y = 0.$

Solution For this equation, $p(x) = x/2$ and $q(x) = 1/2$. Both of these functions are analytic for all real or complex values of x. Since equation (3) has no singular points, the distance between the ordinary point $x = 0$ and the nearest singular point is infinite. Hence the radius of convergence is infinite. ∎ End

The next example helps to answer the questions posed at the end of the last section.

EXAMPLE 2 Find a minimum value for the radius of convergence of a power series solution about $x = 0$ to

(4) $(1 + x^2)y'' - y' + y = 0.$

Solution Here $p(x) = -1/(1 + x^2)$, $q(x) = 1/(1 + x^2)$, and so the singular points of equation (4) occur when $1 + x^2 = 0$; that is, when $x = \pm\sqrt{-1} = \pm i$. Since the only singular points of equation (4) are the complex numbers $\pm i$, we see that $x = 0$ is an ordinary point. Moreover, the distance[‡] from 0 to either $\pm i$ is 1. Thus the radius of convergence of a power series solution about $x = 0$ is at least 1. ∎ End

In equation (28) of Section 8.3, we found the first few terms of a power series solution to equation (4). Because we now know that this series has radius of convergence at least 1, the partial sums of this series will converge to the solution for $|x| < 1$. However, when $|x| \geq 1$, we have no basis on which to decide whether we can use the series to approximate the solution.

[†] See, for example, *Intermediate Differential Equations*, Second Edition, by Earl D. Rainville (Macmillan, New York, 1964), Chapter 3, Section 22.

[‡] Recall that the distance between the two complex numbers $z = a + bi$ and $w = c + di$ is given by $\sqrt{(a - c)^2 + (b - d)^2}$.

Power series expansions about $x_0 = 0$ are somewhat easier to manipulate than expansions about nonzero points. As the next example shows, a simple shift in variable enables us always to expand about the origin.

EXAMPLE 3 Find the first few terms in a power series expansion about $x = 1$ for a general solution to

$$(5) \qquad 2y'' + xy' + y = 0.$$

Also determine the radius of convergence of the series.

Solution As seen in Example 1, there are no singular points for equation (5). Thus $x = 1$ is an ordinary point, and, as a consequence of Theorem 5, equation (5) has a general solution of the form

$$(6) \qquad y(x) = \sum_{n=0}^{\infty} a_n(x - 1)^n.$$

Moreover, the radius of convergence of the series in (6) must be infinite.

We can simplify the computation of the coefficients a_n by shifting the center of the expansion (6) from $x_0 = 1$ to $t_0 = 0$. This is accomplished by the substitution $t = x - 1$. Since $y(x) = y(t + 1)$, we find via the chain rule

$$\frac{dy}{dx} = \frac{dy}{dt}, \qquad \frac{d^2y}{dx^2} = \frac{d^2y}{dt^2},$$

and hence equation (5) is changed into

$$(7) \qquad 2\frac{d^2y}{dt^2} + (t + 1)\frac{dy}{dt} + y = 0.$$

We now seek a general solution of the form

$$(8) \qquad y(t) = \sum_{n=0}^{\infty} a_n t^n,$$

where the a_n's in equations (6) and (8) are the same. Proceeding as usual, we substitute the power series for $y(t)$ into (7), derive a recurrence relation for the coefficients, and ultimately find that

$$y(t) = a_0 \left\{ 1 - \frac{1}{4}t^2 + \frac{1}{24}t^3 + \cdots \right\} + a_1 \left\{ t - \frac{1}{4}t^2 - \frac{1}{8}t^3 + \cdots \right\}$$

(the details are left for the reader). Thus

$$(9) \qquad y(x) = a_0 \left\{ 1 - \frac{1}{4}(x - 1)^2 + \frac{1}{24}(x - 1)^3 + \cdots \right\}$$

$$+ a_1 \left\{ (x - 1) - \frac{1}{4}(x - 1)^2 - \frac{1}{8}(x - 1)^3 + \cdots \right\}. \qquad \blacksquare \text{ End}$$

When the coefficients of a linear equation are not polynomials in x, but are analytic functions, we can still find analytic solutions by essentially the same method.

EXAMPLE 4 Find a power series expansion for the solution to

(10) $y''(x) + e^x y'(x) + (1 + x^2) y(x) = 0; \qquad y(0) = 1, \qquad y'(0) = 0.$

Solution Here $p(x) = e^x$ and $q(x) = 1 + x^2$, both of which are analytic for all x. Thus by Theorem 5, the initial value problem (10) has a power series solution

(11) $\displaystyle y(x) = \sum_{n=0}^{\infty} a_n x^n$

that converges for all $x(\rho = \infty)$. To find the first few terms of this series, we first expand $p(x) = e^x$ in its Maclaurin series:

$$e^x = 1 + x + \frac{x^2}{2!} + \frac{x^3}{3!} + \cdots.$$

Substituting the expansions for $y(x)$, $y'(x)$, $y''(x)$, and e^x into (10) gives

(12) $\displaystyle \sum_{n=2}^{\infty} n(n-1) a_n x^{n-2} + \left(1 + x + \frac{x^2}{2} + \frac{x^3}{6} + \frac{x^4}{24} + \cdots \right) \sum_{n=1}^{\infty} n a_n x^{n-1}$

$\displaystyle + (1 + x^2) \sum_{n=0}^{\infty} a_n x^n = 0.$

Because of the computational difficulties due to the appearance of the product of the power series for e^x and $y'(x)$, we concern ourselves with just those terms up to order 4. Writing out (12) and keeping track of all such terms, we find

(13) $(2a_2 + 6a_3 x + 12a_4 x^2 + 20a_5 x^3 + 30a_6 x^4 + \cdots)$

$\left. \begin{array}{l} + (a_1 + 2a_2 x + 3a_3 x^2 + 4a_4 x^3 + 5a_5 x^4 + \cdots) \\[2mm] \quad + (a_1 x + 2a_2 x^2 + 3a_3 x^3 + 4a_4 x^4 + \cdots) \\[2mm] \qquad + \left(\frac{1}{2} a_1 x^2 + a_2 x^3 + \frac{3}{2} a_3 x^4 + \cdots \right) \\[2mm] \qquad\quad + \left(\frac{1}{6} a_1 x^3 + \frac{1}{3} a_2 x^4 + \cdots \right) \\[2mm] \qquad\qquad + \left(\frac{1}{24} a_1 x^4 + \cdots \right) \end{array} \right\}$
$\quad\begin{array}{l} 1 \cdot \Sigma \, n a_n x^{n-1} \\[2mm] x \cdot \Sigma \, n a_n x^{n-1} \\[2mm] \dfrac{x^2}{2} \cdot \Sigma \, n a_n x^{n-1} \\[2mm] \ \vdots \end{array}$

$+ (a_0 + a_1 x + a_2 x^2 + a_3 x^3 + a_4 x^4 + \cdots) \qquad\qquad 1 \cdot \Sigma \, a_n x^n$

$\quad + (a_0 x^2 + a_1 x^3 + a_2 x^4 + \cdots) = 0. \qquad\qquad x^2 \cdot \Sigma \, a_n x^n$

Grouping the like powers of x in equation (13) (for example, the x^2 terms are shown in color) and then setting the coefficients equal to zero yields the system of equations

$$2a_2 + a_1 + a_0 = 0 \qquad (x^0 \text{ term}),$$

$$6a_3 + 2a_2 + 2a_1 = 0 \qquad (x^1 \text{ term}),$$

$$12a_4 + 3a_3 + 3a_2 + \frac{1}{2}a_1 + a_0 = 0 \qquad (x^2 \text{ term}),$$

$$20a_5 + 4a_4 + 4a_3 + a_2 + \frac{7}{6}a_1 = 0 \qquad (x^3 \text{ term}),$$

$$30a_6 + 5a_5 + 5a_4 + \frac{3}{2}a_3 + \frac{4}{3}a_2 + \frac{1}{24}a_1 = 0 \qquad (x^4 \text{ term}).$$

The initial conditions in (10) imply that $y(0) = a_0 = 1$ and $y'(0) = a_1 = 0$. Using these values for a_0 and a_1, we can solve the above system first for a_2, then a_3, and so on:

$$2a_2 + 0 + 1 = 0 \Rightarrow a_2 = -\frac{1}{2},$$

$$6a_3 - 1 + 0 = 0 \Rightarrow a_3 = \frac{1}{6},$$

$$12a_4 + \frac{1}{2} - \frac{3}{2} + 0 + 1 = 0 \Rightarrow a_4 = 0,$$

$$20a_5 + 0 + \frac{2}{3} - \frac{1}{2} + 0 = 0 \Rightarrow a_5 = -\frac{1}{120},$$

$$30a_6 - \frac{1}{24} + 0 + \frac{1}{4} - \frac{2}{3} + 0 = 0 \Rightarrow a_6 = \frac{11}{720}.$$

Thus the solution to the initial value problem in (10) is

(14) $\qquad y(x) = 1 - \frac{1}{2}x^2 + \frac{1}{6}x^3 - \frac{1}{120}x^5 + \frac{11}{720}x^6 + \cdots .$ ■ End

Thus far we have used the power series method only for homogeneous equations. But the same method applies, with obvious modifications, to nonhomogeneous equations of the form

(15) $\qquad y''(x) + p(x)y'(x) + q(x)y(x) = g(x),$

provided the forcing term $g(x)$ and the coefficient functions are analytic at x_0. For example, to find a power series about $x = 0$ for a general solution to

(16) $\qquad y''(x) - xy'(x) - y(x) = \sin x,$

we use the substitution $y(x) = \Sigma a_n x^n$ to obtain a power series expansion for the left-hand side of (16). We then equate the coefficients of this series with the corresponding coeffi-

cients of the Maclaurin expansion for $\sin x$:

$$\sin x = \sum_{n=0}^{\infty} \frac{(-1)^n}{(2n+1)!} x^{2n+1}.$$

Carrying out the details (see Problem 20), we ultimately find that an expansion for a general solution to (16) is

$$(17) \qquad y(x) = a_0 y_1(x) + a_1 y_2(x) + y_p(x),$$

where

$$(18) \qquad y_1(x) = 1 + \frac{1}{2}x^2 + \frac{1}{8}x^4 + \frac{1}{48}x^6 + \cdots,$$

$$(19) \qquad y_2(x) = x + \frac{1}{3}x^3 + \frac{1}{15}x^5 + \frac{1}{105}x^7 + \cdots$$

are the solutions to the homogeneous equation associated with equation (16) and

$$(20) \qquad y_p(x) = \frac{1}{6}x^3 + \frac{1}{40}x^5 + \frac{19}{5040}x^7 + \cdots$$

is a particular solution to equation (16).

EXERCISES 8.4

In Problems 1 through 6, find a minimum value for the radius of convergence of a power series solution about x_0.

1. $(x+1)y'' - 3xy' + 2y = 0$, $\quad x_0 = 1$.

2. $y'' - xy' - 3y = 0$, $\quad x_0 = 2$.

3. $(1 + x^2)y'' - 3y = 0$, $\quad x_0 = 1$.

4. $(x^2 - 5x + 6)y'' - 3xy' - y = 0$, $\quad x_0 = 0$.

5. $y'' - (\tan x)y' + y = 0$, $\quad x_0 = 0$.

6. $(1 + x^3)y'' - xy' + 3x^2y = 0$, $\quad x_0 = 1$.

In Problems 7 through 12, find at least the first four non-zero terms in a power series expansion about x_0 for a general solution to the given differential equation with the given value for x_0.

7. $y' + 2(x - 1)y = 0$, $\quad x_0 = 1$.

8. $y' - 2xy = 0$, $\quad x_0 = -1$.

9. $(x^2 - 2x)y'' + 2y = 0$, $\quad x_0 = 1$.

10. $x^2y'' - xy' + 2y = 0$, $\quad x_0 = 2$.

11. $x^2y'' - y' + y = 0$, $\quad x_0 = 2$.

12. $y'' + (3x - 1)y' - y = 0$, $\quad x_0 = -1$.

In Problems 13 through 19, find at least the first four non-zero terms in a power series expansion of the solution to the given initial value problem.

13. $x' + (\sin t)x = 0$, $\quad x(0) = 1$.

14. $y' - e^x y = 0$, $\quad y(0) = 1$.

15. $(x^2 + 1)y'' - e^x y' + y = 0$, $\quad y(0) = 1$, $\quad y'(0) = 1$.

16. $y'' + ty' + e^t y = 0$, $\quad y(0) = 0$, $\quad y'(0) = -1$.

17. $y'' - (\sin x)y = 0$, $\quad y(\pi) = 1$, $\quad y'(\pi) = 0$.

18. $y'' - (\cos x)y' - y = 0$, $\quad y(\pi/2) = 1$, $\quad y'(\pi/2) = 1$.

19. $y'' - e^{2x}y' + (\cos x)y = 0$, $\quad y(0) = -1$, $\quad y'(0) = 1$.

20. To derive the general solution given by equations (17)–(20) for the nonhomogeneous equation (16), complete the following steps:

 (a) Substitute $y(x) = \sum_{n=0}^{\infty} a_n x^n$ and the Maclaurin series for $\sin x$ into equation (16) to obtain

$$(2a_2 - a_0) + \sum_{k=1}^{\infty} [(k+2)(k+1)a_{k+2} - (k+1)a_k]x^k$$

$$= \sum_{n=0}^{\infty} \frac{(-1)^n}{(2n+1)!} x^{2n+1}.$$

(b) Equate the coefficients of like powers of x on both sides of the equation in part (a) and thereby deduce the equations

$$a_2 = \frac{a_0}{2}, \quad a_3 = \frac{1}{6} + \frac{a_1}{3}, \quad a_4 = \frac{a_0}{8},$$

$$a_5 = \frac{1}{40} + \frac{a_1}{15}, \quad a_6 = \frac{a_0}{48}, \quad a_7 = \frac{19}{5040} + \frac{a_1}{105}.$$

(c) Show that the relations in part (b) yield the general solution to (16) given in equations (17)–(20).

In Problems 21 through 28, use the procedure illustrated in Problem 20 to find at least the first four nonzero terms in a power series expansion about $x = 0$ of a general solution to the given differential equation.

21. $y' - xy = \sin x$.

22. $w' + xw = e^x$.

23. $z'' + xz' + z = x^2 + 2x + 1$.

24. $y'' - 2xy' + 3y = x^2$.

25. $(1 + x^2)y'' - xy' + y = e^{-x}$.

26. $y'' - xy' + 2y = \cos x$.

27. $(1 - x^2)y'' - y' + y = \tan x$.

28. $y'' - (\sin x)y = \cos x$.

29. The equation

$$(1 - x^2)y'' - 2xy' + n(n + 1)y = 0,$$

where n is an unspecified parameter, is called **Legendre's equation.** This equation occurs in applications of differential equations to physics and engineering.

(a) Find a power series expansion about $x = 0$ for a solution to Legendre's equation.

(b) Show that for n a nonnegative integer, there exists an nth degree polynomial that is a solution to Legendre's equation. These polynomials, up to a constant multiple, are called **Legendre polynomials.**

(c) Determine the first three Legendre polynomials, up to a constant multiple.

30. Aging Spring. As a spring ages, its spring "constant" decreases in value. One such model for a spring-mass system with an aging spring is

$$mx''(t) + bx'(t) + ke^{-\eta t}x(t) = 0,$$

where m is the mass, b the damping constant, k and η positive constants, and $x(t)$ the displacement of the spring from its equilibrium position. Let $m = 1$ kg, $b = 2$ N-sec/m, $k = 1$ N/m, and $\eta = 1$ (sec)$^{-1}$. The system is set in motion by displacing the mass one meter from its equilibrium position and then releasing it ($x(0) = 1$, $x'(0) = 0$). Find at least the first four nonzero terms in a power series expansion about $t = 0$ for the displacement.

31. Aging Spring without Damping. In the spring-mass system for an aging spring discussed in Problem 30, assume there is no damping (i.e., $b = 0$), $m = 1$, and $k = 1$. To see the effect of aging, consider η as a positive parameter.

(a) Redo Problem 30 with $b = 0$ and η arbitrary but fixed.

(b) Set $\eta = 0$ in the expansion obtained in part (a). Does this expansion agree with the expansion for the solution to the problem with $\eta = 0$? [*Hint:* When $\eta = 0$, the solution is $x(t) = \cos t$.]

8.5 CAUCHY-EULER (EQUIDIMENSIONAL) EQUATIONS REVISITED

In the previous sections we considered methods for obtaining power series solutions about an ordinary point for a linear second order equation. However, in certain cases we may want a series expansion about a *singular point* of the equation. (This was the case for the Lane-Emden equation discussed in Section 8.1.) To motivate a procedure for finding such expansions, we return to the class of **Cauchy-Euler equations.** In Chapter 4 we solved these equations by making the change of variables $x = e^t$, which transforms a Cauchy-Euler equation into an equation with constant coefficients. However, it is more instructive for our study of series expansions about singular points to work directly in the variable x.

Recall that a second order homogeneous Cauchy-Euler equation has the form

(1) $ax^2y''(x) + bxy'(x) + cy(x) = 0, \qquad x > 0,$

where a $(\neq 0)$, b, and c are (real) constants. Since here $p(x) = b/ax$ and $q(x) = c/ax^2$, it follows that $x = 0$ is a singular point for (1).

As we found in Chapter 4, equation (1) has solutions of the form $y = x^r$. To determine the values for r, we can proceed as follows. Let L be the differential operator defined by the left-hand side of equation (1), that is,

(2) $L[y](x) := ax^2 y''(x) + bxy'(x) + cy(x),$

and set

(3) $w(r, x) := x^r.$

When we substitute $w(r, x)$ for $y(x)$ in (2), we find

$$L[w](x) = ax^2 r(r - 1)x^{r-2} + bxrx^{r-1} + cx^r$$
$$= \{ar^2 + (b - a)r + c\}x^r.$$

From this, we see that $w = x^r$ is a solution to (1) if and only if r satisfies

(4) $ar^2 + (b - a)r + c = 0.$

Equation (4) is referred to as the **auxiliary**, or **indicial, equation** for (1).

When the indicial equation has two distinct roots, we have

$$L[w](x) = a(r - r_1)(r - r_2)x^r,$$

from which it follows that equation (1) has the two linearly independent solutions

(5) $y_1(x) = w(r_1, x) = x^{r_1}, \qquad x > 0,$

(6) $y_2(x) = w(r_2, x) = x^{r_2}, \qquad x > 0.$

When r_1 and r_2 are complex conjugates, $\alpha \pm i\beta$, we can use Euler's formula to express

$$x^{\alpha + i\beta} = e^{(\alpha + i\beta)\ln x} = e^{\alpha \ln x} \cos(\beta \ln x) + ie^{\alpha \ln x} \sin(\beta \ln x)$$
$$= x^\alpha \cos(\beta \ln x) + ix^\alpha \sin(\beta \ln x).$$

Since the real and imaginary parts of $x^{\alpha + i\beta}$ must also be solutions to (1), we can replace (5) and (6) by the two linearly independent real-valued solutions

$$y_1(x) = x^\alpha \cos(\beta \ln x), \qquad y_2(x) = x^\alpha \sin(\beta \ln x).$$

If the indicial equation (4) has a repeated real root r_0, then it turns out that x^{r_0} and $x^{r_0} \ln x$ are two linearly independent solutions. This can be deduced using the reduction of order approach of Section 4.4. However, it is more instructive for later applications to see how these two linearly independent solutions can be obtained via the operator approach. If r_0 is a repeated root, then

(7) $L[w](x) = a(r - r_0)^2 x^r.$

Setting $r = r_0$ gives the solution

(8) $y_1(x) = w(r_0, x) = x^{r_0}, \qquad x > 0.$

To find a second linearly independent solution, we make the following observation. The right-hand side of (7) has the factor $(r - r_0)^2$, so taking the partial derivative of (7) with

respect to r and setting $r = r_0$, we get *zero*. That is,

(9) $\dfrac{\partial}{\partial r}\{L[w](x)\}\big|_{r=r_0} = \{a(r - r_0)^2 x^r \ln x + 2a(r - r_0)x^r\}\big|_{r=r_0} = 0.$

While it may not appear that we have made any progress toward finding a second solution, a closer look at the expression on the left-hand side of (9) will soon vindicate our efforts.

First note that $w(r, x) = x^r$ has continuous partial derivatives of all orders with respect to both r and x. Hence the mixed partial derivatives are equal:

$$\frac{\partial^3 w}{\partial r \partial x^2} = \frac{\partial^3 w}{\partial x^2 \partial r}, \qquad \frac{\partial^2 w}{\partial r \partial x} = \frac{\partial^2 w}{\partial x \partial r}.$$

Consequently, for the differential operator L, we have

$$\frac{\partial}{\partial r}L[w] = \frac{\partial}{\partial r}\left\{ax^2\frac{\partial^2 w}{\partial x^2} + bx\frac{\partial w}{\partial x} + cw\right\}$$

$$= ax^2\frac{\partial^3 w}{\partial r \partial x^2} + bx\frac{\partial^2 w}{\partial r \partial x} + c\frac{\partial w}{\partial r}$$

$$= ax^2\frac{\partial^3 w}{\partial x^2 \partial r} + bx\frac{\partial^2 w}{\partial x \partial r} + c\frac{\partial w}{\partial r}$$

$$= L\left[\frac{\partial w}{\partial r}\right];$$

that is, the operators $\partial/\partial r$ and L commute. With this fact, (9) can be written as

$$L\left[\frac{\partial w}{\partial r}\right]\bigg|_{r=r_0} = 0.$$

Thus for the case of a repeated root r_0, a second linearly independent solution to (1) is

(10) $y_2(x) = \dfrac{\partial w}{\partial r}(r_0, x) = \dfrac{\partial}{\partial r}(x^r)\big|_{r=r_0} = x^{r_0} \ln x, \qquad x > 0.$

EXAMPLE 1 Find a general solution to

(11) $4x^2 y''(x) + y(x) = 0, \qquad x > 0.$

Solution Let $w(r, x) = x^r$ and let L denote the left-hand side of (11). A short calculation gives

$$L[w](x) = (4r^2 - 4r + 1)x^r.$$

Solving the indicial equation

$$4r^2 - 4r + 1 = (2r - 1)^2 = 4\left(r - \frac{1}{2}\right)^2 = 0$$

yields the repeated root $r_0 = 1/2$. Thus a general solution to (11) is obtained from equations (8) and (10) by setting $r_0 = 1/2$. That is,

$$y(x) = C_1\sqrt{x} + C_2\sqrt{x} \ln x, \qquad x > 0. \quad \blacksquare \text{ End}$$

In Section 8.7 we discuss the problem of finding a second linearly independent series solution to certain differential equations. As we shall see, operator methods similar to those described in this section will lead to the desired second solution.

EXERCISES 8.5

In Problems 1 through 10, use the substitution $y = x^r$ to find a general solution to the given equation for $x > 0$.

1. $x^2 y''(x) + 6xy'(x) + 6y(x) = 0$.

2. $2x^2 y''(x) + 13xy'(x) + 15y(x) = 0$.

3. $x^2 y''(x) - xy'(x) + 17y(x) = 0$.

4. $x^2 y''(x) + 2xy'(x) - 3y(x) = 0$.

5. $\dfrac{d^2 y}{dx^2} = \dfrac{5}{x}\dfrac{dy}{dx} - \dfrac{13}{x^2}y$.

6. $\dfrac{d^2 y}{dx^2} = \dfrac{1}{x}\dfrac{dy}{dx} - \dfrac{4}{x^2}y$.

7. $x^3 y'''(x) + 4x^2 y''(x) + 10xy'(x) - 10y(x) = 0$.

8. $x^3 y'''(x) + 4x^2 y''(x) + xy'(x) = 0$.

9. $x^3 y'''(x) + 3x^2 y''(x) + 5xy'(x) - 5y(x) = 0$.

10. $x^3 y'''(x) + 9x^2 y''(x) + 19xy'(x) + 8y(x) = 0$.

In Problems 11 and 12, use a substitution of the form $y = (x - c)^r$ to find a general solution to the given equation for $x > c$.

11. $2(x - 3)^2 y''(x) + 5(x - 3)y'(x) - 2y(x) = 0$.

12. $4(x + 2)^2 y''(x) + 5y(x) = 0$.

In Problems 13 and 14, use variation of parameters to find a general solution to the given equation for $x > 0$.

13. $x^2 y''(x) - 2xy'(x) + 2y(x) = x^{-1/2}$.

14. $x^2 y''(x) + 2xy'(x) - 2y(x) = 6x^{-2} + 3x$.

In Problems 15 through 17, solve the given initial value problem.

15. $t^2 x''(t) - 12x(t) = 0;$ $x(1) = 3;$ $x'(1) = 5$.

16. $x^2 y''(x) + 5xy'(x) + 4y(x) = 0;$ $y(1) = 3,$ $y'(1) = 7$.

17. $x^3 y'''(x) + 6x^2 y''(x) + 29xy'(x) - 29y(x) = 0;$
$y(1) = 2,$ $y'(1) = -3,$ $y''(1) = 19$.

18. When r_0 is a repeated root of the auxiliary equation $ar^2 + br + c = 0$, then $y_1(t) = e^{r_0 t}$ is a solution to the equation $ay'' + by' + cy = 0$, where a, b, and c are constants. Use a derivation similar to the one given in this section for the case when the indicial equation has a repeated root to show that a second linearly independent solution is $y_2(t) = te^{r_0 t}$.

19. Let $L[y](x) := x^3 y'''(x) + xy'(x) - y(x)$.

 (a) Show that $L[x^r](x) = (r - 1)^3 x^r$.

 (b) Using an extension of the argument given in this section for the case when the indicial equation has a double root, show that $L[y] = 0$ has the general solution

$$y(x) = C_1 x + C_2 x \ln x + C_3 x (\ln x)^2.$$

8.6 METHOD OF FROBENIUS

In the previous section we showed that a homogeneous Cauchy-Euler equation has a solution of the form $y(x) = x^r$, $x > 0$, where r is a certain constant. Cauchy-Euler equations have, of course, a very special form with only one singular point (at $x = 0$). In this section we show how the theory for Cauchy-Euler equations generalizes to other equations that have a special type of singularity.

To motivate the procedure, let's rewrite the Cauchy-Euler equation,

(1) $ax^2 y''(x) + bxy'(x) + cy(x) = 0,$ $x > 0,$

in the standard form

(2) $y''(x) + p(x)y'(x) + q(x)y(x) = 0,$ $x > 0,$

where

$$p(x) = \frac{p_0}{x}, \qquad q(x) = \frac{q_0}{x^2},$$

and p_0, q_0 are the constants b/a and c/a, respectively. When we substitute $w(r, x) = x^r$ for y into equation (2), we get

$$[r(r - 1) + p_0 r + q_0]x^{r-2} = 0,$$

which yields the indicial equation

(3) $r(r - 1) + p_0 r + q_0 = 0.$

Thus, if r_1 is a root of (3), then $w(r_1, x) = x^{r_1}$ is a solution to equations (1) and (2).

Let's now assume, more generally, that (2) is an equation for which $xp(x)$ and $x^2q(x)$, instead of being constants, are *analytic functions*. That is, in some open interval about $x = 0$,

(4) $xp(x) = p_0 + p_1 x + p_2 x^2 + \cdots = \displaystyle\sum_{n=0}^{\infty} p_n x^n,$

(5) $x^2q(x) = q_0 + q_1 x + q_2 x^2 + \cdots = \displaystyle\sum_{n=0}^{\infty} q_n x^n.$

It follows from (4) and (5) that

(6) $\displaystyle\lim_{x\to0} xp(x) = p_0$ and $\displaystyle\lim_{x\to0} x^2q(x) = q_0,$

and hence, for x near 0 we have $xp(x) \approx p_0$ and $x^2q(x) \approx q_0$. Therefore it is reasonable to expect that the solutions to (2) will behave (for x near 0) like the solutions to the Cauchy-Euler equation

$$x^2y'' + p_0 xy' + q_0 y = 0.$$

When $p(x)$ and $q(x)$ satisfy (4) and (5), we say that the singular point at $x = 0$ is regular. More generally we state the following.

REGULAR SINGULAR POINT

Definition 3. A singular point x_0 of

(7) $y''(x) + p(x)y'(x) + q(x)y(x) = 0$

is said to be a **regular singular point** if both $(x - x_0)p(x)$ and $(x - x_0)^2q(x)$ are analytic at x_0.[†] Otherwise x_0 is called an **irregular singular point**.

[†] In the terminology of complex variables, p has a pole of order at most 1 and q has a pole of order at most 2 at x_0.

EXAMPLE 1 Classify the singular points of the equation

(8) $(x^2 - 1)^2 y''(x) + (x + 1)y'(x) - y(x) = 0.$

Solution Here

$$p(x) = \frac{x + 1}{(x^2 - 1)^2} = \frac{1}{(x + 1)(x - 1)^2},$$

$$q(x) = \frac{-1}{(x^2 - 1)^2} = \frac{-1}{(x + 1)^2(x - 1)^2},$$

from which we see that ± 1 are the singular points of (8). For the singularity at 1, we have

$$(x - 1)p(x) = \frac{1}{(x + 1)(x - 1)},$$

which is not analytic at $x = 1$. Therefore $x = 1$ is an irregular singular point.

For the singularity at -1, we have

$$(x + 1)p(x) = \frac{1}{(x - 1)^2}, \qquad (x + 1)^2 q(x) = \frac{-1}{(x - 1)^2},$$

both of which are analytic at $x = -1$. Hence $x = -1$ is a regular singular point. ■ End

Let's assume $x = 0$ is a regular singular point for equation (7) so that $p(x)$ and $q(x)$ satisfy (4) and (5); that is,

(9) $p(x) = \displaystyle\sum_{n=0}^{\infty} p_n x^{n-1}, \qquad q(x) = \sum_{n=0}^{\infty} q_n x^{n-2}.$

The idea of the mathematician Frobenius was that since Cauchy-Euler equations have solutions of the form x^r, then for the regular singular point $x = 0$, there should be solutions to (7) of the form x^r *times an analytic function.*[†] Hence we seek solutions to (7) of the form

(10) $w(r, x) = x^r \displaystyle\sum_{n=0}^{\infty} a_n x^n = \sum_{n=0}^{\infty} a_n x^{n+r}, \qquad x > 0.$

Without loss of generality, we assume a_0 is an arbitrary nonzero constant, so we are left with determining r and the coefficients a_n, $n \geq 1$. Differentiating $w(r, x)$ with respect to x, we have

(11) $w'(r, x) = \displaystyle\sum_{n=0}^{\infty} (n + r)a_n x^{n+r-1},$

(12) $w''(r, x) = \displaystyle\sum_{n=0}^{\infty} (n + r)(n + r - 1)a_n x^{n+r-2}.$

[†] *Historical Footnote:* George Frobenius (1848–1917) developed this method in 1873. He is also known for his research on group theory.

If we substitute the above expansions for $w(r, x)$, $w'(r, x)$, $w''(r, x)$, $p(x)$, and $q(x)$ into (7), we obtain

$$\textbf{(13)}\qquad \sum_{n=0}^{\infty}(n + r)(n + r - 1)a_n x^{n+r-2} + \left(\sum_{n=0}^{\infty}p_n x^{n-1}\right)\left(\sum_{n=0}^{\infty}(n + r)a_n x^{n+r-1}\right)$$

$$+ \left(\sum_{n=0}^{\infty}q_n x^{n-2}\right)\left(\sum_{n=0}^{\infty}a_n x^{n+r}\right) = 0.$$

Now we use the Cauchy product to perform the series multiplications and then group like powers of x, starting with the lowest power, x^{r-2}. This gives

$$\textbf{(14)}\qquad [r(r - 1) + p_0 r + q_0]a_0 x^{r-2}$$

$$+ [(r + 1)ra_1 + (r + 1)p_0 a_1 + p_1 ra_0 + q_0 a_1 + q_1 a_0]x^{r-1} + \cdots = 0.$$

For the expansion on the left-hand side of equation (14) to sum to zero, each coefficient must be zero. Considering the first term, x^{r-2}, we find

$$\textbf{(15)}\qquad [r(r - 1) + p_0 r + q_0]a_0 = 0.$$

We have assumed $a_0 \neq 0$, so the quantity in brackets must be zero. This gives the indicial equation that is the same as the one we derived for Cauchy-Euler equations.

INDICIAL EQUATION

Definition 4. If x_0 is a regular singular point of $y'' + py' + qy = 0$, then the **indicial equation** for this point is

$$\textbf{(16)}\qquad r(r - 1) + p_0 r + q_0 = \mathbf{0},$$

where

$$p_0 := \lim_{x \to x_0} (x - x_0)p(x), \qquad q_0 := \lim_{x \to x_0} (x - x_0)^2 q(x).$$

The roots of the indicial equation are called the **exponents (indices)** of the singularity x_0.

EXAMPLE 2 Find the indicial equation and the exponents at the singularity $x = -1$ of

$$\textbf{(17)}\qquad (x^2 - 1)^2 y''(x) + (x + 1)y'(x) - y(x) = 0.$$

Solution In Example 1, we showed that $x = -1$ is a regular singular point. Since $p(x) = (x + 1)^{-1}(x - 1)^{-2}$ and $q(x) = -(x + 1)^{-2}(x - 1)^{-2}$, we find

$$p_0 = \lim_{x \to -1} (x + 1)p(x) = \lim_{x \to -1} (x - 1)^{-2} = \frac{1}{4},$$

$$q_0 = \lim_{x \to -1} (x + 1)^2 q(x) = \lim_{x \to -1} [-(x - 1)^{-2}] = -\frac{1}{4}.$$

Substituting these values for p_0 and q_0 into (16), we obtain the indicial equation

(18) $r(r - 1) + \dfrac{1}{4}r - \dfrac{1}{4} = 0.$

Multiplying by 4 and factoring gives $(4r + 1)(r - 1) = 0$. Hence $r = 1, -1/4$ are the exponents. ∎ End

As we have seen, we can use the indicial equation to determine those values of r for which the coefficient of x^{r-2} in (14) is zero. If we set the coefficient of x^{r-1} in (14) equal to zero, we have

(19) $[(r + 1)r + (r + 1)p_0 + q_0]a_1 + (p_1r + q_1)a_0 = 0.$

Since a_0 is arbitrary and we know the p_i's, q_i's, and r, we can solve equation (19) for a_1, provided the coefficient of a_1 in (19) is not zero. This will be the case if we take r to be the larger of the two roots of the indicial equation (see Problem 43).[†] Similarly, when we set the coefficient of x^r equal to zero, we can solve for a_2 in terms of the p_i's, q_i's, r, a_0, and a_1. Continuing in this manner, we can recursively solve for the a_n's. The procedure is illustrated in the following example.

EXAMPLE 3 Find a series expansion about the regular singular point $x = 0$ for a solution to

(20) $(x + 2)x^2y'' - xy' + (1 + x)y = 0,$ $x > 0.$

Solution Here $p(x) = -x^{-1}(x + 2)^{-1}$ and $q(x) = x^{-2}(x + 2)^{-1}(1 + x)$, so

$$p_0 = \lim_{x \to 0} xp(x) = \lim_{x \to 0}\left[-(x + 2)^{-1}\right] = -\frac{1}{2},$$

$$q_0 = \lim_{x \to 0} x^2q(x) = \lim_{x \to 0}(x + 2)^{-1}(1 + x) = \frac{1}{2}.$$

Since $x = 0$ is a regular singular point, we seek a solution to (20) of the form

(21) $w(r, x) = x^r \displaystyle\sum_{n=0}^{\infty} a_nx^n = \sum_{n=0}^{\infty} a_nx^{n+r}.$

By the previous discussion, r must satisfy the indicial equation (16). Substituting for p_0 and q_0 in (16), we obtain

$$r(r - 1) - \frac{1}{2}r + \frac{1}{2} = 0,$$

which simplifies to $2r^2 - 3r + 1 = (2r - 1)(r - 1) = 0$. Thus $r = 1$ and $r = 1/2$ are the roots of the indicial equation associated with $x = 0$.

Let's use the larger root $r = 1$ and solve for a_1, a_2, etc., to obtain the solution $w(1, x)$. We can simplify the computations by substituting $w(r, x)$ directly into equation (20), where the coefficients are polynomials in x, rather than dividing by $(x + 2)x^2$ and having to work

[†] Several interesting possibilities arise if we take r to be the smaller root. This will be discussed in the next sections.

with the rational functions $p(x)$ and $q(x)$. Inserting $w(r, x)$ in (20) and recalling the formulas for $w'(r, x)$ and $w''(r, x)$ in (11) and (12) gives (with $r = 1$)

$$(22) \qquad (x + 2)x^2 \sum_{n=0}^{\infty} (n + 1)na_n x^{n-1} - x \sum_{n=0}^{\infty} (n + 1)a_n x^n$$

$$+ (1 + x) \sum_{n=0}^{\infty} a_n x^{n+1} = 0,$$

which we can write as

$$(23) \qquad \sum_{n=0}^{\infty} (n + 1)na_n x^{n+2} + \sum_{n=0}^{\infty} 2(n + 1)na_n x^{n+1} - \sum_{n=0}^{\infty} (n + 1)a_n x^{n+1}$$

$$+ \sum_{n=0}^{\infty} a_n x^{n+1} + \sum_{n=0}^{\infty} a_n x^{n+2} = 0.$$

Next we shift the indices so that each summation in (23) is over x^k. With $k = n + 2$ in the first and last summations and $k = n + 1$ in the rest, (23) becomes

$$(24) \qquad \sum_{k=2}^{\infty} [(k - 1)(k - 2) + 1]a_{k-2} x^k + \sum_{k=1}^{\infty} [2k(k - 1) - k + 1]a_{k-1} x^k = 0.$$

Separating off the $k = 1$ term and combining the rest under one summation yields

$$(25) \qquad [2(1)(0) - 1 + 1]a_0 x + \sum_{k=2}^{\infty} [(k^2 - 3k + 3)a_{k-2} + (2k - 1)(k - 1)a_{k-1}]x^k = 0.$$

Notice that the coefficient of x in (25) is zero. This is because $r = 1$ is a root of the indicial equation, which is the equation we obtained by setting the coefficient of the lowest power of x equal to zero.

We can now determine the a_k's in terms of a_0 by setting the coefficients of x^k in equation (25) equal to zero for $k = 2, 3$, etc. This gives the recurrence relation

$$(26) \qquad (k^2 - 3k + 3)a_{k-2} + (2k - 1)(k - 1)a_{k-1} = 0,$$

or, equivalently,

$$(27) \qquad a_{k-1} = -\frac{k^2 - 3k + 3}{(2k - 1)(k - 1)} a_{k-2}, \qquad k \geq 2.$$

Setting $k = 2, 3$, and 4 in (27), we find

$$a_1 = -\frac{1}{3} a_0 \qquad (k = 2),$$

$$a_2 = -\frac{3}{10} a_1 = \frac{1}{10} a_0 \qquad (k = 3),$$

$$a_3 = -\frac{1}{3} a_2 = -\frac{1}{30} a_0 \qquad (k = 4).$$

Substituting these values for r, a_1, a_2, and a_3 into (21) gives

$$(28) \qquad w(1, x) = a_0 x^1 \left(1 - \frac{1}{3}x + \frac{1}{10}x^2 - \frac{1}{30}x^3 + \cdots \right),$$

where a_0 is arbitrary. In particular, for $a_0 = 1$, we get the solution

$$y_1(x) = x - \frac{1}{3}x^2 + \frac{1}{10}x^3 - \frac{1}{30}x^4 + \cdots , \qquad x > 0.$$

See Figure 8.6. ■ End

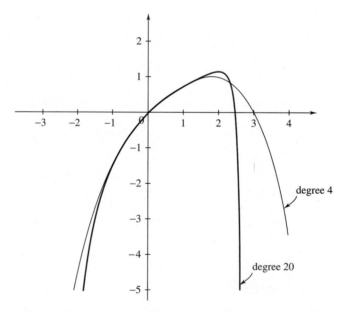

Figure 8.6 Partial sums approximating the solution $y_1(x)$ of Example 3

To find a second linearly independent solution to equation (20), we could try setting $r = 1/2$ and solving for a_1, a_2, ... to obtain a solution $w(1/2, x)$ (see Problem 44). In this particular case, the approach would work. However, if we encounter an indicial equation that has a repeated root, then the method of Frobenius would yield just one solution (apart from constant multiples). To find the desired second solution, we must use another technique, such as the reduction of order procedure discussed in Section 4.4. We tackle the problem of finding a second linearly independent solution in the next section.

The method of Frobenius can be summarized as follows.

METHOD OF FROBENIUS

To derive a series solution about the singular point x_0 of

(29) $a_2(x)y''(x) + a_1(x)y'(x) + a_0(x)y(x) = 0, \qquad x > x_0:$

(a) Set $p(x) := a_1(x)/a_2(x)$, $q(x) := a_0(x)/a_2(x)$. If both $(x - x_0)p(x)$ and $(x - x_0)^2 q(x)$ are analytic at x_0, then x_0 is a regular singular point and the remaining steps apply.

(b) Let

(30) $w(r, x) = (x - x_0)^r \sum_{n=0}^{\infty} a_n(x - x_0)^n = \sum_{n=0}^{\infty} a_n(x - x_0)^{n+r},$

and, using termwise differentiation, substitute $w(r, x)$ into equation (29) to obtain an equation of the form

$$A_0(x - x_0)^{r+J} + A_1(x - x_0)^{r+J+1} + \cdots = 0.$$

(c) Set the coefficients $A_0, A_1, A_2, \ldots$ equal to zero. (Notice that the equation $A_0 = 0$ is just a constant multiple of the indicial equation $r(r - 1) + p_0 r + q_0 = 0$.)

(d) Use the system of equations

$$A_0 = 0, \qquad A_1 = 0, \qquad \ldots, \qquad A_k = 0$$

to find a recurrence relation involving a_k and $a_0, a_1, \ldots, a_{k-1}$.

(e) Take $r = r_1$, the larger root of the indicial equation, and use the relation obtained in step (d) to determine $a_1, a_2, \ldots$ recursively in terms of a_0 and r_1.

(f) A series expansion of a solution to (29) is

(31) $w(r_1, x) = (x - x_0)^{r_1} \sum_{n=0}^{\infty} a_n(x - x_0)^n, \qquad x > x_0,$

where a_0 is arbitrary and the a_n's are defined in terms of a_0 and r_1.

One important question that remains concerns the radius of convergence of the power series that appears in (31). The following theorem contains an answer.[†]

FROBENIUS'S THEOREM

Theorem 6. If x_0 is a regular singular point of equation (29), then there exists at least one series solution of the form (30), where $r = r_1$ is the larger root of the associated indicial equation. Moreover, this series converges for all x such that $0 < x - x_0 < R$, where R is the distance from x_0 to the nearest other singular point (real or complex) of (29).

[†] For a proof of this theorem, see *Ordinary Differential Equations*, by E. L. Ince (Dover Publications, Inc., New York, 1956), Chapter XVI.

For simplicity, in the examples that follow we consider only series expansions about the regular singular point $x = 0$ and only those equations for which the associated indicial equation has real roots.

The following three examples not only illustrate the method of Frobenius but are important examples to which we refer in later sections.

EXAMPLE 4 Find a series solution about the regular singular point $x = 0$ of

$$(32) \qquad x^2 y''(x) - xy'(x) + (1 - x)y(x) = 0, \qquad x > 0.$$

Solution Here $p(x) = -x^{-1}$ and $q(x) = (1 - x)x^{-2}$. It is easy to check that $x = 0$ is a regular singular point of (32), so we compute

$$p_0 = \lim_{x \to 0} xp(x) = \lim_{x \to 0} (-1) = -1,$$

$$q_0 = \lim_{x \to 0} x^2 q(x) = \lim_{x \to 0} (1 - x) = 1.$$

Then the indicial equation is

$$r(r - 1) - r + 1 = r^2 - 2r + 1 = (r - 1)^2 = 0,$$

which has the roots $r_1 = r_2 = 1$.

Next we substitute

$$(33) \qquad w(r, x) = x^r \sum_{n=0}^{\infty} a_n x^n = \sum_{n=0}^{\infty} a_n x^{n+r}$$

into (32) and obtain

$$(34) \qquad x^2 \sum_{n=0}^{\infty} (n + r)(n + r - 1)a_n x^{n+r-2} - x \sum_{n=0}^{\infty} (n + r)a_n x^{n+r-1}$$

$$+ (1 - x) \sum_{n=0}^{\infty} a_n x^{n+r} = 0,$$

which we write as

$$(35) \qquad \sum_{n=0}^{\infty} (n + r)(n + r - 1)a_n x^{n+r} - \sum_{n=0}^{\infty} (n + r)a_n x^{n+r}$$

$$+ \sum_{n=0}^{\infty} a_n x^{n+r} - \sum_{n=0}^{\infty} a_n x^{n+r+1} = 0.$$

Shifting the indices so that each summation in (35) is over x^{k+r}, we take $k = n + 1$ in the last summation and $k = n$ in the rest. This gives

$$(36) \qquad \sum_{k=0}^{\infty} [(k + r)(k + r - 1) - (k + r) + 1]a_k x^{k+r} - \sum_{k=1}^{\infty} a_{k-1} x^{k+r} = 0.$$

Singling out the term corresponding to $k = 0$ and combining the rest under one summation yields

(37) $[r(r - 1) - r + 1]a_0 x^r$

$$+ \sum_{k=1}^{\infty} \{[(k + r)(k + r - 1) - (k + r) + 1]a_k - a_{k-1}\}x^{k+r} = 0.$$

When we set the coefficients equal to zero, we obtain

(38) $[r(r - 1) - r + 1]a_0 = 0,$

and, for $k \geq 1$, the recurrence relation

(39) $[(k + r)^2 - 2(k + r) + 1]a_k - a_{k-1} = 0,$

which reduces to

(40) $(k + r - 1)^2 a_k - a_{k-1} = 0.$

Relation (40) can be used to solve for a_k in terms of a_{k-1}:

(41) $a_k = \dfrac{1}{(k + r - 1)^2} a_{k-1}, \qquad k \geq 1.$

Setting $r = r_1 = 1$ in (38) gives (as expected) $0 \cdot a_0 = 0$, and in (41) gives

(42) $a_k = \dfrac{1}{k^2} a_{k-1}, \qquad k \geq 1.$

For $k = 1, 2,$ and 3, we now find

$$a_1 = \frac{1}{1^2} a_0 = a_0 \qquad\qquad (k = 1),$$

$$a_2 = \frac{1}{2^2} a_1 = \frac{1}{(2 \cdot 1)^2} a_0 = \frac{1}{4} a_0 \qquad (k = 2),$$

$$a_3 = \frac{1}{3^2} a_2 = \frac{1}{(3 \cdot 2 \cdot 1)^2} a_0 = \frac{1}{36} a_0 \qquad (k = 3).$$

In general, we have

(43) $a_k = \dfrac{1}{(k!)^2} a_0.$

Hence equation (32) has a series solution given by

(44) $w(1, x) = a_0 x \left\{ 1 + x + \dfrac{1}{4} x^2 + \dfrac{1}{36} x^3 + \cdots \right\}$

$$= a_0 x \sum_{k=0}^{\infty} \frac{1}{(k!)^2} x^k, \qquad x > 0. \quad \blacksquare \text{ End}$$

Since $x = 0$ is the only singular point for equation (32), it follows from Frobenius's theorem or directly by the ratio test that the series solution (44) converges for all $x > 0$.

In the next two examples, we only outline the method; the reader is invited to furnish the intermediate steps.

EXAMPLE 5 Find a series solution about the regular singular point $x = 0$ of

(45) $xy''(x) + 4y'(x) - xy(x) = 0$, $x > 0$.

Solution Since $p(x) = 4/x$ and $q(x) = -1$, we see that $x = 0$ is indeed a regular singular point and

$$p_0 = \lim_{x \to 0} xp(x) = 4, \qquad q_0 = \lim_{x \to 0} x^2 q(x) = 0.$$

The indicial equation is

$$r(r - 1) + 4r = r^2 + 3r = r(r + 3) = 0,$$

with roots $r_1 = 0$ and $r_2 = -3$.

Now substitute

(46) $w(r, x) = x^r \sum_{n=0}^{\infty} a_n x^n = \sum_{n=0}^{\infty} a_n x^{n+r}$

into (45). After a little algebra and a shift in indices, we get

(47) $[r(r - 1) + 4r]a_0 x^{r-1} + [(r + 1)r + 4(r + 1)]a_1 x^r$

$$+ \sum_{k=1}^{\infty} [(k + r + 1)(k + r + 4)a_{k+1} - a_{k-1}]x^{k+r} = 0.$$

Next we set the coefficients equal to zero and find

(48) $[r(r - 1) + 4r]a_0 = 0$,

(49) $[(r + 1)r + 4(r + 1)]a_1 = 0$,

and, for $k \geq 1$, the recurrence relation

(50) $(k + r + 1)(k + r + 4)a_{k+1} - a_{k-1} = 0$.

For $r = r_1 = 0$, equation (48) becomes $0 \cdot a_0 = 0$ and (49) becomes $4 \cdot a_1 = 0$. Hence, while a_0 is arbitrary, a_1 must be zero. Setting $r = r_1 = 0$ in (50), we find

(51) $a_{k+1} = \dfrac{1}{(k + 1)(k + 4)} a_{k-1}$, $k \geq 1$,

from which it follows (after a few experimental computations) that $a_{2k+1} = 0$ for $k = 0, 1, \ldots$, and

(52) $a_{2k} = \dfrac{1}{[2 \cdot 4 \cdots (2k)][5 \cdot 7 \cdots (2k + 3)]} a_0$

$$= \dfrac{1}{2^k k![5 \cdot 7 \cdots (2k + 3)]} a_0, \qquad k \geq 1.$$

Hence equation (45) has a series solution

(53) $w(0, x) = a_0 \left\{ 1 + \sum_{k=1}^{\infty} \dfrac{1}{2^k k! [5 \cdot 7 \cdots (2k + 3)]} x^{2k} \right\}$, $x > 0$. ■ End

If in Example 5 we had worked with the root $r = r_2 = -3$, then we would actually have obtained *two* linearly independent solutions (see Problem 45).

EXAMPLE 6 Find a series solution about the regular singular point $x = 0$ of

(54) $xy''(x) + 3y'(x) - xy(x) = 0$, $x > 0$.

Solution Since $p(x) = 3/x$ and $q(x) = -1$, we see that $x = 0$ is a regular singular point. Moreover,

$$p_0 = \lim_{x \to 0} xp(x) = 3, \qquad q_0 = \lim_{x \to 0} x^2 q(x) = 0.$$

So the indicial equation is

(55) $r(r - 1) + 3r = r^2 + 2r = r(r + 2) = 0$,

with roots $r_1 = 0$ and $r_2 = -2$.

Substituting

(56) $w(r, x) = x^r \sum_{n=0}^{\infty} a_n x^n = \sum_{n=0}^{\infty} a_n x^{n+r}$

into (54) ultimately gives

(57) $[r(r - 1) + 3r]a_0 x^{r-1} + [(r + 1)r + 3(r + 1)]a_1 x^r$

$$+ \sum_{k=1}^{\infty} [(k + r + 1)(k + r + 3)a_{k+1} - a_{k-1}]x^{k+r} = 0.$$

Setting the coefficients equal to zero, we have

(58) $[r(r - 1) + 3r]a_0 = 0$,

(59) $[(r + 1)r + 3(r + 1)]a_1 = 0$,

and, for $k \geq 1$, the recurrence relation

(60) $(k + r + 1)(k + r + 3)a_{k+1} - a_{k-1} = 0$.

With $r = r_1 = 0$, these equations lead to the following formulas: $a_{2k+1} = 0$, $k = 0, 1, \ldots$, and

(61) $a_{2k} = \dfrac{1}{[2 \cdot 4 \cdots (2k)][4 \cdot 6 \cdots (2k + 2)]} a_0 = \dfrac{1}{2^{2k} k! (k + 1)!} a_0$, $k \geq 0$.

Hence equation (54) has a series solution

(62) $w(0, x) = a_0 \sum_{k=0}^{\infty} \dfrac{1}{2^{2k} k! (k + 1)!} x^{2k}$, $x > 0$. ■ End

Unlike in Example 5, if we work with the second root $r = r_2 = -2$ in Example 6, then we do *not* obtain a second linearly independent solution (see Problem 46).

In the preceding examples we were able to use the method of Frobenius to find a series solution valid to the right ($x > 0$) of the regular singular point $x = 0$. For $x < 0$, we can use the change of variables $x = -t$ and then solve the resulting equation for $t > 0$.

The method of Frobenius also applies to higher-order linear equations (see Problems 35–38).

EXERCISES 8.6

In Problems 1 through 10, classify each singular point (real or complex) of the given equation as regular or irregular.

1. $(x^2 - 1)y'' + xy' + 3y = 0$.

2. $x^2y'' + 8xy' - 3xy = 0$.

3. $(x^2 + 1)z'' + 7x^2z' - 3xz = 0$.

4. $x^2y'' - 5xy' + 7y = 0$.

5. $(x^2 - 1)^2y'' - (x - 1)y' + 3y = 0$.

6. $(x^2 - 4)y'' + (x + 2)y' + 3y = 0$.

7. $(t^2 - t - 2)^2x'' + (t^2 - 4)x' - tx = 0$.

8. $(x^2 - x)y'' + xy' + 7y = 0$.

9. $(x^2 + 2x - 8)^2y'' + (3x + 12)y' - x^2y = 0$.

10. $x^3(x - 1)y'' + (x^2 - 3x)(\sin x)y' - xy = 0$.

In Problems 11 through 18, find the indicial equation and the exponents for the specified singularity of the given differential equation.

11. $x^2y'' - 2xy' - 10y = 0$, at $x = 0$.

12. $x^2y'' + 4xy' + 2y = 0$, at $x = 0$.

13. $(x^2 - x - 2)^2z'' + (x^2 - 4)z' - 6xz = 0$, at $x = 2$.

14. $(x^2 - 4)y'' + (x + 2)y' + 3y = 0$, at $x = -2$.

15. $\theta^3y'' + \theta(\sin \theta)y' - (\tan \theta)y = 0$, at $\theta = 0$.

16. $(x^2 - 1)y'' - (x - 1)y' - 3y = 0$, at $x = 1$.

17. $(x - 1)^2y'' + (x^2 - 1)y' - 12y = 0$, at $x = 1$.

18. $4x(\sin x)y'' - 3y = 0$, at $x = 0$.

In Problems 19 through 24, use the method of Frobenius to find at least the first four nonzero terms in the series expansion about $x = 0$ for a solution to the given equation for $x > 0$.

19. $9x^2y'' + 9x^2y' + 2y = 0$.

20. $2x(x - 1)y'' + 3(x - 1)y' - y = 0$.

21. $x^2y'' + xy' + x^2y = 0$.

22. $xy'' + y' - 4y = 0$.

23. $x^2z'' + (x^2 + x)z' - z = 0$.

24. $3xy'' + (2 - x)y' - y = 0$.

In Problems 25 through 30, use the method of Frobenius to find a general formula for the coefficient a_n in a series expansion about $x = 0$ for a solution to the given equation for $x > 0$.

25. $4x^2y'' + 2x^2y' - (x + 3)y = 0$.

26. $x^2y'' + (x^2 - x)y' + y = 0$.

27. $xw'' - w' - xw = 0$.

28. $3x^2y'' + 8xy' + (x - 2)y = 0$.

29. $xy'' + (x - 1)y' - 2y = 0$.

30. $x(x + 1)y'' + (x + 5)y' - 4y = 0$.

In Problems 31 through 34, first determine a recurrence formula for the coefficients in the (Frobenius) series expansion of the solution about $x = 0$. Use this recurrence formula to determine if there exists a solution to the differential equation that is decreasing for $x > 0$.

31. $xy'' + (1 - x)y' - y = 0$.

32. $x^2y'' - x(1 + x)y' + y = 0$.

33. $3xy'' + 2(1 - x)y' - 4y = 0$.

34. $xy'' + (x + 2)y' - y = 0$.

In Problems 35 through 38, use the method of Frobenius to find at least the first four nonzero terms in the series expansion about $x = 0$ for a solution to the given linear third order equation for $x > 0$.

35. $6x^3y''' + 13x^2y'' + (x + x^2)y' + xy = 0$.

36. $6x^3y''' + 11x^2y'' - 2xy' - (x - 2)y = 0$.

37. $6x^3y''' + 13x^2y'' - (x^2 + 3x)y' - xy = 0$.

38. $6x^3y''' + (13x^2 - x^3)y'' + xy' - xy = 0$.

In Problems 39 and 40, try to use the method of Frobenius to find a series expansion about the irregular singular point $x = 0$ for a solution to the given differential equation. If the method works, give at least the first four nonzero terms in the expansion. If the method does not work, explain what went wrong.

39. $x^2y'' + (3x - 1)y' + y = 0$.

40. $x^2y'' + y' - 2y = 0$.

In certain applications, it is desirable to have an expansion about the point at infinity. To obtain such an expansion, we use the change of variables $z = 1/x$ and expand about $z = 0$. In Problems 41 and 42, show that infinity is a regular singular point of the given differential equation by showing that $z = 0$ is a regular singular point for the transformed equation in z. Also find at least the first four nonzero terms in the series expansion about infinity of a solution to the original equation in x.

41. $x^3y'' - x^2y' - y = 0$.

42. $18(x - 4)^2(x - 6)y'' + 9x(x - 4)y' - 32y = 0$.

43. Show that if r_1 and r_2 are roots of the indicial equation (16), with r_1 the larger root ($\text{Re } r_1 \geq \text{Re } r_2$), then the coefficient of a_1 in equation (19) is not zero when $r = r_1$.

44. To obtain a second linearly independent solution to equation (20):

(a) Substitute $w(r, x)$ given in (21) into (20) and conclude that the coefficients a_k, $k \geq 1$ must satisfy

the recurrence relation

$$(k + r - 1)(2k + 2r - 1)a_k$$
$$+ [(k + r - 1)(k + r - 2) + 1]a_{k-1} = 0.$$

(b) Use the recurrence relation with $r = 1/2$ to derive the second series solution

$$w\left(\frac{1}{2}, x\right) = a_0\left(x^{1/2} - \frac{3}{4}x^{3/2} + \frac{7}{32}x^{7/2}\right.$$
$$\left. - \frac{133}{1920}x^{9/2} + \cdots\right).$$

(c) Use the recurrence relation with $r = 1$ to obtain $w(1, x)$ in (28).

45. In Example 5, show that if we choose $r = r_2 = -3$, then we obtain *two* linearly independent solutions to equation (45). [*Hint:* a_0 and a_3 are arbitrary constants.]

46. In Example 6, show that if we choose $r = r_2 = -2$, then we obtain a solution that is a constant multiple of the solution given in (62). [*Hint:* Show that a_0 and a_1 must be zero while a_2 is arbitrary.]

 47. In applying the method of Frobenius, the following recurrence relation arose: $a_{k+1} = 15^7a_k/(k + 1)^9$, $k = 0, 1, 2, \ldots$.

(a) Show that the coefficients are given by the formula $a_k = 15^{7k}a_0/(k!)^9$, $k = 0, 1, 2, \ldots$.

(b) Use the formula obtained in part (a) with $a_0 = 1$ to compute a_5, a_{10}, a_{15}, a_{20}, and a_{25} on your computer or calculator. What goes wrong?

(c) Now use the recurrence relation to compute a_k for $k = 1, 2, 3, \ldots, 25$, assuming $a_0 = 1$.

(d) What advantage does the recurrence relation have over the formula?

8.7 FINDING A SECOND LINEARLY INDEPENDENT SOLUTION

In the previous section we showed that if $x = 0$ is a regular singular point of

(1) $y''(x) + p(x)y'(x) + q(x)y(x) = 0$, $x > 0$,

then the method of Frobenius can be used to find a series solution valid for x near zero. The first step in the method is to find the roots r_1 and r_2 ($\text{Re } r_1 \geq \text{Re } r_2$) of the associated indicial equation

(2) $r(r - 1) + p_0r + q_0 = 0$.

Then, utilizing the larger root r_1, equation (1) has a series solution of the form

(3) $w(r_1, x) = x^{r_1}\sum_{n=0}^{\infty} a_nx^n = \sum_{n=0}^{\infty} a_nx^{n+r_1}$,

where $a_0 \neq 0$. To find a second linearly independent solution, our first inclination is to set $r = r_2$ and seek a solution of the form

(4) $w(r_2, x) = x^{r_2} \displaystyle\sum_{n=0}^{\infty} a_n x^n = \sum_{n=0}^{\infty} a_n x^{n+r_2}.$

We shall see that this procedure works, *provided $r_1 - r_2$ is not an integer*. However, when $r_1 - r_2$ is an integer, the Frobenius method with $r = r_2$ may just lead to the same solution that we obtained using the root r_1. (This is obviously true when $r_1 = r_2$.)

EXAMPLE 1 Find the first few terms in the series expansion about the regular singular point $x = 0$ for a general solution to

(5) $(x + 2)x^2 y''(x) - xy'(x) + (1 + x)y(x) = 0, \qquad x > 0.$

Solution In Example 3 of Section 8.6, we used the method of Frobenius to find a series solution for (5). In the process we determined that $p_0 = -1/2$, $q_0 = 1/2$ and that the indicial equation has roots $r_1 = 1$, $r_2 = 1/2$. Since these roots do not differ by an integer ($r_1 - r_2 = 1/2$), the method of Frobenius will give two linearly independent solutions of the form

(6) $w(r, x) := x^r \displaystyle\sum_{n=0}^{\infty} a_n x^n = \sum_{n=0}^{\infty} a_n x^{n+r}.$

In Problem 44 of Exercises 8.6, the reader was requested to show that substituting $w(r, x)$ into (5) leads to the recurrence relation

(7) $(k + r - 1)(2k + 2r - 1)a_k + [(k + r - 1)(k + r - 2) + 1]a_{k-1} = 0, \quad k \geq 1.$

With $r = r_1 = 1$ and $a_0 = 1$, we find

(8) $y_1(x) = x - \dfrac{1}{3}x^2 + \dfrac{1}{10}x^3 - \dfrac{1}{30}x^4 + \cdots$

as obtained in the previous section. Moreover, taking $r = r_2 = 1/2$ and $a_0 = 1$ in (7) leads to the second solution,

(9) $y_2(x) = x^{1/2} - \dfrac{3}{4}x^{3/2} + \dfrac{7}{32}x^{7/2} - \dfrac{133}{1920}x^{9/2} + \cdots .$

Consequently a general solution to equation (5) is

(10) $y(x) = c_1 y_1(x) + c_2 y_2(x), \qquad x > 0,$

where $y_1(x)$ and $y_2(x)$ are the series solutions given in equations (8) and (9). See Figure 8.7. ■ End

When the indicial equation has repeated roots, $r_1 = r_2$, obviously substituting $r = r_2$ just gives us back the first solution and gets us nowhere. However, in such a case, we can use the reduction of order formula discussed in Section 4.4 to obtain a second linearly independent solution.

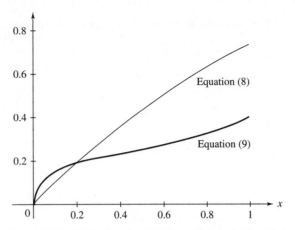

Figure 8.7 Partial sums approximating solutions to Example 1

Recall that if $f(x)$ is a nontrivial solution to equation (1), then a second linearly independent solution is given by the reduction of order formula

(11) $$y(x) = f(x) \int \frac{e^{-\int p(x)\,dx}}{[f(x)]^2}\,dx.$$

In the next example we use this formula to hunt for a second series solution.

EXAMPLE 2 Find the first few terms in the series expansion about the regular singular point $x = 0$ for a general solution to

(12) $$x^2 y''(x) - xy'(x) + (1 - x)y(x) = 0, \qquad x > 0.$$

Solution In Example 4 of Section 8.6, we used the method of Frobenius to obtain a series solution to equation (12). In the process, we found the indicial equation to be $r^2 - 2r + 1 = 0$, which has roots $r_1 = r_2 = 1$. Working with $r_1 = 1$, we derived the series solution

(13) $$y_1(x) = x + x^2 + \frac{1}{4}x^3 + \frac{1}{36}x^4 + \frac{1}{576}x^5 + \cdots = \sum_{k=0}^{\infty} \frac{1}{(k!)^2} x^{k+1}$$

(see equation (44) in Section 8.6 with $a_0 = 1$).

Since $p(x) = -x^{-1}$ for equation (12), we can substitute $f(x) = y_1(x)$ and $p(x) = -x^{-1}$ into the reduction of order formula (11) to obtain

(14) $$y_2(x) = y_1(x) \int \frac{e^{\int x^{-1}\,dx}}{[y_1(x)]^2}\,dx = y_1(x) \int \frac{e^{\ln x}}{[y_1(x)]^2}\,dx$$

$$= y_1(x) \int \frac{x}{\left[x + x^2 + \frac{1}{4}x^3 + \frac{1}{36}x^4 + \frac{1}{576}x^5 + \cdots\right]^2}\,dx.$$

Using the Cauchy product to square the series in the denominator and canceling out an x gives

(15) $$y_2(x) = y_1(x) \int \frac{x}{\left[x^2 + 2x^3 + \frac{3}{2}x^4 + \frac{5}{9}x^5 + \frac{35}{288}x^6 + \cdots\right]} \, dx$$

$$= y_1(x) \int \left\{ \frac{1}{x + 2x^2 + \frac{3}{2}x^3 + \frac{5}{9}x^4 + \frac{35}{288}x^5 + \cdots} \right\} dx.$$

Next, we can use long division to compute the power series for the quotient in the braces, thereby obtaining

(16) $$y_2(x) = y_1(x) \int \left\{ \frac{1}{x} - 2 + \frac{5}{2}x - \frac{23}{9}x^2 + \cdots \right\} dx.$$

Integrating term by term, we then find

(17) $$y_2(x) = y_1(x) \left\{ \ln x - 2x + \frac{5}{4}x^2 - \frac{23}{27}x^3 + \cdots \right\}$$

$$= y_1(x) \ln x + y_1(x) \left(-2x + \frac{5}{4}x^2 - \frac{23}{27}x^3 + \cdots \right).$$

Now we substitute the series for $y_1(x)$ given in (13) and use the Cauchy product to obtain

(18) $$y_2(x) = y_1(x) \ln x + \left(x + x^2 + \frac{1}{4}x^3 + \cdots \right)\left(-2x + \frac{5}{4}x^2 - \frac{23}{27}x^3 + \cdots \right)$$

$$= y_1(x) \ln x + \left(-2x^2 - \frac{3}{4}x^3 - \frac{11}{108}x^4 + \cdots \right).$$

Hence a general solution to equation (12) is

(19) $$y(x) = c_1 y_1(x) + c_2 y_2(x), \qquad x > 0,$$

where c_1 and c_2 are arbitrary constants and $y_1(x)$ and $y_2(x)$ are given by (13) and (18). See Figure 8.8. ∎ End

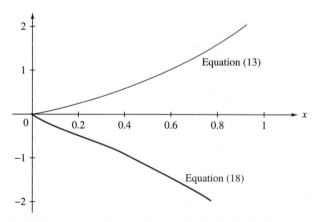

Figure 8.8 Partial sum approximations to solutions of Example 2

Example 2 illustrates the drawback to using the reduction of order formula to find a second linearly independent solution: We must square a series and then compute its reciprocal. As a result, it is very difficult to determine the general term in the series expansion for $y_2(x)$.

In that example, the roots of the indicial equation were equal and we found a second linearly independent solution (18) that involves the first solution $y_1(x)$ multipled by $\ln x$. This should not be too surprising if we recall the analogous situation for a Cauchy-Euler equation having an indicial equation with repeated roots (page 461). As we shall soon see, if an indicial equation has roots that differ by an integer, then the expansion of the second solution may also involve the term $y_1(x)\ln x$. In the following theorem, we give the *general form* of two linearly independent solutions for the three cases where the roots of the indicial equation (a) do not differ by an integer, (b) are equal, or (c) differ by a nonzero integer.

FORM OF SECOND LINEARLY INDEPENDENT SOLUTION

Theorem 7. Let x_0 be a regular singular point for $y'' + py' + qy = 0$ and let r_1 and r_2 be the roots of the associated indicial equation, where $r_1 \geq r_2$ or $\text{Re}\, r_1 \geq \text{Re}\, r_2$.

(a) If $r_1 - r_2$ is not an integer, then there exist two linearly independent solutions of the form

$$(20) \qquad y_1(x) = \sum_{n=0}^{\infty} a_n(x - x_0)^{n+r_1}; \qquad a_0 \neq 0,$$

$$(21) \qquad y_2(x) = \sum_{n=0}^{\infty} b_n(x - x_0)^{n+r_2}, \qquad b_0 \neq 0.$$

(b) If $r_1 = r_2$, then there exist two linearly independent solutions of the form

$$(22) \qquad y_1(x) = \sum_{n=0}^{\infty} a_n(x - x_0)^{n+r_1}, \qquad a_0 \neq 0,$$

$$(23) \qquad y_2(x) = y_1(x)\ln(x - x_0) + \sum_{n=1}^{\infty} b_n(x - x_0)^{n+r_1}.$$

(c) If $r_1 - r_2$ is a positive integer, then there exist two linearly independent solutions of the form

$$(24) \qquad y_1(x) = \sum_{n=0}^{\infty} a_n(x - x_0)^{n+r_1}, \qquad a_0 \neq 0,$$

$$(25) \qquad y_2(x) = Cy_1(x)\ln(x - x_0) + \sum_{n=0}^{\infty} b_n(x - x_0)^{n+r_2}, \qquad b_0 \neq 0,$$

where C is a constant that could be zero.

In each case of the theorem, $y_1(x)$ is just the series solution obtained by the method of Frobenius, with $r = r_1$. When $r_1 - r_2$ is not an integer, the method of Frobenius yields a

second linearly independent solution by taking $r = r_2$. In Section 8.8 we shall prove the formulas stated in Theorem 7 for cases (b) and (c). For now, let's see how knowing the form of the second solution enables us to obtain it. Again, for simplicity, we consider only indicial equations having real roots.

EXAMPLE 3 Find the first few terms in the series expansion about the regular singular point $x = 0$ for two linearly independent solutions to

$$(26) \qquad x^2 y''(x) - xy'(x) + (1 - x)y(x) = 0, \qquad x > 0.$$

Solution This is the same equation we considered in Example 2, where we used the reduction of order formula to find a second linearly independent solution. This time let's try to find a second solution by using the information in Theorem 7. Recall that $r_1 = r_2 = 1$ and so $y_2(x)$ has the form given in (23) (with $x_0 = 0$); that is,

$$(27) \qquad y_2(x) = y_1(x)\ln x + \sum_{n=1}^{\infty} b_n x^{n+1},$$

where (cf. (13))

$$(28) \qquad y_1(x) = \sum_{k=0}^{\infty} \frac{1}{(k!)^2} x^{k+1}.$$

Our goal is to determine the coefficients b_n by substituting $y_2(x)$ directly into equation (26). We begin by differentiating $y_2(x)$ in (27) to obtain

$$y_2'(x) = y_1'(x)\ln x + x^{-1}y_1(x) + \sum_{n=1}^{\infty} (n + 1)b_n x^n,$$

$$y_2''(x) = y_1''(x)\ln x - x^{-2}y_1(x) + 2x^{-1}y_1'(x) + \sum_{n=1}^{\infty} n(n + 1)b_n x^{n-1}.$$

Substituting $y_2(x)$ into (26) yields

$$(29) \qquad x^2 \left\{ y_1''(x)\ln x - x^{-2}y_1(x) + 2x^{-1}y_1'(x) + \sum_{n=1}^{\infty} n(n + 1)b_n x^{n-1} \right\}$$

$$- x \left\{ y_1'(x)\ln x + x^{-1}y_1(x) + \sum_{n=1}^{\infty} (n + 1)b_n x^n \right\}$$

$$+ (1 - x)\left\{ y_1(x)\ln x + \sum_{n=1}^{\infty} b_n x^{n+1} \right\} = 0,$$

which simplifies to

$$(30) \qquad \{x^2 y_1''(x) - xy_1'(x) + (1 - x)y_1(x)\}\ln x - 2y_1(x) + 2xy_1'(x)$$

$$+ \sum_{n=1}^{\infty} n(n + 1)b_n x^{n+1} - \sum_{n=1}^{\infty} (n + 1)b_n x^{n+1} + \sum_{n=1}^{\infty} b_n x^{n+1} - \sum_{n=1}^{\infty} b_n x^{n+2} = 0.$$

Notice that the factor in front of $\ln x$ is just the left-hand side of equation (26) with $y = y_1$. Since y_1 is a solution to (26), this factor is zero. With this observation and a shift in the indices of summation, equation (30) can be rewritten as

(31) $\qquad 2xy_1'(x) - 2y_1(x) + b_1 x^2 + \sum_{k=2}^{\infty} (k^2 b_k - b_{k-1})x^{k+1} = 0.$

To identify the coefficients in (31) so that we can set them equal to zero, we must substitute back in the series expansions for $y_1(x)$ and $y_1'(x)$. From (28), we see that $y_1'(x) = \sum_{k=0}^{\infty} (k+1)x^k/(k!)^2$. Inserting this series together with (28) into (31), we find

(32) $\qquad \sum_{k=0}^{\infty} \dfrac{2(k+1) - 2}{(k!)^2} x^{k+1} + b_1 x^2 + \sum_{k=2}^{\infty} (k^2 b_k - b_{k-1})x^{k+1} = 0.$

Listing separately the $k = 0$ and $k = 1$ terms and combining the remaining terms gives

(33) $\qquad (2 + b_1)x^2 + \sum_{k=2}^{\infty} \left[\dfrac{2k}{(k!)^2} + k^2 b_k - b_{k-1} \right] x^{k+1} = 0.$

Next, we set the coefficients in (33) equal to zero. From the x^2 term we have $2 + b_1 = 0$, so $b_1 = -2$. From the x^{k+1} term, we obtain

$$\dfrac{2k}{(k!)^2} + k^2 b_k - b_{k-1} = 0$$

or

(34) $\qquad b_k = \dfrac{1}{k^2} \left[b_{k-1} - \dfrac{2k}{(k!)^2} \right], \qquad k \geq 2.$

Taking $k = 2$ and 3, we compute

(35) $\qquad b_2 = \dfrac{1}{2^2}[b_1 - 1] = \dfrac{-3}{4}, \qquad b_3 = \dfrac{1}{9}\left[-\dfrac{3}{4} - \dfrac{6}{36} \right] = \dfrac{-11}{108}.$

Hence a second linearly independent solution to (26) is

(36) $\qquad y_2(x) = y_1(x)\ln x - 2x^2 - \dfrac{3}{4}x^3 - \dfrac{11}{108}x^4 + \cdots.$ ■ End

If we compare the different methods used in Examples 2 and 3 for obtaining a second linearly independent solution, we find that the technique of Example 3 has an advantage: It yields a recurrence relation, equation (34), that can be used to solve recursively for the coefficients b_k.

In the next two examples we consider the case when the difference between the roots of the indicial equation is a positive integer. In Example 4, it will turn out that the constant C in formula (25) must be taken to be zero (i.e., no $\ln x$ term is present), while in Example 5, this constant is nonzero (i.e., a $\ln x$ term is present). Since the solutions to these examples require several intermediate computations, we do not display all the details. Rather, we encourage the reader to take an active part by bridging the gaps.

EXAMPLE 4 Find the first few terms in the series expansion about the regular singular point $x = 0$ for a general solution to

(37) $xy''(x) + 4y'(x) - xy(x) = 0, \qquad x > 0.$

Solution In Example 5 of Section 8.6, we used the method of Frobenius to find a series expansion about $x = 0$ for a solution to equation (37). There we found the indicial equation to be $r^2 + 3r = 0$, which has roots $r_1 = 0$ and $r_2 = -3$. Working with $r_1 = 0$, we obtained the series solution

(38) $y_1(x) = 1 + \dfrac{1}{10}x^2 + \dfrac{1}{280}x^4 + \cdots$

(see equation (53) in Section 8.6, with $a_0 = 1$).

Since $r_1 - r_2 = 3$ is a positive integer, it follows from Theorem 7 that equation (37) has a second linearly independent solution of the form

(39) $y_2(x) = Cy_1(x)\ln x + \displaystyle\sum_{n=0}^{\infty} b_n x^{n-3}.$

When we substitute this expression for y_2 into equation (37), we obtain

(40) $x\left\{ Cy_1''(x)\ln x + 2Cx^{-1}y_1'(x) - Cx^{-2}y_1(x) + \displaystyle\sum_{n=0}^{\infty}(n-3)(n-4)b_n x^{n-5} \right\}$

$+ 4\left\{ Cy_1'(x)\ln x + Cx^{-1}y_1(x) + \displaystyle\sum_{n=0}^{\infty}(n-3)b_n x^{n-4} \right\}$

$- x\left\{ Cy_1(x)\ln x + \displaystyle\sum_{n=0}^{\infty} b_n x^{n-3} \right\} = 0,$

which simplifies to

(41) $\{xy_1''(x) + 4y_1'(x) - xy_1(x)\}C\ln x + 3Cx^{-1}y_1(x) + 2Cy_1'(x)$

$+ \displaystyle\sum_{n=0}^{\infty}(n-3)(n-4)b_n x^{n-4} + \sum_{n=0}^{\infty}4(n-3)b_n x^{n-4} - \sum_{n=0}^{\infty} b_n x^{n-2} = 0.$

The factor in braces is zero because $y_1(x)$ satisfies equation (37). After we combine the summations and simplify, equation (41) becomes

(42) $3Cx^{-1}y_1(x) + 2Cy_1'(x) - 2b_1 x^{-3} + \displaystyle\sum_{k=2}^{\infty}[k(k-3)b_k - b_{k-2}]x^{k-4} = 0.$

Substituting in the series for $y_1(x)$ and writing out the first few terms of the summation in (42), we obtain

$-2b_1 x^{-3} + (-2b_2 - b_0)x^{-2} + (3C - b_1)x^{-1} + (4b_4 - b_2)$

$+ \left(\dfrac{7}{10}C + 10b_5 - b_3\right)x + (18b_6 - b_4)x^2 + \left(\dfrac{11}{280}C + 28b_7 - b_5\right)x^3$

$+ \cdots = 0.$

Next, we set the coefficients equal to zero:

$$-2b_1 = 0 \Rightarrow b_1 = 0, \qquad -2b_2 - b_0 = 0 \Rightarrow b_2 = -\frac{1}{2}b_0,$$

$$3C - b_1 = 0 \Rightarrow C = \frac{1}{3}b_1 = 0, \qquad 4b_4 - b_2 = 0 \Rightarrow b_4 = \frac{1}{4}b_2 = -\frac{1}{8}b_0,$$

$$\frac{7}{10}C + 10b_5 - b_3 = 0 \Rightarrow b_5 = \frac{b_3 - \dfrac{7}{10}C}{10} = \frac{1}{10}b_3,$$

$$18b_6 - b_4 = 0 \Rightarrow b_6 = \frac{1}{18}b_4 = -\frac{1}{144}b_0,$$

$$\frac{11}{280}C + 28b_7 - b_5 = 0 \Rightarrow b_7 = \frac{b_5 - \dfrac{11}{280}C}{28} = \frac{1}{280}b_3.$$

Note, in particular, that C must equal zero. Substituting the above values for the b_n's and $C = 0$ back into equation (39) gives

(43) $\quad y_2(x) = b_0 \left\{ x^{-3} - \frac{1}{2}x^{-1} - \frac{1}{8}x - \frac{1}{144}x^3 + \cdots \right\}$

$$+ b_3 \left\{ 1 + \frac{1}{10}x^2 + \frac{1}{280}x^4 + \cdots \right\},$$

where b_0 and b_3 are arbitrary constants. Observe that the expression in braces following b_3 is just the series expansion for $y_1(x)$ given in equation (38). Hence, in order to obtain a second linearly independent solution, we must choose b_0 to be nonzero. Taking $b_0 = 1$ and $b_3 = 0$ gives

(44) $\quad y_2(x) = x^{-3} - \frac{1}{2}x^{-1} - \frac{1}{8}x - \frac{1}{144}x^3 + \cdots .$

Therefore a general solution to equation (37) is

(45) $\quad y(x) = c_1 y_1(x) + c_2 y_2(x), \qquad x > 0,$

where $y_1(x)$ and $y_2(x)$ are given in (38) and (44). (Notice that the right-hand side of (43) coincides with (45) if we identify b_0 as c_2 and b_3 as c_1.) See Figure 8.9 on page 484. ■ End

EXAMPLE 5 Find the first few terms in the series expansion about the regular singular point $x = 0$ for two linearly independent solutions to

(46) $\quad xy''(x) + 3y'(x) - xy(x) = 0, \qquad x > 0.$

Solution In Example 6 of Section 8.6, we used the method of Frobenius to find a series expansion about $x = 0$ for a solution to equation (46). The indicial equation turned out to be $r^2 + 2r = 0$, which has roots $r_1 = 0$ and $r_2 = -2$. Using $r_1 = 0$, we obtained the series

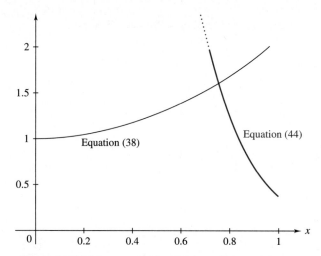

Figure 8.9 Partial sum approximations to solutions of Example 4

solution

(47) $y_1(x) = 1 + \dfrac{1}{8}x^2 + \dfrac{1}{192}x^4 + \dfrac{1}{9216}x^6 + \cdots$

(see equation (62) in Section 8.6 and put $a_0 = 1$).

Because $r_1 - r_2 = 2$ is a positive integer, it follows from Theorem 7 that equation (46) has a second linearly independent solution of the form

(48) $y_2(x) = Cy_1(x)\ln x + \displaystyle\sum_{n=0}^{\infty} b_n x^{n-2}.$

Plugging the expansion for $y_2(x)$ into equation (46) and simplifying yields

(49) $\{xy_1''(x) + 3y_1'(x) - xy_1(x)\}C \ln x + 2Cx^{-1}y_1(x) + 2Cy_1'(x)$

$\quad + \displaystyle\sum_{n=0}^{\infty} (n-2)(n-3)b_n x^{n-3} + \sum_{n=0}^{\infty} 3(n-2)b_n x^{n-3} - \sum_{n=0}^{\infty} b_n x^{n-1} = 0.$

Again the factor in braces is zero because $y_1(x)$ is a solution to equation (46). If we combine the summations and simplify, equation (49) becomes

(50) $2Cx^{-1}y_1(x) + 2Cy_1'(x) - b_1 x^{-2} + \displaystyle\sum_{k=2}^{\infty} [k(k-2)b_k - b_{k-2}]x^{k-3} = 0.$

Substituting in the series expansions for $y_1(x)$ and $y_1'(x)$ and writing out the first few terms of the summation in (50) leads to

(51) $-b_1 x^{-2} + (2C - b_0)x^{-1} + (3b_3 - b_1) + \left(\dfrac{3}{4}C + 8b_4 - b_2\right)x$

$\quad + (15b_5 - b_3)x^2 + \left(\dfrac{5}{96}C + 24b_6 - b_4\right)x^3 + \cdots = 0.$

When we set the coefficients in (51) equal to zero, it turns out that we are free to choose C and b_2 as arbitrary constants:

$$-b_1 = 0 \qquad\qquad \Rightarrow b_1 = 0,$$

$$2C - b_0 = 0 \qquad\qquad \Rightarrow b_0 = 2C \qquad (C \text{ arbitrary}),$$

$$3b_3 - b_1 = 0 \qquad\qquad \Rightarrow b_3 = \frac{1}{3}b_1 = 0,$$

$$8b_4 - b_2 + \frac{3}{4}C = 0 \quad \Rightarrow b_4 = \frac{b_2 - \frac{3}{4}C}{8} = \frac{1}{8}b_2 - \frac{3}{32}C \qquad (b_2 \text{ arbitrary}),$$

$$15b_5 - b_3 = 0 \qquad\qquad \Rightarrow b_5 = \frac{1}{15}b_3 = 0,$$

$$24b_6 - b_4 + \frac{5}{96}C = 0 \Rightarrow b_6 = \frac{b_4 - \frac{5}{96}C}{24} = \frac{1}{192}b_2 - \frac{7}{1152}C.$$

Substituting these values for C and the b_n's back into (48), we obtain the solution

(52) $$y_2(x) = C\left\{ y_1(x)\ln x + 2x^{-2} - \frac{3}{32}x^2 - \frac{7}{1152}x^4 + \cdots \right\}$$

$$+ b_2\left\{ 1 + \frac{1}{8}x^2 + \frac{1}{192}x^4 + \cdots \right\},$$

where C and b_2 are arbitrary constants. Since the factor multiplying b_2 is the first solution $y_1(x)$, we can obtain a second linearly independent solution by choosing $C = 1$ and $b_2 = 0$:

(53) $$y_2(x) = y_1(x)\ln x + 2x^{-2} - \frac{3}{32}x^2 - \frac{7}{1152}x^4 + \cdots .$$

See Figure 8.10. ▪ End

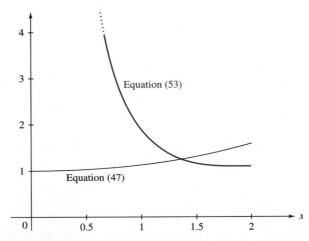

Figure 8.10 Partial sum approximations to the solutions of Example 5

EXERCISES 8.7

In Problems 1 through 14, find at least the first three non-zero terms in the series expansion about $x = 0$ for a general solution to the given equation for $x > 0$. (These are the same equations as in Problems 19 through 32 of Exercises 8.6.)

1. $9x^2y'' + 9x^2y' + 2y = 0$.

2. $2x(x - 1)y'' + 3(x - 1)y' - y = 0$.

3. $x^2y'' + xy' + x^2y = 0$.

4. $xy'' + y' - 4y = 0$.

5. $x^2z'' + (x^2 + x)z' - z = 0$.

6. $3xy'' + (2 - x)y' - y - 0$.

7. $4x^2y'' + 2x^2y' - (x + 3)y = 0$.

8. $x^2y'' + (x^2 - x)y' + y = 0$.

9. $xw'' - w' - xw = 0$.

10. $3x^2y'' + 8xy' + (x - 2)y = 0$.

11. $xy'' + (x - 1)y' - 2y = 0$.

12. $x(x + 1)y'' + (x + 5)y' - 4y = 0$.

13. $xy'' + (1 - x)y' - y = 0$.

14. $x^2y'' - x(1 + x)y' + y = 0$.

In Problems 15 and 16, determine whether the given equation has a solution that is bounded near the origin, all solutions are bounded near the origin, or none of the solutions are bounded near the origin. (These are the same equations as in Problems 33 and 34 of Exercises 8.6.) Note that you need to analyze only the indicial equation in order to answer the question.

15. $3xy'' + 2(1 - x)y' - 4y = 0$.

16. $xy'' + (x + 2)y' - y = 0$.

In Problems 17 through 20, find at least the first three non-zero terms in the series expansion about $x = 0$ for a general solution to the given linear third order equation for $x > 0$. (These are the same equations as in Problems 35 through 38 in Exercises 8.6.)

17. $6x^3y''' + 13x^2y'' + (x + x^2)y' + xy = 0$.

18. $6x^3y''' + 11x^2y'' - 2xy' - (x - 2)y = 0$.

19. $6x^3y''' + 13x^2y'' - (x^2 + 3x)y' - xy = 0$.

20. $6x^3y''' + (13x^2 - x^3)y'' + xy' - xy = 0$.

21. **Buckling Columns.** In the study of the buckling of a column whose cross-section varies, one encounters the equation

$$(54) \qquad x''y''(x) + \alpha^2 y(x) = 0, \qquad x > 0,$$

where x is related to the height above the ground and y is the deflection away from the vertical. The positive constant α depends on the rigidity of the column, its moment of inertia at the top, and the load. The positive integer n depends on the type of column. For example, when the column is a truncated cone (see Figure 8.11(a)), we have $n = 4$.

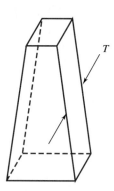

(a) Truncated cone

(b) Truncated pyramid with fixed thickness T and varying width

Figure 8.11 Buckling columns

(a) Use the substitution $x = t^{-1}$ to reduce (54) with $n = 4$ to the form

$$\frac{d^2y}{dt^2} + \frac{2}{t}\frac{dy}{dt} + \alpha^2 y = 0, \qquad t > 0.$$

(b) Find at least the first six nonzero terms in the series expansion about $t = 0$ for a general solution to the equation obtained in part (a).

(c) Use the result of part (b) to give an expansion about $x = 0$ for a general solution to (54).

22. In Problem 21, consider a column with a rectangular cross-section with two sides constant and the other two changing linearly (see Figure 8.11(b)). In this case, $n = 1$. Find at least the first four nonzero terms in the series expansion about $x = 0$ for a general solution to equation (54) when $n = 1$.

23. Use the method of Frobenius and the reduction of order formula, equation (11), to find at least the first three

nonzero terms in the series expansion about the irregular singular point $x = 0$ for a general solution to the differential equation

$$x^2 y'' + y' - 2y = 0.$$

24. The equation

$$xy''(x) + (1 - x)y'(x) + ny(x) = 0,$$

where n is a nonnegative integer, is called **Laguerre's differential equation.** Show that for each n, this equation has a polynomial solution of degree n. These polynomials are denoted by $L_n(x)$ and are called **Laguerre polynomials.** The first few Laguerre polynomials are

$$L_0(x) = 1, \qquad L_1(x) = -x + 1,$$

$$L_2(x) = x^2 - 4x + 2.$$

25. Use the results of Problem 24 and the reduction of order formula, equation (11), to obtain the first few terms in a series expansion about $x = 0$ for a general solution for $x > 0$ to Laguerre's differential equation for $n = 0$ and 1.

8.8 MORE ON THE CASES WHEN ROOTS DIFFER BY AN INTEGER

In this section we discuss another procedure for obtaining a second linearly independent solution when the indicial equation has equal roots or the roots differ by a positive integer. As a consequence of the approach, we will verify the formulas given in parts (b) and (c) of Theorem 7 in the previous section.

Suppose that $x = 0$ is a regular singular point for

(1) $$y''(x) + p(x)y'(x) + q(x)y(x) = 0$$

and that the indicial equation has equal roots, $r_1 = r_2$. As usual, let's try to find a solution to equation (1) of the form

(2) $$w(r, x) = x^r \sum_{n=0}^{\infty} a_n x^n = \sum_{n=0}^{\infty} a_n x^{n+r},$$

where a_0 is an arbitrary nonzero constant. Proceeding as in the method of Frobenius, we substitute the series $p(x) = \sum_{n=0}^{\infty} p_n x^{n-1}$, $q(x) = \sum_{n=0}^{\infty} q_n x^{n-2}$ and the expansions for $w(r, x)$, $w'(r, x)$, and $w''(r, x)$ into equation (1). After some algebra, we arrive at the expansion

(3) $$[r(r - 1) + p_0 r + q_0]a_0 x^{r-2}$$
$$+ [(r + 1)ra_1 + p_0(r + 1)a_1 + p_1 ra_0 + q_0 a_1 + q_1 a_0]x^{r-1} + \cdots = 0$$

(see equations (9)–(14) in Section 8.6). Until now our next step has been to set the coefficient of x^{r-2} equal to zero. This has the effect of forcing r to be a specific number, namely, a root of the indicial equation. As we shall see, there is an advantage to treating r as a *variable*. For the moment, we ignore the x^{r-2} term and set the remaining coefficients in (3) equal to zero. From the x^{r-1} term we get

(4) $$(r + 1)ra_1 + p_0(r + 1)a_1 + p_1 ra_0 + q_0 a_1 + q_1 a_0 = 0.$$

To simplify the computations, we take $a_0 = 1$ and then solve for a_1 in (4):

(5) $$a_1 = -\frac{p_1 r + q_1}{(r + 1)r + p_0(r + 1) + q_0}.$$

For a given differential equation, the quantities p_0, p_1, q_0, and q_1 are known, so (5) can be regarded as a formula for a_1 in terms of the variable r. That is, a_1 is a function of r, and we will denote it by $a_1(r)$. In a similar fashion, we set the successive coefficients in (3) equal to zero and obtain a_n in terms of r. Again we denote this functional dependence by $a_n(r)$.

Since we have chosen the $a_n(r)$'s so that all the coefficients in (3) are equal to zero except for the coefficient of x^{r-2}, substituting $w(r, x)$ into the left-hand side of equation (1) yields

(6) $w''(r, x) + p(x)w'(r, x) + q(x)w(r, x) = [r(r - 1) + p_0 r + q_0]x^{r-2}.$

If we let L denote the differential operator defined by the left-hand side of (1), then we can rewrite (6) as

(7) $L[w(r, x)](x) = [r(r - 1) + p_0 r + q_0]x^{r-2}.$

Now since we are assuming that the indicial equation has roots $r_1 = r_2$, then $r(r - 1) + p_0 r + q_0 = (r - r_1)^2$. Hence equation (7) has the form

(8) $L[w(r, x)](x) = (r - r_1)^2 x^{r-2}.$

Setting $r = r_1$ gives $L[w(r_1, x)](x) = 0$, and therefore, as we know from the method of Frobenius, $w(r_1, x)$ is a solution to equation (1).

To find a second linearly independent solution, let's revive the procedure that effectively dealt with repeated roots associated with a Cauchy-Euler equation (see Section 8.5). Namely, we take the partial derivative with respect to r of both sides of equation (8). This gives

(9) $\dfrac{\partial}{\partial r} L[w(r, x)](x) = 2(r - r_1)x^{r-2} + (r - r_1)^2 x^{r-2} \ln x.$

Setting $r = r_1$, we find

(10) $\dfrac{\partial}{\partial r} L[w(r_1, x)](x) = 0.$

Assuming $w(r, x)$ has continuous third order partial derivatives with respect to both r and x, we can then interchange the order of differentiation and obtain

(11) $L\left[\dfrac{\partial w}{\partial r}(r_1, x)\right](x) = 0.$

Hence $\partial w/\partial r$ evaluated at $r = r_1$ is a second solution to equation (1). We are almost done! Let's find the series expansion for $\partial w/\partial r$.

Since

(12) $w(r, x) = x^r \displaystyle\sum_{n=0}^{\infty} a_n(r)x^n,$

then

(13) $\dfrac{\partial w}{\partial r}(r, x) = x^r \ln x \displaystyle\sum_{n=0}^{\infty} a_n(r)x^n + x^r \sum_{n=0}^{\infty} a'_n(r)x^n.$

Here $a'_n(r)$ is the derivative of $a_n(r)$ with respect to r. Moreover, since we have taken $a_0(r) \equiv 1$, then $a'_0(r) \equiv 0$. Putting $r = r_1$ in (13), we then get

$$\frac{\partial w}{\partial r}(r_1, x) = \left\{ x^{r_1} \sum_{n=0}^{\infty} a_n(r_1)x^n \right\} \ln x + x^{r_1} \sum_{n=1}^{\infty} a'_n(r_1)x^n,$$

which simplifies to

(14) $\dfrac{\partial w}{\partial r}(r_1, x) = w(r_1, x) \ln x + \displaystyle\sum_{n=1}^{\infty} a'_n(r_1)x^{n+r_1}.$

This is the formula for the second linearly independent solution that we have been seeking. Notice that with $y_1(x) = w(r_1, x)$ and $b_n = a'_n(r_1)$, the right-hand side of (14) is the same as the formula for the second linearly independent solution claimed in Theorem 7, equation (23).

EXAMPLE 1 Find series expansions about the regular singular point $x = 0$ for two linearly independent solutions to

(15) $x^2 y''(x) - xy'(x) + (1 - x)y(x) = 0, \qquad x > 0.$

Solution We considered this same equation in Example 4 of Section 8.6 and in Examples 2 and 3 of Section 8.7. The indicial equation is $r^2 - 2r + 1 = 0$, which has roots $r_1 = r_2 = 1$. We now use our new procedure to find two linearly independent solutions.

Let

(16) $w(r, x) = x^r \displaystyle\sum_{n=0}^{\infty} a_n x^n = \sum_{n=0}^{\infty} a_n x^{n+r}.$

As we saw in Section 8.6, taking the derivatives of $w(r, x)$ with respect to x, substituting them into (15), and setting the coefficients equal to zero yields the recurrence relation

(17) $a_k = \dfrac{1}{(r + k - 1)^2} a_{k-1}, \qquad k \geq 1.$

Taking $a_0 = 1$ and solving equation (17) recursively for the a_k's in terms of r, we find

$$a_1 = \frac{1}{r^2} a_0 = \frac{1}{r^2}, \qquad a_2 = \frac{1}{(r+1)^2} a_1 = \frac{1}{[(r+1)r]^2},$$

$$a_3 = \frac{1}{(r+2)^2} a_2 = \frac{1}{[(r+2)(r+1)r]^2},$$

and, in general,

(18) $a_k(r) = \dfrac{1}{[(r + k - 1)(r + k - 2)\cdots r]^2}, \qquad k \geq 1.$

Setting $r = r_1 = 1$ in equation (18) gives

(19) $$a_k(1) = \frac{1}{[k(k-1)\cdots 2 \cdot 1]^2} = \frac{1}{(k!)^2}, \qquad k \geq 1.$$

Hence one solution is

(20) $$y_1(x) = w(1, x) = \sum_{n=0}^{\infty} a_n(1) x^{n+1} = \sum_{n=0}^{\infty} \frac{1}{(n!)^2} x^{n+1}.$$

Here we used the fact that $a_0 = 1 = (0!)^{-2}$.

To get a second linearly independent solution to (15), we must differentiate $a_k(r)$ in equation (18) with respect to r. This is easily done using logarithmic differentiation. We begin by taking the logarithm of both sides of (18), then implicitly differentiate with respect to r. That is,

(21) $$\ln a_k(r) = \ln[(r + k - 1)(r + k - 2)\cdots r]^{-2}$$
$$= -2 \ln(r + k - 1) - 2 \ln(r + k - 2) - \cdots - 2 \ln r,$$

and, differentiating with respect to r, we get

$$\frac{a_k'(r)}{a_k(r)} = -\frac{2}{r + k - 1} - \frac{2}{r + k - 2} - \cdots - \frac{2}{r},$$

which yields

$$a_k'(r) = -2a_k(r)\left\{ \frac{1}{r + k - 1} + \frac{1}{r + k - 2} + \cdots + \frac{1}{r} \right\}.$$

Setting $r = 1$ and using formula (19) for $a_k(1)$, we obtain

(22) $$a_k'(1) = \frac{-2}{(k!)^2}\left\{ \frac{1}{k} + \frac{1}{k - 1} + \cdots + \frac{1}{1} \right\}, \qquad k \geq 1.$$

We now insert these values into formula (14), which we have shown gives a second linearly independent solution:

(23) $$y_2(x) = \frac{\partial w}{\partial r}(1, x) = w(1, x)\ln x + \sum_{n=1}^{\infty} a_n'(1) x^{n+1}$$
$$= w(1, x)\ln x + \sum_{n=1}^{\infty} \frac{-2}{(n!)^2}\left\{ \frac{1}{n} + \frac{1}{n - 1} + \cdots + \frac{1}{1} \right\} x^{n+1}.$$

Thus two linearly independent solutions to equation (15) are given by the series for $y_1(x)$ in (20) and the series for $y_2(x)$ in (23) ■ End

The previous procedure can be modified to accommodate the case when the roots of the indicial equation differ by a positive integer. For this purpose, the new ingredient is to

treat a_0 also as a function of r. Proceeding as in the case of a repeated root, we must now include a_0 in equation (7):

$$(24) \qquad L[w(r, x)](x) = [r(r - 1) + p_0 r + q_0]a_0 x^{r-2}$$

$$= (r - r_1)(r - r_2)a_0 x^{r-2},$$

where r_1, r_2 are the roots of the indicial equation. As we shall soon see, it is convenient to put

$$(25) \qquad a_0 = r - r_2$$

so that equation (24) becomes

$$(26) \qquad L[w(r, x)](x) = (r - r_1)(r - r_2)^2 x^{r-2}.$$

We now exploit the fact that the right-hand side of (26) has the repeated factor $(r - r_2)^2$. Taking the partial derivative with respect to r of both sides of equation (26) gives

$$(27) \qquad \frac{\partial}{\partial r} L[w(r, x)](x) = (r - r_2)^2 x^{r-2} + 2(r - r_1)(r - r_2)x^{r-2}$$

$$+ (r - r_1)(r - r_2)^2 x^{r-2} \ln x.$$

Assuming, as before, that $w(r, x)$ has continuous third order partial derivatives with respect to r and x, then

$$(28) \qquad \frac{\partial}{\partial r} L[w(r, x)](x) = L\left[\frac{\partial w}{\partial r}(r, x)\right](x).$$

Hence on setting $r = r_2$ in (27), we see that

$$(29) \qquad L\left[\frac{\partial w}{\partial r}(r_2, x)\right](x) = 0.$$

Thus the partial derivative $\partial w / \partial r$ evaluated at $r = r_2$ is a solution to equation (1). To write out the series expansion for this solution, we first express $w(r, x)$ in the form

$$w(r, x) = x^r \sum_{n=0}^{\infty} a_n(r)x^n,$$

where we have indicated the fact that all the coefficients a_n (including a_0) are functions of r. Then

$$(30) \qquad \frac{\partial w}{\partial r}(r, x) = x^r \ln x \sum_{n=0}^{\infty} a_n(r)x^n + x^r \sum_{n=0}^{\infty} a_n'(r)x^n$$

$$= w(r, x)\ln x + x^r \sum_{n=0}^{\infty} a_n'(r)x^n,$$

where $a_n'(r)$ is the derivative of $a_n(r)$ with respect to r. Setting $r = r_2$, we obtain the solution

$$(31) \qquad \frac{\partial w}{\partial r}(r_2, x) = w(r_2, x)\ln x + \sum_{n=0}^{\infty} a_n'(r_2)x^{n+r_2},$$

where $a_0 = r - r_2$.

It is clear that if the $a_n(r)$'s are defined for $r = r_2$ and $w(r_2, x) \neq 0$, then $w(r_2, x)$ and $(\partial w/\partial r)(r_2, x)$ are linearly independent solutions because of the factor ln x. What is not so obvious, but nevertheless true, is that $a_R(r)$ where $R = r_1 - r_2$, may fail to be defined for $r = r_2$, so that the preceding method cannot be employed. Fortunately, in this situation two linearly independent solutions can be obtained from $w(r_2, x)$ by treating a_0 and a_R as arbitrary constants (see Problem 9).

The preceding analysis suggests the following procedure when the roots of the indicial equation for (1) differ by a positive integer J. First determine $w(r, x)$, where the coefficients $a_n(r)$ depend on r. If substituting $r = r_2$ leaves a_0 *and a_J* arbitrary, then $w(r_2, x)$ is a general solution, with a_0 and a_J the arbitrary constants. If this substitution leaves a_J arbitrary but $a_i \equiv 0$ for $i = 0, 1, \ldots, J - 1$, then $w(r_2, x) = w(r_1, x)$ is one solution and the second solution is given by $(\partial w/\partial r)(r_2, x)$ (see Problem 10).[†]

The preceding methods are particularly useful in determining the form of the general term in the series expansion of a second linearly independent solution when the recurrence relation is a two-term relation. With suitable modifications, these methods can also be used for higher-order equations.

EXERCISES 8.8

In Problems 1 through 8, use the method discussed in this section to obtain the series expansion about the regular singular point $x = 0$ for two linearly independent solutions to the given equation whose indicial equation has repeated roots.

1. $xy'' + y' - 4y = 0$.

2. $xy'' + (1 - x)y' - y = 0$.

3. $x^2 y'' + (x^2 - x)y' + y = 0$.

4. $x^2 y'' + xy' + x^2 y = 0$.

5. $x^2 y'' + 3x(1 + x)y' + y = 0$.

6. $x^2 y'' - xy' + (1 - x^3)y = 0$.

7. $xy'' + y' - 3y = 0$.

8. $x^2 y'' + (3x - x^2)y' + y = 0$.

9. To find a series expansion for a general solution to
$$xy''(x) + 4y'(x) - xy(x) = 0, \qquad x > 0:$$

(a) Show that the coefficients of $w(r, x)$ satisfy the recurrence relation
$$(k + r + 1)(k + r + 4)a_{k+1} - a_{k-1} = 0, \quad k \geq 1.$$

(b) Setting $r = r_2$, show that not only is a_0 arbitrary, but so is a_3.

(c) Use the recurrence relation to obtain a series expansion for a general solution with a_0 and a_3 as the arbitrary constants. Compare your answer with Example 4 in Section 8.7.

10. To find a series expansion for a general solution to
$$xy''(x) + 3y'(x) - xy(x) = 0, \qquad x > 0:$$

(a) Show that the coefficients of $w(r, x)$ satisfy the recurrence relation
$$(k + r + 1)(k + r + 3)a_{k+1} - a_{k-1} = 0, \quad k \geq 1.$$

(b) Show that $a_1 \equiv 0$ and hence
$$a_{2n+1}(r) \equiv 0, \qquad n = 0, 1, 2, \ldots .$$

(c) Set $r = -2$ and use the recurrence relation to show that $a_0(-2)$ must be zero.

(d) Let $a_0(r) = r + 2$ so that $a_0(-2) = 0$. Show that $a_2(r) = (r + 4)^{-1}$ and, for $n \geq 2$,
$$a_{2n}(r) = \frac{1}{(r + 2n + 2)[(r + 2n)^2 \cdots (r + 6)^2 (r + 4)^2]}.$$

[†] Further details are given in *Intermediate Differential Equations*, Second Edition, by Earl D. Rainville (Macmillan, New York, 1964), Sections 30–32.

(e) Find the series solution $w(-2, x)$.

(f) Let $a_0(r) = r + 2$ and show that $a_0'(r) = 1$, $a_2'(r) = -(r + 4)^{-2}$, and, for $n \geq 2$,

$$a_{2n}'(r) = \frac{(-1)}{(r + 2n + 2)[(r + 2n)^2 \cdots (r + 4)^2]}$$
$$\times \left\{ \frac{1}{r + 2n + 2} + \frac{2}{r + 2n} + \cdots + \frac{2}{r + 4} \right\}.$$

[*Hint:* Use logarithmic differentiation.]

(g) Find the series solution $(\partial w/\partial r)(-2, x)$. Compare your answers with Example 5 in Section 8.7.

In Problems 11 through 16, use the methods discussed in this section to obtain the series expansion about the regular singular point $x = 0$ for two linearly independent solutions for $x > 0$ to the given equation whose roots of the indicial equation differ by a positive integer.

11. $x(1 - x)y'' - 3y' + 2y = 0$.

12. $4x^2y'' + 2x^2y' - (x + 3)y = 0$.

13. $xy'' + (x + 2)y' - y = 0$.

14. $xy'' - y' - xy = 0$.

15. $xy'' + (x - 1)y' - 2y = 0$.

16. $xy'' + xy' + y = 0$.

In Problems 17 through 20, use the method discussed in this section to obtain the series expansion about the regular singular point $x = 0$ for three linearly independent solutions to the given linear third order equation for $x > 0$. [Hint: If the indicial equation has a root of multiplicity three, it is helpful to consider $\partial^2 w/\partial r^2$.]

17. $2x^3y''' + (5x^2 - x^3)y'' + xy' - xy = 0$.

18. $3x^3y''' + 5x^2y'' + (x^2 - x)y' + xy = 0$.

19. $x^2y''' + 3xy'' + y' - y = 0$.

20. $x^3y''' + (x^3 + 6x^2)y'' - (x^2 - 7x)y' + (x + 1)y = 0$.

21. In Problem 10, let $a_0 = r - r_1 = r$ (instead of $r - r_2$), then solve for $a_n(r)$. Show that if you now set $r = r_1 = 0$, then you obtain $y_3(x) = w(0, x) \equiv 0$, and

$$y_4(x) = \frac{\partial w}{\partial r}(0, x) = 0 \cdot \ln x + \sum_{n=0}^{\infty} b_n x^n.$$

Hence you get only one linearly independent solution.

8.9 SPECIAL FUNCTIONS

In advanced work in applied mathematics, engineering, and physics, a few special second order equations arise with amazing frequency. These equations have been extensively studied, and volumes have been written on the properties of their solutions. Three of these equations are the hypergeometric equation, Bessel's equation, and Legendre's equation. The solutions to these and other equations that occur in applications are often referred to as **special functions.** We shall briefly consider the three equations mentioned above. For a more detailed study of special functions, we refer the reader to *Basic Hypergeometric Series*, by G. Gasper and M. Rahman (Cambridge University Press, Cambridge, 1990); *Special Functions*, by E. D. Rainville (Macmillan, New York, 1960); and *Higher Transcendental Functions*, by A. Erdelyi (ed.) (McGraw-Hill, New York, 1953), 3 volumes.

Hypergeometric Equation

The linear second order differential equation

(1) $x(1 - x)y'' + [\gamma - (\alpha + \beta + 1)x]y' - \alpha\beta y = 0,$

where α, β, and γ are fixed parameters, is called the **hypergeometric equation.** This equation has singular points at $x = 0$ and 1, both of which are regular. Thus a series expansion about $x = 0$ for a solution to (1) obtained by the method of Frobenius will converge at least for $0 < x < 1$ (see Theorem 6, page 469). To find this expansion, observe

that the indicial equation associated with $x = 0$ is

$$r(r - 1) + \gamma r = r(r - (1 - \gamma)) = 0,$$

which has roots 0 and $1 - \gamma$. If γ is not an integer, then we can use the root $r = 0$ to obtain a solution to (1) of the form

(2) $$y_1(x) = \sum_{n=0}^{\infty} a_n x^n.$$

Substituting $y_1(x)$ given in (2) into (1), shifting indices, and simplifying ultimately leads to the equation

(3) $$\sum_{n=1}^{\infty} [n(n + \gamma - 1)a_n - (n + \alpha - 1)(n + \beta - 1)a_{n-1}]x^{n-1} = 0.$$

Setting the series coefficients equal to zero yields the recurrence relation

(4) $$n(n + \gamma - 1)a_n - (n + \alpha - 1)(n + \beta - 1)a_{n-1} = 0, \qquad n \geq 1.$$

Since $n \geq 1$ and γ is not an integer, there is no fear of dividing by zero when we rewrite (4) as

(5) $$a_n = \frac{(n + \alpha - 1)(n + \beta - 1)}{n(n + \gamma - 1)} a_{n-1}.$$

Solving recursively for a_n, we obtain

(6) $$a_n = \frac{\alpha(\alpha + 1) \cdots (\alpha + n - 1)\beta(\beta + 1) \cdots (\beta + n - 1)}{n!\gamma(\gamma + 1) \cdots (\gamma + n - 1)} a_0, \qquad n \geq 1.$$

If we employ the **factorial function** $(t)_n$, which is defined for nonnegative integers n by

(7) $$(t)_n := t(t + 1)(t + 2) \cdots (t + n - 1), \qquad n \geq 1,$$
$$(t)_0 := 1, \qquad t \neq 0,$$

then we can express a_n more compactly as

(8) $$a_n = \frac{(\alpha)_n(\beta)_n}{n!(\gamma)_n} a_0, \qquad n \geq 1.$$

(In fact, we can write $n!$ as $(1)_n$.) If we take $a_0 = 1$ and then substitute the expression for a_n in (8) into (2), we obtain the following solution to the hypergeometric equation:

(9) $$y_1(x) = 1 + \sum_{n=1}^{\infty} \frac{(\alpha)_n(\beta)_n}{n!(\gamma)_n} x^n.$$

The solution given in (9) is called a **Gaussian hypergeometric function** and is denoted by the symbol $F(\alpha, \beta; \gamma; x)$.[†] That is,

$$(10) \qquad F(\alpha, \beta; \gamma; x) := 1 + \sum_{n=1}^{\infty} \frac{(\alpha)_n (\beta)_n}{n!(\gamma)_n} x^n.$$

Hypergeometric functions are generalizations of the geometric series. To see this, observe that for any constant β that is not zero or a negative integer,

$$F(1, \beta; \beta; x) = 1 + x + x^2 + x^3 + \cdots .$$

It is interesting to note that many other familiar functions can be expressed in terms of the hypergeometric function. For example,

$$(11) \qquad F(\alpha, \beta; \beta; x) = (1 - x)^{-\alpha},$$

$$(12) \qquad F(1, 1; 2; x) = -x^{-1} \ln(1 - x),$$

$$(13) \qquad F\left(\frac{1}{2}, 1; \frac{3}{2}; x^2\right) = \frac{1}{2} x^{-1} \ln\left(\frac{1 + x}{1 - x}\right),$$

$$(14) \qquad F\left(\frac{1}{2}, 1; \frac{3}{2}; -x^2\right) = x^{-1} \arctan x.$$

We leave the verification of these formulas for the reader.

To obtain a second linearly independent solution to (1) when γ is not an integer, we use the other root, $1 - \gamma$, of the indicial equation and seek a solution of the form

$$(15) \qquad y_2(x) = \sum_{n=0}^{\infty} b_n x^{n+1-\gamma}.$$

Substituting $y_2(x)$ into equation (1) and solving for b_n, we eventually arrive at

$$(16) \qquad y_2(x) = x^{1-\gamma} + \sum_{n=1}^{\infty} \frac{(\alpha + 1 - \gamma)_n (\beta + 1 - \gamma)_n}{n!(2 - \gamma)_n} x^{n+1-\gamma}.$$

Factoring out $x^{1-\gamma}$, we see that the second solution $y_2(x)$ can be expressed in terms of a hypergeometric function. That is,

$$(17) \qquad y_2(x) = x^{1-\gamma} F(\alpha + 1 - \gamma, \beta + 1 - \gamma; 2 - \gamma; x).$$

When γ is an integer, one of the formulas given in (9) or (16) (corresponding to the larger root, 0 or $1 - \gamma$) still gives a solution. We then use the techniques of this chapter to obtain a second linearly independent solution, which may or may not involve a logarithmic term. We omit a discussion of these solutions.

In many books, the hypergeometric function is expressed in terms of the gamma function $\Gamma(x)$ instead of the factorial function. Recall that in Section 7.6, we defined

$$(18) \qquad \Gamma(x) := \int_0^{\infty} e^{-u} u^{x-1} \, du, \qquad x > 0$$

[†] **Historical Footnote:** A detailed study of this function was done by Carl Friedrich Gauss in 1813. The mathematical historian E. T. Bell refers to Gauss as the Prince of Mathematicians.

and showed that

(19) $\Gamma(x + 1) = x\Gamma(x), \qquad x > 0.$

It follows from repeated use of relation (19) that

(20) $(t)_n = \dfrac{\Gamma(t + n)}{\Gamma(t)}$

for $t > 0$ and n any nonnegative integer. Using relation (20), we can express the hypergeometric function as

(21) $F(\alpha, \beta; \gamma; x) = \dfrac{\Gamma(\gamma)}{\Gamma(\alpha)\Gamma(\beta)} \displaystyle\sum_{n=0}^{\infty} \dfrac{\Gamma(\alpha + n)\Gamma(\beta + n)}{n!\,\Gamma(\gamma + n)} x^n.$

Bessel's Equation

The linear second order differential equation

(22) $x^2 y'' + xy' + (x^2 - \nu^2)y = 0,$

where $\nu \geq 0$ is a fixed parameter, is called **Bessel's equation of order** ν. This equation has a regular singular point at $x = 0$ and no other singular points in the complex plane. Hence a series solution for (22) obtained by the method of Frobenius will converge for $0 < x < \infty$. The indicial equation for (22) is

$$r(r - 1) + r - \nu^2 = (r - \nu)(r + \nu) = 0,$$

which has roots $r_1 = \nu$ and $r_2 = -\nu$. If 2ν is not an integer, then the method of Frobenius yields two linearly independent solutions, given by

(23) $y_1(x) = a_0 \displaystyle\sum_{n=0}^{\infty} \dfrac{(-1)^n}{2^{2n} n!(1 + \nu)_n} x^{2n + \nu},$

(24) $y_2(x) = b_0 \displaystyle\sum_{n=0}^{\infty} \dfrac{(-1)^n}{2^{2n} n!(1 - \nu)_n} x^{2n - \nu}.$

If in (23) we take

$$a_0 = \dfrac{1}{2^{\nu}\Gamma(1 + \nu)},$$

then it follows from relation (20) that the function

(25) $J_{\nu}(x) := \displaystyle\sum_{n=0}^{\infty} \dfrac{(-1)^n}{n!\,\Gamma(1 + \nu + n)} \left(\dfrac{x}{2}\right)^{2n + \nu}$

is a solution to (22). We call $J_\nu(x)$ the **Bessel function of the first kind of order** ν.[†] Similarly, taking

$$b_0 = \frac{1}{2^{-\nu}\Gamma(1 - \nu)}$$

in equation (24) gives the solution

(26)　　$J_{-\nu}(x) := \displaystyle\sum_{n=0}^{\infty} \frac{(-1)^n}{n!\Gamma(1 - \nu + n)}\left(\frac{x}{2}\right)^{2n - \nu}$,

which is the Bessel function of the first kind of order $-\nu$. When $r_1 - r_2 = 2\nu$ is not an integer, we know by Theorem 7 of Section 8.7 that $J_\nu(x)$ and $J_{-\nu}(x)$ are linearly independent. Moreover, it can be shown that if ν is not an integer, even though 2ν is, then $J_\nu(x)$ and $J_{-\nu}(x)$ are still linearly independent.

　　What happens in the remaining case when ν is a nonnegative integer, say $\nu = m$? While $J_m(x)$ is still a solution, the function $J_{-m}(x)$ is not even properly defined because formula (26) will involve the gamma function evaluated at a nonpositive integer. From a more in-depth study of the gamma function, it turns out that $1/\Gamma(k) = 0$, for $k = 0, -1, -2, \ldots$. Hence (26) becomes

(27)　　$J_{-m}(x) = \displaystyle\sum_{n=0}^{\infty} \frac{(-1)^n}{n!\Gamma(1 - m + n)}\left(\frac{x}{2}\right)^{2n - m} = \sum_{n=m}^{\infty} \frac{(-1)^n}{n!\Gamma(1 - m + n)}\left(\frac{x}{2}\right)^{2n - m}$.

Comparing (27) with the formula (25) for $J_m(x)$, we see (after a shift in index) that

(**28**)　　$J_{-m}(x) = (-1)^m J_m(x)$,

which means that $J_{-m}(x)$ and $J_m(x)$ are linearly *dependent*. To resolve the problem of finding a second linearly independent solution in the case when ν is a nonnegative integer, we can use the method discussed in Section 8.8 (see Problem 27). There is, however, another approach to this problem, which we now describe.

　　For ν not an integer, we can take linear combinations of $J_\nu(x)$ and $J_{-\nu}(x)$ to obtain other solutions to (22). In particular, let

(29)　　$Y_\nu(x) := \dfrac{\cos(\nu\pi)J_\nu(x) - J_{-\nu}(x)}{\sin(\nu\pi)}$,　　ν not an integer,

for $x > 0$. The function $Y_\nu(x)$ is called the **Bessel function of the second kind of order** ν, and, as can be verified, $J_\nu(x)$ and $Y_\nu(x)$ are linearly independent. Notice that when ν is an integer, the denominator in (29) is zero; but, by formula (28), so is the numerator! Hence it is reasonable to hope that, in a limiting sense, formula (29) is still meaningful. In fact, using l'Hôpital's rule, it is possible to show that for m a nonnegative integer, the function defined by

(30)　　$Y_m(x) := \displaystyle\lim_{\nu \to m} \frac{\cos(\nu\pi)J_\nu(x) - J_{-\nu}(x)}{\sin(\nu\pi)}$

[†] *Historical Footnote:* Frederic Wilhelm Bessel (1784–1846) started his career in commercial navigation and later became an astronomer. In 1817, Bessel introduced the functions $J_\nu(x)$ in his study of planetary orbits.

for $x > 0$ is a solution to (22) with $\nu = m$. Furthermore, $J_m(x)$ and $Y_m(x)$ are linearly independent. We again call $Y_m(x)$ the Bessel function of the second kind of order m. In the literature the func$Y_m(x)$ is sometimes denoted by $N_m(x)$; it has been called the **Neumann function** and the **Weber function.**

Figure 8.12 shows the graphs of $J_0(x)$ and $J_1(x)$ and Figure 8.13 the graphs of $Y_0(x)$ and $Y_1(x)$. Notice that the curves for $J_0(x)$ and $J_1(x)$ behave like damped sine waves and have zeros that interlace (see Problem 28). The fact that the Bessel functions J_n have an infinite number of zeros will be helpful in certain applications (see Project A in Chapter 10).

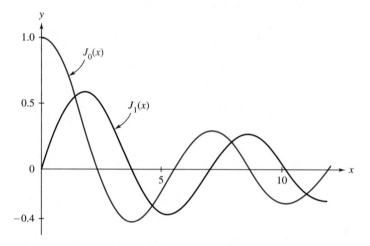

Figure 8.12 Graphs of the Bessel functions $J_0(x)$ and $J_1(x)$

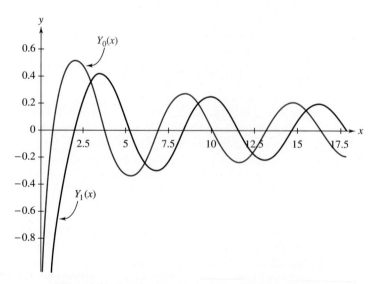

Figure 8.13 Graphs of the Bessel functions $Y_0(x)$ and $Y_1(x)$

There are several useful recurrence relations involving Bessel functions. For example,

(31) $\quad \dfrac{d}{dx}[x^\nu J_\nu(x)] = x^\nu J_{\nu-1}(x),$

(32) $\quad \dfrac{d}{dx}[x^{-\nu} J_\nu(x)] = -x^{-\nu} J_{\nu+1}(x),$

(33) $\quad J_{\nu+1}(x) = \dfrac{2\nu}{x} J_\nu(x) - J_{\nu-1}(x),$

(34) $\quad J_{\nu+1}(x) = J_{\nu-1}(x) - 2J_\nu'(x).$

Furthermore, analogous equations hold for Bessel functions of the second kind.

To illustrate the techniques involved in proving the recurrence relations, let's verify relation (31). We begin by substituting series (25) for $J_\nu(x)$ into the left-hand side of (31). Differentiating, we get

(35) $\quad \dfrac{d}{dx}[x^\nu J_\nu(x)] = \dfrac{d}{dx}\left\{ x^\nu \displaystyle\sum_{n=0}^{\infty} \dfrac{(-1)^n}{n!\Gamma(1+\nu+n)} \left(\dfrac{x}{2}\right)^{2n+\nu} \right\}$

$\qquad\qquad\qquad\;\; = \dfrac{d}{dx}\left\{ \displaystyle\sum_{n=0}^{\infty} \dfrac{(-1)^n x^{2n+2\nu}}{n!\Gamma(1+\nu+n)2^{2n+\nu}} \right\}$

$\qquad\qquad\qquad\;\; = \displaystyle\sum_{n=0}^{\infty} \dfrac{(-1)^n(2n+2\nu)x^{2n+2\nu-1}}{n!\Gamma(1+\nu+n)2^{2n+\nu}}.$

Since $\Gamma(1+\nu+n) = (\nu+n)\Gamma(\nu+n)$, we have from (35)

$$\dfrac{d}{dx}[x^\nu J_\nu(x)] = \sum_{n=0}^{\infty} \dfrac{(-1)^n 2 x^{2n+2\nu-1}}{n!\Gamma(\nu+n)2^{2n+\nu}}.$$

Factoring out an x^ν gives

$$\dfrac{d}{dx}[x^\nu J_\nu(x)] = x^\nu \sum_{n=0}^{\infty} \dfrac{(-1)^n}{n!\Gamma(1+(\nu-1)+n)} \left(\dfrac{x}{2}\right)^{2n+\nu-1} = x^\nu J_{\nu-1}(x),$$

as claimed in equation (31). We leave the verifications of the remaining relations as exercises (see Problems 22–24).

The following approximations are useful when analyzing the behavior of Bessel functions for large arguments:

(36) $\quad J_\nu(x) \approx \sqrt{\dfrac{2}{\pi x}} \cos\left(x - \dfrac{\nu\pi}{2} - \dfrac{\pi}{4}\right), \quad Y_\nu(x) \approx \sqrt{\dfrac{2}{\pi x}} \sin\left(x - \dfrac{\nu\pi}{2} - \dfrac{\pi}{4}\right), \quad x \gg 1.$

This oscillatory behavior is demonstrated in Figures 8.12 and 8.13. Problem 41 provides some justification of formula (36). Of course, for small x the leading terms of the power series expansion dominate and we have

(37) $\quad J_\nu(x) \approx x^\nu/2^\nu\Gamma(1+\nu), \quad Y_0(x) \approx (2\ln x)/\pi, \quad Y_{\nu>0}(x) \approx -\Gamma(\nu)2^\nu/\pi x^\nu, \quad 0 < x \ll 1.$

Legendre's Equation

The linear second order differential equation

$$(38) \qquad (1 - x^2)y'' - 2xy' + n(n + 1)y = 0,$$

where n is a fixed parameter, is called **Legendre's equation.** This equation has a regular singular point at 1, and hence a series solution for (38) about $x = 1$ may be obtained by the method of Frobenius. By setting $z = x - 1$, equation (38) is transformed into

$$(39) \qquad z(z + 2)\frac{d^2y}{dz^2} + 2(z + 1)\frac{dy}{dz} - n(n + 1)y = 0.$$

The indicial equation for (39) at $z = 0$ is

$$r(r - 1) + r = r^2 = 0,$$

which has roots $r_1 = r_2 = 0$. Upon substituting

$$y(z) = \sum_{k=0}^{\infty} a_k z^k$$

into (39) and proceeding as usual, we arrive at the solution

$$(40) \qquad y_1(x) = 1 + \sum_{k=1}^{\infty} \frac{(-n)_k(n + 1)_k}{k!(1)_k}\left(\frac{1 - x}{2}\right)^k,$$

where we have expressed y_1 in terms of the original variable x and have taken $a_0 = 1$. We have written the solution in the above form because it is now obvious from (40) that

$$(41) \qquad y_1(x) = F\left(-n, n + 1; 1; \frac{1 - x}{2}\right),$$

where F is the Gaussian hypergeometric function defined in (10).

For n a nonnegative integer, the factor

$$(-n)_k = (-n)(-n + 1)(-n + 2)\cdots(-n + k - 1)$$

will be zero for $k \geq n + 1$. Hence the solution given in (40) and (41) is a *polynomial* of degree n. Moreover, $y_1(1) = 1$. These polynomial solutions of equation (38) are called the **Legendre polynomials** or **spherical polynomials** and are traditionally denoted by $P_n(x)$. That is,

$$(42) \qquad P_n(x) := 1 + \sum_{k=1}^{\infty} \frac{(-n)_k(n + 1)_k}{k!(1)_k}\left(\frac{1 - x}{2}\right)^k.$$

If we expand about $x = 0$, then $P_n(x)$ takes the form

$$(43) \qquad P_n(x) = 2^{-n} \sum_{m=0}^{[n/2]} \frac{(-1)^m(2n - 2m)!}{(n - m)!m!(n - 2m)!}x^{n-2m},$$

[†] ***Historical Footnote:*** Solutions to this equation were obtained by Adrien Marie Legendre (1752–1833) in 1785 and are referred to as Legendre functions.

where $[n/2]$ is the greatest integer less than or equal to $n/2$ (see Problem 34). The first three Legendre polynomials are

$$P_0(x) = 1, \qquad P_1(x) = x, \qquad P_2(x) = \frac{3}{2}x^2 - \frac{1}{2}.$$

The Legendre polynomials satisfy the **orthogonality condition**

(44) $\qquad \displaystyle\int_{-1}^{1} P_m(x)P_n(x)\,dx = 0 \quad \text{for} \quad n \neq m.$

To see this, we first rewrite equation (38) in what is called the **selfadjoint form:**

(45) $\qquad [(1 - x^2)y']' + n(n + 1)y = 0.$

Since $P_n(x)$ and $P_m(x)$ satisfy (45) with parameters n and m respectively, we have

(46) $\qquad [(1 - x^2)P_n'(x)]' + n(n + 1)P_n(x) = 0,$

(47) $\qquad [(1 - x^2)P_m'(x)]' + m(m + 1)P_m(x) = 0.$

Multiplying (46) by $P_m(x)$ and (47) by $P_n(x)$ and then subtracting, we find

$$P_m(x)[(1 - x^2)P_n'(x)]' - P_n(x)[(1 - x^2)P_m'(x)]'$$
$$+ [n(n + 1) - m(m + 1)]P_m(x)P_n(x) = 0,$$

which can be rewritten in the form

(48) $\qquad (n^2 - m^2 + n - m)P_m(x)P_n(x) = P_n(x)[(1 - x^2)P_m'(x)]' - P_m(x)[(1 - x^2)P_n'(x)]'.$

It is a straightforward calculation to show that the right-hand side of (48) is just

$$\{(1 - x^2)[P_n(x)P_m'(x) - P_n'(x)P_m(x)]\}'.$$

Using this fact and the identity $n^2 - m^2 + n - m = (n - m)(n + m + 1)$, we rewrite equation (48) as

(49) $\qquad (n - m)(n + m + 1)P_m(x)P_n(x) = \{(1 - x^2)[P_n(x)P_m'(x) - P_n'(x)P_m(x)]\}'.$

Integrating both sides of (49) from $x = -1$ to $x = 1$ yields

(50) $\qquad (n - m)(n + m + 1)\displaystyle\int_{-1}^{1} P_m(x)P_n(x)\,dx$

$$= \int_{-1}^{1} \{(1 - x^2)[P_n(x)P_m'(x) - P_n'(x)P_m(x)]\}'\, dx$$
$$= \{(1 - x^2)[P_n(x)P_m'(x) - P_n'(x)P_m(x)]\}\Big|_{-1}^{1}$$
$$= 0$$

because $1 - x^2 = 0$ for $x = \pm 1$. Since n and m are nonnegative integers with $n \neq m$, then $(n - m)(n + m + 1) \neq 0$, and so equation (44) follows from (50).

Legendre polynomials also satisfy the **recurrence formula**

(51) $(n + 1)P_{n+1}(x) = (2n + 1)xP_n(x) - nP_{n-1}(x)$

and **Rodrigues's formula**

(52) $P_n(x) = \dfrac{1}{2^n n!} \dfrac{d^n}{dx^n}\{(x^2 - 1)^n\}$

(see Problems 32 and 33).

The Legendre polynomials are generated by the function $(1 - 2xz + z^2)^{-1/2}$ in the sense that

(53) $(1 - 2xz + z^2)^{-1/2} = \displaystyle\sum_{n=0}^{\infty} P_n(x)z^n, \qquad |z| < 1, \qquad |x| < 1.$

That is, if we expand $(1 - 2xz + z^2)^{-1/2}$ in a Taylor series about $z = 0$, treating x as a fixed parameter, then the coefficients of z^n are the Legendre polynomials $P_n(x)$. The function $(1 - 2xz + z^2)^{-1/2}$ is called a **generating function** for $P_n(x)$ and can be derived from the recurrence formula (51) (see Problem 35).

The Legendre polynomials are an example of a class of special functions called **classical orthogonal polynomials.**[†] The latter includes, for example, the **Jacobi polynomials,** $P_n^{(\alpha,\beta)}(x)$; the **Gegenbauer** or **ultraspherical polynomials,** $C_n^\lambda(x)$; the **Chebyshev** (**Tchebichef**) **polynomials,** $T_n(x)$ and $U_n(x)$; the **Laguerre polynomials,** $L_n^\alpha(x)$; and the **Hermite polynomials,** $H_n(x)$. The properties of the classical orthogonal polynomials can be found in the books mentioned earlier in this section or in the *Handbook of Mathematical Functions with Formulas, Graphs, and Mathematical Tables*, by M. Abramowitz and I. A. Stegun (eds.) (Dover, New York, 1965), Chapter 22.

EXERCISES 8.9

In Problems 1 through 4, express a general solution to the given equation using Gaussian hypergeometric functions.

1. $x(1 - x)y'' + \left(\dfrac{1}{2} - 4x\right)y' - 2y = 0.$

2. $3x(1 - x)y'' + (1 - 27x)y' - 45y = 0.$

3. $2x(1 - x)y'' + (1 - 6x)y' - 2y = 0.$

4. $2x(1 - x)y'' + (3 - 10x)y' - 6y = 0.$

In Problems 5 through 8, verify the following formulas by expanding each function in a power series about $x = 0$.

5. $F(1, 1; 2; x) = -x^{-1}\ln(1 - x).$

6. $F(\alpha, \beta; \beta; x) = (1 - x)^{-\alpha}.$

7. $F\left(\dfrac{1}{2}, 1; \dfrac{3}{2}; x^2\right) = \dfrac{1}{2}x^{-1}\ln\left(\dfrac{1 + x}{1 - x}\right).$

8. $F\left(\dfrac{1}{2}, 1; \dfrac{3}{2}; -x^2\right) = x^{-1}\arctan x.$

In Problems 9 and 10, use one of the methods discussed in Section 8.7 to obtain two linearly independent solutions to the given hypergeometric equation.

9. $x(1 - x)y'' + (1 - 3x)y' - y = 0.$

[†] Here orthogonality is used in the more general sense that $\int_a^b p_n(x)p_m(x)w(x)\,dx = 0$ for $n \neq m$, where $w(x)$ is a weight function on the interval (a, b).

10. $x(1 - x)y'' + (2 - 2x)y' - \dfrac{1}{4}y = 0.$

11. Show that the **confluent hypergeometric equation**

$$xy'' + (\gamma - x)y' - \alpha y = 0,$$

where α and γ are fixed parameters and γ is not an integer, has two linearly independent solutions

$$y_1(x) = {}_1F_1(\alpha; \gamma; x) := 1 + \sum_{n=1}^{\infty} \frac{(\alpha)_n}{n!(\gamma)_n} x^n$$

and

$$y_2(x) = x^{1-\gamma}\,{}_1F_1(\alpha + 1 - \gamma; 2 - \gamma; x).$$

12. Use the property of the gamma function given in (19) to derive relation (20).

In Problems 13 through 18, express a general solution to the given equation using Bessel functions of either the first or second kind.

13. $4x^2y'' + 4xy' + (4x^2 - 1)y = 0.$

14. $9x^2y'' + 9xy' + (9x^2 - 16)y = 0.$

15. $x^2y'' + xy' + (x^2 - 1)y = 0.$

16. $x^2y'' + xy' + x^2y = 0.$

17. $9t^2x'' + 9tx' + (9t^2 - 4)x = 0.$

18. $x^2z'' + xz' + (x^2 - 16)z = 0.$

In Problems 19 and 20, a Bessel equation is given. For the appropriate choice of ν, the Bessel function $J_\nu(x)$ is one solution. Use one of the methods discussed in Section 8.7 to obtain a second linearly independent solution.

19. $x^2y'' + xy' + (x^2 - 1)y = 0.$

20. $x^2y'' + xy' + (x^2 - 4)y = 0.$

21. Show that $x^\nu J_\nu(x)$ satisfies the equation

$$xy'' + (1 - 2\nu)y' + xy = 0, \qquad x > 0$$

and use this result to find a solution for the equation

$$xy'' - 2y' + xy = 0, \qquad x > 0.$$

In Problems 22 through 24, derive the indicated recurrence formulas.

22. Formula (32). **23.** Formula (33).

24. Formula (34).

25. Show that $J_{1/2}(x) = (2/\pi x)^{1/2} \sin x$ and

$J_{-1/2}(x) = (2/\pi x)^{1/2} \cos x.$

26. The Bessel functions of order $\nu = n + 1/2$, n any integer, are related to the **spherical Bessel functions.** Use

relation (33) and the results of Problem 25 to show that such Bessel functions can be represented in terms of $\sin x$, $\cos x$, and powers of x. Demonstrate this by determining a closed form for $J_{-3/2}(x)$ and $J_{5/2}(x)$.

27. Use the method discussed in Section 8.8 to determine a second linearly independent solution to Bessel's equation of order zero in terms of the Bessel function $J_0(x)$.

28. Show that between two consecutive positive roots (zeros) of $J_1(x)$, there is a root of $J_0(x)$. This interlacing property of the roots of Bessel functions is illustrated in Figure 8.12. [*Hint:* Use relation (31) and Rolle's theorem from calculus.]

29. Use formula (43) to determine the first five Legendre polynomials.

30. Show that the Legendre polynomials of even degree are even functions of x, while those of odd degree are odd functions.

31. **(a)** Show that the orthogonality condition (44) for Legendre polynomials implies that

$$\int_{-1}^{1} P_n(x)q(x)\,dx = 0$$

for *any* polynomial $q(x)$ of degree at most $n - 1$. [*Hint:* The polynomials $P_0, P_1, \ldots, P_{n-1}$ are linearly independent and hence span the space of all polynomials of degree at most $n - 1$. Thus $q(x) = a_0P_0(x) + a_1P_1(x) + \cdots + a_{n-1}P_{n-1}(x)$ for suitable constants a_k.]

(b) Prove that if $Q_n(x)$ is a polynomial of degree n such that

$$\int_{-1}^{1} Q_n(x)P_k(x)\,dx = 0 \quad \text{for} \quad k = 0, 1, \ldots, n - 1,$$

then $Q_n(x) = cP_n(x)$ for some constant c. [*Hint:* Select c so that the coefficient of x^n for $Q_n(x) - cP_n(x)$ is zero. Then, since $P_0, \ldots, P_{n-1}$ is a basis,

$$Q_n(x) - cP_n(x) = a_0P_0(x) + \cdots + a_{n-1}P_{n-1}(x).$$

Multiply the last equation by $P_k(x)(0 \le k \le n - 1)$ and integrate from $x = -1$ to $x = 1$ to show that each a_k is zero.]

32. Deduce the recurrence formula (51) for Legendre polynomials by completing the following steps:

(a) Show that the function $Q_{n-1}(x) := (n + 1)P_{n+1}(x) - (2n + 1)xP_n(x)$ is a polynomial of degree $n - 1$. [*Hint:* Compute the coefficient of the x^{n+1} term using

the representation (42). The coefficient of x^n is also zero because $P_{n+1}(x)$ and $xP_n(x)$ are both odd or both even functions, a consequence of Problem 30.]

(b) Using the result of Problem 31(a), show that

$$\int_{-1}^{1} Q_{n-1}(x)P_k(x)\,dx = 0 \quad \text{for} \quad k = 0, 1, \dots, n-2.$$

(c) From Problem 31(b), conclude that $Q_{n-1}(x) = cP_{n-1}(x)$ and, by taking $x = 1$, show that $c = -n$. [*Hint:* Recall that $P_m(1) = 1$ for all m.] From the definition of $Q_{n-1}(x)$ in part (a), the recurrence formula now follows.

33. To prove Rodrigues's formula (52) for Legendre polynomials, complete the following steps:

(a) Let $v_n := (d^n/dx^n)\{(x^2 - 1)^n\}$ and show that $v_n(x)$ is a polynomial of degree n with the coefficient of x^n equal to $(2n)!/n!$.

(b) Use integration by parts n times to show that, for any polynomial $q(x)$ of degree less than n,

$$\int_{-1}^{1} v_n(x)q(x)\,dx = 0.$$

Hint: For example, when $n = 2$,

$$\int_{-1}^{1} \frac{d^2}{dx^2}\{(x^2 - 1)^2\}q(x)\,dx$$

$$= q(x)\frac{d}{dx}\{(x^2 - 1)^2\}\Big|_{-1}^{1} - \{q'(x)(x^2 - 1)^2\}\Big|_{-1}^{1}$$

$$+ \int_{-1}^{1} q''(x)(x^2 - 1)^2\,dx.$$

Since $n = 2$, the degree of $q(x)$ is at most 1, and so $q''(x) \equiv 0$. Thus

$$\int_{-1}^{1} \frac{d^2}{dx^2}\{(x^2 - 1)^2\}q(x)\,dx = 0.$$

(c) Use the result of Problem 31(b) to conclude that $P_n(x) = cv_n(x)$ and show that $c = 1/2^n n!$ by comparing the coefficients of x^n in $P_n(x)$ and $v_n(x)$.

34. Use Rodrigues's formula (52) to obtain the representation (43) for the Legendre polynomials $P_n(x)$. *Hint:* From the binomial formula,

$$P_n(x) = \frac{1}{2^n n!}\frac{d^n}{dx^n}\{(x^2 - 1)^n\}$$

$$= \frac{1}{2^n n!}\frac{d^n}{dx^n}\left\{\sum_{m=0}^{n} \frac{n!(-1)^m}{(n-m)!m!}x^{2n-2m}\right\}.$$

35. The generating function in (53) for Legendre polynomials can be derived from the recurrence formula (51) as follows. Let x be fixed and set $f(z) := \sum_{n=0}^{\infty} P_n(x)z^n$. The goal is to determine an explicit formula for $f(z)$.

(a) Show that multiplying each term in the recurrence formula (51) by z^n and summing the terms from $n = 1$ to ∞ leads to the differential equation

$$\frac{df}{dz} = \frac{x - z}{1 - 2xz + z^2}f.$$

Hint:

$$\sum_{n=1}^{\infty} (n+1)P_{n+1}(x)z^n = \sum_{n=0}^{\infty} (n+1)P_{n+1}(x)z^n - P_1(x)$$

$$= \frac{df}{dz} - x.$$

(b) Solve the differential equation derived in part (a) and use the initial condition $f(0) = P_0(x) \equiv 1$ to obtain $f(z) = (1 - 2xz + z^2)^{-1/2}$.

36. Find a general solution about $x = 0$ for the equation

$$(1 - x^2)y'' - 2xy' + 2y = 0$$

by first finding a polynomial solution and then using the reduction of order formula given in equation (11) in Section 8.7, page 477 (or refer to the inside back cover) to find a second (series) solution.

37. The **Hermite polynomials** $H_n(x)$ are polynomial solutions to Hermite's equation

$$y'' - 2xy' + 2ny = 0.$$

The Hermite polynomials are generated by

$$e^{2tx - t^2} = \sum_{n=0}^{\infty} \frac{H_n(x)}{n!}t^n.$$

Use this equation to determine the first four Hermite polynomials.

38. The **Chebyshev** (Tchebichef) **polynomials** $T_n(x)$ are polynomial solutions to Chebyshev's equation

$$(1 - x^2)y'' - xy' + n^2 y = 0.$$

The Chebyshev polynomials satisfy the recurrence relation

$$T_{n+1}(x) = 2xT_n(x) - T_{n-1}(x),$$

with $T_0(x) = 1$ and $T_1(x) = x$. Use this recurrence relation to determine the next three Chebyshev polynomials.

39. The **Laguerre polynomials** $L_n(x)$ are polynomial solutions to Laguerre's equation

$$xy'' + (1 - x)y' + ny = 0.$$

The Laguerre polynomials satisfy Rodrigues's formula,

$$L_n(x) = \frac{e^x}{n!} \frac{d^n}{dx^n} (x^n e^{-x}).$$

Use this formula to determine the first four Laguerre polynomials.

40. Reduction to Bessel's Equation. The class of equations of the form

(54) $\quad y''(x) + cx^n y(x) = 0, \qquad x > 0,$

where c and n are positive constants, can be solved by transforming the equation into Bessel's equation.

(a) First, use the substitution $y = x^{1/2} z$ to transform (54) into an equation involving x and z.

(b) Second, use the substitution

$$s = \frac{2\sqrt{c}}{n+2} x^{n/2+1}$$

to transform the equation obtained in part (a) into the Bessel equation

$$s^2 \frac{d^2 z}{ds^2} + s \frac{dz}{ds} + \left(s^2 - \frac{1}{(n+2)^2} \right) z = 0, \quad s > 0.$$

(c) A general solution to the equation in part (b) can be given in terms of Bessel functions of the first and second kind. Substituting back in for s and z, obtain a general solution for equation (54).

41. (a) Show that the substitution $z(x) = \sqrt{x}\, y(x)$ renders Bessel's equation (22) in the form

(55) $\qquad z'' + \left(1 + \frac{1 - 4\nu^2}{4x^2} \right) z = 0.$

(b) For $x \gg 1$, equation (55) would apparently be approximated by the equation

(56) $\qquad z'' + z = 0.$

Write down the general solution to (56), reset $y(x) = z(x)/\sqrt{x}$, and argue the plausibility of formula (36).

(c) For $\nu = \pm 1/2$, equation (55) reduces to equation (56) exactly. Relate this observation to Problem 25.

CHAPTER SUMMARY

Power series solutions to differential equations are useful alternatives when explicit solutions involving elementary functions cannot be found.

Power Series

Every power series $\sum_{n=0}^{\infty} a_n (x - x_0)^n$ has a **radius of convergence** ρ, $0 \le \rho \le \infty$, such that the series converges absolutely for $|x - x_0| < \rho$ and diverges when $|x - x_0| > \rho$. By the ratio test,

$$\frac{1}{\rho} = \lim_{n \to \infty} \frac{|a_{n+1}|}{|a_n|},$$

provided that this limit exists as an extended real number. A function $f(x)$ that is the sum of a power series in some open interval about x_0 is said to be **analytic at x_0**. If f is analytic at x_0, its power series representation about x_0 is the **Taylor series**

$$f(x) = \sum_{n=0}^{\infty} \frac{f^{(n)}(x_0)}{n!} (x - x_0)^n.$$

Taylor Series Method

If it is known that the solution $y(x)$ to the initial value problem

$$y' = F(x, y), \qquad y(x_0) = y_0$$

is analytic at the initial point x_0, then the first few terms of its Taylor expansion about x_0 can be computed by successively differentiating the equation:

$$y(x_0) = y_0, \qquad y'(x_0) = F(x_0, y_0),$$

$$y''(x_0) = \frac{\partial F}{\partial x}(x_0, y_0) + \frac{\partial F}{\partial y}(x_0, y_0)F(x_0, y_0), \qquad \text{etc.}$$

Power Series Method for an Ordinary Point

In the case of a linear equation of the form

(1) $y'' + p(x)y' + q(x)y = 0,$

where p and q are analytic at x_0, the point x_0 is called an **ordinary point,** and the equation has a pair of linearly independent solutions expressible as power series about x_0. The radii of convergence of these series solutions are at least as large as the distance from x_0 to the nearest singularity (real or complex) of the equation. To find power series solutions to (1), we substitute $y(x) = \sum_{n=0}^{\infty} a_n(x - x_0)^n$ into (1), group like terms, and set the coefficients of the resulting power series equal to zero. This leads to a recurrence relation for the coefficients a_n, which, in some cases, may even yield a general formula for the a_n. The same method applies to the nonhomogeneous version of (1), provided the forcing function is also analytic at x_0.

Regular Singular Points

If, in equation (1), either p or q fails to be analytic at x_0, then x_0 is a **singular point** of (1). If x_0 is a singular point for which $(x - x_0)p(x)$ and $(x - x_0)^2 q(x)$ are both analytic at x_0, then x_0 is a **regular singular point.** The Cauchy-Euler equation,

(2) $ax^2\dfrac{d^2y}{dx^2} + bx\dfrac{dy}{dx} + cy = 0, \qquad x > 0$

has a regular singular point at $x = 0$, and a general solution to (2) can be obtained by substituting $y = x^r$ and examining the roots of the resulting indicial equation $ar^2 + (b - a)r + c = 0$.

Method of Frobenius

For an equation of the form (1) with a regular singular point at x_0, a series solution can be found by the **method of Frobenius.** This is obtained by substituting

$$w(r, x) = (x - x_0)^r \sum_{n=0}^{\infty} a_n(x - x_0)^n$$

into (1), finding a recurrence relation for the coefficients, and choosing $r = r_1$, the larger root of the **indicial equation**

(3) $r(r - 1) + p_0 r + q_0 = 0,$

where $p_0 := \lim_{x \to x_0}(x - x_0)p(x)$, $q_0 := \lim_{x \to x_0}(x - x_0)^2 q(x)$.

Finding a Second Linearly Independent Solution

If the two roots r_1, r_2 of the indicial equation (3) do not differ by an integer, then a second linearly independent solution to (1) can be found by taking $r = r_2$ in the method of Frobenius. However, if $r_1 = r_2$ or $r_1 - r_2$ is a positive integer, then discovering a second solution requires a different approach. This may be a reduction of order procedure, the utilization of Theorem 7 which gives the *forms* of the solutions, or an operator approach in which, in the method of Frobenius, the coefficients a_n are treated as functions of r.

Special Functions

Some special functions in physics and engineering that arise from series solutions to linear second order equations with polynomial coefficients are Gaussian hypergeometric functions, $F(\alpha, \beta; \gamma; x)$; Bessel functions $J_\nu(x)$; and orthogonal polynomials such as those of Legendre, Chebyshev, Laguerre, and Hermite.

REVIEW PROBLEMS

1. Use the Taylor series method to determine the first four nonzero terms of a series solution for the given initial value problem.
 (a) $y' = xy - y^2$;　　$y(0) = 1$.
 (b) $z'' - x^3 z' + xz^2 = 0$;　　$z(0) = -1$,　　$z'(0) = 1$.

2. Determine all the singular points of the given equation and classify them as regular or irregular.
 (a) $(x^2 - 4)^2 y'' + (x - 4)y' + xy = 0$.
 (b) $(\sin x)y'' + y = 0$.

3. Find at least the first four nonzero terms in a power series expansion about $x = 0$ for a general solution to the given equation.
 (a) $y'' + x^2 y' - 2y = 0$.
 (b) $y'' + e^{-x} y' - y = 0$.

4. Find a general formula for the coefficient a_n in a power series expansion about $x = 0$ for a general solution to the given equation.
 (a) $(1 - x^2)y'' + xy' + 3y = 0$.
 (b) $(x^2 - 2)y'' + 3y = 0$.

5. Find at least the first four nonzero terms in a power series expansion about $x = 2$ for a general solution to
 $$w'' + (x - 2)w' - w = 0.$$

6. Use the substitution $y = x^r$ to find a general solution to the given equation for $x > 0$.
 (a) $2x^2 y''(x) + 5xy'(x) - 12y(x) = 0$.

 (b) $x^3 y'''(x) + 3x^2 y''(x) - 2xy'(x) - 2y(x) = 0$.

7. Use the method of Frobenius to find at least the first four nonzero terms in the series expansion about $x = 0$ for a solution to the given equation for $x > 0$.
 (a) $x^2 y'' - 5xy' + (9 - x)y = 0$.
 (b) $x^2 y'' + (x^2 + 2x)y' - 2y = 0$.

8. Find the indicial equation and its roots and state (but do not compute) the form of the series expansion about $x = 0$ (as in Theorem 7 on page 479) for two linearly independent solutions of the given equation for $x > 0$.
 (a) $x^2 y'' + (\sin x)y' - 4y = 0$.
 (b) $2xy'' + 5y' + xy = 0$.
 (c) $(x \sin x)y'' + xy' + (\tan x)y = 0$.

9. Find at least the first three nonzero terms in the series expansion about $x = 0$ for a general solution to the given equation for $x > 0$.
 (a) $x^2 y'' - x(1 + x)y' + y = 0$.
 (b) $xy'' + y' - 2y = 0$.
 (c) $2xy'' + 6y' + y = 0$.
 (d) $x^2 y'' + (x - 2)y = 0$.

10. Express a general solution to the given equation using Gaussian hypergeometric functions or Bessel functions.
 (a) $x(1 - x)y'' + \left(\dfrac{1}{2} - 6x\right)y' - 6y = 0$.

 (b) $9\theta^2 y'' + 9\theta y' + (9\theta^2 - 1)y = 0$.

TECHNICAL WRITING EXERCISES

1. Knowing that a general solution to a nonhomogeneous linear second order equation can be expressed as a particular solution plus a general solution to the corresponding homogeneous equation, what can you say about the form of a general power series solution to the nonhomogeneous equation about an ordinary point?

2. Discuss advantages and disadvantages of power series solutions over numerical solutions.

3. Explain why it is not very helpful to know that $y(x) = \sum_{n=0}^{\infty} a_n x^n$ when you are solving nonlinear equations with variable coefficients.

4. Give an overview of the method of Frobenius that can be understood by a typical calculus student.

A. Spherically Symmetric Solutions to Schrödinger's Equation for the Hydrogen Atom

In quantum mechanics one is interested in determining the wave function and energy states of an atom. These are determined from Schrödinger's equation. In the case of the hydrogen atom, it is possible to find wave functions ψ that are functions only of r, the distance from the proton to the electron. Such functions are called **spherically symmetric** and satisfy the simpler equation

(1) $$\frac{1}{r}\frac{d^2}{dr^2}(r\psi) = \frac{-8m\pi^2}{h^2}\left(E + \frac{e_0^2}{r}\right)\psi,$$

where e_0^2, m, and h are constants, and E, also a constant, represents the energy of the atom, which we assume here to be negative.

(a) Show that with the substitutions

$$r = \frac{h^2}{4\pi^2 me_0^2}\rho, \qquad E = \frac{2\pi^2 me_0^4}{h^2}\varepsilon,$$

equation (1) reduces to

$$\frac{d^2(\rho\psi)}{d\rho^2} = -\left(\varepsilon + \frac{2}{\rho}\right)\rho\psi.$$

(b) If $f := \rho\psi$, then the preceding equation becomes

(2) $$\frac{d^2f}{d\rho^2} = -\left(\varepsilon + \frac{2}{\rho}\right)f.$$

Show that the substitution $f(\rho) = e^{-\alpha\rho}g(\rho)$, where α is a positive constant, transforms (2) into

(3) $$\frac{d^2g}{d\rho^2} - 2\alpha\frac{dg}{d\rho} + \left(\frac{2}{\rho} + \varepsilon + \alpha^2\right)g = 0.$$

(c) If we choose $\alpha^2 = -\varepsilon$ (ε negative), then (3) becomes

(4) $$\frac{d^2g}{d\rho} - 2\alpha\frac{dg}{d\rho} + \frac{2}{\rho}g = 0.$$

Show that a power series solution $g(\rho) = \sum_{k=1}^{\infty} a_k\rho^k$ (starting with $k = 1$) for (4) must have coefficients a_k that satisfy the recurrence relation

(5) $$a_{k+1} = \frac{2(\alpha k - 1)}{k(k + 1)}a_k, \qquad k \geq 1.$$

(d) Now for $a_1 = 1$ and k very large, $a_{k+1} \approx (2\alpha/k)a_k$ and so $a_{k+1} \approx (2\alpha)^k/k!$, which are the coefficients for $\rho e^{2\alpha\rho}$. Hence g acts like $\rho e^{2\alpha\rho}$, so $f(\rho) = e^{-\alpha\rho}g(\rho)$ is like $\rho e^{\alpha\rho}$. Going back further, we then see that $\psi \approx e^{\alpha\rho}$. Therefore when $r = h^2\rho/4\pi^2 me_0^2$ is large, so is ψ. Roughly speaking, $\psi^2(r)$ is proportional to the probability of finding an electron a distance r from the proton. Thus the above argument would imply that the electron in a hydrogen atom is more likely to be found at a very large distance from the proton! Since this makes no sense physically, we ask: Do there exist positive values for α for which ψ remains bounded as r becomes large?

Show that when $\alpha = 1/n$, $n = 1, 2, 3, \ldots$, then $g(\rho)$ is a polynomial of degree n and argue that ψ is therefore bounded.

(e) Let E_n and $\psi_n(\rho)$ denote, respectively, the energy state and wave function corresponding to $\alpha = 1/n$. Find E_n (in terms of the constants e_0^2, m, and h) and $\psi_n(\rho)$ for $n = 1, 2$, and 3.

B. Airy's Equation

 In aerodynamics one encounters the following initial value problem for **Airy's equation:**

$$y'' + xy = 0; \qquad y(0) = 1, \qquad y'(0) = 0.$$

(a) Find the first five nonzero terms in a power series expansion about $x = 0$ for the solution and graph this polynomial for $-10 \le x \le 10$.

(b) Using the Runge-Kutta subroutine (see Section 5.6) with $h = 0.05$, approximate the solution on the interval $[0, 10]$, i.e., at the points 0.05, 0.1, 0.15, etc.

(c) Using the Runge-Kutta subroutine with $h = 0.05$, approximate the solution on the interval $[-10, 0]$. [*Hint:* With the change of variables $z = -x$, it suffices to approximate the solution to $y'' - zy = 0$; $y(0) = 1$, $y'(0) = 0$, on the interval $[0, 10]$.]

(d) Using your knowledge of constant coefficient equations as a basis for guessing the behavior of the solutions to Airy's equation, decide whether the power series approximation obtained in part (a) or the numerical approximation obtained in parts (b) and (c) better describes the true behavior of the solution on the interval $[-10, 10]$.

C. Buckling of a Tower

A tower is constructed of four angle beams connected by diagonals (see Figure 8.14). The deflection curve $y(x)$ for the tower is governed by the equation

(6) $$x^2\frac{d^2y}{dx^2} + \frac{Pa^2}{EI}y = 0, \qquad a < x < a + L,$$

where x is the vertical coordinate measured down from the extended top of the tower, y is the deflection from the vertical passing through the center of the unbuckled tower, L is the tower height, a is length of the truncation, P is the load, E is the modulus of elasticity, and I is the moment of inertia. The appropriate boundary conditions for this design are

(7) $$y(a) = 0,$$

(8) $$y'(a + L) = 0.$$

Clearly the solution $y(x) \equiv 0$ is always at hand. However, when the load P is heavy enough, the tower can *buckle* and a nontrivial solution $y(x)$ is also possible. We want to predict this phenomenon.

(a) Unbuckled

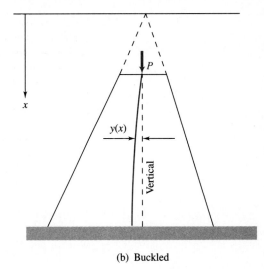

(b) Buckled

Figure 8.14 Buckling tower

(a) Solve equation (6). [*Hint:* Equation (6) is a Cauchy-Euler equation.]

(b) Show that the first boundary condition (7) implies

$$y = Ax^{1/2} \sin[\beta \ln(x/a)],$$

where A is an arbitrary constant and $\beta := \sqrt{Pa^2/EI - 1/4}$.

(c) Show that the second boundary condition (8) gives

$$A\left\{\tan\left[\beta \ln\left(\frac{a+L}{a}\right)\right] + 2\beta\right\} = 0.$$

(d) Use the result of part (c) to argue that *no* buckling takes place (i.e., the only possibility is $A = 0$) if $0 < \beta < \beta_c$, where β_c is the smallest positive real number that makes the expression in braces zero.

(e) The value of the load corresponding to β_c is called the **critical load** P_c. Solve for P_c in terms of β_c, a, E, and I.

(f) Find the critical load P_c if $a = 10$, $L = 40$, and $EI = 1000$.

D. *Aging Spring and Bessel Functions*

In Problems 30 and 31 in Exercises 8.4, page 459, we discussed a model for a spring-mass system with an aging spring. Without damping, the displacement $x(t)$ at time t is governed by the equation

(9) $mx''(t) + ke^{-\eta t}x(t) = 0,$

where m, k, and η are positive constants. The general solution to this equation can be expressed using Bessel functions.

(a) The coefficient of x suggests a change of variables of the form $s = \alpha e^{\beta t}$. Show that (9) transforms into

(10) $$\beta^2 s^2 \frac{d^2 x}{ds^2} + \beta^2 s \frac{dx}{ds} + \frac{k}{m} \left(\frac{s}{\alpha} \right)^{-\eta/\beta} x = 0.$$

(b) Show that choosing α and β to satisfy

$$\frac{-\eta}{\beta} = 2 \quad \text{and} \quad \frac{k}{m\beta^2 \alpha^{-\eta/\beta}} = 1$$

transforms (10) into the Bessel equation of order zero in s and x.

(c) Using the result of part (b), show that a general solution to (9) is given by

$$x(t) = c_1 J_0 \left(\frac{2}{\eta} \sqrt{k/m} \, e^{-\eta t/2} \right) + c_2 Y_0 \left(\frac{2}{\eta} \sqrt{k/m} \, e^{-\eta t/2} \right),$$

where J_0 and Y_0 are the Bessel functions of order zero of the first and second kind, respectively.

(d) Discuss the behavior of the displacement $x(t)$ for c_2 positive, negative, and zero.

(e) Compare the behavior of the displacement $x(t)$ for η a small positive number and η a large positive number.

Chapter 9

Matrix Methods for Linear Systems

Remark: *Chapter 9 uses matrix-algebra techniques to solve systems of differential equations. After covering the first six sections of this chapter, the reader can return to Chapter 5 to study applications of systems in Sections 5.4–5.5, numerical methods in Section 5.6, phase plane analysis in Section 5.7, or dynamical systems and chaos in Section 5.8. For those readers who have already studied Chapter 5, several application problems are dispersed throughout Chapter 9.*

9.1 INTRODUCTION

In Sections 4.10 and 7.6, we presented some elementary techniques for solving linear systems of differential equations, and in Chapter 5, we discussed their application to electric networks, mechanical systems with more than one degree of freedom, and multiple compartment systems. These techniques work quite well when the number of variables is small. However, for large electric networks or mechanical systems with many degrees of freedom, these methods are cumbersome and disguise the underlying theory. One way out of the maze is to use matrix methods.

The kinds of systems we shall deal with are those in the **normal form**:

$$
\begin{aligned}
x_1'(t) &= a_{11}(t)x_1(t) + \cdots + a_{1n}(t)x_n(t) + f_1(t), \\
x_2'(t) &= a_{21}(t)x_1(t) + \cdots + a_{2n}(t)x_n(t) + f_2(t), \\
&\;\;\vdots \qquad\qquad\quad \vdots \qquad\qquad \vdots \qquad\quad \vdots \\
x_n'(t) &= a_{n1}(t)x_1(t) + \cdots + a_{nn}(t)x_n(t) + f_n(t).
\end{aligned}
\tag{1}
$$

The structure of such a system suggests there might be a more convenient way of expressing it. This is in fact the case. Using matrices and vectors, one can write system (1) in the compact form

$$
\mathbf{x}'(t) = \mathbf{A}(t)\mathbf{x}(t) + \mathbf{f}(t),
\tag{2}
$$

where $\mathbf{x}(t) = \text{col}(x_1(t), \ldots, x_n(t))$ and $\mathbf{f}(t) = \text{col}(f_1(t), \ldots, f_n(t))$ are vectors and $\mathbf{A}(t) = [a_{ij}(t)]$ is a matrix. Not only is (2) easier to write and remember, but it also serves as a reminder that first order linear systems behave much like linear first order differential equations, as we shall see.

In this chapter we use matrix methods to study systems of first order differential equations in normal form. We assume the reader is acquainted with the computational aspects of linear algebra usually taught in a basic linear algebra course[†]—in particular, matrix addition and multiplication, solving systems of linear equations, evaluating determinants, and computing the inverse of a matrix. For the reader's convenience, we give a brief review of matrices in the next section. There we also introduce vector and matrix functions and discuss their calculus.

The advantages of the matrix approach become evident in Section 9.3, where we develop the theory for linear systems in normal form. This development parallels very closely the discussion of linear second order equations in Section 4.3 and the theory of linear higher-order equations in Section 6.1. In particular, we discuss linear dependence of vector functions, fundamental solution sets, and the Wronskian of a set of solutions.

In Sections 9.4 and 9.5, we give a procedure for determining a fundamental solution set for a homogeneous normal system with constant coefficients. The approach is analogous to that used for scalar equations. For example, the role previously played by the auxiliary equation and its roots is now taken by the characteristic equation for the matrix and its eigenvalues and eigenvectors. In Section 9.6, we give extensions of the methods of undetermined coefficients and variation of parameters.

The connection between first order normal systems and first order linear equations alluded to earlier is discussed in Section 9.7. There the concept of a matrix exponential function $e^{\mathbf{A}t}$ is discussed. Moreover, we demonstrate how to use the matrix exponential function to obtain a general solution for a homogeneous system. A brief example will illustrate the power of this formalism. For the moment, we presume the reader is familiar with matrix algebra.

EXAMPLE 1 A disk is rotating at an angular speed of ω radians per second. If a point is fixed on the disk at a distance a from its center, its velocity is given by the vector cross product

$$(3) \qquad \mathbf{v} = \mathbf{\Omega} \times \mathbf{r},$$

where $\mathbf{r}$ is its position vector $(x, y, 0)$ in an inertial frame and the vector $\mathbf{\Omega}$ has magnitude ω and points along the z-axis. Describe the motion of the rotating point in the inertial frame if it starts on the x-axis at $t = 0$.

Solution It is clear from Figure 9.1 that the coordinates of the point after a rotation of ωt radians are given by

$$(4) \qquad x = a \cos \omega t, \qquad y = a \sin \omega t, \qquad z = 0.$$

[†] A suitable reference is *Linear Algebra and Its Applications*, by David C. Lay (Addison-Wesley, Reading, Mass., 1994), or almost any other introductory linear algebra text.

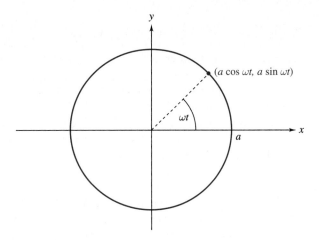

Figure 9.1 Coordinates on a rotating disk

Let's see how we can derive (4) from the differential equation. The coordinate expansion of (3) for $\mathbf{v} = (v_1, v_2, v_3)$ reads

(5) $v_1 = -\omega y,$ $v_2 = \omega x,$ $v_3 = 0.$

There is no z motion, so we reexpress (5) in matrix notation as

$$\begin{bmatrix} v_1 \\ v_2 \end{bmatrix} = \frac{d}{dt}\begin{bmatrix} x \\ y \end{bmatrix} = \begin{bmatrix} 0 & -\omega \\ \omega & 0 \end{bmatrix}\begin{bmatrix} x \\ y \end{bmatrix} = \mathbf{A}\begin{bmatrix} x \\ y \end{bmatrix}.$$

Now the (scalar) differential equation $du/dt = au$ has the general solution

(6) $u = e^{at}c$

(Section 2.3). Can we make any sense out of the matrix version of this:

(7) $\begin{bmatrix} x \\ y \end{bmatrix} = e^{\mathbf{A}t}\mathbf{c} = e^{\mathbf{A}t}\begin{bmatrix} c_1 \\ c_2 \end{bmatrix}$?

Since the exponential has the Taylor expansion

$$e^{at} = 1 + at + \frac{(at)^2}{2!} + \frac{(at)^3}{3!} + \frac{(at)^4}{4!} + \frac{(at)^5}{5!} + \cdots,$$

let's experiment with

(8) $e^{\mathbf{A}t} = \mathbf{I} + \mathbf{A}t + \frac{(\mathbf{A}t)^2}{2!} + \frac{(\mathbf{A}t)^3}{3!} + \frac{(\mathbf{A}t)^4}{4!} + \frac{(\mathbf{A}t)^5}{5!} + \cdots$

(where we have replaced 1 with the identity matrix). Recalling that $\mathbf{A} = \begin{bmatrix} 0 & -\omega \\ \omega & 0 \end{bmatrix}$, we compute

$$\mathbf{I} = \begin{bmatrix} 1 & 0 \\ 0 & 1 \end{bmatrix}, \qquad \mathbf{A}t = \begin{bmatrix} 0 & -\omega t \\ \omega t & 0 \end{bmatrix},$$

$$(\mathbf{A}t)^2 = \begin{bmatrix} -(\omega t)^2 & 0 \\ 0 & -(\omega t)^2 \end{bmatrix}, \qquad (\mathbf{A}t)^3 = \begin{bmatrix} 0 & (\omega t)^3 \\ -(\omega t)^3 & 0 \end{bmatrix},$$

$$(\mathbf{A}t)^4 = \begin{bmatrix} (\omega t)^4 & 0 \\ 0 & (\omega t)^4 \end{bmatrix}, \qquad (\mathbf{A}t)^5 = \begin{bmatrix} 0 & -(\omega t)^5 \\ (\omega t)^5 & 0 \end{bmatrix}, \quad \text{etc.,}$$

and assemble (8) to find

$$e^{\mathbf{A}t} = \begin{bmatrix} 1 - \dfrac{(\omega t)^2}{2!} + \dfrac{(\omega t)^4}{4!} - \cdots & -(\omega t) + \dfrac{(\omega t)^3}{3!} - \dfrac{(\omega t)^5}{5!} + \cdots \\ (\omega t) - \dfrac{(\omega t)^3}{3!} + \dfrac{(\omega t)^5}{5!} - \cdots & 1 - \dfrac{(\omega t)^2}{2!} + \dfrac{(\omega t)^4}{4!} - \cdots \end{bmatrix}.$$

If we recognize the Taylor series for the trigonometric functions

$$\cos \omega t = 1 - \frac{(\omega t)^2}{2!} + \frac{(\omega t)^4}{4!} + \cdots \,,$$

$$\sin \omega t = (\omega t) - \frac{(\omega t)^3}{3!} + \frac{(\omega t)^5}{5!} - \cdots \,,$$

then we identify (8) as

$$e^{\mathbf{A}t} = \begin{bmatrix} \cos \omega t & -\sin \omega t \\ \sin \omega t & \cos \omega t \end{bmatrix}$$

and the general solution (6) would be interpreted as

(9) $$\begin{bmatrix} x \\ y \end{bmatrix} = \begin{bmatrix} \cos \omega t & -\sin \omega t \\ \sin \omega t & \cos \omega t \end{bmatrix} \begin{bmatrix} c_1 \\ c_2 \end{bmatrix}.$$

At $t = 0$, the point has coordinates $(a, 0)$; thus

$$\begin{bmatrix} a \\ 0 \end{bmatrix} = \begin{bmatrix} 1 & 0 \\ 0 & 1 \end{bmatrix} \begin{bmatrix} c_1 \\ c_2 \end{bmatrix} = \begin{bmatrix} c_1 \\ c_2 \end{bmatrix},$$

or $c_1 = a$, $c_2 = 0$. Finally, expressing (9) componentwise recovers the solution (4):

$$\begin{bmatrix} x \\ y \end{bmatrix} = \begin{bmatrix} a \cos \omega t \\ a \sin \omega t \end{bmatrix}. \quad \blacksquare \text{ End}$$

9.2 A BRIEF REVIEW OF MATRICES AND VECTORS

A **matrix** is a rectangular array of numbers arranged in rows and columns. An $m \times n$ matrix—that is, a matrix with m rows and n columns—is usually denoted by

$$\mathbf{A} := \begin{bmatrix} a_{11} & a_{12} & a_{13} & \cdots & a_{1n} \\ a_{21} & a_{22} & a_{23} & \cdots & a_{2n} \\ \vdots & \vdots & \vdots & \cdots & \vdots \\ a_{m1} & a_{m2} & a_{m3} & \cdots & a_{mn} \end{bmatrix},$$

where the element in the ith row and jth column is a_{ij}. The notation $[a_{ij}]$ is also used to designate $\mathbf{A}$. The matrices we will work with usually consist of real numbers, but in certain instances we allow complex number entries.

The matrix obtained from $\mathbf{A}$ by interchanging its rows and columns is called the **transpose** of $\mathbf{A}$ and is denoted by $\mathbf{A}^T$. For example, if

$$\mathbf{A} = \begin{bmatrix} 1 & 2 & 6 \\ -1 & 2 & -1 \end{bmatrix}, \quad \text{then} \quad \mathbf{A}^T = \begin{bmatrix} 1 & -1 \\ 2 & 2 \\ 6 & -1 \end{bmatrix}.$$

Some matrices of special interest are **square matrices,** which have the same number of rows and columns; **diagonal matrices,** which are square matrices with only zero entries off the main diagonal (that is, $a_{ij} = 0$ if $i \neq j$); and (column) **vectors,** which are $n \times 1$ matrices. For example, if

$$\mathbf{A} = \begin{bmatrix} 3 & 4 & -1 \\ 2 & 6 & 5 \\ 0 & 1 & 4 \end{bmatrix}, \quad \mathbf{B} = \begin{bmatrix} 3 & 0 & 0 \\ 0 & 0 & 0 \\ 0 & 0 & 7 \end{bmatrix}, \quad \mathbf{x} = \begin{bmatrix} 4 \\ 2 \\ 1 \end{bmatrix},$$

then $\mathbf{A}$ is a square matrix, $\mathbf{B}$ is a diagonal matrix, and $\mathbf{x}$ is a vector. An $m \times n$ matrix whose entries are all zero is called a **zero matrix** and is denoted by $\mathbf{0}$. The $n \times n$ diagonal matrix with ones down the main diagonal is called the **identity matrix** and is denoted by $\mathbf{I}$, or by $\mathbf{I}_n$ when the size of the matrix is important.

For consistency, we denote matrices by boldfaced capitals, such as $\mathbf{A}$, $\mathbf{B}$, $\mathbf{C}$, $\mathbf{I}$, $\mathbf{X}$, and $\mathbf{Y}$, and reserve boldfaced lowercase letters, such as $\mathbf{c}$, $\mathbf{x}$, $\mathbf{y}$, and $\mathbf{z}$, for vectors. We also write $\mathbf{x} = \text{col}(x_1, \ldots, x_n)$ to mean the vector whose transpose is $\mathbf{x}^T = [x_1 \, x_2 \ldots x_n]$.

Algebra of Matrices

We need the following properties of matrices.

Matrix Addition. Addition is performed by adding corresponding elements; that is, the *sum* of two $m \times n$ matrices is

$$\mathbf{A} + \mathbf{B} = [a_{ij}] + [b_{ij}] = [a_{ij} + b_{ij}].$$

Scalar Multiplication. If $\mathbf{A}$ is an $m \times n$ matrix and r is a number (scalar), then the **scalar multiple** of $\mathbf{A}$ by r is the matrix $r\mathbf{A}$ obtained from $\mathbf{A}$ by multiplying each element in $\mathbf{A}$ by the number r. That is, $r\mathbf{A} = r[a_{ij}] = [ra_{ij}]$. The special case of $(-1)\mathbf{A}$ is written $-\mathbf{A}$.

Properties of Matrix Addition and Scalar Multiplication. If $\mathbf{A}$, $\mathbf{B}$, and $\mathbf{C}$ are $m \times n$ matrices and r, s are scalars, then

$$\mathbf{A} + (\mathbf{B} + \mathbf{C}) = (\mathbf{A} + \mathbf{B}) + \mathbf{C}, \qquad \mathbf{A} + \mathbf{B} = \mathbf{B} + \mathbf{A},$$

$$\mathbf{A} + \mathbf{0} = \mathbf{A}, \qquad\qquad\qquad \mathbf{A} + (-\mathbf{A}) = \mathbf{0},$$

$$r(\mathbf{A} + \mathbf{B}) = r\mathbf{A} + r\mathbf{B}, \qquad\qquad (r + s)\mathbf{A} = r\mathbf{A} + s\mathbf{A},$$

$$r(s\mathbf{A}) = (rs)\mathbf{A} = s(r\mathbf{A}).$$

Matrix Multiplication. If $\mathbf{A}$ is an $m \times n$ matrix and $\mathbf{B}$ is an $n \times p$ matrix, then the **product** of $\mathbf{A}$ and $\mathbf{B}$ is the $m \times p$ matrix $\mathbf{AB}$ defined by

$$\mathbf{AB} := [c_{ij}], \quad \text{where} \quad c_{ij} := \sum_{k=1}^{n} a_{ik}b_{kj}$$

for $i = 1, \ldots, m$ and $j = 1, \ldots, p$. Observe that the ijth entry in $\mathbf{AB}$ is the dot product of the ith row of $\mathbf{A}$ and the jth column of $\mathbf{B}$. For example,

$$\begin{bmatrix} 1 & 0 & 1 \\ 3 & -1 & 2 \end{bmatrix} \begin{bmatrix} 1 & 2 & x \\ -1 & -1 & y \\ 4 & 1 & z \end{bmatrix} = \begin{bmatrix} 1 + 0 + 4 & 2 + 0 + 1 & x + 0 + z \\ 3 + 1 + 8 & 6 + 1 + 2 & 3x - y + 2z \end{bmatrix}$$

$$= \begin{bmatrix} 5 & 3 & x + z \\ 12 & 9 & 3x - y + 2z \end{bmatrix}.$$

It is important to keep in mind that matrix multiplication does not commute; that is, $\mathbf{AB}$ may not equal $\mathbf{BA}$. This is certainly true if $\mathbf{A}$ is 2×3 and $\mathbf{B}$ is 3×2, since then $\mathbf{AB}$ is 2×2, while $\mathbf{BA}$ is 3×3. In some cases, $\mathbf{AB}$ may be defined, while $\mathbf{BA}$ is not. Even for square matrices, $\mathbf{AB}$ is not necessarily equal to $\mathbf{BA}$ (see Problem 8).

Properties of Matrix Multiplication

$(\mathbf{AB})\mathbf{C} = \mathbf{A}(\mathbf{BC})$	(Associativity)
$(\mathbf{A} + \mathbf{B})\mathbf{C} = \mathbf{AC} + \mathbf{BC}$	(Distributivity)
$\mathbf{A}(\mathbf{B} + \mathbf{C}) = \mathbf{AB} + \mathbf{AC}$	(Distributivity)
$(r\mathbf{A})\mathbf{B} = r(\mathbf{AB}) = \mathbf{A}(r\mathbf{B})$	(Associativity)

Let $\mathbf{A}$ be an $m \times n$ matrix and let $\mathbf{x}$ and $\mathbf{y}$ be $n \times 1$ vectors. Then $\mathbf{Ax}$ is an $m \times 1$ vector, and so we can think of multiplication by $\mathbf{A}$ as defining an operator that maps $n \times 1$ vectors into $m \times 1$ vectors. A consequence of the distributivity and associativity properties is that multiplication by $\mathbf{A}$ defines a **linear operator**, since $\mathbf{A}(\mathbf{x} + \mathbf{y}) = \mathbf{Ax} + \mathbf{Ay}$ and $\mathbf{A}(r\mathbf{x}) = r\mathbf{Ax}$. Moreover, if $\mathbf{A}$ is an $m \times n$ matrix and $\mathbf{B}$ is an $n \times p$ matrix, then the $m \times p$ matrix $\mathbf{AB}$ defines a linear operator that is the composition of the linear operator defined by $\mathbf{B}$ with the linear operator defined by $\mathbf{A}$. That is, $(\mathbf{AB})\mathbf{x} = \mathbf{A}(\mathbf{Bx})$, where $\mathbf{x}$ is a $p \times 1$ vector.

Inverse of a Matrix. For square matrices we say that $\mathbf{B}$ is the **inverse** of $\mathbf{A}$, denoted $\mathbf{B} = \mathbf{A}^{-1}$, if $\mathbf{AB} = \mathbf{BA} = \mathbf{I}$. Obviously, if $\mathbf{B}$ is the inverse of $\mathbf{A}$, then $\mathbf{A}$ is the inverse of $\mathbf{B}$. A matrix that has an inverse is called **invertible**, or **nonsingular**. If no inverse exists, the matrix is said to be **singular**. For example, the identity matrix is invertible ($\mathbf{I}^{-1} = \mathbf{I}$), but the zero matrix is singular.

Finding the Inverse of a Matrix. By a **row operation**, we mean any one of the following:

 (a) Interchanging two rows of the matrix
 (b) Multiplying a row of the matrix by a nonzero scalar
 (c) Adding a scalar multiple of one row of the matrix to another row.

If the $n \times n$ matrix $\mathbf{A}$ has an inverse, then $\mathbf{A}^{-1}$ can be determined by performing row operations on the $n \times 2n$ matrix $[\mathbf{A} \mid \mathbf{I}]$ obtained by writing $\mathbf{A}$ and $\mathbf{I}$ side by side. In particular, we perform row operations on the matrix $[\mathbf{A} \mid \mathbf{I}]$ until the first n rows and columns form the identity matrix; that is, the new matrix is $[\mathbf{I} \mid \mathbf{B}]$. Then $\mathbf{A}^{-1} = \mathbf{B}$. We remark that if this procedure fails to produce a matrix of the form $[\mathbf{I} \mid \mathbf{B}]$, then $\mathbf{A}$ has no inverse.

EXAMPLE 1 Find the inverse of $\mathbf{A} = \begin{bmatrix} 1 & 2 & 1 \\ 1 & 3 & 2 \\ 1 & 0 & 1 \end{bmatrix}$ and use it to solve the system $\mathbf{Ax} = \begin{bmatrix} 1 \\ -1 \\ 0 \end{bmatrix}$.

Solution We first form the matrix $[\mathbf{A} \mid \mathbf{I}]$ and *row-reduce* the matrix to $[\mathbf{I} \mid \mathbf{A}^{-1}]$. Computing, we find the following:

The matrix $[\mathbf{A} \mid \mathbf{I}]$
$$\left[\begin{array}{ccc|ccc} 1 & 2 & 1 & 1 & 0 & 0 \\ 1 & 3 & 2 & 0 & 1 & 0 \\ 1 & 0 & 1 & 0 & 0 & 1 \end{array}\right].$$

Subtract the first row from the second and third to obtain
$$\left[\begin{array}{ccc|ccc} 1 & 2 & 1 & 1 & 0 & 0 \\ 0 & 1 & 1 & -1 & 1 & 0 \\ 0 & -2 & 0 & -1 & 0 & 1 \end{array}\right].$$

Add 2 times the second row to the third row to obtain
$$\left[\begin{array}{ccc|ccc} 1 & 2 & 1 & 1 & 0 & 0 \\ 0 & 1 & 1 & -1 & 1 & 0 \\ 0 & 0 & 2 & -3 & 2 & 1 \end{array}\right].$$

Subtract 2 times the second row from the first to obtain
$$\left[\begin{array}{ccc|ccc} 1 & 0 & -1 & 3 & -2 & 0 \\ 0 & 1 & 1 & -1 & 1 & 0 \\ 0 & 0 & 2 & -3 & 2 & 1 \end{array}\right].$$

Multiply the third row by 1/2 to obtain
$$\left[\begin{array}{ccc|ccc} 1 & 0 & -1 & 3 & -2 & 0 \\ 0 & 1 & 1 & -1 & 1 & 0 \\ 0 & 0 & 1 & -\dfrac{3}{2} & 1 & \dfrac{1}{2} \end{array}\right].$$

Add the third row to the first and then subtract the third row from the second to obtain

$$\left[\begin{array}{ccc|ccc} 1 & 0 & 0 & \frac{3}{2} & -1 & \frac{1}{2} \\ 0 & 1 & 0 & \frac{1}{2} & 0 & -\frac{1}{2} \\ 0 & 0 & 1 & -\frac{3}{2} & 1 & \frac{1}{2} \end{array}\right].$$

The matrix shown in color is $\mathbf{A}^{-1}$.

Finally, the solution to the given system is just

$$\mathbf{x} = \mathbf{A}^{-1}\begin{bmatrix} 1 \\ -1 \\ 0 \end{bmatrix} = \begin{bmatrix} \frac{3}{2} & -1 & \frac{1}{2} \\ \frac{1}{2} & 0 & -\frac{1}{2} \\ -\frac{3}{2} & 1 & \frac{1}{2} \end{bmatrix}\begin{bmatrix} 1 \\ -1 \\ 0 \end{bmatrix} = \begin{bmatrix} \frac{5}{2} \\ \frac{1}{2} \\ -\frac{5}{2} \end{bmatrix}. \quad \blacksquare \text{ End}$$

Determinants. For a 2×2 matrix $\mathbf{A}$, the **determinant of** $\mathbf{A}$, denoted $\det \mathbf{A}$ or $|\mathbf{A}|$, is defined by

$$\det \mathbf{A} := \begin{vmatrix} a_{11} & a_{12} \\ a_{21} & a_{22} \end{vmatrix} = a_{11}a_{22} - a_{12}a_{21}.$$

We can define the determinant of a 3×3 matrix $\mathbf{A}$ in terms of its cofactor expansion about the first row; that is,

$$\det \mathbf{A} := \begin{vmatrix} a_{11} & a_{12} & a_{13} \\ a_{21} & a_{22} & a_{23} \\ a_{31} & a_{32} & a_{33} \end{vmatrix} = a_{11}\begin{vmatrix} a_{22} & a_{23} \\ a_{32} & a_{33} \end{vmatrix} - a_{12}\begin{vmatrix} a_{21} & a_{23} \\ a_{31} & a_{33} \end{vmatrix} + a_{13}\begin{vmatrix} a_{21} & a_{22} \\ a_{31} & a_{32} \end{vmatrix}.$$

For example,

$$\begin{vmatrix} 1 & 2 & 1 \\ 0 & 3 & 5 \\ 2 & 1 & -1 \end{vmatrix} = 1\begin{vmatrix} 3 & 5 \\ 1 & -1 \end{vmatrix} - 2\begin{vmatrix} 0 & 5 \\ 2 & -1 \end{vmatrix} + 1\begin{vmatrix} 0 & 3 \\ 2 & 1 \end{vmatrix}$$

$$= 1(-3 - 5) - 2(0 - 10) + 1(0 - 6) = 6.$$

The determinant of an $n \times n$ matrix can be similarly defined by a cofactor expansion involving $(n - 1)$st order determinants. However, a more practical way to evaluate the determinant when n is large involves the row-reduction of the matrix to upper triangular form. Since here we deal mainly with 2×2 and 3×3 matrices, we refer the reader to an elementary linear algebra text for a further discussion of evaluating determinants.[†]

The next theorem summarizes many of the results from elementary linear algebra that we need.

[†] See *Linear Algebra and Its Applications*, by David C. Lay (Addison-Wesley, Reading, Mass., 1994).

MATRICES AND SYSTEMS OF EQUATIONS

Theorem 1. Let **A** be an $n \times n$ matrix. The following statements are equivalent:

(a) A is singular (does not have an inverse).
(b) The determinant of **A** is zero.
(c) Ax = 0 has nontrivial solutions ($\mathbf{x} \neq \mathbf{0}$).
(d) The columns (rows) of **A** form a linearly dependent set.

In part (d), the statement that the n columns of **A** are linearly dependent means that there exist scalars $c_1, \ldots, c_n$ *not all zero* such that

$$c_1 \mathbf{a}_1 + c_2 \mathbf{a}_2 + \cdots + c_n \mathbf{a}_n = \mathbf{0},$$

where $\mathbf{a}_j$ is the vector forming the jth column of **A**.

If **A** is a nonsingular square matrix, then the homogeneous system $\mathbf{Ax} = \mathbf{0}$ has $\mathbf{x} = \mathbf{0}$ as its only solution. However, if **A** is singular, then $\mathbf{Ax} = \mathbf{0}$ has infinitely many solutions. Indeed, Theorem 1 asserts that for any singular matrix **A**, there is a vector $\mathbf{x}_0 \neq \mathbf{0}$ such that $\mathbf{Ax}_0 = \mathbf{0}$, and we can get infinitely many other solutions by multiplying $\mathbf{x}_0$ by any scalar, i.e., taking $\mathbf{x} = c\mathbf{x}_0$.

More generally, if **A** is nonsingular then the nonhomogeneous system $\mathbf{Ax} = \mathbf{b}$, where **b** is any given vector (of the appropriate size), has a unique solution (namely $\mathbf{x} = \mathbf{A}^{-1}\mathbf{b}$).

The situation for singular matrices, however, is more involved. If $\det \mathbf{A} = 0$, then either $\mathbf{Ax} = \mathbf{b}$ has no solutions or it has infinitely many of them (see Problem 16). In the latter case, all of the solutions are of the form

$$\mathbf{x} = \mathbf{x}_p + \mathbf{x}_h,$$

where $\mathbf{x}_p$ is a particular solution to $\mathbf{Ax} = \mathbf{b}$ and $\mathbf{x}_h$ is any of the infinity of solutions to the corresponding homogeneous system $\mathbf{Ax} = \mathbf{0}$ (see Problem 17). The resemblance of this situation to that of solving nonhomogeneous linear differential equations should be quite apparent.

Calculus of Matrices

If we allow the entries $a_{ij}(t)$ in a matrix $\mathbf{A}(t)$ to be functions of the variable t, then $\mathbf{A}(t)$ is a **matrix function of** t. Similarly, if the entries $x_i(t)$ of a vector $\mathbf{x}(t)$ are functions of t, then $\mathbf{x}(t)$ is a **vector function of** t.

These matrix and vector functions have a calculus much like that of real-valued functions. A matrix $\mathbf{A}(t)$ is said to be **continuous at** t_0 if each entry $a_{ij}(t)$ is continuous at t_0. Moreover, $\mathbf{A}(t)$ is **differentiable at** t_0 if each entry $a_{ij}(t)$ is differentiable at t_0, and we write

(1) $$\frac{d\mathbf{A}}{dt}(t_0) = \mathbf{A}'(t_0) := [a'_{ij}(t_0)].$$

Similarly, we define

(2) $$\int_a^b \mathbf{A}(t)\,dt := \left[\int_a^b a_{ij}(t)\,dt \right].$$

EXAMPLE 2 Let $A(t) = \begin{bmatrix} t^2 + 1 & \cos t \\ e^t & 1 \end{bmatrix}$.

Find: (a) $A'(t)$. (b) $\int_0^1 A(t)\,dt$.

Solution Using formulas (1) and (2), we compute

(a) $A'(t) = \begin{bmatrix} 2t & -\sin t \\ e^t & 0 \end{bmatrix}$. (b) $\int_0^1 A(t)\,dt = \begin{bmatrix} \frac{4}{3} & \sin 1 \\ e-1 & 1 \end{bmatrix}$. ■ End

The basic properties of differentiation are valid for matrix functions.

Differentiation Formulas for Matrix Functions

$$\frac{d}{dt}(CA) = C\frac{dA}{dt}, \quad C \text{ a constant matrix.}$$

$$\frac{d}{dt}(A + B) = \frac{dA}{dt} + \frac{dB}{dt}.$$

$$\frac{d}{dt}(AB) = A\frac{dB}{dt} + \frac{dA}{dt}B.$$

In the last formula, the order in which the matrices are written is very important because matrix multiplication does not always commute.

EXERCISES 9.2

1. Let $A := \begin{bmatrix} 2 & 1 \\ 3 & 5 \end{bmatrix}$ and $B := \begin{bmatrix} -1 & 0 \\ 2 & -3 \end{bmatrix}$.

 Find: (a) A^T. (b) $A + B$. (c) $3A - B$.

2. Let $A := \begin{bmatrix} 2 & 0 & 5 \\ 2 & 1 & 1 \end{bmatrix}$ and $B := \begin{bmatrix} 1 & -1 & 2 \\ 0 & 3 & -2 \end{bmatrix}$.

 Find: (a) B^T. (b) $A + B$. (c) $7A - 4B$.

3. Let $A := \begin{bmatrix} 2 & 4 \\ 1 & 1 \end{bmatrix}$ and $B := \begin{bmatrix} -1 & 3 \\ 5 & 2 \end{bmatrix}$.

 Find: (a) AB. (b) $A^2 = AA$. (c) $B^2 = BB$.

4. Let $A := \begin{bmatrix} 2 & 1 \\ 0 & 4 \\ -1 & 3 \end{bmatrix}$ and $B := \begin{bmatrix} 1 & 1 & -1 \\ 0 & 3 & 1 \end{bmatrix}$.

 Find: (a) AB. (b) BA. (c) A^TA.

5. Let $A := \begin{bmatrix} 1 & -2 \\ 2 & -3 \end{bmatrix}$, $B := \begin{bmatrix} 1 & 0 \\ 1 & 1 \end{bmatrix}$, and $C := \begin{bmatrix} -1 & 1 \\ 2 & 1 \end{bmatrix}$.

 Find: (a) AB. (b) AC. (c) $A(B + C)$.

6. Let $A := \begin{bmatrix} 1 & 2 \\ 1 & 1 \end{bmatrix}$, $B := \begin{bmatrix} 0 & 3 \\ 1 & 2 \end{bmatrix}$, and $C := \begin{bmatrix} 1 & -4 \\ 1 & 1 \end{bmatrix}$.

 Find: (a) AB. (b) $(AB)C$. (c) $(A + B)C$.

7. Let $x := \begin{bmatrix} 1 \\ 2 \\ 4 \end{bmatrix}$ and $y := \begin{bmatrix} 2 \\ 4 \\ -1 \end{bmatrix}$.

 Find: (a) x^Tx. (b) y^Ty. (c) x^Ty. (d) y^Tx.

8. Let $\mathbf{A} := \begin{bmatrix} 2 & -1 \\ -3 & 4 \end{bmatrix}$ and $\mathbf{B} := \begin{bmatrix} 1 & 2 \\ 3 & 2 \end{bmatrix}$.

Verify that $\mathbf{AB} \neq \mathbf{BA}$.

In Problems 9 through 14, compute the inverse of the given matrix, if it exists.

9. $\begin{bmatrix} 2 & 1 \\ -1 & 4 \end{bmatrix}$.

10. $\begin{bmatrix} 4 & 1 \\ 5 & 9 \end{bmatrix}$.

11. $\begin{bmatrix} 1 & 1 & 1 \\ 1 & 2 & 1 \\ 2 & 3 & 2 \end{bmatrix}$.

12. $\begin{bmatrix} 1 & 1 & 1 \\ 1 & 2 & 3 \\ 0 & 1 & 1 \end{bmatrix}$.

13. $\begin{bmatrix} -2 & -1 & 1 \\ 2 & 1 & 0 \\ 3 & 1 & -1 \end{bmatrix}$.

14. $\begin{bmatrix} 1 & 1 & 1 \\ 1 & -1 & 2 \\ 1 & 1 & 4 \end{bmatrix}$.

15. Find *all* solutions to the following homogeneous systems ($\mathbf{Ax} = \mathbf{0}$), where

(a) $\mathbf{A} = \begin{bmatrix} -1 & 2 \\ 2 & 3 \end{bmatrix}$.

(b) $\mathbf{A} = \begin{bmatrix} -2 & 2 & -1 \\ 1 & -3 & 1 \\ 4 & -4 & 2 \end{bmatrix}$.

(c) $\mathbf{A} = \begin{bmatrix} -1 & 3 \\ -3 & 9 \end{bmatrix}$.

(d) $\mathbf{A} = \begin{bmatrix} 1 & 1 & -1 \\ -1 & -1 & 1 \\ 1 & 1 & -1 \end{bmatrix}$.

16. Let $\mathbf{A} = \begin{bmatrix} 2 & -1 & 1 \\ -1 & 2 & 1 \\ 1 & 1 & 2 \end{bmatrix}$.

(a) Show that $\mathbf{A}$ is singular.

(b) Show that $\mathbf{Ax} = \begin{bmatrix} 3 \\ 1 \\ 3 \end{bmatrix}$ has no solutions.

(c) Show that $\mathbf{Ax} = \begin{bmatrix} 3 \\ 0 \\ 3 \end{bmatrix}$ has infinitely many solutions.

17. Prove that if $\mathbf{x}_p$ satisfies $\mathbf{Ax}_p = \mathbf{b}$, then every solution to the nonhomogeneous system $\mathbf{Ax} = \mathbf{b}$ is of the form $\mathbf{x} = \mathbf{x}_p + \mathbf{x}_h$, where $\mathbf{x}_h$ is a solution to the corresponding homogeneous system $\mathbf{Ax} = \mathbf{0}$.

18. Find all solutions (if any) to the following nonhomogeneous systems $\mathbf{Ax} = \mathbf{b}$, where

(a) $\mathbf{A} = \begin{bmatrix} 2 & -1 \\ -6 & 3 \end{bmatrix}$, $\mathbf{b} = \begin{bmatrix} 2 \\ 3 \end{bmatrix}$.

(b) $\mathbf{A} = \begin{bmatrix} 1 & 1 & 1 \\ 2 & 3 & 0 \\ 1 & 2 & 1 \end{bmatrix}$, $\mathbf{b} = \begin{bmatrix} 1 \\ 0 \\ 1 \end{bmatrix}$.

(c) $\mathbf{A} = \begin{bmatrix} 2 & 0 & 1 \\ -3 & 1 & 4 \\ -1 & 1 & 5 \end{bmatrix}$, $\mathbf{b} = \begin{bmatrix} -1 \\ 1 \\ 0 \end{bmatrix}$.

Describe the solution set in the form suggested by Problem 17. That is, display a particular solution and express all the solutions to the corresponding homogeneous system.

In Problems 19 through 22, find the matrix function $\mathbf{X}^{-1}(t)$ whose value at t is the inverse of the given matrix $\mathbf{X}(t)$.

19. $\mathbf{X}(t) = \begin{bmatrix} e^t & e^{4t} \\ e^t & 4e^{4t} \end{bmatrix}$.

20. $\mathbf{X}(t) = \begin{bmatrix} \sin 2t & \cos 2t \\ 2\cos 2t & -2\sin 2t \end{bmatrix}$.

21. $\mathbf{X}(t) = \begin{bmatrix} e^t & e^{-t} & e^{2t} \\ e^t & -e^{-t} & 2e^{2t} \\ e^t & e^{-t} & 4e^{2t} \end{bmatrix}$. 22. $\mathbf{X}(t) = \begin{bmatrix} e^{3t} & 1 & t \\ 3e^{3t} & 0 & 1 \\ 9e^{3t} & 0 & 0 \end{bmatrix}$.

In Problems 23 through 28, evaluate the given determinant.

23. $\begin{vmatrix} 4 & 3 \\ -1 & 2 \end{vmatrix}$.

24. $\begin{vmatrix} 12 & 8 \\ 3 & 2 \end{vmatrix}$.

25. $\begin{vmatrix} 1 & 0 & 0 \\ 3 & 1 & 2 \\ 1 & 5 & -2 \end{vmatrix}$.

26. $\begin{vmatrix} 1 & 0 & 2 \\ 0 & 3 & -1 \\ -1 & 2 & 1 \end{vmatrix}$.

27. $\begin{vmatrix} 1 & 4 & 3 \\ -1 & -1 & 2 \\ 4 & 5 & 2 \end{vmatrix}$.

28. $\begin{vmatrix} 1 & 4 & 4 \\ 3 & 0 & -3 \\ 1 & 6 & 2 \end{vmatrix}$.

In Problems 29 through 31, determine the values of r for which $\det(\mathbf{A} - r\mathbf{I}) = 0$.

29. $\mathbf{A} = \begin{bmatrix} 1 & 1 \\ -2 & 4 \end{bmatrix}$.

30. $\mathbf{A} = \begin{bmatrix} 3 & 3 \\ 2 & 4 \end{bmatrix}$.

31. $\mathbf{A} = \begin{bmatrix} 0 & 0 & 0 \\ 0 & 1 & 0 \\ 1 & 0 & 1 \end{bmatrix}$.

32. Illustrate the equivalence of the assertions (a)–(d) in Theorem 1 for the matrix

$$\mathbf{A} = \begin{bmatrix} 4 & -2 & 2 \\ -2 & 4 & 2 \\ 2 & 2 & 4 \end{bmatrix}$$

as follows.

(a) Show that the row-reduction procedure applied to $[\mathbf{A} \mid \mathbf{I}]$ fails to produce the inverse of $\mathbf{A}$.

(b) Calculate det $\mathbf{A}$.

(c) Determine a nontrivial solution $\mathbf{x}$ to $\mathbf{A}\mathbf{x} = \mathbf{0}$.

(d) Find scalars c_1, c_2, and c_3, *not all zero*, so that $c_1\mathbf{a}_1 + c_2\mathbf{a}_2 + c_3\mathbf{a}_3 = \mathbf{0}$, where $\mathbf{a}_1$, $\mathbf{a}_2$, and $\mathbf{a}_3$ are the columns of $\mathbf{A}$.

*In Problems 33 and 34, find d**x**/dt for the given vector functions.*

33. $\mathbf{x}(t) = \begin{bmatrix} e^{3t} \\ 2e^{3t} \\ -e^{3t} \end{bmatrix}$ **34.** $\mathbf{x}(t) = \begin{bmatrix} e^{-t}\sin 3t \\ 0 \\ -e^{-t}\sin 3t \end{bmatrix}$.

*In Problems 35 and 36, find d**X**/dt for the given matrix functions.*

35. $\mathbf{X}(t) = \begin{bmatrix} e^{5t} & 3e^{2t} \\ -2e^{5t} & -e^{2t} \end{bmatrix}$.

36. $\mathbf{X}(t) = \begin{bmatrix} \sin 2t & \cos 2t & e^{-2t} \\ -\sin 2t & 2\cos 2t & 3e^{-2t} \\ 3\sin 2t & \cos 2t & e^{-2t} \end{bmatrix}$.

In Problems 37 and 38, verify that the given vector satisfies the given system.

37. $\mathbf{x}' = \begin{bmatrix} 1 & 1 \\ -2 & 4 \end{bmatrix}\mathbf{x}, \quad \mathbf{x} = \begin{bmatrix} e^{3t} \\ 2e^{3t} \end{bmatrix}$.

38. $\mathbf{x}' = \begin{bmatrix} 0 & 0 & 0 \\ 0 & 1 & 0 \\ 1 & 0 & 1 \end{bmatrix}\mathbf{x}, \quad \mathbf{x} = \begin{bmatrix} 0 \\ e^t \\ -3e^t \end{bmatrix}$.

In Problems 39 and 40, verify that the given matrix satisfies the given matrix differential equation.

39. $\mathbf{X}' = \begin{bmatrix} 1 & -1 \\ 2 & 4 \end{bmatrix}\mathbf{X}, \quad \mathbf{X} = \begin{bmatrix} e^{2t} & e^{3t} \\ -e^{2t} & -2e^{3t} \end{bmatrix}$.

40. $\mathbf{X}' = \begin{bmatrix} 1 & 0 & 0 \\ 0 & 3 & -2 \\ 0 & -2 & 3 \end{bmatrix}\mathbf{X}, \quad \mathbf{X} = \begin{bmatrix} e^t & 0 & 0 \\ 0 & e^t & e^{5t} \\ 0 & e^t & -e^{5t} \end{bmatrix}$.

*In Problems 41 and 42, the matrices **A**(t) and **B**(t) are given. Find*

(a) $\displaystyle\int \mathbf{A}(t)\,dt.$ (b) $\displaystyle\int_0^1 \mathbf{B}(t)\,dt.$ (c) $\dfrac{d}{dt}[\mathbf{A}(t)\mathbf{B}(t)].$

41. $\mathbf{A}(t) = \begin{bmatrix} t & e^t \\ 1 & e^t \end{bmatrix}, \quad \mathbf{B}(t) = \begin{bmatrix} \cos t & -\sin t \\ \sin t & \cos t \end{bmatrix}.$

42. $\mathbf{A}(t) = \begin{bmatrix} 1 & e^{-2t} \\ 3 & e^{-2t} \end{bmatrix}, \quad \mathbf{B}(t) = \begin{bmatrix} e^{-t} & e^{-t} \\ -e^{-t} & 3e^{-t} \end{bmatrix}.$

43. An $n \times n$ matrix $\mathbf{A}$ is called **symmetric** if $\mathbf{A}^T = \mathbf{A}$; that is, if $a_{ij} = a_{ji}$, for all $i, j = 1, \ldots, n$. Show that if $\mathbf{A}$ is an $n \times n$ matrix, then $\mathbf{A} + \mathbf{A}^T$ is a symmetric matrix.

44. Let $\mathbf{A}$ be an $m \times n$ matrix. Show that $\mathbf{A}^T\mathbf{A}$ is a symmetric $n \times n$ matrix and $\mathbf{A}\mathbf{A}^T$ is a symmetric $m \times m$ matrix (see Problem 43).

45. The **inner product** of two vectors is a generalization of the dot product for vectors with complex entries. It is defined by

$$(\mathbf{x}, \mathbf{y}) := \sum_{i=1}^n x_i \bar{y}_i,$$

where $\mathbf{x} = \text{col}(x_1, x_2, \ldots, x_n)$, $\mathbf{y} = \text{col}(y_1, y_2, \ldots, y_n)$ are complex numbers and the overbar denotes complex conjugation.

(a) Show that $(\mathbf{x}, \mathbf{y}) = \mathbf{x}^T\bar{\mathbf{y}}$, where $\bar{\mathbf{y}} = \text{col}(\bar{y}_1, \bar{y}_2, \ldots, \bar{y}_n)$.

(b) Prove that for any $n \times 1$ vectors $\mathbf{x}$, $\mathbf{y}$, $\mathbf{z}$ and any complex number λ, we have

$$(\mathbf{x}, \mathbf{y}) = \overline{(\mathbf{y}, \mathbf{x})}, \quad (\mathbf{x}, \mathbf{y} + \mathbf{z}) = (\mathbf{x}, \mathbf{y}) + (\mathbf{x}, \mathbf{z}),$$

$$(\lambda\mathbf{x}, \mathbf{y}) = \lambda(\mathbf{x}, \mathbf{y}), \quad (\mathbf{x}, \lambda\mathbf{y}) = \bar{\lambda}(\mathbf{x}, \mathbf{y}).$$

9.3 LINEAR SYSTEMS IN NORMAL FORM

A system of n linear differential equations is in **normal form** if it is expressed as

$$(1) \qquad \mathbf{x}'(t) = \mathbf{A}(t)\mathbf{x}(t) + \mathbf{f}(t),$$

where $\mathbf{x}(t) = \text{col}(x_1(t), \ldots, x_n(t))$, $\mathbf{f}(t) = \text{col}(f_1(t), \ldots, f_n(t))$, and $\mathbf{A}(t) = [a_{ij}(t)]$ is an $n \times n$ matrix. As with a scalar linear differential equation, a system is called **homogeneous** when $\mathbf{f}(t) \equiv \mathbf{0}$; otherwise, it is called **nonhomogeneous**. When the elements of $\mathbf{A}$ are all constants, the system is said to have **constant coefficients**.

Notice that a linear differential equation

(2) $$y^{(n)}(t) + p_{n-1}(t)y^{(n-1)}(t) + \cdots + p_0(t)y(t) = g(t)$$

can be rewritten as a first order system in normal form using the substitution $x_1(t) := y(t)$, $x_2(t) := y'(t), \ldots, x_n(t) := y^{(n-1)}(t)$ (see Section 4.10). Indeed, equation (2) is equivalent to $\mathbf{x}'(t) = \mathbf{A}(t)\mathbf{x}(t) + \mathbf{f}(t)$, where $\mathbf{x}(t) = \text{col}(x_1(t), \ldots, x_n(t))$, $\mathbf{f}(t) := \text{col}(0, \ldots, 0, g(t))$, and

$$\mathbf{A}(t) := \begin{bmatrix} 0 & 1 & 0 & \cdots & 0 & 0 \\ 0 & 0 & 1 & & 0 & 0 \\ \vdots & \vdots & \vdots & & \vdots & \vdots \\ 0 & 0 & 0 & \cdots & 0 & 1 \\ -p_0(t) & -p_1(t) & -p_2(t) & \cdots & -p_{n-2}(t) & -p_{n-1}(t) \end{bmatrix}.$$

The theory for systems in normal form parallels very closely the theory of linear differential equations presented in Chapters 4 and 6. In many cases the proofs for scalar linear differential equations carry over to normal systems with appropriate modifications. Conversely, results for normal systems apply to scalar linear equations since, as we showed, any scalar linear equation can be expressed as a normal system. This is the case with the existence and uniqueness theorems for linear differential equations.

The **initial value problem** for the normal system (1) is the problem of finding a differentiable vector function $\mathbf{x}(t)$ that satisfies the system on an interval I and also satisfies the **initial condition** $\mathbf{x}(t_0) = \mathbf{x}_0$, where t_0 is a given point of I and $\mathbf{x}_0 = \text{col}(x_{1,0}, \ldots, x_{n,0})$ is a given vector.

EXISTENCE AND UNIQUENESS

Theorem 2. Suppose $\mathbf{A}(t)$ and $\mathbf{f}(t)$ are continuous on an open interval I that contains the point t_0. Then, for any choice of the initial vector $\mathbf{x}_0 = \text{col}(x_{1,0}, \ldots, x_{n,0})$, there exists a unique solution $\mathbf{x}(t)$ on the whole interval I to the initial value problem

$$\mathbf{x}'(t) = \mathbf{A}(t)\mathbf{x}(t) + \mathbf{f}(t), \qquad \mathbf{x}(t_0) = \mathbf{x}_0.$$

We give a proof of this result in Chapter 13[†] and obtain as corollaries the existence and uniqueness theorems for second order equations (Theorem 2, Section 4.2) and higher-order linear equations (Theorem 1, Section 6.1).

If we rewrite system (1) as $\mathbf{x}' - \mathbf{A}\mathbf{x} = \mathbf{f}$ and define the operator $L[\mathbf{x}] := \mathbf{x}' - \mathbf{A}\mathbf{x}$, then we can express system (1) in the operator form $L[\mathbf{x}] = \mathbf{f}$. Here the operator L maps vector

[†] All references to Chapters 11–13 refer to the expanded text *Fundamentals of Differential Equations and Boundary Value Problems*, Second Edition.

functions into vector functions. Moreover, L is a *linear* operator in the sense that for any scalars a, b and differentiable vector functions $\mathbf{x}$, $\mathbf{y}$, we have

$$L[a\mathbf{x} + b\mathbf{y}] = aL[\mathbf{x}] + bL[\mathbf{y}].$$

The proof of this linearity follows from the properties of matrix multiplication (see Problem 25).

As a consequence of the linearity of L, if $\mathbf{x}_1, \ldots, \mathbf{x}_n$ are solutions to the homogeneous system $\mathbf{x}' = \mathbf{A}\mathbf{x}$, or $L[\mathbf{x}] = \mathbf{0}$ in operator notation, then any linear combination of these vectors, $c_1\mathbf{x}_1 + \cdots + c_n\mathbf{x}_n$, is also a solution. Moreover, we will see that if the solutions $\mathbf{x}_1, \ldots, \mathbf{x}_n$ are linearly independent, then *every* solution to $L[\mathbf{x}] = \mathbf{0}$ can be expressed as $c_1\mathbf{x}_1 + \cdots + c_n\mathbf{x}_n$ for an appropriate choice of the constants $c_1, \ldots, c_n$.

LINEAR DEPENDENCE OF VECTOR FUNCTIONS

Definition 1. The m vector functions $\mathbf{x}_1, \ldots, \mathbf{x}_m$ are said to be **linearly dependent on an interval** I if there exist constants $c_1, \ldots, c_m$, not all zero, such that

$$c_1\mathbf{x}_1(t) + \cdots + c_m\mathbf{x}_m(t) = \mathbf{0}$$

for all t in I. If the vectors are not linearly dependent, they are said to be **linearly independent on** I.

EXAMPLE 1 Show that the vector functions $\mathbf{x}_1(t) = \text{col}(e^t, 0, e^t)$, $\mathbf{x}_2(t) = \text{col}(3e^t, 0, 3e^t)$, and $\mathbf{x}_3(t) = \text{col}(t, 1, 0)$ are linearly dependent on $(-\infty, \infty)$.

Solution Notice that $\mathbf{x}_2$ is just 3 times $\mathbf{x}_1$ and therefore $3\mathbf{x}_1(t) - \mathbf{x}_2(t) + 0 \cdot \mathbf{x}_3(t) = \mathbf{0}$ for all t. Hence, $\mathbf{x}_1$, $\mathbf{x}_2$, and $\mathbf{x}_3$ are linearly dependent on $(-\infty, \infty)$. ∎ End

EXAMPLE 2 Show that the vector functions $\mathbf{x}_1(t) = \text{col}(e^{2t}, 0, e^{2t})$, $\mathbf{x}_2(t) = \text{col}(e^{2t}, e^{2t}, -e^{2t})$, and $\mathbf{x}_3(t) = \text{col}(e^t, 2e^t, e^t)$ are linearly independent on $(-\infty, \infty)$.

Solution To prove independence, we *assume* c_1, c_2, and c_3 are constants for which

$$(3) \qquad c_1\mathbf{x}_1(t) + c_2\mathbf{x}_2(t) + c_3\mathbf{x}_3(t) = \mathbf{0}$$

holds at every t in $(-\infty, \infty)$ and show that this forces $c_1 = c_2 = c_3 = 0$. Let's try $t = 0$ in (3). We have

$$c_1 \begin{bmatrix} 1 \\ 0 \\ 1 \end{bmatrix} + c_2 \begin{bmatrix} 1 \\ 1 \\ -1 \end{bmatrix} + c_3 \begin{bmatrix} 1 \\ 2 \\ 1 \end{bmatrix} = \mathbf{0},$$

which is equivalent to the system of linear equations

$$\begin{aligned} c_1 + c_2 + c_3 &= 0, \\ (4) \qquad c_2 + 2c_3 &= 0, \\ c_1 - c_2 + c_3 &= 0. \end{aligned}$$

Either by solving (4) or by checking that the determinant of its coefficients is non-zero (recall Theorem 1 on page 521), we can verify that (4) has only the trivial solution $c_1 = c_2 = c_3 = 0$. Therefore the vector functions $\mathbf{x}_1$, $\mathbf{x}_2$, and $\mathbf{x}_3$ are linearly independent on $(-\infty, \infty)$ (in fact, on any interval containing $t = 0$). ■ End

As the previous example illustrates, if $\mathbf{x}_1(t), \ldots, \mathbf{x}_n(t)$ are n vector functions each having n components, then these vectors will be linearly independent on an interval I if the determinant

$$\det[\mathbf{x}_1(t) \quad \cdots \quad \mathbf{x}_n(t)]$$

is not zero at some point t in I. Because of the analogy with scalar equations, we call this determinant the **Wronskian**.

WRONSKIAN

Definition 2. The **Wronskian** of n vector functions $\mathbf{x}_1(t) = \mathrm{col}(x_{1,1}, \ldots, x_{n,1}), \ldots,$ $\mathbf{x}_n(t) = \mathrm{col}(x_{1,n}, \ldots, x_{n,n})$ is defined to be the real-valued function

$$W[\mathbf{x}_1, \ldots, \mathbf{x}_n](t) := \begin{vmatrix} x_{1,1}(t) & x_{1,2}(t) & \cdots & x_{1,n}(t) \\ x_{2,1}(t) & x_{2,2}(t) & \cdots & x_{2,n}(t) \\ \vdots & \vdots & & \vdots \\ x_{n,1}(t) & x_{n,2}(t) & \cdots & x_{n,n}(t) \end{vmatrix}.$$

We now show that if $\mathbf{x}_1, \mathbf{x}_2, \ldots, \mathbf{x}_n$ are linearly independent *solutions* on I to the homogeneous system $\mathbf{x}' = \mathbf{A}\mathbf{x}$, where $\mathbf{A}$ is an $n \times n$ matrix of continuous functions, then the Wronskian $W(t) := \det[\mathbf{x}_1 \ \mathbf{x}_2 \ \ldots \ \mathbf{x}_n]$ is never zero on I. For suppose to the contrary that $W(t_0) = 0$ at some point t_0. Then by Theorem 1, page 521, the vanishing of the determinant implies that the column vectors $\mathbf{x}_1(t_0), \mathbf{x}_2(t_0), \ldots, \mathbf{x}_n(t_0)$ are linearly dependent. Thus there exist scalars $c_1, \ldots, c_n$, *not all zero*, such that

$$c_1\mathbf{x}_1(t_0) + \cdots + c_n\mathbf{x}_n(t_0) = \mathbf{0}.$$

However, $c_1\mathbf{x}_1(t) + \cdots + c_n\mathbf{x}_n(t)$ and the vector function $\mathbf{z}(t) \equiv \mathbf{0}$ are both solutions to $\mathbf{x}' = \mathbf{A}\mathbf{x}$ on I, and they agree at the point t_0. So these solutions must be identical on I according to the existence–uniqueness theorem (Theorem 2). That is,

$$c_1\mathbf{x}_1(t) + \cdots + c_n\mathbf{x}_n(t) = \mathbf{0}$$

for *all* t in I. But this contradicts the given information that $\mathbf{x}_1, \ldots, \mathbf{x}_n$ are linearly independent on I. We have shown that $W(t_0) \neq 0$, and since t_0 is an arbitrary point, it follows that $W(t) \neq 0$ for all $t \in I$.

The preceding argument has two important implications that parallel the scalar case. First, *the Wronskian of solutions to $\mathbf{x}' = \mathbf{A}\mathbf{x}$ is either identically zero or never zero on I* (see also Problem 31). Second, *a set of n solutions $\mathbf{x}_1, \ldots, \mathbf{x}_n$ to $\mathbf{x}' = \mathbf{A}\mathbf{x}$ on I is linearly independent on I if and only if their Wronskian is never zero on I*. With these facts in hand, we can imitate the proof given for the scalar case in Section 6.1 (Theorem 2) to obtain the following representation theorem for the solutions to $\mathbf{x}' = \mathbf{A}\mathbf{x}$.

REPRESENTATION OF SOLUTIONS (HOMOGENEOUS CASE)

Theorem 3. Let $\mathbf{x}_1, \ldots, \mathbf{x}_n$ be n linearly independent solutions to the homogeneous system

(5) $\mathbf{x}'(t) = \mathbf{A}(t)\mathbf{x}(t)$

on the interval I, where $\mathbf{A}(t)$ is an $n \times n$ matrix function continuous on I. Then every solution to (5) on I can be expressed in the form

(6) $\mathbf{x}(t) = c_1\mathbf{x}_1(t) + \cdots + c_n\mathbf{x}_n(t),$

where $c_1, \ldots, c_n$ are constants.

A set of solutions $\{\mathbf{x}_1, \ldots, \mathbf{x}_n\}$ that are linearly independent on I or, equivalently, whose Wronskian does not vanish on I, is called a **fundamental solution set** for (5) on I. The linear combination in (6), written with arbitrary constants, is referred to as a **general solution** to (5).

If we take the vectors in a fundamental solution set and let them form the columns of a matrix $\mathbf{X}(t)$, that is,

$$\mathbf{X}(t) = [\mathbf{x}_1(t) \ \mathbf{x}_2(t) \ \ldots \ \mathbf{x}_n(t)] = \begin{bmatrix} x_{1,1}(t) & x_{1,2}(t) & \cdots & x_{1,n}(t) \\ x_{2,1}(t) & x_{2,2}(t) & \cdots & x_{2,n}(t) \\ \vdots & \vdots & & \vdots \\ x_{n,1}(t) & x_{n,2}(t) & \cdots & x_{n,n}(t) \end{bmatrix},$$

then the matrix $\mathbf{X}(t)$ is called a **fundamental matrix** for (5). We can use it to express the general solution (6) as

$$\mathbf{x}(t) = \mathbf{X}(t)\mathbf{c},$$

where $\mathbf{c} = \mathrm{col}(c_1, \ldots, c_n)$ is an arbitrary constant vector. Since $\det \mathbf{X} = W[\mathbf{x}_1, \ldots, \mathbf{x}_n]$ is never zero on I, it follows from Theorem 1 in Section 9.2 that $\mathbf{X}(t)$ is invertible for every t in I.

EXAMPLE 3 Verify that the set

$$S = \left\{ \begin{bmatrix} e^{2t} \\ e^{2t} \\ e^{2t} \end{bmatrix}, \begin{bmatrix} -e^{-t} \\ 0 \\ e^{-t} \end{bmatrix}, \begin{bmatrix} 0 \\ e^{-t} \\ -e^{-t} \end{bmatrix} \right\}$$

is a fundamental solution set for the system

(7) $\mathbf{x}'(t) = \begin{bmatrix} 0 & 1 & 1 \\ 1 & 0 & 1 \\ 1 & 1 & 0 \end{bmatrix} \mathbf{x}(t)$

on the interval $(-\infty, \infty)$ and find a fundamental matrix for (7). Also determine a general solution for (7).

Solution Substituting the first vector in the set S into the right-hand side of (7) gives

$$\mathbf{Ax} = \begin{bmatrix} 0 & 1 & 1 \\ 1 & 0 & 1 \\ 1 & 1 & 0 \end{bmatrix} \begin{bmatrix} e^{2t} \\ e^{2t} \\ e^{2t} \end{bmatrix} = \begin{bmatrix} 2e^{2t} \\ 2e^{2t} \\ 2e^{2t} \end{bmatrix} = \mathbf{x}'(t).$$

Hence this vector satisfies system (7) for all t. Similar computations verify that the remaining vectors in S are also solutions to (7) on $(-\infty, \infty)$. For us to show that S is a fundamental solution set, it is enough to observe that the Wronskian

$$W(t) = \begin{vmatrix} e^{2t} & -e^{-t} & 0 \\ e^{2t} & 0 & e^{-t} \\ e^{2t} & e^{-t} & -e^{-t} \end{vmatrix} = e^{2t} \begin{vmatrix} 0 & e^{-t} \\ e^{-t} & -e^{-t} \end{vmatrix} + e^{-t} \begin{vmatrix} e^{2t} & e^{-t} \\ e^{2t} & -e^{-t} \end{vmatrix} = -3$$

is never zero.

A fundamental matrix $\mathbf{X}(t)$ for (7) is just the matrix we used to compute the Wronskian; that is,

$$(8) \qquad \mathbf{X}(t) := \begin{bmatrix} e^{2t} & -e^{-t} & 0 \\ e^{2t} & 0 & e^{-t} \\ e^{2t} & e^{-t} & -e^{-t} \end{bmatrix}.$$

A general solution to (7) can now be expressed as

$$\mathbf{x}(t) = \mathbf{X}(t)\mathbf{c} = c_1 \begin{bmatrix} e^{2t} \\ e^{2t} \\ e^{2t} \end{bmatrix} + c_2 \begin{bmatrix} -e^{-t} \\ 0 \\ e^{-t} \end{bmatrix} + c_3 \begin{bmatrix} 0 \\ e^{-t} \\ -e^{-t} \end{bmatrix}. \quad \blacksquare \text{ End}$$

It is easy to check that the fundamental matrix in (8) satisfies the equation

$$\mathbf{X}'(t) = \begin{bmatrix} 0 & 1 & 1 \\ 1 & 0 & 1 \\ 1 & 1 & 0 \end{bmatrix} \mathbf{X}(t);$$

indeed, this is equivalent to showing that $\mathbf{x}' = \mathbf{Ax}$ for each column $\mathbf{x}$ in S. In general, a fundamental matrix for a system $\mathbf{x}' = \mathbf{Ax}$ satisfies the corresponding **matrix differential equation** $\mathbf{X}' = \mathbf{AX}$.

Another consequence of the linearity of the operator L defined by $L[\mathbf{x}] := \mathbf{x}' - \mathbf{Ax}$ is the **superposition principle** for linear systems. It states that if $\mathbf{x}_1$ and $\mathbf{x}_2$ are solutions, respectively, to the nonhomogeneous systems

$$L[\mathbf{x}] = \mathbf{g}_1 \quad \text{and} \quad L[\mathbf{x}] = \mathbf{g}_2,$$

then $c_1\mathbf{x}_1 + c_2\mathbf{x}_2$ is a solution to

$$L[\mathbf{x}] = c_1\mathbf{g}_1 + c_2\mathbf{g}_2.$$

Using this superposition principle and the representation theorem for homogeneous systems, we can prove the following theorem.

REPRESENTATION OF SOLUTIONS (NONHOMOGENEOUS CASE)

Theorem 4. Let $\mathbf{x}_p$ be a particular solution to the nonhomogeneous system

(9) $\qquad \mathbf{x}'(t) = \mathbf{A}(t)\mathbf{x}(t) + \mathbf{f}(t)$

on the interval I, and let $\{\mathbf{x}_1, \ldots, \mathbf{x}_n\}$ be a fundamental solution set on I for the corresponding homogeneous system $\mathbf{x}'(t) = \mathbf{A}(t)\mathbf{x}(t)$. Then every solution to (9) on I can be expressed in the form

(10) $\qquad \mathbf{x}(t) = \mathbf{x}_p(t) + c_1\mathbf{x}_1(t) + \cdots + c_n\mathbf{x}_n(t),$

where $c_1, \ldots, c_n$ are constants.

The proof of this theorem is almost identical to the proofs of Theorem 7 in Section 4.7 and Theorem 4 in Section 6.1. We leave the proof as an exercise.

The linear combination of $\mathbf{x}_p, \mathbf{x}_1, \ldots, \mathbf{x}_n$ in (10) written with arbitrary constants $c_1, \ldots, c_n$ is called a **general solution** of (9). This general solution can also be expressed as $\mathbf{x} = \mathbf{x}_p + \mathbf{Xc}$, where $\mathbf{X}$ is a fundamental matrix for the homogeneous system and $\mathbf{c}$ is an arbitrary constant vector.

We now summarize the results of this section as they apply to the problem of finding a general solution to a system of n linear first order differential equations in normal form.

APPROACH TO SOLVING NORMAL SYSTEMS

1. To determine a general solution to the $n \times n$ homogeneous system $\mathbf{x}' = \mathbf{Ax}$:

 (a) Find a fundamental solution set $\{\mathbf{x}_1, \ldots, \mathbf{x}_n\}$ that consists of n linearly independent solutions to the homogeneous system.

 (b) Form the linear combination

 $\qquad \mathbf{x} = \mathbf{Xc} = c_1\mathbf{x}_1 + \cdots + c_n\mathbf{x}_n,$

 where $\mathbf{c} = \text{col}(c_1, \ldots, c_n)$ is any constant vector and $\mathbf{X} = [\mathbf{x}_1, \ldots, \mathbf{x}_n]$ is the fundamental matrix, to obtain a general solution.

2. To determine a general solution to the nonhomogeneous system $\mathbf{x}' = \mathbf{Ax} + \mathbf{f}$:

 (a) Find a particular solution $\mathbf{x}_p$ to the nonhomogeneous system.

 (b) Form the sum of the particular solution and the general solution $\mathbf{Xc} = c_1\mathbf{x}_1 + \cdots + c_n\mathbf{x}_n$ to the corresponding homogeneous system in part 1,

 $\qquad \mathbf{x} = \mathbf{x}_p + \mathbf{Xc} = \mathbf{x}_p + c_1\mathbf{x}_1 + \cdots + c_n\mathbf{x}_n,$

 to obtain a general solution to the given system.

We shall devote the rest of this chapter to methods for finding fundamental solution sets for homogeneous systems and particular solutions for nonhomogeneous systems.

EXERCISES 9.3

In Problems 1 through 4, write the given system in the matrix form $\mathbf{x}' = \mathbf{Ax} + \mathbf{f}$.

1. $x'(t) = 3x(t) - y(t) + t^2$,
$\quad y'(t) = -x(t) + 2y(t) + e^t$.

2. $r'(t) = 2r(t) + \sin t$,
$\quad \theta'(t) = r(t) - \theta(t) + 1$.

3. $\dfrac{dx}{dt} = t^2 x - y - z + t$,

$\quad \dfrac{dy}{dt} = e^t z + 5$,

$\quad \dfrac{dz}{dt} = tx - y + 3z - e^t$.

4. $\dfrac{dx}{dt} = x + y + z$,

$\quad \dfrac{dy}{dt} = 2x - y + 3z$,

$\quad \dfrac{dz}{dt} = x + 5z$.

In Problems 5 through 8, rewrite the given scalar equation as a first order system in normal form. Express the system in the matrix form $\mathbf{x}' = \mathbf{Ax} + \mathbf{f}$.

5. $y''(t) - 3y'(t) - 10y(t) = \sin t$.

6. $x''(t) + x(t) = t^2$.

7. $\dfrac{d^4 w}{dt^4} + w = t^2$.

8. $\dfrac{d^3 y}{dt^3} - \dfrac{dy}{dt} + y = \cos t$.

In Problems 9 through 12, write the given system as a set of scalar equations.

9. $\mathbf{x}' = \begin{bmatrix} 5 & 0 \\ -2 & 4 \end{bmatrix} \mathbf{x} + e^{-2t} \begin{bmatrix} 2 \\ -3 \end{bmatrix}$.

10. $\mathbf{x}' = \begin{bmatrix} 2 & 1 \\ -1 & 3 \end{bmatrix} \mathbf{x} + e^t \begin{bmatrix} t \\ 1 \end{bmatrix}$.

11. $\mathbf{x}' = \begin{bmatrix} 1 & 0 & 1 \\ -1 & 2 & 5 \\ 0 & 5 & 1 \end{bmatrix} \mathbf{x} + e^t \begin{bmatrix} 1 \\ 0 \\ 0 \end{bmatrix} + t \begin{bmatrix} 0 \\ 1 \\ 0 \end{bmatrix}$.

12. $\mathbf{x}' = \begin{bmatrix} 0 & 1 & 0 \\ 0 & 0 & 1 \\ -1 & 1 & 2 \end{bmatrix} \mathbf{x} + t \begin{bmatrix} 1 \\ -1 \\ 2 \end{bmatrix} + \begin{bmatrix} 3 \\ 1 \\ 0 \end{bmatrix}$.

In Problems 13 through 18, determine whether the given vector functions are linearly dependent (LD) or linearly independent (LI) on the interval $(-\infty, \infty)$.

13. $\begin{bmatrix} t \\ 3 \end{bmatrix}$, $\begin{bmatrix} 4 \\ 1 \end{bmatrix}$.

14. $\begin{bmatrix} te^{-t} \\ e^{-t} \end{bmatrix}$, $\begin{bmatrix} e^{-t} \\ e^{-t} \end{bmatrix}$.

15. $\begin{bmatrix} \sin t \\ \cos t \end{bmatrix}$, $\begin{bmatrix} \sin 2t \\ \cos 2t \end{bmatrix}$.

16. $e^t \begin{bmatrix} 1 \\ 5 \end{bmatrix}$, $e^t \begin{bmatrix} -3 \\ -15 \end{bmatrix}$.

17. $e^{2t} \begin{bmatrix} 1 \\ 0 \\ 5 \end{bmatrix}$, $e^{2t} \begin{bmatrix} 1 \\ 1 \\ -1 \end{bmatrix}$, $e^{3t} \begin{bmatrix} 0 \\ 1 \\ 0 \end{bmatrix}$.

18. $\begin{bmatrix} 1 \\ 0 \\ 1 \end{bmatrix}$, $\begin{bmatrix} t \\ 0 \\ t \end{bmatrix}$, $\begin{bmatrix} t^2 \\ 0 \\ t^2 \end{bmatrix}$.

In Problems 19 through 22, the given vector functions are solutions to the system $\mathbf{x}'(t) = \mathbf{Ax}(t)$. Determine whether they form a fundamental solution set. If they do, find a fundamental matrix for the system and give a general solution.

19. $\mathbf{x}_1 = e^{2t} \begin{bmatrix} 1 \\ -2 \end{bmatrix}$, $\quad \mathbf{x}_2 = e^{2t} \begin{bmatrix} -2 \\ 4 \end{bmatrix}$.

20. $\mathbf{x}_1 = e^{-t} \begin{bmatrix} 3 \\ 2 \end{bmatrix}$, $\quad \mathbf{x}_2 = e^{4t} \begin{bmatrix} 1 \\ -1 \end{bmatrix}$.

21. $\mathbf{x}_1 = \begin{bmatrix} e^{-t} \\ 2e^{-t} \\ e^{-t} \end{bmatrix}$, $\quad \mathbf{x}_2 = \begin{bmatrix} e^t \\ 0 \\ e^t \end{bmatrix}$, $\quad \mathbf{x}_3 = \begin{bmatrix} e^{3t} \\ -e^{3t} \\ 2e^{3t} \end{bmatrix}$.

22. $\mathbf{x}_1 = \begin{bmatrix} e^t \\ e^t \\ e^t \end{bmatrix}$, $\quad \mathbf{x}_2 = \begin{bmatrix} \sin t \\ \cos t \\ -\sin t \end{bmatrix}$, $\quad \mathbf{x}_3 = \begin{bmatrix} -\cos t \\ \sin t \\ \cos t \end{bmatrix}$.

23. Verify that the vector functions

$$\mathbf{x}_1 = \begin{bmatrix} e^t \\ e^t \end{bmatrix} \quad \text{and} \quad \mathbf{x}_2 = \begin{bmatrix} e^{-t} \\ 3e^{-t} \end{bmatrix}$$

are solutions to the homogeneous system

$$\mathbf{x}' = \mathbf{Ax} = \begin{bmatrix} 2 & -1 \\ 3 & -2 \end{bmatrix} \mathbf{x},$$

on $(-\infty, \infty)$, and that

$$\mathbf{x}_p = \frac{3}{2} \begin{bmatrix} te^t \\ te^t \end{bmatrix} - \frac{1}{4} \begin{bmatrix} e^t \\ 3e^t \end{bmatrix} + \begin{bmatrix} t \\ 2t \end{bmatrix} - \begin{bmatrix} 0 \\ 1 \end{bmatrix}$$

is a particular solution to the nonhomogeneous system $\mathbf{x}' = \mathbf{Ax} + \mathbf{f}(t)$, where $\mathbf{f}(t) = \text{col}(e^t, t)$. Find a general solution to $\mathbf{x}' = \mathbf{Ax} + \mathbf{f}(t)$.

24. Verify that the vector functions

$$\mathbf{x}_1 = \begin{bmatrix} e^{3t} \\ 0 \\ e^{3t} \end{bmatrix}, \quad \mathbf{x}_2 = \begin{bmatrix} -e^{3t} \\ e^{3t} \\ 0 \end{bmatrix}, \quad \mathbf{x}_3 = \begin{bmatrix} -e^{-3t} \\ -e^{-3t} \\ e^{-3t} \end{bmatrix}$$

are solutions to the homogeneous system

$$\mathbf{x}' = \mathbf{Ax} = \begin{bmatrix} 1 & -2 & 2 \\ -2 & 1 & 2 \\ 2 & 2 & 1 \end{bmatrix} \mathbf{x},$$

on $(-\infty, \infty)$, and that

$$\mathbf{x}_p = \begin{bmatrix} 5t + 1 \\ 2t \\ 4t + 2 \end{bmatrix}$$

is a particular solution to $\mathbf{x}' = \mathbf{A}\mathbf{x} + \mathbf{f}(t)$, where $\mathbf{f}(t) = \text{col}(-9, 0, -18)$. Find a general solution to $\mathbf{x}' = \mathbf{A}\mathbf{x} + \mathbf{f}(t)$.

25. Prove that the operator defined by $L[\mathbf{x}] := \mathbf{x}' - \mathbf{A}\mathbf{x}$, where $\mathbf{A}$ is an $n \times n$ matrix function and $\mathbf{x}$ is an $n \times 1$ differentiable vector function, is a linear operator.

26. Let $\mathbf{X}(t)$ be a fundamental matrix for the system $\mathbf{x}' = \mathbf{A}\mathbf{x}$. Show that $\mathbf{x}(t) = \mathbf{X}(t)\mathbf{X}^{-1}(t_0)\mathbf{x}_0$ is the solution to the initial value problem $\mathbf{x}' = \mathbf{A}\mathbf{x}$, $\mathbf{x}(t_0) = \mathbf{x}_0$.

In Problems 27 and 28, verify that $\mathbf{X}(t)$ is a fundamental matrix for the given system and compute $\mathbf{X}^{-1}(t)$. Use the result of Problem 26 to find the solution to the given initial value problem.

27. $\mathbf{x}' = \begin{bmatrix} 0 & 6 & 0 \\ 1 & 0 & 1 \\ 1 & 1 & 0 \end{bmatrix} \mathbf{x}, \qquad \mathbf{x}(0) = \begin{bmatrix} -1 \\ 0 \\ 1 \end{bmatrix};$

$$\mathbf{X}(t) = \begin{bmatrix} 6e^{-t} & -3e^{-2t} & 2e^{3t} \\ -e^{-t} & e^{-2t} & e^{3t} \\ -5e^{-t} & e^{-2t} & e^{3t} \end{bmatrix}.$$

28. $\mathbf{x}' = \begin{bmatrix} 2 & 3 \\ 3 & 2 \end{bmatrix} \mathbf{x}, \qquad \mathbf{x}(0) = \begin{bmatrix} 3 \\ -1 \end{bmatrix};$

$$\mathbf{X}(t) = \begin{bmatrix} e^{-t} & e^{5t} \\ -e^{-t} & e^{5t} \end{bmatrix}.$$

29. Show that

$$\begin{vmatrix} t^2 & t|t| \\ 2t & 2|t| \end{vmatrix} \equiv 0$$

on $(-\infty, \infty)$, but that the two column vectors

$$\begin{bmatrix} t^2 \\ 2t \end{bmatrix}, \qquad \begin{bmatrix} t|t| \\ 2|t| \end{bmatrix}$$

are linearly independent on $(-\infty, \infty)$.

30. Abel's Formula. If $\mathbf{x}_1, \ldots, \mathbf{x}_n$ are any n solutions to the $n \times n$ system $\mathbf{x}'(t) = \mathbf{A}(t)\mathbf{x}(t)$, then Abel's formula gives a representation for the Wronskian $W(t) := W[\mathbf{x}_1, \ldots, \mathbf{x}_n](t)$. Namely,

$$W(t) = W(t_0)\exp\left(\int_{t_0}^{t} \{a_{11}(s) + \cdots + a_{nn}(s)\} ds \right),$$

where $a_{11}(s), \ldots, a_{nn}(s)$ are the main diagonal elements of $\mathbf{A}(s)$. Prove this formula in the special case when $n = 3$. [*Hint:* Follow the outline in Problem 31 of Exercises 6.1.]

31. Using Abel's formula, prove that the Wronskian of n solutions to $\mathbf{x}' = \mathbf{A}\mathbf{x}$ on the interval I is either identically zero on I or never zero on I.

32. Prove that a fundamental solution set for the homogeneous system $\mathbf{x}'(t) = \mathbf{A}(t)\mathbf{x}(t)$ always exists on an interval I, provided $\mathbf{A}(t)$ is continuous on I. [*Hint:* Use the existence and uniqueness theorem (Theorem 2) and make judicious choices for $\mathbf{x}_0$.]

33. Prove Theorem 3 on the representation of solutions of the homogeneous system.

34. Prove Theorem 4 on the representation of solutions of the nonhomogeneous system.

35. To illustrate the connection between a higher-order equation and the equivalent first order system, consider the equation

(11) $y'''(t) - 6y''(t) + 11y'(t) - 6y(t) = 0.$

(a) Show that $\{e^t, e^{2t}, e^{3t}\}$ is a fundamental solution set for (11).

(b) Using the definition of Section 6.1, compute the Wronskian of $\{e^t, e^{2t}, e^{3t}\}$.

(c) Setting $x_1 = y$, $x_2 = y'$, $x_3 = y''$, show that equation (11) is equivalent to the first order system

(12) $\mathbf{x}' = \mathbf{A}\mathbf{x},$

where

$$\mathbf{A} := \begin{bmatrix} 0 & 1 & 0 \\ 0 & 0 & 1 \\ 6 & -11 & 6 \end{bmatrix}.$$

(d) The substitution used in part (c) suggests that

$$S := \left\{ \begin{bmatrix} e^t \\ e^t \\ e^t \end{bmatrix}, \begin{bmatrix} e^{2t} \\ 2e^{2t} \\ 4e^{2t} \end{bmatrix}, \begin{bmatrix} e^{3t} \\ 3e^{3t} \\ 9e^{3t} \end{bmatrix} \right\}$$

is a fundamental solution set for system (12). Verify that this is the case.

(e) Compute the Wronskian of S. How does it compare with the Wronskian computed in part (b)?

9.4 HOMOGENEOUS LINEAR SYSTEMS WITH CONSTANT COEFFICIENTS

In this section we discuss a procedure for obtaining a general solution for the homogeneous system

(1) $\mathbf{x}'(t) = \mathbf{A}\mathbf{x}(t),$

where $\mathbf{A}$ is a (real) *constant* $n \times n$ matrix. The general solution we seek will be defined for all t because the elements of $\mathbf{A}$ are just constant functions, which are continuous on $(-\infty, \infty)$ (recall Theorem 2, page 525). In Section 9.3 we showed that a general solution to (1) can be constructed from a fundamental solution set consisting of n linearly independent solutions to (1). Thus our goal is to find n such vector solutions.

In Chapter 4 we were successful in solving homogeneous linear equations with constant coefficients by guessing that the equation had a solution of the form e^{rt}. Because any scalar linear equation can be expressed as a system, it is reasonable to expect system (1) to have solutions of the form

$$\mathbf{x}(t) = e^{rt}\mathbf{u},$$

where r is a constant and $\mathbf{u}$ is a constant vector, both of which must be determined. Substituting $e^{rt}\mathbf{u}$ for $\mathbf{x}(t)$ in (1) gives

$$re^{rt}\mathbf{u} = \mathbf{A}e^{rt}\mathbf{u} = e^{rt}\mathbf{A}\mathbf{u}.$$

Canceling the factor e^{rt} and rearranging terms, we find that

(2) $(\mathbf{A} - r\mathbf{I})\mathbf{u} = \mathbf{0},$

where $r\mathbf{I}$ denotes the diagonal matrix with r's along its main diagonal.

The preceding calculation shows that $\mathbf{x}(t) = e^{rt}\mathbf{u}$ is a solution to (1) if and only if r and $\mathbf{u}$ satisfy equation (2). Since the trivial case, $\mathbf{u} = \mathbf{0}$, is of no help in finding linearly independent solutions to (1), we require that $\mathbf{u} \neq \mathbf{0}$. Such vectors are given a special name, as follows.

EIGENVALUES AND EIGENVECTORS

Definition 3. Let $\mathbf{A} = [a_{ij}]$ be an $n \times n$ constant matrix. The **eigenvalues** of $\mathbf{A}$ are those (real or complex) numbers r for which $(\mathbf{A} - r\mathbf{I})\mathbf{u} = \mathbf{0}$ has at least one nontrivial solution[†] $\mathbf{u}$. The corresponding nontrivial solutions $\mathbf{u}$ are called the **eigenvectors** of $\mathbf{A}$ associated with r.

As stated in Theorem 1 of Section 9.2, a linear homogeneous system of n *algebraic* equations in n unknowns has a nontrivial solution if and only if the determinant of its coefficients is zero. Hence a necessary and sufficient condition for (2) to have a nontrivial

[†] We will allow $\mathbf{u}$ to have complex-number entries.

solution is that

(3) $|\mathbf{A} - r\mathbf{I}| = 0.$

Expanding the determinant of $\mathbf{A} - r\mathbf{I}$ in terms of its cofactors, we find that it is an nth degree polynomial in r; that is,

(4) $|\mathbf{A} - r\mathbf{I}| = p(r).$

Therefore *finding the eigenvalues of a matrix* $\mathbf{A}$ *is equivalent to finding the zeros of the polynomial* $p(r)$. Equation (3) is called the **characteristic equation** of $\mathbf{A}$, and $p(r)$ in (4) is the **characteristic polynomial** of $\mathbf{A}$. The characteristic equation plays a role for systems similar to the role played by the auxiliary equation for scalar equations.

There are many commercially available software packages that can be used to compute the eigenvalues and eigenvectors for a given matrix. Four such packages are DERIVE, available from Soft Warehouse, Inc.; MATLAB, available from The MathWorks, Inc.; MATHEMATICA, available from Wolfram Research; and MAPLE, available from the University of Waterloo. Although the reader is encouraged to make use of such packages, the examples and most exercises in this text can be easily carried out without them. Those exercises for which a computer package is desirable are flagged with the icon ▓ .

EXAMPLE 1 Find the eigenvalues and eigenvectors of the matrix

$$\mathbf{A} := \begin{bmatrix} 2 & -3 \\ 1 & -2 \end{bmatrix}.$$

Solution The characteristic equation for $\mathbf{A}$ is

$$|\mathbf{A} - r\mathbf{I}| = \begin{vmatrix} 2 - r & -3 \\ 1 & -2 - r \end{vmatrix} = (2 - r)(-2 - r) + 3 = r^2 - 1 = 0.$$

Hence the eigenvalues of $\mathbf{A}$ are $r_1 = 1$, $r_2 = -1$. To find the eigenvectors corresponding to $r_1 = 1$, we must solve $(\mathbf{A} - r_1\mathbf{I})\mathbf{u} = \mathbf{0}$. Substituting for $\mathbf{A}$ and r_1 gives

(5) $$\begin{bmatrix} 1 & -3 \\ 1 & -3 \end{bmatrix}\begin{bmatrix} u_1 \\ u_2 \end{bmatrix} = \begin{bmatrix} 0 \\ 0 \end{bmatrix}.$$

Notice that this matrix equation is equivalent to the single scalar equation $u_1 - 3u_2 = 0$. Therefore the solutions to (5) are obtained by assigning an arbitrary value for u_2, say $u_2 = s$, and setting $u_1 = 3u_2 = 3s$. Consequently, the eigenvectors associated with $r_1 = 1$ can be expressed as

(6) $$\mathbf{u}_1 = s\begin{bmatrix} 3 \\ 1 \end{bmatrix}.$$

For $r_2 = -1$, the equation $(\mathbf{A} - r_2\mathbf{I})\mathbf{u} = \mathbf{0}$ becomes

$$\begin{bmatrix} 3 & -3 \\ 1 & -1 \end{bmatrix}\begin{bmatrix} u_1 \\ u_2 \end{bmatrix} = \begin{bmatrix} 0 \\ 0 \end{bmatrix}.$$

Solving, we obtain $u_1 = s$ and $u_2 = s$, with s arbitrary. Therefore the eigenvectors associated with the eigenvalue $r_2 = -1$ are

(7) $\mathbf{u}_2 = s \begin{bmatrix} 1 \\ 1 \end{bmatrix}.$ ∎ End

We remark that in the above example the collection (6) of all eigenvectors associated with $r_1 = 1$ forms a one-dimensional subspace when the zero vector is adjoined. The same is true for $r_2 = -1$. These subspaces are called **eigenspaces.**

EXAMPLE 2 Find the eigenvalues and eigenvectors of the matrix

$$\mathbf{A} := \begin{bmatrix} 1 & 2 & -1 \\ 1 & 0 & 1 \\ 4 & -4 & 5 \end{bmatrix}.$$

Solution The characteristic equation for $\mathbf{A}$ is

$$|\mathbf{A} - r\mathbf{I}| = \begin{vmatrix} 1-r & 2 & -1 \\ 1 & -r & 1 \\ 4 & -4 & 5-r \end{vmatrix} = 0,$$

which simplifies to $(r - 1)(r - 2)(r - 3) = 0$. Hence the eigenvalues of $\mathbf{A}$ are $r_1 = 1$, $r_2 = 2$, and $r_3 = 3$. To find the eigenvectors corresponding to $r_1 = 1$, we set $r = 1$ in $(\mathbf{A} - r\mathbf{I})\mathbf{u} = \mathbf{0}$. This gives

(8) $$\begin{bmatrix} 0 & 2 & -1 \\ 1 & -1 & 1 \\ 4 & -4 & 4 \end{bmatrix} \begin{bmatrix} u_1 \\ u_2 \\ u_3 \end{bmatrix} = \begin{bmatrix} 0 \\ 0 \\ 0 \end{bmatrix}.$$

Using elementary row operations (Gaussian elimination), we see that (8) is equivalent to the two equations

$$u_1 - u_2 + u_3 = 0,$$
$$2u_2 - u_3 = 0.$$

Thus we can obtain the solutions to (8) by assigning an arbitrary value to u_2, say $u_2 = s$, solving $2u_2 - u_3 = 0$ for u_3 to get $u_3 = 2s$, and then solving $u_1 - u_2 + u_3 = 0$ for u_1 to get $u_1 = -s$. Hence the eigenvectors associated with $r_1 = 1$ are

(9) $\mathbf{u}_1 = s \begin{bmatrix} -1 \\ 1 \\ 2 \end{bmatrix}.$

For $r_2 = 2$, we solve

$$\begin{bmatrix} -1 & 2 & -1 \\ 1 & -2 & 1 \\ 4 & -4 & 3 \end{bmatrix} \begin{bmatrix} u_1 \\ u_2 \\ u_3 \end{bmatrix} = \begin{bmatrix} 0 \\ 0 \\ 0 \end{bmatrix}$$

in a similar fashion to obtain the eigenvectors

$$(10) \quad \mathbf{u}_2 = s \begin{bmatrix} -2 \\ 1 \\ 4 \end{bmatrix}.$$

Finally, for $r_3 = 3$, we solve

$$\begin{bmatrix} -2 & 2 & -1 \\ 1 & -3 & 1 \\ 4 & -4 & 2 \end{bmatrix} \begin{bmatrix} u_1 \\ u_2 \\ u_3 \end{bmatrix} = \begin{bmatrix} 0 \\ 0 \\ 0 \end{bmatrix}$$

and get the eigenvectors

$$(11) \quad \mathbf{u}_3 = s \begin{bmatrix} -1 \\ 1 \\ 4 \end{bmatrix}. \quad \blacksquare \text{ End}$$

Let's return to the problem of finding a general solution to a homogeneous system of differential equations. We have already shown that $e^{rt}\mathbf{u}$ is a solution to (1) if r is an eigenvalue and $\mathbf{u}$ a corresponding eigenvector. The question is: Can we obtain n linearly independent solutions to the homogeneous system by finding all the eigenvalues and eigenvectors of $\mathbf{A}$?

n LINEARLY INDEPENDENT EIGENVECTORS

Theorem 5. Suppose the $n \times n$ constant matrix $\mathbf{A}$ has n linearly independent eigenvectors $\mathbf{u}_1, \mathbf{u}_2, \ldots, \mathbf{u}_n$. Let r_i be the eigenvalue[†] corresponding to $\mathbf{u}_i$. Then

$$(12) \quad \{e^{r_1 t}\mathbf{u}_1, e^{r_2 t}\mathbf{u}_2, \ldots, e^{r_n t}\mathbf{u}_n\}$$

is a fundamental solution set on $(-\infty, \infty)$ for the homogeneous system $\mathbf{x}' = \mathbf{A}\mathbf{x}$. Consequently, a general solution of $\mathbf{x}' = \mathbf{A}\mathbf{x}$ is

$$(13) \quad \mathbf{x}(t) = c_1 e^{r_1 t}\mathbf{u}_1 + c_2 e^{r_2 t}\mathbf{u}_2 + \cdots + c_n e^{r_n t}\mathbf{u}_n,$$

where $c_1, \ldots, c_n$ are arbitrary constants.

Proof. As we have seen, the vector functions listed in (12) are solutions to the homogeneous system. Moreover, their Wronskian is

$$W(t) = \det[e^{r_1 t}\mathbf{u}_1, \ldots, e^{r_n t}\mathbf{u}_n] = e^{(r_1 + \cdots + r_n)t} \det[\mathbf{u}_1, \ldots, \mathbf{u}_n].$$

Since the eigenvectors are assumed to be linearly independent, it follows from Theorem 1 in Section 9.2 that $\det[\mathbf{u}_1, \ldots, \mathbf{u}_n]$ is not zero. Hence the Wronskian $W(t)$ is never zero. This shows that (12) is a fundamental solution set, and consequently a general solution is given by (13). $\blacksquare$ End

[†] The eigenvalues $r_1, \ldots, r_n$ may be real or complex and need not be distinct. In this section the cases we discuss have real eigenvalues. We consider complex eigenvalues in Section 9.5.

An application of Theorem 5 is given in the next example.

EXAMPLE 3 Find a general solution of

(14) $\mathbf{x}'(t) = \mathbf{A}\mathbf{x}(t)$, where $\mathbf{A} = \begin{bmatrix} 2 & -3 \\ 1 & -2 \end{bmatrix}$.

Solution In Example 1 we showed that the matrix **A** has eigenvalues $r_1 = 1$ and $r_2 = -1$. Taking, say, $s = 1$ in equations (6) and (7), we get the corresponding eigenvectors

$$\mathbf{u}_1 = \begin{bmatrix} 3 \\ 1 \end{bmatrix} \quad \text{and} \quad \mathbf{u}_2 = \begin{bmatrix} 1 \\ 1 \end{bmatrix}.$$

Because $\mathbf{u}_1$ and $\mathbf{u}_2$ are linearly independent, it follows from Theorem 5 that a general solution to (14) is

(15) $\mathbf{x}(t) = c_1 e^t \begin{bmatrix} 3 \\ 1 \end{bmatrix} + c_2 e^{-t} \begin{bmatrix} 1 \\ 1 \end{bmatrix}.$ ■ End

If we sum the vectors on the right-hand side of equation (15) and then write out the expressions for the components of $\mathbf{x}(t) = \text{col}(x_1(t), x_2(t))$, we get

$$x_1(t) = 3c_1 e^t + c_2 e^{-t},$$
$$x_2(t) = c_1 e^t + c_2 e^{-t}.$$

This is the familiar form of a general solution for a system, as discussed in Section 4.10.

Example 3 nicely illustrates the geometric role played by the eigenvectors $\mathbf{u}_1$ and $\mathbf{u}_2$. If the initial vector $\mathbf{x}(0)$ is a scalar multiple of $\mathbf{u}_1$ (i.e., $\mathbf{x}(0) = c_1\mathbf{u}_1$), then the vector solution to the system, $\mathbf{x}(t) = c_1 e^t \mathbf{u}_1$, will always have the same or opposite direction as $\mathbf{u}_1$. That is, it will lie along the straight line determined by $\mathbf{u}_1$ (see Figure 9.2). Furthermore the

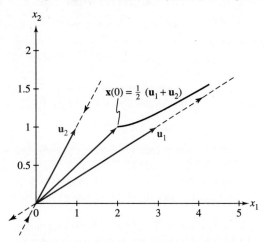

Figure 9.2 Trajectories of solutions for Example 3

trajectory of this solution, as t increases, will tend to infinity, since the corresponding eigenvalue $r_1 = 1$ is positive (observe the e^t term). A similar assertion holds if the initial vector is a scalar multiple of $\mathbf{u}_2$, except that since $r_2 = -1$ is negative, the trajectory $\mathbf{x}(t) = c_2 e^{-t} \mathbf{u}_2$ will approach the origin as t increases (because of e^{-t}). For an initial vector that involves both $\mathbf{u}_1$ and $\mathbf{u}_2$, such as $\mathbf{x}(0) = \frac{1}{2}(\mathbf{u}_1 + \mathbf{u}_2)$, the resulting trajectory is a blend of the above motions, with the contribution due to the larger eigenvalue $r_1 = 1$ dominating as t increases; see Figure 9.2.

A useful property of eigenvectors that concerns their linear independence is stated in the next theorem.

LINEAR INDEPENDENCE OF EIGENVECTORS

Theorem 6. If $r_1, \dots, r_m$ are *distinct* eigenvalues for the matrix $\mathbf{A}$ and $\mathbf{u}_i$ is an eigenvector associated with r_i, then $\mathbf{u}_1, \dots, \mathbf{u}_m$ are linearly independent.

Proof. Let's first treat the case $m = 2$. Suppose, to the contrary, that $\mathbf{u}_1$ and $\mathbf{u}_2$ are linearly dependent so that

(16) $\mathbf{u}_1 = c\mathbf{u}_2$

for some constant c. Multiplying both sides of (16) by $\mathbf{A}$ and using the fact that $\mathbf{u}_1$ and $\mathbf{u}_2$ are eigenvectors with corresponding eigenvalues r_1 and r_2, we obtain

(17) $r_1\mathbf{u}_1 = cr_2\mathbf{u}_2.$

Next we multiply (16) by r_2 and then subtract from (17) to get

$(r_1 - r_2)\mathbf{u}_1 = \mathbf{0}.$

Since $\mathbf{u}_1$ is not the zero vector, we must have $r_1 = r_2$. But this violates the assumption that the eigenvalues are distinct! Hence $\mathbf{u}_1$ and $\mathbf{u}_2$ are linearly independent.

The cases $2 < m$ follow by induction. The details of the proof are left as an exercise. ■ End

Combining Theorems 5 and 6, we get the following corollary.

n DISTINCT EIGENVALUES

Corollary 1. If the $n \times n$ constant matrix $\mathbf{A}$ has n distinct eigenvalues $r_1, \dots, r_n$ and $\mathbf{u}_i$ is an eigenvector associated with r_i, then

$$\{e^{r_1 t}\mathbf{u}_1, \dots, e^{r_n t}\mathbf{u}_n\}$$

is a fundamental solution set for the homogeneous system $\mathbf{x}' = \mathbf{A}\mathbf{x}$.

EXAMPLE 4 Solve the initial value problem

(18) $$\mathbf{x}'(t) = \begin{bmatrix} 1 & 2 & -1 \\ 1 & 0 & 1 \\ 4 & -4 & 5 \end{bmatrix} \mathbf{x}(t), \qquad \mathbf{x}(0) = \begin{bmatrix} -1 \\ 0 \\ 0 \end{bmatrix}.$$

Solution In Example 2 we showed that the 3×3 coefficient matrix $\mathbf{A}$ has the three distinct eigenvalues $r_1 = 1$, $r_2 = 2$, and $r_3 = 3$. If we set $s = 1$ in equations (9), (10), and (11), we obtain the corresponding eigenvectors

$$\mathbf{u}_1 = \begin{bmatrix} -1 \\ 1 \\ 2 \end{bmatrix}, \qquad \mathbf{u}_2 = \begin{bmatrix} -2 \\ 1 \\ 4 \end{bmatrix}, \qquad \mathbf{u}_3 = \begin{bmatrix} -1 \\ 1 \\ 4 \end{bmatrix},$$

whose linear independence is guaranteed by Theorem 6. Hence a general solution to (18) is

(19) $$\mathbf{x}(t) = c_1 e^t \begin{bmatrix} -1 \\ 1 \\ 2 \end{bmatrix} + c_2 e^{2t} \begin{bmatrix} -2 \\ 1 \\ 4 \end{bmatrix} + c_3 e^{3t} \begin{bmatrix} -1 \\ 1 \\ 4 \end{bmatrix}$$

$$= \begin{bmatrix} -e^t & -2e^{2t} & -e^{3t} \\ e^t & e^{2t} & e^{3t} \\ 2e^t & 4e^{2t} & 4e^{3t} \end{bmatrix} \begin{bmatrix} c_1 \\ c_2 \\ c_3 \end{bmatrix}.$$

To satisfy the initial condition in (18), we solve

$$\mathbf{x}_0 = \begin{bmatrix} -1 & -2 & -1 \\ 1 & 1 & 1 \\ 2 & 4 & 4 \end{bmatrix} \begin{bmatrix} c_1 \\ c_2 \\ c_3 \end{bmatrix} = \begin{bmatrix} -1 \\ 0 \\ 0 \end{bmatrix}$$

and find that $c_1 = 0$, $c_2 = 1$, and $c_3 = -1$. Inserting these values into (19) gives the desired solution. ■ End

There is a special class of $n \times n$ matrices that *always* have real eigenvalues and *always* have n linearly independent eigenvectors. These are the real symmetric matrices.

REAL SYMMETRIC MATRICES

Definition 4. A **real symmetric matrix** $\mathbf{A}$ is a matrix with real entries that satisfies $\mathbf{A}^T = \mathbf{A}$.

Taking the transpose of a matrix interchanges its rows and columns. Doing this is equivalent to "flipping" the matrix about its main diagonal. Consequently $\mathbf{A}^T = \mathbf{A}$ if and only if $\mathbf{A}$ is symmetric about its main diagonal.

If $\mathbf{A}$ is an $n \times n$ real symmetric matrix, it is known[†] that there always exist n linearly independent eigenvectors. Thus Theorem 5 applies, and a general solution to $\mathbf{x}' = \mathbf{A}\mathbf{x}$ is given by (13).

[†] See *Linear Algebra and Its Applications*, by David C. Lay (Addison-Wesley, Reading, Mass., 1994).

EXAMPLE 5 Find a general solution of

(20) $\mathbf{x}'(t) = \mathbf{A}\mathbf{x}(t)$, where $\mathbf{A} = \begin{bmatrix} 1 & -2 & 2 \\ -2 & 1 & 2 \\ 2 & 2 & 1 \end{bmatrix}$.

Solution $\mathbf{A}$ is symmetric, so we are assured that $\mathbf{A}$ has three linearly independent eigenvectors. To find them, we first compute the characteristic equation for $\mathbf{A}$:

$$|\mathbf{A} - r\mathbf{I}| = \begin{vmatrix} 1-r & -2 & 2 \\ -2 & 1-r & 2 \\ 2 & 2 & 1-r \end{vmatrix} = -(r-3)^2(r+3) = 0.$$

Thus the eigenvalues of $\mathbf{A}$ are $r_1 = r_2 = 3$ and $r_3 = -3$.

Notice that the eigenvalue $r = 3$ has multiplicity 2 when considered as a root of the characteristic equation. Therefore we must find *two* linearly independent eigenvectors associated with $r = 3$. Substituting $r = 3$ in $(\mathbf{A} - r\mathbf{I})\mathbf{u} = \mathbf{0}$ gives

$$\begin{bmatrix} -2 & -2 & 2 \\ -2 & -2 & 2 \\ 2 & 2 & -2 \end{bmatrix} \begin{bmatrix} u_1 \\ u_2 \\ u_3 \end{bmatrix} = \begin{bmatrix} 0 \\ 0 \\ 0 \end{bmatrix}.$$

This system is equivalent to the single equation $-u_1 - u_2 + u_3 = 0$, so we can obtain its solutions by assigning an arbitrary value to u_2, say $u_2 = v$, and an arbitrary value of u_3, say $u_3 = s$. Solving for u_1, we find $u_1 = u_3 - u_2 = s - v$. Therefore the eigenvectors associated with $r_1 = r_2 = 3$ can be expressed as

$$\mathbf{u} = \begin{bmatrix} s-v \\ v \\ s \end{bmatrix} = s\begin{bmatrix} 1 \\ 0 \\ 1 \end{bmatrix} + v\begin{bmatrix} -1 \\ 1 \\ 0 \end{bmatrix}.$$

By first taking $s = 1$, $v = 0$ and then taking $s = 0$, $v = 1$, we get the two linearly independent eigenvectors

(21) $\mathbf{u}_1 = \begin{bmatrix} 1 \\ 0 \\ 1 \end{bmatrix}$, $\mathbf{u}_2 = \begin{bmatrix} -1 \\ 1 \\ 0 \end{bmatrix}$.

For $r_3 = -3$, we solve

$$(\mathbf{A} + 3\mathbf{I})\mathbf{u} = \begin{bmatrix} 4 & -2 & 2 \\ -2 & 4 & 2 \\ 2 & 2 & 4 \end{bmatrix} \begin{bmatrix} u_1 \\ u_2 \\ u_3 \end{bmatrix} = \begin{bmatrix} 0 \\ 0 \\ 0 \end{bmatrix}$$

to obtain the eigenvectors col$(-s, -s, s)$. Taking $s = 1$ gives

$$\mathbf{u}_3 = \begin{bmatrix} -1 \\ -1 \\ 1 \end{bmatrix}.$$

Since the eigenvectors $\mathbf{u}_1$, $\mathbf{u}_2$, and $\mathbf{u}_3$ are linearly independent, a general solution to (20) is

$$\mathbf{x}(t) = c_1 e^{3t} \begin{bmatrix} 1 \\ 0 \\ 1 \end{bmatrix} + c_2 e^{3t} \begin{bmatrix} -1 \\ 1 \\ 0 \end{bmatrix} + c_3 e^{-3t} \begin{bmatrix} -1 \\ -1 \\ 1 \end{bmatrix}. \quad \blacksquare \text{ End}$$

If a matrix $\mathbf{A}$ *is not symmetric, it is possible for* $\mathbf{A}$ *to have a repeated eigenvalue but not have two linearly independent corresponding eigenvectors.* In particular, the matrix

(22) $$\mathbf{A} = \begin{bmatrix} 1 & -1 \\ 4 & -3 \end{bmatrix}$$

has the repeated eigenvalue $r_1 = r_2 = -1$, but Problem 35 shows that all the eigenvectors associated with $r = -1$ are of the form $\mathbf{u} = s\,\text{col}(1, 2)$. Consequently, no two eigenvectors are linearly independent.

A procedure for finding a general solution in such a case is illustrated in Problems 35–40, but the underlying theory is deferred to Section 9.7, where we discuss the matrix exponential.

EXERCISES 9.4

In Problems 1 through 8, find the eigenvalues and eigenvectors of the given matrix.

1. $\begin{bmatrix} -4 & 2 \\ 2 & -1 \end{bmatrix}$.

2. $\begin{bmatrix} 6 & -3 \\ 2 & 1 \end{bmatrix}$.

3. $\begin{bmatrix} 1 & -1 \\ 2 & 4 \end{bmatrix}$.

4. $\begin{bmatrix} 1 & 5 \\ 1 & -3 \end{bmatrix}$.

5. $\begin{bmatrix} 1 & 0 & 0 \\ 0 & 0 & 2 \\ 0 & 2 & 0 \end{bmatrix}$.

6. $\begin{bmatrix} 0 & 1 & 1 \\ 1 & 0 & 1 \\ 1 & 1 & 0 \end{bmatrix}$.

7. $\begin{bmatrix} 1 & 0 & 0 \\ 2 & 3 & 1 \\ 0 & 2 & 4 \end{bmatrix}$.

8. $\begin{bmatrix} -3 & 1 & 0 \\ 0 & -3 & 1 \\ 4 & -8 & 2 \end{bmatrix}$.

In Problems 9 and 10, some of the eigenvalues of the given matrix are complex. Find all the eigenvalues and eigenvectors.

9. $\begin{bmatrix} 0 & -1 \\ 1 & 0 \end{bmatrix}$.

10. $\begin{bmatrix} 1 & 2 & -1 \\ 0 & 1 & 1 \\ 0 & -1 & 1 \end{bmatrix}$.

In Problems 11 through 16, find a general solution of the system $\mathbf{x}'(t) = \mathbf{A}\mathbf{x}(t)$ *for the given matrix* $\mathbf{A}$.

11. $\begin{bmatrix} -1 & \frac{3}{4} \\ -5 & 3 \end{bmatrix}$.

12. $\mathbf{A} = \begin{bmatrix} 1 & 3 \\ 12 & 1 \end{bmatrix}$.

13. $\mathbf{A} = \begin{bmatrix} 1 & 2 & 2 \\ 2 & 0 & 3 \\ 2 & 3 & 0 \end{bmatrix}$.

14. $\mathbf{A} = \begin{bmatrix} -1 & 1 & 0 \\ 1 & 2 & 1 \\ 0 & 3 & -1 \end{bmatrix}$.

15. $\mathbf{A} = \begin{bmatrix} 1 & 2 & 3 \\ 0 & 1 & 0 \\ 2 & 1 & 2 \end{bmatrix}$.

16. $\mathbf{A} = \begin{bmatrix} -7 & 0 & 6 \\ 0 & 5 & 0 \\ 6 & 0 & 2 \end{bmatrix}$.

17. Consider the system $\mathbf{x}'(t) = \mathbf{A}\mathbf{x}(t)$, $t \geq 0$, with

$$\mathbf{A} = \begin{bmatrix} 1 & \sqrt{3} \\ \sqrt{3} & -1 \end{bmatrix}.$$

 (a) Show that the matrix $\mathbf{A}$ has eigenvalues $r_1 = 2$ and $r_2 = -2$ with corresponding eigenvectors $\mathbf{u}_1 = \text{col}(\sqrt{3}, 1)$ and $\mathbf{u}_2 = \text{col}(1, -\sqrt{3})$.
 (b) Sketch the trajectory of the solution having initial vector $\mathbf{x}(0) = -\mathbf{u}_1$.
 (c) Sketch the trajectory of the solution having initial vector $\mathbf{x}(0) = \mathbf{u}_2$.
 (d) Sketch the trajectory of the solution having initial vector $\mathbf{x}(0) = \mathbf{u}_2 - \mathbf{u}_1$.

18. Consider the system $\mathbf{x}'(t) = \mathbf{A}\mathbf{x}(t)$, $t \geq 0$, with

$$\mathbf{A} = \begin{bmatrix} -2 & 1 \\ 1 & -2 \end{bmatrix}.$$

(a) Show that the matrix $\mathbf{A}$ has eigenvalues $r_1 = -1$ and $r_2 = -3$ with corresponding eigenvectors $\mathbf{u}_1 = \text{col}(1, 1)$ and $\mathbf{u}_2 = \text{col}(1, -1)$.

(b) Sketch the trajectory of the solution having initial vector $\mathbf{x}(0) = \mathbf{u}_1$.

(c) Sketch the trajectory of the solution having initial vector $\mathbf{x}(0) = -\mathbf{u}_2$.

(d) Sketch the trajectory of the solution having initial vector $\mathbf{x}(0) = \mathbf{u}_1 - \mathbf{u}_2$.

In Problems 19 through 24, find a fundamental matrix for the system $\mathbf{x}'(t) = \mathbf{A}\mathbf{x}(t)$ for the given matrix $\mathbf{A}$.

19. $\mathbf{A} = \begin{bmatrix} -1 & 1 \\ 8 & 1 \end{bmatrix}$.

20. $\mathbf{A} = \begin{bmatrix} 5 & 4 \\ -1 & 0 \end{bmatrix}$.

21. $\mathbf{A} = \begin{bmatrix} 0 & 1 & 0 \\ 0 & 0 & 1 \\ 8 & -14 & 7 \end{bmatrix}$.

22. $\mathbf{A} = \begin{bmatrix} 3 & 1 & -1 \\ 1 & 3 & -1 \\ 3 & 3 & -1 \end{bmatrix}$.

23. $\mathbf{A} = \begin{bmatrix} 2 & 1 & 1 & -1 \\ 0 & -1 & 0 & 1 \\ 0 & 0 & 3 & 1 \\ 0 & 0 & 0 & 7 \end{bmatrix}$.

24. $\mathbf{A} = \begin{bmatrix} 4 & -1 & 0 & 0 \\ 0 & 0 & 0 & 0 \\ 0 & 0 & 2 & -3 \\ 0 & 0 & 1 & -2 \end{bmatrix}$.

25. Using matrix algebra techniques, find a general solution of the system

$$x' = x + 2y - z,$$
$$y' = x + z,$$
$$z' = 4x - 4y + 5z.$$

26. Using matrix algebra techniques, find a general solution of the system

$$x' = 3x - 4y,$$
$$y' = 4x - 7y.$$

Compare your solution with the solution to Example 2 in Section 4.10.

In Problems 27 through 30, use a linear algebra software package such as DERIVE, MATLAB, MAPLE, or MATHE-MATICA to compute the required eigenvalues and eigen-

vectors and then give a fundamental matrix for the system $\mathbf{x}'(t) = \mathbf{A}\mathbf{x}(t)$ for the given matrix $\mathbf{A}$.

27. $\mathbf{A} = \begin{bmatrix} 0 & 1.1 & 0 \\ 0 & 0 & 1.3 \\ 0.9 & 1.1 & -6.9 \end{bmatrix}$

28. $\mathbf{A} = \begin{bmatrix} 2 & 1 & 1 \\ -1 & 1 & 0 \\ 3 & 3 & 3 \end{bmatrix}$.

29. $\mathbf{A} = \begin{bmatrix} 0 & 1 & 0 & 0 \\ 0 & 0 & 1 & 0 \\ 0 & 0 & 0 & 1 \\ 2 & -6 & 3 & 3 \end{bmatrix}$.

30. $\mathbf{A} = \begin{bmatrix} 0 & 1 & 0 & 0 \\ 1 & -1 & 0 & 0 \\ 0 & 0 & 0 & 1 \\ 0 & 0 & -2 & 4 \end{bmatrix}$.

In Problems 31 through 34, solve the given initial value problem.

31. $\mathbf{x}'(t) = \begin{bmatrix} 1 & 3 \\ 3 & 1 \end{bmatrix} \mathbf{x}(t), \qquad \mathbf{x}(0) = \begin{bmatrix} 3 \\ 1 \end{bmatrix}$.

32. $\mathbf{x}'(t) = \begin{bmatrix} 6 & -3 \\ 2 & 1 \end{bmatrix} \mathbf{x}(t), \qquad \mathbf{x}(0) = \begin{bmatrix} -10 \\ -6 \end{bmatrix}$.

33. $\mathbf{x}'(t) = \begin{bmatrix} 1 & -2 & 2 \\ -2 & 1 & -2 \\ 2 & -2 & 1 \end{bmatrix} \mathbf{x}(t), \qquad \mathbf{x}(0) = \begin{bmatrix} -2 \\ -3 \\ 2 \end{bmatrix}$.

34. $\mathbf{x}'(t) = \begin{bmatrix} 0 & 1 & 1 \\ 1 & 0 & 1 \\ 1 & 1 & 0 \end{bmatrix} \mathbf{x}(t), \qquad \mathbf{x}(0) = \begin{bmatrix} -1 \\ 4 \\ 0 \end{bmatrix}$.

35. (a) Show that the matrix

$$\mathbf{A} = \begin{bmatrix} 1 & -1 \\ 4 & -3 \end{bmatrix}$$

has the repeated eigenvalue $r = -1$ and that all the eigenvectors are of the form $\mathbf{u} = s \, \text{col}(1, 2)$.

(b) Use the result of part (a) to obtain a nontrivial solution $\mathbf{x}_1(t)$ to the system $\mathbf{x}' = \mathbf{A}\mathbf{x}$.

(c) To obtain a second linearly independent solution to $\mathbf{x}' = \mathbf{A}\mathbf{x}$, try $\mathbf{x}_2(t) = te^{-t}\mathbf{u}_1 + e^{-t}\mathbf{u}_2$. [*Hint:* Substitute $\mathbf{x}_2$ into the system $\mathbf{x}' = \mathbf{A}\mathbf{x}$ and derive the relations

$$(\mathbf{A} + \mathbf{I})\mathbf{u}_1 = \mathbf{0}, \qquad (\mathbf{A} + \mathbf{I})\mathbf{u}_2 = \mathbf{u}_1.$$

Since $\mathbf{u}_1$ must be an eigenvector, set $\mathbf{u}_1 = \text{col}(1, 2)$ and solve for $\mathbf{u}_2$.]

(d) What is $(\mathbf{A} + \mathbf{I})^2\mathbf{u}_2$? (In Section 9.7, $\mathbf{u}_2$ will be identified as a *generalized eigenvector*.)

36. Use the method discussed in Problem 35 to find a general solution to the system

$$x'(t) = \begin{bmatrix} 5 & -3 \\ 3 & -1 \end{bmatrix} x(t).$$

37. (a) Show that the matrix

$$A = \begin{bmatrix} 2 & 1 & 6 \\ 0 & 2 & 5 \\ 0 & 0 & 2 \end{bmatrix}$$

has the repeated eigenvalue $r = 2$ with multiplicity 3 and that all the eigenvectors of A are of the form $u = s\,col(1, 0, 0)$.

(b) Use the result of part (a) to obtain a solution to the system $x' = Ax$ of the form $x_1(t) = e^{2t}u_1$.

(c) To obtain a second linearly independent solution to $x' = Ax$, try $x_2(t) = te^{2t}u_1 + e^{2t}u_2$. [*Hint:* Show that u_1 and u_2 must satisfy

$$(A - 2I)u_1 = 0, \qquad (A - 2I)u_2 = u_1.]$$

(d) To obtain a third linearly independent solution to $x' = Ax$, try

$$x_3(t) = \frac{t^2}{2}e^{2t}u_1 + te^{2t}u_2 + e^{2t}u_3.$$

[*Hint:* Show that u_1, u_2, and u_3 must satisfy

$$(A - 2I)u_1 = 0, \qquad (A - 2I)u_2 = u_1,$$
$$(A - 2I)u_3 = u_2.]$$

(e) Show that $(A - 2I)^2u_2 = (A - 2I)^3u_3 = 0$.

38. Use the method discussed in Problem 37 to find a general solution to the system

$$x'(t) = \begin{bmatrix} 3 & -2 & 1 \\ 2 & -1 & 1 \\ -4 & 4 & 1 \end{bmatrix} x(t).$$

39. (a) Show that the matrix

$$A = \begin{bmatrix} 2 & 1 & 1 \\ 1 & 2 & 1 \\ -2 & -2 & -1 \end{bmatrix}$$

has the repeated eigenvalue $r = 1$ of multiplicity 3 and that all the eigenvectors of A are of the form $u = s\,col(-1, 1, 0) + v\,col(-1, 0, 1)$.

(b) Use the result of part (a) to obtain two linearly independent solutions to the system $x' = Ax$ of the form

$$x_1(t) = e^t u_1 \quad \text{and} \quad x_2(t) = e^t u_2.$$

(c) To obtain a third linearly independent solution to $x' = Ax$, try

$$x_3(t) = te^t u_3 + e^t u_4.$$

[*Hint:* Show that u_3 and u_4 must satisfy

$$(A - I)u_3 = 0, \qquad (A - I)u_4 = u_3. \ (A-I)$$

Choose u_3, an eigenvector of A, so that you can solve for u_4.]

(d) What is $(A - I)^2 u_4$?

40. Use the method discussed in Problem 39 to find a general solution to the system

$$x'(t) = \begin{bmatrix} 1 & 3 & -2 \\ 0 & 7 & -4 \\ 0 & 9 & -5 \end{bmatrix} x(t).$$

41. Use the substitution $x_1 = y$, $x_2 = y'$ to convert the linear equation $ay'' + by' + cy = 0$, where a, b, and c are constants, into a normal system. Show that the characteristic equation for this system is the same as the auxiliary equation for the original equation.

42. Show that the **Cauchy-Euler system**

$$tx'(t) = Ax(t),$$

where A is a constant matrix, has nontrivial solutions of the form $x(t) = t^r u$ if and only if r is an eigenvalue of A and u is a corresponding eigenvector.

In Problems 43 and 44, use the result of Problem 42 to find a general solution of the given system.

43. $tx'(t) = \begin{bmatrix} 1 & 3 \\ -1 & 5 \end{bmatrix} x(t), \quad t > 0.$

44. $tx'(t) = \begin{bmatrix} -4 & 2 \\ 2 & -1 \end{bmatrix} x(t), \quad t > 0.$

45. Mixing Between Interconnected Tanks. Two tanks, each holding 50 L of liquid, are interconnected by pipes with liquid flowing from tank A into tank B at a rate of 4 L/min and from tank B into tank A at 1 L/min (see Figure 9.3). The liquid inside each tank is kept well stirred. Pure water flows into tank A at a rate of 3 L/min, and the solution flows out of tank B at 3 L/min. If, initially, tank A contains 25 kg of salt and tank B contains no salt (only water), determine the mass of salt in each tank at time $t \geq 0$. Graph on the same axes the two quantities $x_1(t)$ and $x_2(t)$, where $x_1(t)$ is the mass of salt in tank A and $x_2(t)$ the mass in tank B.

Figure 9.3 Mixing problem for interconnected tanks

46. Mixing with a Common Drain. Two tanks, each holding 1 L of liquid, are connected by a pipe through which liquid flows from tank A into tank B at a rate of $3 - \alpha$ L/min ($0 < \alpha < 3$). The liquid inside each tank is kept well stirred. Pure water flows into tank A at a rate of 3 L/min. Solution flows out of tank A at α L/min and out of tank B at $3 - \alpha$ L/min. If, initially, tank B contains no salt (only water) and tank A contains 1 kg of salt, determine the mass of salt in each tank at time $t \geq 0$. How does the mass of salt in tank A depend on the choice of α? What is the maximum mass of salt in tank B? See Figure 9.4.

Figure 9.4 Mixing problem for a common drain, $0 < \alpha < 3$

47. To find a general solution to the system

$$\mathbf{x}' = \mathbf{A}\mathbf{x} = \begin{bmatrix} 1 & 3 & -1 \\ 3 & 0 & 1 \\ -1 & 1 & 2 \end{bmatrix} \mathbf{x},$$

proceed as follows:

(a) Use a numerical root-finding procedure to approximate the eigenvalues.

(b) If r is an eigenvalue, then let $\mathbf{u} = \mathrm{col}(u_1, u_2, u_3)$ be an eigenvector associated with r. To solve for $\mathbf{u}$, assume $u_1 = 1$. (If not u_1, then either u_2 or u_3 may be chosen to be 1. Why?) Now solve the system

$$(\mathbf{A} - r\mathbf{I}) \begin{bmatrix} 1 \\ u_2 \\ u_3 \end{bmatrix} = \begin{bmatrix} 0 \\ 0 \\ 0 \end{bmatrix}$$

for u_2 and u_3. Use this procedure to find approximations for three linearly independent eigenvectors for $\mathbf{A}$.

(c) Use these approximations to give a general solution to the system.

48. To complete the proof of Theorem 6, assume the induction hypothesis that $\mathbf{u}_1, \ldots, \mathbf{u}_k$, $2 \leq k$, are linearly independent.

(a) Show that if

$$c_1 \mathbf{u}_1 + \cdots + c_k \mathbf{u}_k + c_{k+1} \mathbf{u}_{k+1} = \mathbf{0},$$

then

$$c_1(r_1 - r_{k+1})\mathbf{u}_1 + \cdots + c_k(r_k - r_{k+1})\mathbf{u}_k = \mathbf{0}.$$

(b) Use the result of part (a) and the induction hypothesis to conclude that $\mathbf{u}_1, \ldots, \mathbf{u}_{k+1}$ are linearly independent. The theorem follows by induction.

49. Stability. A homogeneous system $\mathbf{x}' = \mathbf{A}\mathbf{x}$ with constant coefficients is **stable** if it has a fundamental matrix whose entries all remain bounded as $t \to +\infty$. (It will follow from Lemma 1 in Section 9.7 that if one fundamental matrix of the system has this property, then all fundamental matrices for the system do.) Otherwise, the system is **unstable**. A stable system is **asymptotically stable** if all solutions approach the zero solution as $t \to +\infty$. Stability is discussed in more detail in Chapter 12.[†]

(a) Show that if $\mathbf{A}$ has all distinct real eigenvalues, then $\mathbf{x}'(t) = \mathbf{A}\mathbf{x}(t)$ is stable if and only if all eigenvalues are nonpositive.

(b) Show that if $\mathbf{A}$ has all distinct real eigenvalues, then $\mathbf{x}'(t) = \mathbf{A}\mathbf{x}(t)$ is asymptotically stable if and only if all eigenvalues are negative.

(c) Argue that in parts (a) and (b), we can replace "has distinct real eigenvalues" by "is symmetric" and the statements are still true.

[†] All references to Chapters 11–13 refer to the expanded text *Fundamentals of Differential Equations and Boundary Value Problems*, Second Edition.

9.5 COMPLEX EIGENVALUES

In the previous section we showed that the homogeneous system

(1) $\mathbf{x}'(t) = \mathbf{A}\mathbf{x}(t),$

where $\mathbf{A}$ is a constant $n \times n$ matrix, has a solution of the form $\mathbf{x}(t) = e^{rt}\mathbf{u}$ if and only if r is an eigenvalue of $\mathbf{A}$ and $\mathbf{u}$ is a corresponding eigenvector. In this section we show how to obtain two real vector solutions to system (1) when $\mathbf{A}$ is real and has a pair[†] of complex conjugate eigenvalues $\alpha + i\beta$ and $\alpha - i\beta$.

Suppose $r_1 = \alpha + i\beta$ (α and β real numbers) is an eigenvalue of $\mathbf{A}$ with corresponding eigenvector $\mathbf{z} = \mathbf{a} + i\mathbf{b}$, where $\mathbf{a}$ and $\mathbf{b}$ are real constant vectors. We first observe that the complex conjugate of $\mathbf{z}$, namely, $\bar{\mathbf{z}} := \mathbf{a} - i\mathbf{b}$, is an eigenvector associated with the eigenvalue $r_2 = \alpha - i\beta$. To see this, note that taking the complex conjugate of $(\mathbf{A} - r_1\mathbf{I})\mathbf{z} = \mathbf{0}$ yields $(\mathbf{A} - \bar{r}_1\mathbf{I})\bar{\mathbf{z}} = \mathbf{0}$ because the conjugate of the product is the product of the conjugates and $\mathbf{A}$ and $\mathbf{I}$ have real entries ($\bar{\mathbf{A}} = \mathbf{A}$, $\bar{\mathbf{I}} = \mathbf{I}$). Since $r_2 = \bar{r}_1$, we see that $\bar{\mathbf{z}}$ is an eigenvector associated with r_2. Therefore two linearly independent complex vector solutions to (1) are

(2) $\mathbf{w}_1(t) = e^{r_1 t}\mathbf{z} = e^{(\alpha + i\beta)t}(\mathbf{a} + i\mathbf{b}),$

(3) $\mathbf{w}_2(t) = e^{r_2 t}\bar{\mathbf{z}} = e^{(\alpha - i\beta)t}(\mathbf{a} - i\mathbf{b}).$

As in Section 4.6, where we handled complex roots to the auxiliary equation, let's use one of these complex solutions and Euler's formula to obtain two real vector solutions. With the aid of Euler's formula, we rewrite $\mathbf{w}_1(t)$ as

$$\mathbf{w}_1(t) = e^{\alpha t}(\cos \beta t + i \sin \beta t)(\mathbf{a} + i\mathbf{b})$$
$$= e^{\alpha t}\{(\cos \beta t\, \mathbf{a} - \sin \beta t\, \mathbf{b}) + i(\sin \beta t\, \mathbf{a} + \cos \beta t\, \mathbf{b})\}.$$

We have thereby expressed $\mathbf{w}_1(t)$ in the form $\mathbf{w}_1(t) = \mathbf{x}_1(t) + i\mathbf{x}_2(t)$, where $\mathbf{x}_1(t)$ and $\mathbf{x}_2(t)$ are the two *real* vector functions

(4) $\mathbf{x}_1(t) := e^{\alpha t}\cos \beta t\, \mathbf{a} - e^{\alpha t}\sin \beta t\, \mathbf{b},$

(5) $\mathbf{x}_2(t) := e^{\alpha t}\sin \beta t\, \mathbf{a} + e^{\alpha t}\cos \beta t\, \mathbf{b}.$

Since $\mathbf{w}_1(t)$ is a solution to (1), then

$$\mathbf{w}_1'(t) = \mathbf{A}\mathbf{w}_1(t),$$
$$\mathbf{x}_1' + i\mathbf{x}_2' = \mathbf{A}\mathbf{x}_1 + i\mathbf{A}\mathbf{x}_2.$$

Equating the real and imaginary parts yields

$$\mathbf{x}_1'(t) = \mathbf{A}\mathbf{x}_1(t) \quad \text{and} \quad \mathbf{x}_2'(t) = \mathbf{A}\mathbf{x}_2(t).$$

[†] Recall that the complex roots of a polynomial equation with real coefficients must occur in complex conjugate pairs.

Hence $\mathbf{x}_1(t)$ and $\mathbf{x}_2(t)$ are real vector solutions to (1) associated with the complex conjugate eigenvalues $\alpha \pm i\beta$. Because $\mathbf{a}$ and $\mathbf{b}$ are not both the zero vector, it can be shown that $\mathbf{x}_1(t)$ and $\mathbf{x}_2(t)$ are linearly independent vector functions on $(-\infty, \infty)$ (see Problem 15).

Let's summarize our findings.

COMPLEX EIGENVALUES

If the real matrix $\mathbf{A}$ has complex conjugate eigenvalues $\alpha \pm i\beta$ with corresponding eigenvectors $\mathbf{a} \pm i\mathbf{b}$, then two linearly independent real vector solutions to $\mathbf{x}'(t) = \mathbf{A}\mathbf{x}(t)$ are

(6) $e^{\alpha t} \cos \beta t\, \mathbf{a} - e^{\alpha t} \sin \beta t\, \mathbf{b},$

(7) $e^{\alpha t} \sin \beta t\, \mathbf{a} + e^{\alpha t} \cos \beta t\, \mathbf{b}.$

EXAMPLE 1 Find a general solution of

(8) $\mathbf{x}'(t) = \begin{bmatrix} -1 & 2 \\ -1 & -3 \end{bmatrix} \mathbf{x}(t).$

Solution The characteristic equation for $\mathbf{A}$ is

$$|\mathbf{A} - r\mathbf{I}| = \begin{vmatrix} -1 - r & 2 \\ -1 & -3 - r \end{vmatrix} = r^2 + 4r + 5 = 0.$$

Hence $\mathbf{A}$ has eigenvalues $r = -2 \pm i$.

To find a general solution, we need only find an eigenvector associated with the eigenvalue $r = -2 + i$. Substituting $r = -2 + i$ into $(\mathbf{A} - r\mathbf{I})\mathbf{z} = \mathbf{0}$ gives

$$\begin{bmatrix} 1 - i & 2 \\ -1 & -1 - i \end{bmatrix} \begin{bmatrix} z_1 \\ z_2 \end{bmatrix} = \begin{bmatrix} 0 \\ 0 \end{bmatrix}.$$

The solutions are $z_1 = 2s$ and $z_2 = (-1 + i)s$, with s arbitrary. Hence the eigenvectors associated with $r = -2 + i$ are $\mathbf{z} = s\,\mathrm{col}(2, -1 + i)$. Taking $s = 1$ gives the eigenvector

$$\mathbf{z} = \begin{bmatrix} 2 \\ -1 + i \end{bmatrix} = \begin{bmatrix} 2 \\ -1 \end{bmatrix} + i \begin{bmatrix} 0 \\ 1 \end{bmatrix}.$$

We have found that $\alpha = -2$, $\beta = 1$, and $\mathbf{z} = \mathbf{a} + i\mathbf{b}$ with $\mathbf{a} = \mathrm{col}(2, -1)$, and $\mathbf{b} = \mathrm{col}(0, 1)$, so a general solution to (8) is

$$\mathbf{x}(t) = c_1 \left\{ e^{-2t} \cos t \begin{bmatrix} 2 \\ -1 \end{bmatrix} - e^{-2t} \sin t \begin{bmatrix} 0 \\ 1 \end{bmatrix} \right\}$$

$$+ c_2 \left\{ e^{-2t} \sin t \begin{bmatrix} 2 \\ -1 \end{bmatrix} + e^{-2t} \cos t \begin{bmatrix} 0 \\ 1 \end{bmatrix} \right\}$$

$$= c_1 \begin{bmatrix} 2e^{-2t} \cos t \\ -e^{-2t}(\cos t + \sin t) \end{bmatrix} + c_2 \begin{bmatrix} 2e^{-2t} \sin t \\ e^{-2t}(\cos t - \sin t) \end{bmatrix}. \qquad \blacksquare \text{ End}$$

Figure 9.5 Coupled spring-mass system with fixed ends

Complex eigenvalues occur in modeling coupled spring-mass systems. For example, the motion of the spring-mass system illustrated in Figure 9.5 is governed by the second order system

(9)
$$m_1 x_1'' = -k_1 x_1 + k_2(x_2 - x_1),$$
$$m_2 x_2'' = -k_2(x_2 - x_1) - k_3 x_2,$$

where x_1 and x_2 represent the displacement of the masses m_1 and m_2 to the right of their equilibrium positions and k_1, k_2, k_3 are the spring constants of the three springs (see the discussion in Section 5.4). If we introduce the new variables $y_1 := x_1$, $y_2 := x_1'$, $y_3 := x_2$, $y_4 := x_2'$, then we can rewrite the system in the normal form

(10) $$\mathbf{y}'(t) = \mathbf{A}\mathbf{y}(t) = \begin{bmatrix} 0 & 1 & 0 & 0 \\ -(k_1 + k_2)/m_1 & 0 & k_2/m_1 & 0 \\ 0 & 0 & 0 & 1 \\ k_2/m_2 & 0 & -(k_2 + k_3)/m_2 & 0 \end{bmatrix} \mathbf{y}(t).$$

For such a system, it turns out that $\mathbf{A}$ has only imaginary eigenvalues and they occur in complex conjugate pairs: $\pm i\beta_1$, $\pm i\beta_2$. Hence any solution will consist of sums of sine and cosine functions. The frequencies of these functions

$$f_1 := \frac{\beta_1}{2\pi} \quad \text{and} \quad f_2 := \frac{\beta_2}{2\pi}$$

are called the **normal** or **natural frequencies** of the system (β_1 and β_2 are the **angular frequencies** of the system).

EXAMPLE 2 Determine the normal frequencies for the coupled spring-mass system governed by system (10) when $m_1 = m_2 = 1$ kg, $k_1 = 1$ N/m, $k_2 = 2$ N/m, and $k_3 = 3$ N/m.

Solution To find the eigenvalues of $\mathbf{A}$, we must solve the characteristic equation

$$|\mathbf{A} - r\mathbf{I}| = \begin{vmatrix} -r & 1 & 0 & 0 \\ -3 & -r & 2 & 0 \\ 0 & 0 & -r & 1 \\ 2 & 0 & -5 & -r \end{vmatrix} = r^4 + 8r^2 + 11 = 0.$$

From the quadratic formula we find $r^2 = -4 \pm \sqrt{5}$, so the four eigenvalues of $\mathbf{A}$ are $\pm i\sqrt{4 - \sqrt{5}}$ and $\pm i\sqrt{4 + \sqrt{5}}$. Hence the two normal frequencies for this system are

$$\frac{\sqrt{4 - \sqrt{5}}}{2\pi} \approx 0.211 \quad \text{and} \quad \frac{\sqrt{4 + \sqrt{5}}}{2\pi} \approx 0.397 \text{ cycles per second.} \quad \blacksquare \text{ End}$$

EXERCISES 9.5

In Problems 1 through 4, find a general solution of the system
$\mathbf{x}'(t) = \mathbf{Ax}(t)$ *for the given matrix* $\mathbf{A}$.

1. $\mathbf{A} = \begin{bmatrix} 2 & -4 \\ 2 & -2 \end{bmatrix}$.

2. $\mathbf{A} = \begin{bmatrix} -2 & -5 \\ 1 & 2 \end{bmatrix}$.

3. $\mathbf{A} = \begin{bmatrix} 1 & 2 & -1 \\ 0 & 1 & 1 \\ 0 & -1 & 1 \end{bmatrix}$.

4. $\mathbf{A} = \begin{bmatrix} 5 & -5 & -5 \\ -1 & 4 & 2 \\ 3 & -5 & -3 \end{bmatrix}$.

In Problems 5 through 8, find a fundamental matrix for the system $\mathbf{x}'(t) = \mathbf{Ax}(t)$ *for the given matrix* $\mathbf{A}$.

5. $\mathbf{A} = \begin{bmatrix} -1 & -2 \\ 8 & -1 \end{bmatrix}$.

6. $\mathbf{A} = \begin{bmatrix} -2 & -2 \\ 4 & 2 \end{bmatrix}$.

7. $\mathbf{A} = \begin{bmatrix} 0 & 0 & 1 \\ 0 & 0 & -1 \\ 0 & 1 & 0 \end{bmatrix}$.

8. $\mathbf{A} = \begin{bmatrix} 0 & 1 & 0 & 0 \\ 1 & 0 & 0 & 0 \\ 0 & 0 & 0 & 1 \\ 0 & 0 & -13 & 4 \end{bmatrix}$.

In Problems 9 through 12, use a linear algebra software package to compute the required eigenvalues and eigenvectors for the given matrix $\mathbf{A}$ *and then give a fundamental matrix for the system* $\mathbf{x}'(t) = \mathbf{Ax}(t)$.

9. $\mathbf{A} = \begin{bmatrix} 0 & 1 & 1 \\ -1 & 0 & 1 \\ -1 & -1 & 0 \end{bmatrix}$.

10. $\mathbf{A} = \begin{bmatrix} 0 & 1 & 0 & 0 \\ 0 & 0 & 1 & 0 \\ 0 & 0 & 0 & 1 \\ 13 & -4 & -12 & 4 \end{bmatrix}$.

11. $\mathbf{A} = \begin{bmatrix} 0 & 1 & 0 & 0 \\ 0 & 0 & 1 & 0 \\ 0 & 0 & 0 & 1 \\ -2 & 2 & -3 & 2 \end{bmatrix}$.

12. $\mathbf{A} = \begin{bmatrix} 1 & 0 & 0 & 0 & 0 \\ 0 & 0 & 1 & 0 & 0 \\ 0 & 1 & 0 & 0 & 0 \\ 0 & 0 & 0 & 0 & 1 \\ 0 & 0 & 0 & -29 & -4 \end{bmatrix}$.

In Problems 13 and 14, find the solution to the given system that satisfies the given initial condition.

13. $\mathbf{x}'(t) = \begin{bmatrix} -3 & -1 \\ 2 & -1 \end{bmatrix} \mathbf{x}(t)$,

(a) $\mathbf{x}(0) = \begin{bmatrix} -1 \\ 0 \end{bmatrix}$.

(b) $\mathbf{x}(\pi) = \begin{bmatrix} 1 \\ -1 \end{bmatrix}$.

(c) $\mathbf{x}(-2\pi) = \begin{bmatrix} 2 \\ 1 \end{bmatrix}$.

(d) $\mathbf{x}(\pi/2) = \begin{bmatrix} 0 \\ 1 \end{bmatrix}$.

14. $\mathbf{x}'(t) = \begin{bmatrix} 1 & 0 & -1 \\ 0 & 2 & 0 \\ 1 & 0 & 1 \end{bmatrix} \mathbf{x}(t)$,

(a) $\mathbf{x}(0) = \begin{bmatrix} -2 \\ 2 \\ -1 \end{bmatrix}$.

(b) $\mathbf{x}(-\pi) = \begin{bmatrix} 0 \\ 1 \\ 1 \end{bmatrix}$.

15. Show that $\mathbf{x}_1(t)$ and $\mathbf{x}_2(t)$ given by equations (4) and (5) are linearly independent on $(-\infty, \infty)$, provided $\beta \neq 0$ and $\mathbf{a}$ and $\mathbf{b}$ are not both the zero vector.

16. Show that $\mathbf{x}_1(t)$ and $\mathbf{x}_2(t)$ given by equations (4) and (5) can be obtained as linear combinations of $\mathbf{w}_1(t)$ and $\mathbf{w}_2(t)$ given by equations (2) and (3). [*Hint:* Show that

$$\mathbf{x}_1(t) = \frac{\mathbf{w}_1(t) + \mathbf{w}_2(t)}{2}, \qquad \mathbf{x}_2(t) = \frac{\mathbf{w}_1(t) - \mathbf{w}_2(t)}{2i}.]$$

In Problems 17 and 18, use the results of Problem 42 in Exercises 9.4 to find a general solution to the given Cauchy-Euler system for $t > 0$.

17. $t\mathbf{x}'(t) = \begin{bmatrix} -1 & -1 & 0 \\ 2 & -1 & 1 \\ 0 & 1 & -1 \end{bmatrix} \mathbf{x}(t).$

18. $t\mathbf{x}'(t) = \begin{bmatrix} -1 & -1 \\ 9 & -1 \end{bmatrix} \mathbf{x}(t).$

19. For the coupled spring-mass system governed by system (9), assume $m_1 = m_2 = 1$ kg, $k_1 = k_2 = 2$ N/m, and $k_3 = 3$ N/m. Determine the normal frequencies for this coupled spring-mass system.

20. For the coupled spring-mass system governed by system (9), assume $m_1 = m_2 = 1$ kg, $k_1 = k_2 = k_3 = 1$ N/m, and assume initially that $x_1(0) = 0$ m, $x_1'(0) = 0$ m/sec, $x_2(0) = 2$ m, and $x_2'(0) = 0$ m/sec. Using matrix algebra techniques, solve this initial value problem.

21. RLC Network. The currents in the RLC network given by the schematic diagram in Figure 9.6 are governed by the following equations:

$$4I_2'(t) + 52q_1(t) = 10,$$

$$13I_3(t) + 52q_1(t) = 10,$$

$$I_1(t) = I_2(t) + I_3(t),$$

Figure 9.6 RLC network for Problem 21

where $q_1(t)$ is the charge on the capacitor, $I_1(t) = q_1'(t)$, and initially $q_1(0) = 0$ coulombs and $I_1(0) = 0$ amps. Solve for the currents I_1, I_2, and I_3. [*Hint:* Differentiate the first two equations, eliminate I_1, and form a normal system with $x_1 = I_2$, $x_2 = I_2'$, and $x_3 = I_3$.]

22. RLC Network. The currents in the RLC network given by the schematic diagram in Figure 9.7 are governed by the following equations:

$$50I_1'(t) + 80I_2(t) = 160,$$

$$50I_1'(t) + 800q_3(t) = 160,$$

$$I_1(t) = I_2(t) + I_3(t),$$

Figure 9.7 RLC network for Problem 22

where $q_3(t)$ is the charge on the capacitor, $I_3(t) = q_3'(t)$, and initially $q_3(0) = 0.5$ coulombs and $I_3(0) = 0$ amps. Solve for the currents I_1, I_2, and I_3. [*Hint:* Differentiate the first two equations, use the third equation to eliminate I_3, and form a normal system with $x_1 = I_1$, $x_2 = I_1'$, and $x_3 = I_2$.]

23. Stability. In Problem 49 of Exercises 9.4, we discussed the idea of stability and asymptotic stability for a linear system of the form $\mathbf{x}'(t) = \mathbf{A}\mathbf{x}(t)$. Assume $\mathbf{A}$ has all distinct eigenvalues (real or complex).

(a) Show that the system is stable if and only if all the eigenvalues of $\mathbf{A}$ have nonpositive real part.

(b) Show that the system is asymptotically stable if and only if all the eigenvalues of $\mathbf{A}$ have negative real part.

9.6 NONHOMOGENEOUS LINEAR SYSTEMS

The techniques discussed in Chapter 4 for finding a particular solution to the nonhomogeneous equation $y'' + p(x)y' + q(x)y = g(x)$ have natural extensions to nonhomogeneous linear systems.

Undetermined Coefficients

The method of undetermined coefficients can be used to find a particular solution to the nonhomogeneous linear system.

$$\mathbf{x}'(t) = \mathbf{A}\mathbf{x}(t) + \mathbf{f}(t),$$

when $\mathbf{A}$ is an $n \times n$ *constant* matrix and the entries of $\mathbf{f}(t)$ are polynomials, exponential functions, sines and cosines, or finite sums and products of these functions. We can use Table 4.1 (discussed in Section 4.8 and reproduced on the inside back cover) as a *guide* in choosing the form of a particular solution $\mathbf{x}_p(t)$. Some exceptions are discussed in the exercises (see Problems 25–28).

EXAMPLE 1 Find a general solution of

$$(1) \qquad \mathbf{x}'(t) = \mathbf{A}\mathbf{x}(t) + t\mathbf{g}, \quad \text{where} \quad \mathbf{A} = \begin{bmatrix} 1 & -2 & 2 \\ -2 & 1 & 2 \\ 2 & 2 & 1 \end{bmatrix} \quad \text{and} \quad \mathbf{g} = \begin{bmatrix} -9 \\ 0 \\ -18 \end{bmatrix}.$$

Solution In Example 5 in Section 9.4, we found that a general solution to the corresponding homogeneous system $\mathbf{x}' = \mathbf{A}\mathbf{x}$ is

$$(2) \qquad \mathbf{x}_h(t) = c_1 e^{3t} \begin{bmatrix} 1 \\ 0 \\ 1 \end{bmatrix} + c_2 e^{3t} \begin{bmatrix} -1 \\ 1 \\ 0 \end{bmatrix} + c_3 e^{-3t} \begin{bmatrix} -1 \\ -1 \\ 1 \end{bmatrix}.$$

Since the entries in $\mathbf{f}(t) := t\mathbf{g}$ are just linear functions of t, we are inclined to seek a particular solution of the form

$$\mathbf{x}_p(t) = t\mathbf{a} + \mathbf{b} = t \begin{bmatrix} a_1 \\ a_2 \\ a_3 \end{bmatrix} + \begin{bmatrix} b_1 \\ b_2 \\ b_3 \end{bmatrix},$$

where the constant vectors $\mathbf{a}$ and $\mathbf{b}$ are to be determined. Substituting this expression for $\mathbf{x}_p(t)$ into system (1) yields

$$\mathbf{a} = \mathbf{A}(t\mathbf{a} + \mathbf{b}) + t\mathbf{g},$$

which can be written as

$$t(\mathbf{A}\mathbf{a} + \mathbf{g}) + (\mathbf{A}\mathbf{b} - \mathbf{a}) = \mathbf{0}.$$

Setting the "coefficients" of this vector polynomial equal to zero yields the two systems

$$(3) \qquad \mathbf{A}\mathbf{a} = -\mathbf{g},$$

$$(4) \qquad \mathbf{A}\mathbf{b} = \mathbf{a}.$$

By Gaussian elimination or by using a linear algebra software package, we can solve (3) for $\mathbf{a}$ and we find $\mathbf{a} = \text{col}(5, 2, 4)$. Next we substitute for $\mathbf{a}$ in (4) and solve for $\mathbf{b}$ to obtain $\mathbf{b} = \text{col}(1, 0, 2)$. Hence a particular solution for (1) is

$$(5) \qquad \mathbf{x}_p(t) = t\mathbf{a} + \mathbf{b} = t \begin{bmatrix} 5 \\ 2 \\ 4 \end{bmatrix} + \begin{bmatrix} 1 \\ 0 \\ 2 \end{bmatrix} = \begin{bmatrix} 5t + 1 \\ 2t \\ 4t + 2 \end{bmatrix}.$$

A general solution for (1) is $\mathbf{x}(t) = \mathbf{x}_h(t) + \mathbf{x}_p(t)$, where $\mathbf{x}_h(t)$ is given in (2) and $\mathbf{x}_p(t)$ in (5). ■ End

In the preceding example, the nonhomogeneous term $\mathbf{f}(t)$ was a vector polynomial. If, instead, $\mathbf{f}(t)$ has the form

$$\mathbf{f}(t) = \text{col}(1, \ t, \ \sin t),$$

then, using the superposition principle, we would seek a particular solution of the form

$$\mathbf{x}_p(t) = t\mathbf{a} + \mathbf{b} + \sin t\mathbf{c} + \cos t\mathbf{d}.$$

Similarly, if

$$\mathbf{f}(t) = \text{col}(t, \ e^t, \ t^2),$$

we would take

$$\mathbf{x}_p(t) = t^2\mathbf{a} + t\mathbf{b} + \mathbf{c} + e^t\mathbf{d}.$$

Of course, we must modify our guess should one of the terms be a solution to the corresponding homogeneous system. Problems 25–28 illustrate the strategies.

Variation of Parameters

In Section 4.9, we discussed the method of variation of parameters for a general (not necessarily constant-coefficient) second order linear equation. Simply put, the idea is that if a general solution to the homogeneous equation has the form $x_h(t) = c_1 x_1(t) + c_2 x_2(t)$, where $x_1(t)$ and $x_2(t)$ are linearly independent solutions to the homogeneous equation, then a particular solution to the nonhomogeneous equation would have the form $x_p(t) = v_1(t)x_1(t) + v_2(t)x_2(t)$, where $v_1(t)$ and $v_2(t)$ are certain functions of t. A similar idea can be used for systems.

Let $\mathbf{X}(t)$ be a fundamental matrix for the homogeneous system

$$(6) \qquad \mathbf{x}'(t) = \mathbf{A}(t)\mathbf{x}(t),$$

where now *the entries of* $\mathbf{A}$ *may be any continuous functions of* t. Because a general solution to (6) is given by $\mathbf{X}(t)\mathbf{c}$, where $\mathbf{c}$ is a constant $n \times 1$ vector, we seek a particular solution to the nonhomogeneous system

$$(7) \qquad \mathbf{x}'(t) = \mathbf{A}(t)\mathbf{x}(t) + \mathbf{f}(t)$$

of the form

$$(8) \qquad \mathbf{x}_p(t) = \mathbf{X}(t)\mathbf{v}(t),$$

where $\mathbf{v}(t) = \text{col}(v_1(t), \ \ldots, \ v_n(t))$ is a vector function of t to be determined.

To derive a formula for $\mathbf{v}(t)$, we first differentiate (8) using the matrix version of the product rule to obtain

$$\mathbf{x}_p'(t) = \mathbf{X}(t)\mathbf{v}'(t) + \mathbf{X}'(t)\mathbf{v}(t).$$

Substituting the expressions for $\mathbf{x}_p(t)$ and $\mathbf{x}_p'(t)$ into (7) yields

(9) $\mathbf{X}(t)\boldsymbol{v}'(t) + \mathbf{X}'(t)\boldsymbol{v}(t) = \mathbf{A}(t)\mathbf{X}(t)\boldsymbol{v}(t) + \mathbf{f}(t).$

Since $\mathbf{X}(t)$ satisfies the matrix equation $\mathbf{X}'(t) = \mathbf{A}(t)\mathbf{X}(t)$, equation (9) becomes

$$\mathbf{X}\boldsymbol{v}' + \mathbf{A}\mathbf{X}\boldsymbol{v} = \mathbf{A}\mathbf{X}\boldsymbol{v} + \mathbf{f},$$
$$\mathbf{X}\boldsymbol{v}' = \mathbf{f}.$$

Multiplying both sides of the last equation by $\mathbf{X}^{-1}(t)$ (which exists since the columns of $\mathbf{X}(t)$ are linearly independent) gives

$$\boldsymbol{v}'(t) = \mathbf{X}^{-1}(t)\mathbf{f}(t).$$

Integrating, we obtain

$$\boldsymbol{v}(t) = \int \mathbf{X}^{-1}(t)\mathbf{f}(t)\,dt.$$

Hence a particular solution to (7) is

(10) $\mathbf{x}_p(t) = \mathbf{X}(t)\boldsymbol{v}(t) = \mathbf{X}(t)\displaystyle\int \mathbf{X}^{-1}(t)\mathbf{f}(t)\,dt.$

Combining (10) with the solution $\mathbf{X}(t)\mathbf{c}$ to the homogeneous system yields the following general solution to (7):

(11) $\mathbf{x}(t) = \mathbf{X}(t)\mathbf{c} + \mathbf{X}(t)\displaystyle\int \mathbf{X}^{-1}(t)\mathbf{f}(t)\,dt.$

The elegance of the derivation of the variation of parameters formula (10) for systems becomes evident when one compares it with the more lengthy derivations for the scalar case in Sections 4.9 and 6.4.

Given an initial value problem of the form

(12) $\mathbf{x}'(t) = \mathbf{A}(t)\mathbf{x}(t) + \mathbf{f}(t), \qquad \mathbf{x}(t_0) = \mathbf{x}_0,$

we can use the initial condition $\mathbf{x}(t_0) = \mathbf{x}_0$ to solve for $\mathbf{c}$ in (11). Expressing $\mathbf{x}(t)$ using a definite integral, we have

$$\mathbf{x}(t) = \mathbf{X}(t)\mathbf{c} + \mathbf{X}(t)\int_{t_0}^{t} \mathbf{X}^{-1}(s)\mathbf{f}(s)\,ds.$$

Using the initial condition $\mathbf{x}(t_0) = \mathbf{x}_0$, we find

$$\mathbf{x}_0 = \mathbf{x}(t_0) = \mathbf{X}(t_0)\mathbf{c} + \mathbf{X}(t_0)\int_{t_0}^{t_0} \mathbf{X}^{-1}(s)\mathbf{f}(s)\,ds = \mathbf{X}(t_0)\mathbf{c}.$$

Solving for **c**, we have $\mathbf{c} = \mathbf{X}^{-1}(t_0)\mathbf{x}_0$. Thus the solution to (12) is given by the formula

$$(13) \qquad \mathbf{x}(t) = \mathbf{X}(t)\mathbf{X}^{-1}(t_0)\mathbf{x}_0 + \mathbf{X}(t) \int_{t_0}^{t} \mathbf{X}^{-1}(s)\mathbf{f}(s)\,ds.$$

To apply the variation of parameters formulas, we first must determine a fundamental matrix $\mathbf{X}(t)$ for the homogeneous system. In the case when the coefficient matrix $\mathbf{A}$ is constant, we have discussed methods for finding $\mathbf{X}(t)$. However, if the entries of $\mathbf{A}$ depend on t, the determination of $\mathbf{X}(t)$ may be extremely difficult.

EXAMPLE 2 Find the solution to the initial value problem

$$(14) \qquad \mathbf{x}'(t) = \begin{bmatrix} 2 & -3 \\ 1 & -2 \end{bmatrix} \mathbf{x}(t) + \begin{bmatrix} e^{2t} \\ 1 \end{bmatrix}, \qquad \mathbf{x}(0) = \begin{bmatrix} -1 \\ 0 \end{bmatrix}.$$

Solution In Example 3 in Section 9.4, we found two linearly independent solutions to the corresponding homogeneous system; namely,

$$\mathbf{x}_1(t) = \begin{bmatrix} 3e^t \\ e^t \end{bmatrix} \quad \text{and} \quad \mathbf{x}_2(t) = \begin{bmatrix} e^{-t} \\ e^{-t} \end{bmatrix}.$$

Hence a fundamental matrix for the homogeneous system is

$$\mathbf{x}(t) = \begin{bmatrix} 3e^t & e^{-t} \\ e^t & e^{-t} \end{bmatrix}.$$

Although the solution to (14) can be found via the method of undetermined coefficients, we shall find it directly from formula (13). For this purpose, we need $\mathbf{X}^{-1}(t)$. One way[†] to obtain $\mathbf{X}^{-1}(t)$ is to form the augmented matrix

$$\begin{bmatrix} 3e^t & e^{-t} & \vdots & 1 & 0 \\ e^t & e^{-t} & \vdots & 0 & 1 \end{bmatrix}$$

and row-reduce this matrix to the matrix $[\mathbf{I} \mid \mathbf{X}^{-1}(t)]$. This gives

$$\mathbf{X}^{-1}(t) = \begin{bmatrix} \dfrac{1}{2}e^{-t} & -\dfrac{1}{2}e^{-t} \\ -\dfrac{1}{2}e^t & \dfrac{3}{2}e^t \end{bmatrix}.$$

[†] This procedure works for an invertible matrix of any dimension. For an arbitrary 2×2 invertible matrix $\mathbf{U}(t)$, a formula for $\mathbf{U}^{-1}(t)$ is derived in Problem 32.

Substituting into formula (13), we obtain the solution

$$\mathbf{x}(t) = \begin{bmatrix} 3e^t & e^{-t} \\ e^t & e^{-t} \end{bmatrix} \begin{bmatrix} \dfrac{1}{2} & -\dfrac{1}{2} \\ -\dfrac{1}{2} & \dfrac{3}{2} \end{bmatrix} \begin{bmatrix} -1 \\ 0 \end{bmatrix}$$

$$+ \begin{bmatrix} 3e^t & e^{-t} \\ e^t & e^{-t} \end{bmatrix} \int_0^t \begin{bmatrix} \dfrac{1}{2}e^{-s} & -\dfrac{1}{2}e^{-s} \\ -\dfrac{1}{2}e^s & \dfrac{3}{2}e^s \end{bmatrix} \begin{bmatrix} e^{2s} \\ 1 \end{bmatrix} ds$$

$$= \begin{bmatrix} -\dfrac{3}{2}e^t + \dfrac{1}{2}e^{-t} \\ -\dfrac{1}{2}e^t + \dfrac{1}{2}e^{-t} \end{bmatrix} + \begin{bmatrix} 3e^t & e^{-t} \\ e^t & e^{-t} \end{bmatrix} \int_0^t \begin{bmatrix} \dfrac{1}{2}e^s - \dfrac{1}{2}e^{-s} \\ \dfrac{1}{2}e^{3s} + \dfrac{3}{2}e^s \end{bmatrix} ds$$

$$= \begin{bmatrix} -\dfrac{3}{2}e^t + \dfrac{1}{2}e^{-t} \\ -\dfrac{1}{2}e^t + \dfrac{1}{2}e^{-t} \end{bmatrix} + \begin{bmatrix} 3e^t & e^{-t} \\ e^t & e^{-t} \end{bmatrix} \begin{bmatrix} \dfrac{1}{2}e^t + \dfrac{1}{2}e^{-t} - 1 \\ \dfrac{3}{2}e^t - \dfrac{1}{6}e^{3t} - \dfrac{4}{3} \end{bmatrix}$$

$$= \begin{bmatrix} -\dfrac{9}{2}e^t - \dfrac{5}{6}e^{-t} + \dfrac{4}{3}e^{2t} + 3 \\ -\dfrac{3}{2}e^t - \dfrac{5}{6}e^{-t} + \dfrac{1}{3}e^{2t} + 2 \end{bmatrix} . \quad \blacksquare \text{ End}$$

EXERCISES 9.6

In Problems 1 through 4, use the method of undetermined coefficients to find a general solution to the system $\mathbf{x}'(t) = \mathbf{A}\mathbf{x}(t) + \mathbf{f}(t)$, where $\mathbf{A}$ and $\mathbf{f}(t)$ are given.

1. $\mathbf{A} = \begin{bmatrix} 6 & 1 \\ 4 & 3 \end{bmatrix}$, $\mathbf{f}(t) = \begin{bmatrix} -11 \\ -5 \end{bmatrix}$.

2. $\mathbf{A} = \begin{bmatrix} 1 & 1 \\ 4 & 1 \end{bmatrix}$, $\mathbf{f}(t) = \begin{bmatrix} -t - 1 \\ -4t - 2 \end{bmatrix}$.

3. $\mathbf{A} = \begin{bmatrix} 1 & -2 & 2 \\ -2 & 1 & 2 \\ 2 & 2 & 1 \end{bmatrix}$, $\mathbf{f}(t) = \begin{bmatrix} 2e^t \\ 4e^t \\ -2e^t \end{bmatrix}$.

4. $\mathbf{A} = \begin{bmatrix} 2 & 2 \\ 2 & 2 \end{bmatrix}$, $\mathbf{f}(t) = \begin{bmatrix} -4\cos t \\ -\sin t \end{bmatrix}$.

In Problems 5 through 10, use the method of undetermined coefficients (Table 4.1 on inside back cover) to determine only the form of a particular solution for the system $\mathbf{x}'(t) = \mathbf{A}\mathbf{x}(t) + \mathbf{f}(t)$, where $\mathbf{A}$ and $\mathbf{f}(t)$ are given.

5. $\mathbf{A} = \begin{bmatrix} 1 & 1 \\ 0 & 2 \end{bmatrix}$, $\mathbf{f}(t) = e^{-2t}\begin{bmatrix} t \\ 3 \end{bmatrix}$.

6. $\mathbf{A} = \begin{bmatrix} 1 & -1 \\ 0 & -1 \end{bmatrix}$, $\mathbf{f}(t) = \begin{bmatrix} t \\ e^{3t} \end{bmatrix}$.

7. $\mathbf{A} = \begin{bmatrix} 0 & 1 \\ 2 & 0 \end{bmatrix}$, $\mathbf{f}(t) = \begin{bmatrix} \sin 3t \\ t \end{bmatrix}$.

8. $\mathbf{A} = \begin{bmatrix} -1 & 0 \\ 2 & 2 \end{bmatrix}$, $\mathbf{f}(t) = \begin{bmatrix} t^2 \\ t + 1 \end{bmatrix}$.

9. $\mathbf{A} = \begin{bmatrix} 1 & 1 \\ 0 & 1 \end{bmatrix}$, $\mathbf{f}(t) = \begin{bmatrix} e^{2t} \\ e^{3t} \end{bmatrix}$.

10. $\mathbf{A} = \begin{bmatrix} 2 & -1 \\ 1 & 5 \end{bmatrix}$, $\mathbf{f}(t) = \begin{bmatrix} te^{-t} \\ 3e^{-t} \end{bmatrix}$.

In Problems 11 through 16, use the variation of parameters formula (11) to find a general solution of the system $\mathbf{x}'(t) = \mathbf{A}\mathbf{x}(t) + \mathbf{f}(t)$, where $\mathbf{A}$ and $\mathbf{f}(t)$ are given.

11. $\mathbf{A} = \begin{bmatrix} 0 & 1 \\ -1 & 0 \end{bmatrix}$, $\mathbf{f}(t) = \begin{bmatrix} 1 \\ 0 \end{bmatrix}$.

12. $\mathbf{A} = \begin{bmatrix} 1 & 2 \\ 3 & 2 \end{bmatrix}$, $\mathbf{f}(t) = \begin{bmatrix} 1 \\ -1 \end{bmatrix}$.

13. $\mathbf{A} = \begin{bmatrix} 2 & 1 \\ -3 & -2 \end{bmatrix}$, $\quad \mathbf{f}(t) = \begin{bmatrix} 2e^t \\ 4e^t \end{bmatrix}$.

14. $\mathbf{A} = \begin{bmatrix} 0 & -1 \\ 1 & 0 \end{bmatrix}$, $\quad \mathbf{f}(t) = \begin{bmatrix} t^2 \\ 1 \end{bmatrix}$.

15. $\mathbf{A} = \begin{bmatrix} -4 & 2 \\ 2 & -1 \end{bmatrix}$, $\quad \mathbf{f}(t) = \begin{bmatrix} t^{-1} \\ 4 + 2t^{-1} \end{bmatrix}$.

16. $\mathbf{A} = \begin{bmatrix} 0 & 1 \\ -1 & 0 \end{bmatrix}$, $\quad \mathbf{f}(t) = \begin{bmatrix} 8\sin t \\ 0 \end{bmatrix}$.

In Problems 17 through 20, use the variation of parameters formula (11) and possibly a linear algebra software package to find a general solution of the system $\mathbf{x}'(t) = \mathbf{A}\mathbf{x}(t) + \mathbf{f}(t)$, *where* $\mathbf{A}$ *and* $\mathbf{f}(t)$ *are given.*

17. $\mathbf{A} = \begin{bmatrix} 0 & 1 & 1 \\ 1 & 0 & 1 \\ 1 & 1 & 0 \end{bmatrix}$, $\quad \mathbf{f}(t) = \begin{bmatrix} 3e^t \\ -e^t \\ -e^t \end{bmatrix}$.

18. $\mathbf{A} = \begin{bmatrix} 1 & -1 & 1 \\ 0 & 0 & 1 \\ 0 & -1 & 2 \end{bmatrix}$, $\quad \mathbf{f}(t) = \begin{bmatrix} 0 \\ e^t \\ e^t \end{bmatrix}$.

19. $\mathbf{A} = \begin{bmatrix} 0 & 1 & 0 & 0 \\ -1 & 0 & 0 & 0 \\ 0 & 0 & 0 & 1 \\ 0 & 0 & 1 & 0 \end{bmatrix}$ $\quad \mathbf{f}(t) = \begin{bmatrix} t \\ 0 \\ e^{-t} \\ t \end{bmatrix}$.

20. $\mathbf{A} = \begin{bmatrix} 0 & 1 & 0 & 0 \\ 0 & 0 & 1 & 0 \\ 0 & 0 & 0 & 1 \\ 8 & -4 & -2 & -1 \end{bmatrix}$, $\quad \mathbf{f}(t) = \begin{bmatrix} e^t \\ 0 \\ 1 \\ 0 \end{bmatrix}$.

In Problems 21 and 22, find the solution to the given system that satisfies the given initial condition.

21. $\mathbf{x}'(t) = \begin{bmatrix} 0 & 2 \\ -1 & 3 \end{bmatrix} \mathbf{x}(t) + \begin{bmatrix} e^t \\ -e^t \end{bmatrix}$,

(a) $\mathbf{x}(0) = \begin{bmatrix} 5 \\ 4 \end{bmatrix}$. (b) $\mathbf{x}(1) = \begin{bmatrix} 0 \\ 1 \end{bmatrix}$.

(c) $\mathbf{x}(5) = \begin{bmatrix} 1 \\ 0 \end{bmatrix}$. (d) $\mathbf{x}(-1) = \begin{bmatrix} -4 \\ 5 \end{bmatrix}$.

22. $\mathbf{x}'(t) = \begin{bmatrix} 0 & 2 \\ 4 & -2 \end{bmatrix} \mathbf{x}(t) + \begin{bmatrix} 4t \\ -4t - 2 \end{bmatrix}$,

(a) $\mathbf{x}(0) = \begin{bmatrix} 4 \\ -5 \end{bmatrix}$. (b) $\mathbf{x}(2) = \begin{bmatrix} 1 \\ 1 \end{bmatrix}$.

23. Using matrix algebra techniques and the method of undetermined coefficients, find a general solution for
$$x''(t) + y'(t) - x(t) + y(t) = -1,$$
$$x'(t) + y'(t) - x(t) = t^2.$$
Compare your solution with the solution in Example 3 in Section 4.10.

24. Using matrix algebra techniques and the method of undetermined coefficients, solve the initial value problem
$$x'(t) - 2y(t) = 4t, \qquad x(0) = 4,$$
$$y'(t) + 2y(t) - 4x(t) = -4t - 2, \qquad y(0) = -5.$$
Compare your solution with the solution in Example 1 in Section 7.9.

25. To find a general solution to the system
$$\mathbf{x}'(t) = \begin{bmatrix} 0 & 1 \\ -2 & 3 \end{bmatrix} \mathbf{x}(t) + \mathbf{f}(t), \quad \text{where} \quad \mathbf{f}(t) = \begin{bmatrix} e^t \\ 0 \end{bmatrix},$$
proceed as follows:

(a) Find a fundamental solution set for the corresponding homogeneous system.
(b) The obvious choice for a particular solution would be a vector function of the form $\mathbf{x}_p(t) = e^t\mathbf{a}$; however, the homogeneous system has a solution of this form. The next choice would be $\mathbf{x}_p(t) = te^t\mathbf{a}$. Show that this choice does *not* work.
(c) For systems, multiplying by t is not always sufficient. The proper guess is
$$\mathbf{x}_p(t) = te^t\mathbf{a} + e^t\mathbf{b}.$$
Use this guess to find a particular solution of the given system.
(d) Use the results of parts (a) and (c) to find a general solution of the given system.

26. For the system of Problem 25, we found that a proper guess for a particular solution is $\mathbf{x}_p(t) = te^t\mathbf{a} + e^t\mathbf{b}$. In some cases $\mathbf{a}$ or $\mathbf{b}$ may be zero.

(a) Find a particular solution for the system of Problem 25 if $\mathbf{f}(t) = \text{col}(3e^t, 6e^t)$.
(b) Find a particular solution for the system of Problem 25 if $\mathbf{f}(t) = \text{col}(e^t, e^t)$.

27. Find a general solution of the system
$$\mathbf{x}'(t) = \begin{bmatrix} 0 & 1 & 1 \\ 1 & 0 & 1 \\ 1 & 1 & 0 \end{bmatrix} \mathbf{x}(t) + \begin{bmatrix} -1 \\ -1 - e^{-t} \\ -2e^{-t} \end{bmatrix}.$$
[*Hint:* Use superposition to determine the particular solution.]

28. Find a particular solution for the system

$$\mathbf{x}'(t) = \begin{bmatrix} 1 & -1 \\ -1 & 1 \end{bmatrix} \mathbf{x}(t) + \begin{bmatrix} -3 \\ 1 \end{bmatrix}.$$

[*Hint:* Try $\mathbf{x}_p(t) = t\mathbf{a} + \mathbf{b}$.]

In Problems 29 and 30, find a general solution to the given Cauchy-Euler system for t > 0. Remember to express the system in the form $\mathbf{x}'(t) = \mathbf{A}(t)\mathbf{x}(t) + \mathbf{f}(t)$ *before using the variation-of-parameters formula.*

29. $t\mathbf{x}'(t) = \begin{bmatrix} 2 & -1 \\ 3 & -2 \end{bmatrix} \mathbf{x}(t) + \begin{bmatrix} t^{-1} \\ 1 \end{bmatrix}.$

30. $t\mathbf{x}'(t) = \begin{bmatrix} 4 & -3 \\ 8 & -6 \end{bmatrix} \mathbf{x}(t) + \begin{bmatrix} t \\ 2t \end{bmatrix}.$

31. Use the variation-of-parameters formula (10) to derive a formula for a particular solution y_p to the scalar equation $y'' + p(t)y' + q(t)y = g(t)$ in terms of two linearly independent solutions $y_1(t)$, $y_2(t)$ of the corresponding homogeneous equation. Show that your answer agrees with the formulas derived in Section 4.9. [*Hint:* First write the scalar equation in system form.]

32. Let $\mathbf{U}(t)$ be the invertible 2×2 matrix

$$\mathbf{U}(t) := \begin{bmatrix} a(t) & b(t) \\ c(t) & d(t) \end{bmatrix}.$$

Show that

$$\mathbf{U}^{-1}(t) = \frac{1}{[a(t)d(t) - b(t)c(t)]} \begin{bmatrix} d(t) & -b(t) \\ -c(t) & a(t) \end{bmatrix}.$$

33. RL Network. The currents in the RL network given by the schematic diagram in Figure 9.8 are governed by the following equations:

$$2I_1'(t) + 90I_2(t) = 9,$$

$$I_3'(t) + 30I_4(t) - 90I_2(t) = 0,$$

$$60I_5(t) - 30I_4(t) = 0,$$

$$I_1(t) = I_2(t) + I_3(t),$$

$$I_3(t) = I_4(t) + I_5(t).$$

Figure 9.8 RL network for Problem 33

Assume the currents are initially zero. Solve for the five currents $I_1, \ldots, I_5$. [*Hint:* Eliminate all unknowns except I_2 and I_5, and form a normal system with $x_1 = I_2$ and $x_2 = I_5$.]

34. Conventional Combat Model. When one models a pair of conventional forces in combat, the following system arises

$$\mathbf{x}' = \begin{bmatrix} -a & -b \\ -c & -d \end{bmatrix} \mathbf{x} + \begin{bmatrix} p \\ q \end{bmatrix},$$

where $\mathbf{x} = \text{col}(x_1, x_2)$. The variables $x_1(t)$ and $x_2(t)$ represent the strengths of opposing forces at time t. The terms $-ax_1$ and $-dx_2$ represent the *operational loss rates*, and the terms $-bx_2$ and $-cx_1$ represent the *combat loss rates* for the troops x_1 and x_2, respectively. The constants p and q represent the respective rates of reinforcement. Let $a = 1$, $b = 4$, $c = 3$, $d = 2$, and $p = q = 5$. By solving the appropriate initial value problem, determine which forces will win if

(a) $x_1(0) = 20$, $x_2(0) = 20$.
(b) $x_1(0) = 21$, $x_2(0) = 20$.
(c) $x_1(0) = 20$, $x_2(0) = 21$.

35. Mixing Problem. Two tanks A and B, each holding 50 L of liquid, are interconnected by pipes. The liquid flows from tank A into tank B at a rate of 4 L/min and from B into A at a rate of 1 L/min (see Figure 9.9). The liquid inside each tank is kept well stirred. A brine solution that has a concentration of 2 kg/L of salt flows into tank A at a rate of 4 L/min. A brine solution that has a concentration of 1 kg/L of salt flows into tank B at a rate of 1 L/min. The solutions flow out of the system from both tanks—from tank A at 1 L/min and from tank B at 4 L/min. If, initially, tank A contains only water and tank B contains 5 kg of salt, determine the mass of salt in each tank at time $t \geq 0$. After several minutes have elapsed, which tank has the higher concentration of salt? What is its limiting concentration?

Figure 9.9 Mixing problem for interconnected tanks

9.7 THE MATRIX EXPONENTIAL FUNCTION

In this chapter we have developed various ways to extend techniques for scalar differential equations to systems. In this section we take a substantial step further by showing that with the right notation, the formulas for solving normal systems with constant coefficients are identical to the formulas for solving first order equations with constant coefficients. For example, we know that a general solution to the equation $x'(t) = ax(t)$, where a is a constant, is $x(t) = ce^{at}$. Analogously, we shall show that a general solution to the normal system

(1) $\qquad \mathbf{x}'(t) = \mathbf{A}\mathbf{x}(t),$

where $\mathbf{A}$ is a constant $n \times n$ matrix, is $\mathbf{x}(t) = e^{\mathbf{A}t}\mathbf{c}$. Our first task is to define the matrix exponential $e^{\mathbf{A}t}$.

If $\mathbf{A}$ is a constant $n \times n$ matrix, we define $e^{\mathbf{A}t}$ by taking the series expansion for e^{at} and replacing a by $\mathbf{A}$; that is,

(2) $\qquad e^{\mathbf{A}t} := \mathbf{I} + \mathbf{A}t + \mathbf{A}^2\dfrac{t^2}{2!} + \cdots + \mathbf{A}^n\dfrac{t^n}{n!} + \cdots.$

(Note that we also replace 1 by $\mathbf{I}$.) By the right-hand side of (2), we mean the $n \times n$ matrix whose elements are power series with coefficients given by the corresponding entries in the matrices $\mathbf{I}$, $\mathbf{A}$, $\mathbf{A}^2/2!$, $\dots$.

If $\mathbf{A}$ is a diagonal matrix, then the computation of $e^{\mathbf{A}t}$ is straightforward. For example, if

$$\mathbf{A} = \begin{bmatrix} -1 & 0 \\ 0 & 2 \end{bmatrix},$$

then

$$\mathbf{A}^2 = \mathbf{AA} = \begin{bmatrix} 1 & 0 \\ 0 & 4 \end{bmatrix}, \qquad \mathbf{A}^3 = \begin{bmatrix} -1 & 0 \\ 0 & 8 \end{bmatrix}, \qquad \dots, \qquad \mathbf{A}^n = \begin{bmatrix} (-1)^n & 0 \\ 0 & 2^n \end{bmatrix},$$

and so

$$e^{\mathbf{A}t} = \sum_{n=0}^{\infty} \mathbf{A}^n\frac{t^n}{n!} = \begin{bmatrix} \displaystyle\sum_{n=0}^{\infty} (-1)^n\frac{t^n}{n!} & 0 \\ 0 & \displaystyle\sum_{n=0}^{\infty} 2^n\frac{t^n}{n!} \end{bmatrix} = \begin{bmatrix} e^{-t} & 0 \\ 0 & e^{2t} \end{bmatrix}.$$

More generally, if $\mathbf{A}$ is an $n \times n$ *diagonal* matrix with $r_1, r_2, \dots, r_n$ down its main diagonal, then $e^{\mathbf{A}t}$ is the diagonal matrix with $e^{r_1 t}, e^{r_2 t}, \dots, e^{r_n t}$ down its main diagonal (see Problem 26). If $\mathbf{A}$ is not a diagonal matrix, the computation of $e^{\mathbf{A}t}$ is more involved. We deal with this important problem later in this section.

It can be shown that the series (2) converges for all t and has many of the same properties[†] as the scalar exponential e^{at}.

[†] For proofs of these and other properties of the matrix exponential function, see *Matrices and Linear Transformations*, by Charles G. Cullen (Addison-Wesley Publishing Co., Reading, Mass., 1972), Chapter 8. See also the amusing article ''Nineteen Dubious Ways to Compute the Exponential of a Matrix,'' by Cleve Moler and Charles van Loan, *SIAM Review*, Vol. 20, No. 4 (Oct. 1978).

PROPERTIES OF THE MATRIX EXPONENTIAL FUNCTION

Theorem 7. Let $\mathbf{A}$ and $\mathbf{B}$ be $n \times n$ constant matrices and r, s, and t be real (or complex) numbers. Then,

(a) $e^{\mathbf{A}0} = e^{\mathbf{0}} = \mathbf{I}$.

(b) $e^{\mathbf{A}(t+s)} = e^{\mathbf{A}t}e^{\mathbf{A}s}$.

(c) $(e^{\mathbf{A}t})^{-1} = e^{-\mathbf{A}t}$.

(d) $e^{(\mathbf{A}+\mathbf{B})t} = e^{\mathbf{A}t}e^{\mathbf{B}t}$, provided that $\mathbf{A}\mathbf{B} = \mathbf{B}\mathbf{A}$.

(e) $e^{r\mathbf{I}t} = e^{rt}\mathbf{I}$.

Property (c) has profound implications. First, it asserts that for any matrix $\mathbf{A}$, the matrix $e^{\mathbf{A}t}$ *has* an inverse for all t. Moreover, this inverse is obtained by simply replacing t by $-t$. In applying property (d) (the law of exponents), one must exercise care because of the stipulation that the matrices $\mathbf{A}$ and $\mathbf{B}$ commute (see Problem 25).

Another important property of the matrix exponential arises from the fact that we can differentiate the series in (2) term by term. This gives

$$\frac{d}{dt}(e^{\mathbf{A}t}) = \frac{d}{dt}\left(\mathbf{I} + \mathbf{A}t + \mathbf{A}^2\frac{t^2}{2} + \cdots + \mathbf{A}^n\frac{t^n}{n!} + \cdots\right)$$

$$= \mathbf{A} + \mathbf{A}^2t + \mathbf{A}^3\frac{t^2}{2} + \cdots + \mathbf{A}^n\frac{t^{n-1}}{(n-1)!} + \cdots$$

$$= \mathbf{A}\left[\mathbf{I} + \mathbf{A}t + \mathbf{A}^2\frac{t^2}{2} + \cdots + \mathbf{A}^{n-1}\frac{t^{n-1}}{(n-1)!} + \cdots\right].$$

Hence

$$\frac{d}{dt}(e^{\mathbf{A}t}) = \mathbf{A}e^{\mathbf{A}t},$$

and so $e^{\mathbf{A}t}$ *is a solution to the matrix differential equation* $\mathbf{X}' = \mathbf{A}\mathbf{X}$. Since $e^{\mathbf{A}t}$ is invertible (property (c)), it follows that the columns of $e^{\mathbf{A}t}$ are linearly independent solutions to system (1). Combining these facts we have the following.

$e^{\mathbf{A}t}$ IS A FUNDAMENTAL MATRIX ·

Theorem 8. If $\mathbf{A}$ is an $n \times n$ constant matrix, then the columns of the matrix exponential $e^{\mathbf{A}t}$ form a fundamental solution set for the system $\mathbf{x}'(t) = \mathbf{A}\mathbf{x}(t)$. Therefore $e^{\mathbf{A}t}$ is a fundamental matrix for the system, and a general solution is $\mathbf{x}(t) = e^{\mathbf{A}t}\mathbf{c}$.

Knowing that $e^{\mathbf{A}t}$ is a fundamental matrix is of practical use provided we can calculate $e^{\mathbf{A}t}$. As we observed, if $\mathbf{A}$ is a diagonal matrix, then we simply exponentiate the diagonal elements (times t) to obtain $e^{\mathbf{A}t}$. Also, if $\mathbf{B}$ is a **nilpotent** matrix, that is, $\mathbf{B}^k = \mathbf{0}$ for

some positive integer k, then the series for $e^{\mathbf{B}t}$ has only a finite number of terms, since $\mathbf{B}^k = \mathbf{B}^{k+1} = \cdots = \mathbf{0}$. In such cases, $e^{\mathbf{B}t}$ reduces to

$$e^{\mathbf{B}t} = \mathbf{I} + \mathbf{B}t + \cdots + \mathbf{B}^{k-1}\frac{t^{k-1}}{(k-1)!}.$$

Using the law of exponents and this fact about nilpotent matrices, we can determine $e^{\mathbf{A}t}$ for a special class of matrices. Let r be a scalar. Since

$$e^{\mathbf{A}t} = e^{r\mathbf{I}t}e^{(\mathbf{A}-r\mathbf{I})t} = e^{rt}e^{(\mathbf{A}-r\mathbf{I})t},$$

we get a finite representation for $e^{\mathbf{A}t}$ if $\mathbf{B} = \mathbf{A} - r\mathbf{I}$ is nilpotent for some r. In fact, when the characteristic polynomial for $\mathbf{A}$ has the form $p(r) = (r_1 - r)^n$, that is, when $\mathbf{A}$ has one eigenvalue r_1 of multiplicity n, it is a consequence of the Cayley-Hamilton theorem[†] that $(r_1\mathbf{I} - \mathbf{A})^n = \mathbf{0}$. Hence $\mathbf{A} - r_1\mathbf{I}$ is nilpotent and

$$e^{\mathbf{A}t} = e^{r_1 t}\left\{\mathbf{I} + (\mathbf{A} - r_1\mathbf{I})t + \cdots + (\mathbf{A} - r_1\mathbf{I})^{n-1}\frac{t^{n-1}}{(n-1)!}\right\}.$$

EXAMPLE 1 Find the fundamental matrix $e^{\mathbf{A}t}$ for the system

(3) $\mathbf{x}' = \mathbf{A}\mathbf{x}$, where $\mathbf{A} = \begin{bmatrix} 2 & 1 & 1 \\ 1 & 2 & 1 \\ -2 & -2 & -1 \end{bmatrix}$.

Solution We begin by computing the characteristic polynomial for $\mathbf{A}$:

$$p(r) = |\mathbf{A} - r\mathbf{I}| = \begin{vmatrix} 2-r & 1 & 1 \\ 1 & 2-r & 1 \\ -2 & -2 & -1-r \end{vmatrix} = -r^3 + 3r^2 - 3r + 1 = -(r-1)^3.$$

Thus $r = 1$ is an eigenvalue of $\mathbf{A}$ with multiplicity 3. By the Cayley-Hamilton theorem, $(\mathbf{A} - \mathbf{I})^3 = \mathbf{0}$, and so

(4) $e^{\mathbf{A}t} = e^t e^{(\mathbf{A}-\mathbf{I})t} = e^t\left\{\mathbf{I} + (\mathbf{A} - \mathbf{I})t + (\mathbf{A} - \mathbf{I})^2\frac{t^2}{2}\right\}.$

Computing, we find

$$\mathbf{A} - \mathbf{I} = \begin{bmatrix} 1 & 1 & 1 \\ 1 & 1 & 1 \\ -2 & -2 & -2 \end{bmatrix} \quad \text{and} \quad (\mathbf{A} - \mathbf{I})^2 = \begin{bmatrix} 0 & 0 & 0 \\ 0 & 0 & 0 \\ 0 & 0 & 0 \end{bmatrix}.$$

Substituting into (4) yields

(5) $e^{\mathbf{A}t} = e^t\begin{bmatrix} 1 & 0 & 0 \\ 0 & 1 & 0 \\ 0 & 0 & 1 \end{bmatrix} + te^t\begin{bmatrix} 1 & 1 & 1 \\ 1 & 1 & 1 \\ -2 & -2 & -2 \end{bmatrix} = \begin{bmatrix} e^t + te^t & te^t & te^t \\ te^t & e^t + te^t & te^t \\ -2te^t & -2te^t & e^t - 2te^t \end{bmatrix}.$ ∎ End

[†] The Cayley-Hamilton theorem states that a matrix satisfies its own characteristic equation, that is, $p(\mathbf{A}) = \mathbf{0}$. For a discussion of this theorem, see *Matrices and Linear Transformations*, by Charles G. Cullen (Addison-Wesley, Reading, Mass., 1972), Chapter 5.

In the preceding example we used the nilpotency of the matrix $\mathbf{A} - r\mathbf{I}$ to compute $e^{\mathbf{A}t}$ directly. In general, we cannot expect nilpotency to hold, but we can take advantage of the following relationship between fundamental matrices to help compute $e^{\mathbf{A}t}$.

RELATIONSHIP BETWEEN FUNDAMENTAL MATRICES

Lemma 1. Let $\mathbf{X}(t)$ and $\mathbf{Y}(t)$ be two fundamental matrices for the same system $\mathbf{x}' = \mathbf{A}\mathbf{x}$. Then, there exists a constant matrix $\mathbf{C}$ such that $\mathbf{X}(t) = \mathbf{Y}(t)\mathbf{C}$.

Proof. Let $\mathbf{x}_1(t), \ldots, \mathbf{x}_n(t)$ be the columns of $\mathbf{X}(t)$ and let $\mathbf{y}_1(t), \ldots, \mathbf{y}_n(t)$ be the columns of $\mathbf{Y}(t)$. Since $\{\mathbf{y}_1(t), \ldots, \mathbf{y}_n(t)\}$ is a fundamental solution set and $\mathbf{x}_j(t)$, $j = 1, \ldots, n$ are solutions to $\mathbf{x}' = \mathbf{A}\mathbf{x}$, there exist constants $c_{1j}, c_{2j}, \ldots, c_{nj}$ such that

$$\mathbf{x}_j(t) = c_{1j}\mathbf{y}_1(t) + c_{2j}\mathbf{y}_2(t) + \cdots + c_{nj}\mathbf{y}_n(t)$$

for each $j = 1, \ldots, n$. But this is equivalent to writing $\mathbf{X}(t) = \mathbf{Y}(t)\mathbf{C}$, where $\mathbf{C} = [c_{ij}]$. ■ End

We now use Lemma 1 to find a formula for $e^{\mathbf{A}t}$ when a fundamental matrix $\mathbf{X}(t)$ for $\mathbf{x}' = \mathbf{A}\mathbf{x}$ is known. Since $e^{\mathbf{A}t}$ is also a fundamental matrix for the system, Lemma 1 asserts that $e^{\mathbf{A}t} = \mathbf{X}(t)\mathbf{C}$ for some constant matrix $\mathbf{C}$. Setting $t = 0$ yields $\mathbf{I} = \mathbf{X}(0)\mathbf{C}$, and solving for $\mathbf{C}$ we get $\mathbf{C} = \mathbf{X}^{-1}(0)$. Hence

(6) $e^{\mathbf{A}t} = \mathbf{X}(t)\mathbf{X}^{-1}(0)$.

Although this formula is useful, it does place the burden of determining $e^{\mathbf{A}t}$ on finding a fundamental matrix $\mathbf{X}(t)$. Fortunately, we can use the *properties* of the matrix exponential $e^{\mathbf{A}t}$ to help simplify this task. Because the columns of a fundamental matrix must have the form $e^{\mathbf{A}t}\mathbf{u}$, let's try to find n vectors $\mathbf{u}$ for which the computation of $e^{\mathbf{A}t}\mathbf{u}$ is tractable.

We begin by using the relation $e^{\mathbf{A}t} = e^{rt(\mathbf{A}-r\mathbf{I})t}$ to express $e^{\mathbf{A}t}\mathbf{u}$ as

(7) $e^{\mathbf{A}t}\mathbf{u} = e^{rt}e^{(\mathbf{A}-r\mathbf{I})t}\mathbf{u}$

$$= e^{rt}\left\{\mathbf{u} + t(\mathbf{A} - r\mathbf{I})\mathbf{u} + \cdots + \frac{t^k}{k!}(\mathbf{A} - r\mathbf{I})^k\mathbf{u} + \cdots\right\}.$$

Now we know that if r is an eigenvalue of $\mathbf{A}$ and $\mathbf{u}$ is a corresponding eigenvector, then $e^{rt}\mathbf{u}$ is a solution to (1). Indeed, in this situation $(\mathbf{A} - r\mathbf{I})\mathbf{u} = (\mathbf{A} - r\mathbf{I})^2\mathbf{u} = \cdots = \mathbf{0}$. Hence the series in (7) reduces to the first term, $e^{rt}\mathbf{u}$. While it is too much to expect $\mathbf{A} - r\mathbf{I}$ to be nilpotent, it is not too much to ask that $(\mathbf{A} - r\mathbf{I})^k\mathbf{u} = \mathbf{0}$ for some nontrivial vector $\mathbf{u}$ and some positive integer k.

GENERALIZED EIGENVECTORS

Definition 5. Let $\mathbf{A}$ be an $n \times n$ constant matrix and r be an eigenvalue of $\mathbf{A}$. A nontrivial vector $\mathbf{u}$ that satisfies

$$(\mathbf{A} - r\mathbf{I})^k\mathbf{u} = \mathbf{0}$$

for some positive integer k is called a **generalized eigenvector** associated with r.

A consequence of the primary decomposition theorem in advanced linear algebra[†] is that if the characteristic polynomial for **A** is

$$p(r) = (r_1 - r)^{m_1} \cdots (r_k - r)^{m_k},$$

where the r_i's are distinct eigenvalues of **A** and m_i is the multiplicity of the eigenvalue r_i, then for each i there exist m_i linearly independent generalized eigenvectors associated with r_i, and the combined set of $n = m_1 + \cdots + m_k$ generalized eigenvectors is linearly independent. Moreover, if **u** is a generalized eigenvector associated with r_i, then $(\mathbf{A} - r_i\mathbf{I})^{m_i}\mathbf{u} = \mathbf{0}$. This leads to the following procedure for finding n linearly independent solutions to system (1).

SOLVING SYSTEMS USING GENERALIZED EIGENVECTORS

To obtain a fundamental solution set for $\mathbf{x}' = \mathbf{A}\mathbf{x}$:

(a) Compute the characteristic polynomial $p(t) = |\mathbf{A} - r\mathbf{I}|$ and find the distinct eigenvalues $r_1, \ldots, r_k$.

(b) For each eigenvalue r_i, find m_i linearly independent generalized eigenvectors, where m_i is the multiplicity of the eigenvalue r_i.

(c) Use the n linearly independent generalized eigenvectors obtained in (b) to compute the n linearly independent solutions to $\mathbf{x}' = \mathbf{A}\mathbf{x}$ of the form

$$(8) \qquad e^{\mathbf{A}t}\mathbf{u} = e^{rt}\left\{ \mathbf{u} + t(\mathbf{A} - r\mathbf{I})\mathbf{u} + \frac{t^2}{2}(\mathbf{A} - r\mathbf{I})^2\mathbf{u} + \cdots \right\},$$

where r is an eigenvalue and **u** is a corresponding generalized eigenvector. If r has multiplicity m_i, then the above series reduces to the first m_i terms.

EXAMPLE 2 Find the fundamental matrix $e^{\mathbf{A}t}$ for the system

$$(9) \qquad \mathbf{x}' = \mathbf{A}\mathbf{x}, \quad \text{where} \quad \mathbf{A} = \begin{bmatrix} 1 & 0 & 0 \\ 1 & 3 & 0 \\ 0 & 1 & 1 \end{bmatrix}.$$

Solution We begin by finding the characteristic polynomial for **A**:

$$p(r) = |\mathbf{A} - r\mathbf{I}| = \begin{vmatrix} 1 - r & 0 & 0 \\ 1 & 3 - r & 0 \\ 0 & 1 & 1 - r \end{vmatrix} = -(r - 1)^2(r - 3).$$

Hence the eigenvalues of **A** are $r = 1$ with multiplicity 2 and $r = 3$ with multiplicity 1.

Since $r = 1$ has multiplicity 2, we must determine two linearly independent associated generalized eigenvectors. We begin with $(\mathbf{A} - \mathbf{I})\mathbf{u} = \mathbf{0}$; that is,

$$\begin{bmatrix} 0 & 0 & 0 \\ 1 & 2 & 0 \\ 0 & 1 & 0 \end{bmatrix}\begin{bmatrix} u_1 \\ u_2 \\ u_3 \end{bmatrix} = \begin{bmatrix} 0 \\ 0 \\ 0 \end{bmatrix}.$$

[†] *Matrices and Linear Transformations*, by Charles G. Cullen, Ibid., Theorem 5.11, page 196.

Solving, we obtain $u_1 = u_2 = 0$ and $u_3 = s$, where s is arbitrary. Thus there is only one linearly independent eigenvector corresponding to $r = 1$, and with $s = 1$ we choose $\mathbf{u}_1 = \text{col}(0, 0, 1)$. Hence one solution to (9) is

$$(10) \qquad \mathbf{x}_1(t) = e^t \mathbf{u}_1 = e^t \begin{bmatrix} 0 \\ 0 \\ 1 \end{bmatrix} = \begin{bmatrix} 0 \\ 0 \\ e^t \end{bmatrix}.$$

Next we solve $(\mathbf{A} - \mathbf{I})^2 \mathbf{u} = \mathbf{0}$. From

$$(\mathbf{A} - \mathbf{I})^2 \mathbf{u} = \begin{bmatrix} 0 & 0 & 0 \\ 2 & 4 & 0 \\ 1 & 2 & 0 \end{bmatrix} \begin{bmatrix} u_1 \\ u_2 \\ u_3 \end{bmatrix} = \begin{bmatrix} 0 \\ 0 \\ 0 \end{bmatrix},$$

we find $u_2 = s$, $u_1 = -2u_2 = -2s$, and $u_3 = v$, where s and v are arbitrary. Taking $s = 1$ and $v = 0$, we obtain the generalized eigenvector $\mathbf{u}_2 = \text{col}(-2, 1, 0)$, which is linearly independent of $\mathbf{u}_1$. Now we use $\mathbf{u}_2$ to obtain a second solution to (9). Since $(\mathbf{A} - \mathbf{I})^2 \mathbf{u}_2 = \mathbf{0}$, formula (8) reduces to

$$(11) \qquad \mathbf{x}_2(t) = e^{\mathbf{A}t} \mathbf{u}_2 = e^t \{\mathbf{u}_2 + t(\mathbf{A} - \mathbf{I})\mathbf{u}_2\}$$

$$= e^t \begin{bmatrix} -2 \\ 1 \\ 0 \end{bmatrix} + te^t \begin{bmatrix} 0 & 0 & 0 \\ 1 & 2 & 0 \\ 0 & 1 & 0 \end{bmatrix} \begin{bmatrix} -2 \\ 1 \\ 0 \end{bmatrix}$$

$$= e^t \begin{bmatrix} -2 \\ 1 \\ 0 \end{bmatrix} + te^t \begin{bmatrix} 0 \\ 0 \\ 1 \end{bmatrix} = \begin{bmatrix} -2e^t \\ e^t \\ te^t \end{bmatrix}. ^\dagger$$

For the eigenvalue $r = 3$, we solve $(\mathbf{A} - 3\mathbf{I})\mathbf{u} = \mathbf{0}$, that is,

$$\begin{bmatrix} -2 & 0 & 0 \\ 1 & 0 & 0 \\ 0 & 1 & -2 \end{bmatrix} \begin{bmatrix} u_1 \\ u_2 \\ u_3 \end{bmatrix} = \begin{bmatrix} 0 \\ 0 \\ 0 \end{bmatrix},$$

to obtain the eigenvector $\mathbf{u}_3 = \text{col}(0, 2, 1)$. Hence a third linearly independent solution to (9) is

$$(12) \qquad \mathbf{x}_3(t) = e^{3t} \mathbf{u}_3 = e^{3t} \begin{bmatrix} 0 \\ 2 \\ 1 \end{bmatrix} = \begin{bmatrix} 0 \\ 2e^{3t} \\ e^{3t} \end{bmatrix}.$$

The matrix $\mathbf{X}(t)$ whose columns are the vectors $\mathbf{x}_1(t)$, $\mathbf{x}_2(t)$, and $\mathbf{x}_3(t)$ given in equations (10), (11), and (12),

$$\mathbf{X}(t) = \begin{bmatrix} 0 & -2e^t & 0 \\ 0 & e^t & 2e^{3t} \\ e^t & te^t & e^{3t} \end{bmatrix},$$

†Note that $\mathbf{x}_2(t)$ has been expressed in the format discussed in Problem 35 of Exercises 9.4, page 542.

is a fundamental matrix for (9). Setting $t = 0$ and then computing $\mathbf{X}^{-1}(0)$, we find

$$
\mathbf{X}(0) = \begin{bmatrix} 0 & -2 & 0 \\ 0 & 1 & 2 \\ 1 & 0 & 1 \end{bmatrix} \quad \text{and} \quad \mathbf{X}^{-1}(0) = \begin{bmatrix} -\dfrac{1}{4} & -\dfrac{1}{2} & 1 \\ -\dfrac{1}{2} & 0 & 0 \\ \dfrac{1}{4} & \dfrac{1}{2} & 0 \end{bmatrix}.
$$

It now follows from formula (6) that

$$
e^{\mathbf{A}t} = \mathbf{X}(t)\mathbf{X}^{-1}(0) = \begin{bmatrix} 0 & -2e^{t} & 0 \\ 0 & e^{t} & 2e^{3t} \\ e^{t} & te^{t} & e^{3t} \end{bmatrix} \begin{bmatrix} -\dfrac{1}{4} & -\dfrac{1}{2} & 1 \\ -\dfrac{1}{2} & 0 & 0 \\ \dfrac{1}{4} & \dfrac{1}{2} & 0 \end{bmatrix}
$$

$$
= \begin{bmatrix} e^{t} & 0 & 0 \\ -\dfrac{1}{2}e^{t} + \dfrac{1}{2}e^{3t} & e^{3t} & 0 \\ -\dfrac{1}{4}e^{t} - \dfrac{1}{2}te^{t} + \dfrac{1}{4}e^{3t} & -\dfrac{1}{2}e^{t} + \dfrac{1}{2}e^{3t} & e^{t} \end{bmatrix}. \quad \blacksquare \text{ End}
$$

Use of the fundamental matrix $e^{\mathbf{A}t}$ simplifies many computations. For example, the properties $e^{\mathbf{A}t}e^{-\mathbf{A}s} = e^{\mathbf{A}(t-s)}$ and $(e^{\mathbf{A}t_0})^{-1} = e^{-\mathbf{A}t_0}$ enable us to rewrite the variation of parameters formula (13) in Section 9.6 in a simpler form. Namely, the solution to the initial value problem $\mathbf{x}' = \mathbf{A}\mathbf{x} + \mathbf{f}(t)$, $\mathbf{x}(t_0) = \mathbf{x}_0$ is given by

$$
(13) \qquad \mathbf{x}(t) = e^{\mathbf{A}(t-t_0)}\mathbf{x}_0 + \int_{t_0}^{t} e^{\mathbf{A}(t-s)}\mathbf{f}(s)\,ds,
$$

which is a system version of the formula for the solution to the scalar initial value problem $x' = ax + f(t)$, $x(t_0) = x_0$.

In closing, we remark that the software packages for eigenvalue computation listed in Section 9.4 also contain subroutines for computing the matrix exponential.

EXERCISES 9.7

In Problems 1 through 6, (a) show that the given matrix $\mathbf{A}$ satisfies $(\mathbf{A} - r\mathbf{I})^{k} = \mathbf{0}$ for some number r and some positive integer k and (b) use this fact to determine the matrix $e^{\mathbf{A}t}$. [Hint: Compute the characteristic polynomial and use the Cayley-Hamilton theorem.]

1. $\mathbf{A} = \begin{bmatrix} 3 & -2 \\ 0 & 3 \end{bmatrix}$.

2. $\mathbf{A} = \begin{bmatrix} 1 & -1 \\ 1 & 3 \end{bmatrix}$.

3. $\mathbf{A} = \begin{bmatrix} 2 & 1 & -1 \\ -3 & -1 & 1 \\ 9 & 3 & -4 \end{bmatrix}$.

4. $\mathbf{A} = \begin{bmatrix} 2 & 1 & 3 \\ 0 & 2 & -1 \\ 0 & 0 & 2 \end{bmatrix}$.

5. $\mathbf{A} = \begin{bmatrix} -2 & 0 & 0 \\ 4 & -2 & 0 \\ 1 & 0 & -2 \end{bmatrix}$. **6.** $\mathbf{A} = \begin{bmatrix} 0 & 1 & 0 \\ 0 & 0 & 1 \\ -1 & -3 & -3 \end{bmatrix}$.

In Problems 7 through 10, determine $e^{\mathbf{A}t}$ by first finding a fundamental matrix $\mathbf{X}(t)$ for $\mathbf{x}' = \mathbf{A}\mathbf{x}$ and then using formula (6).

7. $\mathbf{A} = \begin{bmatrix} 0 & 1 \\ -1 & 0 \end{bmatrix}$. **8.** $\mathbf{A} = \begin{bmatrix} 1 & 1 \\ 4 & 1 \end{bmatrix}$.

9. $\mathbf{A} = \begin{bmatrix} 0 & 1 & 0 \\ 0 & 0 & 1 \\ 1 & -1 & 1 \end{bmatrix}$. **10.** $\mathbf{A} = \begin{bmatrix} 0 & 2 & 2 \\ 2 & 0 & 2 \\ 2 & 2 & 0 \end{bmatrix}$.

In Problems 11 and 12, determine $e^{\mathbf{A}t}$ by using generalized eigenvectors to find a fundamental matrix and then using formula (6).

11. $\mathbf{A} = \begin{bmatrix} 5 & -4 & 0 \\ 1 & 0 & 2 \\ 0 & 2 & 5 \end{bmatrix}$. **12.** $\mathbf{A} = \begin{bmatrix} 1 & 1 & 1 \\ 2 & 1 & -1 \\ 0 & -1 & 1 \end{bmatrix}$.

In Problems 13 through 16, use a linear algebra software package to help in determining $e^{\mathbf{A}t}$.

13. $\mathbf{A} = \begin{bmatrix} 0 & 1 & 0 & 0 & 0 \\ 0 & 0 & 1 & 0 & 0 \\ 1 & -3 & 3 & 0 & 0 \\ 0 & 0 & 0 & 0 & 1 \\ 0 & 0 & 0 & -1 & 0 \end{bmatrix}$.

14. $\mathbf{A} = \begin{bmatrix} 1 & 0 & 0 & 0 & 0 \\ 0 & 0 & 1 & 0 & 0 \\ 0 & -1 & -2 & 0 & 0 \\ 0 & 0 & 0 & 0 & 1 \\ 0 & 0 & 0 & -1 & 0 \end{bmatrix}$.

15. $\mathbf{A} = \begin{bmatrix} 0 & 1 & 0 & 0 & 0 \\ 0 & 0 & 1 & 0 & 0 \\ -1 & -3 & -3 & 0 & 0 \\ 0 & 0 & 0 & 0 & 1 \\ 0 & 0 & 0 & -4 & -4 \end{bmatrix}$.

16. $\mathbf{A} = \begin{bmatrix} -1 & 0 & 0 & 0 & 0 \\ 0 & 0 & 1 & 0 & 0 \\ 0 & -1 & -2 & 0 & 0 \\ 0 & 0 & 0 & 0 & 1 \\ 0 & 0 & 0 & -4 & -4 \end{bmatrix}$.

In Problems 17 through 20, use the generalized eigenvectors of $\mathbf{A}$ to find a general solution to the system $\mathbf{x}'(t) = \mathbf{A}\mathbf{x}(t)$, where $\mathbf{A}$ is given.

17. $\mathbf{A} = \begin{bmatrix} 0 & 1 & 0 \\ 0 & 0 & 1 \\ -2 & -5 & -4 \end{bmatrix}$. **18.** $\mathbf{A} = \begin{bmatrix} 0 & 0 & 1 \\ 0 & 1 & 2 \\ 0 & 0 & 1 \end{bmatrix}$.

19. $\mathbf{A} = \begin{bmatrix} 1 & 0 & 1 & 2 \\ 1 & 1 & 2 & 1 \\ 0 & 0 & 2 & 0 \\ 0 & 0 & 1 & 1 \end{bmatrix}$. **20.** $\mathbf{A} = \begin{bmatrix} -1 & -8 & 1 \\ -1 & -3 & 2 \\ -4 & -16 & 7 \end{bmatrix}$.

21. Use the results of Problem 5 to find the solution to the initial value problem

$$\mathbf{x}'(t) = \begin{bmatrix} -2 & 0 & 0 \\ 4 & -2 & 0 \\ 1 & 0 & -2 \end{bmatrix} \mathbf{x}(t), \quad \mathbf{x}(0) = \begin{bmatrix} 1 \\ 1 \\ -1 \end{bmatrix}.$$

22. Use your answer to Problem 12 to find the solution to the initial value problem

$$\mathbf{x}'(t) = \begin{bmatrix} 1 & 1 & 1 \\ 2 & 1 & -1 \\ 0 & -1 & 1 \end{bmatrix} \mathbf{x}(t), \quad \mathbf{x}(0) = \begin{bmatrix} -1 \\ 0 \\ 3 \end{bmatrix}.$$

23. Use the results of Problem 3 and the variation of parameters formula (13) to find the solution to the initial value problem

$$\mathbf{x}'(t) = \begin{bmatrix} 2 & 1 & -1 \\ -3 & -1 & 1 \\ 9 & 3 & -4 \end{bmatrix} \mathbf{x}(t) + \begin{bmatrix} 0 \\ t \\ 0 \end{bmatrix}, \quad \mathbf{x}(0) = \begin{bmatrix} 0 \\ 3 \\ 0 \end{bmatrix}.$$

24. Use your answer to Problem 9 and the variation of parameters formula (13) to find the solution to the initial value problem

$$\mathbf{x}'(t) = \begin{bmatrix} 0 & 1 & 0 \\ 0 & 0 & 1 \\ 1 & -1 & 1 \end{bmatrix} \mathbf{x}(t) + \begin{bmatrix} 0 \\ 0 \\ t \end{bmatrix}, \quad \mathbf{x}(0) = \begin{bmatrix} 1 \\ -1 \\ 0 \end{bmatrix}.$$

25. Let

$$\mathbf{A} = \begin{bmatrix} 1 & 2 \\ -1 & 3 \end{bmatrix} \quad \text{and} \quad \mathbf{B} = \begin{bmatrix} 2 & 1 \\ 0 & 1 \end{bmatrix}.$$

(a) Show that $\mathbf{AB} \neq \mathbf{BA}$.

(b) Show that property (d) in Theorem 7 does not hold for these matrices. That is, show that $e^{(\mathbf{A}+\mathbf{B})t} \neq e^{\mathbf{A}t}e^{\mathbf{B}t}$.

26. Let $\mathbf{A}$ be a diagonal $n \times n$ matrix with entries $r_1, \ldots, r_n$ down its main diagonal. To compute $e^{\mathbf{A}t}$, proceed as follows:

(a) Show that $\mathbf{A}^k$ is the diagonal matrix with entries $r_1^k, \ldots, r_n^k$ down its main diagonal.

(b) Use the result of part (a) to show that $e^{\mathbf{A}t}$ is the diagonal matrix with entries $e^{r_1 t}, \ldots, e^{r_n t}$ down its main diagonal.

27. In Problems 35–40 of Exercises 9.4, page 542, some *ad hoc* formulas were invoked to find general solutions to the system $\mathbf{x}' = \mathbf{A}\mathbf{x}$ when $\mathbf{A}$ had repeated eigenvalues. Using the generalized eigenvector procedure (and

especially equation (8)), justify the *ad hoc* formulas proposed in

(a) Problem 35, Exercises 9.4.
(b) Problem 37, Exercises 9.4
(c) Problem 39, Exercises 9.4.

CHAPTER SUMMARY

In this chapter we discussed the theory of linear systems in normal form and presented methods for solving such systems. The theory and methods are natural extensions of the development for second order and higher-order linear equations. The important properties and techniques are listed below.

Homogeneous Normal Systems

$$\mathbf{x}'(t) = \mathbf{A}(t)\mathbf{x}(t)$$

The $n \times n$ matrix function $\mathbf{A}(t)$ is assumed to be continuous on an interval I.

Fundamental Solution Set: $\{\mathbf{x}_1, \ldots, \mathbf{x}_n\}$. The n vector solutions $\mathbf{x}_1(t), \ldots, \mathbf{x}_n(t)$ of the homogeneous system on the interval I form a **fundamental solution set,** provided they are linearly independent on I or, equivalently, their **Wronskian**

$$W[\mathbf{x}_1, \ldots, \mathbf{x}_n](t) := \det[\mathbf{x}_1, \ldots, \mathbf{x}_n] = \begin{vmatrix} x_{1,1}(t) & x_{1,2}(t) & \cdots & x_{1,n}(t) \\ x_{2,1}(t) & x_{2,2}(t) & \cdots & x_{2,n}(t) \\ \vdots & \vdots & & \vdots \\ x_{n,1}(t) & x_{n,2}(t) & \cdots & x_{n,n}(t) \end{vmatrix}$$

is never zero on I.

Fundamental Matrix: $\mathbf{X}(t)$. An $n \times n$ matrix function $\mathbf{X}(t)$ whose column vectors form a fundamental solution set for the homogeneous system is called a **fundamental matrix.** The determinant of $\mathbf{X}(t)$ is the Wronskian of the fundamental solution set. Since the Wronskian is never zero on the interval I, then $\mathbf{X}^{-1}(t)$ exists for t in I.

General Solution to Homogeneous System: $\mathbf{Xc} = c_1\mathbf{x}_1 + \cdots + c_n\mathbf{x}_n$. If $\mathbf{X}(t)$ is a fundamental matrix whose column vectors are $\mathbf{x}_1, \ldots, \mathbf{x}_n$, then a general solution to the homogeneous system is

$$\mathbf{x}(t) = \mathbf{X}(t)\mathbf{c} = c_1\mathbf{x}_1(t) + c_2\mathbf{x}_2(t) + \cdots + c_n\mathbf{x}_n(t),$$

where $\mathbf{c} = \mathrm{col}(c_1, \ldots, c_n)$ is an arbitrary constant vector.

Homogeneous Systems with Constant Coefficients. The form of a general solution for a homogeneous system with constant coefficients depends on the eigenvalues and eigenvectors of the $n \times n$ constant matrix $\mathbf{A}$. An **eigenvalue** of $\mathbf{A}$ is a number r such that the system $\mathbf{Au} = r\mathbf{u}$ has a nontrivial solution $\mathbf{u}$, called an **eigenvector** of $\mathbf{A}$ associated with the eigenvalue r. Finding the eigenvalues of $\mathbf{A}$ is equivalent to finding the roots of the

characteristic equation

$$|\mathbf{A} - r\mathbf{I}| = 0.$$

The corresponding eigenvectors are found by solving the system $(\mathbf{A} - r\mathbf{I})\mathbf{u} = \mathbf{0}$.

If the matrix $\mathbf{A}$ has n linearly independent eigenvectors $\mathbf{u}_1, \ldots, \mathbf{u}_n$ and r_i is the eigenvalue corresponding to the eigenvector $\mathbf{u}_i$, then

$$\{e^{r_1 t}\mathbf{u}_1, \ e^{r_2 t}\mathbf{u}_2, \ \ldots, \ e^{r_n t}\mathbf{u}_n\}$$

is a fundamental solution set for the homogeneous system. A class of matrices that always has n linearly independent eigenvectors is the set of **symmetric** matrices—that is, matrices that satisfy $\mathbf{A} = \mathbf{A}^T$.

If $\mathbf{A}$ has complex conjugate eigenvalues $\alpha \pm i\beta$ and associated eigenvectors $\mathbf{z} = \mathbf{a} \pm i\mathbf{b}$, where $\mathbf{a}$ and $\mathbf{b}$ are real vectors, then two linearly independent real vector solutions to the homogeneous system are

$$e^{\alpha t}\cos \beta t\,\mathbf{a} - e^{\alpha t}\sin \beta t\,\mathbf{b}, \qquad e^{\alpha t}\sin \beta t\,\mathbf{a} + e^{\alpha t}\cos \beta t\,\mathbf{b}.$$

When $\mathbf{A}$ has a repeated eigenvalue r of multiplicity m, then it is possible that $\mathbf{A}$ does *not* have n linearly independent eigenvectors. However, associated with r are m linearly independent solutions to the homogeneous system of the form

$$\mathbf{x}_1(t) = e^{rt}\mathbf{k}_{1,1}, \qquad \mathbf{x}_2(t) = e^{rt}\mathbf{k}_{2,1} + te^{rt}\mathbf{k}_{2,2}, \quad \ldots,$$

$$\mathbf{x}_m(t) = e^{rt}\mathbf{k}_{m,1} + te^{rt}\mathbf{k}_{m,2} + \cdots + \frac{t^{m-1}}{(m-1)!}e^{rt}\mathbf{k}_{m,m}.$$

Nonhomogeneous Normal Systems

$$\mathbf{x}'(t) = \mathbf{A}(t)\mathbf{x}(t) + \mathbf{f}(t)$$

The $n \times n$ matrix function $\mathbf{A}(t)$ and the vector function $\mathbf{f}(t)$ are assumed continuous on an interval I.

General Solution to Nonhomogeneous System: $\mathbf{x}_p + \mathbf{Xc}$. If $\mathbf{x}_p(t)$ is any particular solution for the nonhomogeneous system and $\mathbf{X}(t)$ is a fundamental matrix for the associated homogeneous system, then a general solution for the nonhomogeneous system is

$$\mathbf{x}(t) = \mathbf{x}_p(t) + \mathbf{X}(t)\mathbf{c} = \mathbf{x}_p(t) + c_1\mathbf{x}_1(t) + \cdots + c_n\mathbf{x}_n(t),$$

where $\mathbf{x}_1(t), \ldots, \mathbf{x}_n(t)$ are the column vectors of $\mathbf{X}(t)$ and $\mathbf{c} = \mathrm{col}(c_1, \ldots, c_n)$ is an arbitrary constant vector.

Undetermined Coefficients. If the nonhomogeneous term $\mathbf{f}(t)$ is a vector whose components are polynomials, or exponential or sinusoidal functions, and $\mathbf{A}$ is a constant matrix, then one can use an extension of the method of undetermined coefficients to decide the form of a particular solution to the nonhomogeneous system.

Variation of Parameters: $\mathbf{X}(t)v(t)$. Let $\mathbf{X}(t)$ be a fundamental matrix for the homogeneous system. A particular solution to the nonhomogeneous system is given by the

variation of parameters formula

$$\mathbf{x}_p(t) = \mathbf{X}(t)\mathbf{v}(t) = \mathbf{X}(t) \int \mathbf{X}^{-1}(t)\mathbf{f}(t)\,dt.$$

Matrix Exponential Function

If $\mathbf{A}$ is a constant $n \times n$ matrix, then the matrix exponential function

$$e^{\mathbf{A}t} := \mathbf{I} + \mathbf{A}t + \mathbf{A}^2\frac{t^2}{2!} + \cdots + \mathbf{A}^n\frac{t^n}{n!} + \cdots$$

is a fundamental matrix for the homogeneous system $\mathbf{x}'(t) = \mathbf{A}\mathbf{x}(t)$. The matrix exponential has some of the same properties as the scalar exponential e^{at}. In particular,

$$e^{\mathbf{0}} = \mathbf{I}, \qquad e^{\mathbf{A}(t+s)} = e^{\mathbf{A}t}e^{\mathbf{A}s}, \qquad (e^{\mathbf{A}t})^{-1} = e^{-\mathbf{A}t}.$$

If $(\mathbf{A} - r\mathbf{I})^k = \mathbf{0}$ for some r and k, then the series for $e^{\mathbf{A}t}$ has only a finite number of terms:

$$e^{\mathbf{A}t} = e^{rt}\left\{ \mathbf{I} + (\mathbf{A} - r\mathbf{I})t + \cdots + (\mathbf{A} - r\mathbf{I})^{k-1}\frac{t^{k-1}}{(k-1)!} \right\}.$$

The matrix exponential function $e^{\mathbf{A}t}$ can also be computed from any fundamental matrix $\mathbf{X}(t)$ via the formula

$$e^{\mathbf{A}t} = \mathbf{X}(t)\mathbf{X}^{-1}(0).$$

Generalized Eigenvectors

If r is an eigenvalue of $\mathbf{A}$, then a **generalized eigenvector** associated with r is a nonzero vector $\mathbf{u}$ that satisfies $(\mathbf{A} - r\mathbf{I})^k\mathbf{u} = \mathbf{0}$ for some positive integer k. *Every* matrix $\mathbf{A}$ has a set of n linearly independent generalized eigenvectors that can be used to compute a fundamental solution set. In particular, associated with each generalized eigenvector $\mathbf{u}$ and corresponding eigenvalue r is a solution to the homogeneous system $\mathbf{x}'(t) = \mathbf{A}\mathbf{x}(t)$ of the form

$$e^{\mathbf{A}t}\mathbf{u} = e^{rt}\left\{ \mathbf{u} + t(\mathbf{A} - r\mathbf{I})\mathbf{u} + \frac{t^2}{2!}(\mathbf{A} - r\mathbf{I})^2\mathbf{u} + \cdots \right\}.$$

If r has multiplicity m, then the above series reduces to the first m terms.

REVIEW PROBLEMS

In Problems 1 through 4, find a general solution for the system $\mathbf{x}'(t) = \mathbf{A}\mathbf{x}(t)$, *where* $\mathbf{A}$ *is given.*

1. $\mathbf{A} = \begin{bmatrix} 6 & -3 \\ 2 & 1 \end{bmatrix}$.

2. $\mathbf{A} = \begin{bmatrix} 3 & 2 \\ -5 & 1 \end{bmatrix}$.

3. $\mathbf{A} = \begin{bmatrix} 1 & 2 & 0 & 0 \\ 2 & 1 & 0 & 0 \\ 0 & 0 & 1 & 2 \\ 0 & 0 & 2 & 1 \end{bmatrix}$.

4. $\mathbf{A} = \begin{bmatrix} 1 & 1 & 0 \\ 0 & 1 & 0 \\ 0 & 0 & 2 \end{bmatrix}$.

In Problems 5 and 6, find a fundamental matrix for the system $\mathbf{x}'(t) = \mathbf{A}\mathbf{x}(t)$, *where* $\mathbf{A}$ *is given.*

5. $\mathbf{A} = \begin{bmatrix} 1 & -1 \\ 2 & 4 \end{bmatrix}$.

6. $\mathbf{A} = \begin{bmatrix} 5 & 0 & 0 \\ 0 & -4 & 3 \\ 0 & 3 & 4 \end{bmatrix}$.

In Problems 7 through 10, find a general solution for the system $\mathbf{x}'(t) = \mathbf{A}\mathbf{x}(t) + \mathbf{f}(t)$, *where* $\mathbf{A}$ *and* $\mathbf{f}(t)$ *are given.*

7. $\mathbf{A} = \begin{bmatrix} 1 & 1 \\ 4 & 1 \end{bmatrix}$, $\mathbf{f}(t) = \begin{bmatrix} 5 \\ 6 \end{bmatrix}$.

8. $\mathbf{A} = \begin{bmatrix} -4 & 2 \\ 2 & -1 \end{bmatrix}$, $\mathbf{f}(t) = \begin{bmatrix} e^{4t} \\ 3e^{4t} \end{bmatrix}$.

9. $\mathbf{A} = \begin{bmatrix} 2 & 1 & -1 \\ -3 & -1 & 1 \\ 9 & 3 & -4 \end{bmatrix}$, $\mathbf{f}(t) = \begin{bmatrix} t \\ 0 \\ 1 \end{bmatrix}$.

10. $\mathbf{A} = \begin{bmatrix} 2 & -2 & 3 \\ 0 & 3 & 2 \\ 0 & -1 & 2 \end{bmatrix}$, $\mathbf{f}(t) = \begin{bmatrix} e^{-t} \\ 2 \\ 1 \end{bmatrix}$.

In Problems 11 and 12, solve the given initial value problem.

11. $\mathbf{x}'(t) = \begin{bmatrix} 0 & 1 \\ -2 & 3 \end{bmatrix} \mathbf{x}(t)$, $\mathbf{x}(0) = \begin{bmatrix} 1 \\ -1 \end{bmatrix}$.

12. $\mathbf{x}'(t) = \begin{bmatrix} 2 & 1 \\ -4 & 2 \end{bmatrix} \mathbf{x}(t) + \begin{bmatrix} te^{2t} \\ e^{2t} \end{bmatrix}$, $\mathbf{x}(0) = \begin{bmatrix} 2 \\ 2 \end{bmatrix}$.

In Problems 13 and 14, find a general solution for the Cauchy-Euler system $t\mathbf{x}'(t) = \mathbf{A}\mathbf{x}(t)$, *where* $\mathbf{A}$ *is given.*

13. $\mathbf{A} = \begin{bmatrix} 0 & 3 & 1 \\ 1 & 2 & 1 \\ 1 & 3 & 0 \end{bmatrix}$.

14. $\mathbf{A} = \begin{bmatrix} 1 & 2 & -1 \\ 2 & 1 & 1 \\ -1 & 1 & 0 \end{bmatrix}$.

In Problems 15 and 16, find the fundamental matrix $e^{\mathbf{A}t}$ *for the system* $\mathbf{x}'(t) = \mathbf{A}\mathbf{x}(t)$, *where* $\mathbf{A}$ *is given.*

15. $\mathbf{A} = \begin{bmatrix} 4 & 2 & 3 \\ 2 & 1 & 2 \\ -1 & 2 & 0 \end{bmatrix}$.

16. $\mathbf{A} = \begin{bmatrix} 0 & 1 & 4 \\ 0 & 0 & 2 \\ 0 & 0 & 0 \end{bmatrix}$.

TECHNICAL WRITING EXERCISES

1. Explain how the theory of homogeneous linear differential equations (as described in Sections 4.3 and 6.1) follows from the theory of linear systems in normal form (as described in Section 9.3).

2. Discuss the similarities and differences between the method for finding solutions to a linear constant coefficient differential equation (see Sections 4.5, 4.6, and 6.2) and the method for finding solutions to a linear system in normal form that has constant coefficients (see Sections 9.4 and 9.5).

3. Explain how the variation of parameters formulas for linear second order equations derived in Section 4.9 follow from the formulas derived in Section 9.6 for linear systems in normal form.

4. Explain how you would define the matrix functions $\sin \mathbf{A}t$ and $\cos \mathbf{A}t$, where $\mathbf{A}$ is a constant $n \times n$ matrix. How are these functions related to the matrix exponential and how are they connected to the solutions of the system $\mathbf{x}'' + \mathbf{A}^2\mathbf{x} = \mathbf{0}$? The reader will find it informative to consider the cases when $\mathbf{A}$ is

$$\begin{bmatrix} 0 & 0 \\ 0 & 0 \end{bmatrix}, \begin{bmatrix} 0 & 1 \\ 0 & 0 \end{bmatrix}, \begin{bmatrix} 0 & 0 \\ 1 & 0 \end{bmatrix}, \text{ and } \begin{bmatrix} 1 & -1 \\ 1 & -1 \end{bmatrix}.$$

GROUP PROJECTS FOR CHAPTER 9

A. Uncoupling Normal Systems

The easiest normal systems to solve are systems of the form

(1) $\qquad \mathbf{x}'(t) = \mathbf{D}\mathbf{x}(t),$

where $\mathbf{D}$ is an $n \times n$ diagonal matrix. Such a system actually consists of n uncoupled equations

(2) $\qquad x_i'(t) = d_{ii}x_i(t), \qquad i = 1, \ldots, n,$

whose solution is

$$x_i(t) = c_i e^{d_{ii}t},$$

where the c_i's are arbitrary constants. This raises the following question: When can we *uncouple* a normal system?

To answer this question, we need the following result from linear algebra. An $n \times n$ matrix $\mathbf{A}$ is diagonalizable if and only if $\mathbf{A}$ has n linearly independent eigenvectors $\mathbf{p}_1, \ldots, \mathbf{p}_n$. Moreover, if $\mathbf{P}$ is the matrix whose columns are $\mathbf{p}_1, \ldots, \mathbf{p}_n$, then

(3) $\qquad \mathbf{P}^{-1}\mathbf{A}\mathbf{P} = \mathbf{D},$

where $\mathbf{D}$ is the diagonal matrix whose entry d_{ii} is the eigenvalue associated with the vector $\mathbf{p}_i$.

(a) Use the above result to show that the system

(4) $\qquad \mathbf{x}'(t) = \mathbf{A}\mathbf{x}(t),$

where $\mathbf{A}$ is an $n \times n$ diagonalizable matrix, is equivalent to an uncoupled system

(5) $\qquad \mathbf{y}'(t) = \mathbf{D}\mathbf{y}(t),$

where $\mathbf{y} = \mathbf{P}^{-1}\mathbf{x}$ and $\mathbf{D} = \mathbf{P}^{-1}\mathbf{A}\mathbf{P}$.

(b) Solve system (5).

(c) Use the results of parts (a) and (b) to show that a general solution to (4) is given by

$$\mathbf{x}(t) = c_1 e^{d_{11}t}\mathbf{p}_1 + c_2 e^{d_{22}t}\mathbf{p}_2 + \cdots + c_n e^{d_{nn}t}\mathbf{p}_n.$$

(d) Use the procedure discussed in parts (a)–(c) to obtain a general solution for the system

$$\mathbf{x}'(t) = \begin{bmatrix} 5 & -4 & 4 \\ 1 & 0 & 1 \\ -1 & 2 & 1 \end{bmatrix} \mathbf{x}(t).$$

Specify $\mathbf{P}$, $\mathbf{D}$, $\mathbf{P}^{-1}$, and $\mathbf{y}$.

B. Matrix Laplace Transform Method

The Laplace transform method for solving systems of linear differential equations with constant coefficients was discussed in Chapter 7. To apply the procedure for equations given in matrix form, we first extend the definition of the Laplace operator $\mathscr{L}$ to a column vector of functions $\mathbf{x} = \mathrm{col}(x_1(t), \ldots, x_n(t))$ by taking the transform of each component:

$$\mathscr{L}\{\mathbf{x}\}(s) := \mathrm{col}(\mathscr{L}\{x_1\}(s), \ldots, \mathscr{L}\{x_n\}(s)).$$

With this notation, the vector analogue of the important property relating the Laplace transform of the derivative of a function (see Theorem 4, Chapter 7, page 365) becomes

(6) $\mathscr{L}\{\mathbf{x}'\}(s) = s\mathscr{L}\{\mathbf{x}\}(s) - \mathbf{x}(0).$

Now suppose we are given the initial value problem

(7) $\mathbf{x}' = \mathbf{A}\mathbf{x} + \mathbf{f}(t), \qquad \mathbf{x}(0) = \mathbf{x}_0,$

where $\mathbf{A}$ is a constant $n \times n$ matrix. Let $\hat{\mathbf{x}}(s)$ denote the Laplace transform of $\mathbf{x}(t)$ and $\hat{\mathbf{f}}(s)$ denote the transform of $\mathbf{f}(t)$. Then, taking the transform of the system and using the relation (6), we get

$$\mathscr{L}\{\mathbf{x}'\} = \mathscr{L}\{\mathbf{A}\mathbf{x} + \mathbf{f}\},$$
$$s\hat{\mathbf{x}} - \mathbf{x}_0 = \mathbf{A}\hat{\mathbf{x}} + \hat{\mathbf{f}}.$$

Next we collect the $\hat{\mathbf{x}}$ terms and solve for $\hat{\mathbf{x}}$ by premultiplying by $(s\mathbf{I} - \mathbf{A})^{-1}$:

$$(s\mathbf{I} - \mathbf{A})\hat{\mathbf{x}} = \hat{\mathbf{f}} + \mathbf{x}_0$$
$$\hat{\mathbf{x}} = (s\mathbf{I} - \mathbf{A})^{-1}(\hat{\mathbf{f}} + \mathbf{x}_0).$$

Finally, we obtain the solution $\mathbf{x}(t)$ by taking the inverse Laplace transform:

$$\mathbf{x} = \mathscr{L}^{-1}\{\hat{\mathbf{x}}\} = \mathscr{L}^{-1}\{(s\mathbf{I} - \mathbf{A})^{-1}(\hat{\mathbf{f}} + \mathbf{x}_0)\}.$$

In applying the matrix Laplace transform method, it is straightforward (but possibly tedious) to compute $(s\mathbf{I} - \mathbf{A})^{-1}$, but the computation of the inverse transform may require some of the special techniques (such as partial fractions) discussed in Chapter 7.

(a) In the above procedure, we used the property that $\mathscr{L}\{\mathbf{A}\mathbf{x}\} = \mathbf{A}\mathscr{L}\{\mathbf{x}\}$ for any constant $n \times n$ matrix $\mathbf{A}$. Show that this property follows from the linearity of the transform in the scalar case.

(b) Use the matrix Laplace transform method to solve the following initial value problems:

(i) $\mathbf{x}'(t) = \begin{bmatrix} 0 & 2 \\ -1 & 3 \end{bmatrix} \mathbf{x}(t), \qquad \mathbf{x}(0) = \begin{bmatrix} -1 \\ 3 \end{bmatrix}.$

(ii) $\mathbf{x}'(t) = \begin{bmatrix} 3 & -2 \\ 4 & -1 \end{bmatrix} \mathbf{x}(t) + \begin{bmatrix} \sin t \\ -\cos t \end{bmatrix}, \qquad \mathbf{x}(0) = \begin{bmatrix} 0 \\ 0 \end{bmatrix}.$

C. Undamped Second Order Systems

We have seen that the coupled spring-mass system depicted in Figure 9.5, page 547, is governed by equations (9) of Section 9.5, which we reproduce here:

$$m_1 x_1'' = -k_1 x_1 + k_2(x_2 - x_1),$$

$$m_2 x_2'' = -k_2(x_2 - x_1) - k_3 x_2.$$

This system was rewritten in normal form as equation (10) in Section 9.5; however, there is some advantage in expressing it as a second order system in the form

$$\begin{bmatrix} m_1 & 0 \\ 0 & m_2 \end{bmatrix} \begin{bmatrix} x_1 \\ x_2 \end{bmatrix}'' = -\begin{bmatrix} k_1 + k_2 & -k_2 \\ -k_2 & k_2 + k_3 \end{bmatrix} \begin{bmatrix} x_1 \\ x_2 \end{bmatrix}.$$

This structure,

(8) $\mathbf{M}\mathbf{x}'' = -\mathbf{K}\mathbf{x},$

with a diagonal *mass matrix* $\mathbf{M}$ and a symmetric *stiffness matrix* $\mathbf{K}$, is typical for most undamped vibrating systems. Our experience with other spring-mass systems (Sections 5.1–5.4) suggests that we seek solutions to (8) of the form

(9) $\mathbf{x} = \cos \omega t \, \mathbf{v}$ or $\mathbf{x} = \sin \omega t \, \mathbf{v},$

where $\mathbf{v}$ is a constant vector and ω is a positive constant.

(a) Show that the system (8) has a nontrivial solution of the form (9) if and only if ω and $\mathbf{v}$ satisfy the "generalized eigenvalue problem" $\mathbf{K}\mathbf{v} = \omega^2 \mathbf{M}\mathbf{v}$.

(b) By employing the inverse of the mass matrix, one can rewrite (8) as

$\mathbf{x}'' = -\mathbf{M}^{-1}\mathbf{K}\mathbf{x} =: \mathbf{B}\mathbf{x}.$

Show that $-\omega^2$ must be an eigenvalue of $\mathbf{B}$ if (9) is a nontrivial solution.

(c) If $\mathbf{B}$ is an $n \times n$ constant matrix, then $\mathbf{x}'' = \mathbf{B}\mathbf{x}$ can be written as a system of $2n$ first order equations in normal form. Thus a general solution can be formed from $2n$ linearly independent solutions. Use the observation in part (b) to find a general solution to the following second order systems:

(i) $\mathbf{x}'' = \begin{bmatrix} 0 & 1 \\ -4 & -5 \end{bmatrix} \mathbf{x}.$

(ii) $\mathbf{x}'' = \begin{bmatrix} -2 & 2 \\ 2 & -5 \end{bmatrix} \mathbf{x}.$

(iii) $\mathbf{x}'' = \begin{bmatrix} -5 & 4 & -4 \\ -1 & 0 & -1 \\ 1 & -2 & -1 \end{bmatrix} \mathbf{x}.$

(iv) $\mathbf{x}'' = \begin{bmatrix} 2 & -1 & -1 \\ -1 & 2 & 0 \\ -1 & 0 & 2 \end{bmatrix} \mathbf{x}.$

D. Strange Behavior of Competing Species—Part II

In Project F in Chapter 5, we used numerical experimentation to study the behavior of solutions to the system

$$p_1' = p_1(1 - p_1 - ap_2 - bp_3),$$

(10) $\quad p_2' = p_2(1 - bp_1 - p_2 - ap_3),$

$$p_3' = p_3(1 - ap_1 - bp_2 - p_3),$$

where a and b are positive constants and p_i is the population of species S_i, $i = 1, 2, 3$. We found that when $a = b = 0.5$, the populations approach the equilibrium solution $p_1 = p_2 = p_3 = 1/2$ as $t \to \infty$. However, when $a = 0.5$ and $b = 2$, the populations cycled between the three critical points —(1, 0, 0), (0, 1, 0), (0, 0, 1)—getting closer and closer to these points. Moreover, the populations remained near each point for longer and longer periods of time before moving on to the next point. To analyze what is happening near these equilibrium points, we will study the corresponding linearized systems.

(a) Let $f_i(p_1, p_2, p_3)$ denote the function on the right-hand side of the equation for p_i' in (10); in particular, $f_1(p_1, p_2, p_3) = p_1(1 - p_1 - ap_2 - bp_3)$. Now $\mathbf{f} = \text{col}(f_1, f_2, f_3)$ is a mapping from $\mathbf{R}^3$ into $\mathbf{R}^3$. Show that the Jacobian matrix for this mapping is

$$\mathbf{A} := \begin{bmatrix} 1 - 2p_1 - ap_2 - bp_3 & -ap_1 & -bp_1 \\ -bp_2 & 1 - 2p_2 - ap_3 - bp_1 & -ap_2 \\ -ap_3 & -bp_3 & 1 - 2p_3 - ap_1 - bp_2 \end{bmatrix}.$$

(b) Show that at the critical point (δ, δ, δ) where $\delta = (1 + a + b)^{-1}$, we have $\mathbf{A} = -\delta\mathbf{B}$, where

$$\mathbf{B} := \begin{bmatrix} 1 & a & b \\ b & 1 & a \\ a & b & 1 \end{bmatrix}.$$

(c) Show that the eigenvalues of $\mathbf{B}$ are

$$1 + a + b, \quad \left(1 - \frac{a + b}{2}\right) \pm i\frac{\sqrt{3}(a - b)}{2}.$$

(d) Now, near the critical point (δ, δ, δ), the solutions to (10) should behave like the solutions to $\mathbf{p}' = -\delta\mathbf{Bp}$. Use the results of part (c) to argue that when $a + b < 2$, the solutions to (10) should approach the critical point (δ, δ, δ) and when $a + b > 2$, they should move away.

(e) Using a similar analysis, conclude that when either $a < 1$ or $b < 1$, the solutions to (10) should move away from the critical points (1, 0, 0), (0, 1, 0), and (0, 0, 1) because at least one of the eigenvalues of the linearized system (about each critical point) is positive.

(f) How do the analyses performed in parts (d) and (e) support the conclusions reached from numerical experimentations conducted in Project F in Chapter 5?

Chapter 10

Partial Differential Equations

10.1 INTRODUCTION: A MODEL FOR HEAT FLOW

> Develop a model for the flow of heat through a thin, insulated wire whose ends are kept at a constant temperature of 0°C and whose initial temperature distribution is to be specified.

Suppose the wire is placed along the x-axis with $x = 0$ at the left end of the wire and $x = L$ at the right end (see Figure 10.1). If we let u denote the temperature of the wire, then u depends on the time t and on the position x within the wire. (We will assume the wire is thin and hence u is constant throughout a cross-section of the wire corresponding to a fixed value of x.) Because the wire is insulated, we assume no heat enters or leaves through the sides of the wire.

To develop a model for heat flow through the thin wire, let's consider the small volume element V of wire between the two cross-sectional planes A and B that are perpendicular to the x-axis, with plane A located at x and plane B located at $x + \Delta x$ (see Figure 10.1).

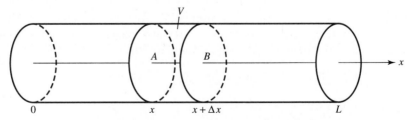

Figure 10.1 Heat flow through a thin piece of wire

The temperature on plane A at time t is $u(x, t)$ and on plane B is $u(x + \Delta x, t)$. We will need the following principles of physics that describe heat flow.[†]

1. **Heat Conduction:** The rate of heat flow (the amount of heat per unit time flowing through a unit of cross-sectional area at A) is proportional to $\partial u / \partial x$, the temperature gradient at A. The proportionality constant k is called the **thermal conductivity** of the material.

2. **Direction of Heat Flow:** The direction of heat flow is always from points of higher temperature to points of lower temperature.

3. **Specific Heat Capacity:** The amount of heat necessary to raise the temperature of an object of mass m by an amount Δu is $cm\,\Delta u$, where the constant c is the **specific heat capacity** of the material.

If we let H represent the amount of heat flowing from left to right through the surface A during an interval of time Δt, then the formula for heat conduction becomes

$$H(x) = -ka\,\Delta t\,\frac{\partial u}{\partial x}(x, t),$$

where a is the cross-sectional area of the wire. The negative sign follows from the second principle—if $\partial u / \partial x$ is positive, then heat flows from right to left (hotter to colder).

Similarly, the amount of heat flowing from left to right across plane B during an interval of time Δt is

$$H(x + \Delta x) = -ka\,\Delta t\,\frac{\partial u}{\partial x}(x + \Delta x, t).$$

The net change in the heat ΔE in volume V is the amount entering at end A minus the amount leaving at end B, plus any heat generated by sources (such as electric currents, chemical reactions, heaters, etc.). The latter is modeled by a term $Q(x, t)\Delta x a\,\Delta t$, where Q is the energy rate (power) density. Therefore

(1) $\quad \Delta E = H(x) - H(x + \Delta x) + Q(x, t)\Delta x a\,\Delta t$

$$= ka\,\Delta t\left[\frac{\partial u}{\partial x}(x + \Delta x, t) - \frac{\partial u}{\partial x}(x, t)\right] + Q(x, t)\Delta x a\,\Delta t.$$

Now by the third principle, the net change in heat is given by $\Delta E = cm\,\Delta u$, where Δu is the change in temperature and c is the specific heat capacity. If we assume the change in temperature in the volume V is essentially equal to the change in temperature at x, that is $\Delta u = u(x, t + \Delta t) - u(x, t)$, and that the mass of the volume V of wire is $a\rho\,\Delta x$, where ρ is the linear density of the wire, then

(2) $\quad \Delta E = ca\rho\,\Delta x[u(x, t + \Delta t) - u(x, t)].$

[†]For a discussion of heat transfer, see *University Physics*, Ninth Edition, by H. D. Young and R. A. Freedman (Addison-Wesley, Reading, Mass., 1996).

Equating the two expressions for ΔE given in equations (1) and (2) yields

$$ka\,\Delta t\left[\frac{\partial u}{\partial x}(x + \Delta x, t) - \frac{\partial u}{\partial x}(x, t)\right] + Q(x, t)\Delta x a\,\Delta t = ca\rho\,\Delta x[u(x, t + \Delta t) - u(x, t)].$$

Now dividing both sides by Δx and Δt and then taking the limits as Δx and Δt approach zero, we obtain

$$k\frac{\partial^2 u}{\partial x^2}(x, t) + Q(x, t) = c\rho\frac{\partial u}{\partial t}(x, t)$$

or

$$(3) \qquad \frac{\partial u}{\partial t}(x, t) = \beta\frac{\partial^2 u}{\partial x^2}(x, t) + P(x, t),$$

where the positive constant $\beta := k/c\rho$ is the **diffusivity** of the material and $P(x, t) := Q(x, t)/c\rho$. Equation (3) the **one-dimensional heat flow equation.**

Equation (3) governs the flow of heat in the wire. We have two other constraints in our original problem. First, we are keeping the ends of the wire at $0°C$. Thus we require that

$$(4) \qquad u(0, t) = u(L, t) = 0$$

for all t. These are called **boundary conditions.** Second, we must be given the initial temperature distribution $f(x)$. That is, we require

$$(5) \qquad u(x, 0) = f(x), \qquad 0 < x < L.$$

Equation (5) is referred to as the **initial condition** on u.

Combining equations (3), (4), and (5), we have the following mathematical model for the heat flow in a wire without internal sources ($P = 0$) whose ends are kept at the constant temperature $0°C$:

$$(6) \qquad \frac{\partial u}{\partial t}(x, t) = \beta\frac{\partial^2 u}{\partial x^2}(x, t), \qquad 0 < x < L, \qquad t > 0,$$

$$(7) \qquad u(0, t) = u(L, t) = 0, \qquad t > 0,$$

$$(8) \qquad u(x, 0) = f(x), \qquad 0 < x < L.$$

This model is an example of an **initial-boundary value problem.** Intuitively we expect that equations (6)–(8) completely and unambiguously specify the temperature in the wire. Once we have found a function $u(x, t)$ that meets all three of these conditions, we can be assured that u is the temperature. (Theorem 7 in Section 10.5 will bear this out.)

In higher dimensions, the heat flow equation (or just **heat equation**) has the form

$$\frac{\partial u}{\partial t} = \beta\,\Delta u + P(x, y, z, t)$$

where Δu, in this context, is the **Laplacian** of u.[†] In two and three dimensions the Laplacian is defined, respectively, as

$$\Delta u := \frac{\partial^2 u}{\partial x^2} + \frac{\partial^2 u}{\partial y^2} \quad \text{and} \quad \Delta u := \frac{\partial^2 u}{\partial x^2} + \frac{\partial^2 u}{\partial y^2} + \frac{\partial^2 u}{\partial z^2}.$$

When the temperature reaches a steady state, that is when u does not depend on time, and there are no sources, then $\partial u/\partial t = 0$ and the temperature satisfies **Laplace's equation**

$$\Delta u = 0.$$

One classical technique for solving the initial-boundary value problem for the heat equation (6)–(8) is the method of *separation of variables*, which effectively allows us to replace the partial derivatives by ordinary derivatives. This technique is discussed in the next section. In using separation of variables, one is often required to express a given function as a trigonometric series. Such series are called *Fourier series*; their properties are discussed in Sections 10.3 and 10.4. We devote the remaining three sections to the three basic partial differential equations that arise most commonly in applications: the heat equation, the wave equation, and Laplace's equation.

10.2 METHOD OF SEPARATION OF VARIABLES

The method of separation of variables is a classical technique that is effective in solving several types of partial differential equations. The idea is roughly the following. We think of a solution, say $u(x, t)$, to a partial differential equation as being an infinite linear combination of simple component functions $u_n(x, t)$, $n = 0, 1, 2, \ldots$, which also satisfy the equation and certain boundary conditions. (This is a reasonable assumption provided the partial differential equation and the boundary conditions are linear and homogeneous.) To determine a component solution, $u_n(x, t)$, we assume it can be written with its variables separated; that is, as

$$u_n(x, t) = X_n(x)T_n(t).$$

Substituting this form for a solution into the partial differential equation and using the boundary conditions leads, in many circumstances, to two *ordinary* differential equations for the unknown functions $X_n(x)$ and $T_n(t)$. In this way we have reduced the problem of solving a partial differential equation to the more familiar problem of solving a differential equation that involves only one variable. In this section we will illustrate this technique for the heat equation and the wave equation.

In the previous section we derived the following initial-boundary value problem as a mathematical model for the sourceless heat flow in a wire whose ends are kept at the

[†] Regrettably, it is the same notation as we just used for $u(x, t + \Delta t) - u(x, t)$.

constant temperature zero:

$$\textbf{(1)} \qquad \frac{\partial u}{\partial t}(x, t) = \beta \frac{\partial^2 u}{\partial x^2}(x, t), \qquad 0 < x < L, \qquad t > 0,$$

$$\textbf{(2)} \qquad u(0, t) = u(L, t) = 0, \qquad t > 0,$$

$$\textbf{(3)} \qquad u(x, 0) = f(x), \qquad 0 < x < L.$$

To solve this problem by the method of separation of variables, we begin by assuming that equation (1) has solutions of the form

$$u(x, t) = X(x)T(t),$$

where X is a function of x alone and T is a function of t alone. To determine X and T, we first compute the partial derivatives of u to obtain

$$\frac{\partial u}{\partial t} = X(x)T'(t) \quad \text{and} \quad \frac{\partial^2 u}{\partial x^2} = X''(x)T(t).$$

Substituting these expressions into (1) gives

$$X(x)T'(t) = \beta X''(x)T(t),$$

and separating variables yields

$$\textbf{(4)} \qquad \frac{T'(t)}{\beta T(t)} = \frac{X''(x)}{X(x)}.$$

We now observe that the functions on the left-hand side of (4) depend only on t, while those on the right-hand side depend only on x. Since x and t are variables that are independent of one another, the two ratios in (4) must then be *constant*. Thus

$$\frac{X''(x)}{X(x)} = K \quad \text{and} \quad \frac{T'(t)}{\beta T(t)} = K,$$

or

$$\textbf{(5)} \qquad X''(x) - KX(x) = 0 \quad \text{and} \quad T'(t) - \beta KT(t) = 0.$$

Consequently, for separable solutions, we have reduced the problem of solving the partial differential equation (1) to solving the two *ordinary* differential equations in (5).

Before proceeding, let's consider the boundary conditions in (2). Since $u(x, t) = X(x)T(t)$, these conditions are

$$X(0)T(t) = 0 \quad \text{and} \quad X(L)T(t) = 0, \qquad t > 0.$$

Hence either $T(t) = 0$ for all $t > 0$, which implies that $u(x, t) \equiv 0$, or

$$\textbf{(6)} \qquad X(0) = X(L) = 0.$$

Ignoring the trivial solution $u(x, t) \equiv 0$, we combine the boundary conditions in (6) with the differential equation for X in (5) and obtain the *boundary value problem*

$$\textbf{(7)} \qquad X''(x) - KX(x) = 0, \qquad X(0) = X(L) = 0,$$

where K can be any constant.

Notice that the function $X(x) \equiv 0$ is a solution for every K. Depending on the choice of K, this may be the *only* solution to the boundary value problem (7). Thus if we seek a nontrivial solution $u(x, t) = X(x)T(t)$ to (1)–(2), we must first determine those values of K for which the boundary value problem (7) has nontrivial solutions. These special values of K are called **eigenvalues,** and the corresponding nontrivial solutions to (7) are the **eigenfunctions.** This nomenclature arises from the similarity of the operator form of the differential equation, $(D^2 - K)[X] = 0$, with the matrix eigenvector equation $(\mathbf{A} - r\mathbf{I})\mathbf{u} = \mathbf{0}$ (Section 9.4, page 533).

To solve the (constant coefficient) equation in (7), we try $X(x) = e^{rx}$, derive the auxiliary equation $r^2 - K = 0$, and consider three cases.

Case 1. $K > 0$. In this case, the roots of the auxiliary equation are $\pm\sqrt{K}$, so a general solution to the differential equation in (7) is

$$X(x) = C_1 e^{\sqrt{K}x} + C_2 e^{-\sqrt{K}x}.$$

To determine C_1 and C_2, we appeal to the boundary conditions:

$$X(0) = C_1 + C_2 = 0,$$
$$X(L) = C_1 e^{\sqrt{K}L} + C_2 e^{-\sqrt{K}L} = 0.$$

From the first equation, we see that $C_2 = -C_1$. The second equation can then be written as $C_1(e^{\sqrt{K}L} - e^{-\sqrt{K}L}) = 0$ or $C_1(e^{2\sqrt{K}L} - 1) = 0$. Since $K > 0$, it follows that $(e^{2\sqrt{K}L} - 1) > 0$. Therefore C_1, and hence C_2, is zero. Consequently, there is *no* nontrivial solution to (7) for $K > 0$.

Case 2. $K = 0$. Here $r = 0$ is a repeated root to the auxiliary equation, and a general solution to the differential equation is

$$X(x) = C_1 + C_2 x.$$

The boundary conditions in (7) yield $C_1 = 0$ and $C_1 + C_2 L = 0$, which imply that $C_1 = C_2 = 0$. Thus for $K = 0$, there is no nontrivial solution to (7).

Case 3. $K < 0$. In this case the roots of the auxiliary equation are $\pm i\sqrt{-K}$. (Note that $-K > 0$ since $K < 0$.) Thus a general solution to $X'' - KX = 0$ is

(8) $X(x) = C_1 \cos \sqrt{-K}x + C_2 \sin \sqrt{-K}x.$

This time the boundary conditions $X(0) = X(L) = 0$ give the system

$$C_1 = 0,$$
$$C_1 \cos \sqrt{-K}L + C_2 \sin \sqrt{-K}L = 0.$$

Because $C_1 = 0$, the system reduces to solving $C_2 \sin \sqrt{-K}L = 0$. Hence, either $\sin \sqrt{-K}L = 0$ or $C_2 = 0$. Now $\sin \sqrt{-K}L = 0$ only when $\sqrt{-K}L = n\pi$, where n is an integer. Therefore (7) has a nontrivial solution ($C_2 \neq 0$) when $\sqrt{-K}L = n\pi$ or $K = -(n\pi/L)^2$, $n = 1, 2, 3, \ldots$ (we exclude $n = 0$, since it makes $K = 0$). Furthermore, the nontrivial solutions

(eigenfunctions) X_n corresponding to the eigenvalue $K = -(n\pi/L)^2$ are given by (cf. (8))

(9) $X_n(x) = a_n \sin\left(\dfrac{n\pi x}{L}\right),$

where the a_n's are arbitrary nonzero constants.

Having determined that $K = -(n\pi/L)^2$ for any positive integer n, let's consider the second equation in (5):

$$T'(t) + \beta\left(\frac{n\pi}{L}\right)^2 T(t) = 0.$$

For each $n = 1, 2, 3, \ldots,$ a general solution to this linear first order equation is

$$T_n(t) = b_n e^{-\beta(n\pi/L)^2 t}.$$

Combining this with equation (9) we obtain, for each $n = 1, 2, 3, \ldots,$ the function

(10) $u_n(x, t) := X_n(x)T_n(t) = a_n \sin(n\pi x/L) b_n e^{-\beta(n\pi/L)^2 t}$

$\qquad\qquad\qquad = c_n e^{-\beta(n\pi/L)^2 t} \sin(n\pi x/L),$

where c_n is an arbitrary constant.

We would like to conclude that each $u_n(x, t)$ is a solution to (1)–(2). But, we have shown *only* that *if* (1)–(2) has a solution of the form $u(x, t) = X(x)T(t)$, then u must be one of the functions given in (10). We leave it for the reader to verify, by direct substitution, that the functions in (10) are indeed solutions to (1)–(2). They satisfy the differential equation and the boundary conditions (but not the initial condition).

A simple computation also shows that if u_n and u_m are solutions to (1)–(2), then so is any linear combination $au_n + bu_m$. (This is a consequence of the fact that the operator $\mathcal{L} := \partial/\partial t - \beta \partial^2/\partial x^2$ is a *linear* operator and the boundary conditions in (2) are *homogeneous*.) Furthermore, if we take an *infinite* sum of these functions, that is,

(11) $u(x, t) = \displaystyle\sum_{n=1}^{\infty} u_n(x, t) = \sum_{n=1}^{\infty} c_n e^{-\beta(n\pi/L)^2 t} \sin(n\pi x/L),$

then this formal series will again be a solution to (1)–(2), provided the infinite series has the proper convergence behavior.

For a solution u of the form (11) to satisfy the initial condition (3), we must have

$$u(x, 0) = \sum_{n=1}^{\infty} c_n \sin\left(\frac{n\pi x}{L}\right) = f(x), \qquad 0 < x < L.$$

The problem (1)–(3) of heat flow in a thin wire has thus been reduced to the problem of determining an expansion for $f(x)$ of the form

(12) $f(x) = \displaystyle\sum_{n=1}^{\infty} c_n \sin\left(\frac{n\pi x}{L}\right).$

Such an expansion is called a **Fourier sine series** and will be discussed in the next two sections. If we can choose the c_n's so that equation (12) holds, then the expansion for

$u(x, t)$ in (11) is called a **formal solution** to the heat flow problem (1)–(3). If this expansion converges to a function with continuous second partial derivatives, then the formal solution is an actual (genuine) solution. Moreover, the solution is unique.

EXAMPLE 1 Find the solution to the heat flow problem

(13) $$\frac{\partial u}{\partial t} = 7\frac{\partial^2 u}{\partial x^2},\qquad\qquad 0 < x < \pi,\qquad t > 0,$$

(14) $u(0,\ t) = u(\pi,\ t) = 0,\qquad\qquad t > 0,$

(15) $u(x,\ 0) = 3\sin 2x - 6\sin 5x,\qquad 0 < x < \pi.$

Solution Comparing equation (13) with (1), we see that $\beta = 7$ and $L = \pi$. Hence we need only determine the values of c_n in formula (12). That is, we must have

$$u(x,\ 0) = 3\sin 2x - 6\sin 5x = \sum_{n=1}^{\infty} c_n \sin nx.$$

Equating the coefficients of like terms, we find that

$$c_2 = 3 \quad \text{and} \quad c_5 = -6,$$

and the remaining c_n's are zero. Hence, from (11), the solution to the heat flow problem (13)–(15) is

$$u(x,\ t) = c_2 e^{-\beta(2\pi/L)^2 t} \sin(2\pi x/L) + c_5 e^{-\beta(5\pi/L)^2 t} \sin(5\pi x/L)$$

$$= 3e^{-28t}\sin 2x - 6e^{-175t}\sin 5x. \quad \blacksquare \text{ End}$$

Another situation in which the separation of variables approach applies occurs in the study of a vibrating string. This concerns the transverse vibrations of a string stretched between two points, such as a guitar string or piano wire. The goal is to find a function $u(x, t)$ that gives the displacement (deflection) of the string at any point x ($0 \leq x \leq L$) and any time $t \geq 0$ (see Figure 10.2). In developing the mathematical model, we assume the string is perfectly flexible and has constant linear density, the tension on the string is constant, gravity is negligible, and no other forces are acting on the string. Under these conditions and the additional assumption that the displacements $u(x, t)$ are small in com-

Figure 10.2 Displacement of string at time t

parison to the length of the string, it turns out that the motion of the string is governed by the following initial-boundary value problem.[†]

$$(16) \qquad \frac{\partial^2 u}{\partial t^2} = \alpha^2 \frac{\partial^2 u}{\partial x^2}, \qquad 0 < x < L, \quad t > 0,$$

$$(17) \qquad u(0, t) = u(L, t) = 0, \qquad t \geq 0,$$

$$(18) \qquad u(x, 0) = f(x), \qquad 0 \leq x \leq L,$$

$$(19) \qquad \frac{\partial u}{\partial t}(x, 0) = g(x), \qquad 0 \leq x \leq L.$$

The constant α^2 appearing in (16) is strictly positive and equals the ratio of the tension to the (linear) density of the string. The physical significance of α, which has units of velocity, will be revealed in Section 10.6. The boundary conditions in (17) reflect the fact that the string is held fixed at the two endpoints $x = 0$ and $x = L$. Equations (18) and (19), respectively, specify the initial displacement of the string and the initial velocity of each point on the string. For the initial and boundary conditions to be consistent, we assume $f(0) = f(L) = 0$ and $g(0) = g(L) = 0$.

Let's apply the method of separation of variables to the initial-boundary value problem for the vibrating string (16)–(19). Thus we begin by assuming equation (16) has a solution of the form

$$u(x, t) = X(x)T(t),$$

where X is a function of x alone and T is a function of t alone. Differentiating u, we obtain

$$\frac{\partial^2 u}{\partial t^2} = X(x)T''(t), \qquad \frac{\partial^2 u}{\partial x^2} = X''(x)T(t).$$

Substituting these expressions into (16), we have

$$X(x)T''(t) = \alpha^2 X''(x)T(t),$$

and separating variables gives

$$\frac{T''(t)}{\alpha^2 T(t)} = \frac{X''(x)}{X(x)}.$$

Just as before, these ratios must equal some constant K:

$$(20) \qquad \frac{X''(x)}{X(x)} = K \quad \text{and} \quad \frac{T''(t)}{\alpha^2 T(t)} = K.$$

Furthermore, with $u(x, t) = X(x)T(t)$, the boundary conditions in (17) give

$$X(0)T(t) = 0, \qquad X(L)T(t) = 0, \qquad t \geq 0.$$

[†]For a derivation of this mathematical model, see *Applied Differential Equations*, Third Edition, by Murray R. Spiegel (Prentice-Hall, Inc., Englewood Cliffs, New Jersey, 1980).

In order for these equations to hold for all $t \geq 0$, either $T(t) \equiv 0$, which implies that $u(x, t) \equiv 0$, or

$$X(0) = X(L) = 0.$$

Ignoring the trivial solution, we combine these boundary conditions with the differential equation for X in (20) and obtain the boundary value problem

(21) $X''(x) - KX(x) = 0,$ $X(0) = X(L) = 0,$

where K can be any constant.

This is the same boundary value problem that we encountered earlier while solving the heat equation. There we found that the eigenvalues are

$$K = -\left(\frac{n\pi}{L}\right)^2, \qquad n = 1, 2, 3, \ldots,$$

with corresponding eigenfunctions (nontrivial solutions)

(22) $X_n(x) = c_n \sin\left(\dfrac{n\pi x}{L}\right),$

where the c_n's are arbitrary nonzero constants.

Having determined that $K = -(n\pi/L)^2$ for some positive integer n, let's consider the second equation in (20) for such K:

$$T''(t) + \frac{\alpha^2 n^2 \pi^2}{L^2} T(t) = 0.$$

For each $n = 1, 2, 3, \ldots$, a general solution is

$$T_n(t) = c_{n,1} \cos\frac{n\pi\alpha}{L}t + c_{n,2} \sin\frac{n\pi\alpha}{L}t.$$

Combining this with equation (22), we obtain, for each $n = 1, 2, 3, \ldots$, the function

$$u_n(x, t) = X_n(x)T_n(t) = \left(c_n \sin\frac{n\pi x}{L}\right)\left(c_{n,1} \cos\frac{n\pi\alpha}{L}t + c_{n,2} \sin\frac{n\pi\alpha}{L}t\right),$$

or, reassembling the constants,

(23) $u_n(x, t) = \left(a_n \cos\dfrac{n\pi\alpha}{L}t + b_n \sin\dfrac{n\pi\alpha}{L}t\right) \sin\dfrac{n\pi x}{L}.$

Using the fact that linear combinations of solutions to (16)–(19) are again solutions, we consider the infinite sum (superposition) of the functions in (23):

(24) $u(x, t) = \displaystyle\sum_{n=1}^{\infty} \left[a_n \cos\dfrac{n\pi\alpha}{L}t + b_n \sin\dfrac{n\pi\alpha}{L}t\right] \sin\dfrac{n\pi x}{L}.$

For a solution of the form (24), substitution into the initial conditions (18)–(19) gives

(25) $$u(x,\ 0) = \sum_{n=1}^{\infty} a_n \sin \frac{n\pi x}{L} = f(x), \qquad 0 \le x \le L,$$

(26) $$\frac{\partial u}{\partial t}(x,\ 0) = \sum_{n=1}^{\infty} \frac{n\pi\alpha}{L} b_n \sin \frac{n\pi x}{L} = g(x), \qquad 0 \le x \le L.$$

We have now reduced the vibrating string problem (16)–(19) to the problem of determining the Fourier sine series expansions for $f(x)$ and $g(x)$:

(27) $$f(x) = \sum_{n=1}^{\infty} a_n \sin \frac{n\pi x}{L}, \qquad g(x) = \sum_{n=1}^{\infty} B_n \sin \frac{n\pi x}{L},$$

where $B_n = (n\pi\alpha/L)b_n$. If we choose the a_n's and b_n's so that the equations in (25) and (26) hold, then the expansion for $u(x,\ t)$ in (24) is a **formal solution** to the vibrating string problem (16)–(19). If this expansion converges to a function with continuous second partial derivatives, then the formal solution is an actual (genuine) solution.

EXAMPLE 2 Find the solution to the vibrating string problem

(28) $$\frac{\partial^2 u}{\partial t^2} = 4 \frac{\partial^2 u}{\partial x^2}, \qquad 0 < x < \pi, \qquad t > 0,$$

(29) $$u(0,\ t) = u(\pi,\ t) = 0, \qquad t \ge 0,$$

(30) $$u(x,\ 0) = \sin 3x - 4 \sin 10x, \qquad 0 \le x \le \pi,$$

(31) $$\frac{\partial u}{\partial t}(x,\ 0) = 2 \sin 4x + \sin 6x, \qquad 0 \le x \le \pi.$$

Solution Comparing equation (28) with equation (16), we see that $\alpha = 2$ and $L = \pi$. Hence we need only determine the values of the coefficients a_n and b_n in formula (24). The a_n's are chosen so that equation (25) holds; that is,

$$u(x,\ 0) = \sin 3x - 4 \sin 10x = \sum_{n=1}^{\infty} a_n \sin nx.$$

Equating coefficients of like terms, we see that

$$a_3 = 1, \qquad a_{10} = -4,$$

and the remaining a_n's are zero. Similarly, referring to equation (26), we must choose the b_n's so that

$$\frac{\partial u}{\partial t}(x,\ 0) = 2 \sin 4x + \sin 6x = \sum_{n=1}^{\infty} n2b_n \sin nx.$$

Comparing coefficients, we find

$$2 = (4)(2)b_4 \quad \text{or} \quad b_4 = \frac{1}{4},$$

$$1 = (6)(2)b_6 \quad \text{or} \quad b_6 = \frac{1}{12},$$

and the remaining b_n's are zero. Hence, from formula (24), the solution to the vibrating string problem (28)–(31) is

$$(32) \qquad u(x,\ t) = \cos 6t \sin 3x + \frac{1}{4} \sin 8t \sin 4x$$

$$+ \frac{1}{12} \sin 12t \sin 6x - 4 \cos 20t \sin 10x. \quad \blacksquare \text{ End}$$

In later sections the method of separation of variables is used to study a wide variety of problems for the heat, wave, and Laplace's equations. However, to use the method effectively, one must be able to compute trigonometric series (or, more generally, eigenfunction expansions) such as the Fourier sine series that we encountered here. These expansions are discussed in the next two sections.

EXERCISES 10.2

In Problems 1 through 8, determine all the solutions, if any, to the given boundary value problem by first finding a general solution to the differential equation.

1. $y'' - y = 0; \quad 0 < x < 1, \quad y(0) = 0, \quad y(1) = -4.$

2. $y'' - 6y' + 5y = 0; \quad 0 < x < 2,$
$y(0) = 1, \quad y(2) = 1.$

3. $y'' + 4y = 0; \quad 0 < x < \pi, \quad y(0) = 0, \quad y'(\pi) = 0.$

4. $y'' + 9y = 0; \quad 0 < x < \pi,$
$y(0) = 0, \quad y'(\pi) = -6.$

5. $y'' - y = 1 - 2x; \quad 0 < x < 1,$
$y(0) = 0, \quad y(1) = 1 + e.$

6. $y'' + y = 0; \quad 0 < x < 2\pi, \quad y(0) = 0, \quad y(2\pi) = 1.$

7. $y'' + y = 0; \quad 0 < x < 2\pi, \quad y(0) = 1, \quad y(2\pi) = 1.$

8. $y'' - 2y' + y = 0; \quad -1 < x < 1,$
$y(-1) = 0, \quad y(1) = 2.$

In Problems 9 through 14, find the values of λ (eigenvalues) for which the given problem has a nontrivial solution. Also determine the corresponding nontrivial solutions (eigenfunctions).

9. $y'' + \lambda y = 0; \quad 0 < x < \pi, \quad y(0) = 0, \quad y'(\pi) = 0.$

10. $y'' + \lambda y = 0; \quad 0 < x < \pi, \quad y'(0) = 0, \quad y(\pi) = 0.$

11. $y'' + \lambda y = 0; \quad 0 < x < 2\pi,$
$y(0) = y(2\pi), \quad y'(0) = y'(2\pi).$

12. $y'' + \lambda y = 0; \quad 0 < x < \pi/2,$
$y'(0) = 0, \quad y'(\pi/2) = 0.$

13. $y'' + \lambda y = 0; \quad 0 < x < \pi,$
$y(0) - y'(0) = 0, \quad y(\pi) = 0.$

14. $y'' - 2y' + \lambda y = 0; \quad 0 < x < \pi,$
$y(0) = 0, \quad y(\pi) = 0.$

In Problems 15 through 18, solve the heat flow problem (1)–(3) with $\beta = 3$, $L = \pi$, and the given initial function $f(x)$.

15. $f(x) = \sin x - 6 \sin 4x.$

16. $f(x) = \sin 3x + 5 \sin 7x - 2 \sin 13x.$

17. $f(x) = \sin x - 7 \sin 3x + \sin 5x.$

18. $f(x) = \sin 4x + 3 \sin 6x - \sin 10x.$

In Problems 19 through 22, solve the vibrating string problem (16)–(19) with $\alpha = 3$, $L = \pi$, and the given initial functions $f(x)$ and $g(x)$.

19. $f(x) = 3 \sin 2x + 12 \sin 13x, \quad g(x) \equiv 0.$

20. $f(x) \equiv 0$, $g(x) = -2 \sin 3x + 9 \sin 7x - \sin 10x$.

21. $f(x) = 6 \sin 2x + 2 \sin 6x$,
$g(x) = 11 \sin 9x - 14 \sin 15x$.

22. $f(x) = \sin x - \sin 2x + \sin 3x$,
$g(x) = 6 \sin 3x - 7 \sin 5x$.

23. Find the formal solution to the heat flow problem (1)–(3) with $\beta = 2$ and $L = 1$ if

$$f(x) = \sum_{n=1}^{\infty} \frac{1}{n^2} \sin n\pi x.$$

24. Find the formal solution to the vibrating string problem (16)–(19) with $\alpha = 4$, $L = \pi$, and

$$f(x) = \sum_{n=1}^{\infty} \frac{1}{n^2} \sin nx,$$

$$g(x) = \sum_{n=1}^{\infty} \frac{(-1)^{n+1}}{n} \sin nx.$$

25. By considering the behavior of the solutions of the equation

$$T'(t) - \beta KT(t) = 0, \qquad t > 0,$$

give an argument that is based on physical grounds to rule out the case where $K > 0$ in equation (5).

26. Verify that $u_n(x, t)$ given in equation (10) satisfies equation (1) and the boundary conditions in (2) by substituting $u_n(x, t)$ directly into the equations involved.

In Problems 27 through 30, a partial differential equation (PDE) is given along with the form of a solution having separated variables. Show that such a solution must satisfy the indicated set of ordinary differential equations.

27. $\dfrac{\partial^2 u}{\partial r^2} + \dfrac{1}{r}\dfrac{\partial u}{\partial r} + \dfrac{1}{r^2}\dfrac{\partial^2 u}{\partial \theta^2} = 0$

with $u(r, \theta) = R(r)T(\theta)$ yields

$$r^2 R''(r) + rR'(r) - \lambda R(r) = 0,$$
$$T''(\theta) + \lambda T(\theta) = 0,$$

where λ is a constant.

28. $\dfrac{\partial^2 u}{\partial t^2} + \dfrac{\partial u}{\partial t} + u = \alpha^2 \dfrac{\partial^2 u}{\partial x^2}$

with $u(x, t) = X(x)T(t)$ yields

$$X''(x) - \lambda X(x) = 0,$$
$$T''(t) + T'(t) + (1 - \lambda\alpha^2)T(t) = 0,$$

where λ is a constant.

29. $\dfrac{\partial u}{\partial t} = \beta \left\{ \dfrac{\partial^2 u}{\partial x^2} + \dfrac{\partial^2 u}{\partial y^2} \right\}$

with $u(x, y, t) = X(x)Y(y)T(t)$ yields

$$T'(t) - \beta KT(t) = 0,$$
$$X''(x) - JX(x) = 0,$$
$$Y''(y) + (J - K)Y(y) = 0,$$

where J, K are constants.

30. $\dfrac{\partial^2 u}{\partial r^2} + \dfrac{1}{r}\dfrac{\partial u}{\partial r} + \dfrac{1}{r^2}\dfrac{\partial^2 u}{\partial \theta^2} + \dfrac{\partial^2 u}{\partial z^2} = 0$

with $u(r, \theta, z) = R(r)T(\theta)Z(z)$ yields

$$T''(\theta) + \mu T(\theta) = 0,$$
$$Z''(z) + \lambda Z(z) = 0,$$
$$r^2 R''(r) + rR'(r) - (r^2\lambda + \mu)R(r) = 0,$$

where μ, λ are constants.

31. For the PDE in Problem 27, assume the following boundary conditions are imposed:

$$u(r, 0) = u(r, \pi) = 0,$$
$$u(r, \theta) \text{ remains bounded as } r \to 0^+.$$

Show that a nontrivial solution of the form $u(r, \theta) = R(r)T(\theta)$ must satisfy the boundary conditions

$$T(0) = T(\pi) = 0,$$
$$R(r) \text{ remains bounded as } r \to 0^+.$$

32. For the PDE in Problem 29, assume the following boundary conditions are imposed:

$$u(0, y, t) = u(a, y, t) = 0, \quad 0 \le y \le b, \quad t \ge 0,$$
$$\dfrac{\partial u}{\partial y}(x, 0, t) = \dfrac{\partial u}{\partial y}(x, b, t) = 0, \quad 0 \le x \le a, \quad t \ge 0.$$

Show that a nontrivial solution of the form $u(x, y, t) = X(x)Y(y)T(t)$ must satisfy the boundary conditions

$$X(0) = X(a) = 0,$$
$$Y'(0) = Y'(b) = 0.$$

33. When the temperature in a wire reaches a steady state, that is, when u depends only on x, then $u(x)$ satisfies Laplace's equation $\partial^2 u / \partial x^2 = 0$.

(a) Find the steady-state solution when the ends of the wire are kept at a constant temperature of 50°C, that is, when $u(0) = u(L) = 50$.

(b) Find the steady-state solution when one end of the wire is kept at 10°C, while the other is kept at 40°C, that is, when $u(0) = 10$ and $u(L) = 40$.

10.3 FOURIER SERIES

While analyzing heat flow and vibrating strings in the previous section, we encountered the problem of expressing a function in a trigonometric series (cf. equations (12) and (25) in Section 10.2). In the next two sections, we discuss the theory of Fourier series, which deals with trigonometric series expansions. First, however, we review some function properties that are particularly relevant to this study: piecewise continuity, periodicity, and even and odd symmetry.

In Section 7.2 we defined a **piecewise continuous** function on $[a, b]$ as a function f that is continuous at every point in $[a, b]$, except possibly for a finite number of points at which f has a *jump discontinuity.* Such functions are necessarily integrable over any finite interval on which they are piecewise continuous.

Recall also that a function is **periodic of period** T if $f(x + T) = f(x)$ for all x in the domain of f. The smallest positive value of T is called the **fundamental period.** The trigonometric functions $\sin x$ and $\cos x$ are examples of periodic functions with fundamental period 2π and $\tan x$ is periodic with fundamental period π. A constant function is a periodic function with arbitrary period T.

There are two symmetry properties of functions that will be useful in the study of Fourier series. A function f that satisfies $f(-x) = f(x)$ for all x in the domain of f has a graph that is symmetric with respect to the y-axis (see Figure 10.3(a)). We say that such a function is an **even** function. A function f that satisfies $f(-x) = -f(x)$ for all x in the domain of f has a graph that is symmetric with respect to the origin (see Figure 10.3(b)). It is said to be an **odd** function. The functions $1, x^2, x^4, \ldots$ are examples of even functions, while the functions $x, x^3, x^5, \ldots$ are odd. The trigonometric functions $\sin x$ and $\tan x$ are odd functions and $\cos x$ is an even function.

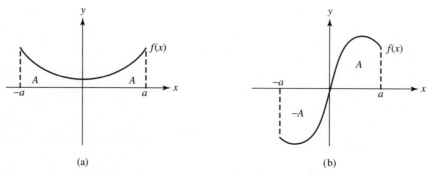

(a) (b)

Figure 10.3 (a) Even function $\int_{-a}^{a} f = A + A = 2\int_{0}^{a} f$
(b) Odd function $\int_{-a}^{a} f = A - A = 0$

EXAMPLE 1 Determine whether the given function is even, odd, or neither.

(a) $f(x) = \sqrt{1 + x^2}$. **(b)** $g(x) = x^{1/3} - \sin x$. **(c)** $h(x) = e^x$.

Solution (a) Since $f(-x) = \sqrt{1 + (-x)^2} = \sqrt{1 + x^2} = f(x)$, then $f(x)$ is an even function.

(b) Because $g(-x) = (-x)^{1/3} - \sin(-x) = -x^{1/3} + \sin x = -(x^{1/3} - \sin x) = -g(x)$, it follows that $g(x)$ is an odd function.

(c) Here $h(-x) = e^{-x}$. Since $e^{-x} = e^x$ only when $x = 0$ and e^{-x} is never equal to $-e^x$, then $h(x)$ is neither an even nor an odd function. ■ End

Knowing that a function is even or odd can be useful in evaluating definite integrals.

PROPERTIES OF SYMMETRIC FUNCTIONS

Theorem 1. If f is an even piecewise continuous function on $[-a, a]$, then

(1) $$\int_{-a}^{a} f(x)\,dx = 2\int_{0}^{a} f(x)\,dx.$$

If f is an odd piecewise continuous function on $[-a, a]$, then

(2) $$\int_{-a}^{a} f(x)\,dx = 0.$$

Proof. If f is an even function, then $f(-x) = f(x)$. Hence

$$\int_{-a}^{a} f(x)\,dx = \int_{-a}^{0} f(x)\,dx + \int_{0}^{a} f(x)\,dx$$

$$= -\int_{a}^{0} f(u)\,du + \int_{0}^{a} f(x)\,dx = 2\int_{0}^{a} f(x)\,dx,$$

where we used the change of variables $u = -x$. This is illustrated in Figure 10.3(a). Formula (2) can be proved by a similar argument and is illustrated in Figure 10.3(b). ■ End

The next example deals with certain integrals that are crucial in Fourier series.

EXAMPLE 2 Evaluate the following integrals when m and n are positive integers:

(a) $\displaystyle\int_{-T}^{T} \sin\frac{m\pi x}{T} \cos\frac{n\pi x}{T}\,dx.$ (b) $\displaystyle\int_{-T}^{T} \sin\frac{m\pi x}{T} \sin\frac{n\pi x}{T}\,dx.$

(c) $\displaystyle\int_{-T}^{T} \cos\frac{m\pi x}{T} \cos\frac{n\pi x}{T}\,dx.$

Solution **(a)** Since $\sin\dfrac{m\pi(-x)}{T}\cos\dfrac{n\pi(-x)}{T} = -\sin\dfrac{m\pi x}{T}\cos\dfrac{n\pi x}{T}$, the integrand in part (a) is an odd function. Thus, by Theorem 1,

(3) $$\int_{-T}^{T} \sin\frac{m\pi x}{T}\cos\frac{n\pi x}{T}\,dx = 0.$$

(b) Using the fact that the integrand is even (check this!) and the trigonometric identity

$$2\sin A\sin B = \cos(A - B) - \cos(A + B),$$

we find for $m \neq n$

$$\int_{-T}^{T} \sin\frac{m\pi x}{T}\sin\frac{n\pi x}{T}\,dx = 2\int_{0}^{T}\sin\frac{m\pi x}{T}\sin\frac{n\pi x}{T}\,dx$$

$$= \int_{0}^{T}\left\{\cos\frac{(m-n)\pi x}{T} - \cos\frac{(m+n)\pi x}{T}\right\}dx$$

$$= \frac{T}{\pi}\left[\frac{\sin\dfrac{(m-n)\pi x}{T}}{m-n} - \frac{\sin\dfrac{(m+n)\pi x}{T}}{m+n}\right]\Bigg|_{0}^{T} = 0.$$

When $m = n$, the integral in (b) becomes

$$\int_{-T}^{T}\left(\sin\frac{n\pi x}{T}\right)^{2}dx = 2\int_{0}^{T}\left(\sin\frac{n\pi x}{T}\right)^{2}dx = \int_{0}^{T}\left\{1 - \cos\frac{2n\pi x}{T}\right\}dx$$

$$= \left[x - \frac{T}{2n\pi}\sin\frac{2n\pi x}{T}\right]\Bigg|_{0}^{T} = T.$$

Hence

(4) $$\int_{-T}^{T}\sin\frac{m\pi x}{T}\sin\frac{n\pi x}{T}\,dx = \begin{cases} 0, & m \neq n, \\ T, & m = n. \end{cases}$$

The integrand in part (c) is also even and can be treated with a similar trigonometric identity. We leave it as an exercise for the reader to verify that (see Problem 8)

(5) $$\int_{-T}^{T}\cos\frac{m\pi x}{T}\cos\frac{n\pi x}{T}\,dx = \begin{cases} 0, & m \neq n, \\ T, & m = n. \end{cases}$$ ■ End

Equations (3)–(5) express an **orthogonality condition**[†] satisfied by the set of trigonometric functions $\{\cos x, \sin x, \cos 2x, \sin 2x, \ldots\}$, where $T = \pi$. We will say more about this later in this section.

It is easy to verify that if each of the functions $f_1, \ldots, f_n$ is periodic of period T, then so is any linear combination

$$c_1 f_1(x) + \cdots + c_n f_n(x).$$

For example, the sum $7 + 3 \cos \pi x - 8 \sin \pi x + 4 \cos 2\pi x - 6 \sin 2\pi x$ has period 2, since each term has period 2. Furthermore, if the infinite series

$$\frac{a_0}{2} + \sum_{n=1}^{\infty} \left(a_n \cos \frac{n\pi x}{T} + b_n \sin \frac{n\pi x}{T} \right)$$

consisting of $2T$-periodic functions converges for all x, then the function to which it converges will be periodic of period $2T$.

Just as we can associate a Taylor series with a function that has derivatives of all orders at a fixed point, we can identify a particular trigonometric series with a piecewise continuous function. To illustrate this, let's assume that $f(x)$ has the series expansion[††]

(6) $$f(x) = \frac{a_0}{2} + \sum_{n=1}^{\infty} \left\{ a_n \cos \frac{n\pi x}{T} + b_n \sin \frac{n\pi x}{T} \right\},$$

where the a_n's and b_n's are constants. (Necessarily, f has period $2T$.)

To determine the coefficients $a_0, a_1, b_1, a_2, b_2, \ldots$, we proceed as follows. Let's integrate $f(x)$ from $-T$ to T, assuming we can integrate term by term:

$$\int_{-T}^{T} f(x)\,dx = \int_{-T}^{T} \frac{a_0}{2}\,dx + \sum_{n=1}^{\infty} a_n \int_{-T}^{T} \cos \frac{n\pi x}{T}\,dx + \sum_{n=1}^{\infty} b_n \int_{-T}^{T} \sin \frac{n\pi x}{T}\,dx.$$

Since $\sin(n\pi x/T)$ is an odd function, $\int_{-T}^{T} \sin(n\pi x/T)\,dx = 0$. Further, for $n = 1, 2, \ldots$, it is easy to check that $\int_{-T}^{T} \cos(n\pi x/T)\,dx = 0$. Hence

$$\int_{-T}^{T} f(x)\,dx = \int_{-T}^{T} \frac{a_0}{2}\,dx = a_0 T,$$

and so

$$a_0 = \frac{1}{T} \int_{-T}^{T} f(x)\,dx.$$

[†] This nomenclature is suggested by the fact that the formulas for the Riemann sums approximating the integrals (3)–(5) look like dot products (Exercises 9.2, page 524) of higher-dimensional vectors. In most calculus texts, it is shown that the dot product of orthogonal vectors in two dimensions is zero.

[††] The choice of a constant $a_0/2$ instead of just a_0 will make the formulas easier to remember.

(Notice that $a_0/2$ is the average value of f over one period $2T$.) Next, to find the coefficient a_m when $m \geq 1$, we multiply (6) by $\cos(m\pi x/T)$ and integrate:

$$(7) \qquad \int_{-T}^{T} f(x) \cos \frac{m\pi x}{T} \, dx = \frac{a_0}{2} \int_{-T}^{T} \cos \frac{m\pi x}{T} \, dx + \sum_{n=1}^{\infty} a_n \int_{-T}^{T} \cos \frac{n\pi x}{T} \cos \frac{m\pi x}{T} \, dx$$

$$+ \sum_{n=1}^{\infty} b_n \int_{-T}^{T} \sin \frac{n\pi x}{T} \cos \frac{m\pi x}{T} \, dx.$$

The orthogonality conditions (3)–(5) render the integrals on the right-hand side quite immediate. We have already observed that

$$\int_{-T}^{T} \cos \frac{m\pi x}{T} \, dx = 0, \qquad m \geq 1,$$

and, by formulas (3) and (5), we have

$$\int_{-T}^{T} \sin \frac{n\pi x}{T} \cos \frac{m\pi x}{T} \, dx = 0,$$

$$\int_{-T}^{T} \cos \frac{n\pi x}{T} \cos \frac{m\pi x}{T} \, dx = \begin{cases} 0, & n \neq m, \\ T, & n = m. \end{cases}$$

Hence, in (7) we see that only one term on the right survives the integration:

$$\int_{-T}^{T} f(x) \cos \frac{m\pi x}{T} \, dx = a_m T.$$

Thus we have a formula for the coefficient a_m:

$$a_m = \frac{1}{T} \int_{-T}^{T} f(x) \cos \frac{m\pi x}{T} \, dx.$$

Similarly, multiplying (6) by $\sin(m\pi x/T)$ and integrating yields

$$\int_{-T}^{T} f(x) \sin \frac{m\pi x}{T} \, dx = b_m T$$

so that the formula for b_m is

$$b_m = \frac{1}{T} \int_{-T}^{T} f(x) \sin \frac{m\pi x}{T} \, dx.$$

Motivated by the above computations, we now make the following definition.

FOURIER SERIES

Definition 1. Let f be a piecewise continuous function on the interval $[-T, T]$. The **Fourier series**[†] of f is the trigonometric series

$$(8) \qquad f(x) \sim \frac{a_0}{2} + \sum_{n=1}^{\infty} \left\{ a_n \cos \frac{n\pi x}{T} + b_n \sin \frac{n\pi x}{T} \right\},$$

where the a_n's and b_n's are given by the formulas[††]

$$(9) \qquad a_n = \frac{1}{T} \int_{-T}^{T} f(x) \cos \frac{n\pi x}{T} \, dx, \qquad n = 0, 1, 2, \ldots,$$

$$(10) \qquad b_n = \frac{1}{T} \int_{-T}^{T} f(x) \sin \frac{n\pi x}{T} \, dx, \qquad n = 1, 2, 3, \ldots.$$

Formulas (9) and (10) are called the **Euler formulas.** We use the symbol $\sim$ in (8) to remind us that this series is associated with $f(x)$ but may not converge to $f(x)$. We will return to the question of convergence later in this section. First let's consider a few examples of Fourier series.

EXAMPLE 3 Compute the Fourier series for

$$f(x) = \begin{cases} 0, & -\pi < x < 0, \\ x, & 0 < x < \pi. \end{cases}$$

Solution Here $T = \pi$. Using formulas (9) and (10), we have

$$a_n = \frac{1}{\pi} \int_{-\pi}^{\pi} f(x) \cos nx \, dx = \frac{1}{\pi} \int_{0}^{\pi} x \cos nx \, dx$$

$$= \frac{1}{\pi n^2} \int_{0}^{\pi n} u \cos u \, du = \frac{1}{\pi n^2} [\cos u + u \sin u] \Big|_{0}^{\pi n}$$

$$= \frac{1}{\pi n^2} (\cos n\pi - 1) = \frac{1}{\pi n^2} [(-1)^n - 1], \qquad n = 1, 2, 3, \ldots,$$

$$a_0 = \frac{1}{\pi} \int_{-\pi}^{\pi} f(x) \, dx = \frac{1}{\pi} \int_{0}^{\pi} x \, dx = \frac{x^2}{2\pi} \Big|_{0}^{\pi} = \frac{\pi}{2},$$

[†]*Historical Footnote:* Joseph B. J. Fourier (1768–1830) developed his series for solving heat flow problems. Lagrange expressed doubts about the validity of the representation, but Dirichlet devised conditions that assured its convergence.

[††]Notice that $f(x)$ need not be defined for every x in $[-T, T]$; we need only that the integrals in (9) and (10) exist.

$$b_n = \frac{1}{\pi}\int_{-\pi}^{\pi} f(x)\sin nx\,dx = \frac{1}{\pi}\int_{0}^{\pi} x\sin nx\,dx$$

$$= \frac{1}{\pi n^2}\int_{0}^{\pi n} u\sin u\,du = \frac{1}{\pi n^2}\left[\sin u - u\cos u\right]\Big|_{0}^{\pi n}$$

$$= \frac{-\cos n\pi}{n} = \frac{(-1)^{n+1}}{n}, \qquad n = 1, 2, 3, \ldots.$$

Therefore,

(11) $$f(x) \sim \frac{\pi}{4} + \sum_{n=1}^{\infty}\left\{\frac{1}{\pi n^2}[(-1)^n - 1]\cos nx + \frac{(-1)^{n+1}}{n}\sin nx\right\}$$

$$= \frac{\pi}{4} - \frac{2}{\pi}\left\{\cos x + \frac{1}{9}\cos 3x + \frac{1}{25}\cos 5x + \cdots\right\}$$

$$+ \left\{\sin x - \frac{1}{2}\sin 2x + \frac{1}{3}\sin 3x + \cdots\right\}.$$

Some partial sums of this series are displayed in Figure 10.4. ■ End

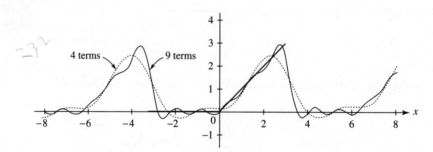

Figure 10.4 Partial sums of Fourier series in Example 3

EXAMPLE 4 Compute the Fourier series for

$$f(x) = \begin{cases} -1 & -\pi < x < 0, \\ 1 & 0 < x < \pi. \end{cases}$$

Solution Again, $T = \pi$. Notice that f is an odd function. Since the product of an odd function and an even function is odd (see Problem 7), $f(x)\cos nx$ is also an odd function. Thus

$$a_n = \frac{1}{\pi}\int_{-\pi}^{\pi} f(x)\cos nx\,dx = 0, \qquad n = 0, 1, 2, \ldots.$$

Furthermore, $f(x) \sin nx$ is the product of two odd functions and therefore is an even function, so

$$b_n = \frac{1}{\pi} \int_{-\pi}^{\pi} f(x) \sin nx\,dx = \frac{2}{\pi} \int_0^{\pi} \sin nx\,dx$$

$$= \frac{2}{\pi} \left[\frac{-\cos nx}{n} \right]\Bigg|_0^{\pi} = \frac{2}{\pi} \left[\frac{1}{n} - \frac{(-1)^n}{n} \right], \qquad n = 1, 2, 3, \ldots,$$

$$= \begin{cases} 0, & n \text{ even}, \\ \dfrac{4}{\pi n}, & n \text{ odd}. \end{cases}$$

Thus

$$(12) \qquad f(x) \sim \frac{2}{\pi} \sum_{n=1}^{\infty} \frac{[1 - (-1)^n]}{n} \sin nx = \frac{4}{\pi} \left[\sin x + \frac{1}{3} \sin 3x + \frac{1}{5} \sin 5x + \cdots \right].$$

Some partial sums of (12) are sketched in Figure 10.5. ■ End

Figure 10.5 Partial sums of Fourier series in Example 4

In Example 4, the odd function f has a Fourier series consisting only of sine functions. It is easy to see that, in general, if f is any odd function, then its Fourier series consists only of sine terms.

EXAMPLE 5 Compute the Fourier series for $f(x) = |x|$, $-1 < x < 1$.

Solution Here $T = 1$. Since f is an even function, $f(x) \sin n\pi x$ is an odd function. Therefore

$$b_n = \int_{-1}^{1} f(x) \sin n\pi x\,dx = 0, \qquad n = 1, 2, 3, \ldots.$$

Since $f(x) \cos n\pi x$ is an even function, we have

$$a_0 = \int_{-1}^{1} f(x)\,dx = 2\int_0^1 x\,dx = x^2 \Big|_0^1 = 1,$$

$$a_n = \int_{-1}^{1} f(x) \cos n\pi x\,dx = 2\int_0^1 x \cos n\pi x\,dx$$

$$= \frac{2}{\pi^2 n^2}\int_0^{\pi n} u \cos u\,du = \frac{2}{\pi^2 n^2}[\cos u + u \sin u]\Big|_0^{\pi n} = \frac{2}{\pi^2 n^2}(\cos n\pi - 1)$$

$$= \frac{2}{\pi^2 n^2}[(-1)^n - 1], \qquad n = 1, 2, 3, \dots .$$

Therefore

$$(13) \qquad f(x) \sim \frac{1}{2} + \sum_{n=1}^{\infty} \frac{2}{\pi^2 n^2}[(-1)^n - 1] \cos n\pi x$$

$$= \frac{1}{2} - \frac{4}{\pi^2}\left\{\cos \pi x + \frac{1}{9}\cos 3\pi x + \frac{1}{25}\cos 5\pi x + \cdots\right\}.$$

The partial sums for (13) are displayed in Figure 10.6. ■ End

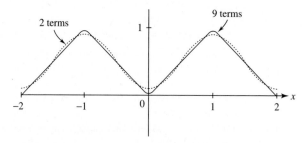

Figure 10.6 Partial sums of Fourier series in Example 5

Notice that the even function f of Example 5 has a Fourier series consisting only of cosine functions and the constant function $1 = \cos 0\pi x$. In general, if f is an even function, then its Fourier series consists only of cosine functions (including $\cos 0\pi x$).

Orthogonal Expansions

Fourier series are examples of orthogonal expansions.[†] A set of functions $\{f_n(x)\}_{n=1}^{\infty}$ is said to be an **orthogonal system** or just **orthogonal** with respect to the nonnegative weight

[†] Orthogonality is also discussed in Section 8.9 on pages 501–502.

function $w(x)$ on the interval $[a, b]$ if

(14) $\qquad \displaystyle\int_a^b f_m(x) f_n(x) w(x)\, dx = 0, \quad$ whenever $\quad m \ne n.$

As we have seen, the set of trigonometric functions

(15) $\qquad \{1,\ \cos x,\ \sin x,\ \cos 2x,\ \sin 2x,\ \dots\}$

is orthogonal on $[-\pi, \pi]$ with respect to the weight function $w(x) \equiv 1$. If we define the **norm** of f as

(16) $\qquad \|f\| := \left[\displaystyle\int_a^b f^2(x) w(x)\, dx \right]^{1/2},$

then we say that a set of functions $\{f_n(x)\}_{n=1}^\infty$ (or $\{f_n(x)\}_{n=1}^N$) is an **orthonormal system with respect to** $w(x)$ if (14) holds and also $\|f_n\| = 1$ for each n. Equivalently, we say the set is an orthonormal system if

(17) $\qquad \displaystyle\int_a^b f_m(x)\, f_n(x) w(x)\, dx = \begin{cases} 0, & m \ne n, \\ 1, & m = n. \end{cases}$

We can always obtain an orthonormal system from an orthogonal system just by dividing each function by its norm. In particular, since

$$\int_{-\pi}^\pi \cos^2 nx\, dx = \int_{-\pi}^\pi \sin^2 nx\, dx = \pi, \qquad n = 1, 2, 3, \dots$$

and

$$\int_{-\pi}^\pi 1\, dx = 2\pi,$$

then the orthogonal system (15) gives rise on $[-\pi, \pi]$ to the orthonormal system

$$\{(2\pi)^{-1/2},\ \pi^{-1/2} \cos x,\ \pi^{-1/2} \sin x,\ \pi^{-1/2} \cos 2x,\ \pi^{-1/2} \sin 2x,\ \dots\}.$$

If $\{f_n(x)\}_{n=1}^\infty$ is an orthogonal system with respect to $w(x)$ on $[a, b]$, we might ask if we can expand a function $f(x)$ in terms of these functions; that is, can we express f in the form

(18) $\qquad f(x) = c_1 f_1(x) + c_2 f_2(x) + \cdots$

for a suitable choice of constants $c_1, c_2, \dots$? Such an expansion is called an **orthogonal expansion,** or a **generalized Fourier series.**

To determine the constants in (18), we can proceed as we did in deriving Euler's formulas for the coefficients of a Fourier series, this time using the orthogonality of the system. Presuming the representation (18) is valid, we multiply by $f_m(x) w(x)$ and integrate

to obtain

(19)
$$\int_a^b f(x) f_m(x) w(x)\, dx = c_1 \int_a^b f_1(x) f_m(x) w(x)\, dx + c_2 \int_a^b f_2(x) f_m(x) w(x)\, dx + \cdots$$

$$= \sum_{n=1}^{\infty} c_n \int_a^b f_n(x) f_m(x) w(x)\, dx.$$

(Here we have also assumed that we can integrate term by term.) Because the system is orthogonal with respect to $w(x)$, every integral on the right-hand side of (19) is zero except when $n = m$. Solving for c_m gives

(20)
$$c_m = \frac{\displaystyle\int_a^b f(x) f_m(x) w(x)\, dx}{\displaystyle\int_a^b f_m^2(x) w(x)\, dx} = \frac{\displaystyle\int_a^b f(x) f_m(x) w(x)\, dx}{\|f_m\|^2}, \qquad n = 1, 2, 3, \ldots .$$

The derivation of the formula for c_m was only *formal*, since the question of the convergence of the expansion in (18) was not answered. If the series $\sum_{n=1}^{\infty} c_n f_n(x)$ converges *uniformly* to $f(x)$ on $[a, b]$, then each step can be justified, and indeed, the coefficients are given by formula (20). The notion of uniform convergence is discussed in the next subsection and in Section 13.2.[†]

Convergence of Fourier Series

Let's turn to the question of the convergence of a Fourier series. Figures 10.4–10.6 provide some clues. In Example 5 it is possible to use a comparison or limit comparison test to show that the series is absolutely dominated by a p-series of the form $\sum_{n=1}^{\infty} 1/n^2$, which converges. However, this is much harder to do in Example 4, since the terms go to zero like $1/n$. Matters can be even worse, since there exist Fourier series that *diverge*.[††] We state two theorems that deal with the convergence of a Fourier series and two dealing with the properties of termwise differentiation and integration. For proofs of these results, see *Partial Differential Equations of Mathematical Physics*, by Tyn Myint-U (Elsevier North Holland, Inc., New York, 1980), Chapter 5; *Advanced Calculus with Applications*, by N. J. DeLillo (Macmillan, New York, 1982), Chapter 9; or an advanced text on the theory of Fourier series.

Before proceeding, we need the following notation. Let

$$f(x^+) := \lim_{h \to 0^+} f(x + h) \quad \text{and} \quad f(x^-) := \lim_{h \to 0^+} f(x - h).$$

[†] All references to Chapters 11–13 refer to the expanded text *Fundamentals of Differential Equations and Boundary Value Problems*, Second Edition.

[††] In fact, there are trigonometric series that converge but are *not* Fourier series; an example is $\displaystyle\sum_{n=1}^{\infty} \frac{\sin nx}{\ln(n + 1)}$.

POINTWISE CONVERGENCE Of FOURIER SERIES

Theorem 2. If f and f' are piecewise continuous on $[-T, T]$, then for any x in $(-T, T)$

(21) $$\frac{a_0}{2} + \sum_{n=1}^{\infty}\left\{a_n\cos\frac{n\pi x}{T} + b_n\sin\frac{n\pi x}{T}\right\} = \frac{1}{2}[f(x^+) + f(x^-)],$$

where the a_n's and b_n's are given by the Euler formulas (9) and (10). For $x = \pm T$, the series converges to $\frac{1}{2}[f(-T^+) + f(T^-)].$[†]

In other words, when f and f' are piecewise continuous on $[-T, T]$, the Fourier series converges to $f(x)$ whenever f is continuous at x and converges to the average of the left- and right-hand limits at points where f is discontinuous.

Observe that the left-hand side of (21) is periodic of period $2T$. This means that if we extend $f(x)$ from the interval $(-T, T)$ to the entire real line using $2T$-periodicity, then equation (21) holds for all x for the $2T$ periodic extension of $f(x)$.

EXAMPLE 6 To which function does the Fourier series for

$$f(x) = \begin{cases} -1, & -\pi < x < 0, \\ 1, & 0 < x < \pi, \end{cases}$$

converge?

Solution In Example 4 we found that the Fourier series for $f(x)$ is given by (12), and in Figure 10.5, we sketched the graphs of two of its partial sums. Now $f(x)$ and $f'(x)$ are piecewise continuous in $[-\pi, \pi]$. Moreover, f is continuous except at $x = 0$. Thus by Theorem 2, the Fourier series of f in (12) converges to the 2π periodic function $g(x)$, where $g(x) = f(x) = -1$ for $-\pi < x < 0$, $g(x) = f(x) = 1$ for $0 < x < \pi$, $g(0) = 1/2[f(0^+) + f(0^-)] = 0$, and at $\pm\pi$ we have $g(\pm\pi) = [f(-\pi^+) + f(\pi^-)]/2 = (-1 + 1)/2 = 0$. The graph of $g(x)$ is given in Figure 10.7. ■ End

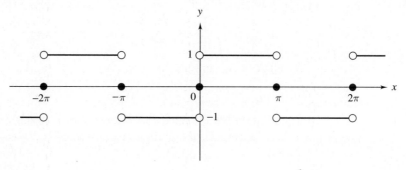

Figure 10.7 The limit function of the Fourier series for $f(x) = \begin{cases} -1, & -\pi < x < 0, \\ 1, & 0 < x < \pi \end{cases}$

[†] From formula (21), we see that it doesn't matter how we define $f(x)$ at its points of discontinuity, since only the left- and right-hand limits are involved. The derivative $f'(x)$, of course, is undefined at such points.

When f is a $2T$-periodic function that is continuous on $(-\infty, \infty)$ and has a piecewise continuous derivative, its Fourier series not only converges at each point—it converges **uniformly on** $(-\infty, \infty)$. This means that for any prescribed tolerance $\varepsilon > 0$, the graph of the partial sum

$$s_N(x) := \frac{a_0}{2} + \sum_{n=1}^{N} \left\{ a_n \cos \frac{n\pi x}{T} + b_n \sin \frac{n\pi x}{T} \right\}$$

will, for all N large, lie in an ε-corridor about the graph of f on $(-\infty, \infty)$ (see Figure 10.8). The property of uniform convergence of Fourier series is particularly helpful when one needs to verify that a formal solution to a partial differential equation is an actual (genuine) solution.

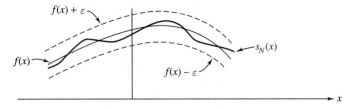

Figure 10.8 An ε-corridor about f

UNIFORM CONVERGENCE OF FOURIER SERIES

Theorem 3. Let f be a continuous function on $(-\infty, \infty)$ and periodic of period $2T$. If f' is piecewise continuous on $[-T, T]$, then the Fourier series for f converges uniformly to f on $[-T, T]$ and hence on any interval. That is, for each $\varepsilon > 0$, there exists an integer N_0 (that depends on ε) such that

$$\left| f(x) - \left\{ \frac{a_0}{2} + \sum_{n=1}^{N} \left\{ a_n \cos \frac{n\pi x}{T} + b_n \sin \frac{n\pi x}{T} \right\} \right\} \right| < \varepsilon,$$

for all $N \geq N_0$, and all $x \in (-\infty, \infty)$.

In Example 5 we obtained the Fourier series expansion given in (13) for $f(x) = |x|$, $-1 < x < 1$. Since $g(x)$, the periodic extension of $f(x)$ (see Figure 10.9) is continuous on $(-\infty, \infty)$ and

$$f'(x) = \begin{cases} -1, & -1 < x < 0, \\ 1, & 0 < x < 1, \end{cases}$$

is piecewise continuous on $[-1, 1]$, the Fourier series expansion (13) converges uniformly to $|x|$ on $[-1, 1]$. Compare Figure 10.6.

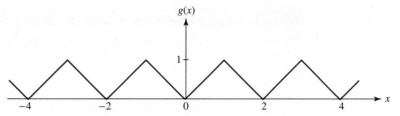

Figure 10.9 Periodic extension of $f(x) = |x|$, $-1 < x < 1$

The term-by-term differentiation of a Fourier series is not always permissible. For example, the Fourier series for $f(x) = x$, $-\pi < x < \pi$ (see Problem 9) is

(22) $$f(x) \sim 2 \sum_{n=1}^{\infty} (-1)^{n+1} \frac{\sin nx}{n},$$

which converges for all x, whereas its derived series

$$2 \sum_{n=1}^{\infty} (-1)^{n+1} \cos nx$$

diverges for every x. The following theorem gives sufficient conditions for using termwise differentiation.

DIFFERENTIATION OF FOURIER SERIES

Theorem 4. Let $f(x)$ be continuous on $(-\infty, \infty)$ and $2T$ periodic. Let $f'(x)$ and $f''(x)$ be piecewise continuous on $[-T, T]$. Then, the Fourier series of $f'(x)$ can be obtained from the Fourier series for $f(x)$ by termwise differentiation. In particular, if

$$f(x) = \frac{a_0}{2} + \sum_{n=1}^{\infty} \left\{ a_n \cos \frac{n \pi x}{T} + b_n \sin \frac{n \pi x}{T} \right\},$$

then

$$f'(x) \sim \sum_{n=1}^{\infty} \frac{\pi n}{T} \left\{ -a_n \sin \frac{n \pi x}{T} + b_n \cos \frac{n \pi x}{T} \right\}.$$

Notice that Theorem 4 does not apply to the example function $f(x)$ and its Fourier series expansion shown in (22), since the 2π-periodic extension of this $f(x)$ fails to be continuous on $(-\infty, \infty)$.

Termwise integration of a Fourier series is permissible under much weaker conditions.

INTEGRATION OF FOURIER SERIES

Theorem 5. Let $f(x)$ be piecewise continuous on $[-T, T]$ with Fourier series

$$f(x) \sim \frac{a_0}{2} + \sum_{n=1}^{\infty} \left\{ a_n \cos \frac{n\pi x}{T} + b_n \sin \frac{n\pi x}{T} \right\}.$$

Then, for any x in $[-T, T]$, we have

$$\int_{-T}^{x} f(t)\,dt = \int_{-T}^{x} \frac{a_0}{2}\,dt + \sum_{n=1}^{\infty} \int_{-T}^{x} \left\{ a_n \cos \frac{n\pi t}{T} + b_n \sin \frac{n\pi t}{T} \right\} dt.$$

EXERCISES 10.3

In Problems 1 through 6, determine whether the given function is even, odd, or neither.

1. $f(x) = x^3 + \sin 2x$. **2.** $f(x) = \sin^2 x$.

3. $f(x) = (1 - x^2)^{-1/2}$. **4.** $f(x) = \sin(x + 1)$.

5. $f(x) = e^{-x} \cos 3x$. **6.** $f(x) = x^{1/5} \cos x^2$.

7. Prove the following properties:

 (a) If f and g are even functions, then so is the product fg.

 (b) If f and g are odd functions, then fg is an even function.

 (c) If f is an even function and g is an odd function, then fg is an odd function.

8. Verify formula (5). [*Hint:* use the identity $2 \cos A \cos B = \cos(A + B) + \cos(A - B)$.]

In Problems 9 through 16, compute the Fourier series for the given function f on the specified interval. Use a computer or graphing calculator to plot a few partial sums of the Fourier series.

9. $f(x) = x$, $-\pi < x < \pi$.

10. $f(x) = |x|$, $-\pi < x < \pi$.

11. $f(x) = \begin{cases} 1, & -2 < x < 0, \\ x, & 0 < x < 2. \end{cases}$

12. $f(x) = \begin{cases} 0, & -\pi < x < 0, \\ x^2, & 0 < x < \pi. \end{cases}$

13. $f(x) = x^2$, $-1 < x < 1$.

14. $f(x) = \begin{cases} x, & 0 < x < \pi, \\ x + \pi, & -\pi < x < 0. \end{cases}$

15. $f(x) = e^x$, $-\pi < x < \pi$.

16. $f(x) = \begin{cases} 0, & -\pi < x < -\pi/2, \\ -1, & -\pi/2 < x < 0, \\ 1, & 0 < x < \pi/2, \\ 0, & \pi/2 < x < \pi. \end{cases}$

In Problems 17 through 24, determine the function to which the Fourier series for $f(x)$, given in the indicated problem, converges.

17. Problem 9. **18.** Problem 10.

19. Problem 11. **20.** Problem 12.

21. Problem 13. **22.** Problem 14.

23. Problem 15. **24.** Problem 16.

25. Find the functions represented by the series obtained by the termwise integration of the given series from $-\pi$ to x:

 (a) $2 \sum_{n=1}^{\infty} \frac{(-1)^{n+1}}{n} \sin nx = x$, $-\pi < x < \pi$.

 (b) $-\frac{4}{\pi} \sum_{n=0}^{\infty} \frac{\sin(2n + 1)x}{(2n + 1)} \sim f(x)$,

 $$f(x) = \begin{cases} -1, & -\pi < x < 0, \\ 1, & 0 < x < \pi. \end{cases}$$

26. Show that the set of functions

$$\left\{ \cos\frac{\pi}{2}x,\ \sin\frac{\pi}{2}x,\ \cos\frac{3\pi}{2}x,\ \sin\frac{3\pi}{2}x,\ \ldots, \right.$$

$$\left. \cos\frac{(2n-1)\pi}{2}x,\ \sin\frac{(2n-1)\pi}{2}x,\ \ldots \right\}$$

is an orthonormal system on $[-1, 1]$ with respect to the weight function $w(x) \equiv 1$.

27. Find the orthogonal expansion (generalized Fourier series) for

$$f(x) = \begin{cases} 0, & -1 < x < 0, \\ 1, & 0 < x < 1, \end{cases}$$

in terms of the orthonormal system of Problem 26.

28. (a) Show that the function $f(x) = x^2$, $-\pi < x < \pi$ has the Fourier series

$$f(x) \sim \frac{\pi^2}{3} + 4\sum_{n=1}^{\infty}\frac{(-1)^n}{n^2}\cos nx.$$

(b) Use the result of part (a) and Theorem 2 to show that

$$\sum_{n=1}^{\infty}\frac{(-1)^{n+1}}{n^2} = \frac{\pi^2}{12}.$$

(c) Use the result of part (a) and Theorem 2 to show that

$$\sum_{n=1}^{\infty}\frac{1}{n^2} = \frac{\pi^2}{6}.$$

29. In Section 8.9, it was shown that the Legendre polynomials $P_n(x)$ are orthogonal on the interval $[-1, 1]$ with respect to the weight function $w(x) \equiv 1$. Using the fact that the first three Legendre polynomials are

$$P_0(x) \equiv 1, \quad P_1(x) = x, \quad P_2(x) = (3/2)x^2 - (1/2),$$

find the first three coefficients in the expansion

$$f(x) = a_0 P_0(x) + a_1 P_1(x) + a_2 P_2(x) + \cdots,$$

where $f(x)$ is the function

$$f(x) := \begin{cases} -1, & -1 < x < 0, \\ 1, & 0 < x < 1. \end{cases}$$

30. As in Problem 29, find the first three coefficients in the expansion

$$f(x) = a_0 P_0(x) + a_1 P_1(x) + a_2 P_2(x) + \cdots,$$

when $f(x) = |x|$, $-1 < x < 1$.

31. The Hermite polynomials $H_n(x)$ are orthogonal on the interval $(-\infty, \infty)$ with respect to the weight function $W(x) = e^{-x^2}$. Verify this fact for the first three Hermite polynomials:

$$H_0(x) \equiv 1, \quad H_1(x) = 2x, \quad H_2(x) = 4x^2 - 2.$$

32. The Chebyshev (Tchebichef) polynomials $T_n(x)$ are orthogonal on the interval $[-1, 1]$ with respect to the weight function $w(x) = (1 - x^2)^{-1/2}$. Verify this fact for the first three Chebyshev polynomials:

$$T_0(x) \equiv 1, \quad T_1(x) = x, \quad T_2(x) = 2x^2 - 1.$$

33. Let $\{f_n(x)\}$ be an orthogonal set of functions on the interval $[a, b]$ with respect to the weight function $w(x)$. Show that they satisfy the **Pythagorean property**

$$\|f_m + f_n\|^2 = \|f_m\|^2 + \|f_n\|^2$$

if $m \neq n$.

34. Norm. The norm of a function $\|f\|$ is like the length of a vector in R^n. In particular, show that the norm defined in (16) satisfies the following properties associated with length (assume f and g are continuous and $w(x) > 0$ on $[a, b]$):

(a) $\|f\| \geq 0$, and $\|f\| = 0$ if and only if $f \equiv 0$.
(b) $\|cf\| = |c| \|f\|$, where c is any real number.
(c) $\|f + g\| \leq \|f\| + \|g\|$.

35. Inner Product. The integral in the orthogonality condition (14) is like the dot product of two vectors in R^n. In particular, show that the **inner product** of two functions defined by

$$(23) \qquad \langle f, g \rangle := \int_a^b f(x)g(x)w(x)\,dx,$$

where $w(x)$ is a positive weight function, satisfies the following properties associated with the dot product (assume f, g, and h are continuous on $[a, b]$):

(a) $\langle f + g, h \rangle = \langle f, h \rangle + \langle g, h \rangle$.
(b) $\langle cf, h \rangle = c\langle f, h \rangle$ where c is any real number.
(c) $\langle f, g \rangle = \langle g, f \rangle$.

36. Complex Form of the Fourier Series

(a) Using the Euler formula $e^{i\theta} = \cos\theta + i\sin\theta$, $i = \sqrt{-1}$, prove that

$$\cos nx = \frac{e^{inx} + e^{-inx}}{2} \quad \text{and} \quad \sin nx = \frac{e^{inx} - e^{-inx}}{2i}.$$

(b) Show that the Fourier series

$$f(x) \sim \frac{a_0}{2} + \sum_{n=1}^{\infty}\{a_n \cos nx + b_n \sin nx\}$$

$$= c_0 + \sum_{n=1}^{\infty}\{c_n e^{inx} + c_{-n}e^{-inx}\},$$

where

$$c_0 = \frac{a_0}{2}, \qquad c_n = \frac{a_n - ib_n}{2}, \qquad c_{-n} = \frac{a_n + ib_n}{2}.$$

(c) Finally, use the results of part (b) to show that

$$f(x) \sim \sum_{n=-\infty}^{\infty} c_n e^{inx}, \qquad -\pi < x < \pi,$$

where

$$c_n = \frac{1}{2\pi} \int_{-\pi}^{\pi} f(x) e^{-inx} \, dx.$$

37. Least Squares Approximation Property. The Nth partial sum of the Fourier series gives the best mean square approximation of f by a trigonometric polynomial. To prove this, proceed as follows. Let

$$f(x) \sim \frac{a_0}{2} + \sum_{n=1}^{\infty} \{a_n \cos nx + b_n \sin nx\},$$

and let $F_N(x)$ denote an arbitrary trigonometric polynomial of degree N:

$$F_N(x) = \frac{\alpha_0}{2} + \sum_{n=1}^{N} \{\alpha_n \cos nx + \beta_n \sin nx\}.$$

Define

$$E := \int_{-\pi}^{\pi} [f(x) - F_N(x)]^2 \, dx,$$

which is the *total square error*. Expanding the integrand, we get

$$E = \int_{-\pi}^{\pi} f^2(x) \, dx - 2 \int_{-\pi}^{\pi} f(x) F_N(x) \, dx$$

$$+ \int_{-\pi}^{\pi} F_N^2(x) \, dx.$$

(a) Show that

$$\int_{-\pi}^{\pi} F_N^2(x) \, dx = \pi \left(\frac{\alpha_0^2}{2} + \alpha_1^2 + \cdots + \alpha_N^2 \right.$$

$$\left. + \beta_1^2 + \cdots + \beta_N^2 \right)$$

and

$$\int_{-\pi}^{\pi} f(x) F_N(x) \, dx = \pi \left(\frac{\alpha_0 a_0}{2} + \alpha_1 a_1 + \cdots + \alpha_N a_N \right.$$

$$\left. + \beta_1 b_1 + \cdots + \beta_N b_N \right).$$

(b) Let E^* be the error when we approximate f by the Nth partial sum of its Fourier series; that is, when we choose $\alpha_n = a_n$ and $\beta_n = b_n$. Show that

$$E^* = \int_{-\pi}^{\pi} f^2(x) \, dx - \pi \left(\frac{a_0^2}{2} + a_1^2 + \cdots + a_N^2 \right.$$

$$\left. + b_1^2 + \cdots + b_N^2 \right).$$

(c) Using the results of parts (a) and (b), show that $E - E^* \geq 0$, that is, $E \geq E^*$, by proving that

$$E - E^* = \pi \left\{ \frac{(\alpha_0 - a_0)^2}{2} + (\alpha_1 - a_1)^2 \right.$$

$$+ \cdots + (\alpha_N - a_N)^2 + (\beta_1 - b_1)^2$$

$$\left. + \cdots + (\beta_N - b_N)^2 \right\}.$$

Hence the Nth partial sum of the Fourier series gives the least total square error, since $E \geq E^*$.

38. Bessel's Inequality. Use the fact that E^*, defined in part (b) of Problem 37, is nonnegative to prove **Bessel's inequality**

$$(24) \qquad \frac{a_0^2}{2} + \sum_{n=1}^{\infty} \{a_n^2 + b_n^2\} \leq \frac{1}{\pi} \int_{-\pi}^{\pi} f^2(x) \, dx.$$

(If f is piecewise continuous on $[-\pi, \pi]$, then we have equality in (24). This result is called **Parseval's identity**.)

39. Gibbs Phenomenon.[†] The American mathematician Josiah Willard Gibbs (1839–1903) observed that near points of discontinuity of f, the partial sums of the Fourier series for f may overshoot by approximately 9% of the jump, regardless of the number of terms. This is illustrated in Figure 10.10 for the function

$$f(x) = \begin{cases} -1, & -\pi < x < 0, \\ 1, & 0 < x < \pi, \end{cases}$$

whose Fourier series has the partial sums

$$f_{2n-1}(x) = \frac{4}{\pi} \left[\sin x + \frac{1}{3} \sin 3x + \cdots + \frac{\sin (2n-1)x}{(2n-1)} \right].$$

[†] *Historical Footnote:* Actually, H. Wilbraham discovered this phenomenon some fifty years earlier than Gibbs did. It is more appropriately called the *Gibbs-Wilbraham* phenomenon.

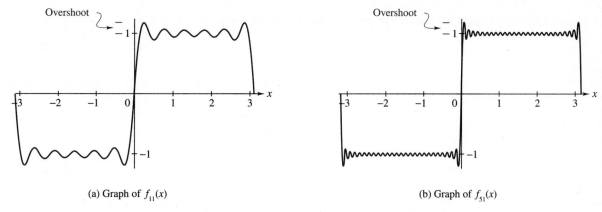

(a) Graph of $f_{11}(x)$ (b) Graph of $f_{51}(x)$

Figure 10.10 Gibbs's phenomenon for partial sums of Fourier series

To verify this for $f(x)$, proceed as follows:

(a) Show that

$$\pi \sin x f_{2n-1}'(x) = 4 \sin x[\cos x + \cos 3x$$

$$+ \cdots + \cos(2n - 1)x]$$

$$= 2 \sin 2nx.$$

(b) Infer from part (a) and the figure that the maximum occurs at $x = \pi/(2n)$ and has the value

$$f_{2n-1}\left(\frac{\pi}{2n}\right) = \frac{4}{\pi}\left[\sin\frac{\pi}{2n} + \frac{1}{3}\sin\frac{3\pi}{2n}\right.$$

$$\left. + \cdots + \frac{1}{2n-1}\sin\frac{(2n-1)\pi}{2n}\right].$$

(c) Show that if one approximates

$$\int_0^\pi \frac{\sin x}{x} dx$$

using the partition $x_k := (2k - 1)(\pi/2n)$, $k = 1$, $2, \ldots, n$, $\Delta x_k = \pi/n$ and choosing the midpoint of each interval as the place to evaluate the integrand,

then

$$\int_0^\pi \frac{\sin x}{x} dx \approx \frac{\sin(\pi/2n)}{\pi/2n}\frac{\pi}{n} + \cdots + \frac{\sin[(2n - 1)\pi/2n]}{(2n - 1)\pi/2n}\frac{\pi}{n}$$

$$= \frac{\pi}{2}f_{2n-1}\left(\frac{\pi}{2n}\right).$$

(d) Use the result of part (c) to show that the overshoot satisfies

$$\lim_{n\to\infty} f_{2n-1}\left(\frac{\pi}{2n}\right) = \frac{2}{\pi}\int_0^\pi \frac{\sin x}{x} dx.$$

(e) Using the result of part (d) and a numerical integration algorithm (such as Simpson's rule, Appendix B) for the **sine integral function**

$$Si(z) := \int_0^z \frac{\sin x}{x} dx,$$

show that $\lim_{n\to\infty} f_{2n-1}(\pi/(2n)) \approx 1.18$. Thus the approximations overshoot the true value of $f(0^+) = 1$ by 0.18 or 9% of the jump from $f(0^-)$ to $f(0^+)$.

10.4 FOURIER COSINE AND SINE SERIES

A typical problem encountered in using separation of variables to solve a partial differential equation is the problem of representing a function defined on some finite interval by a trigonometric series consisting of only sine functions or only cosine functions. For example, in Section 10.2, equation (12), we needed to express the initial values $u(x, 0) = f(x)$, $0 < x < L$, of the solution to the initial-boundary value problem associated with the problem

of heat flow as a trigonometric series of the form

$$\textbf{(1)} \qquad f(x) = \sum_{n=1}^{\infty} c_n \sin\left(\frac{n\pi x}{L}\right).$$

Recalling that the Fourier series for an odd function defined on $[-L, L]$ consists entirely of sine terms, we might try to achieve (1) by extending the function $f(x)$, $0 < x < L$, to the interval $(-L, L)$ in such a way that the extended function is odd. This is accomplished by defining the function

$$f_o(x) := \begin{cases} f(x), & 0 < x < L, \\ -f(-x), & -L < x < 0. \end{cases}$$

and extending $f_o(x)$ to all x using $2L$-periodicity. Since $f_o(x)$ is an odd function, it has a Fourier series consisting entirely of sine terms. Moreover, $f_o(x)$ is an extension of $f(x)$, since $f_o(x) = f(x)$ on $(0, L)$. This extension is called the **odd $2L$-periodic extension** of $f(x)$. The resulting Fourier series expansion is called a half-range expansion for $f(x)$, since it represents the function $f(x)$ on $(0, L)$, which is half of the interval $(-L, L)$ where it represents $f_o(x)$.

In a similar fashion, we can define the even **$2L$-periodic extension** of $f(x)$ as the function

$$f_e(x) := \begin{cases} f(x), & 0 < x < L, \\ f(-x), & -L < x < 0, \end{cases}$$

with $f_e(x + 2L) = f_e(x)$.

To illustrate the various extensions, let's consider the function $f(x) = x$, $0 < x < \pi$. If we extend $f(x)$ to the interval $(-\pi, \pi)$ using π periodicity, then the extension $\tilde{f}$ is given by

$$\tilde{f}(x) = \begin{cases} x, & 0 < x < \pi, \\ x + \pi, & -\pi < x < 0, \end{cases}$$

with $\tilde{f}(x + 2\pi) = \tilde{f}(x)$. In Problem 14 of Exercise 10.3, the reader computed the Fourier series for $\tilde{f}(x)$ and found

$$\tilde{f}(x) \sim \frac{\pi}{2} - \sum_{n=1}^{\infty} \frac{1}{n} \sin 2nx,$$

which consists of both odd functions (the sine terms) and even functions (the constant term), since the π-periodic extension $\tilde{f}(x)$ is neither an even nor an odd function. The odd 2π-periodic extension of $f(x)$ is just $f_o(x) = x$, $-\pi < x < \pi$, which has the Fourier series expansion

$$\textbf{(2)} \qquad f_o(x) \sim 2 \sum_{n=1}^{\infty} \frac{(-1)^{n+1}}{n} \sin nx$$

(see Problem 9 in Exercises 10.3). Because $f_o(x) = f(x)$ on the interval $(0, \pi)$, the expansion in (2) is a half-range expansion for $f(x)$. The even 2π-periodic extension of $f(x)$ is the function $f_e(x) = |x|$, $-\pi < x < \pi$, which has the Fourier series expansion

$$\textbf{(3)} \qquad f_e(x) = \frac{\pi}{2} - \frac{4}{\pi} \sum_{n=0}^{\infty} \frac{1}{(2n+1)^2} \cos (2n+1)x$$

(see Problem 10 in Exercises 10.3).

The preceding three extensions, the π-periodic function $\tilde{f}(x)$, the odd 2π-periodic function $f_o(x)$, and the even 2π-periodic function $f_e(x)$, are natural extensions of $f(x)$. There are many other ways of extending $f(x)$. For example, the function

$$g(x) = \begin{cases} x, & 0 < x < \pi, \\ 0, & -\pi < x < 0, \end{cases} \qquad g(x + 2\pi) = g(x),$$

which we studied in Example 3 of Section 10.3 is also an extension of $f(x)$. However, its Fourier series contains both sine and cosine terms and hence is not as useful as previous extensions. The graphs of these extensions of $f(x)$ are given in Figure 10.11 on page 606.

The Fourier series expansions for $f_o(x)$ and $f_e(x)$ given in (2) and (3) represent $f(x)$ on the interval $(0, \pi)$ (actually, they equal $f(x)$ on $(0, \pi)$). This motivates the following definitions.

FOURIER COSINE AND SINE SERIES

Definition 2. Let $f(x)$ be piecewise continuous on the interval $[0, T]$. The **Fourier cosine series** of $f(x)$ on $[0, T]$ is

(4) $\dfrac{a_0}{2} + \displaystyle\sum_{n=1}^{\infty} a_n \cos \dfrac{n\pi x}{T},$

where

(5) $a_n = \dfrac{2}{T} \displaystyle\int_0^T f(x) \cos \dfrac{n\pi x}{T}\,dx, \qquad n = 0, 1, \ldots .$

The **Fourier sine series** of $f(x)$ on $[0, T]$ is

(6) $\displaystyle\sum_{n=1}^{\infty} b_n \sin \dfrac{n\pi x}{T},$

where

(7) $b_n = \dfrac{2}{T} \displaystyle\int_0^T f(x) \sin \dfrac{n\pi x}{T}\,dx, \qquad n = 1, 2, \ldots .$

The trigonometric series in (4) is just the Fourier series for $f_e(x)$, the even $2T$-periodic extension of $f(x)$, and that in (6) is the Fourier series for $f_o(x)$, the odd $2T$-periodic extension of $f(x)$. These are called **half-range expansions** for $f(x)$.

EXAMPLE 1 Compute the Fourier sine series for

$$f(x) = \begin{cases} x, & 0 \le x \le \pi/2, \\ \pi - x, & \pi/2 \le x \le \pi. \end{cases}$$

Figure 10.11 Extensions of $f(x) = x$, $0 < x < \pi$

Solution Using formula (7) with $T = \pi$, we find

$$
b_n = \frac{2}{\pi} \int_0^{\pi} f(x) \sin nx\, dx = \frac{2}{\pi} \int_0^{\pi/2} x \sin nx\, dx + \frac{2}{\pi} \int_{\pi/2}^{\pi} (\pi - x) \sin nx\, dx
$$

$$
= \frac{2}{\pi n^2} \int_0^{\pi n/2} u \sin u\, du + 2 \int_{\pi/2}^{\pi} \sin nx\, dx - \frac{2}{\pi n^2} \int_{\pi n/2}^{\pi n} u \sin u\, du
$$

$$
= \frac{2}{\pi n^2} [\sin u - u \cos u] \Big|_0^{\pi n/2} - \frac{2}{n} \left[\cos \pi n - \cos \frac{n\pi}{2} \right]
$$

$$
\quad - \frac{2}{\pi n^2} [\sin u - u \cos u] \Big|_{\pi n/2}^{\pi n}
$$

$$
= \frac{4}{\pi n^2} \sin \frac{n\pi}{2} = \begin{cases} 0, & n \text{ even,} \\ \dfrac{4(-1)^{(n-1)/2}}{\pi n^2}, & n \text{ odd.} \end{cases}
$$

So on letting $n = 2k + 1$, we find the Fourier sine series for $f(x)$ to be

(8) $\dfrac{4}{\pi} \displaystyle\sum_{k=0}^{\infty} \dfrac{(-1)^k}{(2k+1)^2} \sin(2k+1)x = \dfrac{4}{\pi} \left\{ \sin x - \dfrac{1}{9} \sin 3x + \dfrac{1}{25} \sin 5x + \cdots \right\}.$ ■ End

Since, in Example 1, the function $f(x)$ is continuous and $f'(x)$ is piecewise continuous on $(0, \pi)$, it follows from Theorem 2 on pointwise convergence of Fourier series that

$$f(x) = \frac{4}{\pi} \left\{ \sin x - \frac{1}{9} \sin 3x + \frac{1}{25} \sin 5x - \frac{1}{49} \sin 7x + \cdots \right\}$$

for all x in $[0, \pi]$.

Let's return to the problem of heat flow in one dimension.

EXAMPLE 2 Find the solution to the heat flow problem

(9) $$\frac{\partial u}{\partial t} = 2 \frac{\partial^2 u}{\partial x^2}, \qquad 0 < x < \pi, \qquad t > 0,$$

(10) $$u(0, t) = u(\pi, t) = 0, \qquad t > 0,$$

(11) $$u(x, 0) = \begin{cases} x, & 0 < x \leq \pi/2, \\ \pi - x, & \pi/2 \leq x < \pi. \end{cases}$$

Solution Comparing equation (9) with equation (1) in Section 10.2, we see that $\beta = 2$ and $L = \pi$. Hence we need only represent $u(x, 0) = f(x)$ in a Fourier sine series (cf. equation (12) on page 579) of the form

$$\sum_{n=1}^{\infty} c_n \sin nx.$$

In Example 1 we obtained this expansion and showed that

$$c_n = \frac{4}{\pi n^2} \sin \frac{n\pi}{2} = \begin{cases} 0, & n \text{ even} \\ \dfrac{4(-1)^{(n-1)/2}}{\pi n^2}, & n \text{ odd}. \end{cases}$$

Hence, from equation (11) on page 579, the solution to the heat flow problem (9)–(11) is

$$\begin{aligned} u(x, t) &= \sum_{n=1}^{\infty} c_n e^{-2n^2 t} \sin nx \\ &= \frac{4}{\pi} \sum_{k=0}^{\infty} \frac{(-1)^k}{(2k+1)^2} e^{-2(2k+1)^2 t} \sin(2k+1)x \\ &= \frac{4}{\pi} \left\{ e^{-2t} \sin x - \frac{1}{9} e^{-18t} \sin 3x + \frac{1}{25} e^{-50t} \sin 5x + \cdots \right\}. \end{aligned}$$

A sketch of a partial sum for $u(x, t)$ is displayed in Figure 10.12. ∎ End

Figure 10.12 Partial sum for $u(x, t)$ in Example 2

EXERCISES 10.4

In Problems 1 through 4, determine (a) the π-periodic extension $\tilde{f}$, (b) the odd 2π-periodic extension f_o, and (c) the even 2π-periodic extension f_e for the given function f and sketch their graphs.

1. $f(x) = x^2$, $0 < x < \pi$.

2. $f(x) = \sin 2x$, $0 < x < \pi$.

3. $f(x) = \begin{cases} 0, & 0 < x < \pi/2, \\ 1, & \pi/2 < x < \pi. \end{cases}$

4. $f(x) = \pi - x$, $0 < x < \pi$.

In Problems 5 through 10, compute the Fourier sine series for the given function.

5. $f(x) = -1$, $0 < x < 1$.

6. $f(x) = \cos x$, $0 < x < \pi$.

7. $f(x) = x^2$, $0 < x < \pi$.

8. $f(x) = \pi - x$, $0 < x < \pi$.

9. $f(x) = x - x^2$, $0 < x < 1$.

10. $f(x) = e^x$, $0 < x < 1$.

In Problems 11 through 16, compute the Fourier cosine series for the given function.

11. $f(x) = \pi - x$, $0 < x < \pi$.

12. $f(x) = 1 + x$, $0 < x < \pi$.

13. $f(x) = e^x$, $0 < x < 1$.

14. $f(x) = e^{-x}$, $0 < x < 1$.

15. $f(x) = \sin x$, $0 < x < \pi$.

16. $f(x) = x - x^2$, $0 < x < 1$.

In Problems 17 through 19, find the solution to the heat flow problem

$$\frac{\partial u}{\partial t} = 5\frac{\partial^2 u}{\partial x^2}, 0 < x < \pi, t > 0,$$

$$u(0, t) = u(\pi, t) = 0, t > 0,$$

$$u(x, 0) = f(x), 0 < x < \pi,$$

where $f(x)$ is given.

17. $f(x) = 1 - \cos 2x$. **18.** $f(x) = x(\pi - x)$.

19. $f(x) = \begin{cases} -x, & 0 < x \leq \pi/2, \\ x - \pi, & \pi/2 \leq x < \pi. \end{cases}$

10.5 THE HEAT EQUATION

In Section 10.1 we developed a model for heat flow in an insulated wire whose ends are kept at the constant temperature 0°C. In particular, we found that the temperature $u(x, t)$ in the wire is governed by the initial-boundary value problem

(1) $$\frac{\partial u}{\partial t} = \beta \frac{\partial^2 u}{\partial x^2}, \qquad 0 < x < L, \qquad t > 0,$$

(2) $u(0, t) = u(L, t) = 0, \qquad t > 0,$

(3) $u(x, 0) = f(x), \qquad 0 < x < L$

(see equations (6)–(8) in Section 10.1). Here equation (2) specifies that the temperature at the ends of the wire is zero, whereas equation (3) specifies the initial temperature distribution.

In Section 10.2 we also derived a formal solution to (1)–(3) using separation of variables. There we found the solution to (1)–(3) to have the form

(4) $$u(x, t) = \sum_{n=1}^{\infty} c_n e^{-\beta(n\pi/L)^2 t} \sin \frac{n\pi x}{L},$$

where the c_n's are the coefficients in the Fourier sine series for $f(x)$:

(5) $$f(x) = \sum_{n=1}^{\infty} c_n \sin \frac{n\pi x}{L}.$$

In other words, solving (1)–(3) reduces to computing the Fourier sine series for the initial value function $f(x)$.

In this section we discuss heat flow problems where the ends of the wire are insulated or kept at a constant, but nonzero, temperature. (The latter involves nonhomogeneous boundary conditions.) We will also discuss the problem in which a heat source is adding heat to the wire. (This results in a nonhomogeneous partial differential equation.) The problem of heat flow in a rectangular plate is also discussed and leads to the topic of double Fourier series. We conclude this section with a discussion of the existence and uniqueness of solutions to the heat flow problem.

In the model of heat flow in a wire, let's replace the assumption that the ends of the wire are kept at a constant temperature zero and instead assume that the ends of the wire are *insulated*, that is, no heat flows out (or in) at the ends of the wire. It follows from the principle of heat conduction (see Section 10.1) that the temperature gradient must be zero at these end points, that is,

$$\frac{\partial u}{\partial x}(0, t) = \frac{\partial u}{\partial x}(L, t) = 0, \qquad t > 0.$$

In the next example we obtain the formal solution to the heat flow problem with these boundary conditions.

EXAMPLE 1 Find a formal solution to the heat flow problem governed by the initial-boundary value problem

(6)
$$\frac{\partial u}{\partial t} = \beta \frac{\partial^2 u}{\partial x^2}, \qquad 0 < x < L, \qquad t > 0,$$

(7) $\dfrac{\partial u}{\partial x}(0, t) = \dfrac{\partial u}{\partial x}(L, t) = 0, \qquad t > 0,$

(8) $u(x, 0) = f(x), \qquad 0 < x < L.$

Solution Using the method of separation of variables, we first assume

$$u(x, t) = X(x)T(t).$$

Substituting into equation (6) and separating variables as was done in Section 10.2 (cf. equation (5) on page 577), we get the two equations

(9) $X''(x) - KX(x) = 0,$

(10) $T'(t) - \beta KT(t) = 0,$

where K is some constant. The boundary conditions in (7) become

$$X'(0)T(t) = 0 \quad \text{and} \quad X'(L)T(t) = 0.$$

For these equations to hold for all $t > 0$, either $T(t) \equiv 0$, which implies that $u(x, t) \equiv 0$, or

(11) $X'(0) = X'(L) = 0.$

Combining the boundary conditions in (11) with equation (9) gives the boundary value problem

(12) $X''(x) - KX(x) = 0; \qquad X'(0) = X'(L) = 0,$

where K can be any constant.

To solve for the nontrivial solutions to (12), we begin with the auxiliary equation $r^2 - K = 0$. When $K > 0$, arguments similar to those used in Section 10.2 show that there are no nontrivial solutions to (12).

When $K = 0$, the auxiliary equation has the repeated root 0 and a general solution to the differential equation is

$$X(x) = A + Bx.$$

The boundary conditions in (12) reduce to $B = 0$ with A arbitrary. Thus, for $K = 0$, the nontrivial solutions to (12) are of the form

$$X(x) = c_0,$$

where $c_0 = A$ is an arbitrary nonzero constant.

When $K < 0$, the auxiliary equation has the roots $r = \pm i\sqrt{-K}$. Thus a general solution to the differential equation in (12) is

$$X(x) = C_1 \cos\sqrt{-K}\, x + C_2 \sin\sqrt{-K}\, x.$$

The boundary conditions in (12) lead to the system

$$\sqrt{-K}C_2 = 0,$$

$$-\sqrt{-K}C_1 \sin\sqrt{-K}L + \sqrt{-K}C_2 \cos\sqrt{-K}L = 0.$$

Hence $C_2 = 0$ and the system reduces to solving $C_1 \sin\sqrt{-K}L = 0$. Since $\sin\sqrt{-K}L = 0$ only when $\sqrt{-K}L = n\pi$, where n is an integer, we obtain a nontrivial solution only when $\sqrt{-K} = n\pi/L$ or $K = -(n\pi/L)^2$, $n = 1, 2, 3, \ldots$. Furthermore, the nontrivial solutions (eigenfunctions) X_n corresponding to the eigenvalue $K = -(n\pi/L)^2$ are given by

$$(13) \qquad X_n(x) = c_n \cos\frac{n\pi x}{L},$$

where the c_n's are arbitrary nonzero constants. In fact, formula (13) also holds for $n = 0$, since $K = 0$ has the eigenfunctions $X_0(x) = c_0$.

Having determined that $K = -(n\pi/L)^2$, $n = 0, 1, 2, \ldots$, let's consider equation (10) for such K:

$$T'(t) + \beta(n\pi/L)^2 T(t) = 0.$$

For $n = 0, 1, 2, \ldots$, the general solution is

$$T_n(t) = b_n e^{-\beta(n\pi/L)^2 t},$$

where the b_n's are arbitrary constants. Combining this with equation (13), we obtain the functions

$$u_n(x, t) = X_n(x)T_n(t) = \left[c_n \cos\frac{n\pi x}{L}\right][b_n e^{-\beta(n\pi/L)^2 t}],$$

$$u_n(x, t) = a_n e^{-\beta(n\pi/L)^2 t} \cos\frac{n\pi x}{L},$$

where $a_n = b_n c_n$ is, again, an arbitrary constant.

If we take an infinite series of these functions, we obtain

$$(14) \qquad u(x, t) = \frac{a_0}{2} + \sum_{n=1}^{\infty} a_n e^{-\beta(n\pi/L)^2 t} \cos\frac{n\pi x}{L},$$

which will be a solution to (6)–(7) provided the series has the proper convergence behavior. Notice that in (14) we have removed the constant term and written it as $a_0/2$, thus producing the standard form for cosine expansions.

Assuming a solution to (6)–(7) is given by the series in (14) and substituting into the initial condition (8), we get

$$(15) \qquad u(x, 0) = \frac{a_0}{2} + \sum_{n=1}^{\infty} a_n \cos\frac{n\pi x}{L} = f(x), \qquad 0 < x < L.$$

This means that if we choose the a_n's as the coefficients in the Fourier cosine series for f,

$$a_n = \frac{2}{L}\int_0^L f(x) \cos\frac{n\pi x}{L}\,dx, \qquad n = 0, 1, 2, \ldots,$$

then $u(x, t)$ given in (14) will be a **formal solution** to the heat flow problem (6)–(8). Again, if this expansion converges to a continuous function with continuous second partial derivatives, then the formal solution is an actual solution. ∎ End

Note that the individual terms in the series (4), (14) are genuine solutions to the heat equation and the associated boundary conditions (but not the initial condition, of course). These solutions are called the **modes** of the system. Observe that the time factor $e^{-\beta(n\pi/L)^2 t}$ decays faster for the more oscillatory (higher n) modes. This makes sense physically; a temperature pattern of many alternating hot and cold strips arranged closely together will equilibrate faster than a more uniform pattern. See Figure 10.13.

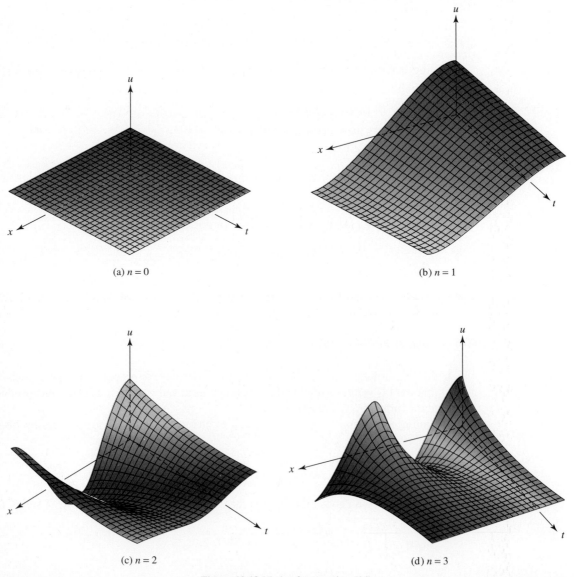

(a) $n = 0$

(b) $n = 1$

(c) $n = 2$

(d) $n = 3$

Figure 10.13 Modes for equation (14)

A good way of expressing how separation of variables "works" is as follows: The initial temperature profile $f(x)$ is decomposed into modes through Fourier analysis—equations (5) or (15). Then each mode is matched with a time decay factor—equations (4) or (14)—and it evolves in time accordingly.

When the ends of the wire are kept at 0°C or when the ends are insulated, the boundary conditions are said to be **homogeneous.** But, when the ends of the wire are kept at constant temperatures different from zero, that is,

(16) $u(0, t) = U_1$ and $u(L, t) = U_2,$ $t > 0,$

then the boundary conditions are called **nonhomogeneous.**

From our experience with vibration problems in Chapter 5 we expect that the solution to the heat flow problem with nonhomogeneous boundary conditions will consist of a **steady-state solution** $v(x)$ that satisfies the nonhomogeneous boundary conditions in (16) plus a **transient solution** $w(x, t)$. That is,

$$u(x, t) = v(x) + w(x, t),$$

where $w(x, t)$ and its partial derivatives tend to zero as $t \to \infty$. The function $w(x, t)$ will then satisfy homogeneous boundary conditions, as illustrated in the next example.

EXAMPLE 2 Find a formal solution to the heat flow problem governed by the initial-boundary value problem

(17) $\dfrac{\partial u}{\partial t} = \beta \dfrac{\partial^2 u}{\partial x^2},$ $0 < x < L,$ $t > 0,$

(18) $u(0, t) = U_1,$ $u(L, t) = U_2,$ $t > 0,$

(19) $u(x, 0) = f(x),$ $0 < x < L.$

Solution Let's assume the solution $u(x, t)$ consists of a steady-state solution $v(x)$ and a transient solution $w(x, t)$, that is,

(20) $u(x, t) = v(x) + w(x, t).$

Substituting for $u(x, t)$ in equations (17)–(19) leads to

(21) $\dfrac{\partial u}{\partial t} = \dfrac{\partial w}{\partial t} = \beta v''(x) + \beta \dfrac{\partial^2 w}{\partial x^2};$ $0 < x < L,$ $t > 0,$

(22) $v(0) + w(0, t) = U_1,$ $v(L) + w(L, t) = U_2,$ $t > 0,$

(23) $v(x) + w(x, 0) = f(x),$ $0 < x < L.$

If we allow $t \to \infty$ in (21)–(22), assuming $w(x, t)$ is a transient solution, we obtain the steady-state boundary value problem

$$v''(x) = 0, \qquad 0 < x < L,$$
$$v(0) = U_1, \qquad v(L) = U_2.$$

Solving for v, we obtain $v(x) = Ax + B$, and choosing A and B so that the boundary conditions are satisfied yields

(24) $v(x) = U_1 + \dfrac{(U_2 - U_1)x}{L}$

as the steady-state solution.

With this choice for $v(x)$, the initial-boundary value problem (21)–(23) reduces to the following initial-boundary value problem for $w(x, t)$:

(25) $\dfrac{\partial w}{\partial t} = \beta \dfrac{\partial^2 w}{\partial x^2}, \qquad 0 < x < L, \qquad t > 0,$

(26) $w(0, t) = w(L, t) = 0, \qquad t > 0,$

(27) $w(x, 0) = f(x) - U_1 - \dfrac{(U_2 - U_1)x}{L}, \qquad 0 < x < L.$

Recall that a formal solution to (25)–(27) is given by equation (4). Hence

$$w(x, t) = \sum_{n=1}^{\infty} c_n e^{-\beta(n\pi/L)^2 t} \sin \frac{n\pi x}{L},$$

where the c_n's are the coefficients of the Fourier sine series expansion

$$f(x) - U_1 - \frac{(U_2 - U_1)x}{L} = \sum_{n=1}^{\infty} c_n \sin \frac{n\pi x}{L}.$$

Therefore the formal solution to (17)–(19) is

(28) $u(x, t) = U_1 + \dfrac{(U_2 - U_1)x}{L} + \displaystyle\sum_{n=1}^{\infty} c_n e^{-\beta(n\pi/L)^2 t} \sin \dfrac{n\pi x}{L},$

with

$$c_n = \frac{2}{L} \int_0^L \left[f(x) - U_1 - \frac{(U_2 - U_1)x}{L} \right] \sin \frac{n\pi x}{L} \, dx. \quad \blacksquare \text{ End}$$

In the next example, we consider the heat flow problem when a heat source is present but is independent of time. (Recall the derivation in Section 10.1.)

EXAMPLE 3 Find a formal solution to the heat flow problem governed by the initial-boundary value problem

(29) $\dfrac{\partial u}{\partial t} = \beta \dfrac{\partial^2 u}{\partial x^2} + P(x), \qquad 0 < x < L, \qquad t > 0,$

(30) $u(0, t) = U_1, \qquad u(L, t) = U_2, \qquad t > 0,$

(31) $u(x, 0) = f(x), \qquad 0 < x < L.$

Solution We begin by assuming that the solution consists of a steady-state solution $v(x)$ and a transient solution $w(x, t)$, namely,

$$u(x, t) = v(x) + w(x, t),$$

where $w(x, t)$ and its partial derivatives tend to zero as $t \to \infty$. Substituting for $u(x, t)$ in (29)–(31) yields

$$(32) \qquad \frac{\partial u}{\partial t} = \frac{\partial w}{\partial t} = \beta v''(x) + \beta \frac{\partial^2 w}{\partial x^2} + P(x), \qquad 0 < x < L, \qquad t > 0,$$

$$(33) \qquad v(0) + w(0, t) = U_1, \qquad v(L) + w(L, t) = U_2, \qquad t > 0,$$

$$(34) \qquad v(x) + w(x, 0) = f(x), \qquad 0 < x < L.$$

Letting $t \to \infty$ in (32)–(33), we obtain the steady-state boundary value problem

$$v''(x) = -\frac{1}{\beta} P(x), \qquad 0 < x < L,$$

$$v(0) = U_1, \qquad v(L) = U_2.$$

The solution to this boundary value problem can be obtained by two integrations using the boundary conditions to determine the constants of integration. The reader can verify that the solution is given by the formula

$$(35) \qquad v(x) = \left[U_2 - U_1 + \int_0^L \left(\int_0^z \frac{1}{\beta} P(s)\,ds \right) dz \right] \frac{x}{L} + U_1 - \int_0^x \left(\int_0^z \frac{1}{\beta} P(s)\,ds \right) dz.$$

With this choice for $v(x)$, we find that the initial-boundary value problem (32)–(34) reduces to the following initial-boundary value problem for $w(x, t)$:

$$(36) \qquad \frac{\partial w}{\partial t} = \beta \frac{\partial^2 w}{\partial x^2}, \qquad 0 < x < L, \qquad t > 0,$$

$$(37) \qquad w(0, t) = w(L, t) = 0, \qquad t > 0,$$

$$(38) \qquad w(x, 0) = f(x) - v(x), \qquad 0 < x < L,$$

where $v(x)$ is given by formula (35). As before, the solution to this initial-boundary value problem is

$$(39) \qquad w(x, t) = \sum_{n=1}^{\infty} c_n e^{-\beta(n\pi/L)^2 t} \sin \frac{n\pi x}{L},$$

where the c_n's are determined from the Fourier sine series expansion of $f(x) - v(x)$:

$$(40) \qquad f(x) - v(x) = \sum_{n=1}^{\infty} c_n \sin \frac{n\pi x}{L}.$$

Thus the formal solution to (29)–(31) is given by

$$u(x, t) = v(x) + w(x, t),$$

where $v(x)$ is given in (35) and $w(x, t)$ is prescribed by (39)–(40). ∎ End

The method of separation of variables is also applicable to problems in higher dimensions. For example, consider the problem of heat flow in a rectangular plate with sides $x = 0$, $x = L$, $y = 0$, and $y = W$. Assuming the two sides $y = 0$, $y = W$ are kept at a constant temperature of $0°C$ and the two sides $x = 0$, $x = L$ are perfectly insulated, then heat flow is governed by the initial-boundary value problem in the following example (see Figure 10.14).

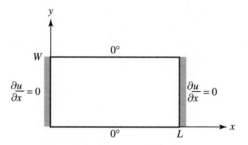

Figure 10.14 Plate with insulated sides

EXAMPLE 4 Find a formal solution $u(x, y, t)$ to the initial-boundary value problem

(41) $$\frac{\partial u}{\partial t} = \beta \left\{ \frac{\partial^2 u}{\partial x^2} + \frac{\partial^2 u}{\partial y^2} \right\}, \qquad 0 < x < L, \qquad 0 < y < W, \qquad t > 0,$$

(42) $$\frac{\partial u}{\partial x}(0, y, t) = \frac{\partial u}{\partial x}(L, y, t) = 0, \qquad 0 < y < W, \qquad t > 0,$$

(43) $u(x, 0, t) = u(x, W, t) = 0, \qquad 0 < x < L, \qquad t > 0,$

(44) $u(x, y, 0) = f(x, y), \qquad 0 < x < L, \qquad 0 < y < W.$

Solution If we assume a solution of the form $u(x, y, t) = V(x, y)T(t)$, then equation (41) separates into the two equations

(45) $T'(t) - \beta K T(t) = 0,$

(46) $$\frac{\partial^2 V}{\partial x^2}(x, y) + \frac{\partial^2 V}{\partial y^2}(x, y) - KV(x, y) = 0,$$

where K can be any constant. To solve equation (46), we again use separation of variables. Here we assume $V(x, y) = X(x)Y(y)$. This allows us to separate equation (46) into the two equations

(47) $X''(x) - JX(x) = 0,$

(48) $Y''(y) + (J - K)Y(y) = 0,$

where J can be any constant (see Problem 29 in Exercises 10.2).

To solve for $X(x)$, we observe that the boundary conditions in (42), in terms of the separated variables, become

$$X'(0)Y(y)T(t) = X'(L)Y(y)T(t) = 0, \qquad 0 < y < W, \qquad t > 0.$$

Hence in order to get a nontrivial solution, we must have

(49) $X'(0) = X'(L) = 0$.

The boundary value problem for X given in equations (47) and (49) was solved in Example 1 (cf. equations (12) and (13)). Here $J = -(m\pi/L)^2$, $m = 0, 1, 2, \ldots$, and

$$X_m(x) = c_m \cos \frac{m\pi x}{L},$$

where the c_m's are arbitrary.

To solve for $Y(y)$, we first observe that the boundary conditions in (43) become

(50) $Y(0) = Y(W) = 0$.

Next, substituting $J = -(m\pi/L)^2$ into equation (48) yields

$$Y''(y) - (K + (m\pi/L)^2)Y(y) = 0,$$

which we can rewrite as

(51) $Y''(y) - EY(y) = 0$,

where $E = K + (m\pi/L)^2$. The boundary value problem for Y consisting of (50)–(51) has also been solved before. In Section 10.2 (cf. equations (7) and (9)) we showed that $E = -(n\pi/W)^2$, $n = 1, 2, 3, \ldots$, and the nontrivial solutions are given by

$$Y_n(y) = a_n \sin \frac{n\pi y}{W},$$

where the a_n's are arbitrary.

Since $K = E - (m\pi/L)^2$, we have

$$K = -(n\pi/W)^2 - (m\pi/L)^2, \qquad m = 0, 1, 2, \ldots, \qquad n = 1, 2, 3, \ldots.$$

Substituting K into equation (45), we can solve for $T(t)$ and obtain

$$T_{mn}(t) = b_{mn}e^{-(m^2/L^2 + n^2/W^2)\beta\pi^2 t}.$$

Substituting in for X_m, Y_n, and T_{mn}, we get

$$u_{mn}(x, y, t) = \left(c_m \cos \frac{m\pi x}{L} \right) \left(a_n \sin \frac{n\pi y}{W} \right) \left(b_{mn}e^{-(m^2/L^2 + n^2/W^2)\beta\pi^2 t} \right),$$

$$u_{mn}(x, y, t) = a_{mn}e^{-(m^2/L^2 + n^2/W^2)\beta\pi^2 t} \cos \frac{m\pi x}{L} \sin \frac{n\pi y}{W},$$

where $a_{mn} := a_n b_{mn} c_m$ ($m = 0, 1, 2, \ldots$, $n = 1, 2, 3, \ldots$) are arbitrary constants.

If we now take a doubly infinite series of such functions, then we obtain the formal series

(52) $$u(x, y, t) = \sum_{m=0}^{\infty} \sum_{n=1}^{\infty} a_{mn}e^{-(m^2/L^2 + n^2/W^2)\beta\pi^2 t} \cos \frac{m\pi x}{L} \sin \frac{n\pi y}{W}.$$

We are now ready to apply the initial conditions (44). Setting $t = 0$, we obtain

(53) $$u(x, y, 0) = f(x, y) = \sum_{m=0}^{\infty} \sum_{n=1}^{\infty} a_{mn} \cos \frac{m\pi x}{L} \sin \frac{n\pi y}{W}.$$

This is a **double Fourier Series.**[†] The formulas for the coefficients a_{mn} are obtained by exploiting the orthogonality conditions twice. Presuming (53) is valid and permits term by term integration, we multiply each side by $\cos (p\pi x/L) \sin (q\pi y/W)$ and integrate over x and y:

$$\int_0^L \int_0^W f(x, y) \cos \frac{p\pi x}{L} \sin \frac{q\pi y}{W} \, dy \, dx$$

$$= \sum_{m=0}^{\infty} \sum_{n=1}^{\infty} a_{mn} \int_0^L \int_0^W \cos \frac{m\pi x}{L} \sin \frac{n\pi y}{W} \cos \frac{p\pi x}{L} \sin \frac{q\pi y}{W} \, dy \, dx.$$

According to the orthogonality conditions, each integral on the right is zero, except when $m = p$ and $n = q$; thus

$$\int_0^L \int_0^W f(x, y) \cos \frac{p\pi x}{L} \sin \frac{q\pi y}{W} \, dy \, dx$$

$$= a_{pq} \int_0^L \cos^2 \frac{p\pi x}{L} \, dx \int_0^W \sin^2 \frac{q\pi y}{W} \, dy = \begin{cases} \dfrac{LW}{4} a_{pq}, & p \neq 0 \\[2mm] \dfrac{LW}{2} a_{pq}, & p = 0. \end{cases}$$

Hence

(54) $$a_{0q} = \frac{2}{LW} \int_0^L \int_0^W f(x, y) \sin \frac{q\pi y}{W} \, dy \, dx, \qquad q = 1, 2, 3, \dots,$$

and for $p \geq 1$, $q \geq 1$,

(55) $$a_{pq} = \frac{4}{LW} \int_0^L \int_0^W f(x, y) \cos \frac{p\pi x}{L} \sin \frac{q\pi y}{W} \, dy \, dx.$$

Finally, the solution to the initial-boundary value problem (41)–(44) is given by equation (52) where the coefficients are prescribed by equations (54) and (55). ■ End

Existence and Uniqueness of Solutions

In the examples that we have studied in this section and in Section 10.2, we were able to obtain formal solutions in the sense that we could express the solution in terms of a series expansion consisting of exponentials, sines, and cosines. To prove that these series converge to actual solutions requires results on the convergence of Fourier series and results from real analysis on uniform convergence. We will not go into these details here, but refer the

[†] For a discussion of double Fourier series, see *Partial Differential Equations of Mathematical Physics*, by Tyn Myint-U (Elsevier North Holland, Inc., New York, 1980), Section 5.14.

reader to Section 6.5 of the text by Tyn Myint-U (see footnote on page 618) for a proof of the existence of a solution to the heat flow problem discussed in Sections 10.1 and 10.2. (A proof of uniqueness is also given there.)

As might be expected, by using Fourier series and the method of separation of variables one can also obtain "solutions" when the initial data are discontinuous, since the formal solutions only require the existence of a convergent Fourier series. This allows one to study idealized problems in which the initial conditions do not agree with the boundary conditions or the initial conditions involve a jump discontinuity. For example, we may assume that initially one half of the wire is at one temperature, whereas the other half is at a different temperature, that is,

$$f(x) = \begin{cases} U_1, & 0 < x < L/2, \\ U_2, & L/2 < x < L. \end{cases}$$

The formal solution that we obtained will make sense for $0 < x < L$, $t > 0$, but we must take care in interpreting the results near the points of discontinuity $x = 0$, $L/2$, and L.

The question of the uniqueness of the solution to the heat flow problem can be answered in various ways. One is tempted to argue that the method of separation of variables yields *formulas* for the solutions and therefore a unique solution. However, this does *not* exclude the possibility of solutions existing that cannot be obtained by the method of separation of variables.

From physical considerations, we know that the *peak* temperature along a wire does not increase spontaneously if there are no heat sources to drive it up. Indeed if hot objects could draw heat from colder objects without external intervention, we would have violations of the second law of thermodynamics. Thus no point on an unheated wire will ever reach a higher temperature than the initial peak temperature. Such statements are called **maximum principles** and can be proved mathematically for the heat equation and Laplace's equation. One such result is the following.[†]

MAXIMUM PRINCIPLE FOR THE HEAT EQUATION

Theorem 6. Let $u(x, t)$ be a continuously differentiable function that satisfies the heat equation

(56) $\dfrac{\partial u}{\partial t} = \beta \dfrac{\partial^2 u}{\partial x^2}, \qquad 0 < x < L, \qquad t > 0,$

and the boundary conditions

(57) $u(0, t) = u(L, t) = 0, \qquad t > 0.$

Then $u(x, t)$ attains its maximum value at $t = 0$, for some x in $[0, L]$, that is,

$$\max_{\substack{t \geq 0 \\ 0 \leq x \leq L}} u(x, t) = \max_{0 \leq x \leq L} u(x, 0).$$

[†] For a discussion of maximum principles and their applications, see *Maximum Principles in Differential Equations*, by M. H. Protter and H. F. Weinberger (Prentice Hall, Inc., Englewood Cliffs, New Jersey, 1967).

We can use the maximum principle to show that the heat flow problem has a unique solution.

UNIQUENESS OF SOLUTION

Theorem 7. The initial-boundary value problem

$$(58) \qquad \frac{\partial u}{\partial t} = \beta \frac{\partial^2 u}{\partial x^2}, \qquad 0 < x < L, \qquad t > 0,$$

$$(59) \qquad u(0, t) = u(L, t) = 0, \qquad t \geq 0,$$

$$(60) \qquad u(x, 0) = f(x), \qquad 0 \leq x \leq L,$$

has at most one continuously differentiable solution.

Proof. Assume $u(x, t)$ and $v(x, t)$ are continuously differentiable functions that satisfy the initial-boundary value problem (58)–(60). Let $w = u - v$. Now w is a continuously differentiable solution to the boundary value problem (56)–(57). By the maximum principle, w must attain its maximum at $t = 0$, and since

$$w(x, 0) = u(x, 0) - v(x, 0) = f(x) - f(x) = 0,$$

we have $w(x, t) \leq 0$. Hence $u(x, t) \leq v(x, t)$ for all $0 \leq x \leq L, t \geq 0$. A similar argument using $\hat{w} = v - u$ yields $v(x, t) \leq u(x, t)$. Therefore we have $u(x, t) = v(x, t)$ for all $0 \leq x \leq L, t \geq 0$. Thus there is at most one continuously differentiable solution to the problem (58)–(60). ∎ End

EXERCISES 10.5

In Problems 1 through 10, find a formal solution to the given initial-boundary value problem.

1. $\dfrac{\partial u}{\partial t} = 5 \dfrac{\partial^2 u}{\partial x^2}, \qquad 0 < x < 1, \qquad t > 0,$

$u(0, t) = u(1, t) = 0, \qquad t > 0,$
$u(x, 0) = (1 - x)x^2, \qquad 0 < x < 1.$

2. $\dfrac{\partial u}{\partial t} = \dfrac{\partial^2 u}{\partial x^2}, \qquad 0 < x < \pi, \qquad t > 0,$

$u(0, t) = u(\pi, t) = 0, \qquad t > 0,$
$u(x, 0) = x^2, \qquad 0 < x < \pi.$

3. $\dfrac{\partial u}{\partial t} = 3 \dfrac{\partial^2 u}{\partial x^2}, \qquad 0 < x < \pi, \qquad t > 0,$

$\dfrac{\partial u}{\partial x}(0, t) = \dfrac{\partial u}{\partial x}(\pi, t) = 0, \qquad t > 0,$
$u(x, 0) = x, \qquad 0 < x < \pi.$

4. $\dfrac{\partial u}{\partial t} = 2 \dfrac{\partial^2 u}{\partial x^2}, \qquad 0 < x < 1, \qquad t > 0,$

$\dfrac{\partial u}{\partial x}(0, t) = \dfrac{\partial u}{\partial x}(1, t) = 0, \qquad t > 0,$
$u(x, 0) = x(1 - x), \qquad 0 < x < 1.$

5. $\dfrac{\partial u}{\partial t} = \dfrac{\partial^2 u}{\partial x^2}, \qquad 0 < x < \pi, \qquad t > 0,$

$\dfrac{\partial u}{\partial x}(0, t) = \dfrac{\partial u}{\partial x}(\pi, t) = 0, \qquad t > 0,$
$u(x, 0) = e^x, \qquad 0 < x < \pi.$

6. $\dfrac{\partial u}{\partial t} = 7 \dfrac{\partial^2 u}{\partial x^2}, \qquad 0 < x < \pi, \qquad t > 0,$

$\dfrac{\partial u}{\partial x}(0, t) = \dfrac{\partial u}{\partial x}(\pi, t) = 0, \qquad t > 0,$
$u(x, 0) = 1 - \sin x, \qquad 0 < x < \pi.$

7. $\dfrac{\partial u}{\partial t} = 2\dfrac{\partial^2 u}{\partial x^2}$, $0 < x < \pi$, $t > 0$,

$u(0, t) = 5$, $u(\pi, t) = 10$, $t > 0$,
$u(x, 0) = \sin 3x - \sin 5x$, $0 < x < \pi$.

8. $\dfrac{\partial u}{\partial t} = \dfrac{\partial^2 u}{\partial x^2}$, $0 < x < \pi$, $t > 0$,

$u(0, t) = 0$, $u(\pi, t) = 3\pi$, $t > 0$,
$u(x, 0) = 0$, $0 < x < \pi$.

9. $\dfrac{\partial u}{\partial t} = \dfrac{\partial^2 u}{\partial x^2} + e^{-x}$, $0 < x < \pi$, $t > 0$,

$u(0, t) = u(\pi, t) = 0$, $t > 0$,
$u(x, 0) = \sin 2x$, $0 < x < \pi$.

10. $\dfrac{\partial u}{\partial t} = 3\dfrac{\partial^2 u}{\partial x^2} + x$, $0 < x < \pi$, $t > 0$,

$u(0, t) = u(\pi, t) = 0$, $t > 0$,
$u(x, 0) = \sin x$, $0 < x < \pi$.

11. Find a formal solution to the initial-boundary value problem

$$\frac{\partial u}{\partial t} = 4\frac{\partial^2 u}{\partial x^2}, \qquad 0 < x < \pi, \qquad t > 0,$$

$$\frac{\partial u}{\partial x}(0, t) = 0, \qquad u(\pi, t) = 0, \quad t > 0,$$

$$u(x, 0) = f(x), \qquad 0 < x < \pi.$$

12. Find a formal solution to the initial-boundary value problem

$$\frac{\partial u}{\partial t} = \frac{\partial^2 u}{\partial x^2}, \qquad 0 < x < \pi, \quad t > 0,$$

$$u(0, t) = 0, \qquad u(\pi, t) + \frac{\partial u}{\partial x}(\pi, t) = 0, \qquad t > 0,$$

$$u(x, 0) = f(x), \qquad 0 < x < \pi.$$

13. Find a formal solution to the initial-boundary value problem

$$\frac{\partial u}{\partial t} = 2\frac{\partial^2 u}{\partial x^2} + 4x, \qquad 0 < x < \pi, \qquad t > 0,$$

$$u(0, t) = u(\pi, t) = 0, \qquad t > 0,$$

$$u(x, 0) = \sin x, \qquad 0 < x < \pi.$$

14. Find a formal solution to the initial-boundary value problem

$$\frac{\partial u}{\partial t} = 3\frac{\partial^2 u}{\partial x^2} + 5, \qquad 0 < x < \pi, \qquad t > 0,$$

$$u(0, t) = u(\pi, t) = 1, \qquad t > 0,$$

$$u(x, 0) = 1, \qquad 0 < x < \pi.$$

In Problems 15 through 18, find a formal solution to the initial-boundary value problem

$$\frac{\partial u}{\partial t} = \frac{\partial^2 u}{\partial x^2} + \frac{\partial^2 u}{\partial y^2}, \quad 0 < x < \pi, \;\; 0 < y < \pi, \;\; t > 0,$$

$$\frac{\partial u}{\partial x}(0, y, t) = \frac{\partial u}{\partial x}(\pi, y, t) = 0, \quad 0 < y < \pi, \;\; t > 0,$$

$$u(x, 0, t) = u(x, \pi, t) = 0, \quad 0 < x < \pi, \;\; t > 0,$$

$$u(x, y, 0) = f(x, y), \quad 0 < x < \pi, \;\; 0 < y < \pi,$$

for the given function f(x, y).

15. $f(x, y) = \cos 6x \sin 4y - 3 \cos x \sin 11y$.

16. $f(x, y) = \cos x \sin y + 4 \cos 2x \sin y - 3 \cos 3x \sin 4y$.

17. $f(x, y) = y$.

18. $f(x, y) = x \sin y$.

19. Chemical Diffusion. Chemical diffusion through a thin layer is governed by the equation

$$\frac{\partial C}{\partial t} = k\frac{\partial^2 C}{\partial x^2} - LC,$$

where $C(x, t)$ is the concentration in moles/cm^3, the diffusivity k is a positive constant with units cm^2/sec, and $L > 0$ is a consumption rate with units sec^{-1}. Assume the boundary conditions are

$$C(0, t) = C(a, t) = 0, \qquad t > 0$$

and the initial concentration is given by

$$C(x, 0) = f(x), \qquad 0 < x < a.$$

Use the method of separation of variables to solve formally for the concentration $C(x, t)$. What happens to the concentration as $t \to +\infty$?

10.6 THE WAVE EQUATION

In Section 10.2 we presented a model for the motion of a vibrating string. If $u(x, t)$ represents the displacement (deflection) of the string and the ends of the string are held fixed, then the motion of the string is governed by the initial-boundary value problem

(1)
$$\frac{\partial^2 u}{\partial t^2} = \alpha^2 \frac{\partial^2 u}{\partial x^2}, \qquad 0 < x < L, \qquad t > 0,$$

(2) $\qquad u(0, t) = u(L, t) = 0, \qquad t > 0,$

(3) $\qquad u(x, 0) = f(x), \qquad 0 < x < L,$

(4)
$$\frac{\partial u}{\partial t}(x, 0) = g(x), \qquad 0 < x < L.$$

Equation (1) is called the **wave equation.**

The constant α^2 appearing in (1) is strictly positive and depends on the linear density and tension of the string. The boundary conditions in (2) reflect the fact that the string is held fixed at the two end points $x = 0$ and $x = L$.

Equations (3) and (4) specify, respectively, the initial displacement and the initial velocity of each point on the string. For the initial and boundary conditions to be consistent, we assume $f(0) = f(L) = 0$ and $g(0) = g(L) = 0$.

Using the method of separation of variables, we found in Section 10.2 that a formal solution to (1)–(4) is given by (cf. equations (24)–(26) on page 582)

(5)
$$u(x, t) = \sum_{n=1}^{\infty} \left[a_n \cos \frac{n\pi\alpha}{L} t + b_n \sin \frac{n\pi\alpha}{L} t \right] \sin \frac{n\pi x}{L},$$

where the a_n's and b_n's are determined from the Fourier sine series

(6)
$$f(x) = \sum_{n=1}^{\infty} a_n \sin \frac{n\pi x}{L},$$

(7)
$$g(x) = \sum_{n=1}^{\infty} b_n \left(\frac{n\pi\alpha}{L} \right) \sin \frac{n\pi x}{L}.$$

See Figure 10.15 for a sketch of partial sums for (5).

Each term (or mode) in expansion (5) can be viewed as a **standing wave** (a wave that vibrates in place without lateral motion along the string). For example, the first term,

$$\left(a_1 \cos \frac{\pi\alpha}{L} t + b_1 \sin \frac{\pi\alpha}{L} t \right) \sin \frac{\pi x}{L},$$

consists of a sinusoidal shape function $\sin(\pi x/L)$ multiplied a time-varying amplitude. The second term is also a sinusoid $\sin(2\pi x/L)$ with a time-varying amplitude. In the latter case, there is a *node* in the middle at $x = L/2$ that never moves. For the nth term, we have a sinusoid $\sin(n\pi x/L)$ with a time-varying amplitude and $(n - 1)$ nodes. This is illustrated in Figure 10.16. Thus separation of variables decomposes the initial data into sinusoids or modes (equations (6)–(7)), assigns each mode a frequency at which to vibrate (equation

Figure 10.15 Partial sums for equation (5) with $f(x) = \begin{cases} x, & 0 < x \le \pi/2 \\ \pi - x, & \pi/2 \le x < \pi \end{cases}$ and $g(x) = 0$

(5)), and represents the solution as the superposition of infinitely many standing waves. Modes with more nodes vibrate at higher frequencies. By choosing the initial configuration of the string to have the same shape as one of the individual terms in the solution, we can activate only that mode.

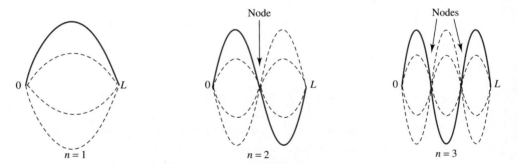

Figure 10.16 Standing waves. Time-varying amplitudes are shown by dashed curves

The different modes of vibration of a guitar string are distinguishable to the human ear by the frequency of the sound they generate; different frequencies are discerned as different "pitches." The *fundamental* is the frequency of the lowest mode, and integer multiples of the fundamental frequency are called "harmonics." A cello and a trombone sound different even if they are playing the same fundamental, say F$^{\#}$, because of the difference in the intensities of the harmonics they generate.

Many engineering devices operate better in some modes than in others. For example, certain modes propagate down a waveguide or an optical fiber with less attenuation than the others. In such cases, engineers design the systems to suppress the undesirable modes, either by shaping the initial excitation or by introducing devices (like resistor cards in a waveguide) which damp out certain modes preferentially.

The fundamental mode on a guitar string can be suppressed by holding the finger lightly in contact with the midpoint, thereby creating a stationary point or node there. According to equation (5), the pitch of the note then doubles, thus producing an "octave." This style of fingering, called "playing harmonics," is used frequently in musical performance.

An important characteristic of an engineering device is the set of angular frequencies supported by its eigenmodes—its "eigenfrequencies." According to equation (5), the eigenfrequencies of the vibrating string are the harmonics $\omega_n = n\pi a/L$, $n = 1, 2, 3, \ldots$. These blend together pleasantly to the human ear, and we enjoy melodies played on string instruments. The eigenfrequencies of a drum are not harmonics (see Problem 21). Therefore drums are used for rhythm, not for melody.

As we have seen in the preceding section, the method of separation of variables can be used to solve problems with nonhomogeneous boundary conditions and nonhomogeneous equations where the forcing term is time independent. In the next example, we will consider a problem with a time-dependent forcing term $h(x, t)$.

EXAMPLE 1 For given functions f, g, and h, find a formal solution to the initial-boundary value problem

$$(8) \qquad \frac{\partial^2 u}{\partial t^2} = \alpha^2 \frac{\partial^2 u}{\partial x^2} + h(x, t), \qquad 0 < x < L, \qquad t > 0,$$

$$(9) \qquad u(0, t) = u(L, t) = 0, \qquad t > 0,$$

$$(10) \qquad u(x, 0) = f(x), \qquad 0 < x < L,$$

$$(11) \qquad \frac{\partial u}{\partial t}(x, 0) = g(x), \quad 0 < x < L.$$

Solution The boundary conditions in (9) certainly require that the solution be zero for $x = 0$ and $x = L$. Motivated by the fact that the solution to the corresponding homogeneous system (1)–(4) consists of a superposition of standing waves, let's try to find a solution to (8)–(11) of the form

$$(12) \qquad u(x, t) = \sum_{n=1}^{\infty} u_n(t) \sin \frac{n\pi x}{L},$$

where the $u_n(t)$'s are functions of t to be determined.

For each fixed t, we can compute a Fourier sine series for $h(x, t)$. If we assume the series is convergent to $h(x, t)$, then

$$(13) \qquad h(x, t) = \sum_{n=1}^{\infty} h_n(t) \sin \frac{n\pi x}{L},$$

where the coefficient $h_n(t)$ is given by (recall equation (6) on page 605)

$$h_n(t) = \frac{2}{L} \int_0^L h(x, t) \sin\frac{n\pi x}{L}\, dx, \qquad n = 1, 2, \ldots .$$

If the series in (13) has the proper convergence properties, then we can substitute (12) and (13) into equation (8) and obtain

$$\sum_{n=1}^{\infty} \left[u_n''(t) + \left(\frac{n\pi\alpha}{L}\right)^2 u_n(t) \right] \sin\frac{n\pi x}{L} = \sum_{n=1}^{\infty} h_n(t) \sin\frac{n\pi x}{L} .$$

Equating the coefficients in each series (why?), we have

$$u_n''(t) + \left(\frac{n\pi\alpha}{L}\right)^2 u_n(t) = h_n(t).$$

This is a nonhomogeneous, constant coefficient equation that can be solved using variation of parameters. The reader should verify that

$$u_n(t) = a_n \cos\frac{n\pi\alpha}{L}t + b_n \sin\frac{n\pi\alpha}{L}t + \frac{L}{n\pi\alpha} \int_0^t h_n(s) \sin\left[\frac{n\pi\alpha}{L}(t - s)\right] ds$$

(cf. Problem 28 of Exercises 4.9 or Project A, Chapter 4, page 231.) Hence, with this choice of $u_n(t)$, the series in (12) is a formal solution to the partial differential equation (8).

Since[†]

$$u_n(0) = a_n \quad \text{and} \quad u_n'(0) = b_n\left(\frac{n\pi\alpha}{L}\right),$$

substituting (12) into the initial conditions (10)–(11) yields

(14) $$u(x, 0) = f(x) = \sum_{n=1}^{\infty} a_n \sin\frac{n\pi x}{L},$$

(15) $$\frac{\partial u}{\partial t}(x, 0) = g(x) = \sum_{n=1}^{\infty} b_n\left(\frac{n\pi\alpha}{L}\right) \sin\frac{n\pi x}{L} .$$

Thus, if we choose the a_n's and b_n's so that equations (14) and (15) are satisfied, a formal solution to (8)–(11) is given by

(16) $$u(x, t) = \sum_{n=1}^{\infty} \left\{ a_n \cos\frac{n\pi\alpha}{L}t + b_n \sin\frac{n\pi\alpha}{L}t \right.$$

$$\left. + \frac{L}{n\pi\alpha} \int_0^t h_n(s) \sin\left[\frac{n\pi\alpha}{L}(t - s)\right] ds \right\} \sin\frac{n\pi x}{L} . \quad \blacksquare \text{ End}$$

The method of separation of variables can also be used to solve initial-boundary value problems for the wave equation in higher dimensions. For example, a vibrating rectangular

[†] To compute $u_n'(0)$, we use the fact that $\dfrac{d}{dt}\displaystyle\int_0^t G(s, t)\, ds = G(t, t) + \displaystyle\int_0^t \dfrac{\partial G}{\partial t}(s, t)\, ds.$

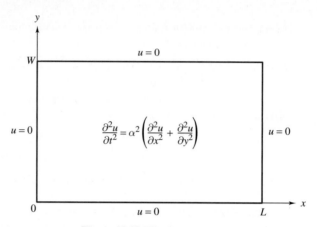

Figure 10.17 Vibrating membrane

membrane of length L and width W (see Figure 10.17) is governed by the following initial-boundary value problem for $u(x,\, y,\, t)$:

$$(17) \qquad \frac{\partial^2 u}{\partial t^2} = \alpha^2 \left(\frac{\partial^2 u}{\partial x^2} + \frac{\partial^2 u}{\partial y^2} \right), \qquad 0 < x < L, \qquad 0 < y < W, \qquad t > 0,$$

$$(18) \qquad u(0,\, y,\, t) = u(L,\, y,\, t) = 0, \qquad 0 < y < W, \qquad t > 0,$$

$$(19) \qquad u(x,\, 0,\, t) = u(x,\, W,\, t) = 0, \qquad 0 < x < L, \qquad t > 0,$$

$$(20) \qquad u(x,\, y,\, 0) = f(x,\, y), \qquad 0 < x < L, \qquad 0 < y < W,$$

$$(21) \qquad \frac{\partial u}{\partial t}(x,\, y,\, 0) = g(x,\, y), \qquad 0 < x < L, \qquad 0 < y < W.$$

Using an argument similar to the one given for the problem of heat flow in a rectangular plate (Example 4 in Section 10.5), we find that the initial-boundary value problem (17)–(21) has a formal solution

$$(22) \qquad u(x,\, y,\, t) = \sum_{m=1}^{\infty} \sum_{n=1}^{\infty} \left\{ a_{mn} \cos\left(\sqrt{\frac{m^2}{L^2} + \frac{n^2}{W^2}}\, \alpha \pi t \right) \right.$$

$$\left. + b_{mn} \sin\left(\sqrt{\frac{m^2}{L^2} + \frac{n^2}{W^2}}\, \alpha \pi t \right) \right\} \sin\frac{m\pi x}{L} \sin\frac{n\pi y}{W},$$

where the constants a_{mn} and b_{mn} are determined from the double Fourier series

$$f(x,\, y) = \sum_{m=1}^{\infty} \sum_{n=1}^{\infty} a_{mn} \sin\frac{m\pi x}{L} \sin\frac{n\pi y}{W},$$

$$g(x,\, y) = \sum_{m=1}^{\infty} \sum_{n=1}^{\infty} \alpha\pi \sqrt{\frac{m^2}{L^2} + \frac{n^2}{W^2}}\, b_{mn} \sin\frac{m\pi x}{L} \sin\frac{n\pi y}{W}.$$

In particular,

$$(23) \qquad a_{mn} = \frac{4}{LW} \int_0^L \int_0^W f(x, y) \sin \frac{m\pi x}{L} \sin \frac{n\pi y}{W} \, dy \, dx,$$

$$(24) \qquad b_{mn} = \frac{4}{LW\pi\alpha \sqrt{\dfrac{m^2}{L^2} + \dfrac{n^2}{W^2}}} \int_0^L \int_0^W g(x, y) \sin \frac{m\pi x}{L} \sin \frac{n\pi y}{W} \, dy \, dx.$$

We leave the derivation of this solution as an exercise (see Problem 19).

We mentioned earlier that the solution to the vibrating string problem (1)–(4) consisted of a superposition of standing waves. There are also "traveling waves" associated with the wave equation. Traveling waves arise naturally out of d'Alembert's solution to the wave equation for an "infinite" string.

To obtain d'Alembert's solution to the wave equation

$$\frac{\partial^2 u}{\partial t^2} = \alpha^2 \frac{\partial^2 u}{\partial x^2},$$

we use the change of variables

$$\psi = x + \alpha t, \qquad \eta = x - \alpha t.$$

If u has continuous second partial derivatives, then $\partial u/\partial x = \partial u/\partial \psi + \partial u/\partial \eta$ and $\partial u/\partial t = \alpha(\partial u/\partial \psi - \partial u/\partial \eta)$, from which we obtain

$$\frac{\partial^2 u}{\partial x^2} = \frac{\partial^2 u}{\partial \psi^2} + 2 \frac{\partial^2 u}{\partial \psi \partial \eta} + \frac{\partial^2 u}{\partial \eta^2},$$

$$\frac{\partial^2 u}{\partial t^2} = \alpha^2 \left\{ \frac{\partial^2 u}{\partial \psi^2} - 2 \frac{\partial^2 u}{\partial \psi \partial \eta} + \frac{\partial^2 u}{\partial \eta^2} \right\}.$$

Substituting these expressions into the wave equation and simplifying yields

$$\frac{\partial^2 u}{\partial \psi \partial \eta} = 0.$$

We can solve this equation directly by first integrating with respect to ψ to obtain

$$\frac{\partial u}{\partial \eta} = b(\eta),$$

where $b(\eta)$ is an arbitrary function of η, and then integrating with respect to η to find

$$u(\psi, \eta) = A(\psi) + B(\eta),$$

where $A(\psi)$ and $B(\eta) = \int b(\eta) \, d\eta$ are arbitrary functions. Substituting the original variables x and t gives **d'Alembert's solution**

$$(25) \qquad u(x, t) = A(x + \alpha t) + B(x - \alpha t).$$

It is easy to check by direct substitution that $u(x, t)$, defined by formula (25), is indeed a solution to the wave equation, provided A and B are twice differentiable functions.

EXAMPLE 2 Using d'Alembert's formula (25), find a solution to the initial value problem

(26) $$\frac{\partial^2 u}{\partial t^2} = \alpha^2 \frac{\partial^2 u}{\partial x^2}, \qquad -\infty < x < \infty, \qquad t > 0,$$

(27) $$u(x, 0) = f(x), \qquad -\infty < x < \infty,$$

(28) $$\frac{\partial u}{\partial t}(x, 0) = g(x), \qquad -\infty < x < \infty.$$

Solution A solution to (26) is given by formula (25), so we need only choose the functions A and B so that the initial conditions (27)–(28) are satisfied. For this we need

(29) $$u(x, 0) = A(x) + B(x) = f(x),$$

(30) $$\frac{\partial u}{\partial t}(x, 0) = \alpha A'(x) - \alpha B'(x) = g(x).$$

Integrating equation (30) from x_0 to x (x_0 arbitrary) and dividing by α gives

(31) $$A(x) - B(x) = \frac{1}{\alpha} \int_{x_0}^{x} g(s)\,ds + C,$$

where C is also arbitrary. Solving the system (29) and (31), we obtain

$$A(x) = \frac{1}{2} f(x) + \frac{1}{2\alpha} \int_{x_0}^{x} g(s)\,ds + \frac{C}{2},$$

$$B(x) = \frac{1}{2} f(x) - \frac{1}{2\alpha} \int_{x_0}^{x} g(s)\,ds - \frac{C}{2}.$$

Using these functions in formula (25) gives

$$u(x, t) = \frac{1}{2} [f(x + \alpha t) + f(x - \alpha t)] + \frac{1}{2\alpha} \left[\int_{x_0}^{x + \alpha t} g(s)\,ds - \int_{x_0}^{x - \alpha t} g(s)\,ds \right],$$

which simplifies to

(32) $$u(x, t) = \frac{1}{2} [f(x + \alpha t) + f(x - \alpha t)] + \frac{1}{2\alpha} \int_{x - \alpha t}^{x + \alpha t} g(s)\,ds. \quad \blacksquare \text{ End}$$

EXAMPLE 3 Find the solution to the initial value problem

(33) $$\frac{\partial^2 u}{\partial t^2} = 4 \frac{\partial^2 u}{\partial x^2}, \qquad -\infty < x < \infty, \qquad t > 0,$$

(34) $$u(x, 0) = \sin x, \qquad -\infty < x < \infty,$$

(35) $$\frac{\partial u}{\partial t}(x, 0) = 1, \qquad -\infty < x < \infty.$$

Solution This is just a special case of the preceding example where $\alpha = 2$, $f(x) = \sin x$, and $g(x) = 1$. Substituting into (32), we obtain the solution

(36) $$u(x, t) = \frac{1}{2}[\sin(x + 2t) + \sin(x - 2t)] + \frac{1}{4}\int_{x-2t}^{x+2t} ds$$

$$= \sin x \cos 2t + t. \quad \blacksquare \text{ End}$$

We now use d'Alembert's formula to show that the solution to the "infinite" string problem consists of traveling waves.

Let $h(x)$ be a function defined on $(-\infty, \infty)$. The function $h(x + a)$, where $a > 0$, is a translation of the function $h(x)$ in the sense that its "shape" is the same as $h(x)$, but its position has been shifted to the left by an amount a. This is illustrated in Figure 10.18 for a function $h(x)$ whose graph consists of a triangular "bump." If we let $t \geq 0$ be a parameter (say time), then the functions $h(x + \alpha t)$ represent a family of functions with the same shape but shifted further and further to the left as $t \to \infty$ (Figure 10.19). We say that $h(x + \alpha t)$ is a **traveling wave** moving to the left with speed α. In a similar fashion, $h(x - \alpha t)$ is a traveling wave moving to the right with speed α.

If we refer to formula (25), we find that the solution to $\partial^2 u/\partial t^2 = \alpha^2 \partial^2 u/\partial x^2$ consists of traveling waves $A(x + \alpha t)$ moving to the left with speed α and $B(x - \alpha t)$ moving to the right at the same speed.

In the special case when the initial velocity $g(x) \equiv 0$, we have

$$u(x, t) = \frac{1}{2}[f(x + \alpha t) + f(x - \alpha t)].$$

Hence $u(x, t)$ is the sum of the traveling waves

$$\frac{1}{2}f(x + \alpha t) \quad \text{and} \quad \frac{1}{2}f(x - \alpha t).$$

Figure 10.18 Graphs of $h(x)$ and $h(x + a)$

Figure 10.19 Traveling wave

These waves are initially superimposed, since

$$u(x, 0) = \frac{1}{2} f(x) + \frac{1}{2} f(x) = f(x).$$

As t increases, the two waves move away from each other with speed 2α. This is illustrated in Figure 10.20 for a triangular wave.

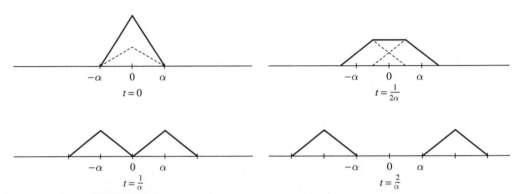

Figure 10.20 Decomposition of initial displacement into traveling waves

EXAMPLE 4 Express the standing wave $\cos \dfrac{\pi\alpha}{L} t \sin \dfrac{\pi x}{L}$ as a superposition of traveling waves.

Solution By a familiar trigonometric identity,

$$\cos \frac{\pi\alpha}{L} t \sin \frac{\pi x}{L} = \frac{1}{2} \sin \frac{\pi}{L}(x - \alpha t) + \frac{1}{2} \sin \frac{\pi}{L}(x + \alpha t)$$

$$= h(x - \alpha t) + h(x + \alpha t),$$

where

$$h(x) = \frac{1}{2} \sin \frac{\pi x}{L}.$$

This standing wave results from adding a wave traveling to the right to the same waveform traveling to the left. ∎ End

Existence and Uniqueness of Solutions

In Example 1, the method of separation of variables was used to derive a formal solution to the given initial-boundary value problem. To show that these series converge to an actual solution requires results from real analysis just as was the case for the formal solutions to the heat equation in Section 10.5. In Examples 2 and 3, we can establish the validity of d'Alembert's solution by direct substitution into the initial value problem, assuming sufficient differentiability of the initial functions. We leave it as an exercise for the reader to show that if f has a continuous second derivative and g has a continuous first derivative, then d'Alembert's solution is a true solution (see Problem 12).

The question of the uniqueness of the solution to the initial-boundary value problem (1)–(4) can be answered using an **energy argument**.

UNIQUENESS OF THE SOLUTION TO THE VIBRATING STRING PROBLEM

Theorem 8. The initial-boundary value problem

$$(37) \qquad \frac{\partial^2 u}{\partial t^2} = \alpha^2 \frac{\partial^2 u}{\partial x^2}, \qquad 0 < x < L, \qquad t > 0,$$

$$(38) \qquad u(0, t) = u(L, t) = 0, \qquad t \geq 0,$$

$$(39) \qquad u(x, 0) = f(x), \qquad 0 \leq x \leq L,$$

$$(40) \qquad \frac{\partial u}{\partial t}(x, 0) = g(x), \qquad 0 \leq x \leq L,$$

has at most one twice continuously differentiable solution.

Proof. Assume both $u(x, t)$ and $v(x, t)$ are twice continuously differentiable solutions to (37)–(40) and let $w(x, t) := u(x, t) - v(x, t)$. It is easy to check that $w(x, t)$ satisfies the initial-boundary value problem (37)–(40) with zero initial data; that is, for $0 \leq x \leq L$,

$$(41) \qquad w(x, 0) = 0 \quad \text{and} \quad \frac{\partial w}{\partial t}(x, 0) = 0.$$

We now show that $w(x, t) \equiv 0$ for $0 \leq x \leq L, t \geq 0$.

If $w(x, t)$ is the displacement of the vibrating string at location x for time t, then with the appropriate units, the total energy $E(t)$ of the vibrating string at time t is defined by the integral

$$(42) \qquad E(x) := \frac{1}{2} \int_0^L \left[\alpha^2 \left(\frac{\partial w}{\partial x} \right)^2 + \left(\frac{\partial w}{\partial t} \right)^2 \right] dx.$$

(The first term in the integrand relates to the stretching of the string at location x and represents the potential energy. The second term is the square of the velocity of the vibrating string at x and represents the kinetic energy.)

We now consider the derivative of $E(t)$:

$$\frac{dE}{dt} = \frac{d}{dt} \left\{ \frac{1}{2} \int_0^L \left[\alpha^2 \left(\frac{\partial w}{\partial x} \right)^2 + \left(\frac{\partial w}{\partial t} \right)^2 \right] dx \right\}.$$

Since w has continuous second partial derivatives (because u and v do), we can interchange the order of integration and differentiation. This gives

$$(43) \qquad \frac{dE}{dt} = \int_0^L \left[\alpha^2 \frac{\partial w}{\partial x} \frac{\partial^2 w}{\partial t \partial x} + \frac{\partial w}{\partial t} \frac{\partial^2 w}{\partial t^2} \right] dx.$$

Again the continuity of the second partials of w guarantees that the mixed partials are equal; that is,

$$\frac{\partial^2 w}{\partial t \partial x} = \frac{\partial^2 w}{\partial x \partial t}.$$

Combining this fact with integration by parts, we obtain

(44)
$$\int_0^L \alpha^2 \frac{\partial w}{\partial x} \frac{\partial^2 w}{\partial t \partial x} dx = \int_0^L \alpha^2 \frac{\partial w}{\partial x} \frac{\partial^2 w}{\partial x \partial t} dx$$

$$= \alpha^2 \frac{\partial w}{\partial x}(L, t) \frac{\partial w}{\partial t}(L, t) - \alpha^2 \frac{\partial w}{\partial x}(0, t) \frac{\partial w}{\partial t}(0, t)$$

$$- \int_0^L \alpha^2 \frac{\partial^2 w}{\partial x^2} \frac{\partial w}{\partial t} dx.$$

The boundary conditions $w(0, t) = w(L, t) = 0$, $t \geq 0$, imply that $(\partial w/\partial t)(0, t) = (\partial w/\partial t)(L, t) = 0$, $t \geq 0$. This reduces equation (44) to

$$\int_0^L \alpha^2 \frac{\partial w}{\partial x} \frac{\partial^2 w}{\partial t \partial x} dx = - \int_0^L \alpha^2 \frac{\partial w}{\partial t} \frac{\partial^2 w}{\partial x^2} dx.$$

Substituting this in for the first integrand in (43), we find

$$\frac{dE}{dt} = \int_0^L \frac{\partial w}{\partial t} \left[\frac{\partial^2 w}{\partial t^2} - \alpha^2 \frac{\partial^2 w}{\partial x^2} \right] dx.$$

Since w satisfies equation (37), the integrand is zero for all x. Thus $dE/dt = 0$, and so $E(t) \equiv C$, where C is a constant. This means that the total energy is conserved within the vibrating string.

The first boundary condition in (41) states that $w(x, 0) = 0$ for $0 \leq x \leq L$. Hence $(\partial w/\partial x)(x, 0) = 0$ for $0 < x < L$. Combining this with the second boundary condition in (41), we find that when $t = 0$, the integrand in (42) is zero for $0 < x < L$. Therefore $E(0) = 0$. Since $E(t) \equiv C$, we must have $C = 0$. Hence

(45)
$$E(t) = \frac{1}{2} \int_0^L \left[\alpha^2 \left(\frac{\partial w}{\partial x} \right)^2 + \left(\frac{\partial w}{\partial t} \right)^2 \right] dx \equiv 0.$$

That is, the total energy of the system is zero.

Because the integrand in (45) is nonnegative and continuous and the integral is zero, the integrand must be zero for $0 \leq x \leq L$. Moreover, the integrand is the sum of two squares and so each term must be zero. Hence

$$\frac{\partial w}{\partial x}(x, t) = 0 \quad \text{and} \quad \frac{\partial w}{\partial t}(x, t) = 0$$

for all $0 \leq x \leq L$, $t \geq 0$. Thus $w(x, t) = K$, where K is a constant. Physically, this says that there is no motion in the string.

Finally, since w is constant and w is zero when $t = 0$, then $w(x, t) \equiv 0$. Consequently, $u(x, t) = v(x, t)$ and the initial-boundary value problem has at most one solution. ■ End

EXERCISES 10.6

In Problems 1 through 4, find a formal solution to the vibrating string problem governed by the given initial-boundary value problem.

1. $\dfrac{\partial^2 u}{\partial t^2} = \dfrac{\partial^2 u}{\partial x^2}$, $0 < x < 1$, $t > 0$,

$u(0, t) = u(1, t) = 0$, $t > 0$,
$u(x, 0) = x(1 - x)$, $0 < x < 1$,
$\dfrac{\partial u}{\partial t}(x, 0) = \sin 7\pi x$, $0 < x < 1$.

2. $\dfrac{\partial^2 u}{\partial t^2} = 16 \dfrac{\partial^2 u}{\partial x^2}$, $0 < x < \pi$, $t > 0$,

$u(0, t) = u(\pi, t) = 0$, $t > 0$,
$u(x, 0) = \sin^2 x$, $0 < x < \pi$,
$\dfrac{\partial u}{\partial t}(x, 0) = 1 - \cos x$, $0 < x < \pi$.

3. $\dfrac{\partial^2 u}{\partial t^2} = 4 \dfrac{\partial^2 u}{\partial x^2}$, $0 < x < \pi$, $t > 0$,

$u(0, t) = u(\pi, t) = 0$, $t > 0$,
$u(x, 0) = x^2(\pi - x)$, $0 < x < \pi$,
$\dfrac{\partial u}{\partial t}(x, 0) = 0$, $0 < x < \pi$.

4. $\dfrac{\partial^2 u}{\partial t^2} = 9 \dfrac{\partial^2 u}{\partial x^2}$, $0 < x < \pi$, $t > 0$,

$u(0, t) = u(\pi, t) = 0$, $t > 0$,
$u(x, 0) = \sin 4x + 7 \sin 5x$, $0 < x < \pi$,
$\dfrac{\partial u}{\partial t}(x, 0) = \begin{cases} x, & 0 < x < \pi/2, \\ \pi - x, & \pi/2 < x < \pi. \end{cases}$

5. The Plucked String. A vibrating string is governed by the initial-boundary value problem (1)–(4). If the string is lifted to a height h_0 at $x = a$ and released, then the initial conditions are

$$f(x) = \begin{cases} h_0 x/a, & 0 < x \le a, \\ h_0(L - x)/(L - a), & a < x < L, \end{cases}$$

and $g(x) \equiv 0$. Find a formal solution.

6. The Struck String. A vibrating string is governed by the initial-boundary value problem (1)–(4). If the

string is struck at $x = a$, then the initial conditions may be approximated by $f(x) \equiv 0$ and

$$g(x) = \begin{cases} v_0 x/a, & 0 < x \le a, \\ v_0(L - x)/(L - a), & a < x < L, \end{cases}$$

where v_0 is a constant. Find a formal solution.

In Problems 7 and 8, find a formal solution to the vibrating string problem governed by the given nonhomogeneous initial-boundary value problem.

7. $\dfrac{\partial^2 u}{\partial t^2} = \dfrac{\partial^2 u}{\partial x^2} + tx$, $0 < x < \pi$, $t > 0$,

$u(0, t) = u(\pi, t) = 0$, $t > 0$,
$u(x, 0) = \sin x$, $0 < x < \pi$,
$\dfrac{\partial u}{\partial t}(x, 0) = 5 \sin 2x - 3 \sin 5x$, $0 < x < \pi$.

8. $\dfrac{\partial^2 u}{\partial t^2} = \dfrac{\partial^2 u}{\partial x^2} + x \sin t$, $0 < x < \pi$, $t > 0$,

$u(0, t) = u(\pi, t) = 0$, $t > 0$,
$u(x, 0) = 0$, $0 < x < \pi$,
$\dfrac{\partial u}{\partial t}(x, 0) = 0$, $0 < x < \pi$.

9. If one end of a string is held fixed while the other is free, then the motion of the string is governed by the initial-boundary value problem

$$\dfrac{\partial^2 u}{\partial t^2} = \alpha^2 \dfrac{\partial^2 u}{\partial x^2}, \quad 0 < x < L, \quad t > 0,$$

$$u(0, t) = 0 \quad \text{and} \quad \dfrac{\partial u}{\partial x}(L, t) = 0, \quad t > 0,$$

$$u(x, 0) = f(x), \quad 0 < x < L,$$

$$\dfrac{\partial u}{\partial t}(x, 0) = g(x), \quad 0 < x < L.$$

Derive a formula for a formal solution.

10. Derive a formula for the solution to the following initial-boundary value problem involving nonhomoge-

neous boundary conditions

$$\frac{\partial^2 u}{\partial t^2} = \alpha^2 \frac{\partial^2 u}{\partial x^2}, \qquad 0 < x < L, \qquad t > 0,$$

$$u(0, t) = U_1, \qquad u(L, t) = U_2, \qquad t > 0,$$

$$u(x, 0) = f(x), \qquad 0 < x < L,$$

$$\frac{\partial u}{\partial t}(x, 0) = g(x), \qquad 0 < x < L,$$

where U_1 and U_2 are constants.

11. The Telegraph Problem.[†] Use the method of separation of variables to derive a formal solution to the telegraph problem

$$\frac{\partial^2 u}{\partial t^2} + \frac{\partial u}{\partial t} + u = \alpha^2 \frac{\partial^2 u}{\partial x^2}, \qquad 0 < x < L, \qquad t > 0,$$

$$u(0, t) = u(L, t) = 0, \qquad t > 0,$$

$$u(x, 0) = f(x), \qquad 0 < x < L,$$

$$\frac{\partial u}{\partial t}(x, 0) = 0, \qquad 0 < x < L.$$

12. Verify d'Alembert's solution (32) to the initial value problem (26)–(28) when $f(x)$ has a continuous second derivative and $g(x)$ has a continuous first derivative by substituting it directly into the equations.

In Problems 13 through 18, find the solution to the initial value problem

$$\frac{\partial^2 u}{\partial t^2} = \alpha^2 \frac{\partial^2 u}{\partial x^2}, \qquad -\infty < x < \infty, \qquad t > 0,$$

$$u(x, 0) = f(x), \qquad -\infty < x < \infty,$$

$$\frac{\partial u}{\partial t}(x, 0) = g(x), \qquad -\infty < x < \infty,$$

for the given functions $f(x)$ and $g(x)$.

13. $f(x) \equiv 0, \qquad f(x) = \cos x.$

14. $f(x) = x^2, \qquad g(x) \equiv 0.$

15. $f(x) = x, \qquad g(x) = x.$

16. $f(x) = \sin 3x, \qquad g(x) \equiv 1.$

17. $f(x) = e^{-x^2}, \qquad g(x) = \sin x.$

18. $f(x) = \cos 2x, \qquad g(x) = 1 - x.$

19. Derive the formal solution given in equations (22)–(24) to the vibrating membrane problem governed by the initial-boundary value problem (17)–(21).

20. Long Water Waves. The motion of long water waves in a channel of constant depth is governed by the **linearized Korteweg and de Vries (KdV) equation**

(46) $\qquad u_t + \alpha u_x + \beta u_{xxx} = 0,$

where $u(x, t)$ is the displacement of the water from its equilibrium depth at location x and at time t, and α and β are positive constants.

(a) Show that equation (46) has a solution of the form

(47) $\qquad \begin{aligned} u(x, t) &= V(z), \\ z &= kx - w(k)t, \end{aligned}$

where k is a fixed constant and $w(k)$ is a function of k, provided V satisfies

(48) $\qquad -w\frac{dV}{dz} + \alpha k\frac{dV}{dz} + \beta k^3 \frac{d^3 V}{dz^3} = 0.$

These solutions, defined by (47), are called **uniform waves.**

(b) Physically, we are interested only in solutions $V(z)$ that are bounded and nonconstant on the infinite interval $(-\infty, \infty)$. Show that such solutions exist only if $\alpha k - w > 0$.

(c) Let $\lambda^2 = (\alpha k - w)/(\beta k^3)$. Show that the solutions from part (b) can be expressed in the form

$$V(x, t) = A \sin [\lambda k x - (\alpha \lambda k - \beta \lambda^3 k^3)t + B],$$

where A and B are arbitrary constants. [*Hint:* Solve for w in terms of λ and k and use (47).]

(d) Since both λ and k can be chosen arbitrarily and they always appear together as the product λk, we can set $\lambda = 1$ without loss of generality. Hence we have

$$V(x, t) = A \sin [kx - (\alpha k - \beta k^3)t + B],$$

as a uniform wave solution to (46). The defining relation

$$w(k) = \alpha k - \beta k^3$$

is called the **dispersion relation,** the ratio $w(k)/k = \alpha - \beta k^2$ is called the **phase velocity,** and the derivative $dw/dk = \alpha - 3\beta k^2$ is called the **group velocity.** When the group velocity is not constant, the waves are called **dispersive.** Show that the standard wave equation $u_{tt} = \alpha^2 u_{xx}$ has only nondispersive waves.

21. Vibrating Drum. A vibrating circular membrane of unit radius whose edges are held fixed in a plane and

[†] For a discussion of the telegraph problem, see *Methods of Mathematical Physics*, by R. Courant and D. Hibert, Vol. II (Wiley-Interscience, New York, 1962).

whose displacement $u(r, t)$ depends only on the radial distance r from the center and on the time t is governed by the initial-boundary value problem

$$\frac{\partial^2 u}{\partial t^2} = \alpha^2 \left(\frac{\partial^2 u}{\partial r^2} + \frac{1}{r} \frac{\partial u}{\partial r} \right), \quad 0 < r < 1, \quad t > 0,$$

$$u(1, t) = 0, \quad t > 0,$$

$u(r, t)$ remains finite as $r \to 0^+$,

$$u(r, 0) = f(r), \quad 0 < r < 1,$$

$$\frac{\partial u}{\partial t}(r, 0) = g(r), \quad 0 < r < 1,$$

where f and g are the initial displacements and velocities, respectively. Use the method of separation of variables to derive a formal solution to the vibrating drum problem. [*Hint:* Show that there is a family of solutions of the form

$$u_n(r, t) = [a_n \cos(k_n \alpha t) + b_n \sin(k_n \alpha t)] J_0(k_n r),$$

where J_0 is the Bessel function of the first kind of order zero and $0 < k_1 < k_2 < \cdots < k_n < \cdots$ are the positive zeros of J_0. Now use superposition.] See Figure 10.21.

(a) (b) (c)

Figure 10.21 Mode shapes for vibrating drum (a) $J_0(2.405r)$, (b) $J_0(5.520r)$, (c) $J_0(8.654r)$

10.7 LAPLACE'S EQUATION

In Section 10.1 we showed how Laplace's equation,

$$\frac{\partial^2 u}{\partial x^2} + \frac{\partial^2 u}{\partial y^2} = 0,$$

arises in the study of steady-state or time-independent solutions to the heat equation. Because these solutions do not depend on time, initial conditions are irrelevant and only boundary conditions are specified. Other applications include the static displacement $u(x, y)$ of a stretched membrane fastened in space along the boundary of a region (here u must satisfy Laplace's equation inside the region); the electrostatic and gravitational potentials in certain force fields (here u must satisfy Laplace's equation in any region that is free of electrical charges or mass); and, in fluid mechanics for an idealized fluid, the stream function $u(x, y)$ whose level curves (stream lines), $u(x, y)$ = constant, represent the path of particles in the fluid (again u satisfies Laplace's equation in the flow region).

There are two basic types of boundary conditions that are usually associated with Laplace's equation: **Dirichlet boundary conditions,** where the solution $u(x, y)$ to Laplace's equation in a domain D is required to satisfy

$$u(x, y) = f(x, y) \qquad \text{on } \partial D,$$

with $f(x, y)$ a specified function defined on the boundary ∂D of D; and **Neumann boundary conditions,** where the directional derivative $\partial u/\partial n$ along the outward normal to the boundary is required to satisfy

$$\frac{\partial u}{\partial n}(x, y) = g(x, y) \qquad \text{on } \partial D,$$

with $g(x, y)$ a specified function defined on ∂D. We say that the boundary conditions are **mixed** if the solution is required to satisfy $u(x, y) = f(x, y)$ on part of the boundary and $(\partial u/\partial n)(x, y) = g(x, y)$ on the remaining portion of the boundary.

In this section we use the method of separation of variables to find solutions to Laplace's equation with various boundary conditions for rectangular, circular, and cylindrical domains. We also discuss the existence and uniqueness of such solutions.

EXAMPLE 1 Find a solution to the following mixed boundary value problem for a rectangle (see Figure 10.22):

(1) $$\frac{\partial^2 u}{\partial x^2} + \frac{\partial^2 u}{\partial y^2} = 0, \qquad 0 < x < a, \qquad 0 < y < b,$$

(2) $$\frac{\partial u}{\partial x}(0, y) = \frac{\partial u}{\partial x}(a, y) = 0, \qquad 0 \le y \le b,$$

(3) $$u(x, b) = 0, \qquad 0 \le x \le a,$$

(4) $$u(x, 0) = f(x), \qquad 0 \le x \le a.$$

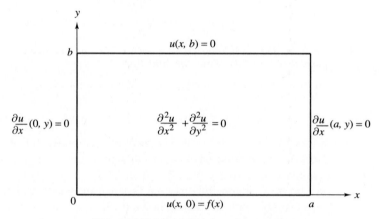

Figure 10.22 Mixed boundary value problem

Solution Separating variables, we first let $u(x, y) = X(x)Y(y)$. Substituting into equation (1), we have

$$X''(x)Y(y) + X(x)Y''(y) = 0,$$

which separates into

$$\frac{X''(x)}{X(x)} = -\frac{Y''(y)}{Y(y)} = K,$$

where K is some constant. This leads to the two ordinary differential equations

(5) $$X''(x) - KX(x) = 0,$$

(6) $$Y''(y) + KY(y) = 0.$$

From the boundary condition (2), we observe that

(7) $$X'(0) = X'(a) = 0.$$

We have encountered the eigenvalue problem in (5) and (7) before (see Example 1 in Section 10.5). The eigenvalues are $K = K_n = -(n\pi/a)^2$, $n = 0, 1, 2, \ldots$, with corresponding solutions

$$\textbf{(8)} \qquad X_n(x) = a_n \cos\left(\frac{n\pi x}{a}\right),$$

where the a_n's are arbitrary constants.

Setting $K = K_n = -(n\pi/a)^2$ in equation (6) and solving for Y gives[†]

$$\textbf{(9)} \qquad \begin{aligned} Y_0(y) &= A_0 + B_0 y, \\ Y_n(y) &= A_n \cosh\left(\frac{n\pi y}{a}\right) + B_n \sinh\left(\frac{n\pi y}{a}\right), \qquad n = 1, 2, \ldots. \end{aligned}$$

Now the boundary condition $u(x, b) = 0$ in (3) will be satisfied if $Y(b) = 0$. Setting $y = b$ in (9), we see that we want $A_0 = -bB_0$ and

$$A_n = -B_n \tanh\left(\frac{n\pi b}{a}\right), \quad \text{or equivalently} \quad Y_n(y) = C_n \sinh\left[\frac{n\pi}{a}(y - b)\right], \quad n = 1, 2, \ldots$$

(see Problem 18). Thus we find that there are solutions to (1)–(3) of the form

$$\begin{aligned} u_0(x, y) &= X_0(x)Y_0(y) = a_0 B_0(y - b) = E_0(y - b), \\ u_n(x, y) &= X_n(x)Y_n(y) = a_n \cos\left(\frac{n\pi x}{a}\right) C_n \sinh\left[\frac{n\pi}{a}(y - b)\right] \\ &= E_n \cos\left(\frac{n\pi x}{a}\right) \sinh\left[\frac{n\pi}{a}(y - b)\right], \qquad n = 1, 2, \ldots, \end{aligned}$$

where the E_n's are constants. In fact, by the superposition principle,

$$\textbf{(10)} \qquad u(x, y) = E_0(y - b) + \sum_{n=1}^{\infty} E_n \cos\left(\frac{n\pi x}{a}\right) \sinh\left[\frac{n\pi}{a}(y - b)\right]$$

is a formal solution to (1)–(3).

Applying the remaining nonhomogeneous boundary condition in (4), we have

$$u(x, 0) = f(x) = -E_0 b + \sum_{n=1}^{\infty} E_n \sinh\left(-\frac{n\pi b}{a}\right) \cos\left(\frac{n\pi x}{a}\right).$$

This is a Fourier cosine series for $f(x)$ and hence the coefficients are given by the formulas

$$\textbf{(11)} \qquad \begin{aligned} E_0 &= \frac{1}{(-ba)} \int_0^a f(x)\, dx, \\ E_n &= \frac{2}{a \sinh\left(-\dfrac{n\pi b}{a}\right)} \int_0^a f(x) \cos\left(\frac{n\pi x}{a}\right) dx, \qquad n = 1, 2, \ldots. \end{aligned}$$

[†] We usually express $Y_n(y) = a_n e^{n\pi y/a} + b_n e^{-n\pi y/a}$. However, computation is simplified in this case by using the hyperbolic functions $\cosh z = (e^z + e^{-z})/2$ and $\sinh z = (e^z - e^{-z})/2$.

Figure 10.23 Partial sums for Example 1 with $f(x) = \begin{cases} x, & 0 < x \leq \pi/2 \\ \pi - x, & \pi/2 \leq x < \pi \end{cases}$

Thus a formal solution is given by (10) with the constants E_n given by (11). See Figure 10.23 for a sketch of partial sums for (10). ■ End

In Example 1 the boundary conditions were homogeneous on three sides of the rectangle and nonhomogeneous on the fourth side, $\{(x, y): y = 0, 0 \leq x \leq a\}$. It is important to note that the method used in Example 1 can also be used to solve problems for which the boundary conditions are nonhomogeneous on all sides. This is accomplished by solving four separate boundary value problems in which three sides have homogeneous boundary conditions and only one side is nonhomogeneous. The solution is then obtained by summing these four solutions (see Problem 5).

For problems involving circular domains, it is usually more convenient to use polar coordinates. In rectangular coordinates the Laplacian has the form

$$\Delta u = \frac{\partial^2 u}{\partial x^2} + \frac{\partial^2 u}{\partial y^2}.$$

In polar coordinates (r, θ), we let

$$x = r \cos \theta, \qquad y = r \sin \theta$$

so that

$$r = \sqrt{x^2 + y^2}, \qquad \tan \theta = y/x.$$

With patience and a little care in applying the chain rule, one can show that the Laplacian in polar coordinates is

(12) $$\Delta u = \frac{\partial^2 u}{\partial r^2} + \frac{1}{r} \frac{\partial u}{\partial r} + \frac{1}{r^2} \frac{\partial^2 u}{\partial \theta^2}$$

(see Problem 6). In the next example we obtain a solution to the **Dirichlet problem** in a disk of radius a.

EXAMPLE 2 A circular metal disk of radius a has its top and bottom insulated. The edge ($r = a$) of the disk is kept at a specified temperature that depends on its location (varies with θ). The steady-state temperature inside the disk satisfies Laplace's equation. Determine the temperature distribution $u(r, \theta)$ inside the disk by finding the solution to the following Dirichlet boundary value problem, depicted in Figure 10.24:

(13) $$\frac{\partial^2 u}{\partial r^2} + \frac{1}{r}\frac{\partial u}{\partial r} + \frac{1}{r^2}\frac{\partial^2 u}{\partial \theta^2} = 0; \qquad 0 \le r < a, \qquad -\pi \le \theta \le \pi,$$

(14) $$u(a, \theta) = f(\theta), \qquad -\pi \le \theta \le \pi.$$

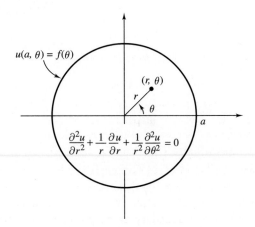

Figure 10.24 Steady-state temperature distribution in a disk

Solution To use the method of separation of variables, we first set

$$u(r, \theta) = R(r)T(\theta),$$

where $0 \le r < a$ and $-\pi \le \theta \le \pi$. Substituting into (13) and separating variables give

$$\frac{r^2 R''(r) + rR'(r)}{R(r)} = -\frac{T''(\theta)}{T(\theta)} = \lambda,$$

where λ is any constant. This leads to the two ordinary differential equations

(15) $$r^2 R''(r) + rR'(r) - \lambda R(r) = 0,$$

(16) $$T''(\theta) + \lambda T(\theta) = 0.$$

For $u(r, \theta)$ to be continuous in the disk $0 \le r < a$, we need $T(\theta)$ to be 2π-periodic; in particular, we require

(17) $$T(-\pi) = T(\pi) \quad \text{and} \quad T'(-\pi) = T'(\pi).$$

Therefore we seek nontrivial solutions to the eigenvalue problem (16)–(17).

When $\lambda < 0$, the general solution to (16) is the sum of two exponentials. Hence we have only trivial 2π-periodic solutions.

When $\lambda = 0$, we find $T(\theta) = A\theta + B$ to be the solution to (16). This linear function is periodic only when $A = 0$, that is, $T_0(\theta) = B$ is the only 2π-periodic solution corresponding to $\lambda = 0$.

When $\lambda > 0$, the general solution to (16) is

$$T(\theta) = A \cos \sqrt{\lambda}\theta + B \sin \sqrt{\lambda}\theta.$$

Here we get a nontrivial 2π-periodic solution only when $\sqrt{\lambda} = n$, $n = 1, 2, \ldots$. (The reader can check this using (17).) Hence we obtain the nontrivial 2π-periodic solutions

(18) $T_n(\theta) = A_n \cos n\theta + B_n \sin n\theta$

corresponding to $\sqrt{\lambda} = n$, $n = 1, 2, \ldots$.

Now for $\lambda = n^2$, $n = 0, 1, 2, \ldots$, equation (15) is the Cauchy-Euler equation

(19) $r^2 R''(r) + r R'(r) - n^2 R(r) = 0$

(see Section 4.5, page 178). When $n = 0$, the general solution is

$$R_0(r) = C + D \ln r.$$

Since $\ln r \to \infty$ as $r \to 0^+$, this solution is unbounded near $r = 0$ when $D \neq 0$. Therefore we must choose $D = 0$ if $u(r, \theta)$ is to be continuous at $r = 0$. We now have $R_0(r) = C$ and so $u_0(r, \theta) = R_0(r)T(\theta) = CB$, which for convenience we write in the form

(20) $u_0(r, \theta) = \dfrac{A_0}{2}$,

where A_0 is an arbitrary constant.

When $\lambda = n^2$, $n = 1, 2, \ldots$, the reader should verify that equation (19) has the general solution

$$R_n(r) = C_n r^n + D_n r^{-n}.$$

Since $r^{-n} \to \infty$ as $r \to 0^+$, we must set $D_n = 0$ in order for $u(r, \theta)$ to be bounded at $r = 0$. Thus

$$R_n(r) = C_n r^n.$$

Now for each $n = 1, 2, \ldots$, we have the solutions

(21) $u_n(r, \theta) = R_n(r)T_n(\theta) = C_n r^n (A_n \cos n\theta + B_n \sin n\theta),$

and by forming an infinite series from the solutions in (20) and (21), we get the following formal solution to (13):

$$u(r, \theta) = \frac{A_0}{2} + \sum_{n=1}^{\infty} C_n r^n (A_n \cos n\theta + B_n \sin n\theta).$$

It is more convenient to write this series in the equivalent form

(22) $u(r, \theta) = \dfrac{a_0}{2} + \sum_{n=1}^{\infty} \left(\dfrac{r}{a}\right)^n (a_n \cos n\theta + b_n \sin n\theta),$

where the a_n's and b_n's are constants. These constants can be determined from the boundary condition; indeed with $r = a$ in (22), condition (14) becomes

$$f(\theta) = \frac{a_0}{2} + \sum_{n=1}^{\infty} (a_n \cos n\theta + b_n \sin n\theta).$$

Hence if $f(\theta)$ is 2π-periodic, we recognize that a_n, b_n are Fourier coefficients. Thus

(23) $a_n = \frac{1}{\pi} \int_{-\pi}^{\pi} f(\theta) \cos n\theta \, d\theta, \qquad n = 0, 1, \ldots,$

(24) $b_n = \frac{1}{\pi} \int_{-\pi}^{\pi} f(\theta) \sin n\theta \, d\theta, \qquad n = 1, 2, \ldots.$

To summarize, if a_n and b_n are defined by formulas (23) and (24), then $u(r, \theta)$ given in (22) is a formal solution to the Dirichlet problem (13)–(14). ▪ End

The procedure in Example 2 can also be used to study the **Neumann problem** in a disk:

(25) $\Delta u = 0, \qquad 0 \le r < a, \qquad -\pi \le \theta \le \pi,$

(26) $\dfrac{\partial u}{\partial r}(a, \theta) = f(\theta), \qquad -\pi \le \theta \le \pi.$

For this problem there is no unique solution, since if u is a solution, then the function u plus a constant is also a solution. Moreover f must satisfy the **consistency condition**

(27) $\displaystyle\int_{-\pi}^{\pi} f(\theta)d\theta = 0.$

If we interpret the solution $u(r, \theta)$ of equation (25) as the steady-state temperature distribution inside a circular disk that does not contain either a heat source or heat sink, then equation (26) specifies the flow of heat across the boundary of the disk. Here the consistency condition (27) is simply the requirement that the net flow of heat across the boundary is zero. (If we keep pumping heat in, the temperature won't reach equilibrium!) We leave the solution of the Neumann problem and the derivation of the consistency condition for the exercises.

The technique used in Example 2 also applies to annular domains, $\{(r, \theta): 0 < a < r < b\}$, and to exterior domains, $\{(r, \theta): a < r\}$. We also leave these applications as exercises.

Laplace's equation in cylindrical coordinates arises in the study of steady-state temperature distributions in a solid cylinder and in determining the electric potential inside a cylinder. In the cylindrical coordinates,

$$x = r \cos \theta, \qquad y = r \sin \theta, \qquad z = z,$$

Laplace's equation becomes

(28) $\Delta u = \dfrac{\partial^2 u}{\partial r^2} + \dfrac{1}{r} \dfrac{\partial u}{\partial r} + \dfrac{1}{r^2} \dfrac{\partial^2 u}{\partial \theta^2} + \dfrac{\partial^2 u}{\partial z^2} = \mathbf{0.}$

The Dirichlet problem for the cylinder $\{(r,\ \theta,\ z): 0 \leq r \leq a,\ 0 \leq z \leq b\}$ has the boundary conditions

(29) $u(a,\ \theta,\ z) = f(\theta,\ z),$ $-\pi \leq \theta \leq \pi,$ $0 \leq z \leq b,$

(30) $u(r,\ \theta,\ 0) = g(r,\ \theta),$ $0 \leq r \leq a,$ $-\pi \leq \theta \leq \pi,$

(31) $u(r,\ \theta,\ b) = h(r,\ \theta),$ $0 \leq r \leq a,$ $-\pi \leq \theta \leq \pi.$

To solve the Dirichlet boundary value problem (28)–(31), we first solve the three boundary value problems corresponding to: (i) $g \equiv 0$ and $h \equiv 0$; (ii) $f \equiv 0$ and $h \equiv 0$; and (iii) $f \equiv 0$ and $g \equiv 0$. Then by the superposition principle, the solution to (28)–(31) will be the sum of these three solutions. This is the same method that was discussed in dealing with Dirichlet problems on rectangular domains. (See the remarks following Example 1.) In the next example we solve the Dirichlet problem when $g \equiv 0$ and $h \equiv 0$.

EXAMPLE 3 The base $(z = 0)$ and the top $(z = b)$ of a charge-free cylinder are grounded and therefore are at zero potential. The potential on the lateral surface $(r = a)$ of the cylinder is given by $u(a,\ \theta,\ z) = f(\theta,\ z)$, where $f(\theta,\ 0) = f(\theta,\ b) = 0$. Inside the cylinder, the potential $u(r,\ \theta,\ z)$ satisfies Laplace's equation. Determine the potential u inside the cylinder by finding a solution to the Dirichlet boundary value problem

(32) $\dfrac{\partial^2 u}{\partial r^2} + \dfrac{1}{r}\dfrac{\partial u}{\partial r} + \dfrac{1}{r^2}\dfrac{\partial^2 u}{\partial \theta^2} + \dfrac{\partial^2 u}{\partial z^2} = 0,$ $0 \leq r < a,$ $-\pi \leq \theta \leq \pi,$ $0 < z < b,$

(33) $u(a,\ \theta,\ z) = f(\theta,\ z),$ $-\pi \leq \theta \leq \pi,$ $0 \leq z \leq b,$

(34) $u(r,\ \theta,\ 0) = u(r,\ \theta,\ b) = 0,$ $0 \leq r < a,$ $-\pi \leq \theta \leq \pi.$

Solution Using the method of separation of variables, we first assume

$$u(r,\ \theta,\ z) = R(r)T(\theta)Z(z).$$

Substituting into equation (32) and separating out the Z's, we find

$$\frac{R''(r) + (1/r)R'(r)}{R(r)} + \frac{1}{r^2}\frac{T''(\theta)}{T(\theta)} = -\frac{Z''(z)}{Z(z)} = \lambda,$$

where λ can be any constant. Separating further the R's and T's gives

$$\frac{r^2 R''(r) + rR'(r)}{R(r)} - r^2\lambda = -\frac{T''(\theta)}{T(\theta)} = \mu,$$

where μ can also be any constant. We now have the three ordinary differential equations:

(35) $r^2 R''(r) + rR'(r) - (r^2\lambda + \mu)R(r) = 0,$

(36) $T''(\theta) + \mu T(\theta) = 0,$

(37) $Z''(z) + \lambda Z(z) = 0.$

For u to be continuous in the cylinder, $T(\theta)$ must be 2π-periodic. Thus let's begin with the eigenvalue problem

$$T''(\theta) + \mu T(\theta) = 0,\quad -\pi < \theta < \pi,$$
$$T(-\pi) = T(\pi)\quad \text{and}\quad T'(-\pi) = T'(\pi).$$

In Example 2 we showed that this problem has nontrivial solutions for $\mu = n^2$, $n = 0, 1, 2, \ldots$, that are given by

(38) $T_n(\theta) = A_n \cos n\theta + B_n \sin n\theta,$

where the A_n's and B_n's are arbitrary constants.

The boundary conditions in (34) imply that $Z(0) = Z(b) = 0$. Therefore Z must satisfy the eigenvalue problem

$$Z''(z) + \lambda Z(z) = 0, \qquad 0 < z < b,$$
$$Z(0) = Z(b) = 0.$$

We have seen this eigenvalue problem several times before. Nontrivial solutions exist for $\lambda = (m\pi/b)^2$, $m = 1, 2, 3, \ldots$ and are given by

(39) $Z_m(z) = C_m \sin\left(\dfrac{m\pi z}{b}\right),$

where the C_m's are arbitrary constants.

Substituting for μ and λ in equation (35) gives

$$r^2 R''(r) + rR'(r) - \left(r^2\left(\frac{m\pi}{b}\right)^2 + n^2\right)R(r) = 0, \qquad 0 \le r < a.$$

The reader should verify that the change of variables $s = (m\pi r/b)$ transforms this equation into the **modified Bessel's equation of order n**[†]

(40) $s^2 R''(s) + sR'(s) - (s^2 + n^2)R(s) = 0, \qquad 0 \le s < \dfrac{m\pi a}{b}.$

The modified Bessel's equation of order n has two linearly independent solutions—the **modified Bessel function of the first kind:**

$$I_n(s) = \sum_{k=0}^{\infty} \frac{(s/2)^{2k+n}}{k!\,\Gamma(k + n + 1)},$$

which remains bounded near zero, and the **modified Bessel function of the second kind:**

$$K_n(s) = \lim_{v \to n} \frac{\pi}{2} \frac{I_{-v}(s) - I_v(s)}{\sin v\pi},$$

which becomes unbounded as $s \to 0$. (Recall that Γ is the gamma function discussed in Section 7.6.) A general solution to (40) has the form $CK_n + DI_n$, where C and D are constants. Since u must remain bounded near $s = 0$, we must take $C = 0$. Thus the desired

[†] The modified Bessel's equation of order n arises in many applications and has been studied extensively. We refer the reader to the text *Special Functions*, by E. D. Raineville (Chelsea Publishing Co., Inc., New York, 1972) for details about its solution.

solutions to (40) have the form

(41) $R_{mn}(r) = D_{mn}I_n\left(\dfrac{m\pi r}{b}\right),$ $n = 0, 1, \dots,$ $m = 1, 2, \dots,$

where the D_{mn}'s are arbitrary constants.

If we multiply the functions in (38), (39), and (41) and then sum over m and n, we obtain the following series solution to (32) and (34):

(42) $u(r, \theta, z) = \displaystyle\sum_{m=1}^{\infty} a_{m0}I_0\left(\frac{m\pi r}{b}\right)\sin\left(\frac{m\pi z}{b}\right)$

$$+ \sum_{n=1}^{\infty}\sum_{m=1}^{\infty}(a_{mn}\cos n\theta + b_{mn}\sin n\theta)I_n\left(\frac{m\pi r}{b}\right)\sin\left(\frac{m\pi z}{b}\right),$$

where the a_{mn}'s and b_{mn}'s are arbitrary constants.

The constants in (42) can be obtained by imposing boundary conditions (33). Setting $r = a$ and rearranging terms, we have

(43) $f(\theta, z) = \displaystyle\sum_{m=1}^{\infty} a_{m0}I_0\left(\frac{m\pi a}{b}\right)\sin\left(\frac{m\pi z}{b}\right)$

$$+ \sum_{n=1}^{\infty}\left[\sum_{m=1}^{\infty} a_{mn}I_n\left(\frac{m\pi a}{b}\right)\sin\left(\frac{m\pi z}{b}\right)\right]\cos n\theta$$

$$+ \sum_{n=1}^{\infty}\left[\sum_{m=1}^{\infty} b_{mn}I_n\left(\frac{m\pi a}{b}\right)\sin\left(\frac{m\pi z}{b}\right)\right]\sin n\theta.$$

Treating this double Fourier series like the one in Section 10.5 (page 618), we find

(44) $a_{m0} = \displaystyle\int_0^b\int_{-\pi}^{\pi} f(\theta, z)\sin\left(\frac{m\pi z}{b}\right)d\theta\,dz \Big/ \pi b I_0\left(\frac{m\pi a}{b}\right),$

(45) $a_{mn} = 2\displaystyle\int_0^b\int_{-\pi}^{\pi} f(\theta, z)\sin\left(\frac{m\pi z}{b}\right)\cos n\theta\,d\theta\,dz \Big/ \pi b I_n\left(\frac{m\pi a}{b}\right),$ $n \geq 1,$

(46) $b_{mn} = 2\displaystyle\int_0^b\int_{-\pi}^{\pi} f(\theta, z)\sin\left(\frac{m\pi z}{b}\right)\sin n\theta\,d\theta\,dz \Big/ \pi b I_n\left(\frac{m\pi a}{b}\right).$

Consequently, a formal solution to (32)–(34) is given by equation (42) with the constants a_{mn} and b_{mn} determined by equations (44)–(46). ■ End

Existence and Uniqueness of Solutions

The existence of solutions to the boundary value problems for Laplace's equation can be established by studying the convergence of the formal solutions that we obtained using the method of separation of variables.

To answer the question of the uniqueness of the solution to a Dirichlet boundary value problem for Laplace's equation, recall that Laplace's equation arises in the search for steady-state solutions to the heat equation. Just as there are maximum principles for the

heat equation, there are also maximum principles for Laplace's equation. We state one such result here.[†]

MAXIMUM PRINCIPLE FOR LAPLACE'S EQUATION

Theorem 9. Let $u(x, y)$ be a solution to Laplace's equation in a bounded domain D with $u(x, y)$ continuous in $\overline{D}$, the closure of D. ($\overline{D} = D \cup \partial D$, where ∂D denotes the boundary of D.) Then $u(x, y)$ attains its maximum value on ∂D.

The uniqueness of the solution to the Dirichlet boundary value problem follows from the maximum principle. We state this result in the next theorem but leave its proof as an exercise (see Problem 19).

UNIQUENESS OF SOLUTION

Theorem 10. Let D be a bounded domain. If there is a continuous solution to the Dirichlet boundary value problem

$$\Delta u(x, y) = 0 \qquad \text{in } D,$$

$$u(x, y) = f(x, y) \quad \text{on } \partial D,$$

then the solution is unique.

Solutions to Laplace's equation in two variables are called **harmonic** functions. These functions arise naturally in the study of analytic functions of a single complex variable. Moreover, complex analysis provides many useful results about harmonic functions. For a discussion of this interaction, we refer the reader to an introductory text on complex analysis such as *Fundamentals of Complex Analysis*, Second Edition, by E. B. Saff and A. D. Snider (Prentice Hall, Inc., Englewood Cliffs, New Jersey, 1993).

EXERCISES 10.7

In Problems 1 through 5, find a formal solution to the given boundary value problem.

1. $\dfrac{\partial^2 u}{\partial x^2} + \dfrac{\partial^2 u}{\partial y^2} = 0, \qquad 0 < x < \pi, \qquad 0 < y < 1,$

$\dfrac{\partial u}{\partial x}(0, y) = \dfrac{\partial u}{\partial x}(\pi, y) = 0, \qquad 0 \le y \le 1,$

$u(x, 0) = 4 \cos 6x + \cos 7x, \qquad 0 \le x \le \pi,$

$u(x, 1) = 0, \qquad 0 \le x \le \pi.$

2. $\dfrac{\partial^2 u}{\partial x^2} + \dfrac{\partial^2 u}{\partial y^2} = 0, \qquad 0 < x < \pi, \qquad 0 < y < \pi,$

$\dfrac{\partial u}{\partial x}(0, y) = \dfrac{\partial u}{\partial x}(\pi, y) = 0, \qquad 0 \le y \le \pi,$

$u(x, 0) = \cos x - 2 \cos 4x, \qquad 0 \le x \le \pi,$

$u(x, \pi) = 0, \qquad 0 \le x \le \pi.$

[†] A proof can be found in Section 8.2 of the text *Partial Differential Equations of Mathematical Physics*, Second Edition, by Tyn Myint-U (Elsevier North Holland, Inc., New York, 1980).

3. $\dfrac{\partial^2 u}{\partial x^2} + \dfrac{\partial^2 u}{\partial y^2} = 0,$ $0 < x < \pi,$ $0 < y < \pi,$

$u(0, y) = u(\pi, y) = 0,$ $0 \le y \le \pi,$
$u(x, 0) = f(x),$ $0 \le x \le \pi,$
$u(x, \pi) = 0,$ $0 \le x \le \pi.$

4. $\dfrac{\partial^2 u}{\partial x^2} + \dfrac{\partial^2 u}{\partial y^2} = 0,$ $0 < x < \pi,$ $0 < y < \pi,$

$u(0, y) = u(\pi, y) = 0,$ $0 \le y \le \pi,$
$u(x, 0) = \sin x + \sin 4x,$ $0 \le x \le \pi,$
$u(x, \pi) = 0,$ $0 \le x \le \pi.$

5. $\dfrac{\partial^2 u}{\partial x^2} + \dfrac{\partial^2 u}{\partial y^2} = 0,$ $0 < x < \pi,$ $0 < y < 1,$

$\dfrac{\partial u}{\partial x}(0, y) = \dfrac{\partial u}{\partial x}(\pi, y) = 0,$ $0 \le y \le 1,$

$u(x, 0) = \cos x - \cos 3x,$ $0 \le x \le \pi,$
$u(x, 1) = \cos 2x,$ $0 \le x \le \pi.$

6. Derive the polar coordinate form of the Laplacian given in equation (12).

In Problems 7 and 8, find a solution to the Dirichlet boundary value problem for a disk:

$$\dfrac{\partial^2 u}{\partial r^2} + \dfrac{1}{r}\dfrac{\partial u}{\partial r} + \dfrac{1}{r^2}\dfrac{\partial^2 u}{\partial \theta^2} = 0, \quad 0 \le r < 2, -\pi \le \theta \le \pi,$$

$$u(2, \theta) = f(\theta), -\pi \le \theta \le \pi$$

for the given function $f(\theta)$.

7. $f(\theta) = |\theta|,$ $-\pi \le \theta \le \pi.$

8. $f(\theta) = \cos^2 \theta,$ $-\pi \le \theta \le \pi.$

9. Find a solution to the Neumann boundary value problem for a disk:

$$\dfrac{\partial^2 u}{\partial r^2} + \dfrac{1}{r}\dfrac{\partial u}{\partial r} + \dfrac{1}{r^2}\dfrac{\partial^2 u}{\partial \theta^2} = 0, \quad 0 \le r < a, -\pi \le \theta \le \pi,$$

$$\dfrac{\partial u}{\partial r}(a, \theta) = f(\theta), -\pi \le \theta \le \pi.$$

10. A solution to the Neumann problem (25)–(26) must also satisfy the consistency condition in (27). To show this, use **Green's second formula**

$$\iint_D (v\Delta u - u\Delta v)\, dx\, dy = \int_{\partial D}\left(v\dfrac{\partial u}{\partial n} - u\dfrac{\partial v}{\partial n}\right) ds,$$

where $\partial/\partial n$ is the outward normal derivative and ds is the differential of arc length. [*Hint:* Take $v \equiv 1$ and observe that $\partial u/\partial n = \partial u/\partial r$.]

11. Find a solution to the following Dirichlet problem for an annulus:

$$\dfrac{\partial^2 u}{\partial r^2} + \dfrac{1}{r}\dfrac{\partial u}{\partial r} + \dfrac{1}{r^2}\dfrac{\partial^2 u}{\partial \theta^2} = 0, \quad 1 < r < 2, -\pi \le \theta \le \pi,$$

$$u(1, \theta) = \sin 4\theta - \cos \theta, \quad -\pi \le \theta \le \pi,$$

$$u(2, \theta) = \sin \theta, \quad -\pi \le \theta \le \pi.$$

12. Find a solution to the following Dirichlet problem for an annulus:

$$\dfrac{\partial^2 u}{\partial r^2} + \dfrac{1}{r}\dfrac{\partial u}{\partial r} + \dfrac{1}{r^2}\dfrac{\partial^2 u}{\partial \theta^2} = 0, \quad 1 < r < 3, -\pi \le \theta \le \pi,$$

$$u(1, \theta) = 0, \quad -\pi \le \theta \le \pi,$$

$$u(3, \theta) = \cos 3\theta + \sin 5\theta, \quad -\pi \le \theta \le \pi.$$

13. Find a solution to the following Dirichlet problem for an exterior domain:

$$\dfrac{\partial^2 u}{\partial r^2} + \dfrac{1}{r}\dfrac{\partial u}{\partial r} + \dfrac{1}{r^2}\dfrac{\partial^2 u}{\partial \theta^2} = 0, \quad 1 < r, -\pi \le \theta \le \pi,$$

$$u(1, \theta) = f(\theta), \quad -\pi \le \theta \le \pi,$$

$u(r, \theta)$ remains bounded as $r \to \infty.$

14. Find a solution to the following Neumann problem for an exterior domain:

$$\dfrac{\partial^2 u}{\partial r^2} + \dfrac{1}{r}\dfrac{\partial u}{\partial r} + \dfrac{1}{r^2}\dfrac{\partial^2 u}{\partial \theta^2} = 0, \quad 1 < r, -\pi \le \theta \le \pi,$$

$$\dfrac{\partial u}{\partial r}(1, \theta) = f(\theta), -\pi \le \theta \le \pi,$$

$u(r, \theta)$ remains bounded as $r \to \infty.$

15. Find a solution to the following Dirichlet problem for a half disk:

$$\dfrac{\partial^2 u}{\partial r^2} + \dfrac{1}{r}\dfrac{\partial u}{\partial r} + \dfrac{1}{r^2}\dfrac{\partial^2 u}{\partial \theta^2} = 0, \quad 0 < r < 1, 0 < \theta < \pi,$$

$$u(r, 0) = 0, \quad 0 \le r \le 1,$$

$$u(r, \pi) = 0, \quad 0 \le r \le 1,$$

$$u(1, \theta) = \sin 3\theta, \quad 0 \le \theta \le \pi,$$

$u(0, \theta)$ bounded.

16. Find a solution to the following Dirichlet problem for a half annulus:

$$\dfrac{\partial^2 u}{\partial r^2} + \dfrac{1}{r}\dfrac{\partial u}{\partial r} + \dfrac{1}{r^2}\dfrac{\partial^2 u}{\partial \theta^2} = 0, \quad \pi < r < 2\pi, 0 < \theta < \pi,$$

$$u(r, 0) = \sin r, \quad \pi \le r \le 2\pi,$$

$$u(r, \pi) = 0, \quad \pi \le r \le 2\pi,$$

$$u(\pi, \theta) = u(2\pi, \theta) = 0, \quad 0 \le \theta \le \pi.$$

17. Find a solution to the mixed boundary value problem

$$\frac{\partial^2 u}{\partial r^2} + \frac{1}{r}\frac{\partial u}{\partial r} + \frac{1}{r^2}\frac{\partial^2 u}{\partial \theta^2} = 0, \quad 1 < r < 3, \quad -\pi \le \theta \le \pi,$$

$$u(1, \theta) = f(\theta), \qquad -\pi \le \theta \le \pi,$$

$$\frac{\partial u}{\partial r}(3, \theta) = g(\theta), \quad -\pi \le \theta \le \pi.$$

18. Show that

$$-B_n \tanh\left(\frac{n\pi b}{a}\right) \cosh\left(\frac{n\pi y}{a}\right) + B_n \sinh\left(\frac{n\pi y}{a}\right)$$

$$= C_n \sinh\left[\frac{n\pi}{a}(y - b)\right]$$

where

$$C_n = B_n / \cosh\left(\frac{n\pi b}{a}\right).$$

19. Prove Theorem 10 on the uniqueness of the solution to the Dirichlet problem.

20. Stability. Use the maximum principle to prove the following theorem on the continuous dependence of the solution on the boundary conditions:

Theorem. Let u_1 and u_2 be continuous functions in $\overline{D}$, where D is a bounded domain. For $i = 1$ and 2, let u_i be the solution to the Dirichlet problem

$$\Delta u = 0, \quad \text{in } D,$$

$$u = f_i, \quad \text{on } \partial D.$$

If the boundary values satisfy

$$|f_1(x, y) - f_2(x, y)| \le \varepsilon \quad \text{for all } (x, y) \text{ on } \partial D,$$

where $\varepsilon > 0$ is some constant, then

$$|u_1(x, y) - u_2(x, y)| \le \varepsilon \quad \text{for all } (x, y) \text{ in } D.$$

21. For the Dirichlet problem described in Example 3, let $a = b = \pi$ and assume the potential on the lateral side $(r = \pi)$ of the cylinder is $f(\theta, z) = \sin z$. Use equations (44)–(46) to compute the solution given by equation (42).

22. Invariance of Laplace's Equation. A complex-valued function $f(z)$ of the complex variable $z = x + iy$ can be written in the form $f(z) = u(x, y) + iv(x, y)$, where u and v are real-valued functions. If $f(z)$ is analytic in a planar region D, then its real and imaginary parts satisfy the Cauchy-Riemann equations in D; that is, in D

$$\frac{\partial u}{\partial x} = \frac{\partial v}{\partial y}, \qquad \frac{\partial u}{\partial y} = -\frac{\partial v}{\partial x}.$$

Let f be analytic and one-to-one in D and assume its inverse f^{-1} is analytic in D', where D' is the image of D under f. Then

$$\frac{\partial x}{\partial u} = \frac{\partial y}{\partial v}, \qquad \frac{\partial x}{\partial v} = -\frac{\partial y}{\partial u}$$

are just the Cauchy-Riemann equations for f^{-1}. Show that if $\phi(x, y)$ satisfies Laplace's equation for (x, y) in D, then $\psi(u, v) := \phi(x(u, v), y(u, v))$ satisfies Laplace's equation for (u, v) in D'.

23. Fluid Flow Around a Corner. The stream lines that describe the fluid flow around a corner (see Figure 10.25) are given by $\phi(x, y) = k$, where k is a constant and ϕ, the stream function, satisfies the boundary value problem

$$\frac{\partial^2 \phi}{\partial x^2} + \frac{\partial^2 \phi}{\partial y^2} = 0, \quad x > 0, \quad y > 0,$$

$$\phi(x, 0) = 0, \quad 0 \le x,$$

$$\phi(0, y) = 0, \quad 0 \le y.$$

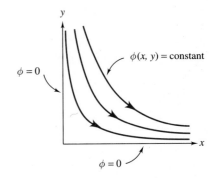

Figure 10.25 Flow around a corner

(a) Using the results of Problem 22, show that this problem can be reduced to finding the flow above a flat plate (see Figure 10.26 on page 684). That is, show that the problem reduces to finding the solution to

$$\frac{\partial^2 \psi}{\partial u^2} + \frac{\partial^2 \psi}{\partial v^2} = 0, \quad v > 0, \quad -\infty < u < \infty,$$

$$\psi(u, 0) = 0, \quad -\infty < u < \infty,$$

where ψ and ϕ are related as follows: $\psi(u, v) = \phi(x(u, v), y(u, v))$ with the mapping between (u, v) and (x, y) given by the analytic function $f(z) = z^2$.

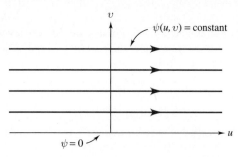

Figure 10.26 Flow above a flat surface

(b) Verify that a nonconstant solution to the problem in part (a) is given by $\psi(u, v) = v$.

(c) Using the result of part (b), find a stream function $\phi(x, y)$ for the original problem.

24. Unbounded Domain. Using separation of variables, find a solution of Laplace's equation in the infinite rectangle $0 < x < \pi$, $0 < y < \infty$ that is zero on the sides $x = 0$ and $x = \pi$, approaches zero as $y \to \infty$, and equals $f(x)$ for $y = 0$.

CHAPTER SUMMARY

Separation of Variables: A classical technique that is effective in solving boundary value problems for partial differential equations is the **method of separation of variables**. Briefly the idea is to assume first that there exists a solution that can be written with the variables separated: e.g., $u(x, t) = X(x)T(t)$. Substituting $u = XT$ into the partial differential equation and then imposing the boundary conditions leads to the problem of finding the eigenvalues and eigenfunctions for a boundary value problem for an ordinary differential equation. Solving for the eigenvalues and eigenfunctions, one eventually obtains solutions $u_n(x, t) = X_n(x)T_n(t)$, $n = 1, 2, 3, \ldots$ that solve the partial differential equation and the boundary conditions. Taking infinite linear combinations of the $u_n(x, t)$ yields solutions to the partial differential equation and boundary conditions of the form

$$u(x, t) = \sum_{n=1}^{\infty} c_n u_n(x, t).$$

The coefficients c_n are then obtained using the initial conditions or some other boundary conditions.

Fourier Series: Let $f(x)$ be a piecewise continuous function on the interval $[-T, T]$. The *Fourier series* of f is the trigonometric series

$$f(x) \sim \frac{a}{2} + \sum_{n=1}^{\infty} \left\{ a_n \cos \frac{n\pi x}{T} + b_n \sin \frac{n\pi x}{T} \right\},$$

where the a_n's and b_n's are given by **Euler's formulas:**

$$a_n = \frac{1}{T} \int_{-T}^{T} f(x) \cos \frac{n\pi x}{T} \, dx, \qquad n = 0, 1, 2, \ldots,$$

$$b_n = \frac{1}{T} \int_{-T}^{T} f(x) \sin \frac{n\pi x}{T} \, dx, \qquad n = 1, 2, 3, \ldots.$$

Let f be piecewise continuous on the half-interval $[0, T]$. The **Fourier cosine series** of $f(x)$ on $[0, T]$ is

$$f(x) \sim \frac{a_0}{2} + \sum_{n=1}^{\infty} a_n \cos \frac{n\pi x}{T},$$

where

$$a_n = \frac{2}{T} \int_0^T f(x) \cos \frac{n\pi x}{T}\, dx.$$

The **Fourier sine series** of $f(x)$ on $[0, T]$ is

$$f(x) \sim \sum_{n=1}^{\infty} b_n \sin \frac{n\pi x}{T},$$

where

$$b_n = \frac{2}{T} \int_0^T f(x) \sin \frac{n\pi x}{T}\, dx.$$

Fourier series and the method of separation of variables are used to solve boundary value problems and initial-boundary value problems for the three classical equations:

Heat equation
$$\frac{\partial u}{\partial t} = \beta \frac{\partial^2 u}{\partial x^2}.$$

Wave equation
$$\frac{\partial^2 u}{\partial t^2} = \alpha^2 \frac{\partial^2 u}{\partial x^2}.$$

Laplace's equation
$$\frac{\partial^2 u}{\partial x^2} + \frac{\partial^2 u}{\partial y^2} = 0.$$

For the heat equation, there is a unique solution $u(x, t)$ when the boundary values at the ends of a conducting wire are specified along with the initial temperature distribution. The wave equation yields the displacement $u(x, t)$ of a vibrating string when the ends are held fixed and the initial displacement and initial velocity at each point of the string are given. In such a case, separation of variables yields a solution that is the sum of standing waves. For an infinite string with specified initial displacement and velocity, d'Alembert's solution yields traveling waves. Laplace's equation arises in the study of steady-state solutions to the heat equations where either Dirichlet or Neumann boundary conditions are specified.

TECHNICAL WRITING EXERCISES

1. The method of separation of variables is an important technique in solving initial-boundary value problems and boundary value problems for *linear* partial differential equations. Explain where the linearity of the differential equation plays a critical role in the method of separation of variables.

2. In applying the method of separation of variables, we have encountered a variety of *special functions*, such as sines, cosines, Bessel functions, and modified Bessel functions. Describe three or four examples of partial differential equations that involve other special functions, such as Legendre polynomials, Hermite polynomials, and Laguerre polynomials. (Some exploring in the library may be needed.)

3. A constant coefficient second order partial differential equation of the form

$$a \frac{\partial^2 u}{\partial x^2} + b \frac{\partial^2 u}{\partial x \partial y} + c \frac{\partial^2 u}{\partial y^2} = 0$$

can be classified using the discriminant $D := b^2 - 4ac$. In particular, the equation is called

parabolic	if $D = 0$,
hyperbolic	if $D > 0$,
elliptic	if $D < 0$.

Verify that the heat equation is parabolic, the wave equation is hyperbolic, and Laplace's equation is elliptic. It can be shown that such parabolic (hyperbolic, elliptic) equations can be transformed by a linear change of variables into the heat (wave, Laplace's) equation. Based on your knowledge of the heat equation, describe which types of problems (initial value, boundary value, etc.) are appropriate for parabolic equations. Do the same for hyperbolic equations and elliptic equations.

A. Steady-State Temperature Distribution in a Circular Cylinder

When the temperature u inside a circular cylinder reaches a steady state, it satisfies Laplace's equation $\Delta u = 0$. If the temperature on the lateral surface ($r = a$) is kept at zero, the temperature on the top ($z = b$) is kept at zero, and the temperature on the bottom ($z = 0$) is given by $u(r, \theta, 0) = f(r, \theta)$, then the steady-state temperature satisfies the boundary value problem

$$\frac{\partial^2 u}{\partial r^2} + \frac{1}{r}\frac{\partial u}{\partial r} + \frac{1}{r^2}\frac{\partial^2 u}{\partial \theta^2} + \frac{\partial^2 u}{\partial z^2} = 0, \qquad 0 \leq r < a, \qquad -\pi \leq \theta \leq \pi, \qquad 0 < z < b,$$

$$u(a, \theta, z) = 0, \qquad -\pi \leq \theta \leq \pi, \qquad 0 \leq z \leq b,$$

$$u(r, \theta, b) = 0, \qquad 0 \leq r < a, \qquad -\pi \leq \theta \leq \pi,$$

$$u(r, \theta, 0) = f(r, \theta), \qquad 0 \leq r < a, \qquad -\pi \leq \theta \leq \pi,$$

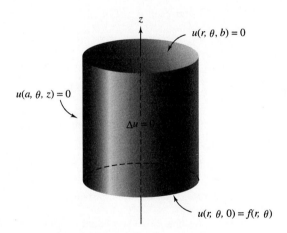

Figure 10.27 Dirichlet problem for cylinder

where $f(a, \theta) = 0$ for $-\pi \leq \theta \leq \pi$ (see Figure 10.27). To find a solution to this boundary value problem, proceed as follows:

(a) Let $u(r, \theta, z) = R(r)T(\theta)Z(z)$. Show that R, T, and Z must satisfy the three ordinary differential equations

$$r^2 R'' + rR' - (r^2\lambda + \mu)R = 0,$$

$$T'' + \mu T = 0,$$

$$Z'' + \lambda Z = 0.$$

(b) Show that $T(\theta)$ has the form

$$T(\theta) = A \cos n\theta + B \sin n\theta$$

for $\mu = n^2$, $n = 0, 1, 2, \ldots$.

(c) Show that $Z(z)$ has the form

$$Z(z) = C \sinh \beta(b - z)$$

for $\lambda = -\beta^2$, where $\beta > 0$.

(d) Show that $R(r)$ has the form, for each n,

$$R_n(r) = DJ_n(\beta r),$$

where J_n is the Bessel function of the first kind.

(e) Show that the boundary conditions require $R(a) = 0$, and so

$$J_n(\beta a) = 0.$$

Hence, for each n, if $0 < \alpha_{n1} < \alpha_{n2} < \cdots < \alpha_{nm} < \cdots$ are the zeros of J_n, then

$$\beta_{nm} = \alpha_{nm}/a.$$

Moreover,

$$R_n(r) = DJ_n(\alpha_{nm}r/a).$$

(f) Use the preceding results to show that $u(r, \theta, z)$ has the form

$$u(r, \theta, z) = \sum_{n=0}^{\infty} \sum_{m=1}^{\infty} J_n\left(\frac{\alpha_{nm}r}{a}\right) (a_{nm} \cos n\theta + b_{nm} \sin n\theta) \sinh\left(\frac{\alpha_{nm}(b - z)}{a}\right),$$

where a_{nm} and b_{nm} are constants.

(g) Use the final boundary condition and a double orthogonal expansion involving Bessel functions and trigonometric functions to derive the formulas

$$a_{0m} = \frac{1}{\pi a^2 \sinh(\alpha_{0m}b/a)[J_1(\alpha_{0m})]^2} \int_0^a \int_0^{2\pi} f(r, \theta)J_0\left(\frac{\alpha_{0m}r}{a}\right) r\, dr\, d\theta,$$

for $m = 1, 2, 3, \ldots$, and for $n, m = 1, 2, 3, \ldots$,

$$a_{nm} = \frac{2}{\pi a^2 \sinh(\alpha_{nm}b/a)[J_{n+1}(\alpha_{nm})]^2} \int_0^a \int_0^{2\pi} f(r, \theta)J_n\left(\frac{\alpha_{nm}r}{a}\right) \cos(n\theta) r\, dr\, d\theta,$$

$$b_{nm} = \frac{2}{\pi a^2 \sinh(\alpha_{nm}b/a)[J_{n+1}(\alpha_{nm})]^2} \int_0^a \int_0^{2\pi} f(r, \theta)J_n\left(\frac{\alpha_{nm}r}{a}\right) \sin(n\theta) r\, dr\, d\theta.$$

B. A Laplace Transform Solution of the Wave Equation

Laplace transforms can be used to solve certain partial differential equations. To illustrate this technique, consider the initial-boundary value problem

(1) $$\frac{\partial^2 u}{\partial t^2} = \alpha^2 \frac{\partial^2 u}{\partial x^2}; \qquad 0 < x < \infty, \qquad t > 0,$$

(2) $u(0, t) = h(t), \qquad t > 0,$

(3) $u(x, 0) = 0, \qquad 0 < x < \infty,$

(4) $\dfrac{\partial u}{\partial t}(x, 0) = 0, \qquad 0 < x < \infty,$

(5) $\lim_{x \to \infty} u(x, t) = 0, \qquad t \geq 0.$

This problem arises in studying a semi-infinite string that is initially horizontal and at rest and where one end is being moved vertically. Let $u(x, t)$ be the solution to (1)–(5). For each x, let

$$U(x, s) := \mathscr{L}\{u(x, t)\}(x, s) = \int_0^\infty e^{-st} u(x, t)\, dt.$$

(a) Using the fact that

$$\mathscr{L}\left\{\frac{\partial^2 u}{\partial x^2}\right\} = \frac{\partial^2}{\partial x^2} \mathscr{L}\{u\},$$

show that $U(x, s)$ satisfies the equation

(6) $$s^2 U(x, s) = \alpha^2 \frac{\partial^2 U}{\partial x^2}, \qquad 0 < x < \infty.$$

(b) Show that the general solution to (6) is

$$U(x, s) = A(s) e^{-sx/\alpha} + B(s) e^{sx/\alpha},$$

where $A(s)$ and $B(s)$ are arbitrary functions of s.

(c) Since $u(x, t) \to 0$ as $x \to \infty$ for all $0 \leq t < \infty$, we have $U(x, s) \to 0$ as $x \to \infty$. Use this fact to show that the $B(s)$ in part (b) must be zero.

(d) Using equation (2), show that

$$A(s) = H(s) = \mathscr{L}\{h\}(s),$$

where $A(s)$ is given in part (b).

(e) Use the results of parts (b), (c), and (d) to obtain a formal solution to (1)–(5).

C. Green's Function

Let Ω be a region in the xy-plane having a smooth boundary $\partial\Omega$. Associated with Ω is a **Green's function** $G(x,y;\xi,\eta)$ defined for pairs of distinct points (x,y), (ξ,η) in Ω. The function $G(x,y;\xi,\eta)$ has the following property.

Let $\Delta u := \partial^2 u/\partial x^2 + \partial^2 u/\partial y^2$ denote the Laplacian operator on Ω; let $h(x, y)$ be a given continuous function on Ω; and let $f(x, y)$ be a given continuous function on $\partial\Omega$. Then, a continuous solution to

the Dirichlet boundary value problem

$$\Delta u(x, y) = h(x, y), \quad \text{in } \Omega,$$
$$u(x, y) = f(x, y), \quad \text{on } \partial\Omega,$$

is given by

(7)
$$u(x, y) = \iint_\Omega G(x, y; \xi, \eta) h(\xi, \eta) \, d\xi \, d\eta$$

$$+ \int_{\partial\Omega} f(\xi, \eta) \frac{\partial G}{\partial n}(x, y; \xi, \eta) \, d\sigma(\xi, \eta),$$

where n is the outward normal to the boundary $\partial\Omega$ of Ω and the second integral is the line integral around the boundary of Ω with the interior of Ω on the left as the boundary is traversed. In (7) we assume that $\partial\Omega$ is sufficiently smooth so that the integrands and integrals exist.

When Ω is the upper half-plane, the Green's function is[†]

$$G(x, y; \xi, \eta) = \frac{1}{4\pi} \ln\left[\frac{(x - \xi)^2 + (y - \eta)^2}{(x - \xi)^2 + (y + \eta)^2} \right].$$

(a) Using (7), show that a solution to

(8) $\Delta u(x, y) = h(x, y), \quad -\infty < x < \infty, \quad 0 < y,$

(9) $u(x, 0) = f(x, 0), \quad -\infty < x < \infty,$

is given by

$$u(x, y) = \frac{y}{\pi} \int_{-\infty}^{\infty} \frac{f(\xi)}{(x - \xi)^2 + y^2} \, d\xi$$

$$+ \frac{1}{4\pi} \int_0^\infty \int_{-\infty}^{\infty} \ln\left[\frac{(x - \xi)^2 + (y - \eta)^2}{(x - \xi)^2 + (y + \eta)^2} \right] h(\xi, \eta) \, d\xi \, d\eta.$$

(b) Use the result of part (a) to determine a solution to (8)–(9) when $h \equiv 0$ and $f \equiv 1$.

(c) When Ω is the interior of the unit circle $x^2 + y^2 = 1$ the Green's function, in polar coordinates (r, θ) and (ρ, ϕ), is given by

$$G(r, \theta; \rho, \phi) = \frac{1}{4\pi} \ln[r^2 + \rho^2 - 2r\rho \cos(\phi - \theta)]$$

$$- \frac{1}{4\pi} \ln[r^2 + \rho^{-2} - 2r\rho^{-1} \cos(\phi - \theta)] - \frac{1}{4\pi} \ln \rho.$$

Using (7), show that the solution to

$$\frac{\partial^2 u}{\partial r^2} + \frac{1}{r} \frac{\partial u}{\partial r} + \frac{1}{r^2} \frac{\partial^2 u}{\partial \theta^2} = 0, \quad 0 \le r < 1, \quad 0 \le \theta \le 2\pi,$$

$$u(1, \theta) = f(\theta), \quad 0 \le \theta \le 2\pi,$$

[†] Techniques for determining Green's functions can be found in texts on partial differential equations such as *Partial Differential Equations of Mathematical Physics*, Second Edition, by Tyn Myint-U (Elsevier North Holland, Inc., New York, 1980). Chapter 10.

is given by

$$u(r, \theta) = \frac{1}{2\pi} \int_0^{2\pi} \frac{1 - r^2}{1 + r^2 - 2r \cos(\phi - \theta)} f(\phi) d\phi.$$

This is known as **Poisson's integral formula.**

(d) Use Poisson's integral formula to derive the following theorem for solutions to Laplace's equation.

MEAN VALUE PROPERTY

Theorem 11. Let $u(x, y)$ satisfy $\Delta u = 0$ in a bounded domain Ω in $\mathbf{R}^2$ and let (x_0, y_0) lie in Ω. Then

(10) $$u(x_0, y_0) = \frac{1}{2\pi} \int_0^{2\pi} u(x_0 + r \cos \theta, y_0 + r \sin \theta) d\theta$$

for all $r > 0$ for which the disk $\bar{B}(x_0, y_0; r) = \{(x, y): (x - x_0)^2 + (y - y_0)^2 \le r^2\}$ lies entirely in Ω.

[*Hint:* Use a change of variables that maps the disk $\bar{B}(x_0, y_0; r)$ to the unit disk $\bar{B}(0, 0; 1)$.]

D. Numerical Method for $\Delta u = f$ on a Rectangle

Let R denote the open rectangle

$$R = \{(x, y): a < x < b, \ c < y < d\}$$

and ∂R be its boundary. Here we describe a numerical technique for solving the generalized Dirichlet problem

(11)
$$\frac{\partial^2 u}{\partial x^2} + \frac{\partial^2 u}{\partial y^2} = f(x, y), \qquad \text{for } (x, y) \text{ in } R,$$

$$u(x, y) = g(x, y), \qquad \text{for } (x, y) \text{ on } \partial R.$$

The method is similar to the finite-difference technique discussed in Chapter 4. We begin by selecting positive integers m and n and step sizes h and k so that $b - a = hm$ and $d - c = kn$. The interval $[a, b]$ is now partitioned into m equal subintervals and $[c, d]$ into n equal subintervals. The partition points are

$$x_i = a + ih, \qquad 0 \le i \le m,$$
$$y_j = c + jk, \qquad 0 \le j \le n$$

(see Figure 10.28). The (dashed) lines $x = x_i$ and $y = y_j$ are called **grid lines** and their intersections (x_i, y_j) are the **mesh points** of the partition. Our goal is to obtain approximations to the solution $u(x, y)$ of problem (11) at each *interior* mesh point, i.e., at (x_i, y_j) where $1 \le i \le m - 1$, $1 \le j \le n - 1$. (Of course, from (11), we are given the values of $u(x, y)$ at the boundary mesh points, e.g., $u(x_0, y_j) = g(x_0, y_j)$, $0 \le j \le n$.)

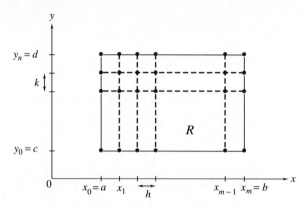

Figure 10.28 Rectangular grid

The next step is to approximate the partial derivatives $\partial^2 u/\partial x^2$ and $\partial^2 u/\partial y^2$ using the *centered-difference approximations*

$$(12) \qquad \frac{\partial^2 u}{\partial x^2}(x_i, y_j) \approx \frac{1}{h^2}\left[u(x_{i+1}, y_j) - 2u(x_i, y_j) + u(x_{i-1}, y_j)\right] \quad \text{for } 1 \le i \le m - 1$$

and

$$(13) \qquad \frac{\partial^2 u}{\partial y^2}(x_i, y_j) \approx \frac{1}{k^2}\left[u(x_i, y_{j+1}) - 2u(x_i, y_j) + u(x_i, y_{j-1})\right] \quad \text{for } 1 \le j \le n - 1.$$

These approximations are based on the generic formula

$$y''(x) = [y(x + h) - 2y(x) + y(x - h)]/h^2 + [\text{terms involving } h^2, h^3, \ldots],$$

which follows from the Taylor series expansion for $y(x)$.

(a) Show that substituting the approximations (12) and (13) into the Dirichlet problem (11) yields the following system:

$$(14) \qquad 2\left[\left(\frac{h}{k}\right)^2 + 1\right]u_{i,j} - (u_{i+1,j} + u_{i-1,j}) - \left(\frac{h}{k}\right)^2(u_{i,j+1} + u_{i,j-1}) = -h^2 f(x_i, y_j)$$

for $i = 1, 2, \ldots, m - 1$, and $j = 1, 2, \ldots, n - 1$;

$$u_{0,j} = g(a, y_j), \qquad u_{m,j} = g(b, y_j), \qquad j = 0, 1, \ldots, n,$$
$$u_{i,0} = g(x_i, c), \qquad u_{i,n} = g(x_i, d), \qquad i = 1, 2, \ldots, m - 1,$$

where $u_{i,j}$ approximates $u(x_i, y_j)$.

Notice that each equation in (14) involves approximations to the solution that appear in a cross centered at a mesh point (see Figure 10.29).

(b) Show that the system in part (a) is a linear system of $(m - 1)(n - 1)$ unknowns in $(m - 1)(n - 1)$ equations.

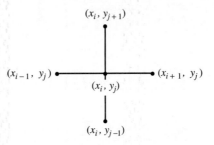

Figure 10.29 Approximate solution $u_{i,j}$ at (x_i, y_j) is obtained from values at the ends of the cross

(c) For Laplace's equation where $f(x, y) \equiv 0$, show that when $h = k$, equation (14) yields

$$u_{i,j} = \frac{1}{4}(u_{i-1,j} + u_{i+1,j} + u_{i,j-1} + u_{i,j+1}).$$

Compare this averaging formula with the mean value property of Theorem 11 (Project C).

(d) For the square plate

$$R = \{(x, y): \quad 0 \le x \le 0.4, \quad 0 \le y \le 0.4\},$$

the boundary is maintained at the following temperatures:

$u(0, y) = 0,$ $u(0.4, y) = 10y,$ $0 \le y \le 0.4.$

$u(x, 0) = 0,$ $u(x, 0.4) = 10x,$ $0 \le x \le 0.4.$

Using the system in part (a) with $f(x, y) \equiv 0$, $m = n = 4$, and $h = k = 0.1$, find approximations to the steady-state temperatures at the mesh points of the plate. [*Hint:* It is helpful to label these grid points with a single index, say $p_1, p_2, \ldots, p_q$, choosing the ordering in a book-reading sequence.]

Chapter 11

Eigenvalue Problems and Sturm-Liouville Equations

11.1 INTRODUCTION: HEAT FLOW IN A NONUNIFORM WIRE

> Develop a model for the flow of heat through a thin, insulated wire whose ends are kept at a constant temperature of 0°C. Assume the density, thermal conductivity, and specific heat capacity of the wire are not uniform but vary along the length of the wire.

In Section 10.1 we developed a model for heat flow assuming the density ρ, the thermal conductivity k, and the specific heat capacity c were constant. If these quantities depend on their location x along the wire (see Figure 10.1, page 573), then we must modify the model developed in Section 10.1 as follows.

The net change in the heat ΔE is now (cf. equation (1) in Section 10.1)

(1)
$$\Delta E = H(x) - H(x + \Delta x) + Q(x, t)\Delta x a\Delta t$$
$$= a\Delta t \left[k(x + \Delta x)\frac{\partial u}{\partial x}(x + \Delta x, t) - k(x)\frac{\partial u}{\partial x}(x, t) \right] + Q(x, t)\Delta x a\Delta t.$$

Using the principle concerning specific heat capacity, we also have (cf. equation (2) in Section 10.1)

(2)
$$\Delta E = c(x)\rho(x)a\Delta x[u(x, t + \Delta t) - u(x, t)].$$

Equating the change in heat given in equations (1) and (2) and dividing by $a\Delta x\Delta t$, we find

(3)
$$c(x)\rho(x)\left[\frac{u(x, t + \Delta t) - u(x, t)}{\Delta t} \right]$$
$$= \frac{k(x + \Delta x)\frac{\partial u}{\partial x}(x + \Delta x, t) - k(x)\frac{\partial u}{\partial x}(x, t)}{\Delta x} + Q(x, t).$$

Letting Δt and Δx approach zero gives

(4) $$c(x)\rho(x)\frac{\partial u}{\partial t}(x, t) = \frac{\partial}{\partial x}\left[k(x)\frac{\partial u}{\partial x}(x, t)\right] + Q(x, t).$$

Since the ends of the wire remain at 0°C, we require $u(x, t)$ to satisfy the boundary conditions

(5) $$u(0, t) = u(L, t) = 0, \qquad t > 0,$$

where L is the length of the wire. If we let $f(x)$ be the initial temperature distribution in the wire, then $u(x, t)$ must also satisfy the initial condition

(6) $$u(x, 0) = f(x), \qquad 0 < x < L.$$

Thus we see that a mathematical model for the flow of heat through a nonuniform wire is given by the initial-boundary value problem consisting of equations (4), (5), and (6).

In certain cases, such as heat given off by a chemical reaction, the internal source/sink expression $Q(x, t)$ can be modeled by a term proportional to the temperature, $q(x)u(x, t)$. Most other types of sources are independent of temperature and are modeled by an explicit function $h(x, t)$. Hence a general equation for heat flow in a nonuniform wire is given by

(7) $$c(x)\rho(x)\frac{\partial u}{\partial t}(x, t) = \frac{\partial}{\partial x}\left[k(x)\frac{\partial u}{\partial x}(x, t)\right] + q(x)u(x, t) + h(x, t).$$

In the original problem the temperatures at the ends of the wire were held constant. More generally, the ends of the wire could be insulated, $\partial u/\partial x = 0$, or just partially insulated, $\partial u/\partial x + cu = 0$. To include these possibilities, we assume $u(x, t)$ satisfies the boundary conditions

(8) $$a_1u(a, t) + a_2\frac{\partial u}{\partial x}(a, t) = 0, \qquad b_1u(b, t) + b_2\frac{\partial u}{\partial x}(b, t) = 0$$

for all $t > 0$. These are called **separated boundary conditions**, since the condition imposed on $u(x, t)$ at $x = a$ does not depend on the values of $u(x, t)$ at $x = b$ and vice versa.

From (7), (8), and (6), we get the following general model for heat flow in a nonuniform wire

(9) $$r(x)\frac{\partial u}{\partial t}(x, t) = \frac{\partial}{\partial x}\left[p(x)\frac{\partial u}{\partial x}(x, t)\right] + q(x)u(x, t) + h(x, t), \qquad a < x < b, \qquad t > 0,$$

(10) $$a_1u(a, t) + a_2\frac{\partial u}{\partial x}(a, t) = 0, \qquad b_1u(b, t) + b_2\frac{\partial u}{\partial x}(b, t) = 0, \qquad t > 0,$$

(11) $$u(x, 0) = f(x), \qquad a < x < b,$$

where $r(x) = c(x)\rho(x)$ and $p(x) = k(x)$.

In Chapter 10 we used the method of separation of variables to solve these types of problems. In particular, we discussed techniques for handling the nonhomogeneous term $h(x, t)$ (see Example 3 in Section 10.5 and Example 1 in Section 10.6). For now, we will

assume $h(x, t) \equiv 0$. Applying the method of separation of variables to equations (9) and (10), that is, setting $u(x, t) = y(x)T(t)$, leads to the **eigenvalue problem for $y(x)$:**

$$(12) \qquad \frac{d}{dx}\left[p(x)\frac{dy}{dx}\right] + q(x)y + \lambda r(x)y = 0, \qquad a < x < b,$$

$$(13) \qquad a_1 y(a) + a_2 y'(a) = 0, \qquad b_1 y(b) + b_2 y'(b) = 0,$$

where λ is a number to be determined (the reader should verify this). The system (12)–(13) is called a **Sturm-Liouville boundary value problem.**[†] We exclude the trivial boundary conditions where $a_1 = a_2 = 0$ or $b_1 = b_2 = 0$. When $p(x)$, $q(x)$, and $r(x)$ are continuous on $[a, b]$ and $p(x) > 0$ and $r(x) > 0$ on $[a, b]$, we refer to this system as a **regular** Sturm-Liouville boundary value problem.

Most of the eigenvalue problems encountered in Chapter 10 were regular. However, when the assumptions on p, p', q, and r hold only on the open interval (a, b) and not the closed, we say that the equation is **singular**, provided p is zero at a or b; $p(x)$, $q(x)$, or $r(x)$ becomes unbounded as x approaches either a or b; or the interval (a, b) is unbounded. An example of a singular Sturm-Liouville boundary value problem is the problem of finding nontrivial solutions for Bessel's equation

$$(14) \qquad \frac{d}{dx}\left[x\frac{dy}{dx}\right] + \lambda xy = 0, \qquad 0 < x < b,$$

that satisfy the boundary conditions

$$(15) \qquad \lim_{x \to 0^+} y(x) \quad \text{and} \quad \lim_{x \to 0^+} y'(x) \quad \text{exist and are finite and } y(b) = 0.$$

This problem arises in the study of heat flow in a cylinder. Here $p(x) = r(x) = x$, which is zero at $x = 0$. The boundary conditions in (15) are typical of singular Sturm-Liouville boundary value problems.

Another important class of boundary conditions is the class of periodic conditions that we previously encountered in solving Laplace's equation in a disk (see Section 10.7). If we seek P-periodic solutions to equation (12), then we require the solution $y(x)$ to satisfy the **periodic boundary conditions**

$$(16) \qquad y(0) = y(P), \qquad y'(0) = y'(P).$$

As we shall see, the theory of periodic Sturm-Liouville problems is closely related to the theory of regular Sturm-Liouville boundary value problems.

The type of boundary value problem that arises from the application of the method of separation of variables depends on the geometry of the region. Of course, a coordinate system must be chosen that allows the variables to be separated. In Chapter 10, we made use of rectangular and polar coordinates.[††]

[†] *Historical Footnote:* These systems are also called **Sturm-Liouville systems** or **Sturm-Liouville problems.** They are named after the Swiss mathematician Jacques Charles François Sturm (1803–1855) and the French mathematician Joseph Liouville (1809–1882), who studied this class of boundary value problems in the 1830s.

[††] Other coordinate systems that are used for special geometric regions are discussed in the text by G. Arfken, *Mathematical Methods for Physicists*, Third Edition (Academic Press, New York, 1985), Chapter 2.

In this chapter we study the properties of Sturm-Liouville boundary value problems. We begin, in Section 11.2, with a discussion of eigenvalues and eigenfunctions. In Section 11.3 we analyze regular Sturm-Liouville boundary value problems. Nonhomogeneous boundary value problems and the Fredholm alternative are discussed in Section 11.4. We give two methods for solving nonhomogeneous problems; the method of eigenfunction expansions (Section 11.5) and Green's functions (Section 11.6). In Section 11.7 we discuss three examples of singular Sturm-Liouville boundary value problems: Bessel's equation, Legendre's equation, and Hermite's equation. Finally, in Section 11.8, we describe the location of the zeros for solutions to Sturm-Liouville equations as given by the Sturm separation theorem, the Sturm fundamental theorem, and the Picone comparison theorem.

11.2 EIGENVALUES AND EIGENFUNCTIONS

The problem of finding a solution to a second order linear differential equation

(1) $y'' + p(x)y' + q(x)y = f(x), \qquad a < x < b,$

that satisfies the boundary conditions

(2) $a_{11}y(a) + a_{12}y'(a) + b_{11}y(b) + b_{12}y'(b) = c_1,$
 $a_{21}y(a) + a_{22}y'(a) + b_{21}y(b) + b_{22}y'(b) = c_2,$

is called a **two-point boundary value problem.** The conditions in (2) are examples of linear boundary conditions. When $c_1 = c_2 = 0$, we say that the boundary conditions are **homogeneous;** otherwise, we refer to them as **nonhomogeneous.**

There are special types of boundary conditions that occur frequently in applications. They are

Separated: $a_1 y(a) + a_2 y'(a) = c_1,$
 $b_1 y(b) + b_2 y'(b) = c_2.$

Dirichlet[†]**:** $y(a) = c_1, \qquad y(b) = c_2.$

Neumann[††]**:** $y'(a) = c_1, \qquad y'(b) = c_2.$

Periodic: or $y(-T) = y(T), \qquad y'(-T) = y'(T),$
 $y(0) = y(2T), \qquad y'(0) = y'(2T),$

where the period is $2T$.

Notice that the Dirichlet and Neumann conditions are special types of separated boundary conditions.

[†] ***Historical Footnote:*** Named for the German mathematician Peter Gustav Lejeune Dirichlet (1805–1859) as a result of his work in the calculus of variations. He was also the first to give a rigorous proof of the convergence of Fourier series.
[††] ***Historical Footnote:*** Named for the German mathematician Karl Gottfried Neumann (1832–1925), who studied differential and integral equations as well as complex variables.

For boundary value problems involving third or higher-order equations, one usually specifies the same number of boundary conditions as the order of the equation. When the order is even, these boundary conditions are often separated with half the conditions at the endpoint $x = a$ and half at $x = b$. We give an example of a fourth order problem later in this section.

For the homogeneous equation with homogeneous boundary conditions

(3) $y'' + p(x)y' + q(x)y = 0, \qquad a < x < b,$

(4) $a_{11}y(a) + a_{12}y'(a) + b_{11}y(b) + b_{12}y'(b) = 0,$
$a_{21}y(a) + a_{22}y'(a) + b_{21}y(b) + b_{22}y'(b) = 0,$

it is easy to verify that if $\phi(x)$ is a nontrivial solution to (3)–(4), then so is $A\phi$ for any choice of the constant A. This gives a *one-parameter family* of solutions. If (3)–(4) has two linearly independent solutions $\phi_1(x)$ and $\phi_2(x)$, then $A_1\phi_1(x) + A_2\phi_2(x)$ is also a solution for any choice of A_1 and A_2; hence we have a *two-parameter family* of solutions. The remaining possibility is that $\phi(x) \equiv 0$ is the only solution to (3)–(4), in which case there is a unique solution. Consequently three situations can arise: the boundary value problem (3)–(4) has a unique solution, a one-parameter family of solutions, or a two-parameter family of solutions (see Problem 31).

The situation for the nonhomogeneous problem (1)–(2) is similar. However, there is an additional possibility: *the nonhomogeneous problem may have no solutions*. The following examples illustrate the various situations.

EXAMPLE 1 Find all the solutions to the boundary value problem

(5) $y'' + 2y' + 5y = 0,$

(6) $y(0) = 2, \qquad y(\pi/4) = 1.$

Solution The auxiliary equation for (5) is $r^2 + 2r + 5 = 0$, which has roots $r = -1 \pm 2i$. Hence a general solution for (5) is $y(x) = c_1 e^{-x}\cos 2x + c_2 e^{-x}\sin 2x$. We will now try to determine c_1 and c_2 so that the boundary conditions in (6) are satisfied. Setting $x = 0$ and $\pi/4$, we find

$$y(0) = c_1 = 2, \qquad y(\pi/4) = c_2 e^{-\pi/4} = 1.$$

Hence $c_1 = 2$ and $c_2 = e^{\pi/4}$, and so problem (5)–(6) has the unique solution

$$y(x) = 2e^{-x}\cos 2x + e^{\pi/4}e^{-x}\sin 2x. \quad \blacksquare \text{ End}$$

EXAMPLE 2 Find all the solutions to the boundary value problem

(7) $y'' + y = \cos 2x,$

(8) $y'(0) = 0, \qquad y'(\pi) = 0.$

Solution The auxiliary equation for (7) is $r^2 + 1 = 0$, so a general solution for the corresponding homogeneous equation is $y_h(x) = c_1\cos x + c_2\sin x$. As we know from the method of undetermined coefficients, a particular solution to (7) has the form $y_p(x) = A\cos 2x +$

$B \sin 2x$. Substituting y_p into (7) and solving for A and B, we find $A = -1/3$ and $B = 0$. Hence $y_p(x) = -(1/3) \cos 2x$. Therefore a general solution for (7) is

(9) $y(x) = c_1 \cos x + c_2 \sin x - (1/3) \cos 2x.$

To determine the values c_1 and c_2, we substitute the general solution in (9) into (8) and obtain

$$y'(0) = c_2 = 0, \quad y'(\pi) = -c_2 = 0.$$

Thus $c_2 = 0$ and c_1 is arbitrary. So problem (7)–(8) has a one-parameter family of solutions; namely,

$$y(x) = c_1 \cos x - (1/3) \cos 2x,$$

where c_1 is any real number. ■ End

EXAMPLE 3 Find the solutions to

(10) $y'' + 4y = 0,$

(11) $y(-\pi) = y(\pi), \qquad y'(-\pi) = y'(\pi).$

Solution The auxiliary equation for (10) is $r^2 + 4 = 0$, so a general solution is

(12) $y(x) = c_1 \cos 2x + c_2 \sin 2x.$

Since $\cos 2x$ and $\sin 2x$ are π (and hence 2π)-periodic functions, it follows that $y(-\pi) = y(\pi)$. Moreover, $y'(x) = -2c_1 \sin 2x + 2c_2 \cos 2x$ is also π-periodic and so $y'(-\pi) = y'(\pi)$. Thus (10)–(11) has a two-parameter family of solutions given by (12), where c_1 and c_2 are the parameters. ■ End

Let's see what happens if, in the preceding example, we replace equation (10) by the nonhomogeneous equation

(13) $y'' + 4y = 4x.$

A general solution for (13) is

(14) $y(x) = x + c_1 \cos 2x + c_2 \sin 2x.$

Since $y(-\pi) = -\pi + c_1$ and $y(\pi) = \pi + c_1$, we find that there are no solutions to (13) that satisfy $y(-\pi) = y(\pi)$. Thus the nonhomogeneous boundary value problem (13)–(11) has *no solutions*.

The Sturm-Liouville boundary value problems introduced in the preceding section are examples of two-point boundary problems that involve a parameter λ. Our goal is to determine for which values of λ the boundary value problem

(15) $\dfrac{d}{dx}\left(p(x)\dfrac{dy}{dx}\right) + q(x)y + \lambda r(x)y = 0, \qquad a < x < b,$

(16) $a_1 y(a) + a_2 y'(a) = 0, \qquad b_1 y(b) + b_2 y'(b) = 0,$

has nontrivial solutions. Such problems are called **eigenvalue problems.** The nontrivial solutions are called **eigenfunctions** and the corresponding number λ an **eigenvalue.**[†]

One reason eigenvalue problems are important is that they arise in using the method of separation of variables to solve partial differential equations (see Section 10.2). It is therefore worthwhile to keep in mind some simple examples of these problems that were solved in Chapter 10.

Dirichlet boundary conditions:

(17) $y'' + \lambda y = 0; \qquad y(0) = y(L) = 0,$

which has eigenvalues $\lambda_n = (n\pi/L)^2$, $n = 1, 2, 3, \ldots$, and corresponding eigenfunctions

$$\phi_n(x) = a_n \sin\left(\frac{n\pi x}{L}\right),$$

where the a_n's are arbitrary nonzero constants (see Section 10.2).

Neumann boundary conditions:

(18) $y'' + \lambda y = 0; \qquad y'(0) = y'(L) = 0,$

which has eigenvalues $\lambda_n = (n\pi/L)^2$, $n = 0, 1, 2, \ldots$, and corresponding eigenfunctions

$$\phi_n(x) = c_n \cos\left(\frac{n\pi x}{L}\right),$$

where the c_n's are arbitrary nonzero constants (see Example 1 in Section 10.5).

Periodic boundary conditions:

(19) $y'' + \lambda y = 0; \qquad y(-\pi) = y(\pi), \qquad y'(-\pi) = y'(\pi),$

which has eigenvalues $\lambda_n = n^2$, $n = 0, 1, 2, \ldots$, and corresponding eigenfunctions

$$\phi_0(x) = A_0, \qquad \phi_n(x) = A_n \cos nx + B_n \sin nx, \qquad n \geq 1,$$

where the A_n's and B_n's are arbitrary constants, not both zero (see Example 2 in Section 10.7). Notice that for the eigenvalue problem (19) involving periodic boundary conditions, there are two linearly independent eigenfunctions for each eigenvalue λ_n, $n = 1, 2, 3, \ldots$.

In the next example we illustrate that even for the simple equation $y'' + \lambda y = 0$, eigenvalues may be difficult to determine exactly.

EXAMPLE 4 Find all the real eigenvalues and eigenfunctions for

(20) $y'' + \lambda y = 0,$

(21) $y(0) = 0, \qquad 3y(\pi) - y'(\pi) = 0.$

[†] Readers familiar with linear algebra will recognize these eigenvalues and eigenfunctions as the eigenvalues and eigenvectors for the linear operator defined by $L[y] := -(1/r)(py')' - (q/r)y$, where the domain of L consists of those twice-differentiable functions satisfying (16).

Solution The auxiliary equation for (20) is $r^2 + \lambda = 0$. The nature of the roots varies for λ negative, zero, and positive, so we will consider these cases separately.

Case 1. $\boldsymbol{\lambda < 0}$. Let $\lambda = -\mu^2$ for $\mu > 0$. The roots of the auxiliary equation are $r = \pm\mu$, so a general solution to (20) is $y(x) = c_1 e^{\mu x} + c_2 e^{-\mu x}$. It is more convenient to express the general solution in terms of hyperbolic functions. Thus we write $y(x)$ in the equivalent form $y(x) = C_1 \cosh \mu x + C_2 \sinh \mu x$. Substituting in the boundary conditions in (21) yields

$$C_1 = 0, \quad 3(C_1 \cosh \mu\pi + C_2 \sinh \mu\pi) - (\mu C_1 \sinh \mu\pi + \mu C_2 \cosh \mu\pi) = 0.$$

With $C_1 = 0$, the last equation becomes

$$C_2(3 \sinh \mu\pi - \mu \cosh \mu\pi) = 0.$$

To obtain a nontrivial solution, we must have $C_2 \neq 0$, so $3 \sinh \mu\pi - \mu \cosh \mu\pi = 0$; that is, μ must satisfy

(22) $$\tanh \mu\pi = \frac{1}{3}\mu.$$

It is easy to see that in the μy-plane, the line $y = \mu/3$ intersects the curve $y = \tanh \mu\pi$ only once for $\mu > 0$ (see Figure 11.1). Therefore there is only one positive solution to (22) which we denote by μ_0.

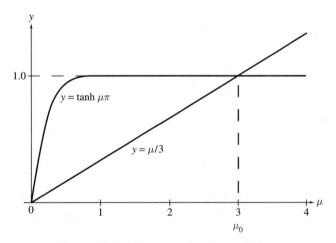

Figure 11.1 Graphs of $y = \mu/3$ and $y = \tanh \mu\pi$

To summarize, the boundary value problem (20)–(21) has one negative eigenvalue $\lambda_0 = -\mu_0^2$, where $\tanh \mu_0\pi = \mu_0/3$, and the corresponding eigenfunctions are $y_0(x) = c_0 \sinh \mu_0 x$, with $c_0 \neq 0$ arbitrary.

Case 2. $\boldsymbol{\lambda = 0}$. Here zero is a double root of the auxiliary equation, so a general solution to (20) is $y(x) = c_1 x + c_2$. Substituting into (21) gives

$$c_2 = 0, \quad 3\pi c_1 + 3c_2 - c_1 = 0,$$

which has the solution $c_1 = c_2 = 0$. Thus we have only the trivial solution.

Case 3. $\lambda > 0$. Let $\lambda = \mu^2$ for $\mu > 0$. The roots of the auxiliary equation are $r = \pm \mu i$, and a general solution to (20) is $y(x) = c_1 \cos \mu x + c_2 \sin \mu x$. Substituting into (21) gives

$$c_1 = 0, \qquad 3(c_1 \cos \mu\pi + c_2 \sin \mu\pi) - (-\mu c_1 \sin \mu\pi + \mu c_2 \cos \mu\pi) = 0.$$

Setting $c_1 = 0$ in the last equation, we get

$$c_2(3 \sin \mu\pi - \mu \cos \mu\pi) = 0.$$

Hence either c_2 is zero or $3 \sin \mu\pi - \mu \cos \mu\pi = 0$. That is, for nontrivial solutions to exist, μ must satisfy the equation

(23) $$\tan \mu\pi = \frac{1}{3}\mu.$$

To see that there exist values for $\mu > 0$ that satisfy (23), we examine the graphs of $y = \mu/3$ and $y = \tan \mu\pi$. For those values of $\mu > 0$ where the two graphs intersect, we have a solution to (23). From Figure 11.2, we see that there are infinitely many such values of μ. Let μ_1 denote the smallest value, μ_2 the next larger, and so on. Now for each μ_n, we get that $y_n(x) = c_2 \sin \mu_n x$ with $c_2 \neq 0$ is a nontrivial solution to (20)–(21) with $\lambda = \mu_n^2$, $n = 1, 2, 3, \ldots$.

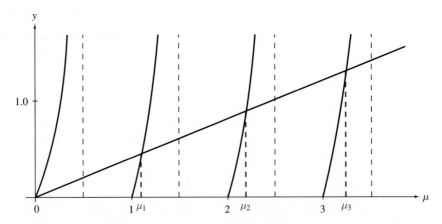

Figure 11.2 Graphs of $y = \mu/3$ and $y = \tan \mu\pi$

From the three cases above, we see that the eigenvalues of (20)–(21) are $\lambda_0 = -\mu_0^2$ where $\tanh \mu_0\pi = \mu_0/3$ and $\lambda_n = \mu_n^2$, $n = 1, 2, 3, \ldots$, where $\tan \mu_n\pi = \mu_n/3$. The corresponding eigenfunctions are $y_0 = c_0 \sinh \mu_0 x$ and $y_n = c_n \sin \mu_n x$, $n = 1, 2, 3, \ldots$, with the c_n's being arbitrary nonzero constants. It is interesting to note from Figure 11.2 that as n becomes large, μ_n approaches an odd multiple of $1/2$. That is, $\lambda_n \approx (2n + 1)^2/4$ when n is large. Using a numerical scheme such as Newton's method to solve $\tanh \mu_0\pi = \mu_0/3$ and $\tan \mu_1\pi = \mu_1/3$, we can show that $\lambda_0 \approx -9.00000$ and $\lambda_1 \approx 1.23897$. ■ End

In the above example we dealt only with the *real* eigenvalues. In the next section we show that the eigenvalues for regular Sturm-Liouville boundary value problems are necessarily real, not complex numbers.

EXAMPLE 5 Determine all real eigenvalues and eigenfunctions for

$$(24) \qquad (xy')' + \frac{\lambda}{x}y = 0, \qquad 1 < x < e,$$

$$(25) \qquad y(1) = 0, \qquad y(e) = 0.$$

Solution Equation (24) can be rewritten as the Cauchy-Euler equation

$$(26) \qquad x^2y'' + xy' + \lambda y = 0.$$

(Cauchy-Euler equations were studied in Sections 4.5 and 8.5.) Substituting $y = x^r$ gives $r^2 + \lambda = 0$ as the auxiliary equation for (26). We again consider the three cases $\lambda < 0$, $\lambda = 0$, and $\lambda > 0$.

Case 1. $\lambda < 0$. Let $\lambda = -\mu^2$ for $\mu > 0$. The roots of the auxiliary equation are $r = \pm\mu$, so a general solution to (26) is $y(x) = c_1 x^\mu + c_2 x^{-\mu}$. Substituting into the boundary conditions (25) gives

$$c_1 + c_2 = 0,$$

$$c_1 e^\mu + c_2 e^{-\mu} = 0,$$

whose only solution for $\mu > 0$ is $c_1 = c_2 = 0$. Thus there are no eigenvalues $\lambda > 0$.

Case 2. $\lambda = 0$. Here zero is a double root of the auxiliary equation and a general solution to (26) is $y(x) = c_1 + c_2 \ln x$. Substituting into (25), we have

$$c_1 = 0, \qquad c_1 + c_2 = 0.$$

Again the solution is $c_1 = c_2 = 0$, so $\lambda = 0$ is not an eigenvalue.

Case 3. $\lambda > 0$. Let $\lambda = \mu^2$ for $\mu > 0$. The roots of the auxiliary equation are now $r = \pm\mu i$, and a general solution to (26) is $y(x) = c_1 \cos(\mu \ln x) + c_2 \sin(\mu \ln x)$. Substituting into (25) yields

$$c_1 = 0, \qquad c_1 \cos \mu + c_2 \sin \mu = 0.$$

With $c_1 = 0$, the last equation becomes $c_2 \sin \mu = 0$. For $\mu = n\pi$, $n = 1, 2, 3, \ldots$, $\sin \mu = \sin n\pi = 0$ and we get the nontrivial solutions $y = c_2 \sin(n\pi \ln x)$ to (24)–(25).

To summarize, the problem (24)–(25) has eigenvalues $\lambda_n = (n\pi)^2$, $n = 1, 2, 3, \ldots$, with corresponding eigenfunctions $y_n = c_n \sin(n\pi \ln x)$, where c_n is an arbitrary nonzero constant. ■ End

Eigenvalue problems for higher-order linear equations occur in continuum mechanics and, in particular, in vibration and bending problems for beams. One such equation is

$$(27) \qquad \frac{d^4 y}{dx^4} + \lambda y = 0, \qquad 0 < x < L,$$

where y is the transverse displacement of a uniform elastic beam of length L. If the beam has linear density ρ, cross-sectional moment of inertia I, and Young's modulus of elasticity E, then $\lambda = K\rho/EI$, where K is a constant to be determined. If the beam is simply supported at its ends, then one imposes the boundary conditions

$$y(0) = y''(0) = 0, \qquad y(L) = y''(L) = 0,$$

(see Figure 11.3(a)). If we have a cantilever beam where one end is clamped while the other is free to move, then the boundary conditions are

$$y(0) = y'(0) = 0, \qquad y''(L) = y'''(L) = 0$$

(see Figure 11.3(b)). In both of these cases, it can be shown that the eigenvalues must be negative. We leave the computation of the eigenvalues and eigenfunctions as exercises (see Problems 28 and 29).

(a) (b)

Figure 11.3 Simply supported beam (a) and a cantilever beam (b)

The eigenvalue problems for differential equations that we have studied here and the eigenvalue problems for matrices that were discussed in Chapter 9 are both examples of eigenvalue problems for linear operators. Such problems are studied in the field of mathematics known as *spectral theory*. In the next section, we will see that there is a very close connection between the properties of the eigenvalues and eigenvectors for real symmetric (or Hermitian) matrices and for regular Sturm-Liouville boundary value problems.

EXERCISES 11.2

In Problems 1 through 12, determine the solutions, if any, to the given boundary value problem.

1. $y'' + 2y' + 26y = 0$; $y(0) = 1$, $y(\pi) = -e^{-\pi}$.

2. $y'' + y = x + 1$; $y(0) = 0$, $y'(\pi) = 0$.

3. $y'' - 4y' + 13y = 0$; $y(0) = 0$, $y(\pi) = 0$.

4. $y'' + 4y' + 8y = 0$; $y'(0) = 0$, $y'(\pi) = 0$.

5. $y'' + y = \sin 2x$; $y(0) = y(2\pi)$, $y'(0) = y'(2\pi)$.

6. $y'' - y = x$; $y(0) = 3$, $y'(1) = 2e - e^{-1} - 1$.

7. $y'' + y = e^x$; $y(0) = 0$, $y(\pi) + y'(\pi) = 0$.

8. $y'' + 4y = 4$; $y(-\pi) = y(\pi)$, $y'(-\pi) = y'(\pi)$.

9. $y'' + 4y = \sin 2x$; $y(-\pi) = y(\pi)$, $y'(-\pi) = y'(\pi)$.

10. $y'' + 2y' + 2y = e^x \cos x$; $y(0) = 0$, $y(\pi) = 0$.

11. $x^2 y'' + xy' + y = 4x^{-1}$; $y(1) = 2$, $y(e^\pi) = 2e^{-\pi}$.

12. $y'' + y = x$; $y'(0) = 0$, $y'(\pi) = 0$.

In Problems 13 through 20, find all the real eigenvalues and eigenfunctions for the given eigenvalue problem.

13. $y'' + \lambda y = 0$; $y(0) = 0$, $y'(1) = 0$.

14. $y'' + \lambda y = 0$; $y(-3) = y(3)$, $y'(-3) = y'(3)$.

15. $y'' + 3y + \lambda y = 0$; $y'(0) = 0$, $y'(\pi) = 0$.

16. $y'' - 2y + \lambda y = 0$; $y(0) = 0$, $y'(\pi) = 0$.

17. $y'' + \lambda y = 0$; $2y(0) + y'(0) = 0$, $y(\pi) = 0$.

18. $y'' + \lambda y = 0$; $y(0) = 0$, $y(\pi) - 2y'(\pi) = 0$.

19. $(xy')' + \lambda x^{-1}y = 0;$ $y'(1) = 0,$ $y(e^{\pi}) = 0.$

20. $(xy')' + \lambda x^{-1}y = 0;$ $y'(1) = 0,$ $y'(e) = 0.$

In Problems 21 and 22, use Newton's method (Appendix A) to determine to three decimal places the first three eigenvalues for the given eigenvalue problem.

21. $y'' + \lambda y = 0;$ $y(0) + 2y'(0) = 0,$ $y(\pi) = 0.$

22. $y'' + \lambda y = 0;$ $y(0) = 0,$ $y(1) + y'(1) = 0.$

In Problems 23 through 26, find all real values of λ for which the given boundary value problem has a nontrivial solution. Also find the corresponding solutions.

23. $y'' + y' + \lambda(y' + y) = 0;$ $y'(0) = 0,$ $y(\pi) = 0.$

24. $x^2y'' + \lambda(xy' - y) = 0;$ $y(1) = 0,$ $y'(e) = 0.$

25. $y'' + \lambda y = 0;$ $y(0) = 0,$ $y(1 + \lambda^2) = 0.$

26. $y'' + \lambda y = 0;$ $y(0) = 0,$ $\lambda y(\pi) - y'(\pi) = 0.$

In Problems 27 through 29, find all the nonpositive real eigenvalues and eigenfunctions of the given fourth order boundary value problems. [Hint: Let $\lambda = -\mu^4,$ $\mu > 0$ or $\lambda = 0.$]

27. $y^{(4)} + \lambda y = 0;$ $y(0) = y'(0) = 0,$ $y(L) = y'(L) = 0.$

28. $y^{(4)} + \lambda y = 0;$ $y(0) = y'(0) = 0,$ $y''(L) = y'''(L) = 0.$

29. $y^{(4)} + \lambda y = 0;$ $y(0) = y''(0) = 0,$ $y(L) = y''(L) = 0.$

30. Prove that the nonhomogeneous boundary value problem (1)–(2) has a unique solution for each continuous function f and numbers c_1 and c_2 if and only if the corresponding homogeneous boundary value problem, with $f(x) \equiv 0$ and $c_1 = c_2 = 0,$ has only the trivial solution $y(x) \equiv 0.$

31. Prove that the homogeneous boundary value problem (3)–(4) has either the trivial solution, a one-parameter family of solutions, or a two-parameter family of solutions.

32. Prove that the nonhomogeneous boundary value problem (1)–(2) has either no solution, a unique solution, a one-parameter family of solutions, or a two-parameter family of solutions. [Hint: If y_p is a solution to (1)–

(2) and y is another solution to (1)–(2), use the superposition principle to show that $y - y_p$ is a solution to the homogeneous problem (3)–(4). The conclusion follows from the result in Problem 31.]

33. Partial Differential Equation. Assume the problem

$$u_{tt} = u_{xx} + 2u_x; \quad 0 < x < \pi, \quad t > 0,$$

$$u(0, t) = 0, \quad t > 0,$$

$$u(\pi, t) = 0, \quad t > 0,$$

has a solution of the form $u(x, t) = X(x)T(t).$

(a) Use the method of separation of variables discussed in Section 10.2 to derive the following eigenvalue problem for $X(x)$:

$$X''(x) + 2X'(x) + \lambda X(x) = 0; \quad X(0) = X(\pi) = 0.$$

(b) Find the eigenvalues λ and the corresponding eigenfunctions for the problem given in part (a).

34. Assume the problem

$$u_t = u_{xx} + u_x; \quad 0 < x < 1, \quad t > 0,$$

$$u_x(0, t) = 0, \quad t > 0,$$

$$u_x(1, t) = 0, \quad t > 0.$$

has a solution of the form $u(x, t) = X(x)T(t).$

(a) Use the method of separation of variables discussed in Section 10.2 to derive the eigenvalue problem

$$X''(x) + (1 + \lambda)X(x) = 0; \quad X'(0) = X'(1) = 0.$$

(b) Find the eigenvalues λ and the corresponding eigenfunctions for the problem given in part (a).

35. Helmholtz Equation. In solving the vibrating membrane problem, one encounters the Helmholtz equation

$$u_{xx} + u_{yy} + \alpha^2 u = 0, \quad 0 < x < a, \quad 0 < y < b,$$

with boundary conditions

$$u(0, y) = u(a, y) = 0, \quad 0 < y < b,$$

$$u(x, 0) = u(x, b) = 0, \quad 0 < x < a.$$

Use the method of separation of variables discussed in Section 11.2 to find the eigenvalues α^2 and corresponding eigenfunctions.

11.3 REGULAR STURM-LIOUVILLE BOUNDARY VALUE PROBLEMS

In this section we analyze regular Sturm-Liouville boundary value problems, that is, eigenvalue problems of the form

(1) $\qquad (p(x)y'(x))' + q(x)y(x) + \lambda r(x)y(x) = 0, \qquad a < x < b,$

(2) $\qquad \begin{aligned} &a_1 y(a) + a_2 y'(a) = 0, \\ &b_1 y(b) + b_2 y'(b) = 0, \end{aligned}$

where $p(x)$, $p'(x)$, $q(x)$, and $r(x)$ are real-valued continuous functions on $[a, b]$ and $p(x) > 0$ and $r(x) > 0$ on $[a, b]$. We exclude those cases when $a_1 = a_2 = 0$ or $b_1 = b_2 = 0$.

The following question immediately comes to mind. Can a general second order homogeneous linear differential equation be written in the form of a Sturm-Liouville equation? That is, can we convert the equation

(3) $\qquad A_2(x)y''(x) + A_1(x)y'(x) + A_0(x)y(x) + \lambda \rho(x)y(x) = 0$

into an equation of the same type as (1)? To do this, we would like to determine p so that $(py')' = py'' + p'y' = A_2 y'' + A_1 y'$. This means $p = A_2$ and $p' = A_1$. But, in general, $A_2' \neq A_1$, so this straightforward approach is not always possible. We can remedy this situation by finding an *integrating factor* $\mu(x)$ such that when (3) is multiplied by μ, we do get $(\mu A_2)' = \mu A_1$. Setting $p = \mu A_2$, the last equation becomes $p' = \mu A_1 = pA_1/A_2$. When we solve this separable equation for p, we find

$$p(x) = Ce^{\int A_1(x)/A_2(x)\,dx},$$

where C is arbitrary. Taking $C = 1$ and recalling that $\mu = p/A_2$, we get

(4) $\qquad \mu(x) = [1/A_2(x)]e^{\int A_1(x)/A_2(x)\,dx}.$

Multiplying (3) by this choice for μ gives

$$\mu A_2 y'' + \mu A_1 y' + \mu A_0 y + \lambda \mu \rho y = 0,$$

which simplifies to

(5) $\qquad (py')' + qy + \lambda ry = 0,$

where $p = \mu A_2$, $q = \mu A_0$, and $r = \mu \rho$. Notice that we need the coefficients continuous and $A_2(x)$ not equal to zero. This procedure is illustrated in the following example.

EXAMPLE 1 Convert the following equation into the form of a Sturm-Liouville equation

(6) $\qquad 3x^2 y''(x) + 4xy'(x) + 6y(x) + \lambda y(x) = 0, \qquad x > 0.$

Solution Here $A_2(x) = 3x^2$ and $A_1(x) = 4x$. Using formula (4), we find

$$\mu(x) = \frac{1}{3x^2}\, e^{\int A_1(x)/A_2(x)\,dx} = \frac{1}{3x^2}\, e^{\int (4x)/(3x^2)\,dx}$$

$$= \frac{1}{3x^2}\, e^{(4/3)\int x^{-1}\,dx} = \frac{1}{3x^2}\, e^{(4/3)\ln x} = \frac{x^{4/3}}{3x^2} = \frac{1}{3x^{2/3}}.$$

Multiplying equation (6) by $\mu(x) = 1/(3x^{2/3})$, we get

$$x^{4/3}y''(x) + \frac{4}{3}x^{1/3}y'(x) + 2x^{-2/3}y(x) + \lambda(3x^{2/3})^{-1}y(x) = 0,$$

(7) $(x^{4/3}y'(x))' + 2x^{-2/3}y(x) + \lambda(3x^{2/3})^{-1}y(x) = 0.$ ■ End

Throughout this section we set

(8) $L[y](x) := (p(x)y'(x))' + q(x)y(x),$

so that equation (1) can be written in the compact form

$L[y](x) + \lambda r(x)y(x) = 0.$

We shall use this notation to express the various identities that arise in our study of regular Sturm-Liouville boundary value problems.

LAGRANGE'S IDENTITY FOR $L[y] = (py')' + qy$

Theorem 1. Let u and v be functions having continuous second derivatives on the interval $[a, b]$. Then,

(9) $uL[v] - vL[u] = \dfrac{d}{dx}(pW[u, v]),$

where $W[u, v] = uv' - vu'$ is the Wronskian of u and v.

Proof. Using the product rule and adding and subtracting $pu'v'$, we have

$$\begin{aligned}
uL[v] - vL[u] &= u[(pv')' + qv] - v[(pu')' + qu] \\
&= u(pv')' + quv - v(pu')' - quv \\
&= u(pv')' + u'(pv') - v'(pu') - v(pu')' \\
&= [u(pv')]' - [v(pu')]' \\
&= [p(uv' - vu')]' \\
&= \frac{d}{dx}[pW[u, v]]. \text{ ■ End}
\end{aligned}$$

By integrating Lagrange's identity, we obtain the following equation called **Green's formula.**

GREEN'S FORMULA FOR $L[y] = (py')' + qy$

Corollary 1. Let u and v be functions having continuous second derivatives on the interval $[a, b]$. Then,

(10) $\displaystyle\int_a^b (uL[v] - vL[u])(x)\,dx = (pW[u, v])(x)\,\Big|_a^b.$

If u and v also satisfy the boundary conditions in (2), then Green's formula simplifies to

(11) $\displaystyle\int_a^b (uL[v] - vL[u])(x)\,dx = 0.$

To see this, observe that when $a_2 \neq 0$ and $b_2 \neq 0$ in (2), then

$$u'(a) = -(a_1/a_2)u(a), \qquad u'(b) = -(b_1/b_2)u(b),$$
$$v'(a) = -(a_1/a_2)v(a), \qquad v'(b) = -(b_1/b_2)v(b).$$

Substituting these values into the right-hand side of (10) gives

$$p(b)W[u, v](b) - p(a)W[u, v](a)$$
$$= p(b)[u(b)v'(b) - u'(b)v(b)] - p(a)[u(a)v'(a) - u'(a)v(a)]$$
$$= p(b)[u(b)(-b_1/b_2)v(b) - (-b_1/b_2)u(b)v(b)]$$
$$\qquad - p(a)[u(a)(-a_1/a_2)v(a) - (-a_1/a_2)u(a)v(a)]$$
$$= 0.$$

When $a_2 = 0$ (so that $a_1 \neq 0$), it follows from (2) that both $u(a)$ and $v(a)$ are zero. Hence $[u(a)v'(a) - u'(a)v(a)] = 0$. Similarly when $b_2 = 0$, we get $[u(b)v'(b) - u'(b)v(b)] = 0$. Therefore, if u and v satisfy the boundary conditions in (2), then (11) holds.

Equation (11) takes a particularly useful and elegant form when expressed in terms of an inner product. If we set

(12) $\displaystyle (f, g) = \int_a^b f(x)g(x)\,dx,$

then $(\cdot,\cdot)$ is an **inner product** on the set of continuous real-valued functions on $[a, b]$.[†] Using this notation, we can rewrite (11) as

(13) $(u, L[v]) = (L[u], v).$

A linear differential operator L that satisfies (13) for all functions u and v in its domain is called a **selfadjoint operator.** We have just shown that if L is defined by (8) and the domain of L is the set of functions that have continuous second derivatives on $[a, b]$ and satisfy the boundary conditions in (2), then L is a selfadjoint operator. Selfadjoint operators are like symmetric matrices in that their eigenvalues are real-valued. We now prove this for regular Sturm-Liouville boundary value problems.

EIGENVALUES AND EIGENFUNCTIONS ARE REAL-VALUED

Theorem 2. The eigenvalues for the regular Sturm-Liouville boundary value problem (1)–(2) are real and have real-valued eigenfunctions.

[†] For further discussion on inner products, see the text by B. Kolman, *Elementary Linear Algebra*, Fifth Edition (Macmillan Publishing Co., New York, 1991), Chapter 3.

Proof. Assume λ, possibly a complex number, is an eigenvalue for (1)–(2) with eigenfunction $\phi(x)$, which may be complex-valued. That is,

(14) $L[\phi](x) + \lambda r(x)\phi(x) = 0,$

and $\phi(\not\equiv 0)$ satisfies the boundary conditions in (2). Taking the complex conjugate of both sides of (14) and using the fact that p, q, and r are real-valued, we obtain

(15) $\overline{L[\phi](x) + \lambda r(x)\phi(x)} = L[\bar{\phi}](x) + \bar{\lambda} r(x)\bar{\phi}(x) = 0.$

Further, since a_1, a_2, b_1, and b_2 are real, a similar argument shows that $\bar{\phi}$ also satisfies the boundary conditions in (2). Hence $\bar{\lambda}$ is an eigenvalue with eigenfunction $\bar{\phi}$.

From equation (14), we get

(16) $\displaystyle\int_a^b L[\phi](x)\bar{\phi}(x)\,dx = -\lambda \int_a^b r(x)\phi(x)\bar{\phi}(x)\,dx = -\lambda \int_a^b r(x)\big|\phi(x)\big|^2 dx.$

In addition, using identity (11) and the fact that $\bar{\lambda}$ is an eigenvalue with eigenfunction $\bar{\phi}$ we find

(17) $\displaystyle\int_a^b L[\phi](x)\bar{\phi}(x)\,dx = \int_a^b \phi(x)L[\bar{\phi}](x)\,dx = -\bar{\lambda} \int_a^b r(x)\big|\phi(x)\big|^2 dx.$

But the left-hand sides of (16) and (17) are the same, so equating the right-hand sides yields

$$-\lambda \int_a^b r(x)\big|\phi(x)\big|^2 dx = -\bar{\lambda} \int_a^b r(x)\big|\phi(x)\big|^2 dx.$$

Since $r(x) > 0$ and ϕ is a nontrivial solution to (1)–(2), then $\int_a^b r(x)\big|\phi(x)\big|^2 dx > 0$. Thus we can divide out this integral to obtain $\lambda = \bar{\lambda}$, which means that λ is a real number.

If ϕ itself is not real-valued, we can obtain real-valued eigenfunctions corresponding to λ simply by taking the real or imaginary parts of ϕ (see Problem 16). This completes the proof of Theorem 2. ■ End

If all the eigenfunctions associated with a particular eigenvalue are just scalar multiples of each other, then the eigenvalue is called **simple.**

EIGENVALUES ARE SIMPLE

Theorem 3. All the eigenvalues of the regular Sturm-Liouville boundary value problem (1)–(2) are simple.

Proof. Assume $\phi(x)$ and $\psi(x)$ are two eigenfunctions corresponding to the eigenvalue λ (thus ϕ and ψ are solutions to the same differential equation). They satisfy (2), so we have

$$a_1\phi(a) + a_2\phi'(a) = 0, \qquad a_1\psi(a) + a_2\psi'(a) = 0.$$

Let's assume $a_2 \neq 0$. (A similar argument applies if $a_2 = 0$ and $a_1 \neq 0$.) Solving for $\phi'(a)$ and $\psi'(a)$ gives

$$\phi'(a) = (-a_1/a_2)\phi(a), \qquad \psi'(a) = (-a_1/a_2)\psi(a).$$

If we now compute the Wronskian of ϕ and ψ at $x = a$, we find

(18) $W[\phi, \psi](a) = \phi(a)\psi'(a) - \phi'(a)\psi(a)$

$\qquad\qquad\qquad = \phi(a)(-a_1/a_2)\psi(a) - (-a_1/a_2)\phi(a)\psi(a) = 0.$

Recall that in Section 4.3, we proved that if the Wronskian of two solutions to a second order linear homogeneous equation is zero at a point, then the two solutions are linearly dependent. Consequently, ϕ and ψ are linearly dependent, so one must be a multiple of the other. Thus λ is a simple eigenvalue. ■ End

We now show that the eigenfunctions associated with different eigenvalues enjoy the property of orthogonality. Recall that in the discussion of orthogonal expansions in Section 10.3, we said that two real-valued functions f and g are **orthogonal** with respect to a positive weight function $w(x)$ on the interval $[a, b]$ if

(19) $\displaystyle\int_a^b f(x)g(x)w(x)\,dx = 0.$

ORTHOGONALITY OF EIGENFUNCTIONS

Theorem 4. Eigenfunctions that correspond to distinct eigenvalues of the regular Sturm-Liouville boundary value problem (1)–(2) are orthogonal with respect to the weight function $r(x)$ on $[a, b]$.

Proof. Let λ and μ be distinct eigenvalues with corresponding eigenfunctions ϕ and ψ, respectively; that is, $L[\phi] = -\lambda r\phi$ and $L[\psi] = -\mu r\psi$. We can assume, by Theorem 2, that ϕ and ψ are real-valued functions. Setting $u = \phi$ and $v = \psi$ in (11) and substituting for $L[\phi]$ and $L[\psi]$, we find

$$0 = \int_a^b (\phi L[\psi] - \psi L[\phi])(x)\,dx$$

$$= \int_a^b (-\phi\mu r\psi + \psi\lambda r\phi)(x)\,dx$$

$$= (\lambda - \mu)\int_a^b \phi(x)\psi(x)r(x)\,dx.$$

Since $\lambda \neq \mu$, we see that

$$\int_a^b \phi(x)\psi(x)r(x)\,dx = 0.$$

Therefore the eigenfunctions ϕ and ψ are orthogonal with respect to $r(x)$ on $[a, b]$. ∎ End

To illustrate the properties of regular Sturm-Liouville boundary value problems, let's consider the problem

(20) $y'' + \lambda y = 0;$ $y(0) = y(\pi) = 0.$

In Section 11.2, we mentioned that the eigenvalues of (20) are $\lambda_n = n^2$, $n = 1, 2, 3, \ldots$, with corresponding eigenfunctions $\phi_n(x) = c_n \sin nx$. Notice that the eigenvalues are real and simple as predicted by Theorems 2 and 3. Moreover, in Example 2 in Section 10.3, we proved that

(21) $$\int_0^\pi \sin nx \sin mx\,dx = 0, \quad \text{for} \quad m \neq n.$$

Thus the eigenfunctions corresponding to distinct eigenvalues are orthogonal with respect to $r(x) \equiv 1$ on $[0, \pi]$, which illustrates Theorem 4.

Notice further that the eigenvalues for problem (20) form an increasing sequence

$$\lambda_1 = 1^2 < \lambda_2 = 2^2 < \lambda_3 = 3^2 < \cdots,$$

with $\lim_{n\to\infty} \lambda_n = \lim_{n\to\infty} n^2 = +\infty$. A similar ordering is true for any regular Sturm-Liouville boundary value problem.

SEQUENCE OF EIGENVALUES

Theorem 5. The eigenvalues for the regular Sturm-Liouville boundary value problem (1)–(2) form a countable, increasing sequence

(22) $\lambda_1 < \lambda_2 < \lambda_3 < \cdots,$

with $\lim_{n\to\infty} \lambda_n = +\infty$.

This theorem tells us that a regular Sturm-Liouville boundary value problem has infinitely many eigenvalues and that they can be arranged in a sequence that increases to infinity. The proof of this fact is beyond the scope of this book, so we refer the reader to more advanced texts.[†]

[†] See for example the texts by G. Birkhoff and G. Rota, *Ordinary Differential Equations*, Third Edition (John Wiley and Sons, New York, 1978), Chapter 10, or E. Coddington and N. Levinson, *Theory of Ordinary Differential Equations* (Krieger, New York, 1984), Chapter 8.

Eigenfunction Expansions

Having in hand a sequence of orthogonal functions, we can use these functions to construct series expansions for a general class of functions. For example, the eigenfunctions $\phi_n(x) = \sin nx$ for problem (20) lead to the Fourier sine series expansions discussed in Section 10.4:

$$f(x) \sim \sum_{n=1}^{\infty} b_n \sin nx.$$

In fact, if f is continuous on $[0, \pi]$, f' is piecewise continuous on $[0, \pi]$, and $f(0) = f(\pi)$, then

$$f(x) = \sum_{n=1}^{\infty} b_n \sin nx, \qquad 0 \le x \le \pi,$$

where

$$b_n = \frac{2}{\pi} \int_0^{\pi} f(x) \sin nx\, dx, \qquad n = 1, 2, 3, \dots .$$

General orthogonal expansions were discussed in Section 10.3. There we showed that if the set $\{\phi_n(x)\}_{n=1}^{\infty}$ is orthonormal with respect to a positive weight function $w(x)$ on $[a, b]$, that is,

$$(23) \qquad \int_a^b \phi_n(x)\phi_m(x)w(x)\,dx = \begin{cases} 0, & m \ne n, \\ 1, & m = n, \end{cases}$$

then with any piecewise continuous function f, we can identify an orthogonal expansion

$$(24) \qquad f(x) \sim \sum_{n=1}^{\infty} c_n\phi_n(x),$$

where

$$(25) \qquad c_n = \int_a^b f(x)\phi_n(x)w(x)\,dx.$$

The eigenfunctions corresponding to distinct eigenvalues are orthogonal with respect to $r(x)$ for a regular Sturm-Liouville boundary value problem. Hence we can use the eigenfunctions to form an *orthonormal* system with respect to $r(x)$ simply by normalizing (scaling) each eigenfunction ϕ_n so that

$$(26) \qquad \int_a^b \phi_n^2(x)r(x)\,dx = 1.$$

EXAMPLE 2 Construct an orthonormal system of eigenfunctions corresponding to the regular Sturm-Liouville boundary value problem

$$(27) \qquad y'' + \lambda y = 0; \quad y(0) = 0, \quad y'(\pi) = 0.$$

Solution In Exercises 10.2, Problem 9, we found that the eigenvalues of (27) are $\lambda_n = (2n - 1)^2/4$, $n = 1, 2, 3, \ldots$, with corresponding eigenfunctions

$$\phi_n(x) = a_n \sin\left(\frac{2n - 1}{2}\right)x.$$

The eigenfunctions are orthogonal with respect to $r(x) \equiv 1$ on $[0, \pi]$, so we need only choose the a_n's so that (26) holds. We compute

$$\int_0^\pi a_n^2 \sin^2\left(\frac{2n - 1}{2}x\right) dx = a_n^2 \int_0^\pi \left[\frac{1}{2} - \frac{1}{2}\cos(2n - 1)x\right] dx$$

$$= a_n^2 \left[\frac{x}{2} - \frac{1}{2(2n - 1)}\sin(2n - 1)x\right]\Bigg|_0^\pi$$

$$= a_n^2(\pi/2).$$

Hence we can take $a_n = \sqrt{2/\pi}$, which gives

$$(28) \qquad \left\{\sqrt{2/\pi}\sin\left(\frac{2n - 1}{2}x\right)\right\}_{n=1}^\infty$$

as an orthonormal system of eigenfunctions. ∎ End

If $\{\phi_n\}_{n=1}^\infty$ is an orthonormal system of eigenfunctions, then we can use (24) and (25) to obtain an eigenfunction expansion for a function f. But, does the eigenfunction expansion converge? If so, to what function does it converge? These questions are partially answered in the next theorem.

UNIFORM CONVERGENCE OF EIGENFUNCTION EXPANSIONS

Theorem 6. Let $\{\phi_n\}_{n=1}^\infty$ be an orthonormal system of eigenfunctions for the regular Sturm-Liouville boundary value problem (1)–(2). Let f be a continuous function on $[a, b]$ with f' piecewise continuous[†] on $[a, b]$. If f satisfies the boundary conditions in (2), then

$$(29) \qquad f(x) = \sum_{n=1}^\infty c_n \phi_n(x), \qquad a \leq x \leq b,$$

where c_n is given by formula (25) with $w(x) = r(x)$. Moreover, the eigenfunction expansion in (29) converges uniformly on $[a, b]$.

The proof of Theorem 6 can be found in more advanced texts.[††]

[†] A function $g(x)$ is piecewise continuous on $[a, b]$ if it is continuous except for a finite number of points at which g has jump discontinuities (see Section 7.2).

[††] See for example the text by E. Coddington and N. Levinson mentioned in the footnote on page 675.

EXAMPLE 3 Express

$$f(x) = \begin{cases} 2x/\pi, & 0 \le x \le \pi/2, \\ 1, & \pi/2 \le x \le \pi, \end{cases}$$

in an eigenfunction expansion using the orthonormal system of eigenfunctions given in (28).

Solution To obtain the eigenfunction expansion for $f(x)$, we use formula (25). Thus

$$c_n = \int_0^\pi f(x)\sqrt{2/\pi} \sin\left(\frac{2n-1}{2}x\right) dx$$

$$= \left(\frac{2}{\pi}\right)^{3/2} \int_0^{\pi/2} x \sin\left(\frac{2n-1}{2}x\right) dx + \left(\frac{2}{\pi}\right)^{1/2} \int_{\pi/2}^\pi \sin\left(\frac{2n-1}{2}\right) dx$$

$$= \left(\frac{2}{\pi}\right)^{3/2} \left[\frac{-2x}{2n-1}\cos\left(\frac{2n-1}{2}x\right) + \frac{4}{(2n-1)^2}\sin\left(\frac{2n-1}{2}x\right)\right]\Bigg|_0^{\pi/2}$$

$$- \left(\frac{2}{\pi}\right)^{1/2} \left[\frac{2}{2n-1}\cos\left(\frac{2n-1}{2}x\right)\right]\Bigg|_{\pi/2}^{\pi},$$

and so, for $n = 1, 2, 3, \ldots,$

$$(30) \qquad c_n = \frac{2^{7/2}}{\pi^{3/2}(2n-1)^2}\sin\left(\frac{n\pi}{2} - \frac{\pi}{4}\right).$$

Hence

$$(31) \qquad f(x) = \sum_{n=1}^\infty c_n\sqrt{2/\pi}\sin\left(\frac{2n-1}{2}x\right)$$

$$= \frac{2^{7/2}}{\pi^2}\sin(x/2) + \frac{2^{7/2}}{9\pi^2}\sin(3x/2) - \frac{2^{7/2}}{25\pi^2}\sin(5x/2) - \frac{2^{7/2}}{49\pi^2}\sin(7x/2) + \cdots,$$

where the c_n's are as given in equation (30). Since $f(0) = 0$ and $f'(\pi) = 0$, then f satisfies the boundary conditions in (27). Moreover, f is continuous and f' piecewise continuous on $[0, \pi]$. Therefore, by Theorem 6, the series in (31) converges uniformly to f on $[0, \pi]$. ■ End

There are further results concerning the pointwise convergence of the eigenfunction expansions that are similar to the results for Fourier series (cf. Theorem 2, Section 10.3). We state without proof one such result.

POINTWISE CONVERGENCE OF EIGENFUNCTION EXPANSIONS

Theorem 7. Let $\{\phi_n\}_{n=1}^{\infty}$ be an orthonormal sequence of eigenfunctions for the regular Sturm-Liouville boundary value problem (1)–(2). Let f and f' be piecewise continuous on $[a, b]$. Then, for any x in (a, b),

$$(32) \qquad \sum_{n=1}^{\infty} c_n \phi_n(x) = \frac{1}{2}[f(x^+) + f(x^-)],$$

where the c_n's are given by formula (25) with $w(x) = r(x)$.

Periodic Sturm-Liouville Problems

Periodic Sturm-Liouville problems have properties similar to those of regular Sturm-Liouville boundary value problems. For example, the periodic Sturm-Liouville problem

$$(33) \qquad y'' + \lambda y = 0; \qquad y(0) = y(2\pi), \qquad y'(0) = y'(2\pi),$$

has eigenvalues $\lambda_n = n^2$, $n = 0, 1, 2, \ldots$, and eigenfunctions

$$\phi_0 = a_0/2, \quad \phi_{n,1}(x) = a_n \cos nx \text{ and } \phi_{n,2}(x) = b_n \sin nx \text{ for } n = 1, 2, 3, \ldots.$$

The set of eigenfunctions

$$\{1/2, \cos x, \sin x, \cos 2x, \sin 2x, \ldots\}$$

form an orthogonal set. The usual Fourier series expansions discussed in Section 10.3 are just eigenfunction expansions using this set; that is,

$$f(x) \sim a_0/2 + \sum_{n=1}^{\infty} [a_n \cos nx + b_n \sin nx].$$

The main distinction between the properties of regular and periodic Sturm-Liouville problems is the fact that the eigenvalues for the periodic Sturm-Liouville problem do not have to be simple. For example, both $\cos nx$ and $\sin nx$ are eigenfunctions corresponding to the eigenvalue $\lambda_n = n^2$ for the periodic problem (33). In the following theorem we give, without proof, the properties of the **periodic Sturm-Liouville problem:**

$$(34) \qquad (py')' + qy + \lambda ry = 0; \qquad y(0) = y(2T), \qquad y'(0) = y'(2T),$$

where p, p', q, and r are continuous $2T$-periodic functions with p and r positive.

PROPERTIES OF PERIODIC STURM-LIOUVILLE PROBLEMS

Theorem 8. For the periodic Sturm-Liouville problem (34):

1. The eigenvalues are real and form a countable, increasing sequence
 $\lambda_1 < \lambda_2 < \lambda_3 \ldots$ with $\lim_{n \to \infty} \lambda_n = +\infty$.
2. The eigenfunctions can be chosen to be real-valued.
3. The eigenfunctions corresponding to distinct eigenvalues are orthogonal with respect to $r(x)$ on $[0, 2T]$.
4. Let $\{\phi_n\}_{n=1}^{\infty}$ be an orthonormal sequence of all the eigenfunctions and let f be a $2T$-periodic function such that f is continuous and f' is piecewise continuous. Then,

$$(35) \qquad f(x) = \sum_{n=1}^{\infty} a_n \phi_n(x),$$

where

$$(36) \qquad a_n = \int_0^{2T} f(x)\phi_n(x)r(x)\,dx.$$

Moreover, the series in (35) converges uniformly on $[0, 2T]$ (and hence on $(-\infty, \infty)$).

The reader may have observed from the examples that eigenfunctions appear to be oscillatory; for example, $\sin nx$ and $\sin(2n - 1)x/2$. On closer examination, we see that the number of oscillations in $[a, b]$ actually increases as n increases. This behavior, which is typical of solutions to Sturm-Liouville equations, is discussed in Section 11.8.

EXERCISES 11.3

In Problems 1 through 6, convert the given equation into the form of a Sturm-Liouville equation.

1. $y'' + 6y' + \lambda y = 0$.

2. $xy'' + \lambda y = 0, \qquad x > 0$.

3. $x(1 - x)y'' - 2xy' + \lambda y = 0, \qquad 0 < x < 1$.

4. $x^2 y'' + xy' + (x^2 - \lambda)y = 0, \qquad x > 0$.

5. $x^2 y'' + xy' + \lambda y = 0, \qquad x > 0$.

6. $(1 - x^2)y'' - 2xy' + \lambda y = 0, \qquad -1 < x < 1$.

In Problems 7 through 11, determine whether the given linear differential operator $L[y]$, whose domain consists of all functions that have continuous second derivatives on the interval $[0, \pi]$ and satisfy the given boundary conditions, is selfadjoint.

7. $L[y] := y'' + 2y' + 5y; \qquad y(0) = 0, \qquad y(\pi) = 0$.

8. $L[y] := y'' + \lambda y; \qquad y(0) = y(\pi), \qquad y'(0) = y'(\pi)$.

9. $L[y] := y'' + \lambda y;$
 $y(0) = -y(\pi), \qquad y'(0) = -y'(\pi)$.

10. $L[y] := y'' + \lambda y;$
 $y(0) + y'(\pi) = 0, \qquad y(\pi) + y'(0) = 0$.

11. $L[y] := (1 + x^2)y'' + 2xy' + \lambda r y;$
 $y'(0) = 0, \qquad y(\pi) - (1 + \pi^2)y'(\pi) = 0$.

12. **Inner Product.** The notion of inner product $(\cdot, \cdot)$ is important in the study of general vector spaces and generalizes the dot product of vectors in $\mathbf{R}^n$. Recall we encountered inner products in our study of Fourier series and orthogonal expansions in Section 10.3. Ver-

ify that (f, g) defined in (12) is an inner product for f and g continuous on $[a, b]$; that is, show for f, g, and h continuous on $[a, b]$ and c a real number that (f, g) satisfies the following:

(a) $(f, f) \geq 0$ and equals zero only when $f \equiv 0$.
(b) $(f, g) = (g, f)$.
(c) $(f + h, g) = (f, g) + (h, g)$.
(d) $(cf, g) = c(f, g)$.

13. Let $L[y] := y^{(4)}$. Let the domain of L be the set of functions that have four continuous derivatives on $[0, \pi]$ and satisfy

$$y(0) = y'(0) = 0 \quad \text{and} \quad y(\pi) = y'(\pi) = 0.$$

(a) Show that L is selfadjoint.
(b) Use the result of part (a) to show that eigenfunctions corresponding to distinct eigenvalues of the problem

$$(37) \qquad \begin{matrix} y^{(4)} + \lambda y = 0; \\ y(0) = y'(0) = 0, \qquad y(\pi) = y'(\pi) = 0, \end{matrix}$$

are orthogonal with respect to $w(x) \equiv 1$ on $[0, \pi]$.

14. Let λ be an eigenvalue and ϕ a corresponding eigenfunction for the regular Sturm-Liouville boundary value problem (1)–(2).

(a) Show that

$$(38) \qquad \lambda \int_a^b r(x)\phi^2(x)\,dx$$

$$= \int_a^b \{p(x)[\phi'(x)]^2 - q(x)\phi^2(x)\}\,dx$$

$$- [p(x)\phi(x)\phi'(x)]\Big|_a^b.$$

(b) Use equation (38) to show that if $q(x) \leq 0$ and $a_1 = b_1 = 0$ in (2), then λ is nonnegative.
(c) Show that if $q(x) \leq 0$ and $a_2 = b_2 = 0$ in (2), then λ is nonnegative.
(d) Show that if $q(x) \leq 0$, $a_1/a_2 < 0$, and $b_1/b_2 > 0$, then λ is nonnegative.

15. Show that the eigenvalues of the eigenvalue problem (37) in Problem 13 are nonpositive. [*Hint:* Derive an equation similar to equation (38) in Problem 14.]

16. Show that if $\phi = u + iv$ is an eigenfunction corresponding to λ for the regular Sturm-Liouville problem (1)–(2), then the real part u and the imaginary part v each satisfy (1)–(2).

In Problems 17 through 24, (a) determine the normalized eigenfunctions for the given problem and (b) use the results of part (a) to express $f(x) = x$ in an eigenfunction expansion. (These are the same equations as in Problems 13 through 20 of Exercises 11.2.)

17. $y'' + \lambda y = 0;$ $\quad y(0) = 0,$ $\quad y'(1) = 0.$
18. $y'' + \lambda y = 0;$ $\quad y(-3) = y(3),$ $\quad y'(-3) = y'(3).$
19. $y'' + 3y + \lambda y = 0;$ $\quad y'(0) = 0,$ $\quad y'(\pi) = 0.$
20. $y'' - 2y + \lambda y = 0;$ $\quad y(0) = 0,$ $\quad y'(\pi) = 0.$
21. $y'' + \lambda y = 0;$ $\quad 2y(0) + y'(0) = 0,$ $\quad y(\pi) = 0.$
22. $y'' + \lambda y = 0;$ $\quad y(0) = 0,$ $\quad y(\pi) - 2y'(\pi) = 0.$
23. $(xy')' + \lambda x^{-1}y = 0;$ $\quad y'(1) = 0,$ $\quad y(e^\pi) = 0.$
24. $(xy')' + \lambda x^{-1}y = 0;$ $\quad y'(1) = 0,$ $\quad y'(e) = 0.$

25. **Bessel's Inequality.** Let $\{\phi_n\}_{n=1}^\infty$ be an orthonormal system with respect to a positive weight function $w(x)$ on $[a, b]$ and let f be continuous on $[a, b]$. Show that if

$$f(x) \sim \sum_{n=1}^\infty c_n\phi_n(x),$$

then

$$(39) \qquad \sum_{n=1}^\infty c_n^2 \leq \int_a^b f^2(x)w(x)\,dx.$$

[*Hint:* Model your proof after that of Bessel's inequality for Fourier series (cf. Problem 38 in Exercises 10.3).]

26. Prove that the linear differential operator associated with the periodic Sturm-Liouville problem (34) is selfadjoint.

11.4 NONHOMOGENEOUS BOUNDARY VALUE PROBLEMS AND THE FREDHOLM ALTERNATIVE

Unlike homogeneous boundary value problems (which always have $y \equiv 0$ as a solution), nonhomogeneous problems need not have any solution. In this section our goal is to develop

a test to determine whether the nonhomogeneous problem

(1) $L[y](x) := A_2(x)y''(x) + A_1(x)y'(x) + A_0(x)y(x) = h(x), \qquad a < x < b;$

(2)
$$B_1[y] := a_{11}y(a) + a_{12}y'(a) + b_{11}y(b) + b_{12}y'(b) = 0,$$
$$B_2[y] := a_{21}y(a) + a_{22}y'(a) + b_{21}y(b) + b_{22}y'(b) = 0,$$

has a solution. Before we proceed, we need to introduce the concept of an adjoint.

FORMAL ADJOINT

Definition 1. Let L be the differential operator given by the left-hand side of equation (1), i.e., $L[y] = A_2y'' + A_1y' + A_0y$. The **formal adjoint** (or Lagrange adjoint) of L is the differential operator L^+ defined by

(3) $L^+[y] := (A_2y)'' - (A_1y)' + A_0y,$

provided A_2'', A_1', and A_0 are continuous on $[a, b]$.

Although at this stage the peculiar definition of L^+ is unmotivated, we shall soon uncover its relationship to the operator L. The reader should take a minute to verify that if $A_1 = A_2'$, i.e., if equation (1) is in Sturm-Liouville form, then $L^+ = L$. In the following examples, we compute the formal adjoint for an operator with constant coefficients and for a Cauchy-Euler type operator.

EXAMPLE 1 Determine the formal adjoint of

(4) $L[y] = y'' + 2y' + 2y.$

Solution Here $A_0 = A_1 = 2$ and $A_2 = 1$. Substituting into formula (3) gives

(5) $L^+[y] = y'' - 2y' + 2y.$ ■ End

EXAMPLE 2 Determine the formal adjoint of

(6) $L[y](x) = x^2y''(x) + xy'(x) + 4y(x).$

Solution Here $A_2(x) = x^2$, $A_1(x) = x$, and $A_0(x) = 4$. Substituting into formula (3) and simplifying gives

(7) $L^+[y](x) = [x^2y(x)]'' - [xy(x)]' + 4y(x)$
$$= x^2y''(x) + 3xy'(x) + 5y(x). \quad ■ \text{ End}$$

The differential operator L and its formal adjoint are related by a Lagrange identity. This is a generalization of the Lagrange identity we proved for Sturm-Liouville operators (see page 671).

LAGRANGE'S IDENTITY

Theorem 9. Let L be the differential operator defined in (1) and let L^+ be its formal adjoint. Then,

$$(8) \qquad L[u]v - uL^+[v] = \frac{d}{dx}[P(u, v)],$$

where $P(u, v)$, the **bilinear concomitant** associated with L, is defined as

$$(9) \qquad P(u, v) := uA_1v - u(A_2v)' + u'A_2v.$$

Proof. Expanding the left-hand side of (8), we get

$$(10) \qquad L[u]v - uL^+[v] = (A_2u'' + A_1u' + A_0u)v - u[(A_2v)'' - (A_1v)' + A_0v]$$

$$= A_2u''v + A_1u'v - A_2uv'' - 2A_2'uv' - A_2''uv + A_1'uv + A_1uv'.$$

From the definition of $P(u, v)$ in (9), we compute

$$(11) \qquad \frac{d}{dx}[P(u, v)] = (u'A_1v + uA_1'v + uA_1v') - (uA_2v' + uA_2'v)'$$

$$+ (u''A_2v + u'A_2'v + u'A_2v')$$

$$= A_2u''v + A_1u'v - A_2uv'' - 2A_2'uv' - A_2''uv + A_1'uv + A_1uv',$$

which is the same as the right-hand side of (10). ∎ End

Integrating Lagrange's identity leads to Green's formula.

GREEN'S[†] FORMULA

Corollary 2. Let L be the differential operator defined in (1) and let L^+ be its formal adjoint. Then,

$$(12) \qquad \int_a^b (L[u]v - uL^+[v])(x)\,dx = P(u, v)(x)\Big|_a^b.$$

In Section 11.3 we defined an inner product for continuous real-valued functions on $[a, b]$ by

$$(13) \qquad (f, g) = \int_a^b f(x)g(x)\,dx.$$

[†] **Historical Footnote:** George Green (1793–1841) was an English mathematician who made the first attempt at a mathematical theory of electromagnetism (1828).

With this notation, Green's formula becomes

(14) $(L[u], v) = (u, L^+[v]) + P(u, v)\Big|_a^b .$

The elegance of formula (14) serves as a justification for the peculiar definitions of L^+ and $P(u, v)$. Indeed, it shows us how to transfer the operator L from the function u to the other function v, when computing inner products.

While the operators L and L^+ are meaningful when applied to any function y having a continuous second derivative on $[a, b]$ (i.e., for y in $C^2[a, b]$), in applications we need to further restrict the collection of functions y on which L and L^+ act. For example, in solving boundary value problems, we are interested in applying L only to those functions in $C^2[a, b]$ that satisfy the given boundary conditions. Therefore, when we consider the operators L and L^+, we will hereafter assume their domains $D(L)$ and $D(L^+)$, respectively, to be certain subspaces of $C^2[a, b]$.

Given the linear differential operator L with domain $D(L)$, we shall see that it is advantageous to restrict the formal adjoint L^+ to a domain $D(L^+)$ so that

(15) $(L[u], v) = (u, L^+[v])$

holds for all u in $D(L)$ and all v in $D(L^+)$. Of course, simply taking $D(L^+)$ to consist of the single function $v \equiv 0$ will do the job, but this is of no interest. What we would really like is to choose $D(L^+)$ as the largest possible set of functions in $C^2[a, b]$ for which (15) holds. With this optimal choice for $D(L^+)$, we say that L^+ is the **adjoint operator**[†] of L (i.e., we drop the adjective "formal"). For those special cases when $L^+ = L$ and (15) holds with $D(L^+) = D(L)$, we say that L is **selfadjoint.** For example, in the preceding section, we showed that for a regular Sturm-Liouville boundary value problem, the associated operator L is selfadjoint (the domain of L consisting of those functions in $C^2[a, b]$ that satisfy the boundary conditions).

Thanks to Green's formula, the adjoint will simply be L^+ restricted to a suitable class of functions v so that

$$P(u, v)\Big|_a^b = 0, \qquad \text{for } u \text{ in } D(L), \quad v \text{ in } D(L^+).$$

EXAMPLE 3 Find the adjoint operator for

(16) $L[y] = y'' + 2y' + 2y,$

where the domain of L consists of those functions that have continuous second derivatives on $[0, \pi]$ and satisfy the boundary conditions

(17) $y(0) = y(\pi) = 0.$

[†]The adjoint of an operator is usually studied in connection with a Hilbert space. For a further discussion of Hilbert spaces and adjoint operators, see a text on functional analysis such as *Functional Analysis*, by W. Rudin (McGraw-Hill, New York, 1973).

Solution In Example 1 we found that $L^+[v] = v'' - 2v' + 2v$. We must now determine the domain of L^+. We first observe that v'' must be continuous on $[0, \pi]$. Further, in order for equation (15) to hold, we must have

(18) $P(u, v)(x)\Big|_0^\pi = 0$

for all u in $D(L)$ and v in $D(L^+)$. From the formula for $P(u, v)$ in (9) with $A_1 = 2$ and $A_2 = 1$, condition (18) becomes

$$\{2u(\pi)v(\pi) - u(\pi)v'(\pi) + u'(\pi)v(\pi)\} - \{2u(0)v(0) - u(0)v'(0) + u'(0)v(0)\} = 0.$$

Since u is in $D(L)$, we know that $u(0) = u(\pi) = 0$. Thus the above equation simplifies to

(19) $u'(\pi)v(\pi) - u'(0)v(0) = 0.$

Now observe that $u'(\pi)$ and $u'(0)$ can take on any value, so we must have $v(0) = v(\pi) = 0$ in order for equation (19) to hold for all u in $D(L)$. Hence the domain of L^+ consists of those functions v that have continuous second derivatives on $[0, \pi]$ and that satisfy the boundary conditions

(20) $v(0) = v(\pi) = 0.$ ∎ End

ADJOINT BOUNDARY VALUE PROBLEM

Definition 2. Let $B[u] = 0$ represent the boundary conditions required for a function to be in the domain of L and let $B^+[u] = 0$ be the boundary conditions required for a function to be in the domain of its adjoint operator L^+. Then, the **adjoint boundary value problem** for

(21) $L[u] = 0,$ $B[u] = 0$

is the boundary value problem

(22) $L^+[u] = 0,$ $B^+[u] = 0.$

For example, as seen from Example 3, the adjoint boundary value problem for

(23) $y'' + 2y' + 2y = 0;$ $y(0) = y(\pi) = 0,$

is the boundary value problem

(24) $y'' - 2y' + 2y = 0;$ $y(0) = y(\pi) = 0.$

Two other examples follow.

EXAMPLE 4 Find the adjoint boundary value problem for

(25) $x^2 y''(x) + xy'(x) + 4y(x) = 0;$ $y(1) = y(e^\pi) = 0.$

Solution In Example 2 we found that the formal adjoint for $L[y](x) = x^2 y''(x) + xy'(x) + 4y(x)$ is

(26) $L^+[y](x) = x^2 y''(x) + 3xy'(x) + 5y(x).$

To determine the adjoint boundary conditions $B^+(y) = 0$, we use Green's formula and require that

$$P(u, v)(x) \Big|_1^{e^\pi} = 0,$$

for u in $D(L)$ and v in $D(L^+)$. Since $A_2(x) = x^2$ and $A_1(x) = x$, this condition becomes

$$xu(x)v(x) - x^2u(x)v'(x) - 2xu(x)v(x) + x^2u'(x)v(x) \Big|_1^{e^\pi} = 0,$$

that is,

$$e^\pi u(e^\pi)v(e^\pi) - e^{2\pi}u(e^\pi)v'(e^\pi) - 2e^\pi u(e^\pi)v(e^\pi) + e^{2\pi}u'(e^\pi)v(e^\pi)$$
$$- u(1)v(1) + u(1)v'(1) + 2u(1)v(1) - u'(1)v(1) = 0.$$

For u in $D(L)$, we have $u(1) = u(e^\pi) = 0$, so the above equation simplifies to

(27) $e^{2\pi}u'(e^\pi)v(e^\pi) - u'(1)v(1) = 0.$

Because $u'(e^\pi)$ and $u'(1)$ can take on any values, we must have $v(e^\pi) = v(1) = 0$ in order for equation (27) to hold for all u in $D(L)$. Therefore the adjoint boundary value problem is

(28) $x^2y''(x) + 3xy'(x) + 5y(x) = 0;$ $y(1) = y(e^\pi) = 0.$ ∎ End

EXAMPLE 5 Find the adjoint boundary value problem for

(29) $y'' + y = 0;$ $y(0) + y(\pi) = 0,$ $y'(0) = 0.$

Solution Here $L^+[y] = L[y] = y'' + y$. To determine the adjoint boundary value conditions, we again require

$$P(u, v) \Big|_0^\pi = 0$$

for u in $D(L)$ and v in $D(L^+)$. Since $A_1 = 0$ and $A_2 = 1$, we want

$$u'(\pi)v(\pi) - u(\pi)v'(\pi) - u'(0)v(0) + u(0)v'(0) = 0.$$

Now for u in $D(L)$, we have $u(\pi) = -u(0)$ and $u'(0) = 0$. Thus the above equation reduces to

(30) $u'(\pi)v(\pi) + u(0)[v'(\pi) + v'(0)] = 0.$

Since $u'(\pi)$ and $u(0)$ are arbitrary, we require that $v(\pi) = 0$ and $v'(0) + v'(\pi) = 0$ so that equation (30) holds for all u in $D(L)$. Therefore the adjoint boundary value problem is

(31) $y'' + y = 0;$ $y'(0) + y'(\pi) = 0,$ $y(\pi) = 0.$ ∎ End

Let's return to the question of when does the nonhomogeneous boundary value problem

(32) $L[y] = h,$ $B[y] = 0,$

have a solution. If it does have a solution u, then

$$(L[u],\ v) = (h,\ v)$$

for any v. Choosing v to be a solution to the (homogeneous) adjoint problem

(33) $L^+[v] = 0,\qquad B^+[v] = 0,$

we find that

$$(L[u],\ v) = (u,\ L^+[v]) = (u,\ 0) = 0.$$

Hence a necessary condition for the existence of a solution to (32) is that

$$(h,\ v) = 0,$$

for all v that satisfy (33). In orther words, for the nonhomogeneous problem to have a solution, the nonhomogeneous term h must be orthogonal to all solutions of the adjoint problem. It turns out that this condition is also sufficient for the existence of solutions.[†] This test is known as the Fredholm alternative.[††]

FREDHOLM ALTERNATIVE

Theorem 10. Let L be a linear differential operator and let B represent a set of linear boundary conditions. The nonhomogeneous boundary value problem

(34)
$$L[y](x) = h(x),\qquad a < x < b,$$
$$B[y] = 0$$

has a solution if and only if

(35)
$$\int_a^b h(x)z(x)\,dx = 0$$

for every solution z of the adjoint boundary value problem

(36)
$$L^+[z](x) = 0,\qquad a < x < b,$$
$$B^+[z] = 0.$$

Remark. According to Theorem 10, if the adjoint boundary value problem (36) has only the trivial solution, then the nonhomogeneous boundary value problem (34) has a solution for each h. In such a case, if the boundary conditions in (2) are either separated or periodic, then it can be shown that there is a *unique* solution for each h. For more

[†]For a proof that the condition is sufficient, we refer the reader to Chapter 11 of the text by E. Coddington and N. Levinson referenced in the footnote on page 675.

[††]*Historical Footnote:* I. Fredholm (1866–1927), in his study of linear integral equations (circa 1900), realized that an analogy exists between linear integral equations and systems of n equations in n unknowns. This analogy also extends to differential equations and, more generally, to linear operators.

general boundary conditions, it is possible to have a one-parameter family of solutions (see Problem 28).

In the following examples, we illustrate how the Fredholm alternative is used to determine whether a nonhomogeneous boundary value problem has a solution.

EXAMPLE 6 Determine conditions on h that guarantee that the problem

(37) $y'' + 2y' + 2y = h;$ $y(0) = y(\pi) = 0,$

has a solution.

Solution The operator $L[y] = y'' + 2y' + 2y$ is the same as that in Example 3. Hence (see (23) and (24)), the adjoint problem is

(38) $y'' - 2y' + 2y = 0;$ $y(0) = y(\pi) = 0.$

The auxiliary equation for (38) is $r^2 - 2r + 2 = 0$, whose roots are $r = 1 \pm i$. Hence a general solution to the differential equation in (38) is

$$y(x) = c_1 e^x \cos x + c_2 e^x \sin x.$$

Using the boundary conditions in (38) to determine c_1 and c_2, we find

$$y(0) = c_1 = 0 \quad \text{and} \quad y(\pi) = -c_1 e^\pi = 0.$$

Thus $c_1 = 0$ and c_2 is arbitrary. Therefore every solution to the adjoint problem (38) has the form

$$z(x) = c_2 e^x \sin x.$$

It now follows from the Fredholm alternative that if h is continuous, then the nonhomogeneous problem (37) has a solution if and only if

$$\int_0^\pi h(x) e^x \sin x \, dx = 0. \quad \blacksquare \text{ End}$$

Recall from Section 11.2 that if (37) does have a solution, say when $h(x) = e^{-x} \cos x$, then it has a one-parameter family of solutions, since the corresponding homogeneous problem ((37) with $h = 0$) has a one-parameter family of solutions, $y(x) = ce^{-x} \sin x$.

EXAMPLE 7 Determine conditions on h that guarantee that the problem

(39) $x^2 y''(x) + xy'(x) + 4y(x) = h(x);$ $y(1) = y(e^\pi) = 0$

has a solution.

Solution In Example 4 we determined that the corresponding adjoint problem is

(40) $x^2 y''(x) + 3xy'(x) + 5y(x) = 0;$ $y(1) = y(e^\pi) = 0.$

Now the auxiliary equation for the Cauchy-Euler equation in (40) is $r(r - 1) + 3r + 5 = 0$, which has roots $r = -1 \pm 2i$. Hence a general solution to the differential equation in

(40) is

$$y(x) = c_1 x^{-1} \cos(2 \ln x) + c_2 x^{-1} \sin(2 \ln x).$$

Using the boundary conditions in (40) to determine c_1 and c_2, we find

$$y(1) = c_1 = 0 \quad \text{and} \quad y(e^\pi) = e^{-\pi} c_1 = 0.$$

Hence $c_1 = 0$ and c_2 is arbitrary. Therefore any solution to the adjoint problem (40) has the form

$$z(x) = c_2 x^{-1} \sin(2 \ln x).$$

It now follows from the Fredholm alternative that for h continuous, the nonhomogeneous problem (39) has a solution if and only if

(41) $\displaystyle\int_1^{e^\pi} h(x) x^{-1} \sin(2 \ln x)\, dx = 0.$ ∎ End

Applying the Fredholm alternative to a nonhomogeneous regular Sturm-Liouville boundary value problem is easy because the operator is selfadjoint.

EXAMPLE 8 Give conditions on h that guarantee that the nonhomogeneous regular Sturm-Liouville boundary value problem

(42) $y'' + \mu y = h; \qquad y(0) = 0, \qquad y'(\pi) = 0,$

has a solution.

Solution Since (42) is selfadjoint, we must have h orthogonal to every solution to the corresponding homogeneous problem. From Example 2 of Section 11.3 (cf. page 676), we know that this homogeneous problem has nontrivial solutions when and only when $\mu = \lambda_n = (2n - 1)^2/4$ for some $n = 1, 2, 3, \ldots$; that is, μ is an eigenvalue. Therefore we must consider two cases.

Case 1. If $\mu \neq (2n - 1)^2/4$ for $n = 1, 2, 3, \ldots$, then the only solution to the homogeneous problem is the trivial solution. It then follows from the Fredholm alternative that the nonhomogeneous problem (42) has a unique solution for every h.

Case 2. If $\mu = (2n - 1)^2/4$ for some $n = 1, 2, 3, \ldots$, then the solutions of the homogeneous problem (eigenfunctions) are of the form $\phi_n(x) = b_n \sin((2n - 1)x/2)$, where b_n is arbitrary. The Fredholm alternative tells us that the nonhomogeneous problem (42) has a solution if and only if

(43) $\displaystyle\int_0^{\pi} h(x) \sin\left(\frac{2n - 1}{2} x\right) dx = 0.$

Moreover, if this condition is satisfied, there is a one-parameter family of solutions, since we can always add a multiple of $\sin((2n - 1)x/2)$ to an existing solution to obtain a new solution. ∎ End

In the next two sections we give procedures for solving nonhomogeneous boundary value problems. The first procedure produces an eigenfunction expansion for the solution. The second procedure involves Green's functions and expresses the solution as a definite integral.

EXERCISES 11.4

In Problems 1 through 6, compute the formal adjoint of the given linear differential operator L.

1. $L[y](x) = xy''(x) + (2 - x^2)y'(x) + 3y(x)$.

2. $L[y](x) = x^2y''(x) + (\sin x)y'(x) + 2xy(x)$.

3. $L[y](x) = (1 + x^2)y''(x) + 2xy'(x) + y(x)$.

4. $L[y](x) = x^2y''(x) - 2xy'(x) + 3y(x)$.

5. $L[y](x) = 2x^2y''(x) + xy'(x) - 5y(x)$.

6. $L[y](x) = (\sin x)y''(x) - e^xy'(x) + y(x)$.

In Problems 7 through 10, find the adjoint operator and its domain for the given linear differential operator L whose domain consists of functions that have continuous second derivatives and satisfy the indicated boundary conditions.

7. $L[y] = y'' - 2y' + 10y$; $y(0) = 0$, $y(\pi) = 0$.

8. $L[y] = y'' - 4y' + 5y$; $y(0) = 0$, $y(2\pi) = 0$.

9. $L[y](x) = 2x^2y''(x) + 3xy'(x) - y(x)$;
 $y(1) = 0$, $y(4) = 0$.

10. $L[y](x) = x^2y''(x) + 2xy'(x) + (5/4)y(x)$;
 $y(1) = 0$, $y(e^\pi) = 0$.

In Problems 11 through 16, find the adjoint boundary value problem for the given boundary value problem.

11. $y'' + 6y' + 10y = 0$; $y'(0) = 0$, $y'(\pi) = 0$.

12. $y'' + y' + y = 0$; $y(0) = y(\pi)$, $y'(0) = y'(\pi)$.

13. $y'' + 4y = 0$; $y(0) = y(\pi)$, $y'(0) = y'(\pi)$.

14. $y'' = 0$; $y(0) = y(\pi/2)$, $y'(0) = y'(\pi/2)$.

15. $x^2y''(x) + 3xy'(x) + 2y(x) = 0$;
 $y'(1) = 0$, $y'(e^\pi) = 0$.

16. $x^2y''(x) + 2xy'(x) = 0$;
 $y(1) = y(2)$, $y'(1) = y'(2)$.

In Problems 17 through 26, determine conditions on h that guarantee that the given nonhomogeneous boundary value problem has a solution. (These are nonhomogeneous problems associated with the homogeneous Problems 7 through 16.)

17. $y'' - 2y' + 10y = h$; $y(0) = 0$, $y(\pi) = 0$.

18. $y'' - 4y' + 5y = h$; $y(0) = 0$, $y(2\pi) = 0$.

19. $2x^2y''(x) + 3xy'(x) - y(x) = h(x)$;
 $y(1) = 0$, $y(4) = 0$.

20. $x^2y''(x) + 2xy'(x) + (5/4)y(x) = h(x)$;
 $y(1) = 0$, $y(e^\pi) = 0$.

21. $y'' + 6y' + 10y = h$; $y'(0) = 0$, $y'(\pi) = 0$.

22. $y'' + y' + y = h$; $y(0) = y(\pi)$, $y'(0) = y'(\pi)$.

23. $y'' + 4y = h$; $y(0) = y(\pi)$, $y'(0) = y'(\pi)$.

24. $y'' = h$; $y(0) = y(\pi/2)$, $y'(0) = y'(\pi/2)$.

25. $x^2y''(x) + 3xy'(x) + 2y(x) = h(x)$;
 $y'(1) = 0$, $y'(e^\pi) = 0$.

26. $x^2y''(x) + 2xy'(x) = h(x)$;
 $y(1) = y(2)$, $y'(1) = y'(2)$.

27. Prove Green's formula by starting with

$$(L[u], v) = \int_a^b (A_2u'' + A_1u' + A_0u)v \, dx$$

and then, performing integration by parts, obtain

$$(L[u], v) = (u, L^+[v]) + P(u, v)\Big|_a^b.$$

28. Show that the "boundary value problem"

$$y'' + y = h(x), \quad 0 < x < \pi, \quad y(0) = 0,$$

has a one-parameter family of solutions for each h continuous on $[0, \pi]$. [*Hint:* Show that the homogeneous equation has a one-parameter family of solutions, while the adjoint problem has only the trivial solution.]

29. Show that the boundary value problem

$$y'' + y = h(x);$$
$$y(0) = 0, \quad y(\pi) = 0, \quad y'(\pi) = 0,$$

has a unique solution for each h that is continuous on $[0, \pi]$ and satisfies

$$\int_0^\pi h(x) \sin x \, dx = 0.$$

[*Hint:* Show that the homogeneous equation has only the trivial solution.]

11.5 SOLUTION BY EIGENFUNCTION EXPANSION

In the previous section we found that the Fredholm alternative provides a test to determine whether a given nonhomogeneous boundary value problem has a solution. But how do we find such a solution? Of course, if we can find a general solution to the nonhomogeneous equation, then we can simply substitute into the boundary conditions and determine a solution (see Example 2 in Section 11.2). However, most regular Sturm-Liouville boundary value problems are not constant coefficient equations. Therefore it may be difficult to find such a general solution.

In this section, we take advantage of the fact that the eigenfunctions associated with a regular Sturm-Liouville boundary value problem form an orthogonal system. Our goal is to find an **eigenfunction expansion** for a solution to the nonhomogeneous regular Sturm-Liouville boundary value problem

(1) $L[y] + \mu r y = f,$

(2) $a_1 y(a) + a_2 y'(a) = 0, \qquad b_1 y(b) + b_2 y'(b) = 0,$

where μ is a fixed real number and

(3) $L[y] := (p y')' + q y.$

Let $\{\lambda_n\}_{n=1}^{\infty}$ be the eigenvalues with corresponding eigenfunctions $\{\phi_n\}_{n=1}^{\infty}$ for the homogeneous eigenvalue problem associated with (1)–(2), with μ replaced by the parameter λ; that is, $L[\phi_n] + \lambda_n r \phi_n = 0$ and ϕ_n satisfies (2). We know (recall Theorem 4 in Section 11.3) that $\{\phi_n\}_{n=1}^{\infty}$ forms an orthogonal system with respect to the weight function $r(x)$ on $[a, b]$.

Let

(4) $\displaystyle \phi(x) = \sum_{n=1}^{\infty} c_n \phi_n(x).$

Our aim is to determine the coefficients c_n so that ϕ is a solution to (1)–(2).

For the moment, we shall ignore the important issues of convergence and differentiability of this series; that is, we shall treat ϕ as a *formal* series expansion. First notice that since each ϕ_n satisfies the boundary conditions in (2), so does ϕ. Further, substituting the expansion for ϕ into the left-hand side of (1) and using the fact that $L[\phi_n] = -\lambda_n r \phi_n$, we find

(5) $\displaystyle L[\phi] + \mu r \phi = L\left[\sum_{n=1}^{\infty} c_n \phi_n \right] + \mu r \sum_{n=1}^{\infty} c_n \phi_n$

$\displaystyle \qquad\qquad\qquad = \sum_{n=1}^{\infty} c_n L[\phi_n] + \mu r \sum_{n=1}^{\infty} c_n \phi_n$

$\displaystyle \qquad\qquad\qquad = \sum_{n=1}^{\infty} c_n (-\lambda_n r \phi_n) + \mu r \sum_{n=1}^{\infty} c_n \phi_n$

$\displaystyle \qquad\qquad\qquad = r \sum_{n=1}^{\infty} (\mu - \lambda_n) c_n \phi_n.$

If we express f as the product $r(f/r)$ and expand f/r in an eigenfunction expansion

$$f/r = \sum_{n=1}^{\infty} \gamma_n \phi_n, \qquad \gamma_n = \frac{\int_a^b (f/r)\, \phi_n r\, dx}{\int_a^b \phi_n^2 r\, dx} = \frac{\int_a^b f \phi_n\, dx}{\int_a^b \phi_n^2 r\, dx},$$

then the right-hand side of (1) becomes

(6) $\qquad r \sum_{n=1}^{\infty} \gamma_n \phi_n.$

Equating the left- and right-hand sides of (1) given in (5) and (6), we find

(7) $\qquad r \sum_{n=1}^{\infty} (\mu - \lambda_n) c_n \phi_n = r \sum_{n=1}^{\infty} \gamma_n \phi_n.$

Since $r > 0$, we can simplify (7) to

$$\sum_{n=1}^{\infty} [(\mu - \lambda_n) c_n - \gamma_n] \phi_n = 0.$$

This expansion sums to zero if and only if the coefficients are all zero. Therefore we set

(8) $\qquad (\mu - \lambda_n) c_n - \gamma_n = 0, \qquad n = 1, 2, 3, \dots .$

We now consider two cases.

Case 1. $\mu \neq \lambda_n$ for every $n = 1, 2, 3, \dots .$ In this case, we can solve (8) for c_n:

$$c_n = \frac{\gamma_n}{\mu - \lambda_n}.$$

The unique solution to (1)–(2) is thus given formally by

(9) $\qquad \phi = \sum_{n=1}^{\infty} \frac{\gamma_n}{\mu - \lambda_n} \phi_n.$

We have referred to ϕ as the unique solution, since if $\mu \neq \lambda_n$ for every $n = 1, 2, 3, \dots ,$ the corresponding homogeneous equation has only the trivial solution and hence the nonhomogeneous boundary value problem has a unique solution.

Case 2. $\mu = \lambda_N$ for some N. In the case when $n = N$, equation (8) reduces to

$$\gamma_N = 0.$$

There are now two possibilities. If $\gamma_N \neq 0$, we have a contradiction and so the nonhomogeneous equation does not have a solution expressible as an eigenfunction expansion. If $\gamma_N = 0$, then there is no restriction on c_N. Thus we get a one-parameter family of solutions to the nonhomogeneous boundary value problem (1)–(2), with c_N being the parameter.

But what does it mean for $\gamma_N = 0$? From the definition of γ_N, we see that since

$$\gamma_N = \int_a^b f(x)\phi_N(x)\,dx \Big/ \int_a^b \phi_N^2(x)r(x)\,dx,$$

the function f must be orthogonal to ϕ_N when $\gamma_N = 0$. But this is just the Fredholm alternative condition for the nonhomogeneous problem (1)–(2).

METHOD OF EIGENFUNCTION EXPANSIONS

To obtain an eigenfunction expansion for the solution ϕ to the nonhomogeneous regular Sturm-Liouville boundary value problem

$$L[y] + \mu ry = f, \qquad B[y] = 0:$$

(a) Find an orthogonal system of eigenfunctions $\{\phi_n\}_{n=1}^{\infty}$ and corresponding eigenvalues $\{\lambda_n\}_{n=1}^{\infty}$ for

$$L[y] + \lambda ry = 0, \qquad B[y] = 0.$$

(b) Compute the eigenfunction expansion for f/r; that is, determine the coefficients γ_n so that

$$f/r = \sum_{n=1}^{\infty} \gamma_n\phi_n.$$

(c) If $\mu \neq \lambda_n$ for $n = 1, 2, 3, \ldots$, then the solution ϕ is given by

$$\phi = \sum_{n=1}^{\infty} \frac{\gamma_n}{\mu - \lambda_n}\,\phi_n.$$

(d) If $\mu = \lambda_N$ for some N and $\gamma_N \neq 0$, then there is no solution.

(e) If $\mu = \lambda_N$ for some N and $\gamma_N = 0$, then

$$\phi = c_N\phi_N + \sum_{\substack{n=1 \\ n \neq N}}^{\infty} \frac{\gamma_n}{\mu - \lambda_n}\,\phi_n$$

is a solution for any choice of the parameter c_N.

In the following examples we illustrate this procedure.

EXAMPLE 1 Find an eigenfunction expansion for the solution to the regular Sturm-Liouville boundary value problem

(10) $y'' + 3y = 5\sin 3x + 2\sin 8x; \qquad y(0) = 0, \qquad y(\pi) = 0.$

Solution Recall from Section 11.2 that the boundary value problem

$$y'' + \lambda y = 0; \qquad y(0) = 0, \qquad y(\pi) = 0,$$

has eigenvalues $\lambda_n = n^2$, $n = 1, 2, \ldots$, with corresponding eigenfunctions

$$\phi_n(x) = \sin nx, \qquad n = 1, 2, \ldots.$$

Here $r(x) \equiv 1$, so we need to determine coefficients γ_n such that

$$f(x)/r(x) = f(x) = \sum_{n=1}^{\infty} \gamma_n \sin nx$$

$$= 5 \sin 3x + 2 \sin 8x.$$

Clearly $\gamma_3 = 5$, $\gamma_8 = 2$, and the remaining γ_n's are zero. Since $\mu = 3 \neq n^2 = \lambda_n$ for $n = 1, 2, \ldots$, it follows that the unique solution to (10) is

$$\textbf{(11)} \qquad \phi(x) = \sum_{n=1}^{\infty} \frac{\gamma_n}{\mu - \lambda_n} \phi_n(x) = \frac{5}{3 - 9} \sin 3x + \frac{2}{3 - 64} \sin 8x$$

$$= -\frac{5}{6} \sin 3x - \frac{2}{61} \sin 8x. \quad \blacksquare \text{ End}$$

As a check on our computations and as a verification of the theory, the reader is invited to show directly by substitution that ϕ given in (11) is the solution to problem (10).

EXAMPLE 2 Find a formal eigenfunction expansion for the solution to the regular Sturm-Liouville boundary value problem

$$\textbf{(12)} \qquad y'' + y = f, \qquad y(0) = y'(\pi) = 0,$$

where

$$\textbf{(13)} \qquad f(x) = \begin{cases} 2x/\pi, & 0 \leq x \leq \pi/2, \\ 1, & \pi/2 \leq x \leq \pi. \end{cases}$$

Solution In Example 3 in Section 11.3, we determined that the sequence

$$\left\{ \sqrt{\frac{2}{\pi}} \sin \left(\frac{2n - 1}{2} x \right) \right\}_{n=1}^{\infty}$$

is an orthonormal system of eigenfunctions for the boundary value problem

$$y'' + \lambda y = 0; \qquad y(0) = y'(\pi) = 0,$$

where the associated eigenvalues are $\lambda_n = (2n - 1)^2/4$. We also found that $f(x)$ $(= f(x)/r(x)$, since $r(x) \equiv 1)$ given in (13) has the eigenfunction expansion

$$\textbf{(14)} \qquad f(x) = \sum_{n=1}^{\infty} \gamma_n \sqrt{\frac{2}{\pi}} \sin \left(\frac{2n - 1}{2} x \right),$$

where

(15) $\gamma_n = \dfrac{2^{7/2}}{\pi^{3/2}(2n-1)^2} \sin\left(\dfrac{n\pi}{2} - \dfrac{\pi}{4}\right), \qquad n = 1, 2, 3, \ldots.$

Since $\mu = 1$ is *not* an eigenvalue, the nonhomogeneous problem (12) has a unique solution ϕ given formally by the expansion

(16) $\phi(x) = \displaystyle\sum_{n=1}^{\infty} \dfrac{\gamma_n}{1 - \lambda_n}\, \phi_n(x) = \sum_{n=1}^{\infty} \dfrac{\gamma_n}{1 - \dfrac{(2n-1)^2}{4}} \sqrt{\dfrac{2}{\pi}} \sin\left(\dfrac{2n-1}{2}\, x\right),$

where the γ_n's are given by equation (15). ■ End

A solution to the nonhomogeneous boundary value problem in (12) can also be obtained by finding two general solutions to the differential equation in (12): one valid on the interval $[0, \pi/2]$ and the other on $[\pi/2, \pi]$. The *four* arbitrary constants in these general solutions are selected to assure the boundary conditions in (12) are satisfied and that the solution and its derivative are continuous on $[0, \pi]$. We leave it to the reader to carry out this procedure and obtain the solution

(17) $\phi(x) = \begin{cases} \dfrac{2}{\pi}x, & 0 \le x \le \pi/2, \\[2ex] 1 - \dfrac{2}{\pi}\cos x, & \pi/2 \le x \le \pi. \end{cases}$

In advanced texts, it is shown that if f is continuous on $[a, b]$, then the formal solution in (9) converges uniformly to the genuine solution. For example, the expansion in (16) converges uniformly to the function ϕ in (17).

If one has an eigenfunction expansion for the solution to a nonhomogeneous boundary value problem, then one can obtain an *approximate solution* by truncating the expansion after N terms. This truncated expansion is an example of a **Galerkin approximation** for the solution. Galerkin's method and **collocation**[†] are two widely used procedures for approximating the solution to a boundary value problem.

EXERCISES 11.5

In Problems 1 through 8, find a formal eigenfunction expansion for the solution to the given nonhomogeneous boundary value problem, if it exists.

1. $y'' + 2y = 5\sin x - 7\sin 3x;\qquad y(0) = 0,\qquad y(\pi) = 0.$

2. $y'' - y = \sin 4x - \sin 11x;\qquad y(0) = 0,\quad y(\pi) = 0.$

3. $y'' + 4y = \sin 2x + \sin 8x;\qquad y(0) = 0,\qquad y(\pi) = 0.$

4. $y'' + \pi y = \cos 7x + 5\cos 10x;\qquad y'(0) = 0,\qquad y'(\pi) = 0.$

5. $y'' + y = \cos 4x + \cos 7x;\qquad y'(0) = 0,\qquad y'(\pi) = 0.$

6. $y'' + 6y = \cos 4x - \cos 5x;\qquad y'(0) = 0,\qquad y'(\pi) = 0.$

7. $y'' + y = x^2 - 2\pi x;\qquad y(0) = 0,\qquad y'(\pi) = 0.$

8. $y'' + 2y = \pi x - x^2;\qquad y(0) = 0,\qquad y(\pi) = 0.$

[†] For more information on numerical methods for solving boundary value problems, see *Numerical Methods for Two-Point Boundary Value Problems*, by H. B. Keller (Krieger, New York, 1992). The **shooting method**, in particular, is illustrated in Project D.

In Problems 9 through 14, find a formal eigenfunction expansion for the solution to the given nonhomogeneous boundary value problem. Assume f satisfies the conditions of the Fredholm alternative.

9. $y'' + y = f;$ $y'(0) = 0,$ $y(\pi) = 0.$

10. $y'' + 7y = f;$ $y(0) = 0,$ $y'(\pi) = 0.$

11. $(xy')' + 5x^{-1}y = f;$ $y'(1) = 0,$ $y(e^\pi) = 0.$

12. $xy'' + y' + x^{-1}y = f;$ $y(1) = 0,$ $y(e^\pi) = 0.$

13. $y'' + 6y = f;$ $y(0) = 0,$ $3y(\pi) - y'(\pi) = 0.$
 [*Hint*: See Example 4 in Section 11.2.]

14. $(xy')' - x^{-1}y = f;$ $y'(1) = 0,$ $y'(e) = 0.$

15. Derive the solution to problem (12) given in equation (17).

11.6 GREEN'S FUNCTIONS

In this section we derive an integral representation for the solution to a nonhomogeneous boundary value problem. Namely, we show that the solution can be expressed in the form

$$(1) \qquad y(x) = \int_a^b G(x, s)f(s)\,ds,$$

where the kernel function, $G(x, s)$, is called a *Green's function*, or an *influence function*. This representation is known to exist for very general problems.[†] However, we will consider only the regular Sturm-Liouville boundary value problems

$$(2) \qquad L[y](x) = -f(x), \qquad a < x < b,$$

$$(3) \qquad a_1 y(a) + a_2 y'(a) = 0,$$

$$(4) \qquad b_1 y(b) + b_2 y'(b) = 0,$$

where

$$(5) \qquad L[y] := \frac{d}{dx}\left(p\frac{dy}{dx}\right) + qy$$

and p, p', and q are continuous on $[a, b]$ with $p(x) > 0$ on $[a, b]$. We will assume f is continuous on $[a, b]$. While it is not necessary, it is customary in the present context to write $-f(x)$, rather than $f(x)$, as the nonhomogeneous term in equation (2).

Throughout this section **we assume that the corresponding homogeneous problem** $(f(x) \equiv 0)$ **has only the trivial solution,** or equivalently, we assume that $\lambda = 0$ is not an eigenvalue. It then follows from the Fredholm alternative that the nonhomogeneous problem (2)–(4) has a solution. This solution is unique, since the corresponding homogeneous problem has a unique solution.

We will use the method of variation of parameters to obtain an integral representation for the solution to (2)–(4) (see Section 4.9). To begin, choose two nontrivial solutions z_1 and z_2 to the homogeneous equation $L[y] = 0$. Pick z_1 so that it satisfies (3) and z_2 so that

[†] For a discussion of Green's functions for boundary value problems for nth order linear ordinary differential equations, see the text by E. Coddington and N. Levinson, *Theory of Ordinary Differential Equations* (Krieger, New York, 1984), Chapter 7.

it satisfies (4). The existence of z_1 and z_2 follows from the existence and uniqueness theorem for initial value problems (see Theorem 2 in Section 4.2), since one can choose initial conditions at a (respectively b) so that the boundary condition (3) (respectively (4)) is satisfied and the solution is nontrivial.

Before we can use z_1 and z_2 with the method of variation of parameters, we must verify that z_1 and z_2 are linearly independent. Note that if z_1 and z_2 were linearly *dependent*, then one would be a constant multiple of the other, say $z_1 = cz_2$. Since z_2 satisfies (4), it follows that z_1 also satisfies (4). Recall that z_1 also satisfies (3). Thus z_1 satisfies $L[z_1] = 0$ and (3)–(4). But we have assumed that the corresponding homogeneous problem has *only* the trivial solution. This contradiction implies that z_1 and z_2 are linearly independent.

To apply the formulas from the variation of parameters method, we first write the nonhomogeneous equation (2) in the standard form

$$y'' + \frac{p'}{p}y' + \frac{q}{p}y = -\frac{f}{p}.$$

A particular solution to this equation is then given by (cf. formulas (4) and (12) in Section 4.9)

(6) $y(x) = c_1(x)z_1(x) + c_2(x)z_2(x),$

where

(7) $c_1'(x) = \dfrac{f(x)z_2(x)}{p(x)W[z_1,\, z_2](x)}, \qquad c_2'(x) = \dfrac{-f(x)z_1(x)}{p(x)W[z_1,\, z_2](x)},$

and $W[z_1,\, z_2] = z_1z_2' - z_1'z_2$ is the Wronskian of z_1 and z_2. Since we are free to pick the constants in the antiderivatives for c_1' and c_2', it will turn out to be convenient to choose

(8) $c_1(x) = \displaystyle\int_x^b \dfrac{-z_2(s)f(s)}{p(s)W[z_1,\, z_2](s)}\,ds$

and

(9) $c_2(x) = \displaystyle\int_a^x \dfrac{-z_1(s)f(s)}{p(s)W[z_1,\, z_2](s)}\,ds.$

Substituting into (6), we obtain the following solution to (2):

(10) $y(x) = z_1(x)\displaystyle\int_x^b \dfrac{-z_2(s)f(s)}{p(s)W[z_1,\, z_2](s)}\,ds + z_2(x)\displaystyle\int_a^x \dfrac{-z_1(s)f(s)}{p(s)W[z_1,\, z_2](s)}\,ds$

$\qquad\qquad = \displaystyle\int_a^b G(x,\, s)f(s)\,ds,$

where

$$G(x, s) := \begin{cases} \dfrac{-z_1(s)z_2(x)}{p(s)W[z_1, z_2](s)}, & a \leq s \leq x, \\[3mm] \dfrac{-z_1(x)z_2(s)}{p(s)W[z_1, z_2](s)}, & x \leq s \leq b. \end{cases}$$

Using the fact that z_1 and z_2 satisfy the equation $L[y] = 0$, one can show that

(11) $p(x)W[z_1, z_2](x) = C, \quad x \text{ in } [a, b],$

where C is a constant (see Problem 31). Hence $G(x, s)$ has the simpler form

(12) $G(x, s) = \begin{cases} -z_1(s)z_2(x)/C, & a \leq s \leq x, \\ -z_1(x)z_2(s)/C, & x \leq s \leq b, \end{cases}$

where C is given by equation (11). The function $G(x, s)$ is called the **Green's function** for the problem (2)–(4).

INTEGRAL REPRESENTATION OF SOLUTION

Theorem 11. Let $G(x, s)$ be the Green's function for a regular Sturm-Liouville boundary value problem. Then, the function

(13) $$y(x) = \int_a^b G(x, s)f(s)\,ds$$

is the solution to the regular Sturm-Liouville boundary value problem with $L[y](x) = -f(x)$ for any function f that is continuous on $[a, b]$.

Proof. We used the method of variation of parameters to derive $y(x)$, so we know that it satisfies (2). Hence we need only show that $y(x)$ given in (13) satisfies the boundary conditions (3) and (4). We first compute $y'(x)$ from equation (10) (with $pW[z_1, z_2] = C$):

(14) $$y'(x) = z_1'(x) \int_x^b \frac{-z_2(s)f(s)}{C}\,ds + \frac{z_1(x)z_2(x)f(x)}{C}$$

$$+ z_2'(x) \int_a^x \frac{-z_1(s)f(s)}{C}\,ds - \frac{z_1(x)z_2(x)f(x)}{C}$$

$$= z_1'(x) \int_x^b (-z_2(s)f(s)/C)\,ds + z_2'(x) \int_a^x (-z_1(s)f(s)/C)\,ds.$$

Using equations (10) and (14), we have

$$a_1 y(a) + a_2 y'(a) = a_1 z_1(a) \int_a^b (-z_2(s)f(s)/C)\, ds + a_2 z_1'(a) \int_a^b (-z_2(s)f(s)/C)\, ds$$

$$= [a_1 z_1(a) + a_2 z_1'(a)] \int_a^b (-z_2(s)f(s)/C)\, ds$$

$$= 0,$$

since z_1 satisfies (3). Thus y given in (13) satisfies (3). A similar computation shows that y also satisfies (4). Hence y solves the problem (2)–(4). ■End

We now summarize the above procedure for solving the nonhomogeneous boundary value problem (2)–(4) when the homogeneous problem has only the trivial solution.

SOLUTION USING GREEN'S FUNCTIONS

To obtain the solution to

$$L[y] = -f; \qquad B_1[y] = 0, \qquad B_2[y] = 0,$$

where $L[y] = (py')' + qy$ and f is continuous on $[a, b]$:

(a) Find two nontrivial solutions z_1 and z_2 to $L[y] = 0$ that satisfy $B_1[z_1] = 0$ and $B_2[z_2] = 0$.

(b) Compute $C = p(x)W[z_1, z_2](x)$, where $W[z_1, z_2] = z_1 z_2' - z_1' z_2$.

(c) Construct the Green's function:

$$(15) \qquad G(x, s) = \begin{cases} -z_1(s)z_2(x)/C, & a \le s \le x, \\ -z_1(x)z_2(s)/C, & x \le s \le b. \end{cases}$$

The desired solution is given by

$$(16) \qquad y(x) = \int_a^b G(x, s)f(s)\, ds.$$

EXAMPLE 1 Find the Green's function $G(x, s)$ for the boundary value problem

$$(17) \qquad y''(x) = -f(x); \qquad y(0) = 0, \qquad y(\pi) = 0.$$

Use the Green's function to obtain the solution when $f(x) = x$.

Solution A general solution to the homogeneous problem $y'' = 0$ is $y_h(x) = Ax + B$, so $z_1(x)$ and $z_2(x)$ must be of this form. To get $z_1(x)$ we want to choose A and B so that $z_1(0) = B = 0$. Since A is arbitrary, we can set it equal to 1. Hence we take $z_1(x) = x$. To get $z_2(x)$, we want to choose A and B so that $z_2(\pi) = A\pi + B = 0$. Thus $B = -A\pi$. Taking $A = -1$, we get $z_2(x) = \pi - x$.

We now compute

$$C = p(x)W[z_1, z_2](x) = (1)[(x)(-1) - (1)(\pi - x)] = -\pi.$$

(This step also serves as a check on our work, since all appearances of the variable x must cancel to obtain a constant.) Thus the Green's function is

$$
\textbf{(18)} \qquad G(x,\, s) = \begin{cases} -z_1(s)z_2(x)/C, & 0 \le s \le x, \\ -z_1(x)z_2(s)/C, & x \le s \le \pi, \end{cases}
$$

$$
= \begin{cases} s(\pi - x)/\pi, & 0 \le s \le x, \\ x(\pi - s)/\pi, & x \le s \le \pi. \end{cases}
$$

When $f(x) = x$, the solution to (17) is given by equation (16). Substituting for $f(x)$ and $G(x,\, s)$, we find

$$
y(x) = \int_a^b G(x,\, s)f(s)\, ds
$$

$$
= \int_0^x [s(\pi - x)s/\pi]\, ds + \int_x^\pi [x(\pi - s)s/\pi]\, ds
$$

$$
= \frac{(\pi - x)}{\pi}\left[\frac{s^3}{3}\right]\Big|_0^x + \frac{x}{\pi}\left[\frac{\pi s^2}{2} - \frac{s^3}{3}\right]\Big|_x^\pi
$$

$$
= x(\pi^2 - x^2)/6. \quad \blacksquare \text{ End}
$$

In the next theorem we give some basic properties of Green's functions.

PROPERTIES OF GREEN'S FUNCTIONS

Theorem 12. Let $G(x,\, s)$ be the Green's function defined by equation (12) for the regular Sturm-Liouville boundary value problem (2)–(4). Then

(a) $G(x,\, s)$ is continuous on the square $[a,\, b] \times [a,\, b]$. For each fixed s, the partial derivatives $(\partial G/\partial x)(x,\, s)$ and $(\partial^2 G/\partial x^2)(x,\, s)$ are continuous functions of x for $x \ne s$.

(b) At $x = s$, the partial derivative $\partial G/\partial x$ has a jump discontinuity:

$$
\textbf{(19)} \qquad \lim_{x \to s^+} \frac{\partial G}{\partial x}(x,\, s) - \lim_{x \to s^-} \frac{\partial G}{\partial x}(x,\, s) = \frac{-1}{p(s)}.
$$

(c) For each fixed s, the function $G(x,\, s)$ satisfies the corresponding homogeneous problem for $x \ne s$; that is,

$$
\textbf{(20)} \qquad L[G(\cdot,\, s)] = 0, \qquad \text{for } x \ne s.
$$

$$
\textbf{(21)} \qquad a_1 G(a,\, s) + a_2 \frac{\partial G}{\partial x}(a,\, s) = 0, \qquad b_1 G(b,\, s) + b_2 \frac{\partial G}{\partial x}(b,\, s) = 0.
$$

(d) There is only one function satisfying properties (a)–(c).

(e) $G(x,\, s)$ is symmetric; that is,

$$
G(x,\, s) = G(s,\, x).
$$

Proof

(a) Since z_1 and z_2 are both solutions to $L[y] = 0$ on $[a, b]$, they are continuous and have continuous first and second derivatives. Hence it follows from the formula for $G(x, s)$ that $G(x, s)$ is continuous on the square $[a, b] \times [a, b]$. Furthermore, for fixed s, it follows that $\partial G/\partial x$ and $\partial^2 G/\partial x^2$ are continuous functions of x for $x \neq s$.

(b) Using the fact that $C = p(s)W[z_1, z_2](s)$, we find

$$\lim_{x \to s^+} \frac{\partial G}{\partial x}(x, s) - \lim_{x \to s^-} \frac{\partial G}{\partial x}(x, s) = \frac{-z_1(s)z_2'(s)}{C} - \frac{-z_1'(s)z_2(s)}{C}$$

$$= -\frac{z_1(s)z_2'(s) - z_1'(s)z_2(s)}{C}$$

$$= -\frac{W[z_1, z_2](s)}{C} = \frac{-1}{p(s)}.$$

(c) For fixed s, the function $G(x, s)$ is a constant multiple of $z_1(x)$ when $x < s$ and a constant multiple of $z_2(x)$ when $s < x$. Since both z_1 and z_2 satisfy $L[y] = 0$, then so does $G(\cdot, s)$ for $x \neq s$. Next we verify (21). When $x < s$, we have $G(x, s) = (-z_2(s)/C)z_1(x)$. Since $z_1(x)$ satisfies (3), we have

$$a_1 G(a, s) + a_2 \frac{\partial G}{\partial x}(a, s) = (-z_2(s)/C)[a_1 z_1(a) + a_2 z_1'(a)] = 0.$$

Furthermore, when $s < x$, we have $G(x, s) = (-z_1(s)/C)z_2(x)$. Since $z_2(x)$ satisfies (4), we find

$$b_1 G(b, s) + b_2 \frac{\partial G}{\partial x}(b, s) = (-z_1(s)/C)[b_1 z_2(b) + b_2 z_2'(b)] = 0.$$

Thus, for each fixed s, $G(\cdot, s)$ satisfies (20)–(21).

(d) Assume that both $G(x, s)$ and $H(x, s)$ satisfy properties (a), (b), and (c). We shall show that

$$K(x, s) := G(x, s) - H(x, s)$$

is identically zero on $[a, b] \times [a, b]$. For this purpose, fix $s = s_0$. By property (a), $K(x, s_0)$ is continuous on $[a, b]$, and $(\partial K/\partial x)(x, s_0)$ and $(\partial^2 K/\partial x^2)(x, s_0)$ are continuous for $x \neq s_0$. Moreover, since property (b) says that $(\partial G/\partial x)(x, s_0)$ and $(\partial H/\partial x)(x, s_0)$ have equal jumps at $x = s_0$, we see that $(\partial K/\partial x)(x, s_0) = (\partial G/\partial x)(x, s_0) - (\partial H/\partial x)(x, s_0)$ exists and is continuous (has no jump) at $x = s_0$. Hence $K(x, s_0)$ and $(\partial K/\partial x)(x, s_0)$ are continuous on $[a, b]$. From property (c), equation (20), it follows that for $x \neq s_0$, $L[K(\cdot, s_0)] = 0$. Solving this equation for $(\partial^2 K/\partial x^2)(x, s_0)$, we find that $(\partial^2 K/\partial x^2)(x, s_0)$ is equal to a function that is continuous at $x = s_0$. Consequently $(\partial^2 K/\partial x^2)(x, s_0)$ is continuous on $[a, b]$. Thus $K(x, s_0)$ is a solution to the homogeneous boundary value problem on the whole interval $[a, b]$. But this problem has *only* the trivial solution, so

$K(x, s_0) \equiv 0$. Finally, since s_0 is arbitrary, $K(x, s) \equiv 0$ on $[a, b] \times [a, b]$, that is, the Green's function is unique.

(e) The symmetry is obvious from the definition of $G(x, s)$ in (12); simply interchange x and s in the formulas. ■ End

Green's functions satisfying properties (a)–(d) also exist for general nonselfadjoint boundary value problems. Furthermore, it can be shown that the existence of a *symmetric* Green's function is equivalent to the linear operator L being selfadjoint.[†]

Because of the uniqueness of the Green's function, we could have defined $G(x, s)$ to be the function that satisfies properties (a)–(c) in Theorem 12 and then used these properties to derive formula (12). We leave this derivation as an exercise (see Problem 32).

The Green's function can also be characterized using the Dirac delta function (see Section 7.8). If we assume that there exists a function $G(x, s)$ such that

$$y(x) = \int_a^b G(x, s) f(s) \, ds$$

is the solution to (2)–(4), then operating on both sides with L, and assuming we can interchange differentiation and integration, gives

$$L[y](x) = \int_a^b L[G(\cdot, s)](s) f(s) \, ds = -f(x).$$

That is, $L[G(\cdot, s)](x)$ must act like the Dirac delta $-\delta(x - s)$. Moreover, $G(\cdot, s)$ must satisfy the boundary conditions (3) and (4). If we define $G(x, s)$ as the solution to

(22) $L[G(\cdot, s)](x) = -\delta(x - s),$

(23) $a_1 G(a, s) + a_2 \dfrac{\partial G}{\partial x}(a, s) = 0,$

(24) $b_1 G(b, s) + b_2 \dfrac{\partial G}{\partial x}(b, s) = 0,$

then, using the properties of the Dirac delta function, one can formally show that $G(x, s)$ satisfies properties (a)–(c) in Theorem 12 (see Problem 33).

The characterization of the Green's function given in (22)–(24) supports the following physical interpretation. We view $G(x, s)$ as the deflection of a string (beam, etc.) at a point x as a result of the action of a point force of unit intensity at some other point s. That is, $G(x, s)$ represents the *influence* of the force at s on the string at the point x. Hence the name, **influence function.** Furthermore, $G(x, s)f(s)$ is the influence of a point force of intensity $f(s)$. Taking a superposition of *all* the point forces for s between a and b, we obtain

$$y(x) = \int_a^b G(x, s) f(s) \, ds$$

as the resultant deflection of a string under an external force of $-f(x)$.

[†] See E. Coddington and N. Levinson, *Ibid.*

EXERCISES 11.6

In Problems 1 through 10, find the Green's function $G(x, s)$ for the given boundary value problem.

1. $y'' = -f(x)$; $y(0) = 0$, $y'(\pi) = 0$.

2. $y'' - y = -f(x)$; $y(0) = 0$, $y(1) = 0$.

3. $y'' = -f(x)$; $y(0) = 0$, $y(\pi) + y'(\pi) = 0$.

4. $y'' + y = -f(x)$; $y'(0) = 0$, $y(\pi) = 0$.

5. $y'' + 4y = -f(x)$; $y(0) = 0$, $y'(\pi) = 0$.

6. $y'' + y = -f(x)$; $y(0) + y'(0) = 0$, $y(\pi) = 0$.

7. $(x^2 y')' - 2y = -f(x)$; $y(1) = 0$, $y'(2) = 0$.

8. $(xy')' - (4/x)y = -f(x)$; $y'(1) = 0$, $y(2) = 0$.

9. $(e^{2x} y')' + e^{2x} y = -f(x)$; $y(0) = 0$, $y(1) = 0$.

10. $(e^{-4x} y')' - 5e^{-4x} y = -f(x)$; $y(0) = 0$, $y'(1) = 0$.

In Problems 11 through 20, use Green's functions to solve the given boundary value problem.

11. $y'' = -e^x$; $y(0) = 0$, $y(\pi) = 0$.

12. $y'' = x^2$; $y(0) = 0$, $y(\pi) = 0$.

13. $y'' = -x$; $y(0) = 0$, $y(\pi) + y'(\pi) = 0$.

14. $y'' = -x^2$; $y(0) = 0$, $y'(\pi) = 0$.

15. $y'' + y = 1$; $y(0) = 0$, $y(1) = 0$.

16. $y'' + y = -12$; $y(0) = 0$, $y(2) = 0$.

17. $y'' - y = x$; $y(0) = 0$, $y(1) = 0$.

18. $y'' - y = 24$; $y'(0) = 0$, $y'(1) = 0$.

19. $(xy')' - (4/x)y = -x$; $y(1) = 0$, $y(2) = 0$.

20. $(x^2 y')' = 1$; $y(1) = 0$, $y(2) = 0$.

21. Derive a formula involving a Green's function for the solution to the following nonhomogeneous equation with nonhomogeneous boundary conditions

(25) $y''(x) = -f(x)$; $y(0) = \alpha$, $y(1) = \beta$.

[*Hint:* Let $y = u + v$, where $u'' = 0$ and u satisfies the nonhomogeneous boundary conditions in (25), while $v'' = -f$ and v satisfies the corresponding homogeneous boundary conditions where $\alpha = \beta = 0$.]

22. Nonselfadjoint Equations. Assume the homogeneous problem associated with the boundary value problem

$$A_2(x)y''(x) + A_1(x)y'(x) + A_0(x)y(x) = -f(x),$$
$$a < x < b,$$

$$a_1 y(a) + a_2 y'(a) = 0 \quad \text{and} \quad b_1 y(b) + b_2 y'(b) = 0$$

has only the trivial solution. Using the method of variation of parameters and proceeding in a manner similar to that used in this section, show that the solution to this problem is given by the formula

$$y(x) = \int_a^b K(x, s)f(s)\,ds,$$

where

$$K(x, s) = \begin{cases} \dfrac{-z_1(s)z_2(x)}{W[z_1, z_2](s)A_2(s)}, & a \leq s \leq x, \\[3mm] \dfrac{-z_1(x)z_2(s)}{W[z_1, z_2](s)A_2(s)}, & x \leq s \leq b. \end{cases}$$

The function $K(x, s)$ is called the **Green's function** for this nonselfadjoint problem.

In Problems 23 through 26, use the results of Problem 22 to (a) find the Green's function $K(x, s)$ for the given (non-selfadjoint) boundary value problem and (b) solve the given boundary value problem using Green's functions.

23. $y'' + 3y' + 2y = -x$; $y(0) = 0$, $y(1) = 0$.

24. $y'' - 2y' + y = 1$; $y(0) = 0$, $y(1) = 0$.

25. $x^2 y'' - 2xy' + 2y = x$; $y(1) = 0$, $y(2) = 0$.

26. $x^2 y'' + 3xy' - 3y = x$; $y(1) = 0$, $y'(2) = 0$.

In Problems 27 through 30, find the Green's function $H(x, s)$ for the fourth order equation

$$y^{(4)}(x) = -f(x),$$

with given boundary conditions. [*Hint: Solve the boundary value problem for $f(x) = \delta(x - s)$. Observe that the jump condition is*

$$\lim_{x \to s^+} \frac{\partial^3 H(x, s)}{\partial x^3} - \lim_{x \to s^-} \frac{\partial^3 H(x, s)}{\partial x^3} = -1.]$$

27. $y(0) = y'(0) = 0$, $y(\pi) = y'(\pi) = 0$.

28. $y(0) = y''(0) = 0$, $y(\pi) = y''(\pi) = 0$.

29. $y(0) = y'(0) = 0$, $y''(\pi) = y'''(\pi) = 0$.

30. $y(0) = y'(0) = 0$, $y(\pi) = y''(\pi) = 0$.

31. Let y_1 and y_2 be solutions to

$$y'' + \frac{p'}{p}y' + \frac{q}{p}y = 0,$$

on (a, b). Use Abel's identity (equation (17) in Section 4.3) to show that

$$p(x)W[y_1, y_2](x) = C$$

for all x in $[a, b]$, where C is a constant.

32. Define $G(x, s)$ to be the (unique) function that satisfies properties (a)–(c) in Theorem 12. To show that $G(x, s)$ is given by formula (12), proceed as follows:

 (i) Using property (c), show that $G(x, s)$ has the form

 $$G(x, s) = \begin{cases} u(s)z_1(x), & a \le x < s, \\ v(s)z_2(x), & s < x \le b, \end{cases}$$

 where $L[z_1] = 0$ and z_1 satisfies the boundary conditions in (21) at $x = a$, while $L[z_2] = 0$ and z_2 satisfies the boundary conditions in (21) at $x = b$.

 (ii) Imposing conditions (a) and (b) on $G(x, s)$, show that u and v must satisfy the system

 $$v(s)z_2(s) - u(s)z_1(s) = 0,$$

 $$v(s)z_2'(s) - u(s)z_1'(s) = -\frac{1}{p(s)}.$$

 (iii) Solving this system for u and v, derive formula (12) for $G(x, s)$.

33. Define $G(x, s)$ to be the solution to equations (22)–(24). To formally show that $G(x, s)$ has properties (a)–(c) of Theorem 12, proceed as follows:

 (i) Use the fact that $\delta(x - s) = 0$ for $x \ne s$ to show that $G(x, s)$ satisfies property (c). Moreover, for each fixed s, the partial derivatives $(\partial G/\partial x)(x, s)$ and $(\partial^2 G/\partial x^2)(x, s)$ are continuous functions of x for $x \ne s$.

 (ii) To show that $G(x, s)$ is continuous on $[a, b] \times [a, b]$ and complete the proof of property (a), use the fact that

 $$y(x) = \int_a^b G(x, s)f(s)\,ds,$$

the solution to (2)–(4), is continuous and hence, for any f,

$$0 = \lim_{\varepsilon \to 0^+} [y(s + \varepsilon) - y(s - \varepsilon)]$$

$$= \int_a^b \lim_{\varepsilon \to 0^+} [G(s + \varepsilon, s) - G(s - \varepsilon, s)]f(s)\,ds.$$

 (iii) Fix s and let $x_1 < s < x_2$. Formally integrate both sides of (22) with respect to x from $x = x_1$ to $x = x_2$, to obtain

 $$p(x)\frac{\partial G}{\partial x}(x, s)\Big|_{x_1}^{x_2} + \int_{x_1}^{x_2} q(x)G(x, s)\,dx$$

 $$= -\int_{x_1}^{x_2} \delta(x - s)\,dx.$$

 From this equation, use the continuity of p, q, and G and the property of the Dirac delta function that

 $$\int_{-\infty}^{\infty} \delta(x - s)\,dx = 1$$

 to derive property (b).

34. Let $G(x, s)$ be the Green's function for the regular Sturm-Liouville boundary value problem

 $$L[y] + \mu y = -f; \qquad B[y] = 0.$$

 Let $f(x) = \delta(x - s)$. Using the method of eigenfunction expansions discussed in Section 11.5, derive the eigenfunction formula for $G(x, s)$,

 $$G(x, s) = \sum_{n=1}^{\infty} \frac{\phi_n(x)\phi_n(s)}{\lambda_n - \mu},$$

 where $\{\phi_n\}_{n=1}^{\infty}$ is an orthonormal system of eigenfunctions with corresponding eigenvalues $\{\lambda_n\}_{n=1}^{\infty}$ for

 $$L[y] + \lambda y = 0; \qquad B[y] = 0.$$

 Assume μ is *not* an eigenvalue.

35. Show that property (b) in Theorem 12 is equivalent to

 $$\lim_{s \to x^+} \frac{\partial G}{\partial x}(x, s) - \lim_{s \to x^-} \frac{\partial G}{\partial x}(x, s) = \frac{1}{p(x)}.$$

11.7 SINGULAR STURM-LIOUVILLE BOUNDARY VALUE PROBLEMS

In applications to heat flow on a cylinder, one encounters Sturm-Liouville boundary value problems that do not satisfy *all* the conditions required for the problem to be regular (see Section 11.3).

Let

(1) $$L[y](x) := \frac{d}{dx}\left[p(x)\frac{dy}{dx}\right] + q(x)y$$

and consider the equation

(2) $$L[y](x) + \lambda r(x)y(x) = 0, \qquad a < x < b.$$

We assume $p(x)$, $p'(x)$, $q(x)$, and $r(x)$ are real-valued continuous functions on (a, b) and $p(x)$ and $r(x)$ are positive on (a, b). Notice that our assumptions apply only on the *open* interval (a, b) and *not* the closed interval $[a, b]$, which is the case for a regular Sturm-Liouville boundary value problem. We will call equation (2) a **singular Sturm-Liouville equation**[†] if one or more of the following situations hold:

 (a) $\lim_{x \to a^+} p(x) = 0$ or $\lim_{x \to b^-} p(x) = 0$.
 (b) $p(x)$, $q(x)$, or $r(x)$ becomes unbounded as x approaches either a or b.
 (c) The interval (a, b) is unbounded (e.g., (a, ∞), $(-\infty, b)$, or $(-\infty, \infty)$).

Three important examples of singular Sturm-Liouville equations are Bessel's equation, Legendre's equation, and Hermite's equation.

Bessel's Equation of Order ν

(3) $$t^2 y''(t) + t y'(t) + (t^2 - \nu^2)y = 0, \qquad 0 < t < \sqrt{\lambda}b$$

is transformed by the substitution $t = \sqrt{\lambda}x$ into the singular Sturm-Liouville equation

(4) $$\frac{d}{dx}\left[x\frac{dy}{dx}\right] - \frac{\nu^2}{x}y + \lambda xy = 0, \qquad 0 < x < b.$$

Here $p(x) = x$, which is zero for $x = 0$. Moreover, for $\nu \neq 0$, the function $q(x) = -\nu^2/x$ becomes unbounded as $x \to 0^+$.

Legendre's Equation

(5) $$(1 - x^2)y'' - 2xy' + n(n + 1)y = 0, \qquad -1 < x < 1$$

can be written as the singular Sturm-Liouville equation

(6) $$\frac{d}{dx}\left[(1 - x^2)\frac{dy}{dx}\right] + \lambda y = 0, \qquad -1 < x < 1,$$

where $\lambda = n(n + 1)$. Here $p(x) = 1 - x^2$ is zero at both endpoints ± 1.

[†] For a more rigorous treatment of general singular Sturm-Liouville problems, see "Orthogonal Polynomials and Singular Sturm-Liouville Systems, I", by Lance L. Littlejohn and Allan M. Krall, *Rocky Mountain Journal of Mathematics*, Vol. 16, No. 3, 1986, 435–479.

Hermite's Equation[†]

(7) $y'' - 2xy' + 2ny = 0, \qquad -\infty < x < \infty,$

when multiplied by e^{-x^2}, becomes the singular Sturm-Liouville equation

(8) $\dfrac{d}{dx}\left[e^{-x^2} \dfrac{dy}{dx} \right] + \lambda e^{-x^2} y = 0, \qquad -\infty < x < \infty,$

where $\lambda = 2n$. Here the interval $(-\infty, \infty)$ is unbounded.

In Section 11.3 we observed that the linear operator associated with a regular Sturm-Liouville boundary value problem was a *selfadjoint* operator. Since many of the properties of regular Sturm-Liouville boundary value problems are consequences of this fact, it is useful to know which boundary conditions make a singular Sturm-Liouville equation into a selfadjoint problem. To investigate this question, we return to Green's formula (see equation (10) in Section 11.3). Since p and q are continuous on (a, b), the only possible change in Green's formula for a singular Sturm-Liouville equation is that the integral may be improper at one or both endpoints. In this case, we have

(9) $\displaystyle\int_a^b (uL[v] - vL[u])(x)\,dx = \lim_{x \to b^-} p(x)W[u,\,v](x) - \lim_{x \to a^+} p(x)W[u,\,v](x).$

Now the operator L will be selfadjoint if the right-hand side of (9) is zero (cf. equation (13) in Section 11.3). This would be the case if each limit, taken separately, were zero. So let's consider just one of the limits on the right-hand side of (9) and ask what boundary conditions on u and v ensure that

(10) $\displaystyle\lim_{x \to a^+} p(x)[u(x)v'(x) - u'(x)v(x)] = 0.$

One type of boundary condition for which (10) holds is the separated boundary condition. This is the type of boundary condition we imposed on a regular Sturm-Liouville boundary value problem. Furthermore, it occurs naturally in such applications as the flow of heat in a wire (see Section 11.1). Another type of boundary condition that we encountered in our study of Laplace's equation is the requirement that both $y(x)$ and $y'(x)$ remain bounded as x approaches the endpoint (see Examples 2 and 3 in Section 10.7). This condition is more general than the separated boundary condition. A third condition arising in applications is the requirement that $\sqrt{p(x)}y(x)$ and $\sqrt{p(x)}y'(x)$ go to zero as x approaches the boundary (see Project A). These three types of boundary conditions, with suitable assumptions on $p(x)$, guarantee that equation (10) holds at the endpoint a, as we now show.

[†] *Historical Footnote:* Named after the French mathematician Charles Hermite (1822–1901), who contributed significantly to algebra and analysis.

> ## SINGULAR BOUNDARY CONDITIONS
>
> **Lemma 1.** Any one of the following three conditions guarantees that equation (10) holds:
>
> **(i)** $\lim_{x \to a^+} p(x) = p(a)$ exists and u, v satisfy the boundary condition
>
> **(11)** $a_1 y(a) + a_2 y'(a) = 0,$ a_1, a_2 not both zero.
>
> **(ii)** $\lim_{x \to a^+} p(x) = 0$ and u, v satisfy the boundary condition
>
> **(12)** $y(x)$, $y'(x)$ remain bounded as $x \to a^+$.
>
> **(iii)** The functions u, v satisfy the boundary condition
>
> **(13)** $\lim\limits_{x \to a^+} \sqrt{p(x)}\, y(x) = 0$ and $\lim\limits_{x \to a^+} \sqrt{p(x)}\, y'(x) = 0.$

Proof

(i) Without loss of generality, assume that $a_1 \neq 0$. Then from (11), we have $u(a) = -(a_2/a_1)u'(a)$ and $v(a) = -(a_2/a_1)v'(a)$. Substituting into the left-hand side of (10) yields

(14) $p(a)[u(a)v'(a) - u'(a)v(a)] = p(a)[-(a_2/a_1)u'(a)v'(a) + (a_2/a_1)u'(a)v'(a)]$

$$= 0.$$

(ii) If u, u', v, and v' remain bounded as x approaches a from the right, then the factor in brackets on the left-hand side of (10) remains bounded. Since $\lim_{x \to a^+} p(x) = 0$, the left-hand side is just a bounded factor times a factor that goes to zero. Hence the left-hand side goes to zero and (10) holds.

(iii) If u and v satisfy (13), then substituting into the left-hand side of (10) gives

$$\lim_{x \to a^+} p(x)[u(x)v'(x) - u'(x)v(x)] = \left(\lim_{x \to a^+} \sqrt{p(x)}\, u(x) \right)\left(\lim_{x \to a^+} \sqrt{p(x)}\, v'(x) \right)$$
$$- \left(\lim_{x \to a^+} \sqrt{p(x)}\, u'(x) \right)\left(\lim_{x \to a^+} \sqrt{p(x)}\, v(x) \right)$$
$$= 0 - 0 = 0.$$

(iii) Hence equation (10) is satisfied. ∎ End

Although Lemma 1 deals with the left endpoint a, an analogous lemma can be stated for the right endpoint b that ensures

(15) $\lim\limits_{x \to b^-} p(x)[u(x)v'(x) - u'(x)v(x)] = 0.$

When both (10) and (15) hold, the operator L will be selfadjoint (recall equation (9)). For example, if condition (i) of Lemma 1 holds for a and the analogue of condition (ii) holds for b, then L is selfadjoint.

Even when boundary conditions imposed on the singular Sturm-Liouville equation give a linear operator that is selfadjoint, other difficulties can arise. For example, the eigenvalues may *not* be discrete (isolated). Indeed, there may be an entire interval I such that if λ is any number in I, then the problem has a nontrivial solution (see Projects B and C). In such a case, the problem is said to have a **continuous spectrum.** It is also possible to have both discrete eigenvalues and a continuous spectrum (see Figure 11.4). When the singular Sturm-Liouville boundary value problem has no continuous spectra but has an infinite number of discrete eigenvalues, the arguments given in Section 11.3 for regular Sturm-Liouville boundary value problems can be modified to show that the eigenvalues must be real and that eigenfunctions corresponding to distinct eigenvalues are orthogonal with respect to $r(x)$ on (a, b) (see Problem 7). Moreover, the formulas for the coefficients of an eigenfunction expansion remain the same. However, the convergence of the eigenfunction expansion must be examined on an individual basis. For an in-depth discussion of singular Sturm-Liouville boundary value problems, we refer the reader to an advanced text on the subject.[†]

Figure 11.4 Set of eigenvalues containing discrete points as well as an interval

To illustrate the behavior of solutions to singular Sturm-Liouville boundary value problems, let's study separately the boundary value problems associated with Bessel's equation, Legendre's equation, and Hermite's equation.

Bessel's Equation

A typical singular Sturm-Liouville boundary value problem associated with Bessel's equation of order ν is

(16) $\quad \dfrac{d}{dx}\left[x\dfrac{dy}{dx}\right] - \dfrac{\nu^2}{x}y + \lambda x y = 0, \qquad 0 < x < 1,$

(17) $\quad y(x),\ y'(x)$ remain bounded as $x \to 0^+$,

(18) $\quad y(1) = 0.$

This problem is selfadjoint, since condition (ii) of Lemma 1 holds at $a = 0$ and the analogue of condition (i) of Lemma 1 holds at $b = 1$.

If $\lambda > 0$, the substitution $t = \sqrt{\lambda}\,x$ transforms (16) into the usual Bessel's equation

(19) $\quad t^2 y'' + t y' + (t^2 - \nu^2)y = 0$

[†] See for example the book by K. Yosida, *Lectures on Differential and Integral Equations* (Dover, New York, 1991).

(the reader should check this!). Using power series methods, we found in Section 8.9 that a general solution to (19) is

$$y(t) = c_1 J_\nu(t) + c_2 Y_\nu(t),$$

or

$$y(x) = c_1 J_\nu(\sqrt{\lambda}x) + c_2 Y_\nu(\sqrt{\lambda}x).$$

It follows from the power series expansion for $J_\nu(\sqrt{\lambda}x)$ that when $\nu = 0$ or $\nu \geq 1$, both $J_\nu(\sqrt{\lambda}x)$ and $\sqrt{\lambda}J_\nu'(\sqrt{\lambda}x)$ remain bounded as $x \to 0^+$ (see Problem 9). However, the function $Y_\nu(\sqrt{\lambda}x)$ does not. Hence we must have $c_2 = 0$ for (17) to be satisfied. (The case when $0 < \nu < 1$ is discussed in Problem 10.)

For (18) to be satisfied for $y(x) = c_1 J_\nu(\sqrt{\lambda}x)$, we need $c_1 = 0$ or

$$J_\nu(\sqrt{\lambda}) = 0.$$

That is, $\sqrt{\lambda}$ must be a zero of J_ν in order for $J_\nu(\sqrt{\lambda}x)$ to be a nontrivial solution to (16)–(18). It is known that the Bessel function J_ν has an increasing sequence of real zeros:

$$0 < \alpha_{\nu 1} < \alpha_{\nu 2} < \alpha_{\nu 3} < \cdots.$$

Hence $\lambda_n = \alpha_{\nu n}^2$, $n = 1, 2, 3, \ldots$ are eigenvalues of (16)–(18) with corresponding eigenfunctions

(20) $$y_n(x) = c_1 J_\nu(\alpha_{\nu n} x).$$

These are the only positive eigenvalues. Moreover, one can show that there are no nonpositive eigenvalues of (16)–(18) (see Problem 11).

The eigenfunctions in (20) are orthogonal with respect to the weight function $r(x) = x$:

(21) $$\int_0^1 J_\nu(\alpha_{\nu n} x) J_\nu(\alpha_{\nu m} x) x \, dx = 0, \qquad n \neq m.$$

If $f(x)$ is a given function, then an eigenfunction expansion associated with f is

(22) $$f \sim \sum_{n=1}^\infty a_n J_\nu(\alpha_{\nu n} x),$$

where

$$a_n = \frac{\displaystyle\int_0^1 f(x) J_\nu(\alpha_{\nu n} x) x \, dx}{\displaystyle\int_0^1 J_\nu^2(\alpha_{\nu n} x) x \, dx}, \qquad n = 1, 2, 3, \ldots.$$

When f and f' are piecewise continuous on $[0, 1]$, the expansion in (22) converges for x in $(0, 1)$ to $[f(x^+) + f(x^-)]/2$.

Legendre's Equation

A typical singular Sturm-Liouville boundary value problem associated with Legendre's equation is

$$(23) \qquad \frac{d}{dx}\left[(1 - x^2)\frac{dy}{dx}\right] + \lambda y = 0, \quad -1 < x < 1,$$

$$(24) \qquad y(x), \; y'(x) \text{ remain bounded as } x \to \pm 1.$$

This problem is selfadjoint, since condition (ii) of Lemma 1 holds at each endpoint. In Section 8.9 we found that (23) has polynomial solutions for $\lambda = n(n + 1)$, $n = 0, 1, 2, \ldots$ These solutions are just constant multiples of the Legendre polynomials

$$(25) \qquad P_n(x) = 2^{-n} \sum_{m=0}^{[n/2]} \frac{(-1)^m(2n - 2m)!}{(n - m)!m!(n - 2m)!} x^{n-2m},$$

where $[n/2]$ is the greatest integer less than or equal to $n/2$ (cf. (41) on page 500, Section 8.9). The first four Legendre polynomials are

$$P_0(x) = 1, \qquad P_1(x) = x, \qquad P_2(x) = \frac{3}{2}x^2 - \frac{1}{2}, \qquad P_3(x) = \frac{5}{2}x^3 - \frac{3}{2}x.$$

It is left to the reader to show that the only eigenvalues for (23)–(24) are $\lambda = n(n + 1)$, $n = 0, 1, 2, \ldots$, with corresponding eigenfunctions $P_n(x)$, the Legendre polynomials (see Problem 12). That the Legendre polynomials are (up to constant multiples) the only eigenfunctions corresponding to these eigenvalues follows from the fact that a second linear independent solution to (23) involves a $\ln(1 - x)$ term that becomes unbounded as $x \to 1^-$ (see Problem 13).

The Legendre polynomials $P_n(x)$ are orthogonal on $[-1, 1]$ with respect to $r(x) = 1$:

$$(26) \qquad \int_{-1}^{1} P_n(x)P_m(x)\,dx = 0, \qquad n \neq m.$$

If $f(x)$ is a given function, then an eigenfunction expansion for $f(x)$ in terms of Legendre polynomials is

$$(27) \qquad f \sim \sum_{n=0}^{\infty} a_n P_n(x),$$

where

$$a_n = \frac{\displaystyle\int_{-1}^{1} f(x)P_n(x)\,dx}{\displaystyle\int_{-1}^{1} P_n^2(x)\,dx}, \qquad n = 0, 1, 2, \ldots.$$

Convergence results similar to those for Fourier series hold for the expansion in (27) (cf. Section 10.3).

Hermite's Equation

A singular Sturm-Liouville boundary value problem associated with Hermite's equation is

(28) $\dfrac{d}{dx}\left[e^{-x^2}\dfrac{dy}{dx}\right] + \lambda e^{-x^2}y = 0,\qquad -\infty < x < \infty,$

(29) $\lim\limits_{x\to\pm\infty} e^{-x^2/2}y(x) = 0\quad\text{and}\quad \lim\limits_{x\to\pm\infty} e^{-x^2/2}y'(x) = 0.$

By condition (iii) of Lemma 1, the linear operator associated with (28)–(29) is selfadjoint.

In Section 8.9 we observed that Hermite's equation

(30) $y'' - 2xy' + 2ny = 0$

(which is equivalent to (28)) has polynomial solutions that are constant multiples of the Hermite polynomials $H_n(x)$, $n = 0, 1, 2, \dots$. Since polynomial solutions satisfy the boundary conditions (29), we find that the singular Sturm-Liouville boundary value problem (28)–(29) has eigenvalues $\lambda_n = 2n$, $n = 0, 1, 2, \dots$, with corresponding eigenfunctions $H_n(x)$. In fact, it can be shown that these are the only eigenvalues. The first four Hermite polynomials are

$$H_0(x) = 1,\qquad H_1(x) = 2x,\qquad H_2(x) = 4x^2 - 2,\qquad H_3(x) = 8x^3 - 12x$$

(see Exercises 8.9, Problem 37).

The Hermite polynomials $H_n(x)$ are orthogonal on $(-\infty, \infty)$ with respect to the weight function $r(x) = e^{-x^2}$:

(31) $\displaystyle\int_{-\infty}^{\infty} H_n(x)H_m(x)e^{-x^2}\,dx = 0,\qquad n \neq m.$

If f is a given function, then an eigenfunction expansion associated with f is given by

(32) $\displaystyle f \sim \sum_{n=0}^{\infty} a_n H_n(x),$

where

$$a_n = \frac{\displaystyle\int_{-\infty}^{\infty} f(x)H_n(x)e^{-x^2}\,dx}{\displaystyle\int_{-\infty}^{\infty} H_n^2(x)e^{-x^2}\,dx},\qquad n = 0, 1, 2, \dots.$$

A detailed study of the eigenvalues and eigenfunctions for Hermite's boundary value problem (28)–(29) along with conditions for the convergence of the eigenfunction expansion in (32) is beyond the scope of this book. We refer the reader to an advanced book on these topics.[†]

[†] See for example the books by F. V. Atkinson, *Discrete and Continuous Boundary Value Problems* (Academic Press, Orlando, Florida, 1964) and E. C. Titchmarsh, *Eigenfunction Expansions Associated with Second-Order Differential Equations*, Vols. I and II (Clarendon, Oxford, 1962).

In Problems 1 through 6, first verify that the corresponding homogeneous problem is selfadjoint. Then, find a formal eigenfunction expansion for the solution to the given nonhomogeneous boundary value problem where f is a continuous function and μ is not an eigenvalue of the corresponding homogeneous problem

1. $(xy')' - \dfrac{4}{x}y + \mu xy = f(x), \qquad 0 < x < 1,$

 $y(x), y'(x)$ remain bounded as $x \to 0^+$, $y(1) = 0.$

2. $(xy')' - \dfrac{9}{x}y + \mu xy = f(x), \qquad 0 < x < 1,$

 $y(x), y'(x)$ remain bounded as $x \to 0^+$, $y(1) = 0.$

3. $(xy')' + \mu xy = f(x), \qquad 0 < x < 1,$

 $y(x), y'(x)$ remain bounded as $x \to 0^+$, $y'(1) = 0.$

4. $[(1 - x^2)y']' + \mu y = f(x), \qquad -1 < x < 1,$

 $y(x), y'(x)$ remain bounded as $x \to \pm 1.$

5. $[(1 - x^2)y']' + \mu y = f(x), \qquad 0 < x < 1,$

 $y(0) = 0$, $y(x), y'(x)$ remain bounded as $x \to 1^-.$

 [*Hint:* Eigenfunctions are odd degree Legendre polynomials.]

6. $[(1 - x^2)y']' + \mu y = f(x), \qquad 0 < x < 1,$

 $y'(0) = 0$, $y(x), y'(x)$ remain bounded as $x \to 1^-.$

 [*Hint:* Eigenfunctions are even degree Legendre polynomials.]

7. Let (2) be a singular Sturm-Liouville equation. Let $B[y] = 0$ represent linear homogeneous boundary conditions such that, together with (1), they give rise to a selfadjoint operator L. Assume this singular Sturm-Liouville boundary value problem has no continuous spectra but has an infinite number of discrete eigenvalues λ_n, $n = 1, 2, 3, \ldots$. Using arguments similar to those in Section 11.3 for regular Sturm-Liouville boundary value problems:

 (a) Show that the eigenvalues must be real.

 (b) Show that eigenfunctions corresponding to distinct eigenvalues are orthogonal with respect to $r(x)$ on (a, b).

8. Verify that the change of variables $t = \sqrt{\lambda}x$ transforms equation (3) into equation (4).

9. The power series expansion for $J_\nu(\sqrt{\lambda}x)$ has the form $(\sqrt{\lambda}x)^\nu g(\sqrt{\lambda}x)$, where g is analytic about $x = 0$. Show that for $\nu = 0$ or $\nu \geq 1$, both $J_\nu(\sqrt{\lambda}x)$ and $J'_\nu(\sqrt{\lambda}x)$ remain bounded as $x \to 0^+$.

10. For Bessel's equation of order ν where $0 < \nu < 1$, we replace the boundary condition (17) by

 (33) $\qquad \lim_{x \to 0^+} y(x) = 0, \qquad \lim_{x \to 0^+} xy'(x) = 0.$

 (a) Show that if u and v satisfy the boundary conditions in (33), then equation (10) holds at $a = 0$ and hence the singular Sturm-Liouville boundary value problem (16), (18), and (33) gives rise to a selfadjoint operator.

 (b) Using the representation for $J_\nu(\sqrt{\lambda}x)$ given in Problem 9, show that $J_\nu(\sqrt{\lambda}x)$ satisfies the boundary conditions in (33).

11. To show that the eigenvalues for Bessel's equation must be positive, proceed as follows:

 (a) Let $\phi(x)$ be an eigenfunction for (16)–(18) with corresponding eigenvalue λ. Substitute $\phi(x)$ into (16). Now multiply both sides of (16) by $\phi(x)$ and integrate from 0 to 1. Using integration by parts, obtain

 (34) $\qquad -\displaystyle\int_0^1 x[\phi'(x)]^2\,dx - \nu^2 \int_0^1 x^{-1}[\phi(x)]^2\,dx$

 $$+ \lambda \int_0^1 x[\phi(x)]^2\,dx = 0.$$

 (b) For $\nu > 0$, use equation (34) to conclude that $\lambda > 0.$

 (c) For $\nu = 0$, use equation (34) to conclude that $\lambda \geq 0$. To prove $\lambda = 0$ is not an eigenvalue for $\nu = 0$, find a general solution to $xy'' + y' = 0$ and show that only the trivial solution satisfies the boundary conditions in (17)–(18).

12. (a) Show that the only nonnegative eigenvalues for (23)–(24) are $\lambda_n = n(n + 1)$, $n = 0, 1, 2, \ldots$. [*Hint:* Find a power series expansion for a general solution to (23).]

 (b) Show that (23)–(24) does not have any negative eigenvalues. [*Hint:* Use a procedure similar to the one outlined in Problem 11.]

13. Show that the only eigenfunction of (23)–(24) corresponding to the eigenvalues $\lambda_n = n(n + 1)$, $n = 0, 1, 2, \ldots$ are multiples of the Legendre polynomials $P_n(x)$. [*Hint:* In Section 8.9, we showed that exponents of the singularity $x_0 = 1$ for Legendre's equation were both zero. Use Theorem 7 in Section 8.7 to show that any

other solution has a logarithmic term that becomes unbounded as $x \to 1^-$.]

14. (a) Use formula (25) to show that $P_n(x)$ is an odd function when n is odd and is an even function when n is even.

(b) Show that the odd degree Legendre polynomials are orthogonal on $(0, 1)$, that is,

$$\int_0^1 P_{2m+1}(x)P_{2n+1}(x)\,dx = 0, \qquad m \neq n.$$

(c) Show that the even degree Legendre polynomials are orthogonal on $(0, 1)$, that is,

$$\int_0^1 P_{2m}(x)P_{2n}(x)\,dx = 0, \qquad m \neq n.$$

15. Chebyshev's (Tchebichef's) Equation

(a) Show that Chebyshev's equation

$$(1 - x^2)y'' - xy' + \lambda y = 0, \qquad -1 < x < 1,$$

can be written in the Sturm-Liouville form

$$(35) \qquad [(1 - x^2)^{1/2}y']' + \lambda(1 - x^2)^{-1/2}y = 0,$$
$$-1 < x < 1.$$

(b) It can be shown that Chebyshev's equation (35) with the boundary conditions

$$y(x), \ y'(x) \text{ remain bounded as } x \to \pm 1,$$

has eigenvalues $\lambda_n = n^2$, $n = 0, 1, 2, \ldots$. The corresponding eigenfunctions are the Chebyshev polynomials $T_n(x)$. The first four are

$$T_0(x) = 1, \qquad T_1(x) = x, \qquad T_2(x) = 2x^2 - 1,$$
$$T_3(x) = 4x^3 - 3x.$$

Show that

$$\int_{-1}^1 \frac{T_m(x)T_n(x)}{(1 - x^2)^{1/2}}\,dx = 0, \qquad m \neq n.$$

(c) Find a formal eigenfunction expansion for the solution to the nonhomogeneous boundary value problem

$$[(1 - x^2)^{1/2}y']' + \mu(1 - x^2)^{-1/2}y = f(x),$$
$$-1 < x < 1,$$

$$y(x), \ y'(x) \text{ remain bounded as } x \to \pm 1,$$

where f is continuous on $[-1, 1]$ and μ is not an eigenvalue of the corresponding homogeneous problem.

16. Laguerre's Equation

(a) Show that Laguerre's equation

$$xy'' + (1 - x)y' + \lambda y = 0, \qquad 0 < x < \infty$$

can be written in the Sturm-Liouville form

$$(36) \qquad (xe^{-x}y')' + \lambda e^{-x}y = 0, \qquad 0 < x < \infty.$$

(b) It can be shown that Laguerre's equation (36) with boundary conditions

$$y(x), \ y'(x) \text{ remain bounded as } x \to 0^+,$$
$$x^{1/2}e^{-x/2}y(x) \text{ and } x^{1/2}e^{-x/2}y'(x) \to 0 \text{ as } x \to \infty$$

has eigenvalues $\lambda_n = n$, $n = 0, 1, 2, \ldots$. The corresponding eigenfunctions are the Laguerre polynomials $L_n(x)$. The first four are

$$L_0(x) = 1, \ L_1(x) = 1 - x, \ L_2(x) = (2 - 4x + x^2)/2,$$
$$L_3(x) = (6 - 18x + 9x^2 - x^3)/6.$$

Show that

$$\int_0^\infty L_m(x)L_n(x)e^{-x}\,dx = 0, \qquad m \neq n.$$

(c) Find a formal eigenfunction expansion for the solution to the nonhomogeneous equation

$$(xe^{-x}y')' + \mu e^{-x}y = f(x), \qquad 0 < x < \infty,$$

satisfying the boundary conditions in part (b), where μ is not an eigenvalue of the corresponding homogeneous problem and f is continuous on $[0, \infty)$ and satisfies

$$\lim_{s \to \infty} f(s)e^{-s/2} = 0.$$

11.8 OSCILLATION AND COMPARISON THEORY

In Chapter 5 we developed a model for a simple spring-mass system involving a second order linear differential equation. We found that when the damping constant is sufficiently small, the mass vibrates back and forth or **oscillates.** In fact, the solution was a product of a damping factor $Ae^{-(b/2m)t}$ and a sine function $\sin(\beta t + \theta)$ (cf. equation (6) in Section 5.2).

Functions that oscillate cannot generally be expressed in such a nice fashion. In fact the concept of oscillation is difficult to capture mathematically. Clearly (nonconstant) periodic functions oscillate, but many other types of functions also exhibit what we would call oscillatory behavior. For our purposes, we will identify oscillatory behavior of $y = f(x)$ with repeated contact with the x-axis. Thus the more frequently the graph of a function intersects the x-axis, the more it oscillates. If a function has an *infinite* number of zeros on an interval $[a, \infty)$, we say that the function is **oscillatory**. Hence studying the oscillatory behavior of a function means we are investigating the number and locations of its zeros.

In this section we investigate the zeros of solutions to the Sturm-Liouville equation

(1) $$\frac{d}{dx}\left(p(x)\frac{dy}{dx} \right) + q(x)y = 0, \qquad a < x < b,$$

where p, p', and q are continuous on $[a, b]$ and p is positive on $[a, b]$.[†] The first question one might ask is: Can a solution have an infinite number of zeros in (a, b)? The following theorem shows that only the trivial solution may have an infinite number of zeros.

FINITE NUMBER OF ZEROS

Theorem 13. Let ϕ be a solution to (1) on (a, b). If ϕ has an infinite number of zeros in *any* closed, bounded interval $[a', b'] \subset (a, b)$, then $\phi(x) \equiv 0$ in (a, b).

Proof. To prove this theorem, we need a result from real analysis called the *Bolzano-Weierstrass* theorem. It states that a bounded infinite set has at least one limit point. That is, there is a sequence of distinct points $\{x_n\}_{n=1}^{\infty}$ in the set and a point x_0 such that $\lim_{n\to\infty} x_n = x_0$. Here we are assuming an infinite number of zeros of ϕ lie in $[a', b']$, a bounded interval. By the Bolzano-Weierstrass theorem, there is a sequence $\{x_n\}_{n=1}^{\infty}$ of (distinct) zeros of ϕ that converges to some point x_0. Since $[a', b']$ is a closed interval, the point x_0 must be in $[a', b']$. We claim

The point x_0 is a zero of ϕ. Since x_n is a zero of ϕ — that is, $\phi(x_n) = 0$, $n = 1$, $2, \ldots$ — it follows from the continuity of ϕ that

$$\phi(x_0) = \lim_{n\to\infty} \phi(x_n) = \lim_{n\to\infty} 0 = 0.$$

The point x_0 is a zero of ϕ'. By assumption, ϕ' exists for x in $[a', b']$, and we have

$$\phi'(x_0) = \lim_{x\to x_0} \frac{\phi(x) - \phi(x_0)}{x - x_0} = \lim_{x\to x_0} \frac{\phi(x)}{x - x_0}.$$

Since this limit must hold for *any* sequence of distinct points tending to x_0, we can choose the sequence $\{x_n\}_{n=1}^{\infty}$ of zeros of ϕ. Thus

$$\phi'(x_0) = \lim_{n\to\infty} \frac{\phi(x_n)}{x_n - x_0} = 0.$$

[†] For an unbounded interval such as $(0, \infty)$, we assume the conditions on p and q hold on the open interval $(0, \infty)$.

Finally, since $\phi(x_0) = 0$ and $\phi'(x_0) = 0$, it follows from the existence and uniqueness theorem for second order linear equations that $\phi(x) \equiv 0$ on (a, b) (see Theorem 2 in Section 4.2). ■ End

Remark. Theorem 13 shows that nontrivial oscillatory solutions (solutions with infinitely many zeros) can occur only on an unbounded interval if p is everywhere positive.

Another question one might ask is: Can two solutions to equation (1) have a common zero on $[a, b]$? The answer is yes, but if they do, one function must be a constant times the other.

COMMON ZEROS

Theorem 14. If ϕ and ψ are two solutions to (1) and $\phi(x_0) = \psi(x_0) = 0$ for some x_0 in $[a, b]$, then ϕ and ψ are linearly dependent on $[a, b]$.

Proof. Computing the Wronskian of ϕ and ψ and using the fact that $\phi(x_0) = \psi(x_0) = 0$, we have

$$(2) \qquad W[\phi, \psi](x_0) = \phi(x_0)\psi'(x_0) - \phi'(x_0)\psi(x_0) = 0.$$

But when the Wronskian of two solutions vanishes at some point in $[a, b]$, these solutions must be linearly dependent on $[a, b]$ according to Corollary 1 in Section 4.3. Thus ϕ and ψ are linearly dependent. ■ End

An immediate consequence of Theorem 14 is that if ϕ and ψ are two linearly independent solutions, then they must have *different zeros*. But, is there any relationship between their respective zeros? One answer is given by the Sturm separation theorem.

STURM SEPARATION THEOREM

Theorem 15. Let ϕ and ψ be two linearly independent solutions to (1). Then, between any two consecutive zeros of ϕ in $[a, b]$, there is precisely one zero of ψ (see Figure 11.5).

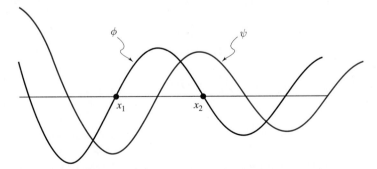

Figure 11.5 Interlacing of zeros of linearly independent solutions

Proof. Let x_1 and x_2 be two consecutive zeros of ϕ in $[a, b]$. Since ϕ and ψ are linearly independent, it follows from the preceding theorem that $\psi(x_1) \neq 0$ and $\psi(x_2) \neq 0$.

Assume the conclusion is false, that is, $\psi(x) \neq 0$ for x in $[x_1, x_2]$. Then the function $f(x) := \phi(x)/\psi(x)$ is well defined and continuous on $[x_1, x_2]$ and continuously differentiable on (x_1, x_2). Since $\phi(x_1) = \phi(x_2) = 0$, we have $f(x_1) = f(x_2) = 0$. It now follows from Rolle's theorem that there exists a number z in (x_1, x_2) such that $f'(z) = 0$.

Computing $f'(z)$, we find

$$(3) \qquad 0 = f'(z) = \frac{\phi'(z)\psi(z) - \phi(z)\psi'(z)}{[\psi(z)]^2} = \frac{W[\psi, \phi](z)}{[\psi(z)]^2}.$$

But this means that the Wronskian of ϕ and ψ vanishes at z, which contradicts the fact that ϕ and ψ are linearly independent. Thus $\psi(x)$ must have a zero in (x_1, x_2).

By interchanging the roles of ϕ and ψ, we also see that between any two consecutive zeros of ψ is a zero of ϕ. Hence there cannot be more than one zero of ψ between two consecutive zeros of ϕ. ∎ End

One consequence of this theorem is the fact that between two consecutive zeros of a Bessel function of the first kind J_ν, there is a zero of the Bessel function of the second kind Y_ν (and vice versa). This is because these two functions are linearly independent solutions to the Sturm-Liouville equation

$$(4) \qquad \frac{d}{dx}\left(x\frac{dy}{dx}\right) + \left(x - \frac{\nu^2}{x}\right)y = 0, \qquad x > 0.$$

Since J_ν has infinitely many zeros on $(0, \infty)$, we conclude from this interlacing property of the zeros that Y_ν also has infinitely many zeros. This illustrates a general phenomenon.

OSCILLATORY SOLUTIONS

Corollary 3. Let the interval in Theorem 15 be the interval $[a, \infty)$. If one (nontrivial) solution to (1) is oscillatory, then all solutions to (1) are oscillatory.

Proof. Let ϕ be a nontrivial solution to (1) that has an infinite number of zeros on $[a, \infty)$. By the Sturm separation theorem, any other solution must have a zero between each successive pair of zeros of ϕ and hence must also have an infinite number of zeros on $[a, \infty)$. ∎ End

When we apply Theorem 15 to Hermite's equation in Sturm-Liouville form

$$(5) \qquad \frac{d}{dx}\left(e^{-x^2}\frac{dy}{dx}\right) + 2ne^{-x^2}y = 0, \qquad -\infty < x < \infty,$$

where n is a nonnegative integer, we find that since one solution is the Hermite polynomial $H_n(x)$ of degree n which has n real zeros, then the other solutions must have one zero between consecutive zeros of $H_n(x)$ for a minimum of $n - 1$ zeros. Moreover, any other (nontrivial) solution cannot have more than $n + 1$ zeros because if the contrary were true, it would imply that $H_n(x)$ has more zeros than its degree permits.

Thus far we have compared the zeros of two solutions to the *same* equation. For solutions of two different equations, it still may be possible to relate their zeros. For example, for the equations

$$y'' + m^2 y = 0 \quad \text{and} \quad y'' + n^2 y = 0,$$

we see that the first has a general solution of the form $\phi_1(x) = A_1 \sin(m(x - \theta_1))$ and the second, $\phi_2(x) = A_2 \sin(n(x - \theta_2))$. The distance between the consecutive zeros of the first is π/m and the second is π/n. Therefore, when $m < n$, the distance between the zeros of ϕ_1 is greater than the distance between the zeros of ϕ_2. In other words, for $n^2 > m^2$, between any two consecutive zeros of $\phi_1(x)$ there is a zero of $\phi_2(x)$. A similar result holds when the constants m^2 and n^2 are replaced by functions of x. The next result is a version of the Sturm fundamental theorem.

STURM FUNDAMENTAL THEOREM

Theorem 16. Let ϕ_1 be a nontrivial solution to

$$(6) \qquad y'' + q_1(x)y = 0, \qquad a < x < b,$$

and let ϕ_2 be a nontrivial solution to

$$(7) \qquad y'' + q_2(x)y = 0, \qquad a < x < b.$$

Assume $q_2(x) \geq q_1(x)$ for x in (a, b). If x_1 and x_2 are two consecutive zeros of $\phi_1(x)$ on (a, b), then there exists a zero of ϕ_2 between x_1 and x_2 unless $q_2(x) = q_1(x)$ on $[x_1, x_2]$ and ϕ_1 and ϕ_2 are linearly dependent on $[x_1, x_2]$.

Notice that the conclusion tells us that either ϕ_2 has a zero between x_1 and x_2 or $\phi_2(x_1) = \phi_2(x_2) = 0$. See Figure 11.6.

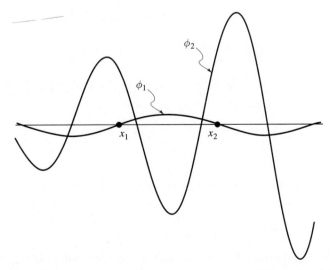

Figure 11.6 Sturm fundamental theorem: zeros of ϕ_2 separate zeros of ϕ_1

Proof of Theorem 16. Let x_1 and x_2 be two successive zeros of ϕ_1. Assume that $\phi_2(x) \neq 0$ in (x_1, x_2). Our goal is to show that $q_2 = q_1$ and that ϕ_1, ϕ_2 are linearly dependent on $[x_1, x_2]$. Without loss of generality, we further assume that $\phi_1(x) > 0$ and $\phi_2(x) > 0$ in (x_1, x_2). Combining this with the fact that $q_2(x) \geq q_1(x)$, we find

$$(8) \qquad \frac{d}{dx}\left(W[\phi_2, \phi_1]\right) = \frac{d}{dx}\left(\phi_2\phi_1' - \phi_2'\phi_1\right) = \phi_2'\phi_1' + \phi_2\phi_1'' - \phi_2'\phi_1' - \phi_1\phi_2''$$

$$= \phi_2\phi_1'' - \phi_2''\phi_1 = \phi_2(-q_1\phi_1) - (-q_2\phi_2)\phi_1$$

$$= \phi_1\phi_2(q_2 - q_1) \geq 0$$

for x in (x_1, x_2). Hence $W[\phi_2, \phi_1](x)$ is nondecreasing on (x_1, x_2). However, since $\phi_1(x_1) = \phi_1(x_2) = 0$ and ϕ_1 is positive on (x_1, x_2), we must have $\phi_1'(x_1) \geq 0$ and $\phi_1'(x_2) \leq 0$. Therefore, since $\phi_2(x) \geq 0$,

$$W[\phi_2, \phi_1](x_1) = \phi_2(x_1)\phi_1'(x_1) \geq 0$$

and

$$W[\phi_2, \phi_1](x_2) = \phi_2(x_2)\phi_1'(x_2) \leq 0.$$

Because $W[\phi_2, \phi_1]$ is nondecreasing, the only way for it to be nonnegative at x_1 and nonpositive at x_2 is for

$$(9) \qquad W[\phi_2, \phi_1](x) \equiv 0, \qquad x \text{ in } [x_1, x_2].$$

Clearly equation (9) implies that $\dfrac{d}{dx}W[\phi_2, \phi_1](x) = 0$ on $[x_1, x_2]$. From equation (8), we deduce that $q_2 = q_1$ on $[x_1, x_2]$. But then ϕ_1, ϕ_2 satisfy the same differential equation on $[x_1, x_2]$ and their Wronskian vanishes on this interval. Therefore ϕ_1 and ϕ_2 are linearly dependent on $[x_1, x_2]$. ∎ End

One immediate consequence of the Sturm fundamental theorem is the following.

NONOSCILLATING SOLUTIONS

Corollary 4. If $q(x) \leq 0$ on $[a, b]$, then a nontrivial solution ϕ of the equation

$$(10) \qquad y'' + q(x)y = 0, \qquad a < x < b,$$

can have at most one zero in $[a, b]$.

Proof. Since $\psi(x) \equiv 1$ is a solution to the equation $y'' + 0 \cdot y = 0$ and $q(x) \leq 0$, it follows from the Sturm fundamental theorem (with $q_1 = q$, $q_2 \equiv 0$) that if ϕ has two or more zeros in $[a, b]$, then ψ must have a zero between them. Since ψ is never zero, ϕ can have at most one zero in $[a, b]$. ∎ End

The Sturm fundamental theorem tells us that the oscillatory behavior of a solution to equation (10) is determined by the sign and magnitude of $q(x)$. For example, when $q(x) \leq 0$, a solution can have at most one zero or, when $q(x) > m^2$, a solution must have a zero between

any two zeros of a solution to $y'' + m^2 y = 0$. In the latter case, the zeros of the solutions to (10) must lie within a distance π/m of each other because $y'' + m^2 y = 0$ has a solution with zeros that are π/m distance apart (see Problem 12).

In the next example, we consider the distance between consecutive zeros of the solutions to Bessel's equation.

EXAMPLE 1 For $a > 0$, discuss the number of zeros in the interval $[a, a + \pi)$ of the solutions to Bessel's equation of order ν:

(11) $x^2 y'' + xy' + (x^2 - \nu^2)y = 0, \qquad x > 0.$

Solution The substitution $y = ux^{-1/2}$ transforms equation (11) into

(12) $u'' + \left[1 - \dfrac{4\nu^2 - 1}{4x^2}\right] u = 0, \qquad x > 0,$

(see Problem 13). We will now compare the solutions of (12) with those of $u'' + u = 0$. Observe that $\phi(x) = A \sin(x - a)$ is a solution to $u'' + u = 0$ and has zeros at a and $a + \pi$.

Case 1. $\nu > 1/2$. In this case, $4\nu^2 - 1 > 0$, so

$$1 - \frac{4\nu^2 - 1}{4x^2} < 1$$

for x in $[a, a + \pi)$. Thus, by the Sturm fundamental theorem, a solution to (12) cannot have more than one zero in $[a, a + \pi)$ because ϕ does not have a zero in $(a, a + \pi)$. Since $y = ux^{-1/2}$, we have shown that when $\nu > 1/2$ any solution to Bessel's solution (11) can have *at most one* zero in $[a, a + \pi)$.

Case 2. $0 \le \nu < 1/2$. In this case, $4\nu^2 - 1 < 0$, so

$$1 - \frac{4\nu^2 - 1}{4x^2} > 1$$

for x in $[a, a + \pi)$. Again, by the Sturm fundamental theorem, a solution to (12) must have a zero in $(a, a + \pi)$, since a and $a + \pi$ are consecutive zeros of $\phi(x) = A \sin(x - a)$. Because $y = ux^{-1/2}$, we have shown that when $0 \le \nu < 1/2$, every solution to Bessel's equation (11) *must have a zero* in $[a, a + \pi)$.

Case 3. $\nu = 1/2$. In this case, (12) reduces to $u'' + u = 0$, which has the general solution $u(x) = c_1 \cos x + c_2 \sin x$. Thus

$$y(x) = c_1 x^{-1/2} \cos x + c_2 x^{-1/2} \sin x = Ax^{-1/2} \sin(x - \theta)$$

for a suitable constant θ. Consequently any solution to Bessel's equation (11) with $\nu = 1/2$ has *exactly one zero* in $[a, a + \pi)$.

Summarizing, we have shown that a solution to Bessel's equation (11) has

(a) at most one zero in $[a, a + \pi)$ when $\nu > 1/2$;
(b) at least one zero in $[a, a + \pi)$ when $0 \le \nu < 1/2$; and
(c) exactly one zero in $[a, a + \pi)$ when $\nu = 1/2$. ∎ End

The Sturm fundamental theorem was extended to more general Sturm-Liouville equations by M. Picone.

PICONE COMPARISON THEOREM

Theorem 17. Let $\phi_1(x)$ be a nontrivial solution to the Sturm-Liouville equation

$$(13) \qquad \frac{d}{dx}\left(p_1 \frac{dy}{dx}\right) + q_1 y = 0, \qquad a < x < b,$$

and let $\phi_2(x)$ be a nontrivial solution to

$$(14) \qquad \frac{d}{dx}\left(p_2 \frac{dy}{dx}\right) + q_2 y = 0, \qquad a < x < b.$$

Assume $p_1(x) \geq p_2(x) > 0$ and $q_1(x) \leq q_2(x)$ for x in $[a, b]$. Then, between any two consecutive zeros x_1 and x_2 of ϕ_1 in $[a, b]$, there is a zero of ϕ_2, unless ϕ_1 and ϕ_2 are linearly dependent on $[x_1, x_2]$, in which case $q_1(x) \equiv q_2(x)$ on $[x_1, x_2]$.

A proof of the Picone comparison theorem using the **Picone identity** is outlined in Project C. It can also be proved using the Prüfer substitution.[†]

If we apply the Picone comparison theorem to a Sturm-Liouville equation, we can obtain estimates on the distance between consecutive zeros of its solutions. In the next corollary, let f_M and f_m denote, respectively, the maximum and minimum values of f over the interval $[a, b]$.

DISTANCE BETWEEN ZEROS

Corollary 5. Let

$$(15) \qquad \frac{d}{dx}\left(p \frac{dy}{dx}\right) + qy + \lambda r y = 0, \qquad a < x < b,$$

be a Sturm-Liouville equation, where p, p', q, and r are continuous on $[a, b]$ and p and r are positive on $[a, b]$. Let ϕ be a nontrivial solution to (15) with consecutive zeros x_1 and x_2 in $[a, b]$. If

$$\lambda > \max\left\{\frac{-q_M}{r_M}, \frac{-q_m}{r_m}, 0\right\},$$

then $x_2 - x_1$, the distance between the zeros of ϕ, is bounded by

$$(16) \qquad \pi \sqrt{\frac{p_m}{q_M + \lambda r_M}} \leq x_2 - x_1 \leq \pi \sqrt{\frac{p_M}{q_m + \lambda r_m}}.$$

[†] For a discussion of the Prüfer substitution and a proof of the Picone comparison theorem, also referred to as the Sturm comparison theorem, see the text by G. Birkhoff and G. C. Rota, *Ordinary Differential Equations*, Third Edition (John Wiley and Sons, New York, 1978), Chapter 10.

Proof. The constant coefficient equation

(17) $p_M y'' + (q_m + \lambda r_m)y = 0$

has the solution $\psi(x) = \sin[K(x - x_1)]$, where $K^2 = (q_m + \lambda r_m)/p_M$ (here we have used the hypothesis $\lambda > -q_m/r_m$ so that $q_m + \lambda r_m > 0$). Moreover, $\psi(x)$ has a zero at $x = x_1$ and its next zero is at $x_1 + (\pi/K)$, that is, the distance to its next zero is

$$\pi\sqrt{\frac{p_M}{q_m + \lambda r_m}} .$$

Since $p_M \geq p(x)$, $\lambda > 0$, and $q_m + \lambda r_m \leq q(x) + \lambda r(x)$ for all x in (a, b), it follows from the Picone comparison theorem that ϕ must have a zero between any two successive zeros of $\psi(x)$ (or $\phi(x_1 + (\pi/K)) = 0$ if ϕ and ψ are linearly dependent). Therefore, $x_2 \leq x_1 + (\pi/K)$, so

$$x_2 - x_1 \leq \pi\sqrt{\frac{p_M}{q_m + \lambda r_m}}$$

(see Figure 11.7).

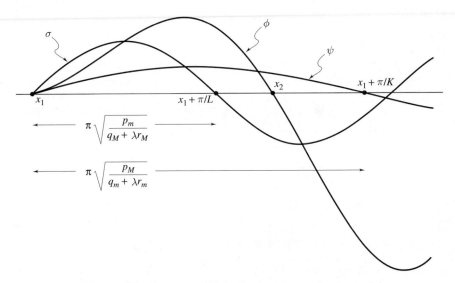

Figure 11.7 Estimating the distance between consecutive zeros of ϕ

To obtain a lower estimate for the distance between consecutive zeros, we compare the given equation with the constant coefficient equation

(18) $p_m y'' + (q_M + \lambda r_M)y = 0,$

which has the solution $\sigma(x) = \sin[L(x - x_1)]$, where $L^2 = (q_M + \lambda r_M)/p_m$. Notice that $\sigma(x)$ has consecutive zeros x_1 and $x_1 + (\pi/L)$. Since $p(x) \geq p_m$ and $q(x) + \lambda r(x) \leq q_M + \lambda r_M$ for x in (a, b), it follows from the Picone comparison theorem that $\sigma(x)$ must have a zero

between any two consecutive zeros of ϕ or have the same zeros (see Figure 11.7). Thus $x_1 + (\pi/L) \leq x_2$, so

$$\pi \sqrt{\frac{p_m}{q_M + \lambda r_M}} \leq x_2 - x_1. \quad \blacksquare \text{ End}$$

An immediate consequence of Corollary 5 is that as $\lambda \to +\infty$, the distance between the zeros of a solution to a Sturm-Liouville equation approaches zero. This means that since the eigenvalues λ_n associated with a regular Sturm-Liouville boundary value problem increase to infinity, the number of zeros in (a, b) of an eigenfunction ϕ_n increases as n increases. In fact, one can show using the Prüfer substitution that if a regular Sturm-Liouville boundary value problem has eigenvalues $\lambda_1 < \lambda_2 < \cdots$, then the eigenfunction ϕ_n belonging to λ_n has *exactly n* $-$ 1 zeros in the open interval (a, b).[†]

EXERCISES 11.8

1. Let $\phi(x)$ be a solution to the equation

$$\frac{d}{dx}\left(e^x \frac{dy}{dx}\right) + xy = 0$$

on $(-\infty, \infty)$. Given that $\phi(1/n) = 0$ for $n = 1, 2, \ldots,$ determine $\phi(x)$.

2. The function $\phi(x) = \sin x$ is a solution to $y'' + y = 0$ on $(-\infty, \infty)$ and has an infinite number of zeros. Does this contradict Theorem 13? Explain.

3. Let $\phi(x)$ be a solution on $(-\infty, \infty)$ to

$$\frac{d}{dx}\left[(x^2 + 1)\frac{dy}{dx}\right] - 2y = 0.$$

If $\phi(0) = 0$, prove that $\phi(x) = cx$ for some constant c.

4. Can the function $\phi(x) = x^4 \sin(1/x)$ be a solution on $(-1, 1)$ to some differential equation of the form (1), where p, p', and q are continuous on $[-1, 1]$ and p is positive on $[-1, 1]$? Explain.

5. Use the Sturm fundamental theorem to prove that any nontrivial solution to $y'' + (1 - e^x)y = 0$ has at most one zero in $(0, \infty)$ but has infinitely many zeros in $(-\infty, 0)$.

6. Use the Sturm fundamental theorem to prove that any nontrivial solution to $y'' + (x^2 - 1)y = 0$ has at most one zero in $(-1, 1)$ but has infinitely many zeros on $(-\infty, -1)$ and on $(1, \infty)$.

7. The Legendre polynomial $P_n(x)$ of degree n satisfies the equation

$$\frac{d}{dx}\left[(1 - x^2)\frac{dy}{dx}\right] + n(n + 1)y = 0, \quad -1 < x < 1,$$

and has n distinct zeros in $(-1, 1)$. Use the Picone comparison theorem to show that between any two zeros of $P_{m-1}(x)$, there is a zero of $P_m(x)$ for $m \geq 2$.

8. Let $\phi(x)$ be a nontrivial solution to

$$y'' + 9y + e^{-3x}y = 0$$

on $(0, \infty)$. Use the Sturm fundamental theorem to estimate the distance between consecutive zeros of $\phi(x)$ when x is large.

9. Let $\phi(x)$ be a nontrivial solution to

$$y'' + x^{-2}y' + (4 - e^{-x})y = 0$$

on $(0, \infty)$. Use the Picone comparison theorem to estimate the distance between consecutive zeros of $\phi(x)$ when x is large. [*Hint:* First express the equation in Sturm-Liouville form.]

10. Use Corollary 5 to estimate the distance between consecutive zeros in $[0, 5]$ of a nontrivial solution to

$$\frac{d}{dx}\left[e^{-x^2}\frac{dy}{dx}\right] + (\sin x)y + \lambda(1 + x^2)^{-1}y = 0.$$

[†] For a proof of this result, see the text by G. Birkhoff and G. C. Rota, *Ibid.*

11. Use Corollary 5 to estimate the distance between consecutive zeros in $[0, 5]$ of a nontrivial solution to

$$\frac{d}{dx}\left[(1 + x)\frac{dy}{dx}\right] + e^{-x}y + \lambda y = 0.$$

12. In equation (10), assume $q(x) \geq m^2$ on $[a, b]$. Prove that the distance between two consecutive zeros of a solution ϕ to (10) is less than or equal to π/m. [*Hint:* If x_1 is a zero of ϕ, compare ϕ with the function $\psi(x) = \sin(m(x - x_1))$, which is a solution to $y'' + m^2 y = 0$.]

13. Show that the substitution $y = ux^{-1/2}$ transforms equation (11) into equation (12).

14. Show that if $q(x) \geq m^2 > 0$ on $[a, \infty)$, then every solution to $y'' + q(x)y = 0$ is oscillatory.

15. Assume $0 < q_m \leq q(x) \leq q_M$ on $[a, b]$. Prove the following:

 (a) If $q_m > k^2\pi^2/(b - a)^2$, a positive integer, then any solution to equation (10) has at least k zeros in $[a, b]$.

 (b) If $q_M < k^2\pi^2/(b - a)^2$, then any nontrivial solution to equation (10) has at most k zeros in $[a, b]$.

16. **Sturm Convexity Theorem.** Prove that if $q(x)$ is a continuous, increasing function on $[0, \infty)$ and $\phi(x)$ is a nontrivial solution to

$$y'' + q(x)y = 0, \qquad 0 < x < \infty,$$

then $t_k - t_{k-1} > t_{k+1} - t_k$, where $\{t_n\}_{n=1}^\infty$ is the sequence of consecutive zeros in $(0, \infty)$ of the solution $\phi(x)$.

17. Show that if $w(x)$ and $q(x)$ are continuous on $[a, b]$ and $q(x) < 0$ on $[a, b]$, then a nontrivial solution ϕ of the equation

$$y'' + w(x)y' + q(x)y = 0$$

can have at most one zero in $[a, b]$.

CHAPTER SUMMARY

In this chapter we discussed the theory of regular Sturm-Liouville boundary value problems, the Fredholm alternative and nonhomogeneous boundary value problems, examples of singular Sturm-Liouville boundary value problems, and the oscillation and comparison theory for Sturm-Liouville equations. The important properties and concepts are given below.

Eigenvalue Problems

The problem of finding those values of λ for which the boundary value problem

$$\textbf{(1)} \qquad \begin{array}{ll} (py')' + qy + \lambda ry = 0, & a < x < b, \\ a_1 y(a) + a_2 y'(a) = 0, & b_1 y(b) + b_2 y'(b) = 0, \end{array}$$

has nontrivial solutions is called an **eigenvalue problem**. The number λ is an **eigenvalue** and the corresponding nontrivial solutions are **eigenfunctions.**

Sturm-Liouville Equation

A second order linear differential equation of the form

$$(p(x)y'(x))' + q(x)y(x) + \lambda r(x)y(x) = 0, \qquad a < x < b,$$

is called a **Sturm-Liouville equation.** Any homogeneous second order linear equation can be rewritten as a Sturm-Liouville equation.

Regular Sturm-Liouville Boundary Value Problems

An eigenvalue problem of the form (1) where $p(x)$, $p'(x)$, $q(x)$, and $r(x)$ are real-valued continuous functions on $[a, b]$ and $p(x) > 0$ and $r(x) > 0$ on $[a, b]$ (we do not allow $a_1 = a_2 = 0$ or $b_1 = b_2 = 0$) is called a **regular Sturm-Liouville boundary value problem.** These boundary value problems have the following properties:

1. The eigenvalues are real, simple, and form a countable, increasing sequence $\lambda_1 < \lambda_2 < \cdots$ with $\lim_{n \to \infty} \lambda_n = +\infty$.
2. The eigenfunctions can be chosen to be real-valued.
3. The eigenfunctions corresponding to distinct eigenvalues are orthogonal with respect to $r(x)$ on $[a, b]$.
4. Let $\{\phi_n\}_{n=1}^{\infty}$ be an orthonormal sequence of eigenfunctions and let f be a continuous function on $[a, b]$ with f' piecewise continuous on $[a, b]$. If f satisfies the two boundary conditions, then

$$f(x) = \sum_{n=1}^{\infty} a_n \phi_n(x), \qquad a \leq x \leq b,$$

where

$$a_n = \int_a^b f(x) \phi_n(x) r(x) \, dx.$$

Moreover, the series converges uniformly on $[a, b]$.

Periodic Sturm-Liouville problems have similar properties.

Fredholm Alternative

Let $L[y](x) := A_2(x)y''(x) + A_1(x)y'(x) + A_0(x)y(x)$ and let $B[y] = 0$ represent a set of separated boundary conditions. Let $L^+[y](x) = (A_2(x)y(x))'' - (A_1(x)y(x))' + A_0(x)y(x)$ be the adjoint of L with adjoint boundary conditions $B^+[y] = 0$. (The adjoint boundary conditions are chosen so that $(Lu, v) = (u, L^+v)$ for all u and v satisfying $B[u] = 0$ and $B^+[v] = 0$.) The nonhomogeneous boundary value problem

$$L[y](x) = h(x), \qquad a < x < b, \qquad B[y] = 0,$$

has a solution if and only if

$$\int_a^b h(x)z(x) \, dx = 0$$

for every solution z of the adjoint boundary value problem

$$L^+[z](x) = 0, \qquad a < x < b, \qquad B^+[z] = 0.$$

Solution by Eigenfunction Expansion

Let $L[y] := (py')' + qy$ and assume $L[y] + \mu r y = 0$, $B[y] = 0$ is a regular Sturm-Liouville boundary value problem with eigenvalues $\{\lambda_n\}_{n=1}^{\infty}$ and corresponding eigenfunctions $\{\phi_n\}_{n=1}^{\infty}$.

If $\mu \neq \lambda_n$ for $n = 1, 2, \ldots$, then the solution to the nonhomogeneous problem

$$L[y] + \mu r y = h, \qquad B[y] = 0,$$

is given by

$$\phi = \sum_{n=1}^{\infty} \frac{\gamma_n}{\mu - \lambda_n} \phi_n,$$

where the constants γ_n are determined from the eigenfunction expansion of h/r:

$$\frac{h}{r} = \sum_{n=1}^{\infty} \gamma_n \phi_n.$$

If $\mu = \lambda_N$ for some N and $\gamma_N \neq 0$, then there is no solution to the nonhomogeneous problem.

If $\mu = \lambda_N$ for some N and $\gamma_N = 0$, then

$$\phi = c_N \phi_N + \sum_{\substack{n=1 \\ n \neq N}}^{\infty} \frac{\gamma_n}{\mu - \lambda_n} \phi_n$$

is a solution to the nonhomogeneous problem for any choice of the parameter c_N.

Green's Functions

The solution to

$$L[y] = -f, \qquad B_1[y] = 0, \qquad B_2[y] = 0,$$

where we assume that $L[y] := (py')' + qy$, the function f is continuous on $[a, b]$, and the corresponding homogeneous problem has only the trivial solution, is given by

$$\phi(x) = \int_a^b G(x, s) f(s) \, ds,$$

where $G(x, s)$ is the **Green's function**

$$G(x, s) = \begin{cases} -z_1(s) z_2(x)/C, & a \leq s \leq x, \\ -z_1(x) z_2(s)/C, & x \leq s \leq b. \end{cases}$$

Here z_1 and z_2 are two nontrivial solutions to $L[y] = 0$ satisfying $B_1[z_1] = 0$ and $B_2[z_2] = 0$. The constant C is given by $C = p(x)W[z_1, z_2](x)$ for any $x \in [a, b]$.

Singular Sturm-Liouville Problems

Certain *non*regular Sturm-Liouville boundary value problems, called **singular** Sturm-Liouville boundary value problems, arise in solving Bessel's, Legendre's, and Hermite's equations. These singular Sturm-Liouville problems also have increasing sequences of simple eigenvalues. Moreover, their corresponding sequences of eigenfunctions are orthogonal and can be used to form eigenfunction expansions with convergence properties similar to those associated with regular Sturm-Liouville problems.

Oscillation and Comparison Theory

Several results are known about the location of the zeros of solutions to Sturm-Liouville equations. The **Sturm separation theorem** says that the zeros of two linearly independent solutions are intertwined. The **Sturm fundamental theorem** and the **Picone comparison theorem** allow one to compare the location of the zeros of a solution to one Sturm-Liouville equation with those of a related equation.

REVIEW PROBLEMS

1. Find all the real eigenvalues and eigenfunctions for the given eigenvalue problem.

 (a) $y'' + 6y' + \lambda y = 0$; $y(0) = 0,$ $y(1) = 0.$
 (b) $y'' + \lambda y = 0$; $y(0) = 0,$ $y(\pi) + 2y'(\pi) = 0.$

2. Convert the given equation into the form of a Sturm-Liouville equation.

 (a) $y'' + 7y' + \lambda y = 0.$
 (b) $y'' - 3xy' + \lambda y = 0.$
 (c) $xy'' + (1 - x)y' + \lambda y = 0,$ $x > 0.$

3. **(a)** Determine the eigenfunctions, which are normalized with respect to the weight function $r(x) \equiv 1$ on $[0, \pi]$, for the problem

 $$y'' + \lambda y = 0; \qquad y'(0) = 0, \qquad y(\pi) = 0.$$

 (b) Express $f(x) = 1 - x$ in an eigenfunction expansion using the result of part (a).

4. Find the adjoint boundary value problem for the given boundary value problem.

 (a) $y'' + xy' + y = 0$; $y'(0) = 0,$ $y(1) + y'(1) = 0.$
 (b) $x^2 y'' + 2xy' - 3y = 0$; $y(1) = 0,$ $y'(e) = 0.$

5. Use the Fredholm alternative to determine conditions on h that guarantee that the given nonhomogeneous boundary value problem has a solution.

 (a) $y'' - 6y' + 13y = h$; $y(0) = 0,$ $y(2\pi) = 0.$
 (b) $xy'' + y' = h$; $y'(1) = 0,$ $y'(2) = 0.$

6. Find a formal eigenfunction expansion for the solution to the nonhomogeneous boundary value problem

 $$y'' + 3y = x - 4 \cos 2x;$$
 $$y'(0) = 0, \qquad y'(\pi) = 0.$$

7. Find the Green's function $G(x, s)$ and use it to solve the given boundary value problem.

 (a) $y'' = x + 1,$ $y(0) = 0,$ $y'(1) = 0.$
 (b) $y'' - y = 4,$ $y'(0) = 0,$ $y(1) = 0.$

8. Find a formal eigenfunction expansion for the solution to the given nonhomogeneous boundary value problem where f is a continuous function and μ is not an eigenvalue of the corresponding homogeneous problem.

 (a) $(xy')' - (49/x)y + \mu xy = f(x),$ $0 < x < 1,$
 $y(x), y'(x)$ remain bounded as $x \to 0^+,$ $y(1) = 0.$

 (b) $[(1 - x^2)y']' + \mu y = f(x),$ $-1 < x < 1,$
 $y(x), y'(x)$ remain bounded as $x \to \pm 1.$

9. Let $\phi(x)$ be a nontrivial solution to

 $$[(1 - e^{-x})y']' + \left(\frac{x + 1}{x}\right)y = 0$$

 on $(0, \infty)$. Use the Picone comparison theorem to estimate the distance between consecutive zeros of $\phi(x)$ when x is large.

10. Use Corollary 5 in Section 11.8 to estimate the distance between consecutive zeros in $[0, \pi]$ of a nontrivial solution to

 $$[(2 + \cos x)y']' + (\sin x)y + 5y = 0.$$

TECHNICAL WRITING EXERCISES

1. The method of solving a nonhomogeneous regular Sturm-Liouville boundary value problem using an eigenfunction expansion as discussed in Section 11.5 can be extended to periodic Sturm-Liouville problems. Describe this extension and illustrate it with an example.

2. Green's functions also exist for certain periodic Sturm-Liouville problems. Explain how to construct a Green's function for the problem

$$y''(x) + 2y(x) = -f(x);$$
$$y(0) = y(2\pi), \qquad y'(0) = y'(2\pi).$$

3. Formulate a Fredholm alternative for the system $\mathbf{Ax} = \mathbf{b}$, where $\mathbf{A}$ is an $n \times n$ real-valued constant matrix, $\mathbf{b}$ is an $n \times 1$ constant vector, and $\mathbf{x}$ is the $n \times 1$ unknown vector. Illustrate your formulation by examples.

A. Hermite Polynomials and the Harmonic Oscillator

In quantum mechanics the motion of a particle is described by Schrödinger's equation. In one-dimensional space, the equation has the form

(1) $\psi'' + (2m/\hbar^2)[E - V(x)]\psi = 0,$

where the function $V(x)$ represents the potential energy, m is the mass of the particle, $\hbar$ is Planck's constant h divided by 2π, the constant E represents the energy, and $\underline{\psi}$ is the **wave function.** The wave function ψ may be real- or complex-valued. The quantity $\psi(x)\overline{\psi(x)}\,dx = |\psi(x)|^2\,dx$ gives the probability that the particle will be found in the interval $(x, x + dx)$. Physicists are interested in the solutions to (1) that satisfy the boundary conditions

(2) $\psi(x)$ **and** $\psi'(x)$ **remain bounded as** $x \to \pm\infty$.

The eigenvalues $E = E_n$ of (1)–(2) are called the **energy levels** of the system.

 (a) For the **harmonic oscillator,** the potential energy is $V(x) = \alpha x^2$, where $\alpha > 0$ is a constant. Show that for the harmonic oscillator, the boundary value problem (1)–(2) simplifies to

(3) $y''(t) + (\lambda - t^2)y(t) = 0,$

(4) $y(t)$ and $y'(t)$ remain bounded at $t \to \pm\infty$,

 where $t = \beta x$, $\beta^4 = 2m\alpha/\hbar^2$, $y(t) = \psi(x)$, and $\lambda = 2mE/(\beta^2\hbar^2)$.

 (b) Verify that the change of variables $y(t) = e^{-t^2/2}u(t)$ transforms (3)–(4) into

(5) $(e^{-t^2}u')' + (\lambda - 1)e^{-t^2}u = 0,$

(6) $e^{-t^2/2}u(t)$ and $e^{-t^2/2}u'(t) - te^{-t^2/2}u(t)$ remain bounded as $t \to \pm\infty$.

 (c) Use the result of part (b) to show that $\lambda = 2n + 1$, $n = 0, 1, 2, \ldots$, are eigenvalues of the singular Sturm-Liouville boundary value problem (3)–(4) with corresponding eigenfunctions $y_n(t) = e^{-t^2/2}H_n(t)$, where H_n is the **Hermite polynomial** of degree n (cf. Section 11.7 or Section 8.9).

 (d) Sketch the first three wave functions $y_0(t)$, $y_1(t)$, and $y_2(t)$.

B. Continuous and Mixed Spectra

Free Particles. When the potential energy $V(x)$ in Schrödinger's equation (1) is identically zero, the particle is said to be **free.** Thus the motion of a **free particle** is governed by the singular Sturm-Liouville boundary value problem

(7) $\psi'' + \lambda\psi = 0,$

(8) $\psi(x)$ **and** $\psi'(x)$ **remain bounded as** $x \to \pm\infty$,

where $\lambda = 2mE/\hbar^2$.

(a) Show that every nonnegative number λ is an eigenvalue of (7)–(8) and that there are no negative eigenvalues. Hence the spectrum of the free particle is continuous and consists of the interval $[0, \infty)$.

Square-Well Potential. A **square-well** potential is a function of the form

$$V(x) = \begin{cases} -\alpha^2, & |x| < a, \\ 0, & |x| > a, \end{cases}$$

where α and a are positive constants. For such a potential, Schrödinger's equation has the form

(9) $\psi'' + \lambda\psi = \begin{cases} -\beta^2\psi, & |x| < a, \\ 0, & |x| > a, \end{cases}$

where $\lambda = 2mE/\hbar^2$ and $\beta^2 = 2m\alpha^2/\hbar^2$. We also impose the boundary conditions

(10) $\psi(x)$ and $\psi'(x)$ remain bounded as $x \to \pm\infty$.

Equation (9) can be solved in the sense that we can find solutions on the three intervals $(-\infty, -a)$, $(-a, a)$, and (a, ∞) and then patch them together to get a function ψ such that ψ and ψ' are continuous on $(-\infty, \infty)$.

(b) Show that each $\lambda > 0$ is an eigenvalue for the singular Sturm-Liouville boundary value problem (9)–(10) and has two linearly independent eigenfunctions associated with it.

(c) Show that $\lambda = 0$ is an eigenvalue for the boundary value problem (9)–(10) if and only if $\beta a = k\pi/a$ for some integer k. Moreover, when $\beta a = k\pi/a$, there is only one linearly independent eigenfunction associated with $\lambda = 0$.

(d) Show that for $\lambda < -\beta^2$ the only solution to (9)–(10) is the trivial solution and hence there are no eigenvalues less than $-\beta^2$.

(e) Show that for $-\beta^2 < \lambda < 0$, there can exist a finite number of discrete eigenvalues λ_n of (9)–(10). [*Hint:* Let $\gamma = \sqrt{\beta^2 + \lambda}$. The function $A\cos(\gamma x)$ is part of an eigenfunction for $x > a$ if and only if $\gamma\tan(\gamma a) = \sqrt{-\lambda}$. Further, $B\sin(\gamma x)$ is part of an eigenfunction for $x > a$ if and only if $\gamma\cot(\gamma a) = -\sqrt{-\lambda}$.]

Consequently the spectrum of a particle under the influence of a square-well potential may have a discrete spectrum made up of a finite number of negative eigenvalues in addition to its continuous spectrum $[0, \infty)$.

C. *Picone Comparison Theorem*

The purpose of this project is to give a proof of the Picone comparison theorem (see Theorem 17, page 720). Using the notation of that theorem, we proceed as follows.

Let x_1 and x_2 be two consecutive zeros of ϕ_1 and assume $\phi_2 \neq 0$ in $[x_1, x_2]$.

(a) Show that

(11) $(q_2 - q_1)\phi_1^2\phi_2^2 + (p_2\phi_2')'\,\phi_1^2\phi_2 - (p_1\phi_1')'\,\phi_1\phi_2^2 = 0.$

(b) Integrating equation (11) from x_1 to x_2 and using integration by parts, obtain

(12) $\displaystyle\int_{x_1}^{x_2} (q_2 - q_1)\phi_1^2\,dx + \int_{x_1}^{x_2} p_1(\phi_1')^2\,dx - \int_{x_1}^{x_2} \frac{2p_2\phi_2'\phi_1\phi_1'}{\phi_2}\,dx + \int_{x_1}^{x_2} \frac{p_2(\phi_2')^2\phi_1^2}{\phi_2^2}\,dx = 0.$

(c) From equation (12), derive the **Picone identity**

$$(13) \qquad \int_{x_1}^{x_2} (q_2 - q_1)\phi_1^2 \, dx + \int_{x_1}^{x_2} (p_1 - p_2)(\phi_1')^2 \, dx + \int_{x_1}^{x_2} p_2 \left[\phi_1' - \frac{\phi_1 \phi_2'}{\phi_2} \right]^2 dx = 0.$$

(d) If $\phi_2(x) \neq 0$ for x in (x_1, x_2), but $\phi_2(x)$ is zero at either $x = x_1$ or $x = x_2$, then some of the preceding integrals must be treated as improper integrals. Verify that the Picone identity still holds. [*Hint:* If $\phi_2(x_1) = 0$, then $\phi_2'(x_1) \neq 0$. Use this fact and l'Hôpital's rule to show that the boundary terms, which arise from using integration by parts, are still zero.]

(e) Use the Picone identity and the hypotheses of Theorem 17, page 720 to show that if $q_1(x) \not\equiv q_2(x)$ on $[x_1, x_2]$, then ϕ_2 must have a zero in (x_1, x_2).

(f) Use the Picone identity and the hypotheses of Theorem 17, page 720, to show that if $q_1(x) \equiv q_2(x)$ on $[x_1, x_2]$, then ϕ_1 and ϕ_2 are linearly dependent on $[x_1, x_2]$. [*Hint:* On the intervals where $p_1 \equiv p_2$, show that ϕ_1 and ϕ_2 satisfy the same equation and their Wronskian is zero. On the intervals where $p_1(x) \neq p_2(x)$, show that $\phi_1' = 0$ and $\phi_2' = 0$. Combine to conclude ϕ_1 and ϕ_2 are linearly dependent.]

This completes the proof of the Picone comparison theorem.

D. Shooting Method

The shooting method for approximating the solution to the second order boundary value problem

$$(14) \qquad y''(x) = f(x, y, y'), \qquad a \leq x \leq b, \qquad y(a) = \alpha, \qquad y(b) = \beta,$$

involves solving a sequence of initial value problems of the form

$$(15) \qquad y''(x) = f(x, y, y'), \qquad a \leq x \leq b, \qquad y(a) = \alpha, \qquad y'(a) = t,$$

where t is a parameter. The idea is to "shoot" from a starting location, $y(a) = \alpha$, with an initial "elevation," $y'(a) = t$, at a target located at $y(b) = \beta$. (See Figure 11.8.) If we miss on the first attempt, we adjust the "elevation" t and try again.

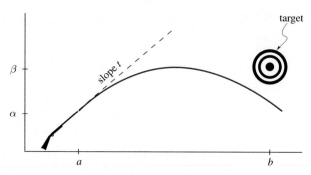

Figure 11.8 Shooting method

The shooting method generates a sequence of values t_k that (we hope) converges to a value $\hat{t}$ for which the solution to the initial value problem (15) with $y'(a) = \hat{t}$ satisfies $y(b) = \beta$. If we let $y(x, t)$ denote the solution to (15) with $y'(a) = t$, then our problem is to find t such that $y(b, t) = \beta$ or

(16) $y(b, t) - \beta = 0.$

We can use a root-finding procedure such as the **bisection method** to generate a sequence t_k that converges to the solution to (16). To begin, we need two guesses t_0 and t_1 such that $y(b, t_0) - \beta$ and $y(b, t_1) - \beta$ are of opposite signs (one positive and one negative). We then pick $t_2 = (t_0 + t_1)/2$. If $y(b, t_2) - \beta = 0$, we are done. Otherwise, $y(b, t_2) - \beta$ will have the opposite sign of either $y(b, t_0) - \beta$ or $y(b, t_1) - \beta$. If $y(b, t_2) - \beta$ and $y(b, t_0) - \beta$ have opposite signs, then we pick $t_3 = (t_0 + t_2)/2$. Otherwise we pick $t_3 = (t_1 + t_2)/2$. We continue until t_n and t_{n-1} are sufficiently close together (that is, within a prescribed tolerance) and recall that the desired root of (16) is located somewhere between them. Notice that this procedure requires solving an initial value problem for each t_k (see Section 5.6).

(a) Use the shooting method to approximate the solution to the nonlinear boundary value problem

$$y''(x) = \frac{1}{2} y^3(x); \qquad 1 \le x \le 3, \qquad y(1) = 2/3, \qquad y(3) = 2/5.$$

Compare your approximate solution with the actual solution $y(x) = 2/(x + 2)$.

(b) For linear boundary value problems of the form

(17) $y''(x) = p(x)y'(x) + q(x)y(x) + g(x), \qquad a \le x \le b, \qquad y(a) = \alpha, \qquad y(b) = \beta,$

the shooting method requires only two "shots." Using the superposition principle, show that the solution to (17) is given by

$$y(x) = y_1(x) + \frac{\beta - y_1(b)}{y_2(b)} y_2(x),$$

where $y_1(x)$ is the solution to the initial value problem

$$y''(x) = p(x)y'(x) + q(x)y(x) + g(x); \qquad a \le x \le b, \qquad y(a) = \alpha, \qquad y'(a) = 0,$$

and $y_2(x)$ is the solution to the initial value problem

$$y''(x) = p(x)y'(x) + q(x)y(x); \qquad a \le x \le b, \qquad y(a) = 0, \qquad y'(a) = 1.$$

(We assume $y_2(b) \ne 0$; otherwise the boundary value problem (17) does not have a unique solution.)

(c) Use the result of part (b) to approximate the solution to the linear boundary value problem

$$y''(x) = -\frac{11}{3x} y'(x) + \frac{1}{x^2} y(x) + \frac{8}{3x^2} - \frac{\ln x}{x^2}; \qquad 1 \le x \le 4;$$

$$y(1) = 1, \qquad y(4) = 4^{1/3} + \ln x \approx 2.973695413.$$

Compare your approximate solution with the actual solution

$$y(x) = x^{1/3} + \ln x.$$

E. Finite-Difference Method for Boundary Value Problems

A classical method for numerically approximating a solution to a boundary value problem for a linear second order differential equation consists of replacing each derivative in the differential equation by a difference quotient that approximates that derivative. To illustrate this procedure, consider the linear boundary value problem

(18) $y''(x) + p(x)y'(x) + q(x)y(x) = g(x); \quad a < x < b,$

(19) $y(a) = \alpha, \quad y(b) = \beta,$

where p, q, and g are continuous on $[a, b]$ and α, β are given constants.

We begin by partitioning the interval $[a, b]$ into $N + 1$ subintervals of equal length h. That is, let $h := (b - a)/(N + 1)$ and for $k = 0, 1, \ldots, N + 1$, set

$$x_k := a + kh.$$

Then $a = x_0 < x_1 < \cdots < x_N < x_{N+1} = b$.

Assuming a solution $y(x)$ to (1)–(2) exists, our aim is to approximate $y(x)$ at the interior mesh points x_k, $k = 1, 2, \ldots, N$. At these points, we know that

(20) $y''(x_k) + p(x_k)y'(x_k) + q(x_k)y(x_k) = g(x_k).$

The essence of the finite-difference method is to find difference quotients that closely approximate the derivatives $y'(x_k)$ and $y''(x_k)$ when h is small.

(a) Assuming y has a Taylor series representation of the form

(21) $y(x) = y(x_k) + y'(x_k)(x - x_k) + \dfrac{y''(x_k)}{2}(x - x_k)^2 + \dfrac{y'''(x_k)}{3!}(x - x_k)^3 + \cdots$

$$= \sum_{j=0}^{\infty} \frac{y^{(j)}(x_k)}{j!}(x - x_k)^j,$$

derive the following **centered-difference formula** for $y'(x_k)$:

(22) $y'(x_k) = \dfrac{1}{2h}[y(x_{k+1}) - y(x_{k-1})] + [\text{terms involving } h^2, h^3, \ldots].$

[*Hint:* Work with the Taylor representations for $y(x_{k+1})$ and $y(x_{k-1})$.]

(b) Again using (21), derive the following centered-difference formula for $y''(x_k)$:

(23) $y''(x_k) = \dfrac{1}{h^2}[y(x_{k+1}) - 2y(x_k) + y(x_{k-1})] + [\text{terms involving } h^2, h^3, \ldots].$

(c) If we let w_k denote the approximation (to be computed) for $y(x_k)$, $k = 1, 2, \ldots, N$ and set $w_0 = y(x_0) = \alpha$, $w_{N+1} = y(x_{N+1}) = \beta$, then from (23) and (22) we obtain the approximations

(24) $y''(x_k) \approx \dfrac{1}{h^2}(w_{k+1} - 2w_k + w_{k-1}),$

(25) $y'(x_k) \approx \dfrac{1}{2h}(w_{k+1} - w_{k-1}).$

Using the centered-difference approximations (24) and (25) in place of $y''(x_k)$ and $y'(x_k)$ in (20) yields the system of equations

$$(26) \qquad \left(\frac{w_{k+1} - 2w_k + w_{k-1}}{h^2}\right) + p(x_k)\left(\frac{w_{k+1} - w_{k-1}}{2h}\right) + q(x_k)w_k = g(x_k),$$

$k = 1, 2, \ldots, N$. The equation (26) represent N linear equations in the N unknowns $w_1, \ldots, w_N$. Thus on solving (26), we obtain approximations for $y(x)$ at the interior mesh points x_k, $k = 1, \ldots, N$.

Use this method to approximate the solution to the boundary value problem

$$y''(x) - 9y(x) = 0; \qquad y(0) = 1, \qquad y(1) = e^{-3},$$

taking $h = 1/4$ ($N = 3$).

(d) Compare the approximations obtained in part (c) with the actual solution $y = e^{-3x}$.

(e) Let E denote the electrostatic potential between two concentric metal spheres of radii r_1 and r_2 ($r_1 < r_2$), where the potential of the inner sphere is kept constant at 100 volts and the potential of the outer sphere at 0 volts. The potential E will then depend only on the distance r from the center of the spheres and must satisfy Laplace's equation. This leads to the following boundary value problem for E:

$$\frac{d^2E}{dr^2} + \frac{2}{r}\frac{dE}{dr} = 0; \qquad E(r_1) = 100, \qquad E(r_2) = 0,$$

where $r_1 < r < r_2$.

For $r_1 = 1$ and $r_2 = 3$, obtain finite-difference approximations for $E(2)$ using $h = 1/2$ ($N = 3$) and $h = 1/4$ ($N = 7$).

(f) Compare your approximations in part (e) with the actual solution $E(r) = 150r^{-1} - 50$.

Chapter 12

Stability of Autonomous Systems

12.1 INTRODUCTION: COMPETING SPECIES

> The populations $x(t)$ and $y(t)$ at time t of two species that compete for the same food supply are governed by the system
>
> $$(1) \qquad \frac{dx}{dt} = x(a_1 - b_1 x - c_1 y),$$
>
> $$(2) \qquad \frac{dy}{dt} = y(a_2 - b_2 y - c_2 x),$$
>
> where a_1, a_2, b_1, b_2, c_1, and c_2 are positive constants that depend on the species and environment under investigation. Describe the long-term behavior of these populations.

That (1)–(2) is a plausible model for competing species can be seen as follows. Using the logistic model of Section 3.2 for the growth of a single species, let's assume that in the absence of species y, population x is governed by an equation of the form

$$\frac{dx}{dt} = a_1 x - b_1 x^2 = x(a_1 - b_1 x)$$

and, similarly, that in the absence of species x, population y is governed by

$$\frac{dy}{dt} = a_2 y - b_2 y^2 = y(a_2 - b_2 y),$$

where a_1, a_2, b_1, and b_2 are positive constants. When both species are present and competing for the same food, the effect is to decrease their rate of growth. Because this effect depends on the size of both populations, it is reasonable to assume it is proportional to the product xy. Consequently we replace the growth rate for species x by $a_1 x - b_1 x^2 - c_1 xy$ and for species y by $a_2 y - b_2 y^2 - c_2 xy$, which yields the model (1)–(2).

Generally, we cannot explicitly solve the nonlinear system (1)–(2) for x and y in terms of t. Further, if we attempt a phase plane analysis (without the aid of numerical methods) as done in Section 5.7, we will encounter insurmountable difficulties—the integral curves cannot be found in a closed form and therefore we cannot analytically describe the trajectories. However, by studying the qualitative properties of the solutions $x(t)$ and $y(t)$, we can obtain useful information about certain trajectories. Specifically, we will try to determine how the solutions behave near a critical point.

Let's first determine the critical points for the system (1)–(2). Setting the right-hand sides equal to zero, we have

$$x(a_1 - b_1 x - c_1 y) = 0,$$

$$y(a_2 - b_2 y - c_2 x) = 0.$$

This system has (at most) four solutions: $(0, 0)$, $(0, a_2/b_2)$, $(a_1/b_1, 0)$, and the point of intersection of the two lines $a_1 - b_1 x - c_1 y = 0$, $a_2 - b_2 y - c_2 x = 0$, which, if it exists, is the point

$$(3) \qquad P = \left(\frac{a_1 b_2 - a_2 c_1}{b_1 b_2 - c_1 c_2}, \frac{a_2 b_1 - a_1 c_2}{b_1 b_2 - c_1 c_2} \right).$$

Notice that the first three equilibrium solutions lie on the nonnegative axes that border the first quadrant in the xy-plane. Because populations must be nonnegative, we will confine the rest of our analysis to the first quadrant.

To determine the flow of trajectories in the first quadrant, let's start with equation (1) involving dx/dt:

$$\frac{dx}{dt} = x(a_1 - b_1 x - c_1 y).$$

From this equation, we see that in the first quadrant, the sign of dx/dt is positive when $a_1 - b_1 x - c_1 y > 0$ and negative when $a_1 - b_1 x - c_1 y < 0$. Consequently $x(t)$ increases along a trajectory that lies below the line $a_1 - b_1 x - c_1 y = 0$ and decreases along trajectories above this line (see Figure 12.1(a)). Similarly, $y(t)$ increases along a trajectory that lies below the line $a_2 - b_2 y - c_2 x = 0$ and decreases when it is above this line (see Figure 12.1(b)).

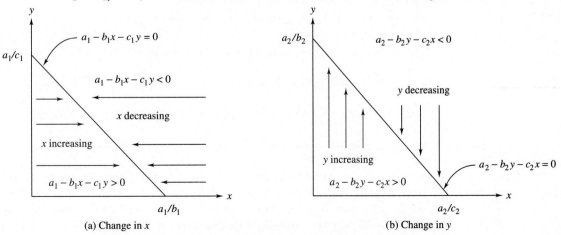

(a) Change in x (b) Change in y

Figure 12.1 Regions where the populations are increasing or decreasing

It turns out that there are four cases to analyze, depending on the relative positions of the two lines and whether they intersect in the first quadrant. These four cases are illustrated in Figure 12.2. The arrow systems (e.g., ↘) indicate whether the variables x and y are increasing or decreasing and are based on the analysis illustrated in Figure 12.1. The colored dots are the critical points in the first quadrant. The black lines denote where $dx/dt = 0$ and the colored lines where $dy/dt = 0$. Typical trajectories are also shown in color.

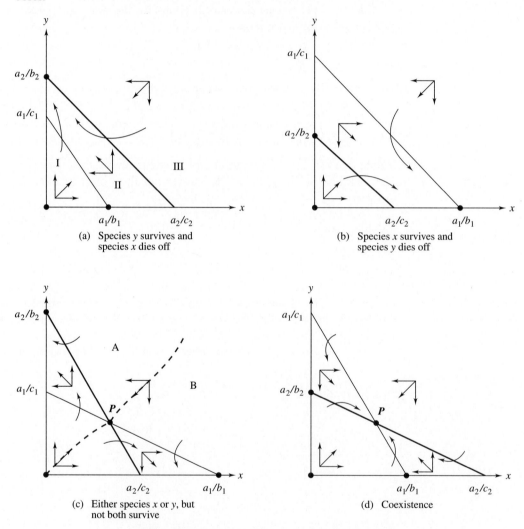

(a) Species y survives and species x dies off

(b) Species x survives and species y dies off

(c) Either species x or y, but not both survive

(d) Coexistence

Figure 12.2 Phase plane diagrams for competing species

Let's begin with the case illustrated in Figure 12.2(a). If we start in region I, that is, $(x(0), y(0)) = (x_0, y_0)$ lies in region I, then populations x and y are increasing (↗). At some point in time, the trajectory $(x(t), y(t))$ will move into region II. In region II, the population

$x(t)$ decreases while $y(t)$ continues to increase. Observe that the trajectory *cannot* cross the colored line bounding region II because this would require either $x(t)$ to be increasing, which it is not in region II, or $y(t)$ to increase *across* the colored line, which is impossible because $dy/dt = 0$ on that line. Consequently the trajectories in region II must move up and to the left ($\nwarrow$). Since $dx/dt = 0$ on the y-axis, the trajectory cannot cross it and so must approach the critical point $(0, a_2/b_2)$ in the upper corner of region II. When the populations start in region III, both $x(t)$ and $y(t)$ are decreasing ($\swarrow$). Since the trajectory cannot leave the first quadrant, the trajectory must cross the colored line and enter region II, where it then moves up toward the critical point $(0, a_2/b_2)$. To summarize, we have shown that when the initial populations $x(0)$ and $y(0)$ are positive, the species x dies off ($x(t) \to 0$), while the population of species y approaches the limiting population a_2/b_2. Graphically, all trajectories beginning in the first quadrant approach the critical point $(0, a_2/b_2)$ as $t \to +\infty$. Typical trajectories are illustrated by colored curves in Figure 12.2(a).

A similar analysis of the case illustrated in Figure 12.2(b) shows that species y dies off, while the population of species x approaches the limiting population a_1/b_1. Graphically, all trajectories beginning in the first quadrant approach the critical point $(a_1/b_1, 0)$ as $t \to +\infty$. Typical trajectories are again illustrated by colored curves.

The analysis of the case illustrated in Figure 12.2(c) yields a different result. There exists a dividing curve called a *separatrix* (illustrated by a dashed colored curve) that separates the first quadrant into two regions A and B. Trajectories beginning in region A approach the critical point $(0, a_2/b_2)$ as $t \to +\infty$, and those beginning in region B approach the critical point $(a_1/b_1, 0)$. In other words, for some initial populations, species x dies off, while the population of species y approaches a_2/b_2. For other initial populations, species y dies off, while the population of species x approaches a_1/b_1. Unfortunately the separatrix is usually difficult to determine.

Finally, an analysis of the case illustrated in Figure 12.2(d) shows that every trajectory in the first quadrant heads toward the critical point P given in (3). That is, the populations $x(t)$ and $y(t)$ approach the equilibrium populations

$$x = \frac{a_1 b_2 - a_2 c_1}{b_1 b_2 - c_1 c_2}, \qquad y = \frac{a_2 b_1 - a_1 c_2}{b_1 b_2 - c_1 c_2}.$$

This is the only case that allows for the coexistence of the two species. The critical point P is an example of an *asymptotically stable* critical point, since all the nearby trajectories approach P as $t \to +\infty$. By comparison the origin is an *unstable* critical point because nearby trajectories move away from it.

In Section 12.2, we study the stability of critical points for general linear systems. Namely, we analyze the behavior of trajectories near the critical point of a linear system in the plane and classify the type of behavior as $t \to +\infty$. In Section 12.3, we discuss how the stability of a critical point for an ''almost linear system'' can be determined by studying the corresponding linear system. In Section 12.4 we introduce the energy method and use it to study conservative systems. Section 12.5 describes a generalization of the energy method known as Lyapunov's direct method, which is used to analyze the stability of a critical point. In Section 12.6, we present results of Poincaré and Bendixson, which concern the existence of periodic solutions for autonomous systems in the plane. Finally, in Section 12.7, we discuss stability for higher-dimensional systems using results from Chapter 9.

12.2 LINEAR SYSTEMS IN THE PLANE

An autonomous system in the plane has the form

(1) $x'(t) = f(x, y)$,

(2) $y'(t) = g(x, y)$,

where the essential feature is that f and g are real-valued functions that do not depend explicitly on t. In Section 5.7 we studied such systems using a phase plane analysis. However, that approach depended heavily on our being able to find the integral curves for the system. In this chapter we try to determine the behavior of the solutions to an autonomous system when integral curves cannot be found or are difficult to graph. (The reader is advised to review the material in Section 5.7 before proceeding.)

We begin by studying the behavior of trajectories in a neighborhood of a critical point. Recall that a **critical point** of system (1)–(2) is a point (x_0, y_0) where both f and g are zero. If (x_0, y_0) is a critical point, then the constant functions $x(t) \equiv x_0$, $y(t) \equiv y_0$ form a solution to (1)–(2). This is called an **equilibrium solution**, since the functions do not change with t.

Before turning to nonlinear systems, we will study the easier problem of stability of critical points for linear systems. By a simple translation, we can always transform a given linear system into one of the form

(3) $x'(t) = ax + by$,

(4) $y'(t) = cx + dy$,

where the origin $(0, 0)$ is now the critical point (see Problem 21). So, without loss of generality, we analyze the linear system (3)–(4), but under the additional assumption that $ad - bc \neq 0$. This last condition makes $(0, 0)$ an *isolated* critical point; that is, there is a disk about the origin that does *not* contain any other critical point of (3)–(4) (see Problem 22). (In fact, $(0, 0)$ is the only critical point of (3)–(4) when $ad - bc \neq 0$.)

System (3)–(4) can be solved using the elimination method discussed in Section 4.10 or by the matrix methods of Sections 9.4 and 9.5. Either approach yields solutions of the form $x(t) = ue^{rt}$, $y(t) = ve^{rt}$, where u, v, and r are constants to be determined. Substituting these expressions for x and y into system (3)–(4), we have

$$rue^{rt} = aue^{rt} + bve^{rt},$$
$$rve^{rt} = cue^{rt} + dve^{rt},$$

which simplifies to

$$(r - a)u - bv = 0,$$
$$-cu + (r - d)v = 0.$$

This pair of equations has a nontrivial solution for u, v if and only if the determinant of the coefficient matrix is zero; that is,

(5) $(r - a)(r - d) - bc = r^2 - (a + d)r + (ad - bc) = \mathbf{0}.$

This is called the **characteristic equation** for the system (3)–(4).[†]

[†] This is also the condition for r to be an *eigenvalue* for the matrix $A = \begin{bmatrix} a & b \\ c & d \end{bmatrix}$.

The asymptotic (long-term) behavior of the solutions to (3)–(4) is linked to the nature of the roots r_1 and r_2 of the characteristic equation; namely, whether these roots are both real or both complex, both positive, both negative, of opposite signs, etc. We consider each case in detail by first discussing a special example and then indicating what happens in general. These general results (whose proofs we omit) can be obtained using a linear change of variables[†] (see Problem 23).

For the reader's convenience, at the end of this section, we summarize all these results in Theorem 1.

Case 1. r_1, r_2 real, distinct, and positive. In this case, the general solution to system (3)–(4) has the form

$$\textbf{(6)} \qquad \begin{aligned} x(t) &= A_1 e^{r_1 t} + A_2 e^{r_2 t}, \\ y(t) &= B_1 e^{r_1 t} + B_2 e^{r_2 t}, \end{aligned}$$

where the A_i's and B_i's are interrelated. As a simple example, we discuss the system

$$\frac{dx}{dt} = x, \qquad \frac{dy}{dt} = 3y.$$

Here $r_1 = 1$ and $r_2 = 3$ and a general solution is

$$x(t) = A e^t, \qquad y(t) = B e^{3t}.$$

Consequently $y = B(e^t)^3 = (B/A^3)x^3$ defines integral curves for the system. A phase plane diagram for this system is given in Figure 12.3(a). Notice that the trajectories move *away* from the origin and that all the integral curves are tangent to the x-axis at the origin, except for the two integral curves along the y-axis. In this case, we say that the origin is an **unstable improper node**. It is unstable because the trajectories move *away* from the origin, and it is an improper node because (almost all) integral curves have the same tangent line at the origin. (A proper node occurs in Case 4; see Figures 12.3 and 12.8 for a comparison.)

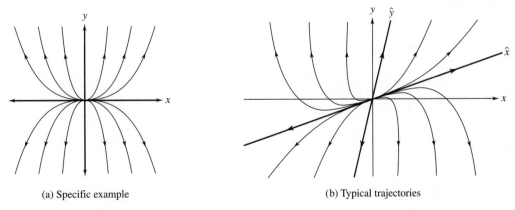

(a) Specific example (b) Typical trajectories

Figure 12.3 Unstable improper node; roots are real, distinct, and positive

[†] Such change of variables procedures are studied in linear algebra as "change of basis," where it is shown that the structure of the eigenvalues and eigenspaces remains unchanged when a matrix is diagonalized. See the book by Charles G. Cullen, *Matrices and Linear Transformations* (Dover Publishing, New York, 1992), Chapter 5.

In general, when the roots are distinct, real, and positive, the origin is an *unstable improper node* whose phase plane diagram near the origin resembles Figure 12.3(b). The colored lines labeled $\hat{x}$ and $\hat{y}$ represent the images of the x- and y-axes under a change of variables. We make two key observations. First, (most) trajectories in Figures 12.3(a) and 12.3(b) are becoming parallel to the y- and $\hat{y}$-axes, respectively, as they move away from the origin. Second, (most) trajectories are leaving the origin along integral curves that are tangent to the x- or $\hat{x}$-axes. We refer to the $\hat{x}$- and $\hat{y}$-axes as the **transformed axes**. Observe that these transformed axes are themselves integral curves.

Case 2. r_1, r_2 real, distinct, negative. The general solutions to (3)–(4) have the same form as in (6). A simple example is the system

$$\frac{dx}{dt} = -2x, \qquad \frac{dy}{dt} = -y,$$

for which $r_1 = -2$, $r_2 = -1$ and

$$x(t) = Ae^{-2t}, \qquad y(t) = Be^{-t}$$

is a general solution. The reader should verify that the integral curves for this example have the form $x = (A/B^2)y^2$. A phase plane diagram for this system is given in Figure 12.4(a). Notice that in this example, the trajectories are moving *toward* the origin. For this reason, the origin is called an **asymptotically stable** critical point. Again, since (most) trajectories approach the origin along curves that are tangent to the y-axis, the origin is an improper node. Thus the critical point at the origin is an asymptotically stable improper node.

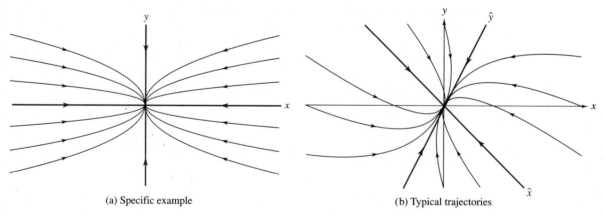

(a) Specific example (b) Typical trajectories

Figure 12.4 Stable improper node; roots are real, distinct, and negative

In general, when the roots are distinct, real, and negative, the origin is an *asymptotically stable improper node* whose phase plane diagram near the origin resembles that in Figure 12.4(b). Cases 1 and 2 are quite similar. They differ mainly in that when the roots are both

positive, the trajectories move away from the origin (the origin is unstable), while when the roots are both negative, the trajectories approach the origin (the origin is stable).

EXAMPLE 1 Classify the critical point at the origin and sketch a phase plane diagram for the system

(7) $$\frac{dx}{dt} = -5x + 2y,$$

(8) $$\frac{dy}{dt} = x - 4y.$$

Solution The characteristic equation for this system is $r^2 + 9r + 18 = (r + 6)(r + 3) = 0$, which has roots $r_1 = -6$, $r_2 = -3$. Thus the origin is an asymptotically stable improper node. To sketch the phase plane diagram, we must first determine the two lines passing through the origin that correspond to the transformed axes. These lines have the form $y = mx$ and must be trajectories for the given system. To determine the values of m, we first divide equations (7) and (8) to obtain

(9) $$\frac{dy}{dx} = \frac{dy/dt}{dx/dt} = \frac{x - 4y}{-5x + 2y}.$$

Substituting in $y = mx$ gives

$$m = \frac{x - 4mx}{-5x + 2mx} = \frac{1 - 4m}{-5 + 2m}.$$

Solving for m yields $2m^2 - m - 1 = (m - 1)(2m + 1) = 0$, so $m = 1, -1/2$. Hence the two axes are $y = x$ and $y = -(1/2)x$. As we know, the trajectories approach one of these lines tangentially. To determine which, let's graph a few of the isoclines for (9). For example, when $dy/dx = 0$, we get $x - 4y = 0$ or $y = (1/4)x$. So, when a trajectory crosses the line $y = (1/4)x$, its slope is zero. Moreover, when dy/dx is infinite or undefined, then $-5x + 2y = 0$. Hence, when a trajectory crosses the line $y = (5/2)x$, the tangent line to the trajectory must be vertical. These isoclines are represented in Figure 12.5 by horizontal and vertical hash marks, respectively. It is now clear that the trajectories must be tangent to the line $y = x$ as they approach the origin. ■ End

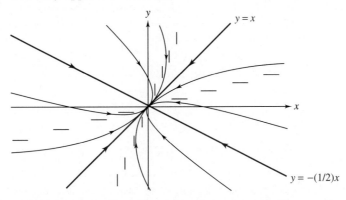

Figure 12.5 Phase plane diagram for $dx/dt = -5x + 2y$; $dy/dt = x - 4y$

Case 3. r_1, r_2 **real, opposite signs.** Again, a general solution to system (3)–(4) has the same form as in (6). A simple example is the system

$$\frac{dx}{dt} = x, \qquad \frac{dy}{dt} = -y,$$

which has a general solution of the form

$$x(t) = Ae^{t}, \qquad y(t) = Be^{-t},$$

with $r_1 = 1$, $r_2 = -1$. The trajectories lie on the integral curves $y = ABx^{-1}$ (see Figure 12.6(a)).

In this case, the trajectories come in along the positive and negative y-axis and go out along the positive and negative x-axis. The origin is unstable, since there are trajectories that pass arbitrarily near to the origin, but then eventually move away. (The two trajectories on the y-axis do approach the origin, but they are the only ones.) For apt reasons, the origin is said to be a **saddle point**. Hence, when the roots are real and of opposite signs, the origin is an *unstable saddle point*. In general, the trajectories come in alongside one of the transformed axes[†] and go out alongside the other. This is illustrated in Figure 12.6(b). Observe that the trajectories have both of these lines as asymptotes and that both lines are themselves trajectories for the system.

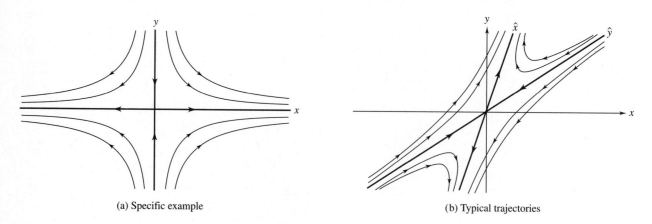

(a) Specific example (b) Typical trajectories

Figure 12.6 Unstable saddle point; roots are real and of opposite signs

[†] The phase plane diagram can be generated using a computer software package such as PHASEPLANE, PHASER, or MACMATH; however, knowing the transformed axes is still helpful.

EXAMPLE 2 Classify the critical point at the origin and sketch a phase plane diagram for the system

(10) $$\frac{dx}{dt} = 5x - 3y,$$

(11) $$\frac{dy}{dt} = 4x - 3y.$$

Solution The characteristic equation for this system is $r^2 - 2r - 3 = (r + 1)(r - 3) = 0$, which has roots $r_1 = -1$, $r_2 = 3$. Thus the origin is an unstable saddle point. To sketch the phase plane diagram, we must first determine the two lines corresponding to the transformed axes. Solving for dy/dx, we have

$$\frac{dy}{dx} = \frac{dy/dt}{dx/dt} = \frac{4x - 3y}{5x - 3y}.$$

Next we substitute $y = mx$ and find

$$m = \frac{4x - 3mx}{5x - 3mx} = \frac{4 - 3m}{5 - 3m}.$$

Solving for m gives $3m^2 - 8m + 4 = (3m - 2)(m - 2) = 0$ so that $m = 2, 2/3$. Hence the two transformed axes are $y = 2x$ and $y = (2/3)x$. Since the origin is a saddle point, the trajectories are moving in toward the origin alongside one of the transformed axes and they are moving out alongside the other. On the line $y = 2x$ itself, we find that

$$\frac{dx}{dt} = 5x - 3y = 5x - 3(2x) = -x.$$

So, when $x > 0$, this trajectory moves toward the origin, since $dx/dt < 0$. The same is true for $x < 0$, since $dx/dt > 0$. Consequently, on the other line, $y = (2/3)x$, the trajectories must be moving away from the origin. Indeed, a quick check shows that on $y = (2/3)x$

$$\frac{dx}{dt} = 5x - 3y = 5x - 3(2/3)x = 3x,$$

so the trajectories do, in fact, recede from the origin. The phase plane diagram is shown in Figure 12.7. ∎ End

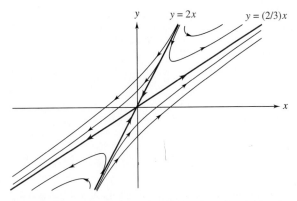

Figure 12.7 Phase plane diagram for $dx/dt = 5x - 3y$; $dy/dt = 4x - 3y$

Case 4. Equal roots, $r_1 = r_2$. In this case, a general solution may or may not[†] involve a factor of t times the exponential e^{rt}. As we shall see, the presence of the factor of t alters the phase plane diagram, but not the stability. We will discuss the two possibilities.

A simple example where the factor of t is not present is

$$\frac{dx}{dt} = rx, \qquad \frac{dy}{dt} = ry,$$

which has the general solution

$$x(t) = Ae^{rt}, \qquad y(t) = Be^{rt}.$$

Here the trajectories lie on the integral curves $y = (B/A)x$. When $r > 0$, these trajectories move away from the origin, so the origin is unstable (see Figure 12.8(a)). When $r < 0$, the trajectories approach the origin and the origin is asymptotically stable (see Figure 12.8(b)). In either case, the trajectories lie on lines passing through the origin and each line is an integral curve. Because of the nonrestrictive directions of these integral curves, the origin is called a **proper node.**

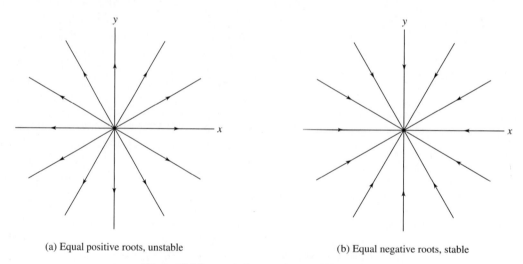

(a) Equal positive roots, unstable (b) Equal negative roots, stable

Figure 12.8 Proper nodes; equal roots with no factor of t

In general, when the roots are equal and a factor of t is not present, the origin is a *proper node* that is *unstable* when the root $r > 0$ and *asymptotically stable* when $r < 0$.

An example where the factor of t arises is

$$\frac{dx}{dt} = rx, \qquad \frac{dy}{dt} = x + ry,$$

[†] In Section 9.7 we found that when we have a repeated eigenvalue and there are not two linearly independent eigenvectors, then the factor of t is present.

which has the general solution

$$x(t) = Ae^{rt}, \qquad y(t) = (At + B)e^{rt}.$$

To find the trajectories, we first solve for t in terms of x to obtain

$$t = \frac{1}{r}\ln(x/A).$$

Substituting for t in the expression for $y(t)$ and simplifying, we find

$$y = \frac{B}{A}x + \frac{x}{r}\ln(x/A).$$

These integral curves are sketched in Figure 12.9(a). Observe that dy/dx is undefined for $x = 0$. When $r > 0$, the trajectories move away from the origin, so it is unstable. Since the trajectories lie on curves that are tangent to the y-axis at the origin, we have an improper node (see Figure 12.9(a)). When $r < 0$, the trajectories approach the origin, so the origin is an asymptotically stable point.

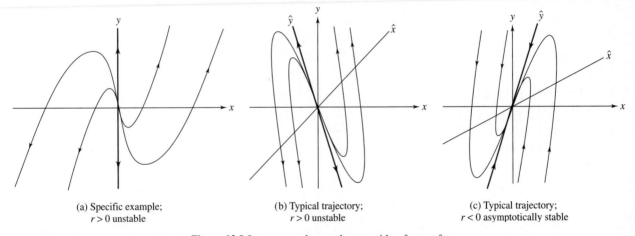

(a) Specific example;
$r > 0$ unstable

(b) Typical trajectory;
$r > 0$ unstable

(c) Typical trajectory;
$r < 0$ asymptotically stable

Figure 12.9 Improper node; equal roots with a factor of t

In summary, when the roots are equal and a factor of t is present in the general solution, the origin is an *improper node* that is *unstable* when $r > 0$ (see Figure 12.9(b)) and *asymptotically stable* when $r < 0$ (see Figure 12.9(c)).

Case 5. Complex roots, $r = \alpha \pm i\beta$. We now consider the case of complex roots $r = \alpha \pm i\beta$, where $\alpha \neq 0$ and $\beta \neq 0$. (Pure imaginary roots, $r = \pm i\beta$, are discussed in Case 6.) For complex roots, a general solution to system (3)–(4) is given by

$$x(t) = e^{\alpha t}[A_1 \cos \beta t + B_1 \sin \beta t],$$
$$y(t) = e^{\alpha t}[A_2 \cos \beta t + B_2 \sin \beta t].$$

It is clear that when $\alpha > 0$, the trajectories will travel far away from the origin. Hence the origin is unstable. But, when $\alpha < 0$, the trajectories approach the origin and the origin is asymptotically stable.

To better describe the behavior of trajectories near the origin, we analyze the system

(12) $\dfrac{dx}{dt} = \alpha x - \beta y,$

(13) $\dfrac{dy}{dt} = \beta x + \alpha y,$

which has characteristic roots $r = \alpha \pm i\beta$. If we set $z(t) = x(t) + iy(t)$ to form a complex-valued function, then

$$\frac{dz}{dt} = \frac{dx}{dt} + i\frac{dy}{dt} = (\alpha x - \beta y) + i(\beta x + \alpha y)$$

$$= \alpha(x + iy) + i\beta(x + iy) = (\alpha + i\beta)z.$$

Using the polar coordinate representation $z(t) = \rho(t)e^{i\theta(t)}$, we obtain

$$\frac{dz}{dt} = \frac{d\rho}{dt}e^{i\theta} + i\rho e^{i\theta}\frac{d\theta}{dt}.$$

Setting the two expressions for dz/dt equal yields

$$\frac{d\rho}{dt}e^{i\theta} + i\rho e^{i\theta}\frac{d\theta}{dt} = (\alpha + i\beta)\rho e^{i\theta},$$

and simplifying, we have

$$\frac{d\rho}{dt} + i\rho\frac{d\theta}{dt} = \alpha\rho + i\beta\rho.$$

By equating the real and imaginary parts, we obtain the simple system

$$\frac{d\rho}{dt} = \rho\alpha, \qquad \frac{d\theta}{dt} = \beta,$$

which has the general solution

$$\rho(t) = \rho_0 e^{\alpha t}, \qquad \theta(t) = \beta t + \theta_0,$$

where ρ_0 and θ_0 are real-valued constants. Again, we see that if $\alpha > 0$, the solution moves away from the origin, since $\rho \to \infty$ as $t \to \infty$. But, because θ is a linear function of t, the trajectories spiral around the origin. Similarly, when $\alpha < 0$, the solutions spiral in toward the origin. We therefore refer to the origin as a **spiral point.** We can determine the direction of rotation of the spiral from the original system. For example, setting $y = 0$ in equation (4), we find that $dy/dt = cx$. Now, if $c < 0$, then as the trajectory crosses the positive x-axis ($y = 0$), we have $dy/dt < 0$. Consequently, the trajectory is spiraling clockwise about the origin (see Figures 12.10(a) and 12.10(b)).

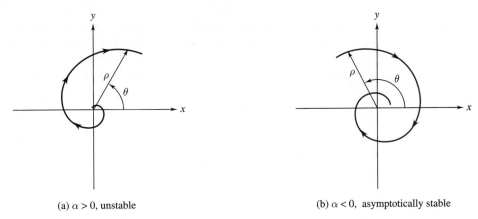

(a) $\alpha > 0$, unstable

(b) $\alpha < 0$, asymptotically stable

Figure 12.10 Spiral point; $r = \alpha \pm i\beta$ with clockwise rotation

In general, the spirals may be skewed, but stability and orientation remain the same. So when the roots $r = \alpha \pm i\beta$ are complex numbers, the origin is a *spiral point* that is *unstable* when $\alpha > 0$ (see Figure 12.11(a)) and *asymptotically stable* when $\alpha < 0$ (see Figure 12.11(b)).

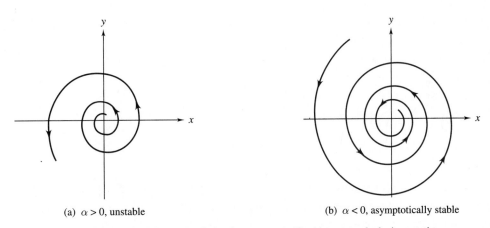

(a) $\alpha > 0$, unstable

(b) $\alpha < 0$, asymptotically stable

Figure 12.11 General spiral point; $r = \alpha \pm i\beta$ with counterclockwise rotation

EXAMPLE 3 Classify the critical point at the origin and sketch the phase plane diagram for the system

(14) $\dfrac{dx}{dt} = 2x - y,$

(15) $\dfrac{dy}{dt} = x + 2y.$

Solution The characteristic equation for this system is $(r - 2)^2 + 1 = 0$, which has complex roots $r = 2 \pm i$. Thus the origin is an unstable spiral point, since $\alpha = 2 > 0$. Setting $y = 0$ in the equation (15), we obtain $dy/dt = x$. Hence $dy/dt > 0$ when a trajectory crosses the positive x-axis. Thus the trajectories are spiraling with a counterclockwise rotation about the origin. See Figure 12.12. ■ End

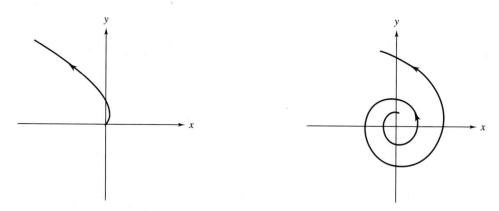

Figure 12.12 Trajectories for $dx/dt = 2x - y$; $dy/dt = x + 2y$

Case 6. Pure imaginary roots, $r = \pm i\beta$. When the roots are pure imaginary, $r = \pm i\beta$, a general solution has the form

$$x(t) = A_1 \cos \beta t + B_1 \sin \beta t,$$

$$y(t) = A_2 \cos \beta t + B_2 \sin \beta t.$$

These solutions all have period $2\pi/\beta$ and remain bounded. A simple example is the system

$$\frac{dx}{dt} = -\beta y, \qquad \frac{dy}{dt} = \beta x,$$

which is just system (12)–(13) with $\alpha = 0$. Hence, taking $\alpha = 0$ in the analysis of Case 5, we find that, in polar coordinates,

$$\frac{d\rho}{dt} = 0, \qquad \frac{d\theta}{dt} = \beta.$$

Hence $\rho = \rho_0$ and $\theta(t) = \beta t + \theta_0$. Since ρ is constant (but arbitrary), the trajectories are concentric circles about the origin. Thus the motion is a periodic rotation around a circle centered at the origin. Appropriately the origin is called a **center** and is a stable critical point.

 In general, when the roots are pure imaginary, the origin is a *stable center*. The trajectories are ellipses centered at the origin (see Problem 24). A typical phase diagram is shown in Figure 12.13.

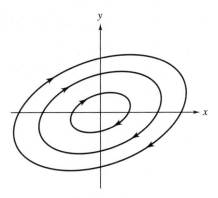

Figure 12.13 Stable center; pure imaginary roots, clockwise rotation

We now summarize the analysis of the behavior of trajectories of a linear system near an isolated critical point at the origin.

STABILITY OF LINEAR SYSTEMS

Theorem 1. Assume the origin $(0, 0)$ is an isolated critical point for the linear system

$$x'(t) = ax + by,$$
$$y'(t) = cx + dy,$$

where a, b, c, and d are real and $ad - bc \neq 0$. Let r_1 and r_2 be the roots of the characteristic equation

$$r^2 - (a + d)r + (ad - bc) = 0.$$

The stability of the origin and the classification of the origin as a critical point depends on the roots r_1 and r_2 as follows:

Roots	Type of critical point	Stability
distinct, positive	improper node	unstable
distinct, negative	improper node	asymptotically stable
opposite signs	saddle point	unstable
equal, positive	proper or improper node	unstable
equal, negative	proper or improper node	asymptotically stable
complex-valued:		
positive real part	spiral point	unstable
negative real part	spiral point	asymptotically stable
pure imaginary	center	stable

By a simple change of variables, Theorem 1 can be used to classify critical points that are not located at the origin.

EXAMPLE 4 Find and classify the critical point of the linear system

$$(16) \qquad \frac{dx}{dt} = 2x + y + 3,$$

$$(17) \qquad \frac{dy}{dt} = -3x - 2y - 4.$$

Solution The critical point is the solution to the system

$$2x + y + 3 = 0,$$
$$-3x - 2y - 4 = 0.$$

Solving this system, we obtain the critical point $(-2, 1)$. Now we will use the change of variables

$$x = u - 2, \qquad y = v + 1,$$

to translate the critical point $(-2, 1)$ to the origin $(0, 0)$. Substituting into system (16)–(17) and simplifying, we obtain a system of differential equations in u and v:

$$\frac{du}{dt} = \frac{dx}{dt} = 2(u - 2) + (v + 1) + 3 = 2u + v,$$

$$\frac{dv}{dt} = \frac{dy}{dt} = -3(u - 2) - 2(v + 1) - 4 = -3u - 2v.$$

The characteristic equation for this linear system is $r^2 - 1 = 0$, which has roots $r = \pm 1$. Hence the origin is an unstable saddle point for the new system. Therefore, the critical point $(-2, 1)$ is an unstable saddle point for the original system (16)–(17). ■ End

EXERCISES 12.2

In Problems 1 through 6, classify the critical point at the origin of the given linear system.

In Problems 7 through 12, find and classify the critical point of the given linear system.

1. $\dfrac{dx}{dt} = 5x + 6y,$

$\dfrac{dy}{dt} = -5x - 8y.$

2. $\dfrac{dx}{dt} = 3x,$

$\dfrac{dy}{dt} = 3y.$

3. $\dfrac{dx}{dt} = 3x + 5y,$

$\dfrac{dy}{dt} = -5x - 5y.$

4. $\dfrac{dx}{dt} = 6x - y,$

$\dfrac{dy}{dt} = 8y.$

5. $\dfrac{dx}{dt} = -6x + 3y,$

$\dfrac{dy}{dt} = x - 4y.$

6. $\dfrac{dx}{dt} = 2x + 13y,$

$\dfrac{dy}{dt} = -x - 2y.$

7. $\dfrac{dx}{dt} = -4x + 2y + 8,$

$\dfrac{dy}{dt} = x - 2y + 1.$

8. $\dfrac{dx}{dt} = -3x - y - 4,$

$\dfrac{dy}{dt} = 29x + y + 30.$

9. $\dfrac{dx}{dt} = 2x + y + 9,$

$\dfrac{dy}{dt} = -5x - 2y - 22.$

10. $\dfrac{dx}{dt} = 2x + 4y + 4,$

$\dfrac{dy}{dt} = 3x + 5y + 4.$

11. $\dfrac{dx}{dt} = -x + y + 8,$

$\dfrac{dy}{dt} = -x - 4y - 17.$

12. $\dfrac{dx}{dt} = -x + y + 4,$

$\dfrac{dy}{dt} = -y + 1.$

In Problems 13 through 20, classify the critical point at the origin and either by hand or using a computer software package, sketch a phase plane diagram for the given system.

13. $\dfrac{dx}{dt} = x - 6y,$

$\dfrac{dy}{dt} = 2x - 7y.$

14. $\dfrac{dx}{dt} = 5x,$

$\dfrac{dy}{dt} = 5y.$

15. $\dfrac{dx}{dt} = x + 2y,$

$\dfrac{dy}{dt} = 5x - 2y.$

16. $\dfrac{dx}{dt} = -3x - 5y,$

$\dfrac{dy}{dt} = 2x - y.$

17. $\dfrac{dx}{dt} = -8x + y,$

$\dfrac{dy}{dt} = 2x + 2y.$

18. $\dfrac{dx}{dt} = 3x - y,$

$\dfrac{dy}{dt} = -2x + 4y.$

19. $\dfrac{dx}{dt} = -2x + y,$

$\dfrac{dy}{dt} = -2y.$

20. $\dfrac{dx}{dt} = -3x - y,$

$\dfrac{dy}{dt} = 13x + 3y.$

21. Show that when the system

$$x'(t) = ax + by + p,$$
$$y'(t) = cx + dy + q$$

has a critical point at (h, k), then the change of variables $x = u + h$, $y = v + k$ yields the new system

$$u'(t) = au + bv,$$
$$v'(t) = cu + dv,$$

with the origin as the critical point.

22. Show that if $ad - bc = 0$, then there exist critical points for the system (3)–(4) that are arbitrarily close to $(0, 0)$. [*Hint:* If a or b is not zero, then the points on the line $ax + by = 0$ are all critical points.]

23. Show that the change of variables

$$x = 2u + v, \qquad y = 3u + v$$

transforms the linear system

$$\frac{dx}{dt} = x - 2y, \qquad \frac{dy}{dt} = 3x - 4y$$

into the linear system

$$\frac{du}{dt} = -2u, \qquad \frac{dv}{dt} = -v,$$

which is the example studied in Case 2. Verify that the u-axis maps to the line $y = (3/2)x$ and the v-axis maps to the line $y = x$.

24. Show that the set of points

$$C := \{(x, y): x^2 + y^2 = r^2\}$$

is mapped by the change of variables

$$x = au + bv, \qquad y = cu + dv,$$

where $ad - bc \neq 0$, into the set of points

$$E := \{(u, v): (a^2 + c^2)u^2 + (2ab + 2cd)uv + (b^2 + d^2)v^2 = r^2\}.$$

Verify that E is an ellipse in the uv-plane.

12.3 ALMOST LINEAR SYSTEMS

Having in mind the discussion of the preceding section concerning stability for linear autonomous systems, we now turn to the general situation of a *nonlinear* system and ask: Exactly what is meant by stability of a critical point? Let's say we are dealing with the autonomous system

(1) $\qquad \dfrac{dx}{dt} = f(x, y),$

(2) $\qquad \dfrac{dy}{dt} = g(x, y),$

where f and g are real-valued functions that do not depend explicitly on t.

STABILITY OF A CRITICAL POINT

Definition 1. A critical point (x_0, y_0) of the autonomous system (1)–(2) is **stable** if, given any $\varepsilon > 0$, there exists a $\delta > 0$ such that every solution $x = \phi(t)$, $y = \psi(t)$ of the system that satisfies

$$\sqrt{(\phi(0) - x_0)^2 + (\psi(0) - y_0)^2} < \delta$$

at $t = 0$ also satisfies

$$\sqrt{(\phi(t) - x_0)^2 + (\psi(t) - y_0)^2} < \varepsilon$$

for all $t \geq 0$.

 If (x_0, y_0) is stable and there exists an $\eta > 0$ such that every solution $x = \phi(t)$, $y = \psi(t)$ that satisfies

$$\sqrt{(\phi(0) - x_0)^2 + (\psi(0) - y_0)^2} < \eta$$

at $t = 0$ also satisfies

$$\lim_{t \to \infty} \phi(t) = x_0, \qquad \lim_{t \to \infty} \psi(t) = y_0,$$

then the critical point is **asymptotically stable.**
 A critical point that is not stable is called **unstable.**

In other words, a critical point (x_0, y_0) is stable if any trajectory that begins near (within δ of) the point (x_0, y_0) remains near (within ε of) this point. This definition is illustrated in Figures 12.14(a) and 12.14(b). When all the trajectories that start close to a stable critical point actually approach it as $t \to \infty$, then the critical point is asymptotically stable (see Figure 12.14(a)). As we have seen in the case of a linear system whose characteristic equation has pure imaginary roots, it is possible for trajectories to remain close, but not actually approach a critical point (see Figure 12.14(b)); that is, there are indeed critical points that are stable but not asymptotically stable.

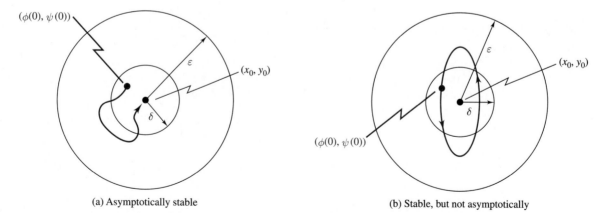

(a) Asymptotically stable (b) Stable, but not asymptotically

Figure 12.14 Stability of a critical point

A remark is in order. The essential property of asymptotic stability, that $\phi(t) \to x_0$ and $\psi(t) \to y_0$ as $t \to \infty$, is *not* sufficient for a critical point to be stable. There are, in fact, peculiar systems with the property that trajectories starting arbitrarily close to a critical point first travel away from the critical point before turning back.[†] This is shown schematically in Figure 12.15. For such a system, (x_0, y_0) is not stable and hence not asymptotically stable.

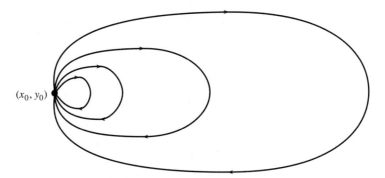

Figure 12.15 A peculiar unstable critical point

Our analysis in the preceding section concerning the behavior of trajectories near a critical point for a linear system suggests that our classification of the critical point as stable, asymptotically stable, and unstable is consistent with the definitions just given. To rigorously verify this, we would have to explain, for example, how to select the δ's given the ε's. However, we will not pursue such details here.

Our intuition leads us to believe that if we were to make a small perturbation of a linear system near a critical point, then the stability properties of the critical point for the perturbed system would be the same as those for the original linear system. In many cases this is true, but there are certain exceptions. These exceptions can be observed even when the perturbed system is linear. Since the classification of a critical point depends only on the roots of the characteristic equation (see equation (5) in Section 12.2), this classification can be determined by answering the question: How do the roots of the characteristic equation change under small perturbations of the coefficients? If the roots are real, distinct, and both positive (negative), then under a small perturbation they would still be positive (negative). Consequently the nature of the critical point would remain the same. This is also true if the roots are of opposite signs or are complex (not pure imaginary). However, there are two cases where small changes can substantially alter the nature of the roots and hence possibly alter the classification of the critical point.

The first is the case of pure imaginary roots. A small perturbation of such a system could result in any of three possible situations: pure imaginary roots, complex conjugate

[†]For a detailed example, see the book by L. Cesari, *Asymptotic Behavior and Stability Problems in Ordinary Differential Equations*, Third Edition (Springer-Verlag, New York, 1991), page 96.

roots with positive real parts ($\alpha > 0$), or complex conjugate roots with negative real parts ($\alpha < 0$). For example, the system

$$\frac{dx}{dt} = \alpha x - (1 + \gamma)y,$$

$$\frac{dy}{dt} = (1 + \gamma)x + \alpha y$$

has the characteristic equation

$$r^2 - 2\alpha r + [\alpha^2 + (1 + \gamma)^2] = 0,$$

which has roots $r = \alpha \pm (1 + \gamma)i$. Let the unperturbed system be the system with $\alpha = \gamma = 0$, for which the characteristic equation has pure imaginary roots $r = \pm i$. If we perturb this system, keeping α and γ small, we can have

(a) pure imaginary roots $r = \pm(1 + \gamma)i$ for $\alpha = 0$, or
(b) complex roots $r = \alpha \pm (1 + \gamma)i$ with positive real parts when $\alpha > 0$ and negative real parts when $\alpha < 0$.

Consequently if the critical point for the unperturbed system is a center, then the critical point for the perturbed linear system can be a center, an unstable spiral point, or an asymptotically stable spiral point. For arbitrarily small perturbations, these are the only possibilities.

A second exceptional situation is that of (real) equal roots. Here an arbitrarily small perturbation may result in distinct real roots or complex conjugate roots (depending on whether the double root separates horizontally or vertically in the plane; see Problem 18). If the equal roots were positive (negative), then under small perturbations, we would have real positive (negative) roots or complex conjugate roots whose real parts are positive (negative). In either case, the stability is not affected.

The situation is similar when the perturbed system is "almost linear."

ALMOST LINEAR SYSTEMS

Definition 2. Let the origin $(0, 0)$ be a critical point of the autonomous system

(3) $x'(t) = ax + by + F(x, y),$
(4) $y'(t) = cx + dy + G(x, y),$

where a, b, c, d are constants and F, G are continuous in a neighborhood (disk) about the origin. Assume that $ad - bc \neq 0$ so that the origin is an isolated critical point for the **corresponding linear system,** which we obtain by setting $F = G = 0$ in (3) and (4). Then, system (3)–(4) is said to be **almost linear** near the origin if

(5) $\dfrac{F(x, y)}{\sqrt{x^2 + y^2}} \to 0$ and $\dfrac{G(x, y)}{\sqrt{x^2 + y^2}} \to 0,$ as $\sqrt{x^2 + y^2} \to 0.$

Condition (5) can be written in a simpler fashion using the notation $r = \sqrt{x^2 + y^2}$ for the distance from (x, y) to the origin. The requirement now becomes $F(x, y)/r \to 0$ and $G(x, y)/r \to 0$ as $r \to 0^+$.

Almost linear systems are often encountered in applications, where F and G involve higher-order powers of x and y. For example, the system

$$\frac{dx}{dt} = 3x + 2y - xy + y^3,$$

$$\frac{dy}{dt} = -x + y + x^2 - y^2$$

is almost linear since $ad - bc = (3)(1) - (2)(-1) \neq 0$, and $F(x, y) = -xy + y^3$ and $G(x, y) = x^2 - y^2$ involve only higher-order terms in x and y. In fact, if the function $f(x, y)$ in equation (1) can be expanded in a Taylor series in x and y about $(0, 0)$ with $f(0, 0) = 0$, and if $g(x, y)$ in equation (2) can also be expanded in a Taylor series about $(0, 0)$ with $g(0, 0) = 0$, then system (1)–(2) will be almost linear, provided the origin is an isolated critical point for the corresponding linear system.

We now state a classical result due to Poincaré that relates the stability of the origin for an almost linear system with the stability of the origin in the corresponding linear system. The situation turns out to be the same as we found for perturbations that result in a linear system.[†]

STABILITY OF ALMOST LINEAR SYSTEMS

Theorem 2. Let r_1 and r_2 be the roots of the characteristic equation

(6) $r^2 - (a + d)r + (ad - bc) = 0$

for the linear system corresponding to the almost linear system (3)–(4). Then, the stability properties of the critical point at the origin for the almost linear system are the same as the stability properties of the origin for the corresponding linear system with one exception: When the roots are pure imaginary, the stability properties for the almost linear system cannot be deduced from the linear system (a different approach is required).

In addition to the stability, the behavior of the trajectories near the origin is essentially the same for the almost linear system and its corresponding linear system. The exceptions discussed for linear perturbations are the same as in the almost linear case. These are summarized in Table 12.1 on page 756.

[†] We do not give a proof here but refer the reader to the book by D. W. Jordan and P. Smith, *Nonlinear Ordinary Differential Equations,* Second Edition (Clarendon Press, Oxford, 1987), Chapter 10, where a detailed analysis using Lyapunov functions is given.

TABLE 12.1 STABILITY OF ALMOST LINEAR AND LINEAR SYSTEMS

Roots	Linear Systems		Almost Linear Systems	
r_1 and r_2	Type	Stability	Type	Stability
distinct, positive	improper node	unstable	improper node	unstable
distinct, negative	improper node	asymptotically stable	improper node	asymptotically stable
opposite signs	saddle point	unstable	saddle point	unstable
equal, positive	improper or proper node	unstable	improper or proper node or spiral point	unstable
equal, negative	improper or proper node	asymptotically stable	improper or proper node or spiral point	asymptotically stable
complex; real part				
positive	spiral point	unstable	spiral point	unstable
negative	spiral point	asymptotically stable	spiral point	asymptotically stable
pure imaginary	center	stable	center or spiral point	indeterminant

Here r_1 and r_2 denote the roots of the characteristic equation (6) for the linear system corresponding to the almost linear system (3)–(4).

Competing Species

In Section 12.1, we presented a model for two competing species with populations $x(t)$ and $y(t)$. These populations were assumed to satisfy the system

$$(7) \qquad \frac{dx}{dt} = x(a_1 - b_1 x - c_1 y),$$

$$(8) \qquad \frac{dy}{dt} = y(a_2 - b_2 y - c_2 x),$$

where the positive constants a_1, a_2, b_1, b_2, c_1, c_2 depend on the species and environment under investigation. Using Theorem 2 and Table 12.1, we present a detailed analysis of the phase plane diagram and discuss the stability for a particular competing species problem. The analytical methods presented here combined with the use of a computer can produce a more detailed phase plane diagram (see Project C).

Assume the populations satisfy the almost linear system

$$(9) \qquad \frac{dx}{dt} = x(7 - x - 2y) = 7x - x^2 - 2xy,$$

$$(10) \qquad \frac{dy}{dt} = y(5 - y - x) = 5y - y^2 - xy.$$

Then the critical points for this system are solutions to the pair of equations

$$x(7 - x - 2y) = 0,$$
$$y(5 - y - x) = 0.$$

One solution is $(0, 0)$. Setting $x = 0$, we also obtain $y = 5$ and setting $y = 0$, we get $x = 7$. So three of the critical points are $(0, 0)$, $(0, 5)$ and $(7, 0)$. While these points are not strictly inside the first quadrant, they do lie on its border. Therefore knowing the behavior of trajectories near these points will help us to obtain a phase diagram for the first quadrant. A fourth critical point occurs at the intersection of the lines

$$7 - x - 2y = 0,$$
$$5 - y - x = 0,$$

which is the point $(3, 2)$. We begin with an analysis of the trajectories near the origin $(0, 0)$.

Critical point $(0, 0)$. We get the corresponding linear system by deleting the second order terms in (9)–(10). This yields

$$\frac{dx}{dt} = 7x, \qquad \frac{dy}{dt} = 5y.$$

The characteristic equation for this system is $(r - 7)(r - 5) = 0$, which has roots $r_1 = 7$, $r_2 = 5$. Thus the origin is an unstable improper node. Furthermore, this system is easy to solve (the variables are uncoupled!). We find that $x(t) = Ae^{7t}$ and $y(t) = Be^{5t}$, so we can directly compute the trajectories. They are $y = B(e^t)^5 = (B/A^{5/7})x^{5/7}$. Observe that the trajectories are tangent to the y-axis and move away from the origin. In Figure 12.16, we have sketched the phase plane diagram in the first quadrant near the origin.

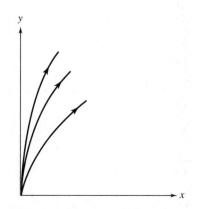

Figure 12.16 Phase plane diagram near $(0, 0)$

Critical point (0, 5). We begin by translating the point $(0, 5)$ to the origin by making the change of variables $x = u$, $y = v + 5$. Substituting into system (9)–(10) and simplifying, we obtain the almost linear system

$$\frac{du}{dt} = -3u - u^2 - 2uv,$$

$$\frac{dv}{dt} = -5u - 5v - v^2 - uv.$$

The corresponding linear system is

$$\frac{du}{dt} = -3u, \qquad \frac{dv}{dt} = -5u - 5v,$$

which has the characteristic equation $(r + 3)(r + 5) = 0$ with roots $r_1 = -3$, $r_2 = -5$. Hence the point $(0, 5)$ is an asymptotically stable improper node. To find the transformed axes $\hat{u}$ and $\hat{v}$, we set $v = mu$. Substituting into

$$\frac{dv}{du} = \frac{dv/dt}{du/dt} = \frac{-5u - 5v}{-3u},$$

we obtain

$$m = \frac{-5 - 5m}{-3},$$

and solving for m, we find $m = -5/2$. Thus one transformed axis is $v = -(5/2)u$. What about the other axis? By substituting $v = mu$, we allow for all lines through the origin except the vertical line $u = 0$. Thus we suspect that the v-axis ($u = 0$) is the missing transformed axis. We can confirm this by solving for du/dv:

$$\frac{du}{dv} = \frac{-3u}{-5u - 5v},$$

which is indeed satisfied by $u = 0$. Therefore the transformed axes are $v = -(5/2)u$ and $u = 0$. If we check the direction field along the u-axis, we see that $dv/du = 5/3$. Moreover, $dv/du = 0$ along the line $v = -u$. Consequently the trajectories approach the transformed axis $v = -(5/2)u$ tangentially. The phase plane diagram for the linearized system is sketched in Figure 12.17(a), and the diagram near the critical point $(0, 5)$ of the original system is shown in Figure 12.17(b).

Critical point (7, 0). To translate the point $(7, 0)$ to the origin, we set $x = u + 7$, $y = v$. Substituting into (9)–(10) and simplifying yields the almost linear system

$$\frac{du}{dt} = -7u - 14v - u^2 - 2uv,$$

$$\frac{dv}{dt} = -2v - v^2 - uv.$$

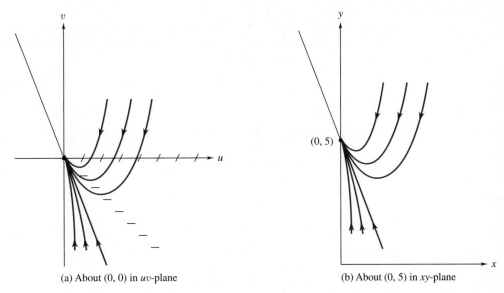

(a) About (0, 0) in uv-plane (b) About (0, 5) in xy-plane

Figure 12.17 Phase plane diagram near the critical point (0, 5)

The corresponding linear system

$$\frac{du}{dt} = -7u - 14v, \qquad \frac{dv}{dt} = -2v$$

has the characteristic equation $(r + 7)(r + 2) = 0$ with roots $r_1 = -7$, $r_2 = -2$. Thus the point $(7, 0)$ is an asymptotically stable improper node. To get the transformed axes, we set $v = mu$. For the linear system, substituting into

$$\frac{dv}{du} = \frac{dv/dt}{du/dt} = \frac{-2v}{-7u - 14v},$$

we obtain

$$m = \frac{-2m}{-7 - 14m},$$

which yields $m = 0$, $-5/14$. Therefore the two transformed axes are $v = 0$ and $v = -(5/14)u$. We now check the direction field along a few lines. When $u = 0$, $dv/du = 1/7$, and when $v = -u$, $dv/du = 2/7$. Moreover, dv/du is undefined when $v = -(1/2)u$. Consequently the trajectories approach the transformed axis $v = -(5/14)u$ tangentially. The phase plane diagrams in the uv- and xy-planes are sketched in Figures 12.18(a) and (b) on page 760.

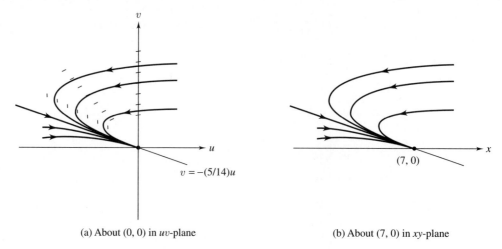

(a) About (0, 0) in uv-plane (b) About (7, 0) in xy-plane

Figure 12.18 Phase plane diagram near the critical point (7, 0)

Critical point (3, 2). Setting $x = u + 3$, $y = v + 2$ and substituting into system (9)–(10), we obtain the almost linear system

$$\frac{du}{dt} = -3u - 6v - u^2 - 2uv,$$

$$\frac{dv}{dt} = -2u - 2v - v^2 - uv.$$

The corresponding linear system has the characteristic equation $r^2 + 5r - 6 = (r + 6)(r - 1) = 0$ with roots $r = -6, 1$. Hence the point (3, 2) is an unstable saddle point. To obtain the transformed axes, we set $v = mu$. For the linear system, substituting into

$$\frac{dv}{du} = \frac{dv/dt}{du/dt} = \frac{-2u - 2v}{-3u - 6v},$$

we obtain

$$m = \frac{-2 - 2m}{-3 - 6m}.$$

Solving for m yields $m = 1/2, -2/3$. Thus the two transformed axes are $v = (1/2)u$ and $v = -(2/3)u$. For a saddle point, the trajectories move toward the critical point alongside one transformed axis and away alongside the other. If we set $v = -(2/3)u$, then for the linear system, we have $du/dt = -3u - 6(-2/3)u = u$. The trajectories along this transformed axis are moving away from the origin because $du/dt > 0$ if $u > 0$ and $du/dt < 0$ for $u < 0$. Checking the line $v = (1/2)u$, we see that $du/dt = -3u - 6(1/2)u = -6u$, so the trajectories are moving toward the origin. The phase plane diagrams in the uv- and xy-plane are sketched in Figure 12.19.

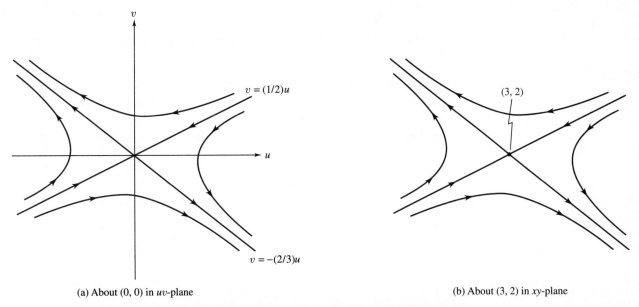

(a) About (0, 0) in uv-plane

(b) About (3, 2) in xy-plane

Figure 12.19 Phase plane diagram near the critical point (3, 2)

In Figure 12.20, we have combined the results for each of the four critical points and sketched the phase plane diagram in the first quadrant for the system (9)–(10). We have included four trajectories that separate the first quadrant into four regions labeled I, II, III, and IV. (These trajectories are difficult to find and the curves sketched represent only rough

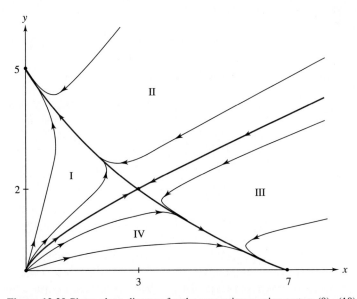

Figure 12.20 Phase plane diagram for the competing species system (9)–(10)

approximations of the true curves; see Project C.) These curves are called *separatrices*. Observe that a trajectory in region I remains in that region and approaches the critical point $(0, 5)$ as $t \to +\infty$. Similarly, trajectories in region II also approach $(0, 5)$ as $t \to +\infty$. The trajectories in regions III and IV approach the critical point $(7, 0)$ as $t \to +\infty$. Regions I and II form part of the **regions of asymptotic stability for the critical point** $(0, 5)$; that is, a trajectory beginning in that region is attracted to $(0, 5)$. Similarly, regions III and IV are part of the regions of asymptotic stability for the critical point $(7, 0)$.[†]

What does the phase plane diagram tell us about the competing species problem? If the initial population vector $(x(0), y(0))$ lies in either region I or II, then species x dies off and species y approaches the limiting population of 5. (Units may be thousands, millions, etc.) If, however, $(x(0), y(0))$ lies in region III or IV, then species y dies off and the population x will approach the limiting value 7. Mathematically, there can be coexistence only if the initial population vector $(x(0), y(0))$ lies on either separatrix that approaches the critical point $(3, 2)$ or lies precisely at the critical point $(3, 2)$. However, this situation does not arise in practice because the model, no matter how good, is just an approximation and any error can move the true trajectory off a separatrix.

EXERCISES 12.3

In Problems 1 through 8, show that the given system is almost linear near the origin and discuss the type and stability of the critical point at the origin.

1. $\dfrac{dx}{dt} = 3x + 2y - y^2,$

$\dfrac{dy}{dt} = -2x - 2y + xy.$

2. $\dfrac{dx}{dt} = -2x + 2xy,$

$\dfrac{dy}{dt} = x - y + x^2.$

3. $\dfrac{dx}{dt} = x - y,$

$\dfrac{dy}{dt} = 3x + 5y - x^3.$

4. $\dfrac{dx}{dt} = -x + 4y + y^2,$

$\dfrac{dy}{dt} = -x - y - xy.$

5. $\dfrac{dx}{dt} = x + 5y - y^2,$

$\dfrac{dy}{dt} = -x - y - y^2.$

6. $\dfrac{dx}{dt} = 4x + 5y + xy,$

$\dfrac{dy}{dt} = x + x^2 - y^2.$

7. $\dfrac{dx}{dt} = e^{x+y} - \cos x,$

$\dfrac{dy}{dt} = \cos y + x - 1.$

8. $\dfrac{dx}{dt} = \sin(y - 3x),$

$\dfrac{dy}{dt} = \cos x - e^y.$

In Problems 9 through 12, find all the critical points for the given system and discuss the type and stability of the critical points.

9. $\dfrac{dx}{dt} = 16 - xy,$

$\dfrac{dy}{dt} = x - y^3.$

10. $\dfrac{dx}{dt} = x^2 - y,$

$\dfrac{dy}{dt} = x + y^2.$

11. $\dfrac{dx}{dt} = x + y,$

$\dfrac{dy}{dt} = 5y - xy + 6.$

12. $\dfrac{dx}{dt} = 4 - xy,$

$\dfrac{dy}{dt} = x - y.$

 In Problems 13 through 16, find all the critical points for the given system, discuss the type and stability of each critical point, and either by hand or using a computer software package, sketch the phase plane diagrams near each of the critical points.

13. $\dfrac{dx}{dt} = 1 - xy,$

$\dfrac{dy}{dt} = x - y^3.$

14. $\dfrac{dx}{dt} = -x + xy - 6,$

$\dfrac{dy}{dt} = x - y.$

[†] Regions of asymptotic stability are also called "basins of attraction." When *every* trajectory approaches a critical point, the critical point is said to be globally stable. This happens in the case of an asymptotically stable critical point for an autonomous *linear* system. But, as we see here, an asymptotically stable critical point need not be globally stable.

15. $\dfrac{dx}{dt} = 6x - 2x^2 - xy,$ **16.** $\dfrac{dx}{dt} = x + y^2,$

$\dfrac{dy}{dt} = 6y - 2y^2 - xy.$ $\dfrac{dy}{dt} = x - 2y.$

17. van der Pol's equation

(a) Show that van der Pol's equation

$$\frac{d^2x}{dt^2} + \varepsilon(x^2 - 1)\frac{dx}{dt} + x = 0$$

is equivalent to the system

$$\frac{dx}{dt} = y,$$

$$\frac{dy}{dt} = -x + \varepsilon y - \varepsilon x^2 y,$$

where $y := dx/dt$.

(b) Describe how the stability of the zero solution of van der Pol's equation (equivalently, the origin of the system) depends on the parameter ε.

18. Consider the system

$$\frac{dx}{dt} = (\lambda + \alpha)x + \beta y,$$

$$\frac{dy}{dt} = -\beta x + (\lambda + \gamma)y,$$

where α, β, γ, and λ are real numbers. The characteristic equation for the unperturbed system, where $\alpha = \beta = \gamma = 0$, has a repeated root λ. Show that if we perturb the system, keeping α, β, and γ small, then the characteristic equation for the perturbed system can have the following:

(a) Equal roots for $\beta = 0$ and $\alpha = \gamma$.

(b) Distinct real roots for $\beta = 0$ and $\alpha \neq \gamma$.

(c) Complex roots for $\beta \neq 0$ and $\alpha = \gamma = 0$.

(d) What does this suggest about the type of critical point for an almost linear system if the characteristic equation for the corresponding linear system has equal real roots?

19. Let $x(t)$ and $y(t)$ represent the populations at time t of two species of insects that compete with each other for

food. Assume their populations are governed by the system (7)–(8) (cf. equations (1) and (2) in Section 12.1).

(a) Show that when $y \equiv 0$, species x has a stable equilibrium population of a_1/b_1. This is called the *carrying capacity* for x.

(b) What is the carrying capacity for y?

(c) Let A and B denote the carrying capacities for species x and y respectively. Show that the system can be rewritten in the equivalent form

$$\frac{dx}{dt} = a_1 x \left(1 - \frac{x}{A} - \frac{d_1 y}{A}\right),$$

$$\frac{dy}{dt} = a_2 y \left(1 - \frac{y}{B} - \frac{d_2 x}{B}\right),$$

where $d_i = c_i/b_i$, $i = 1, 2$.

(d) In Section 12.1 we observed that when $a_1/c_1 > a_2/b_2$ and $a_1/b_1 < a_2/c_2$, the two species can coexist. Assume this is initially the case but now an insecticide is being applied that kills only species x. Describe what happens to the competing populations if the application of the insecticide is increased (that is, a_1 is decreased).

20. Show that coexistence occurs in the competing species model governed by system (7)–(8) if and only if there is an asymptotically stable critical point in the first quadrant.

21. When one of the populations in a competing species model is being harvested, the phase plane diagram is completely changed. For example, consider the competition model

$$\frac{dx}{dt} = x(1 - 4x - y) - h,$$

$$\frac{dy}{dt} = y(1 - 2y - 5x),$$

where h is the constant harvest rate. For $h = 0$, $1/32$, and $5/32$, determine the critical points, discuss the type and stability of the critical points, and sketch the phase plane diagrams. In each case, discuss the implication of the competition model.

12.4 ENERGY METHODS

Many differential equations arise from problems in mechanics and for these equations it is natural to study the effect that energy in the system has on the solutions to the system. By analyzing the energy, we can often determine the stability of a critical point (even in the more elusive case when it is a center of the corresponding linear equation). As we shall see in the next section, these ideas can be generalized to systems that are unrelated to problems in mechanics.

Conservative Systems

A **conservative system** is one in which there are no dissipative forces acting. Two examples are a spring-mass system with no damping or friction and free fall without air resistance. In the absence of a dissipative force, the energy of the system must remain constant.

The total energy E in a mechanical system is the sum of the kinetic energy T and the potential energy U. If we let $x(t)$ denote the position coordinate, then the kinetic energy is given by

$$T = \frac{1}{2} m \left(\frac{dx}{dt} \right)^2,$$

where m is the mass, which we assume is constant. We also assume the potential energy is a function only of the position coordinate x. (For a spring-mass system, $U(x) = cx^2$ and for the simple pendulum, $U(x) = c(1 - \cos x)$.) The total energy is

$$E = T + U = \frac{1}{2} m \left(\frac{dx}{dt} \right)^2 + U(x).$$

For a conservative system, the total energy is constant during motion, so differentiating E with respect to t, we obtain

$$\frac{dE}{dt} = m \frac{dx}{dt} \frac{d^2x}{dt^2} + U'(x) \frac{dx}{dt} = \frac{dx}{dt} \left[m \frac{d^2x}{dt^2} + U'(x) \right] = 0.$$

Now either $dx/dt = 0$, in which case x is a constant, or

$$m \frac{d^2x}{dt^2} + U'(x) = 0.$$

Dividing by m, we obtain the standard form[†] for a conservative system:

(1) $$\frac{d^2x}{dt^2} + g(x) = 0,$$

where $g(x) := U'(x)/m$.

[†] Also called a **Newton equation.**

The Poincaré phase plane for these conservative systems will provide some insight. If we let $v = dx/dt$, then equation (1) is equivalent to the system

(2) $$\frac{dx}{dt} = v,$$

(3) $$\frac{dv}{dt} = -g(x).$$

Solving $dv/dx = -g(x)/v$, we obtain the integral curves

(4) $$\frac{v^2}{2} + G(x) = k,$$

where $G(x) := \int g(x)\,dx = U(x)/m + C$. Observe that these integral curves correspond to curves of constant total energy. We refer to

$$E(x, v) := \frac{v^2}{2} + G(x)$$

as an **energy function** and $G(x)$ as a **potential function**. A level curve $E(x, v) = k$ is called an **energy integral**.

EXAMPLE 1 Find the energy function $E(x, v)$ for the following equations. Select E so that the energy at the critical point $(0, 0)$ is zero, that is, $E(0, 0) = 0$.

(a) $\dfrac{d^2x}{dt^2} + x - x^3 = 0.$

(b) $\dfrac{d^2x}{dt^2} + \sin x = 0$ (pendulum equation).

(c) $\dfrac{d^2x}{dt^2} + x - x^4 = 0.$

Solution (a) Here $g(x) = x - x^3$. By integrating g, we obtain the potential function $G(x) = (1/2)x^2 - (1/4)x^4 + C$. Thus $E(x, v) = (1/2)v^2 + (1/2)x^2 - (1/4)x^4 + C$. Now $E(0, 0) = 0$ implies that $C = 0$. Hence, the energy function we seek is $E(x, v) = (1/2)v^2 + (1/2)x^2 - (1/4)x^4$.

(b) Since $g(x) = \sin x$, the potential function is $G(x) = -\cos x + C$, which gives $E(x, v) = (1/2)v^2 - \cos x + C$. Setting $E(0, 0) = 0$, we find $C = 1$. Thus the energy function is $E(x, v) = (1/2)v^2 - \cos x + 1$.

(c) Here $g(x) = x - x^4$ and the potential function is $G(x) = (1/2)x^2 - (1/5)x^5 + C$. Setting $E(0, 0) = 0$, we get $C = 0$, so the energy function is $E(x, v) = (1/2)v^2 + (1/2)x^2 - (1/5)x^5$. ∎ End

Notice that the pendulum equation in part (b) also has a critical point at $(\pi, 0)$. Setting the energy to be zero at $(\pi, 0)$, that is, $E(\pi, 0) = 0$, we arrive at the energy function $E(x, v) = (1/2)v^2 - \cos x - 1$.

For conservative systems, much useful information about the phase plane can be gained directly from the potential function $G(x)$. A convenient way of doing this is to place the graph of the potential function $z = G(x)$ (the potential plane) directly above the phase plane diagram with the z-axis aligned with the v-axis (see Figure 12.21). This places the relative extrema for $G(x)$ directly over the critical points for the system, since $G'(x_0) = g(x_0) = 0$ implies that $(x_0, 0)$ is a critical point for (2)–(3).

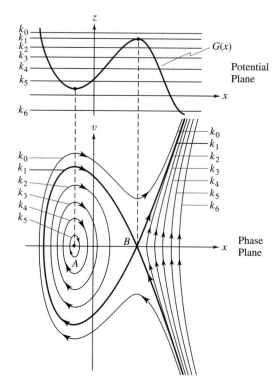

Figure 12.21 Relationship between potential plane (above) and phase plane (below)

To understand the relationship between the potential plane and the phase plane, let's look at the level curves for the energy function. In general, we have

$$\frac{v^2}{2} + G(x) = k,$$

where $G(x)$ is the potential function with $G'(x) = g(x)$. Solving for v gives

$$v = \pm\sqrt{2[k - G(x)]}.$$

Thus the velocity v exists (is real-valued) only when $k - G(x) \geq 0$. This generally results in three types of behavior.

When $G(x)$ has a strict local minimum at x_0 (above the point A) as illustrated in Figure 12.21, the integral curves $E(x, v) = k$, where k is slightly greater than $G(x_0)$, are closed curves encircling the critical point $A = (x_0, 0)$. To see this, observe that there is an interval (a, b) containing x_0 with $G(a) = G(b) = k_5$ and $G(x) < k_5$ for x in (a, b). Since $v = \sqrt{2[k_5 - G(x)]}$ is defined on (a, b) with $v = 0$ at $x = a$ and b, the two curves $v = \pm\sqrt{2[k_5 - G(x)]}$ join up at a and b to produce the closed curve. As this occurs for any k satisfying $G(x_0) < k \le k_5$, the critical point A is encircled by closed trajectories corresponding to periodic solutions. Thus A is a center.

When $G(x)$ has a strict local maximum at x_1 (above the point B), the integral curves $E(x, v) = k$ are *not* closed curves encircling the critical point $B = (x_1, 0)$. For $k = G(x_1)$, the curves touch the x-axis at x_1, but for $k > G(x_1)$, $v = \sqrt{2[k - G(x)]}$ is strictly positive and the level curves do not touch the x-axis. Finally, when $k < G(x_1)$, there is an interval about x_1 above and below which there are *no* points of the curve $E(x, v) = k$ (the quantity under the radical in the formula for v is negative). Hence the critical point B is a saddle point, and the level curves near B resemble hyperbolas. This is illustrated in Figure 12.21 by the level curves corresponding to k_0, k_1, and k_2 near B.

Finally, away from the extreme points of $G(x)$, the level curves may be part of a closed trajectory, such as the integral curve corresponding to $k = k_2$, or may be unbounded, such as the integral curves corresponding to $k = k_1$ or k_6.

EXAMPLE 2 Use the potential plane to help sketch the phase plane diagram for the equations:

$$\textbf{(a)}\;\; \frac{d^2x}{dt^2} + x - x^3 = 0. \qquad \textbf{(b)}\;\; \frac{d^2x}{dt^2} + \sin x = 0. \qquad \textbf{(c)}\;\; \frac{d^2x}{dt^2} + x - x^4 = 0.$$

Solution In Example 1 we found that potential functions for these equations are
(a) $G(x) = (1/2)x^2 - (1/4)x^4$; (b) $G(x) = 1 - \cos x$; (c) $G(x) = (1/2)x^2 - (1/5)x^5$.

(a) The local maxima and minima of $G(x)$ occur when $G'(x) = g(x) = x - x^3 = 0$; namely, at $x = 0, -1, 1$. Consequently the phase plane diagram for equation (a) has critical points at $(0, 0)$, $(-1, 0)$, and $(1, 0)$. Since $G(x)$ has a strict local minimum at $x = 0$, the critical point $(0, 0)$ is a center. Furthermore, $G(x)$ has local maxima at $x = \pm 1$, so $(-1, 0)$ and $(1, 0)$ are saddle points. The potential plane and the phase plane are displayed in Figure 12.22 on page 768.

(b) The local maxima and minima of $G(x) = 1 - \cos x$ occur when $G'(x) = g(x) = \sin x = 0$. Here $G(x)$ has local minima for $x = 0, \pm 2\pi, \pm 4\pi, \ldots$ and local maxima for $x = \pm\pi, \pm 3\pi, \pm 5\pi, \ldots$. Consequently the critical points $(2n\pi, 0)$, where n is an integer, are centers, while the critical points $((2n + 1)\pi, 0)$ are saddle points. The potential and phase planes are given in Figure 12.23 on page 768.

(c) The extreme values for $G(x) = (1/2)x^2 - (1/5)x^5$ occur when $G'(x) = g(x) = x - x^4 = 0$; that is, for $x = 0$ and 1. Here G has a minimum at $x = 0$ and a maximum at $x = 1$. Hence the critical point $(0, 0)$ is a center and the critical point $(1, 0)$ is a saddle point (see Figure 12.24 on page 769). ∎ End

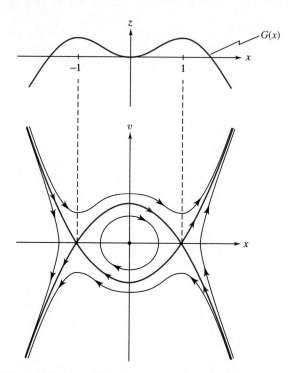

Figure 12.22 Potential and phase plane for $d^2x/dt^2 + x - x^3 = 0$

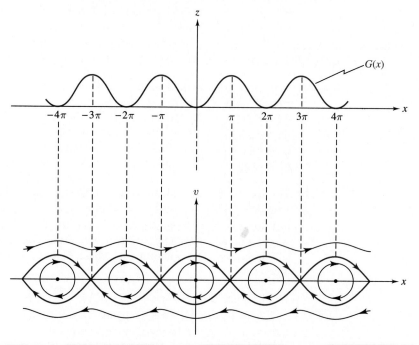

Figure 12.23 Potential and phase plane for $d^2x/dt^2 + \sin x = 0$

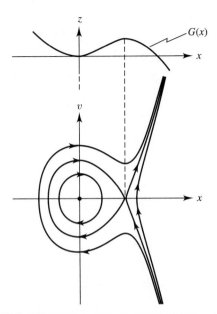

Figure 12.24 Potential and phase plane for $d^2x/dt^2 + x - x^4 = 0$

The reader may have already observed that the three conservative systems in Example 2 have the *same* corresponding linear equation $d^2x/dt^2 + x = 0$. As we know, the origin is a stable center for this linear equation. Thus Theorem 2 of the preceding section gives us no information about the stability of the origin for the three almost linear equations. However, using the energy method approach, we were able to determine that for these three conservative systems the origin is indeed a stable center.

Nonconservative Systems

For nonconservative systems, the energy of the system does *not* remain constant along some trajectory in the Poincaré phase plane. Consequently the level curves of the energy function are *not* integral curves for the system. However, if we know that the energy is decreasing along a trajectory, then it follows that the trajectory is moving toward a state that has a lower total energy. This state may correspond to a critical point or a limit cycle (see Definition 5 on page 782). We will demonstrate this behavior in the following examples.

A nonconservative system with nonlinear damping has the form

$$\frac{d^2x}{dt^2} + h(x,\, dx/dt) + g(x) = 0,$$

where $G(x) := \int g(x)\, dx$ corresponds to the potential energy and $h(x,\, dx/dt)$ represents damping. Let

$$E(x,\, dx/dt) := \frac{1}{2}\left(\frac{dx}{dt}\right)^2 + G(x)$$

as in the case of a conservative system ($h \equiv 0$). If we now compute dE/dt, we find

$$\frac{dE}{dt} = \frac{dx}{dt}\frac{d^2x}{dt^2} + G'(x)\frac{dx}{dt} = \frac{dx}{dt}\left(\frac{d^2x}{dt^2} + g(x)\right)$$

$$= -\frac{dx}{dt}h(x,\ dx/dt),$$

where we have used $d^2x/dt^2 + g(x) = -h(x,\ dx/dt)$. Consequently the *energy E* is decreasing when $vh(x,\ v) > 0$ and is increasing when $vh(x,\ v) < 0$.

EXAMPLE 3 Sketch the phase plane for the damped pendulum equation

$$\frac{d^2x}{dt^2} + b\frac{dx}{dt} + \sin x = 0,$$

where b is a positive constant.

Solution We first observe that $vh(x,\ v) = v(bv) = bv^2 > 0$ for $v \neq 0$. Hence the energy is continually decreasing along a trajectory. The level curves for the energy function $E(x,\ v) = (1/2)v^2 + 1 - \cos x$ are just the integral curves for the (undamped) simple pendulum equation and are sketched in Figure 12.23. (Recall that the integral curves $E(x,\ v) = k$ are symmetric with respect to the x-axis and shrink to points on the x-axis as k decreases.) Hence a trajectory must move toward the x-axis. The reader can verify that the critical points for the damped pendulum are the same as for the simple pendulum. Moreover, they are of the same type. The resulting phase plane diagram is sketched in Figure 12.25. In this case, the separatrices (colored curves) divide the plane into strips. A typical trajectory in a strip decays and spirals into an asymptotically stable spiral point (colored dots) located at multiples of 2π along the x-axis. ■ End

Figure 12.25 Phase plane diagram for damped pendulum

In the next example, the trajectories approach a limit cycle.

EXAMPLE 4 Sketch the phase plane diagram for the equation

$$\frac{d^2x}{dt^2} + \left(x^2 + \left(\frac{dx}{dt}\right)^2 - 1\right)\frac{dx}{dt} + x = 0.$$

Solution Here $h(x, v) = (x^2 + v^2 - 1)v$. Hence $vh(x, v) = v^2(x^2 + v^2 - 1)$, which is zero on the unit circle $x^2 + v^2 = 1$, positive outside the unit circle, and negative inside. Consequently trajectories beginning inside the unit circle spiral out and trajectories beginning outside the circle spiral in (see Figure 12.26). The unit circle (in color) is the trajectory for the solution $x(t) = \cos t$ and is called a **limit cycle**. ∎ End

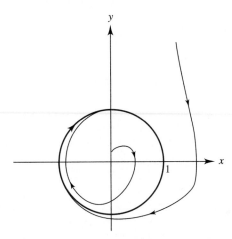

Figure 12.26 Phase plane diagram for $d^2x/dt^2 + [x^2 + (dx/dt)^2 - 1]dx/dt + x = 0$

EXERCISES 12.4

In Problems 1 through 6, find the potential energy function $G(x)$ and the energy function $E(x, v)$ for the given equations. Select E so that $E(0, 0) = 0$.

1. $\dfrac{d^2x}{dt^2} + x^2 - 3x + 1 = 0.$

2. $\dfrac{d^2x}{dt^2} + \cos x = 0.$

3. $\dfrac{d^2x}{dt^2} + \dfrac{x^2}{x-1} = 0.$

4. $\dfrac{d^2x}{dt^2} + x - \dfrac{x^3}{6} + \dfrac{x^5}{120} = 0.$

5. $\dfrac{d^2x}{dt^2} + x^3 + x^2 - x = 0.$

6. $\dfrac{d^2x}{dt^2} + e^x - 1 = 0.$

In Problems 7 through 12, use the potential plane to help sketch the phase plane diagrams for the given equations.

7. $\dfrac{d^2x}{dt^2} + x^3 - x = 0.$

8. $\dfrac{d^2x}{dt^2} + 9x = 0.$

9. $\dfrac{d^2x}{dt^2} + 2x^2 + x - 1 = 0.$

10. $\dfrac{d^2x}{dt^2} - \sin x = 0.$

11. $\dfrac{d^2x}{dt^2} + \dfrac{x}{x-2} = 0.$

12. $\dfrac{d^2x}{dt^2} + (x-1)^3 = 0.$

In Problems 13 through 16, use the energy function to assist in sketching the phase plane diagrams for the given nonconservative systems.

13. $\dfrac{d^2x}{dt^2} + \dfrac{dx}{dt} + x - x^3 = 0.$

14. $\dfrac{d^2x}{dt^2} + \dfrac{dx}{dt} + x - x^4 = 0.$

15. $\dfrac{d^2x}{dt^2} + \dfrac{dx}{dt} + \dfrac{x}{x-2} = 0.$

16. $\dfrac{d^2x}{dt^2} + \dfrac{dx}{dt} + 2x^2 + x - 1 = 0.$

17. Nonlinear Spring. The general nonlinear spring equation

$$\frac{d^2x}{dt^2} + \alpha x + \beta x^3 = 0,$$

where $\alpha > 0$ and β are parameters, is used to model a variety of physical phenomena. Holding $\alpha > 0$ fixed,

sketch the potential function and the phase plane diagram for $\beta > 0$ and also for $\beta < 0$. Describe how the behavior of solutions to the equation differs in these cases.

18. Cusps. We have observed that where the potential energy function $G(x)$ for a conservative system has a strict maximum (minimum), the corresponding critical point is a saddle point (center). At a critical point $(x_0, 0)$ for which $G(x)$ also has a point of inflection (e.g., $G'(x_0) = g(x_0) = 0$, $G''(x_0) = g'(x_0) = 0$, and $G'''(x_0) = g''(x_0) \neq 0$), the curve in the phase plane has a **cusp**. Demonstrate this by sketching the potential plane and the phase plane diagrams for the equation

$$\frac{d^2x}{dt^2} + x^2 = 0.$$

19. General Relativity. In studying the relativistic motion of a particle moving in the gravitational field of a larger body, one encounters the equation

$$\frac{d^2u}{d\theta^2} + u - \lambda u^2 - 1 = 0,$$

where u is inversely proportional to the distance of the particle from the body, θ is an angle in the plane of motion, and λ is a parameter with $0 < \lambda < 1$.

(a) Sketch the phase plane diagram for $0 < \lambda < 1/4$.
(b) Sketch the phase plane diagram for $1/4 < \lambda < 1$.
(c) What observations can be made about the motion of the particle in these two cases?

12.5 LYAPUNOV'S DIRECT METHOD

In the preceding section we used the energy of the system, $E(x, v)$, to determine the qualitative behavior of the trajectories of solutions of conservative as well as nonconservative systems. An advantage of that method is that we do not actually have to solve the equation but can deduce the behavior directly. As we shall see, this approach can be extended to the study of the stability of an isolated critical point for a general autonomous system. The technique is called **Lyapunov's direct method** (also known as **Lyapunov's second method**).[†]

[†] *Historical Footnote:* A. M. Lyapunov (Liapunov, Liapounov) (1857–1918) was a Russian mathematician whose research included not only nonlinear differential equations and stability, but probability theory and celestial mechanics.

Before we present the results that make up Lyapunov's direct method, let's analyze the system

(1) $\qquad \dfrac{dx}{dt} = y - xy^2 - x^3,$

(2) $\qquad \dfrac{dy}{dt} = -x - x^2 y - y^3.$

This is almost linear with the origin as an isolated critical point. The corresponding linear system

(3) $\qquad \dfrac{dx}{dt} = y, \qquad \dfrac{dy}{dt} = -x$

has a *center* at the origin, and therefore the results of Section 12.3 do not tell us whether the origin is stable, asymptotically stable, or unstable for the almost linear system.

Let's try to apply the energy method. We begin by observing that the linear system (3) is equivalent to the second order equation

(4) $\qquad \dfrac{d^2 x}{dt^2} + x = 0,$

where $y = x'$. This suggests that we investigate the "energy" function

$$V(x, y) = (1/2)x^2 + (1/2)y^2$$

that is associated with the mechanical system governed by equation (4). The function $V(x, y)$ is positive except at the origin where $V(0, 0) = 0$. Moreover, the level curves $V(x, y) = k$ are concentric circles about the origin. Again, following the energy method, we now consider how $V(x, y)$ changes along trajectories of the almost linear system. Let $x = \phi(t)$ and $y = \psi(t)$ be a solution to system (1)–(2). The partial derivatives V_x, V_y are just $V_x = x$, $V_y = y$, so we can compute dV/dt using the chain rule:

$$\frac{d}{dt} V(\phi(t), \psi(t)) = V_x(\phi(t), \psi(t))\frac{d\phi}{dt} + V_y(\phi(t), \psi(t))\frac{d\psi}{dt}$$

$$= \phi(t)\frac{d\phi}{dt} + \psi(t)\frac{d\psi}{dt}$$

$$= \phi(t)[\psi(t) - \phi(t)\psi^2(t) - \phi^3(t)] + \psi(t)[-\phi(t) - \phi^2(t)\psi(t) - \psi^3(t)]$$

$$= -[\phi^2(t) + \psi^2(t)]^2.$$

Hence dV/dt is negative everywhere except at the origin, where it is zero. Consequently $V(\phi(t), \psi(t))$ is decreasing along any trajectory of (1)–(2). Geometrically this means that as a trajectory intersects the level curve $V(x, y) = k$ (a circle), it then has to move into the region enclosed by that level curve (see Figure 12.27 on page 774). Consequently every trajectory must approach the origin as $t \to \infty$, which means the origin is asymptotically stable.

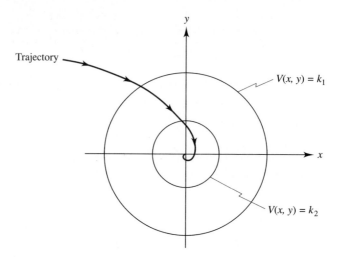

Figure 12.27 Trajectory entering regions bounded by level curves $V(x, y) = k$

In this example, the function $V(x, y)$ was selected based on our knowledge of the total energy for the spring-mass system that was governed by the corresponding linear system. Actually, all that is needed is a function $V(x, y)$ whose level curves encircle the critical point and whose values decrease along the trajectories of the almost linear system. To help describe such a function, we introduce the following definitions.

DEFINITE AND SEMIDEFINITE FUNCTIONS

Definition 3. Let $V(x, y)$ be continuous on some domain D containing the origin and assume that $V(0, 0) = 0$. We say $V(x, y)$ is **positive definite** on D if $V(x, y) > 0$ for all (x, y) in D except the origin.

 If $V(x, y) \geq 0$ for all (x, y) in D, we say that $V(x, y)$ is **positive semidefinite** on D.

 Similarly, we define **negative definite** and **negative semidefinite** by reversing the appropriate signs in the above inequalities.

The energy functions

$$V(x, y) = \frac{y^2}{2} + \frac{x^2}{2} - \frac{x^4}{4},$$

(5) $$V(x, y) = \frac{y^2}{2} + 1 - \cos x,$$

$$V(x, y) = \frac{y^2}{2} + \frac{x^2}{2} - \frac{x^5}{5}$$

that we encountered in Example 1 of Section 12.4 are all positive definite near the origin (see Problem 16).

To study the effect on $V(x, y)$ as we move along a trajectory, we introduce the notion of the derivative of $V(x, y)$ with respect to a system.

THE DERIVATIVE $\dot{V}(x, y)$

Definition 4. Suppose the first order partial derivatives of $V(x, y)$ are continuous in a domain D containing the origin. The **derivative of V with respect to the system**

$$\frac{dx}{dt} = f(x, y), \qquad \frac{dy}{dt} = g(x, y)$$

is the function

$$\dot{V}(x, y) := V_x(x, y)f(x, y) + V_y(x, y)g(x, y).$$

Remark. As we saw in the preceding example, the derivative of $V(x, y)$ with respect to the system, when evaluated at a solution $x = \phi(t)$, $y = \psi(t)$ of the system, is just the total derivative of $V(\phi(t), \psi(t))$ with respect to t; that is

$$\frac{dV}{dt}(\phi, \psi) = V_x(\phi, \psi)\frac{d\phi}{dt} + V_y(\phi, \psi)\frac{d\psi}{dt}$$

$$= V_x(\phi, \psi)f(\phi, \psi) + V_y(\phi, \psi)g(\phi, \psi)$$

$$= \dot{V}(\phi, \psi).$$

We now state and prove Lyapunov's theorem, which gives sufficient conditions for the origin to be stable.

LYAPUNOV'S STABILITY THEOREM

Theorem 3. Let $V(x, y)$ be a positive definite function on a domain D containing the origin. Suppose the origin is an isolated critical point for the system

(6) $$\frac{dx}{dt} = f(x, y),$$

(7) $$\frac{dy}{dt} = g(x, y)$$

and let $\dot{V}(x, y)$ be the derivative of $V(x, y)$ with respect to this system.

(a) If $\dot{V}(x, y)$ is negative semidefinite on D, then the origin is stable.
(b) If $\dot{V}(x, y)$ is negative definite on D, then the origin is asymptotically stable.

Proof

(a) Let B_r denote the open disk

$$B_r := \{(x, y): x^2 + y^2 < r^2\}$$

and $\bar{B}_r$ be the closed disk of radius r. For a given $\varepsilon > 0$, we need to determine $\delta > 0$ such that for any solution $x = \phi(t)$, $y = \psi(t)$ of (6)–(7) satisfying $(\phi(t_0), \psi(t_0)) \in B_\delta$, we have $(\phi(t), \psi(t)) \in B_\varepsilon$ for all $t \geq t_0$. To specify δ, we proceed as follows.

Choose $\alpha > 0$ so that $\bar{B}_\alpha \subset D$ and, for $\varepsilon < \alpha$, let $\bar{A}_{\varepsilon,\alpha}$ denote the closed annulus

$$\bar{A}_{\varepsilon,\alpha} := \{(x, y): \varepsilon^2 \leq x^2 + y^2 \leq \alpha^2\}$$

with inner radius ε and outer radius α. Let μ denote the minimum value of V on $\bar{A}_{\varepsilon,\alpha}$. Since V is continuous and is positive everywhere except at the origin, we have $\mu > 0$. Also, since V is zero at the origin, we can select a $\delta > 0$ such that $\delta < \varepsilon$ and $V(x, y) < \mu$ for all (x, y) in B_δ. We now show that for this δ, the stability condition is satisfied.

Let $0 < \varepsilon < \alpha$ and $x = \phi(t)$, $y = \psi(t)$ be a nontrivial solution to (6)–(7) such that $(\phi(t_0), \psi(t_0)) \in B_\delta$. If $(\phi(t), \psi(t)) \in B_\varepsilon$ for all $t \geq t_0$, then we are done. If this is not the case, then there is a first value of t, say $t_1 > t_0$, where the trajectory touches the boundary of B_ε (see Figure 12.28). Then on the interval $[t_0, t_1]$, we have $(\phi(t), \psi(t)) \in \bar{B}_\varepsilon \subset B_\alpha$. Since $\dot{V}$ is negative semidefinite, we have for $t \in [t_0, t_1]$,

$$V(\phi(t), \psi(t)) - V(\phi(t_0), \psi(t_0)) = \int_{t_0}^{t} \frac{dV}{ds}(\phi(s), \psi(s))\,ds$$

$$= \int_{t_0}^{t} \dot{V}(\phi(s), \psi(s))\,ds$$

$$\leq 0.$$

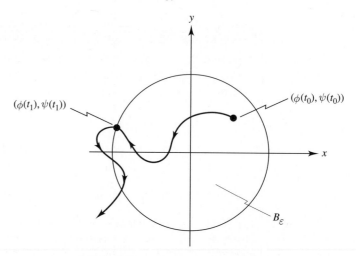

Figure 12.28 A trajectory leaving B_ε at $t = t_1$

Thus

(8) $V(\phi(t), \psi(t)) \leq V(\phi(t_0), \psi(t_0))$, for $t \in [t_0, t_1]$.

But $V(\phi(t_0), \psi(t_0)) < \mu$ because $(\phi(t_0), \psi(t_0)) \in B_\delta$, and $V(\phi(t_1), \psi(t_1)) \geq \mu$, since $(\phi(t_1), \psi(t_1)) \in \bar{A}_{\varepsilon, \alpha}$. Consequently, from (8),

$$\mu \leq V(\phi(t_1), \psi(t_1)) \leq V(\phi(t_0), \psi(t_0)) < \mu,$$

which is a contradiction. Therefore $(\phi(t), \psi(t)) \in B_\varepsilon$ for all $t \geq t_0$. Hence the stability is established.

(b) Here $\dot{V}$ is negative definite on D, so we already know the origin is stable from part (a). It remains to be shown that there exists an $\eta > 0$ such that if $x = \phi(t)$, $y = \psi(t)$ is a solution to (6)–(7) for which $(\phi(t_0), \psi(t_0))$ lies in B_η, then

$$\lim_{t \to \infty} \phi(t) = 0, \qquad \lim_{t \to \infty} \psi(t) = 0.$$

Now if we can prove that $V(\phi(t), \psi(t)) \to 0$ as $t \to +\infty$, then it will follow from the positive definiteness of V that $\phi(t) \to 0$ and $\psi(t) \to 0$ as $t \to +\infty$.

To specify the constant η, take $\varepsilon = \alpha/2$ and for this ε choose δ as in the proof of part (a).[†] We will show that with $\eta = \delta$ the asymptotic stability criterion is satisfied.

Suppose to the contrary that there exists a solution $x = \phi(t)$, $t = \psi(t)$ to (6)–(7) such that $(\phi(t_0), \psi(t_0))$ lies in B_η but $(\phi(t), \psi(t))$ does not approach the origin as $t \to +\infty$. Since $V(\phi(t), \psi(t))$ is decreasing for $t \geq t_0$ (see (8)), it follows that there is an $L > 0$ such that

$$V(\phi(t), \psi(t)) \geq L > 0 \quad \text{for all} \quad t \geq t_0.$$

Also, since V is continuous and $V(0, 0) = 0$, there exists an $h > 0$ such that

$$V(x, y) < L \quad \text{for all} \quad (x, y) \in B_h.$$

Hence the trajectory $(\phi(t), \psi(t))$ must lie in the annulus $\bar{A}_{h, \alpha/2}$ for $t \geq t_0$ (see Figure 12.29 on page 778).

Let $\gamma(<0)$ be the maximum value of $\dot{V}$ on the annulus $\bar{A}_{h, \alpha/2}$. Then, since the trajectory $(\phi(t), \psi(t))$ lies in $\bar{A}_{h, \alpha/2}$, we have

$$\frac{dV}{dt}(\phi(t), \psi(t)) \leq \gamma.$$

So, integrating we find

$$V(\phi(t), \psi(t)) \leq V(\phi(t_0), \psi(t_0)) + \gamma(t - t_0) \quad \text{for all} \quad t \geq t_0.$$

Since $\gamma < 0$, it now follows that, for t sufficiently large, $V(\phi(t), \psi(t)) < 0$, which contradicts the fact that V is positive definite. Hence we must have $V(\phi(t), \psi(t)) \to 0$ as $t \to +\infty$. ■ End

[†] There is nothing special about $\alpha/2$. Any fixed choice of ε less than α would work as well.

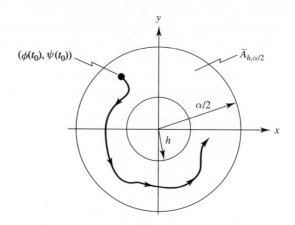

Figure 12.29 The trajectory lies in the annulus $\bar{A}_{h,\alpha/2}$

The function $V(x, y)$ in Theorem 3 is called a **Lyapunov function.** It is clear from the theorem that Lyapunov functions play a key role in determining the stability of an isolated critical point. *Unfortunately, there is no clear strategy available for choosing $V(x, y)$. However, for those systems that can be identified with a mechanical problem, the total energy function is a good candidate and, for many other systems, expressions of the form*

$$V(x, y) = ax^2 + by^2, \qquad V(x, y) = ax^2 + by^4,$$
$$V(x, y) = ax^4 + by^2, \qquad V(x, y) = ax^4 + by^4,$$

may work for an appropriate choice of the constants a, b.[†]

In the next example, we use Lyapunov's direct method to determine the stability of the origin for an autonomous system whose corresponding linear system does *not* have the origin as an isolated critical point; such a system cannot be analyzed using the results of Section 12.3.

EXAMPLE 1 Use Lyapunov's direct method to determine the stability for the origin for the system

(9) $$\frac{dx}{dt} = -2y^3,$$

(10) $$\frac{dy}{dt} = x - 3y^3.$$

Solution The origin is an isolated critical point for this system, since the pair of equations

$$-2y^3 = 0, \qquad x - 3y^3 = 0$$

[†] For further discussion, see the book by J. LaSalle and S. Lefschetz, *Stability by Liapunov's Direct Method with Applications* (Academic Press, New York, 1961).

has only $x = y = 0$ as a solution. (Note, however, that for the corresponding linear system $dx/dt = 0$, $dy/dt = x$ the origin is *not* an isolated critical point ($ad - bc = 0$). Hence (9)–(10) is not an almost linear system.)

Since $f(x, y) = -2y^3$ and $g(x, y) = x - 3y^3$, let's try to find a Lyapunov function of the form $V(x, y) = ax^2 + by^4$, with $a, b > 0$. Clearly V is positive definite on any domain containing the origin.

Next, computing $\dot{V}(x, y)$, we find

$$\dot{V}(x, y) = V_x(x, y)f(x, y) + V_y(x, y)g(x, y)$$
$$= (2ax)(-2y^3) + (4by^3)(x - 3y^3)$$
$$= 4(b - a)xy^3 - 12by^6.$$

So, if we take $a = b = 1$, then

$$\dot{V}(x, y) = -12y^6 \leq 0,$$

and we get that $\dot{V}(x, y)$ is negative semidefinite. It now follows from Theorem 3 that the origin is a stable critical point for the system (9)–(10). ■ End

The next result will give conditions for the instability of a system.

LYAPUNOV'S INSTABILITY THEOREM

Theorem 4. Supppose the origin is an isolated critical point for the autonomous system (6)–(7) and let $V(x, y)$ with $V(0, 0) = 0$ be defined and continuous on a domain D containing the origin. Suppose further that $\dot{V}(x, y)$ is positive definite on D. If, in every disk centered at the origin, there exists some point (a, b) such that $V(a, b)$ is positive, then the origin is unstable.

Proof. Fix $t_0 > 0$ and let $\alpha > 0$ be chosen so that $\bar{B}_\alpha \subset D$, where $\bar{B}_\alpha$ is the closed disk with center $(0, 0)$ and radius α. Since V is continuous, there exists an $L > 0$ such that $|V(x, y)| \leq L$ for all (x, y) in $\bar{B}_\alpha$. We will show that no matter how small we take $\varepsilon > 0$, there is a solution $x = \phi(t)$, $y = \psi(t)$ with $(\phi(t_0), \psi(t_0))$ in B_ε such that the trajectory $(\phi(t), \psi(t))$ eventually leaves the disk B_α. Thereby it follows that the origin is unstable.

Now, by hypothesis, given ε, with $0 \leq \varepsilon < \alpha$, there exists a point (a, b) in B_ε for which $V(a, b) > 0$. Let $x = \phi(t)$, $y = \psi(t)$ be the solution to (6)–(7) satisfying the initial condition

$$\phi(t_0) = a, \quad \psi(t_0) = b.$$

Also let t_1 be the first value of $t > t_0$ for which the trajectory $(\phi(t), \psi(t))$ leaves B_α ($t_1 = \infty$ means $(\phi(t), \psi(t))$ remains in B_α). Since $\dot{V}$ is positive definite on B_α, we have

$$V(\phi(t), \psi(t)) - V(\phi(t_0), \psi(t_0)) = \int_{t_0}^{t} \frac{dV}{ds}(\phi(s), \psi(s))\,ds$$

$$\geq 0$$

for all t in $[t_0, t_1)$. If we let V_0 denote $V(\phi(t_0), \psi(t_0))$, then $V_0 > 0$ and, for t in $[t_0, t_1)$,

$$V(\phi(t), \psi(t)) \geq V_0.$$

Now since V is continuous, there exists a $\lambda > 0$ such that $|V(x, y)| < V_0$ for (x, y) in B_λ. Therefore the trajectory $(\phi(t), \psi(t))$ must lie in the annulus $\bar{A}_{\lambda, \alpha}$ for $t \in [t_0, t_1)$. Let $\mu > 0$ be the minimum value of $\dot{V}$ on $\bar{A}_{\lambda, \alpha}$. Then, for t in $[t_0, t_1)$, we have

$$V(\phi(t), \psi(t)) - V(\phi(t_0), \psi(t_0)) = \int_{t_0}^{t} \frac{dV}{ds}(\phi(s), \psi(s))\,ds$$

$$\geq \mu(t - t_0),$$

that is,

$$V(\phi(t), \psi(t)) \geq V_0 + \mu(t - t_0) \quad \text{for} \quad t \in [t_0, t_1).$$

If $t_1 = +\infty$, then it follows that $V(\phi(t), \psi(t)) \to +\infty$ as $t \to +\infty$. But this is absurd, since $|V(x, y)| \leq L$ for (x, y) in B_α. Hence $t_1 < \infty$; that is, the trajectory $(\phi(t), \psi(t))$ leaves the disk B_α. ∎ End

EXAMPLE 2 Determine the stability of the critical point at the origin for the system

(11) $$\frac{dx}{dt} = -y^3,$$

(12) $$\frac{dy}{dt} = -x^3.$$

Solution The origin is not an isolated critical point for the corresponding linear system, so we cannot use the comparison approach of Section 12.3. Suspecting that the origin is unstable, we will attempt to apply Theorem 4.

Notice that $f(x, y) = -y^3$ is a function only of y and $g(x, y) = -x^3$ a function only of x. So let's try $V(x, y) = -xy$. Then V is continuous, $V(0, 0) = 0$, and in every disk centered at the origin, V is positive at some point (namely, those points of the disk lying in the second and fourth quadrants). Moreover,

$$\dot{V}(x, y) = (-y)(-y^3) + (-x)(-x^3) = y^4 + x^4,$$

which is positive definite in any domain containing the origin. Hence, by Lyapunov's instability theorem, the system (11)–(12) is unstable. ∎ End

EXERCISES 12.5

In Problems 1 through 8, use Lyapunov's direct method to determine the stability of the origin for the given system.

1. $\dfrac{dx}{dt} = -3x^3 + 2xy^2,$

$\dfrac{dy}{dt} = -y^3.$

[*Hint:* Try $V(x, y) = x^2 + y^2$.]

2. $\dfrac{dx}{dt} = -y - x^3 - xy^2,$

$\dfrac{dy}{dt} = x - x^2y - y^3.$

[*Hint:* Try $V(x, y) = x^2 + y^2$.]

3. $\dfrac{dx}{dt} = y^2 + xy^2 - x^3,$

$\dfrac{dy}{dt} = -xy + x^2y - y^3.$

[*Hint:* Try $V(x, y) = x^2 + y^2$.]

4. $\dfrac{dx}{dt} = x^3y + x^2y^3 - x^5,$

$\dfrac{dy}{dt} = -2x^4 - 6x^3y^2 - 2y^5.$

[*Hint:* Try $V(x, y) = 2x^2 + y^2$.]

5. $\dfrac{dx}{dt} = 2x^3,$

$\dfrac{dy}{dt} = 2x^2y - y^3.$

[*Hint:* Try $V(x, y) = x^2 - y^2$.]

6. $\dfrac{dx}{dt} = x + y + y^3,$

$\dfrac{dy}{dt} = x + y + x^5.$

[*Hint:* Try $V(x, y) = xy$.]

7. $\dfrac{dx}{dt} = 2y - x^3,$

$\dfrac{dy}{dt} = -x^3 - y^5.$

[*Hint:* Try $V(x, y) = ax^4 + by^2$.]

8. $\dfrac{dx}{dt} = y^3 - 2x^3,$

$\dfrac{dy}{dt} = -3x - y^3.$

[*Hint:* Try $V(x, y) = ax^2 + by^4$.]

In Problems 9 through 14, use Lyapunov's direct method to determine the stability of the zero solution for the given equation.

9. $\dfrac{d^2x}{dt^2} + |x|\dfrac{dx}{dt} + x^3 = 0.$

10. $\dfrac{d^2x}{dt^2} + (1 - x^2)\dfrac{dx}{dt} + x = 0.$

11. $\dfrac{d^2x}{dt^2} + \left(1 - \left(\dfrac{dx}{dt}\right)^2\right)\dfrac{dx}{dt} + x = 0.$

12. $\dfrac{d^2x}{dt^2} + x^2\dfrac{dx}{dt} + x^5 = 0.$

13. $\dfrac{d^2x}{dt^2} + \dfrac{dx}{dt} + \sin x = 0.$

[*Hint:* Use $V(x, y) = (1/2)(x + y)^2 + x^2 + (1/2)y^2$, where $y = dx/dt$.]

14. $\dfrac{d^2x}{dt^2} + \cos x\dfrac{dx}{dt} + \sin x = 0.$

15. Let f be continuous and have continuous first partial derivatives in some domain D containing the origin. Show that the system

$$\dfrac{dx}{dt} = -y + xf(x, y),$$

$$\dfrac{dy}{dt} = x + yf(x, y)$$

(a) is asymptotically stable if $f(x, y) < 0$ in D.
(b) is unstable if $f(x, y) > 0$ in D.

16. Verify that the three energy functions given in (5) are positive definite in some domain containing the origin.

17. Prove that the zero solution for a conservative system

$$\dfrac{d^2x}{dt^2} + g(x) = 0,$$

where g is continuous in an interval containing zero, is stable provided $g(0) = 0$ and, for some $k > 0$, $sg(s) > 0$ for $0 < |s| < k$.

18. Assume $V(x, y)$ is positive definite in a domain containing the origin and the level curves $V(x, y) = k$ are simple closed curves that expand out from the origin such that the outward normal **N** to the level curve at the point (a, b) is given by

$$\mathbf{N} = \nabla V(a, b) = V_x(a, b)\mathbf{i} + V_y(a, b)\mathbf{j},$$

where **i** and **j** are unit vectors in the positive x and y directions, respectively. Let $(\phi(t), \psi(t))$, $t \to +\infty$, be a trajectory for the solution to system (6)–(7) that passes through (a, b) at time $t = t_0$. Recall that the tangent **T** to the trajectory at (a, b) is

$$\mathbf{T} = \frac{d\phi}{dt}(t_0)\mathbf{i} + \frac{d\phi}{dt}(t_0)\mathbf{j}.$$

(a) Verify that

$$\dot{V}(\phi, \psi) = \mathbf{N} \cdot \mathbf{T} = |\mathbf{N}||\mathbf{T}| \cos \theta,$$

where θ is the angle between **N** and **T** (see Figure 12.30).

(b) Show that if $\dot{V}$ is negative definite, then a trajectory

(c) Show that if $\dot{V}$ is negative semidefinite, then a trajectory that enters the region bounded by $V(x, y) = k$ cannot leave this region.

(d) Show that if $\dot{V}$ is positive definite, then a trajectory $(\phi(t), \psi(t))$ that touches the level curve $V(x, y) = k$ must leave the region bounded by this level curve.

$(\phi(t), \psi(t))$ that touches the level curve $V(x, y) = k$ must enter the region bounded by this level curve.

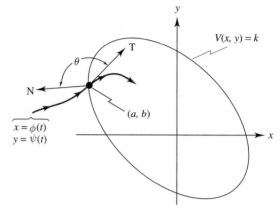

Figure 12.30 Intersection of a trajectory with a level curve

12.6 LIMIT CYCLES AND PERIODIC SOLUTIONS

In Example 4 in Section 12.4, we used the energy method to study a nonconservative system involving nonlinear damping. Our analysis led us to the existence of a closed trajectory that encircled the unstable spiral point at the origin. What was interesting about this closed trajectory was that all trajectories beginning inside of it spiraled out toward it, while those beginning outside spiraled into it. Since closed trajectories correspond to periodic solutions, we have in hand an example of when all solutions approach the periodic solution as $t \to \infty$ (except for the zero solution corresponding to the critical point at the origin). This closed trajectory is an example of a **limit cycle.** In this section we discuss the criteria for the existence and nonexistence of a limit cycle in the plane.

LIMIT CYCLE

Definition 5. A nontrivial[†] closed trajectory that is isolated is called a **limit cycle.**

[†] Critical points are technically closed trajectories, but we exclude these trivial ones. The closed trajectories that surround a *center* are also excluded, since they are not isolated. (Recall Section 12.2, page 748.)

The next example illustrates the behavior of trajectories near a limit cycle. Let $r^2 = x^2 + y^2$ and consider the system

(1) $\dfrac{dx}{dt} = x - y - x(3r - r^2 - 1),$

(2) $\dfrac{dy}{dt} = x + y - y(3r - r^2 - 1),$

for which the origin is the only critical point. It is helpful to use polar coordinates $x = r \cos \theta$, $y = r \sin \theta$ to study this system. We need two identities

$$x\frac{dx}{dt} + y\frac{dy}{dt} = r\frac{dr}{dt},$$

$$x\frac{dy}{dt} - y\frac{dx}{dt} = r^2\frac{d\theta}{dt},$$

which follow from the polar equations (see Problem 2). Now multiplying equation (1) by x and equation (2) by y and adding, we find

$$
\begin{aligned}
r\frac{dr}{dt} = x\frac{dx}{dt} + y\frac{dy}{dt} &= x(x - y - x(3r - r^2 - 1)) + y(x + y - y(3r - r^2 - 1)) \\
&= (x^2 + y^2) - (x^2 + y^2)(3r - r^2 - 1) \\
&= r^2(r - 1)(r - 2).
\end{aligned}
$$

Thus we have

(3) $\dfrac{dr}{dt} = r(r - 1)(r - 2).$

We now multiply equation (2) by x and equation (1) by y and subtract the latter from the former to get

$$
\begin{aligned}
r^2\frac{d\theta}{dt} = x\frac{dy}{dt} - y\frac{dx}{dt} &= x(x + y - y(3r - r^2 - 1)) - y(x - y - x(3r - r^2 - 1)) \\
&= x^2 + y^2 = r^2.
\end{aligned}
$$

Hence $d\theta/dt = 1$, which tells us that the trajectories revolve counterclockwise about the origin. From equation (3), we observe that dr/dt is zero when $r = 0$, 1, and 2. The critical point (0, 0) is represented by $r = 0$. When $r = 1$ or 2, the corresponding trajectories are just circles of radius 1 or 2, respectively; hence these circles are limit cycles. Now when r lies in (0, 1), we see from (3) that $dr/dt > 0$, so a trajectory in this region spirals out toward the limit cycle $r = 1$. When r lies in (1, 2), $dr/dt < 0$ and the trajectories spiral in

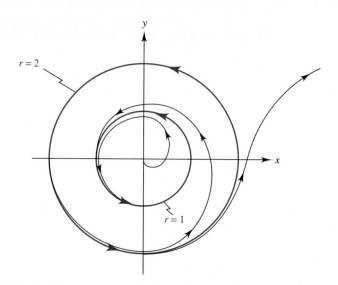

Figure 12.31 Phase plane diagram with limit cycles and spiraling trajectories

toward the limit cycle $r = 1$ and away from the limit cycle $r = 2$. Finally, when $r > 2$, $dr/dt > 0$ and the trajectories spiral out away from the limit cycle $r = 2$ (see Figure 12.31).

When nearby trajectories approach a limit cycle, we say that the limit cycle is **stable.** When they recede, the limit cycle is **unstable.** It is possible for trajectories to approach one side of the cycle while receding from the other. Such a limit cycle is called **semistable** (see Problem 1). In the previous example, the limit cycle $r = 2$ is unstable while the limit cycle $r = 1$ is stable. Clearly the origin is an unstable spiral point.

In applications when limit cycles do exist, they are seldom circles and, unfortunately, finding them may be difficult. In this regard, certain theoretical results, which we now present, may be helpful. The first theorem certainly makes sense intuitively; however, its proof is beyond the scope of this text.[†]

LIMIT CYCLES SURROUND CRITICAL POINTS

Theorem 5. A limit cycle in the plane must enclose at least one critical point. Moreover, any critical point enclosed by a limit cycle cannot be a saddle point.

Observe that in the previous example, both limit cycles encircled the origin, which was an unstable spiral point.

The next result is due to I. Bendixson. It tells us when a limit cycle does not exist.

[†] For a proof, see the book by P. W. Jordan and P. Smith, *Nonlinear Ordinary Differential Equations,* Second Edition (Clarendon Press, Oxford, 1987), Section 3.4.

BENDIXSON NEGATIVE CRITERION

Theorem 6. Let $f(x, y)$ and $g(x, y)$ have continuous first partial derivatives in the simply connected domain[†] D and assume

$$f_x(x, y) + g_y(x, y)$$

is of one sign in D. Then there are no (nonconstant) periodic solutions to

$$(4) \qquad \frac{dx}{dt} = f(x, y),$$

$$(5) \qquad \frac{dy}{dt} = g(x, y)$$

that lie entirely in D.

Proof. Assume $x = \phi(t)$, $y = \psi(t)$ is a (nonconstant) T-periodic solution to system (4)–(5) whose trajectory $\Gamma(t) := (\phi(t), \psi(t))$ lies entirely in D. Let Ω denote the region enclosed by Γ. Green's theorem in the plane says that[††]

$$(6) \qquad \oint_\Gamma [f(x, y)dy - g(x, y)dx] = \iint_\Omega [f_x(x, y) + g_y(x, y)]dxdy.$$

The assumption that $f_x + g_y$ is of one sign in D implies the integral on the right-hand side is not zero. Using the parametric representation $\Gamma(t) = (\phi(t), \psi(t))$, $0 \le t \le T$, and the fact that ϕ, ψ are solutions to (4)–(5), we find the integral on the left-hand side of (6) to be equal to

$$\int_0^T \left[f(\phi(t), \psi(t)) \frac{d\psi}{dt} - g(\phi(t), \psi(t)) \frac{d\phi}{dt} \right] dt = \int_0^T \left[\frac{d\phi}{dt} \frac{d\psi}{dt} - \frac{d\psi}{dt} \frac{d\phi}{dt} \right] dt = 0.$$

This yields a contradiction, since the right-hand side of (6) is not zero. Thus there is no (nonconstant) periodic solution whose trajectory lies entirely in D. ∎ End

We use the Bendixson criterion in the next example to show that limit cycles do not exist for the given systems.

EXAMPLE 1 Show that the system

$$(7) \qquad \frac{dx}{dt} = \alpha(y)x + \beta(y),$$

$$(8) \qquad \frac{dy}{dt} = \gamma(x)y + \delta(x),$$

[†] A simply connected domain is a connected region that does not have any ''holes'' in it.
[††] If ϕ, ψ is a nonconstant periodic solution of (4)–(5), then the assumptions on f and g will guarantee that Γ and Ω satisfy the conditions of Green's theorem.

where β, δ are continuous and α, γ are both continuous and positive (negative) in some simply connected domain, has no nonconstant periodic solution.

Solution Here $f(x, y) = \alpha(y)x + \beta(y)$ and $g(x, y) = \gamma(x)y + \delta(x)$. Now computing $f_x + g_y$, we find

$$f_x + g_y = \alpha(y) + \gamma(x),$$

which is always positive (negative). Hence, by the Bendixson negative criterion, there can be no nonconstant periodic solutions. ■ End

A mechanical system with positive (negative) damping is governed by a second order equation of the form

(9) $$\frac{d^2x}{dt^2} + \gamma(x)\frac{dx}{dt} + \delta(x) = 0,$$

where $\gamma(x)$ is positive (negative). This equation can be written in the form of system (7)–(8). It then follows from the Bendixson negative criterion that this mechanical system has no nonconstant periodic solution (see Problem 23). This conclusion is consistent with the results of the energy method that show that positive damping causes trajectories to move in toward the origin and negative damping causes them to move away from the origin.

What happens if there is negative damping near the origin and positive damping away from the origin? Is there a nonconstant periodic solution? The following result gives sufficient conditions for the existence of a nonconstant periodic solution. Its proof is left for more advanced texts.[†]

POINCARÉ-BENDIXSON THEOREM

Theorem 7. Let f and g have continuous first partial derivatives on the closed bounded region R and assume the system

$$\frac{dx}{dt} = f(x, y), \quad \frac{dy}{dt} = g(x, y)$$

has no critical points in R. If a solution $x = \phi(t)$, $y = \psi(t)$ to the system exists for all $t \geq t_0$ and its trajectory $\Gamma(t) := (\phi(t), \psi(t))$ remains inside R for $t \geq t_0$, then either Γ is a limit cycle or it spirals toward a limit cycle in R. In either case, the system has a nonconstant periodic solution.

To apply this theorem, we must find a closed bounded region R that contains a trajectory Γ, but not a critical point of the system. Since limit cycles encircle critical points (Theorem 5), it follows that the region R has a hole in it.

A typical strategy for determining R is as follows. First find a curve C_1 encircling a critical point and show that all trajectories crossing C_1 must head out. Second, find a curve C_2 encircling C_1 and show that all trajectories crossing C_2 must head in. If R is the region

[†] For a proof, see the book by J. K. Hale, *Ordinary Differential Equations*, Second Edition (Krieger, New York, 1980), Chapter II.

between C_1 and C_2, including C_1 and C_2 as its boundary, then any trajectory beginning in R must remain in R. Otherwise it must cross either C_1 headed in or C_2 headed out. Both situations yield a contradiction. We illustrate this approach in the next example.

EXAMPLE 2 Show that the equation

(10) $$\frac{d^2x}{dt^2} + \left(4x^2 + \left(\frac{dx}{dt}\right)^2 - 4\right)\frac{dx}{dt} + x^3 = 0$$

has a nonconstant periodic solution.

Solution Let $y = dx/dt$ and rewrite equation (10) as the equivalent system

$$\frac{dx}{dt} = y,$$

$$\frac{dy}{dt} = -x^3 - (4x^2 + y^2 - 4)y.$$

Notice that the origin is the only critical point for this system.

Motivated by the energy method, let's consider the Lyapunov function

$$V(x, y) := \frac{1}{2}y^2 + \frac{1}{4}x^4.$$

The level curves $V(x, y) = k$, $k > 0$, consist of simple closed curves encircling the origin. Moreover,

$$\dot{V}(x, y) = V_x \frac{dx}{dt} + V_y \frac{dy}{dt}$$

$$= x^3 y + y(-x^3 - y(4x^2 + y^2 - 4))$$

$$= -y^2(4x^2 + y^2 - 4).$$

Now if (x, y) lies inside the ellipse $4x^2 + y^2 = 4$, then $\dot{V}(x, y) \geq 0$, and if it lies outside, then $\dot{V}(x, y) \leq 0$.

To apply the Poincaré-Bendixson theorem, we must determine a region R that contains a trajectory but no critical point. To define R, select k_1 sufficiently small so that the level curve $V(x, y) = k_1$ lies inside the ellipse $4x^2 + y^2 = 4$. On this level curve, $\dot{V}(x, y) \geq 0$, so trajectories cannot cross into the region enclosed by $V(x, y) = k_1$. Next select k_2 sufficiently large so that the level curve $V(x, y) = k_2$ encircles the ellipse $4x^2 + y^2 = 4$. Here, on this level curve, $\dot{V}(x, y) \leq 0$, so trajectories cannot leave the region enclosed by $V(x, y) = k_2$. Let R be the region between $V(x, y) = k_1$ and $V(x, y) = k_2$ (see Figure 12.32 on page 788). It now follows from the Poincaré-Bendixson theorem that there exists a nonconstant periodic solution inside R. ■ End

In the above example, the ellipse $4x^2 + y^2 = 4$ is *not* the limit cycle, but we can conclude that the limit cycle is intertwined with it. Indeed, when the limit cycle lies inside the ellipse, its "energy" V is increasing, and when it is outside, its "energy" is decreasing.

Example 2 is a situation where the damping is negative near the origin and positive away from the origin. In the analysis of the triode vacuum tube, a slightly different behavior

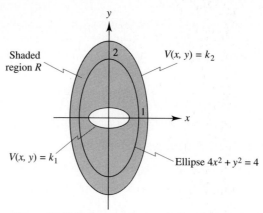

Figure 12.32 Region R where periodic solution exists

arises. The model governing the tube is the van der Pol equation[†]

(11) $$\frac{d^2x}{dt^2} + \varepsilon(x^2 - 1)\frac{dx}{dt} + x = 0,$$

where ε is a positive constant. Here the damping coefficient $\varepsilon(x^2 - 1)$ is positive for $|x| > 1$ but negative for $|x| < 1$. Therefore this damping coefficient is never positive on any curve encircling the origin. While it is not obvious, it turns out that the van der Pol equation has a unique periodic solution for any $\varepsilon > 0$. The limit cycles corresponding to these periodic solutions are illustrated in Figure 12.33 for $\varepsilon = 0.1$, 1, and 5.

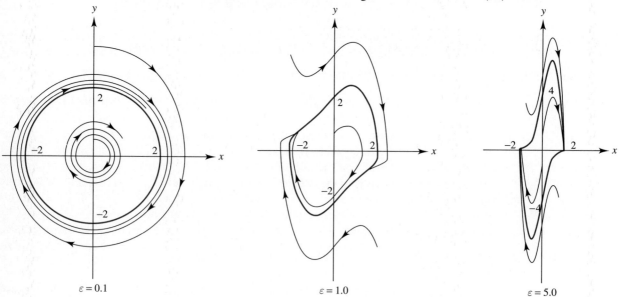

Figure 12.33 Limit cycles for van der Pol equations

[†] For a detailed discussion, see the book by L. Cesari, *Asymptotic Behavior and Stability Problems in Ordinary Differential Equations*, Third Edition (Springer-Verlag, New York, 1971), pages 154–156.

We conclude this section with a statement of a theorem on **Lienard equations** due to N. Levinson and O. K. Smith,[†] which can be used to show that van der Pol's equation has a unique nonconstant periodic solution (see Problem 24).

LEVINSON AND SMITH'S THEOREM

Theorem 8. Let $f(x)$ and $g(x)$ be continuous functions and let

$$F(x) := \int_0^x f(s)\,ds, \qquad G(x) := \int_0^x g(s)\,ds.$$

The Lienard equation

$$\frac{d^2x}{dt^2} + f(x)\frac{dx}{dt} + g(x) = 0$$

has a unique nonconstant periodic solution whenever all of the following conditions hold:

(a) $f(x)$ is even.
(b) $F(x) < 0$, for $0 < x < a$;
$\quad\ F(x) > 0$, for $x > a$, for some a.
(c) $F(x) \to +\infty$ as $x \to +\infty$, monotonically for $x > a$.
(d) $g(x)$ is an odd function with $g(x) > 0$ for $x > 0$.
(e) $G(x) \to +\infty$ as $x \to +\infty$.

EXERCISES 12.6

1. Semistable Limit Cycle. For the system

$$\frac{dx}{dt} = x - y - x(2r - r^2),$$

$$\frac{dy}{dt} = x + y - y(2r - r^2),$$

where $x = r\cos\theta$ and $y = r\sin\theta$:

(a) Sketch the phase plane diagram and identify each limit cycle and critical point.
(b) Explain why the limit cycle $r = 1$ is semistable.

2. For $x = r\cos\theta$ and $y = r\sin\theta$, prove that

$$r\frac{dr}{dt} = x\frac{dx}{dt} + y\frac{dy}{dt},$$

$$r^2\frac{d\theta}{dt} = x\frac{dy}{dt} - y\frac{dx}{dt}.$$

[*Hint:* Use the facts that $r^2 = x^2 + y^2$ and $\tan\theta = y/x$.]

3. For systems of the form

$$\frac{dx}{dt} = x - y + xf(r)/r,$$

$$\frac{dy}{dt} = x + y + yf(r)/r,$$

where $x = r\cos\theta$ and $y = r\sin\theta$:

(a) Show that the system has periodic solutions corresponding to the values of r such that $r + f(r) = 0$.
(b) Determine the direction of rotation for the trajectories of this system.

4. For systems of the form

$$\frac{dx}{dt} = y + xf(r)/r,$$

$$\frac{dy}{dt} = -x + yf(r)/r,$$

where $x = r\cos\theta$ and $y = r\sin\theta$:

[†] See *Duke Mathematics Journal*, Vol. 9, 1942, pages 382–403.

(a) Show that the system has periodic solutions corresponding to the zeros of $f(r)$.

(b) Determine the direction of rotation for the trajectories of this system.

▪ In Problems 5 through 12, either by hand or using a computer software package sketch the phase plane diagrams for the given system. Discuss the stability of each limit cycle and critical point. Here $x = r \cos \theta$ and $y = r \sin \theta$.

5. $\dfrac{dx}{dt} = x - y + x(r^3 - 4r^2 + 5r - 3),$

$\dfrac{dy}{dt} = x + y + y(r^3 - 4r^2 + 5r - 3).$

6. $\dfrac{dx}{dt} = y + x(r^2 - 7r + 10),$

$\dfrac{dy}{dt} = -x + y(r^2 - 7r + 10).$

7. $\dfrac{dx}{dt} = x - y + x(r^3 - r - 1),$

$\dfrac{dy}{dt} = x + y + y(r^3 - r - 1).$

8. $\dfrac{dx}{dt} = x - y + x(x^2 + y^2),$

$\dfrac{dy}{dt} = x + y + y(x^2 + y^2).$

9. $\dfrac{dx}{dt} = x + y + x(r^2 - 3r + 1),$

$\dfrac{dy}{dt} = -x + y + y(r^2 - 3r + 1).$

10. $\dfrac{dx}{dt} = -y + x(r^2 - 5r + 6),$

$\dfrac{dy}{dt} = x + y(r^2 - 5r + 6).$

11. $\dfrac{dx}{dt} = y + x \sin(1/r),$

$\dfrac{dy}{dt} = -x + y \sin(1/r).$

12. $\dfrac{dx}{dt} = y + \dfrac{x}{r} \sin r,$

$\dfrac{dy}{dt} = -x + \dfrac{y}{r} \sin r.$

In Problems 13 through 18, show that the given system or equation has only constant periodic solutions in the specified domain D.

13. $\dfrac{dx}{dt} = x - y + x^2 y,$

$\dfrac{dy}{dt} = x + x^2 y + \dfrac{1}{3} y^3.$

(D is entire plane.)

14. $\dfrac{dx}{dt} = 2x + y - xy^2,$

$\dfrac{dy}{dt} = x - 3y - x^2 y.$

(D is entire plane.)

15. $\dfrac{dx}{dt} = -8x + y + x^3,$

$\dfrac{dy}{dt} = 2x - 7y + y^3.$

(D is the disk $x^2 + y^2 < 5$.)

16. $\dfrac{dx}{dt} = 3x + 2y - xy^2,$

$\dfrac{dy}{dt} = x + 4y - 2x^2 y.$

(D is interior of ellipse $2x^2 + y^2 = 7$.)

17. $\dfrac{d^2 x}{dt^2} + (1 - x^2) \dfrac{dx}{dt} + x^3 = 0.$

(D is the strip $-1 < x < 1$ in the xv-plane.)

18. $\dfrac{d^2 x}{dt^2} + \left(3 - \left(\dfrac{dx}{dt} \right)^2 \right) \dfrac{dx}{dt} + x = 0.$

(D is the slab $-1 < v < 1$ in the xv-plane).

In Problems 19 through 22, show that the given system or equation has a nonconstant periodic solution.

19. $\dfrac{dx}{dt} = 2x - y - 2x^3 - 3xy^2,$

$\dfrac{dy}{dt} = 2x + 4y - 4y^3 - 2x^2 y.$

[*Hint:* Use the Lyapunov function $V(x, y) = 2x^2 + y^2$.]

20. $\dfrac{dx}{dt} = 8x - 2y - 4x^3 - 2xy^2,$

$\dfrac{dy}{dt} = x + 4y - 2y^3 - 3x^2y.$

[*Hint:* Use the Lyapunov function $V(x, y) = x^2 + 2y^2$.]

21. $\dfrac{d^2x}{dt^2} + \left(x^4 + \left(\dfrac{dx}{dt}\right)^2 - 1\right)\dfrac{dx}{dt} + x = 0.$

22. $\dfrac{d^2x}{dt^2} + \ln\left(x^2 + \left(\dfrac{dx}{dt}\right)^2 + \dfrac{1}{2}\right)\dfrac{dx}{dt} + x^3 = 0.$

23. Strictly Positive Damping. Use the Bendixson negative criterion to show that a mechanical system governed by equation (9) with γ and δ continuous on $(-\infty, \infty)$ and γ strictly positive (negative) has only constant periodic solutions.

24. Show that van der Pol's equation (11) satisfies the conditions of Levinson-Smith's theorem and therefore has a unique nonconstant periodic solution.

In Problems 25 through 28, show that the given Lienard equations have unique nonconstant periodic solutions.

25. $\dfrac{d^2x}{dt^2} + x^2(x^2 - 1)\dfrac{dx}{dt} + x = 0.$

26. $\dfrac{d^2x}{dt^2} + (x^2 - 4)\dfrac{dx}{dt} + x^3 = 0.$

27. $\dfrac{d^2x}{dt^2} + (x^2 - \cos x - 1)\dfrac{dx}{dt} + x + \sin x = 0.$

28. $\dfrac{d^2x}{dt^2} + (x^4 - 1)\dfrac{dx}{dt} + x^5 = 0.$

12.7 STABILITY OF HIGHER-DIMENSIONAL SYSTEMS

Many of the results we have obtained on the stability of linear and almost linear systems in the plane can be extended to higher-dimensional systems (a notable exception is the Poincaré-Bendixson theorem discussed in Section 12.6). To give these extensions, we must first introduce some definitions and notation. We assume the reader is familiar with the material in Chapter 9.

Let $\mathbf{z} = \mathrm{col}(z_1, z_2, \ldots, z_n)$ be a vector in $\mathbf{R}^n$. We define the **norm**[†] of $\mathbf{z}$ to be

$$\|\mathbf{z}\| := \max_{i=1,\ldots,n} |z_i|.$$

In a similar fashion, if $\mathbf{A}$ is an $n \times n$ matrix with entries a_{ij}, $i, j = 1, \ldots n$, we define the norm of $\mathbf{A}$ to be

$$\||\mathbf{A}\|| := \max_{i,j=1,\ldots,n} |a_{ij}|.$$

A simple analysis using the triangle inequality shows that

$$\|\mathbf{Az}\| = \max_{i=1,\ldots,n} \left|\sum_{k=1}^{n} a_{ik}z_k\right|$$

$$\leq \max_{i=1,\ldots,n} \left(\sum_{k=1}^{n} |a_{ik}||z_k|\right)$$

$$\leq \sum_{k=1}^{n} \||\mathbf{A}\|| \, |z_k|$$

$$\leq \||\mathbf{A}\|| \sum_{k=1}^{n} \|\mathbf{z}\|,$$

[†] There are other norms that can be used in $\mathbf{R}^n$, such as the Euclidean norm $\sqrt{z_1^2 + z_2^2 + \cdots + z_n^2}$. In most of the results that we give, the choice of norm is not essential.

which reduces to the inequality

(1) $\|\mathbf{Az}\| \leq n\|\|\mathbf{A}\|\| \|\mathbf{z}\|.$

We can now define stability for a solution to the system of differential equations

(2) $\mathbf{x}'(t) = \mathbf{f}(t, \mathbf{x}),$

where $\mathbf{x}'(t) = \mathrm{col}(x_1'(t), \ldots, x_n'(t))$ and $\mathbf{f}(t, \mathbf{x}) = \mathrm{col}(f_1(t, \mathbf{x}), \ldots, f_n(t, \mathbf{x}))$. We assume each $f_i(t, \mathbf{x})$, $i = 1, \ldots, n$, is a continuous function of $t, x_1, \ldots, x_n$ with continuous partial derivatives with respect to $x_1, \ldots, x_n$.

LYAPUNOV STABILITY

Definition 6. A solution $\boldsymbol{\phi}(t)$ to system (2) is **stable** (also called **Lyapunov stable**) for $t \geq t_0$ if for any $\varepsilon > 0$ there exists $\delta(t_0, \varepsilon) > 0$ such that whenever $\|\mathbf{x}(t_0) - \boldsymbol{\phi}(t_0)\| < \delta$, where $\mathbf{x}(t)$ is any solution to (2), then $\|\mathbf{x}(t) - \boldsymbol{\phi}(t)\| < \varepsilon$ for all $t \geq t_0$. In addition, if for any such $\mathbf{x}$ we have $\mathbf{x}(t) - \boldsymbol{\phi}(t) \to \mathbf{0}$ as $t \to \infty$, then $\boldsymbol{\phi}$ is **asymptotically stable.**
 If $\boldsymbol{\phi}$ not stable, then it is **unstable.**

The relationship between Lyapunov stability of a solution and the stability of a critical point is the following. A solution $\boldsymbol{\phi}$ of (2) is stable if and only if the equilibrium solution $\mathbf{z}(t) \equiv \mathbf{0}$ is a stable solution to the system

(3) $\mathbf{z}'(t) = \mathbf{g}(t, \mathbf{z}),$

where

$$\mathbf{g}(t, \mathbf{z}) := \mathbf{f}(t, \mathbf{z} + \boldsymbol{\phi}) - \mathbf{f}(t, \boldsymbol{\phi}).$$

This, in turn, is equivalent to the origin $\mathbf{0} = \mathrm{col}(0, 0, \ldots, 0)$ being a stable critical point for (3). Indeed, if $\mathbf{x}$ and $\boldsymbol{\phi}$ are solutions to (2), then $\mathbf{z} = \mathbf{x} - \boldsymbol{\phi}$ is a solution to (3). So if $\boldsymbol{\phi}$ is stable and $\|\mathbf{x}(t) - \boldsymbol{\phi}(t)\| < \varepsilon$, then $\|\mathbf{z}(t)\| < \varepsilon$ and the zero solution is stable. The converse is also true.

For linear systems it is sufficient to study the stability of the zero solution to the homogeneous system.

STABILITY AND THE ZERO SOLUTION

Theorem 9. A solution $\boldsymbol{\phi}(t)$ to the linear system

(4) $\mathbf{x}'(t) = \mathbf{A}(t)\mathbf{x}(t) + \mathbf{h}(t),$

where $\mathbf{A}(t)$ is an $n \times n$ matrix whose entries are functions of t and $\mathbf{h}(t) = \mathrm{col}(h_1(t), \ldots, h_n(t))$ is stable (asymptotically stable) if and only if the zero solution $\mathbf{z}(t) \equiv \mathbf{0}$ is a stable (an asymptotically stable) solution to

$$\mathbf{x}'(t) = \mathbf{A}(t)\mathbf{x}(t).$$

We leave the proof as an exercise (Problem 7).

The next result relates the stability of a linear system to properties of a fundamental matrix for the system.

FUNDAMENTAL MATRIX AND STABILITY

Theorem 10. Let $\mathbf{A}(t)$ be an $n \times n$ matrix-valued function whose entries are continuous functions for $t \geq t_0$ and let $\mathbf{X}(t)$ be a fundamental matrix for the homogeneous system

$$(5) \qquad \mathbf{x}'(t) = \mathbf{A}(t)\mathbf{x}(t), \qquad t \geq t_0.$$

If there exists a constant $K > 0$ such that $|||\mathbf{X}(t)||| \leq K$ for $t \geq t_0$, then the zero solution $\mathbf{x}(t) = \mathbf{0}$ is stable. Moreover, if $\lim_{t \to \infty} |||\mathbf{X}(t)||| = 0$, then the zero solution is asymptotically stable.

Proof. Without loss of generality, assume $\mathbf{X}(t_0) = \mathbf{I}$. (Otherwise, use the fundamental matrix $\mathbf{Y}(t) = \mathbf{X}(t)\mathbf{X}^{-1}(t_0)$.) If $\mathbf{x}(t)$ is any solution to (5), then we can express $\mathbf{x}(t)$ in the form

$$\mathbf{x}(t) = \mathbf{X}(t)\mathbf{x}(t_0)$$

(cf. Theorem 3 on page 528 and the subsequent discussion). Taking the norm of both sides and using inequality (1), we find

$$(6) \qquad \|\mathbf{x}(t)\| = \|\mathbf{X}(t)\mathbf{x}(t_0)\| \leq n|||\mathbf{X}(t)||| \, \|\mathbf{x}(t_0)\|.$$

If $|||\mathbf{X}(t)||| \leq K$ for $t \geq t_0$, then we have

$$\|\mathbf{x}(t)\| \leq nK\|\mathbf{x}(t_0)\|.$$

So, given $\varepsilon > 0$, we select $\delta = \varepsilon/(nK)$. Then, whenever $\|\mathbf{x}(t_0)\| < \delta = \varepsilon/(nK)$, it follows that

$$\|\mathbf{x}(t)\| < nK\delta = \varepsilon \quad \text{for} \quad t \geq t_0,$$

and the origin is stable.

Moreover, if $\lim_{t \to \infty} |||\mathbf{X}(t)||| = 0$, then it follows from (6) that $\mathbf{x}(t) \to \mathbf{0}$ as $t \to \infty$, and so the zero solution is asymptotically stable. ■ End

As a corollary, we have the following result for homogeneous linear systems with constant coefficients.

HOMOGENEOUS SYSTEMS WITH CONSTANT COEFFICIENTS

Corollary 1. Let $\mathbf{A}$ be an $n \times n$ constant matrix. If all eigenvalues of $\mathbf{A}$ have negative real parts, then for any t_0 the zero solution $\mathbf{x}(t) \equiv \mathbf{0}$ to

$$\mathbf{x}'(t) = \mathbf{A}\mathbf{x}(t), \qquad t \geq t_0,$$

is asymptotically stable.

However, if at least *one* eigenvalue of $\mathbf{A}$ has positive real part, then the zero solution is unstable.

Proof. Recall from Chapter 9 that the entries of a fundamental matrix consist of functions of the form

(7) $e^{\alpha t}[p(t)\cos\beta t + q(t)\sin\beta t]$,

where $r = \alpha + i\beta$ is an eigenvalue of **A** and $p(t)$ and $q(t)$ are polynomials in t. (*Note:* If r is real, then $\beta = 0$ and the function reduces to $e^{\alpha t}p(t)$.) Now if all eigenvalues have negative real parts, that is, all α's satisfy $\alpha < 0$, then each of the entries in the fundamental matrix $\mathbf{X}(t)$ is not only bounded, but also goes to zero as $t \to \infty$, since

$$\lim_{t\to\infty} e^{\alpha t}p(t) = 0, \quad \text{for} \quad \alpha < 0.$$

Hence the zero solution is asymptotically stable.

If some eigenvalue $r = \alpha + i\beta$ has a positive real part, $\alpha > 0$, then the function in (7) becomes unbounded as $t \to \infty$, since

$$\lim_{t\to\infty} \left| e^{\alpha t}p(t) \right| = +\infty, \quad \text{for} \quad \alpha > 0.$$

We can multiply a solution to a linear system by an arbitrarily small constant and still have a solution, so we conclude that starting arbitrarily close to the zero solution is a solution that becomes unbounded. Thus the zero solution is unstable. ∎ End

Notice that, in the corollary, we omitted the case when one of the eigenvalues has zero real part. In this case, the zero solution may be stable or unstable, depending on whether the dimension of the eigenspace is the same as the multiplicity of the eigenvalue (see Problem 8).

Corollary 1 gives us an easy test for stability.

EXAMPLE 1 Determine the stability of the zero solution to the system

(8) $\mathbf{x}'(t) = \begin{bmatrix} -1 & 0 & 0 \\ 0 & -1 & 4 \\ 0 & -1 & -1 \end{bmatrix} \mathbf{x}(t), \qquad t \geq 0.$

Solution From the characteristic equation

$$(r + 1)[(r + 1)^2 + 4] = 0,$$

we find that the eigenvalues are -1, $-1 \pm 2i$. Since all the eigenvalues have negative real parts, the zero solution is asymptotically stable. ∎ End

EXAMPLE 2 Determine the stability of the zero solution to the system

(9) $\mathbf{x}'(t) = \begin{bmatrix} 1 & -1 & 0 \\ 3 & 0 & 0 \\ 0 & 0 & -2 \end{bmatrix} \mathbf{x}(t), \qquad t \geq 0.$

Solution From the characteristic equation

$$(r + 2)(r^2 - r + 3) = 0,$$

we find that the eigenvalues are -2, $(1/2) \pm (\sqrt{11}/2)i$. Because at least one eigenvalue, $(1/2) + (\sqrt{11}/2)i$, has positive real part, the zero solution is unstable. ∎ End

EXAMPLE 3 Determine the stability of the zero solution to the system

$$(10) \qquad \mathbf{x}'(t) = \begin{bmatrix} -2 & 1 & 0 \\ 0 & -2 & 1 \\ 0 & 0 & -2 \end{bmatrix} \mathbf{x}(t), \qquad t \geq 0.$$

Solution Here there is just one eigenvalue -2, which has multiplicity 3. Even though the eigenspace is one dimensional, the zero solution is asymptotically stable, since the eigenvalue is negative. ∎ End

Certain results for almost linear systems in the plane can be extended to higher-dimensional systems and to nonautonomous systems.

ALMOST LINEAR SYSTEMS

Definition 7. Let $\mathbf{A}$ be an $n \times n$ constant matrix whose determinant is nonzero. Let $\mathbf{f}(t, \mathbf{x})$ and $(\partial \mathbf{f}/\partial x_i)(t, \mathbf{x})$, $i = 1, \ldots, n$, be continuous functions for $t \geq 0$ and for $\|\mathbf{x}\| < K$. Assume $\mathbf{f}(t, \mathbf{0}) = \mathbf{0}$ for $t \geq 0$. If

$$\lim_{\|\mathbf{z}\| \to 0} \frac{\|\mathbf{f}(t, \mathbf{z})\|}{\|\mathbf{z}\|} = 0,$$

uniformly[†] in t, then we say that

$$(11) \qquad \mathbf{x}'(t) = \mathbf{A}\mathbf{x}(t) + \mathbf{f}(t, \mathbf{x})$$

is an **almost linear system.**

We conclude with the following result due to Poincaré and Perron, which gives conditions on $\mathbf{A}$ for the zero solution to the almost linear system to be asymptotically stable.[††]

POINCARÉ AND PERRON'S THEOREM

Theorem 11. If all the eigenvalues of $\mathbf{A}$ have negative real parts, then the zero solution of the almost linear system (11) is asymptotically stable. If at least one eigenvalue of $\mathbf{A}$ has positive real part, then the zero solution of (11) is unstable.

[†] That is, for every $\varepsilon > 0$, there exists a $\delta > 0$ such that whenever $0 < \|\mathbf{z}\| < \delta$, then $\|\mathbf{f}(t, \mathbf{z})\|/\|\mathbf{z}\| < \varepsilon$ for all $t \geq 0$.

[††] We refer the reader to the book by F. Baurer and J. A. Nohel, *The Qualitative Theory of Ordinary Differential Equations* (Dover Publications, Inc., New York, 1969), Chapter 4, for a proof.

EXERCISES 12.7

In Problems 1 through 6, determine the stability of the zero solution to the system $\mathbf{x}' = \mathbf{Ax}$ for the given matrix $\mathbf{A}$.

1. $\begin{bmatrix} -1 & 2 & 1 \\ 0 & -3 & 2 \\ 0 & 0 & -4 \end{bmatrix}$.

2. $\begin{bmatrix} -3 & 0 & 0 \\ 0 & -2 & -1 \\ 0 & 1 & -2 \end{bmatrix}$.

3. $\begin{bmatrix} -1 & -2 & 1 \\ -1 & 0 & -1 \\ -4 & 4 & -5 \end{bmatrix}$.

4. $\begin{bmatrix} 1 & 1 & 0 \\ 0 & 1 & 0 \\ 0 & 0 & 2 \end{bmatrix}$.

5. $\begin{bmatrix} 0 & 1 & 0 \\ 0 & 0 & 1 \\ 1 & -1 & 1 \end{bmatrix}$.

6. $\begin{bmatrix} 2 & 1 & -1 \\ -3 & -1 & 1 \\ 9 & 3 & -4 \end{bmatrix}$.

7. Prove Theorem 9.

8. **Eigenvalues with Zero Real Part.** To illustrate the relationship between stability and the degeneracy of the eigenspace for an eigenvalue with zero real part, consider the system

$$\mathbf{x}' = \begin{bmatrix} 0 & \alpha \\ 0 & 0 \end{bmatrix} \mathbf{x},$$

where α is a constant.

(a) Show that when $\alpha = 0$, zero is an eigenvalue of multiplicity two and has two linearly independent eigenvectors. Find the general solution and show directly that the zero solution is stable.

(b) Show that when $\alpha = 1$, zero is an eigenvalue of multiplicity two and has only one linearly independent eigenvector. Find the general solution and show directly that the zero solution is unstable.

(c) When an eigenspace is degenerate, then one of the polynomials, $p(t)$ or $q(t)$, in (7) is at least of order one. Use this fact to prove that if an eigenvalue for the matrix $\mathbf{A}$ has zero real part and is degenerate, then the zero solution to $\mathbf{x}' = \mathbf{Ax}$ is unstable.

(d) When an eigenspace is nondegenerate, then the polynomials $p(t)$ and $q(t)$ in (7) are constant. Use this fact to prove that if all the eigenvalues of the matrix $\mathbf{A}$ have nonpositive real parts and those with zero real parts are nondegenerate, then the zero solution to $\mathbf{x}' = \mathbf{Ax}$ is stable.

In Problems 9 and 10, use the results of Problem 8 to determine the stability of the zero solution to the system $\mathbf{x}' = \mathbf{Ax}$ for the given matrix $\mathbf{A}$.

9. $\begin{bmatrix} 0 & 1 & 1 & 0 \\ -1 & 0 & 0 & 1 \\ 0 & 0 & 0 & 1 \\ 0 & 0 & -1 & 0 \end{bmatrix}$.

10. $\begin{bmatrix} 0 & 4 & 0 & 0 \\ -4 & 0 & 0 & 0 \\ 0 & 0 & 0 & 4 \\ 0 & 0 & -4 & 0 \end{bmatrix}$.

In Problems 11 through 14, verify that the given system is almost linear and discuss the stability of the zero solution.

11. $\mathbf{x}' = \begin{bmatrix} 0 & 1 & 0 \\ 0 & 0 & 1 \\ -2 & -5 & -4 \end{bmatrix} \mathbf{x} + \begin{bmatrix} x_1^3 - x_2 x_3 \\ x_3^2 \\ -x_1 x_2 x_3 \end{bmatrix}$.

12. $\mathbf{x}' = \begin{bmatrix} -1 & -1 & 0 \\ 2 & -1 & 1 \\ 0 & 1 & -1 \end{bmatrix} \mathbf{x} + (2 - \cos t) \begin{bmatrix} x_1^2 \\ x_2^2 \\ x_3^2 \end{bmatrix}$.

13. $\dfrac{dx_1}{dt} = e^{-x_1} + \cos x_2 - 2$,

$\dfrac{dx_2}{dt} = -x_2 + \sin x_3$,

$\dfrac{dx_3}{dt} = 1 - e^{x_2 + x_3}$.

14. $\dfrac{dx_1}{dt} = -x_1 + x_2 + e^{x_3} - 1$,

$\dfrac{dx_2}{dt} = \sin x_3 - x_2$,

$\dfrac{dx_3}{dt} = -2x_3$.

In Problems 15 through 18, find all the critical points for the given system and discuss the stability of the corresponding equilibrium solutions.

15. $\dfrac{dx_1}{dt} = -x_1 + 1$,

$\dfrac{dx_2}{dt} = -2x_1 - x_2 + 2x_3 - 4$,

$\dfrac{dx_3}{dt} = -3x_1 - 2x_2 - x_3 + 1$.

16. $\dfrac{dx_1}{dt} = x_2 + x_3 - 1,$

$\dfrac{dx_2}{dt} = x_1 + x_3 + 2,$

$\dfrac{dx_3}{dt} = x_1 + x_2 + 3.$

17. $\dfrac{dx_1}{dt} = x_1\left(1 - x_1 - \dfrac{1}{2}x_2 - x_3\right),$

$\dfrac{dx_2}{dt} = x_2\left(1 - x_1 - x_2 - \dfrac{1}{2}x_3\right),$

$\dfrac{dx_3}{dt} = x_3\left(1 - \dfrac{1}{2}x_1 - x_2 - x_3\right).$

18. $\dfrac{dx_1}{dt} = x_3 - x_3^2,$

$\dfrac{dx_2}{dt} = -x_1,$

$\dfrac{dx_3}{dt} = x_2.$

CHAPTER SUMMARY

In this chapter we discussed the stability of critical points, limit cycles, and solutions for autonomous systems. While the discussion dealt primarily with systems in the plane, many of the results can be extended to higher-order systems. The techniques can loosely be grouped into two categories. First, there is an analytic method dating back to Poincaré that uses information about a related linear system to deduce information about the nonlinear system. Second, there is a geometric method mainly due to Lyapunov that uses a generalization of the energy of a physical system.

Linear Systems in the Plane

If the origin is an isolated critical point, then the stability of the origin and the classification of the origin as a critical point depends on the roots of the characteristic equation. The origin may be stable, asymptotically stable, or unstable and may be a node, saddle point, spiral point, or center. The results are summarized in Theorem 1, page 749, and again in Table 12.1, page 756.

Almost Linear Systems

If the origin is an isolated critical point for the corresponding linear system, then the stability and classification of the critical point at the origin for the linear and nonlinear systems are closely related, as shown in Table 12.1. In particular, the stability of the origin for the linear system and for the nonlinear system is the same, except when the origin of the associated linear system is a center. In this case, the nonlinear system may have either a center or spiral point at the origin and the stability must be decided by a different method.

Conservative and Nonconservative Systems

For second order differential equations that arise from problems in mechanics, the **energy method** is very helpful in sketching the phase plane diagram for the equation. For a conservative system governed by an equation of the form

$$\frac{d^2x}{dt^2} + g(x) = 0,$$

the function $G(x) = \int g(x)\,dx$ is called a **potential function** and $E(x, v) = (1/2)v^2 + G(x)$ is an **energy function.** The level curves for the energy function are integral curves for the system. Moreover, the critical points for the system are located at the critical values of the potential function. Graphing the potential function above the phase plane diagram can be very helpful in sketching the phase plane diagram.

Lyapunov's Direct Method

The energy method generalizes to arbitrary systems. This generalization is called Lyapunov's direct method. If $V(x, y)$ is positive definite and

$$\dot{V}(x, y) = V_x(x, y)f(x, y) + V_y(x, y)g(x, y)$$

is the derivative of $V(x, y)$ with respect to the system

$$\frac{dx}{dt} = f(x, y), \qquad \frac{dy}{dt} = g(x, y),$$

then the origin is stable if $\dot{V}$ is negative semidefinite, asymptotically stable if $\dot{V}$ is negative definite, and unstable if $\dot{V}$ is positive definite. The function $V(x, y)$ is called a **Lyapunov function.**

Limit Cycles and Periodic Solutions

A **limit cycle** is an isolated nontrivial closed trajectory. Limit cycles correspond to non-constant periodic solutions. The Bendixson negative criterion is a simple test to show that no limit cycles exist in a certain region of the plane, while the Poincaré-Bendixson theorem gives sufficient conditions for the existence of a limit cycle.

Higher-Dimensional Systems

The concept of stability extends to higher-dimensional systems. For linear systems, the stability of the zero solution (the origin) follows from properties of a fundamental matrix for the system. If the norm of the fundamental matrix $|||\mathbf{X}(t)|||$ is bounded, then the zero solution is stable. If $|||\mathbf{X}(t)||| \to 0$ as $t \to +\infty$, then the zero solution is asymptotically stable. When the linear system has constant coefficients, the stability is determined by the eigenvalues of the coefficient matrix $\mathbf{A}$. If all eigenvalues have negative real parts, then the zero solution is asymptotically stable. If at least one eigenvalue has positive real part, then the zero solution is unstable.

For almost linear systems, if the eigenvalues of the corresponding linear system all have negative real parts, then the zero solution of the almost linear system is asymptotically stable.

REVIEW PROBLEMS

In Problems 1 through 6, find all the critical points for the given system, discuss the type and stability of each critical point, and sketch the phase plane diagrams near each of the critical points.

1. $\dfrac{dx}{dt} = -x - 4y,$

 $\dfrac{dy}{dt} = -2x - 7y.$

2. $\dfrac{dx}{dt} = -x + 3y,$

 $\dfrac{dy}{dt} = 5x - 3y.$

3. $\dfrac{dx}{dt} = x + 5y + xy,$

 $\dfrac{dy}{dt} = -5x - 5y - xy.$

4. $\dfrac{dx}{dt} = x + 13y,$

 $\dfrac{dy}{dt} = -2x - y.$

5. $\dfrac{dx}{dt} = x - 3x^2 - xy,$

 $\dfrac{dy}{dt} = 2y - y^2 - 4xy.$

6. $\dfrac{dx}{dt} = 3x + 2y - y^3,$

 $\dfrac{dy}{dt} = x + 2y + y^3.$

In Problems 7 and 8, use the potential plane to help sketch the phase plane diagrams for the given equations.

7. $\dfrac{d^2x}{dt^2} + x^3 - 2x^2 - x + 2 = 0.$

8. $\dfrac{d^2x}{dt^2} + x^3 - x^2 = 0.$

In Problems 9 through 12, use Lyapunov's direct method to determine the stability for the origin of the given system.

9. $\dfrac{dx}{dt} = -x^3 - xy^2 + xy^5,$

 $\dfrac{dy}{dt} = -y^3 + x^2y - 3x^2y^4.$

 [*Hint:* Try $V(x, y) = 3x^2 + y^2.$]

10. $\dfrac{dx}{dt} = x^2y + y^3,$

 $\dfrac{dy}{dt} = -xy^2 + x^3.$

 [*Hint:* Try $V(x, y) = xy.$]

11. $\dfrac{dx}{dt} = x^3 + xy^2,$

 $\dfrac{dy}{dt} = -y^3 - x^2y.$

 [*Hint:* Try $V(x, y) = y^2 - x^2.$]

12. $\dfrac{dx}{dt} = -x^3 + y^3,$

 $\dfrac{dy}{dt} = -y^5.$

 [*Hint:* Try $V(x, y) = x^4 + y^2.$]

In Problems 13 and 14, sketch the phase plane diagrams for the given system. Discuss the stability of each limit cycle and critical point. Here $x = r \cos \theta$ and $y = r \sin \theta$.

13. $\dfrac{dx}{dt} = -y + x(r^3 - 5r^2 + 8r - 4),$

 $\dfrac{dy}{dt} = x + y(r^3 - 5r^2 + 8r - 4).$

14. $\dfrac{dx}{dt} = x - y + x(r^3 - 9r^2 + 26r - 25),$

 $\dfrac{dy}{dt} = x + y + y(r^3 - 9r^2 + 26r - 25).$

In Problems 15 and 16, determine whether the given system has a nonconstant periodic solution in the entire plane.

15. $\dfrac{dx}{dt} = 2x - 8x^3 - 3xy^2,$

 $\dfrac{dy}{dt} = \dfrac{1}{2}y - \dfrac{1}{2}y^3 - x^2y.$

 [*Hint:* Use the Lyapunov function $V(x, y) = x^2 + y^2.$]

16. $\dfrac{dx}{dt} = x - x^3 + 2x^3y,$

 $\dfrac{dy}{dt} = -2y - y^3 - 3x^2y^2.$

In Problems 17 and 18, determine the stability of the zero solution to the given system.

17. $\mathbf{x}' = \begin{bmatrix} -1 & 1 & 0 \\ 0 & -1 & 0 \\ 0 & 0 & -2 \end{bmatrix} \mathbf{x} + \begin{bmatrix} x_1^2 x_2 \\ x_2^3 \\ x_3^3 \end{bmatrix}.$

18. $\mathbf{x}' = \begin{bmatrix} -3 & 0 & 0 \\ 0 & -1 & 2 \\ 0 & -2 & -1 \end{bmatrix} \mathbf{x} + \begin{bmatrix} -x_1 x_2^2 \\ x_1 x_3 \\ -x_2 x_3 \end{bmatrix}.$

TECHNICAL WRITING EXERCISES

1. Discuss the relationship between the stability of the almost linear system and the stability of the corresponding linear system.

2. What are the advantages and disadvantages of the two general approaches (analytic vs geometric) in determining the stability of a critical point?

3. Periodic solutions of autonomous systems in the plane and limit cycles are closely identified. Describe this relationship and discuss the existence and stability of limit cycles.

4. Much of the chapter deals with almost linear systems in the plane. Discuss the extension of the results to almost linear systems in higher dimensions.

A. Solitons and Korteweg-de Vries Equation

The Korteweg-de Vries (KdV) equation

$$\frac{\partial u}{\partial t} + \frac{\partial^3 u}{\partial x^3} + \frac{1}{2}\frac{d}{dx}(u^2) = 0,$$

where $u(x, t)$ is the vertical displacement at position x and time t, is used to model long water waves. Of particular interest are **traveling wave** solutions, which are solutions of the form $u(x, t) = v(x - ct)$, where c is the constant wave speed.

(a) Show that $v(x - ct)$ is a solution to the KdV equation provided $v(z)$, $z = x - ct$, satisfies

$$-c\frac{dv}{dz} + \frac{d^3 v}{dz^3} + \frac{1}{2}\frac{d}{dz}(v^2) = 0.$$

(b) Of the traveling waves, we are most interested in **solitons,** traveling waves that are positive and decay to zero at $\pm\infty$. Therefore integrate once and set the constant of integration equal to zero to obtain

$$\frac{d^2 v}{dz^2} + \frac{1}{2}v^2 - cv = 0.$$

(c) Sketch the phase plane diagram for this conservative system.

(d) Use the results of parts (a) and (b) to argue that there exist solitons for the KdV equation. [*Hint:* The existence of a positive function $v(z)$ that decays to zero as $z \rightarrow \pm\infty$ corresponds to a trajectory in the vv'-plane that begins at the origin and returns to the origin.]

(e) Does increasing the speed of the soliton produce a shorter or taller wave?

B. Burger's Equation

A simple model for diffusive waves in fluid mechanics that incorporates nonlinearity and diffusion is **Burger's equation:**

$$\frac{\partial u}{\partial t} + u\frac{\partial u}{\partial x} = c\frac{\partial^2 u}{\partial x^2}.$$

The existence of shock waves is associated with the existence of traveling waves $u(x, t) = v(x - ct)$.

(a) Using an approach similar to that used in Project A, find a second order equation for $v(z)$, $z = x - ct$, that corresponds to a traveling wave solution to Burger's equation.

(b) Sketch the phase plane diagram for the equation obtained in part (a).

(c) Describe the type of traveling waves that exist for Burger's equation.

C. Computing Phase Plane Diagrams

In this chapter our emphasis was on analytic techniques for determining the stability of a critical point or a limit cycle. The phase plane diagrams that we constructed, although somewhat crude, provided us with a geometric representation for the behavior of the trajectories. Naturally, having a detailed phase plane diagram can improve our understanding of the behavior of the trajectories.

Using a numerical method such as Runge-Kutta for systems discussed in Section 5.6 or a computer software package such as PHASEPLANE, PHASER, or MACMATH, and a little experimentation—*keeping in mind the results of the analytical methods*—one can obtain a detailed phase plane diagram. In addition to knowing the type and location of each critical point and where limit cycles might be located, we offer a few practical suggestions.

(i) In computing trajectories near an unstable critical point, replace the step size h by $-h$. This has the effect of reversing "time" and makes unstable nodes and spiral points into stable nodes and spiral points. It can also transform an unstable limit cycle into a stable limit cycle.

(ii) In general there is no way of finding an initial point lying on a limit cycle. Hence, it is customary to locate limit cycles by following two trajectories—one approaching from the outside and one from the inside—until they appear to converge together and jointly map out the limit cycle. (The limit cycle lies between these two trajectories.) If the limit cycle is not stable, then the trajectories are computed using $-h$.

(a) Using a numerical method such as Runge-Kutta for systems or a computer software package generate a detailed phase plane diagram for the system

$$\frac{dx}{dt} = x(7 - x - 2y),$$

$$\frac{dy}{dt} = y(5 - y - x).$$

Compare with Figure 12.20.

(b) Generate a detailed phase plane diagram for the equation

$$\frac{d^2x}{dt^2} + x - x^3 = 0.$$

Compare with Figure 12.22.

(c) Generate a detailed phase plane diagram for the van der Pol equation

$$\frac{d^2x}{dt^2} + 5(x^2 - 1)\frac{dx}{dt} + x = 0.$$

Compare with Figure 12.33(c).

D. Ecosystem on Planet GLIA-2

Alfred Clark, Jr., University of Rochester

You are the chief mathematical biologist on a scientific expedition to the starsystem Glia. For the last week, your spaceship has been orbiting the small planet Glia-2, gathering data on what appears to be a unique biological system. Although most of the planet is barren, a small portion contains a

thriving ecosystem consisting of just three species. There is a flowering plant, Jancor, a small mouse-like mammal, Murrot, which feeds on Jancor, and a ferret-like predator, Vekton, which feeds on Murrot. After studying the data brought back by landing parties, you have concluded that this system can be modeled by the following equations for the Murrot population M and the Vekton population V, both in units of millions:

$$\frac{dM}{dt} = rM\left(1 - \frac{M}{A}\right) - \frac{\beta MV}{M + B}, \qquad \frac{dV}{dt} = \frac{bVM}{M + B} - cV.$$

Thus the Murrots obey a logistic law with a net birth rate r and a maximum sustainable population A. They suffer a loss to the Vektons, represented by the term $-\beta MV/(M + B)$. The Vektons have a natural death rate c—hence the loss term cV—and a birth rate $bVM/(M + B)$, which is strongly dependent on the Murrot population. From the landing party observations, the following parameter values are established:

$r = 12 \text{ yr}^{-1}$

$A = 14$ (units of one million)

$\beta = 20 \text{ yr}^{-1}$

$B = 4$ (units of one million)

$b = 4 \text{ yr}^{-1}$

$c = 1.6 \text{ yr}^{-1}$

In the following calculations, you will need to make use of a software package such as PHASER, PHASEPLANE, MACMATH, or XFUNCTIONS in order to integrate the equations.

(a) Determine whether the population is in stable equilibrium or in a limit cycle. You may assume the system has been going for some time with the above parameter values.

(b) Consider the possibility of intervening. If the population of either species drops below 1 million, either in equilibrium or during a limit cycle, the ecosystem will be susceptible to destruction by random natural events, such as extreme weather. Show that the system with the above parameter values is, by this criterion, in danger. Hence intervention, even though contrary to general Federation guidelines, is warranted. The only intervention tool you have is the ability to clear-cut some of the Jancor, thereby reducing the maximum sustainable population A. Assume you have the power to make A take on any value ≤ 14. Is it actually feasible to increase the minimum population by restricting the Murrot food supply (that is, reducing A)?

■ E. *A Growth Model for Phytoplankton—Part II*

Olivier Bernard, Station Zoologique de Villefranche-sur-Mer, France, and Jean-Luc Gouzé, INRIA Sophia-Antipolis, France

In the first part of this project in Chapter 5 (page 318), we used experimental data to identify the parameters of the following model for the chemostat:

(1)
$$\begin{cases} \dfrac{dX}{dt} = \rho\, \dfrac{SX}{S+k} - \alpha X, \\[2mm] \dfrac{dS}{dt} = \alpha(s_i - S) - \dfrac{\rho}{y}\, \dfrac{SX}{S + k}. \end{cases}$$

Suppose $\rho > \alpha$, as in the experiments.

(a) Compute the two equilibria (critical points) of the system.

(b) Determine the corresponding linear systems for the two equilibria. Study their stability and show that there is one saddle point and one stable node.

(c) Show that $S + X/y$ obeys a simple equation. Write the analytical solution and show that

(2) $$\lim_{t \to \infty} (S + X/y) = s_i.$$

What does it mean on the phase diagram?

(d) For $t > 4$, plot the experimental data (Table 5.6 on page 319) in the (X, S)-plane.

(e) Using (c) and the fact that at the experimental equilibrium S is very small (cf. Table 5.6), estimate s_i from the data and equation (2) (take $y = 0.13$).

(f) In the phase plane, draw the isoclines $dS/dX = 0$ and $dX/dS = 0$ corresponding to the estimated parameters $\rho = 2$, $k = 0.5$, $y = 0.13$, $\alpha = 1.06$, and the value of s_i obtained in (e). Compare with (d).

(g) Sketch a phase diagram for the system (1).

(h) Do you think that (1) is a reasonable model?

Chapter 13

Existence and Uniqueness

Theory

13.1 INTRODUCTION: SUCCESSIVE APPROXIMATIONS

In nearly every chapter of this book, the theory, application, or procedure discussed depended on a fundamental theorem concerning the existence and uniqueness of the solution to an initial value problem. In this chapter, we provide the proofs of these basic theorems.

How does one know that a solution exists? When possible, the simplest way is actually to exhibit the solution in an explicit form. While we were able to do this for the entire class of linear first order differential equations (see Section 2.3) and also for homogeneous equations (or homogeneous normal systems) that have constant coefficients (see Chapters 4, 6, and 9), most differential equations do not fall into these categories.

Another approach to proving the existence of a solution is to construct a sequence of **approximate solutions** to the initial value problem and then show that these approximations converge to the desired solution. This is the strategy we adopt in this chapter.

Some methods for constructing approximations are already familiar to us. For example, we have discussed numerical procedures such as Euler's method (see Sections 1.4 and 3.5) and analytic techniques such as Taylor's method for generating the partial sums of a power series solution (see Section 8.2). For the purpose of handling a very general class of differential equations, our procedure will be to construct a sequence of **successive approximations (iterations)** and show that these converge to the desired solution.

To describe the procedure, we begin by expressing the initial value problem

(1) $\qquad y'(x) = f(x, y(x)), \qquad y(x_0) = y_0$

as an **integral equation.** This is accomplished by integrating both sides of equation (1) from x_0 to x:

(2) $\qquad \displaystyle\int_{x_0}^{x} y'(t)\,dt = y(x) - y(x_0) = \int_{x_0}^{x} f(t, y(t))\,dt.$

Setting $y(x_0) = y_0$ and solving for $y(x)$, we have

(3) $$y(x) = y_0 + \int_{x_0}^{x} f(t, y(t))\,dt.$$

Our derivation shows that a solution to (1) must also be a solution to (3). Conversely, using the fundamental theorem of calculus, we can show that if f and y are continuous, then a solution to (3) satisfies the differential equation in (1) (see Problem 17). Furthermore, such a solution must satisfy the initial condition $y(x_0) = y_0$, as can be seen by substituting $x = x_0$ in (3). Therefore, with certain continuity assumptions, the initial value problem (1) and the integral equation (3) are equivalent in the sense that they have the same solutions.

We can view the right-hand side of (3) as defining an operator that maps a function y to a function $T[y]$, where

(4) $$T[y](x) := y_0 + \int_{x_0}^{x} f(t, y(t))\,dt.$$

(In general, T is a nonlinear operator.) With this notation, we can express equation (3) in operator form as

(5) $$y = T[y].$$

In (5), equality is meant in the sense that the functions y and $T[y]$ are equal on some interval I. A solution to (5) is called a **fixed point** of T because for equation (5) to be satisfied, T must map y to itself, and y therefore remains fixed. Of course, the term ''point'' refers to the *function* $y(t)$, in this context.

To gain some insight into the procedure for determining approximate solutions to (5), let's consider the simpler but analogous problem of finding the fixed points of a function $g(x)$; that is, we seek the roots of the equation

(6) $$x = g(x).$$

Geometrically, solving equation (6) is equivalent to finding the points where the graph of $y = g(x)$ intersects the line $y = x$. If we consider the particular graphs in Figure 13.1, then we can devise the following geometric procedure that leads to the point of intersection P. Starting with some point P_0 on the curve $y = g(x)$, we follow a horizontal line until we reach a point Q_0 on the line $y = x$. From Q_0, we follow a vertical line down to a point P_1 on the graph of $y = g(x)$. Repeating the procedure, we follow a horizontal line to the point Q_1 on $y = x$, then a vertical line to the point P_2. Continuing in this staircase fashion, we draw closer and closer to the point of intersection P.

To describe this procedure algebraically, let (x_0, y_0) be the coordinates of the point P_0, (x_1, y_1) be the coordinates of the point P_1, and so on. Each point P_n lies on the graph of the curve $y = g(x)$, so we have $y_n = g(x_n)$. Recall that the points P_0 and Q_0 lie on the same horizontal line; thus the coordinates of Q_0 are (y_0, y_0). Also the points P_1 and Q_0 are on the same vertical line, so $x_1 = y_0$. Since $y_0 = g(x_0)$, we conclude that $x_1 = g(x_0)$. By the same reasoning, we find that $x_2 = g(x_1)$, $x_3 = g(x_2)$, and so forth. In general, we have

(7) $$x_{n+1} = g(x_n), \qquad n = 0, 1, 2, \dots.$$

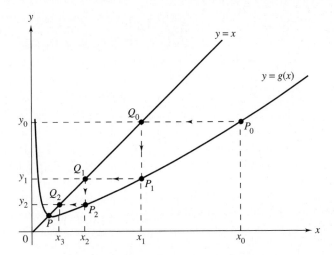

Figure 13.1 Method of successive substitutions

To summarize, we start with an initial approximation (a guess) x_0, substitute it into (7), and obtain a new approximation x_1. Continuing this iteration, we obtain a sequence of approximations $\{x_n\}$ that we hope will converge to a solution of (6). This procedure is called the method of **successive substitutions.**

EXAMPLE 1 Use the method of successive substitutions to approximate a solution to the equation

(8) $x = \dfrac{x}{2} + \dfrac{1}{x}.$

Solution The graph of the function $g(x) = x/2 + 1/x$ is shown in Figure 13.1. Using formula (7), the recurrence relation is

(9) $x_{n+1} = \dfrac{x_n}{2} + \dfrac{1}{x_n}.$

With $x_0 = 1$ as an initial approximation, we find

$$x_1 = \dfrac{1}{2} + \dfrac{1}{1} = 1.5, \qquad x_2 = \dfrac{1.5}{2} + \dfrac{1}{1.5} = 1.4166667,$$

and so on. In Table 13.1 on page 808 we give x_n for $n = 0, 1, \ldots, 5$ as computed on a pocket calculator. We stopped iterating after x_5 because x_4 and x_5 agree to seven decimal places, which is the accuracy of our calculator.[†] Hence

$$x_4 \approx x_5 = \dfrac{x_4}{2} + \dfrac{1}{x_4},$$

and 1.4142136 is the desired approximation for a solution to (8). ∎ End

[†] Stopping procedures for numerical approximations are discussed in Section 3.5.

TABLE 13.1 APPROXIMATIONS FOR A SOLUTION OF $x = x/2 + 1/x$	
$x_0 = 1$	$x_3 = 1.4142157$
$x_1 = 1.5$	$x_4 = 1.4142136$
$x_2 = 1.4166667$	$x_5 = 1.4142136$

If the value of x_5 seems familiar, that is because x_5 agrees with the number $\sqrt{2}$ to eight digits. Indeed, the reader can verify that equation (8) is equivalent to the equation $x^2 = 2$.

While the method of successive substitutions worked quite well (converged) for the problem of Example 1, there are situations in which it will not converge to a fixed point (see Problem 19). The following theorem[†] gives a set of conditions that ensure convergence.

BANACH FIXED POINT THEOREM

Theorem 1. Suppose g is a differentiable function on the closed interval I and $g(x)$ lies in I for all x in I. If there exists a constant K such that $|g'(x)| \le K < 1$ for all x in I, then the equation $x = g(x)$ has a unique solution in I. Moreover, the sequence of **successive approximations** defined by

(10) $x_{n+1} = g(x_n), \quad n = 0, 1, 2, \dots,$

converges to this solution for any choice of the starting value x_0 in I.

Proof. While it is not difficult to prove directly that g has a fixed point in I, we shall establish this fact by analyzing the sequence of successive approximations. We divide the argument into four parts.

The Approximations Are Well Defined. Choose an x_0 in I. Since x_0 lies in the domain of g, then $x_1 = g(x_0)$ is well defined. Moreover, x_1 is a point of I, so $x_2 = g(x_1)$ is well defined. By induction, we see that the entire sequence $\{x_n\}$ is well defined.

The Approximations Converge to an Element of I. To establish convergence, observe that we can write

(11) $x_n = x_0 + (x_1 - x_0) + (x_2 - x_1) + \cdots + (x_n - x_{n-1}).$

That is, x_n is the $(n + 1)$th partial sum of the series

(12) $x_0 + \displaystyle\sum_{j=0}^{\infty} (x_{j+1} - x_j).$

[†] ***Historical Footnote:*** This theorem is a special case of a theorem proved by Stephen Banach in 1922.

Therefore, to prove that the sequence $\{x_n\}$ converges, we just have to show that the series in (12) converges. We can prove this using the comparison test, but first we must obtain an estimate on the absolute value of each term in the series.

Let x and y be two numbers in I. By the mean value theorem, there exists some number z between x and y such that

$$g(x) - g(y) = g'(z)(x - y).$$

Using the assumption that $|g'(z)| \leq K$ for all z in I, we have

(13) $\qquad |g(x) - g(y)| \leq K|x - y|, \qquad (K < 1)$

for all x, y in I. In particular, from formula (10), we have

$$|x_2 - x_1| = |g(x_1) - g(x_0)| \leq K|x_1 - x_0|.$$

In a similar manner, we find

$$|x_3 - x_2| \leq K|x_2 - x_1| \leq K^2|x_1 - x_0|$$

and, in general,

(14) $\qquad |x_{j+1} - x_j| \leq K^j|x_1 - x_0|, \qquad j = 0, 1, 2, \ldots.$

Since $K \leq 1$, the geometric series

$$\sum_{j=0}^{\infty} K^j|x_1 - x_0| = |x_1 - x_0| \sum_{j=0}^{\infty} K^j$$

converges. Thus, by the comparison test, the series in (12) converges absolutely, which means that $\{x_n\}$ converges. Let $x_\infty := \lim x_n$. Since $\{x_n\}$ lies in I and I is a closed interval, then x_∞ is in I.

x_∞ *Is a Fixed Point.* To see that x_∞ is a solution of the equation $x = g(x)$, we take the limit as $n \to \infty$ of both sides of the equation $x_{n+1} = g(x_n)$. Since g is continuous, we have

$$x_\infty = \lim_{n \to \infty} x_{n+1} = \lim_{n \to \infty} g(x_n) = g\left(\lim_{n \to \infty} x_n\right) = g(x_\infty).$$

Hence x_∞ is a fixed point of g.

Uniqueness. To show that x_∞ is the unique fixed point in I, assume z is some other fixed point in I. Then, by inequality (13),

$$|x_\infty - z| = |g(x_\infty) - g(z)| \leq K|x_\infty - z|.$$

Simplifying, we find $(1 - K)|x_\infty - z| \leq 0$, and since $0 \leq K < 1$, we must have $x_\infty = z$. Therefore there is only one fixed point of g in I, and it is the limit of the successive approximations defined in (10) for any starting value x_0 in I. ∎ End

We leave it for the reader to verify that the function $g(x)$ in Example 1 satisfies the hypotheses of Theorem 1 on the interval $[1, 2]$ (see Problem 20).

Several times in the proof of Theorem 1 we used inequality (13). Functions that satisfy such an inequality (with $K < 1$) for all x, y in an interval I are called **contractions.** Contractions map two points x, y into two points $g(x)$, $g(y)$ that are closer together than were the original points. If we view the interval I as pliable, then we can visualize the Banach fixed point theorem as stating that if g contracts the interval I into a subset of itself, then some point remains fixed. Not surprisingly, Theorem 1 is also called the **contraction mapping principle.**

In the next section we use another form of the Banach fixed point theorem to prove the existence of a unique solution to the initial value problem (1). For now, let's apply the method of successive substitutions to the integral equation (3). That is, we define the sequence of functions $\{y_n\}_{n=0}^{\infty}$ by

(15) $$y_{n+1}(x) := T[y_n](x) = y_0 + \int_{x_0}^{x} f(t, y_n(t))\,dt,$$

where we must specify the starting function $y_0(x)$. The simplest choice for $y_0(x)$ is to let $y_0(x) \equiv y_0$. The successive approximations y_n of the solution to the integral equation (3) are called the **Picard iterations.**[†]

EXAMPLE 2 Compute the Picard iterations for the initial value problem

(16) $y'(x) = 2y(x)$, $y(0) = 1$,

starting with $y_0(x) \equiv 1$.

Solution The initial value problem (16) is equivalent to the integral equation

(17) $$y(x) = 1 + \int_{0}^{x} 2y(t)\,dt.$$

Computing the Picard iterations via (15), we find

$$y_1(x) = 1 + \int_{0}^{x} 2y_0(t)\,dt = 1 + \int_{0}^{x} 2\,dt = 1 + 2x,$$

$$y_2(x) = 1 + \int_{0}^{x} 2(1 + 2t)\,dt = 1 + 2x + \frac{(2x)^2}{2!},$$

$$y_3(x) = 1 + \int_{0}^{x} 2\left[1 + 2t + \frac{(2t)^2}{2!}\right]dt = 1 + 2x + \frac{(2x)^2}{2!} + \frac{(2x)^3}{3!},$$

[†] *Historical Footnote:* The method can be traced back to J. Liouville, who used it in 1838 in his study of linear second order equations. It was also used by J. Caque in 1864, L. Fuchs in 1870, and G. Peano in 1888. But it was E. Picard whose name became associated with the method because of his work on successive approximation in 1890 and 1896.

and so forth. The pattern seems clear: The nth iteration is given by

$$(18) \qquad y_n(x) = 1 + 2x + \frac{(2x)^2}{2!} + \cdots + \frac{(2x)^n}{n!}.$$

Here we recognize $y_n(x)$ as the partial sum of the Maclaurin series for e^{2x}:

$$e^{2x} = 1 + 2x + \frac{(2x)^2}{2!} + \cdots + \frac{(2x)^n}{n!} + \cdots.$$

Hence we see that the sequence of Picard iterations converges to e^{2x}, which is the solution to the initial value problem (16). ■ End

In Example 2, we were very fortunate to be able to decipher the general form of the sequence of Picard iterations. This enabled us to determine that the sequence converged to the solution to the initial value problem. In most cases, it is impossible to determine a simple form for the Picard iterations. As we show in the next section, however, there are conditions on f that guarantee that the sequence of iterations converges to a solution.

EXERCISES 13.1

In Problems 1 through 4, express the given initial value problem as an integral equation.

1. $y'(x) = x^2 - y(x)$, $y(1) = 4$.

2. $y'(x) = \sin(x + y(x))$, $y(\pi) = 0$.

3. $y'(x) = [y(x) - x]^2$, $y(1) = -3$.

4. $y'(x) = e^{y(x)}$, $y(0) = 1$.

In Problems 5 through 10, use the method of successive substitutions to approximate a solution of the given equation, starting with the given value for x_0.

5. $x = \dfrac{1}{2}\left(x + \dfrac{3}{x}\right)$, $x_0 = 3$,

6. $x = \dfrac{3}{8 - x^3}$, $x_0 = 0$.

7. $x = \dfrac{1}{x^2 + 4}$, $x_0 = 0.5$.

8. $x = \dfrac{1}{2}\left(x + \dfrac{5}{x}\right)$, $x_0 = 2$.

9. $x = \sqrt[4]{\dfrac{5 - x}{3}}$, $x_0 = 1$.

10. $x = 1 + \sin x$, $x_0 = 2$.

In Problems 11 through 16, compute the Picard iterations $y_1(x)$ and $y_2(x)$ for the given initial value problem, starting with $y_0(x) \equiv y_0$, where y_0 is the initial value for y.

11. $y'(x) = -y(x)$, $y(0) = 2$.

12. $y'(x) = y^2(x)$, $y(0) = 1$.

13. $y'(x) = 3x^2$, $y(1) = 2$.

14. $y'(x) = \cos x$, $y(\pi) = 0$.

15. $y'(x) = y(x) - e^x$, $y(0) = 0$.

16. $y'(x) = x - y(x)$, $y(1) = 1$.

17. Assuming f and y in equation (3) are continuous, use the fundamental theorem of calculus to show that a solution to (3) is a solution to the initial value problem (1).

18. Show that equation (8) is equivalent to $x^2 - 2 = 0$ in the sense that a solution to one is a solution to the other.

19. Show that the method of successive substitutions for the equation $x = (x^2 + 1)/2$ converges to the solution $x = 1$ when $x_0 = 0$ but diverges when x_0 is chosen to be 2 or even 1.1. [*Hint:* Sketch the curve $y = (x^2 + 1)/2$.]

20. To verify that $g(x) = x/2 + 1/x$ satisfies the hypotheses of Theorem 1 on the interval $[1, 2]$, proceed as follows:

(a) Show that for x in [1, 2],

$$-\frac{1}{2} \le g'(x) \le \frac{1}{4},$$

so $|g'(x)| \le 1/2$.

(b) Show that for x in [1, 2],

$$\sqrt{2} \le g(x) \le \frac{3}{2},$$

so $g(x)$ lies in [1, 2].

13.2 PICARD'S EXISTENCE AND UNIQUENESS THEOREM

In this section we show that, under mild assumptions, the successive approximations obtained using Picard's method converge to the solution to the initial value problem.

What do we mean when we say a sequence of functions $\{y_n\}$ converges to a function y on an interval [a, b]? Intuitively, we want the functions y_n to get close to the function y as n approaches ∞. We therefore need to define what is meant by *close*.

Let $C[a, b]$ denote the set of all functions that are continuous on the interval [a, b]. For y in $C[a, b]$, we define the **norm** of y to be

(1) $$\|y\| := \max_{x \in [a, b]} |y(x)|.$$

When we wish to emphasize the interval I over which the norm is taken, we write $\|y\|_I$. The norm of y can be thought of as the distance from the function y to the identically zero function. This is like saying that $|x|$ is the distance from x to the origin. Analogous to the formula for the distance between two real numbers, we define the **distance between two functions** y and z in $C[a, b]$ to be $\|y - z\|$.

Using the properties of the absolute value, it can be shown that the norm satisfies the following three properties:

(2) $\|y - z\| = 0$ if and only if $y(x) = z(x)$ for all x in [a, b].

(3) $\|cy\| = |c| \|y\|$, where c is any real number.

(4) $\|y + z\| \le \|y\| + \|z\|$ (triangle inequality).

Here we have assumed that y and z are elements of $C[a, b]$.

With this concept of distance, we say a sequence of functions $\{y_n\}$ **converges uniformly** to y, denoted $y_n \xrightarrow{\text{unif}} y$, if

(5) $$\lim_{n \to \infty} \|y_n - y\| = 0.$$

That is, the distance between y_n and y goes to zero as n approaches ∞. Geometrically we can think of uniform convergence as requiring that the sequence $\{y_n\}$ eventually gets inside an arbitrarily small ε-**corridor** about $y(x)$ (see Figure 13.2).

EXAMPLE 1 Show that the sequence of functions $y_n(x) = x^2 + x/n$, $n = 1, 2, 3, \ldots$, converges uniformly on [0, 1] to the function $y(x) = x^2$.

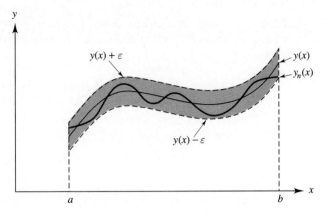

Figure 13.2 An ε-corridor about $y(x)$

Solution We first compute

$$\|y_n - y\| = \max_{x \in [0,1]} |y_n(x) - y(x)|$$

$$= \max_{x \in [0,1]} \left| x^2 + \frac{x}{n} - x^2 \right|$$

$$= \max_{x \in [0,1]} \left| \frac{x}{n} \right| = \frac{1}{n}.$$

Thus

$$\lim_{n \to \infty} \|y_n - y\| = \lim_{n \to \infty} \frac{1}{n} = 0,$$

and so y_n converges uniformly on $[0, 1]$ to y. ■ End

The reader is probably familiar with a related type of convergence for functions that is called pointwise convergence. We say that a sequence of functions $\{y_n\}$ **converges pointwise** on $[a, b]$ to the function y if for every x in $[a, b]$,

(6) $$\lim_{n \to \infty} y_n(x) = y(x).$$

Uniform convergence is stronger than pointwise convergence because if $\|y_n - y\| \to 0$, then certainly

$$\lim_{n \to \infty} [y_n(x) - y(x)] = 0$$

for every x in $[a, b]$; hence y_n converges pointwise to y. However, pointwise convergence does *not* imply uniform convergence (see Problem 7).

In the previous section we proved that under suitable conditions, the sequence of approximations obtained by the method of successive substitutions converged to the fixed point of the function. This was a consequence of the Banach fixed point theorem for

functions defined on a closed interval I. To prove that the Picard iterations converge, we need an extension of this theorem that applies to *operators* defined on a closed subset of functions in $C[a, b]$.

BANACH FIXED POINT THEOREM FOR OPERATORS

Theorem 2. Let S denote the set of continuous functions on $[a, b]$ that lie within a distance α of a given function y_0; that is,

$$S = \{y \in C[a, b]: \|y - y_0\| \leq \alpha\}.$$

Suppose G is an operator mapping S into S and that G is a **contraction** on S in the sense that there exists a constant K, $0 \leq K < 1$, such that

(7) $\|G[w] - G[z]\| \leq K\|w - z\|$ for all w, z in S.

Then the equation $y = G[y]$ has a unique solution in S. Moreover, the sequence of successive approximations defined by

(8) $y_{n+1} := G[y_n]$, $n = 0, 1, 2, \ldots,$

converges uniformly to this solution for any choice of the starting function y_0 in S.

Proof. The proof of Theorem 2 is essentially the same as the proof of Theorem 1 in Section 13.1, so we only outline the argument.

The Approximations Are Well Defined. As in the proof of Theorem 1, the y_n's are well defined because of the hypothesis that G maps S into S.

The Approximations Converge Uniformly to an Element of S. Using the fact that G is a contraction on S, we can derive the inequalities

(9) $\|y_{j+1} - y_j\| \leq K^j\|y_1 - y_0\|$, $j = 0, 1, 2, \ldots,$

in the same way that we established inequalities (14) in Section 13.1. Furthermore, y_n is the $(n + 1)$th partial sum of the series

$$y_0 + \sum_{j=0}^{\infty} (y_{j+1} - y_j),$$

and from (9), we know that the norm of each term in this series is bounded by the constant $M_j := K^j\|y_1 - y_0\|$. Since the series $\Sigma\, M_j$ is convergent (because $K < 1$), we can appeal to an extension of the comparison test, called the **Weierstrass M test,**[†] to conclude that the sequence $\{y_n\}$ converges uniformly to a function y_∞ in S.

[†] This test is also called the **Weierstrass comparison test** and is discussed in most advanced calculus or real analysis texts; for example, *Advanced Calculus*, Fourth Edition, by Wilfred Kaplan (Addison-Wesley, Reading, Mass., 1991), or *Principles of Mathematical Analysis*, Third Edition, by Walter Rudin (McGraw-Hill, New York, 1976).

y_∞ Is a Fixed Point. Because G is a contraction, $\|G[y_\infty] - G[y_n]\| \le K\|y_\infty - y_n\|$ for all n. We already know that $\|y_\infty - y_n\| \to 0$. Thus $\|G[y_\infty] - G[y_n]\| \to 0$ as $n \to \infty$; that is, $G[y_n] \xrightarrow{\text{unif}} G[y_\infty]$. But $y_{n+1} = G[y_n]$ and $y_{n+1} \xrightarrow{\text{unif}} y_\infty$. Therefore, by the uniqueness of the limit, we have $y_\infty = G[y_\infty]$. Hence y_∞ is a fixed point of G.

Uniqueness. That y_∞ is the unique fixed point of G again follows from the fact that G is a contraction, just as in the proof of Theorem 1. ■ End

In Section 13.1 we showed that when f and y are continuous functions, the initial value problem

(10) $y'(x) = f(x, y(x)), \qquad y(x_0) = y_0,$

is equivalent to the operator equation

(11) $y = T[y],$

where

(12) $T[y](x) := y_0 + \displaystyle\int_{x_0}^{x} f(t, y(t))\,dt.$

Our goal is to apply the Banach fixed point theorem to this operator T and thereby show (with certain conditions on f) that equation (11) has a unique solution, which is the limit of the **Picard iterations**

(13) $y_{n+1} := T[y_n], \qquad y_0(x) \equiv y_0.$

PICARD'S EXISTENCE AND UNIQUENESS THEOREM

Theorem 3. Assume f and $\partial f/\partial y$ are continuous functions in a rectangle $R = \{(x, y): a < x < b, c < y < d\}$ that contains the point (x_0, y_0). Then the initial value problem (10) has a unique solution in some interval $[x_0 - h, x_0 + h]$, where h is a positive number. Moreover, the Picard iterations defined in (13) converge uniformly on $[x_0 - h, x_0 + h]$ to this solution.

Proof. To apply Theorem 2 to the operator T in (12), we must show that there exist an interval $I := [x_0 - h, x_0 + h]$ and a positive constant α such that T maps the set

$$S := \{g \in C(I): \|g - y_0\|_I \le \alpha\}$$

into itself and T is a contraction on S.

First we choose positive constants h_1 and α_1 so that

$$R_1 := \{(x, y): |x - x_0| \le h_1,\ |y - y_0| \le \alpha_1\} \subset R.$$

Since R_1 is a closed rectangle and f as well as $\partial f/\partial y$ are continuous on R_1, there exist constants $M > 0$ and $L > 0$ such that for all (x, y) in R_1,

$$|f(x, y)| \le M \quad \text{and} \quad \left|\frac{\partial f}{\partial y}(x, y)\right| \le L.$$

Let g be a continuous function on $I_1 := [x_0 - h_1, x_0 + h_1]$ that satisfies $|g(x) - y_0| \leq \alpha_1$ for x in I_1. For such a g, we have

$$T[g](x) = y_0 + \int_{x_0}^{x} f(t, g(t)) \, dt.$$

Hence for $|x - x_0| \leq h_1$,

$$(14) \qquad |T[g](x) - y_0| = \left| \int_{x_0}^{x} f(t, g(t)) \, dt \right| \leq \left| \int_{x_0}^{x} |f(t, g(t))| \, dt \right|$$

$$\leq M \left| \int_{x_0}^{x} dt \right| = M|x - x_0|.$$

Geometrically we have shown that if the function g satisfies $|g(x) - y_0| \leq \alpha_1$ for x in I_1, then the graph of the function $T[g]$ must lie in the sectors bounded by the lines $y = y_0 \pm M(x - x_0)$ (see Figure 13.3). To keep $T[g]$ in the rectangle R_1, we must choose h less than $h_2 := \min\{h_1, \alpha_1/M\}$. We will see later that we also need h to be less than $1/L$ in order for T to be a contraction. Therefore we choose h so that

$$0 < h < \min\{h_1, \alpha_1/M, 1/L\}.$$

Now let $S := \{g \in C(I) : \|g - y_0\|_I \leq \alpha\}$, where $\alpha = \alpha_1$ and $I = [x_0 - h, x_0 + h]$.

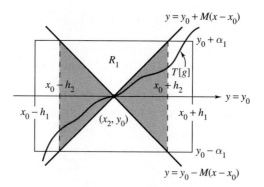

Figure 13.3 Graph of $T[g]$

T Maps S into S. Clearly $T[g](x)$ is a continuous function on I. For g in S, inequality (14) holds, so for x in I, we have

$$|T[g](x) - y_0| \leq M|x - x_0| \leq Mh < M\left(\frac{\alpha_1}{M}\right) = \alpha_1.$$

Hence $\|T[g] - y_0\|_I < \alpha_1$ (see the shaded region in Figure 13.3). Therefore, if g lies in S, then $T[g]$ does also. Thus T maps S into S.

T Is a Contraction on S. Let u and v be functions in S. On R_1, we have $|\partial f/\partial y| \leq L$. Thus, using the mean value theorem,

$$|T[u](x) - T[v](x)| = \left| \int_{x_0}^{x} \{ f(t, u(t)) - f(t, v(t)) \} \, dt \right|$$

$$= \left| \int_{x_0}^{x} \frac{\partial f}{\partial y}(t, z(t))[u(t) - v(t)] \, dt \right|$$

$$\leq L \left| \int_{x_0}^{x} |u(t) - v(t)| \, dt \right|$$

$$\leq L \|u - v\|_I |x - x_0| \leq Lh \|u - v\|_I,$$

where $z(t)$ is a function that lies between $u(t)$ and $v(t)$. Setting $K := Lh$, we have shown that

$$\|T[u] - T[v]\|_I \leq K \|u - v\|_I,$$

and since (by assumption) $h < 1/L$, we have $K < 1$. Hence T is a contraction.

It now follows from the Banach fixed point theorem (page 814) that T has a unique fixed point in S, so the initial value problem (10) has a unique solution in S. Moreover, this solution is the uniform limit of the sequence of successive approximations defined in (13).

To complete the proof, we need to resolve one subtle but important point. That is, although we have shown that problem (10) has a unique solution in S, we have not shown that every solution to (10) on the interval $[x_0 - h, x_0 + h]$ must lie in S. To see this, suppose $u(x)$ is a solution to (10) on $[x_0 - h, x_0 + h]$ and recall that f is bounded by M on the rectangle R_1 of Figure 13.3. Since $u(x_0) = y_0$, the graph of $u(x)$ must lie in R_1 for x near x_0. For such values of x, we have $|f(x, u(x))| \leq M$. Thus $|u'(x)| = |f(x, u(x))| \leq M$. Therefore, for x near x_0, the graph of $u = T[u]$ must be in the shaded sector shown in Figure 13.3. In fact, the graph of u cannot escape from this sector on the interval $|x - x_0| \leq h \; (\leq h_2)$ because this would require that $|u'(x)| = |f(x, u(x))| > M$ at some point of the sector. Hence $|u(x) - y_0| \leq \alpha_1$ for all $|x - x_0| \leq h$; that is, $u \in S$. ∎ End

The preceding discussion carries over to first order systems if we define the distance between two vector functions properly. To be specific, let $C_n[a, b]$ denote the set of all vector functions $\mathbf{y} = (y_1, y_2, \ldots, y_n)$ with n components that are continuous on $[a, b]$. For $\mathbf{y}$ in $C_n[a, b]$, define

$$\|\mathbf{y}\| := \max_{x \in [a,b]} |\mathbf{y}(x)|,$$

where $|\mathbf{y}(x)| := [|y_1(x)|^2 + |y_2(x)|^2 + \cdots + |y_n(x)|^2]^{1/2}$.

By extending the Banach fixed point theorem to operators defined on closed subsets of $C_n[a, b]$, we obtain the following version of Picard's existence and uniqueness theorem.

PICARD'S THEOREM FOR SYSTEMS

Theorem 4. Assume $\mathbf{f}$ and $\partial \mathbf{f}/\partial x_i$, $i = 1, \ldots, n$, are continuous functions in a rectangle

$$R = \{(t, x_1, \ldots, x_n) : a < t < b, \, c_i < x_i < d_i, \, i = 1, \ldots, n\}$$

that contains the point $(t_0, \mathbf{x}_0)$. Then the initial value problem

$$\mathbf{x}'(t) = \mathbf{f}(t, \mathbf{x}), \qquad \mathbf{x}(t_0) = \mathbf{x}_0,$$

where $\mathbf{x}_0 = \text{col}(x_{1,0}, \ldots, x_{n,0})$, has a unique solution in some interval $[t_0 - h, t_0 + h]$, where h is a positive constant.

EXERCISES 13.2

*In Problems 1 through 6, determine whether the given sequence of functions $\{y_n\}$ converges **uniformly** to the given function y on the interval specified.*

1. $y_n(x) = 1 - \dfrac{x}{n}$ and $y(x) \equiv 1$ on $[-1, 1]$.

2. $y_n(x) = \left(\dfrac{x}{2}\right)^n$ and $y(x) \equiv 0$ on $[0, 2]$.

3. $y_n(x) = \dfrac{nx}{1 + n^2 x^2}$ and $y(x) \equiv 0$ on $[0, 1]$.

4. $y_n(x) = \dfrac{1}{2x + n}$ and $y(x) \equiv 0$ on $[0, 1]$.

5. $y_n(x) = \displaystyle\sum_{k=0}^{n} x^k$ and $y(x) = \dfrac{1}{1 - x}$ on $\left[0, \dfrac{1}{2}\right]$.

6. $y_n(x) = x + \dfrac{x^2}{n}$ and $y(x) = x$ on $[0, 1]$.

7. Show that the sequence of functions

$$y_n(x) = \begin{cases} n^2 x, & 0 \le x \le 1/n, \\ 2n - n^2 x, & 1/n < x < 2/n, \\ 0, & 2/n \le x \le 1 \end{cases} \quad n = 2, 3, 4, \ldots,$$

converges pointwise on $[0, 1]$ to $y(x) \equiv 0$ but does not converge uniformly to y on $[0, 1]$. [*Hint:* Show that $\|y_n - y\| = n$ for all n.]

8. Prove inequality (9) in the proof of Theorem 2.

9. In the proof of Theorem 3, h is chosen to be strictly less than $\min\{h_1, \alpha_1/M, 1/L\}$, where h_1, α_1, M, and L are defined in the proof. Find an h for which the initial

value problem

$$y'(x) = y^2(x) - x, \qquad y(1) = 0,$$

has a unique solution on $[1 - h, 1 + h]$ if

$$R_1 = \{(x, y) : |x - 1| \le 1, |y| \le 1\};$$

that is, $h_1 = 1$, $\alpha_1 = 1$.

10. The initial value problem

$$(15) \qquad y''(x) + y(x) = 6x, \qquad y(0) = 1, \qquad y'(0) = -1$$

is equivalent to the integral equation

$$(16) \qquad y(x) = x^3 - x + 1 + \int_0^x (u - x) y(u)\,du.$$

To prove this, proceed as follows:

(a) Integrate equation (15) from 0 to x using the initial condition $y'(0) = -1$ to obtain

$$(17) \qquad y'(x) = 3x^2 - 1 - \int_0^x y(u)\,du.$$

(b) Integrate equation (17) from 0 to x using the initial condition $y(0) = 1$ to get

$$(18) \qquad y(x) = x^3 - x + 1 - \int_0^x \left(\int_0^t y(u)\,du\right) dt.$$

(c) Inverting the order of integration and integrating with respect to t, obtain equation (16).

11. Let

$$T[y](x) := x^3 - x + 1 + \int_0^x (u - x) y(u)\,du.$$

Equation (16) in Problem 10 can now be written $y = T[y]$, and we can therefore try to approximate the solution to the initial value problem (15) using the Picard iterations $y_{n+1} = T[y_n]$. Setting $y_0(x) = x^3 - x + 1$, compute the approximations $y_1(x)$ and $y_2(x)$.

12. Let T be the operator defined in Problem 11. We can use the Banach fixed point theorem for operators to show that T has a fixed point in the set

$$S = \{g \in C[0, 1]: \|g - y_0\| \le 2\},$$

where $y_0(x) = x^3 - x + 1$ and the norm is taken over $[0, 1]$.

(a) Show that if g, f lie in S, then

$$\|T[g] - T[f]\| \le \frac{1}{2}\|g - f\|,$$

and hence T is a contraction on S.

(b) Show that $\|y_0\| = 1$.

(c) Show that if $\|g - y_0\| \le 2$, then $\|g\| \le 2 + \|y_0\| = 3$.

(d) Using the results of parts (a), (b), and (c), show that if g is in S, then

$$\|T[g] - y_0\| \le \frac{1}{2}\|g\| \le \frac{3}{2}.$$

Hence T maps S into S.

(e) Apply the Banach fixed point theorem for operators to prove that T has a fixed point in S.

(f) Why is $y_0(x) = x^3 - x + 1$ a good starting point for the Picard iterations in Problem 11?

13. Let $\{y_n\}$ be a sequence of functions in $C[a, b]$ that converge uniformly to a function y in $C[a, b]$. Prove that

$$\lim_{n \to \infty} \int_a^b y_n(x)\,dx = \int_a^b y(x)\,dx.$$

14. Is the assertion of Problem 13 true if we replace uniform convergence by pointwise convergence? [*Hint:* See Problem 7.]

15. **Picard Iterations for Systems**

(a) Show that the initial value problem

$$\begin{aligned} x'(t) &= -y^2(t) & x(0) &= 0, \\ y'(t) &= z(t), & y(0) &= 1, \\ z'(t) &= x(t)y(t), & z(0) &= 1 \end{aligned}$$

is equivalent to the system of integral equations

$$x(t) = -\int_0^t y^2(s)\,ds,$$

$$y(t) = 1 + \int_0^t z(s)\,ds,$$

$$z(t) = 1 + \int_0^t x(s)y(s)\,ds.$$

(b) The Picard iterations for this system consist of the three sequences $\{x_n\}$, $\{y_n\}$, and $\{z_n\}$ defined by

$$x_{n+1}(t) := -\int_0^t y_n^2(s)\,ds, \qquad x_0(t) \equiv 0,$$

$$y_{n+1}(t) := 1 + \int_0^t z_n(s)\,ds, \qquad y_0(t) \equiv 1,$$

$$z_{n+1}(t) := 1 + \int_0^t x_n(s)y_n(s)\,ds, \qquad z_0(t) \equiv 1.$$

Compute the first two iterations: $x_1(t)$, $y_1(t)$, $z_1(t)$ and $x_2(t)$, $y_2(t)$, $z_2(t)$.

13.3 EXISTENCE OF SOLUTIONS OF LINEAR EQUATIONS

In Chapter 2 we established an important property of first order *linear* differential equations. Namely, if the functions $P(t)$ and $Q(t)$ are continuous in an interval I containing t_0, then there exists a solution to

(1) $y'(t) + P(t)y(t) = Q(t), \qquad y(t_0) = y_0,$

on the *entire* interval I. In the preceding section the best we could say was that if f and $\partial f/\partial y$ are continuous in a rectangle

$$R = \{(t, y): a < t < b, c < y < d\}$$

containing (t_0, y_0), then the solution to

$$y'(t) = f(t, y(t)), \qquad y(t_0) = y_0,$$

exists on *some* subinterval $[t_0 - h, t_0 + h] \subset [a, b]$.

For example, the seemingly simple nonlinear problem

(2) $y'(t) = y^2(t), \qquad y(0) = a(\neq 0),$

has the solution $y(t) = 1/(a^{-1} - t)$, which is undefined at $t = 1/a$ even though $f(t, y) = y^2$ is very well behaved.

In this section we show that for *linear* systems in normal form, a solution exists over the entire interval. This fact is a consequence of the following theorem.

CONTINUATION OF SOLUTION

Theorem 5. Let $\mathbf{x} = (x_1, \ldots, x_n)$ and $\mathbf{f}(t, \mathbf{x})$ denote the vector function $\mathbf{f}(t, \mathbf{x}) = (f_1(t, x_1, \ldots, x_n), \ldots, f_n(t, x_1, \ldots, x_n))$. Suppose $\mathbf{f}$ and $\partial \mathbf{f}/\partial x_i$, $i = 1, \ldots, n$, are continuous on the strip

$$R = \{(t, \mathbf{x}) \in \mathbf{R}^{n+1} : a \leq t \leq b, \ \mathbf{x} \text{ arbitrary}\}$$

containing the point $(t_0, \mathbf{x}_0)$. Assume further that there exists a positive constant L such that, for $i = 1, \ldots, n$,

(3) $\left| \dfrac{\partial \mathbf{f}}{\partial x_i}(t, \mathbf{x}) \right| \leq L$

for all $(t, \mathbf{x})$ in R. Then the initial value problem

(4) $\mathbf{x}'(t) = \mathbf{f}(t, \mathbf{x}), \qquad \mathbf{x}(t_0) = \mathbf{x}_0,$

has a unique solution on the entire interval $a \leq t \leq b$.

Proof. We prove Theorem 5 for scalar equations ($n = 1$). The proof for systems is essentially the same.

The proof when $n = 1$ proceeds along the same line of reasoning as the proofs of the Banach fixed point theorems. We show that the Picard iterations $\{x_n\}$, where

(5) $x_{n+1}(t) = T[x_n](t) := x_0 + \displaystyle\int_{t_0}^{t} f(s, x_n(s)) \, ds, \qquad x_0(t) \equiv x_0,$

converge uniformly to a unique solution on the interval $[a, b]$.

The Approximations Are Well Defined. Since $x_0(t) \equiv x_0$ is continuous on $[a, b]$ and f is continuous on the strip R, it follows that $x_1(t) = T[x_0](t)$ is well defined and continuous on $[a, b]$. By induction, we find that $x_n(t)$ is well defined and continuous on $[a, b]$.

The Approximations Converge Uniformly on [*a, b*]. Recall that the sequence of successive approximations $\{x_n\}$ consists of the partial sums of the series

(6) $$x_0 + \sum_{j=0}^{\infty} (x_{j+1}(t) - x_j(t)).$$

Thus, if this series converges uniformly on [*a, b*], then so does the sequence. But to show convergence on the entire interval [*a, b*], we need more precise estimates for the terms in (6) than we obtained in the proofs of the Banach fixed point theorems.

Since $f(t, x_0)$ is continuous on [*a, b*], it is bounded; that is, there is some constant M such that $|f(t, x_0)| \le M$ on [*a, b*]. To simplify the computations, let's consider t in the interval $[t_0, b]$. (The case when $t \in [a, t_0]$ can be handled in a similar fashion.) For such t,

(7) $$|x_1(t) - x_0| = \left| \int_{t_0}^{t} f(s, x_0(s))\,ds \right| \le \int_{t_0}^{t} |f(s, x_0)|\,ds \le M(t - t_0).$$

Let $w(t)$ and $v(t)$ be two continuous functions on $[t_0, b]$. Using the mean value theorem and the hypothesis that $|\partial f/\partial x|$ is bounded by L on R, we find

(8) $$|f(t, w(t)) - f(t, v(t))| = \left| \frac{\partial f}{\partial x}(t, z(t))[w(t) - v(t)] \right| \le L|w(t) - v(t)|$$

for all t in $[t_0, b]$, where $z(t)$ is some function between $w(t)$ and $v(t)$. Using inequality (8) and then (7), we find that $x_2 = T[x_1]$ satisfies

$$|x_2(t) - x_1(t)| \le \int_{t_0}^{t} |f(s, x_1(s)) - f(s, x_0(s))|\,ds$$

$$\le L \int_{t_0}^{t} |x_1(s) - x_0(s)|\,ds$$

$$\le LM \int_{t_0}^{t} (s - t_0)\,ds = LM \frac{(t - t_0)^2}{2!}$$

for t in $[t_0, b]$. Similarly, for $n = 1, 2, 3, \ldots$, we find by induction that

$$|x_{n+1}(t) - x_n(t)| \le \int_{t_0}^{t} |f(s, x_n(s)) - f(s, x_{n-1}(s))|\,ds$$

(9) $$\le L \int_{t_0}^{t} |x_n(s) - x_{n-1}(s)|\,ds$$

$$\le L^n M \int_{t_0}^{t} \frac{(s - t_0)^n}{n!}\,ds = L^n M \frac{(t - t_0)^{n+1}}{(n + 1)!}.$$

From inequalities (9), we see that the terms in (6) are bounded in absolute value by the terms in the positive series

$$\left(\frac{M}{L}\right)\sum_{j=0}^{\infty}\frac{[L(b-t_0)]^{j+1}}{(j+1)!},$$

which converges to $(M/L)(e^{L(b-t_0)}-1)$. Therefore, by the Weierstrass M test,[†] the series in (6) converges uniformly on $[t_0, b]$ to a continuous function, $x_\infty(t)$. Hence the sequence $\{x_n\}$ converges uniformly on $[t_0, b]$ to x_∞.

A similar argument establishes convergence on $[a, t_0]$, so $\{x_n\}$ converges uniformly on $[a, b]$ to x_∞.

x_∞ *Is a Solution.* Since $\{x_n\}$ converges uniformly to x_∞, we can take the limit inside the integral (see Problem 13 in Exercises 13.2). Thus, using equation (5), we obtain

$$x_\infty(t) = \lim_{n\to\infty} x_{n+1}(t) = \lim_{n\to\infty}\left[x_0 + \int_{t_0}^{t} f(s, x_n(s))ds\right]$$

$$= x_0 + \int_{t_0}^{t} f(s, x_\infty(s))ds.$$

Since x_∞ satisfies the equivalent integral equation, it must satisfy the initial value problem (4).

Uniqueness. Let z be a solution to (4) on $[a, b]$. Then z must satisfy the integral equation

(10) $$z(t) = x_0 + \int_{t_0}^{t} f(s, z(s))ds,$$

just as x_∞ does. Therefore, for t in $[t_0, b]$,

$$|z(t) - x_\infty(t)| \leq \int_{t_0}^{t}|f(s, z(s)) - f(s, x_\infty(s))|ds$$

(11) $$\leq L\int_{t_0}^{t}|z(s) - x_\infty(s)|ds$$

$$\leq L\|z - x_\infty\|\int_{t_0}^{t}ds = L\|z - x_\infty\|(t - t_0),$$

[†] See footnote on page 814.

where the norm is taken over $[t_0, b]$. Using (11), we next show by a similar string of inequalities that

$$\left| z(t) - x_\infty(t) \right| \le \int_{t_0}^t \left| f(s, z(s)) - f(s, x_\infty(s)) \right| ds$$

$$\le L \int_{t_0}^t \left| z(s) - x_\infty(s) \right| ds$$

$$\le L^2 \| z - x_\infty \| \int_{t_0}^t (s - t_0) ds = L^2 \| z - x_\infty \| \frac{(t - t_0)^2}{2!}.$$

By induction, we find

$$\left| z(t) - x_\infty(t) \right| \le L^n \frac{(t - t_0)^n}{n!} \| z - x_\infty \|, \qquad n = 1, 2, 3, \ldots,$$

for t in $[t_0, b]$. Taking the limit as $n \to \infty$, we deduce that $\left| z(t) - x_\infty(t) \right| \equiv 0$. That is, z and x_∞ agree on $[t_0, b]$. A similar argument shows that they agree on $[a, t_0]$, so the initial value problem has a unique solution on the interval $[a, b]$. ∎ End

Theorem 5 is called a **continuation theorem** because it specifies conditions that guarantee that the solution to initial value problem (4) can be **continued** from the small interval $[t_0 - h, t_0 + h]$ to the entire interval $[a, b]$. As we now show, these conditions are satisfied by linear systems.

EXISTENCE AND UNIQUENESS THEOREM FOR LINEAR SYSTEMS

Theorem 6. Suppose the $n \times n$ matrix function $\mathbf{A}(t) = [a_{i\,j}(t)]$ and the vector function $\mathbf{f}(t)$ are continuous on an open interval (a, b) that contains the point t_0. Then, for any choice of the initial vector $\mathbf{x}_0$, there exists a unique solution on the whole interval (a, b) to the initial value problem

(12) $\mathbf{x}'(t) = \mathbf{A}(t)\mathbf{x}(t) + \mathbf{f}(t), \qquad \mathbf{x}(t_0) = \mathbf{x}_0.$

Proof. Let $[\tilde{a}, \tilde{b}]$ be a closed subinterval of (a, b) containing t_0. For system (12), we let $\mathbf{g}(t, \mathbf{x}) := \mathbf{A}(t)\mathbf{x} + \mathbf{f}(t)$, which is continuous on $\tilde{R} := \{(t, \mathbf{x}): \tilde{a} \le t \le \tilde{b}, \mathbf{x} \text{ arbitrary}\}$. Furthermore,

$$\frac{\partial \mathbf{g}}{\partial x_i} = \text{col}\,(a_{1i}(t), a_{2i}(t), \ldots, a_{ni}(t))$$

for $i = 1, \ldots, n$. Since each a_{ij} is continuous on the closed interval $[\tilde{a}, \tilde{b}]$, we can find a positive constant L such that $\left| \partial \mathbf{g}/\partial x_i \right| \le L$ in $\tilde{R}$. Therefore, by Theorem 5, there exists a unique solution to (12) on the interval $[\tilde{a}, \tilde{b}]$. Since $[\tilde{a}, \tilde{b}]$ is an arbitrary interval in (a, b), we must have a unique solution on the whole interval (a, b). ∎ End

EXAMPLE 1 Over what interval are we assured that there is a unique solution to

$$(13) \quad \mathbf{x}'(t) = \begin{bmatrix} t^2 & \tan t \\ 1 & \sin t \end{bmatrix} \mathbf{x}(t) + \begin{bmatrix} \sqrt{1-t} \\ 0 \end{bmatrix}, \quad \mathbf{x}(0) = \begin{bmatrix} -1 \\ 1 \end{bmatrix}?$$

Solution The matrix $\mathbf{A}(t)$ is continuous on the interval $(-\pi/2, \pi/2)$, but on no larger interval containing 0 because $a_{12}(t) = \tan t$ has vertical asymptotes at $t = \pm \pi/2$. Now $\mathbf{f}(t)$ is continuous on the interval $(-\infty, 1]$ but not for $t > 1$, since $f_1(t) = \sqrt{1-t}$ is not defined for $t > 1$. Hence $\mathbf{A}(t)$ and $\mathbf{f}(t)$ are simultaneously continuous on the interval $(-\pi/2, 1)$, which contains 0. Therefore, by Theorem 6, the initial value problem (13) has a unique solution on the interval $(-\pi/2, 1)$. ∎ End

A careful examination of the proof of Theorem 6 reveals that the initial value problem (13) actually has a unique solution on the half-closed interval $(-\pi/2, 1]$. We leave the verification as an exercise.

In Section 4.10 we demonstrated how a higher-order linear equation can be reduced to a first order system in normal form. Using this reduction, we prove the following.

EXISTENCE AND UNIQUENESS THEOREM FOR LINEAR EQUATIONS

Theorem 7. Suppose $p_1(t), \ldots, p_n(t)$ and $g(t)$ are continuous on an interval (a, b) containing the point t_0. Then, for every choice of the initial values $y_0, y_1, \ldots, y_{n-1}$, there exists a unique solution on the whole interval (a, b) to the initial value problem

$$(14) \quad \begin{aligned} &y^{(n)}(t) + p_1(t)y^{(n-1)}(t) + \cdots + p_n(t)y(t) = g(t); \\ &y(t_0) = y_0, \quad y'(t_0) = y_1, \quad \ldots, \quad y^{(n-1)}(t_0) = y_{n-1}. \end{aligned}$$

Proof. Let $x_1 = y$, $x_2 = y'$, $\ldots$, $x_n = y^{(n-1)}$. Then the initial value problem (14) is reduced to the system

$$(15) \quad \begin{aligned} x_1'(t) &= x_2(t), & x_1(t_0) &= y_0, \\ &\vdots \quad \vdots & &\vdots \\ x_{n-1}'(t) &= x_n(t), & x_{n-1}(t_0) &= y_{n-2}, \\ x_n'(t) &= -p_n(t)x_1(t) - \cdots - p_1(t)x_n(t) + g(t), & x_n(t_0) &= y_{n-1}. \end{aligned}$$

When we express (15) in the matrix form $\mathbf{x}'(t) = \mathbf{A}(t)\mathbf{x}(t) + \mathbf{f}(t)$, we find that the entries in $\mathbf{A}(t)$ are either $-p_1(t), \ldots, -p_n(t)$, 0, or 1 and the entries in $\mathbf{f}(t)$ are either 0 or $g(t)$. Hence, if $p_1(t), \ldots, p_n(t)$ and $g(t)$ are continuous on an interval (a, b), then so are $\mathbf{A}(t)$ and $\mathbf{f}(t)$. Therefore, by Theorem 6, the initial value problem in (15) has a unique solution on (a, b). But since (15) is equivalent to (14), there is a unique solution to (14) on (a, b). ∎ End

EXAMPLE 2 Over what interval are we assured that there is a unique solution to

$$(16) \quad \begin{aligned} &y'''(t) + t^2 y''(t) + (\ln t)y'(t) - (\cos t)y(t) = \frac{1}{2-t}; \\ &y(1) = 3, \quad y'(1) = -2, \quad y''(1) = -6? \end{aligned}$$

Solution Here $p_1(t) = t^2$ is continuous everywhere, $p_2(t) = \ln t$ is continuous on $(0, \infty)$, $p_3(t) = -\cos t$ is continuous everywhere, and $g(t) = (2 - t)^{-1}$ is continuous for $t \neq 2$. Thus $p_1(t)$, $p_2(t)$, $p_3(t)$, and $g(t)$ are simultaneously continuous on $(0, 2)$, which contains $t = 1$. Hence, by Theorem 7, there is a unique solution to (16) on $(0, 2)$. ∎ End

There are many other theorems dealing with the existence and uniqueness of solutions to initial value problems. For these and other results, we refer the reader to an advanced text that studies the theory of ordinary differential equations.[†]

EXERCISES 13.3

In Problems 1 through 6, determine over what interval we are assured that there is a unique solution to the given initial value problem.

1. $\mathbf{x}'(t) = \begin{bmatrix} \cos t & \sqrt{t} \\ t^3 & -1 \end{bmatrix} \mathbf{x}(t) + \begin{bmatrix} \tan t \\ e^t \end{bmatrix}, \qquad \mathbf{x}(2) = \begin{bmatrix} 0 \\ 4 \end{bmatrix}.$

2. $\mathbf{x}'(t) = \begin{bmatrix} e^t & 5 \\ 0 & \sqrt{2+t} \end{bmatrix} \mathbf{x}(t) + \begin{bmatrix} (1-t)^{-1} \\ t \end{bmatrix}, \qquad \mathbf{x}(0) = \begin{bmatrix} 1 \\ 2 \end{bmatrix}.$

3. $y'''(t) - (\ln t)y''(t) + (\tan t)y(t) = e^{2t}$;
$y(1) = 1, \qquad y'(1) = -2, \qquad y''(1) = 0.$

4. $ty^{(4)}(t) - y(t) = \sqrt{3 - t}$;
$y(1) = y'(1) = y''(1) = 0, \qquad y'''(1) = -2.$

5. $\mathbf{x}'(t) = \begin{bmatrix} x_1'(t) \\ x_2'(t) \end{bmatrix} = \begin{bmatrix} \sin x_2(t) \\ 3x_1(t) \end{bmatrix}, \qquad \mathbf{x}(0) = \begin{bmatrix} 1 \\ 2 \end{bmatrix}.$

6. $x'(t) = \arctan(x(t)), \qquad x(0) = 2.$

7. Find the unique solution to the initial value problem
$$y'''(t) - (\sin t)y'(t) + e^{-t}y(t) = 0;$$
$$y(0) = y'(0) = y''(0) = 0.$$

8. Prove that the initial value problem (13) in Example 1 actually has a unique solution on the half-closed interval $(-\pi/2, 1]$.

13.4 CONTINUOUS DEPENDENCE OF SOLUTIONS

In the preceding sections we discussed the existence and uniqueness of solutions to the initial value problem

(1) $\dfrac{dy}{dx} = f(x, y), \qquad y(x_0) = y_0.$

In this section we investigate properties of the solution. Namely, we give conditions for the solution to be a continuous function of the initial value y_0 or of the function f. We state and prove these results for a scalar first order equation. Similar results hold for systems.[††]

To show that the solution to equation (1) is a continuous function of the initial value y_0, we need the following inequality due to T. H. Gronwall.

[†] For example, *Ordinary Differential Equations*, by G. Birkhoff and G. C. Rota (John Wiley and Sons, New York, 1978).

[††] For a discussion of the results of this section for systems, see an advanced text such as *Ordinary Differential Equations*, G. Birkhoff and G. C. Rota, Third Edition (John Wiley and Sons, New York, 1978), Chapter 6.

GRONWALL'S INEQUALITY

Theorem 8. Let u and v be continuous functions satisfying $u(x) > 0$ and $v(x) \geq 0$ on $[a, b]$. Let $c \geq 0$ be a constant. If

$$(2) \qquad v(x) \leq c + \int_a^x u(t)v(t)\,dt, \qquad a \leq x \leq b,$$

then

$$(3) \qquad v(x) \leq c \exp\left\{\int_a^x u(t)\,dt\right\}, \qquad a \leq x \leq b.$$

Proof. First assume $c > 0$ and define

$$V(x) := c + \int_a^x u(t)v(t)\,dt, \qquad a \leq x \leq b.$$

Then $V(x) \geq v(x)$ by (2), and since u and v are nonnegative, $V(x) \geq V(a) = c > 0$ on $[a, b]$. Moreover, $V'(x) = u(x)v(x) \leq u(x)V(x)$. Dividing by $V(x)$, we get

$$(4) \qquad u(x) \geq \frac{V'(x)}{V(x)}, \qquad a \leq x \leq b.$$

When we integrate both sides of (4) from a to x, we find

$$(5) \qquad \int_0^x u(s)\,ds \geq \int_a^x \frac{V'(s)}{V(s)}\,ds = \ln V(s)\,\Big|_a^x = \ln V(x) - \ln V(a) = \ln\left(V(x)/c\right).$$

Solving for $V(x)$ (recall that the exponential function is increasing), we obtain

$$(6) \qquad V(x) \leq c \exp\left\{\int_a^x u(s)\,ds\right\}, \qquad a \leq x \leq b.$$

Since $v(x) \leq V(x)$, inequality (3) follows from (6).

If inequality (2) holds with $c = 0$, then it certainly holds for all $c > 0$. Hence, as we have shown in the preceding discussion, inequality (3) is valid for all $c > 0$. Taking the limit as $c \to 0^+$, we get $v(x) \leq 0$, which is the same as (3) with $c = 0$. ■ End

As we have just seen, when $c = 0$ in inequality (2), it follows that $v(x) \leq 0$. But since $v(x)$ is nonnegative, we must have $v(x) \equiv 0$. This observation can be used to prove that an initial value problem has a unique solution (see Problem 12).

The next theorem is of practical importance. Suppose we need to solve the initial value problem (1), but we know the value of y_0 only to within $\pm 10^{-3}$ (for example, if y_0 is obtained by some measurement, its value will be known only up to the accuracy of the measuring instrument). Let's denote the approximate value of y_0 by $\tilde{y}_0$ so that $|y_0 - \tilde{y}_0| \leq 10^{-3}$. When we solve

(7) $y' = f(x, y), \qquad y(x_0) = \tilde{y}_0,$

we obtain a solution $\phi(x, \tilde{y}_0)$, where we have displayed the initial value for emphasis. The question is: Will $\phi(x, \tilde{y}_0)$ be close to the solution $\phi(x, y_0)$ of (1) for x in an interval about x_0? In other words, if we slightly perturb the initial value y_0, will the solution change significantly? As we now show, under suitable assumptions on f, small changes in y_0 lead to small changes in the solution.

CONTINUOUS DEPENDENCE ON INITIAL VALUE

Theorem 9. Let f and $\partial f/\partial y$ be continuous functions on an open rectangle

$$R = \{(x, y): a < x < b, c < y < d\}$$

containing the point (x_0, y_0). Assume that for all $\tilde{y}_0$ sufficiently close to y_0, the solution $\phi(x, \tilde{y}_0)$ to (7) exists on the interval $[x_0 - h, x_0 + h]$ and its graph lies within a fixed closed rectangle $R_0 \subset R$. Then, for $|x - x_0| \leq h$,

(8) $|\phi(x, y_0) - \phi(x, \tilde{y}_0)| \leq |y_0 - \tilde{y}_0|e^{Lh},$

where L is any positive constant such that $|(\partial f/\partial y)(x, y)| \leq L$ for all (x, y) in R_0. Moreover, as $\tilde{y}_0$ approaches y_0, the solution $\phi(x, \tilde{y}_0)$ approaches $\phi(x, y_0)$ uniformly on $[x_0 - h, x_0 + h]$.

Notice that (8) implies that, for each $\varepsilon > 0$,

$$|\phi(x, y_0) - \phi(x, \tilde{y}_0)| \leq \varepsilon \quad \text{on} \quad [x_0 - h, x_0 + h],$$

whenever $|y_0 - \tilde{y}_0| \leq \varepsilon e^{-Lh}$.

Proof of Theorem 9. We know that the solution $\phi(x, y_0)$ to the initial value problem (1) satisfies the integral equation

(9) $\phi(x, y_0) = y_0 + \displaystyle\int_{x_0}^{x} f(s, \phi(s, y_0))\,ds, \qquad |x - x_0| \leq h.$

Similarly, for $\tilde{y}$ close enough to y_0,

(10) $\phi(x, \tilde{y}_0) = \tilde{y}_0 + \displaystyle\int_{x_0}^{x} f(s, \phi(s, \tilde{y}_0))\,ds, \qquad |x - x_0| \leq h.$

Subtracting the second of these equations from the first gives

(11) $\phi(x, y_0) - \phi(x, \tilde{y}_0) = y_0 - \tilde{y}_0 + \int_{x_0}^{x} [f(s, \phi(s, y_0)) - f(s, \phi(s, \tilde{y}_0))]\, ds.$

Assume first that $x \geq x_0$. Taking the absolute value of both sides of (11) and using the triangle inequality, we obtain

(12) $|\phi(x, y_0) - \phi(x, \tilde{y}_0)| \leq |y_0 - \tilde{y}_0| + \int_{x_0}^{x} |f(s, \phi(s, y_0)) - f(s, \phi(s, \tilde{y}_0))|\, ds.$

Since $|\partial f/\partial y|$ is bounded by L on R_0, it follows from the mean value theorem that

(13) $|f(s, w) - f(s, z)| \leq L|w - z|$

for (s, w) and (s, z) in R_0. Using this last inequality in (12), we get

(14) $|\phi(x, y_0) - \phi(x, \tilde{y}_0)| \leq |y_0 - \tilde{y}_0| + \int_{x_0}^{x} L|\phi(s, y_0) - \phi(s, \tilde{y}_0)|\, ds.$

Now we can apply Gronwall's inequality (with $v(x) = |\phi(x, y_0) - \phi(x, \tilde{y}_0)|$) to (14) and obtain

$$|\phi(x, y_0) - \phi(x, \tilde{y}_0)| \leq |y_0 - \tilde{y}_0| \exp\left\{ \int_{x_0}^{x} L\, dt \right\}$$
$$= |y_0 - \tilde{y}_0| e^{L(x - x_0)}$$
$$\leq |y_0 - \tilde{y}_0| e^{Lh}, \qquad x_0 \leq x \leq x_0 + h.$$

For $x \leq x_0$, we can use the substitution $x = x_0 - t$ to show that inequality (8) continues to hold in this range (see Problem 13). ■ End

Geometrically speaking. Theorem 9 says that for initial values that are close together, the graphs of the solutions passing through them will remain close, at least on the interval $|x - x_0| \leq h$.

In the next theorem we show that solutions to different equations, but with the same initial conditions, remain close if the differential equations are nearly the same.

CONTINUOUS DEPENDENCE ON $f(x, y)$

Theorem 10. Let f and $\partial f/\partial y$ be continuous functions on an open rectangle $R = \{(x, y): a < x < b, c < y < d\}$ containing the point (x_0, y_0). Let F be continuous on R and assume

(15) $\left| F(x, y) - f(x, y) \right| \leq \varepsilon,$ for (x, y) in R.

Let ϕ be the solution to the initial value problem

(16) $\dfrac{dy}{dx} = f(x, y), \qquad y(x_0) = y_0,$

and let ψ be a solution to

(17) $\dfrac{dy}{dx} = F(x, y), \qquad y(x_0) = y_0.$

Assume both solutions $\phi(x)$, $\psi(x)$ exist on $[x_0 - h, x_0 + h]$ and their graphs lie in a closed rectangle $R_0 \subset R$. Then, for $|x - x_0| \leq h$,

(18) $\left| \phi(x) - \psi(x) \right| \leq \varepsilon h e^{Lh},$

where L is any positive constant such that $\left| (\partial f/\partial y)(x, y) \right| \leq L$ for all (x, y) in R_0.
 Moreover, as F approaches f uniformly on R, that is, as $\varepsilon \to 0^+$ in (15), the solution $\psi(x)$ approaches $\phi(x)$ uniformly on $[x_0 - h, x_0 + h]$.

Proof. We begin with the integral representations for $\phi(x)$ and $\psi(x)$; namely,

$$\phi(x) = y_0 + \int_{x_0}^{x} f(s, \phi(s))\,ds, \qquad \psi(x) = y_0 + \int_{x_0}^{x} F(s, \psi(s))\,ds.$$

Subtracting these equations gives

$$\phi(x) - \psi(x) = \int_{x_0}^{x} [f(s, \phi(s)) - F(s, \psi(s))]\,ds$$

and inserting $0 = -f(s, \psi(s)) + f(s, \psi(s))$ in the integrand yields

(19) $\phi(x) - \psi(x) = \displaystyle\int_{x_0}^{x} [f(s, \phi(s)) - f(s, \psi(s))]\,ds + \int_{x_0}^{x} [f(s, \psi(s)) - F(s, \psi(s))]\,ds.$

For $x_0 \leq x \leq x_0 + h$, the triangle inequality applied to (19) gives

(20) $\left| \phi(x) - \psi(x) \right| \leq \displaystyle\int_{x_0}^{x} \left| f(s, \phi(s)) - f(s, \psi(s)) \right|\,ds + \int_{x_0}^{x} \left| f(s, \psi(s)) - F(s, \psi(s)) \right|\,ds.$

The first integral in (20) can be estimated just as in the proof of Theorem 9. That is, since

$$\left|\frac{\partial f}{\partial y}(s,\, t)\right| \leq L \quad \text{for} \quad (s,\, t) \text{ in } R_0,$$

we have

$$|f(s,\, \phi(s)) - f(s,\, \psi(s))| \leq L|\phi(s) - \psi(s)|.$$

Hence, for $x_0 \leq x \leq x_0 + h$,

(21) $$\int_{x_0}^{x} |f(s,\, \phi(s)) - f(s,\, \psi(s))|\,ds \;\leq\; \int_{x_0}^{x} L|\phi(s) - \psi(s)|\,ds.$$

To estimate the second integral in (20), we use the hypothesis (15). This gives

(22) $$\int_{x_0}^{x} |f(s,\, \psi(s)) - F(s,\, \psi(s))|\,ds \leq \varepsilon(x - x_0) \leq \varepsilon h.$$

Using the estimates (21) and (22) in inequality (20), we obtain

$$|\phi(x) - \psi(x)| \leq \varepsilon h + \int_{x_0}^{x} L|\phi(s) - \psi(s)|\,ds.$$

Now we are in position to apply Gronwall's inequality (Theorem 8). It yields

$$|\phi(x) - \psi(x)| \leq \varepsilon h \exp\left\{\int_{x_0}^{x} L\,ds\right\}$$

$$\leq \varepsilon h e^{Lh} \quad \text{for} \quad x_0 \leq x \leq x_0 + h.$$

Since the last inequality can also be verified for $x_0 - h \leq x \leq x_0$, the conclusion (18) of the theorem is proved.

Finally, by letting $\varepsilon \to 0^+$ in (18), we see that $\psi(x)$ approaches $\phi(x)$ uniformly on $[x_0 - h,\, x_0 + h]$. That is, as F comes uniformly close to f on R, then ψ comes uniformly close to ϕ on $[x_0 - h,\, x_0 + h]$. ∎ End

In the next example, we use inequality (18) to bound the difference between the solutions to two different equations that are nearly the same.

EXAMPLE 1 Let $\phi(x)$ be the solution to

(23) $$y' = e^y, \quad y(0) = y_0,$$

and let $\psi(x)$ be the solution to

(24) $$y' = 1 + y, \quad y(0) = y_0.$$

Assume the solutions exist on $[-1/2, 1/2]$ and their graphs remain in the rectangle $R_0 = \{(x, y): -1/2 \leq x \leq 1/2, -2 \leq y \leq 2\}$. Use Theorem 10 to find a bound on the difference between $\phi(x)$ and $\psi(x)$ for x in $[-1/2, 1/2]$.

Solution We can use inequality (18) in Theorem 10 to obtain the bound, but first we must determine the constant L and the constant ε. Here $x_0 = 0$ and $h = 1/2$.

Since e^y and $1 + y$ are both continuous and have continuous partial derivatives with respect to y in R_0, we will use the one with the smaller bound. If we set $F(x, y) := e^y$, then $(\partial F/\partial y)(x, y) = e^y$. Hence $|(\partial F/\partial y)(x, y)| \leq e^2$ for (x, y) in R_0. For $f(x, y) = 1 + y$, we find that $|(\partial f/\partial y)(x, y)| = 1$ for (x, y) in R_0. Since $1 \leq e^2$, we will let $f(x, y) = 1 + y$ in Theorem 10 and take $L = 1$.

We must now measure the distance between $1 + y$ and e^y on R_0. Recall that Taylor's theorem with remainder gives the representation

(25) $$g(y) = g(0) + g'(0)y + \frac{g''(0)y^2}{2!} + \cdots + \frac{g^{(k)}(0)y^k}{k!} + \frac{g^{(k+1)}(\xi)y^{k+1}}{(k+1)!},$$

where ξ, which depends on y, lies between 0 and y. For $g(y) = e^y$, we have

$$e^y = 1 + y + \frac{e^\xi y^2}{2}, \qquad \xi \text{ between 0 and } y.$$

Thus

$$\left|e^y - (1 + y)\right| = \left|\frac{e^\xi y^2}{2}\right| \leq \frac{e^2(2)^2}{2} = 2e^2$$

for y in $[-2, 2]$. Consequently

$$\left|F(x, y) - f(x, y)\right| = \left|e^y - (1 + y)\right| \leq 2e^2 =: \varepsilon$$

for (x, y) in R_0.

It now follows from inequality (18) in Theorem 10 that

(26) $$\left|\phi(x) - \psi(x)\right| \leq (2e^2)\left(\frac{1}{2}\right)e^{(1)(1/2)} = e^{2.5}$$

for x in $[-1/2, 1/2]$. ■ End

EXERCISES 13.4

In Problems 1 through 6, let $\phi(x, y_0)$ be the solution to the given differential equation that satisfies $y(0) = y_0$ and let $\phi(x, \tilde{y}_0)$ be the solution satisfying $y(0) = \tilde{y}_0$. Assume the solutions exist on $[-1, 1]$ and their graphs remain in the rectangle $R_0 = \{(x, y): -1 \leq x \leq 1, -4 \leq y \leq 4\}$. Use Theorem 9 to find a bound on the differences between $\phi(x, y_0)$ and $\phi(x, \tilde{y}_0)$ for x in $[-1, 1]$ if $|y_0 - \tilde{y}_0| < 10^{-2}$.

1. $y' = \sin(x + y)$.

2. $y' = x + y$.

3. $y' = e^{\cos y} + x^2$.

4. $y' = e^{-y^2}$.

5. $y' = \dfrac{x^2 - y^2}{1 + x^2 + y^2}$.

6. $y' = \arctan(xy)$.

In Problems 7 through 10, let $\phi(x)$ be the solution to $y' = f(x, y)$, $y(0) = y_0$ and let $\psi(x)$ be the solution to $y' = F(x, y)$, $y(0) = y_0$, where f and F are given func-

tions. *Assume the solutions exist on* $[-1, 1]$ *and their graphs remain in the rectangle* $R_0 = \{(x, y): -1 \le x \le 1, -1 \le y \le 1\}$. *Use Theorem 10 to find a bound on the difference between* $\phi(x)$ *and* $\psi(x)$ *for x in* $[-1, 1]$.

7. $f(x, y) = e^{-y}$, $\quad F(x, y) = 1 - y + y^2/2 - y^3/6$.

8. $f(x, y) = x - \cos y$, $\quad F(x, y) = x - 1 + y^2/2$.

9. $f(x, y) = \sin x + \dfrac{1}{1 + y^2}$, $\quad F(x, y) = x + 1 - y^2$.

10. $f(x, y) = \sin(xy)$, $\quad F(x, y) = xy$.

11. Let $f(x, y) = y^2$.

(a) Solve explicitly for $\phi(x, 1)$, the solution to $y' = f(x, y)$, $y(0) = 1$, and for $\phi(x, 1 + \varepsilon)$, the solution to $y' = f(x, y)$, $y(0) = 1 + \varepsilon$, where $\varepsilon > 0$.

(b) Show that $|\partial f/\partial y| \le 4$ on the rectangle
$R = \{(x, y): -1 < x < 1, -2 < y < 2\}$.

(c) Using the result of part (b) in inequality (8) in Theorem 9, it appears that
$$|\phi(x, 1) - \phi(x, 1 + \varepsilon)| \le \varepsilon e^4, \qquad x \text{ in } (-1, 1).$$
Show that this is *not* true (even for $\varepsilon > 0$ small) by using the results of part (a) to verify that $\phi(x, 1) \to 1 + \varepsilon^{-1}$ as $x \to 1/(1 + \varepsilon)$, while $|\phi(x, 1 + \varepsilon)| \to +\infty$ as $x \to 1/(1 + \varepsilon)$.

(d) Which assumption in Theorem 9 is *not* satisfied so that inequality (8) is not applicable to this situation?

12. Let f and $\partial f/\partial y$ be continuous functions on an open rectangle R containing the point (x_0, y_0). Let $\phi(x)$ and $\psi(x)$ be two solutions to $y' = f(x, y)$, $y(x_0) = y_0$, on the interval $[x_0 - h, x_0 + h]$. Assume their graphs lie in a fixed closed rectangle $R_0 \subset R$. Use Gronwall's inequality to show that $\phi(x) \equiv \psi(x)$ for x in $[x_0 - h, x_0 + h]$. [*Hint:* Proceed as in the proof of Theorem 9.]

13. To verify that inequality (8) holds for $x \le x_0$, let $x = x_0 - t$ and define $\psi(t, y_0) := \phi(x_0 - t, y_0)$ and $\psi(t, \tilde{y}_0) := \phi(x_0 - t, \tilde{y}_0)$.

(a) Show that $\psi(t, y_0)$ is a solution to the initial value problem
$$\frac{dz}{dt} = g(t, z), \qquad z(0) = y_0,$$
where $g(t, z) := -f(x_0 - t, z)$.

(b) Use inequality (8) to show that for $0 \le t \le h$,
$$|\psi(t, y_0) - \psi(t, \tilde{y}_0)| \le |y_0 - \tilde{y}_0| e^{Lh}.$$

(c) Use the result of part (b) to show that for $x_0 - h \le x \le x_0$,
$$|\phi(x, y_0) - \phi(x, \tilde{y}_0)| \le |y_0 - \tilde{y}_0| e^{Lh}.$$

14. Assume the constant L in Theorem 9 is replaced by a positive, continuous function $\ell(x)$, that is
$$\left| \frac{\partial f}{\partial y}(x, y) \right| \le \ell(x) \quad \text{for } (x, y) \text{ in } R_0.$$
Show that Theorem 9 remains valid if inequality (8) is replaced by
$$|\phi(x, y_0) - \phi(x, \tilde{y}_0)| \le |y_0 - \tilde{y}_0| \exp\left\{\left|\int_{x_0}^{x} \ell(x)\, dx\right|\right\}.$$

15. Let $f(x, y, \lambda)$ and $(\partial f/\partial y)(x, y, \lambda)$ be continuous for $|x - x_0| < A$, $|y - y_0| < B$, $|\lambda - \lambda_0| < C$. Let $\phi(x)$ be the solution to
$$dy/dx = f(x, y, \lambda_0), \qquad y(x_0) = y_0,$$
and let $\psi(x)$ be the solution to
$$dy/dx = f(x, y, \lambda), \qquad y(x_0) = y_0.$$
Assume both solutions $\phi(x)$ and $\psi(x)$ exist on $[x_0 - h, x_0 + h]$ and their graphs lie in a closed rectangle
$$R_0 \subset R = \{(x, y): |x - x_0| < A, |y - y_0| < B\}.$$
Show that for each $\varepsilon > 0$, there exists a $\delta > 0$ such that whenever $|\lambda - \lambda_0| < \delta$, the solutions $\phi(x)$ and $\psi(x)$ satisfy
$$|\phi(x) - \psi(x)| < \varepsilon, \quad x_0 - h \le x \le x_0 + h.$$

16. Prove the following combination of Theorems 9 and 10: Let f and $\partial f/\partial y$ be continuous functions on an open rectangle $R = \{(x, y): a < x < b, c < y < d\}$ containing the point (x_0, y_0). Let F be continuous on R and assume that
$$|F(x, y) - f(x, y)| \le \varepsilon \quad \text{for } (x, y) \text{ in } R.$$
Let ϕ be the solution to the initial value problem
$$dy/dx = f(x, y), \qquad y(x_0) = y_0,$$
and let ψ be a solution to
$$dy/dx = F(x, y), \qquad y(x_0) = \tilde{y}_0.$$
Assume both solutions $\phi(x)$, $\psi(x)$ exist on $[x_0 - h, x_0 + h]$ and their graphs lie in a closed rectangle $R_0 \subset R$. Then, for $|x - x_0| \le h$,
$$|\phi(x) - \psi(x)| \le \{|y_0 - \tilde{y}_0| + \varepsilon h\} e^{Lh},$$
where L is a positive constant such that $|(\partial f/\partial y)(x, y)| \le L$ for (x, y) in R_0.

CHAPTER SUMMARY

In this chapter we discussed the existence and uniqueness of solutions to initial value problems. Our approach was to construct a sequence of **successive approximations** that converge uniformly to the desired solution.

For first order equations, we showed that the initial value problem

(1) $y'(x) = f(x, y(x))$, $y(x_0) = y_0$

is equivalent to the **integral equation**

(2) $y(x) = y_0 + \displaystyle\int_{x_0}^{x} f(t, y(t))\,dt.$

If we let the right-hand side of (2) define the operator

(3) $T[y](x) := y_0 + \displaystyle\int_{x_0}^{x} f(t, y(t))\,dt,$

then the integral equation (2) is equivalent to the **operator equation**

(4) $y = T[y]$.

Under certain restrictions on f, we used the **Banach fixed point theorem** to show that the operator equation (4) has a unique solution and consequently the initial value problem (1) has a unique solution on some interval $[x_0 - h, x_0 + h]$, where h is some positive constant. Moreover, this solution can be obtained as the uniform limit of the **Picard iterations** $\{y_n\}$, where

(5) $y_{n+1}(x) := T[y_n](x) = y_0 + \displaystyle\int_{x_0}^{x} f(t, y_n(t))\,dt,$ $y_0(x) \equiv y_0.$

These results have natural generalizations to first order systems in normal form.

For linear systems and linear higher-order equations, we showed that the solutions can be extended beyond the (possibly) small interval $[x_0 - h, x_0 + h]$. For example, if $\mathbf{A}(t)$ and $\mathbf{f}(t)$ are continuous on the interval (a, b) containing t_0, then there exists a unique solution on the *entire* interval (a, b) to the initial value problem

(6) $\mathbf{x}'(t) = \mathbf{A}(t)\mathbf{x}(t) + \mathbf{f}(t)$, $\mathbf{x}(t_0) = \mathbf{x}_0.$

When $f(x, y)$ and $f_y(x, y)$ are continuous functions on an open rectangle containing the point (x_0, y_0), we proved that the solution to the initial value problem (1) depends continuously on the initial value y_0 and on the function f. That is, if y_0 or f are "slightly changed," then the solution to the perturbed initial value problem will be close to the solution of the original problem. Both of these properties follow from **Gronwall's inequality.**

REVIEW PROBLEMS

In Problems 1 and 2, use the method of successive substitutions to approximate a solution of the given equation starting with the given value for x_0.

1. $x = \dfrac{2x^3 + 1}{3x^2 - 1}$, $x_0 = 1$.

2. $x = \cos x$, $x_0 = 0$.

In Problems 3 and 4, express the given initial value problem as an integral equation.

3. $y'(x) = y^2(x) + \sin x$, $y(-1) = -2$.

4. $y'(x) = x^2 y^3(x) - y^2(x)$, $y(0) = 9$.

In Problems 5 and 6, compute the Picard iterations $y_1(x)$ and $y_2(x)$ for the given initial value problem, starting with $y_0(x) \equiv y_0$, where y_0 is the initial value for y.

5. $y'(x) = y^3(x)$, $y(0) = 1$.

6. $y'(x) = -2y(x)$, $y(0) = -1$.

In Problems 7 and 8, determine whether the given sequence of functions $\{y_n\}_{n=1}^{\infty}$ converges uniformly to the given limit function y on the specified interval.

7. $y_n(x) = \dfrac{\sin x}{1 + nx}$ and $y(x) \equiv 0$ on $[0, 1]$.

 [Hint: $|\sin x| \le 2x/\pi$ on $[0, 1]$.]

8. $y_n(x) = \dfrac{1}{1 + nx^2}$ and $y(x) \equiv 0$ on $[-1, 1]$.

In Problems 9 and 10, determine over what interval we are assured that there exists a unique solution to the given initial value problem.

9. $\begin{bmatrix} x_1'(t) \\ x_2'(t) \\ x_3'(t) \end{bmatrix} = \begin{bmatrix} \exp(-x_2^2(t)) \\ (\sin t)x_3(t) \\ x_1(t) - x_3(t) \end{bmatrix}$, $\mathbf{x}(0) = \begin{bmatrix} 0 \\ 1 \\ 1 \end{bmatrix}$.

10. $y'''(t) + (\tan t)y'(t) - e^t y(t) = t^2 - t + 1$;
 $y(1) = 0$, $y'(1) = 1$, $y''(1) = 5$.

11. Let $\phi(x, y_0)$ be the solution to $y' = x \tan y$ satisfying $y(0) = y_0$ and let $\phi(x, \tilde{y}_0)$ be the solution satisfying $y(0) = \tilde{y}_0$. Assume these solutions exist on $[0, 2]$ and their graphs remain in the rectangle $R_0 = \{(x, y): 0 \le x \le 2, -\pi/4 \le y \le \pi/4\}$. Find a bound on the difference between $\phi(x, y_0)$ and $\phi(x, \tilde{y}_0)$ for x on $[0, 2]$ if $|y_0 - \tilde{y}_0| < 10^{-3}$.

12. Let $\phi(x)$ be the solution to $y' = x \sin y$, $y(0) = y_0$, and let $\psi(x)$ be the solution to $y' = xy$, $y(0) = y_0$. Assume the solutions exist on $[-1, 1]$ and their graphs remain in the rectangle $R_0 = \{(x, y): -1 \le x \le 1, -\pi \le y \le \pi\}$. Find a bound on the difference between $\phi(x)$ and $\psi(x)$ for x in $[-1, 1]$.

TECHNICAL WRITING EXERCISES

1. The Banach fixed point theorem for operators (Theorem 2, page 814) gives conditions for an operator mapping a set of continuous functions into itself to have a fixed point. A similar result holds for an operator mapping a set of functions analytic in a disk into itself. What implications does such a result have on the existence of a power series solution to an initial value problem (see Sections 8.3 and 8.4)?

2. Given the initial value problem

$$y' = f(x, y), y(x_0) = y_0,$$

suppose you know that f and $\partial f/\partial y$ are continuous in an open rectangle $R = \{(x, y): a < x < b, c < y < d\}$ that contains the point (x_0, y_0). Give a step-by-step procedure (an algorithm) that can be used to determine a constant h so that the initial value problem has a unique solution on $[x_0 - h, x_0 + h]$. Pretend that your algorithm will be converted to a computer program and discuss how you intend to implement each step.

3. The results of Theorems 9 and 10 on the continuous dependence of solutions discussed in Section 13.4 are stated only for scalar first order equations. How should these results be stated for first order systems?

Appendices

A NEWTON'S METHOD

To solve an equation $g(x) = 0$, we must find the point or points where the graph of $y = g(x)$ meets the x-axis. One procedure for approximating a solution is **Newton's method.**

To motivate Newton's method geometrically, we let $\tilde{x}$ be a root of $g(x) = 0$ and let x_1 be our guess at the value of $\tilde{x}$. If $g(x_1) = 0$, we are done. If $g(x_1) \neq 0$, then we are off by some amount that we call dy (see Figure A.1). Then

$$\frac{dy}{dx} = g'(x_1),$$

and so

(1) $$dx = \frac{dy}{g'(x_1)}.$$

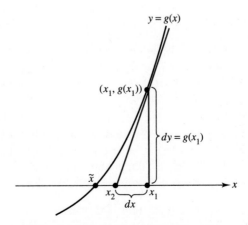

Figure A.1 Tangent line approximation of root

Now $dx = x_1 - x_2$ or $x_2 = x_1 - dx$, where x_2 is at the point where the tangent line through $(x_1, g(x_1))$ intersects the x-axis (Figure A.1). Using equation (1) and the fact that $dy = g(x_1)$, we obtain

$$x_2 = x_1 - \frac{dy}{g'(x_1)} = x_1 - \frac{g(x_1)}{g'(x_1)},$$

which we use as the next approximation to the root $\tilde{x}$.

Repeating this process with x_2 in place of x_1, we obtain the next approximation x_3 to the root $\tilde{x}$. In general, we find the next approximation x_{n+1} by the formula

$$(2) \qquad x_{n+1} = x_n - \frac{g(x_n)}{g'(x_n)}, \qquad n = 1, 2, \ldots.$$

The process is illustrated in Figure A.2.

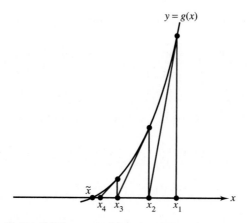

Figure A.2 Sequence of iterations converging to root

If the initial guess x_1 is sufficiently close to a root $\tilde{x}$, then the sequence of iterations $\{x_n\}_{n=1}^{\infty}$ usually converges to the root $\tilde{x}$. However, if we make a bad guess for x_1, then the process may lead away from $\tilde{x}$.

EXAMPLE 1 Find a root to four decimal places of the equation

$$(3) \qquad x^3 + 2x - 4 = 0.$$

Solution Setting $g(x) = x^3 + 2x - 4$, we find that $g'(x) = 3x^2 + 2$ is positive for all x. Hence g is increasing and has at most one zero. Furthermore, since $g(1) = -1$ and $g(2) = 8$, this zero must lie between 1 and 2. Thus we begin the procedure with the initial guess $x_1 = 1.5$. For $g(x) = x^3 + 2x - 4$, equation (2) becomes

$$(4) \qquad x_{n+1} = x_n - \frac{x_n^3 + 2x_n - 4}{3x_n^2 + 2}, \qquad n = 1, 2, \ldots.$$

With $x_1 = 1.5$, equation (4) gives

$$x_2 = 1.5 - \frac{(1.5)^3 + 2(1.5) - 4}{3(1.5)^2 + 2} = 1.5 - \frac{2.375}{8.75} \approx 1.22857.$$

Using x_2 to compute x_3 and so on, we find

$$x_3 = 1.18085,$$
$$x_4 = 1.17951,$$
$$x_5 = 1.17951,$$

where we have rounded off the computations to five decimal places. Since x_4 and x_5 agree to four decimal places and we are uncertain of the fifth decimal place because of roundoff, we surmise that the root $\tilde{x}$ of (3) agrees with 1.1795 to four decimal places. Indeed

$$g(1.1795) = -0.00005 \ldots \quad \text{and} \quad g(1.1796) = 0.00056 \ldots,$$

so $1.1795 < \tilde{x} < 1.1796$. Consequently $\tilde{x} = 1.1795 \ldots$. ■ End

Observe that Newton's method transforms the problem of finding a root to the equation $g(x) = 0$ into the problem of finding a fixed point for the function $h(x) = x - g(x)/g'(x)$; that is, finding a number x such that $x = h(x)$. (See equation (2).)

Several theorems give conditions that guarantee that the sequence of iterations $\{x_n\}_{n=1}^{\infty}$ defined by (2) will converge to a zero of $g(x)$. We mention one such result.

CONVERGENCE OF NEWTON'S METHOD

Theorem 1. Suppose a zero $\tilde{x}$ of $g(x)$ lies in the interval (a, b) and that in this interval

$$g'(x) > 0 \quad \text{and} \quad g''(x) > 0.$$

If we select x_1 so that $\tilde{x} < x_1 < b$, then the sequence of iterations defined by (2) will decrease to $\tilde{x}$.

We do not give a proof of this theorem, but refer the reader to an introductory numerical analysis text such as *Numerical Analysis*, Fifth Edition, by R. Burden and J. Faires (Prindle, Weber, and Schmidt, Boston, 1993).

B SIMPSON'S RULE

A useful procedure for approximating the value of a definite integral is **Simpson's rule.**

Let the interval $[a, b]$ be divided into $2n$ equal parts and let $x_0, x_1, \ldots, x_{2n}$ be the points of the partition, that is,

$$x_k := a + kh, \qquad k = 0, 1, \ldots, 2n,$$

where $h := (b - a)/(2n)$. If

$$y_k := f(x_k), \quad k = 0, 1, \ldots, 2n,$$

then the Simpson's rule approximation I_s for the value of the definite integral

$$\int_a^b f(x)\,dx$$

is given by

(1) $$I_s = \frac{h}{3}[y_0 + 4y_1 + 2y_2 + 4y_3 + \cdots + 2y_{2n-2} + 4y_{2n-1} + y_{2n}]$$

$$= \frac{h}{3}\sum_{k=1}^{n}(y_{2k-2} + 4y_{2k-1} + y_{2k}).$$

If

$$E := \int_a^b f(x)\,dx - I_s$$

is the error that results from using Simpson's rule to approximate the value of the definite integral, then

(2) $$|E| \leq \frac{(b - a)}{180}h^4 M,$$

where $M := \max|f^{(4)}(x)|$ for all x in $[a, b]$.

EXAMPLE 1 Use Simpson's rule with $n = 4$ to approximate the value of the definite integral

(3) $$\int_0^1 \frac{1}{1 + x^2}\,dx.$$

Solution Here $h = 1/8$, $x_k = k/8$, $k = 0, 1, \ldots, 8$, and

$$y_k = (1 + x_k^2)^{-1} = \frac{1}{1 + \dfrac{k^2}{64}} = \frac{64}{64 + k^2}.$$

By Simpson's rule (1), we find

$$I_s = \frac{\left(\dfrac{1}{8}\right)}{3}\left[1 + 4\left(\frac{64}{64 + 1}\right) + 2\left(\frac{64}{64 + 4}\right) + 4\left(\frac{64}{64 + 9}\right)\right.$$

$$+ 2\left(\frac{64}{64 + 16}\right) + 4\left(\frac{64}{64 + 25}\right) + 2\left(\frac{64}{64 + 36}\right)$$

$$\left. + 4\left(\frac{64}{64 + 49}\right) + \left(\frac{64}{64 + 64}\right)\right] = 0.7854.$$

Hence the value of the definite integral in (3) is approximately $I_s = 0.7854$. ■ End

For a more detailed discussion of Simpson's rule, we refer the reader to a numerical analysis book such as *Numerical Analysis,* Fifth Edition, by R. Burden and J. Faires (Prindle, Weber, and Schmidt, Boston, 1993).

C CRAMER'S RULE

When a system of n linear equations in n unknowns has a unique solution, determinants can be used to obtain a formula for the unknowns. This procedure is called **Cramer's rule.** When n is small, these formulas provide a simple procedure for solving the system.

Suppose that for a system of n linear equations in n unknowns,

(1)
$$
\begin{aligned}
a_{11}x_1 + a_{12}x_2 + \cdots + a_{1n}x_n &= b_1, \\
a_{21}x_1 + a_{22}x_2 + \cdots + a_{2n}x_n &= b_2, \\
&\ \ \vdots \\
a_{n1}x_1 + a_{n2}x_2 + \cdots + a_{nn}x_n &= b_n,
\end{aligned}
$$

the coefficient matrix

(2)
$$
\mathbf{A} := \begin{bmatrix}
a_{11} & a_{12} & \cdots & a_{1n} \\
a_{21} & a_{22} & \cdots & a_{2n} \\
\vdots & \vdots & & \vdots \\
a_{n1} & a_{n2} & \cdots & a_{nn}
\end{bmatrix}
$$

has a nonzero determinant. Then Cramer's rule gives the solutions

(3)
$$
x_i = \frac{\det \mathbf{A}_i}{\det \mathbf{A}}, \qquad i = 1, 2, \ldots, n,
$$

where $\mathbf{A}_i$ is the matrix obtained from $\mathbf{A}$ by replacing the ith column of $\mathbf{A}$ by the column vector

$$
\begin{bmatrix}
b_1 \\
b_2 \\
\vdots \\
b_n
\end{bmatrix}
$$

consisting of the constants on the right-hand side of system (1). Again, we assume $\det \mathbf{A} \neq 0$.

EXAMPLE 1 Use Cramer's rule to solve the system

$$
\begin{aligned}
x_1 + 2x_2 - x_3 &= 0, \\
2x_1 + x_2 + x_3 &= 9, \\
x_1 - x_2 - 2x_3 &= 1.
\end{aligned}
$$

Solution We first compute the determinant of the coefficient matrix:

$$\det \begin{bmatrix} 1 & 2 & -1 \\ 2 & 1 & 1 \\ 1 & -1 & -2 \end{bmatrix} = 12.$$

Using formula (3), we find

$$x_1 = \frac{1}{12}\det \begin{bmatrix} 0 & 2 & -1 \\ 9 & 1 & 1 \\ 1 & -1 & -2 \end{bmatrix} = \frac{48}{12} = 4,$$

$$x_2 = \frac{1}{12}\det \begin{bmatrix} 1 & 0 & -1 \\ 2 & 9 & 1 \\ 1 & 1 & -2 \end{bmatrix} = \frac{-12}{12} = -1,$$

$$x_3 = \frac{1}{12}\det \begin{bmatrix} 1 & 2 & 0 \\ 2 & 1 & 9 \\ 1 & -1 & 1 \end{bmatrix} = \frac{24}{12} = 2. \quad \blacksquare \text{ End}$$

For a more detailed discussion of Cramer's rule, we refer the reader to an introductory linear algebra text such as *Linear Algebra and Its Applications*, by David C. Lay (Addison-Wesley, Reading, Mass., 1993).

D METHOD OF LEAST SQUARES

The **method of least squares** is a procedure for fitting a straight line to a set of measured data. Consider the *scatter diagram* in Figure D.1, consisting of an x,y-plot of some data points $\{(x_i, y_i); i = 1, 2, \ldots, N\}$. It is desired to construct the straight line $y = \alpha + \beta x$ that best fits the data points, in the sense that the sum of the squares of the vertical deviations from the points to the line is minimized.

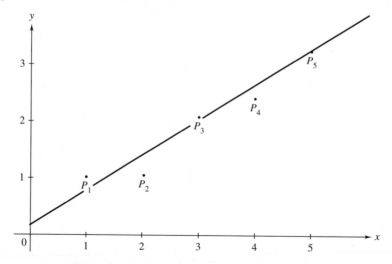

Figure D.1 Scatter diagram and least squares linear fit

If the y-axis intercept α and the slope β of the line are known, then the y value on the line corresponding to the measured x value x_i is given by $\alpha + \beta x_i$. The corresponding measured y value, y_i, thus deviates from the line by $[y_i - (\alpha + \beta x_i)]$, and the total sum of squares of deviations is

$$S := \sum_{i=1}^{N} [y_i - (\alpha + \beta x_i)]^2.$$

The interesting feature of the function S is that the symbols x_i and y_i are constants, while α and β are the (unknown) variables. The values of α and β that minimize S will force its partial derivatives to be zero:

$$0 = \frac{\partial S}{\partial \alpha} = \sum_{i=1}^{N} 2[y_i - (\alpha + \beta x_i)](-1),$$

$$0 = \frac{\partial S}{\partial \beta} = \sum_{i=1}^{N} 2[y_i - (\alpha + \beta x_i)](-x_i).$$

Displaying these conditions in a form that emphasizes the roles of α and β, we rewrite them after a little algebra as

$$\alpha N + \beta \sum_{i=1}^{N} x_i = \sum_{i=1}^{N} y_i,$$

$$\alpha \sum_{i=1}^{N} x_i + \beta \sum_{i=1}^{N} x_i^2 = \sum_{i=1}^{N} x_i y_i,$$

and obtain the formulas for the optimal values of slope and intercept:

$$\alpha = \frac{\left(\sum_{i=1}^{N} x_i^2\right)\left(\sum_{i=1}^{N} y_i\right) - \left(\sum_{i=1}^{N} x_i\right)\left(\sum_{i=1}^{N} x_i y_i\right)}{N \sum_{i=1}^{N} x_i^2 - \left(\sum_{i=1}^{N} x_i\right)^2},$$

$$\beta = \frac{N \sum_{i=1}^{N} x_i y_i - \left(\sum_{i=1}^{N} x_i\right)\left(\sum_{i=1}^{N} y_i\right)}{N \sum_{i=1}^{N} x_i^2 - \left(\sum_{i=1}^{N} x_i\right)^2}.$$

EXAMPLE 1 Find the least-squares linear fit to the data points $P_1(1, 1)$, $P_2(2, 1)$, $P_3(3, 2)$, $P_4(4, 2.5)$, and $P_5(5, 3.1)$, which are plotted in Figure D.1.

Solution Arranging the data as in Table D.1 yields

$$\alpha = \frac{55(9.6) - 15(34.5)}{5(55) - (15)^2} = \frac{10.5}{50} = 0.21,$$

$$\beta = \frac{5(34.5) - 15(9.6)}{5(55) - (15)^2} = \frac{28.5}{50} = 0.57.$$

Thus $y = 0.21 + 0.57x$ is the equation for the best-fitting line, which is graphed in Figure D.1. ■ End

TABLE D.1				
i	x_i	y_i	$x_i y_i$	x_i^2
1	1	1	1	1
2	2	1	2	4
3	3	2	6	9
4	4	2.5	10	16
5	5	3.1	15.5	25
Sums	15	9.6	34.5	55

E RUNGE-KUTTA PROCEDURE FOR *n* EQUATIONS

The following subroutine and algorithm generalize the programs given in Section 5.6 (pages 279–280) to systems of *n* first order ordinary differential equations.

FOURTH ORDER RUNGE-KUTTA SUBROUTINE (*n* EQUATIONS)

Purpose To approximate the solution to the initial value problem

$$x_i' = f_i(t, x_1, \ldots, x_n);$$
$$x_i(t_0) = a_i, \quad i = 1, 2, \ldots, n$$

for $t_0 \leq t \leq c$.

INPUT $n, t_0, a_1, \ldots, a_n, c, N$ (number of steps), PRNTR (=1 to print table).

Step 1 Set step size

$$h = (c - t_0)/N, \quad t = t_0, \quad x_1 = a_1, \quad \ldots, \quad x_n = a_n$$

Step 2 For $j = 1$ to N do Steps 3–5

Step 3 Set

$$k_{1,i} = hf_i(t, x_1, \ldots, x_n),$$
$$i = 1, \ldots, n;$$

$$k_{2,i} = hf_i\left(t + \frac{h}{2}, x_1 + \frac{1}{2}k_{1,1}, \ldots, x_n + \frac{1}{2}k_{1,n}\right),$$
$$i = 1, \ldots, n;$$

$$k_{3,i} = hf_i\left(t + \frac{h}{2}, x_1 + \frac{1}{2}k_{2,1}, \ldots, x_n + \frac{1}{2}k_{2,n}\right),$$
$$i = 1, \ldots, n;$$

$$k_{4,i} = hf_i(t + h, x_1 + k_{3,1}, \ldots, x_n + k_{3,n}),$$
$$i = 1, \ldots, n$$

Step 4 Set

$$t = t + h;$$

$$x_i = x_i + \frac{1}{6}(k_{1,i} + 2k_{2,i} + 2k_{3,i} + k_{4,i}), \quad i = 1, \ldots, n$$

Step 5 If PRNTR = 1, print $t, x_1, x_2, \ldots, x_n$

FOURTH ORDER RUNGE-KUTTA ALGORITHM WITH TOLERANCE (*n* EQUATIONS)

Purpose To approximate the solution to the initial value problem

$$x_i' = f_i(t, x_1, \ldots, x_n);$$
$$x_i(t_0) = a_i, \qquad i = 1, 2, \ldots, n$$

at $t = c$, with tolerance ε.

INPUT $n, t_0, a_1, \ldots, a_n, c$
ε (tolerance)
M (maximum number of iterations)

Step 1 Set $z_i = a_i, \quad i = 1, 2, \ldots, n; \quad$ set PRNTR = 0

Step 2 For $m = 0$ to M do Steps 3–7 (or, to save time, start with $m > 0$)

Step 3 Set $N = 2^m$

Step 4 Call FOURTH ORDER RUNGE-KUTTA SUBROUTINE (*n* EQUATIONS)

Step 5 Print $h, x_1, x_2, \ldots, x_n$

Step 6 If $|z_i - x_i| < \varepsilon$ for $i = 1, \ldots, n$, go to Step 10

Step 7 Set $z_i = x_i, \qquad i = 1, \ldots, n$

Step 8 Print ''$x_i(c)$ is approximately,'' x_i (for $i = 1, \ldots, n$); ''but may not be within the tolerance''; ε

Step 9 Go to Step 11

Step 10 Print ''$x_i(c)$ is approximately,''; x_i (for $i = 1, \ldots, n$); ''with tolerance''; ε

Step 11 STOP

OUTPUT Approximations of the solution to the initial value problem at $t = c$, using 2^m steps.

Answers to Odd-Numbered Problems

CHAPTER 1

Exercises 1.1, page 5

1. ODE, 2nd order, ind. var. t, dep. var. x, linear.
3. ODE, 1st order, ind. var. x, dep. var. y, nonlinear.
5. ODE, 1st order, ind. var. t, dep. var. p, nonlinear.
7. ODE, 1st order, ind. var. x, dep. var. y, nonlinear.
9. ODE, 2nd order, ind. var. x, dep. var. y, linear.
11. PDE, 2nd order, ind. var. t, r, dep. var. N.
13. ODE, 2nd order, ind. var. x, dep. var. y, nonlinear.
15. $dT/dt = k(M - T)$, where k is the proportionality constant.
17. Alison wins by $6\sqrt{3} - 4\sqrt{6} \approx 0.594$ seconds.

Exercises 1.2, page 14

3. Yes. **5.** No. **7.** Yes. **9.** No. **11.** Yes.
13. Yes. **19.** The left-hand side is always ≥ 3.
21. (a) $1/3$, -3. **(b)** $1 \pm \sqrt{6}$. **23.** Yes. **25.** Yes.
27. No. **31. (a)** No.

Exercises 1.3, page 22

1.

Figure B.1
Solution to Problem 1(b)

Figure B.2
Solution to Problem 1(c)

1. (d) The solutions in parts (b) and (c) both become infinite and have the line $y = 2x$ as an asymptote as $x \to \infty$. As $x \to -\infty$, the solution in part (b) becomes infinite and has $y = -2x$ as an asymptote, while the solution in part (c) does not even exist for x negative.

3. All solutions have limiting value 8 as $t \to +\infty$.

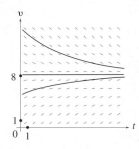

Figure B.3 Solutions to Problem 3 satisfying $v(0) = 5$, $v(0) = 8$, $v(0) = 15$

5. (b) $\lim_{t \to \infty} p(t) = 3/2$. **(c)** $\lim_{t \to \infty} p(t) = 3/2$. **(d)** No.

Figure B.4 Direction field for Problem 5(a)

7. (a)

Figure B.5 Direction field for Problem 7(a)

(b) 2. **(c)** 2. **(d)** 0. **(e)** No.

9. (d) It increases and asymptotically approaches the line $y = x - 1$.

 (f), (g) See Figure B.6.

Figure B.6 Direction field and sketch of $\phi(x)$ for Problem 9(f), (g)

11.

Figure B.7 Solution to Problem 11

13.

Figure B.8 Solution to Problem 13

15.

Figure B.9 Solution to Problem 15

17. It approaches 3.

Exercises 1.4, page 28

1.

x_n	0.1	0.2	0.3	0.4	0.5
y_n	-1.000	-1.010	-1.030	-1.059	-1.097

(rounded to three decimal places).

3.

x_n	0.1	0.2	0.3	0.4	0.5
y_n	2.700	2.511	2.383	2.292	2.225

(rounded to three decimal places).

5.

x_n	1.2	1.4	1.6	1.8
y_n	1.400	1.960	2.789	4.110

7.

N	$\approx x(1)$
1	1
2	1.06185
4	1.13920
8	1.19157

9.

x_n	y_n
1.1	-0.9
1.2	-0.81654
1.3	-0.74572
1.4	-0.68480
1.5	-0.63176
1.6	-0.58511
1.7	-0.54371
1.8	-0.50669
1.9	-0.47335
2.0	-0.44314

11. 1.46635.

15. $T(1) \approx 80.460°$, $T(2) \approx 73.647°$.

CHAPTER 2

Exercises 2.2, page 44

1. Yes. **3.** Yes. **5.** No. **7.** $y = (3x - x^3 + C)^{1/3}$.

9. $x = Ce^{t^3}$. **11.** $2y + \sin(2y) = 4 \arctan x + C$.

13. $x = \dfrac{Ce^t}{Ce^t - 1}$, $x \equiv 1$. **15.** $y = 1/(C - e^{\cos x})$.

17. $y = 2e^{-x^4/4} + 1$. **19.** $y = -3e^{-1 - \cos \theta}$.

21. $y = \sin^2 x + 2 \sin x.$ **23.** $y = \arctan(1 + x^2).$

25. $y = 4e^{x^3/3} - 1.$

27. (a) $y(x) = \int_0^x e^{t^2}\, dt.$ **(b)** $y(x) = \left(1 + 3\int_0^x e^{t^2}\, dt\right)^{1/3}.$

 (c) $y(x) = \tan\left(\int_0^x \sqrt{1 + \sin t}\, dt + \pi/4\right).$

 (d) $y(0.5) \approx 1.381.$

29. (d) $\partial f/\partial y$ is not continuous at $(0, 0)$. **33.** 281 kg.

35. (a) 82.2 min. **(b)** 31.8 min.

 (c) Never attains desired temperature.

37. (a) \$1105.17. **(b)** 27.73 years. **(c)** \$4427.59.

Exercises 2.3, page 52

1. Linear. **3.** Neither. **5.** Both.

7. $y = (1/2)e^{3x} + Ce^x.$ **9.** $r = \sin\theta + C\cos\theta.$

11. $y = -t - 2 + Ce^t.$ **13.** $x = y^3 + Cy^{-2}.$

15. $y = 1 + C(x^2 + 1)^{-1/2}.$ **17.** $y = xe^x - x.$

19. $x = t^{-1}/2 - 2t^{-3}.$ **21.** $y = x^2 \cos x - \pi^2 \cos x.$

23. $y(t) = (38/3)e^{-5t} - (8/3)e^{-20t}.$ **25. (b)** $y(3) \approx 0.183.$

27. (b) 0.6400. **(c)** 0.6158, 0.6283.

29. $x = e^{4y}/2 + Ce^{2y}.$

31. (a) $y = x - 1 + Ce^{-x}.$ **(b)** $y = x - 1 + 2e^{-x}.$

 (c) $y = x/3 - 1/9 + Ce^{-3x}.$

 (d) $y = \begin{cases} x - 1 + 2e^{-x}, & 0 \le x \le 2, \\ x/3 - 1/9 + (4e^6/9 + 2e^4)e^{-3x}, & 2 < x. \end{cases}$

 (e)

Figure B.10 Solution to Problem 31(e)

33. (a) $y = x$ is only solution in a neighborhood of $x = 0$.

 (b) $y = -3x + Cx^2$ satisfies $y(0) = 0$ for any C.

35. (a) 0.281 kg/L. **(b)** 0.446 kg/L.

37. $x(t) = \dfrac{1}{2} - \dfrac{2\cos(\pi t/12)}{4 + (\pi/12)^2} - \dfrac{(\pi/12)\sin(\pi t/12)}{4 + (\pi/12)^2}$
$\qquad + \left(\dfrac{19}{2} + \dfrac{2}{4 + (\pi/12)^2}\right)e^{-2t}.$

Exercises 2.4, page 63

1. Linear with y as dep. var. **3.** Exact.

5. Exact, also linear with x as dep. var.

7. Linear with r as dep. var. **9.** $y = (C - 3x)/(x^2 - 1).$

11. $\sin x \cos y + x^2 - y^2 = C.$ **13.** $t \ln y + t = C.$

15. $r = (C - e^\theta)\sec\theta.$ **17.** Not exact.

19. $x^2 - y^2 + \arctan(xy) = C.$

21. $\ln x + x^2 y^2 - \sin y = \pi^2.$ **23.** $y = -2/(te^t + 2).$

25. $\sin x - x \cos x = \ln y + 1/y + \pi - 1$ (equation is separable, not exact).

27. (a) $-\ln|y| + f(x);$

 (b) $\cos x \sin y - y^2/2 + f(x);$ where f is a function of x only.

29. (c) $y = x^2/(C - x).$ **(d)** Yes, $y \equiv 0.$

33. (a) $x = cy^2;\quad x \equiv 0;\quad y \equiv 0.$ **(b)** $x^2 + 4y^2 = c.$

 (c) $2y^2 \ln y - y^2 + 2x^2 = c.$ **(d)** $2x^2 + y^2 = c.$

Exercises 2.5, page 69

1. Separable, linear with x as dep. var., has integrating factor depending on y alone.

3. Integrating factor depending on y alone.

5. Integrating factor depending on y alone.

7. $\mu = x^{-2};\quad y^2/2 - y/x + 3x = C$ or $x \equiv 0.$

9. $\mu = x;\quad x^2 y^2 + x^2 y + x^4 = C.$

11. $\mu = y^{-2};\quad x^2 y^{-1} + x = C$ or $y \equiv 0.$

13. $\mu = xy;\quad x^2 y^3 - 2x^3 y^2 = C.$

15. $\mu(z) = \exp(\int H(z)\, dz);\quad z = xy.$

17. (b) $\mu = e^y;\quad x = y - 1 + Ce^{-y}.$

Exercises 2.6, page 76

1. $y' = G(ax + by).$ **3.** Linear coefficients.

5. Bernoulli. **7.** Homogeneous.

9. $\ln(y^2/x^6) - y^2/x^2 = C.$

11. $y = x/(\ln|x| + C)$ or $y \equiv 0.$

13. $\sqrt{1 + x^2/t^2} = \ln|t| + C.$ **15.** $(x^2 - 4y^2)^3 x^2 = C.$

17. $y = (x + C)^2/4 - x.$

19. $y = x + (6 + 4Ce^{2x})/(1 + Ce^{2x}).$

21. $y = 2/(Cx - x^3)$ or $y \equiv 0.$

23. $y = 5x^2/(x^5 + C)$ or $y \equiv 0.$

25. $x^{-2} = 2t^2 \ln|t| + Ct^2$ or $x \equiv 0.$

27. $r = \theta^2/(C - \theta)$ or $r \equiv 0.$

29. $(y + 2)^2 + 2(x + 1)(y + 2) - 3(x + 1)^2 = C.$

31. $(2x + 2y - 3)^3 = C(2x + y - 2)^2.$

33. $y = 4x + (3 + Ce^{4x})/(1 - Ce^{4x}).$

35. $\ln[(x + 3)^2 + 3(t - 1)^2]$
$\qquad + (2/\sqrt{3})\arctan[(x + 3)/\sqrt{3}\,(t - 1)] = C.$

37. $y^{-2} = -e^{2x}/2 + Ce^{-2x}$ or $y \equiv 0.$

39. $\theta y^2 = C(\theta + y)^2$ or $y = -\theta.$

41. $(x - y + 2)^2 = Ce^{2x} + 1.$ **45.** $(y - 4x)^2(y + x)^3 = C.$

Review Problems, page 79

1. $e^x + ye^{-y} = C.$ **3.** $x^2 y - x^3 + y^{-2} = C.$

5. $y + x\sin(xy) = C.$ **7.** $y = (7\ln|t| + C)^{-1/7}.$

9. $(x^2 + 4y^2)^3 x^2 = C.$ **11.** $\tan(t - x) + t = C.$

13. $y = -(x^2/2)\cos(2x) + (x/4)\sin(2x) + Cx.$

15. $y = 2x + 3 - (x + C)^2/4$.

17. $y = 2/(1 + Ce^{2\theta})$ or $y \equiv 0$.

19. $y^2 = x^2 + Cx^3$ or $x \equiv 0$.

21. $xy - x^2 - x + y^2/2 - 4y = C$.

23. $y^2 + 2xy - x^2 = C$.

25. $x^2y^{-2} - 2xy^{-1} - 4xy^{-2} = C$ or $y \equiv 0$.

27. $[(y - 4)^2 - 3(x - 3)^2]\{[\sqrt{3}(x - 3) + (y - 4)]$
$/[\sqrt{3}(x - 3) - (y - 4)]\}^{1/\sqrt{3}} = C$.

29. $x^4y^3 - 3x^3y^2 + x^4y^2 = C$. **31.** $y = -x^3/2 + 7x/2$.

33. $x = -t - 2 + 3e^{-t}$. **35.** $y = -2x\sqrt{2x^2 - 1}$.

37. $\ln[(y-2)^2 + 2(x-1)^2] + \sqrt{2}\arctan\left[\dfrac{y-2}{\sqrt{2}(x-1)}\right] = \ln 2$.

39. $y = \sqrt{(19x^4 - 1)/2}$.

CHAPTER 3

Exercises 3.2, page 98

1. $50 - 45e^{-2t/25}$ kg, 5.07 min.

3. $(0.4)(100 - t) - (39 \times 10^{-8})(100 - t)^4$ L; 19.96 min.

5. 0.0097%; 73.24 hr.

7. $(0.2)(1 - e^{-3t/125})$ g/cm^3; 28.88 sec.

9. 4804; 27,000. **13.** 5769; 6000.

15. 1527; 1527.

17. (a)

Year	$\dfrac{1}{p}\dfrac{dp}{dt}$	Logistic (Least Squares)
1790	0.0351	3.06
1800	0.0363	4.23
1810	0.0331	5.86
1820	0.0335	8.08
1830	0.0326	11.12
1840	0.0359	15.22
1850	0.0356	20.72
1860	0.0267	27.99
1870	0.0260	37.42
1880	0.0255	49.37
1890	0.0207	64.06
1900	0.0210	81.49
1910	0.0149	101.29
1920	0.0161	122.72
1930	0.0072	144.71
1940	0.0149	166.09
1950	0.0185	185.82
1960	0.0133	203.16
1970	0.0115	217.76
1980	0.0102	229.61
1990		238.96
2000		246.15

(b) $a = 0.0330615$, $b = 0.000123985$.

(c) $p_0 = 3.05548$ (using all data, including 1990).

(d) See table in part (a).

19. $(1/2)\ln 15 \approx 1.354$ years; 14 million tons per year.

21. 1 hr; 2 hr. **23.** 11.7% **25.** 31,606 yr.

27. e^{-2t} kg of Hh, $2e^{-t} - 2e^{-2t}$ kg of It, and
$1 - 2e^{-t} + e^{-2t}$ kg of Bu.

Exercises 3.3, page 106

1. 20.7 min. **3.** 22.6 min. **5.** 9:08 AM.

7. 28.3°C; 32.5°C; 1:16 PM.

9. 16.3°C; 19.1°C; 31.7°C; 28.9°C. **11.** 30.4 min.

13. 148.6°F.

15. $T - M = C(T + M)\exp[2\arctan(T/M) - 4M^3kt]$;
for T near M, $M^4 - T^4 \approx 4M^3(M - T)$, and so
$dT/dt \approx k_1(M - T)$, where $k_1 = 4M^3k$.

Exercises 3.4, page 114

1. $(0.981)t + (0.0981)e^{-10t} - 0.0981$ m; 1019 sec.

3. 18.6 sec. **5.** $4.91t + 22.55 - 22.55e^{-2t}$ m; 97.3 sec.

7. 241 sec.

9. $95.65t + 956.5e^{-t/10} - 956.5$ m; 13.2 sec.

11. $e^{kv}(kv - mg)^{mg} = e^{v_0k}(kv_0 - mg)^{mg}e^{-k^2x/m}$.

13. 2.69 sec.; 101.19 m. **15.** $(\omega_0 - T/k)e^{-kt/I} + T/k$.

17. 1164 sec.

19. $2636e^{-t/20} + 131.8t - 2636$ m; 1.768 sec.

21. $5e^{-2t}/2 + 6t - 5/2$; 6 m/sec. **23.** Sailboat B.

25. (e) 11.18 km/sec. **(f)** 2.38 km/sec.

Exercises 3.5, page 125

3.

h	"e"
1	3
0.1	2.720551414
0.01	2.718304482
0.001	2.718282082
0.0001	2.718281824

7.

x_n	y_n
1.1	0.10450
1.2	0.21668
1.3	0.33382
1.4	0.45300
1.5	0.57135

9.

x_n	y_n
0.2	0.61784
0.4	1.23864
0.6	1.73653
0.8	1.98111
1.0	1.99705
1.2	1.88461
1.4	1.72447
1.6	1.56184
1.8	1.41732
2.0	1.29779

11. $\phi(1) \approx x(1; 2^{-3}) = 1.25494$.

13. $\phi(1) \approx y(1; 2^{-3}) = 0.71698$.

15. $x = 1.27$.

17.

x_n	$y_n(h = 0.2)$	$y_n(h = 0.1)$	$y_n(h = 0.025)$
0.1		−1	0.06250
0.2	−3	1	0.00391
0.3		−1	0.00024
0.4	9	1	0.00002
0.5		−1	0.00000
0.6	−27	1	0.00000
0.7		−1	0.00000
0.8	81	1	0.00000
0.9		−1	0.00000
1.0	−243	1	0.00000

We conclude that step size can dramatically affect convergence.

19.

x_n	$r = 1.5$	$r = 2$	$r = 3$
0.25	1.58286	1.53125	1.39063
0.5	2.35144	2.04960	1.55347
0.75	3.26750	2.44003	1.62885
1.0	4.25316	2.68675	1.66999
1.25	5.21675	2.82920	1.69406
1.5	6.08340	2.90804	1.70858
1.75	6.81163	2.95080	1.71748
2.0	7.39215	2.97377	1.72298
2.25	7.83709	2.98604	1.72640
2.5	8.16851	2.99257	1.72852
2.75	8.41036	2.99605	1.72985
3.0	8.58432	2.99790	1.73067
3.25	8.70817	2.99889	1.73119
3.5	8.79571	2.99941	1.73151
3.75	8.85729	2.99969	1.73171
4.0	8.90044	2.99983	1.73184
4.25	8.93062	2.99991	1.73192
4.5	8.95168	2.99995	1.73197
4.75	8.96637	2.99997	1.73200
5.0	8.97660	2.99999	1.73202

21.

Time	$K = 0.2$	$K = 0.4$	$K = 0.6$
Midnight	65.0000	65.0000	65.0000
4 AM	69.1639	68.5644	68.1299
8 AM	71.4836	72.6669	73.6678
Noon	72.9089	75.1605	76.9783
4 PM	72.0714	73.5977	74.7854
8 PM	69.8095	69.5425	69.2832
Midnight	68.3852	67.0500	65.9740

T_n (header over the K columns)

Exercises 3.6, page 136

1. $y_{n+1} = y_n + h\cos(x_n + y_n)$
$$- \frac{h^2}{2}\sin(x_n + y_n)[1 + \cos(x_n + y_n)].$$

3. $y_{n+1} = y_n + h(x_n - y_n) + \frac{h^2}{2}(1 - x_n + y_n)$
$$- \frac{h^3}{6}(1 - x_n + y_n) + \frac{h^4}{24}(1 - x_n + y_n).$$

5. Order 2, $\phi(1) \approx 1.3725$; order 4, $\phi(1) \approx 1.3679$.
7. −11.7679. **9.** 1.36789. **11.** $x = 1.41$.
13. $x = 0.50$. **15.**

x_n	y_n
0.5	0.21462
1.0	0.13890
1.5	−0.02668
2.0	−0.81879
2.5	−1.69491
3.0	−2.99510

19. $v(3) \approx 0.24193$ with $h = 0.0625$.
21. $z(1) \approx 2.87083$ with $h = 0.03125$.

CHAPTER 4

Exercises 4.2, page 158

1. Linear, homogeneous, variable coefficients.
3. Nonlinear.
5. Linear, nonhomogeneous, variable coefficients.
7. Linear, nonhomogeneous, constant coefficients.
9. (a) $3x\sin x - (x^2 + 5)\cos x$. (b) 0.
 (c) $(r^2 - 4r - 5)x^r$.
13. (a) $2e^{2x}\cos x - 3e^{2x}\sin x$.
 (b) $-4e^{2x}\cos x + 3e^{2x}\sin x$.
15. Unique solution on $(-\pi/2, \pi/2)$.
17. Unique solution on $(0, 3)$.
19. Does not apply; equation is nonlinear.
21. Unique solution on $(-\infty, 1)$.
23. $D^2 + 7D + 12$. **25.** $D^2 + x$.
27. (a) 2. (b) $x\sin x + \cos x$. **31.** −2; 1; 5.
33. (a) $3x^2 - x^3$. (b) $6x + 3x^2 - 2x^3$. (c) $3x^2 + 2x^3$.
 (d) $6x + 3x^2 - 2x^3$. (e) $D^2 + D - 2$.
 (f) $6x + 3x^2 - 2x^3$.
35. (a) $(D + 4)(D - 1)$. (b) $(D + 3)(D - 2)$.
 (c) $(2D - 1)(D + 5)$. (d) $(D + \sqrt{2})(D - \sqrt{2})$.

Exercises 4.3, page 167

1. Lin. indep.; $2e^{-2x}$. **3.** Lin. indep.; $-e^{4x}$.
5. Lin. dep.; 0.
7. (b) $y = c_1 x^2 + c_2 x^{-1}$. (c) $y = -3x^2 + x^{-1}$.

9. (b) $y = c_1 e^x \cos 2x + c_2 e^x \sin 2x.$
 (c) $y = 2e^x \cos 2x - e^x \sin 2x.$
11. (b) $y = c_1 e^t + c_2(t^2 + 2t + 2).$
 (c) $y = 5e^{t-1} - t^2 - 2t - 2.$
13. (c) $\phi(x) = (1)e^x + (-1)(e^x - e^{-6x}),$
 $\phi(x) = (-3)e^x + (1)(3e^x + e^{-6x}).$
15. The function in (b) cannot be a Wronskian of solutions.
17. (a) True. **(b)** False. **19.** $W = Cxe^{-x}.$
25. (a) Lin. indep. **(b)** Lin. dep. **(c)** Lin. indep.
 (d) Lin. dep.
27. (a) 2. **(b)** 0. **(c)** $2e^{3x}.$

Exercises 4.4, page 172

1. $y = e^{-5x}.$ **3.** $y = x^4.$ **5.** $x = t + 1.$
7. $xw'' + 2xw' + (x + 1)w = 0.$

9. (a) $y = x \int x^{-2} e^{x^2/2} \, dx.$

 (b) $y = \sum_{n=0}^{\infty} \frac{x^{2n}}{n! \, 2^n (2n - 1)} = -1 + \frac{x^2}{2} + \frac{x^4}{24} + \cdots.$

13. (a) $y = (1 - 2x^2) \int (1 - 2x^2)^{-2} e^{x^2} \, dx.$

 (b) $y = (3x - 2x^3) \int (3x - 2x^3)^{-2} e^{x^2} \, dx.$

15. (a) $y = (x - 1) \int e^x x^{-1} (x - 1)^{-2} \, dx.$

 (b) $y = (x^2 - 4x + 2) \int e^x x^{-1} (x^2 - 4x + 2)^{-2} \, dx.$

17. (b) $g(2) \approx 10.7983.$

Exercises 4.5, page 180

1. $c_1 e^{-3x} + c_2 e^{-2x}.$ **3.** $c_1 e^{-4x} + c_2 x e^{-4x}.$
5. $c_1 e^{(-1-\sqrt{5})x/2} + c_2 e^{(-1+\sqrt{5})x/2}.$ **7.** $c_1 e^{x/2} + c_2 e^{-4x}.$
9. $c_1 e^{(1+3\sqrt{5})x/2} + c_2 e^{(1-3\sqrt{5})x/2}.$
11. $c_1 e^{-5x/2} + c_2 x e^{-5x/2}.$ **13.** $3e^{-4x}.$
15. $e^{-x} - 2xe^{-x}.$ **17.** $(\sqrt{3}/2)[e^{(1+\sqrt{3})x} - e^{(1-\sqrt{3})x}].$
19. $2e^{5(x+1)} + e^{-(x+1)}.$
21. (a) $ar + b = 0.$ **(b)** $ce^{-bx/a}.$ **23.** $ce^{-4x/5}.$
25. $ce^{13x/6}.$ **27.** $c_1 e^{-x} + c_2 e^x + c_3 e^{6x}.$
29. $c_1 e^{-x} + c_2 e^{3x} + c_3 e^{5x}.$
31. $c_1 e^x + c_2 e^{(-1+\sqrt{3})x} + c_3 e^{(-1-\sqrt{3})x}.$
33. $c_1 x + c_2 x^{-7}.$ **35.** $c_1 x^{-1} + c_2 x^{-4}.$
37. $c_1 x^{-1/3} + c_2 x^{-1/3} \ln x.$ **39.** $x - 3x^4.$
41. $c_1(x - 2) + c_2(x - 2)^7.$ **43.** No.
45. $c_1 x^{-2} + c_2(-x)^{-1/2}.$ **47.** $c_1 x + c_2 x \ln x + c_3 x^3.$
49. (a) $c_1 e^{r_1 x} + c_2 e^{r_2 x} + c_3 e^{r_3 x},$ where $r_1 = -4.832,$
 $r_2 = -1.869,$ and $r_3 = 0.701.$

 (b) $c_1 e^{r_1 x} + c_2 e^{-r_1 x} + c_3 e^{r_2 x} + c_4 e^{-r_2 x},$
 where $r_1 = 1.176,$ $r_2 = 1.902.$
 (c) $c_1 e^{-x} + c_2 e^{-2x} + c_3 e^x + c_4 e^{2x} + c_5 e^{3x}.$
53. $b = 2000\sqrt{3},$ multiply by the factor $\sqrt{2}.$

Exercises 4.6, page 187

1. $c_1 \cos 3x + c_2 \sin 3x.$ **3.** $c_1 e^{3x} \cos x + c_2 e^{3x} \sin x.$
5. $c_1 e^{-2x} \cos \sqrt{2}x + c_2 e^{-2x} \sin \sqrt{2}x.$
7. $c_1 e^{x/2} \cos(5x/2) + c_2 e^{x/2} \sin(5x/2).$ **9.** $c_1 e^x + c_2 e^{7x}.$
11. $c_1 e^{-5x} + c_2 x e^{-5x}.$
13. $c_1 e^{-x} \cos 2x + c_2 e^{-x} \sin 2x.$
15. $c_1 e^{-5x} \cos 4x + c_2 e^{-5x} \sin 4x.$
17. $c_1 e^{x/2} \cos(3\sqrt{3}x/2) + c_2 e^{x/2} \sin(3\sqrt{3}x/2).$
19. $c_1 e^x + c_2 e^{-x} \cos 2x + c_3 e^{-x} \sin 2x.$
21. $2e^{-x} \cos x + 3e^{-x} \sin x.$
23. $(\sqrt{2}/4)[e^{(2+\sqrt{2})x} - e^{(2-\sqrt{2})x}].$
25. $e^x \sin x - e^x \cos x.$ **27.** $e^{2x} - \sqrt{2}e^x \sin\sqrt{2}x.$
29. (a) $c_1 e^{-x} + c_2 e^x \cos\sqrt{2}x + c_3 e^x \sin\sqrt{2}x.$
 (b) $c_1 e^{2x} + c_2 e^{-2x} \cos 3x + c_3 e^{-2x} \sin 3x.$
 (c) $c_1 \cos 2x + c_2 \sin 2x + c_3 \cos 3x + c_4 \sin 3x.$
33. $c_1 x^2 \cos(\sqrt{2} \ln x) + c_2 x^2 \sin(\sqrt{2} \ln x).$
35. $c_1 x^{-1} \cos(2 \ln x) + c_2 x^{-1} \sin(2 \ln x).$
37. (a) $x(t) = 30e^{-3t} \cos 4t + 20e^{-3t} \sin 4t$ cm.
 (b) $2/\pi.$
 (c) Decreases the frequency of oscillation,
 introduces the factor $e^{-3t},$ causing the solution to
 decay to zero.
39. $b \geq 2\sqrt{Ik}.$
41. (a) $c_1 \cos x + c_2 \sin x + c_3 x \cos x + c_4 x \sin x.$
 (b) $(c_1 + c_2 x)e^{-x} \cos(\sqrt{3}x)$
 $+ (c_3 + c_4 x)e^{-x} \sin(\sqrt{3}x).$

Exercises 4.7, page 191

1. $c_1 e^x + c_2 e^{-x} - x.$ **3.** $c_1 e^{2x} + c_2 e^{-x} + x - 1.$
5. $e^{-x}(c_1 \cos\sqrt{3}x + c_2 \sin\sqrt{3}x) + \sin 2x.$
7. $e^{t/2}[c_1 \cos(\sqrt{3}t/2) + c_2 \sin(\sqrt{3}t/2)] + \cos t.$
9. $c_1 e^x + c_2 x e^x + x^2 e^x.$
11. (a) $5 \cos x.$ **(b)** $\cos x - e^{2x}.$ **(c)** $4 \cos x + 6e^{2x}.$
13. (a) $c_1 e^{\sqrt{2}x} + c_2 e^{-\sqrt{2}x} + (1 + \tan x)/2.$
 (b) $c_1 e^{\sqrt{2}x} + c_2 e^{-\sqrt{2}x} + 1 - x/2.$
 (c) $c_1 e^{\sqrt{2}x} + c_2 e^{-\sqrt{2}x} + \tan x - 3x/2 + 2.$
 (d) $c_1 e^{\sqrt{2}x} + c_2 e^{-\sqrt{2}x} - (x + \tan x)/8.$
17. (a) $c_1 e^x + c_2 e^{2x} + c_3 e^{-5x} - x^3 + 2x^2.$
 (b) $c_1 e^x + c_2 e^{2x} + c_3 e^{-5x} + 2x^3 + x^2.$
19. (a) $v = e^x/2 + c_1 e^{-x}.$
 (b) $y = -e^x/2 + c_1 e^{-x} + c_2 e^{2x}.$

Exercises 4.8, page 200

1. $y_p \equiv -3$. **3.** e^{2x}. **5.** $3t^2 - 2t + 4$.

7. $\cos 3x$. **9.** $re^r/2 + 3e^r/4$. **11.** $2^x[(\ln 2)^2 + 1]^{-1}$.

13. $4x^2e^x$. **15.** $x^2/9 + 4x/27 + 2/27 + e^x/4$.

17. $c_1e^x + c_2e^{-x} + 11x - 1$.

19. $c_1e^x + c_2e^{-2x} - x^2/2 + x/2 - 7/4$.

21. $c_1e^t + c_2e^{2t} + e^t(\cos t - \sin t)/2$.

23. $e^{-\theta}(c_1 \cos \theta + c_2 \sin \theta) + (\theta e^{-\theta} \sin \theta)/2$.

25. $c_1e^{2t} + c_2te^{2t} + t^3e^{2t}/6$.

27. $c_1e^{\sqrt{3}x} + c_2e^{-\sqrt{3}x} - x^2/3 - 2/9 + e^x/2$.

29. $e^{-x/2}[c_1 \cos(\sqrt{3}x/2) + c_2 \sin(\sqrt{3}x/2)] + \sin x$
$\qquad + e^x(-x^2/3 + 2x/3 - 4/9)$.

31. $e^x - 1$. **33.** $e^{-x} + \sin x - \cos x$.

35. $(3/20) \sin 2x - (1/20) \cos 2x - (3/10) \cos x$
$\qquad - (1/10) \sin x$.

37. $x^2/10 + 7x/50 - 161/500 + e^x/4 + 25e^{2x}/4$
$\qquad - 1589e^{5x}/500$.

39. $-(1/2) \sin \theta - e^{2\theta}/3 + 3e^{\theta}/4 + 7e^{-\theta}/12$.

41. $(Ax^2 + Bx) \sin x + (Cx^2 + Dx) \cos x + E10^x$.

43. $e^t(A \cos t + B \sin t) + Ct^2 + Dt + E$.

45. $Ae^{5x} + (Bx + C) \cos 3x + (Dx + E) \sin 3x$.

47. $Aue^u + B + C \sin u + D \cos u$.

49. No. **51.** Yes. **53.** No.

55. $(1/5) \cos x + (2/5) \sin x$.

57. $-(1/2) - (4/25)xe^x + (1/10)x^2e^x$.

59. $c_1x + c_2x^{-3} + x^3/12$.

61. $c_1x^3 + c_2x^{-2} - (1/5)x^{-2} \ln x + 1$.

63. $c_1 \cos(\ln x) + c_2 \sin(\ln x) - (1/4)(\ln x)^2 \cos(\ln x)$
$\qquad + (1/4)(\ln x) \sin(\ln x)$.

65. (a) $-e^{-x} \sin 2x - 2e^{-x} \cos 2x + 2$.

 (b) $e^{-x}(c_1 \sin 2x + c_2 \cos 2x)$.

 (c) $y =$
$$\begin{cases} -e^{-x} \sin 2x - 2e^{-x} \cos 2x + 2, & 0 \le x \le 3\pi/2; \\ (-1 - e^{3\pi/2})e^{-x} \sin 2x + (-2 - 2e^{3\pi/2})e^{-x} \cos 2x, \\ \qquad\qquad\qquad\qquad\qquad\qquad x \ge 3\pi/2. \end{cases}$$

69. (a) $x(t) = A \sin \beta t + B \cos \beta t + x_h$, where
$$A = \frac{k - m\beta^2}{(k - m\beta^2)^2 + b^2\beta^2},$$
$$B = \frac{-b\beta}{(k - m\beta^2)^2 + b^2\beta^2} \text{ and}$$
$$x_h = e^{-bt/2m}\left[c_1 \cos\left(\frac{\sqrt{4mk - b^2}}{2m}t\right)\right.$$
$$\left. + c_2 \sin\left(\frac{\sqrt{4mk - b^2}}{2m}t\right)\right].$$

 (b) In each case, $x_h \to 0$ as $t \to \infty$. So as $t \to \infty$,
 $x(t)$ approaches the function
 $x_p(t) = A \sin \beta t + B \cos \beta t$.

Exercises 4.9, page 206

1. $c_1 \cos 2x + c_2 \sin 2x$
$\qquad - (1/4)(\cos 2x) \ln|\sec 2x + \tan 2x|$.

3. $c_1e^{2t} + c_2e^{-t} + e^{3t}/4$. **5.** $c_1e^x + c_2xe^x + xe^x \ln|x|$.

7. $c_1 \cos 4\theta + c_2 \sin 4\theta + (\theta/4) \sin 4\theta$
$\qquad + (1/16)(\cos 4\theta) \ln|\cos 4\theta|$.

9. $c_1 \cos 2x + c_2 \sin 2x$
$\qquad + [(\cos 2x) \ln|\csc 2x + \cot 2x| - 1]/4$.

11. $c_1 \cos(3 \ln x) + c_2 \sin(3 \ln x)$
$\qquad + (1/9) \cos(3 \ln x) \ln|\sec(3 \ln x) + \tan(3 \ln x)|$.

13. $c_1 \cos x + c_2 \sin x + (\sin x) \ln|\sec x + \tan x| - 2$.

15. $c_1 \cos 2x + c_2 \sin 2x + (1/24) \sec^2 2x - 1/8$
$\qquad + (1/8)(\sin 2x) \ln|\sec 2x + \tan 2x|$.

17. $c_1 \cos x + c_2 \sin x - x^2 + 3 + 3x \sin x$
$\qquad + 3(\cos x) \ln|\cos x|$.

19. $c_1 \cos 2x + c_2 \sin 2x - e^x/5$
$\qquad - (1/2)(\cos 2x) \ln|\sec 2x + \tan 2x|$.

21. $c_1x + c_2x \ln|x| + (1/2)x(\ln|x|)^2$
$\qquad + 3x(\ln|x|)[\ln(\ln |x|)]$.

23. $c_1e^x + c_2(x + 1) - x^2$.

25. $c_1(5x - 1) + c_2e^{-5x} - x^2e^{-5x}/10$.

27. $y = e^{1-x} - e^{x-1} + \dfrac{e^x}{2}\displaystyle\int_1^x \dfrac{e^{-t}}{t}\,dt - \dfrac{e^{-x}}{2}\displaystyle\int_1^x \dfrac{e^t}{t}\,dt$,
$\qquad y(2) \approx -1.93$.

Exercises 4.10, page 214

1. $x_1' = x_2$, $x_2' = 3x_1 - tx_2 + t^2$; $x_1(0) = 3$,
$\qquad x_2(0) = -6$.

3. $x_1' = x_2$, $x_2' = x_3$, $x_3' = x_4$,
$\qquad x_4' = x_4 - 7x_1 + \cos t$; $x_1(0) = x_2(0) = 1$,
$\qquad x_3(0) = 0$, $x_4(0) = 2$.

5. $x = c_2e^{-t}$; $y = c_1e^{-2t} + c_2e^{-t}$.

7. $x = c_1 + c_2e^{-2t}$; $y = c_2e^{-2t}$. **9.** $x \equiv -5$; $y \equiv 1$.

11. $u = c_1 + c_2e^{-t} + (1/2)e^t + (5/3)t$;
$\qquad v = c_1 - 2c_2e^{-t} + (5/3)t$.

13. $x = c_1e^t + (1/4) \cos t - (1/4) \sin t$;
$\qquad y = -3c_1e^t - (3/4) \cos t - (1/4) \sin t$.

15. $u = c_1 \cos 2t + c_2 \sin 2t + c_3e^{\sqrt{3}t} + c_4e^{-\sqrt{3}t}$
$\qquad - (3/10)e^t$;
$\qquad v = c_1 \cos 2t + c_2 \sin 2t - (2/5)c_3e^{\sqrt{3}t} - (2/5)e^{-\sqrt{3}t}$
$\qquad + (1/5)e^t$.

17. $x = 2c_2e^t \cos 2t - 2c_1e^t \sin 2t$;
$\qquad y = c_1e^t \cos 2t + c_2e^t \sin 2t$.

19. $w = -(2/3)c_1e^{2t} + c_2e^{7t} + t + 1$;
$\qquad z = c_1e^{2t} + c_2e^{7t} - 5t - 2$.

21. $x = c_1 \cos t + c_2 \sin t - 4c_3 \cos(\sqrt{6}t)$
$\qquad - 4c_4 \sin(\sqrt{6}t)$;
$\qquad y = c_1 \cos t + c_2 \sin t + c_3 \cos(\sqrt{6}t) + c_4 \sin(\sqrt{6}t)$.

23. For $\lambda < 1$, solutions go to zero as $t \rightarrow +\infty$. For $\lambda = 1$, solutions remain bounded as $t \rightarrow +\infty$. For $\lambda > 1$, solutions become unbounded as $t \rightarrow +\infty$, that is, they go to either $+\infty$ or $-\infty$.

25. Infinitely many solutions satisfying
$x + y = e^t + e^{-2t}$.

27. $x(t) = -c_1 e^t - 2c_2 e^{2t} - c_3 e^{3t}$,
$y(t) = c_1 e^t + c_2 e^{2t} + c_3 e^{3t}$,
$z(t) = 2c_1 e^t + 4c_2 e^{2t} + 4c_3 e^{3t}$.

29. $x = -100 e^{-t/10} + 150$;
$y = -50 e^{-t/10} - 10 t e^{-t/10} + 150$.

31. $x = \left[\dfrac{200 - 100\sqrt{5}}{\sqrt{5}} \right] e^{(-3+\sqrt{5})t/100}$
$\quad - \left[\dfrac{200 + 100\sqrt{5}}{\sqrt{5}} \right] e^{(-3-\sqrt{5})t/100} + 200$;

$y = -\left(\dfrac{100}{\sqrt{5}} \right) e^{(-3+\sqrt{5})t/100} + \left(\dfrac{100}{\sqrt{5}} \right) e^{(-3-\sqrt{5})t/100} + 200$.

33. $90.4°F$. **35.** $460/11 \approx 41.8°F$.

Exercises 4.11, page 224

1. $c_1(2)^n + c_2(4)^n$.

3. $c_1[-(3/2) + (\sqrt{13}/2)]^n + c_2[-(3/2) - (\sqrt{13}/2)]^n$.

5. $c_1 11^{n/2} \cos n\theta + c_2 11^{n/2} \sin n\theta$, where
$\theta = \arctan(\sqrt{2}/3)$.

7. $c_1(-5)^n + c_2 n(-5)^n$.

9. $c_1 2^n \cos(n\theta) + c_2 2^n \sin(n\theta)$,
where $\theta = \arctan\sqrt{15}$.

11. $c_1 \cos(n\pi/2) + c_2 \sin(n\pi/2) + (1/2)n^2 - n$.

13. $c_1 + c_2(-1)^n - (1/2)\cos(3n)$
$\quad + (1/2)[\sin 6/(1 - \cos 6)]\sin(3n)$.

15. $c_1(-1)^n + c_2(-5)^n + (1/21)2^n$.

17. $c_1 2^n + c_2 n 2^n + (1/8)n^2 2^n$.

19. $c_1(-1)^n + c_2 2^n + c_3(-2)^n$.

21. $c_1 + c_2 2^{n/2} \cos(n\pi/4) + c_3 2^{n/2} \sin(n\pi/4)$.

23. $5^{n/2} \cos n\theta$, where $\theta = \pi - \arctan 2$.

25. $(n/6)3^n - (1/4)3^n + 1/4$.

27. $1 + (7/3)n - (1/2)n^2 + (1/6)n^3$.

31. Unstable for $h \geq 1$.

Review Problems, page 229

1. $c_1 e^{-9x} + c_2 e^x$.

3. $c_1 e^{x/2} \cos(3x/2) + c_2 e^{x/2} \sin(3x/2)$.

5. $c_1 e^{3x/2} + c_2 e^{x/3}$.

7. $c_1 e^{-x/3} \cos(x/6) + c_2 e^{-x/3} \sin(x/6)$.

9. $c_1 e^{7x/4} + c_2 x e^{7x/4}$.

11. $t^{1/2}\{c_1 \cos[(\sqrt{19}/2) \ln t] + c_2 \sin[(\sqrt{19}/2) \ln t]\}$.

13. $c_1 \cos 4x + c_2 \sin 4x + (1/17)x e^x - (2/289)e^x$.

15. $c_1 e^{-2x} + c_2 e^{-x} + c_3 e^{-x/3}$.

17. $c_1 e^x + c_2 e^{-x/2} \cos(\sqrt{43}x/2) + c_3 e^{-x/2} \sin(\sqrt{43}x/2)$.

19. $c_1 e^{-3x} + c_2 e^{x/2} + c_3 x e^{x/2}$.

21. $c_1 e^{3x/2} \cos(\sqrt{19}x/2) + c_2 e^{3x/2} \sin(\sqrt{19}x/2) - e^x/5$
$\quad + x^2 + 6x/7 + 4/49$.

23. $c_1 \cos 4\theta + c_2 \sin 4\theta$
$\quad - (1/16)(\cos 4\theta) \ln|\sec 4\theta + \tan 4\theta|$.

25. $c_1 e^{3x/2} + c_2 x e^{3x/2} + e^{3x}/9 + e^{5x}/49$.

27. $c_1 x + c_2 x^{-2} - 2x^{-2} \ln x + x \ln x$.

29. $e^{-2x} \cos(\sqrt{3}x)$.

31. $2e^x \cos 3x - (7/3)e^x \sin 3x - \sin 3x$.

33. $-e^{-x} - 3e^{5x} + e^{8x}$.

35. $\cos \theta + 2 \sin \theta + \theta \sin \theta + (\cos \theta) \ln|\cos \theta|$.

37. $c_1 e^{-x} + c_2(x - 1)$.

39. $c_1 x + c_2 \left[\dfrac{x}{2} \ln\left(\dfrac{1 + x}{1 - x} \right) - 1 \right]$.

41. $x = c_1 e^{\sqrt{3}t} + c_2 e^{-\sqrt{3}t}$; $y = \sqrt{3}c_1 e^{\sqrt{3}t} - \sqrt{3}c_2 e^{-\sqrt{3}t}$.

43. $x = c_1 e^{-t/5} + t - 4$; $y = (3/2)c_1 e^{-t/5} + t - 6$.

45. $x(t) = y(t) \equiv 0$.

CHAPTER 5

Exercises 5.1, page 245

1. $x(t) = (0.05) \cos 7t = (0.05) \sin(7t + \pi/2)$; 0.05;
$2\pi/7$; $7/(2\pi)$.

3. $x(t) = (1/4) \sin 4t - (1/2) \cos 4t =$
$(\sqrt{5}/4) \sin(4t + \phi)$, where
$\phi = -\arctan 2 \approx -1.107$; down at 0.823 ft/sec.

5. $x(t) = (1/10) \cos(7\sqrt{10}t/5)$
$\quad - (1/(14\sqrt{10})) \sin(7\sqrt{10}t/5)$
$\quad \approx (0.103) \sin(4.427t + 1.793)$; 1.369 sec.

7. $x(t) = (1/2) \cos 4t - (1/8) \sin 4t =$
$(\sqrt{17}/8) \sin(4t + \phi)$, where
$\phi = \pi - \arctan 4 \approx 1.816$; 0.331 sec.

Exercises 5.2, page 253

1. $b = 6$: $x(t) = e^{-3t} \cos\sqrt{7}t + (3/\sqrt{7})e^{-3t} \sin\sqrt{7}t =$
$(4/\sqrt{7})e^{-3t} \sin(\sqrt{7}t + \phi)$, where
$\phi = \arctan\sqrt{7}/3 \approx 0.723$.

Figure B.11

$b = 8$: $x(t) = (1 + 4t)e^{-4t}$.

Figure B.12

$b = 10$: $x(t) = (4/3)e^{-2t} - (1/3)e^{-8t}$.

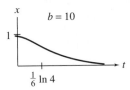

Figure B.13

3. $k = 20$: $x(t) = [(1 + \sqrt{5})/2]e^{(-5+\sqrt{5})t}$
$\qquad + [(1 - \sqrt{5})/2]e^{(-5-\sqrt{5})t}$.

Figure B.14

$k = 25$: $x(t) = (1 + 5t)e^{-5t}$.

Figure B.15

$k = 30$: $x(t) = e^{-5t}\cos\sqrt{5}t + \sqrt{5}e^{-5t}\sin\sqrt{5}t = \sqrt{6}e^{-5t}\sin(\sqrt{5}t + \phi)$, where
$\phi = \arctan(1/\sqrt{5}) \approx 0.421$.

Figure B.16

5. $x(t) = (-3/4)e^{-8t}\cos 8t - e^{-8t}\sin 8t = (5/4)e^{-8t}\sin(8t + \phi)$, where
$\phi = \pi + \arctan(3/4) \approx 3.785$;
damp factor $= (5/4)e^{-8t}$;
quasiperiod $= \pi/4$; quasifreq. $= 4/\pi$.

7. 0.242 m.

9. $(10/\sqrt{9999})\arctan(\sqrt{9999}) \approx 0.156$ sec.

11. Relative extrema at
$t = [\pi/3 + n\pi - \arctan(\sqrt{3}/2)]/(2\sqrt{3})$
for $n = 0, 1, 2, \ldots$; but touches curves $\pm\sqrt{7/12}\,e^{-2t}$
at $t = [\pi/2 + m\pi - \arctan(\sqrt{3}/2)]/(2\sqrt{3})$
for $m = 0, 1, 2, \ldots$.

13. First measure half the quasiperiod P as the time between two successive zero crossings. Then compute the ratio
$x(t + P)/x(t) = e^{-(b/2m)P}$.

Exercises 5.3, page 259

1. $x_p(t) = (0.08)\cos 2t + (0.06)\sin 2t$
$\qquad = (0.1)\sin(2t + \theta)$, where
$\theta = \arctan(4/3) \approx 0.927$.

3. $x(t) = -(18/85)e^{-2t}\cos 6t - (22/255)e^{-2t}\sin 6t$
$\qquad + (2/\sqrt{85})\sin(2t + \theta)$, where
$\theta = \arctan(9/2) \approx 1.352$;
res. freq. $= 2\sqrt{2}/\pi$ cycles/sec.

5. $M(\gamma) = 1/\sqrt{(1 - 4\gamma^2)^2 + 4\gamma^2}$.

Figure B.17

7. $x(t) = \cos 3t + (1/3)t\sin 3t$.

Figure B.18

9. (a) $x(t) = -[F_0/(k - m\gamma^2)]\cos(\sqrt{k/m}\,t)$
$\qquad + [F_0/(k - m\gamma^2)]\cos\gamma t$
$\qquad = (F_0/[m(\omega^2 - \gamma^2)])(\cos\gamma t - \cos\omega t)$.

(c) $x(t) = \sin 8t \sin t$.

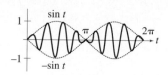

$$\frac{\sin t}{\underbrace{\qquad\qquad}_{-\sin t}}$$

Figure B.19

11. $x(t) = c_1 e^{r_1 t} + c_2 e^{r_2 t} + \dfrac{F_0 \sin(\gamma t + \theta)}{\sqrt{(k - m\gamma^2)^2 + b^2\gamma^2}}$,

where $r_1, r_2 = -(b/2m) \pm (1/2m)\sqrt{b^2 - 4mk}$
and $\tan\theta = (k - m\gamma^2)/(b\gamma)$ as in equation (7).

Exercises 5.4, page 266

1. $m_1 x'' = k_1(y - x)$,
$m_2 y'' = -k_1(y - x) - k_2 y$;
$x(0) = -1,\, x'(0) = 0,\, y(0) = 0,\, y'(0) = 0.$
$x(t) = -(8/17)\cos t - (9/17)\cos(\sqrt{20/3}\,t)$,
$y(t) = -(6/17)\cos t + (6/17)\cos(\sqrt{20/3}\,t)$.

3. $mx'' = -kx + k(y - x)$,
$my'' = -k(y - x) + k(z - y)$,
$mz'' = -k(z - y) - kz$;
The normal frequency, $(1/2\pi)\sqrt{(2 + \sqrt{2})(k/m)}$, has
the mode $x(t) = z(t) = -(1/\sqrt{2})\,y(t)$; the normal
frequency $(1/2\pi)\sqrt{(2 - \sqrt{2})(k/m)}$ has the mode
$x(t) = z(t) = (1/\sqrt{2})\,y(t)$; and the normal frequency
$(1/2\pi)\sqrt{2k/m}$ has the mode $x(t) = -z(t),\, y(t) \equiv 0.$

5. $x(t) = -e^{-t} - te^{-t} - \cos t;\; y(t) = e^{-t} + te^{-t} - \cos t.$

7. $x(t) = (2/5)\cos t + (4/5)\sin t - (2/5)\cos\sqrt{6}\,t$
$+ (\sqrt{6}/5)\sin\sqrt{6}\,t - \sin 2t;$
$y(t) = (4/5)\cos t + (8/5)\sin t + (1/5)\cos\sqrt{6}\,t$
$- (\sqrt{6}/10)\sin\sqrt{6}\,t - (1/2)\sin 2t.$

9. $(1/2\pi)\sqrt{g/l};\; (1/2\pi)\sqrt{(g/l) + (2k/m)}.$

Exercises 5.5, page 274

1. $I(t) = (19/\sqrt{21})\,[e^{(-25-5\sqrt{21})t/2} - e^{(-25+5\sqrt{21})t/2}].$

3. $I_p(t) = (4/51)\cos 20t - (1/51)\sin 20t$; resonance
frequency is $5/\pi$.

5. $M(\gamma) = 1/\sqrt{(100 - 4\gamma^2)^2 + 100\gamma^2}.$

Figure B.20

7. $L = 35$ henrys, $R = 10$ ohms, $C = 1/15$ farads, and
$E(t) = 50\cos 10t$ volts.

11. $I_1 = (3/5)e^{-3t/2} - (8/5)e^{-2t/3} + 1$,
$I_2 = (1/5)e^{-3t/2} - (6/5)e^{-2t/3} + 1$,
$I_3 = (2/5)e^{-3t/2} - (2/5)e^{-2t/3}.$

13. $(1/2)I_1' + 2q_3 = \cos 3t$ (where $I_3 = q_3'$),
$(1/2)I_1' + I_2 = 0,\, I_1 = I_2 + I_3$:
$I_1(0) = I_2(0) = I_3(0) = 0$;
$I_1 = -(36/61)e^{-t}\cos\sqrt{3}\,t$
$\quad - (42\sqrt{3}/61)e^{-t}\sin\sqrt{3}\,t + (36/61)\cos 3t$
$\quad + (30/61)\sin 3t$,
$I_2 = (45/61)e^{-t}\cos\sqrt{3}\,t - (39\sqrt{3}/61)e^{-t}\sin\sqrt{3}\,t$
$\quad - (45/61)\cos 3t + (54/61)\sin 3t$,
$I_3 = -(81/61)e^{-t}\cos\sqrt{3}\,t$
$\quad - (3\sqrt{3}/61)e^{-t}\sin\sqrt{3}\,t + (81/61)\cos 3t$
$\quad - (24/61)\sin 3t.$

Exercises 5.6, page 284

1. i	t_i	$y(t_i)$
1	0.250	0.96924
2	0.500	0.88251
3	0.750	0.75486
4	1.000	0.60656

3. i	t_i	$y(t_i)$
1	1.250	0.80761
2	1.500	0.71351
3	1.750	0.69724
4	2.000	0.74357

5. $y(8) \approx 24.01540.$ **7.** $y(1) \approx x_1(1, 2^{-3}) = 2.77493.$

9. $t_{n+1} = t_n + h,\; n = 0, 1, 2, \ldots;$
$$x_{n+1,i} = x_{n,i} + \frac{h}{2}\,[\,f_i(t_n, x_{n,1}, \ldots, x_{n,m})$$
$$+ f_i(t_n + h, x_{n,1} + hf_1(t_n, x_{n,1}, \ldots, x_{n,m}), \ldots,$$
$$x_{n,m} + hf_m(t_n, x_{n,1}, \ldots, x_{n,m}))],$$
$$i = 1, 2, \ldots, m.$$

11. i	t_i	$x_1(t_i) \approx H(t_i)$
1	0.5	0.09573
2	1.0	0.37389
3	1.5	0.81045
4	2.0	1.37361
5	2.5	2.03111
6	3.0	2.75497
7	3.5	3.52322
8	4.0	4.31970
9	4.5	5.13307
10	5.0	5.95554

13. Yes, yes.

15. $x(1) \approx 127.773;\; y(1) \approx -423.476.$

17. $u(1; 2^{-2}) = v(1; 2^{-2}) = 0.36789.$

19.

		Part (a)		Part (b)		Part (c)	
i	t_i	$x(t_i)$	$y(t_i)$	$x(t_i)$	$y(t_i)$	$x(t_i)$	$y(t_i)$
1	0.5	1.95247	2.25065	1.48118	2.42311	0.91390	2.79704
2	1.0	3.34588	1.83601	2.66294	1.45358	1.63657	1.13415
3	1.5	4.53662	3.36527	5.19629	2.40348	4.49334	1.07811
4	2.0	2.47788	4.32906	3.10706	4.64923	5.96115	5.47788
5	2.5	1.96093	2.71900	1.92574	3.32426	1.51830	5.93110
6	3.0	2.86412	1.96166	2.34143	2.05910	0.95601	2.18079
7	3.5	4.28449	2.77457	3.90106	2.18977	2.06006	0.98131
8	4.0	3.00965	4.11886	3.83241	3.89043	5.62642	1.38072
9	4.5	2.18643	3.14344	2.32171	3.79362	5.10594	5.10462
10	5.0	2.63187	2.25824	2.21926	2.49307	1.74187	5.02491

21.

i	t_i	$x(t_i)$	$y(t_i)$
1	0.250	-0.59733	1.15365
2	0.500	-1.58112	0.48705
3	0.750	-3.41862	-0.19129
4	1.000	-7.01992	-1.13136

23.

i	t_i	$x(t_i)$	$y(t_i)$
1	0.250	0.98661	0.75492
2	0.500	0.91749	1.04259
3	0.750	0.79823	1.38147
4	1.000	0.63303	1.77948

25.

i	t_i	$y(t_i)$
1	0.25	0.75755
2	0.50	0.55876
3	0.75	0.44552
4	1.00	0.46272

27.

i	t_i	$y(t_i)$
1	0.25	-0.72103
2	0.50	-0.39038
3	0.75	-0.01072
4	1.00	0.42498

29. $y(1) \approx x_1(1; 2^{-3}) = 1.25958.$

31. $y(0.1) \approx 0.00647, \ldots, y(2.0) \approx 1.60009.$

33. (a) $P_1(10) \approx 0.567, \ P_2(10) \approx 0.463, \ P_3(10) \approx 0.463.$
 (b) $P_1(10) \approx 0.463, \ P_2(10) \approx 0.567, \ P_3(10) \approx 0.463.$
 (c) $P_1(10) \approx 0.463, \ P_2(10) \approx 0.463, \ P_3(10) \approx 0.567.$
 All populations approach 0.5.

Exercises 5.7, page 297

1. $x = y^3, y > 0.$

Figure B.21

3. $(1/\sqrt{2}, 1/\sqrt{2}), \quad (-1/\sqrt{2}, -1/\sqrt{2}).$ **5.** $(0,0).$
7. $(1, 1)$ and line $y = 2.$ **9.** $e^x + ye^{-y} = c.$
11. $y = x/(\ln|x| + c).$ **13.** $e^x + xy - y^2 = c.$
15. $y^2 - x^2 = c.$ **17.** $(x - 1)^2 + (y - 1)^2 = c.$

Figure B.22

Figure B.23

19. $y = cx^{2/3}.$

Figure B.24

21. $(-2, 1)$ is unstable.

$(-2, 1)$

Figure B.25

23. $(0, 0)$ is stable.

Figure B.26

25. $(0, 0)$ is unstable; $(0, 5)$ is stable;
$(7, 0)$ is stable; $(3, 2)$ is unstable.

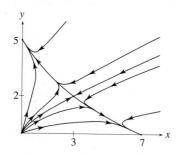

Figure B.27

27. $v^2 - x^2 = c$.

Figure B.28

29. $3v^2 + 3x^2 + x^6 = c$.

Figure B.29

31. $(0, 0)$ is stable; $(1, 0)$ is unstable.

Figure B.30

33. (a)

Figure B.31

(b) 295 persons.

35. (a)

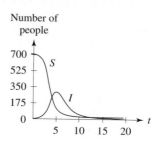

Figure B.32

(b) 4.9 days. **(c)** 18.3 days.

37. Critical point is $(0, 0)$; integral curves are
$2v^2 + 2x^2 + x^4 = c$; motion is periodic about
equilibrium position.

Figure B.33

41. Integral curves are $v^2 + ax^2 + 2ab \ln(C - x) = k$. When $C^2 - 4b < 0$, the bar is pulled over to the wire (see Figure B.34). When $C^2 - 4b > 0$, there is a critical point at $(C - \sqrt{C^2 - 4b}, 0)$, and the bar may oscillate (see closed trajectories in Figure B.35), or may be pulled over to the wire.

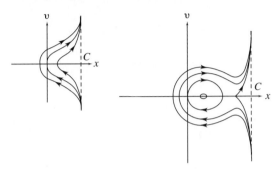

Figure B.34 **Figure B.35**

Exercises 5.8, page 308

1. For $\omega = 3/2$: The Poincaré map alternates between the points $(0.8, 1.5)$ and $(0.8, -1.5)$. There is a subharmonic solution of period 4π. For $\omega = 3/5$: The Poincaré map cycles through the points $(-1.5625, 0.6)$, $(-2.1503, -0.4854)$, $(-0.6114, 0.1854)$, $(-2.5136, 0.1854)$, and $(-0.9747, -0.4854)$. There is a subharmonic solution of period 10π.

3. The Poincaré map converges as $n \to \infty$ to the point $(-1.060065, 0.262366)$.

5. The attractor is the point $(0, 1.092)$.

9. **(a)** $\{1/7, 2/7, 4/7, 1/7, \ldots\}$, $\{3/7, 6/7, 5/7, 3/7, \ldots\}$.
 (b) $\{1/15, 2/15, 4/15, 8/15, 1/15, \ldots\}$, $\{1/5, 2/5, 4/5, 3/5, 1/5, \ldots\}$, $\{1/3, 2/3, 1/3, \ldots\}$, $\{7/15, 14/15, 13/15, 11/15, 7/15, \ldots\}$.
 (c) $x_n = 0$ for $n \geq j$.

13.

Figure B.36

CHAPTER 6

Exercises 6.1, page 328

1. $(-\infty, 0)$. **3.** $(3\pi/2, 5\pi/2)$. **5.** $(0, \infty)$.
7. Lin. indep.; $48e^{7x}$. **9.** Lin. dep.; 0.
11. Lin. indep.; $(3/2)x^{-5/2}$. **13.** Lin. indep.; $12x^4$.
15. $c_1e^{3x} + c_2e^{-x} + c_3e^{-4x}$. **17.** $c_1x + c_2x^2 + c_3x^3$.
19. **(a)** $c_1e^x + c_2e^{-x} \cos 2x + c_3e^{-x} \sin 2x + x^2$.
 (b) $-e^x + e^{-x} \sin 2x + x^2$.
21. **(a)** $c_1x + c_2x \ln x + c_3x(\ln x)^2 + \ln x$.
 (b) $3x - x \ln x + x(\ln x)^2 + \ln x$.
23. **(a)** $2 \sin x - x$. **(b)** $4x - 6 \sin x$.
29. **(b)** Let $f_1(x) = |x - 1|$ and $f_2(x) = x - 1$.

Exercises 6.2, page 334

1. $c_1 + c_2e^{2x} + c_3e^{-4x}$. **3.** $c_1e^{-x} + c_2e^{-2x/3} + c_3e^{x/2}$.
5. $c_1e^{-x} + c_2e^{-x} \cos 5x + c_3e^{-x} \sin 5x$.
9. $c_1e^{3x} + c_2xe^{3x} + c_3x^2e^{3x}$.
11. $c_1e^{-x} + c_2xe^{-x} + c_3x^2e^{-x} + c_4x^3e^{-x}$.
13. $c_1 \cos \sqrt{2}x + c_2x \cos \sqrt{2}x + c_3 \sin \sqrt{2}x + c_4x \sin \sqrt{2}x$.
15. $c_1e^x + c_2xe^x + c_3e^{-3x} + (c_4 + c_5x)e^{-x} \cos 2x$
 $+ (c_6 + c_7x)e^{-x} \sin 2x$.
17. $c_1e^{-4x} + c_2e^{3x} + (c_3 + c_4x + c_5x^2)e^{-2x}$
 $+ (c_6 + c_7x)e^{-2x} \cos x + (c_8 + c_9x)e^{-2x} \sin x + c_{10}$
 $+ c_{11}x + c_{12}x^2 + c_{13}x^3 + c_{14}x^4$.
19. $e^x - 2e^{-2x} - 3e^{2x}$. **21.** $e^{2x} - \sqrt{2}e^x \sin \sqrt{2}x$.
23. $x(t) = c_1 + c_2t + c_3e^t$, $y(t) = c_1 - c_2 + c_2t$.
27. $c_1e^{1.120x} + c_2e^{0.296x} + c_3e^{-0.520x} + c_4e^{-2.896x}$.
29. $c_1e^{-0.5x} \cos(0.866x) + c_2e^{-0.5x} \sin(0.866x)$
 $+ c_3e^{-0.5x} \cos(1.323x) + c_4e^{-0.5x} \sin(1.323x)$.
31. **(a)** $\{x, x^{-1}, x^2\}$. **(b)** $\{x, x^2, x^{-1}, x^{-2}\}$.
 (c) $\{x, x^2 \cos(3 \ln x), x^2 \sin(3 \ln x)\}$.
33. **(b)** $x(t) = c_1 \cos t + c_2 \sin t + c_3 \cos \sqrt{6}t$
 $+ c_4 \sin \sqrt{6}t$.
 (c) $y(t) = 2c_1 \cos t + 2c_2 \sin t - (c_3/2) \cos \sqrt{6}t$
 $- (c_4/2) \sin \sqrt{6}t$.
 (d) $x(t) = (3/5) \cos t + (2/5) \cos \sqrt{6}t$,
 $y(t) = (6/5) \cos t - (1/5) \cos \sqrt{6}t$.
35. $c_1e^{rx} + c_2e^{-rx} + c_3 \cos rx + c_4 \sin rx$, where
 $r^4 = k/(EI)$.

Exercises 6.3, page 341

1. $c_1xe^x + c_2 + c_3x + c_4x^2$. **3.** $c_1x^2e^{-2x}$.
5. $c_1e^x + c_2e^{3x} + c_3e^{-2x} - (1/6)xe^x + (1/6)x^2$
 $+ (5/18)x + 37/108$.
7. $c_1e^x + c_2e^{-2x} + c_3xe^{-2x} - (1/6)x^2e^{-2x}$.
9. $c_1e^x + c_2xe^x + c_3x^2e^x + (1/6)x^3e^x$.

11. D^5. **13.** $D + 7$. **15.** $(D - 2)(D - 1)$.

17. $[(D + 1)^2 + 4]^3$. **19.** $(D + 2)^2[(D + 5)^2 + 9]^2$.

21. $c_3 \cos 2x + c_4 \sin 2x + c_5$.

23. $c_3 x e^{3x} + c_4 x^2 + c_5 x + c_6$.

25. $c_3 + c_4 x + c_5 \cos 2x + c_6 \sin 2x$.

27. $c_3 x e^{-x} \cos x + c_4 x e^{-x} \sin x + c_5 x^2 + c_6 x + c_7$.

29. $c_2 x + c_3 x^2 + c_6 x^2 e^x$.

31. $-2e^{3x} + e^{-2x} + x^2 - 1$. **33.** $x^2 e^{-2x} - x^2 + 3$.

39. $x(t) = -(1/63)e^{3t} + c_1 + c_2 t - c_3 e^{\sqrt{2}t} - c_4 e^{-\sqrt{2}t}$,

 $y(t) = (8/63)e^{3t} + c_1 + c_2 t + c_3 e^{\sqrt{2}t} + c_4 e^{-\sqrt{2}t}$.

Exercises 6.4, page 345

1. $(1/6)x^2 e^{2x}$. **3.** $e^{2x}/16$.

5. $\ln(\sec x) - (\sin x)\ln(\sec x + \tan x)$.

7. $c_1 x + c_2 x^2 + c_3 x^3 - (1/24)x^{-1}$.

9. $-(1/2)e^x \int e^{-x} g(x)\,dx + (1/6)e^{-x} \int e^x g(x)\,dx$

 $+ (1/3)e^{2x} \int e^{-2x} g(x)\,dx$.

11. $c_1 x + c_2 x^{-1} + c_3 x^3 - x \sin x - 3 \cos x + 3x^{-1} \sin x$.

Review Problems, page 348

1. (a) $(0, \infty)$. **(b)** $(-4, -1), (-1, 1), (1, \infty)$.

5. (a) $e^{-5x}(c_1 + c_2 x) + e^{2x}(c_3 + c_4 x + c_5 x^2)$

 $+ (\cos x)(c_6 + c_7 x) + (\sin x)(c_8 + c_9 x)$.

 (b) $c_1 + c_2 x + c_3 x^2 + c_4 x^3 + e^x(c_5 + c_6 x)$

 $+ (e^{-x} \cos \sqrt{3} x)(c_7 + c_8 x)$

 $+ (e^{-x} \sin \sqrt{3} x)(c_9 + c_{10} x)$.

7. (a) D^3. **(b)** $D^2(D - 3)$. **(c)** $[D^2 + 4]^2$.

 (d) $[(D + 2)^2 + 9]^3$.

 (e) $D^3(D + 1)^2(D^2 + 4)(D^2 + 9)$.

9. $c_1 x + c_2 x^5 + c_3 x^{-1} - (1/21)x^{-2}$.

CHAPTER 7

Exercises 7.2, page 363

1. $\dfrac{1}{s^2}, \; s > 0$. **3.** $\dfrac{1}{s - 6}, \; s > 6$. **5.** $\dfrac{s}{s^2 + 4}, \; s > 0$.

7. $\dfrac{s - 2}{(s - 2)^2 + 9}, \; s > 2$. **9.** $e^{-2s}\left(\dfrac{2s + 1}{s^2}\right), \; s > 0$.

11. $\dfrac{e^{-\pi s} + 1}{s^2 + 1}$, all s. **13.** $\dfrac{6}{s + 3} - \dfrac{2}{s^3} + \dfrac{2}{s^2} - \dfrac{8}{s}, \; s > 0$.

15. $\dfrac{6}{s^4} - \dfrac{1}{(s - 1)^2} + \dfrac{s - 4}{(s - 4)^2 + 1}, \; s > 4$.

17. $\dfrac{6}{(s - 3)^2 + 36} - \dfrac{6}{s^4} + \dfrac{1}{s - 1}, \; s > 3$.

19. $\dfrac{24}{(s - 5)^5} - \dfrac{s - 1}{(s - 1)^2 + 7}, \; s > 5$.

21. Continuous (hence piecewise continuous).

23. Piecewise continuous.

25. Continuous (hence piecewise continuous).

27. Neither. **29.** All but functions (c), (e), and (h).

Exercises 7.3, page 369

1. $\dfrac{2}{s^3} + \dfrac{2}{(s - 1)^2 + 4}$. **3.** $\dfrac{s + 1}{(s + 1)^2 + 9} + \dfrac{1}{s - 6} - \dfrac{1}{s}$.

5. $\dfrac{4}{(s + 1)^3} - \dfrac{1}{s^2} + \dfrac{s}{s^2 + 16}$. **7.** $\dfrac{24}{s^5} - \dfrac{24}{s^4} + \dfrac{12}{s^3} - \dfrac{4}{s^2} + \dfrac{1}{s}$.

9. $\dfrac{4(s + 1)}{[(s + 1)^2 + 4]^2}$. **11.** $\dfrac{s}{s^2 - b^2}$.

13. $\dfrac{1}{2s} - \dfrac{s}{2(s^2 + 4)}$. **15.** $\dfrac{3s}{4(s^2 + 1)} + \dfrac{s}{4(s^2 + 9)}$.

17. $\dfrac{s}{2(s^2 + 9)} - \dfrac{s}{2(s^2 + 49)}$.

19. $\dfrac{n + m}{2[s^2 + (n + m)^2]} + \dfrac{m - n}{2[s^2 + (m - n)^2]}$.

21. $\dfrac{s - a}{(s - a)^2 + b^2}$. **23.** $\dfrac{s^2 + 2}{s(s^2 + 4)}$.

25. (a) $\dfrac{s^2 - b^2}{(s^2 + b^2)^2}$. **(b)** $\dfrac{2s^3 - 6sb^2}{(s^2 + b^2)^3}$.

29. $\dfrac{1}{s^2 + 6s + 10}$. **33.** e^{-s}/s^2. **35.** $e^{-(\pi/2)s}/(s^2 + 1)$.

Exercises 7.4, page 379

1. $e^t t^3$. **3.** $e^{-t} \cos 3t$. **5.** $(1/2)e^{-2t} \sin 2t$.

7. $2e^{-2t} \cos 3t + 4e^{-2t} \sin 3t$.

9. $(3/2)e^t \cos 2t - 3e^t \sin 2t$.

11. $\dfrac{6}{s + 5} - \dfrac{1}{s + 2} - \dfrac{4}{s - 1}$. **13.** $\dfrac{1}{(s + 1)^2} - \dfrac{2}{s}$.

15. $-\dfrac{3}{s + 1} + \dfrac{(s - 1) + 2}{(s - 1)^2 + 4}$.

17. $-\dfrac{5}{6s} + \dfrac{11}{10(s - 2)} - \dfrac{4}{15(s + 3)}$.

19. $\dfrac{1}{17}\left[\dfrac{1}{(s - 3)} - \dfrac{s + 1}{(s + 1)^2 + 1} - \dfrac{4}{(s + 1)^2 + 1}\right]$.

21. $(1/3) + e^t + (14/3)e^{6t}$. **23.** $-e^{-3t} + 2te^{-3t} + 6e^{-t}$.

25. $8e^{2t} - e^{-t} \cos 2t + 3e^{-t} \sin 2t$.

27. $-(5/3)e^{-t} + (5/12)e^{2t} + (5/4)e^{-2t}$.

29. $3e^{-2t} + 7e^t \cos t + 11e^t \sin t$.

31. $F_1(s) = F_2(s) = F_3(s) = 1/s^2$, $\mathcal{L}^{-1}\{1/s^2\}(t) = f_3(t) = t$.

33. $e^{5t}/t - e^{-2t}/t$. **35.** $2(\cos t - \cos 3t)/t$.

39. $\dfrac{A}{s} + \dfrac{B}{s-1} + \dfrac{C}{s+2}$, where

$$A = \dfrac{2s+1}{(s-1)(s+2)}\Big|_{s=0} = \dfrac{-1}{2},$$

$$B = \dfrac{2s+1}{s(s+2)}\Big|_{s=1} = 1,$$

$$C = \dfrac{2s+1}{s(s-1)}\Big|_{s=-2} = \dfrac{-1}{2}.$$

41. $2e^{-t} - 4e^{3t} + 5e^{2t}$. **43.** $\dfrac{4}{s+2} + \dfrac{2(s-1)+2(3)}{(s-1)^2+2^2}$.

Exercises 7.5, page 387

1. $2e^{t}\cos 2t + e^{t}\sin 2t$. **3.** $-e^{-3t} + 3te^{-3t}$.

5. $t^2 + \cos t - \sin t$. **7.** $\cos t - 4e^{5t} + 8e^{2t}$.

9. $3te^{t} - e^{-6t}$. **11.** $2 - t + e^{2-t} + 2e^{t-2}$.

13. $(7/5)\sin t + (11/5)\cos t + (3/5)e^{2t-\pi} - e^{-(t-\pi/2)}$.

15. $\dfrac{-s^2+s-1}{(s^2+1)(s-1)(s-2)}$. **17.** $\dfrac{s^5+s^4+6}{s^4(s^2+s-1)}$.

19. $\dfrac{s^3+5s^2-6s+1}{s(s-1)(s^2+5s-1)}$. **21.** $\dfrac{s^3+s^2+2s}{(s^2+1)(s-1)^2}$.

23. $\dfrac{-s^3+1+3se^{-2s}-e^{-2s}}{s^2(s^2+4)}$. **25.** $2e^{t} - \cos t - \sin t$.

27. $(t^2-4)e^{-t}$. **29.** $(3a-b)e^{t}/2 + (b-a)e^{3t}/2$.

31. $5/2 + (a-5/2)e^{-t}\cos t + (a+b-5/2)e^{-t}\sin t$.

35. $t^2/2$.

37. $\cos t + t\sin t + c(\sin t - t\cos t)$, ($c$ arbitrary).

39. $e(t) = -a\cos\sqrt{k/I}\,t$.

41. $e(t) = (-2aI/\sqrt{4Ik-\mu^2})e^{-\mu t/2I}\sin(\sqrt{4Ik-\mu^2}\,t/2I)$.

Exercises 7.6, page 399

1. $2e^{-s}/s^3$. **3.** $e^{-2s}(4s^2+4s+2)/s^3$.

5. $(2e^{-s} - e^{-2s} + 2e^{-3s})/s$. **7.** $[(e^{-s}-e^{-2s})(s+1)]/s^2$.

9. $(e^{-s} - 2e^{-2s} + e^{-3s})/s^2$. **11.** $e^{t-2}u(t-2)$.

13. $e^{-2(t-2)}u(t-2) - 3e^{-2(t-4)}u(t-4)$.

15. $e^{-2(t-3)}[\cos(t-3) - 2\sin(t-3)]u(t-3)$.

17. $(7e^{6-2t} - 6e^{3-t})u(t-3)$.

19. $10 - 10u(t-3\pi)[1 + e^{-(t-3\pi)}(\cos t + \sin t)]$
$+ 10u(t-4\pi)[1 - e^{-(t-4\pi)}(\cos t + \sin t)]$.

21. $\dfrac{1-2se^{-2s}-e^{-2s}}{s^2(1-e^{-2s})}$.

$f(t)$

Figure B.37

23. $\dfrac{1}{1-e^{-2s}}\left[\dfrac{1-e^{-s-1}}{s+1} + \dfrac{e^{-s}-e^{-2s}}{s}\right]$.

$f(t)$

Figure B.38

25. $\dfrac{1}{s(1+e^{-as})}$. **27.** $\dfrac{1-e^{-as}}{as^2(1+e^{-as})}$.

29. $\sin t + [1 - \cos(t-3)]u(t-3)$.

Figure B.39

31. $t + [4 - t + \sin(t-2) - 2\cos(t-2)]u(t-2)$.

Figure B.40

33. $e^{-t}\cos t + 2e^{-t}\sin t$
$+ (1/2)[1 - e^{2\pi-t}(\cos t + \sin t)]u(t-2\pi)$
$- (1/2)[1 - e^{4\pi-t}(\cos t + \sin t)]u(t-4\pi)$.

35. $e^{-t} + e^{-2t} + (1/2)[e^{-3t} + 2e^{-2(t+1)} - e^{-(t+4)}]u(t-2)$.

37. $\cos 2t + (1/3)[1 - u(t-2\pi)]\sin t$
$+ (1/6)[8 + u(t-2\pi)]\sin 2t$.

39. $2e^{-2t} - 2e^{-3t}$
$+ [1/36 + (1/6)(t-1) - (1/4)e^{-2(t-1)}$
$+ (2/9)e^{-3(t-1)}]u(t-1)$
$- [19/36 + (1/6)(t-5)$
$- (7/4)e^{-2(t-5)} + (11/9)e^{-3(t-5)}]u(t-5)$.

45. $\dfrac{e^{t-n}}{6} - \dfrac{e^{-t}}{2}\left[\dfrac{e^{n+1}-1-e^{n+2}+e^2}{e-1}\right]$
$+ \dfrac{e^{-2t}}{3}\left[\dfrac{e^{2n+2}-1-e^{2n+3}+e^3}{e^2-1}\right]$, for $n < t < n+1$.

47. $\displaystyle\sum_{n=1}^{\infty}\dfrac{1}{s^n} = \dfrac{1}{s-1}$. **49.** $\displaystyle\sum_{n=1}^{\infty}\dfrac{(-1)^{n+1}}{2ns^{2n}} = \dfrac{1}{2}\ln\left(1 + \dfrac{1}{s^2}\right)$.

51. (a) $\sqrt{\pi/s}$. **(b)** $105\sqrt{\pi}/(16s^{9/2})$.

59. $e^t - e^{-t} + (1/2)[e^{t-1} + e^{1-t} - 2]u(t - 1)$
 $- (1/2)[e^{t-4} + e^{4-t} - 2]u(t - 4).$

61. $4 - 2e^{-3t/125} + 2u(t - 10)[1 - e^{-3(t-10)/125}].$

63. The resulting differential equation has variable coefficients, so the Laplace transform method will result in a differential equation for the transform.

Exercises 7.7, page 409

1. $2te^t - e^t + \int_0^t e^{t-v}(t - v)g(v)\,dv.$

3. $\int_0^t g(v)e^{2v-2t}\sin(t - v)\,dv + e^{-2t}\cos t + 3e^{-2t}\sin t.$

5. $1 - \cos t.$ **7.** $2e^{5t} - 2e^{-2t}.$ **9.** $(t/2)\sin t.$

11. $(2/3)e^{-2t} + (1/3)e^t.$ **13.** $s^{-2}(s - 3)^{-1}.$

15. $t/4 + (3/8)\sin 2t.$ **17.** $\cos t.$ **19.** $3.$

21. $e^{-t/2}\cos(\sqrt{3}\,t/2) - (1/\sqrt{3})e^{-t/2}\sin(\sqrt{3}\,t/2).$

23. $H(s) = (s^2 + 9)^{-1};$ $h(t) = (1/3)\sin 3t;$
 $y_k(t) = 2\cos 3t - \sin 3t;$
 $y(t) = (1/3)\int_0^t (\sin 3(t - v))g(v)\,dv + 2\cos 3t - \sin 3t.$

25. $H(s) = (s^2 - s - 6)^{-1};$ $h(t) = (e^{3t} - e^{-2t})/5;$
 $y_k(t) = 2e^{3t} - e^{-2t};$
 $y(t) = (1/5)\int_0^t [e^{3(t-v)} - e^{-2(t-v)}]g(v)\,dv + 2e^{3t} - e^{-2t}.$

27. $H(s) = (s^2 - 2s + 5)^{-1};$ $h(t) = (1/2)e^t\sin 2t;$
 $y_k(t) = e^t\sin 2t;$
 $y(t) = (1/2)\int_0^t e^{(t-v)}(\sin 2(t - v))g(v)\,dv + e^t\sin 2t.$

29. $(1/30)\int_0^t e^{-2(t-v)}(\sin 6(t - v))e(v)\,dv - e^{-2t}\cos 6t$
 $+ e^{-2t}\sin 6t.$

31. $t^2/2.$

Exercises 7.8, page 417

1. $-1.$ **3.** $-1.$ **5.** $e^{-2}.$ **7.** $e^{-s} - e^{-3s}.$

9. $e^{-s}.$ **11.** $0.$ **13.** $-(\sin t)u(t - \pi).$

15. $e^t + e^{-3t} + (1/4)(e^{t-1} - e^{3-3t})u(t - 1)$
 $- (1/4)(e^{t-2} - e^{6-3t})u(t - 2).$

17. $2(e^{t-2} - e^{-(t-2)})u(t - 2) + 2e^t - t^2 - 2.$

19. $e^{-t} - e^{-5t} + (e/4)(e^{1-t} - e^{5-5t})u(t - 1).$

21. $\sin t + (\sin t)u(t - 2\pi).$

Figure B.41

23. $\sin t + (\sin t)u(t - \pi) + (\sin t)u(t - 2\pi).$

Figure B.42

25. $(1/2)e^{-2t}\sin 2t.$ **27.** $(1/2)e^t\sin 2t.$

29. The mass remains stopped at $x(t) \equiv 0,$ $t > \pi/2.$

35. $\dfrac{L}{6EI}[3\lambda x^2 - x^3 + (x - \lambda)^3 u(x - \lambda)].$

Exercises 7.9, page 420

1. $x = e^t;$ $y = e^t.$ **3.** $z = e^t;$ $w \equiv 0.$

5. $x = (7/4)e^t + (7/4)e^{-t} - (3/2)\cos t;$
 $y = (7/4)e^t - (7/4)e^{-t} + (1/2)\sin t.$

7. $x = -(150/17)e^{5t/2}\cos(\sqrt{15}\,t/2)$
 $- (334\sqrt{15}/85)e^{5t/2}\sin(\sqrt{15}\,t/2) - (3/17)e^{-3t};$
 $y = (46/17)e^{5t/2}\cos(\sqrt{15}\,t/2)$
 $- (146\sqrt{15}/85)e^{5t/2}\sin(\sqrt{15}\,t/2) + (22/17)e^{-3t}.$

9. $x = 4e^{-2t} - e^{-t} - \cos t;$ $y = 5e^{-2t} - e^{-t}.$

11. $x = (e^t - e^{-t})/2 - (1/2)[e^{t-2} - e^{-(t-2)}]u(t - 2);$
 $y = 1 - (e^t + e^{-t})/2 - [1 - (e^{t-2} + e^{-(t-2)})/2]u(t - 2).$

13. $x = e^{-t} + (1/2)[e^{-(t-\pi)} + \cos t - \sin t]u(t - \pi);$
 $y = e^{-t} - [1 - (1/2)e^{-(t-\pi)} + (1/2)\cos t$
 $+ (1/2)\sin t]u(t - \pi).$

15. $x = t^2;$ $y = t - 1.$

17. $x = (t - 2)e^{t-2};$ $y = e^{t-2}.$

19. $x = -7e^{-t} + e^t;$ $y = 2e^{-t};$ $z = -13e^{-t} + e^t.$

21. $2I_1 + (0.1)I_3' + (0.2)I_1' = 6,$ $(0.1)I_3' - I_2 = 0,$
 $I_1 = I_2 + I_3;$ $I_1(0) = I_2(0) = I_3(0) = 0;$
 $I_1 = -e^{-20t} - 2e^{-5t} + 3,$
 $I_2 = -2e^{-20t} + 2e^{-5t},$
 $I_3 = e^{-20t} - 4e^{-5t} + 3.$

Review Problems, page 423

1. $\dfrac{3}{s} + e^{-2s}\left[\dfrac{1}{s} - \dfrac{1}{s^2}\right].$ **3.** $\dfrac{2}{(s + 9)^3}.$

5. $\dfrac{1}{s - 2} - \dfrac{6}{s^4} + \dfrac{2}{s^3} - \dfrac{5}{s^2 + 25}.$ **7.** $\dfrac{s^2 - 36}{(s^2 + 36)^2}.$

9. $2e^{-4s}\left[\dfrac{1}{s^3} + \dfrac{4}{s^2} + \dfrac{8}{s}\right].$ **11.** $(7/2)t^2e^{-3t}$

13. $2e^t + 2e^{-2t}\cos 3t + e^{-2t}\sin 3t.$

15. $e^{-2t} + e^{-t} - 2te^{-t}.$ **17.** $[2e^{t-2} + 2e^{4-2t}]u(t - 2).$

19. $e^{2t} - e^{5t}.$ **21.** $-(3/2) + t + t^2/2 + (3/2)e^{-t}\cos t.$

23. $(2/\sqrt{7})e^{-3t/2}\sin(\sqrt{7}\,t/2)$
$+\{(1/4)-(3/(4\sqrt{7}))e^{-3(t-1)/2}\sin(\sqrt{7}\,(t-1)/2)$
$-(1/4)e^{-3(t-1)/2}\cos(\sqrt{7}\,(t-1)/2)\}u(t-1)$.

25. $c[t+te^{-2t}+e^{-2t}-1]$.

27. $(9/10)e^{-3t}+(1/10)\cos t-(3/10)\sin t$.

29. $(s^2-5s+6)^{-1}$; $e^{3t}-e^{2t}$.

CHAPTER 8

Exercises 8.2, page 440

1. $[-1,3)$. **3.** $(-4,0)$. **5.** $[1,3]$.

9. $\displaystyle\sum_{n=0}^{\infty}\left[\frac{1}{n+1}+2^{-n-1}\right]x^n$. **11.** $x+x^2+(1/3)x^3+\cdots$.

13. $1-2x+(5/2)x^2+\cdots$.

15. (c) $1-(1/2)x+(1/4)x^2-(1/24)x^3+\cdots$.

17. $\displaystyle\sum_{n=1}^{\infty}(-1)^n nx^{n-1}$. **19.** $\displaystyle\sum_{n=1}^{\infty}a_n nx^{n-1}$.

21. $\ln(x+1)=\displaystyle\sum_{n=0}^{\infty}\frac{(-1)^n}{n+1}x^{n+1}$. **23.** $\displaystyle\sum_{k=0}^{\infty}(k+1)a_{k+1}x^k$.

25. $\displaystyle\sum_{k=1}^{\infty}a_{k-1}x^k$. **29.** $\displaystyle\sum_{n=0}^{\infty}\frac{(-1)^{n+1}(x-\pi)^{2n}}{(2n)!}$.

31. $1+\displaystyle\sum_{n=1}^{\infty}2x^n$. **33.** $6(x-1)+3(x-1)^2+(x-1)^3$.

35. (a) $\displaystyle\sum_{n=0}^{\infty}(-1)^n(x-1)^n$. (b) $\displaystyle\sum_{n=1}^{\infty}\frac{(-1)^{n-1}}{n}(x-1)^n$.

37. $1+x+x^2+\cdots$. **39.** $x+x^2+(1/2)x^3+\cdots$.

41. $1-(1/6)t^3+(1/180)t^6+\cdots$.

43. $(1/6)\theta^3-(1/120)\theta^5+(1/5040)\theta^7+\cdots$.

47. $t+(1/2)t^2-(1/6)t^3+\cdots$.

Exercises 8.3, page 452

1. -1. **3.** $\pm\sqrt{2}$. **5.** $-1,2$.

7. $x=n\pi$, n an integer.

9. $\theta\le 0$ and $\theta=n\pi$, $n=1,2,3,\ldots$.

11. $y=a_0(1-2x+(3/2)x^2-x^3/3+\cdots)$.

13. $a_0(1+x^4/12+\cdots)+a_1(x+x^5/20+\cdots)$.

15. $a_0(1-x^2/2-x^3/6+\cdots)+a_1(x+x^2/2-x^3/6+\cdots)$.

17. $a_0(1-x^2/2+\cdots)+a_1(x-x^3/6+\cdots)$.

19. $a_0\displaystyle\sum_{n=0}^{\infty}\frac{1}{n!}x^{2n}$.

21. $a_0(1-2x^2+x^4/3)$
$+a_1\left(x+\displaystyle\sum_{k=1}^{\infty}\frac{(-3)(-1)\cdots(2k-5)}{(2k+1)!}x^{2k+1}\right)$.

23. $a_{3k+2}=0,\ k=0,1,\ldots$.
$a_0\left(1+\displaystyle\sum_{k=1}^{\infty}\frac{[1\cdot 4\cdot 7\cdots(3k-2)]^2}{(3k)!}x^{3k}\right)$
$+a_1\left(x+\displaystyle\sum_{k=1}^{\infty}\frac{[2\cdot 5\cdot 8\cdots(3k-1)]^2}{(3k+1)!}x^{3k+1}\right)$.

25. $2+x^2-(5/12)x^4+(11/72)x^6+\cdots$.

27. $x+(1/6)x^3-(1/12)x^4+(7/120)x^5+\cdots$.

29. $1-2x+x^2-(1/6)x^3$.

31. $1+2x-(3/4)x^2-(5/6)x^3$.

35. $10-25t^2+(250/3)t^3-(775/4)t^4+\cdots$.

Exercises 8.4, page 458

1. 2. **3.** $\sqrt{2}$. **5.** $\pi/2$.

7. $a_0[1-(x-1)^2+(1/2)(x-1)^4-(1/6)(x-1)^6+\cdots]$.

9. $a_0[1+(x-1)^2+\cdots]$
$+a_1[(x-1)+(1/3)(x-1)^3+\cdots]$.

11. $a_0[1-(1/8)(x-2)^2+(5/96)(x-2)^3+\cdots]$
$+a_1[(x-2)+(1/8)(x-2)^2-(1/96)(x-2)^3+\cdots]$.

13. $1-(1/2)t^2+(1/6)t^4-(31/720)t^5+\cdots$.

15. $1+x+(1/24)x^4+(1/60)x^5+\cdots$.

17. $1-(1/6)(x-\pi)^3+(1/120)(x-\pi)^5$
$+(1/180)(x-\pi)^6+\cdots$.

19. $-1+x+x^2+(1/2)x^3+\cdots$.

21. $a_0[1+(1/2)x^2+(1/8)x^4+(1/48)x^6+\cdots]$
$+[(1/2)x^2+(1/12)x^4+(11/720)x^6+\cdots]$.

23. $a_0[1-(1/2)x^2+\cdots]+a_1[x-(1/3)x^3+\cdots]$
$+[(1/2)x^2+(1/3)x^3+\cdots]$.

25. $a_0[1-(1/2)x^2+\cdots]+a_1[x+\cdots]$
$+[(1/2)x^2-(1/6)x^3+\cdots]$.

27. $a_0[1-(1/2)x^2-(1/6)x^3+\cdots]$
$+a_1[x+(1/2)x^2+\cdots]+[(1/6)x^3+\cdots]$.

29. (a) $a_0\left[1+\displaystyle\sum_{k=1}^{\infty}(-1)^k\frac{n(n-2)(n-4)\cdots(n-2k+2)(n+1)(n+3)\cdots(n+2k-1)}{(2k)!}x^{2k}\right]$
$+a_1\left[x+\displaystyle\sum_{k=1}^{\infty}(-1)^k\frac{(n-1)(n-3)\cdots(n-2k+1)(n+2)(n+4)\cdots(n+2k)}{(2k+1)!}x^{2k+1}\right]$.

(c) $P_0(x) = 1$, $P_1(x) = x$, $P_2(x) = (1/2)(3x^2 - 1)$.

31. (a) $1 - (1/2)t^2 + (\eta/6)t^3 + [(1 - \eta^2)/24]t^4 + \cdots$.

(b) Yes.

Exercises 8.5, page 462

1. $c_1 x^{-2} + c_2 x^{-3}$. **3.** $c_1 x \cos(4 \ln x) + c_2 x \sin(4 \ln x)$.

5. $c_1 x^3 \cos(2 \ln x) + c_2 x^3 \sin(2 \ln x)$.

7. $c_1 x + c_2 x^{-1} \cos(3 \ln x) + c_3 x^{-1} \sin(3 \ln x)$.

9. $c_1 x + c_2 x^{-1/2} \cos[(\sqrt{19}/2) \ln x]$
$+ c_3 x^{-1/2} \sin[(\sqrt{19}/2) \ln x]$.

11. $c_1(x - 3)^{-2} + c_2(x - 3)^{1/2}$.

13. $c_1 x + c_2 x^2 + (4/15)x^{-1/2}$. **15.** $2t^4 + t^{-3}$.

17. $(31/17)x + (3/17)x^{-2} \cos(5 \ln x)$
$- (76/85)x^{-2} \sin(5 \ln x)$.

Exercises 8.6, page 474

1. ± 1 are regular. **3.** $\pm i$ are regular.

5. 1 is regular, -1 is irregular.

7. 2 is regular, -1 is irregular.

9. -4 is regular and 2 is irregular.

11. $r^2 - 3r - 10 = 0$; $r_1 = 5$, $r_2 = -2$.

13. $r^2 - 5r/9 - 4/3 = 0$; $r_1 = (5 + \sqrt{457})/18$,
$r_2 = (5 - \sqrt{457})/18$.

15. $r^2 - 1 = 0$; $r_1 = 1$, $r_2 = -1$.

17. $r^2 + r - 12 = 0$; $r_1 = 3$, $r_2 = -4$.

19. $a_0[x^{2/3} - (1/2)x^{5/3} + (5/28)x^{8/3} - (1/21)x^{11/3} + \cdots]$.

21. $a_0[1 - (1/4)x^2 + (1/64)x^4 - (1/2304)x^6 + \cdots]$.

23. $a_0[x - (1/3)x^2 + (1/12)x^3 - (1/60)x^4 + \cdots]$.

25. $a_0 \sum_{n=0}^{\infty} \dfrac{(-1)^n x^{n+(3/2)}}{2^{n-1}(n+2)!}$. **27.** $a_0 \sum_{n=0}^{\infty} \dfrac{x^{2n+2}}{2^{2n}(n+1)!\,n!}$.

29. $a_0 x^2$. **31.** $a_0 \sum_{n=0}^{\infty} \dfrac{x^n}{n!} = a_0 e^x$; Yes, $a_0 < 0$.

33. $a_0 \left[x^{1/3} + \sum_{n=1}^{\infty} \dfrac{2^{n-2}(3n+4)x^{n+(1/3)}}{3^n n!} \right]$; Yes, $a_0 < 0$.

35. $a_0[x^{5/6} - (1/11)x^{11/6} + (1/374)x^{17/6}$
$- (1/25,806)x^{23/6} + \cdots]$.

37. $a_0[x^{4/3} + (1/17)x^{7/3} + (1/782)x^{10/3}$
$+ (7/88,442)x^{13/3} + \cdots]$.

39. The expansion $\sum_{n=0}^{\infty} n!\,x^n$ diverges for $x \neq 0$.

41. The transformed equation is
$z\,d^2y/dz^2 + 3\,dy/dz - y = 0$. Also $zp(z) = 3$ and
$z^2 q(z) = -z$ are analytic at $z = 0$; hence $z = 0$ is a
regular singular point.
$$y_1(x) = a_0[1 + (1/3)x^{-1} + (1/24)x^{-2} + (1/360)x^{-3} + \cdots].$$

45. $a_0[x^{-3} - (1/2)x^{-1} - (1/8)x - (1/144)x^3 + \cdots]$
$+ a_3[1 + (1/10)x^2 + (1/280)x^4$
$+ (1/15,120)x^6 + \cdots]$.

Exercises 8.7, page 486

1. $c_1 y_1(x) + c_2 y_2(x)$, where
$y_1(x) = x^{2/3} - (1/2)x^{5/3} + (5/28)x^{8/3} + \cdots$ and
$y_2(x) = x^{1/3} - (1/2)x^{4/3} + (1/5)x^{7/3} + \cdots$.

3. $c_1 y_1(x) + c_2 y_2(x)$, where
$y_1(x) = 1 - (1/4)x^2 + (1/64)x^4 + \cdots$ and
$y_2(x) = y_1(x) \ln x + (1/4)x^2 - (3/128)x^4$
$+ (11/13,824)x^6 + \cdots$.

5. $c_1[x - (1/3)x^2 + (1/12)x^3 + \cdots] + c_2[x^{-1} - 1]$.

7. $c_1 y_1(x) + c_2 y_2(x)$, where
$y_1(x) = x^{3/2} - (1/6)x^{5/2} + (1/48)x^{7/2} + \cdots$ and
$y_2(x) = x^{-1/2} - (1/2)x^{1/2}$.

9. $c_1 w_1 + c_2 w_2$, where
$w_1(x) = x^2 + (1/8)x^4 + (1/192)x^6 + \cdots$ and
$w_2(x) = w_1(x) \ln x + 2 - (3/32)x^4 - (7/1728)x^6 + \cdots$.

11. $c_1 y_1(x) + c_2 y_2(x)$, where $y_1(x) = x^2$ and
$y_2(x) = x^2 \ln x - 1 + 2x - (1/3)x^3 + (1/24)x^4$.

13. $c_1 y_1(x) + c_2 y_2(x)$, where
$y_1(x) = 1 + x + (1/2)x^2 + \cdots$ and
$y_2(x) = y_1(x) \ln x - [x + (3/4)x^2 + (11/36)x^3 + \cdots]$.

15. $c_1 y_1(x) + c_2 y_2(x)$, where
$y_1(x) = x^{1/3} + (7/6)x^{4/3} + (5/9)x^{7/3} + \cdots$ and
$y_2(x) = 1 + 2x + (6/5)x^2 + \cdots$; all solutions are
bounded near the origin.

17. $c_1 y_1(x) + c_2 y_2(x) + c_3 y_3(x)$, where
$y_1(x) = x^{5/6} - (1/11)x^{11/6} + (1/374)x^{17/6} + \cdots$,
$y_2(x) = 1 - x + (1/14)x^2 + \cdots$, and
$y_3(x) = y_2(x) \ln x + 7x - (117/196)x^2$
$+ (4997/298116)x^3 + \cdots$.

19. $c_1 y_1(x) + c_2 y_2(x) + c_3 y_3(x)$, where
$y_1(x) = x^{4/3} + (1/17)x^{7/3} + (1/782)x^{10/3} + \cdots$,
$y_2(x) = 1 - (1/3)x - (1/30)x^2 + \cdots$, and
$y_3(x) = x^{-1/2} - (1/5)x^{1/2} - (1/10)x^{3/2} + \cdots$.

21. (b) $c_1 y_1(t) + c_2 y_2(t)$, where
$y_1(t) = 1 - (1/6)(\alpha t)^2 + (1/120)(\alpha t)^4 + \cdots$, and
$y_2(t) = t^{-1}[1 - (1/2)(\alpha t)^2 + (1/24)(\alpha t)^4 + \cdots]$.
(c) $c_1 y_1(x) + c_2 y_2(x)$, where
$y_1(x) = 1 - (1/6)(\alpha/x)^2 + (1/120)(\alpha/x)^4 + \cdots$, and
$y_2(x) = x[1 - (1/2)(\alpha/x)^2 + (1/24)(\alpha/x)^4 + \cdots]$.

23. $c_1 y_1(x) + c_2 y_2(x)$, where
$y_1(x) = 1 + 2x + 2x^2$ and
$y_2(x) = -(1/6)x^{-1} - (1/24)x^{-2} - (1/120)x^{-3} + \cdots$.

25. For $n = 0$, $c_1 + c_2[\ln x + x + (1/4)x^2 + (1/18)x^3 + \cdots]$
and for $n = 1$,
$c_1(1 - x) + c_2[(1 - x) \ln x + 3x - (1/4)x^2$
$- (1/36)x^3 + \cdots]$.

Exercises 8.8, page 492

1. $y_1(x) = \sum_{n=0}^{\infty} \dfrac{4^n x^n}{(n!)^2}$ and

$y_2(x) = y_1(x) \ln x$

$\qquad - \sum_{n=1}^{\infty} \dfrac{2^{2n+1}}{(n!)^2} \left[1 + \dfrac{1}{2} + \dfrac{1}{3} + \cdots + \dfrac{1}{n} \right] x^n.$

3. $y_1(x) = \sum_{n=0}^{\infty} \dfrac{(-1)^n x^{n+1}}{n!}$ and

$y_2(x) = y_1(x) \ln x$

$\qquad - \sum_{n=1}^{\infty} \dfrac{(-1)^n}{n!} \left[1 + \dfrac{1}{2} + \dfrac{1}{3} + \cdots + \dfrac{1}{n} \right] x^{n+1}.$

5. $y_1(x) = x^{-1} + 3$ and

$y_2(x) = y_1(x) \ln x - \left[9 + \sum_{n=2}^{\infty} \dfrac{(-3)^n x^{n-1}}{n(n-1)n!} \right].$

7. $y_1(x) = \sum_{n=0}^{\infty} \dfrac{3^n x^n}{(n!)^2}$ and

$y_2(x) = y_1(x) \ln x - \sum_{n=1}^{\infty} \dfrac{2 \cdot 3^n}{(n!)^2} \left[1 + \dfrac{1}{2} + \dfrac{1}{3} + \cdots + \dfrac{1}{n} \right] x^n.$

9. **(c)** $a_0[x^{-3} - (1/2)x^{-1} - (1/8)x - (1/144)x^3 + \cdots]$
$\qquad + a_3[1 + (1/10)x^2 + (1/280)x^4 + \cdots].$

11. $y_1(x) = 1 + (2/3)x + (1/3)x^2$ and

$y_2(x) = \sum_{n=4}^{\infty} (n-3)x^n.$

13. $y_1(x) = 2 + x$ and

$y_2(x) = y_1(x) \ln x + x^{-1} - 3 - 4x$

$\qquad + \sum_{n=3}^{\infty} \dfrac{2(-1)^n x^{n-1}}{n!(n-1)(n-2)}.$

15. $y_1(x) = x^2$ and

$y_2(x) = y_1(x) \ln x - 1 + 2x + 2 \sum_{n=3}^{\infty} \dfrac{(-x)^n}{n!(n-2)}.$

17. $y_1(x) = x^{1/2} + \sum_{n=1}^{\infty} \dfrac{(4n^2 - 8n + 7) \cdots 39 \cdot 19 \cdot 7 \cdot 3}{[(2n+2)(2n-1) \cdots 5 \cdot 3]^2 2^n n!} x^{n+1/2},$

$y_2(x) = 1 + \sum_{n=1}^{\infty} a_n(0)x^n,$

$y_3(x) = y_2(x) \ln x + \sum_{n=1}^{\infty} a_n(0) \left\{ \left[\dfrac{2n-3}{n^2 - 3n + 3} + \dfrac{2n-5}{(n-1)^2 - 3(n-1) + 3} + \cdots + \dfrac{1}{1} + \dfrac{-1}{1} \right] \right.$

$\qquad \left. - 2 \left[\dfrac{1}{n} + \dfrac{1}{n-1} + \cdots + \dfrac{1}{1} \right] - 2 \left[\dfrac{1}{2n-1} + \dfrac{1}{2n-3} + \cdots + \dfrac{1}{3} + \dfrac{1}{1} \right] \right\} x^n, \quad \text{where}$

$a_n(0) = \dfrac{(n^2 - 3n + 3) \cdots 7 \cdot 3 \cdot 1 \cdot 1}{[(2n-1)(2n-3) \cdots 3 \cdot 1](n!)^2}.$

19. $y_1(x) = \sum_{n=0}^{\infty} \dfrac{x^n}{(n!)^3},$

$y_2(x) = y_1(x) \ln x - 3 \sum_{n=1}^{\infty} \dfrac{1}{(n!)^3} \left[1 + \dfrac{1}{2} + \dfrac{1}{3} + \cdots + \dfrac{1}{n} \right] x^n,$

$y_3(x) = y_1(x) (\ln x)^2 - 6(\ln x) \sum_{n=1}^{\infty} \dfrac{1}{(n!)^3} \left[1 + \dfrac{1}{2} + \dfrac{1}{3} + \cdots + \dfrac{1}{n} \right] x^n$

$\qquad + \sum_{n=1}^{\infty} \dfrac{3}{(n!)^3} \left[1 + \dfrac{1}{2} + \dfrac{1}{3} + \cdots + \dfrac{1}{n} \right] \left[1 + \dfrac{1}{2^2} + \dfrac{1}{3^2} + \cdots + \dfrac{1}{n^2} \right] x^n.$

21. $a_{2n+1} \equiv 0$ and

$a_{2n} = \dfrac{(2n + 2 + r)(2 + r)r}{[(2n + 2 + r)(2n + r) \cdots (2 + r)]^2}.$

Exercises 8.9, page 502

1. $c_1 F\left(1, 2; \dfrac{1}{2}; x\right) + c_2 x^{1/2} F\left(\dfrac{3}{2}, \dfrac{5}{2}; \dfrac{3}{2}; x\right).$

3. $c_1 F\left(1, 1; \dfrac{1}{2}; x\right) + c_2 x^{1/2} F\left(\dfrac{3}{2}, \dfrac{3}{2}; \dfrac{3}{2}; x\right).$

5. $F(1, 1; 2; x) = \sum\limits_{n=0}^{\infty} \dfrac{1}{(n+1)} x^n = -x^{-1} \ln(1-x).$

7. $F\left(\dfrac{1}{2}, 1; \dfrac{3}{2}; x^2\right) = \sum\limits_{n=0}^{\infty} (1/2)_n x^{2n}/(3/2)_n$

$\qquad = \sum\limits_{n=0}^{\infty} x^{2n}/(2n+1)$

$\qquad = \dfrac{1}{2} x^{-1} \ln\left(\dfrac{1+x}{1-x}\right).$

9. $(1-x)^{-1}, \quad (1-x)^{-1} \ln x.$

13. $c_1 J_{1/2}(x) + c_2 J_{-1/2}(x).$ **15.** $c_1 J_1(x) + c_2 Y_1(x).$

17. $c_1 J_{2/3}(x) + c_2 J_{-2/3}(x).$

19. $J_1(x) \ln x - x^{-1} + (3/64)x^3 - (7/2304)x^5 + \cdots.$

21. $x^{3/2} J_{3/2}(x).$

27. $J_0(x) \ln x + \sum\limits_{n=1}^{\infty} \dfrac{(-1)^{n+1}}{2^{2n}(n!)^2} \left[1 + \dfrac{1}{2} + \cdots + \dfrac{1}{n}\right] x^{2n}.$

29. $1, \quad x, \quad (3x^2 - 1)/2,$
$\quad (5x^3 - 3x)/2, \quad (35x^4 - 30x^2 + 3)/8.$

37. $1, \quad 2x, \quad 4x^2 - 2, \quad 8x^3 - 12x.$

39. $1, \quad 1 - x, \quad (2 - 4x + x^2)/2,$
$\quad (6 - 18x + 9x^2 - x^3)/6.$

41. (b) $z = c_1 \cos x + c_2 \sin x, \quad$ so for x large,
$\quad y(x) \approx c_1 x^{-1/2} \cos x + c_2 x^{-1/2} \sin x.$

Review Problems, page 507

1. (a) $1 - x + (3/2)x^2 - (5/3)x^3 + \cdots.$
 (b) $-1 + x - (1/6)x^3 + (1/6)x^4 + \cdots.$

3. (a) $a_0[1 + x^2 + \cdots] + a_1[x + (1/3)x^3 + \cdots].$
 (b) $a_0[1 + (1/2)x^2 - (1/6)x^3 + \cdots]$
 $\quad + a_1[x - (1/2)x^2 + (1/2)x^3 + \cdots].$

5. $a_0[1 + (1/2)(x-2)^2 - (1/24)(x-2)^4 + \cdots]$
 $\quad + a_1(x-2).$

7. (a) $a_0[x^3 + x^4 + (1/4)x^5 + (1/36)x^6 + \cdots].$
 (b) $a_0[x - (1/4)x^2 + (1/20)x^3 - (1/120)x^4 + \cdots].$

9. (a) $c_1 y_1(x) + c_2 y_2(x),$ where
 $\quad y_1(x) = x + x^2 + (1/2)x^3 + \cdots = xe^x$ and
 $\quad y_2(x) = y_1(x) \ln x - x^2 - (3/4)x^3 - (11/36)x^4 + \cdots.$
 (b) $c_1 y_1(x) + c_2 y_2(x),$ where
 $\quad y_1(x) = 1 + 2x + x^2 + \cdots$ and
 $\quad y_2(x) = y_1(x) \ln x - [4x + 3x^2 + (22/27)x^3 + \cdots].$
 (c) $c_1 y_1(x) + c_2 y_2(x),$ where
 $\quad y_1(x) = 1 - (1/6)x + (1/96)x^2 + \cdots$ and
 $\quad y_2(x) = y_1(x) \ln x - 8x^{-2} - 4x^{-1} + (29/36) + \cdots.$

(d) $c_1 y_1(x) + c_2 y_2(x),$ where
 $\quad y_1(x) = x^2 - (1/4)x^3 + (1/40)x^4 + \cdots$ and
 $\quad y_2(x) = x^{-1} + (1/2) + (1/4)x + \cdots.$

CHAPTER 9

Exercises 9.2, page 522

1. (a) $\begin{bmatrix} 2 & 3 \\ 1 & 5 \end{bmatrix}.$ **(b)** $\begin{bmatrix} 1 & 1 \\ 5 & 2 \end{bmatrix}.$ **(c)** $\begin{bmatrix} 7 & 3 \\ 7 & 18 \end{bmatrix}.$

3. (a) $\begin{bmatrix} 18 & 10 \\ 4 & 4 \end{bmatrix}.$ **(b)** $\begin{bmatrix} 8 & 12 \\ 3 & 5 \end{bmatrix}.$ **(c)** $\begin{bmatrix} 16 & 0 \\ 0 & 16 \end{bmatrix}.$

5. (a) $\begin{bmatrix} -1 & -2 \\ -1 & -3 \end{bmatrix}.$ **(b)** $\begin{bmatrix} -5 & -1 \\ -8 & -1 \end{bmatrix}.$ **(c)** $\begin{bmatrix} -6 & -3 \\ -9 & -4 \end{bmatrix}.$

7. (a) 21. **(b)** 21. **(c)** 6. **(d)** 6.

9. $\begin{bmatrix} 4/9 & -1/9 \\ 1/9 & 2/9 \end{bmatrix}.$ **11.** Doesn't exist.

13. $\begin{bmatrix} 1 & 0 & 1 \\ -2 & 1 & -2 \\ 1 & 1 & 0 \end{bmatrix}.$

15. (a) $\mathbf{x} = \mathbf{0}.$ **(b)** $\mathbf{x} = s\begin{bmatrix} -1 \\ 1 \\ 4 \end{bmatrix}, \ s$ arbitrary.

 (c) $\mathbf{x} = s\begin{bmatrix} 3 \\ 1 \end{bmatrix}, \ s$ arbitrary.

 (d) $\mathbf{x} = s\begin{bmatrix} 1 \\ 0 \\ 1 \end{bmatrix} + v\begin{bmatrix} -1 \\ 1 \\ 0 \end{bmatrix}, \ s, v$ arbitrary.

19. $\begin{bmatrix} (4/3)e^{-t} & -(1/3)e^{-t} \\ -(1/3)e^{-4t} & (1/3)e^{-4t} \end{bmatrix}.$

21. $\begin{bmatrix} e^{-t} & (1/2)e^{-t} & -(1/2)e^{-t} \\ (1/3)e^t & -(1/2)e^t & (1/6)e^t \\ -(1/3)e^{-2t} & 0 & (1/3)e^{-2t} \end{bmatrix}.$

23. 11. **25.** $-12.$ **27.** 25. **29.** 2, 3.

31. 0, 1, 1. **33.** $\begin{bmatrix} 3e^{3t} \\ 6e^{3t} \\ -3e^{3t} \end{bmatrix}.$ **35.** $\begin{bmatrix} 5e^{5t} & 6e^{2t} \\ -10e^{5t} & -2e^{2t} \end{bmatrix}.$

41. (a) $\begin{bmatrix} (1/2)t^2 + c_1 & e^t + c_2 \\ t + c_3 & e^t + c_4 \end{bmatrix}.$

 (b) $\begin{bmatrix} \sin 1 & -1 + \cos 1 \\ 1 - \cos 1 & \sin 1 \end{bmatrix}.$

(c) $\begin{bmatrix} (1+e^t)\cos t + (e^t-t)\sin t & (e^t-t)\cos t - (e^t+1)\sin t \\ (e^t-1)\sin t + e^t\cos t & (e^t-1)\cos t - e^t\sin t \end{bmatrix}$.

Exercises 9.3, page 531

1. $\begin{bmatrix} x'(t) \\ y'(t) \end{bmatrix} = \begin{bmatrix} 3 & -1 \\ -1 & 2 \end{bmatrix}\begin{bmatrix} x(t) \\ y(t) \end{bmatrix} + \begin{bmatrix} t^2 \\ e^t \end{bmatrix}$.

3. $\begin{bmatrix} dx/dt \\ dy/dt \\ dz/dt \end{bmatrix} = \begin{bmatrix} t^2 & -1 & -1 \\ 0 & 0 & e^t \\ t & -1 & 3 \end{bmatrix}\begin{bmatrix} x \\ y \\ z \end{bmatrix} + \begin{bmatrix} t \\ 5 \\ -e^t \end{bmatrix}$.

5. $\begin{bmatrix} x_1'(t) \\ x_2'(t) \end{bmatrix} = \begin{bmatrix} 0 & 1 \\ 10 & 3 \end{bmatrix}\begin{bmatrix} x_1(t) \\ x_2(t) \end{bmatrix} + \begin{bmatrix} 0 \\ \sin t \end{bmatrix}$.

7. $\begin{bmatrix} x_1'(t) \\ x_1'(t) \\ x_3'(t) \\ x_4'(t) \end{bmatrix} = \begin{bmatrix} 0 & 1 & 0 & 0 \\ 0 & 0 & 1 & 0 \\ 0 & 0 & 0 & 1 \\ -1 & 0 & 0 & 0 \end{bmatrix}\begin{bmatrix} x_1(t) \\ x_2(t) \\ x_3(t) \\ x_4(t) \end{bmatrix} + \begin{bmatrix} 0 \\ 0 \\ 0 \\ t^2 \end{bmatrix}$.

9. $x_1'(t) = 5x_1(t) + 2e^{-2t}$,
$x_2'(t) = -2x_1(t) + 4x_2(t) - 3e^{-2t}$.

11. $x_1'(t) = x_1(t) + x_3(t) + e^t$;
$x_2'(t) = -x_1(t) + 2x_2(t) + 5x_3(t) + t$;
$x_3'(t) = 5x_2(t) + x_3(t)$.

13. LI. **15.** LI. **17.** LI.
19. Not a fundamental solution set.

21. Yes; $\begin{bmatrix} e^{-t} & e^t & e^{3t} \\ 2e^{-t} & 0 & -e^{3t} \\ e^{-t} & e^t & 2e^{3t} \end{bmatrix}$;

$c_1\begin{bmatrix} e^{-t} \\ 2e^{-t} \\ e^{-t} \end{bmatrix} + c_2\begin{bmatrix} e^t \\ 0 \\ e^t \end{bmatrix} + c_3\begin{bmatrix} e^{3t} \\ -e^{3t} \\ 2e^{3t} \end{bmatrix}$.

23. $\dfrac{3}{2}\begin{bmatrix} te^t \\ te^t \end{bmatrix} - \dfrac{1}{4}\begin{bmatrix} e^t \\ 3e^t \end{bmatrix} + \begin{bmatrix} t \\ 2t \end{bmatrix} - \begin{bmatrix} 0 \\ 1 \end{bmatrix} + c_1\begin{bmatrix} e^t \\ e^t \end{bmatrix} + c_2\begin{bmatrix} e^{-t} \\ 3e^{-t} \end{bmatrix}$.

27. $\mathbf{X}^{-1}(t) = \begin{bmatrix} 0 & (1/4)e^t & -(1/4)e^t \\ -(1/5)e^{2t} & (4/5)e^{2t} & -(2/5)e^{2t} \\ (1/5)e^{-3t} & (9/20)e^{-3t} & (3/20)e^{-3t} \end{bmatrix}$;

$\mathbf{x}(t) = \begin{bmatrix} -(3/2)e^{-t} + (3/5)e^{-2t} - (1/10)e^{3t} \\ (1/4)e^{-t} - (1/5)e^{-2t} - (1/20)e^{3t} \\ (5/4)e^{-t} - (1/5)e^{-2t} - (1/20)e^{3t} \end{bmatrix}$.

35. (b) $2e^{6t}$. **(e)** $2e^{6t}$; same.

Exercises 9.4, page 541

1. Eigenvalues are $r_1 = 0$ and $r_2 = -5$ with associated eigenvectors $\mathbf{u}_1 = \begin{bmatrix} s \\ 2s \end{bmatrix}$ and $\mathbf{u}_2 = \begin{bmatrix} 2s \\ -s \end{bmatrix}$.

3. Eigenvalues are $r_1 = 2$ and $r_2 = 3$ with associated eigenvectors $\mathbf{u}_1 = s\begin{bmatrix} 1 \\ -1 \end{bmatrix}$ and $\mathbf{u}_2 = s\begin{bmatrix} 1 \\ -2 \end{bmatrix}$.

5. Eigenvalues are $r_1 = 1$, $r_2 = 2$, and $r_3 = -2$ with associated eigenvectors $\mathbf{u}_1 = s\begin{bmatrix} 1 \\ 0 \\ 0 \end{bmatrix}$, $\mathbf{u}_2 = s\begin{bmatrix} 0 \\ 1 \\ 1 \end{bmatrix}$,

and $\mathbf{u}_3 = s\begin{bmatrix} 0 \\ -1 \\ 1 \end{bmatrix}$.

7. Eigenvalues are $r_1 = 1$, $r_2 = 2$, and $r_3 = 5$ with associated eigenvectors $\mathbf{u}_1 = s\begin{bmatrix} 2 \\ -3 \\ 2 \end{bmatrix}$, $\mathbf{u}_2 = s\begin{bmatrix} 0 \\ -1 \\ 1 \end{bmatrix}$,

and $\mathbf{u}_3 = s\begin{bmatrix} 0 \\ 1 \\ 2 \end{bmatrix}$.

9. Eigenvalues are $r_1 = i$ and $r_2 = -i$, with associated eigenvectors $\mathbf{u}_1 = s\begin{bmatrix} i \\ 1 \end{bmatrix}$ and $\mathbf{u}_2 = s\begin{bmatrix} -i \\ 1 \end{bmatrix}$.

11. $c_1 e^{3t/2}\begin{bmatrix} 3 \\ 10 \end{bmatrix} + c_2 e^{t/2}\begin{bmatrix} 1 \\ 2 \end{bmatrix}$.

13. $c_1 e^{-t}\begin{bmatrix} -2 \\ 1 \\ 1 \end{bmatrix} + c_2 e^{-3t}\begin{bmatrix} 0 \\ -1 \\ 1 \end{bmatrix} + c_3 e^{5t}\begin{bmatrix} 1 \\ 1 \\ 1 \end{bmatrix}$.

15. $c_1 e^{-t}\begin{bmatrix} 3 \\ 0 \\ -2 \end{bmatrix} + c_2 e^t\begin{bmatrix} 1 \\ -6 \\ 4 \end{bmatrix} + c_3 e^{4t}\begin{bmatrix} 1 \\ 0 \\ 1 \end{bmatrix}$.

17. (b)

Figure B.43

(c)

Figure B.44

(d)

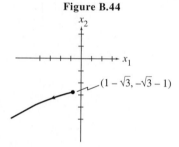

Figure B.45

19. $\begin{bmatrix} e^{3t} & e^{-3t} \\ 4e^{3t} & -2e^{-3t} \end{bmatrix}$.

21. $\begin{bmatrix} e^{t} & e^{2t} & e^{4t} \\ e^{t} & 2e^{2t} & 4e^{4t} \\ e^{t} & 4e^{2t} & 16e^{4t} \end{bmatrix}$.

23. $\begin{bmatrix} e^{2t} & e^{-t} & e^{3t} & -e^{7t} \\ 0 & -3e^{-t} & 0 & e^{7t} \\ 0 & 0 & e^{3t} & 2e^{7t} \\ 0 & 0 & 0 & 8e^{7t} \end{bmatrix}$.

25. $x = -c_1 e^{t} - 2c_2 e^{2t} - c_3 e^{3t}$,
$y = c_1 e^{t} + c_2 e^{2t} + c_3 e^{3t}$,
$z = 2c_1 e^{t} + 4c_2 e^{2t} + 4c_3 e^{3t}$.

27. $\begin{bmatrix} e^{-0.3473t} & e^{0.5237t} & 0.0286e^{-7.0764t} \\ -0.3157e^{-0.3473t} & 0.4761e^{0.5237t} & -0.1837e^{-7.0764t} \\ 0.0844e^{-0.3473t} & 0.1918e^{0.5237t} & e^{-7.0764t} \end{bmatrix}$.

29. $\begin{bmatrix} 0.0251e^{3.4142t} & -0.2361e^{-1.6180t} & e^{0.6180t} & e^{0.5858t} \\ 0.0858e^{3.4142t} & 0.3820e^{-1.6180t} & 0.6180e^{0.6180t} & 0.5858e^{0.5858t} \\ 0.2929e^{3.4142t} & -0.6180e^{-1.6180t} & 0.3820e^{0.6180t} & 0.3431e^{0.5858t} \\ e^{3.4142t} & e^{-1.6180t} & 0.2361e^{0.6180t} & 0.2010e^{0.5858t} \end{bmatrix}$.

31. $\begin{bmatrix} 2e^{4t} + e^{-2t} \\ 2e^{4t} - e^{-2t} \end{bmatrix}$. **33.** $\begin{bmatrix} -3e^{-t} + e^{5t} \\ -2e^{-t} - e^{5t} \\ e^{-t} + e^{5t} \end{bmatrix}$.

35. (b) $\mathbf{x}_1(t) = e^{-t}\begin{bmatrix} 1 \\ 2 \end{bmatrix}$. **(c)** $\mathbf{x}_2(t) = te^{-t}\begin{bmatrix} 1 \\ 2 \end{bmatrix} + e^{-t}\begin{bmatrix} 1 \\ 1 \end{bmatrix}$.

37. (b) $\mathbf{x}_1(t) = e^{2t}\begin{bmatrix} 1 \\ 0 \\ 0 \end{bmatrix}$. **(c)** $\mathbf{x}_2(t) = te^{2t}\begin{bmatrix} 1 \\ 0 \\ 0 \end{bmatrix} + e^{2t}\begin{bmatrix} 0 \\ 1 \\ 0 \end{bmatrix}$.

(d) $\mathbf{x}_3(t) = \dfrac{t^2}{2}e^{2t}\begin{bmatrix} 1 \\ 0 \\ 0 \end{bmatrix} + te^{2t}\begin{bmatrix} 0 \\ 1 \\ 0 \end{bmatrix} + e^{2t}\begin{bmatrix} 0 \\ -6/5 \\ 1/5 \end{bmatrix}$.

39. (b) $\mathbf{x}_1(t) = e^{t}\begin{bmatrix} -1 \\ 1 \\ 0 \end{bmatrix}$; $\mathbf{x}_2(t) = e^{t}\begin{bmatrix} -1 \\ 0 \\ 1 \end{bmatrix}$.

(c) $\mathbf{x}_3(t) = te^{t}\begin{bmatrix} 1 \\ 1 \\ -2 \end{bmatrix} + e^{t}\begin{bmatrix} 1 \\ 0 \\ 0 \end{bmatrix}$. **(d) 0.**

43. $c_1\begin{bmatrix} 3t^2 \\ t^2 \end{bmatrix} + c_2\begin{bmatrix} t^4 \\ t^4 \end{bmatrix}$.

45. $x_1(t) = (25/2)\,(e^{-3t/25} + e^{-t/25})$,
$x_2(t) = (25/4)\,(e^{-t/25} - e^{-3t/25})$.

47. (a) $r_1 = 2.39091$, $r_2 = -2.94338$, $r_3 = 3.55247$.

(b) $\mathbf{u}_1 = \begin{bmatrix} 1 \\ -2.64178 \\ -9.31625 \end{bmatrix}$; $\mathbf{u}_2 = \begin{bmatrix} 1 \\ -1.16825 \\ 0.43862 \end{bmatrix}$;

$\mathbf{u}_3 = \begin{bmatrix} 1 \\ 0.81004 \\ -0.12236 \end{bmatrix}$.

(c) $c_1 e^{r_1 t}\mathbf{u}_1 + c_2 e^{r_2 t}\mathbf{u}_2 + c_3 e^{r_3 t}\mathbf{u}_3$, where the r_i's and the $\mathbf{u}_i$'s are given in parts (a) and (b).

Exercises 9.5, page 548

1. $c_1\begin{bmatrix} 2\cos 2t \\ \cos 2t + \sin 2t \end{bmatrix} + c_2\begin{bmatrix} 2\sin 2t \\ \sin 2t - \cos 2t \end{bmatrix}.$

3. $c_1 e^t \cos t \begin{bmatrix} -1 \\ 1 \\ 0 \end{bmatrix} - c_1 e^t \sin t \begin{bmatrix} -2 \\ 0 \\ 1 \end{bmatrix} + c_2 e^t \sin t \begin{bmatrix} -1 \\ 1 \\ 0 \end{bmatrix} + c_2 e^t \cos t \begin{bmatrix} -2 \\ 0 \\ 1 \end{bmatrix} + c_3 e^t \begin{bmatrix} 1 \\ 0 \\ 0 \end{bmatrix}.$

5. $\begin{bmatrix} e^{-t}\cos 4t & e^{-t}\sin 4t \\ 2e^{-t}\sin 4t & -2e^{-t}\cos 4t \end{bmatrix}.$

7. $\begin{bmatrix} 1 & \cos t & \sin t \\ 0 & -\cos t & -\sin t \\ 0 & -\sin t & \cos t \end{bmatrix}.$

9. $\begin{bmatrix} 1 & \sqrt{3}\sin(\sqrt{3}t) - \cos(\sqrt{3}t) & -\sin(\sqrt{3}t) - \sqrt{3}\cos(\sqrt{3}t) \\ -1 & \sqrt{3}\sin(\sqrt{3}t) + \cos(\sqrt{3}t) & \sin(\sqrt{3}t) - \sqrt{3}\cos(\sqrt{3}t) \\ 1 & 2\cos(\sqrt{3}t) & 2\sin(\sqrt{3}t) \end{bmatrix}.$

11. $\begin{bmatrix} e^t\sin t - e^t\cos t & -e^t\sin t - e^t\cos t & -\cos t & -\sin t \\ 2e^t\sin t & -2e^t\cos t & \sin t & -\cos t \\ 2e^t\sin t + 2e^t\cos t & 2e^t\sin t - 2e^t\cos t & \cos t & \sin t \\ 4e^t\cos t & 4e^t\sin t & -\sin t & \cos t \end{bmatrix}.$

13. (a) $\begin{bmatrix} e^{-2t}(\sin t - \cos t) \\ -2e^{-2t}\sin t \end{bmatrix}.$ **(b)** $\begin{bmatrix} -e^{-2(t-\pi)}\cos t \\ e^{-2(t-\pi)}(\cos t - \sin t) \end{bmatrix}.$ **(c)** $\begin{bmatrix} e^{-2(t+2\pi)}(2\cos t - 3\sin t) \\ e^{-2(t+2\pi)}(\cos t + 5\sin t) \end{bmatrix}.$

(d) $\begin{bmatrix} e^{\pi-2t}\cos t \\ e^{\pi-2t}(\sin t - \cos t) \end{bmatrix}.$

17. $c_1\begin{bmatrix} t^{-1} \\ 0 \\ -2t^{-1} \end{bmatrix} + c_2\begin{bmatrix} t^{-1}\cos(\ln t) \\ t^{-1}\sin(\ln t) \\ -t^{-1}\cos(\ln t) \end{bmatrix} + c_3\begin{bmatrix} t^{-1}\sin(\ln t) \\ -t^{-1}\cos(\ln t) \\ -t^{-1}\sin(\ln t) \end{bmatrix}.$

19. $\dfrac{\sqrt{9 - \sqrt{17}}}{2\sqrt{2}\pi} \approx 0.249;$ $\dfrac{\sqrt{9 + \sqrt{17}}}{2\sqrt{2}\pi} \approx 0.408.$

21. $I_1 = (5/6)e^{-2t}\sin 3t,$
$I_2 = -(10/13)e^{-2t}\cos 3t + (25/78)e^{-2t}\sin 3t,$
$I_3 = (10/13)e^{-2t}\cos 3t + (20/39)e^{-2t}\sin 3t.$

Exercises 9.6, page 554

1. $c_1 e^{7t}\begin{bmatrix} 1 \\ 1 \end{bmatrix} + c_2 e^{2t}\begin{bmatrix} 1 \\ -4 \end{bmatrix} + \begin{bmatrix} 2 \\ -1 \end{bmatrix}.$

3. $c_1 e^{-3t}\begin{bmatrix} 1 \\ 1 \\ -1 \end{bmatrix} + c_2 e^{3t}\begin{bmatrix} 1 \\ 0 \\ 1 \end{bmatrix} + c_3 e^{3t}\begin{bmatrix} -1 \\ 1 \\ 0 \end{bmatrix} + e^t\begin{bmatrix} 1 \\ 0 \\ -1 \end{bmatrix}.$

5. $\mathbf{x}_p = e^{-2t}[\mathbf{a} + t\mathbf{b}].$
7. $\mathbf{x}_p = t\mathbf{a} + \mathbf{b} + \sin 3t\,\mathbf{c} + \cos 3t\,\mathbf{d}.$
9. $\mathbf{x}_p = e^{2t}\mathbf{a} + e^{3t}\mathbf{b}.$

11. $c_1\begin{bmatrix} \sin t \\ \cos t \end{bmatrix} + c_2\begin{bmatrix} \cos t \\ -\sin t \end{bmatrix} + \begin{bmatrix} 0 \\ -1 \end{bmatrix}.$

13. $c_1 e^t\begin{bmatrix} 1 \\ -1 \end{bmatrix} + c_2 e^{-t}\begin{bmatrix} 1 \\ -3 \end{bmatrix} + \begin{bmatrix} 5te^t + (3/2)e^t \\ -5te^t + (9/2)e^t \end{bmatrix}.$

15. $c_1\begin{bmatrix} 1 \\ 2 \end{bmatrix} + c_2 e^{-5t}\begin{bmatrix} -2 \\ 1 \end{bmatrix} + \begin{bmatrix} \ln|t| + (8/5)t - 8/25 \\ 2\ln|t| + (16/5)t + 4/25 \end{bmatrix}.$

17. $c_1 e^{2t}\begin{bmatrix} 1 \\ 1 \\ 1 \end{bmatrix} + c_2 e^{-t}\begin{bmatrix} -1 \\ 0 \\ 1 \end{bmatrix} + c_3 e^{-t}\begin{bmatrix} -1 \\ 1 \\ 0 \end{bmatrix} + e^t\begin{bmatrix} 1 \\ -1 \\ -1 \end{bmatrix}.$

19.
$$\begin{bmatrix} c_1 \cos t + c_2 \sin t + 1 \\ -c_1 \sin t + c_2 \cos t - t \\ c_3 e^t + c_4 e^{-t} - (1/4)e^{-t} - t + (1/2)te^{-t} \\ c_3 e^t - c_4 e^{-t} - (1/4)e^{-t} - 1 - (1/2)te^{-t} \end{bmatrix}.$$

21. (a) $\begin{bmatrix} 4te^t + 5e^t \\ 2te^t + 4e^t \end{bmatrix}.$

(b) $\begin{bmatrix} -2e^{t-1} + 2e^{2(t-1)} - 3e^{2t-1} + (4t-1)e^t \\ -e^{t-1} + 2e^{2(t-1)} - 3e^{2t-1} + (2t+1)e^t \end{bmatrix}.$

(c) $\begin{bmatrix} (-20 + 2e^{-5})e^t + (-3e^{-5} - e^{-10})e^{2t} + (4t+3)e^t \\ (-10 + e^{-5})e^t + (-3e^{-5} - e^{-10})e^{2t} + (2t+3)e^t \end{bmatrix}.$

(d) $\begin{bmatrix} 2e^{t+1} - 3e^{2t+1} + (4t+7)e^t \\ e^{t+1} - 3e^{2t+1} + (2t+5)e^t \end{bmatrix}.$

23. $x = (c_1 + c_2 t)e^t + c_3 e^{-t} - t^2 - 4t - 6,$
$y = -c_2 e^t - 2c_3 e^{-t} - t^2 - 2t - 3.$

25. (a) $\left\{ \begin{bmatrix} e^t \\ e^t \end{bmatrix}, \begin{bmatrix} e^{2t} \\ 2e^{2t} \end{bmatrix} \right\}.$ **(c)** $te^t \begin{bmatrix} 2 \\ 2 \end{bmatrix} + e^t \begin{bmatrix} 0 \\ 1 \end{bmatrix}.$

(d) $c_1 e^t \begin{bmatrix} 1 \\ 1 \end{bmatrix} + c_2 e^{2t} \begin{bmatrix} 1 \\ 2 \end{bmatrix} + te^t \begin{bmatrix} 2 \\ 2 \end{bmatrix} + e^t \begin{bmatrix} 0 \\ 1 \end{bmatrix}.$

27. $c_1 e^{2t} \begin{bmatrix} 1 \\ 1 \\ 1 \end{bmatrix} + c_2 e^{-t} \begin{bmatrix} -1 \\ 0 \\ 1 \end{bmatrix} + c_3 e^{-t} \begin{bmatrix} -1 \\ 1 \\ 0 \end{bmatrix}$

$\qquad + te^{-t} \begin{bmatrix} 1 \\ 0 \\ -1 \end{bmatrix} + e^{-t} \begin{bmatrix} 1 \\ 0 \\ 0 \end{bmatrix} + \begin{bmatrix} 0 \\ 0 \\ 1 \end{bmatrix}.$

29. $c_1 t \begin{bmatrix} 1 \\ 1 \end{bmatrix} + c_2 t^{-1} \begin{bmatrix} 1 \\ 3 \end{bmatrix} + \begin{bmatrix} -(3/4)t^{-1} - (1/2)t^{-1} \ln t + 1 \\ -(3/4)t^{-1} - (3/2)t^{-1} \ln t + 2 \end{bmatrix}.$

33. $I_1 = I_2 + 3I_5,\ I_3 = 3I_5,\ I_4 = 2I_5,$ where
$I_2 = (1/(20\sqrt{817}))[(13 - \sqrt{817})e^{-(31+\sqrt{817})5t/2}$
$\qquad - (13 + \sqrt{817})e^{(-31+\sqrt{817})5t/2}] + 1/10,$
$I_5 = (3/(40\sqrt{817}))[(31 - \sqrt{817})e^{-(31+\sqrt{817})5t/2}$
$\qquad - (31 + \sqrt{817})e^{(-31+\sqrt{817})5t/2}] + 3/20.$

35. $x_1 = (2050 - (5845/4)e^{-3t/50} - (2355/4)e^{-7t/50})/21,$
$x_2 = (1850 - (5845/2)e^{-3t/50} + (2355/2)e^{-7t/50})/21.$
Tank A will have the higher concentration. The limiting concentration in tank A is 41/21 kg/L and in tank B is 37/21 kg/L.

Exercises 9.7, page 563

1. (a) $r = 3;\quad k = 2.$ **(b)** $e^{3t} \begin{bmatrix} 1 & -2t \\ 0 & 1 \end{bmatrix}.$

3. (a) $r = -1;\quad k = 3.$ **(b)** $e^{-t} \begin{bmatrix} 1 + 3t - (3/2)t^2 & t & -t + (1/2)t^2 \\ -3t & 1 & t \\ 9t - (9/2)t^2 & 3t & 1 - 3t + (3/2)t^2 \end{bmatrix}.$

5. (a) $r = -2;\quad k = 2.$ **(b)** $e^{-2t} \begin{bmatrix} 1 & 0 & 0 \\ 4t & 1 & 0 \\ t & 0 & 1 \end{bmatrix}.$ **7.** $\begin{bmatrix} \cos t & \sin t \\ -\sin t & \cos t \end{bmatrix}.$

9. $\dfrac{1}{2} \begin{bmatrix} e^t + \cos t - \sin t & 2 \sin t & e^t - \cos t - \sin t \\ e^t - \sin t - \cos t & 2 \cos t & e^t + \sin t - \cos t \\ e^t - \cos t + \sin t & -2 \sin t & e^t + \cos t + \sin t \end{bmatrix}.$

11. $\dfrac{1}{25} \begin{bmatrix} -4 + 29e^{5t} - 20te^{5t} & 20 - 20e^{5t} & -8 + 8e^{5t} - 40te^{5t} \\ -5 + 5e^{5t} & 25 & -10 + 10e^{5t} \\ 2 - 2e^{5t} + 10te^{5t} & -10 + 10e^{5t} & 4 + 21e^{5t} + 20te^{5t} \end{bmatrix}.$

13. $\begin{bmatrix} (1 - t + t^2/2)e^t & (t - t^2)e^t & (t^2/2)e^t & 0 & 0 \\ (t^2/2)e^t & (1 - t + t^2)e^t & (t + t^2/2)e^t & 0 & 0 \\ (t + t^2/2)e^t & (-3t - t^2)e^t & (1 + 2t + t^2/2)e^t & 0 & 0 \\ 0 & 0 & 0 & \cos t & \sin t \\ 0 & 0 & 0 & -\sin t & \cos t \end{bmatrix}.$

15.
$$\begin{bmatrix} (1+t+t^2/2)e^{-t} & (t+t^2)e^{-t} & (t^2/2)e^{-t} & 0 & 0 \\ (-t^2/2)e^{-t} & (1+t-t^2)e^{-t} & (t-t^2/2)e^{-t} & 0 & 0 \\ (-t+t^2/2)e^{-t} & (-3t+t^2)e^{-t} & (1-2t+t^2/2)e^{-t} & 0 & 0 \\ 0 & 0 & 0 & (1+2t)e^{-2t} & te^{-2t} \\ 0 & 0 & 0 & -4te^{-2t} & (1-2t)e^{-2t} \end{bmatrix}.$$

17. $c_1 e^{-t}\begin{bmatrix} 1 \\ -1 \\ 1 \end{bmatrix} + c_2 e^{-t}\begin{bmatrix} -t \\ -1+t \\ 2-t \end{bmatrix} + c_3 e^{-2t}\begin{bmatrix} 1 \\ -2 \\ 4 \end{bmatrix}.$

19. $c_1 e^{2t}\begin{bmatrix} 3 \\ 6 \\ 1 \\ 1 \end{bmatrix} + c_2 e^{t}\begin{bmatrix} 0 \\ 1 \\ 0 \\ 0 \end{bmatrix} + c_3 e^{t}\begin{bmatrix} 1 \\ t \\ 0 \\ 0 \end{bmatrix} + c_4 e^{t}\begin{bmatrix} 2t \\ t+t^2 \\ 0 \\ 1 \end{bmatrix}.$

21. $\begin{bmatrix} e^{-2t} \\ 4te^{-2t} + e^{-2t} \\ te^{-2t} - e^{-2t} \end{bmatrix}.$ **23.** $e^{-t}\begin{bmatrix} 3t \\ 3 \\ 9t \end{bmatrix} + e^{At}\begin{bmatrix} 2 - e^{t}(t^2 - 2t + 2) \\ 1 + e^{t}(t-1) \\ 6 - 3e^{t}(t^2 - 2t + 2) \end{bmatrix},$ where e^{At} is the matrix in the answer to Problem 3.

Review Problems, page 567

1. $c_1 e^{3t}\begin{bmatrix} 1 \\ 1 \end{bmatrix} + c_2 e^{4t}\begin{bmatrix} 3 \\ 2 \end{bmatrix}.$ **3.** $c_1 e^{-t}\begin{bmatrix} 1 \\ -1 \\ 0 \\ 0 \end{bmatrix} + c_2 e^{-t}\begin{bmatrix} 0 \\ 0 \\ 1 \\ -1 \end{bmatrix} + c_3 e^{3t}\begin{bmatrix} 1 \\ 1 \\ 0 \\ 0 \end{bmatrix} + c_4 e^{3t}\begin{bmatrix} 0 \\ 0 \\ 1 \\ 1 \end{bmatrix}.$ **5.** $\begin{bmatrix} e^{2t} & e^{3t} \\ -e^{2t} & -2e^{3t} \end{bmatrix}.$

7. $c_1 e^{-t}\begin{bmatrix} 1 \\ -2 \end{bmatrix} + c_2 e^{3t}\begin{bmatrix} 1 \\ 2 \end{bmatrix} + \begin{bmatrix} -1/3 \\ -14/3 \end{bmatrix}.$ **9.** $c_1 e^{-t}\begin{bmatrix} 1 \\ 0 \\ 3 \end{bmatrix} + c_2 e^{-t}\begin{bmatrix} t \\ 1 \\ 3t \end{bmatrix} + c_3 e^{-t}\begin{bmatrix} -t + (1/2)t^2 \\ t \\ 1 - 3t + (3/2)t^2 \end{bmatrix} + \begin{bmatrix} 2 + t \\ 7 - 3t \\ 10 \end{bmatrix}.$

11. $\begin{bmatrix} 3e^{t} - 2e^{2t} \\ 3e^{t} - 4e^{2t} \end{bmatrix}.$ **13.** $c_1 t^{-1}\begin{bmatrix} -3 \\ 1 \\ 0 \end{bmatrix} + c_2 t^{-1}\begin{bmatrix} -1 \\ 0 \\ 1 \end{bmatrix} + c_3 t^{4}\begin{bmatrix} 1 \\ 1 \\ 1 \end{bmatrix}.$

15. $\begin{bmatrix} (1/2)e^{t} + (2/3)e^{5t} - (1/6)e^{-t} & -e^{t} + (2/3)e^{5t} + (1/3)e^{-t} & -(1/2)e^{t} + (2/3)e^{5t} - (1/6)e^{-t} \\ (1/3)e^{5t} - (1/3)e^{-t} & (1/3)e^{5t} + (2/3)e^{-t} & (1/3)e^{5t} - (1/3)e^{-t} \\ -(1/2)e^{t} + (1/2)e^{-t} & e^{t} - e^{-t} & (1/2)e^{t} + (1/2)e^{-t} \end{bmatrix}.$

CHAPTER 10

Exercises 10.2, page 584

1. $y = [4/(e - e^{-1})](e^{-x} - e^{x})$. **3.** $y \equiv 0$.

5. $y = e^{x} + 2x - 1$. **7.** $y = \cos x + c \sin x$; c arbitrary.

9. $\lambda_n = (2n-1)^2/4$ and $y_n = c_n \sin[(2n-1)x/2]$, where $n = 1, 2, 3, \ldots$ and the c_n's are arbitrary.

11. $\lambda_n = n^2$, $n = 0, 1, 2, \ldots$; $y_0 = a_0$ and $y_n = a_n \cos nx + b_n \sin nx$, $n = 1, 2, 3, \ldots$, where a_0, a_n, and b_n are arbitrary.

13. The eigenvalues are the roots of $\tan(\sqrt{\lambda_n}\,\pi) + \sqrt{\lambda_n} = 0$, where $\lambda_n > 0$.

For n large, $\lambda_n \approx (2n-1)^2/4$, n is a positive integer. The eigenfunctions are $y_n = c_n[\sin(\sqrt{\lambda_n}x) + \sqrt{\lambda_n}\cos(\sqrt{\lambda_n}x)]$, where the c_n's are arbitrary.

15. $u(x, t) = e^{-3t}\sin x - 6e^{-48t}\sin 4x$.

17. $u(x, t) = e^{-3t}\sin x - 7e^{-27t}\sin 3x + e^{-75t}\sin 5x$.

19. $u(x, t) = 3\cos 6t\sin 2x + 12\cos 39t\sin 13x$.

21. $u(x, t) = 6\cos 6t\sin 2x + 2\cos 18t\sin 6x + (11/27)\sin 27t\sin 9x - (14/45)\sin 45t\sin 15x$.

23. $u(x, t) = \sum_{n=1}^{\infty} n^{-2}e^{-2\pi^2 n^2 t}\sin n\pi x$.

25. If $K > 0$, then $T(t)$ becomes unbounded as $t \to \infty$, and so the temperature $u(x, t) = X(x)T(t)$ becomes unbounded at each position x. Since the temperature must remain bounded for all time, $K \not> 0$.

33. (a) $u(x) \equiv 50$. (b) $u(x) = 30x/L + 10$.

Exercises 10.3, page 600

1. Odd. **3.** Even. **5.** Neither.

9. $f(x) \sim \displaystyle\sum_{n=1}^{\infty} \frac{2(-1)^{n+1}}{n} \sin nx.$

11. $f(x) \sim 1 + \displaystyle\sum_{n=1}^{\infty} \left[\frac{2}{\pi^2 n^2}(-1 + (-1)^n) \cos \frac{n\pi x}{2} \right.$
$\left. + \frac{1}{\pi n}((-1)^{n+1} - 1) \sin \frac{n\pi x}{2} \right].$

13. $f(x) \sim \dfrac{1}{3} + \displaystyle\sum_{n=1}^{\infty} \frac{4(-1)^n}{n^2 \pi^2} \cos n\pi x.$

15. $f(x) \sim [(\sinh \pi)/\pi] \left(1 + \displaystyle\sum_{n=1}^{\infty} \left[\frac{(-1)^n}{1 + n^2} \cos nx \right. \right.$
$\left. \left. + \frac{(-1)^{n+1} n}{1 + n^2} \sin nx \right] \right).$

17. The 2π periodic function $g(x)$, where
$$g(x) = \begin{cases} x, & -\pi < x < \pi, \\ 0, & x = \pm\pi. \end{cases}$$

19. The 4 periodic function $g(x)$, where
$$g(x) = \begin{cases} 1, & -2 < x < 0, \\ x, & 0 < x < 2, \\ 1/2, & x = 0, \\ 3/2, & x = \pm 2. \end{cases}$$

21. The 2 periodic function $g(x)$, where
$$g(x) = x^2, \quad -1 \le x \le 1.$$

23. The 2π periodic function $g(x)$, where
$$g(x) = \begin{cases} e^x, & -\pi < x < \pi, \\ (e^\pi + e^{-\pi})/2, & x = \pm\pi. \end{cases}$$

25. (a) $F(x) = (x^2 - \pi^2)/2$. (b) $F(x) = |x| - \pi$.

27. $f(x) \sim \displaystyle\sum_{n=1}^{\infty} \frac{2}{(2n-1)\pi} \left[(-1)^{n+1} \cos\left(\frac{(2n-1)\pi x}{2}\right) \right.$
$\left. + \sin\left(\frac{(2n-1)\pi x}{2}\right) \right].$

29. $a_0 = 0$; $a_1 = 3/2$; $a_2 = 0$.

Exercises 10.4, page 608

1. (a) The π periodic function $\tilde{f}(x)$, where
$$\tilde{f}(x) = x^2, \quad 0 < x < \pi.$$

(b) The 2π periodic function $f_o(x)$, where
$$f_o(x) = \begin{cases} x^2, & 0 < x < \pi, \\ -x^2, & -\pi < x < 0. \end{cases}$$

(c) The 2π periodic function $f_e(x)$, where
$$f_e(x) = \begin{cases} x^2, & 0 < x < \pi, \\ x^2, & -\pi < x < 0. \end{cases}$$

3. (a) The π periodic function $\tilde{f}(x)$, where
$$\tilde{f}(x) = \begin{cases} 0, & 0 < x < \pi/2, \\ 1, & \pi/2 < x < \pi. \end{cases}$$

(b) The 2π periodic function $f_o(x)$, where
$$f_o(x) = \begin{cases} -1, & -\pi < x < -\pi/2, \\ 0, & -\pi/2 < x < 0, \\ 0, & 0 < x < \pi/2, \\ 1, & \pi/2 < x < \pi. \end{cases}$$

(c) The 2π periodic function $f_e(x)$, where
$$f_e(x) = \begin{cases} 1, & -\pi < x < -\pi/2, \\ 0, & -\pi/2 < x < 0, \\ 0, & 0 < x < \pi/2, \\ 1, & \pi/2 < x < \pi. \end{cases}$$

5. $f(x) \sim -\dfrac{4}{\pi} \displaystyle\sum_{k=1}^{\infty} \frac{1}{2k-1} \sin(2k-1)\pi x.$

7. $f(x) \sim \displaystyle\sum_{n=1}^{\infty} \left[\frac{2\pi(-1)^{n+1}}{n} + \frac{4}{\pi n^3}((-1)^n - 1) \right] \sin nx.$

9. $f(x) \sim \displaystyle\sum_{k=0}^{\infty} \frac{8}{(2k+1)^3 \pi^3} \sin(2k+1)\pi x.$

11. $f(x) \sim \dfrac{\pi}{2} + \dfrac{4}{\pi} \displaystyle\sum_{k=1}^{\infty} \frac{1}{(2k-1)^2} \cos(2k-1)x.$

13. $f(x) \sim e - 1 + 2 \displaystyle\sum_{n=1}^{\infty} \frac{(-1)^n e - 1}{1 + \pi^2 n^2} \cos n\pi x.$

15. $f(x) \sim \dfrac{2}{\pi} + \dfrac{2}{\pi} \displaystyle\sum_{k=1}^{\infty} \left(\frac{1}{2k+1} + \frac{1}{2k-1} \right) \cos 2kx.$

17. $u(x, t) =$
$$\frac{2}{\pi} \sum_{k=1}^{\infty} \left[\frac{2}{2k-1} - \frac{1}{2k+1} - \frac{1}{2k-3} \right] e^{-5(2k-1)^2 t} \sin(2k-1)x.$$

19. $u(x, t) = \dfrac{4}{\pi} \displaystyle\sum_{k=0}^{\infty} \frac{(-1)^{k+1}}{(2k+1)^2} e^{-5(2k+1)^2 t} \sin(2k+1)x.$

Exercises 10.5, page 620

1. $u(x, t) = \displaystyle\sum_{n=1}^{\infty} \frac{8(-1)^{n+1} - 4}{\pi^3 n^3} e^{-5\pi^2 n^2 t} \sin n\pi x.$

3. $u(x, t) = \dfrac{\pi}{2} - \displaystyle\sum_{k=0}^{\infty} \frac{4}{\pi(2k+1)^2} e^{-3(2k+1)^2 t} \cos(2k+1)x.$

5. $u(x,t) = \dfrac{2(e^\pi - 1)}{\pi} + \displaystyle\sum_{n=1}^\infty \dfrac{2e^\pi(-1)^n - 2}{\pi(1+n^2)} e^{-n^2 t} \cos nx.$

7. $u(x,t) = 5 + \dfrac{5}{\pi} x - \dfrac{30}{\pi} e^{-2t} \sin x + \dfrac{5}{\pi} e^{-8t} \sin 2x$

$\qquad + \left(1 - \dfrac{10}{\pi}\right) e^{-18t} \sin 3x$

$\qquad + \dfrac{5}{2\pi} e^{-32t} \sin 4x - \left(1 + \dfrac{6}{\pi}\right) e^{-50t} \sin 5x$

$\qquad + \displaystyle\sum_{n=6}^\infty \dfrac{10}{\pi n} [2(-1)^n - 1] e^{-2n^2 t} \sin nx.$

9. $u(x,t) = \dfrac{e^{-\pi} - 1}{\pi} x - e^{-x} + 1 + \displaystyle\sum_{n=1}^\infty c_n e^{-n^2 t} \sin nx,$

where

$$c_n = \begin{cases} \dfrac{2e^{-\pi} - 2}{\pi n}(-1)^n + \dfrac{2n}{\pi(1+n^2)}((-1)^{n+1}e^{-\pi} + 1) \\ \qquad + \dfrac{2}{\pi n}[(-1)^n - 1], \quad n \neq 2, \\[2mm] \dfrac{e^{-\pi} - 1}{\pi} + \dfrac{4}{5\pi}(1 - e^{-\pi}) + 1, \qquad n = 2. \end{cases}$$

11. $u(x,t) = \displaystyle\sum_{n=0}^\infty a_n e^{-4(n+1/2)^2 t} \cos(n + 1/2)x,$ where

$f(x) = \displaystyle\sum_{n=0}^\infty a_n \cos(n + 1/2)x.$

13. $u(x,t) = \dfrac{\pi^2}{3} x - \dfrac{1}{3} x^3 - 3e^{-2t} \sin x$

$\qquad + \displaystyle\sum_{n=2}^\infty \dfrac{4(-1)^n}{n^3} e^{-2n^2 t} \sin nx.$

15. $u(x,y,t) = e^{-52t} \cos 6x \sin 4y - 3e^{-122t} \cos x \sin 11y.$

17. $u(x,y,t) = 2\displaystyle\sum_{n=1}^\infty \dfrac{(-1)^{n+1}}{n} e^{-n^2 t} \sin ny.$

19. $C(x,t) = \displaystyle\sum_{n=1}^\infty c_n e^{-[L+kn^2\pi^2/a^2]t} \sin\left(\dfrac{n\pi x}{a}\right),$

where

$$c_n = \dfrac{2}{a} \int_0^a f(x) \sin\left(\dfrac{n\pi x}{a}\right) dx.$$

Concentration goes to zero as $t \to +\infty.$

Exercises 10.6, page 633

1. $u(x,t) = \dfrac{1}{7\pi} \sin 7\pi t \sin 7\pi x$

$\qquad + \displaystyle\sum_{k=0}^\infty \dfrac{8}{((2k+1)\pi)^3} \cos(2k+1)\pi t \sin(2k+1)\pi x.$

3. $u(x,t) = \displaystyle\sum_{n=1}^\infty \dfrac{4}{n^2} [2(-1)^{n+1} - 1] \cos 2nt \sin nx.$

5. $u(x,t) = \dfrac{2h_0 L^2}{\pi^2 a(L - a)} \displaystyle\sum_{n=1}^\infty \dfrac{1}{n^2} \sin\dfrac{n\pi a}{L} \sin\dfrac{n\pi x}{L} \cos\dfrac{n\pi at}{L}.$

7. $u(x,t) = \cos t \sin x + \dfrac{5}{2} \sin 2t \sin 2x - \dfrac{3}{5} \sin 5t \sin 5x$

$\qquad + 2\displaystyle\sum_{n=1}^\infty \dfrac{(-1)^{n+1}}{n^3}\left[t - \dfrac{\sin nt}{n}\right] \sin nx.$

9. $u(x,t) = \displaystyle\sum_{n=0}^\infty \left(a_n \cos\dfrac{(2n + 1)\pi\alpha t}{2L}\right.$

$\qquad \left. + b_n \sin\dfrac{(2n + 1)\pi\alpha t}{2L}\right) \sin\dfrac{(2n + 1)\pi x}{2L},$ where

$f(x) = \displaystyle\sum_{n=0}^\infty a_n \sin\dfrac{(2n + 1)\pi x}{2L}$ and

$g(x) = \displaystyle\sum_{n=0}^\infty b_n\dfrac{(2n + 1)\pi\alpha}{2L} \sin\dfrac{(2n + 1)\pi x}{2L}.$

11. $u(x,t) = \displaystyle\sum_{n=1}^\infty a_n T_n(t) \sin\dfrac{n\pi x}{L},$ where

$$a_n = \dfrac{2}{L}\int_0^L f(x) \sin\dfrac{n\pi x}{L}\,dx,$$

and

$$T_n(t) = e^{-t/2}\left(\cos \beta_n t + \dfrac{1}{2\beta_n} \sin \beta_n t\right),$$

where

$$\beta_n = \dfrac{1}{2L}\sqrt{3L^2 + 4\alpha^2\pi^2 n^2}.$$

13. $u(x,t) = \dfrac{1}{2\alpha}[\sin(x + \alpha t) - \sin(x - \alpha t)]$

$\qquad = \dfrac{1}{\alpha} \sin \alpha t \cos x.$

15. $u(x,t) = x + tx.$

17. $u(x,t) = \dfrac{1}{2}\left[e^{-(x+\alpha t)^2} + e^{-(x-\alpha t)^2}\right.$

$\qquad \left. + \dfrac{\cos(x - \alpha t) - \cos(x + \alpha t)}{\alpha}\right].$

21. $u(r,t) = \displaystyle\sum_{n=1}^\infty [a_n \cos(k_n\alpha t) + b_n \sin(k_n\alpha t)]J_0(k_n r),$

where

$$a_n = \dfrac{1}{c_n} \int_0^1 f(r)J_0(k_n r)r\,dr,$$

and

$$b_n = \dfrac{1}{\alpha k_n c_n} \int_0^1 g(r)J_0(k_n r)r\,dr,$$

with

$$c_n = \int_0^1 J_0^2(k_n r)r\,dr.$$

Exercises 10.7, page 645

1. $u(x,y) = \dfrac{4 \cos 6x \sinh(6(y - 1))}{\sinh(-6)} + \dfrac{\cos 7x \sinh(7(y - 1))}{\sinh(-7)}.$

3. $u(x, y) = \sum\limits_{n=1}^{\infty} A_n \sin nx \sinh(ny - n\pi)$, where

$$A_n = \frac{2}{\pi \sinh(-n\pi)} \int_0^\pi f(x) \sin nx \, dx.$$

5. $u(x, y) = \dfrac{\cos x \sinh(y - 1)}{\sinh(-1)} - \dfrac{\cos 3x \sinh(3y - 3)}{\sinh(-3)}$

$\qquad + \dfrac{\cos 2x \sinh 2y}{\sinh(2)}.$

7. $u(r, \theta) = \dfrac{\pi}{2} - \sum\limits_{k=0}^{\infty} \dfrac{r^{2k+1}}{(2k + 1)^2 \pi 2^{2k-1}} \cos(2k + 1)\theta.$

9. $u(r, \theta) = \dfrac{a_0}{2} + \sum\limits_{n=1}^{\infty} \left(\dfrac{r}{a}\right)^n (a_n \cos n\theta + b_n \sin n\theta)$, where

a_0 is arbitrary, and for $n = 1, 2, 3, \ldots$

$$a_n = \frac{a}{\pi n} \int_{-\pi}^{\pi} f(\theta) \cos n\theta \, d\theta,$$

$$b_n = \frac{a}{n\pi} \int_{-\pi}^{\pi} f(\theta) \sin n\theta \, d\theta.$$

11. $u(r, \theta) = \left(\dfrac{1}{3} r - \dfrac{4}{3} r^{-1}\right) \cos \theta + \left(\dfrac{2}{3} r - \dfrac{2}{3} r^{-1}\right) \sin \theta$

$\qquad + \left(-\dfrac{1}{225} r^4 + \dfrac{256}{255} r^{-4}\right) \sin 4\theta.$

13. $u(r, \theta) = \dfrac{a_0}{2} + \sum\limits_{n=1}^{\infty} r^{-n}(a_n \cos n\theta + b_n \sin n\theta)$, where

$$a_n = \frac{1}{\pi} \int_{-\pi}^{\pi} f(\theta) \cos n\theta \, d\theta, \qquad n = 0, 1, 2, \ldots,$$

and

$$b_n = \frac{1}{\pi} \int_{-\pi}^{\pi} f(\theta) \sin n\theta \, d\theta, \qquad n = 1, 2, 3, \ldots.$$

15. $u(r, \theta) = r^3 \sin 3\theta.$

17. $u(r, \theta) = a_0 + b_0 \ln r + \sum\limits_{n=1}^{\infty} [(a_n r^n + b_n r^{-n}) \cos n\theta$

$\qquad + (c_n r^n + d_n r^{-n}) \sin n\theta]$, where

$$a_0 = \frac{1}{2\pi} \int_{-\pi}^{\pi} f(\theta) \, d\theta,$$

$$b_0 = \frac{3}{2\pi} \int_{-\pi}^{\pi} g(\theta) \, d\theta,$$

and for $n = 1, 2, 3, \ldots$

$$a_n + b_n = \frac{1}{\pi} \int_{-\pi}^{\pi} f(\theta) \cos n\theta \, d\theta,$$

$$n3^{n-1} a_n - n3^{-n-1} b_n = \frac{1}{\pi} \int_{-\pi}^{\pi} g(\theta) \cos n\theta \, d\theta,$$

and

$$c_n + d_n = \frac{1}{\pi} \int_{-\pi}^{\pi} f(\theta) \sin n\theta \, d\theta,$$

$$n3^{n-1} c_n - n3^{-n-1} d_n = \frac{1}{\pi} \int_{-\pi}^{\pi} g(\theta) \sin n\theta \, d\theta.$$

21. $u(r, \theta, z) = [I_0(r)/I_0(\pi)] \sin z.$

23. **(c)** $\phi(x, y) = 2xy.$

CHAPTER 11

Exercises 11.2, page 668

1. $y = e^{-x} \cos 5x + ce^{-x} \sin 5x$, where c is any real number.

3. $y = ce^{2x} \sin 3x$, where c is any real number.

5. $y = c_1 \cos x + c_2 \sin x - (1/3) \sin 2x$, where c_1 and c_2 are real numbers.

7. $y = -(1/2) \cos x + (1/2 + e^\pi) \sin x + (1/2)e^x.$

9. No solution.

11. $y = c \sin(\ln x) + 2x^{-1}$, where c is any real number.

13. $\lambda_n = (n + 1/2)^2 \pi^2$, $n = 0, 1, 2, \ldots$; and $y_n = c_n \sin[(n + 1/2)\pi x]$, $n = 0, 1, 2, \ldots$, where the c_n's are arbitrary.

15. $\lambda_n = n^2 - 3$, $n = 0, 1, 2, \ldots$; and $y_n = c_n \cos nx$, $n = 0, 1, 2, \ldots$, where the c_n's are arbitrary.

17. $\lambda_0 = -\mu_0^2$, where $\tanh(\mu_0 \pi) = \mu_0/2$; and $y_0 = c_0(\cosh(\mu_0 x) - (2/\mu_0) \sinh(\mu_0 x))$ and $\lambda_n = \mu_n^2$, where $\tan(\mu_n \pi) = \mu_n/2$, $n = 1, 2, 3, \ldots$; and $y_n = c_n(\cos(\mu_n x) - (2/\mu_n) \sin(\mu_n x))$, where the c_n's are arbitrary.

19. $\lambda_n = (n + 1/2)^2$, $n = 0, 1, 2, \ldots$; and $y_n = c_n \cos[(n + 1/2) \ln x]$, $n = 0, 1, 2, \ldots$, where the c_n's are arbitrary.

21. $\lambda_1 = 1.664$, $\lambda_2 = 5.631$, $\lambda_3 = 11.622.$

23. No nontrivial solutions.

25. λ_n, $n = 1, 2, 3, \ldots$ such that $(\lambda_n^2 + 1)\sqrt{\lambda_n} = n\pi$; and $y_n = c_n \sin(\sqrt{\lambda_n} x)$, $n = 1, 2, 3, \ldots$, where the c_n's are arbitrary.

27. $\lambda_n = -\mu_n^4$, $\mu_n > 0$ and $\cos \mu_n L \cdot \cosh \mu_n L = 1$; $y_n = c_n[(\cos \mu_n L - \cosh \mu_n L)(\sin \mu_n x - \sinh \mu_n x) - (\sin \mu_n L - \sinh \mu_n L)(\cos \mu_n x - \cosh \mu_n x)]$, $n = 1, 2, \ldots$, where the c_n's are arbitrary.

29. $\lambda_n = -(n\pi/L)^4$, $n = 1, 2, 3, \ldots$, and $y_n = c_n \sin(n\pi x/L)$, $n = 1, 2, 3, \ldots$, where the c_n's are arbitrary.

33. **(b)** $\lambda_n = n^2 + 1$, $n = 1, 2, \ldots$, and $y_n = c_n e^{-x} \sin nx$, where the c_n's are arbitrary.

35. $U_{m,n} = c_{n,m}\left(\sin \dfrac{\pi n}{a} x\right)\left(\sin \dfrac{\pi m}{b} y\right)$,

$$\alpha_{m,n}^2 = \left(\frac{\pi n}{a}\right)^2 + \left(\frac{\pi m}{b}\right)^2, \quad n = 1, 2, \ldots,$$

$m = 1, 2, \ldots$, where the $c_{n,m}$'s are arbitrary.

Exercises 11.3, page 680

1. $(e^{6x}y')' + \lambda e^{6x}y = 0.$
3. $((1 - x)^2 y')' + \lambda[(1 - x)/x]y = 0.$
5. $(xy')' + \lambda yx^{-1} = 0.$ **7.** Not self-adjoint.
9. Self-adjoint. **11.** Self-adjoint.
17. $\varphi_n = \sqrt{2}\,\sin[(n + 1/2)\pi x], \quad n = 0, 1, 2, \ldots;$

$$x = \frac{8}{\pi^2} \sum_{n=0}^{\infty} \frac{(-1)^n}{(2n + 1)^2} \sin[(n + 1/2)\pi x].$$

19. $\varphi_n = \sqrt{\dfrac{2}{\pi}}\,\cos nx, \quad n = 0, 1, 2, \ldots;$

$$\sqrt{\frac{2}{\pi}} \sum_{n=0}^{\infty} \frac{(-1)^n - 1}{n^2} \cos nx.$$

21. $\varphi_0 = \gamma_0^{-1}[\cosh(\mu_0 x) - (2/\mu_0)\sinh(\mu_0 x)]$ and
$\varphi_n = \gamma_n^{-1}[-(\mu_n/2)\cos(\mu_n x) + \sin(\mu_n x)],$
$n = 1, 2, 3, \ldots,$ where $\mu_0 = 2\tanh(\mu_0 x),$
$\mu_n = 2\tan(\mu_n \pi),$
$\gamma_0^2 = (4\mu_0^{-2} - 1)[(4\mu_0)^{-1}\sinh(2\mu_0\pi) - \pi/2],$ and
$\gamma_n^2 = (\mu_n^2/4)[\pi/2 + (4\mu_n)^{-1}\sin(2\mu_n\pi)]$
$\quad - (1/2)\sin^2(\mu_n\pi) + \pi/2 - (4\mu_n)^{-1}\sin(2\mu_n\pi);$
$\gamma_0^{-2}[(\pi/\mu_0)\sinh(\mu_0\pi)(1 - 4/\mu_0) + \mu_0^{-2}][\cosh(\mu_0 x)$

$$\quad - (2/\mu_0)\sinh(\mu_0 x)] + \sum_{n=1}^{\infty} \gamma_n^{-2}[(\mu_n^{-2}$$

$\quad - \pi/2)\sin(\mu_n\pi) + \mu_n^{-1}(\pi - 1/2)\cos(\mu_n\pi)$
$\quad + (2/\mu_n)^{-1}][-(\mu_n/2)\cos(\mu_n x) + \sin(\mu_n x)].$

23. $\varphi_n = \sqrt{2/\pi}\,\cos[(n + 1/2)\ln x], \quad n = 0, 1, 2, \ldots;$
$$\frac{2}{\pi} \sum_{n=0}^{\infty} \frac{(-1)^n e^{\pi}(n + 1/2) - 1}{1 + (n + 1/2)^2} \cos[(n + 1/2)\ln x].$$

Exercises 11.4, page 690

1. $L^+[y] = xy'' + x^2 y' + (2x + 3)y.$
3. $L^+[y] = (1 + x^2)y'' + 2xy' + y.$
5. $L^+[y] = 2x^2 y'' + 7xy' - 2y.$
7. $L^+[y] = y'' + 2y' + 10y,$ where
$D(L^+) = \{v \text{ in } C^2[0, \pi]: v(0) = v(\pi) = 0\}.$
9. $L^+[y] = 2x^2 y'' + 5xy',$ where
$D(L^+) = \{v \text{ in } C^2[1, 4]: v(1) = v(4) = 0\}.$
11. $y'' - 6y' + 10y = 0; \quad 6y(0) - y'(0) = 0,$
$6y(\pi) - y'(\pi) = 0.$
13. $y'' + 4y = 0; \quad y(0) = y(\pi), y'(0) = y'(\pi).$
15. $x^2 y'' + xy' + y = 0; \quad y(1) - y'(1) = 0,$
$y(e^{\pi}) - e^{\pi}y'(e^{\pi}) = 0.$

17. $\displaystyle\int_0^{\pi} h(x)e^{-x}\sin 3x\,dx = 0.$

19. Unique solution for each h.

21. $\displaystyle\int_0^{\pi} h(x)e^{3x}(\cos x + 3\sin x)\,dx = 0.$

23. $\displaystyle\int_0^{\pi} h(x)\cos 2x\,dx = 0$ and

$\displaystyle\int_0^{\pi} h(x)\sin 2x\,dx = 0.$

25. $\displaystyle\int_1^{e^{\pi}} h(x)[\cos(\ln x) + \sin(\ln x)]\,dx = 0.$

Exercises 11.5, page 695

1. $5\sin x + \sin 3x.$ **3.** No solution.
5. $C_1 \cos x - (1/15)\cos 4x - (1/48)\cos 7x,$
where C_1 is arbitrary.

7. $\displaystyle\sum_{n=1}^{\infty} \frac{-32}{\pi(2n - 1)^3[1 - (n - 1/2)^2]} \sin[(n - 1/2)x].$

9. $\displaystyle\sum_{n=0}^{\infty} \frac{\gamma_n}{[1 - (n + 1/2)^2]} \sqrt{\frac{2}{\pi}}\cos[(n + 1/2)x],$ where

$\gamma_n = \sqrt{\dfrac{2}{\pi}}\displaystyle\int_0^{\pi} f(x)\cos[(n + 1/2)x]\,dx.$

11. $\displaystyle\sum_{n=0}^{\infty} \frac{\gamma_n}{[5 - (n + 1/2)^2]} \sqrt{\frac{2}{\pi}}\cos[(n + 1/2)\ln x],$ where

$\gamma_n = \sqrt{\dfrac{2}{\pi}}\displaystyle\int_1^{e^{\pi}} f(x)\cos[(n + 1/2)\ln x]\,dx.$

13. $\dfrac{\gamma_0}{6 - \lambda_0}\sinh(\mu_0 x) + \displaystyle\sum_{n=1}^{\infty} \frac{\gamma_n}{6 - \lambda_n}\sin(\mu_n x),$

where $\gamma_0 = \dfrac{\int_0^{\pi} f(x)\sinh(\mu_0 x)\,dx}{\int_0^{\pi} \sinh^2(\mu_0 x)\,dx}$ and

$\gamma_n = \dfrac{\int_0^{\pi} f(x)\sin(\mu_n x)\,dx}{\int_0^{\pi} \sin^2(\mu_n x)\,dx};$

and μ_0 and μ_n satisfy
$$\tanh \mu_0 \pi = \mu_0/3$$
and
$$\tan \mu_n \pi = \mu_n/3, \quad n = 1, 2, 3, \ldots.$$

Exercises 11.6, page 703

1. $G(x, s) = \begin{cases} s, & 0 \leq s \leq x, \\ x, & x \leq s \leq \pi. \end{cases}$

3. $G(x, s) = \begin{cases} -\dfrac{s(x - 1 - \pi)}{1 + \pi}, & 0 \leq s \leq x, \\ -\dfrac{x(s - 1 - \pi)}{1 + \pi}, & x \leq s \leq \pi. \end{cases}$

5. $G(x, s) = \begin{cases} (\sin 2s \cos 2x)/2, & 0 \leq s \leq x, \\ (\sin 2x \cos 2s)/2, & x \leq s \leq \pi. \end{cases}$

7. $G(x, s) = \begin{cases} -(s^{-2} - s)(4x^{-2} + x)/15, & 1 \leq s \leq x, \\ -(x^{-2} - x)(4s^{-2} + s)/15, & x \leq s \leq 2. \end{cases}$

9. $G(x, s) = \begin{cases} -se^{-s}(xe^{-x} - e^{-x}), & 0 \le s \le x, \\ -xe^{-x}(se^{-s} - e^{-s}), & x \le s \le 1. \end{cases}$

11. $G(x, s) = \begin{cases} -s(x - \pi)/\pi, & 0 \le s \le x, \\ -x(s - \pi)/\pi, & x \le s \le \pi, \end{cases}$ and

$y = 1 - e^{-x} - x/\pi + e^{-\pi}x/\pi.$

13. $G(x, s) = \begin{cases} -s(x - 1 - \pi)/(1 + \pi), & 0 \le s \le x, \\ -x(s - 1 - \pi)/(1 + \pi), & x \le s \le \pi, \end{cases}$

and $y = -x^3/6 + (3\pi^2 + \pi^3)x/(6 + 6\pi),$

15. $G(x, s) = \begin{cases} -\sin s \sin(x - 1)/(\sin 1), & 0 \le s \le x, \\ -\sin x \sin(s - 1)/(\sin 1), & x \le s \le 1, \end{cases}$

and $y = [\sin 1 + \sin(x - 1) - \sin x]/(\sin 1).$

17. $G(x, s) =$

$\begin{cases} (e^s - e^{-s})(e^x - e^{2-x})/(2 - 2e^2), & 0 \le s \le x, \\ (e^x - e^{-x})(e^s - e^{2-s})/(2 - 2e^2), & x \le s \le 1, \end{cases}$

and

$y = -x + \dfrac{\sinh x}{\sinh 1} = -x + \dfrac{e^{x+1} - e^{1-x}}{e^2 - 1}.$

19. $G(x, s) =$

$\begin{cases} -(s^2 - s^{-2})(x^2 - 16x^{-2})/60, & 1 \le s \le x, \\ -(x^2 - x^{-2})(s^2 - 16s^{-2})/60, & x \le s \le 2, \end{cases}$

and

$y = [16x^2 \ln 2 - 16x^{-2} \ln 2 - 15x^2 \ln x]/60.$

21. $G(x, s) = \begin{cases} -s(x - 1), & 0 \le s \le x, \\ -x(s - 1), & x \le s \le 1, \end{cases}$ and

$y = \alpha + (\beta - \alpha)x - x\displaystyle\int_0^1 sf(s)\, ds + \int_0^x sf(s)\, ds$

$+ x\displaystyle\int_x^1 f(x)\, ds.$

23. (a) $K(x, s) =$

$\begin{cases} -(e^{-s} - e^{-2s})(e^{-x} - e^{1-2x})/[(e - 1)e^{-3s}], \\ \qquad\qquad 0 \le s \le x, \\ -(e^{-x} - e^{-2x})(e^{-s} - e^{1-2s})/[(e - 1)e^{-3s}], \\ \qquad\qquad x \le s \le 1. \end{cases}$

(b) $y = (1/4)(3 - e^2)e^{-x}/(e - 1)$
$+ (1/4)e(e - 3)e^{-2x}/(e - 1)$
$- (1/2)x + (3/4).$

25. (a) $K(x, s) = \begin{cases} -x(2 - x)(s^{-3} - s^{-2}), & 1 \le s \le x, \\ -x(1 - x)(2s^{-3} - s^{-2}), & x \le s \le 2. \end{cases}$

(b) $y = x^2 \ln 2 - x \ln x - x \ln 2.$

27. $H(x, s) =$

$\begin{cases} -[(s^3/2 - \pi s^2)/\pi^2 + s/2]x^2 - [(-s^3/3 \\ \qquad + \pi s^2/2)/\pi^3 - 1/6]x^3, \\ \qquad\qquad 0 \le s \le x, \\ -[(x^3/2 - \pi x^2)/\pi^2 + x/2]s^2 - [(-x^3/3 \\ \qquad + \pi x^2/2)/\pi^3 - 1/6]s^3, \\ \qquad\qquad x \le s \le \pi. \end{cases}$

29. $H(x, s) = \begin{cases} s^2(s - 3x)/6, & 0 \le s \le x. \\ x^2(x - 3s)/6, & x \le s \le \pi. \end{cases}$

Exercises 11.7, page 712

1. $\displaystyle\sum_{n=1}^{\infty} b_n J_2(\alpha_{2n}x),$ where $\{\alpha_{2n}\}$ is the increasing

sequence of real zeros of J_2 and

$b_n = \dfrac{\int_0^1 f(x)J_2(\alpha_{2n}x)\, dx}{(\mu - \alpha_{2n}^2)\int_0^1 J_2^2(\alpha_{2n}x)x\, dx}.$

3. $\displaystyle\sum_{n=1}^{\infty} a_n J_0(\alpha_{1n}x),$ where $\{\alpha_{1n}\}$ is the increasing

sequence of real zeros of J_0', which are also the zeros

of J_1, and

$a_n = \dfrac{\int_0^1 f(x)J_0(\alpha_{1n}x)\, dx}{(\mu - \alpha_{1n}^2)\int_0^1 J_0^2(\alpha_{1n}x)x\, dx}.$

5. $\displaystyle\sum_{n=0}^{\infty} a_{2n+1} P_{2n+1}(x),$ where

$a_{2n+1} = \dfrac{\int_0^1 f(x)P_{2n+1}(x)\, dx}{[\mu - (2n + 1)(2n + 2)]\int_0^1 P_{2n+1}^2(x)\, dx}.$

15. (c) $\displaystyle\sum_{n=0}^{\infty} a_n T_n(x),$ where

$a_n = \dfrac{\int_{-1}^{1} f(x)T_n(x)(1 - x^2)^{-1/2}\, dx}{(\mu - n^2)\int_{-1}^{1} T_n^2(x)(1 - x^2)^{-1/2}\, dx}.$

Exercises 11.8, page 722

1. $\varphi(x) \equiv 0.$ **9.** $\pi/2.$

11. Between $\pi\sqrt{1/(\lambda + 1)}$ and $\pi\sqrt{6/(\lambda + e^{-5})}.$

Review Problems, page 726

1. (a) $\lambda_n = 9 + n^2\pi^2,$ $n = 1, 2, 3, \ldots;$ and
$y_n = c_n e^{-3x} \sin(n\pi x),$ $n = 1, 2, 3, \ldots,$
where c_n is any real number.

(b) $\lambda_n = \mu_n^2,$ $n = 1, 2, 3, \ldots$ where
$\tan \mu_n \pi = -2\mu_n;$ and $y_n = c_n \sin \mu_n x,$
$n = 1, 2, 3, \ldots,$ where the c_n is any real
number.

3. (a) $\lambda_n = (n - 1/2)^2,$ $n = 1, 2, 3, \ldots,$ and
$y_n = \sqrt{2/\pi} \cos[(n - 1/2)x], n = 1, 2, 3, \ldots.$

(b) $\dfrac{2}{\pi} \displaystyle\sum_{n=1}^{\infty} \left[\dfrac{(-1)^{n-1}(1 - \pi)}{(n - 1/2)} + \dfrac{1}{(n - 1/2)^2}\right]$
$\times \cos[(n - 1/2)x].$

5. (a) $\displaystyle\int_0^{2\pi} h(x)e^{-3x} \sin 2x\, dx = 0.$

(b) $\displaystyle\int_1^2 h(x)\, dx = 0.$

7. (a) $G(x, s) = \begin{cases} s, & 0 \leq s \leq x, \\ x, & x \leq s \leq 1; \end{cases}$

$y = x^3/6 + x^2/2 - 3/2.$

(b) $G(x, s) = \begin{cases} -(e^s + e^{-s})(e^x - e^{2-x})/(2e^2 + 2), \\ \qquad\qquad 0 \leq s \leq x, \\ -(e^x + e^{-x})(e^s - e^{2-s})/(2e^2 + 2), \\ \qquad\qquad x \leq s \leq 1; \end{cases}$

$y = 4(e^{1+x} + e^{1-x} - 1 - e^2)/(e^2 + 1).$

9. $\pi.$

CHAPTER 12

Exercises 12.2, page 750

1. Unstable saddle point.
3. Asymptotically stable spiral point.
5. Asymptotically stable improper node.
7. $(3, 2)$ is an asymptotically stable improper node.
9. $(-4, -1)$ is a stable center.
11. $(3, -5)$ is an asymptotically stable improper node.
13. Asymptotically stable improper node.

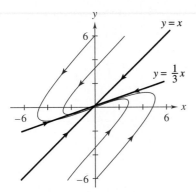

Figure B.46

15. Unstable saddle point.

Figure B.47

17. Unstable saddle point.

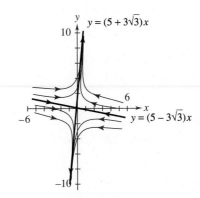

Figure B.48

19. Asymptotically stable improper node.

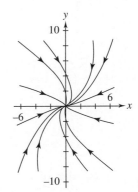

Figure B.49

Exercises 12.3, page 762

1. Unstable saddle point. **3.** Unstable improper node.
5. Center or spiral point of indeterminant stability.
7. Unstable saddle point.
9. $(8, 2)$ is an asymptotically stable improper node; $(-8, -2)$ is an unstable saddle point.
11. $(2, -2)$ is an unstable improper node; $(3, -3)$ is an unstable saddle point.
13. $(1, 1)$ is an asymptotically stable improper node, proper node, or spiral point; $(-1, -1)$ is an unstable saddle point.

Figure B.50

15. $(0, 0)$ is an unstable improper node, proper node, or spiral point; $(0, 3)$ and $(3, 0)$ are unstable saddle points; and $(2, 2)$ is an asymptotically stable improper node.

Figure B.51

17. (b) $\varepsilon < -2$, asymptotically stable improper node;
$\varepsilon = -2$, asymptotically stable improper node, proper node, or spiral point;
$-2 < \varepsilon < 0$, asymptotically stable spiral point;
$\varepsilon = 0$, stable center;
$0 < \varepsilon < 2$, unstable spiral point;
$\varepsilon = 2$, unstable improper node, proper node, or spiral point;
$\varepsilon > 2$, unstable improper node.

19. (b) a_2/b_2. **(d)** As a_1 is decreased, the population of species x approaches zero (reaches zero when $a_1/c_1 = a_2/b_2$) and the population of species y approaches a_2/b_2.

21. $h = 0$: $(0, 0)$ is an unstable improper node;
$(0, 1/2)$ is an unstable saddle point;
$(1/4, 0)$ is an asymptotically stable improper node;
$(1/3, -1/3)$ is an unstable saddle point, but not of interest.
Species x survives while species y dies off.

Figure B.52

$h = 1/32$: $((2 - \sqrt{2})/16, 0)$ is an unstable improper node;
$((2 + \sqrt{2})/16, 0)$ is an asymptotically stable improper node;
$(1/12, 7/24)$ is an unstable saddle point;
$(1/4, -1/8)$ is an unstable saddle point, but not of interest.

Competitive exclusion: One survives, but
the other dies off.

Figure B.53

$h = 5/32$: There are no critical points. Species y
survives while species x dies off.

Figure B.54

7.

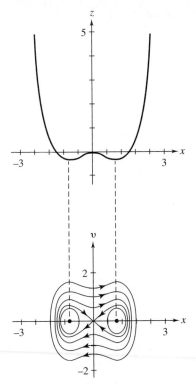

Figure B.55

Exercises 12.4, page 771

1. $G(x) = (1/3)x^3 - (3/2)x^2 + x + C$;
 $E(x,v) = (1/2)v^2 + (1/3)x^3 - (3/2)x^2 + x$.
3. $G(x) = (1/2)x^2 + x + \ln|x - 1| + C$;
 $E(x,v) = (1/2)v^2 + (1/2)x^2 + x + \ln(1 - x)$.
5. $G(x) = (1/4)x^4 + (1/3)x^3 - (1/2)x^2 + C$;
 $E(x,v) = (1/2)v^2 + (1/4)x^4 + (1/3)x^3 - (1/2)x^2$.

9.

Figure B.56

11.

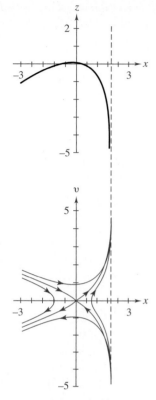

Figure B.57

13. $vh(x,v) = v^2$, so energy is decreasing along a trajectory.

15. $vh(x,v) = v^2$, so energy is decreasing along a trajectory.

Figure B.58

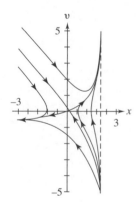

Figure B.59

17. When $\beta > 0$, all trajectories are closed and encircle the origin. Hence, all solutions are periodic. When $\beta < 0$, only trajectories near the origin are closed. Hence, periodic solutions exist, but only for initial values near the origin.

Figure B.60

Figure B.61

19. (a)

Figure B.62

(b)

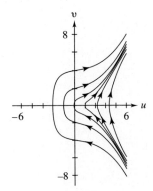

Figure B.63

(c) When $0 < \lambda < 1/4$, closed trajectories exist and hence there can be periodic motion. When $1/4 < \lambda < 1$, there cannot be periodic motion.

Exercises 12.5, page 781

1. Asymptotically stable. **3.** Stable. **5.** Unstable.
7. Asymptotically stable. **9.** Stable. **11.** Stable.
13. Stable.

Exercises 12.6, page 789

1. (a)

Figure B.64

(b) Trajectories inside the unit circle spiral out and approach the unit circle as a limit cycle. Trajectories outside the unit circle also spiral out, away from the unit circle.
3. (b) Counterclockwise.
5. $r = 0$ is an asymptotically stable spiral point. $r = 1$ is a semistable limit cycle. $r = 2$ is an unstable limit cycle. (See Figure B.65.)

Figure B.65

7. $r = 0$ is an asymptotically stable spiral point. $r = 1$ is an unstable limit cycle. (See Figure B.66.)

Figure B.66

9. $r = 0$ is an unstable spiral point. $r = 1$ is a stable limit cycle. $r = 2$ is an unstable limit cycle. (See Figure B.67.)

Figure B.67

11. $r = 0$ is not an isolated critical point, but it is stable. $r = (n\pi)^{-1}$, $n = 1, 2, \ldots$, is a limit cycle that is stable for n even and unstable for n odd. (See Figure B.68.)

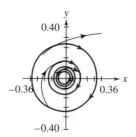

Figure B.68

Exercises 12.7, page 796

1. Asymptotically stable. **3.** Asymptotically stable.
5. Unstable. **9.** Unstable.

11. Asymptotically stable.

13. Asymptotically stable.

15. The equilibrium solution corresponding to the critical point $(1, -2, 2)$ is asymptotically stable.

17. The equilibrium solution corresponding to the critical points $(0,0,0)$, $(1,0,0)$, $(0,1,0)$, and $(0,0,1)$ are unstable. The one corresponding to $(2/5, 2/5, 2/5)$ is asymptotically stable.

Review Problems, page 799

1. $(0, 0)$ is an unstable saddle point.

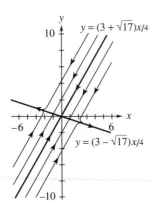

Figure B.69

3. $(0, 0)$ is an asymptotically stable spiral point. (See Figure B.70.)

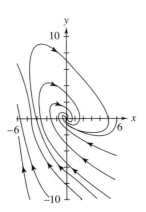

Figure B.70

5. $(0,0)$ is an unstable improper node; $(0, 2)$ is an asymptotically stable improper node; $(1/3, 0)$ is an unstable saddle point; $(1, -2)$ is an asymptotically stable spiral point. (See Figure B.71.)

Figure B.71

7.

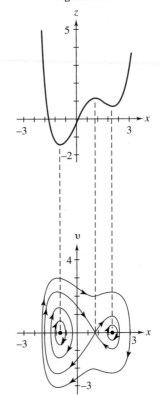

Figure B.72

9. Asymptotically stable. **11.** Unstable.

13. $r = 0$ is an asymptotically stable spiral point.
$r = 1$ is an unstable limit cycle.
$r = 2$ is a semistable limit cycle.
(See Figure B.73.)

Figure B.73

15. Yes. **17.** Asymptotically stable.

CHAPTER 13

Exercises 13.1, page 811

1. $y(x) = 4 + \displaystyle\int_1^x [t^2 - y(t)] \, dt.$

3. $y(x) = -3 + \displaystyle\int_1^x [y(t) - t]^2 \, dt.$ **5.** 1.7320508.

7. 0.2462662. **9.** 1.0698479.

11. $y_1(x) = 2 - 2x, \quad y_2(x) = 2 - 2x + x^2.$

13. $y_1(x) = y_2(x) = 1 + x^3.$

15. $y_1(x) = 1 - e^x, \quad y_2(x) = 2 + x - 2e^x.$

Exercises 13.2, page 818

1. Yes. **3.** No. **5.** Yes.

9. Any h satisfying $0 < h < 1/2$ will work.

11. $y_1(x) = -(1/20)x^5 + (7/6)x^3 - (1/2)x^2 - x + 1;$
$y_2(x) = (1/840)x^7 - (7/120)x^5 + (1/24)x^4 + (7/6)x^3 - (1/2)x^2 - x + 1.$

15. **(b)** $x_1(t) = -t, \quad y_1(t) = 1 + t, \quad z_1(t) = 1;$
$x_2(t) = -t - t^2 - t^3/3, \quad y_2(t) = 1 + t,$
$z_2(t) = 1 - t^2/2 - t^3/3.$

Exercises 13.3, page 825

1. $(\pi/2, 3\pi/2).$ **3.** $(0, \pi/2).$ **5.** $(-\infty, \infty).$

7. $y(t) \equiv 0.$

Exercises 13.4, page 831

1. $10^{-2}e.$ **3.** $10^{-2}e^e.$ **5.** $10^{-2}e^{16}.$

7. $(1/24)e^{7/2}.$ **9.** $(73/6)e^{3\sqrt{3}/8}.$

11. **(a)** $\phi(x, 1) = (1 - x)^{-1};$
$\phi(x, 1 + \varepsilon) = (1 + \varepsilon)(1 - (1 + \varepsilon)x)^{-1}.$

 (d) The graph of $\phi(x, 1 + \varepsilon)$ does not remain in any closed rectangle $R_0 \subset R.$

Review Problems, page 834

1. 1.3247180.

3. $y(x) = -2 + \displaystyle\int_{-1}^x [y^2(t) + \sin t] \, dt.$

5. $y_1(x) = 1 + x;$
$y_2(x) = 1 + x + (3/2)x^2 + x^3 + (1/4)x^4.$

7. Yes. **9.** $(-\infty, \infty).$ **11.** $10^{-3}e^4.$

Index

TABLE 4.1 UNDETERMINED COEFFICIENTS FOR $L[y](x) = g(x)$

Type	$g(x)$	$y_p(x)$
(I) $p_n(x) = a_n x^n + \cdots + a_1 x + a_0$		$x^s P_n(x) = x^s \{A_n x^n + \cdots + A_1 x + A_0\}^\dagger$
(II) $ae^{\alpha x}$		$x^s A e^{\alpha x}$
(III) $a \cos \beta x + b \sin \beta x$		$x^s \{A \cos \beta x + B \sin \beta x\}$
(IV) $p_n(x)e^{\alpha x}$		$x^s P_n(x) e^{\alpha x}$
(V) $p_n(x)\cos \beta x + q_m(x)\sin \beta x$, where $q_m(x) = b_m x^m + \cdots + b_1 x + b_0$		$x^s \{P_N(x)\cos \beta x + Q_N(x)\sin \beta x\}$, where $Q_N(x) = B_N x^N + \cdots + B_1 x + B_0$ and $N = \max(n, m)$
(VI) $ae^{\alpha x} \cos \beta x + be^{\alpha x} \sin \beta x$		$x^s \{A e^{\alpha x} \cos \beta x + B e^{\alpha x} \sin \beta x\}$
(VII) $p_n(x)e^{\alpha x} \cos \beta x + q_m(x)e^{\alpha x} \sin \beta x$		$x^s e^{\alpha x} \{P_N(x)\cos \beta x + Q_N(x)\sin \beta x\}$, where $N = \max(n, m)$

The nonnegative integer s is chosen to be the smallest integer so that no term in the particular solution $y_p(x)$ is a solution to the corresponding homogeneous equation $L[y](x) = 0$.

$^\dagger P_n(x)$ must include all its terms even if $p_n(x)$ has some terms that are zero.

LINEAR FIRST ORDER EQUATIONS

A general solution to the first order linear equation $dy/dx + P(x)y = Q(x)$ is

$$y(x) = [\mu(x)]^{-1} \left(\int \mu(x)Q(x)\, dx + C \right), \quad \text{where} \quad \mu(x) = \exp\left(\int P(x)\, dx \right).$$

REDUCTION OF ORDER FORMULA

Given a nontrivial solution $f(x)$ to $y'' + py' + qy = 0$, a second linearly independent solution is

$$y(x) = f(x) \int \frac{e^{-\int p(x)\, dx}}{[f(x)]^2}\, dx.$$

VARIATION OF PARAMETERS FORMULA

If y_1 and y_2 are two linearly independent solutions to $y'' + py' + qy = 0$, then a particular solution to $y'' + py' + qy = g$ is $y = v_1 y_1 + v_2 y_2$, where

$$v_1(x) = \int \frac{-g(x)y_2(x)}{W[y_1, y_2](x)}\, dx, \qquad v_2(x) = \int \frac{g(x)y_1(x)}{W[y_1, y_2](x)}\, dx,$$

and $W[y_1, y_2](x) = y_1(x)y_2'(x) - y_1'(x)y_2(x).$